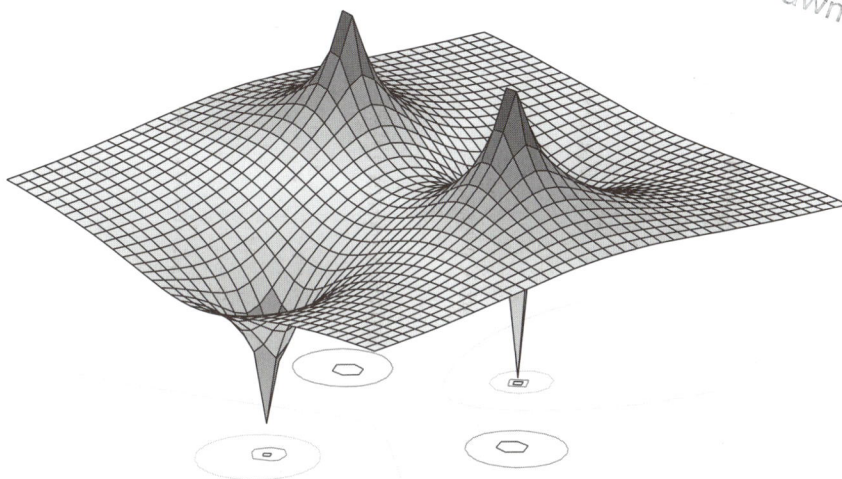

Introduction to Digital Signal Processing Using MATLAB®

Second Edition

Robert J. Schilling and Sandra L. Harris

Clarkson University
Potsdam, NY

CENGAGE
Learning™

Australia • Brazil • Japan • Korea • Mexico • Singapore • Spain • United Kingdom • United States

CENGAGE
Learning™

Introduction to Digital Signal Processing Using MATLAB®

Robert J. Schilling and Sandra L. Harris

Publisher, Global Engineering:
Christopher M. Shortt

Acquisitions Editor: Swati Meherishi

Senior Developmental Editor: Hilda Gowans

Assistant Development Editor: Debarati Roy

Editorial Assistant: Tanya Altieri

Team Assistant: Carly Rizzo

Marketing Manager: Lauren Betsos

Media Editor: Chris Valentine

Content Project Manager:
D. Jean Buttrom

Production Service:
RPK Editorial Services, Inc.

Copyeditor: Shelly Gerger-Knechtl

Proofreader: Becky Taylor

Indexer: Shelly Gerger-Knechtl

Compositor: MPS Limited,
a Macmillan Company

Senior Art Director: Michelle Kunkler

Internal Designer: Carmela Periera

Cover Designer: Patti Hudepohl

Photo Credits:

B/W Image: Digital Vision

Cover Image: Shutterstock Images / jeabjeab

Rights Acquisitions Specialist: John Hill

Text and Image Permissions Researcher:
Kristiina Paul

First Print Buyer: Arethea L. Thomas

Library of Congress Control Number: 2010938463

International Edition:

ISBN-13: 978-1-111-42602-6

ISBN-10: 1-111-42602-3

Cengage Learning International Offices

Asia
www.cengageasia.com
tel: (65) 6410 1200

Australia/New Zealand
www.cengage.com.au
tel: (61) 3 9685 4111

Brazil
www.cengage.com.br
tel: (55) 11 3665 9900

India
www.cengage.co.in
tel: (91) 11 4364 1111

Latin America
www.cengage.com.mx
tel: (52) 55 1500 6000

UK/Europe/Middle East/Africa
www.cengage.co.uk
tel: (44) 0 1264 332 424

Represented in Canada by Nelson Education, Ltd.
tel: (416) 752 9100/(800) 668 0671
www.nelson.com

Cengage Learning is a leading provider of customized learning solutions with office locations around the globe, including Singapore, the United Kingdom, Australia, Mexico, Brazil, and Japan. Locate your local office at: **www.cengage.com/global**

For product information: **www.cengage.com/international**
Visit your local office: **www.cengage.com/global**
Visit our corporate website: **www.cengage.com**

MATLAB is a registered trademark of The MathWorks, 3 Apple Hill Drive, Natick, MA 01760.

Printed in Canada
1 2 3 4 5 6 7 14 13 12 11

In memory of our fathers:

Edgar J. Schilling

and

George W. Harris

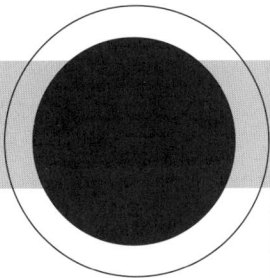

Preface

Digital signal processing, more commonly known as DSP, is a field of study with increasingly widespread applications in the modern technological world. This book focuses on the fundamentals of digital signal processing with an emphasis on practical applications. The text, *Introduction to Digital Signal Processing*, consists of the three parts pictured in Figure 1.

FIGURE 1: Parts of Text

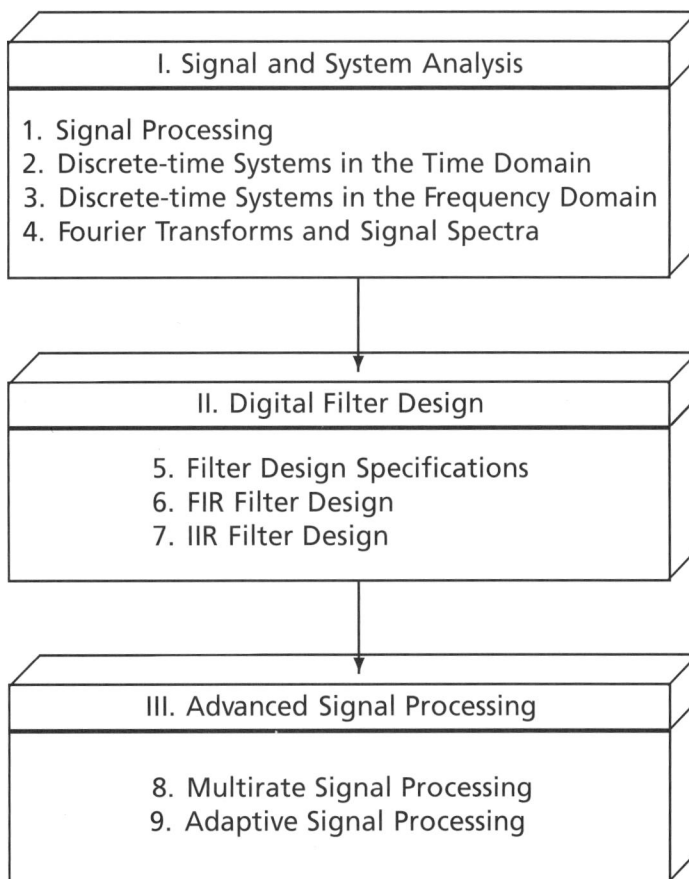

I. Signal and System Analysis

1. Signal Processing
2. Discrete-time Systems in the Time Domain
3. Discrete-time Systems in the Frequency Domain
4. Fourier Transforms and Signal Spectra

II. Digital Filter Design

5. Filter Design Specifications
6. FIR Filter Design
7. IIR Filter Design

III. Advanced Signal Processing

8. Multirate Signal Processing
9. Adaptive Signal Processing

● ● ● ● ● ● ● ● ● ● ● ● ● ● ● ● ● ●

Audience and Prerequisites

This book is targeted primarily toward second-semester juniors, seniors, and beginning graduate students in electrical and computer engineering and related fields that rely on digital signal processing. It is assumed that the students have taken a circuits course, or a signals and systems course, or a mathematics course that includes an introduction to the Fourier transform and the Laplace transform. There is enough material, and sufficient flexibility in the way it can be covered, to provide for courses of different lengths without adding supplementary material. Exposure to MATLAB® programming is useful, but it is not essential. Graphical user interface (GUI) modules are included at the end of each chapter that allow students to interactively explore signal processing concepts and techniques without any need for programming. MATLAB computation problems are supplied for those users who are familiar with MATLAB, and are interested in developing their own programs.

This book is written in an informal style that endeavors to provide motivation for each new topic, and features a careful transition between topics. Significant terms are set apart for convenient reference using Margin Notes and Definitions. Important results are stated as Propositions in order to highlight their significance, and Algorithms are included to summarize the steps used to implement important design procedures. In order to motivate students with examples that are of direct interest, many of the examples feature the processing of speech and music. This theme is also a focus of the course software that includes a facility for recording and playing back speech and sound on a standard PC. This way, students can experience directly the effects of various signal processing techniques.

● ● ● ● ● ● ● ● ● ● ● ● ● ● ● ● ●

Chapter Structure

Each of the chapters of this book follows the template shown in Figure 2. Chapters start with motivation sections that introduce one or more examples of practical problems that can be solved using techniques covered in the chapter. The main body of each chapter is used to

FIGURE 2: Chapter Structure

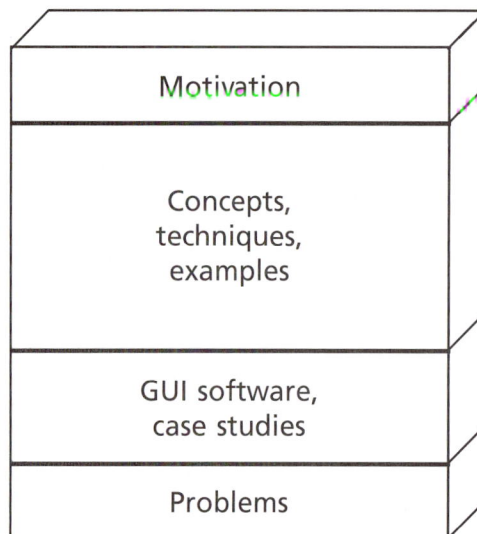

Motivation

Concepts, techniques, examples

GUI software, case studies

Problems

introduce a series of analysis tools and signal processing techniques. Within these sections, the analysis methods and processing techniques evolve from the simple to the more complex. Sections marked with a ∗ near the end of the chapter denote more advanced or specialized material that can be skipped without loss of continuity. Numerous examples are used throughout to illustrate the principles involved.

Near the end of each chapter is a GUI software and case studies section that introduces GUI modules designed to allow the student to interactively explore the chapter concepts and techniques without any need for programming. The GUI modules feature a standard user interface that is simple to use and easy to learn. Data files created as output from one module can be imported as input into other modules. This section also includes case study examples that present complete solutions to practical problems in the form of MATLAB programs. The Chapter Summary section concisely reviews important concepts, and it provides a list of student learning outcomes for each section. The chapter concludes with an extensive set of homework problems separated into three categories and cross referenced to the sections. The Analysis and Design problems can be done by hand or with a calculator. They are used to test student understanding of, and in some cases extend, the chapter material. The GUI Simulation problems allow the student to interactively explore processing and design techniques using the chapter GUI modules. No programming is required for these problems. MATLAB Computation problems are provided that require the user to write programs that apply the signal processing techniques covered in the chapter.

● ● ● ● ● ● ● ● ● ● ● ● ● ● ● ● ● ●
FDSP Toolbox

One of the unique features of this textbook is an integrated software package called the Fundamentals of Digital Signal Processing (FDSP) Toolbox that can be downloaded from the companion web site of the publisher at www.cengage.com/international. Questions and comments concerning the text and the software can be addressed to the authors at: schillin@clarkson.edu.

The FDSP toolbox includes the chapter GUI modules, a library of signal processing functions, all of the MATLAB examples, figures and tables that appear in the text, and on-line help. All of the course software can be accessed easily through a simple menu-based FDSP driver program that is executed with the following command from the MATLAB command prompt.

```
>> f_dsp
```

The FDSP toolbox is self-contained in the sense that only the standard MATLAB interpreter is required. There is no need to for users to have access to optional MATLAB toolboxes such as the Signal Processing and Filter Design toolboxes. Solution PDFs are available to instructors upon request.

● ● ● ● ● ● ● ● ● ● ● ● ● ● ● ● ● ●
Support Material

To access additional course materials please visit www.cengagebrain.com. At the cengagebrain.com home page, search for the ISBN of your title (from the back cover of your book) using the search box at the top of the page. This will take you to the product page where these resources can be found.

● ● ● ● ● ● ● ● ● ● ● ● ● ● ● ⋯

Acknowledgments

This project has been years in the making and many individuals have contributed to its completion. The reviewers commissioned by Brooks/Cole and Cengage Learning made numerous thoughtful and insightful suggestions that were incorporated into the final draft. Thanks to graduate students Joe Tari, Rui Guo, and Lingyun Bai for helping review the initial FDSP toolbox software. We would also like to thank a number of individuals at Brooks/Cole who helped see this project to completion and mold the final product. Special thanks to Bill Stenquist who worked closely with us throughout, and to Rose Kernan. The second edition from Cengage Learning was made possible through the efforts and support of the dedicated group at Global Engineering including Swati Meherishi, Hilda Gowans, Lauren Betsos, Tanya Altieri, and Chris Shortt.

Robert J. Schilling
Sandra L. Harris
Potsdam, NY

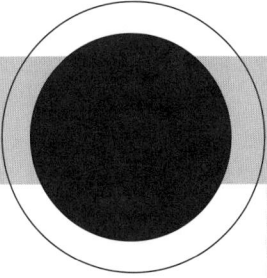

Contents

Margin Contents   xvii

PART I   Signal and System Analysis   1

1   Signal Processing   3

- **1.1**   Motivation   3
  - **1.1.1**   Digital and Analog Processing   4
  - **1.1.2**   Total Harmonic Distortion (THD)   6
  - **1.1.3**   A Notch Filter   7
  - **1.1.4**   Active Noise Control   7
  - **1.1.5**   Video Aliasing   10
- **1.2**   Signals and Systems   11
  - **1.2.1**   Signal Classification   11
  - **1.2.2**   System Classification   16
- **1.3**   Sampling of Continuous-time Signals   21
  - **1.3.1**   Sampling as Modulation   21
  - **1.3.2**   Aliasing   23
- **1.4**   Reconstruction of Continuous-time Signals   26
  - **1.4.1**   Reconstruction Formula   26
  - **1.4.2**   Zero-order Hold   29
- **1.5**   Prefilters and Postfilters   33
  - **1.5.1**   Anti-aliasing Filter   33
  - **1.5.2**   Anti-imaging Filter   37
- *__1.6__   DAC and ADC Circuits   39
  - **1.6.1**   Digital-to-analog Converter (DAC)   39
  - **1.6.2**   Analog-to-digital Converter (ADC)   41
- **1.7**   The FDSP Toolbox   46
  - **1.7.1**   FDSP Driver Module   46
  - **1.7.2**   Toolbox Functions   46
  - **1.7.3**   GUI Modules   49
- **1.8**   GUI Software and Case Studies   52
- **1.9**   Chapter Summary   60

*Sections marked with a * contain more advanced or specialized material that can be skipped without loss of continuity.

1.10 Problems   62
    1.10.1 Analysis and Design   62
    1.10.2 GUI Simulation   67
    1.10.3 MATLAB Computation   68

● ● ● ● ● ● ● ● ● ● ● ● ● ● ● ● ●

√ **2 Discrete-time Systems in the Time Domain   70**

2.1 Motivation   70
    2.1.1 Home Mortgage   71
    2.1.2 Range Measurement with Radar   72
2.2 Discrete-time Signals   74
    2.2.1 Signal Classification   74
    2.2.2 Common Signals   79
2.3 Discrete-time Systems   82
2.4 Difference Equations   86
    2.4.1 Zero-input Response   87
    2.4.2 Zero-state Response   90
2.5 Block Diagrams   94
2.6 The Impulse Response   96
    2.6.1 FIR Systems   97
    2.6.2 IIR Systems   98
2.7 Convolution   100
    2.7.1 Linear Convolution   100
    2.7.2 Circular Convolution   103
    2.7.3 Zero Padding   105
    2.7.4 Deconvolution   108
    2.7.5 Polynomial Arithmetic   109
2.8 Correlation   110
    2.8.1 Linear Cross-correlation   110
    2.8.2 Circular Cross-correlation   114
2.9 Stability in the Time Domain   117
2.10 GUI Software and Case Studies   119
2.11 Chapter Summary   129
2.12 Problems   132
    2.12.1 Analysis and Design   133
    2.12.2 GUI Simulation   139
    2.12.3 MATLAB Computation   141

● ● ● ● ● ● ● ● ● ● ● ● ● ● ● ● ●

√ **3 Discrete-time Systems in the Frequency Domain   145**

3.1 Motivation   145
    3.1.1 Satellite Attitude Control   146
    3.1.2 Modeling the Vocal Tract   148
3.2 Z-transform Pairs   149
    3.2.1 Region of Convergence   150
    3.2.2 Common Z-transform Pairs   153
3.3 Z-transform Properties   157
    3.3.1 General Properties   157
    3.3.2 Causal Properties   162

3.4 Inverse Z-transform 164
    3.4.1 Noncausal Signals 164
    3.4.2 Synthetic Division 164
    3.4.3 Partial Fractions 166
    3.4.4 Residue Method 170
3.5 Transfer Functions 174
    3.5.1 The Transfer Function 174
    3.5.2 Zero-State Response 176
    3.5.3 Poles, Zeros, and Modes 177
    3.5.4 DC Gain 180
3.6 Signal Flow Graphs 181
3.7 Stability in the Frequency Domain 184
    3.7.1 Input-output Representations 184
    3.7.2 BIBO Stability 185
    3.7.3 The Jury Test 188
3.8 Frequency Response 191
    3.8.1 Frequency Response 191
    3.8.2 Sinusoidal Inputs 193
    3.8.3 Periodic Inputs 196
3.9 System Identification 198
    3.9.1 Least-squares Fit 199
    3.9.2 Persistently Exciting Inputs 202
3.10 GUI Software and Case Studies 203
    3.10.1 g_sysfreq: Discrete-time System Analysis
          in the Frequency Domain 203
3.11 Chapter Summary 213
3.12 Problems 215
    3.12.1 Analysis and Design 215
    3.12.2 GUI Simulation 225
    3.12.3 MATLAB Computation 226

4 Fourier Transforms and Spectral Analysis 228
4.1 Motivation 228
    4.1.1 Fourier Series 229
    4.1.2 DC Wall Transformer 230
    4.1.3 Frequency Response 232
4.2 Discrete-time Fourier Transform (DTFT) 233
    4.2.1 DTFT 233
    4.2.2 Properties of the DTFT 236
4.3 Discrete Fourier Transform (DFT) 241
    4.3.1 DFT 241
    4.3.2 Matrix Formulation 243
    4.3.3 Fourier Series and Discrete Spectra 245
    4.3.4 DFT Properties 248
4.4 Fast Fourier Transform (FFT) 256
    4.4.1 Decimation in Time FFT 256
    4.4.2 FFT Computational Effort 260
    4.4.3 Alternative FFT Implementations 262

4.5 Fast Convolution and Correlation   263

   4.5.1 Fast Convolution   263

   *__4.5.2__ Fast Block Convolution   267

   4.5.3 Fast Correlation   270

4.6 White Noise   274

   4.6.1 Uniform White Noise   274

   4.6.2 Gaussian White Noise   278

4.7 Auto-correlation   282

   4.7.1 Auto-correlation of White Noise   282

   4.7.2 Power Density Spectrum   284

   4.7.3 Extracting Periodic Signals from Noise   286

4.8 Zero Padding and Spectral Resolution   291

   4.8.1 Frequency Response Using the DFT   291

   4.8.2 Zero Padding   295

   4.8.3 Spectral Resolution   296

4.9 Spectrogram   299

   4.9.1 Data Windows   299

   4.9.2 Spectrogram   301

4.10 Power Density Spectrum Estimation   304

   4.10.1 Bartlett's Method   304

   4.10.2 Welch's Method   308

4.11 GUI Software and Case Studies   311

4.12 Chapter Summary   319

4.13 Problems   323

   4.13.1 Analysis and Design   323

   4.13.2 GUI Simulation   329

   4.13.3 MATLAB Computation   331

PART II   **Digital Filter Design**   **335**

● ● ● ● ● ● ● ● ● ● ● ● ● ● ● ● ● ●

∨ 5   Filter Design Specifications   337

5.1 Motivation   337

   5.1.1 Filter Design Specifications   338

   5.1.2 Filter Realization Structures   339

5.2 Frequency-selective Filters   342

   5.2.1 Linear Design Specifications   343

   5.2.2 Logarithmic Design Specifications (dB)   348

5.3 Linear-phase and Zero-phase Filters   350

   5.3.1 Linear Phase   350

   5.3.2 Zero-phase Filters   356

5.4 Minimum-phase and Allpass Filters   358

   5.4.1 Minimum-phase Filters   359

   5.4.2 Allpass Filters   362

   5.4.3 Inverse Systems and Equalization   366

5.5 Quadrature Filters   367

   5.5.1 Differentiator   367

   5.5.2 Hilbert Transformer   369

   5.5.3 Digital Oscillator   372

5.6 Notch Filters and Resonators   374
    5.6.1 Notch Filters   374
    5.6.2 Resonators   376
5.7 Narrowband Filters and Filter Banks   378
    5.7.1 Narrowband Filters   378
    5.7.2 Filter Banks   381
5.8 Adaptive Filters   383
5.9 GUI Software and Case Study   386
    5.9.1 g_filters: Evaluation of Digital Filter Characteristics   386
5.10 Chapter Summary   392
5.11 Problems   395
    5.11.1 Analysis and Design   395
    5.11.2 GUI Simulation   403
    5.11.3 MATLAB Computation   404

6 FIR Filter Design   406
6.1 Motivation   406
    6.1.1 Numerical Differentiators   407
    6.1.2 Signal-to-noise Ratio   409
6.2 Windowing Method   411
    6.2.1 Truncated Impulse Response   412
    6.2.2 Windowing   416
6.3 Frequency-sampling Method   423
    6.3.1 Frequency Sampling   424
    6.3.2 Transition-band Optimization   425
6.4 Least-squares Method   430
6.5 Equiripple Filters   434
    6.5.1 Minimax Error Criterion   434
    6.5.2 Parks-McClellan Algorithm   436
6.6 Differentiators and Hilbert Transformers   442
    6.6.1 Differentiators   442
    6.6.2 Hilbert Transformers   445
6.7 Quadrature Filters   448
    6.7.1 Generation of a Quadrature Pair   448
    6.7.2 Quadrature Filter   450
    6.7.3 Equalizer Design   453
6.8 Filter Realization Structures   457
    6.8.1 Direct Forms   457
    6.8.2 Cascade Form   459
    6.8.3 Lattice Form   461
6.9 Finite Word Length Effects   464
    6.9.1 Binary Number Representation   465
    6.9.2 Input Quantization Error   466
    6.9.3 Coefficient Quantization Error   470
    6.9.4 Roundoff Error, Overflow, and Scaling   473
6.10 GUI Software and Case Study   477
6.11 Chapter Summary   484

6.12 Problems   488
    6.12.1 Analysis and Design   488
    6.12.2 GUI Simulation   492
    6.12.3 MATLAB Computation   494

● ● ● ● ● ● ● ● ● ● ● ● ● ● ● ● ● ● ●

√7   IIR Filter Design   499

7.1 Motivation   499
    7.1.1 Tunable Plucked-string Filter   500
    7.1.2 Colored Noise   502
7.2 Filter Design by Pole-zero Placement   504
    7.2.1 Resonator   504
    7.2.2 Notch Filter   508
    7.2.3 Comb Filters   510
7.3 Filter Design Parameters   514
7.4 Classical Analog Filters   517
    7.4.1 Butterworth Filters   517
    7.4.2 Chebyshev-I Filters   522
    7.4.3 Chebyshev-II Filters   525
    7.4.4 Elliptic Filters   526
7.5 Bilinear transformation Method   529
7.6 Frequency Transformations   535
    7.6.1 Analog Frequency Transformations   536
    7.6.2 Digital Frequency Transformations   539
7.7 Filter Realization Structures   541
    7.7.1 Direct Forms   541
    7.7.2 Parallel Form   544
    7.7.3 Cascade Form   547
*__7.8__ Finite Word Length Effects   550
    7.8.1 Coefficient Quantization Error   550
    7.8.2 Roundoff Error, Overflow, and Scaling   553
    7.8.3 Limit Cycles   557
7.9 GUI Software and Case Study   560
7.10 Chapter Summary   567
7.11 Problems   571
    7.11.1 Analysis and Design   571
    7.11.2 GUI Simulation   575
    7.11.3 MATLAB Computation   577

PART III   Advanced Signal Processing   581

● ● ● ● ● ● ● ● ● ● ● ● ● ● ● ● ● ● ●

8   Multirate Signal Processing   583

8.1 Motivation   583
    8.1.1 Narrowband Filter Banks   584
    8.1.2 Fractional Delay Systems   586
8.2 Integer Sampling Rate Converters   587
    8.2.1 Sampling Rate Decimator   587
    8.2.2 Sampling Rate Interpolator   588

8.3   Rational Sampling Rate Converters   591
    8.3.1   Single-stage Converters   591
    8.3.2   Multistage Converters   593
8.4   Multirate Filter Realization Structures   596
    8.4.1   Polyphase Decimator   596
    8.4.2   Polyphase Interpolator   598
8.5   Narrowband Filters and Filter Banks   600
    8.5.1   Narrowband Filters   600
    8.5.2   Filter Banks   601
8.6   A Two-channel QMF Bank   607
    8.6.1   Rate Converters in the Frequency Domain   608
    8.6.2   An Alias-free QMF Bank   610
8.7   Oversampling ADC   612
    8.7.1   Anti-aliasing Filters   612
    8.7.2   Sigma-delta ADC   615
8.8   Oversampling DAC   620
    8.8.1   Anti-imaging Filters   620
    8.8.2   Passband Equalization   621
8.9   GUI Software and Case Study   623
8.10   Chapter Summary   630
8.11   Problems   633
    8.11.1   Analysis and Design   633
    8.11.2   GUI Simulation   640
    8.11.3   MATLAB Computation   641

● ● ● ● ● ● ● ● ● ● ● ● ● ● ● ● ● ⋯⋯

9   Adaptive Signal Processing   645
9.1   Motivation   645
    9.1.1   System Identification   646
    9.1.2   Channel Equalization   647
    9.1.3   Signal Prediction   648
    9.1.4   Noise Cancellation   648
9.2   Mean Square Error   649
    9.2.1   Adaptive Transversal Filters   649
    9.2.2   Cross-correlation Revisited   650
    9.2.3   Mean Square Error   651
9.3   The Least Mean Square (LMS) Method   656
9.4   Performance Analysis of LMS Method   660
    9.4.1   Step Size   660
    9.4.2   Convergence Rate   663
    9.4.3   Excess Mean Square Error   666
9.5   Modified LMS Methods   669
    9.5.1   Normalized LMS Method   669
    9.5.2   Correlation LMS Method   671
    9.5.3   Leaky LMS Method   674
9.6   Adaptive FIR Filter Design   678
    9.6.1   Pseudo-filters   678
    9.6.2   Linear-phase Pseudo-filters   681

9.7   The Recursive Least-Squares (RLS) Method   684
   9.7.1   Performance Criterion   684
   9.7.2   Recursive Formulation   685
9.8   Active Noise Control   690
   9.8.1   The Filtered-x LMS Method   691
   9.8.2   Secondary Path Identification   693
   9.8.3   Signal-synthesis Method   695
9.9   Nonlinear System Identification   700
   9.9.1   Nonlinear Discrete-time Systems   700
   9.9.2   Grid Points   701
   9.9.3   Radial Basis Functions   703
   9.9.4   Adaptive RBF Networks   707
9.10 GUI Software and Case Study   713
9.11 Chapter Summary   718
9.12 Problems   721
   9.12.1   Analysis and Design   721
   9.12.2   GUI Simulation   725
   9.12.3   MATLAB Computation   727

References and Further Reading   734

Appendix 1   Transform Tables   738

1.1   Fourier Series   738
1.2   Fourier Transform   739
1.3   Laplace Transform   741
1.4   Z-transform   743
1.5   Discrete-time Fourier Transform   744
1.6   Discrete Fourier Transform (DFT)   745

Appendix 2   Mathematical Identities   747

2.1   Complex Numbers   747
2.2   Euler's Identity   747
2.3   Trigonometric Identities   748
2.4   Inequalities   748
2.5   Uniform White Noise   749

Appendix 3   FDSP Toolbox Functions   750

3.1   Installation   750
3.2   Driver Module: *f_dsp*   751
3.3   Chapter GUI Modules   751
3.4   FDSP Toolbox Functions   752

Index   755

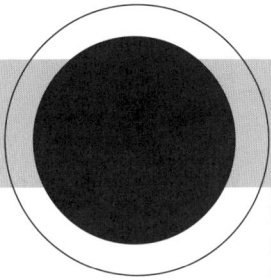

Margin Contents

TABLE I: ▶
Definitions

Number	Term	Symbol	Page
1.1	Causal signal	$x_a(t)$	15
1.2	Linear system	S	17
1.3	Time-invariant system	S	17
1.4	Stable system	S	18
1.5	Frequency response	$H_a(f)$	19
1.6	Impulse response	$h_a(t)$	20
1.7	Bandlimited signal	$x_a(t)$	24
1.8	Transfer function	$H_a(s)$	29
2.1	Impulse response	$h(k)$	97
2.2	FIR and IIR systems	S	97
2.3	Linear convolution	$h(k) \star x(k)$	101
2.4	Circular convolution	$h(k) \circ x(k)$	104
2.5	Linear cross-correlation	$r_{yx}(k)$	111
2.6	Circular cross-correlation	$c_{yz}(k)$	114
2.7	BIBO stable	$\|h\|_1 < \infty$	117
3.1	Z-transform	$X(z)$	149
3.2	Transfer function	$H(z)$	174
3.3	Frequency response	$H(f)$	191
4.1	Discrete-time Fourier transform (DTFT)	$X(f)$	233
4.2	Discrete Fourier transform (DFT)	$X(i)$	242
4.3	Expected value	$E[f(x)]$	275
4.4	Circular auto-correlation	$c_{xx}(k)$	282
4.5	Spectrogram	$G(m, i)$	301
5.1	Group delay	$D(f)$	351
5.2	Linear-phase filter	$H(z)$	351
5.3	Minimum-phase filter	$H(z)$	359
5.4	Allpass Filter	$H(z)$	362
6.1	Signal-to-noise ratio	$SNR(y)$	410
6.2	Quantization operator	$Q_N(x)$	466
9.1	Random cross-correlation	$r_{yx}(i)$	651

xviii Margin Contents

TABLE II: ▶ Propositions

Number	Description	Page
1.1	Signal Sampling	25
1.2	Signal Reconstruction	28
2.1	BIBO Stability: Time Domain	118
3.1	BIBO Stability: Frequency Domain	186
3.2	Frequency Response	194
4.1	Parseval's Identity: DTFT	238
4.2	Parseval's Identity: DFT	254
5.1	Paley-Wiener Theorem	344
5.2	Linear-phase Filter	353
5.3	Minimum-phase Allpass Decomposition	363
6.1	Alternation Theorem	436
6.2	Flow Graph Reversal Theorem	458
9.1	LMS Convergence	661

TABLE III: ▶ Algorithms

Number	Description	Page
1.1	Successive Approximation	43
3.1	Residue Method	172
4.1	Bit Reversal	258
4.2	FFT	260
4.3	Problem Domain	261
4.4	IFFT	262
4.5	Fast Block Convolution	268
5.1	Zero-phase Filter	357
5.2	Minimum-phase Allpass Decomposition	364
6.1	Windowed FIR Filter	421
6.2	Equiripple FIR Filter	438
6.3	Lattice-form Realization	462
7.1	Bilinear Transformation Method	532
9.1	RLS Method	687
9.2	RBF Network Evaluation	708

PART I

Signal and System Analysis

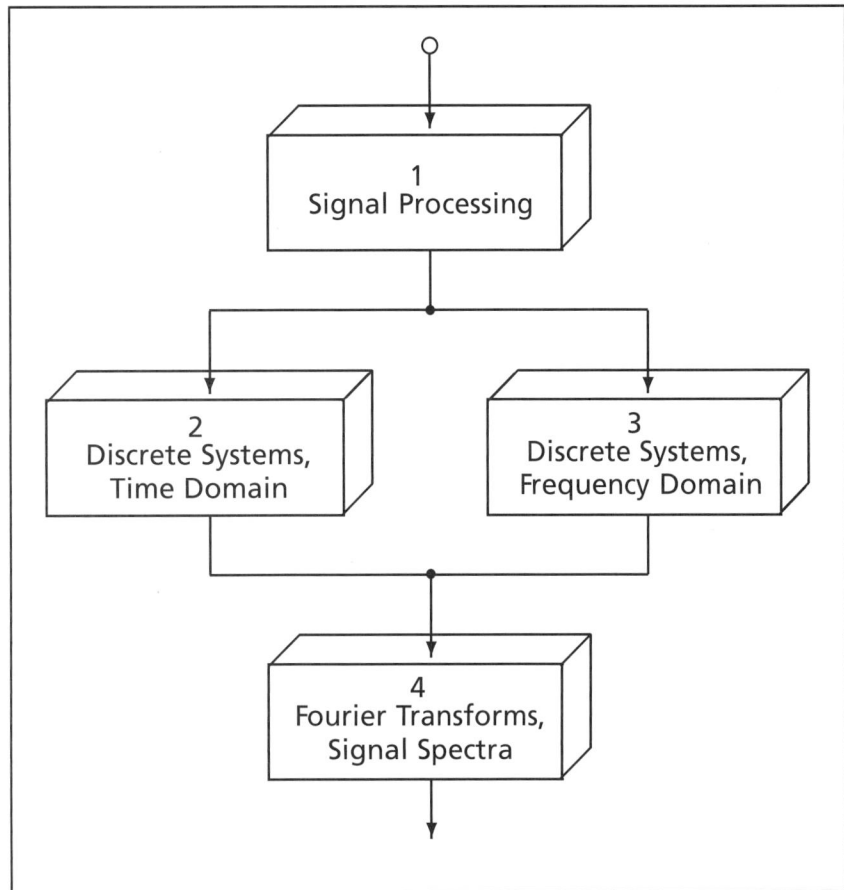

CHAPTER 1

Signal Processing

Chapter Topics

1.1 Motivation

1.2 Signals and Systems

1.3 Sampling of Continuous-time Signals

1.4 Reconstruction of Continuous-time Signals

1.5 Prefilters and Postfilters

1.6 DAC and ADC Circuits

1.7 The FDSP Toolbox

1.8 GUI Software and Case Study

1.9 Chapter Summary

1.10 Problems

1.1 Motivation

Continuous-time signal

Discrete-time signal

A signal is a physical variable whose value varies with time or space. When the value of the signal is available over a continuum of time it is referred to as a *continuous-time* or analog signal. Everyday examples of analog signals include temperature, pressure, liquid level, chemical concentration, voltage and current, position, velocity, acceleration, force, and torque. If the value of the signal is available only at discrete instants of time, it is called a *discrete-time* signal. Although some signals, for example economic data, are inherently discrete-time signals, a more common way to produce a discrete-time signal, $x(k)$, is to take samples of an underlying analog signal, $x_a(t)$.

$$x(k) \triangleq x_a(kT), \quad |k| = 0, 1, 2, \cdots$$

Sampling interval

Digital signal

Here T denotes the *sampling interval* or time between samples, and $\triangleq$ means equals by definition. When finite precision is used to represent the value of $x(k)$, the sequence of quantized values is then called a *digital* signal. A system or algorithm which processes one digital signal $x(k)$ as its input and produces a second digital signal $y(k)$ as its output is a digital signal processor. Digital signal processing (DSP) techniques have widespread applications, and they play an increasingly important role in the modern world. Application areas include speech recognition, detection of targets with radar and sonar, processing of music and video, seismic exploration for oil and gas deposits, medical signal processing including EEG, EKG, and ultrasound, communication channel equalization, and satellite image processing. The focus of this book is the development, implementation, and application of modern DSP techniques.

We begin this introductory chapter with a comparison of digital and analog signal processing. Next, some practical problems are posed that can be solved using DSP techniques. This is followed by characterization and classification of signals. The fundamental notion of the spectrum of a signal is then presented including the concepts of bandlimited and white-noise signals. This leads naturally to the sampling process which takes a continuous-time signal and produces a corresponding discrete-time signal. Simple conditions are presented that ensure that an analog signal can be reconstructed from its samples. When these conditions are not satisfied, the phenomenon of aliasing occurs. The use of guard filters to reduce the effects of aliasing is discussed. Next DSP hardware in the form of analog-to-digital converters (ADCs) and digital-to-analog converters (DACs) is examined. The hardware discussion includes ways to model the quantization error associated with finite precision converters. A custom MATLAB toolbox, called FDSP, is then introduced that facilitates the development of simple DSP pro-

GUI modules grams. The FDSP toolbox also includes a number of graphical user interface (GUI) modules that can be used to browse examples and explore digital signal processing techniques without any need for programming. The GUI module *g_sample* allows the user to investigate the signal sampling process, while the companion module *g_reconstruct* allows the user to explore the signal reconstruction process. The chapter concludes with a case study example, and a summary of continuous-time and discrete-time signal processing.

1.1.1 Digital and Analog Processing

For many years, almost all signal processing was done with analog circuits as shown in Figure 1.1. Here, operational amplifiers, resistors, and capacitors are used to realize frequency selective filters.

With the advent of specialized microprocessors with built-in data conversion circuits (Papamichalis, 1990), it is now commonplace to perform signal processing digitally as shown in Figure 1.2. Digital processing of analog signals is more complex because it requires, at a minimum, the three components shown in Figure 1.2. The analog-to-digital converter or ADC at the front end converts the analog input $x_a(t)$ into an equivalent digital signal $x(k)$. The

FIGURE 1.1: Analog Signal Processing

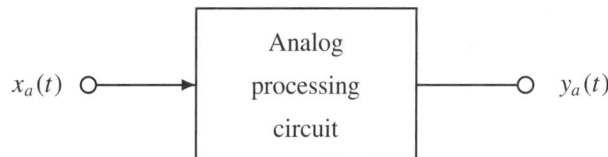

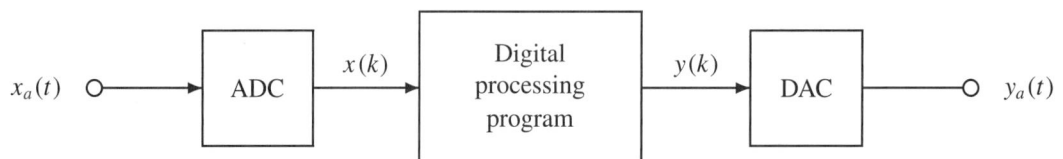

FIGURE 1.2: Digital Signal Processing

	Feature	Analog Processing	Digital Processing
TABLE 1.1: ▶ Comparison of Analog and Digital Signal Processing	Speed	Fast	Moderate
	Cost	Low to moderate	Moderate
	Flexibility	Low	High
	Performance	Moderate	High
	Self-calibration	No	Yes
	Data-logging capability	No	Yes
	Adaptive capability	Limited	Yes

processing of $x(k)$ is then achieved with an algorithm that is implemented in software. For a filtering operation, the DSP algorithm consists of a difference equation, but other types of processing are also possible and are often used. The digital output signal $y(k)$ is then converted back to an equivalent analog signal $y_a(t)$ by the digital-to-analog converter or DAC.

Although the DSP approach requires more steps than analog signal processing, there are many important benefits to working with signals in digital form. A comparison of the relative advantages and disadvantages of the two approaches is summarized in Table 1.1. Although the DSP approach requires more steps than analog signal processing, there are many important benefits to working with signals in digital form. A comparison of the relative advantages and disadvantages of the two approaches is summarized in Table 1.1.

The primary advantages of analog signal processing are *speed* and *cost*. Digital signal processing is not as fast due to the limits on the sampling rates of the converter circuits. In addition, if substantial computations are to be performed between samples, then the clock rate of the processor also can be a limiting factor. Speed can be an issue in *real-time* applications where the kth output sample $y(k)$ must be computed and sent to the DAC as soon as possible after the kth input sample $x(k)$ is available from the ADC. However, there are also applications where the entire input signal is available ahead of time for processing off-line. For this batch mode type of processing, speed is less critical.

Real time

DSP hardware is often somewhat more expensive than analog hardware because analog hardware can consist of as little as a few discrete components on a stand-alone printed circuit board. The cost of DSP hardware varies depending on the performance characteristics required. In some cases, a PC may already be available to perform other functions for a given application, and in these instances the marginal expense of adding DSP hardware is not large.

In spite of these limitations, there are great benefits to using DSP techniques. Indeed, DSP is superior to analog processing with respect to virtually all of the remaining features listed in Table 1.1. One of the most important advantages is the inherent *flexibility* available with a software implementation. Whereas an analog circuit might be tuned with a potentiometer to vary its performance over a limited range, the DSP algorithm can be completely replaced, on the fly, when circumstances warrant.

DSP also offers considerably higher *performance* than analog signal processing. For example, digital filters with arbitrary magnitude responses and linear phase responses can be designed easily whereas this is not feasible with analog filters.

A common problem that plagues analog systems is the fact that the component values tend to *drift* with age and with changes in environmental conditions such as temperature. This leads to a need for periodic calibration or tuning. With DSP there is no drift problem and therefore no need to manually calibrate.

Since data are already available in digital form in a DSP system, with little or no additional expense, one can *log* the data associated with the operation of the system so that its performance can be monitored, either locally of remotely over a network connection. If an unusual operating condition is detected, its exact time and nature can be determined and a higher-level control

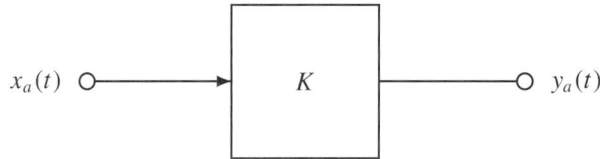

system can be alerted. Although strip chart recorders can be added to an analog system, this substantially increases the expense thereby negating one of its potential advantages.

The flexibility inherent in software can be exploited by having the parameters of the DSP algorithm vary with time and *adapt* as the characteristics of the input signal or the processing task change. Applications, like system identification and active noise control, exploit adaptive signal processing, a topic that is addressed in Chapter 9.

1.1.2 Total Harmonic Distortion (THD)

With the widespread use of digital computers, DSP applications are now commonplace. As a simple initial example, consider the problem of designing an audio amplifier to boost signal strength without distorting the shape of the input signal. For the amplifier shown in Figure 1.3, suppose the input signal $x_a(t)$ is a pure sinusoidal tone of amplitude a and frequency F_0 Hz.

$$x_a(t) = a \cos(2\pi F_0 t) \tag{1.1.1}$$

An ideal amplifier will produce a desired output signal $y_d(t)$ that is a scaled and delayed version of the input signal. For example, if the scale factor or amplifier *gain* is K and the delay is τ, then the desired output is

$$y_d(t) = K x_a(t - \tau)$$
$$= K a \cos[2\pi F_0(t - \tau)] \tag{1.1.2}$$

In a practical amplifier, the relationship between the input and the output is only approximately linear, so some additional terms are present in the actual output y_a.

$$y_a(t) = F[x_a(t)]$$
$$\approx \frac{d_0}{2} + \sum_{i=1}^{M-1} d_i \cos(2\pi i F_0 t + \theta_i) \tag{1.1.3}$$

The presence of the additional harmonics indicates that there is *distortion* in the amplified signal due to nonlinearities within the amplifier. For example, if the amplifier is driven with an input whose amplitude a is too large, then the amplifier will saturate with the result that the output is a *clipped* sine wave that sounds distorted when played through a speaker. To quantify the amount of distortion, the average power contained in the ith harmonic is $d_i^2/2$ for $i \geq 1$ and $d_i^2/4$ for $i = 0$. Thus the *average power* of the signal $y_a(t)$ is

$$P_y = \frac{d_0^2}{4} + \frac{1}{2} \sum_{i=1}^{M-1} d_i^2 \tag{1.1.4}$$

Total harmonic distortion The *total harmonic distortion* or THD of the output signal $y_a(t)$ is defined as the power in the spurious harmonic components, expressed as a percentage of the total power. Thus the following can be used to measure the *quality* of the amplifier output.

$$\text{THD} \triangleq \frac{100(P_y - d_1^2/2)}{P_y} \%$$ (1.1.5)

For an ideal amplifier $d_i = 0$ for $i \neq 1$, and

$$d_1 = Ka$$ (1.1.6a)
$$\theta_1 = -2\pi F_0 \tau$$ (1.1.6b)

Consequently, for a high-quality amplifier, the THD is small, and when no distortion is present THD $= 0$. Suppose the amplifier output is sampled to produce the following digital signal of length $N = 2M$.

$$y(k) = y_a(kT), \quad 0 \leq k < N$$ (1.1.7)

If the sampling interval is set to $T = 1/(N F_0)$, then this corresponds to one period of $y_a(t)$. By processing the digital signal $x(k)$ with the discrete Fourier transform or DFT, it is possible to determine d_i and θ_i for $0 \leq i < M$. In this way the total harmonic distortion can be measured. The DFT is a key analytic tool that is introduced in Chapter 4.

1.1.3 A Notch Filter

As a second example of a DSP application, suppose one is performing sensitive acoustic measurements in a laboratory setting using a microphone. Here, any ambient background sounds in the range of frequencies of interest have the potential to corrupt the measurements with unwanted noise. Preliminary measurements reveal that the overhead fluorescent lights are emitting a 120 Hz hum which corresponds to the second harmonic of the 60 Hz commercial AC power. The problem then is to remove the 120 Hz frequency component while affecting the other nearby frequency components as little as possible. Consequently, you want to process the acoustic data samples with a *notch* filter designed to remove the effects of the fluorescent lights. After some calculations, you arrive at the following digital filter to process the measurements $x(k)$ to produce a filtered signal $y(k)$.

Notch filter

$$y(k) = 1.6466y(k-1) - .9805y(k-2) + .9905x(k)$$
$$-1.6471x(k-1) + .9905x(k-2)$$ (1.1.8)

The filter in (1.1.8) is a notch filter with a bandwidth of 4 Hz, a notch frequency of $F_n = 120$ Hz, and a sampling frequency of $f_s = 1280$ Hz. A plot of the frequency response of this filter is shown in Figure 1.4 where a sharp notch at 120 Hz is apparent. Notice that except for frequencies near F_n, all other frequency components of $x(k)$ are passed through the filter without attenuation. The design of notch filters is discussed in Chapter 7.

1.1.4 Active Noise Control

An application area of DSP that makes use of adaptive signal processing is active control of acoustic noise (Kuo and Morgan, 1996). Examples include industrial noise from rotating machines, propeller and jet engine noise, road noise in an automobile, and noise caused by air flow in heating, ventilation, and air conditioning systems. As an illustration of the latter, consider the active noise control system shown in Figure 1.5 which consists of an air duct with two microphones and a speaker. The basic principle of active noise control is to inject a secondary sound into the environment so as to cancel the primary sound using destructive interference.

FIGURE 1.4:
Magnitude
Response of a
Notch Filter with
$F_n = 120$ Hz

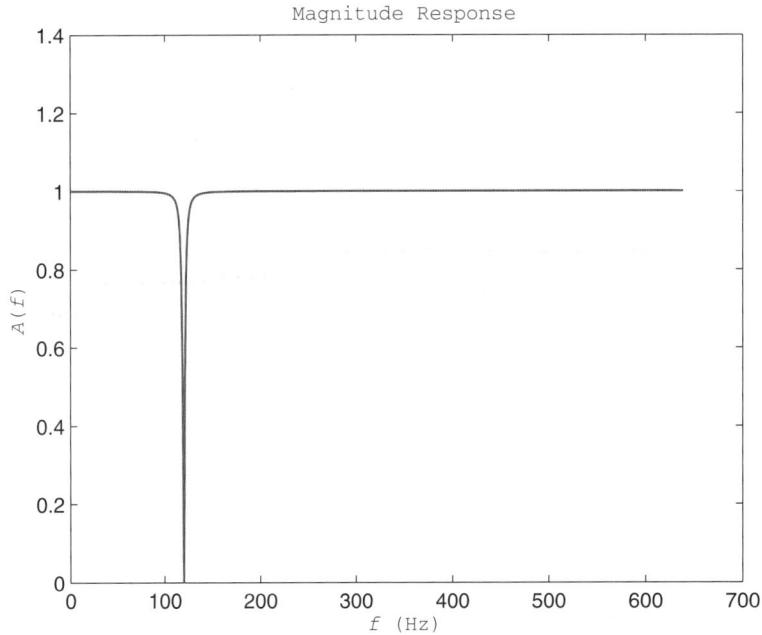

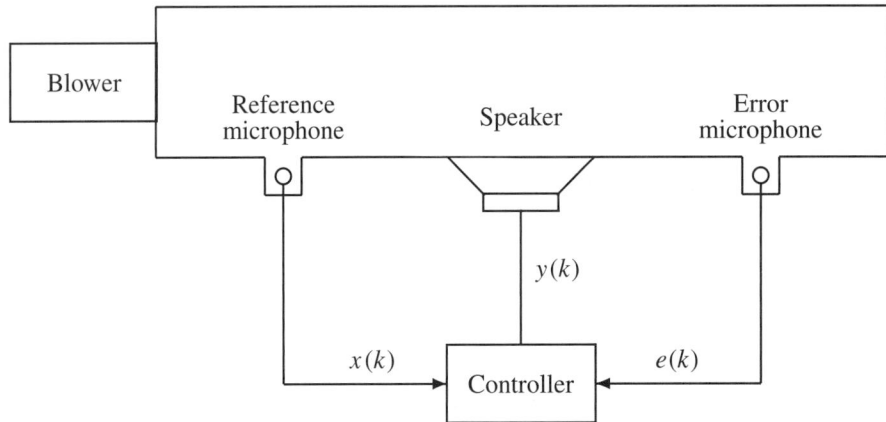

The purpose of the reference microphone in Figure 1.5 is to detect the primary noise $x(k)$ generated by the noise source or blower. The primary noise signal is then passed through a digital filter of the following form.

$$y(k) = \sum_{i=0}^{m} w_i(k)x(k-i) \tag{1.1.9}$$

Antisound

The output of the filter $y(k)$ drives a speaker that creates the secondary sound sometimes called *antisound*. The error microphone, located downstream of the speaker, detects the sum of the primary and secondary sounds and produces an error signal $e(k)$. The objective of the adaptive algorithm is the take $x(k)$ and $e(k)$ as inputs and adjust the filter weights $w(k)$ so as to drive $e^2(k)$ to zero. If zero error can be achieved, then silence is observed at the error microphone. In practical systems, the error or residual sound is significantly reduced by active noise control.

To illustrate the operation of this adaptive DSP system, suppose the blower noise is modeled as a periodic signal with fundamental frequency F_0 and r harmonics plus some random white noise $v(k)$.

blower noise:

$$x(k) = \sum_{i=1}^{r} a_i \cos(2\pi i k F_0 T + \theta_i) + v(k), \quad 0 \le k < p \tag{1.1.10}$$

For example, suppose $F_0 = 100$ Hz and there are $r = 4$ harmonics with amplitudes $a_i = 1/i$ and random phase angles. Suppose the random white noise term is distributed uniformly over the interval $[-.5, .5]$. Let $p = 2048$ samples, suppose the sampling interval is $T = 1/1600$ sec, and the filter order is $m = 40$. The adaptive algorithm used to adjust the filter weights is called the FXLMS method, and it is discussed in detail in Chapter 9. The results of applying this algorithm are shown in Figure 1.6.

Initially the filter weights are set to $w(0) = 0$ which corresponds to no noise control at all. The adaptive algorithm is not activated until sample $k = 512$, so the first quarter of the plot in Figure 1.6 represents the ambient or primary noise detected at the error microphone. When adaptation is activated, the error begins to decrease rapidly and after a short transient period it reaches a steady-state level that is almost two orders of magnitude quieter than the primary noise itself. We can quantify the noise reduction by using the following measure of overall noise cancellation.

$$E = 10 \log_{10} \left(\sum_{i=0}^{p/4-1} e^2(i) \right) - 10 \log_{10} \left(\sum_{i=3p/4}^{p-1} e^2(i) \right) \quad \text{dB} \tag{1.1.11}$$

The overall noise cancellation E is the log of the ratio of the average power of the noise during the first quarter of the samples divided by the average power of the noise during the last quarter of the samples, expressed in units of decibels. Using this measure, the noise cancellation observed in Figure 1.6 is $E = 37.8$ dB.

FIGURE 1.6: Error Signal with Active Noise Control Activated at $k = 512$

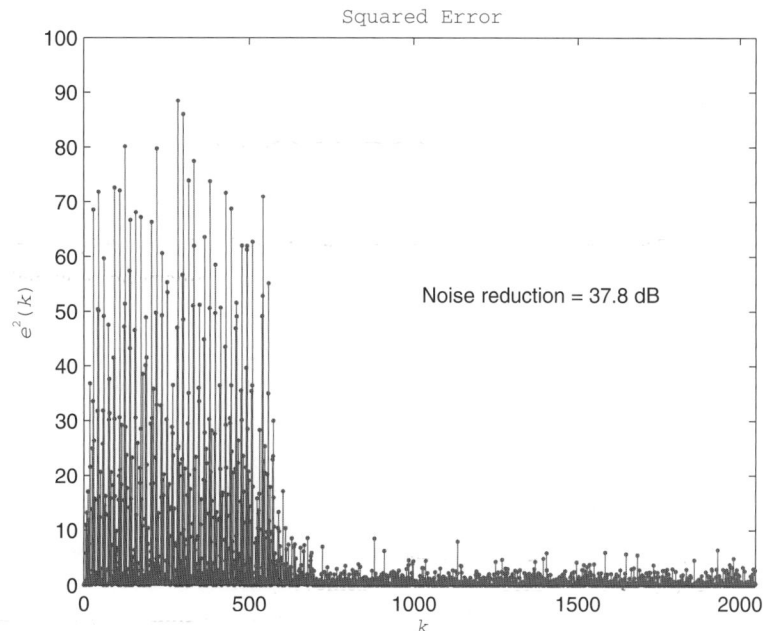

Squared Error

Noise reduction = 37.8 dB

1.1.5 Video Aliasing

Later in Chapter 1 we focus on the problem of sampling a continuous-time signal $x_a(t)$ to produce the following discrete-time signal where $T > 0$ is the sampling interval and $f_s = 1/T$ is the sampling frequency.

$$x(k) = x_a(kT), \quad |k| = 0, 1, 2, \ldots \tag{1.1.12}$$

An important theoretical and practical question that arises in connection with the sampling process is this: under what conditions do the samples $x(k)$ contain all the information needed to reconstruct the signal $x_a(t)$? The Shannon sampling theorem (Proposition 1.1), says that if the signal $x_a(t)$ is bandlimited and the sampling rate f_s is greater than twice the bandwidth or highest frequency present, then it is possible to interpolate between the $x(k)$ to precisely reconstruct $x_a(t)$. However, if the sampling frequency is too low, then the samples become corrupted, a process known as *aliasing*. An easy way to interpret aliasing is to examine a video signal in the form of an $M \times N$ image $I_a(t)$ that varies with time. Here $I_a(t)$ consists of an $M \times N$ array of picture elements or *pixels* where the number of rows M and columns N depends on the video format used. If $I_a(t)$ is sampled with a sampling interval of T then the resulting MN-dimensional discrete-time signal is

Aliasing

Pixels

$$I(k) = I_a(kT), \quad |k| = 0, 1, 2, \ldots \tag{1.1.13}$$

Here, $f_s = 1/T$ is the sampling rate in frames/second. Depending on the content of the image, the sampling rate f_s may or may not be sufficiently high to avoid aliasing.

As a simple illustration, suppose the image consists of a rotating disk with a dark line on it to indicate orientation as shown in Figure 1.7. A casual look at the sequence of frames in Figure 1.7 suggests that the disk appears to be rotating counterclockwise at a rate of 45 degrees per frame. However, this is not the only interpretation possible. For example, an alternative

FIGURE 1.7: Four Video Frames of a Rotating Disk

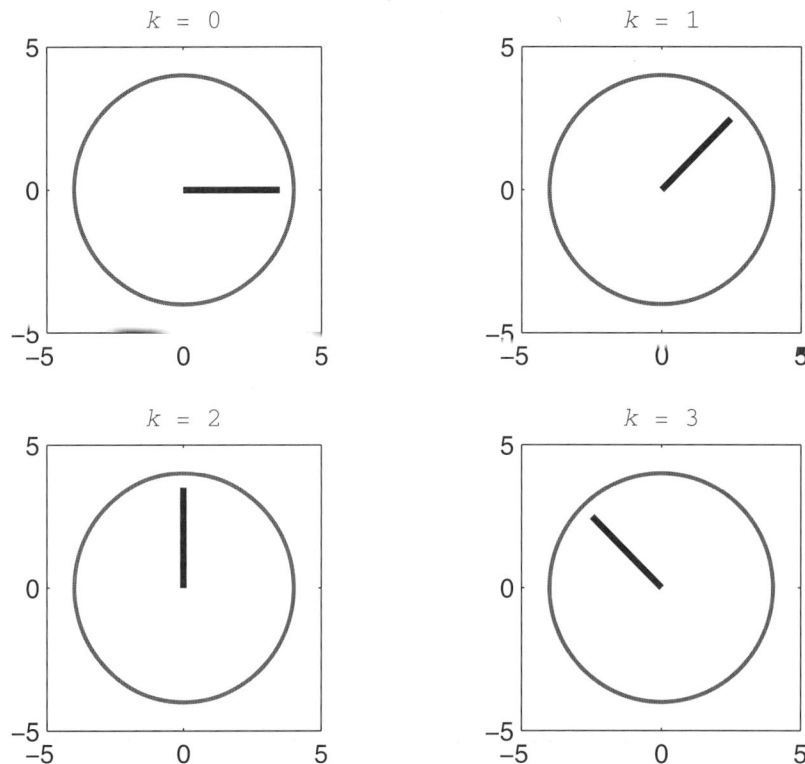

explanation is that the disk is actually rotating clockwise at a rate of 315 degrees/frame. Both interpretations are plausible. Is the motion captured by the snapshots a fast clockwise rotation or a slow counter clockwise rotation? If the disk is in fact rotating clockwise at F_0 revolutions/second, but the sampling rate is $f_s \leq 2F_0$, then aliasing occurs in which case the disk can appear to turn backwards at a slow rate. Interestingly, this manifestation of aliasing was quite common in older western films that featured wagon trains heading west. The spokes on the wagon wheels sometimes appeared to move backwards because of the slow frame rate used to shoot the film and display it on older TVs.

1.2 Signals and Systems

1.2.1 Signal Classification

Recall that a signal is a physical variable whose value varies with respect to time or space. To simplify the notation and terminology, we will assume that, unless noted otherwise, the independent variable denotes time. If the value of the signal, the dependent variable, is available over a continuum of times, $t \in R$, then the signal is referred to as a *continuous-time* signal. An example of a continuous-time signal, $x_a(t)$, is shown in Figure 1.8.

Continuous-time signal

In many cases of practical interest, the value of the signal is only available at discrete instants of time in which case it is referred to as a *discrete-time* signal. That is, signals can be classified into continuous-time or discrete-time depending on whether the independent variable is continuous or discrete, respectively. Common everyday examples of discrete-time signals include economic statistics such as the daily balance in one's savings account, or the monthly inflation rate. In DSP applications, a more common way to produce a discrete-time signal, $x(k)$, is to sample an underlying continuous-time signal, $x_a(t)$, as follows.

Discrete-time signal

$$x(k) = x_a(kT), \quad |k| = 0, 1, 2, \cdots \tag{1.2.1}$$

FIGURE 1.8: A Continuous-time Signal $x_a(t)$

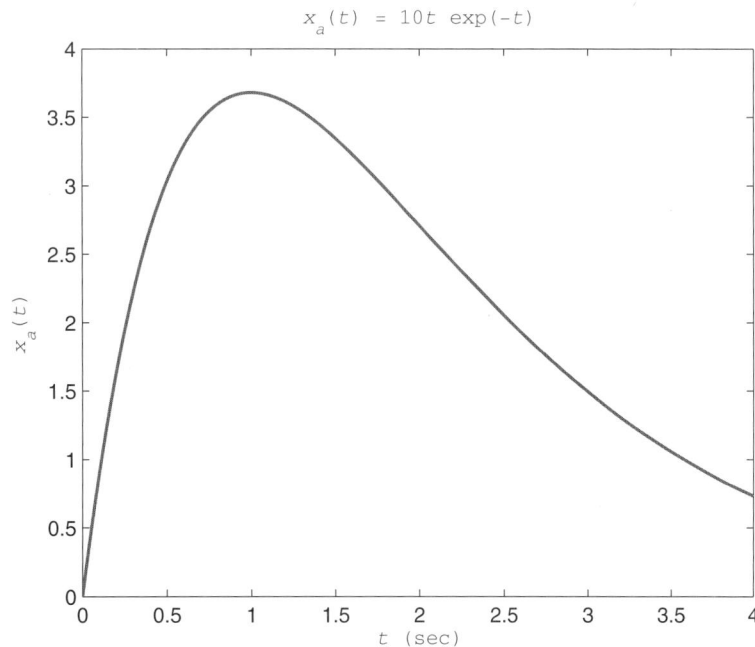

Sampling interval
Sampling frequency

Here, $T > 0$ is the time between samples or *sampling interval* in seconds. The sample spacing also can be specified using the reciprocal of the sampling interval which is called the *sampling frequency*, f_s.

$$f_s \triangleq \frac{1}{T} \text{ Hz} \tag{1.2.2}$$

Discrete-time

Here, the unit of Hz is understood to mean samples/second. Notice that the integer k in (1.2.1) denotes *discrete time* or, more specifically, the sample number. The sampling interval T is left implicit on the left-hand side of (1.2.1) because this simplifies subsequent notation. In those instances where the value of T is important, it will be stated explicitly. An example of a discrete-time signal generated by sampling the continuous-time signal in Figure 1.8 using $T = .25$ seconds is shown in Figure 1.9.

Quantized signal

Just as the independent variable can be continuous or discrete, so can the dependent variable or amplitude of the signal be continuous or discrete. If the number of bits of precision used to represent the value of $x(k)$ is finite, then we say that $x(k)$ is a *quantized* or discrete-amplitude signal. For example, if N bits are used to represent the value of $x(k)$, then there are 2^N distinct values that $x(k)$ can assume. Suppose the value of $x(k)$ ranges over the interval $[x_m, x_M]$. Then the *quantization level*, or spacing between adjacent discrete values of $x(k)$, is

Quantization level

$$q = \frac{x_M - x_m}{2^N} \tag{1.2.3}$$

Quantization operator

The quantization process can be thought of as passing a signal through a piecewise-constant staircase type function. For example, if the quantization is based on rounding to the nearest N bits, then the process can be represented with the following *quantization operator*.

$$Q_N(x) \triangleq q \cdot \text{round}\left(\frac{x}{q}\right) \tag{1.2.4}$$

Digital signal

A graph of $Q_N(x)$ for x ranging over the interval $[-1, 1]$ using $N = 5$ bits is shown in Figure 1.10. A quantized discrete-time signal is called a *digital* signal. That is, a digital signal,

FIGURE 1.9: A Discrete-time Signal $x(k)$ with $T = .25$

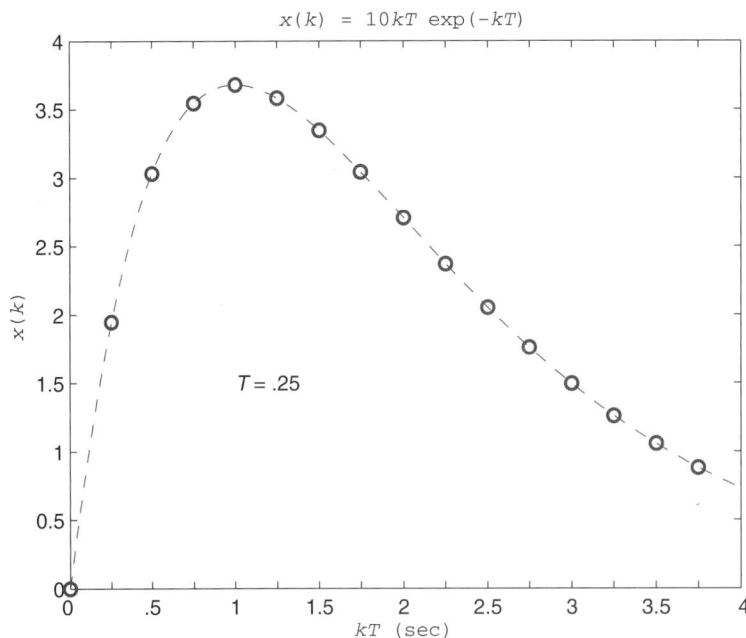

FIGURE 1.10:
Quantization over
$[-1, 1]$ Using $N = 5$
Bits

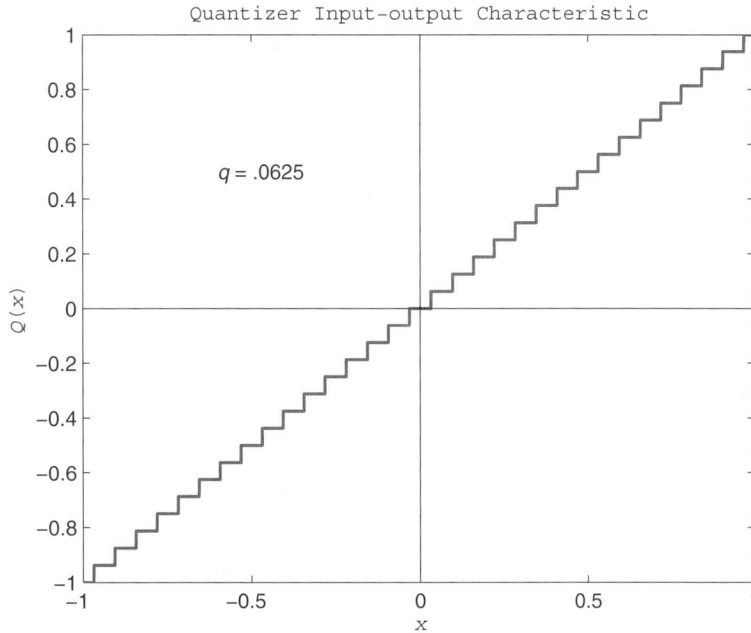

Quantizer Input-output Characteristic

$q = .0625$

$x_q(k)$, is discrete in both time and amplitude with

$$x_q(k) = Q_N[x_a(kT)] \tag{1.2.5}$$

Analog signal By contrast, a signal that is continuous in both time and amplitude is called an *analog* signal. An example of a digital signal obtained by quantizing the amplitude of the discrete-time signal in Figure 1.9 is shown in Figure 1.11. In this case, the 5-bit quantizer in Figure 1.10 is used to

FIGURE 1.11: A
Digital Signal $x_q(k)$

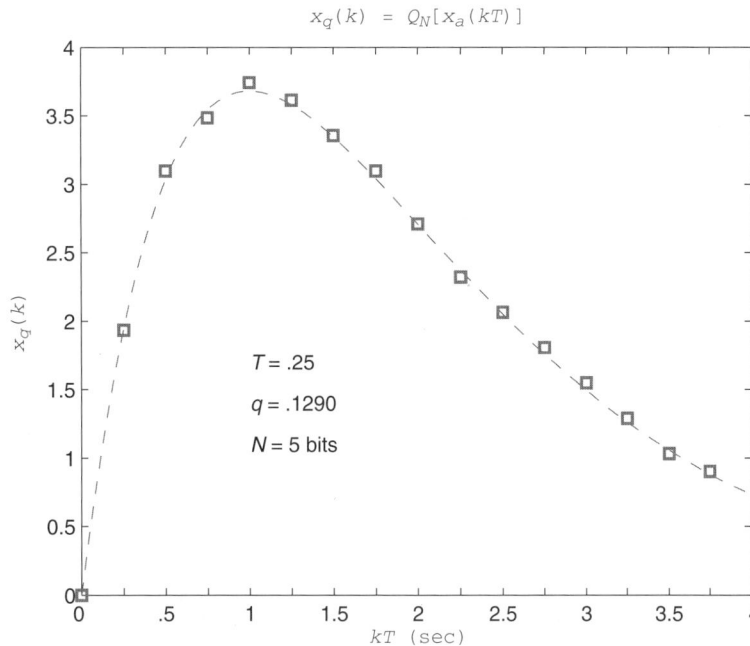

$x_q(k) = Q_N[x_a(kT)]$

$T = .25$

$q = .1290$

$N = 5$ bits

produce $x_q(k)$. Careful inspection of Figure 1.11 reveals that at some of the samples there are noticeable differences between $x_q(k)$ and $x_a(kT)$. If rounding is used, then the magnitude of the error is, at most, $q/2$.

Most of the analysis in this book will be based on discrete-time signals rather than digital signals. That is, infinite precision is used to represent the value of the dependent variable. Finite precision, or finite word length effects, are examined in Chapters 6 and 7 in the context of digital filter design. When digital filter are implemented in MATLAB using the default double-precision arithmetic, this corresponds to 64 bits of precision (16 decimal digits). In most instances this is sufficiently high precision to yield insignificant finite word length effects.

Quantization noise   A digital signal $x_q(k)$ can be modeled as a discrete-time signal $x(k)$ plus random *quantization noise*, $v(k)$, as follows.

$$x_q(k) = x(k) + v(k) \tag{1.2.6}$$

Expected value   An effective way to measure the size or strength of the quantization noise is to use average power defined as the mean, or *expected value*, of $v^2(k)$. Typically, $v(k)$ is modeled as a random variable uniformly distributed over the interval $[-q/2, q/2]$ with probability density $p(x) = 1/q$. In this case, the expected value of $v^2(k)$ is

$$E[v^2] = \int_{-q/2}^{q/2} p(x)x^2 dx$$

$$= \frac{1}{q} \int_{-q/2}^{q/2} x^2 dx \tag{1.2.7}$$

Thus, the average power of the quantization noise is proportional to the square of the quantization level with

$$E[v^2] = \frac{q^2}{12} \tag{1.2.8}$$

Example 1.1   **Quantization Noise**

Suppose the value of a discrete-time signal $x(k)$ is constrained to lie in the interval $[-10, 10]$. Let $x_q(k)$ denote a digital version of $x(k)$ using quantization level q, and consider the following problem. Suppose the average power of the quantization noise, $v(k)$, is to be less than .001. What is the minimum number of bits that are needed to represent the value of $x_q(k)$? The constraint on the average power of the quantization noise is

$$E[v^2] < .001$$

Thus, from (1.2.3) and (1.2.8), we have

$$\frac{(x_M - x_m)^2}{12(2^N)^2} < .001$$

Recall that the signal range is $x_m = -10$ and $x_M = 10$. Multiplying both sides by 12, taking the square root of both sides, and then solving for 2^N yields

$$2^N > \frac{20}{\sqrt{.012}}$$

Finally, taking the natural log of both sides and solving for N we have

$$N > \frac{\ln(182.5742)}{\ln(2)} = 7.5123$$

Since N must be an integer, the minimum number of bits needed to ensure that the average power of the quantization noise is less than .001 is $N = 8$ bits.

Signals can be further classified depending on whether or not they are nonzero for negative values of the independent variable.

DEFINITION
....
1.1: Causal Signal

A signal $x_a(t)$ defined for $t \in R$ is *causal* if and only if it is zero for negative t. Otherwise, the signal is *noncausal*.

$$x_a(t) = 0 \text{ for } t < 0$$

Most of the signals that we work with will be causal signals. A simple, but important, example of a causal signal is the *unit step* which is denoted $\mu_a(t)$ and defined

Unit step

$$\mu_a(t) \overset{\Delta}{=} \begin{cases} 0, & t < 0 \\ 1, & t \geq 0 \end{cases} \qquad (1.2.9)$$

Note that any signal can be made into a causal signal by multiplying by the unit step. For example, $x_a(t) = \exp(-t/\tau)\mu_a(t)$ is a causal decaying exponential with time constant τ.

Unit impulse

Another important example of a causal signal is the *unit impulse* which is denoted $\delta_a(t)$. Strictly speaking, the unit impulse is not a function because it is not defined at $t = 0$. However, the unit impulse can be defined implicitly by the equation

$$\int_{-\infty}^{t} \delta_a(\tau)d\tau = \mu_a(t) \qquad (1.2.10)$$

That is, the unit impulse $\delta_a(t)$ is a signal that, when integrated, produces the unit step $\mu_a(t)$. Consequently, we can loosely think of the unit impulse as the derivative of the unit step function, keeping in mind that the derivative of the unit step is not defined at $t = 0$. The two essential characteristics of the unit impulse that follow from (1.2.10) are

$$\delta_a(t) = 0, \quad t \neq 0 \qquad (1.2.11a)$$

$$\int_{-\infty}^{\infty} \delta_a(t)dt = 1 \qquad (1.2.11b)$$

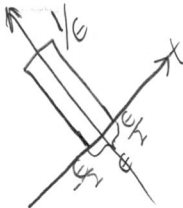

A more informal way to view the unit impulse is to consider a narrow pulse of width ϵ and height $1/\epsilon$ starting at $t = 0$. The unit impulse can be thought of as the limit of this sequence of pulses as the pulse width ϵ goes to zero. By convention, we graph the unit impulse as a vertical arrow with the height of the arrow equal to the *strength*, or area, of the impulse as shown in Figure 1.12.

The unit impulse has an important property that is a direct consequence of (1.2.11). If $x_a(t)$ is a continuous function, then

$$\int_{-\infty}^{\infty} x_a(\tau)\delta_a(\tau - t_0)d\tau = \int_{-\infty}^{\infty} x_a(t_0)\delta_a(\tau - t_0)d\tau$$

$$= x_a(t_0)\int_{-\infty}^{\infty} \delta_a(\tau - t_0)d\tau$$

$$= x_a(t_0)\int_{-\infty}^{\infty} \delta_a(\alpha)d\alpha \qquad (1.2.12)$$

FIGURE 1.12: Unit Impulse, $\delta_a(t)$, and Unit Step, $\mu_a(t)$

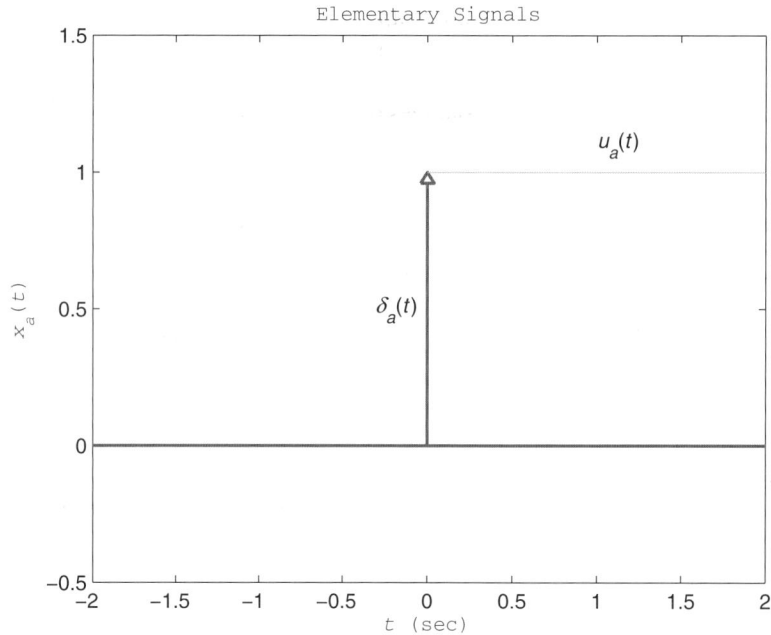

FIGURE 1.13: A System S with Input x and Output y

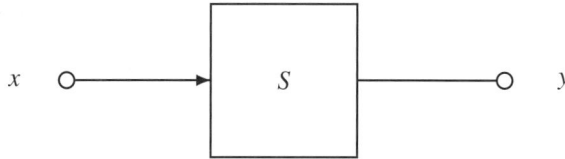

Sifting property

Since the area under the unit impulse is one, we then have the following *sifting property* of the unit impulse

$$\int_{-\infty}^{\infty} x_a(t)\delta_a(t - t_0)dt = x_a(t_0) \qquad (1.2.13)$$

From (1.2.13) we see that when a continuous function of time is multiplied by an impulse and then integrated, the effect is to sift out or sample the value of the function at the time the impulse occurs.

1.2.2 System Classification

Just as signals can be classified, so can the systems that process those signals. Consider a system S with *input x* and *output y* as shown in Figure 1.13. In some instances, for example biomedical systems, the input is referred to as the *stimulus*, and the output is referred to as the *response*. We can think of the system in Figure 1.13 as an operator S that acts on the input signal x to produce the output signal y.

$$y = Sx \qquad (1.2.14)$$

Continuous, discrete systems

If the input and output are continuous-time signals, then the system S is called a *continuous-time* system. A *discrete-time* system is a system S that processes a discrete-time input $x(k)$

to produce a discrete-time output $y(k)$. There are also examples of systems that contain both continuous-time signals and discrete-time signals. These systems are referred to as sampled-data systems.

Almost all of the examples of systems in this book belong to an important class of systems called linear systems.

DEFINITION
....
1.2: Linear System

Let x_1 and x_2 be arbitrary inputs and let a and b be arbitrary scalars. A system S is *linear* if and only if the following holds, otherwise it is a *nonlinear* system.

$$S(ax_1 + bx_2) = aSx_1 + bSx_2$$

Thus a linear system has two distinct characteristics. When $a = b = 1$, we see that the response to a sum of inputs is just the sum of the responses to the individual inputs. Similarly, when $b = 0$, we see that the response to a scaled input is just the scaled response to the original input. Examples of linear discrete-time systems include the notch filter in (1.1.8) and the adaptive filter in (1.1.9). On the other hand, if the analog audio amplifier in Figure 1.3 is over driven and its output saturates to produce harmonics as in (1.1.3), then this is an example of a nonlinear continuous-time system. Another important class of systems is time-invariant systems.

DEFINITION
....
1.3: Time-invariant
System

A system S with input $x_a(t)$ and output $y_a(t)$ is *time-invariant* if and only if whenever the input is translated in time by τ, the output is also translated in time by τ. Otherwise the system is a *time-varying* system.

$$Sx_a(t - \tau) = y_a(t - \tau)$$

For a time-invariant system, delaying or advancing the input delays or advances the output by the same amount, but it does not otherwise affect the shape of the output. Therefore the results of an input-output experiment do not depend on when the experiment is performed. Time-invariant systems described by differential or difference equations have *constant* coefficients. More generally, physical time-invariant systems have constant parameters. The notch filter in (1.1.8) is an example of a discrete-time system that is both linear and time-invariant. On the other hand, the adaptive digital filter in (1.1.9) is a time-varying system because the weights $w(k)$ are coefficients that change with time as the system adapts. The following example shows that the concepts of linearity and time-invariance can sometimes depend on how the system is characterized.

Example 1.2   **System Classification**

Consider the operational amplifier circuit shown in Figure 1.14. Here input resistor R_1 is fixed, but feedback resistor R_2 represents a sensor or transducer whose resistance changes with respect to a sensed environmental variable such as temperature or pressure. For this inverting amplifier configuration, the output voltage $y_a(t)$ is

$$y_a(t) = -\left[\frac{R_2(t)}{R_1}\right] x_1(t)$$

This is an example of a linear continuous-time system that is time-varying because parameter $R_2(t)$ varies as the temperature or pressure changes. However, another way to model this

FIGURE 1.14: An Inverting Amplifier with a Feedback Transducer

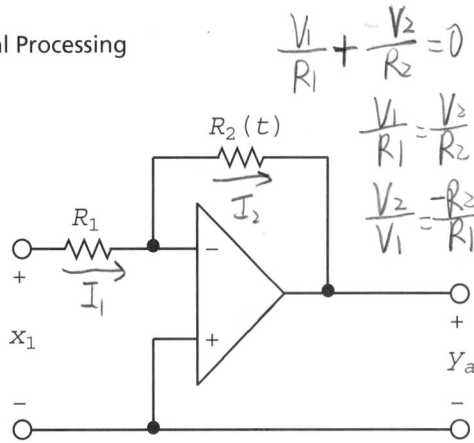

$$\frac{V_1}{R_1} + \frac{-V_2}{R_2} = 0$$

$$\frac{V_1}{R_1} = \frac{V_2}{R_2}$$

$$\frac{V_2}{V_1} = \frac{-R_2}{R_1}$$

system is to consider the variable resistance of the sensor as a second input $x_2(t) = R_2(t)$. Viewing the system in this way, the system output is

$$y_a(t) = -\frac{x_1(t)x_2(t)}{R_1}$$

This formulation of the model is a nonlinear time-invariant system, but with two inputs. Thus, by introducing a second input we have converted a single-input time-varying linear system to a two-input time-invariant nonlinear system.

Another important classification of systems focuses on the question of what happens to the signals as time increases. We say that a signal $x_a(t)$ is *bounded* if and only if there exists a $B_x > 0$ called a *bound* such that

Bounded signal

$$|x_a(t)| \le B_x \quad \text{for} \quad t \in R \tag{1.2.15}$$

DEFINITION
....
1.4: Stable System

A system S is with input $x_a(t)$ and output $y_a(t)$ is *stable*, in a bounded input bounded output (BIBO) sense, if and only if every bounded input produces a bounded output. Otherwise it is an *unstable* system.

Thus an unstable system is a system for which the magnitude of the output grows arbitrarily large with time for a least one bounded input.

Example 1.3   **Stability**

As a simple example of a system that can be stable or unstable depending on its parameter values, consider the following first-order linear continuous-time system where $a \ne 0$.

$$\frac{dy_a(t)}{dt} + ay_a(t) = x_a(t)$$

Suppose the input is the unit step $x_a(t) = \mu_a(t)$ which is bounded with a bound of $B_x = 1$. Direct substitution can be used to verify that for $t \ge 0$, the solution is

$$y_a(t) = y_a(0)\exp(-at) + \frac{1}{a}\left[1 - \exp(-at)\right]$$

If $a > 0$, then the exponential terms grow without bound which means that the bounded input $u_a(t)$ produces an unbounded output $y_a(t)$. Thus this system is unstable, in a BIBO sense, when $a > 0$.

Just as light can be decomposed into a spectrum of colors, signals also contain energy that is distributed over a range of frequencies. To decompose a continuous-time signal $x_a(t)$ into its spectral components, we use the *Fourier transform*.

Fourier transform

$$X_a(f) = F\{x_a(t)\} \triangleq \int_{-\infty}^{\infty} x_a(t)\exp(-j2\pi ft)dt \qquad (1.2.16)$$

It is assumed that the reader is familiar with the basics of continuous-time transforms, specifically the Laplace transform and the Fourier transform. Tables of transform pairs and transform properties for all of the tranforms used in this text can be found in Appendix 1. Here, $f \in R$ denotes frequency in cycles/sec or Hz. In general the Fourier transform, $X_a(f)$, is complex. As such, it can be expressed in *polar form* in terms of its magnitude $A_a(f) = |X_a(f)|$ and phase angle $\phi_a(f) = \angle X_a(f)$ as follows.

Polar form

$$X_a(f) = A_a(f)\exp[j\phi_a(f)] \qquad (1.2.17)$$

Magnitude, phase spectrum

The real-valued function $A_a(f)$ is called the *magnitude spectrum* of $x_a(t)$, while the real-valued function $\phi_a(f)$ is called the *phase spectrum* of $x_a(t)$. More generally, $X_a(f)$ itself is called the *spectrum* of $x_a(t)$. For a real $x_a(t)$, the magnitude spectrum is an even function of f, and the phase spectrum is an odd function of f.

When a signal passes through a linear system, the shape of its spectrum changes. Systems designed to reshape the spectrum in a particular way are called *filters*. The effect that a linear system has on the spectrum of the input signal can be characterized by the frequency response.

Filters

DEFINITION

1.5: Frequency Response

Let S be a stable linear time-invariant continuous-time system with input $x_a(t)$ and output $y_a(t)$. Then the *frequency response* of the system S is denoted $H_a(f)$ and defined

$$H_a(f) \triangleq \frac{Y_a(f)}{X_a(f)}$$

Thus the frequency response of a linear system is just the Fourier transform of the output divided by the Fourier transform of the input. Since $H_a(f)$ is complex, it can be represented by its magnitude $A_a(f) = |H_a(f)|$ and its phase angle $\phi_a(f) = \angle H_a(f)$ as follows

$$H_a(f) = A_a(f)\exp[j\phi_a(f)] \qquad (1.2.18)$$

Magnitude, phase response

The function $A_a(f)$ is called the *magnitude response* of the system, while $\phi_a(f)$ is called the *phase response* of the system. The magnitude response indicates how much each frequency component of $x_a(t)$ is scaled as it passes through the system. That is, $A_a(f)$ is the *gain* of the system at frequency f. Similarly, the phase response indicates how much each frequency component of $x_a(t)$ gets advanced in phase by the system. That is, $\phi_a(f)$ is the *phase shift* of the system at frequency f. Therefore, if the input to the stable system is a pure sinusoidal tone $x_a(t) = \sin(2\pi F_0 t)$, the steady-state output of the stable system is

$$y_a(t) = A_a(F_0)\sin[2\pi F_0 t + \phi_a(F_0)] \qquad (1.2.19)$$

The magnitude response of a real system is an even function of f, while the phase response is an odd function of f. This is similar to the magnitude and phase spectra of a real signal. Indeed, there is a simple relationship between the frequency response of a system and the spectrum of a signal. To see this, consider the impulse response.

DEFINITION

1.6: Impulse Response

Suppose the initial condition of a continuous-time system S is zero. Then the output of the system corresponding to the unit impulse input is denoted $h_a(t)$ and is called the system *impulse response*.

$$h_a(t) = S\delta_a(t)$$

From the sifting property of the unit impulse in (1.2.13) one can show that the Fourier transform of the unit impulse is simply $\Delta_a(f) = 1$. It then follows from Definition 1.5 that when the input is the unit impulse, the Fourier transform of the system output is $Y_a(f) = H_a(f)$. That is, an alternative way to represent the frequency response is as the Fourier transform of the impulse response.

$$H_a(f) = F\{h_a(t)\} \tag{1.2.20}$$

In view of (1.2.17), the magnitude response of a system is just the magnitude spectrum of the impulse response, and the phase response is just the phase spectrum of the impulse response. It is for this reason that the same symbol, $A_a(f)$, is used to denote both the magnitude spectrum of a signal and the magnitude response of a system. A similar remark holds for $\phi_a(f)$ which is used to denote both the phase spectrum of a signal and the phase response of a system.

Example 1.4

Ideal lowpass filter

Ideal Lowpass Filter

An important example of a continuous-time system is the ideal lowpass filter. An *ideal lowpass filter* with cutoff frequency B Hz, is a system whose frequency response is the following pulse of height one and radius B centered at $f = 0$.

$$\rho_B(f) \triangleq \begin{cases} 1, & |f| \le B \\ 0, & |f| > B \end{cases}$$

A plot of the ideal lowpass frequency response is shown in Figure 1.15.

Recall from Definition 1.5 that $Y_a(f) = H_a(f)X_a(f)$. Consequently, the filter in Figure 1.15 passes the frequency components of $x_a(t)$ in the range $[-B, B]$ through the filter without any distortion whatsoever, not even any phase shift. Furthermore, the remaining frequency components of $x_a(t)$ outside the range $[-B, B]$ are completely eliminated by the filter. The idealized nature of the filter becomes apparent when we look at the impulse response of the filter. To compute the impulse response from the frequency response we must apply the inverse Fourier transform. Using the table of Fourier transform pairs in Appendix 1, this yields

$$h_a(t) = 2B \cdot \text{sinc}(2Bt)$$

FIGURE 1.15: Frequency Response of Ideal Lowpass Filter

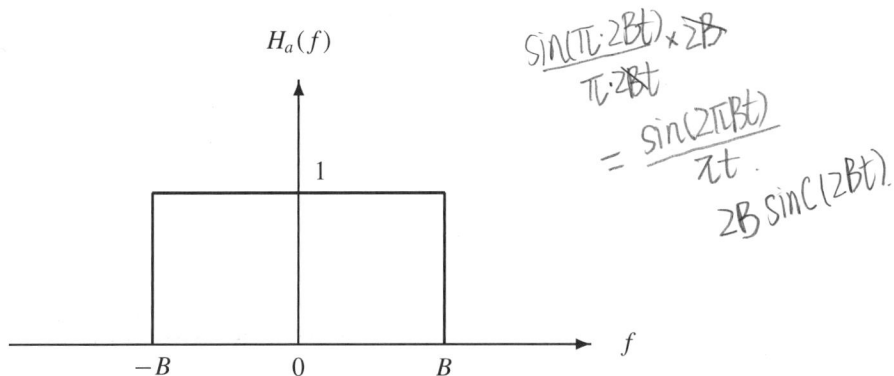

$$\frac{\sin(\pi \cdot 2Bt) \times 2B}{\pi \cdot 2Bt}$$

$$= \frac{\sin(2\pi Bt)}{\pi t}$$

$$2B \, \text{sinc}(2Bt)$$

FIGURE 1.16: Impulse Response of Ideal Lowpass Filter when $B = 100$ Hz

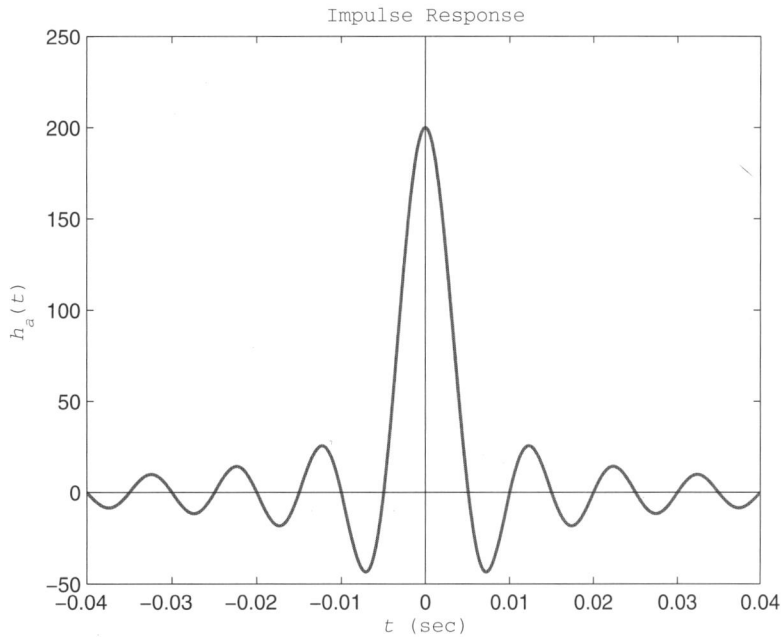

Here the normalized *sinc* function is defined as follows.

$$\checkmark \quad \text{sinc}(x) \overset{\Delta}{=} \frac{\sin(\pi x)}{\pi x}$$

Sinc function

The sinc function is a two-sided decaying sinusoid that is confined to the envelope $1/(\pi x)$. Thus $\text{sinc}(k) = 0$ for $k \neq 0$. The value of $\text{sinc}(x)$ at $x = 0$ is determined by applying L'Hospital's rule which yields $\text{sinc}(0) = 1$. Some authors define the sinc function as $\text{sinc}(x) = \sin(x)/x$. The impulse response of the ideal lowpass filter is $\text{sinc}(2BT)$ scaled by $2B$. A plot of the impulse response for the case $B = 100$ Hz is shown in Figure 1.16.

Notice that the sinc function, and therefore the impulse response, is not a causal signal. But $h_a(t)$ is the filter output when a unit impulse input is applied at time $t = 0$. Consequently, for the ideal filter we have a causal input producing a noncausal output. This is not possible for a physical system. Therefore, the frequency response in Figure 1.15 cannot be realized with physical hardware. In Section 1.4, we examine some lowpass filters that are physically realizable that can be used to approximate the ideal frequency response characteristic.

● ● ● ● ● ● ● ● ● ● ● ● ● ● ● ● ● ●

1.3 Sampling of Continuous-time Signals

1.3.1 Sampling as Modulation

Periodic impulse train

The process of sampling a continuous-time signal $x_a(t)$ to produce a discrete-time signal $x(k)$ can be viewed as a form of amplitude modulation. To see this, let $\delta_T(t)$ denote a periodic train of impulses of period T.

$$\delta_T(t) \overset{\Delta}{=} \sum_{k=-\infty}^{\infty} \delta_a(t - kT) \tag{1.3.1}$$

FIGURE 1.17:
Sampling as
Amplitude
Modulation of an
Impulse Train

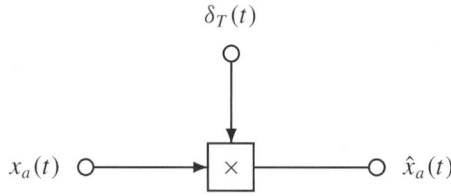

Sampled signal

Thus $\delta_T(t)$ consists of unit impulses at integer multiples of the sampling interval T. The *sampled version* of signal $x_a(t)$ is denoted $\hat{x}_a(t)$, and is defined as the following product.

$$\hat{x}_a(t) \stackrel{\Delta}{=} x_a(t)\delta_T(t) \tag{1.3.2}$$

Amplitude modulation

Since $\hat{x}_a(t)$ is obtained from $x_a(t)$ by multiplication by a periodic signal, this process is a form of *amplitude modulation* of $\delta_T(t)$. In this case $\delta_T(t)$ plays a role similar to the high-frequency carrier wave in AM radio, and $x_a(t)$ represents the low-frequency information signal. A block diagram of the impulse model of sampling is shown in Figure 1.17.

Using the basic properties of the unit impulse in (1.2.11), the sampled version of $x_a(t)$ can be written as follows.

$$\hat{x}_a(t) = x_a(t)\delta_T(t)$$

$$= x_a(t) \sum_{k=-\infty}^{\infty} \delta_a(t - kT)$$

$$= \sum_{k=-\infty}^{\infty} x_a(t)\delta_a(t - kT)$$

$$= \sum_{k=-\infty}^{\infty} x_a(kT)\delta_a(t - kT) \tag{1.3.3}$$

Thus the sampled version of $x_a(t)$ is the following amplitude modulated impulse train.

$$\hat{x}_a(t) = \sum_{k=-\infty}^{\infty} x(k)\delta_a(t - kT) \tag{1.3.4}$$

Whereas $\delta_T(t)$ is a constant-amplitude or uniform train of impulses, $\hat{x}_a(t)$ is a nonuniform impulse train with the area of the kth impulse equal to sample $x(k)$. A graph illustrating the relationship between $\delta_T(t)$ and $\hat{x}_a(t)$ for the case $x_a(t) = 10t \exp(-t)u_a(t)$ is shown in Figure 1.18.

It is useful to note from (1.3.4) that $\hat{x}_a(t)$ is actually a continuous-time signal, rather than a discrete-time signal. However, it is a very special continuous-time signal in that it is zero everywhere except at the samples where it has impulses whose areas correspond to the sample values. Consequently, there is a simple one-to-one relationship between the continuous-time signal $\hat{x}_a(t)$ and the discrete-time signal $x(k)$. If $\hat{x}_a(t)$ is a causal continuous-time signal, we can apply the Laplace transform to it. The *Laplace transform* of a causal continuous-time signal $x_a(t)$ is denoted $X_a(s)$ and is defined

Laplace transform

$$X_a(s) = L\{x_a(t)\} \stackrel{\Delta}{=} \int_0^\infty x_a(t) \exp(-st)\, dt \tag{1.3.5}$$

It is assumed that the reader is familiar with the basics of the Laplace transform. Tables of common Laplace transform pairs and Laplace transform properties can be found in Appendix 1. Comparing (1.3.5) with (1.2.16) it is clear that for causal signals, the Fourier transform is just the Laplace transform, but with the complex variable s replaced by $j2\pi f$. Consequently, the

FIGURE 1.18: Periodic Impulse Train in (a) and Sampled Version of $x_a(t)$ in (b) Using Impulse Sampling

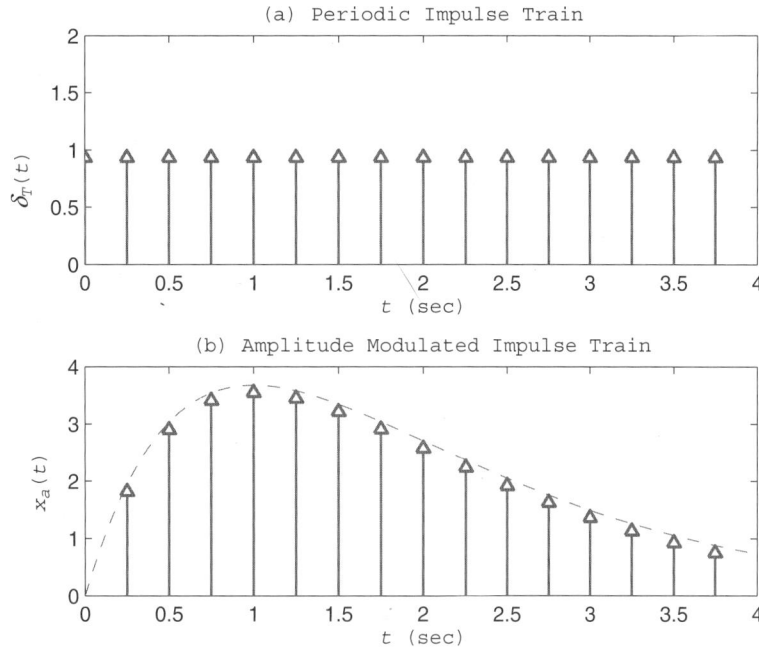

(a) Periodic Impulse Train

(b) Amplitude Modulated Impulse Train

spectrum of a causal signal can be obtained from its Laplace transform as follows.

$$X_a(f) = X_a(s)|_{s=j2\pi f} \tag{1.3.6}$$

note

At this point a brief comment about notation is in order. Note that the same *base* symbol, X_a, is being used to denote both the Laplace transform, $X_a(s)$, in (1.3.5), and the Fourier transform, $X_a(f)$, in (1.2.16). Clearly, an alternative approach would be to introduce distinct symbols for each. However, the need for additional symbols will arise repeatedly in subsequent chapters, so using separate symbols in each case quickly leads to a proliferation of symbols that can be confusing in its own right. Instead, the notational convention adopted here is to rely on the *argument type*, a complex s or a real f, to distinguish between the two cases and dictate the meaning of X_a. The subscript a denotes a continuous-time or *analog* quantity. The less cumbersome X without a subscript, is reserved for discrete-time quantities introduced later.

If the periodic impulse train $\delta_T(t)$ is expanded into a complex Fourier series, the result can be substituted into the definition of $\hat{x}_a(t)$ in (1.3.2). Taking the Laplace transform of $\hat{x}_a(t)$ and converting the result using (1.3.6), we then arrive at the following expression for the spectrum of the sampled version of $x_a(t)$.

$$\hat{X}_a(f) = \frac{1}{T} \sum_{i=-\infty}^{\infty} X_a(f - if_s) . \tag{1.3.7}$$

1.3.2 Aliasing

Aliasing formula

The representation of the spectrum of the sampled version of $x_a(t)$ depicted in (1.3.7) is called the *aliasing formula*. The aliasing formula holds the key to determining conditions under which the samples $x(k)$ contain all the information necessary to completely reconstruct or recover $x_a(t)$ from the samples. To see this, we first consider the notion of a bandlimited signal.

Typically B is chosen to be as small as possible. Thus if $x_a(t)$ is bandlimited to B, then the highest frequency component present in $x_a(t)$ is B Hz. It should be noted that some authors use a slightly different definition of the term bandlimited by replacing the strict inequality in Definition 1.7 with $|f| \geq B$.

DEFINITION

1.7: Bandlimited Signal

A continuous-time signal $x_a(t)$ is *bandlimited* to bandwidth B if and only if its magnitude spectrum satisfies

$$|X_a(f)| = 0 \quad \text{for} \quad |f| > B$$

The aliasing formula in (1.3.7) is quite revealing when it is applied to bandlimited signals. Notice that the aliasing formula says that the spectrum of the sampled version of a signal is just a sum of scaled and shifted spectra of the original signal with the replicated versions of $X_a(f)$ centered at integer multiples of the sampling frequency f_s. This is a characteristic of amplitude modulation in general where the unshifted spectrum ($i = 0$) is called the *base band*

Base, side bands   and the shifted spectra ($i \neq 0$) are called *side bands*. An illustration comparing the magnitude spectra of $x_a(t)$ and $\hat{x}_a(t)$ is shown in Figure 1.19.

Undersampling   The case shown in Figure 1.19 corresponds to $f_s = 3B/2$ and is referred to as *undersampling* because $f_s \leq 2B$. The details of the shape of the even function $|X_a(f)|$ within $[-B, B]$ are not important, so for convenience a triangular spectrum is used. Note how the sidebands in Figure 1.19b overlap with each other and with the baseband. This overlap is an indication of

Aliasing   an undesirable phenomenon called *aliasing*. As a consequence of the overlap, the shape of the spectrum of $\hat{x}_a(t)$ in $[-B, B]$ has been altered and is different from the shape of the spectrum of $x_a(t)$ in Figure 1.19a. The end result is that no amount of *signal-independent* filtering of $\hat{x}_a(t)$ will allow us to recover the spectrum of $x_a(t)$ from the spectrum of $\hat{x}_a(t)$. That is, the overlap or aliasing has caused the samples to be *corrupted* to the point that the original signal $x_a(t)$ can no longer be recovered from the samples. Since $x_a(t)$ is bandlimited, it is evident

FIGURE 1.19:
Magnitude Spectra of $x_a(t)$ in (a) and $\hat{x}_a(t)$ in (b) when $B = 100$, $f_s = 3B/2$

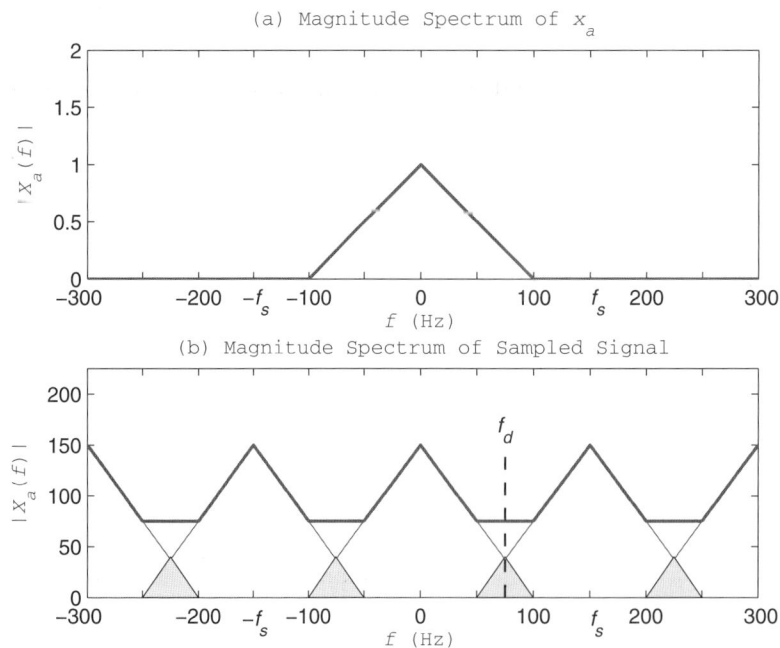

that there will be no aliasing if the sampling rate is sufficiently high. This fundamental result is summarized in the Shannon sampling theorem.

PROPOSITION
....
1.1: Signal Sampling

Suppose a continuous-time signal $x_a(t)$ is bandlimited to B Hz. Let $\hat{x}_a(t)$ denote the sampled version of $x_a(t)$ using impulse sampling with a sampling frequency of f_s. Then the samples $x(k)$ contain all the information necessary to recover the original signal $x_a(t)$ if

$$f_s > 2B$$

In view of the sampling theorem, it should be possible to reconstruct a continuous-time signal from its samples if the signal is bandlimited and the sampling frequency exceeds twice the bandwidth. When $f_s > 2B$, the sidebands of $\hat{X}_a(f)$ do not overlap with each other or the baseband. By properly filtering $\hat{X}_a(f)$ it should be possible to recover the baseband and rescale it to produce $X_a(f)$. Before we consider how to do this, it is of interest to see what happens in the time domain when aliasing occurs due to an inadequate sampling rate. If aliasing occurs, it means that there is another lower-frequency signal that will produce identical samples. Among all signals that generate a given set of samples, there is only one signal that is bandlimited to less than half the sampling rate. All other signals that generate the same samples are high-frequency *impostors* or aliases. The following example illustrates this point.

Impostors

Example 1.5   ### Aliasing

The simplest example of a bandlimited signal is a pure sinusoidal tone that has all its power concentrated at a single frequency F_0. For example, consider the following signal where $F_0 = 90$ Hz.

$$x_a(t) = \sin(180\pi t)$$

From the Fourier transform pair table in Appendix 1, the spectrum of $x_a(t)$ is

$$X_a(f) = \frac{j[\delta(f+90) - \delta(f-90)]}{2}$$

Thus $x_a(t)$ is a bandlimited signal with bandwidth $B = 90$ Hz. From the sampling theorem, we need $f_s > 180$ Hz to avoid aliasing. Suppose $x_a(t)$ is sampled at the rate $f_s = 100$ Hz. In this case $T = .01$ seconds, and the samples are

$$
\begin{aligned}
x(k) &= x_a(kT) \\
&= \sin(180\pi kT) \\
&= \sin(1.8\pi k) \\
&= \sin(2\pi k - .2\pi k) \\
&= \sin(2\pi k)\cos(.2\pi k) - \cos(2\pi k)\sin(.2\pi k) \\
&= -\sin(.2\pi k) \\
&= -\sin(20\pi kT)
\end{aligned}
$$

Thus the samples of the 90 Hz signal $x_a(t) = \sin(180\pi t)$ are identical to the samples of the following lower-frequency signal that has its power concentrated at 10 Hz.

$$x_b(t) = -\sin(20\pi t)$$

A plot comparing the two signals $x_a(t)$ and $x_b(t)$ and their shared samples is shown in Figure 1.20.

FIGURE 1.20:
Common Samples
of Two Bandlimited
Signals

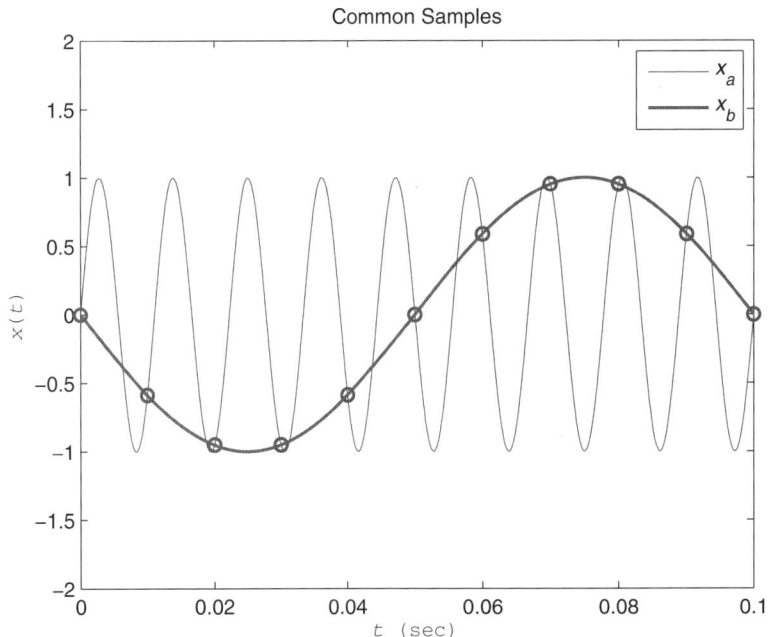

The existence of a lower-frequency signal associated with the samples, $x(k)$, can be predicted directly from the aliasing formula. Indeed, a simple way to interpret (1.3.7) is to introduce something called the *folding* frequency.

Folding frequency

$$f_d \triangleq \frac{f_s}{2} \tag{1.3.8}$$

Thus the folding frequency is simply one half of the sampling frequency. If $x_a(t)$ has any frequency components outside of f_d, then in $\hat{x}_a(t)$ these frequencies get reflected about f_d and *folded back* into the range $[-f_d, f_d]$. For the case in Example 1.5, $f_d = 50$ Hz. Thus the original frequency component at $F_0 = 90$ Hz, gets reflected about f_d to produce a frequency component at 10 Hz. Notice that in Figure 1.19 the folding frequency is at the center of the first region of overlap. The part of the spectrum of $x_a(t)$ that lies outside of the folding frequency gets aliased back into the range $[-f_d, f_d]$ as a result of the overlap.

● ● ● ● ● ● ● ● ● ● ● ● ● ● ● ●

1.4   Reconstruction of Continuous-time Signals

1.4.1 Reconstruction Formula

When the signal $x_a(t)$ is bandlimited and the sampling rate f_s is higher than twice the bandwidth, the samples $x(k)$ contain all the information needed to reconstruct $x_a(t)$. To illustrate the reconstruction process, consider the bandlimited signal whose magnitude spectrum is shown in Figure 1.21a. In this case we have *oversampled* by selecting a sampling frequency that is three times the bandwidth, $B = 100$ Hz. The magnitude spectrum of $\hat{x}_a(t)$ is shown in Figure 1.21b.

FIGURE 1.21:
Magnitude Spectra
of $x_a(t)$ in (a) and
$\hat{x}_a(t)$ in (b) when
$B = 100$, $f_s = 3B$

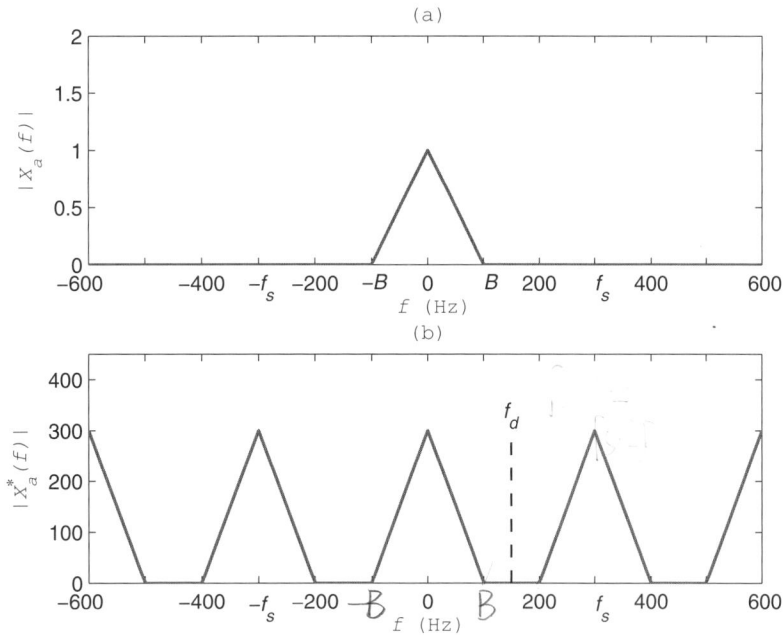

Note how the increase in f_s beyond $2B$ has caused the sidebands to spread out so they no longer overlap with each other or the baseband. In this case there are no spectral components of $x_a(t)$ beyond the folding frequency $f_d = f_s/2$ to be aliased back into the range $[-f_d, f_d]$.

The problem of reconstructing the signal $x_a(t)$ from $\hat{x}_a(t)$ reduces to one of recovering the spectrum $X_a(f)$ from the spectrum $\hat{X}_a(f)$. This can be achieved by passing $\hat{x}_a(t)$ through an ideal lowpass reconstruction filter $H_{\text{ideal}}(f)$ that removes the side bands and rescales the base band. The required frequency response of the reconstruction filter is shown in Figure 1.22. To remove the side bands, the cutoff frequency of the filter should be set to the folding frequency. From the aliasing formula in (1.3.7), the gain of the filter needed to rescale the baseband is T. Thus the required ideal lowpass frequency response is

$$H_{\text{ideal}}(f) \triangleq \begin{cases} T, & |f| \le f_d \\ 0, & |f| > f_d \end{cases} \tag{1.4.1}$$

From the aliasing formula in (1.3.7), one can recover the spectrum of $x_a(t)$ as follows.

$$X_a(f) = H_{\text{ideal}}(f)\hat{X}_a(f) \tag{1.4.2}$$

FIGURE 1.22:
Frequency
Response of Ideal
Lowpass
Reconstruction
Filter

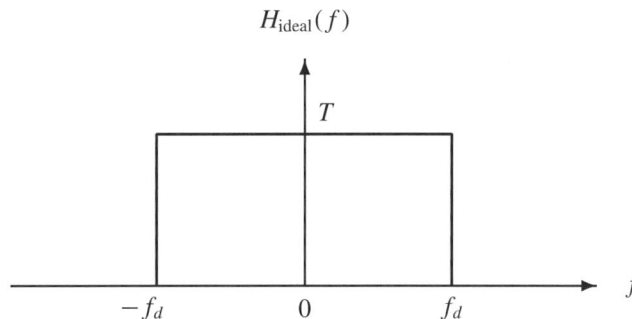

Using (1.2.20) and the table of Fourier transform pairs in Appendix 1, the impulse response of the ideal reconstruction filter is

$$
\begin{aligned}
h_{\text{ideal}}(t) &= F^{-1}\{H_{\text{ideal}}(f)\} \\
&= 2Tf_d \operatorname{sinc}(2f_d t) \\
&= \operatorname{sinc}(f_s t)
\end{aligned}
\tag{1.4.3}
$$

Next, we take the inverse Fourier transform of both sides of (1.4.2). Using (1.3.4), the sifting property of the unit impulse, and the convolution property of the Fourier transform (Appendix 1), we have

$$
\begin{aligned}
x_a(t) &= F^{-1}\{H_{\text{ideal}}(f)\hat{X}_a(f)\} \\
&= \int_{-\infty}^{\infty} h_{\text{ideal}}(t-\tau)\hat{x}_a(\tau)d\tau \\
&= \int_{-\infty}^{\infty} h_{\text{ideal}}(t-\tau)\sum_{k=-\infty}^{\infty} x(k)\delta_a(\tau-kT)d\tau \\
&= \sum_{k=-\infty}^{\infty} x(k)\int_{-\infty}^{\infty} h_{\text{ideal}}(t-\tau)\delta_a(\tau-kT)d\tau \\
&= \sum_{k=-\infty}^{\infty} x(k)h_{\text{ideal}}(t-kT)
\end{aligned}
\tag{1.4.4}
$$

Finally, substituting (1.4.3) into (1.4.4) yields the following formulation called the Shannon reconstruction formula.

PROPOSITION

1.2: Signal Reconstruction

Suppose a continuous-time signal $x_a(t)$ is bandlimited to B Hz. Let $x(k) = x_a(kT)$ be the kth sample of $x_a(t)$ using a sampling frequency of $f_s = 1/T$. If $f_s > 2B$, then $x_a(t)$ can be reconstructed from $x(k)$ as follows.

$$
x_a(t) = \sum_{k=-\infty}^{\infty} x(k)\operatorname{sinc}[f_s(t-kT)]
$$

The Shannon reconstruction formula is an elegant result that is valid as long as $x_a(t)$ is bandlimited to B and $f_s > 2B$. Note that the *sinc* function is used to interpolate between the samples. The importance of the reconstruction formula is that it demonstrates that all the essential information about $x_a(t)$ is contained in the samples $x(k)$ as long as $x_a(t)$ is bandlimited and the sampling rate exceeds twice the bandwidth.

Example 1.6

Signal Reconstruction

Consider the following signal. For what range of values of the sampling interval T can this signal be reconstructed from its samples?

$$
x_a(t) = \sin(5\pi t)\cos(3\pi t)
$$

The signal $x_a(t)$ does not appear in the table of Fourier transform pairs in Appendix 1. However, using the trigonometric identities in Appendix 2 we have

$$
x_a(t) = \frac{\sin(8\pi t) + \sin(2\pi t)}{2}
$$

Since $x_a(t)$ is the sum of two sinusoids and the Fourier transform is linear, it follows that $x_a(t)$ is bandlimited with a bandwidth of $B = 4$ Hz. From the sampling theorem we can reconstruct

$x_a(t)$ from its samples if $f_s > 2B$ or $1/T > 8$. Hence the range of sampling intervals over which $x_a(t)$ can be reconstructed from $x(k)$ is

$$0 < T < .125 \text{ sec}$$

1.4.2 Zero-order Hold

Exact reconstruction of $x_a(t)$ from its samples requires an ideal filter. One can approximately reconstruct $x_a(t)$ using a practical filter. We begin by noting that an efficient way to characterize a linear time-invariant continuous-time system in general is in term of its transfer function.

DEFINITION

1.8: Transfer Function

Let $x_a(t)$ be a causal nonzero input to a continuous-time linear system, and let $y_a(t)$ be the corresponding output assuming zero initial conditions. Then the *transfer function* of the system is defined

$$H_a(s) \triangleq \frac{Y_a(s)}{X_a(s)}$$

Thus the transfer function is just the Laplace transform of the output divided by the Laplace transform of the input assuming zero initial conditions. A table of common Laplace transform pairs can be found in Appendix 1. From the sifting property of the unit impulse in (1.2.13), the Laplace transform of the unit impulse is $\Delta_a(s) = 1$. In view of Definition 1.8, this means that an alternative way to characterize the transfer function is to say that $H_a(s)$ is the Laplace transform of the impulse response, $h_a(t)$.

$$H_a(s) = L\{h_a(t)\} \tag{1.4.5}$$

note

Again note that the same *base* symbol, H_a, is being used to denote both the transfer function, $H_a(s)$, in Definition 1.8, and the frequency response, $H_a(f)$, in Definition 1.5. The notational convention adopted here and throughout the text is to rely on the *argument type*, a complex s or a real f, to distinguish between the two cases and dictate the meaning of H_a.

Example 1.7

Transportation Lag

As an illustration of a continuous-time system and its transfer function, consider a system which delays the input by τ seconds.

$$y_a(t) = x_a(t - \tau)$$

This type of system might be used, for example, to model a transportation lag in a process control system or a signal propagation delay in a telecommunication system. Using the definition of the Laplace transform in (1.3.5), and the fact that $x_a(t)$ is causal, we have

$$\begin{aligned}
Y_a(s) &= L\{x_a(t - \tau)\} \\
&= \int_0^\infty x_a(t - \tau)\exp(-st)dt \\
&= \int_{-\tau}^\infty x_a(\alpha)\exp[-s(\alpha + \tau)]d\alpha \quad \} \alpha = t - \tau \\
&= \exp(-s\tau)\int_0^\infty x_a(\alpha)\exp(-s\alpha)d\alpha \\
&= \exp(-s\tau)X_a(s)
\end{aligned}$$

FIGURE 1.23:
Transfer Function
of Transportation
Lag with Delay τ

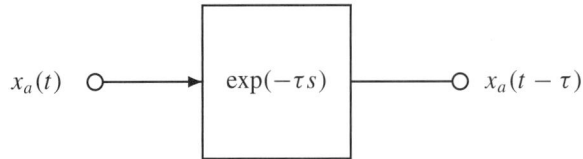

$$x_a(t) \quad\bigcirc\!\!\longrightarrow\!\! \boxed{\exp(-\tau s)} \!\!\longrightarrow\!\!\bigcirc\quad x_a(t-\tau)$$

It then follows from Definition 1.8 that the transfer function of a transportation lag with delay τ is

$$H_a(s) = \exp(-\tau s)$$

A block diagram of the transportation lag is shown in Figure 1.23.

Zero-order hold

The reconstruction of $x_a(t)$ in Proposition 1.2 interpolates between the samples using the sinc function. A simpler form of interpolation is to use a low degree polynomial fitted to the samples. To that end, consider the following linear system with a delay called a *zero-order hold*.

$$y_a(t) = \int_0^t [x_a(\tau) - x_a(\tau - T)]d\tau \tag{1.4.6}$$

Recalling that the integral of the unit impulse is the unit step, we find that the impulse response of this system is

$$\begin{aligned} h_0(t) &= \int_0^t [\delta_a(\tau) - \delta_a(\tau - T)]d\tau \\ &= \int_0^t \delta_a(\tau)d\tau - \int_0^t \delta_a(\tau - T)d\tau \\ &= \mu_a(t) - \mu_a(t - T) \end{aligned} \tag{1.4.7}$$

Thus the impulse response of the zero-order hold is the pulse of unit height and width T starting at $t = 0$ as shown in Figure 1.24.

FIGURE 1.24:
Impulse Response
of Zero-order Hold
Filter

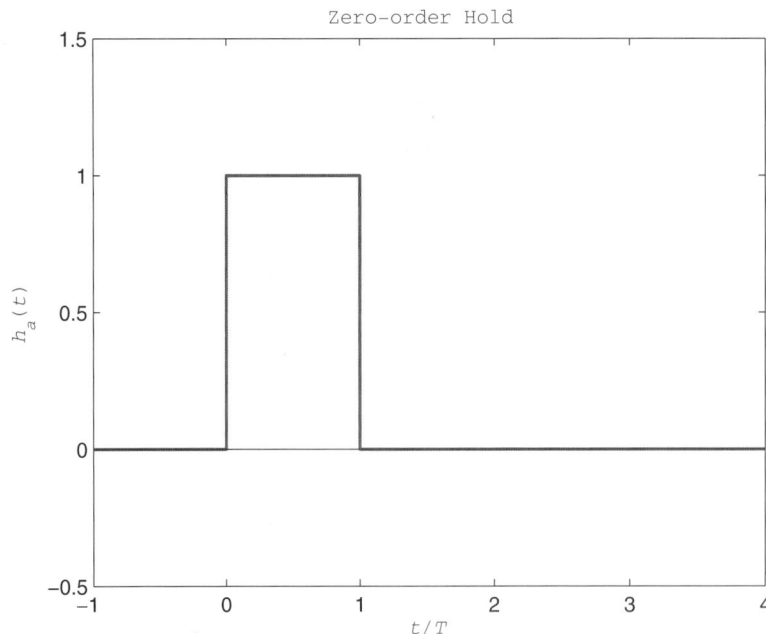

FIGURE 1.25:
Reconstruction of
$x_a(t)$ with a
Zero-order Hold

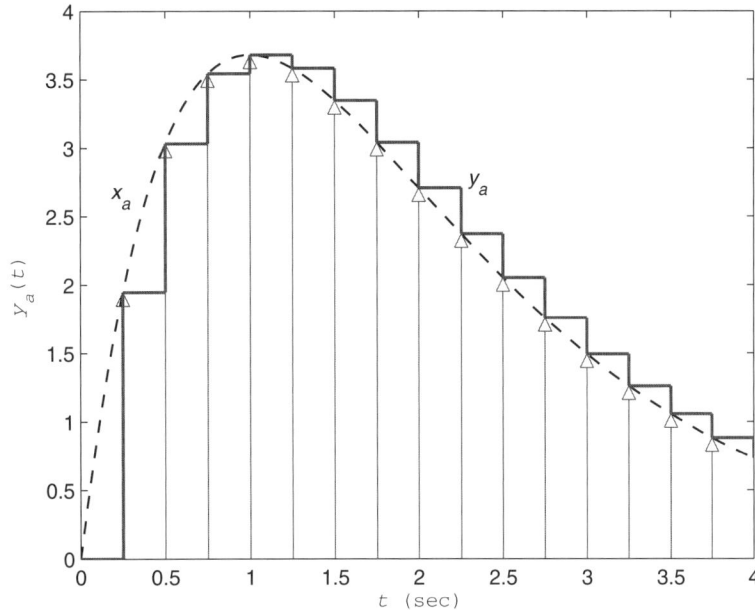

Since the zero-order hold is linear and time-invariant, the response to an impulse of strength $x(k)$ at time $t = kT$ will then be a pulse of height $x(k)$ and width T starting at $t = kT$. When the input is $\hat{x}_a(t)$, we simply add up all the responses to the scaled and shifted impulses to get a *piecewise-constant* approximation to $x_a(t)$ as shown in Figure 1.25. Notice that this is equivalent to interpolating between the samples with a polynomial of degree zero. It is for this reason that the system in (1.4.6) is called a *zero-order* hold. It holds onto the most recent sample and extrapolates to the next one using a polynomial of degree zero. Higher-degree hold filters are also possible (Proakis and Manolakis, 1992), but the zero-order hold is the most popular.

The zero-order hold filter also can be described in terms of its transfer function. Recall from (1.4.5) that the transfer function is just the Laplace transform of the impulse response in (1.4.7). Using the linearity of the Laplace transform, and the results of Example 1.7, we have

$$H_0(s) = L\{h_0(t)\}$$
$$= L\{\mu_a(t)\} - L\{\mu_a(t - T)\}$$
$$= [1 - \exp(-Ts)]L\{\mu_a(t)\} \qquad (1.4.8)$$

Finally, from the table of Laplace transform pairs in Appendix 1, the transfer function for a zero-order hold filter is.

$$H_0(s) = \frac{1 - \exp(-Ts)}{s} \qquad (1.4.9)$$

A typical DSP system was described in Figure 1.2 as an analog-to-digital converter (ADC), followed by a digital signal processing program, followed by a digital-to-analog converter (DAC). We now have mathematical models available for the two converter blocks. The ADC can be modeled by an impulse sampler, while the DAC can be modeled with a zero-order hold filter as shown in Figure 1.26 which is an updated version of Figure 1.2. Note that the impulse sampler is represented symbolically with a switch that opens and closes every T seconds.

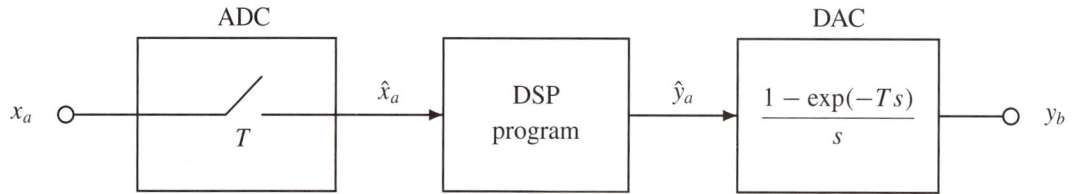

FIGURE 1.26: Mathematical Model of DSP System

It should be emphasized that the formulation depicted in Figure 1.26 uses *mathematical* models of signal sampling and signal reconstruction. With the impulse model of sampling we are able to determine constraints on $X_a(f)$ and T that ensure that all the essential information about $x_a(t)$ is contained in the samples $x(k)$. For signal reconstruction, the piecewise-constant output from the zero-order hold filter is an effective model for the DAC output signal. To complement the mathematical models in Figure 1.26, physical circuit models of the ADC and the DAC blocks are investigated in Section 1.6.

FDSP Functions

The Fundamentals of Digital Signal Processing (FDSP) toolbox supplied with this text, and discussed in Section 1.7, contains the following function for evaluating the frequency response of a linear continuous-time system.

```
% F_FREQS: Compute frequency response of continuous-time system
%
% Usage:
%          [H,f] = f_freqs (b,a,N,fmax);
% Pre:
%          b    = vector of length m+1 containing numerator coefficients
%          a    = vector of length n+1 containing denominator coefficients
%          N    = number of discrete frequencies
%          fmax = maximum frequency (0 <= f <= fmax)
% Post:
%          H = 1 by N complex vector containing the frequency response
%          f = 1 by N vector the containing frequencies at which H is
%              evaluated
% Notes:
%          H(s) must be stable.
```

To plot the magnitude and phase responses on a single screen one can use the following built-in MATLAB functions.

```
A = abs(H);              % magnitude response
phi = angle(H);          % phase response
subplot(2,1,1)           % top half of screen
plot (f,A)               % magnitude response plot
subplot(2,1,2)           % bottom half of screen
plot(f,phi)              % phase response plot
```

1.5 Prefilters and Postfilters

The sampling theorem in Proposition 1.1 tells us that to avoid aliasing during the sampling process, the signal $x_a(t)$ must be bandlimited, and we must sample at a rate that is greater than twice the bandwidth. Unfortunately, a quick glance at the Fourier transform pair table in Appendix 1 reveals that many of those signals are, in fact, not bandlimited. The exceptions are sines and cosines that have all their power concentrated at a single frequency. Of course a general periodic signal can be expressed as a Fourier series, and as long as the series is truncated to a finite number of harmonics, the resulting signal will be bandlimited. For a general nonperiodic signal, if we are to avoid aliasing we must bandlimit the signal explicitly by passing it through an analog lowpass filter as in Figure 1.27.

1.5.1 Anti-aliasing Filter

Anti-aliasing filter

The prefilter in Figure 1.27 is called a *anti-aliasing filter* or a *guard filter*. Its function is to remove all frequency components outside the range $[-F_c, F_c]$ where $F_c < f_d$ so that aliasing does not occur during sampling. The optimal choice for an anti-aliasing filter is an ideal lowpass filter. Since this filter is not physically realizable, we instead approximate the ideal lowpass characteristic. For example, a widely used family of lowpass filters is the set of Butterworth filters (Ludeman, 1986). A lowpass *Butterworth filter* of order n has the following magnitude response.

Butterworth filters

$$|H_a(f)| = \frac{1}{\sqrt{1 + (f/F_c)^{2n}}}, \quad n \geq 1 \tag{1.5.1}$$

Cutoff frequency

Notice that $|H_a(F_c)| = 1/\sqrt{2}$ where F_c is called the 3 dB *cutoff frequency* of the filter. The term arises from the fact that when the filter gain is expressed in units of decibels or dB, we have

$$20 \log_{10}\{|H_a(F_c)|\} \approx -3 \text{ dB} \tag{1.5.2}$$

The magnitude responses of several Butterworth filters are shown in Figure 1.28. Note that as the order n increases, the magnitude response approaches the ideal lowpass characteristic which is shown for comparison. However, unlike the ideal lowpass filter, the Butterworth filters introduce phase shift as well.

The transfer function of a lowpass Butterworth filter of order n with radian cutoff frequency $\Omega_c = 2\pi F_c$ can be expressed as follows.

$$H_a(s) = \frac{\Omega_c^n}{s^n + \Omega_c a_1 s^{n-1} + \Omega_c^2 a_2 s^{n-2} \cdots + \Omega_c^n} \tag{1.5.3}$$

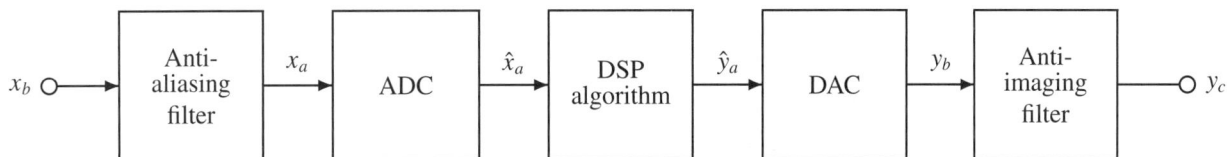

FIGURE 1.27: DSP System with Analog Prefilter and Postfilter

FIGURE 1.28:
Magnitude
Responses of
Lowpass
Butterworth Filters
with $F_c = 1$

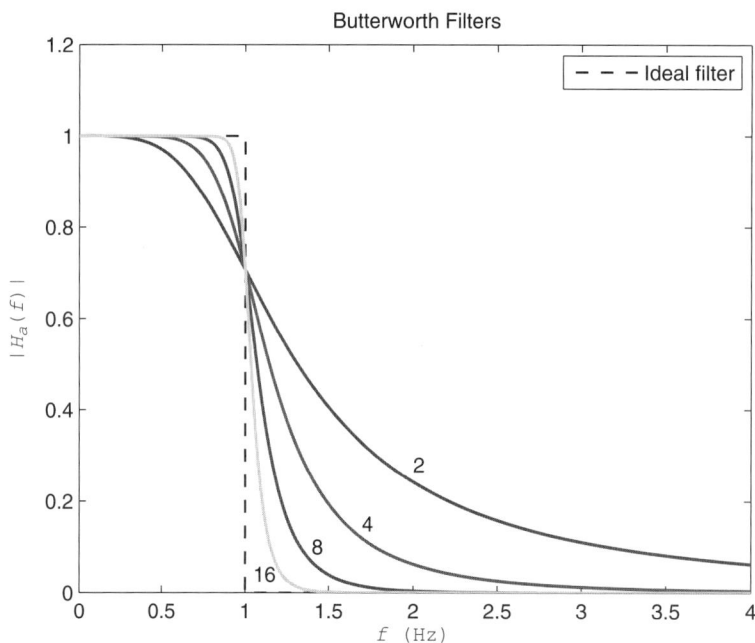

Butterworth Filters

TABLE 1.2: ▶
Second-order
Factors of
Normalized
Butterworth
Lowpass Filters

Order	$a(s)$
1	$(s + 1)$
2	$(s + 1.5142s + 1)$
3	$(s + 1)(s^2 + s + 1)$
4	$(s^2 + 1.8478s + 1)(s^2 + 0.7654s + 1)$
5	$(s + 1)(s^2 + 1.5180s + 1)(s^2 + 0.6180s + 1)$
6	$(s^2 + 1.9318s + 1)(s^2 + 1.5142s + 1)(s^2 + 0.5176s + 1)$
7	$(s + 1)(s^2 + 1.8022s + 1)(s^2 + 1.2456s + 1)(s^2 + 0.4450s + 1)$
8	$(s^2 + 1.9622s + 1)(s^2 + 1.5630s + 1)(s^2 + 1.1110s + 1)(s^2 + 0.3986s + 1)$

Normalized filter

When $\Omega_c = 1$ rad/sec or $F_c = .5/\pi$ Hz, this corresponds to a *normalized* Butterworth filter. A list of the coefficients for the first few normalized Butterworth filters is summarized in Table 1.2. Notice that the denominator polynomials in Table 1.2 are factored into quadratic factors when n is even and quadratic and linear factors when n is odd. This is done because the preferred way to realize $H_a(s)$ with a circuit is as a series or cascade configuration of first and second order blocks. That way, the overall transfer function is less sensitive to the precision of the circuit elements.

Example 1.8

First-order Filter

Consider the problem of realizing a lowpass Butterworth filter of order $n = 1$ with a circuit. From Table 1.2, we have $a_1 = 1$. Thus from (1.5.3) the first-order transfer function is

$$H_1(s) = \frac{\Omega_c}{s + \Omega_c}$$

This transfer function can be realized with a simple RC circuit with $RC = 1/\Omega_c$. However, a passive circuit realization of $H_1(s)$ can experience electrical loading effects when it is connected in series with other blocks to form a more general filter. Therefore, consider instead the active

FIGURE 1.29: Active Circuit Realization of First-order Lowpass Butterworth Filter Block

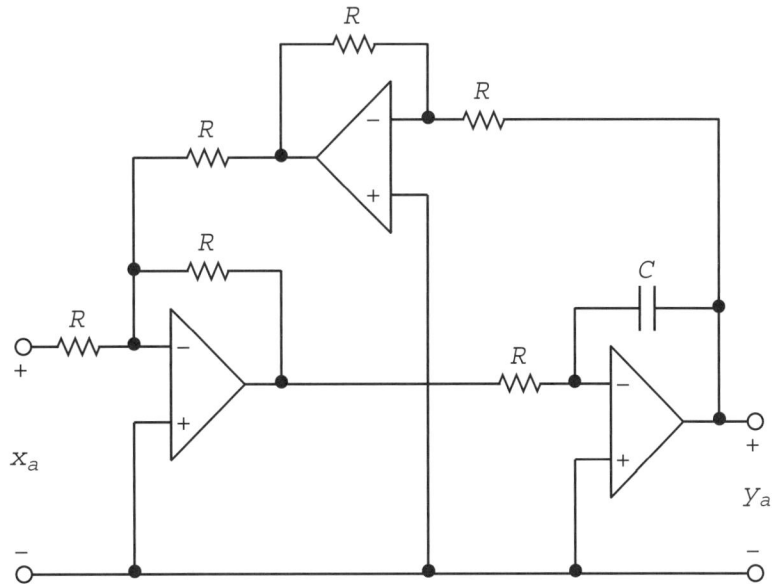

circuit realization shown in Figure 1.29 which uses three operational amplifiers (op amps). This circuit has a high input impedance and a low output impedance which means it will not introduce significant loading effects when connected to other circuits. The transfer function of the circuit (Dorf and Svoboda, 2000) is

$$H_1(s) = \frac{1/(RC)}{s + 1/(RC)}$$

Thus we require $1/(RC) = \Omega_c$. Typically, C is chosen to be a convenient value, and then R is computed using

$$R = \frac{1}{\Omega_c C}$$

As an illustration, suppose a cutoff frequency of $F_c = 1000$ Hz is desired. Then $\Omega_c = 2000\pi$. If $C = 0.01 \ \mu$F, which is a common value, then the required value for R is

$$R = 15.915 \ \text{k}\Omega$$

Integrated circuits typically contain up to four op amps. Therefore, the first-order filter section can be realized with a single integrated circuit and seven discrete components. Since the resistors all have the same resistance R, a resistor network chip can be used.

Since all of the first-order filter sections in Table 1.2 have $a_1 = 1$, the filter in Figure 1.29 can be used for a general first-order filter section. An nth order Butterworth filter also contains second-order filter sections.

Example 1.9

Second-order Filter

Consider the problem of realizing a lowpass Butterworth filter of order $n = 2$ with a circuit. From (1.5.3) the general form of the second-order transfer function is

$$H_2(s) = \frac{\Omega_c^2}{s^2 + \Omega_c a_1 + \Omega_c^2}$$

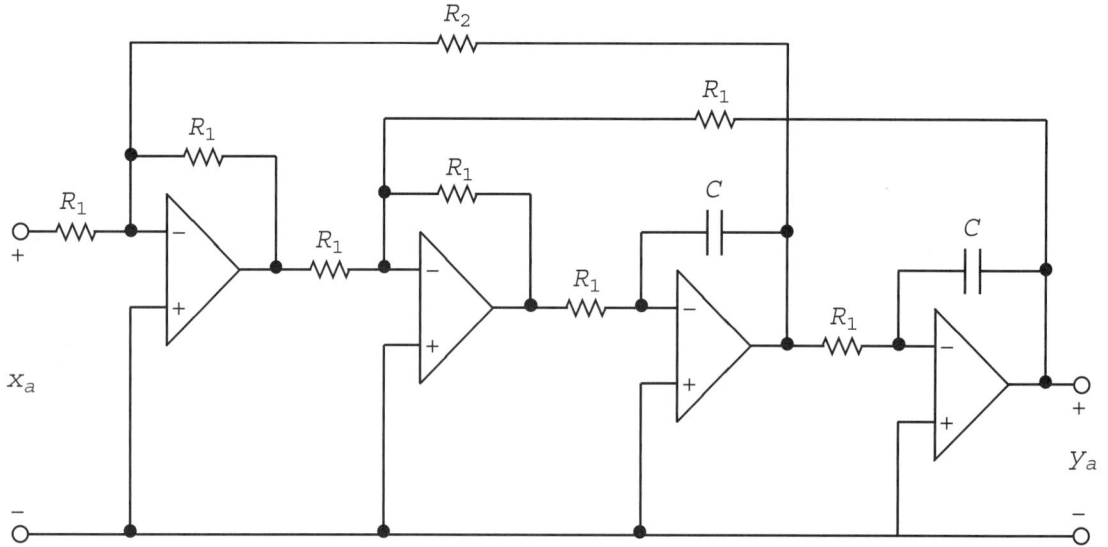

FIGURE 1.30: Active Circuit Realization of Second-order Lowpass Butterworth Filter Block

Using a state-space formulation of $H_2(s)$, this second-order block can be realized with the active circuit shown in Figure 1.30 which uses four op amps. The transfer function of the circuit (Dorf and Svoboda, 2000) is

$$H_2(s) = \frac{1/(R_1 C)^2}{s^2 + s/(R_2 C) + 1/(R_1 C)^2}$$

From the last term of the denominator, we require $1/(R_1 C)^2 = \Omega_c^2$. Again, typically C is chosen to be a convenient value, and then R_1 is computed using

$$R_1 = \frac{1}{\Omega_c C}$$

Next from the linear term of the denominator we have $1/(R_2 C) = \Omega_c a_1$. Solving for R_2 and expressing the final result in terms of R_1 yields

$$R_2 = \frac{R_1}{a_1}$$

As an illustration, suppose a cutoff frequency of $F_c = 5000$ Hz is desired. Then $\Omega_c = 10^4 \pi$, and from Table 1.2 we have $a_1 = 1.5142$. If we pick $C = 0.01\ \mu\text{F}$, then the two resistors are

$$R_1 \approx 3.183\ \text{k}\Omega$$
$$R_2 \approx 2.251\ \text{k}\Omega$$

This filter also can be realized with a single integrated circuit plus 10 discrete components.

Cascade connection A general lowpass Butterworth filter can be realized by using a cascade connection of first and second-order blocks where the output of one block is used as the input to the next block. In this way an anti-aliasing filter of order n can be constructed.

The Butterworth family of lowpass filters is one of several families of classical analog filters (Lam, 1979). Other classical analog filters that could be used for anti-aliasing filters include Chebyshev filters and elliptic filters. Classical analog lowpass filters are considered in detail in Chapter 7 where we investigate a digital design technique that converts an analog filter into an equivalent digital filter.

Classical analog filters are sufficiently popular that they have been implemented as integrated circuits using switched capacitor technology (Jameco, 2010). For example, the National Semiconductor LMF6-100 is a sixth-order lowpass Butterworth filter whose cutoff frequency is tunable with an external clock signal of frequency $f_{clock} = 100F_c$. The switched capacitor technology involves internal sampling or switching at the rate f_{clock}. Therefore, if the signal $x_a(t)$ does not have any significant spectral content beyond 50 times the desired sampling rate f_s, then the switch-capacitor filter can be used as an anti-aliasing filter to remove the frequency content in the range $[f_s/2, 50f_s]$. For example, with a desired sampling rate of 2 kHz, a switched-capacitor filter can remove (i.e. significantly reduce) frequencies in the range from 1 kHz to 100 kHz.

1.5.2 Anti-imaging Filter

Anti-imaging filter The postfilter in Figure 1.27 is called an *anti-imaging filter* or smoothing filter. The function of this filter is to remove the residual high-frequency components of $y_b(t)$. The zero-order hold transfer function of the DAC tends to reduce the size of the sidebands of $\hat{y}_a(t)$ which can be thought of as *images* of the baseband spectrum. However, it does not eliminate them completely. The magnitude response of the zero-order hold, shown in Figure 1.31, reveals a lowpass type of characteristic that is different from the ideal reconstruction filter due to the presence of a series of lobes. For the lowpass filter in Figure 1.31, the output is the piecewise constant signal $y_b(t)$ in Figure 1.27, rather than the ideally reconstructed signal $y_a(t)$.

FIGURE 1.31: Magnitude Response of Zero-order Hold

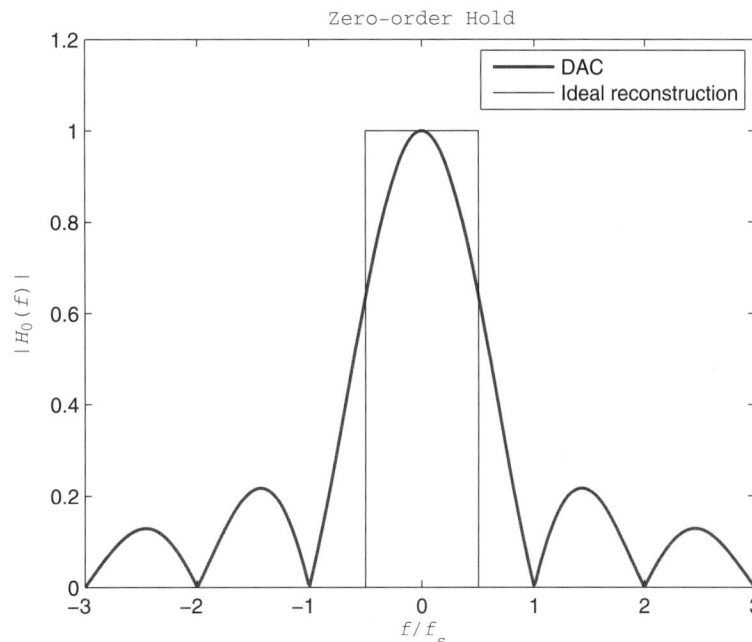

FIGURE 1.32:
Magnitude Spectra
of DAC Input in (a)
and Output in (b)
when $f_s = 2B$

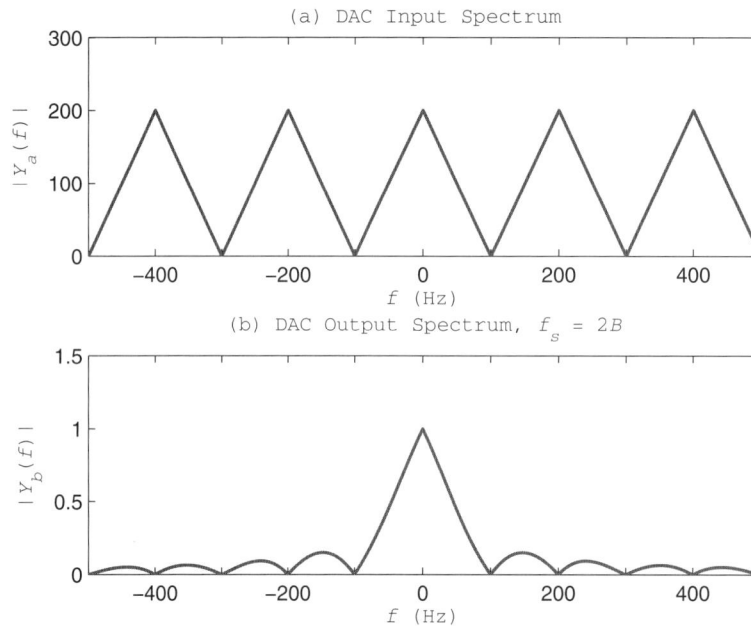

(a) DAC Input Spectrum

(b) DAC Output Spectrum, $f_s = 2B$

Note from Figure 1.31 that the zero-order hold has zero gain at multiples of the sampling frequency where the images of the baseband spectrum are centered. The effect of this low-pass type of characteristics is to reduce the size of the sidebands as can be seen Figure 1.32 which shows a discrete-time DAC input with a triangular spectrum in Figure 1.32a and the corresponding spectrum of the piecewise-constant DAC output in Figure 1.32b.

The purpose of the anti-imaging filter is to further reduce the residual images of the baseband spectrum centered at multiples of the sampling frequency. A lowpass filter similar to the anti-aliasing filter can be used for this purpose. A careful inspection of the baseband spectrum of $y_b(t)$ in Figure 1.32 reveals that it is slightly distorted due to the nonflat passband characteristic of the zero-order hold shown in Figure 1.31. Interestingly enough, this can be compensated for with a digital filter as part of the DSP algorithm by *preprocessing* $\hat{y}_a(t)$ before it is sent into the DAC.

A simple way to reduce the need for an anti-imaging filter is to increase the sampling rate f_s beyond the minimum needed to avoid aliasing. As we shall see in Chapter 8, the DAC sampling rate can be increased as part of the DSP algorithm by inserting samples between those coming from the ADC, a process known as interpolation. The effect of oversampling is to spread out the images along the frequency axis which allows the zero-order hold to more effectively attenuate them. This can be seen in Figure 1.33 which is identical to Figure 1.32, except that the signals have been oversampled by a factor of two. Notice that oversampling also has the beneficial effect of reducing the distortion of the baseband spectrum that is caused by the nonideal passband characteristic of the zero-order hold.

Finally, it is worth noting that there are applications where an anti-imaging filter may not be needed at all. For example, in a digital control application the output of the DAC might be used to drive a relatively slow electro-mechanical device such as a motor. These devices already have a lowpass frequency response characteristic so it is not necessary to filter the DAC output. For some DSP applications, the desired output may be information that can be extracted directly from the discrete-time signal $\hat{y}_a$ in which case there is no need to convert from digital back to analog.

FIGURE 1.33:
Magnitude Spectra
of DAC Input in (a)
and Output in (b)
when $f_s = 4B$

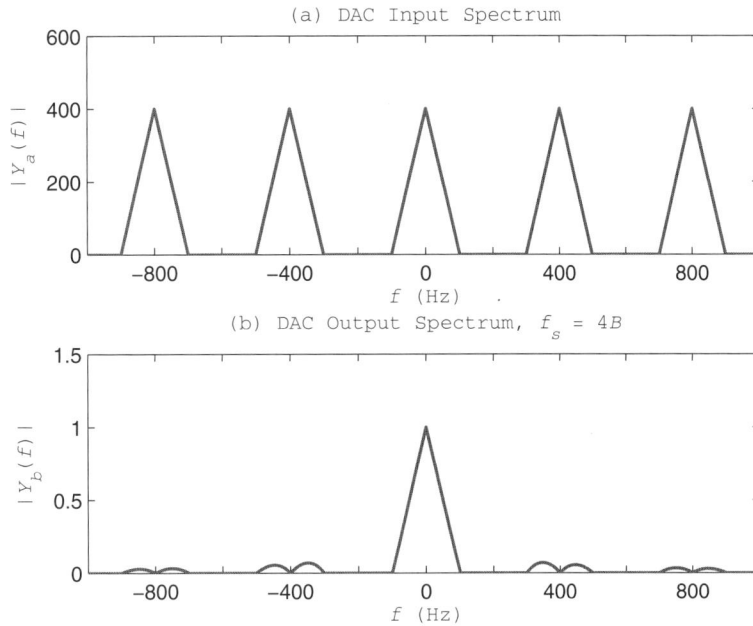

(a) DAC Input Spectrum

(b) DAC Output Spectrum, $f_s = 4B$

● ● ● ● ● ● ● ● ● ● ● ● ● ● ● ● ●

1.6   DAC and ADC Circuits

This optional section presents circuit realizations of digital-to-analog and analog-to-digital converters. This material is included for those readers specifically interested in hardware. This section, and others like it marked with ∗, can be skipped without loss of continuity.

1.6.1   Digital-to-analog Converter (DAC)

DAC

Unipolar, bipolar

Recall that a digital-to-analog converter or DAC can be modeled mathematically as a zero-order hold as in (1.4.9). A physical model of a *DAC* is entirely different because the input is an N-bit binary number, $b = b_{N-1}b_{N-2}\cdots b_1 b_0$, rather than an amplitude-modulated impulse train. A DAC is designed to produce an analog output y_a that is proportional to the decimal equivalent of the binary input b. DAC circuits can be classified as *unipolar* if $y_a \geq 0$ or *bipolar* if y_a is both positive and negative. For the simpler case of a unipolar DAC, the decimal equivalent of the binary input b is

$$x = \sum_{k=0}^{N-1} b_k 2^k \qquad (1.6.1)$$

The binary input of a bipolar DAC represents negative numbers as well using two's complement, offset binary, or a sign-magnitude format (Grover and Deller, 1999). The most common type of DAC is the R-$2R$ ladder circuit shown in Figure 1.34 for the case $N = 4$. The configuration of resistors across the top is the R-$2R$ ladder. On each rung of the ladder is a digitally controlled single pole double throw (SPDT) switch that sends current into either the inverting input (when $b_k = 1$) or the non-inverting input (when $b_k = 0$) of the operational amplifier or *op amp*.

Op amp

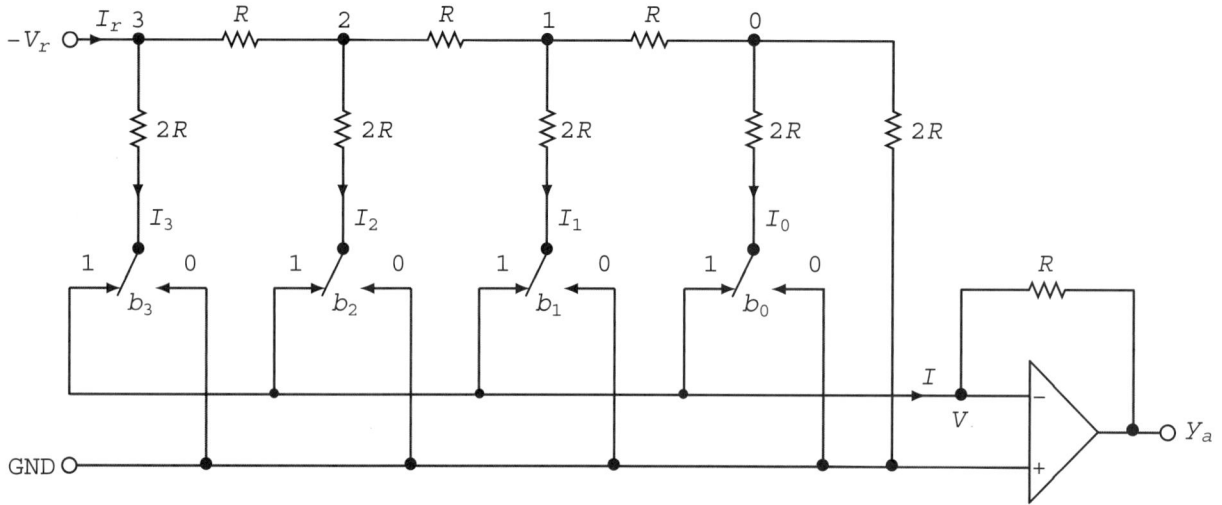

FIGURE 1.34: A 4-bit Unipolar DAC Using an R-2R Ladder

To analyze the operation of the circuit in Figure 1.34 we begin at the end of the ladder and work backwards. First note that for an ideal op amp, the voltage at the inverting input is the same as that at the noninverting input, namely $V = 0$. Consequently, the current I_k through the kth switch does not depend on the switch position. The equivalent resistance looking into node 0 from the left is therefore $R_0 = R$ because it consists of two resistors of resistance $2R$ in parallel. Thus the current entering node 0 from the left is split in half. Next consider node 1. The equivalent resistance to the right of node 1 is $2R$ because it consists of a series combination of R and R_0. This means that the equivalent resistance looking into node 1 from the left is again $R_1 = R$ because it consists of two resistors of resistance $2R$ in parallel. This again means that the current entering node 1 from the left is split in half. This process can be repeated as many times as needed until we conclude that the equivalent resistance looking into node $N - 1$ from the left is R. Consequently, the current drawn by the R-$2R$ ladder, regardless of the switch positions, is as follows where V_r is the *reference voltage*.

Reference voltage

$$I_r = \frac{-V_r}{R} \tag{1.6.2}$$

Since the current is split in half each time it enters another node of the ladder, the current shunted through the kth switch is $I_k = I_r/2^{N-k}$. Consequently, using (1.6.1) the total current entering the op amp section is

$$I = \sum_{k=0}^{N-1} b_k I_k$$

$$= \sum_{k=0}^{N-1} b_k I_r 2^{k-N}$$

$$= \left(\frac{I_r}{2^N}\right) \sum_{k=0}^{N-1} b_k 2^k$$

$$= \left(\frac{I_r}{2^N}\right) x \tag{1.6.3}$$

Finally, an ideal op amp has infinite input impedance which means that the current drawn by the inverting input is zero. Therefore, using (1.6.2) and (1.6.3), the op amp output is

$$y_a = -RI$$

$$= \left(\frac{-RI_r}{2^N}\right) x$$

$$= \left(\frac{V_r}{2^N}\right) x \qquad (1.6.4)$$

Setting the binary input in (1.6.4) to $x = 1$, we see that the quantization level of the DAC is $q = V_r/2^N$. The range of output values for the DAC in Figure 1.34 is

$$0 \leq y_a \leq \left(\frac{2^N - 1}{2^N}\right) V_r \qquad (1.6.5)$$

In view of (1.6.5), the DAC in Figure 1.34 is a unipolar DAC. It can be converted to a bipolar DAC with outputs in the range $-V_r \leq y_a < V_r$ by replacing V_r with $2V_r$ and adding a second op amp circuit at the output that performs level shifting (see Problem 1.15). In this case b is interpreted as an offset binary input whose decimal equivalent is as follows where x is as in (1.6.1).

$$x_{\text{bipolar}} = x - 2^{N-1} \qquad (1.6.6)$$

In general, a *signal conditioning* circuit can be added to the DAC in Figure 1.34 to perform scaling, offset, and impedance matching (Dorf and Svoboda, 2000). It is useful for the DAC circuit to have a low output impedance so that the DAC output is capable of supplying adequate current to drive the desired load.

1.6.2 Analog-to-digital Converter (ADC)

ADC An analog-to-digital converter or *ADC* must take an analog input $-V_r \leq x_a < V_r$ and convert it to a binary output b whose decimal value is equivalent to x_a. The input-output characteristic of an N-bit bipolar ADC is shown in Figure 1.35 for the case $V_r = 5$ and $N = 4$. Note that the staircase is shifted to the left by half a step so that the ADC output is less sensitive to low-level noise when $x_a = 0$. The horizontal length of the step is the quantization level which, for a bipolar ADC, can be expressed

$$q = \frac{V_r}{2^{N-1}} \qquad (1.6.7)$$

Successive-approximation Converters

The most widely used ADC consists of a comparator circuit plus a DAC in the configuration shown in Figure 1.36. The input is the analog voltage x_a to be converted and the output is the equivalent N-bit binary number b.

A very simple form of the ADC can be realized by replacing the block labeled SAR logic with a binary counter that counts the pulses of the periodic pulse train or clock signal, f_{clock}. As the counter output b increases from 0 to $2^N - 1$, the DAC output y_a ranges from $-V_r$ to V_r. When the value of y_a becomes larger than the input x_a, the comparator output u switches from 1 to 0 at which time the counter is disabled. The count b is then the digital equivalent of the analog input x_a using the offset binary code in (1.6.6).

Although an ADC based on the use of a counter is appealing because of its conceptual simplicity, there is a significant practical drawback. The time required to perform a conversion

FIGURE 1.35:
Input-output
Characteristic of
4-bit ADC with
$V_r = 5$

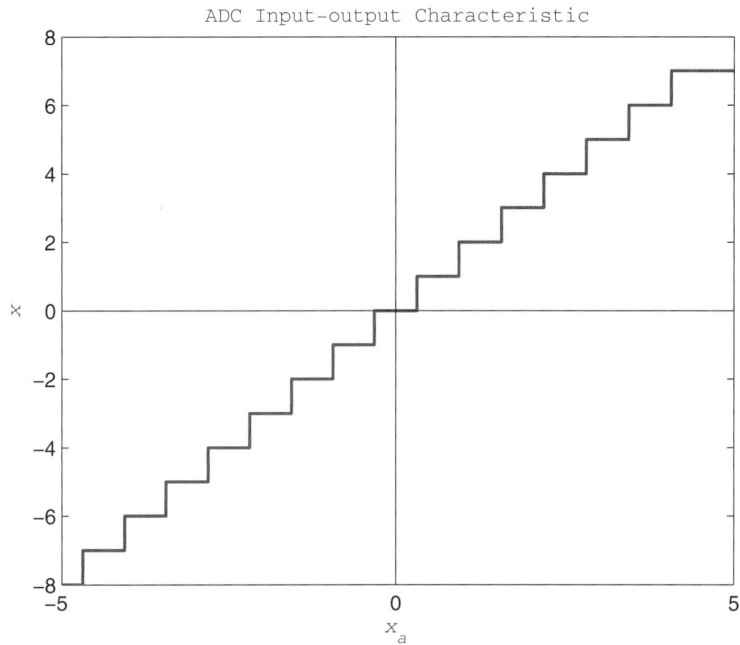

ADC Input-output Characteristic

FIGURE 1.36: An
N-bit Successive
Approximation
ADC

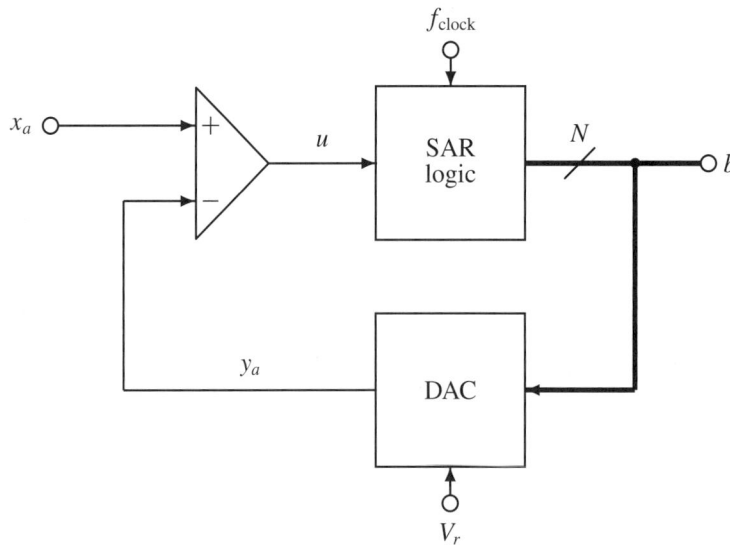

is variable and can be quite long. For random inputs with a mean value of zero, it takes on the average 2^{N-1} clock pulses to perform a conversion, and it can take as long as 2^N clock pulses when $x_a = V_r$. A much more efficient way to perform a conversion is to execute a binary search for the proper value of b by using a successive approximation register (SAR). The basic idea is to start with the most significant bit, b_{N-1}, and determine if it should be 0 or 1 based on the comparator output. This cuts the range of uncertainty for the value of x_a in half and can be done in one clock pulse. Once b_{N-1} is determined, the process is then repeated for bit b_{N-2} and so on until the least significant bit, b_0, is determined. The successive approximation technique is summarized in Algorithm 1.1.

ALGORITHM

1.1: Successive Approximation

1. Set $\Delta y = V_r$, $y_a = -V_r$
2. For $k = N - 1$ down to 0 do
{
(a) If $y_a + \Delta y > x_a$

$$u(k) = 0$$
$$b_k = 0$$

else

$$u(k) = 1$$
$$b_k = 1$$

(b) Set

$$y_a = y_a + b_k \Delta y$$
$$\Delta y = \Delta y/2$$

}

The virtue of the binary search approach is that it takes exactly N clock pulses to perform a conversion, independent of the value of x_a. Thus the conversion time is constant and the process is much faster. For example, for a precision of $N = 12$ bits, the conversion time in comparison with the counter method is reduced, on the average, by a factor of $2^{11}/12 = 170.7$ or two orders of magnitude.

Example 1.10

Successive Approximation

As an illustration of the successive approximation conversion technique, suppose the reference voltage is $V_r = 5$ volts, the converter precision is $N = 10$ bits, and the value to be converted is $x_a = 2.891$ volts. When Algorithm 1.1 is applied, the traces of $y_a(k)$ and $u_a(k)$ for $0 \le k < N$ are as shown in Figure 1.37. Note how the DAC output quickly adjusts in Figure 1.37a to the value of x_a by cutting the interval of uncertainty in half with each clock pulse. From (1.6.7), the quantization level in this case is

$$q = \frac{V_r}{2^{N-1}}$$
$$= \frac{5}{512}$$
$$= 9.8 \text{ mV}$$

Flash Converters

Flash converter

There is another type of ADC, called the *flash* converter, that is used in applications where very high speed conversion is essential, such as in a digital oscilloscope. A simple 2-bit flash converter is shown in Figure 1.38. It consists of a linear resistor array, an array of comparator circuits, and an encoder circuit.

The resistors in the resistor array are selected such that the voltage drop between successive inverting inputs to the comparators is the quantization level q in (1.6.7). The fractional

FIGURE 1.37: DAC Output in (a) and SAR Input in (b) during Successive Approximation Steps

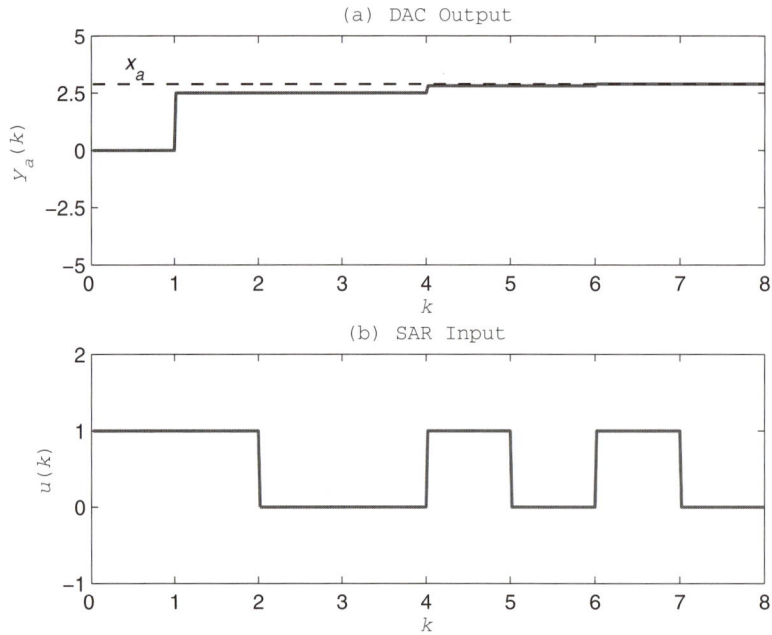

FIGURE 1.38: A 2-bit Flash Converter

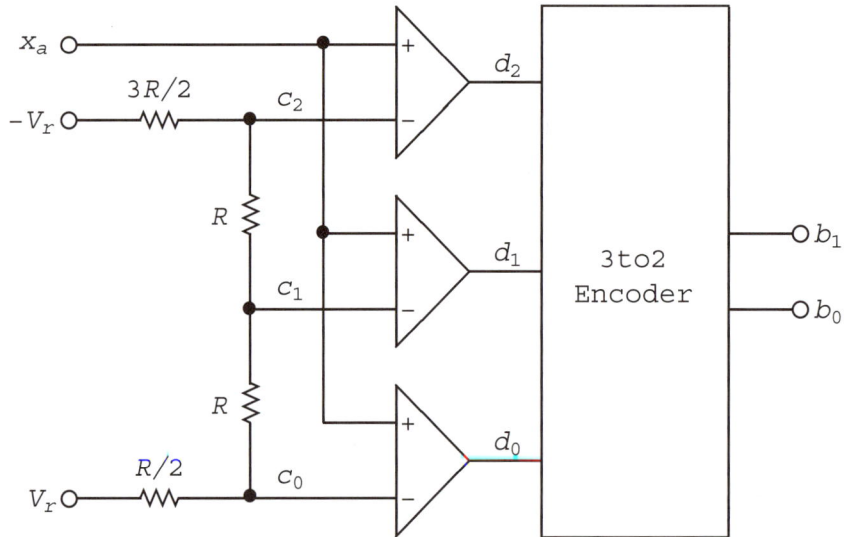

resistances at the ends of the array cause the input-output characteristic to shift to the left by $q/2$ as shown previously in Figure 1.35. Therefore, the voltage at the inverting input of the kth comparator is

$$c_k = -V_r + \left(k + \frac{1}{2}\right) q, \quad 0 \le k < N - 1 \tag{1.6.8}$$

The analog input x_a is compared to each of the threshold voltages c_k. For those k for which $x_a > c_k$, the comparator outputs will be $d_k = 1$, while the remaining comparators will have outputs $d_k = 0$. Thus the comparator outputs d can be thought of as a bar graph code where all the bits to one side of a certain bit are turned on. The encoder circuit takes the $2^N - 1$ comparator outputs d and converts them to an N-bit binary output b. It does so by setting the decimal

["

Input Range	$d = d_2\,d_1\,d_0$	$b = b_1\,b_0$
$-1 \le x_a/V_r < -.75$	000	00
$-.75 \le x_a/V_r < -.25$	001	01
$-.25 \le x_a/V_r < .25$	011	10
$.25 \le x_a/V_r \le 1$	111	11

TABLE 1.3: ▶ Inputs and Outputs of Encoder Circuit when $n = 2$

equivalent of b equal to i where i is the largest subscript such that $d_i = 1$. For example, for the 2-bit converter in Figure 1.38, the four possible inputs and outputs are summarized in Table 1.3.

The beauty of the flash converter is that the entire conversion can be done in a single clock pulse. More specifically, the conversion time is limited only by the settling time of the comparator circuits and the propagation delay of the encoder circuit. Unfortunately, the extremely fast conversion time is achieved at a price. The converter shown in Figure 1.38 has a precision of only 2 bits. In general for an N-bit converter there will be a total of $2^N - 1$ comparator circuits required. Furthermore, the encoder circuit will require $2^N - 1$ inputs. Consequently, as N increases, the flash converter becomes very hardware intensive. As a result, practical high speed flash converters are typically lower precision (6 to 8 bits) in comparison with the medium speed successive approximation converters (8 to 16 bits). There are a number of other types of ADCs as well including the slower, but very high precision, sigma delta converters (see Chapter 8) and dual integrating converters (Grover and Deller, 1999).

FDSP Functions

The FDSP toolbox that accompanies this text contains the following functions for performing analog-to-digital and digital-to-analog conversions.

```
% F_ADC: Perform N-bit analog-to-digital conversion
%
% Usage:
%          [b,d,y] = f_adc (x,N,Vr);
% Pre:
%          x  = analog input
%          N  = number of bits
%          Vr = reference voltage (-Vr <= x < Vr)
% Post:
%          b = 1 by N vector containing binary output
%          d = decimal output (offset binary)
%          y = quantized analog output
% F_DAC: Perform N-bit digital to analog conversion
%
% Usage:
%          y = f_dac (b,N,Vr);
% Pre:
%          b  = string containing N-bit binary input
%          N  = number of bits
%          Vr = reference voltage (-Vr <= y < Vr)
% Post:
%           y = analog output
```

● ● ● ● ● ● ● ● ● ● ● ● ● ● ● ● ●

1.7   The FDSP Toolbox

The software that accompanies this text is available on the publisher's companion web site. It includes a Fundamentals of Digital Signal Processing (FDSP) *toolbox* that contains MATLAB functions that implement the signal processing algorithms developed in the text. This software is provided as an aid to help students solve the Computation problems as well as the GUI Simulation problems that appear at the end of each chapter. The FDSP toolbox also provides the user with a convenient way to run the all of the MATLAB examples and reproduce the MATLAB figures that appear in the text. A novel supplementary component of the toolbox is a collection of graphical user interface (GUI) modules that allow the user to *interactively* explore the signal processing techniques covered in each chapter without any need for programming. In addition, the source for all of the FDSP toolbox functions is provided for users who prefer to write their own MATLAB programs.

The FDSP toolbox is installed using MATLAB itself (Marwan, 2003). When the FDSP zip file is downloaded from the companion web site and unzipped, it places a few files in the default folder at c:\fdsp_net. The FDSP toolbox is then installed by entering the following commands from the MATLAB command prompt.

```
>> cd c:\fdsp_net
>> setup
```

1.7.1   FDSP Driver Module

All of the course software can be conveniently accessed through a driver module called *f_dsp*. The driver module is launched by entering the following command from the MATLAB command prompt:

```
>> f_dsp
```

The startup screen for *f_dsp* is shown in Figure 1.39. Most of the options on the menu toolbar produce submenus of selections. The settings option allows the user to configure f_dsp by selecting default folders for loading, saving, and printing user data. The GUI Modules option is used to run the chapter graphical user interface modules. Using the Examples option, the code for all of the MATLAB examples appearing in the text can be viewed and executed. Similarly, the Figures and Tables options are used to recreate and display the figures and tables from the text. The Help option provides online help for the toolbox functions and the GUI modules. For users interested in obtaining updates and supplementary course material, the Web option connects the user to the companion web site of the publisher. The following author web site also can be used to download updates of the latest FDSP software.

```
www.clarkson.edu/~rschilli/fdsp
```

1.7.2   Toolbox Functions

Algorithms developed in the text are implemented as a library of FDSP toolbox functions supplemented by functions that are already available as part of the standard MATLAB interpreter. These functions are described in detail in the chapters. They fall into two broad categories,

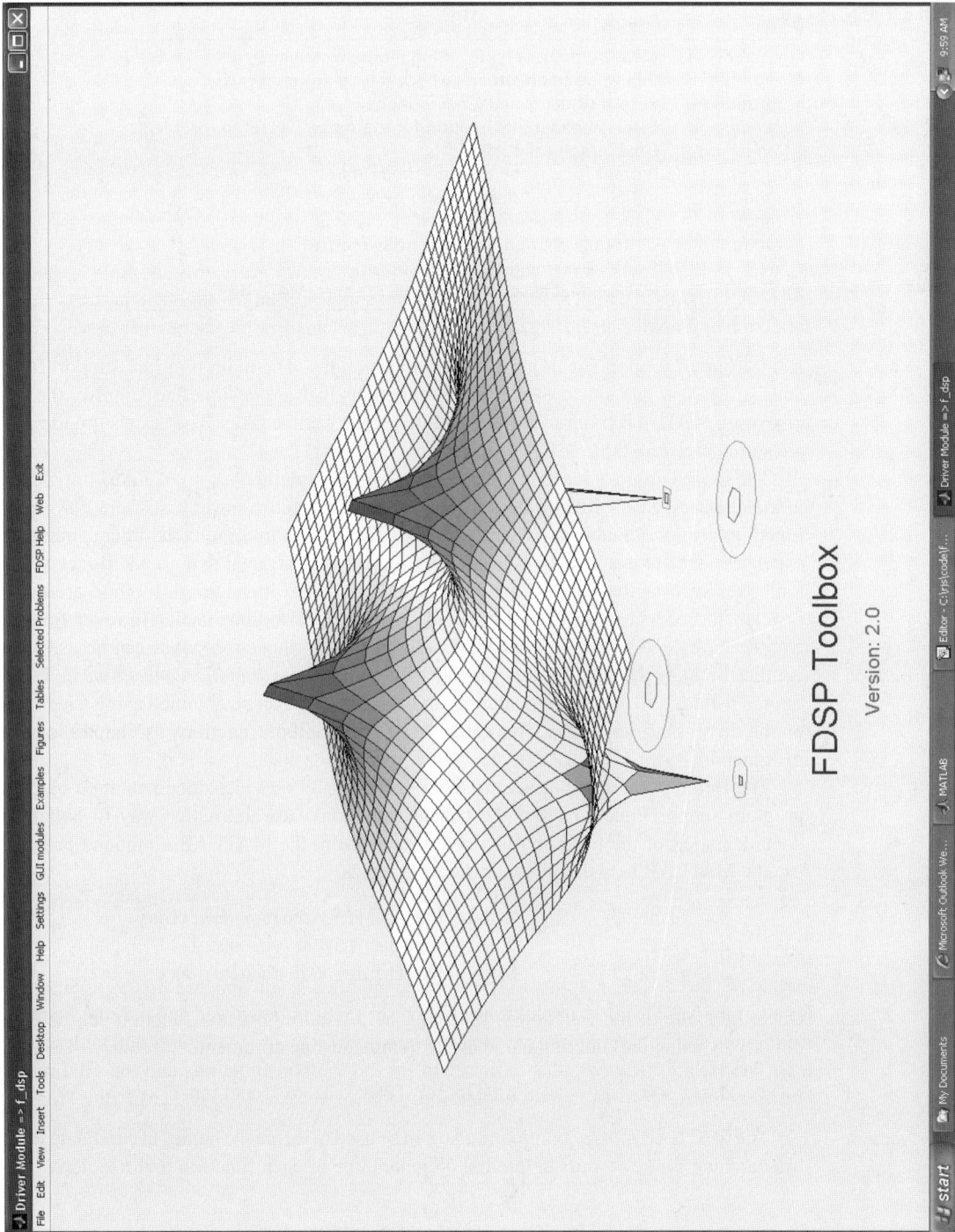

FIGURE 1.39: Driver Module *f_dsp* for FDSP Toolbox

TABLE 1.4: ▶
FDSP Main Program
Support Functions

Name	Description
f_caliper	Measure points on plot using mouse cross hairs
f_clip	Clip value to an interval, check calling arguments
f_getsound	Record signal from PC microphone
f_header	Display header information for examples, figures, or problems
f_labels	Label graphical output
f_prompt	Prompt for a scalar in a specified range
f_randinit	Initialize the random number generator
f_randg	Gaussian white noise matrix
f_randu	Uniformly distributed white noise matrix
f_wait	Pause to examine displayed output
soundsc	Play a signal as sound on the PC speaker (MATLAB)

main program support functions and chapter functions. The main program support functions consist of a few general low-level utility functions that are designed to simplify the process of writing MATLAB programs by performing some routine tasks. The main support functions are summarized in Table 1.4.

The second group of toolbox functions consists of implementations of algorithms developed in the chapters. Specialized functions are developed in those instances where corresponding functions are not available as part of the standard MATLAB interpreter. In order to minimize the expense to the student and maximize accessibility, it is assumed that no additional MATLAB toolboxes such as the Signal Processing toolbox or the Filter Design toolbox are available. Instead, the necessary functions are provided in the FDSP toolbox itself. However, for students who do have access to additional MATLAB toolboxes, those toolboxes can be used without

File name convention
conflict because the FDSP functions all follow the *f_xxx naming convention* in Table 1.4. Source listings and user documentation for each function can be obtained using a menu option on the driver program *f_dsp*. A summary of FDSP toolbox functions by chapter is provided in Appendix 3.

The Help option in the driver module in Figure 1.39 provides documentation on the main program support functions and the chapter functions. An alternative way to obtain online documentation of the toolbox functions directly from the MATLAB command prompt is to use the MATLAB help, helpwin, or *doc* commands.

```
doc fdsp          % Help for all FDSP toolbox functions
doc f_dsp         % Help for the FDSP driver module
doc g_xxx         % Help for chapter GUI module g_xxx
```

Once the name of the function is known, a help window for that function can be obtained directly by using the function name as the command-line argument.

```
doc f_xxx         % Help for FDSP toolbox function f_xxx
```

lookfor
The MATLAB *lookfor* command also can be used to locate the names of candidate functions by searching for key words in the first comment line of each function in the toolbox.

Example 1.11

FDSP Help

To illustrate the type of information provided by the doc command, consider the following example which elicits help documentation for one of the toolbox functions appearing in Chapter 4.

```
doc f_corr
```

The resulting display, shown below, follows a standard format for all toolbox functions. It includes a description of what the function does and how the function is called including definitions of all input and output arguments.

```
F_CORR: Fast cross-correlation of two discrete-time signals

Usage:
        r = f_corr (y,x,circ,norm)
Pre:
        y    = vector of length L containing first signal
        x    = vector of length M <= L containing second signal
        circ = optional correlation type code (default 0):

                0 = linear correlation
                1 = circular correlation

        norm = optional normalization code (default 0):

                0 = no normalization
                1 = normalized cross-correlation
Post:
        r = vector of length L contained selected cross-
            correlation of y with x.
Notes:
        To compute auto-correlation use x = y.

See also: f_corrcoef, f_conv, f_blockconv, conv, corrcoef
```

FDSP Functions

Empty matrix

The standard MATLAB convention for optional input arguments is used for all of the FDSP toolbox functions. If the optional argument appears at the end of the list, it can simply be left off. If it appears in the middle of the list, it can be replaced by the *empty matrix*, [ ]. For example, the following calls of function *f_corr* from Example 1.11 are equivalent.

```
r = f_corr (x,y);        % default values for circ, norm
r = f_corr (x,y,0);
r = f_corr (x,y,[],0);   % [] for default value of circ
r = f_corr (x,y,0,0);
```

1.7.3 GUI Modules

The FDSP toolbox functions are provided to facilitate the development of user programs. Alternatively, a higher level approach (easier but less flexible) is to use the graphical user interface (GUI) modules. When the GUI Modules option is selected from the FDSP driver module, the user is provided with the list of chapter GUI modules summarized in Table 1.5.

TABLE 1.5: ▶
FDSP Toolbox GUI
Modules

Module	Description	Chapter
g_sample	Signal sampling	1
g_reconstruct	Signal reconstruction	1
g_correlate	Signal correlation	2
g_systime	Discrete-time systems in the time domain	2
g_sysfreq	Discrete-time systems in the frequency domain	3
g_spectra	Signal spectral analysis	4
g_filter	Filter specifications and characteristics	5
g_fir	FIR filter design	6
g_iir	IIR filter design	7
g_multirate	Multirate signal processing	8
g_adapt	Adaptive signal processing	9

TABLE 1.6: ▶
User Interface
Features of Chapter
GUI Modules

Feature	Description	Comments
Block diagram	System or algorithm under investigation	Color coded
Parameters	Edit boxes containing simulation parameters	Scalar and vector
Type	Select signal or system type	Radio buttons
View	Select contents of plot window	Radio buttons
Plot	Display selected graphical results	Bottom half of screen
Slider	Adjust scalar design parameter	Fixed range
Menu	Caliper, Save data, Print, Help, Exit	Top of screen

Each of the GUI modules is described in detail near the end of the corresponding chapter. The GUI modules are designed to provide users with a convenient means of investigating the signal processing concepts covered in that chapter without any need for programming. Users who are familiar with MATLAB programming can make use of the FDSP toolbox functions to write their own MATLAB programs.

The GUI modules in Table 1.5 have a standardized user interface that is simple to learn and easy to use. The startup screens consist of a set of tiled windows populated with GUI controls. A typical startup screen for a GUI module is shown in Figure 1.40. A summary of the common user interface features of the chapter GUI modules can be found in Table 1.6. In the
Block diagram   upper left of the screen is a *Block diagram* window showing the system or signal processing
Edit boxes   operation under investigation. Below the block diagram are *edit boxes* that contain simulation parameters. The contents of each edit box can be directly modified by the user with parameter changes activated with the Enter key. Any MATLAB statement or statements that compute the parameters can be entered by the user.
Type options   To the right of the *Block diagram* window is the *Type* window where radio button controls are provided that allow the user to select the signal or system type. In addition to predefined types, there is a User-defined selection that prompts for the name of a user-suppled MATLAB file that supplies the signal or system information. For signal selection, there is also an option to record signals using the PC microphone. To verify that an acceptable recording has been created, a pushbutton control is provided to play the signal on the PC speaker.
View options   To the right of the *Type* window is a *View* window that includes radio button controls that allow the user to select which simulation results to view. The selected graphical output appears in the *Plot* window along the bottom half of the screen. Below the *Type* and *View* windows is a
Slider control   horizontal *Slider bar* that is provided so the user can directly adjust a scalar design parameter such as the sampling frequency, the number of samples, the number of bits, or the filter order. Within the *Parameters*, *Type*, and *View* windows there are also checkbox controls that allow the user to toggle optional features on and off such as dB display, signal clipping, and additive noise.

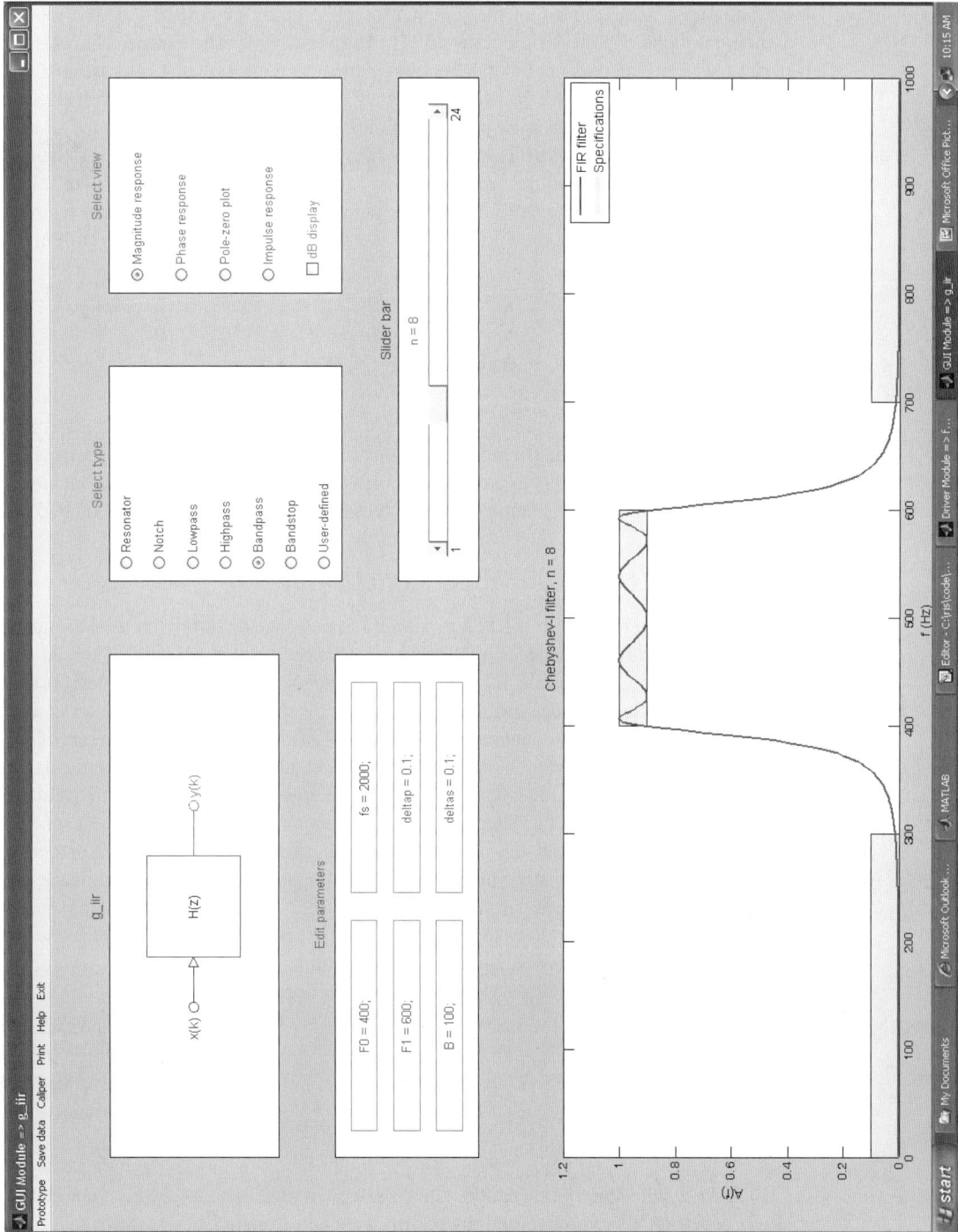

FIGURE 1.40: Example Screen, GUI Module *g_iir*

Menu options   The *Menu* bar at the top of the screen contains options specific to the GUI module. The common options include Caliper, Save, Print, Help, and Exit. The Caliper option is provided so the user can use cross hairs and the mouse to measure a point of interest on the current plot. The Save option allows the user to save data in a user-specified MAT-file. The contents of the MAT-file can be loaded back into the same or other GUI modules using the User-defined

Export   option in the *Type* window. In this way, results can be exported between the GUI modules. MAT-files produced by GUI modules use a common format. The Print option provides a hard copy graphical output of the current plot. The Help option is used to display directions on how to make effective use of the GUI module. Finally, the Exit option returns control to the calling program.

● ● ● ● ● ● ● ● ● ● ● ● ● ● ⋯

1.8   GUI Software and Case Studies

This section focuses on sampling and reconstruction of continuous-time signals. Two GUI modules are presented, one for signal sampling and the other for signal reconstruction. Both allow the user to interactively explore the concepts covered in this chapter without any need for programming. Case study examples are then presented and solved using a MATLAB program.

g_sample: Continuous-time Signal Sampling

The graphical user interface module *g_sample* is designed to allow the user to interactively investigate the sampling process. GUI module *g_sample* features a display screen with tiled windows as shown in Figure 1.41. The *Block diagram* window in the upper-left corner shows a block diagram that includes two signal processing blocks, an anti-aliasing filter and an ADC. Below each block is pair of edit boxes that allow the user to change the parameters of the signal processing blocks. Parameter changes are activated with the Enter key. The anti-aliasing filter is a lowpass Butterworth filter with user-selectable filter order, n, and cutoff frequency, F_c. Selecting $n = 0$ removes the filtering operation completely so $x_b = x_a$. The user-selectable parameters for the ADC are the number of bits of precision, N, and the reference voltage, V_r. Input signals larger in magnitude than V_r get clipped, and when N is small the effects of quantization error become apparent.

The *Type* and *View* windows in the upper-right corner of the screen allow the user to select both the type of input signal x_a, and the viewing mode. The inputs include several common signals plus a user-defined input. For the latter selection, the user must provide the file name (without the .m extension) of a user-supplied M-file function that returns the input vector xa evaluated at the time vector t. For example, if the following file is saved under the name *user1.m*,

User function   then the signal that it generates can be sampled and analyzed with the GUI module *g_sample*.

```
function xa = user1(t)        % Example user function
xa = t .* exp(-t);            % t can be a vector
```

The *View* window options include the color-coded time signals (x_a, x_b, x), and their magnitude spectra. Other viewing options portray the characteristics of the two signal processing blocks. They include the magnitude response of the anti-aliasing filter, and the input-output characteristic of the ADC. The *Plot* window along the bottom half of the screen shows the selected view.

The *Menu* bar at the top of the screen includes several menu options. The *Caliper* option allows the user to measure any point on the current plot by moving the mouse cross hairs to that point and clicking. The *Print* option prints the contents of the plot window to a printer

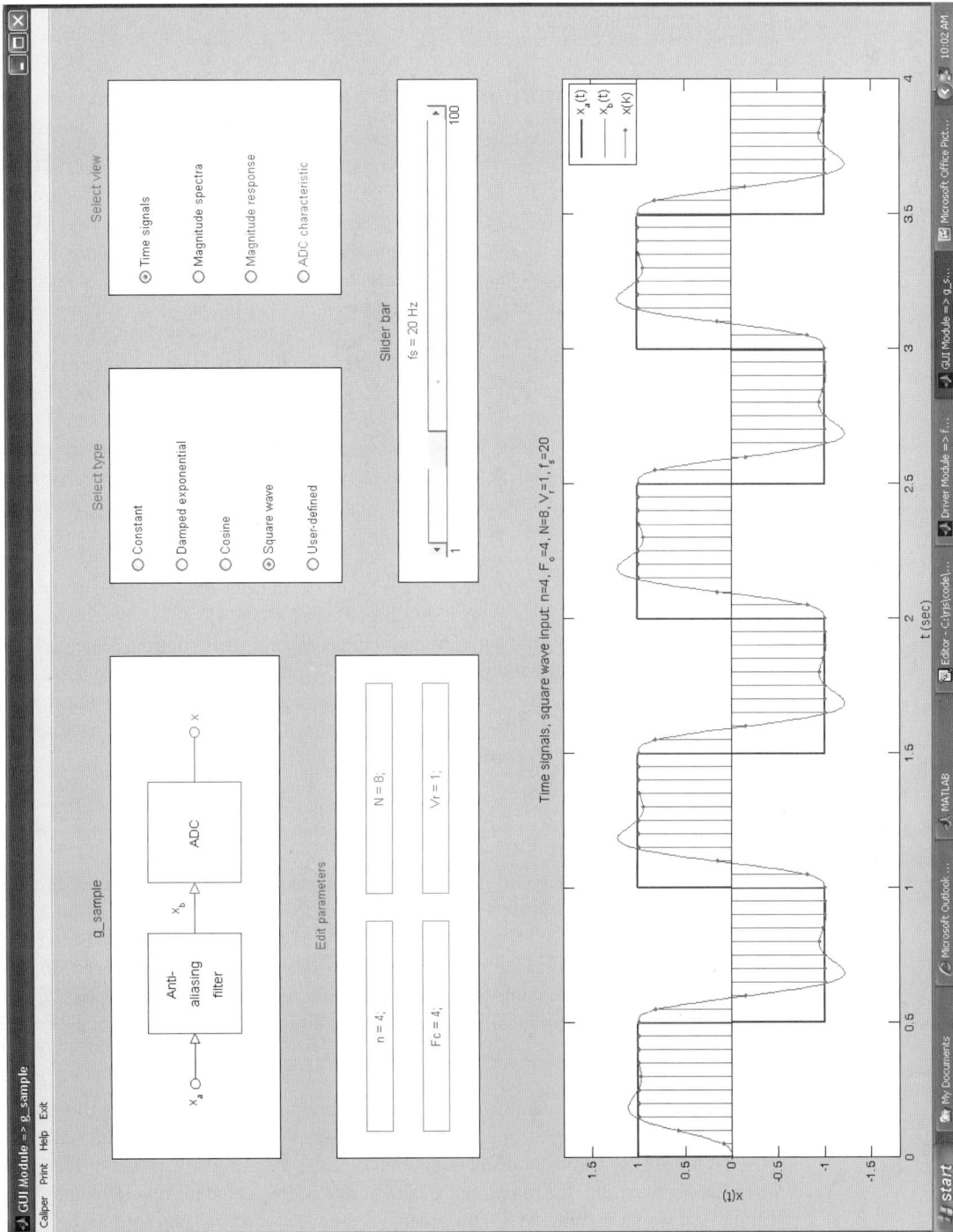

FIGURE 1.41: Display Screen of Chapter GUI Module *g_sample*

or a file. Finally, the *Help* option provides the user with some helpful suggestions on how to effectively use module *g_sample*.

g_reconstruct: Continuous-time Signal Reconstruction

The graphical user interface module *g_reconstruct* is a companion to module *g_sample* that allows the user to interactively investigate the signal reconstruction process. GUI module *g_reconstruct* features a display screen with tiled windows as shown in Figure 1.42. The *Block diagram* window in the upper-left corner shows a block diagram that includes two signal processing blocks, a DAC and an anti-imaging filter. Below each block is pair of edit boxes that allow the user to change the parameters of the signal processing blocks. Parameter changes are activated with the Enter key. For the DAC, the user can select the number of bits of precision, N, and the reference voltage, V_r. When N is small, the effects of quantization error become apparent. The user-selectable parameters for the anti-imaging Butterworth filter are the filter order, n, and the cutoff frequency, F_c. Selecting $n = 0$ removes the filtering operation completely so $y_a = y_b$.

The *Type* and *View* windows in the upper-right corner of the screen allow the user to select both the type of input signal y, and the viewing mode. The inputs include several common signals plus a user-defined input selection where the user supplies the file name of a user-defined M-file function as was described for GUI module *g_sample*. The viewing options

View options   include the color-coded time signals (y_b, y_a, y), and their magnitude spectra. Another viewing option shows the magnitude responses of the DAC and the anti-aliasing filter. The *Plot* window along the bottom half of the screen shows the selected view.

Menu options   The *Menu* bar at the top of the screen includes several menu options. The *Caliper* option allows the user to measure any point on the current plot by moving the mouse cross hairs to that point and clicking. The *Print* option prints the contents of the plot window to a printer or a file. Finally, the *Help* option provides the user with some helpful suggestions on how to effectively use module *g_reconstruct*.

CASE STUDY 1.1 | **Anti-aliasing Filter Design**

An anti-aliasing filter, or guard filter, is an analog lowpass filter that is placed in front of the ADC to reduce the effects of aliasing when the signal to be sampled is not bandlimited. Consider the configuration shown in Figure 1.43 which features an nth-order lowpass Butterworth filter followed by an ADC. Suppose the Butterworth filter has a cutoff frequency of F_c and the ADC is a bipolar N-bit analog-to-digital converter with a reference voltage of V_r. Since the

Over sampling   Butterworth filter is not an ideal lowpass filter, we oversample by a factor of $\alpha > 1$. That is

$$f_s = 2\alpha F_c, \quad \alpha > 1 \tag{1.8.1}$$

The design task is then as follows. Find the minimum filter order, n, that will ensure that the magnitude of the aliasing error is no larger than the quantization error of the ADC.

Given the monotonically decreasing nature of the magnitude response of the Butterworth filter (see Figure 1.28), the maximum aliasing error occurs at the folding frequency $f_d = f_s/2$. The largest signal that the ADC can process has magnitude V_r. Thus from (1.5.1) and (1.8.1) the maximum aliasing error is

$$E_a = V_r |H_a(f_s/2)|$$
$$= V_r |H_a(\alpha F_c)|$$
$$= \frac{V_r}{\sqrt{1 + \alpha^{2n}}} \tag{1.8.2}$$

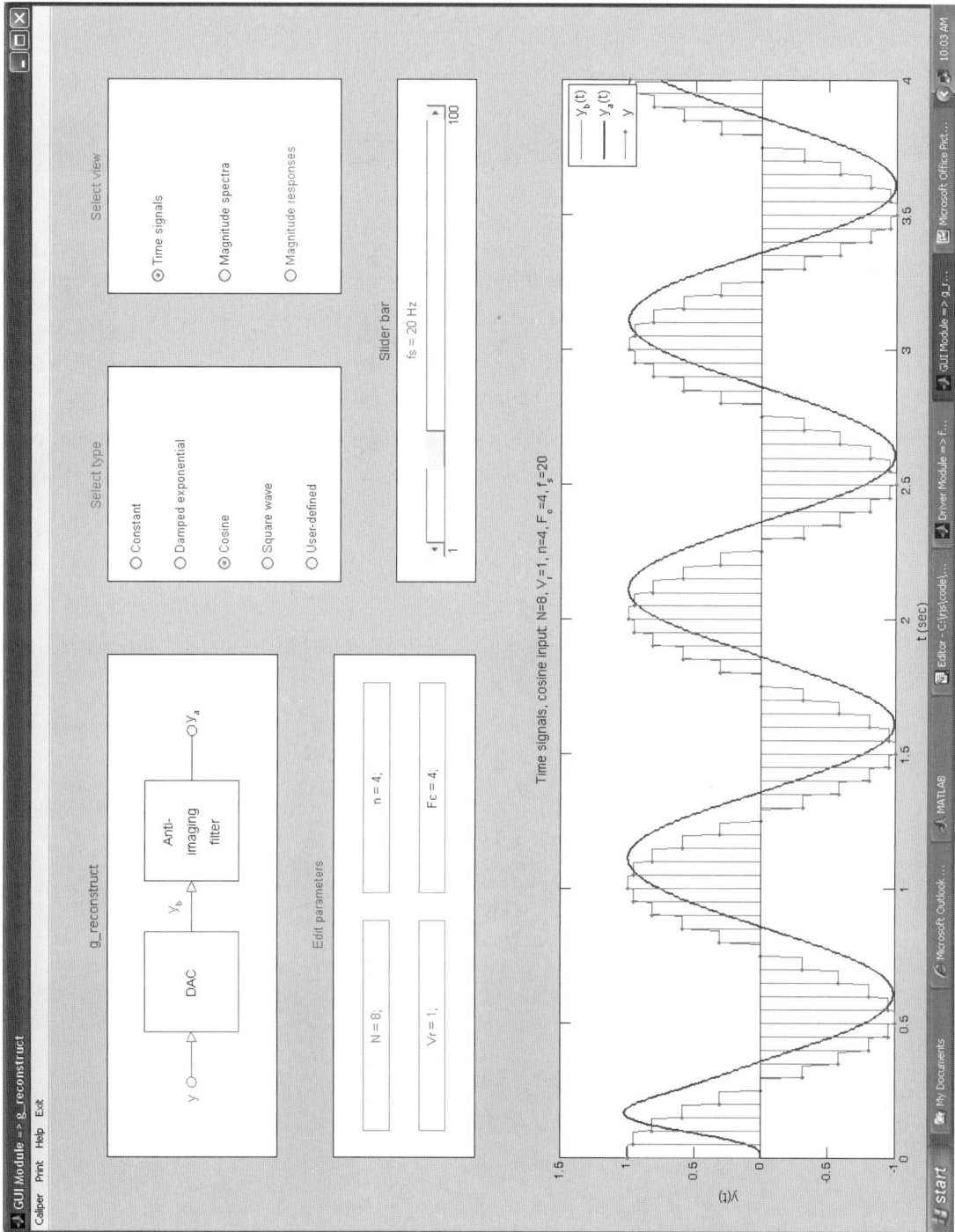

FIGURE 1.42: Display Screen of Chapter GUI Module *g_reconstruct*

FIGURE 1.43:
Preprocessing with
an Anti-Aliasing
Filter

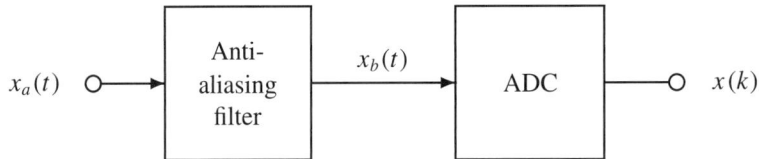

Next, if the bipolar ADC input-output characteristic is offset by $q/2$ as in Figure 1.35, then the size of the quantization error is $|e_q| \leq q/2$ where q is the quantization level. Thus from (1.6.7), the maximum quantization error is

$$E_q = \frac{q}{2}$$

$$= \frac{V_r}{2^N} \tag{1.8.3}$$

Setting $E_a^2 = E_q^2$, we observe that the reference voltage, V_r drops out. Taking reciprocals then yields

$$1 + \alpha^{2n} = 2^{2N} \tag{1.8.4}$$

Finally, solving for n, the required order of the anti-aliasing filter must satisfy

$$n \geq \frac{\ln(2^{2N} - 1)}{2\ln(\alpha)} \tag{1.8.5}$$

Of course the filter order n must be an integer. Consequently, the required order of n is as follows where *ceil* rounds up to the nearest integer.

$$n = \text{ceil}\left[\frac{\ln(4^N - 1)}{2\ln(\alpha)}\right] \tag{1.8.6}$$

CASE STUDY 1.1

MATLAB function *case1_1* computes n using (1.8.6) for oversampling rates in the range $2 \leq \alpha \leq 4$ and ADC precisions in the range $10 \leq N \leq 16$ bits. It can be executed directly from the menu of the FDSP driver program *f_dsp*.

```
function case1_1

% CASE STUDY 1.1: Anti-aliasing filter design

f_header('Case Study 1.1: Anti-aliasing filter design')
F_c = 1;

% Compute minimum filter order

alpha = [2 : 4]
N = [8 : 12]
r = length(N);
n = zeros(r,3);
for i = 1 : r
    for j = 1 : 3
        n(i,j) = ceil(log(4^N(i)-1)/(2*log(alpha(j))));
    end
end
n
```

FIGURE 1.44:
Minimum Order of
Anti-aliasing Filter
Needed to Ensure
that Magnitude of
Aliasing Error
Matches ADC
Quantization Error
for Different Levels
of Oversampling α

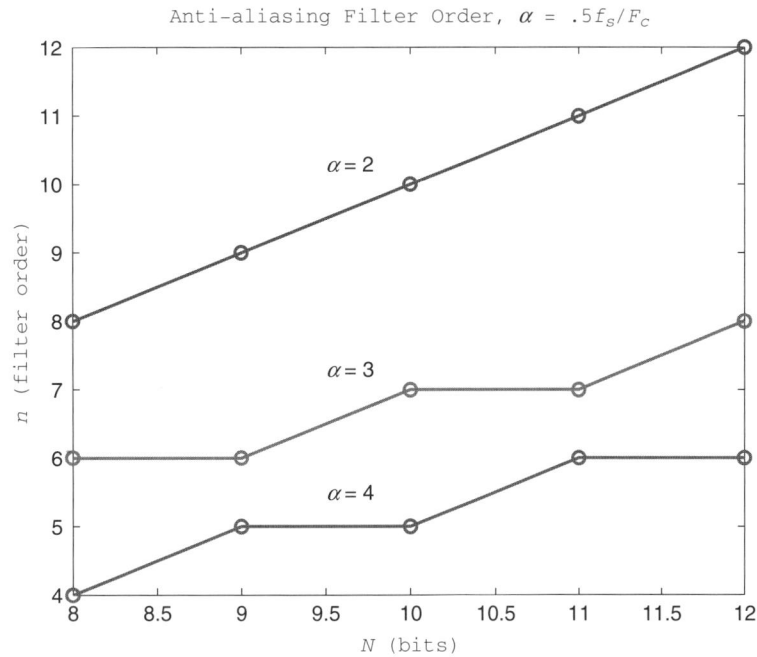

Anti-aliasing Filter Order, $\alpha = .5f_s/F_c$

```
% Display results

figure
plot (N,n,'o-','LineWidth',1.5)
f_labels ('Anti-aliasing filter order, \alpha = .5f_s/F_c','N 'n (bits)',...
          'n (filter order)')
text (9.5,5.5,'\alpha = 4')
text (9.5,7.3,'\alpha = 3')
text (9.5,10.3,'\alpha = 2')
f_wait
```

When *case1_1* is run, it produces the plot shown in Figure 1.44 which displays the required anti-aliasing filter order versus the ADC precision for three different values of oversampling. As expected, as the oversampling factor α increases, the required filter order decreases.

CASE STUDY 1.2

Video Aliasing

Recall from Section 1.1 that a video signal or movie can be represented by an $M \times N$ image $I_a(t)$ that varies with time. Here, $I_a(t)$ consists of an array of picture elements or pixels where the number of rows, M, and columns, N, depends on the video format used. Each pixel is a solid color. For example, if the RGB true color format is used, then the color is specified with 24 bits, 8 for red, 8 for green, and 8 for blue. Consequently each pixel can take on an integer color value in the large, but finite, range $0 \leq c < 2^{24}$. This makes $I_a(t)$ a quantized signal. If

Digital signal $I_a(t)$ is sampled with a sampling interval of T, the resulting signal is discrete in both time and amplitude, in which case it is an MN-dimensional digital signal.

$$I(k) = I_a(kT), \quad k \geq 0 \tag{1.8.7}$$

To illustrate the sampling process and the phenomenon of aliasing, suppose the image $I_a(t)$ features a rotating disk with a dark radial line to indicate orientation as shown previously in Figure 1.7. Suppose the disk is rotating clockwise at a rate of F_0 revolutions/second. If the line on the disk starts out horizontal and to the right, this corresponds to an initial angle of $\theta(0) = 0$. For a clockwise rotation, the angle at time t is

$$\theta_a(t) = -2\pi F_0 t \tag{1.8.8}$$

Next suppose the image $I_a(t)$ is sampled at a rate of f_s frames/sec. The angle of the line, as seen by the viewer of the kth frame, is

$$\theta(k) = \frac{-2\pi F_0 k}{f_s} \tag{1.8.9}$$

Since the disk is rotating at a constant rate of F_0 Hz, a reference point at the end of a line segment of radius r can be thought of as a two-dimensional signal $x_a(t) \in R^2$ with the following rectangular coordinates.

$$x_a(t) = \begin{bmatrix} r\cos[\theta_a(t)] \\ r\sin[\theta_a(t)] \end{bmatrix} \tag{1.8.10}$$

It follows that the signal $x_a(t)$ is bandlimited to F_0 Hz. From the sampling theorem in Proposition 1.1, aliasing will be avoided if $f_s > 2F_0$. Note from (1.8.9) that this corresponds to the line rotating less than π radians. Thus when $f_s > 2F_0$, the sequence of images will show a disk that appears to be rotating clockwise which, in fact, it is. However, for $f_s \leq 2F_0$, aliasing will occur. For the limiting case, $f_s = 2F_0$, the disk rotates exactly half a turn in each image, so it is not possible to tell which direction it is turning. Indeed, when $f_s = F_0$, the disk does not appear to be rotating at all. Since the sampling frequency must satisfy $f_s > 2F_0$ to avoid aliasing, it is convenient to represent the sample rate as a fraction of $2F_0$.

$$f_s = 2\alpha F_0 \tag{1.8.11}$$

Oversampling factor

Here α is the *oversampling factor* where $\alpha > 1$ corresponds to oversampling and $\alpha \leq 1$ represents undersampling. When $\alpha > 1$ the disk appears to turn clockwise, and when $\alpha = .5$ it appears to stop. For values of α near .5 the perceived direction and speed of rotation vary. The following function can be used to interactively view the spinning disk at different sampling

CASE STUDY 1.2

rates to see the effects of aliasing first hand. Like all examples in this text, *case1_2* can be executed from the driver program *f_dsp*. Give it a try and see what you think.

```
function case1_2

% CASE STUDY 1.2: Video aliasing

f_header('Case Study 1.2: Video aliasing')
quit = 0;
tau = 4;
theta = 0;
phi = linspace(0,360,721)/(2*pi);
r1 = 4;
x = r1*cos(phi);
y = r1*sin(phi);
r2 = r1 - .5;
alpha = 2;
F0 = 2;
fs = alpha*2*abs(F0);
T = 1/fs;
```

```
% Main loop

while ~quit

% Select an option

    choice = menu('Case Study 1.2: Video aliasing',...
                  'Enter the oversampling factor, alpha',...
                  'Create and play the video',...
                  'Exit');

% Implement it

    switch (choice)
        case 1,
            alpha = f_prompt ('Enter the oversampling factor, alpha',0,4,alpha);
            fs = alpha*2*abs(F0);
            T = 1/fs;
        case 2,
            k = 1;
            hp = plot(x,y,'b','LineWidth',1.5);
            axis square
            axis([-5 5 -5 5])
            hold on
            caption = sprintf('Oversampling factor \\alpha = %.2f',alpha);
            title(caption)
            frames = fs*tau;
            for i = 1 : frames
                theta = -2*pi*F0*k*T;
                x1 = r2*cos(theta);
                y1 = r2*sin(theta);
                if k > 1
                    plot([0 x0],[0 y0],'w','LineWidth',1.5);
                end
                plot([0 x1],[0 y1],'k','LineWidth',1.5);
                x0 = x1;
                y0 = y1;
                M(k) = getframe;
                k = k + 1;
                tic
                while (toc < T) end
            end
            f_wait
        case 3,
            quit = 1;
    end
end
```

● ● ● ● ● ● ● ● ● ● ● ● ● ● ● ●

1.9   Chapter Summary

This chapter focused on signals, systems, and the sampling and reconstruction process. A signal is a physical variable whose value changes with time or space. Although our focus is on time signals, the DSP techniques introduced also can be applied to two-dimensional spatial signals such as images.

Signals and Systems

Continuous, discrete signals

A *continuous-time* signal, $x_a(t)$, is a signal whose independent variable, t, takes on values over a continuum. A *discrete-time* signal, $x(k)$, is a signal whose independent variable is available only at discrete instants of time $t = kT$ where T is the sampling interval. If $x(k)$ is the sampled version of $x_a(t)$, then

$$x(k) = x_a(kT), \quad |k| = 0, 1, 2, \ldots \tag{1.9.1}$$

Just as the independent variable can be continuous or discrete, so can the dependent variable or amplitude of the signal. When a discrete-time signal is represented with finite precision, the signal is said to be *quantized* because it can only take on discrete values. A quantized discrete-time signal is called a *digital* signal. That is, a digital signal is discrete both in time and in amplitude. The spacing between adjacent discrete values is called the quantization level q.

Quantization level

For an N-bit signal with values ranging over the interval $[x_m, x_M]$, the quantization level is

$$q = \frac{x_M - x_m}{2^N} \tag{1.9.2}$$

Just as light can be decomposed into a spectrum of colors, a signal $x_a(t)$ contains energy that is distributed over a range of frequencies. These spectral components are obtained by applying the Fourier transform to produce a complex-valued frequency domain representation of the signal, $X_a(f)$, called the signal spectrum. The magnitude of $X_a(f)$ is called the *magnitude spectrum*, and the phase angle of $X_a(f)$ is called the *phase spectrum*. The frequency response of a linear continuous-time system with input $x_a(t)$ and output $y_a(t)$ is the spectrum of the output signal divided by the spectrum of the input signal.

Magnitude, phase spectra

$$H_a(f) = \frac{Y_a(f)}{X_a(f)} \tag{1.9.3}$$

Frequency response

The *frequency response* is complex-valued, with the magnitude, $A_a(f) = |H_a(f)|$, called the *magnitude response* of the system, and the phase angle, $\phi_a(f) = \angle H_a(f)$, called the *phase response* of the system. A linear system that is designed to reshape the spectrum of the input signal in some desired way is called a frequency-selective *filter*.

Continuous-time Signal Sampling

The process of creating a discrete-time signal by sampling a continuous-time signal $x_a(t)$ can be modeled mathematically as amplitude modulation of a uniform periodic impulse train, $\delta_T(t)$.

$$\hat{x}_a(t) = x_a(t)\delta_T(t) \tag{1.9.4}$$

The effect of sampling is to scale the spectrum of $x_a(t)$ by $1/T$ where T is the sampling interval, and to replicate the spectrum of $x_a(t)$ at integer multiples of the sampling frequency $f_s = 1/T$. A signal $x_a(t)$ is *bandlimited* to B Hz if the spectrum is zero for $|f| > B$. A

Bandlimited signal

bandlimited signal of bandwidth B can be reconstructed from its samples by passing it through an ideal lowpass filter with gain T and a cutoff frequency at the folding frequency $f_d = f_s/2$ as long as the sampling frequency satisfies:

$$f_s > 2B \qquad (1.9.5)$$

Sampling theorem

Consequently, a signal can be reconstructed from its samples if the signal is bandlimited and the sampling frequency is greater than twice the bandwidth. This fundamental result is known as the Shannon *sampling theorem*. When the signal $x_a(t)$ is not bandlimited, or the sampling rate does not exceed twice the bandwidth, the replicated spectral components of $\hat{x}_a(t)$ overlap and it is not possible recover $x_a(t)$ from its samples. The phenomenon of spectral overlap is known as *aliasing*.

Anti-aliasing filter

Most signals of interest are not bandlimited. To minimize the effects of aliasing, the input signal is preprocessed with a lowpass analog filter called an *anti-aliasing* filter before it is sampled with an analog-to-digital converter (ADC). A practical lowpass filter, such as a Butterworth filter, does not completely remove spectral components above the cutoff frequency F_c. However, the residual aliasing can be further reduced by *oversampling* at a rate $f_s = 2\alpha F_c$ where $\alpha > 1$ is the oversampling factor.

Continuous-time Signal Reconstruction

Once the digital input signal $x(k)$ from the ADC is processed with a DSP algorithm, the resulting output $y(k)$ is typically converted back to an analog signal using a digital-to-analog converter (DAC). Whereas an ADC can be modeled mathematically with an impulse sampler, a DAC is modeled as a *zero-order hold* filter with transfer function

Zero-order hold

$$H_0(s) = \frac{1 - \exp(-Ts)}{s} \qquad (1.9.6)$$

The transfer function of the DAC is the Laplace transform of the output divided by the Laplace transform of the input assuming zero initial conditions. The piecewise-constant output of the DAC contains high-frequency spectral components called images centered at integer multiples of the sampling frequency. These can be reduced by postprocessing with a second lowpass analog filter called an *anti-imaging* filter.

Anti-imaging filter

A circuit realization of a DAC was presented that featured an R-$2R$ resistor network, an operational amplifier, and a bank of digitally-controlled analog switches. Circuit realizations of ADCs that were considered included the successive approximation ADC and the flash ADC. The successive approximation ADC is a widely used, high precision, medium speed converter that requires N clock pulses to perform an N-bit conversion. By contrast, the flash ADC is a very fast, hardware intensive, medium precision converter that requires $2^N - 1$ comparator circuits, but performs a conversion in a single clock pulse.

FDSP Toolbox

The final topic covered in this chapter was the FDSP toolbox, a collection of MATLAB functions and graphical user interface (GUI) modules designed for use with the basic MATLAB interpreter. User programs can make use of the toolbox functions that include general support functions provided to facilitate program development, plus chapter functions that implement the DSP algorithms developed in each chapter. The GUI modules are provided so the user can interactively explore the DSP topics covered in each chapter without any need for programming. For example, the FDSP toolbox includes GUI modules called *g_sample* and *g_reconstruct* that allow the user to interactively investigate the sampling and reconstruction of continuous-time signals including aliasing and quantization effects. The FDSP toolbox software is available from the companion web site of the publisher. It can be conveniently accessed through the

TABLE 1.7: ▶
Learning Outcomes
for Chapter 1

Num.	Learning Outcome	Sec.
1	Understand the advantages and disadvantages of digital signal processing	1.1
2	Know how to classify signals in terms of their independent and the dependent variables	1.2
3	Know how to model quantization error and understand its source	1.2
4	Understand what it means for a signal to be bandlimited, and how to bandlimit a signal	1.5
5	Understand the significance of the Sampling Theorem and how to apply it to avoid aliasing	1.5
6	Know how to reconstruct a signal from its samples	1.6
7	Understand how to specify and use anti-aliasing and anti-imaging filters	1.7
8	Understand the operation and limitations of ADC and DAC converters	1.8
9	Know how to use the FDSP toolbox to write MATLAB scripts	1.9
10	Be able to use the GUI modules *g_sample* and *g_reconstruct* to investigate signal sampling and reconstruction	1.9

driver module, *f_dsp*. This includes running the GUI modules, viewing and executing all of the MATLAB examples and figures that appear throughout the text, and viewing pdf file solutions to selected end of chapter problems. The driver module also provides toolbox help and includes an option for downloading the latest version of the FDSP toolbox from the internet.

Learning Outcomes

This chapter was designed to provide the student with an opportunity to achieve the learning outcomes summarized in Table 1.7.

● ● ● ● ● ● ● ● ● ● ● ● ● ● ● ● ● ●

1.10   Problems

The problems are divided into Analysis and Design problems that can be solved by hand or with a calculator, GUI Simulation problems that are solved using GUI modules *g_sample* and *g_reconstruct*, and MATLAB Computation problems that require user programs.

1.10.1   Analysis and Design

1.1 If a real analog signal $x_a(t)$ is square integrable, then the *energy* that the signal contains within the frequency band $[F_0, F_1]$ where $F_0 \geq 0$ can be computed as follows.

$$E(F_0, F_1) = 2 \int_{F_0}^{F_1} |X_a(f)|^2 df$$

Consider the following two-sided exponential signal with $c > 0$.

$$x_a(t) = \exp(-c|t|)$$

(a) Find the total energy, $E(0, \infty)$.
(b) Find the percentage of the total energy that lies in the frequency range $[0, 2]$ Hz.

1.2 Consider the following *signum* function that returns the sign of its argument.

$$\text{sgn}(t) \triangleq \begin{cases} 1, & t > 0 \\ 0, & t = 0 \\ -1, & t < 0 \end{cases}$$

(a) Using Appendix 1, find the magnitude spectrum
(b) Find the phase spectrum

1.3 Suppose the input to an amplifier is $x_a(t) = \sin(2\pi F_0 t)$ and the steady-state output is

$$y_a(t) = 100\sin(2\pi F_0 t + \phi_1) - 2\sin(4\pi F_0 t + \phi_2) + \cos(6\pi F_0 t + \phi_3)$$

(a) Is the amplifier a linear system or is it a nonlinear system?
(b) What is the gain of the amplifier?
(c) Find the average power of the output signal.
(d) What is the total harmonic distortion of the amplifier?

1.4 Show that the spectrum of a causal signal $x_a(t)$ can be obtained from the Laplace transform $X_a(s)$ be replacing s by $j2\pi f$. Is this also true for noncausal signals?

1.5 Parseval's identity states that a signal and its spectrum are related in the following way.

$$\int_{-\infty}^{\infty} |x_a(t)|^2 dt = \int_{-\infty}^{\infty} |X_a(f)|^2 df$$

Use Parseval's identity to compute the following integral.

$$J = \int_{-\infty}^{\infty} \text{sinc}^2(2Bt) dt$$

1.6 Consider the following discrete-time signal where the samples are represented using N bits.

$$x(k) = \exp(-ckT)\mu(k)$$

(a) How many bits are needed to ensure that the quantization level is less than .001?
(b) Suppose $N = 8$ bits. What is the average power of the quantization noise?

1.7 Consider the following periodic signal.

$$x_a(t) = 1 + \cos(10\pi t)$$

(a) Compute the magnitude spectrum of $x_a(t)$.
(b) Suppose $x_a(t)$ is sampled with a sampling frequency of $f_s = 8$ Hz. Sketch the magnitude spectra of $x_a(t)$ and the sampled signal, $\hat{x}_a(t)$.
(c) Does aliasing occur when $x_a(t)$ is sampled at the rate $f_s = 8$ Hz? What is the folding frequency in this case?
(d) Find a range of values for the sampling interval T that ensures that aliasing does not occur.
(e) Assuming $f_s = 8$ Hz, find an alternative lower-frequency signal, $x_b(t)$, that has the same set of samples as $x_a(t)$.

1.8 Consider the causal exponential signal

$$x_a(t) = \exp(-ct)\mu_a(t)$$

(a) Using Appendix 1, find the magnitude spectrum.
(b) Find the phase spectrum
(c) Sketch the magnitude and phase spectra when $c = 1$.

1.9 Let $x_a(t)$ be a periodic signal with period T_0. The *average power* of $x_a(t)$ can be defined as follows.

$$P_x = \frac{1}{T_0} \int_0^{T_0} |x_a(t)|^2 dt$$

Find the average power of the following periodic continuous-time signals.

(a) $x_a(t) = \cos(2\pi F_0 t)$
(b) $x_a(t) = c$
(c) A periodic train of pulses of amplitude a, duration T, and period T_0.

1.10 Consider the following bandlimited signal.

$$x_a(t) = \sin(4\pi t)[1 + \cos^2(2\pi t)]$$

(a) Using the trigonometric identities in Appendix 2, find the maximum frequency present in $x_a(t)$.

(b) For what range of values for the sampling interval T can this signal be reconstructed from its samples?

1.11 There are special circumstances where it is possible to reconstruct a signal from its samples even when the sampling rate is less than twice the bandwidth. To see this, consider a signal $x_a(t)$ whose spectrum $X_a(f)$ has a hole in it as shown in Figure 1.45.

FIGURE 1.45: A Signal whose Spectrum has a Hole in it

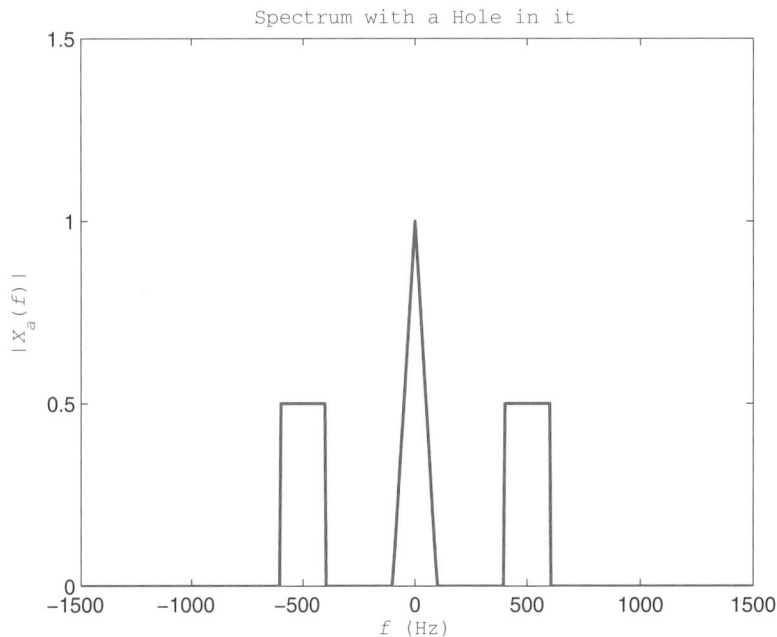

(a) What is the bandwidth of the signal $x_a(t)$ whose spectrum is shown in Figure 1.45? The pulses are of radius 100 Hz.

(b) Suppose the sampling rate is $f_s = 750$ Hz. Sketch the spectrum of the sampled signal $\hat{x}_a(t)$.

(c) Show that $x_a(t)$ can be reconstructed from $\hat{x}_a(t)$ by finding an idealized reconstruction filter with input $\hat{x}_a(t)$ and output $x_a(t)$. Sketch the magnitude response of the reconstruction filter.

(d) For what range of sampling frequencies below $2f_s$ can the signal be reconstructed from the samples using the type of reconstruction filter from part (c)?

1.12 It is not uncommon for students to casually restate the sampling theorem in Proposition 1.1 in the following way. "A signal must be sampled at twice the highest frequency present to avoid aliasing." Interesting enough, this informal formulation is not quite correct. To verify this, consider the following simple signal.

$$x_a(t) = \sin(2\pi t)$$

(a) Find the magnitude spectrum of $x_a(t)$, and verify that the highest frequency present is $F_0 = 1$ Hz.

(b) Suppose $x_a(t)$ is sampled at the rate $f_s = 2$ Hz. Sketch the magnitude spectra of $x_a(t)$ and the sampled signal, $\hat{x}_a(t)$. Do the replicated spectra overlap?

(c) Compute the samples $x(k) = x_a(kT)$ using the sampling rate $f_s = 2$ Hz. Is it possible to reconstruct $x_a(t)$ from $x(k)$ using the reconstruction formula in Proposition 1.2 in this instance?

(d) Restate the sampling theorem in terms of the highest frequency present, but this time correctly.

1.13   Why is it not possible to physically construct an ideal lowpass filter? Use the impulse response, $h_a(t)$, to explain your answer.

1.14   Consider the problem of using an anti-aliasing filter as shown in Figure 1.46. Suppose the anti-aliasing filter is a lowpass Butterworth filter of order $n = 4$ with cutoff frequency $F_c = 2$ kHz.

(a) Find a lower bound f_L on the sampling frequency that ensures that the aliasing error is reduced by a factor of at least .005.

(b) The lower bound f_L represents oversampling by what factor?

FIGURE 1.46:
Preprocessing with an Anti-aliasing Filter

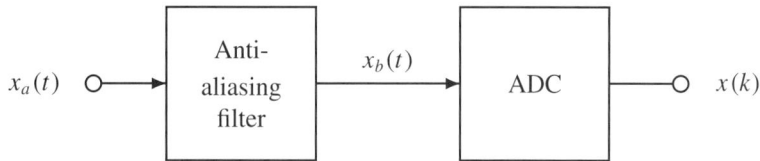

1.15   A bipolar DAC can be constructed from a unipolar DAC by inserting an operational amplifier at the output as shown in Figure 1.47. Note that the unipolar N-bit DAC uses a reference voltage of $2V_R$, rather than $-V_r$ as in Figure 1.34. This means that the unipolar DAC output is $-2y_a$ where y_a is given in (1.6.4). Analysis of the operational amplifier section of the circuit reveals that the bipolar DAC output is then

$$z_a = 2y_a - V_r$$

FIGURE 1.47: A Bipolar N-bit DAC

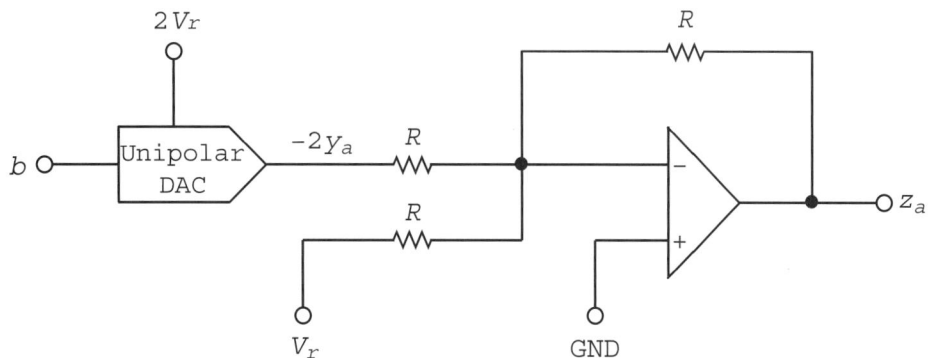

(a) Find the range of values for z_a.

(b) Suppose the binary input is $b = b_{N-1}b_{N-2} \cdots b_0$. For what value of b is $z_a = 0$?

(c) What is the quantization level of this bipolar DAC?

1.16   Show that the transfer function of a linear continuous-time system is the Laplace transform of the impulse response.

1.17 Suppose a bipolar ADC is used with a precision of $N = 12$ bits, and a reference voltage of $V_r = 10$ volts.

(a) What is the quantization level q?
(b) What is the maximum value of the magnitude of the quantization noise assuming the ADC input-output characteristics is offset by $q/2$ as in Figure 1.35?
(c) What is the average power of the quantization noise?

1.18 An alternative to the R-$2R$ ladder DAC is the weighted-resistor DAC shown in Figure 1.48 for the case $N = 4$. Here the switch controlled by bit b_k is open when $b_k = 0$ and closed when $b_k = 1$. Recall that the the decimal equivalent of the binary input b is as follows.

$$x = \sum_{k=0}^{N-1} b_k 2^k$$

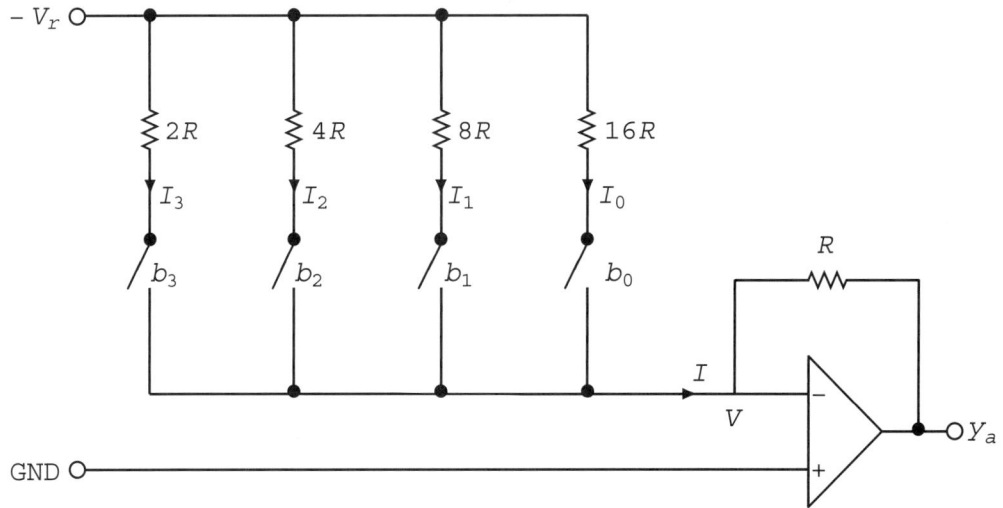

FIGURE 1.48: A 4-bit Weighted-resistor DAC

(a) Show that the current through the kth branch of an N-bit weighted-resistor DAC is

$$I_k = \frac{-V_r b_k}{2^{N-k} R}, \quad 0 \le k < N$$

(b) Show that the DAC output voltage is

$$y_a = \left(\frac{V_r}{2^N}\right) x$$

(c) Find the range of output values for this DAC.
(d) Is this DAC unipolar, or is it bipolar?
(e) Find the quantization level of this DAC.

1.19 Suppose an 8-bit bipolar successive approximation ADC has reference voltage $V_r = 10$ volts.

(a) If the analog input is $x_a = -3.941$ volts, find the successive approximations by filling in the entries in Table 1.8.
(b) If the clock rate is $f_{clock} = 200$ kHz, what is the sampling rate of this ADC?

(c) Find the quantization level of this ADC.

(d) Find the average power of the quantization noise.

TABLE 1.8: ▶	k	b_{n-k}	μ_k	y_k
Successive				
Approximations	0			
	1			
	2			
	3			
	4			
	5			
	6			
	7			

1.10.2 GUI Simulation

1.20 Consider the following exponentially damped sine wave with $c = 1$ and $F_s = 1$.

$$x_a(t) = \exp(-ct)\sin(2\pi F_0 t)\mu_a(t)$$

(a) Write a M-file function called *u-sample2* that returns the value $x_a(t)$.

(b) Use the User-defined option in GUI module *g_sample* to sample this signal at $f_s = 12$ Hz. Plot the time signals.

(c) Adjust the sampling rate to $f_s = 4$ Hz and the cutoff frequency to $F_d = 1$ Hz. Plot the magnitude spectra.

1.21 Use GUI module *g_sample* to plot the magnitude spectra of the User-defined signal in the file, *u_sample1*. Set $F_d = 1$ and do the following two cases. For which ones are there noticeable aliasing?

(a) $f_s = 2$ Hz

(b) $f_s = 10$ Hz

1.22 Consider the exponentially damped sine wave in Problem 1.20.

(a) Write a M-file function called *user1* that returns the value $x_a(t)$.

(b) Use the User-defined option in GUI module *g_reconstruct* to sample this signal at $f_s = 8$ Hz. Plot the time signals.

(c) Adjust the sampling rate to $f_s = 4$ Hz and set $F_d = 2$ Hz. Plot the magnitude spectra.

1.23 Use GUI module *g_reconstruct* to load the User-defined signal in the file, *u_reconstruct1*. Adjust Af_s to 12 Hz and $V_r = 4$.

(a) Plot the time signals, and use the Caliper option to display the amplitude and time of the peak value of the output.

(b) Plot the magnitude spectra.

1.24 Use GUI module *g_sample* to plot the time signals and magnitude spectra of the square wave using $f_s = 10$ Hz. On the magnitude spectra plot, use the Caliper option to display the amplitude and frequency of the third harmonic of $x(k)$. Are there even harmonics present in the square wave?

1.25 Use GUI module *g_reconstruct* to plot the magnitude responses of the following anti-imaging filters. What is the oversampling factor in each case?

(a) $n = 2$, $F_c = 1$

(b) $n = 6$, $F_c = 2$

1.26 Use GUI module *g_reconstruct* with the damped exponential input to plot the time signals using the following DACs. What is the quantization level in each case?

(a) $N = 4$, $V_r = .5$
(b) $N = 12$, $V_r = 2$

1.27 Use GUI module *g_sample* with the damped exponential input to plot the time signals using the following ADCs. For what cases does the ADC output saturate? Write down the quantization level on each time plot.

(a) $N = 4$, $V_r = 1$
(b) $N = 8$, $V_r = .5$
(c) $N = 8$, $V_r = 1$

1.28 Use the GUI module *g_reconstruct* to plot the magnitude responses of a 12-bit DAC with reference voltage $V_r = 10$ volts, and a 6th order Butterworth anti-imaging filter with cutoff frequency $F_c = 2$ Hz. Use oversampling by a factor of two.

1.29 Use GUI module *g_sample* to plot the magnitude response of the following anti-aliasing filters. What is the oversampling factor, α, in each case?

(a) $n = 2$, $F_c = 1$, $f_s = 2$
(b) $n = 6$, $F_c = 2$, $f_s = 12$

1.10.3 MATLAB Computation

1.30 Write a MATLAB function called *u_sinc* that returns the value of the normalized sinc function

$$\text{sinc}(x) = \frac{\sin(\pi x)}{\pi x}$$

Note that, by L'Hospital's rule, $\text{sinc}(0) = 1$. Make sure your function works properly when $x = 0$. Plot $\text{sinc}(2t)$ for $-1 \le t \le 1$ using $N = 401$ samples.

1.31 The purpose of this problem is to numerically verify the signal reconstruction formula in Proposition 1.2. Consider the following bandlimited periodic signal which can be thought of as a truncated Fourier series.

$$x_a(t) = 1 - 2\sin(\pi t) + \cos(2\pi t) + 3\cos(3\pi t)$$

Write a MATLAB program that uses the function *u_sinc* from Problem 1.25 to approximately reconstruct $x_a(t)$ as follows.

$$x_p(t) = \sum_{k=-p}^{p} x_a(kT)\text{sinc}[f_s(t - kT)]$$

Use a sampling rate of $f_s = 6$ Hz. Plot $x_a(t)$ and $x_p(t)$ on the same graph using 101 points equally spaced over the interval $[-2, 2]$. Using *f_prompt*, prompt for the number p and do the following three cases.

(a) $p = 5$
(b) $p = 10$
(c) $p = 20$

1.32 Consider the following elliptic lowpass filter from Chapter 7.

$$H_a(s) = \frac{2.0484s^2 + 171.6597}{s^3 + 6.2717s^2 + 50.0487s + 171.6597}$$

Write a MATLAB program the uses the FDSP toolbox function *f_freqs* to compute the magnitude response of this filter. Plot it over the range [0, 3] Hz. This filter has the sharpest cutoff and is optimal in the sense that the passband ripples and the stopband ripples are all of the same size.

1.33 Consider the following Chebyshev-II lowpass filter from Chapter 7.

$$H_a(s) = \frac{3s^4 + 499s^2 + 15747}{s^5 + 20s^4 + 203s^3 + 1341s^2 + 5150s + 15747}$$

Write a MATLAB program the uses the FDSP toolbox function *f_freqs* to compute the magnitude response of this filter. Plot it over the range [0, 3] Hz. This filter is optimal in the sense that the stopband ripples are all of the same size.

1.34 The Butterworth filter is optimal in the sense that, for a given filter order, the magnitude response is as flat as possible in the passband. If ripples are allowed in the passband, then an analog filter with a sharper cutoff can be achieved. Chebyshev lowpass filters will be discussed in Chapter 7. Consider the following Chebyshev-I lowpass filter from Chapter 7.

$$H_a(s) = \frac{1263.7}{s^5 + 6.1s^4 + 67.8s^3 + 251.5s^2 + 934.3s + 1263.7}$$

Write a MATLAB program that uses the FDSP toolbox function *f_freqs* to compute the magnitude response of this filter. Plot it over the range [0, 3] Hz. This filter is optimal in the sense that the passband ripples are all of the same size.

CHAPTER 2

Discrete-time Systems in the Time Domain

● ● ● ● ● ● ● ● ● ● ● ● ● ● ● ● ● **Chapter Topics**

2.1 Motivation

2.2 Discrete-time Signals

2.3 Discrete-time Systems

2.4 Difference Equations

2.5 Block Diagrams

2.6 The Impulse Response

2.7 Convolution

2.8 Correlation

2.9 Stability in the Time Domain

2.10 GUI Software and Case Studies

2.11 Chapter Summary

2.12 Problems

● ● ● ● ● ● ● ● ● ● ● ● ● ● ●

2.1 Motivation

A discrete-time system is an entity that processes a discrete-time input signal $x(k)$ to produce a discrete-time output signal $y(k)$. Recall from Chapter 1 that if the signals $x(k)$ and $y(k)$ are discrete in amplitude, as well as in time, then they are digital signals and the associated system is a digital signal processor or digital filter. In this chapter we focus our attention on analyzing the input-output behavior of linear time-invariant (LTI) discrete-time systems. This material lays a mathematical foundation for subsequent chapters where we design digital filters and develop digital signal processing (DSP) algorithms. A finite-dimensional LTI discrete-time system can be represented in the time domain by a constant-coefficient *difference equation*.

Difference equation

$$y(k) + \sum_{i=1}^{n} a_i y(k-i) = \sum_{i=0}^{m} b_i x(k-i)$$

There are a number of important DSP applications that require the processing of *pairs* of signals. For example, if $h(k)$ is the response of a linear discrete-time system to a unit impulse

Convolution   input, the response to an arbitrary input $x(k)$ when the initial condition is zero is the *convolution* of the impulse response with the input.

$$y(k) = \sum_{i=0}^{k} h(i)x(k-i), \quad k \geq 0$$

There is a second operation involving a pair of signals that closely resembles convolution. To measure the degree to which an L-point signal $h(k)$ is similar to an M-point signal $x(k)$ where *Cross-correlation* $M \leq L$, one can compute the *cross-correlation* of h with x.

$$r_{hx}(k) = \frac{1}{L} \sum_{i=0}^{L-1} h(i)x(i-k), \quad 0 \leq k < L$$

Comparing the cross-correlation with the convolution, we see that, apart from scaling, the essential difference is in the sign of the independent variable of the second signal. Cross-correlation has a number of important applications. For example, in radar or sonar processing, cross-correlation can be used to determine if a received signal contains an echo of the transmitted signal.

We begin this chapter by examining a number of practical problems that can be modeled as discrete-time systems using difference equations. Techniques for solving difference equations in the time domain are then presented. The complete solution of a linear difference equation can be decomposed into the sum of two parts. The zero-input response is the part of the solution that arises from nonzero initial conditions, and the zero-state solution is the part of the solution that is associated with a nonzero input. The interconnection of simple discrete-time systems to produce more complex systems is illustrated using block diagrams. The zero-state response of an LTI system is completely characterized by the response to a single input, the unit impulse. Discrete-time systems are then classified into two basic groups, those with a finite duration impulse response (FIR) and those with an infinite duration impulse response (IIR). Linear and circular convolution and correlation are then considered. The relationship between them is investigated, and implementations based on matrix multiplication are presented. One of the most important qualitative characteristics of discrete-time systems is stability because almost all practical systems are stable. The stability of an LTI discrete-time system can be determined directly from its impulse response. Finally, GUI modules called *g_systime* and *g_correlate* are presented that allow the user to interactively explore discrete-time systems in the time domain, and perform convolution and correlation operations without any need for programming. The chapter concludes with some case study examples, and a summary of discrete-time system analysis in the time domain.

2.1.1 Home Mortgage

As a simple illustration of a discrete-time system that affects many families, consider the problem of purchasing a home with a mortgage loan. Suppose a fixed-rate mortgage is taken out at an annual interest rate of r compounded monthly where r is expressed as a fraction, rather than a percent. Let $y(k)$ denote the balance owed to the lending agency at the end of month k, and let $x(k)$ denote the monthly mortgage payment. The balance owed at the end of month k is the balance owed at the end of month $k-1$, plus the monthly interest on that balance, minus the monthly payment.

$$y(k) = y(k-1) + \left(\frac{r}{12}\right) y(k-1) - x(k) \tag{2.1.1}$$

Here the initial condition $y(-1) = y_0$ is the size of the mortgage. Given the *difference-equation* representation in (2.1.1), there are a number of practical questions that a potential home buyer

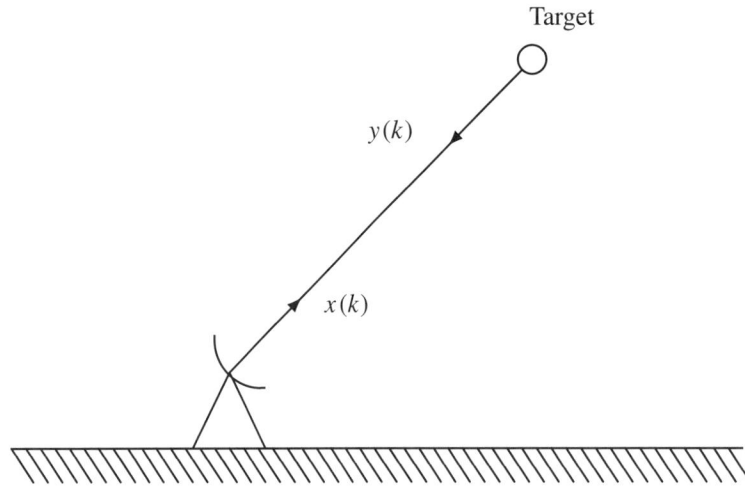

FIGURE 2.1: Radar Illuminating a Target

might want to ask. For example, if the size of the mortgage is D dollars and the duration is N months, what is the required monthly mortgage payment? Another question is how many months does it take before the majority of the monthly payment goes toward reducing principal rather than paying interest? These questions can be easily answered once we develop the tools for solving this system. As we shall see, the mortgage loan system is an example of an unstable discrete-time system.

2.1.2 Range Measurement with Radar

An application where correlation arises is in the measurement of range to a target using radar. Consider the radar installation shown in Figure 2.1.

Here the radar antenna transmits an electromagnetic wave $x(k)$ into space. When a target is illuminated by the radar, some of signal energy is reflected back and returns to the radar receiver. The received signal $y(k)$ can be modeled using the following difference equation.

$$y(k) = cx(k - d) + v(k) \tag{2.1.2}$$

The first term in (2.1.2) represents the *echo* of the transmitted signal. Typically the echo is very faint ($0 < c \ll 1$) because only a small fraction of the original signal is reflected back, with most of the signal energy dissipated through dispersion into space. The delay of d samples accounts for the propagation time required for the signal to travel from the transmitter to the target and back again. Thus d is proportional to the time of flight. The second term in (2.1.2) accounts for random atmospheric measurement noise $v(k)$ that is picked up and amplified by the receiver. In effect, (2.1.2) models of the communication channel from the transmitter to the receiver when an echo is received.

The objective in processing the received signal $y(k)$ is to determine whether or not an echo is present. If an echo is detected, then the distance to the target can be obtained from the delay d. If T is the sampling interval, then the time of flight in seconds is $\tau = dT$. Suppose γ denotes the signal propagation speed. For radar this is the speed of light. Then the distance to the target is computed as follows, where the factor of two arises because the signal must make a round trip.

$$r = \frac{\gamma dT}{2} \tag{2.1.3}$$

FIGURE 2.2: Signal Received at Radar Station

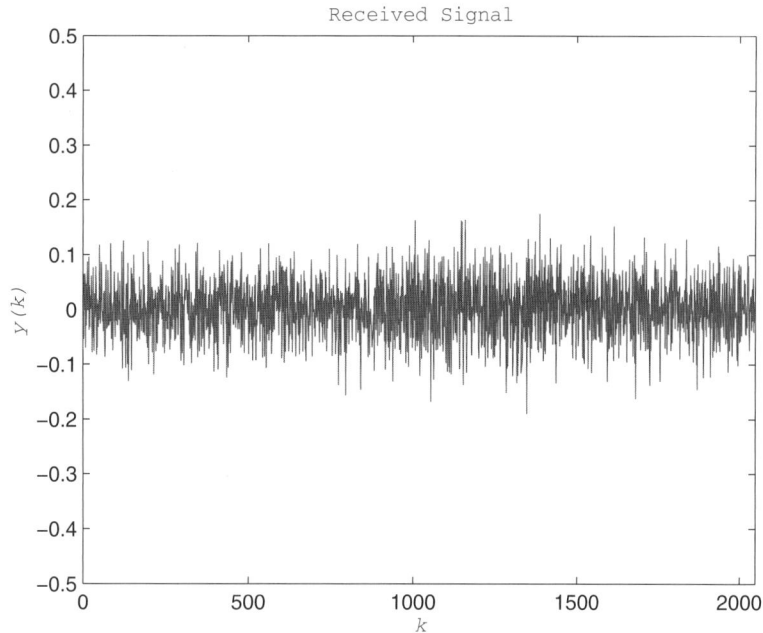

To detect if an echo is present in the noise-corrupted received signal $y(k)$, we compute the normalized cross-correlation of $y(k)$ with $x(k)$. If the measurement noise $v(k)$ is not present, the normalized cross-correlation will have a spike of unit amplitude at $i = d$. When $v(k) \neq 0$, the height of the spike will be reduced.

As an illustration, suppose the transmitted signal $x(k)$ consists of $M = 512$ points of white noise uniformly distributed over $[-1, 1]$. Other broadband signals, for example, a multifrequency chirp, might also be used. Next let the received signal $y(k)$ consist of $N = 2048$ points. Suppose the attenuation factor is $c = 0.05$. Let $v(k)$ consist of zero-mean Gaussian white noise with a standard deviation of $\sigma = 0.05$. A plot of the received signal is shown in Figure 2.2.

Note that from a direct inspection of Figure 2.2, it is not at all apparent whether $y(k)$ contains a delayed and attenuated echo of $x(k)$, much less where it is located. However, if we compute the normalized cross-correlation of $y(k)$ with $x(k)$, then the presence and location of an echo are evident, as can be seen in Figure 2.3. Using the MATLAB *max* function, we find that the time of flight in this case is $d = 939$ samples. The range to the target then can be found using (2.1.3).

MATLAB Functions

There are two built-in MATLAB functions for creating white noise signals. The following code segment creates an N-point white noise signal v that is uniformly distributed between a and b.

```
N = 100;                        % number of points
a = -1;                         % lower limit
b = 1;                          % upper limit
v = a + (b-a)*rand(N,1)         % uniform white noise in [a,b]
```

FIGURE 2.3: Normalized Cross-correlation of Transmitted Signal with Received Signal

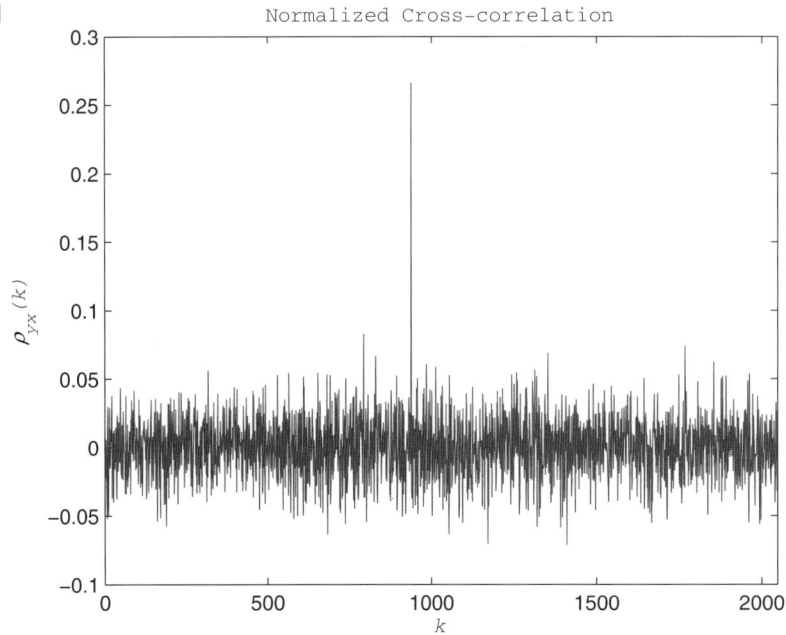

A similar technique can be used to generate normal or Gaussian white noise by replacing *rand* with *randn*. In this case, a will be the mean and $b - a$ will be the standard deviation. White noise is discussed in detail in Chapter 4.

● ● ● ● ● ● ● ● ● ● ● ● ● ● ● ● ● ●

2.2 Discrete-time Signals

The process of sampling a continuous-time signal $x_a(t)$ to produce a discrete-time signal $x(k) = x_a(kT)$ was introduced in Chapter 1. This is the most common way for discrete-time signals to arise in practice because physical variables exist in continuous time. Discrete-time signals can be categorized in a number of useful ways.

2.2.1 Signal Classification

Finite and Infinite Signals

One of the most fundamental ways to characterize a discrete-time signal is to count the number

Finite signal
of samples. A signal $x(k)$ is a finite duration signal, or a *finite signal* , if and only if $x(k)$ is defined over a finite interval $N_1 \leq k \leq N_2$ where $N_2 \geq N_1$. Alternatively, one can think of $x(k)$ as being defined for all integers k but equal to zero outside the interval $[N_1, N_2]$.

$$x(k) = 0, \quad k \notin [N_1, N_2] \tag{2.2.1}$$

Thus a finite signal exists, or can be nonzero, for a finite number of samples N. Otherwise,

Infinite signal
it is an infinite duration signal or an *infinite signal*. When practical computational algorithms such as the fast Fourier transform (FFT) are implemented in software, they are applied to finite signals.

Finite signals are often used to approximate infinite signals. For example, suppose a signal $x(k)$ is defined for $0 \leq k < \infty$ but approaches zero asymptotically as k approaches infinity.

$$|x(k)| \to 0 \quad \text{as} \quad k \to \infty \qquad (2.2.2)$$

This signal can be approximated by a finite signal over the range $[0, N - 1]$ by using a sufficiently large N and truncating or removing the "tail" of the infinite signal.

Causal and Noncausal Signals

Causal signal

When we work with discrete-time systems, it is often convenient to consider input signals whose samples are zero for negative time. The contribution from the part of the input signal that would have come from negative time can be accounted for using initial conditions. A signal $x(k)$ whose samples do not exist or are zero for $k < 0$ is referred to as a *causal signal*.

$$x(k) = 0, \quad k < 0 \qquad (2.2.3)$$

Thus a causal signal $x(k)$ is a signal whose nonzero samples start at $k = 0$. Otherwise, the signal is a *noncausal signal*. Note that a signal that is defined over an interval $[0, N - 1]$ is both finite and causal.

Periodic and Aperiodic Signals

Periodic signal

Among signals $x(k)$ that are defined for all k, there is a subset of signals that are periodic. A signal $x(k)$ is a *periodic signal* if and only if there exists an integer $N > 0$ called the *period* such that

$$x(k + N) \equiv x(k) \qquad (2.2.4)$$

Here the notation $\equiv$, which is pronounced *identically equal*, indicates that the two sides are equal for all values of the independent variable k. A casual observer might think that if a continuous-time signal $x_a(t)$ is periodic, then the discrete-time signal generated by sampling $x_a(t)$ must also be periodic. This is not necessarily the case, as can be seen from the following example.

Example 2.1

Periodic Signals

Suppose a discrete-time signal $x(k)$ is constructed by sampling a continuous-time signal $x_a(t)$ using a sampling rate of $f_s = 1/T$ Hz. For example, consider the following periodic continuous-time signal and its samples.

$$x_a(t) = \cos(2\pi F_0 t)$$
$$x(k) = \cos(2\pi F_0 k T)$$

Here $x_a(t)$ is periodic with a period of $T_0 = 1/F_0$ sec. However, for $x(k)$ to be periodic there must be an integer $N > 0$ such that $x(k + N) \equiv x(k)$. This will occur if there are an integer number of samples per period of $x_a(t)$. That is,

$$f_s = N F_0, \quad N \geq 3$$

Recall from the sampling theorem that to avoid aliasing, it is necessary that the bandlimited signal $x_a(t)$ be sampled at a rate of $f_s > 2F_0$. Thus the minimum sampling rate that will yield a periodic alias-free discrete-time signal is $f_s = 3F_0$. Here the discrete-time period in seconds is $\tau = NT$, and the continuous-time period is $T_0 = 1/F_0$.

Interestingly enough, it is possible to produce a periodic signal $x(k)$ whose period is longer than the period of $x_a(t)$. Consider the case when the sampling rate f_s is related to the fundamental frequency F_0 by a rational number L/M, where L and M are positive integers.

$$f_s = \frac{LF_0}{M}, \quad L > 2M$$

In this case, $Mf_s = LF_0$ or $LT = MT_0$. Thus the time associated with L samples corresponds to M periods of $x_a(t)$. In this instance, $x(k)$ is periodic with period L. Of course, it is possible for the ratio f_s/F_0 to be irrational. When f_s and F_0 are related by an irrational number such as $f_s = \sqrt{5}F_0$, then $x(k)$ is not periodic.

Bounded and Unbounded Signals

Later in the chapter we consider the important concept of stability. Although several different types of stability can be defined, one of the most useful ones is bounded-input bounded-output (BIBO) stability. A signal $x(k)$ is a *bounded signal* if and only if there exists a constant $B_x > 0$, called a *bound*, such that for all k

Bounded signal

$$|x(k)| \le B_x \tag{2.2.5}$$

Unbounded signal

Otherwise, $x(k)$ is an *unbounded signal*. Thus, the graph of a bounded signal lies within a horizontal strip of radius B_x about the time axis where B_x is the bound, as shown in Figure 2.4. Unbounded signals grow arbitrarily large. For example, the periodic signal $x(k) = \sin(2\pi F_0 kT)$ is bounded with bound $B_x = 1$, but the undamped exponential signal $x(k) = 2^k$ is unbounded.

For signals $x(k)$ that are defined for all k, it is possible to think of these signals as vectors in a vector space of sequences. One way to measure the length or *norm* of such a vector is using the ℓ_1 norm.

$$\|x\|_1 \triangleq \sum_{k=-\infty}^{\infty} |x(k)| \tag{2.2.6}$$

FIGURE 2.4: A Bounded Signal $x(k)$

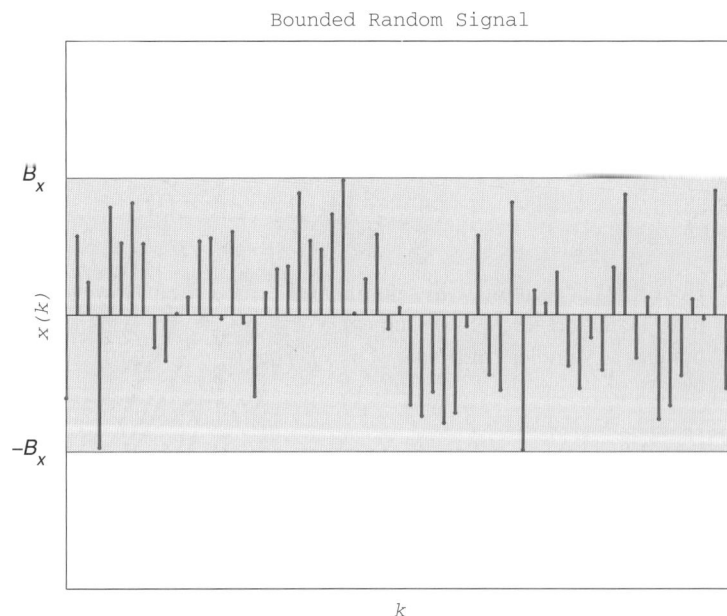

Bounded Random Signal

FIGURE 2.5:
Instantaneous
Signal Power,
$p(k) = x^2(k)/R$

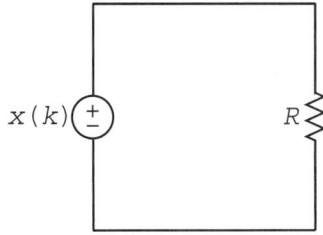

Absolutely summable

If $\|x\|_1 < \infty$, then we say that the signal $x(k)$ is *absolutely summable*. Another way to measure the length of x is by using the ℓ_2 norm.

$$\|x\|_2 \triangleq \left[\sum_{k=-\infty}^{\infty} |x(k)|^2 \right]^{1/2} \tag{2.2.7}$$

Square summable

If $\|x\|_2 < \infty$, then $x(k)$ is said to be *square summable*. Note that $(|a| + |b|)^2 \geq |a|^2 + |b|^2$ due to the presence of cross terms. As a consequence, the absolutely summable signals include the square summable signals as a special case.

$$\|x\|_1 < \infty \Rightarrow \|x\|_2 < \infty \tag{2.2.8}$$

Energy and Power Signals

Closely related to square summable signals are the concepts of energy and power. Suppose $x(k)$ represents the voltage across a one ohm resistor at time k, as shown in Figure 2.5. Then $p(k) = x^2(k)$ can be interpreted as the instantaneous power dissipated by the resistor at time k. For continuous-time signals, the integral of power over time is the total energy. For discrete-time signals, we define the *energy* of a signal $x(k)$ as follows.

Energy

$$E_x \triangleq \sum_{k=-\infty}^{\infty} |x(k)|^2 \tag{2.2.9}$$

Energy signal

Comparing (2.2.9) with (2.2.7), we see that $E_x = \|x\|_2^2$. Therefore the energy E_x is finite if and only if $x(k)$ is square summable. A signal for which $E_x < \infty$ is called an *energy signal*. It is clear that not all signals are energy signals. For example, the constant signal $x(k) = c$ for $c \neq 0$ has infinite energy and therefore is not an energy signal. For signals that do not go to zero fast enough to be square summable, it is useful to compute the average of the instantaneous power $p(x) = |x(k)|^2$. The *average power* of a signal $x(k)$ is denoted P_x and defined as follows.

Average power

$$P_x \triangleq \lim_{N \to \infty} \frac{1}{2N+1} \sum_{k=-N}^{N} |x(k)|^2 \tag{2.2.10}$$

Power signal   A signal for which P_x is nonzero and finite is called a *power signal*. If $x(k)$ is not an energy signal because its energy is infinite, it may qualify as a power signal. For example, the constant nonzero signal $x(k) = c$ has infinite energy but average power $P_x = |c|^2$.

The computation of average power simplifies when $x(k)$ is a causal signal, a periodic signal, or a finite signal. For a causal signal, the average value of the power is

$$P_x = \lim_{N \to \infty} \frac{1}{N+1} \sum_{k=0}^{N} |x(k)|^2 \tag{2.2.11}$$

Similarly, if $x(k) \equiv x(k + N)$ is periodic with a period of N, then to compute the average power, it is necessary to use only one period.

$$P_x = \frac{1}{N} \sum_{k=0}^{N-1} |x(k)|^2 \tag{2.2.12}$$

Periodic extension   Finally, if $x(k)$ is a causal finite signal with N nonzero samples, (2.2.12) can be used to compute its average power. Note that every finite signal $x(k)$ has a *periodic extension* $x_p(k)$ that is defined for all k where $x(k)$ for $0 \leq k < N$ represents one period. The expression in (2.2.12) then represents the average power of one period of $x_p(k)$.

There is a simple relationship between energy and power. For the finite causal signal $x = [x(0), \dots, x(N-1)]^T$, the energy and the average power are both finite and related as follows.

$$E_x = NP_x \tag{2.2.13}$$

This relationship between energy and average power also holds in the limit as $N \to \infty$ in the following sense. For infinite signals, if $x(k)$ is an energy signal, then its average power is $P_x = 0$, and if $x(k)$ is a power signal, then its energy is infinite. Finite nonzero signals are both energy signals and power signals.

Geometric Series

There is a well-known result from mathematics that is helpful in determining the energy of an important class of signals. Let z be real or a complex scalar. The following power series converges if and only if $|z| < 1$ (see Problem 3.17).

$$\sum_{i=0}^{\infty} z^i = \frac{1}{1-z}, \quad |z| < 1 \tag{2.2.14}$$

Geometric series   The power series in (2.2.14) is called the *geometric series*. It is clear that the geometric series will not converge for $|z| \geq 1$ since there are an infinite number of terms, and each one is at least as large as one. Many of the discrete-time signals that appear in practice consist of a sum of terms each of which can be analyzed with the geometric series.

Example 2.2   **Geometric Series**

Consider the following causal signal that includes a gain A and an exponential factor c.

$$x(k) = A(c)^k, \quad k \geq 0$$

First, consider under what conditions $x(k)$ is absolutely summable. Using (2.2.13)

$$\|x\|_1 = \sum_{k=0}^{\infty} |A(c)^k|$$

$$= \sum_{k=0}^{\infty} |A| \cdot |c^k|$$

$$= |A| \sum_{k=0}^{\infty} |c|^k$$

$$= \frac{|A|}{1-|c|}, \quad |c| < 1$$

Thus $x(k)$ is absolutely summable if and only if $|c| < 1$. From (2.2.8), every signal that is absolutely summable is also square summable. Therefore, $x(k)$ is an energy signal. The energy of $x(k)$ is

$$E_x = \sum_{k=0}^{\infty} |A(c)^k|^2$$

$$= \sum_{k=0}^{\infty} |A|^2 \cdot |c^k|^2$$

$$= |A|^2 \sum_{k=0}^{\infty} (|c|^2)^k$$

$$= \frac{|A|^2}{1-|c|^2}, \quad |c| < 1$$

2.2.2 Common Signals

Unit Impulse

Unit impulse

There are a few common signals that arise repeatedly in applications and examples. Perhaps the simplest of these is the *unit impulse* signal which is nonzero at a single sample.

$$\delta(k) \triangleq \begin{cases} 1, & k = 0 \\ 0, & k \neq 0 \end{cases} \tag{2.2.15}$$

The unit impulse $\delta(k)$ plays a role in discrete time that is analogous to the unit impulse or Dirac delta $\delta_a(t)$ in continuous time. Note that the unit impulse is both finite and causal. A graph of the unit impulse is shown in Figure 2.6(a).

Unit Step

Unit Step

Another very common signal that is related to the unit impulse in a simple way is the *unit step* signal which is causal, but not finite.

$$\mu(k) \triangleq \begin{cases} 1, & k \geq 0 \\ 0, & k < 0 \end{cases} \tag{2.2.16}$$

A plot of $\mu(k)$ is shown in Figure 2.6(b). In continuous time, the unit step $\mu_a(t)$ is the integral of the unit impulse $\delta_a(t)$. This is also true in discrete time, where the integral is replaced

FIGURE 2.6: (a) Unit Impulse Signal, (b) Unit Step Signal

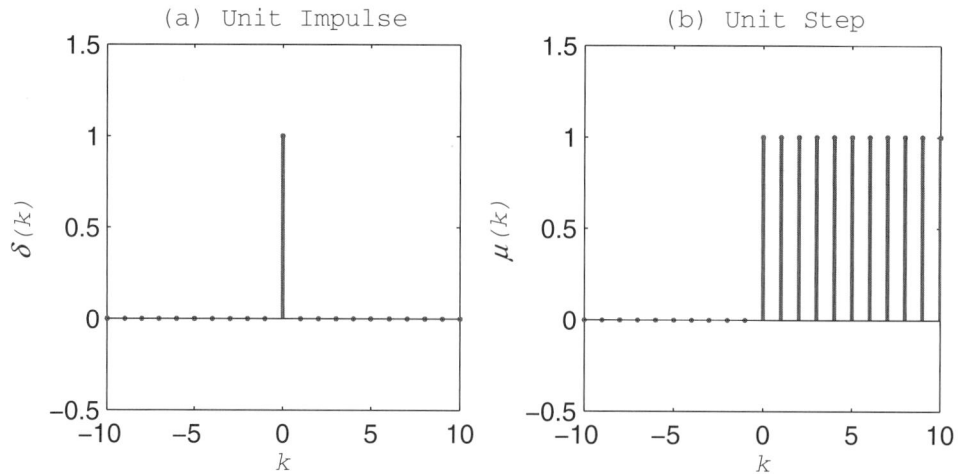

by a sum.

$$\mu(k) = \sum_{i=-\infty}^{k} \delta(i) \tag{2.2.17}$$

Note that if $x(k)$ starts out as a general signal defined for all k, the signal $x(k)\mu(k)$ is a causal signal because multiplication by $\mu(k)$ has the effect of zeroing the samples associated with negative time.

Causal Exponential

Causal exponential

Perhaps the most common discrete-time signal that appears in several different forms and many applications is the *causal exponential* signal, where c is a scalar that may be positive or negative, real or complex.

$$x(k) = c^k \mu(k) \tag{2.2.18}$$

A causal exponential is bounded for $|c| \leq 1$ and unbounded otherwise. When $c < 0$, the value of $x(k)$ oscillates between positive and negative, as can be seen in the plots of $x(k)$ in Figure 2.7. From Example 2.2, the causal exponential is absolutely summable and an energy signal if and only if $|c| < 1$. Note that $x(k)$ in (2.2.18) includes the exponential form $x(k) = \exp(-kT/\tau)$ by setting $c = \exp(-1/\tau)$.

Power Signals

Power signals

The most common examples of power signals include basic periodic signals such as the cosine and sine.

$$x_1(k) = \cos(2\pi F_0 kT) \tag{2.2.19a}$$
$$x_2(k) = \sin(2\pi F_0 kT) \tag{2.2.19b}$$

A general sinusoid can be written as a linear combination of $x_1(k)$ and $x_2(k)$. Interestingly enough, the cosine and sine signals also can be written in terms of exponentials. Suppose $c = \exp(j2\pi F_0 T)$, where $j = \sqrt{-1}$, and c^* denotes the complex conjugate of c. Then, using Euler's identity from Appendix 2, the cosine in (2.2.19a) can be written as $x_1(k) = (c^k + c^{*k})/2$. Similarly the sine in (2.2.19b) can be written as $x_2(k) = (c^k - c^{*k})/(j2)$. Plots of the discrete-time cosine and sine are shown in Figure 2.8.

FIGURE 2.7: Causal Exponentials $x(k) = c^k \mu(k)$. (a) Bounded $c > 0$, (b) Unbounded $c > 0$, (c) Bounded $c < 0$, (d) Unbounded $c < 0$

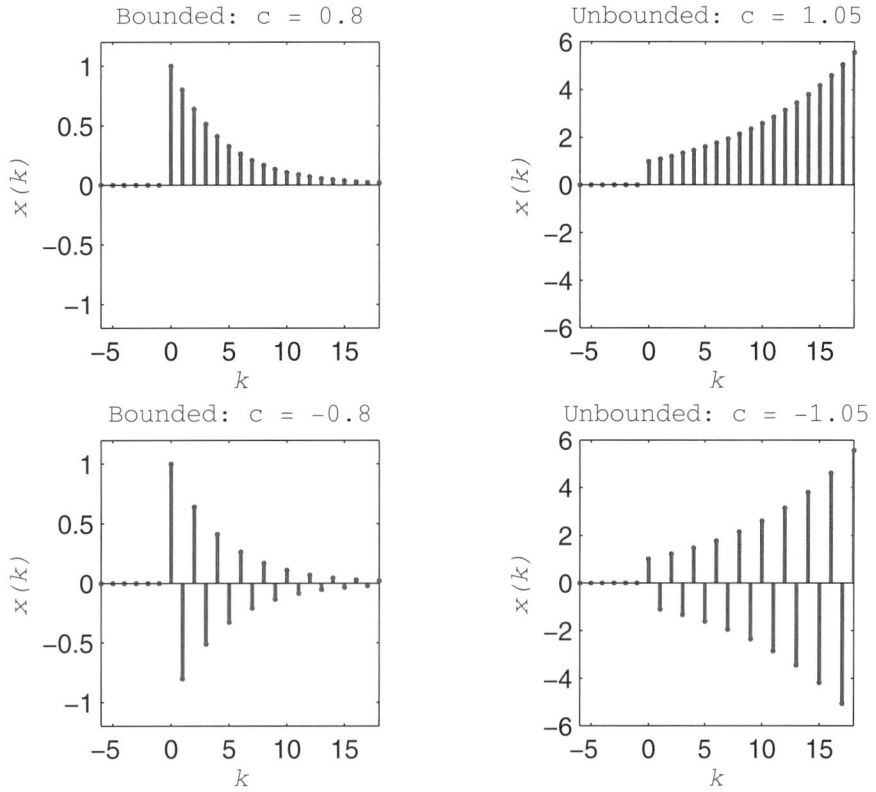

FIGURE 2.8: Periodic Power Signals with $f_s = 20F_0$. (a) $x_1(k) = \cos(2\pi F_0 kT)$, (b) $x_2(k) = \sin(2\pi F_0 kT)$

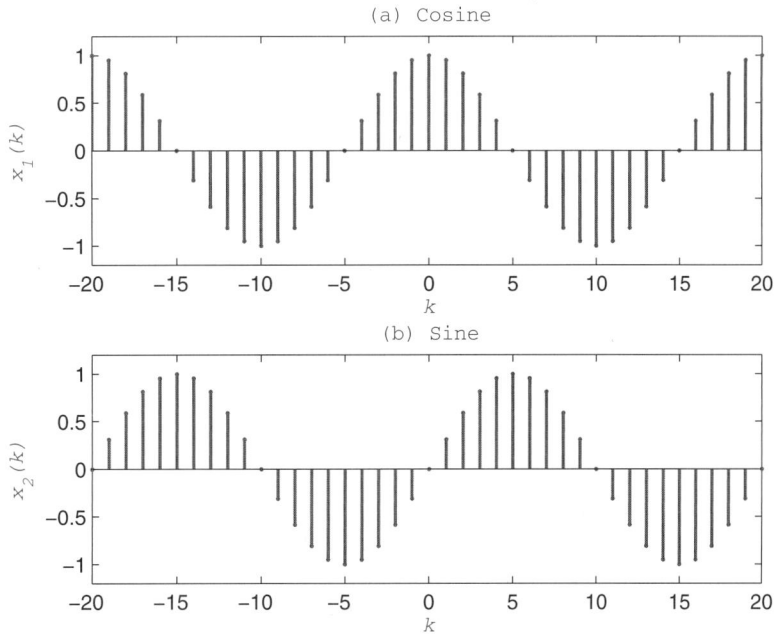

TABLE 2.1: ▶
Common
Discrete-time
Signals and Their
Characteristics

$x(k)$	Finite	Causal	Periodic	Bound	Energy	Average Power
$\delta(k)$	yes	yes	no	$B = 1$	1	1
$\mu(k)$	no	yes	no	$B = 1$	∞	1
$c^k\mu(k),\ |c| < 1$	no	yes	no	$B = |c|$	$\dfrac{1}{1 - |c|^2}$	0
$c^k\mu(k),\ |c| > 1$	no	yes	no	∞	∞	∞
$\cos(2\pi k/N)$	no	no	yes	$B = 1$	∞	.5
$\sin(2\pi k/N)$	no	no	yes	$B = 1$	∞	.5

The common signals introduced in this section all can be classified according to the criteria introduced in Section 2.2.1. Their characteristics are summarized in Table 2.1.

● ● ● ● ● ● ● ● ● ● ● ● ● ● ● ●

2.3 Discrete-time Systems

A discrete-time system S processes a discrete-time input signal $x(k)$ to produce a discrete-time output signal $y(k)$, as shown in Figure 2.9. Discrete-time systems can be categorized in a number of useful ways.

Linear and Nonlinear Systems

Just as with continuous-time systems, discrete-time systems can be linear or nonlinear. Suppose $y_i(k)$ is the output or *response* of the system S to the input $x_i(k)$ for $1 \leq i \leq 2$. If a and b are

Linear system   arbitrary scalars, then S is a *linear system* if and only if

$$x(k) = ax_1(k) + bx_2(k) \Rightarrow y(k) = ay_1(k) + by_2(k) \qquad (2.3.1)$$

Nonlinear system   Otherwise the system S is *nonlinear*. A block diagram of a linear discrete-time system is shown
Superposition   in Figure 2.10. A system that follows (2.3.1) is said to obey the *principle of superposition*. There are two important special cases. If $b = 0$ in (2.3.1), the input $x(k) = ax_1(k)$ produces the output $y(k) = ay_1(k)$. Thus when S is linear, scaling the input by a has the effect of scaling the output by the same constant a. For example, when $a = 0$, this says that a zero input generates a zero output. Similarly, if $a = b = 1$ in (2.3.1), the input $x(k) = x_1(k) + x_2(k)$ produces the output $y(k) = y_1(k) + y_2(k)$. Thus when S is linear, the response to the sum of two inputs is the sum of the responses to the individual inputs. Almost all of the discrete-time systems we consider will be linear.

FIGURE 2.9: A
Discrete-time
System S with Input
$x(k)$ and Output
$y(k)$

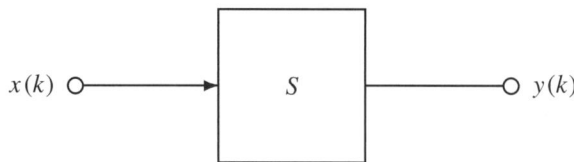

FIGURE 2.10: A
Linear Discrete-time
System

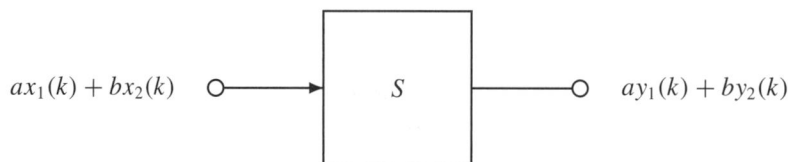

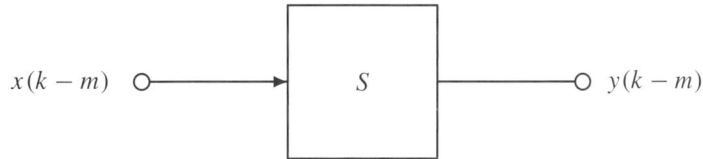

FIGURE 2.11: A Time-invariant Discrete-time System

$x(k - m)$ S $y(k - m)$

Time-invariant and Time-varying Systems

Another important classification has to do with whether or not the input-output behavior of the system S changes with time. Suppose $y(k)$ is the output corresponding to the input $x(k)$.

Time-invariant system   A discrete-time system S is *time-invariant* if and only if for an arbitrary integer m the output produced by the shifted input $x(k - m)$ is $y(k - m)$. Otherwise, S is a *time-varying* system.

For a time-invariant system, shifting the input in time shifts the output by the same amount. It does not otherwise change the output. A block diagram of a time-invariant system is shown in Figure 2.11. Time-invariant systems have parameters that are constant. For example, for a discrete-time system that is implemented as a difference equation, the system is time-invariant if and only if the coefficients of the difference equation are constant. Adaptive filters, which are introduced in Chapter 9, are examples of time-varying systems.

Causal and Noncausal Systems

Physical systems operating in real time have the fundamental property that the output at the present time k can not depend on the future input $x(i)$ for $i > k$ because that input has not yet occurred. Such a system is said to be causal. More specifically, suppose that $y_i(k)$ is the

Causal system   output produced by input $x_i(k)$ for $1 \leq i \leq 2$. A discrete-time system is *causal* if and only if for each time k

$$x_1(i) = x_2(i) \text{ for } i \leq k \Rightarrow y_1(k) = y_2(k) \tag{2.3.2}$$

For a causal system S, if two inputs are identical up to time k, then the corresponding

Noncausal system   outputs also must be identical at time k. Otherwise, S is a *noncausal* system. This implies that for a causal system, the present output $y(k)$ cannot depend on the future input $x(i)$ for $i > k$.

Offline   There are instances where signal processing is not performed online in real time. In these

processing   cases, noncausal systems can be used to process the data *offline* in batch mode, where "future" samples of the input are available. The following example illustrates the two types of systems.

Example 2.3   **Causal and Noncausal Systems**

One of the problems that arises in practice is the need to develop a numerical estimate of the derivative of a continuous-time signal $x_a(t)$ from the samples of the signal $x(k) = x_a(kT)$. For example, one might want to estimate speed from position samples. The simplest technique that can be implemented in real time is to approximate $dx_a(t)/dt$ using the slope of the line connecting the present sample $x(k)$ with the previous sample $x(k - 1)$.

$$y_1(k) = \frac{x(k) - x(k - 1)}{T}$$

This is referred to as a first-order backwards difference approximation to the derivative, also

Backward Euler   called the *backward Euler approximation.* It is first-order because the maximum distance

approximation   between the samples used is one. It is backwards because it goes backwards in time to get the samples used. The backwards difference differentiator is a causal system because the present output $y_1(k)$ does not depend on future values of the input $x(k)$.

Although the backwards difference approximation is simple and can be implemented in real time, it has the disadvantage that the estimate of the derivative ends up being centered between the samples $x(k)$ and $x(k-1)$. Thus it produces an estimate that is delayed by half a sample or $T/2$ seconds. This problem can be rectified by using the following second-order central difference approximate that uses the slope of the line connecting samples $x(k+1)$ and $x(k-1)$.

$$y_2(k) = \frac{x(k+1) - x(k-1)}{2T}$$

This is called a central difference because it is centered at sample $x(k)$. Thus there is no delay in this estimate of the derivative. However, the price paid is that this numerical differentiator can not be implemented in real time because it is a noncausal system. Notice that the present output $y_2(k)$ depends on the future input $x(k+1)$. Consequently, a central difference approximation can be implemented offline only after the samples of $x(k)$ have been recorded and saved. As an illustration of the numerical differentiation, consider the following continuous-time input sampled at a rate of $f_s = 10$ Hz.

$$x_a(t) = \sin(\pi t)$$

Here $dx_a(t)/dt = \pi \cos(\pi t)$. A plot of $dx_a(t)/dt$, $y_1(k)$, and $y_2(k)$ is shown in Figure 2.12. Observe that the backward difference $y_1(k)$ is delayed by half a sample, while $y_2(k)$ is not. The central difference $y_2(k)$ is also somewhat more accurate because it is a second-order approximation. Finally, it should be noted that the process of numerical differentiation is inherently sensitive to the presence of noise, an issue that is of concern for practical signals. Even though the noise itself may be small in amplitude, the slope of the noise signal can still be large. Consequently, differentiation tends to amplify the high frequency component of the noise. The issue of designing digital filters that approximate differentiators is discussed in more detail in Chapter 6.

FIGURE 2.12: Causal and Noncausal Numerical Estimates of the Derivative of $x_a(t) = \sin(\pi t)$

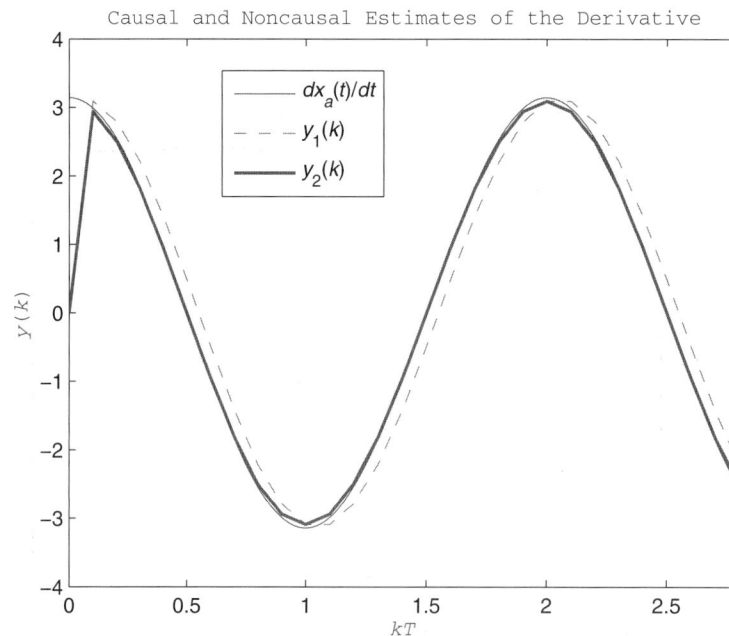

Causal and Noncausal Estimates of the Derivative

Stable and Unstable Systems

Stable system

Another fundamental classification of discrete-time systems is based on the notion of stability. Recall that input $x(k)$ is bounded if and only if there exists a $B_x > 0$, called a bound, such that $|x(k)| \leq B_x$ for all k. We say that a discrete-time systems S is a bounded-input bounded-output (BIBO) *stable* if and only if every bounded input $x(k)$ produces a bounded output $y(k)$.

$$|x(k)| \leq B_x \Rightarrow |y(k)| \leq B_y \qquad (2.3.3)$$

Unstable system

If at least one bounded input can be found that generates an unbounded output, then the system S is BIBO *unstable*.

Example 2.4

Stable and Unstable Systems

As a simple illustration of a discrete-time system that is BIBO stable, consider a running average filter of order $M - 1$.

$$y(k) = \frac{1}{M} \sum_{i=0}^{M-1} x(k-i)$$

Suppose the input $x(k)$ is bounded with bound B_x. Then

$$|y(k)| = \left| \frac{1}{M} \sum_{i=0}^{M-1} x(k-i) \right|$$

$$\leq \frac{1}{M} \sum_{i=0}^{M-1} |x(k-i)|$$

$$= B_x$$

Consequently the output $y(k)$ is bounded with bound $B_y = B_x$. By contrast, consider the home mortgage system introduced in Section 2.1.

$$y(k) = y(k-1) + \left(\frac{r}{12} \right) y(k-1) - x(k)$$

$$= \left(1 + \frac{r}{12} \right) y(k-1) - x(k)$$

Recall that $x(k)$ is the monthly payment, $y(k)$ is the balance owed at the end of month k, and $r > 0$ is the annual interest rate expressed as a fraction. This is an example of a system that is not BIBO stable. Later in this chapter, and also in Chapter 3, we develop direct tests for stability. For now, we can use the definition to check stability. Recall that the initial condition $y(-1)$ represents the amount of the loan. If the first payment is due at $k = 0$, then the balance owed would have grown by $ry(-1)/12$ over the course of the first month. If the payment $x(0)$ is less than the accumulated interest, then the balance owed will increase with $y(0) > y(-1)$. Thus, the following bounded input produces an unbounded output.

$$|x(k)| < \frac{ry(-1)}{12} \Rightarrow |y(k)| \to \infty \text{ as } k \to \infty$$

It follows that the home mortgage system is a BIBO unstable system. Indeed, if the bounded input $x(k) = 0$ is used, then no monthly payments are made, and the balance owed grows without bound due to the accumulated interest.

Passive and Active Systems

Passive system

The final classification of systems that we consider has to do with the energy of the input and output signals. Suppose the input is square summable and is therefore an energy signal with energy E_x, as in (2.2.9). Similarly, suppose the resulting output $y(k)$ is also an energy signal with energy E_y. The discrete-time system S is a *passive* system if and only if the energy does not increase.

$$E_y \leq E_x \tag{2.3.4}$$

Lossless system

Consequently, a passive system does not add energy to the input signal as it propagates through the system to produce the output signal. Otherwise, S is referred to as an *active* system. An active physical system requires a power source, whereas a passive system does not. A special case of a passive system is a *lossless* system which is a discrete-time system S for which the energy stays the same.

$$E_y = E_x \tag{2.3.5}$$

Lossless physical systems contain energy storage elements such as capacitors, inductors, springs, and masses. However, they do not contain energy dissipative elements such as resistors or dampers.

● ● ● ● ● ● ● ● ● ● ● ● ● ● ● ● ●

2.4   Difference Equations

LTI system

Every finite dimensional *linear time-invariant* (LTI) system can be represented in the time domain by a constant-coefficient difference equation with input $x(k)$ and output $y(k)$.

$$y(k) + \sum_{i=1}^{n} a_i y(k-i) = \sum_{i=0}^{m} b_i x(k-i) \tag{2.4.1}$$

System dimension

Here $M = \max\{n, m\}$ is the *dimension* of the system. For convenience, we will refer to the LTI system in (2.4.1) as the system S. Note that the coefficient of the present output $y(k)$ has been normalized to $a_0 = 1$. This can always be done by first dividing both sides of (2.4.1) by a_0 if needed. The output of the system S at time k can be expressed as follows.

$$y(k) = \sum_{i=0}^{m} b_i x(k-i) - \sum_{i=1}^{n} a_i y(k-i) \tag{2.4.2}$$

The system S, in addition to being linear and time-invariant, is also causal because the present output $y(k)$ does not depend on the future input $x(i)$ for $i > k$.

Initial condition

When we consider inputs to discrete-time systems, it is convenient to focus on causal signals where $x(k) = 0$ for $k < 0$. When causal inputs are used, the output or response of a discrete-time system will depend on both the input $x(k)$ and the *initial condition*, which is represented by a vector $y_0 \in R^n$ of past outputs.

$$y_0 \triangleq [y(-1), y(-2), \ldots, y(-n)]^T \tag{2.4.3}$$

In general, the complete response $y(k)$ will depend on both y_0 and the input $x(i)$ for $0 \leq i \leq k$, a shown in Figure 2.13. For the system S, the contributions to the output from initial condition y_0 and the input $x(k)$ can be considered separately. Because the system is linear, they can be added to produce the complete response $y(k)$.

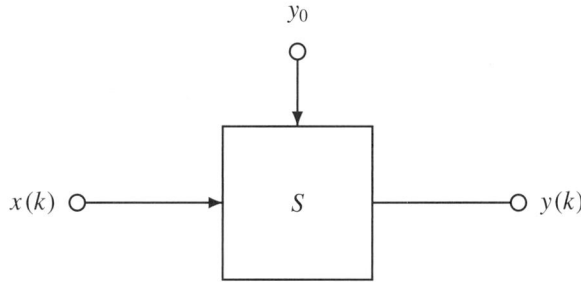

FIGURE 2.13: The Complete Response of the System S Depends on the Initial Condition y_0 and the Input $x(k)$

2.4.1 Zero-input Response

Zero-input response

The output of the discrete-time system S when the input is $x(k) = 0$ is denoted $y_{zi}(k)$ and is referred to as the *zero-input response* of the system. Thus the zero-input response is the solution of the simplified system.

$$y(k) + \sum_{i=1}^{n} a_i y(k - i) = 0, \quad y_0 \neq 0 \tag{2.4.4}$$

The zero-input response $y_{zi}(k)$ is the part of the overall response that arises from the initial condition y_0. To find the zero-input response, consider a generic solution of the form $y(k) = z^k$, where the complex scalar z is yet to be determined. Substituting $y(k) = z^k$ into (2.4.4) and multiplying both sides by z^{n-k} yields

$$a(z) = z^n + a_1 z^{n-1} + \cdots + a_n = 0 \tag{2.4.5}$$

Characteristic polynomial

The polynomial $a(z) = z^n + a_1 z^{n-1} + \cdots + a_n$ is called the *characteristic polynomial* of the system. If we factor $a(z)$ into its n roots $p_1, p_2, \ldots, p_n$, the *factored form* of the characteristic polynomial of S is

$$a(z) = (z - p_1)(z - p_2) \cdots (z - p_n) \tag{2.4.6}$$

One can assume that the roots are *nonzero* because if $a_n = 0$, this makes $a(z)$ a polynomial in z^{-1} of degree less than n. The characteristic polynomial is the key to determining the zero-input response. The signal $y(k) = cz^k$ is a solution of (2.4.4) for an arbitrary scalar c if and only if z is a root of the characteristic polynomial. The simplest case occurs when there are

Simple roots

n distinct roots, in which case we say that the roots *simple*. For n simple roots, individual solutions of the form $c_i p_i^k$ can be added to produce the zero-input response

$$y_{zi}(k) = \sum_{i=1}^{n} c_i p_i^k, \quad k \geq -n \tag{2.4.7}$$

Natural mode

The terms in (2.4.7) of the form $c_i p_i^k$ are called the *natural modes* of the system. Each simple root of the characteristic polynomial $a(z)$ generates a simple natural mode term in the zero-input response.

$$\text{simple mode} = c(p)^k \tag{2.4.8}$$

The weighting coefficients $c = [c_1, c_2, \ldots, c_n]^T$ in (2.4.7) depend on the initial conditions. In particular, setting $y_{zi}(k) = y(k)$ for $-n \leq k \leq -1$ yields n equations in the n unknown elements of the coefficient vector $c \in R^n$.

Example 2.5

Zero-input Response: Simple Roots

To illustrate the process of finding a zero-input response, consider the following two-dimensional discrete-time system.

$$y(k) - .6y(k-1) + .05y(k-2) = 2x(k) + x(k-1)$$

Suppose the initial condition is $y(-1) = 3$ and $y(-2) = 2$, in which case $y_0 = [3, 2]^T$. By inspection, the characteristic polynomial of this system is

$$\begin{aligned} a(z) &= z^2 - .6z + .05 \\ &= (z - .5)(z - .1) \end{aligned}$$

Thus the vector of simple roots is $p = [.5, .1]^T$ and the form of the zero-input response is

$$y_{zi}(k) = c_1(.5)^k + c_2(.1)^k$$

To find the coefficient vector $c \in R^2$, we apply the initial condition $y_{zi}(-1) = 3$ and $y_{zi}(-2) = 2$. In matrix form, these two equations can be written as

$$\begin{bmatrix} 2 & 10 \\ 4 & 100 \end{bmatrix} c = \begin{bmatrix} 3 \\ 2 \end{bmatrix}$$

Subtracting twice the first equation from the second yields $80c_2 = -4$ or $c_2 = -1/20$. Similarly, subtracting ten times the first equation from the second yields $-16c_1 = -28$ or $c_2 = 7/4$. Thus the zero-input response associated with this initial condition is

$$y_{zi}(k) = 1.75(.5)^k - .05(.1)^k$$

MATLAB functions   An alternative way to compute the coefficient vector c using MATLAB is

```
A = [2 10 ; 4 100];
y0 = [3 ; 2];
c = A \ y0
```

The natural modes in (2.4.7) correspond to the simplest and most common special case. A more involved case arises when one of the roots in (2.4.6) is repeated. For example, suppose root p occurs r times as a root of $a(z)$. In this case, p is referred to as a root of *multiplicity* r, *Multiple roots*   and it generates a multiple natural mode term of the following form.

$$\text{multiple mode} = (c_1 + c_2 k + \cdots + c_r k^{r-1}) p^k \qquad (2.4.9)$$

Observe that the coefficient of a root of multiplicity r is a polynomial $c(k)$ of degree $r - 1$. Again the r coefficients of $c(k)$ are determined from the initial conditions. For the special case when $r = 1$, the coefficient polynomial $c(k)$ reduces to a polynomial of degree zero, that is, a constant, as in (2.4.8). The expression in (2.4.9) represents a general natural mode of order r, whereas for simple roots the natural modes are all modes of order one, as in (2.4.8).

Example 2.6

Zero-input Response: Multiple Roots

As an illustration of what happens when the characteristic polynomial has multiple roots, consider the following two-dimensional discrete-time system.

$$y(k) + y(k-1) + .25y(k-2) = 3x(k)$$

Suppose the initial condition is $y(-1) = -1$ and $y(-2) = 6$ or $y_0 = [-1, 6]^T$. The characteristic polynomial of this system is

$$a(z) = z^2 + z + .25$$
$$= (z + .5)^2$$

Thus the vector of roots is $p = [-.5, -.5]^T$, which means that $p = -.5$ is a root of multiplicity 2. Thus the form of the zero-input response is

$$y_{zi}(k) = (c_1 + c_2 k)(-.5)^k$$

To find the coefficient vector $c \in R^2$, we apply the initial condition $y_{zi}(-1) = -1$ and $y_{zi}(-2) = 6$. In matrix form, these two equations can be written as

$$\begin{bmatrix} -2 & 2 \\ 4 & -8 \end{bmatrix} c = \begin{bmatrix} -1 \\ 6 \end{bmatrix}$$

Adding twice the first equation to the second yields $-4c_2 = 4$ or $c_2 = -1$. Adding four times the first equation to the second yields $-4c_1 = 2$ or $c_1 = -.5$. Thus the zero-input response associated with this initial condition is

$$y_{zi}(k) = -(.5 + k)(-.5)^k$$

Complex conjugate roots

The roots of the characteristic polynomial can be real or complex. Since the coefficients of $a(z)$ are real, complex roots always occur in conjugate pairs. The two natural mode terms associated with a complex conjugate pair of roots can be combined by using Euler's identity.

$$\exp(\pm j\theta) = \cos(\theta) \pm j \sin(\theta) \tag{2.4.10}$$

Euler's identity

Note that the left side of Euler's identity is the polar form of a complex variable whose magnitude is $r = 1$, and the right side is the rectangular form. When complex roots occur in conjugate pairs, their coefficients also form complex conjugate pairs. Suppose a pair of complex conjugate roots is expressed in polar form as $p_{1,2} = r \exp(\pm j\theta)$. By using Euler's identity, it is possible to show that the pair of natural mode terms can be combined to form the following damped sinusoid called a complex mode.

$$\text{complex mode} = r^k[c_1 \cos(k\theta) + c_2 \sin(k\theta)] \tag{2.4.11}$$

Example 2.7

Zero-input Response: Complex Roots

As an example of discrete-time systems whose characteristic polynomial has complex roots, consider the following two-dimensional system.

$$y(k) + .49y(k-2) = 3x(k)$$

Suppose the initial condition is $y(-1) = 4$ and $y(-2) = -2$ or $y_0 = [4, -2]^T$. The characteristic polynomial of this system is

$$a(z) = (z^2 + 1) \times z^2 + 0.4p$$
$$= (z - j.7)(z + j.7)$$

Thus the roots are $p_{1,2} = \pm j.7$. If the roots are expressed in polar form, the magnitude is $r = .7$ and the phase angle is $\theta = \pi/2$. That is, $p_{1,2} = .7 \exp(\pm j\pi/2)$. From (2.4.11), the form of the zero-input response is

$$y_{zi}(k) = .7^k[c_1 \cos(\pi k/2) + c_2 \sin(\pi k/2)]$$

Next, applying the initial conditions $y(-1) = 4$ and $y(-2) = -2$ yields

$$-c_2/.7 = 4$$
$$-c_1/.49 = -2$$

Thus $c_1 = .98$, $c_2 = -2.8$, and the zero-input response is

$$y_{zi}(k) = .7^k[.98 \cos(k\pi/2) - 2.8 \sin(k\pi/2)]$$

In the event that some of the roots are simple while others are multiple, the zero-input response consists of a combination of first-order and higher-order modes which can be real or complex. Together, these modes will include n unknown coefficients that are solved for by applying the n initial conditions.

2.4.2 Zero-state Response

Zero-state response

In general, the output of a LTI discrete-time system S also contains a component that is due to the presence of the input $x(k)$. The output of S corresponding to an arbitrary input $x(k)$, when the initial condition vector is zero, is denoted $y_{zs}(k)$ and is referred to as the *zero-state response*.

$$y(k) + \sum_{i=1}^{n} a_i y(k - i) = \sum_{i=0}^{m} b_i x(k - i), \quad y_0 = 0 \qquad (2.4.12)$$

The computation of the zero-state response is more involved than the computation of the zero-input response because there are infinitely many inputs that might be considered. In Section 2.7, a systematic procedure for finding the zero-state response to any input is presented. For now, we illustrate the process of finding the zero-state response by considering the following important class of inputs.

$$x(k) = A p_0^k \mu(k) \qquad (2.4.13)$$

Here $x(k)$ is a causal exponential with amplitude A and exponential factor p_0. To simplify the final result we assume that the roots of the characteristic polynomial $a(z)$ are distinct from one another and from p_0. Just as the characteristic polynomial $a(z)$ can be obtained from inspection of the coefficients of the output in (2.4.12), a second polynomial $b(z)$ called the *input polynomial* can be generated from the coefficients of the input. For the case when $m \le n$, this polynomial is

Input polynomial

$$b(z) = b_0 z^n + b_1 z^{n-1} + \cdots + b_m z^{n-m} \qquad (2.4.14)$$

Given an input as in (2.4.13) and $n + 1$ distinct roots, the zero-state response has a form generally similar to the zero-input response.

$$y_{zs}(k) = \sum_{i=0}^{n} d_i p_i^k \mu(k) \qquad (2.4.15)$$

The weighting coefficient vector $d \in R^{n+1}$ can be computed directly from the polynomials $a(z)$ and $b(z)$ as follows.

$$d_i = \left. \frac{A(z - p_i)b(z)}{(z - p_0)a(z)} \right|_{z=p_i}, \quad 0 \le i \le n \qquad (2.4.16)$$

Note that, in general, $d_i \ne 0$ because p_i is root of the denominator in (2.4.16) for $0 \le i \le n$.

Example 2.8

Zero-state Response

Consider the two-dimensional discrete-time system from Example 2.5 with the following input $x(k)$.

$$y(k) = .6y(k - 1) - .05y(k - 2) + 2x(k) + x(k - 1)$$
$$x(k) = .8^{k+1}\mu(k)$$

The characteristic polynomial of this system is

$$a(z) = z^2 - .6z + .05$$
$$= (z - .5)(z - .1)$$

From inspection of the input terms, the input polynomial $b(z)$ is

$$b(z) = 2z^2 + z$$
$$= 2z(z + .5)$$

In this case we have $A = .8$ and

$$\frac{Ab(z)}{(z - p_0)a(z)} = \frac{1.6z(z + .5)}{(z - .8)(z - .5)(z - .1)}$$

From (2.4.16), the weighting coefficients are

$$z=0.8 \qquad d_0 = \frac{1.6(.8)(1.3)}{.3(.7)} = 7.92$$

$$z=0.5 \qquad d_1 = \frac{1.6(.5)(1.0)}{-.3(.4)} = -6.67$$

$$z=0.1 \qquad d_2 = \frac{1.6(.1)(.6)}{-.7(-.4)} = .343$$

If we apply (2.4.15), the zero-state response associated with the input $x(k)$ is then

$$y_{zs}(k) = [7.92(.8)^k - 6.67(.5)^k + .343(.1)^k]\mu(k)$$

Complete response

Because the system S in (2.4.1) is linear, the *complete response* associated with both a nonzero initial condition y_0 and a nonzero input $x(k)$ is just the sum of the zero-input response and the zero-state response.

$$y(k) = y_{zi}(k) + y_{zs}(k) \qquad (2.4.17)$$

Stable system

Most of the discrete-time systems we investigate later have the property that they are BIBO stable. For a *stable system*, the zero-input response will go to zero as k approaches infinity.

$$|y_{zi}(k)| \to 0 \quad \text{as} \quad k \to \infty \qquad (2.4.18)$$

If the input to a stable system is a power signal, then the complete response will be dominated by the zero-state response. Consequently, we will focus most of our attention on the zero-state response. However, when the transient behavior associated with the initial condition is of interest, then the zero-input response must be computed as well.

Example 2.9   **Complete Response**

Again consider the two-dimensional discrete-time system from Example 2.5 and Example 2.8. Suppose the initial condition is $y_0 = [3, 2]^T$, and the input is

$$x(k) = .8^{k+1}\mu(k)$$

The zero-input response associated with the initial condition was computed in Example 2.5, and the zero-state response associated with the input was computed in Example 2.8. Thus from (2.4.17), the complete response of this system is

$$y(k) = y_{zi}(k) + y_{zs}(k)$$
$$= 1.75(.5)^k - .05(.1)^k + [7.92(.8)^k - 6.67(.5)^k + .343(.1)^k]\mu(k)$$

To verify that $y(k)$ satisfies the initial condition $y_0 = [3, 2]^T$, we have

$$y(-1) = 1.75(.5)^{-1} - .05(.1)^{-1} = 3$$
$$y(-2) = 1.75(.5)^{-2} - .05(.1)^{-2} = 2$$

Plots of the zero-input response, the zero-state response, and the complete response are shown in Figure 2.14. Notice that for $k \gg 1$, the complete response is dominated by the zero-state response because the zero-input response quickly dies out.

FIGURE 2.14: The Output of the System. (a) Zero-input Response with $y_0 = [3, 2]^T$, (b) Zero-state Response with $x(k) = 10(.8)^k\mu(k)$, (c) Complete Response.

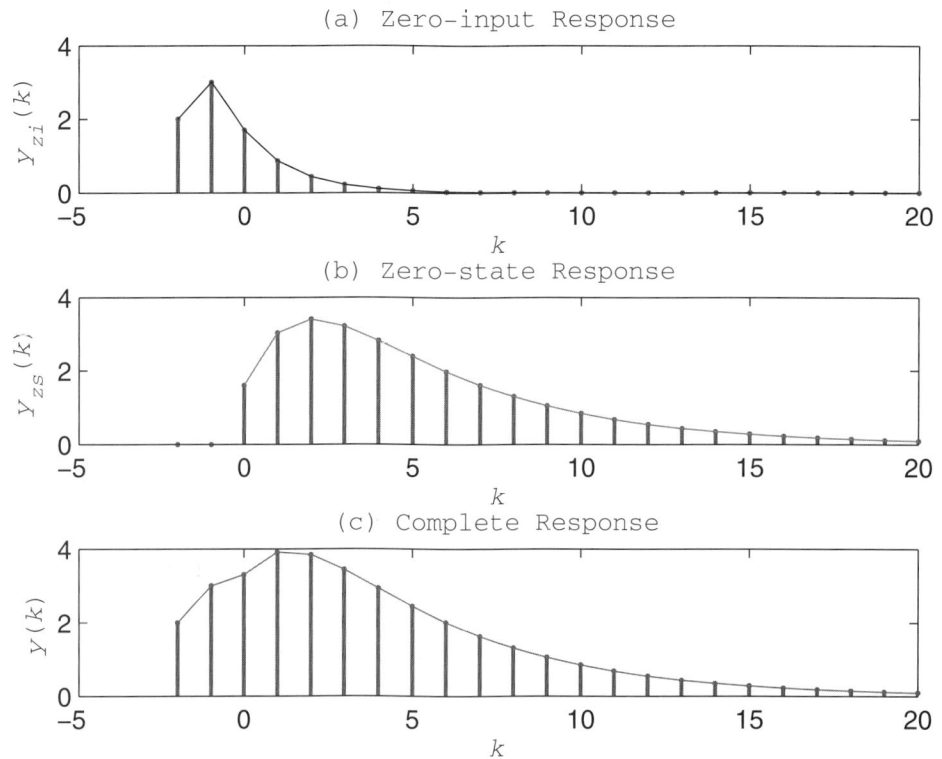

FDSP Functions

A convenient way to find the complete responses is to compute it numerically using the FDSP function *f_filter0*.

```
% F_FILTERO Compute the complete response of a discrete-time system
%
% Usage:
%          y = f_filter0 (b,a,x,y0);
% Pre:
%          b  = vector of length m+1 containing input coefficients
%          a  = vector of length n+1 containing output coefficients
%          x  = vector of N input samples x = [x(0),...,x(N-1)]
%          y0 = optional vector of n past outputs y0 = [y(-1),...,y(-n)]
% Post:
%          y  = vector of length p containing the output samples.  If
%               y0 is present, then y = [y(-n),y(-n+1),...,y(N-1)],
%               otherwise y = [y(0),y(1),...,y(N-1)].
```

MATLAB Functions

If only the zero-state response is needed, then the standard MATLAB function *filter* can be used. Calling *filter* is equivalent to calling *f_filter0* without the optional initial condition argument *y0*.

```
% FILTER: Compute output of a linear discrete-time system
%
% Usage:
%          y = f_filter(b,a,x);
% Pre:
%          b = vector of length m+1 containing input coefficients
%          a = vector of length n+1 containing output coefficients
%          x = vector of length N containing samples of input
% Post:
%          y = vector of length N containing zero state response
```

The advantage of the numerical technique is that it works for any input $x(k)$. Furthermore, the *filter* and *f_filter0* functions are valid for any $n \geq 0$ and any $m \geq 0$. Thus the numerical approach is very general. However, it does not produce a closed-form expression for the system output. Instead, it yields a fixed number of output samples.

Example 2.10

Numerical Zero-state Response

Consider the following four-dimensional discrete-time system driven by input $x(k)$.

$$y(k) = 3y(k-1) - 3.49y(k-2) + 1.908y(k-3) - .4212y(k-4) + x(k) - 2x(k-3)$$
$$x(k) = 2k(.7)^k \sin(2\pi k/10)\mu(k)$$

FIGURE 2.15:
Numerical Solution
of the Zero-state
Response for the
Four-dimensional
System Subject to
the Input $x(k) =$
$2k(.7)^k \sin(2\pi k/10)\mu(k)$

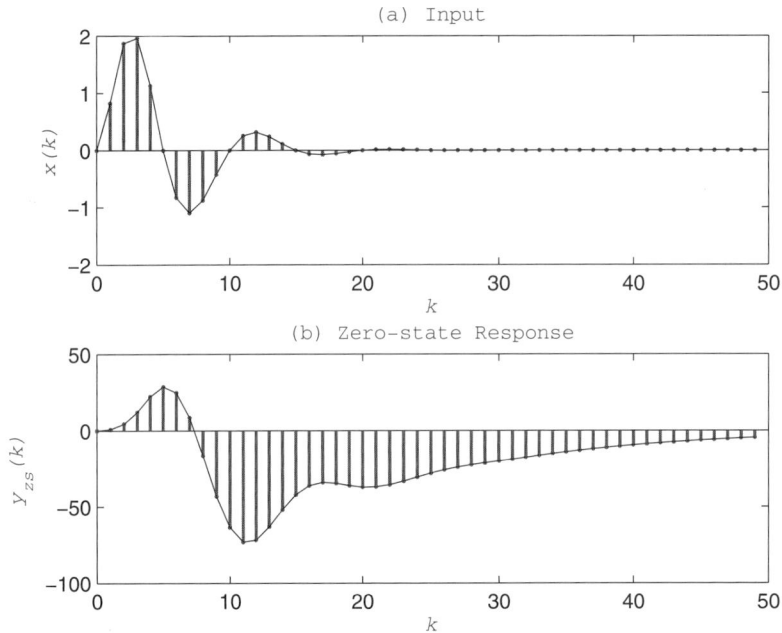

(a) Input

(b) Zero-state Response

The two coefficient vectors are

$$a = [1, 3, -3.49, 1.908, -.4212]^T$$
$$b = [1, 0, 0, -2]^T$$

MATLAB functions   If we use the MATLAB function $p = roots(a)$, the roots of the characteristic polynomial are

$$p = \begin{bmatrix} .9 \\ .9 \\ .6 + j.4 \\ .6 - j.4 \end{bmatrix}$$

Thus there is a double root at $p_{1,2} = .9$ and a complex conjugate pair of roots at $p_{3,4} = .6 \pm j.4$. Plots of the first $N = 50$ points of the input and zero-state response are shown in Figure 2.15, which was generated by running *exam2_10* from the driver program *f_dsp*.

• • • • • • • • • • • • • •

2.5 Block Diagrams

Discrete-time systems can be displayed graphically in the form of block diagrams. With block diagrams, the flow of information and the interconnection between subsystems can be visualized. A block diagram is a set of blocks that represent processing units interconnected by directed line segments that represent signals. There are four basic components, as shown in Figure 2.16. The gain block scales the signal by the constant specified in the block label. The *Delay block*   delay block, which is labeled z^{-1}, delays the signal by one sample. The summer or summing

FIGURE 2.16: Basic Elements of Block Diagrams

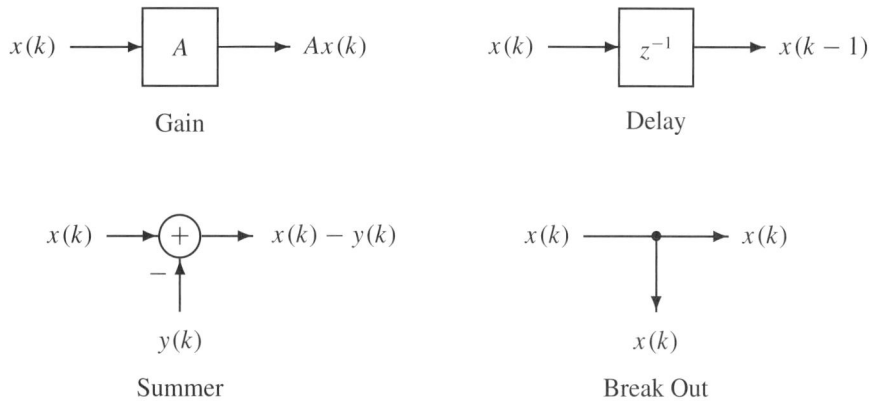

Gain

Delay

Summer

Break Out

junction block adds or subtracts two or more signals, as indicated by the sign labels. Finally, the break out point extracts a copy of the signal.

To develop a block diagram for a linear discrete-time system, first consider the case when the output is a weighted sum of the past inputs.

$$y(k) = \sum_{i=0}^{m} b_i x(k-i) \tag{2.5.1}$$

MA model   This is sometimes referred to as a *moving average* or MA model because, when $b_i = 1/(m+1)$, the output is a moving average of the past $m+1$ samples of the input. More generally, it is a weighted average with the coefficient vector b specifying the weights. An MA model has a simple block diagram, as can be seen in Figure 2.17 which illustrates the case $m = 3$. Since each z^{-1} block delays the input by one sample, the structure in Figure 2.17 is sometimes referred to as a *tapped delay line*.

Next consider a block diagram for a more general LTI system where the current output $y(k)$ depends on both the past inputs and the past outputs.

$$y(k) = \sum_{i=0}^{m} b_i x(k-i) - \sum_{i=1}^{n} a_i y(k-i) \tag{2.5.2}$$

Recall that $M = \max\{m, n\}$ is the dimension of the system. Suppose $m \leq n$. In this case we can pad the coefficient vector $b \in R^{m+1}$ with $n - m$ zeros so that $b \in R^{n+1}$. Alternatively, if $m > n$, we can pad the coefficient vector $a \in R^{n+1}$ with $m - n$ zeros so that $a \in R^{m+1}$. In either case, the zero-padded coefficient vectors a and b will both be of length $M + 1$,

FIGURE 2.17: Block Diagram of a Moving Average (MA) System of Order $m = 3$

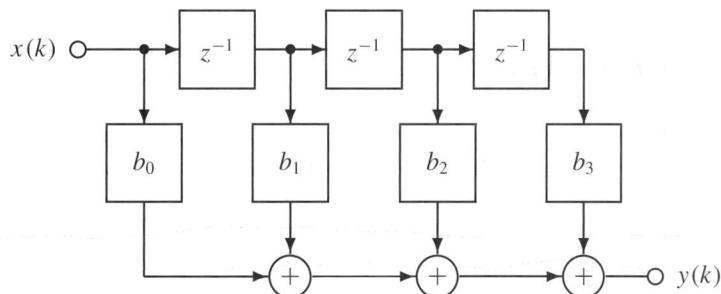

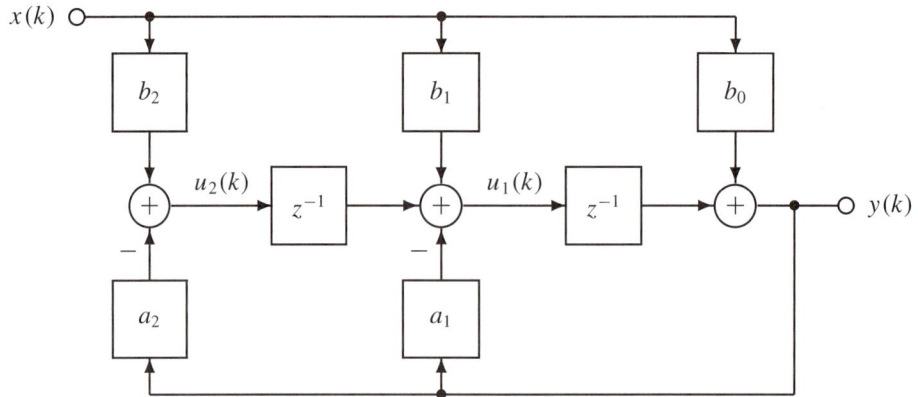

System dimension

where $M = \max\{n, m\}$ is the system dimension. The difference equation in (2.5.2) then can be expressed using a single sum as follows.

$$y(k) = b_0 x(k) + \sum_{i=1}^{M} b_i x(k - i) - a_i y(k - i) \tag{2.5.3}$$

Using this formulation, the system S can be represented in block diagram form, as shown in Figure 2.18 which illustrates the two-dimensional case $M = 2$. Notice that in general the intermediate signals $u_i(k)$ and the output $y(k)$ can be defined recursively as follows.

$$u_M(k) = b_M x(k) - a_M y(k)$$
$$u_{M-1}(k) = b_{M-1} x(k) - a_{M-1} y(k) + u_M(k - 1) \tag{2.5.4}$$
$$\vdots$$
$$u_1(k) = b_1 x(k) - a_1 y(k) + u_2(k - 1)$$
$$y(k) = b_0 x(k) + u_1(k - 1) \tag{2.5.5}$$

The vector $u(-1) \in R^M$ can be thought of as a generalized initial condition vector for the system S. The block diagram shown in Figure 2.18 is called a transformed direct form II realization of the system S. Later, in Chapter 6 and Chapter 7, we examine a number of alternative realizations, each with its own advantages and disadvantages. One of the useful features of the block diagram in Figure 2.18 is that it requires a minimum number of delay elements. Consequently, it is optimal in terms of memory usage. The MATLAB function *filter* is based on the realization in Figure 2.18.

• • • • • • • • • • • • • • •

2.6   **The Impulse Response**

The simplest nonzero input that can be applied to a discrete-time system is the unit impulse input $\delta(k)$. When the initial condition of the system S is zero, the effect of applying an impulse input is to excite the natural modes of the system. The resulting output is called the impulse response.

DEFINITION

2.1: Impulse Response

The *impulse response* of a linear time invariant system S is the zero-state response $h(k)$ produced by the unit impulse input.

$$x(k) = \delta(k) \Rightarrow h(k) = y_{zs}(k)$$

Note that if the system S is a causal system, then its impulse response $h(k)$ will be a causal signal. As we shall see, the impulse response $h(k)$ is the key to computing the zero-state response to any input $x(k)$. Before examining how this is accomplished, it is useful to investigate the impulse response of two important classes of systems.

2.6.1 FIR Systems

Consider a discrete-time system S described by difference equation (2.4.1), but where the output coefficient vector is simply $a = 1$.

$$y(k) = \sum_{i=0}^{m} b_i x(k - i) \tag{2.6.1}$$

The impulse response of this type of system can be computed directly from Definition 2.1, which yields

$$h(k) = \sum_{i=0}^{m} b_i \delta(k - i) \tag{2.6.2}$$

FIR impulse response

Recall that $\delta(k - i) = 0$ for $k \neq i$ and $\delta(0) = 1$. Consequently, $h(k)$ in (2.6.2) can be nonzero only for $0 \leq k \leq m$. More specifically, the impulse response of the FIR system in (2.6.1) is

$$h(k) = \begin{cases} b_k, & 0 \leq k \leq m \\ 0, & m < k < \infty \end{cases} \tag{2.6.3}$$

The first m samples of the impulse response are just the input coefficients b_i, and the remaining samples are all zero. A system whose impulse response contains a finite number of nonzero samples is called a finite impulse response system.

FIR

DEFINITION

2.2: FIR and IIR Systems

A linear system S is a *finite impulse response* or FIR system if and only if the impulse response $h(k)$ has a finite number of nonzero samples. Otherwise, S is an *infinite impulse response* or IIR system.

It follows that the system in (2.6.1) is an FIR system with an impulse response of duration m. FIR systems have a number of useful characteristics that make them good candidates for digital filters. The design of digital FIR filters is presented in Chapter 6, and the design of digital IIR filters is presented in Chapter 7.

Example 2.11

Impulse Response: FIR

As an illustration of an FIR system, consider the running average filter introduced in Example 2.4. Here the filter output at time k is the average of the last M samples of the input.

$$y(k) = \frac{1}{M} \sum_{i=0}^{M-1} x(k - i)$$

FIGURE 2.19:
Impulse Response
of Running Average
Filter of Using
$M = 10$ Samples

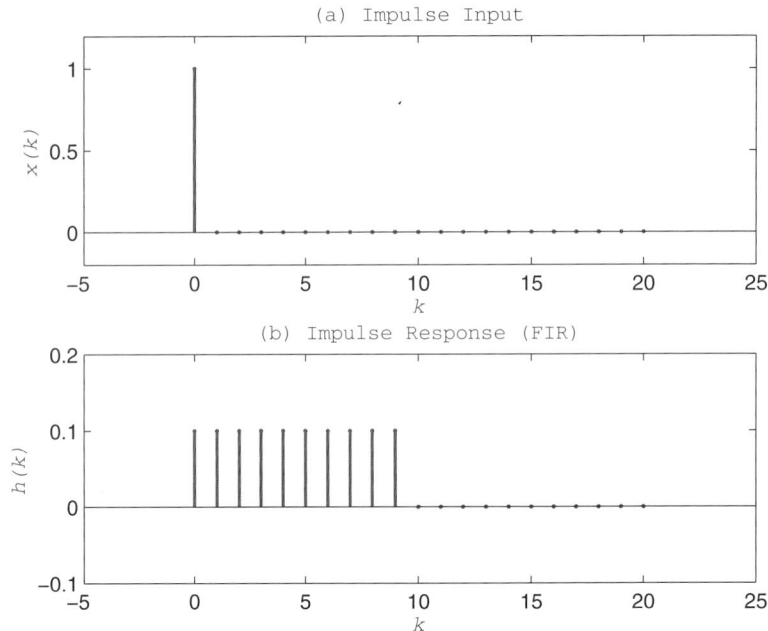

A running average filter tends to smooth out the variation in the input signal depending on the value of M. From (2.6.3), the impulse response of the running average filter is

$$h(k) = \begin{cases} \dfrac{1}{M}, & 0 \le k < M \\ 0, & M \le k < \infty \end{cases}$$

Thus the effect of the running average filter is to take an impulse input of height one and width one sample, and spread it out to form a pulse of height $1/M$ and width M samples. In both cases the height times the width is one. The running average impulse response when $M = 10$ is shown in Figure 2.19.

2.6.2 IIR Systems

Next consider a linear time-invariant system S as in (2.4.1) where the number of output terms is $n \ge 1$.

$$y(k) = \sum_{i=0}^{m} b_i x(k-i) - \sum_{i=1}^{n} a_i y(k-i) \qquad (2.6.4)$$

A systematic technique for finding the impulse response for this more general type of system is introduced in Chapter 3 using the Z-transform. For now, to illustrate the computation of $h(k)$, suppose that $m \le n$ and the roots of the characteristic polynomial $a(z)$ are simple and nonzero. The factored form of $a(z)$ is

$$a(z) = (z - p_1)(z - p_2) \cdots (z - p_n) \qquad (2.6.5)$$

IIR impulse response When $m \leq n$ and the characteristic polynomial has simple nonzero roots, the form of the impulse response is

$$h(k) = d_0 \delta(k) + \sum_{i=1}^{n} d_i (p_i)^k \mu(k) \qquad (2.6.6)$$

The coefficient vector $d \in R^{n+1}$ is computed as follows, where $b(z)$ is the input polynomial defined in (2.4.14) and $p_0 = 0$

$$d_i = \left. \frac{(z - p_i)b(z)}{z a(z)} \right|_{z=p_i}, \quad 0 \leq i \leq n \qquad (2.6.7)$$

If $d_i \neq 0$ for some $i > 0$, then the duration of the impulse response $h(k)$ is infinite, in which case S is an IIR system.

Example 2.12

Impulse Response: IIR

As an example of an IIR system, consider the following two-dimensional discrete-time system S.

$$y(k) = 2x(k) - 3x(k-1) + 4x(k-2) + .2y(k-1) + .8y(k-2)$$

The characteristic polynomial of S is

$$\begin{aligned} a(z) &= z^2 - .2z - .8 \\ &= (z-1)(z+.8) \end{aligned}$$

Thus S has simple nonzero roots at $p = [1, -.8]^T$. From (2.6.6), the form of the impulse response is

$$h(k) = d_0 \delta(k) + [d_1 + d_2(-.5)^k] \mu(k)$$

By inspection, the polynomial associated with the input terms is

$$b(z) = 2z^2 - 3z + 4$$

Applying (2.6.7) with $p_0 = 0$, the coefficient vector $d \in R^3$ is

$$d_0 = \frac{4}{(-1)(.8)} = -5$$

$$d_1 = \frac{2 - 3 + 4}{1.8} = 1.67$$

$$d_2 = \frac{2(.64) - 3(-.8) + 4}{(-.8)(-1.8)} = 5.33$$

Finally, the impulse response of the system S is

$$h(k) = -5\delta(k) + [1.67 - 5.33(-.5)^k]\mu(k)$$

A plot of $h(k)$ is shown in Figure 2.20. Clearly, $h(k)$ has infinite duration and S is an IIR system.

FIGURE 2.20: Impulse
Response of the IIR
System

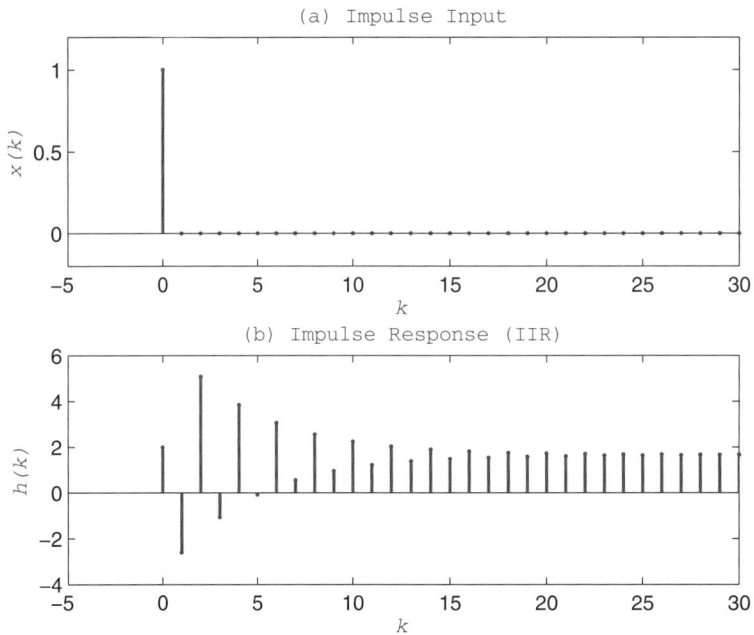

(a) Impulse Input

(b) Impulse Response (IIR)

FDSP Functions

The FDSP toolbox contains the following function for computing the impulse response of
a linear discrete-time system.

```
% F_IMPULSE: Compute impulse response
%
% Usage:
%        [h,k] = f_impulse (b,a,N);
% Pre:
%        b = vector of length m+1 containing input coefficients
%        a = vector of length n+1 containing output coefficients
%        N = number of samples
% Post:
%        h = vector of length N containing the impulse response
%        k = vector of length N containing the discrete solution times
```

2.7 Convolution

2.7.1 Linear Convolution

Once the impulse response is known, the zero-state response to an arbitrary input $x(k)$ can be
computed from it. To see this, first note that a causal signal $x(k)$ can be written as a weighted

Sifting property sum of impulses as follows. This is called the *sifting property* of the unit impulse.

$$x(k) = \sum_{i=0}^{k} x(i)\delta(k - i) \tag{2.7.1}$$

Here the ith term is an impulse of amplitude $x(i)$ occurring at time $k = i$. Recall that the zero-state response to a impulse of unit amplitude occurring at $k = 0$ is $h(k)$. If the system is linear and time-invariant, scaling the input by $x(i)$ and delaying it by i samples produces an output that is also scaled by $x(i)$ and delayed by i samples. That is, the output corresponding to the ith term in (2.7.1) is simply $x(i)h(k - i)$. Furthermore, since S is a linear system, the response to the sum of inputs in (2.7.1) is just the sum of the responses to the individual inputs. Thus the zero-state output produced by $x(k)$ in (2.7.1) can be written as

$$y(k) = \sum_{i=0}^{k} h(k - i)x(i) \tag{2.7.2}$$

The formulation on the right-hand side of (2.7.2) is referred to as the linear convolution of the signal $h(k)$ with the signal $x(k)$. Note that if $x(k)$ is not causal, then the lower limit of the sum becomes $i = -\infty$, and if $h(k)$ is not causal, the upper limit of the sum becomes $i = \infty$.

DEFINITION

2.3: Linear Convolution

Let $h(k)$ and $x(k)$ be causal discrete-time signals. The *linear convolution* of $h(k)$ with $x(k)$ is denoted $h(k) \star x(k)$ and defined

$$h(k) \star x(k) \triangleq \sum_{i=0}^{k} h(k - i)x(i), \quad k \geq 0$$

Convolution operator

The operator $\star$ is called the *convolution operator*. Comparing Definition 2.3 with (2.7.2), we see that the zero-state response of a discrete-time system can be expressed in terms of convolution as

$$y_{zs}(k) = h(k) \star x(k) \tag{2.7.3}$$

Zero-state response

That is, the *zero-state response* is the linear convolution of the impulse response with the input. The convolution operator has a number of useful properties. For example, starting with Definition 2.3 and using a change of variable

$$h(k) \star x(k) = \sum_{i=0}^{k} h(k - i)x(i)$$

$$= \sum_{m=k}^{0} h(m)x(k - m), \quad m = k - i$$

$$= \sum_{m=0}^{k} h(m)x(k - m)$$

$$= x(k) \star h(k) \tag{2.7.4}$$

TABLE 2.2: ▶
Properties of the
Linear Convolution
Operator

Name	Property		
Commutative	$f \star g$	$=$	$g \star f$
Associative	$f \star (g \star h)$	$=$	$(f \star g) \star h$
Distributive	$f \star (g + h)$	$=$	$f \star g + f \star h$

Commutative operator

Thus the convolution operator is *commutative*, and the two arguments can be interchanged without changing the result. Consequently, the zero state response also can be written as

$$y(k) = \sum_{i=0}^{k} h(i)x(k-i) \tag{2.7.5}$$

Zero-state response

When the system S is the FIR system in (2.6.1) with an impulse response of duration m, the upper limit in (2.7.5) can be replaced by the constant m and the impulse response can be replaced with the input coefficients $h(k) = b_k$.

In addition to being commutative, the convolution operator is also associative and distributive, as summarized in Table 2.2. The other two properties can be established using Definition 2.3 and are left as an exercise (Problems 2.29, 2.30).

MATLAB Functions

The linear convolution of two finite discrete-time signals h and x can be computed using the MATLAB function *conv* as follows.

```
% CONV: Perform linear convolution
%
% Usage:
%          y = conv (h,x);
% Pre:
%          h = vector of length L
%          x = vector of length M
% Post:
%          y = vector of length L+M-1 containing convolution of h with x
```

Example 2.13

Linear Convolution

Suppose the input to the system in Example 2.12 is the following causal sine.

$$x(k) = 10 \sin(.1\pi k)\mu(k)$$

If the initial condition is zero, the output of this system using the convolution representation is

$$y(k) = \sum_{i=0}^{k} h(i)x(k-i)$$

$$= 10 \sum_{i=0}^{k} \{-5\delta(i) + [1.67 - 5.33(-.5)^i]\} \sin[.1\pi(k-i)], \quad k \geq 0$$

FIGURE 2.21: Zero-state Response for Example 2.13

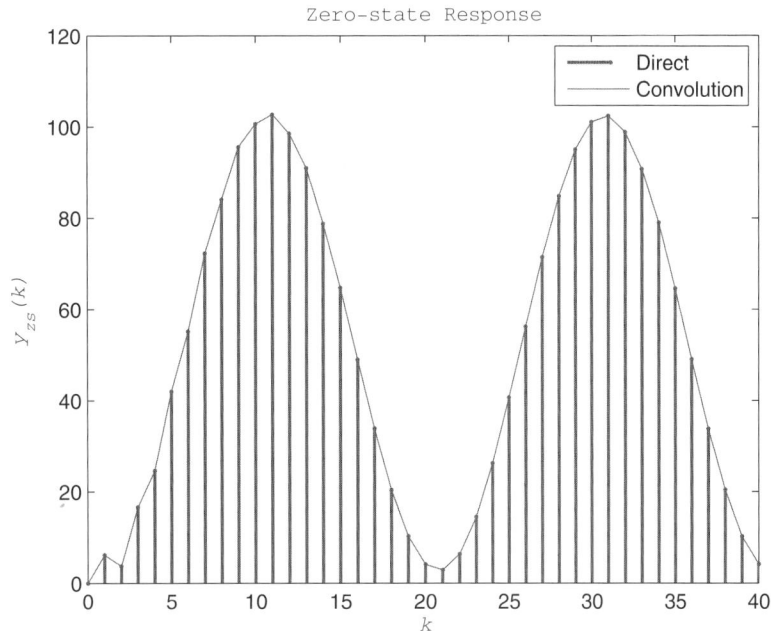

A plot of the zero-state response, obtained by running *exam2_13*, is shown in Figure 2.21. Here the discrete-time signal was computed directly using *filter*, and the interpolated version was computed with convolution using *conv*.

2.7.2 Circular Convolution

For a numerical implementation of the convolution operation to be practical, the signals $h(k)$ and $x(k)$ must be finite. Suppose $h(k)$ is defined for all k, but nonzero only for $0 \leq k < L$. Similarly, let $x(k)$ be a signal that is nonzero for $0 \leq k < M$. Then the linear convolution in (2.7.5) can be expressed as

$$y(k) = \sum_{i=0}^{L-1} h(i)x(k-i), \quad 0 \leq k < L + M - 1 \tag{2.7.6}$$

The upper limit on the summation has been changed from k to $L - 1$ because $h(i) = 0$ for $i \geq L$. Observe that when $i = L - 1$, we have $x(k - i) \neq 0$ for $L - 1 \leq k < L + M$. Consequently, the linear convolution of an L-point signal with an M-point signal is a signal of length $L + M - 1$.

There is an alternative way to define convolution, where the length of the result is the same as the lengths of the two operands. To define this form of convolution, we first introduce the notion of a periodic extension of a finite signal $x(k)$. The *periodic extension* of an N-point signal $x(k)$ is a signal $x_p(k)$ defined for all integers k as follows.

Periodic extension

$$x_p(k) \triangleq x[\text{mod}(k, N)] \tag{2.7.7}$$

Periodic ramp

Here the MATLAB function $\text{mod}(k, N)$ is read as k modulo N. For a fixed integer N, the function $\text{mod}(k, N)$ is a periodic ramp of period N, where $\text{mod}(k, N) = k$ for $0 \leq k < N$. Consequently, the periodic extension $x_p(k) = x(k)$ for $0 \leq k < N$ and $x_p(k)$ extends $x(k)$ periodically in both positive and negative directions. The following is an alternative form of convolution called circular convolution.

DEFINITION

2.4: Circular
Convolution

Let $h(k)$ and $x(k)$ be N-point signals, and let $x_p(k)$ be the periodic extension of $x(k)$. Then the *circular convolution* of $h(k)$ with $x(k)$ is denoted $y_c(k) = h(k) \circ x(k)$ and defined as

$$h(k) \circ x(k) \overset{\Delta}{=} \sum_{i=0}^{N-1} h(i) x_p(k-i), \quad 0 \leq k < N$$

Evaluating the periodic extension $x_p(k - i)$ is equivalent to a counterclockwise circular shift of $x(-i)$ by k samples, as illustrated in Figure 2.22. Note how $h(i)$ is distributed counterclockwise around the outer ring, while $x(-i)$ is distributed clockwise around the inner ring. For different values of k, signal $x(-i)$ gets rotated k samples counterclockwise. The circular convolution is just the sum of the products of the N points distributed around the circle.

It is possible to represent circular convolution using matrix multiplication. Let h and y_c be $N \times 1$ column vectors containing the samples of $h(k)$ and $y_c(k) = h(k) \circ x(k)$, respectively.

$$h = [h(0), h(1), \ldots, h(N-1)]^T \tag{2.7.8a}$$

$$y_c = [y_c(0), y_c(1), \ldots, y_c(N-1)]^T \tag{2.7.8b}$$

FIGURE 2.22:
Counterclockwise
Circular Shift of
$x(-i)$ by $k = 2$ with
$N = 8$

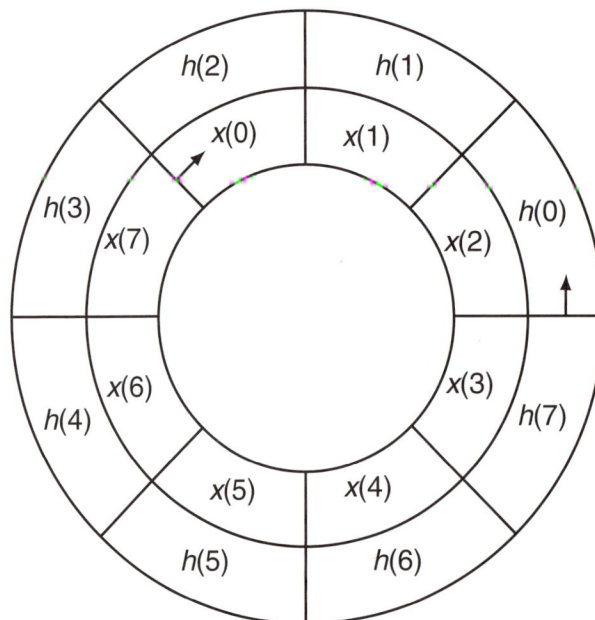

Since circular convolution is a linear transformation from h to y_c, it can be represented by an $N \times N$ matrix $C(x)$. Consider, in particular, the following matrix which corresponds to the case $N = 5$.

$$C(x) = \begin{bmatrix} x(0) & x(4) & x(3) & x(2) & x(1) \\ x(1) & x(0) & x(4) & x(3) & x(2) \\ x(2) & x(1) & x(0) & x(4) & x(3) \\ x(3) & x(2) & x(1) & x(0) & x(4) \\ x(4) & x(3) & x(2) & x(1) & x(0) \end{bmatrix} \tag{2.7.9}$$

Circular convolution matrix

Observe that the columns of the *circular convolution matrix $C(x)$* are just downward rotations of $x(i)$. Using (2.7.8) and (2.7.9), we can express the circular convolution in Definition 2.4 in vector form as

$$y_c = C(x)h \tag{2.7.10}$$

Note from (2.7.9) that if the input $x(k)$ is selected with some care, the circular convolution matrix $C(x)$ will be nonsingular, in which case h can be recovered from y_c using $h = C^{-1}(x)y_c$.

Example 2.14

Circular Convolution

To illustrate how to compute circular convolution using the matrix formulation, consider the following two finite signals of length $N = 3$.

$$h = [2, -1, 6]^T$$
$$x = [5, 3, -4]^T$$

From (2.7.9), the 3×3 circular convolution matrix $C(x)$ is

$$C(x) = \begin{bmatrix} 5 & -4 & 3 \\ 3 & 5 & -4 \\ -4 & 3 & 5 \end{bmatrix}$$

Using (2.7.10), if $y_c(k) = h(k) \circ x(k)$, then

$$\begin{aligned} y_c &= C(x)h \\ &= \begin{bmatrix} 5 & -4 & 3 \\ 3 & 5 & -4 \\ -4 & 3 & 5 \end{bmatrix} \begin{bmatrix} 2 \\ -1 \\ 6 \end{bmatrix} \\ &= [32, -23, 19]^T \end{aligned}$$

2.7.3 Zero Padding

In Chapter 4, a high-speed version of circular convolution is implemented using the fast Fourier transform (FFT). Unfortunately, a direct application of circular convolution will not produce the zero-state response of a linear discrete-time system, even when it is an FIR system with a finite input. To see this, note from the Definition 2.4 that if we evaluate $y_c(k)$ for $k \geq N$, we find that circular convolution is periodic with a period of N.

$$y_c(k + N) = y_c(k) \tag{2.7.11}$$

Since the zero-state response is not, in general, periodic, it follows that circular convolution produces a different response than linear convolution. Fortunately, there is a simple prepro-cessing step that can be performed that allows us to achieve linear convolution using circular

Zero padding

convolution. To keep the formulation general, let $h(k)$ be an L-point signal, and let $x(k)$ be an M-point signal. Suppose we construct two new signals, each of length $N = L + M - 1$, by using *zero padding*. In particular, we can pad $h(k)$ with $M - 1$ zeros and $x(k)$ with $L - 1$ zeros.

$$h_z = [h(0), h(1), \ldots, h(L-1), \overbrace{0, \ldots, 0}^{M-1}]^T \tag{2.7.12a}$$

$$x_z = [x(0), x(1), \ldots, x(M-1), \underbrace{0, \ldots, 0}_{L-1}]^T \tag{2.7.12b}$$

Thus h_z and x_z are zero-padded vectors of length $N = L + M - 1$. Next, consider the circular convolution of $h_z(k)$ with $x_z(k)$. If $x_{zp}(k)$ is the periodic extension of $x_z(k)$, then

$$y_c(k) = h_z(k) \circ x_z(k)$$

$$= \sum_{i=0}^{N-1} h_z(i) x_{zp}(k-i)$$

$$= \sum_{i=0}^{L-1} h_z(i) x_{zp}(k-i), \quad 0 \le k < N \tag{2.7.13}$$

Since $0 \le i < L$ and $0 \le k < N$, the minimum value for $k - i$ at $k = 0$ and $i = L - 1$ is $-(L-1)$. But $x_z(i)$ has $L - 1$ zeros padded to the end of it. Therefore, $x_{zp}(-i) = 0$ for $0 \le i < L$. This means that $x_{zp}(k-i)$ in (2.7.13) can be replaced by $x_z(k-i)$. The result is then the linear convolution of $h_z(k)$ with $x_z(k)$.

$$h_z(k) \circ x_z(k) = h(k) \star x(k), \quad 0 \le k < N \tag{2.7.14}$$

Equivalent convolution

In summary, linear convolution can be achieved by computing the circular convolution of zero-padded versions of the two signals. Consequently, the zero-state response can be computed using circular convolution. The following example illustrates this technique.

Example 2.15   **Zero Padding**

Consider the signals from Example 2.14. In this case $L = 3$, $M = 3$, and $N = L + M - 1 = 5$. The zero-padded signals are

$$h_z = [2, -1, 6, 0, 0]^T$$

$$x_z = [5, 3, -4, 0, 0]^T$$

From (2.7.9), the 5×5 circular convolution matrix $C(x)$ is

$$C(x) = \begin{bmatrix} 5 & 0 & 0 & -4 & 3 \\ 3 & 5 & 0 & 0 & -4 \\ -4 & 3 & 5 & 0 & 0 \\ 0 & -4 & 3 & 5 & 0 \\ 0 & 0 & -4 & 3 & 5 \end{bmatrix}$$

Thus from (2.7.10) and (2.7.14), the linear convolution of $h(k)$ with $x(k)$ is

$$y = C(x_z)h_z$$

$$= \begin{bmatrix} 5 & 0 & 0 & -4 & 3 \\ 3 & 5 & 0 & 0 & -4 \\ -4 & 3 & 5 & 0 & 0 \\ 0 & -4 & 3 & 5 & 0 \\ 0 & 0 & -4 & 3 & 5 \end{bmatrix} \begin{bmatrix} 2 \\ -1 \\ 6 \\ 0 \\ 0 \end{bmatrix}$$

$$= [10, 1, 19, 22, -24]^T$$

FDSP Functions

Both linear and circular convolution of finite discrete-time signals h and x can be computed using the FDSP toolbox function *f_conv* as follows.

```
% F_CONV: Fast linear or circular convolution
%
% Usage:
%           y = f_conv (h,x,circ)
% Pre:
%           h   = vector of length L containing impulse
%                 response signal
%           x   = vector of length M containing input signal
%           circ = optional convolution type code (default: 0)
%
%                 0 = linear convolution
%                 1 = circular convolution (requires M = L)
% Post:
%           y = vector of length L+M-1 containing the
%               convolution of h with x. If circ=1, y
%               is of length L.
% Note:
%           If h is the impulse response of a discrete-time
%           linear system and x is the input, then y is the
%           zero-state response when circ = 0.
```

The function *f_conv* implements a fast form of convolution based on the FFT, as discussed in Chapter 4. The relationship between linear and circular convolution is summarized in Table 2.3.

TABLE 2.3: ▶
Linear and Circular
Convolution

Property	Equation
Linear convolution	$h(k) \star x(k) = \displaystyle\sum_{i=0}^{k} h(i)x(k-i)$
Circular convolution	$h(k) \circ x(k) = \displaystyle\sum_{i=0}^{N-1} h(i)x_p(k-i)$
Zero padding	$h(k) \star x(k) = h_z(k) \circ x_z(i)$

2.7.4 Deconvolution

Deconvolution

There are applications where one knows the impulse response $h(k)$ and the output $y(k)$, and the objective is to reconstruct the input $x(k)$. The process of finding the input $x(k)$ given the impulse response $h(k)$, and the output $y(k)$ is referred to as *deconvolution*. Deconvolution also includes finding the impulse response $h(k)$, given the input $x(k)$ and the output $y(k)$. This is a special case of a more general problem called system identification, a topic that is discussed in Chapter 3 and Chapter 9. When $h(k)$ and $x(k)$ are both causal noise-free signals, recovery of $h(k)$ is relatively simple. Suppose the input $x(k)$ is chosen such that $x(0) \neq 0$. Evaluating (2.7.5) at $k = 0$ then yields $y(0) = h(0)x(0)$ or

$$h(0) = \frac{y(0)}{x(0)} \tag{2.7.15}$$

Once $h(0)$ is known, the remaining samples of $h(k)$ can be obtained recursively. For example, evaluating (2.7.5) at $k = 1$ yields

$$y(1) = h(0)x(1) + h(1)x(0) \tag{2.7.16}$$

Solving (2.7.16) for $h(1)$ we then have

$$h(1) = \frac{y(1) - h(0)x(1)}{x(0)} \tag{2.7.17}$$

This process can be repeated for $2 \leq k < N$. The expression for the general case is obtained by decomposing the sum in (2.7.5) into the $i < k$ terms and the $i = k$ term. Solving for $h(k)$ we then arrive at the following recursive formulation of the impulse response.

$$h(k) = \frac{1}{x(0)} \left[y(k) - \sum_{i=0}^{k-1} h(i)x(k-i) \right], \quad k \geq 1 \tag{2.7.18}$$

With the initialization in (2.7.15), the formulation in (2.7.18) solves the deconvolution problem, thereby reconstructing the impulse response from the input and the output. The following example illustrates this technique.

Example 2.16

Deconvolution

Consider the signals from Example 2.14. In this case, both $h(k)$ and $x(k)$ are of length $N = 3$, with

$$h = [2, -1, 6]^T$$
$$x = [5, 3, -4]^T$$

Let $y(k) = h(k) \star x(k)$. Circular convolution with zero padding was used in Example 2.15 to find the following zero-state response.

$$y = [10, 1, 19, 22, -24]^T$$

To recover $h(k)$ from $x(k)$ and $y(k)$ using deconvolution, we start with (2.7.15). The first sample of the impulse response is $h(0) = y(0)/x(0) = 2$. Next, applying (2.7.18) with $k = 1$ yields

$$h(1) = \frac{y(1) - h(0)x(1)}{x(0)}$$

$$= \frac{1 - 2(3)}{5}$$

$$= -1$$

Finally, applying (2.7.18) with $k = 2$ then yields

$$h(2) = \frac{y(2) - h(0)x(2) - h(1)x(1)}{x(0)}$$

$$= \frac{19 - 2(-4) - (-1)3}{5}$$

$$= 6$$

Thus the impulse response vector is

$$h = [2, -1, 6]^T \checkmark$$

2.7.5 Polynomial Arithmetic

Linear convolution has a simple and useful interpretation in terms of polynomial arithmetic. Suppose $a(z)$ and $b(z)$ are polynomials of degree L and M, respectively.

$$a(z) = a_0 z^L + a_1 z^{L-1} + \cdots + a_L \tag{2.7.19a}$$
$$b(z) = b_0 z^M + b_1 z^{M-1} + \cdots + b_M \tag{2.7.19b}$$

Thus the coefficient vectors a and b are of length $L + 1$ and $M + 1$, respectively. Next let $c(z)$ be the following product polynomial.

$$c(z) = a(z)b(z) \tag{2.7.20}$$

In this case, $c(z)$ is of degree $N = L + M$, and its coefficient vector c is of length $L + M + 1$. That is,

$$c(z) = c_0 z^{L+M} + c_1 z^{L+M-1} + \cdots + c_{L+M} \tag{2.7.21}$$

The coefficient vector of $c(z)$ can be obtained directly from the coefficient vectors of $a(z)$ and $b(z)$ using linear convolution as follows.

$$c(k) = a(k) \star b(k), \quad 0 \leq k < L + M + 2 \tag{2.7.22}$$

Polynomial multiplication, division

Thus linear convolution is equivalent to *polynomial multiplication*. Since deconvolution allows us to recover $a(k)$ from $b(k)$ and $c(k)$, deconvolution is equivalent to *polynomial division*.

Example 2.17

Polynomial Multiplication

To illustrate the relationship between linear convolution and polynomial arithmetic, consider the following two polynomials.

$$a(z) = 2z^2 - z + 6$$
$$b(z) = 5z^2 + 3z - 4$$

In this case the coefficient vectors are $a = [2, -1, 6]^T$ and $b = [5, 3, -4]^T$. Suppose $c(k) = a(k) \star b(k)$. Then, from Example 2.15,

$$c = [10, 1, 19, 22, -24]^T$$

Thus the product of $a(z)$ with $b(z)$ is the polynomial of degree four whose coefficient vector is given by c. That is,

$$b(z) = a(z)b(z)$$
$$= 10z^4 + z^3 + 19z^2 + 22z - 24$$

Using direct multiplication of $a(z)$ times $b(z)$, one can verify this result.

$$c(z) = \left\{ \begin{array}{lllll} 10z^4 & -5z^3 & +30z^2 & & \\ & 6z^3 & -3z^2 & +18z & \\ & & -8z^2 & +4z & -24 \\ \hline 10z^4 & +z^3 & +19z^2 & +22z & -24 \end{array} \right\}$$

MATLAB Functions

The deconvolution of two finite discrete-time signals h and x can be computed using the MATLAB function *deconv* as follows.

```
% DECONV: Perform linear deconvolution
%
% Usage:
%           [h,r] = deconv (y,x);
% Pre:
%           y = vector of length N containing output
%           x = vector of length M < N containing input
% Post:
%           y = vector of length N containing impulse response
% Note:
%           If y and x represent the coefficients of polynomials, then
%           h contains the coefficients of the quotient polynomial and
%           r contains the coefficients of the remainder polynomial.
%
%           y(z) = h(z)x(z) + r(z)
```

2.8   Correlation

2.8.1   Linear Cross-correlation

Next we turn our attention to an operation with finite signals that is closely related to convolution called correlation.

DEFINITION

2.5: Linear Cross-correlation

Let $y(k)$ be an L-point signal and let $x(k)$ be an M-point signal where $M \leq L$. Then the *linear cross-correlation* of $y(k)$ with $x(k)$ is denoted $r_{yx}(i)$ and defined

$$r_{yx}(k) \triangleq \frac{1}{L} \sum_{i=0}^{L-1} y(i)x(i-k), \quad 0 \leq k < L$$

Since $x(k)$ is causal, the lower limit of the sum in Definition 2.5 can be set to $i = k$. The variable k is sometimes called the *lag* variable because it represents the number of samples that $x(i)$ is shifted to the right, or delayed, before the sum of products is computed.

Definition 2.5 indicates how to compute the cross-correlation of two deterministic discrete-time signals. Linear cross-correlation is sometimes defined without the scale factor $1/L$. A scale factor is included in Definition 2.5 because this way Definition 2.5 is consistent with an alternative statistical definition of the cross-correlation of two random signals. The cross-correlation of a pair of random signals is introduced in Chapter 9 using the expected value operator. The formulation in Definition 2.5 applies to finite causal signals. For practical computations, this is the most important special case. However, it is possible to extend the definition of correlation to noncausal signals and to infinite signals (power signals) by extending the lower and upper limits, respectively, of the summation in Definition 2.5.

Just as was the case with convolution, there is a matrix formulation of cross-correlation. Let y and r_{yx} be $L \times 1$ column vectors containing the samples of $y(k)$ and $r_{yx}(k)$, respectively.

$$y = [y(0), y(1), \ldots, y(L-1)]^T \tag{2.8.1a}$$

$$r_{yx} = [r_{yx}(0), r_{yx}(1), \ldots, r_{yx}(L-1)]^T \tag{2.8.1b}$$

Cross-correlation is a linear transformation from y to r_{yx}. Consequently, it can be represented by an $L \times L$ matrix $D(x)$. Consider, in particular, the following matrix which corresponds to the case where $L = 5$ and $M = 3$.

$$D(x) = \frac{1}{5} \begin{bmatrix} x(0) & x(1) & x(2) & 0 & 0 \\ 0 & x(0) & x(1) & x(2) & 0 \\ 0 & 0 & x(0) & x(1) & x(2) \\ 0 & 0 & 0 & x(0) & x(1) \\ 0 & 0 & 0 & 0 & x(0) \end{bmatrix} \tag{2.8.2}$$

Note how the rows of $D(x)$ are constructed by shifting $x(k)$ to the right. However, unlike the circular convolution matrix in (2.7.9), when the samples of x get shifted off the right end of $D(x)$, they do not wrap around and reappear on the left end. Using (2.8.1), (2.8.2) and Definition 2.5, the linear cross-correlation of $y(k)$ with $x(k)$ can be expressed in vector form as

$$r_{yx} = D(x)y \tag{2.8.3}$$

Observe from (2.8.2) that if $x(0) \neq 0$, then the cross-correlation matrix $D(x)$ is nonsingular which means signal $y(k)$ can be recovered from the cross-correlation using $y = D^{-1}(x)r_{yx}$.

Signal shape   Linear cross-correlation can be used to measure the degree to which the *shape* of one signal is similar to the shape of another signal. The following example illustrates this point.

Example 2.18   **Linear Cross-correlation**

To demonstrate how linear cross-correlation can be computed using the matrix formulation, consider the following pair of discrete-time signals.

$$x = [0, 2, 1, 2, 0]^T$$
$$y = [4, -1, -2, 0, 4, 2, 4, 0, -2, 2]^T$$

Here $L = 10$ and $M = 5$. A plot of the two signals $y(k)$ and $x(k)$ is shown in Figure 2.23.

Note how the graph of $x(k)$ is a flattened "M"-shaped signal. Furthermore, the longer signal $y(k)$ contains a scaled and translated version of $x(k)$ with a larger "M" starting at $k = 3$. To verify that $y(k)$ contains a scaled and shifted version of $x(k)$, we compute the linear cross-correlation. From (2.8.2) and (2.8.3), the linear cross-correlation of $y(k)$ with $x(k)$ is

$$r_{yx} = D(x)y$$

$$= \frac{1}{10}\begin{bmatrix} 0 & 2 & 1 & 2 & 0 & 0 & 0 & 0 & 0 & 0 \\ 0 & 0 & 2 & 1 & 2 & 0 & 0 & 0 & 0 & 0 \\ 0 & 0 & 0 & 2 & 1 & 2 & 0 & 0 & 0 & 0 \\ 0 & 0 & 0 & 0 & 2 & 1 & 2 & 0 & 0 & 0 \\ 0 & 0 & 0 & 0 & 0 & 2 & 1 & 2 & 0 & 0 \\ 0 & 0 & 0 & 0 & 0 & 0 & 2 & 1 & 2 & 0 \\ 0 & 0 & 0 & 0 & 0 & 0 & 0 & 2 & 1 & 2 \\ 0 & 0 & 0 & 0 & 0 & 0 & 0 & 0 & 2 & 1 \\ 0 & 0 & 0 & 0 & 0 & 0 & 0 & 0 & 0 & 2 \\ 0 & 0 & 0 & 0 & 0 & 0 & 0 & 0 & 0 & 0 \end{bmatrix}\begin{bmatrix} 4 \\ -1 \\ -2 \\ 0 \\ 4 \\ 2 \\ 4 \\ 0 \\ -2 \\ 2 \end{bmatrix}$$

$$= [-0.4, 0.4, 0.8, 1.8, 0.8, 0.4, 0.2, -0.2, 0.4, 0.0]^T$$

In this case $r_{yx}(k)$ reaches its maximum value at a lag of $k = 3$. This is evident from the plot of $r_{yx}(k)$ shown in Figure 2.24. The fact that $r_{yx}(k)$ has a clear peak at $k = 3$ indicates there

FIGURE 2.23: Two Finite Discrete-time Signals

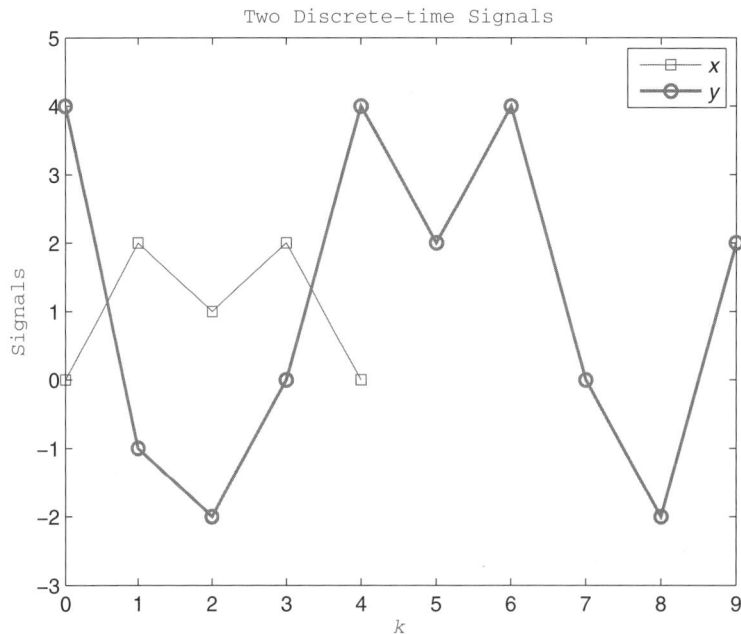

FIGURE 2.24: Linear Cross-correlation of the Signals in Figure 2.23

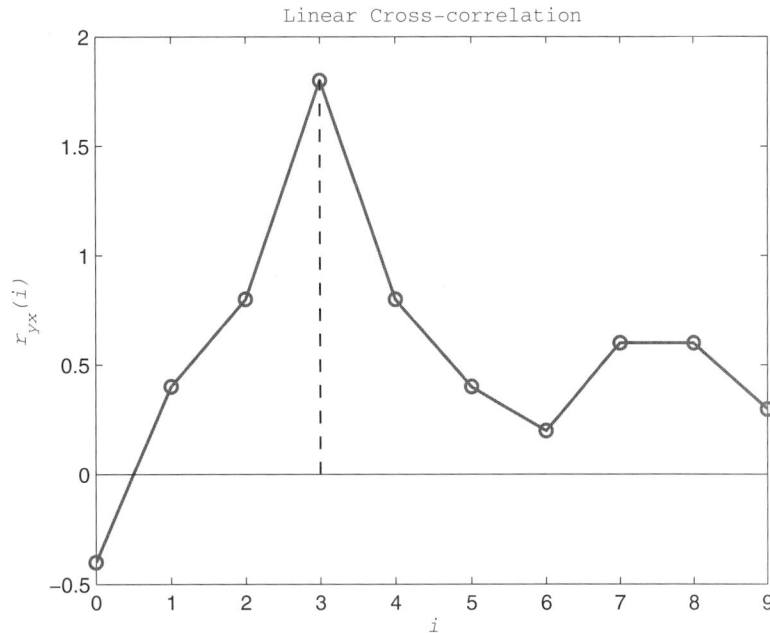

is a strong positive correlation between $y(k)$ and $x(k-3)$. Indeed, except for scaling, the two signals exactly match in this case.

$$y(k) = 2x(k-3), \quad 3 \le k < 3 + M$$

Thus a dominant peak in $r_{yx}(k)$ at $k = p$ is an indication that a signal similar to $x(k)$ is present in signal $y(k)$, starting at $k = p$.

Although the cross-correlation in Definition 2.5 can be used to detect the presence of one signal in another signal, it does suffer from a practical drawback. When $y(k)$ contains a scaled and shifted version of $x(k)$, the cross-correlation will contain a distinct peak. However, the height of the peak will depend on the data in $y(k)$ and $x(k)$. For example, scaling $y(k)$ or $x(k)$ by a will scale the height of the peak by a. Consequently, one is left with the question, how high does the peak have to be for a significant correlation to exist? A simple way to solve this problem is to develop a normalized version of cross-correlation. It can be shown (Proakis and Manolakis, 1988) that the square of the cross-correlation is bounded from above in the following way.

$$r_{yx}^2(k) \le \left(\frac{M}{L}\right) r_{xx}(0) r_{yy}(0), \quad 0 \le k < L \tag{2.8.4}$$

Normalized linear cross-correlation

By making use of (2.8.4), we can introduce the following scaled version of cross-correlation called the *normalized linear cross-correlation* of $y(k)$ with $x(k)$.

$$\rho_{yx}(k) \triangleq \frac{r_{yx}(k)}{\sqrt{(M/L) r_{xx}(0) r_{yy}(0)}}, \quad 0 \le k < L \tag{2.8.5}$$

By construction, the magnitude of the normalized cross-correlation is bounded by $B_\rho = 1$. That is, the normalized cross-correlation is guaranteed to lie within the following interval.

$$-1 \leq \rho_{yx}(k) \leq 1, \quad 0 \leq k < L \tag{2.8.6}$$

If the normalized cross-correlation is used, then any peak that begins to approach the maximum value of one indicates a very strong positive correlation between $y(k)$ and $x(k)$, regardless of whether one of the signals is very small or very large in comparison with the other. For the cross-correlation example shown in Figure 2.24, the peak of the normalized cross-correlation is $\rho_{yx}(3) = 0.744$, indicating there is a strong positive correlation between $y(k)$ and $x(k-3)$.

2.8.2   Circular Cross-correlation

Practical cross-correlations often involve long signals, so it is helpful to develop a numerical implementation of linear cross-correlation that is more efficient than the direct method in Definition 2.5. Recall that linear convolution can be achieved by using circular convolution with zero padding. A similar approach can be used with cross-correlation.

DEFINITION
....
2.6: Circular Cross-correlation

Let $y(k)$ and $x(k)$ be N-point signals, and let $x_p(k)$ be the periodic extension of $x(k)$. The *circular cross-correlation* of $y(k)$ with $x(k)$ is denoted $c_{yx}(k)$ and defined

$$c_{yx}(k) \triangleq \frac{1}{N} \sum_{i=0}^{N-1} y(i)x_p(i-k), \quad 0 \leq k < N$$

Circular cross-correlation operates on two signals of the same length. Comparing Definition 2.6 with Definition 2.5, we see that for circular cross-correlation, $x(k)$ is replaced by its periodic extension $x_p(k)$. The effect of this change is to replace the linear shift of $x(k)$ with a circular shift or rotation of $x(k)$, hence the name circular cross-correlation. A diagram illustrating circular cross-correlation for the case $N = 8$ and $k = 2$ is shown in Figure 2.25. Note that evaluating $x_p(i)$ at $i - k$ is equivalent to a clockwise circular shift of $x(i)$ by k samples, as shown in Figure 2.25. Observe how $y(i)$ is distributed counterclockwise around the outer ring, while $x(i)$ is distributed counterclockwise around the inner ring. For different values of the lag k, the signal $x_p(i - k)$ gets rotated k samples clockwise. The circular cross-correlation is just the sum of the products of the N points distributed around the circle.

Just as was the case with linear cross-correlation, we can scale $c_{yx}(k)$ to produce the following *normalized circular cross-correlation*, whose value is restricted to the interval $[-1, 1]$ (see Problem 2.34).

Normalized circular cross-correlation

$$\sigma_{yx}(k) = \frac{c_{yx}(k)}{\sqrt{c_{xx}(0)c_{yy}(0)}} \tag{2.8.7}$$

There are a number of useful properties of circular cross-correlation. For example, consider the effect of interchanging the roles of $y(k)$ and $x(k)$. Note from Definition 2.6 that $y(k)$ can be replaced by its periodic extension $y_p(k)$ without affecting the result. Consequently, using

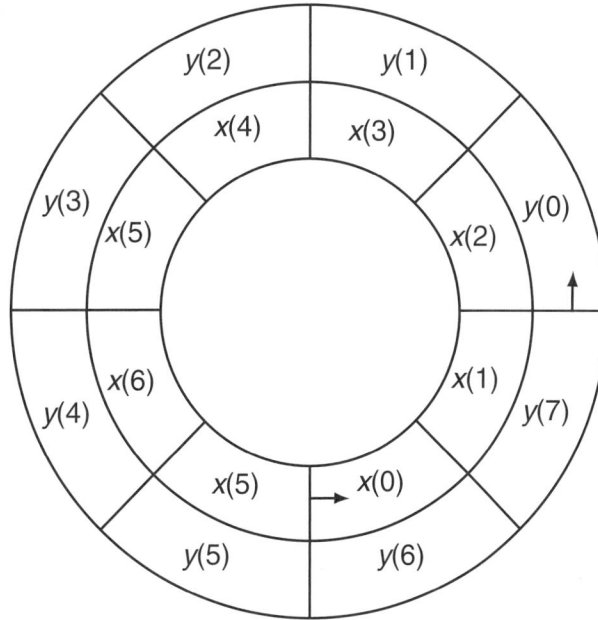

FIGURE 2.25: Clockwise Circular Shift of $x(i)$ by $k = 2$ with $N = 8$

the change of variable $q = i - k$, we have

$$c_{xy}(k) = \frac{1}{N}\sum_{i=0}^{N-1} x(i)y_p(i-k)$$

$$= \frac{1}{N}\sum_{i=0}^{N-1} x_p(i)y_p(i-k)$$

$$= \frac{1}{N}\sum_{q=-k}^{N-1-k} x_p(q+k)y_p(q)\ \}\, q = i-k$$

$$= \frac{1}{N}\sum_{q=0}^{N-1} y_p(q)x_p(q+k)$$

$$= \frac{1}{N}\sum_{q=0}^{N-1} y(q)x_p(q+k) \tag{2.8.8}$$

Observe that the summations in (2.8.8) all extend over one period. Consequently, the starting sample of the sum can be changed from $q = -k$ to $q = 0$ without affecting the result. From Definition 2.6, the last line of (2.8.8) is $c_{yx}(-k)$. Thus we have the following *symmetry property* of circular cross-correlation, which says that changing the order of y and x is equivalent to changing the sign of the lag variable.

Symmetry property

$$c_{xy}(k) = c_{yx}(-k), \quad 0 \le k < N \tag{2.8.9}$$

There is a simple and elegant relationship between circular cross-correlation and circular convolution. Comparing the expression in Definition 2.6 with that in Definition 2.5, observe that the circular cross-correlation of $y(k)$ with $x(k)$ is just a scaled version of the circular convolution of $y(k)$ with $x(-k)$. That is,

$$c_{yx}(k) = \frac{y(k) \circ x(-k)}{N}, \quad 0 \le k < N \tag{2.8.10}$$

With the use of zero padding, linear cross-correlation can be achieved using circular cross-correlation. Suppose $y(k)$ is an L-point signal and $x(k)$ is an M-point signal with $M \le L$. Let $y_z(k)$ be a zero-padded version of $y(k)$ with $M + p$ zeros appended, where $p \ge -1$. Similarly, let $x_z(k)$ be a zero-padded version of $x(k)$ with $L + p$ zeros. Therefore y_z and x_z are both signals of length $N = L + M + p$.

$$x_z = [x(0), \cdots, x(M-1), \overbrace{0, \cdots, 0}^{L+p}]^T \tag{2.8.11a}$$
$$y_z = [y(0), \cdots, y(L-1), \underbrace{0, \cdots, 0}_{M+p}]^T \tag{2.8.11b}$$

Next, let x_{zp} be the periodic extension of $x_z(k)$ as in (2.7.7), and consider the circular cross-correlation of $y_z(k)$ with $x_z(k)$.

$$c_{y_z x_z}(k) = \frac{1}{N} \sum_{i=0}^{N-1} y_z(i) x_{zp}(i - k), \quad 0 \le k < N \tag{2.8.12}$$

If we restrict $c_{y_z x_z}(k)$ to $0 \le k < L$, it can be shown to be proportional to the linear cross-correlation in Definition 2.5. In particular, recalling that $y(k)$ is an L-point signal, we have

$$c_{y_z x_z}(k) = \frac{1}{N} \sum_{i=0}^{L-1} y_z(i) x_{zp}(i - k), \quad 0 \le k < L \tag{2.8.13}$$

Since $0 \le k < L$, the minimum value for $i - k$ is $-(L-1)$. But $x_z(k)$ has $L + p$ zeros padded to the end of it. Therefore $x_{zp}(-k) = 0$ for $0 \le k \le L + p$. It follows that $x_{zp}(i - k)$ in (2.8.13) can be replaced by $x_z(i - k)$ as long as $p \ge -1$. The result is then the linear cross-correlation of $y_z(k)$ with $x_z(k)$. But for $0 \le k < L$, the linear cross-correlation of $y_z(k)$ with $x_z(k)$ is identical to the linear cross-correlation of $y(k)$ with $x(k)$, except for a scale factor of L/N. Consequently,

$$r_{yx}(k) = \left(\frac{N}{L}\right) c_{y_z x_z}(k), \quad 0 \le k < L \tag{2.8.14}$$

The relationship between linear and circular cross-correlation and the properties of cross-correlation are summarized in Table 2.4.

TABLE 2.4: ▶
Linear and Circular
Cross-correlation

Property	Equation
Linear cross-correlation	$r_{yx}(k) = \dfrac{1}{L} \sum_{i=0}^{L-1} y(i) x(i - k)$
Circular cross-correlation	$c_{yx}(k) = \dfrac{1}{N} \sum_{i=0}^{N-1} y(i) x_p(i - k)$
Time reversal	$c_{yx}(-k) = c_{xy}(k)$
Circular convolution	$c_{yx}(k) = \dfrac{y(k) \circ x(-k)}{N}$
Zero padding	$r_{yx}(k) = \left(\dfrac{N}{L}\right) c_{y_z x_z}(k)$

FDSP Functions

If the MATLAB Signal Processing Toolbox is available, then the function *xcorr* can be used to perform linear cross-correlation. Alternatively, the FDSP toolbox contains the following implementation of both linear and circular cross-correlation.

```
% F_CORR: Fast cross-correlation of two discrete-time signals
%
% Usage:
%        r = f_corr (y,x,circ,norm)
% Pre:
%        y    = vector of length L containing first signal
%        x    = vector of length M <= L containing second signal
%        circ = optional correlation type code (default 0):
%
%               0 = linear correlation
%               1 = circular correlation
%
%        norm = optional normalization code (default 0):
%
%               0 = no normalization
%               1 = normalized cross-correlation
% Post:
%        r = vector of length L contained selected cross-
%            correlation of y with x.
% Notes:
%        To compute auto-correlation use x = y.
```

● ● ● ● ● ● ● ● ● ● ● ● ● ● ●

2.9 Stability in the Time Domain

Practical discrete-time systems, particularly digital filters, tend to have one qualitative feature in common: they are stable. Recall from Section 2.2 that a signal $x(k)$ is bounded if and only if $|x(k)| \leq B_x$ for some finite bound $B_x > 0$.

DEFINITION

2.7: BIBO Stable

> A discrete-time system is *bounded-input bounded-output (BIBO) stable* if and only if every bounded input produces a bounded output. Otherwise, the system is *unstable*.

The stability of a discrete-time system can be determined directly from the impulse response $h(k)$. Suppose the input $x(k)$ is bounded by a bound B_x. Recalling that the output is the convolution of the input with the impulse response, from (2.7.5) and the inequalities in

Appendix 2, we have

$$|y(k)| = |\sum_{i=0}^{k} h(i)x(k-i)|$$

$$\leq \sum_{i=0}^{k} |h(i)x(k-i)|$$

$$= \sum_{i=0}^{k} |h(i)| \cdot |x(k-i)|$$

$$\leq B_x \sum_{i=0}^{k} |h(i)|$$

$$\leq B_x \sum_{i=-\infty}^{\infty} |h(i)| \qquad (2.9.1)$$

It follows from (2.9.1) that the system is BIBO stable if the infinite series on the right-hand side converges to a finite value. This condition is not only sufficient for BIBO stability, it is also necessary (see Problem 2.39). This leads to the following fundamental stability result.

PROPOSITION
....
2.1: BIBO Stability: Time
Domain

A linear time-invariant discrete-time system is BIBO stable if and only if the impulse response $h(k)$ is absolutely summable.

$$\|h\|_1 = \sum_{k=-\infty}^{\infty} |h(k)| < \infty$$

There is an important class of discrete-time systems that is always stable. Consider an FIR system.

$$y(k) = \sum_{i=0}^{m} b_i x(k-i) \qquad (2.9.2)$$

Recall from (2.6.3) that for this system the impulse response is $h(k) = b_k$ for $0 \leq k \leq m$, and $h(k) = 0$ for $k > m$. It follows that for an FIR system, the impulse response is absolutely summable.

$$\|h\|_1 = \sum_{i=0}^{m} |b_i| \qquad (2.9.3)$$

Therefore FIR systems are always BIBO stable. This is one of the features that make FIR systems useful candidates for digital filters (Chapter 6), and particularly for adaptive digital filters (Chapter 9). The more general IIR systems, by contrast, can be stable or unstable.

Example 2.19   **BIBO Stability: Time Domain**

Consider the home mortgage system introduced in Section 2.1, where k denotes the month and the fraction $r > 0$ represents the annual interest rate.

$$y(k) = \left(1 + \frac{r}{12}\right) y(k-1) - x(k)$$

First we compute the impulse response using (2.6.6). By inspection, the characteristic polynomial is

$$a(z) = z - p_1$$
$$p_1 = 1 + \frac{r}{12}$$

Hence the form of the impulse response is

$$h(k) = d_0 \delta(k) + d_1 (p_1)^k \mu(k)$$

The polynomial associated with the input terms is $b(z) = -z$. Using (2.6.7) with $p_0 = 0$, we find the coefficient vector $d \in R^2$ to be

$$d_0 = 0$$
$$d_1 = \frac{-1}{p_1}$$

Hence the impulse response of the home mortgage system is

$$h(k) = -(1 + r/12)^{k-1} \mu(k)$$

Since $r > 0$, the samples of $h(k)$ grow with time, so $h(k)$ is clearly not absolutely summable. It follows that the home mortgage system is not BIBO stable.

● ● ● ● ● ● ● ● ● ● ● ● ● ● ● ● ● ●

2.10   GUI Software and Case Studies

This section focuses on applications of discrete-time systems. Graphical user interface modules called *g_systime* and *g_correlate* are introduced that allow the user to explore the input-output behavior of linear discrete-time systems in the time domain and perform cross-correlations and convolutions without any need for programming. Case study examples are then presented and solved using MATLAB programs.

g_systime: Discrete-time System Analysis in the Time Domain

The graphical user interface module *g_systime* allows the user to explore the input-output behavior of linear discrete-time systems in the time domain. GUI module *g_systime* features a display screen with tiled windows, as shown in Figure 2.26. The *Block Diagram* window in the upper-left corner of the screen contains a color-coded block diagram of the system under investigation. Below the block diagram are a number of edit boxes whose contents can be modified by the user. The edit boxes for a and b allow the user to select the coefficients of the characteristic polynomial $a(z)$ and input polynomial $b(z)$ of the following difference equation where $a_0 = 1$.

$$y(k) = \sum_{i=0}^{m} b_i x(k-i) - \sum_{i=1}^{n} a_i y(k-i), \quad y_0 \in R^n \qquad (2.10.1)$$

The initial condition vector $y_0 = [y(-1), \ldots, y(-n)]^T$, and the coefficient vectors a and b can be edited directly by clicking on the shaded area and entering in new values. Any MATLAB statement defining a, b, or y_0 can be entered in the box. For example, the edit box can be

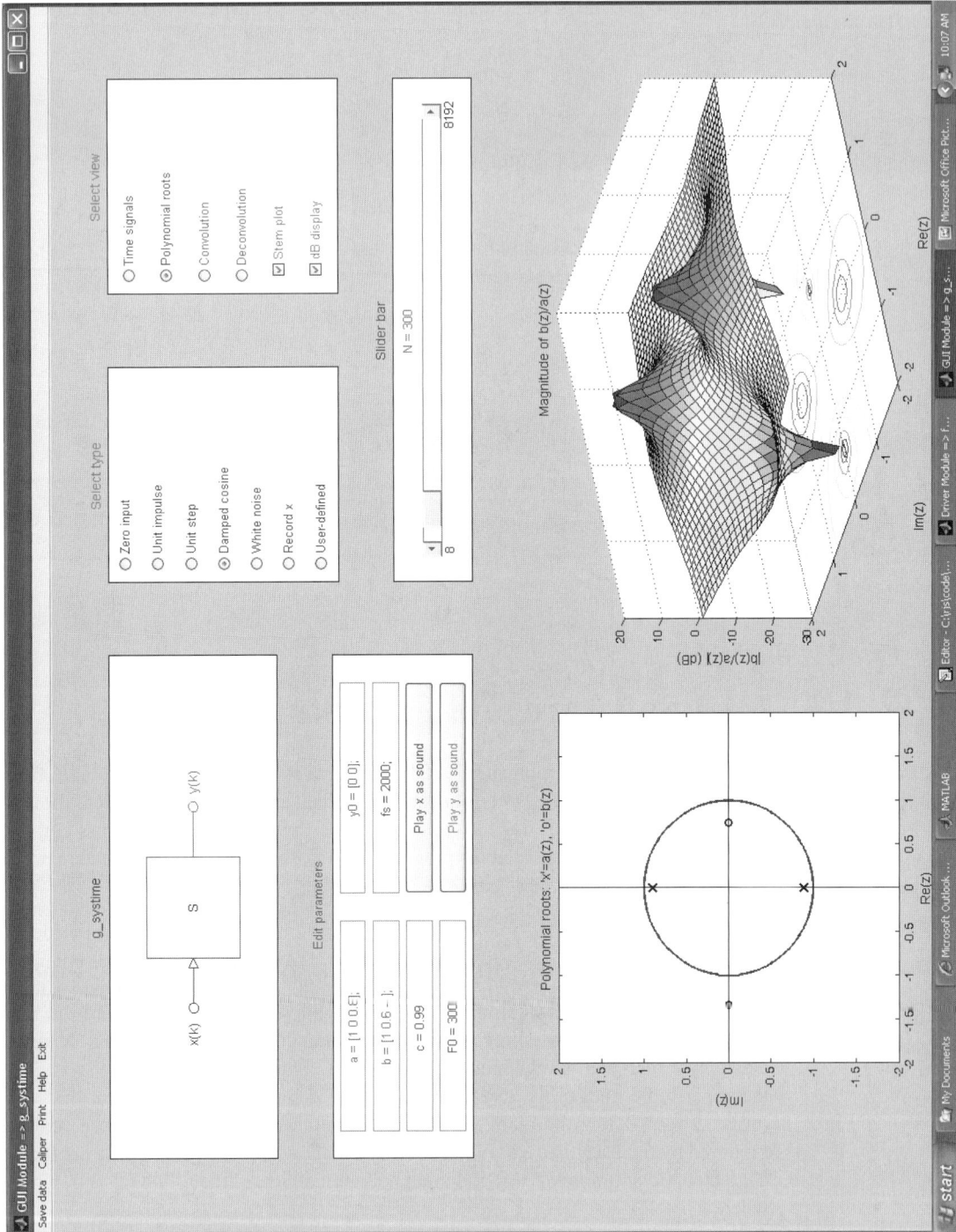

FIGURE 2.26: Display Screen of Chapter GUI Module *g_systime*

cleared and then the MATLAB function *poly* can be used to compute a coefficient vector from a vector of desired roots as follows.

```
a = poly([.7*j,-.7*j,.9,-.4]);
```

The Enter key is used to activate a change to a parameter. Additional scalar parameters that appear in edit boxes are associated with the causal exponential and damped causal cosine inputs.

$$x(k) = c^k \mu(k) \tag{2.10.2}$$
$$x(k) = c^k \cos(2\pi F_0 kT)\mu(k) \tag{2.10.3}$$

Speakers

They include the sampling frequency fs, the exponential damping factor c which is constrained to lie the interval $[-1, 1]$, and the input frequency $0 \le F0 \le fs/2$. The *Parameters* window also contains two push button controls. The push button controls play the signals $x(k)$ and $y(k)$ on the PC speaker using the current sampling rate f_s. This option allows the user to hear the filtering effects of the system S on various types of inputs.

The *Type* and *View* windows in the upper-right corner of the screen allow the user to select both the type of input signal and the viewing mode. The inputs include the zero input, the unit impulse, the unit step, a damped cosine input, white noise uniformly distributed over $[-1, 1]$, recorded sounds from a PC microphone, and user-defined inputs from a MAT file. The Recorded sound option can be used to record up to one second of sound at a sampling rate of $f_s = 8192$ Hz. For the User-defined option, a MAT file containing the input vector x, the sampling frequency fs, and the coefficient vectors a and b must be supplied by the user.

View options include time plots of the input $x(k)$ and output $y(k)$, the roots of the characteristic and input polynomials $a(z)$ and $b(z)$, polynomial multiplication using convolution, and polynomial division using deconvolution. The time plots use continuous time or discrete time depending on the status of the Stem plot check box control. The sketch of the roots of $a(z)$ and $b(z)$ also includes a surface plot of $|b(z)/a(z)|$ which can be linear or in decibels (dB) as $20 \log_{10}(|b(z)/a(z)|)$. The *Plot* window on the bottom half of the screen shows the selected view. The curves are color-coded to match the block diagram labels. The slider bar below the *Type* and *View* window allows the user to change the number of samples N.

The *Menu* bar at the top of the screen includes several menu options. The *Caliper* option allows the user to measure any point on the current plot by moving the mouse cross hairs to that point and clicking. The *Save data* option is used to save the current a, b, x, y, and fs plus the coefficients of the product, quotient, and remainder polynomials in a user-specified MAT file for future use. The User-defined input option can be used to reload these data. The *Print* option prints the contents of the plot window. Finally, the *Help* option provides the user with some helpful suggestions on how to effectively use module *g_systime*.

g_correlate: Correlation and Convolution

GUI Module

The graphical user interface module *g_correlate* is designed to allow the user to explore cross-correlations and convolutions of pairs of signals and auto-correlations of individual signals. GUI module *g_correlate* features a display screen with tiled windows, as shown in Figure 2.27. The upper-left corner of the screen contains a block diagram that specifies the operation being performed. The signals are color-coded, and the labels change depending on the signal processing operation selected.

The *Parameters* window below the block diagram contains three edit boxes. Parameters L, M, and c can be edited directly by the user. Parameters L and M are the lengths of the input signals $x(k)$ and $y(k)$, respectively, with $M \le L$. Parameter c represents a scale factor that is used to add a scaled and shifted version of $x(k)$ to $y(k)$ so that its presence can be

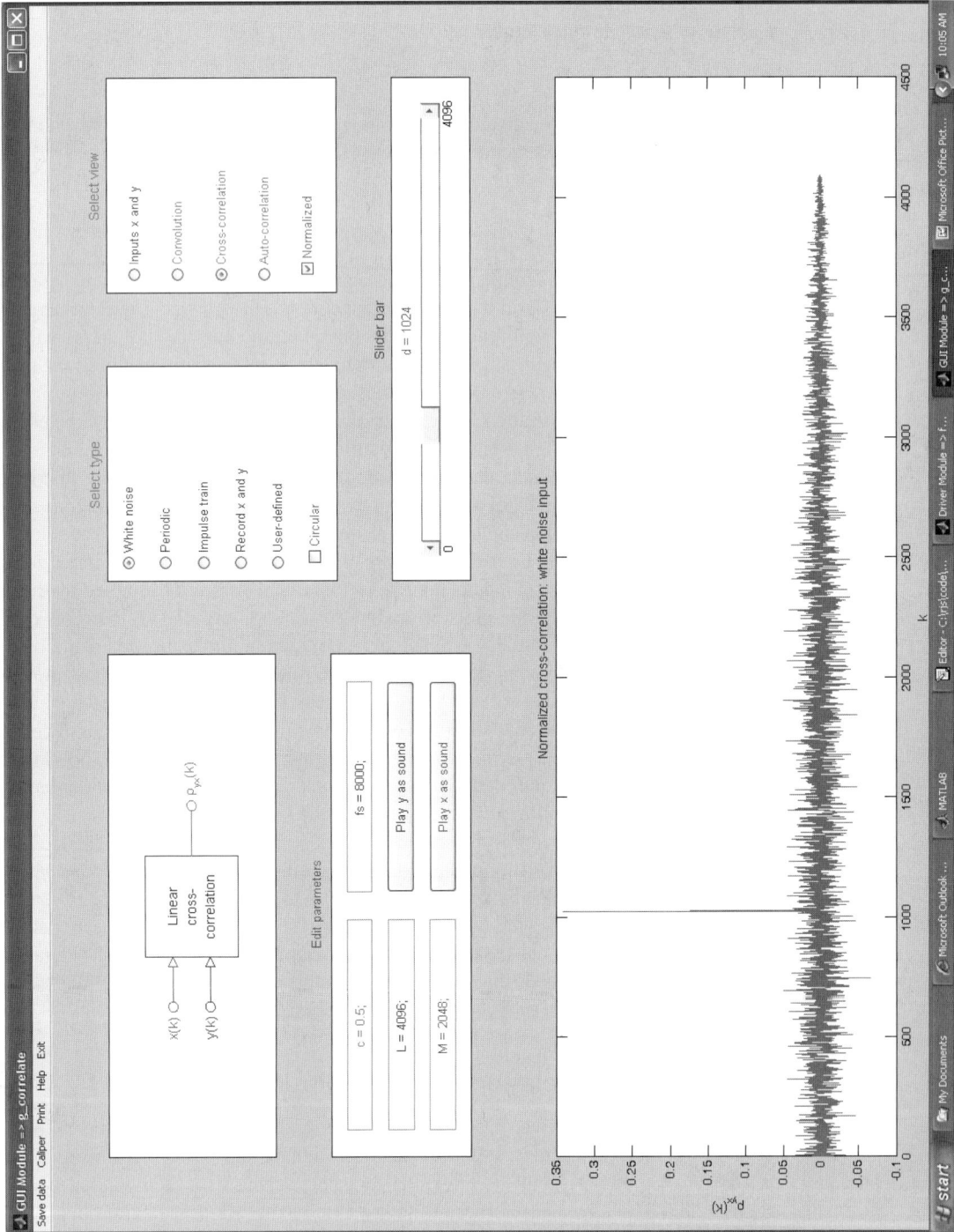

FIGURE 2.27: Display Screen of Chapter GUI Module *g_correlate*

detected using cross-correlation. Changes to parameter values are activated with the Enter key. The *Parameters* window also includes push button and check box controls. The push button *Sound* controls play the signals $x(k)$ and $y(k)$ on the PC speaker using the current sampling rate f_s. The first check box allows the user to toggle between linear correlation or convolution and circular correlation or convolution. The second check box allows the user to choose between regular and normalized cross-correlations and auto-correlations.

The *Type* and *View* windows in the upper-right corner of the screen allow the user to select both the type of input signal and the viewing mode. The inputs include white noise inputs, periodic inputs, a periodic y with a synchronized impulse train for x, inputs recorded from a PC microphone, and user-defined inputs stored in a MAT file. For the recorded inputs, up to .5 second for x and 2 seconds for y can be recorded at $f_s = 8192$ Hz. The user-defined inputs are specified in a user-supplied MAT file containing the vectors x, y, and f_s. For the white noise inputs, the signal $y(k)$ contains a scaled and delayed version of the signal $x(k)$. That is, $y(k)$ is computed as follows, where $x_z(k)$ is a zero-extended version of $x(k)$.

$$y(k) = cx_z(k - d) + v(k), \quad 0 \le k < L \tag{2.10.4}$$

The scale factor c in the *Parameters* window can be modified by the user. The delay d can be set anywhere between 0 and L using the horizontal slider bar that appears below the *Type* and *View* windows.

The *View* options include plots of $x(k)$, $y(k)$, the convolution of $x(k)$ with $y(k)$, the cross-correlation of $y(k)$ with $x(k)$, and the auto-correlation of $y(k)$. The *Plot* window on the bottom half of the screen shows the selected view. The curves are color-coded to match the block diagram labels.

The *Menu* bar at the top of the screen includes several menu options. The *Caliper* option allows the user to measure any point on the current plot by moving the mouse crosshairs to that point and clicking. The *Save Data* option is used to save the current x, y, and f_s in a user-specified MAT file for future use. Files created in this manner can be loaded with the User-defined input option. The *Print* option prints the contents of the *Plot* window. Finally, the *Help* option provides the user with some helpful suggestions on how to effectively use module *g_correlate*.

CASE STUDY 2.1   ## Home Mortgage

Recall from Section 2.1 that the following discrete-time system can be used to model a home mortgage loan.

$$y(k) = y(k - 1) + \left(\frac{r}{12}\right) y(k - 1) - x(k)$$

Here $y(k)$ is the balance owed at the end of month k, r is the annual interest rate expressed as a fraction, and $x(k)$ is the monthly mortgage payment. One of the questions posed in Section 2.1 was the following. If the size of the mortgage is y_0 dollars, and the duration of the mortgage is N months, what is the required monthly payment? Now that we have the necessary tools in place, we can answer this and related questions. To streamline the notation, let

$$p_1 = 1 + \frac{r}{12}$$

Then the difference equation can be simplified to

$$y(k) = -p_1 y(k - 1) - x(k)$$

The size of the mortgage y_0 enters as an initial condition $y(-1) = y_0$. The characteristic and input polynomials of this system are

$$a(z) = z - p_1$$
$$b(z) = -z$$

Thus the form of the zero-input part of the response is

$$y_{zi}(k) = c_1(p_1)^k$$

Setting $y(-1) = y_0$ yields $c_1/p_1 = y_0$. Thus the zero-input response arising from the initial condition is

$$y_{zi}(k) = y_0(p_1)^{k+1}, \quad k \geq -1$$

Next we examine the zero-state response. Let A be the monthly payment. Then the input $x(k)$ is a step of amplitude A. Note that this is a special case of the causal exponential input in (2.4.13) where the exponential factor is $p_0 = 1$. Thus from (2.4.15) the form of the zero-state response is

$$y_{zs}(k) = [d_0 + d_1(p_1)^k]\mu(k)$$

If we applying (2.4.16) with $p_0 = 1$, the coefficient vector $d \in R^2$ is

$$d_0 = \frac{A(-1)}{1 - p_1}$$
$$d_1 = \frac{A(-p_1)}{p_1 - 1}$$

Thus the zero-state response produced to the input $x(k) = A\mu(k)$ is

$$y_{zs}(k) = \frac{A(1 - p_1^{k+1})\mu(k)}{p_1 - 1}$$

Finally, the zero-input and zero-state responses can be combined to produce the complete response.

$$y(k) = y_{zi}(k) + y_{zs}(k)$$
$$= y_0(p_1)^{k+1} + \frac{A(1 - p_1^{k+1})\mu(k)}{p_1 - 1}, \quad k \geq -1$$

The length of the mortgage is N months. Setting $y(N) = 0$ and solving for A yields the required monthly payment.

$$A = \frac{y_0(1 - p_1)p_1^{N+1}}{1 - p_1^{N+1}}$$

CASE
STUDY 2.1

The first part of *case2_1* prompts the user for the interest rate and then computes the required monthly payment for loans of different sizes and durations. The results are shown in Figure 2.28.

```
function case2_1

% CASE STUDY 2.1: Home mortgage

f_header('Case Study 2.1: Home Mortgage')
N = 31;
b = linspace(0,300,N)';        % size of mortgage
d = [15 20 30];                % duration in years
p = zeros(N,3);                % monthly payments
```

FIGURE 2.28: Monthly Mortgage Payments for 6% Interest

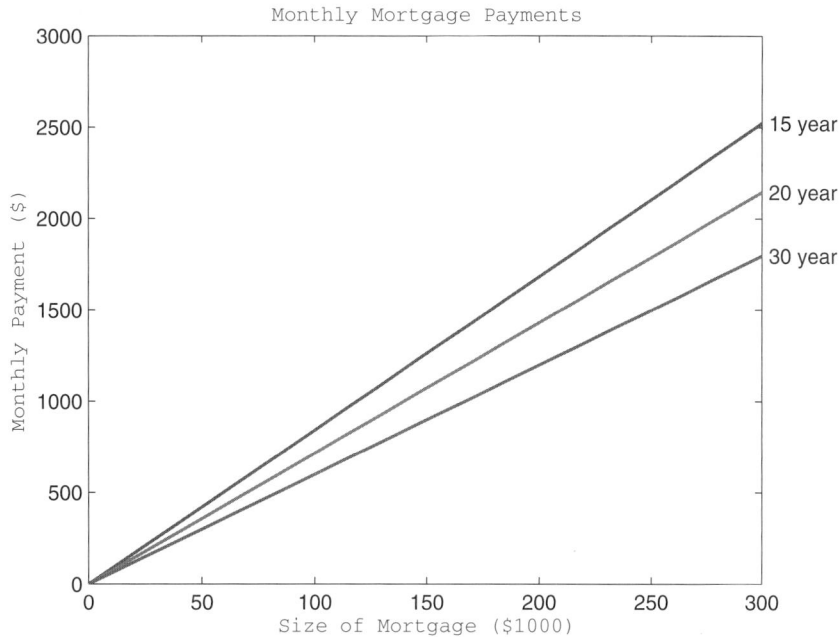

```
% Compute monthly payments

r = f_prompt('Enter interest rate in percent:',0,20,6.0)/100;
p1 = 1 + r/12;
c = p1.^(12*d+1);
for k = 1 : N
    for j = 1 : length(d)
        p(k,j) = 1000*b(k)*c(j)*(p1-1)/(c(j)-1);
    end
end
figure
plot(b,p,'LineWidth',1.5)
f_labels ('Monthly mortgage payments','Size of mortgage ($1000)',...
          'Monthly payment ($)')
hold on
for j = 1 : length(d)
    duration = sprintf ('%d year',d(j));
    text (b(N)+3,p(N,j),duration)
end
f_wait

% Compute balance vs time

q = 12*d(3)+1;                      % number of months
k = [0 : q-1]';                     % discrete times
y_zi = b(21)*p1.^(k+1);            % zero-input response
num = [-1 0];                       % numerator coefficients
den = [1 -p1];                      % denominator coefficients
A = p(21,3);                        % $200,000, 30 years
x = A*ones(q,1);                    % step of amplitude A
```

```
y_zs = filter(num,den,x)/1000;    % zero-state response
y(:,1) = y_zi + y_zs;
y(:,2) = (6*A/(1000*r))*ones(q,1);
figure
h = plot (k,y(:,1),k,y(:,2));
set (h(1),'LineWidth',1.5)
f_labels ('Balance due on 30 year $200,000 mortgage','Month','Balance ($1000)')

% Find crossover month

text (k(q)+5,y(q,2),'6A/r')
crossover = min(find(y(:,1) < y(:,2)))
hold on
plot ([k(crossover),k(crossover)],[0,y(crossover,2)],'k--','LineWidth',1.0)
plot (k(crossover),y(crossover),'.')
legend ('Balance due')
axis ([0 400 0 200])
f_wait
```

Another question that was asked in Section 2.1 was how many months does it take before more than half of the monthly payment goes toward reducing the principal, rather than paying interest? The monthly interest is $(r/12)y(k)$. Thus we require $(r/12)y(k) < A/2$ or

$$y(k) < \frac{6A}{r}$$

The smallest k that satisfies this inequality is the *crossover* month. The second half of *case2_1* computes the zero-input response directly and uses the MATLAB function *filter* to compute the zero-state response corresponding to a $200,000 mortgage over 30 years. The resulting plot of the balance owed is shown in Figure 2.29. Note that the crossover month does not occur until month 250, well beyond the midpoint of the 30-year loan.

FIGURE 2.29:
Balance Owed over
the Duration of the
Mortgage

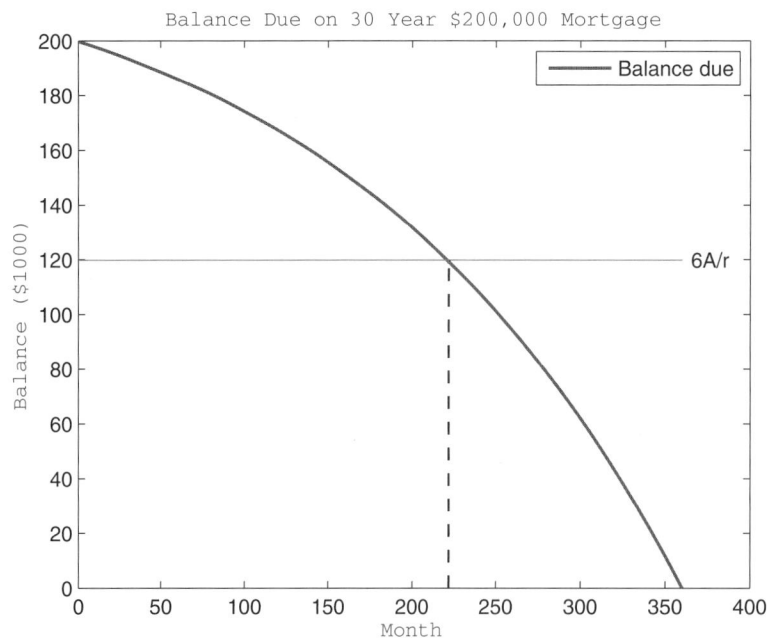

| CASE STUDY 2.2 | **Echo Detection** |

Recall from Section 2.1 that one of the applications of cross-correlation is for radar processing, as shown previously in Figure 2.1. Here, $x(k)$ is the transmitted signal and $y(k)$ is the received signal. First, consider the transmitted signal. Suppose the sampling frequency is $f_s = 1$ MHz, and the number of transmitted samples is $M = 512$. One possible choice for the transmitted signal is a uniformly distributed white noise signal like the one used in Section 2.1. Another possibility is a multi-frequency *chirp*, a sinusoidal signal whose frequency varies with time. For example, let $T = 1/f_s$, and consider the following chirp signal with variable frequency $f(k)$ for $0 \le k < M$.

Chirp

$$f(k) = \frac{kf_s}{2(M-1)}$$
$$y(k) = \sin[2\pi f(k)kT]$$

The received signal $y(k)$ includes a scaled and delayed version of the transmitted signal plus measurement noise. Suppose the received signal consists of $L = 2048$ samples. If $x_z(k)$ denotes the transmitted signal, zero-extended to L points, then the received signal can be expressed as follows.

$$y(k) = cx_z(k - d) + v(k), \quad 0 \le k < L$$

The first term in $y(k)$ is the echo of the transmitted signal that is reflected back from the illuminated target. Typically, the echo will be attenuated due to the dispersion, so $c \ll 1$. In addition, the echo will be delayed by d samples due to the time it takes for the transmitted signal to travel to the target, bounce off, and return. The second term in $y(k)$ is random atmospheric measurement noise picked up by the receiver. For example, suppose $v(k)$ is white noise uniformly distributed over the interval $[-0.1, 0.1]$.

To determine the range to the target, let γ be the propagation speed of the transmitted signal. For radar applications, this corresponds to the speed of light or $\gamma = 1.86 \times 10^8$ miles/sec. The time of flight of the signal is then $\tau = dT$ sec. Multiplying τ by the signal propagation speed, and dividing by two for the round trip, we then arrive at the following expression for the *range* to the target.

$$r = \frac{\gamma dT}{2}$$

CASE STUDY 2.2

Thus the key to finding the distance to the target is to detect the presence, and location, of the echo of the transmitted signal $x(k)$ in the received signal $y(k)$. This can be achieved by running *case2_2* from the FDSP driver program.

```
function case2_2

% CASE STUDY 2.2: Echo detection

f_header('Case Study 2.2: Echo detection')
rand('state',1000)

% Construct transmitted signal x

M = 512;
fs = 1.e7;
T = 1/fs;
k = 0 : M-1;
```

```
freq = (fs/2)*k/(M-1);
x = sin(2*pi*freq.*k*T);
figure
plot (k,x)
set(gca,'Ylim',[-1.5 1.5])
f_labels ('Transmitted chirp signal','k','x(k)')
f_wait

% Construct received signal y

L = 2048;
c = 0.02;
d = 1304;
y = f_randu(1,L,-0.1,0.1);
y(d+1:d+M) = y(d+1:d+M) + c*x;
k = 0 : L-1;
figure
plot (k,y)
f_labels ('Received signal','k','y(k)')
f_wait

% Locate echo and compute range

rho = f_corr (y,x,0,1);
figure
plot (k,rho)
f_labels ('Normalized linear cross-correlation','k','\rho_{yx}(k)')
[rmax,kmax] = max(rho)
delay = kmax - 1
gamma = 1.86e5;
r = gamma*delay*T/2
precision = gamma*T/2
f_wait
```

When *case2_2* is run, it first constructs the chirp signal $x(k)$. The resulting plot is shown in Figure 2.30. Note how the frequency changes with time. Function *case2_2* then constructs the received signal $y(k)$ and computes the normalized linear cross-correlation of $y(k)$ with $x(k)$. The echo of $x(k)$ buried in $y(k)$ becomes evident when we examine the normalized cross-correlation plot shown in Figure 2.31. For this example, $a = .02$ and $d = 1304$. Although the peak at $\rho_{yx}(d) = .147$ is not large due to the noise, it is distinct. If we use the MATLAB function *max* to locate the peak, the value reported for the range to the target is

$$r = 12.13 \text{ miles}$$

The precision with which the range can be measured depends on both the sampling frequency f_s and the propagation speed γ. The smallest increment for r corresponds to a delay of $d = 1$ sample which yields

$$\Delta r = .0093 \text{ miles}$$

FIGURE 2.30: The Transmitted Multi-frequency Chirp Signal $x(k)$

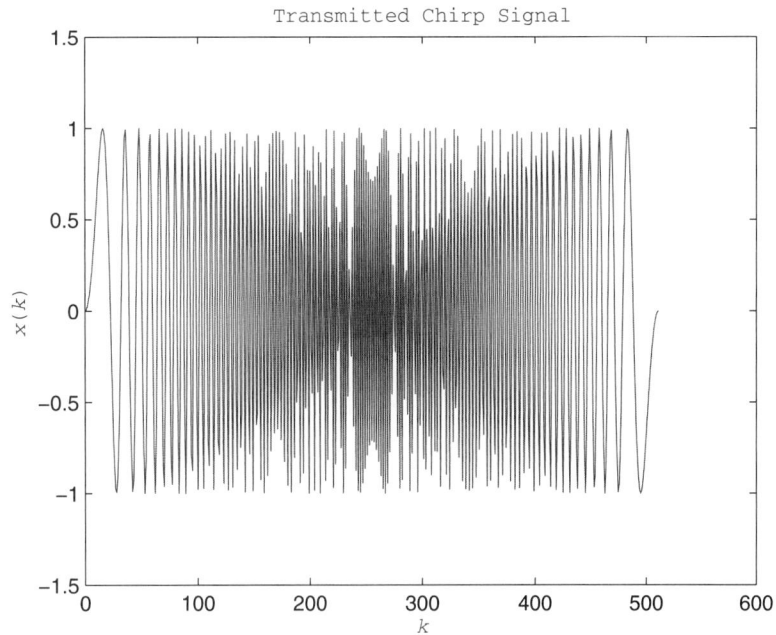

Transmitted Chirp Signal

FIGURE 2.31: Normalized Linear Cross-correlation of Received Signal $y(k)$ with Transmitted Signal $x(k)$

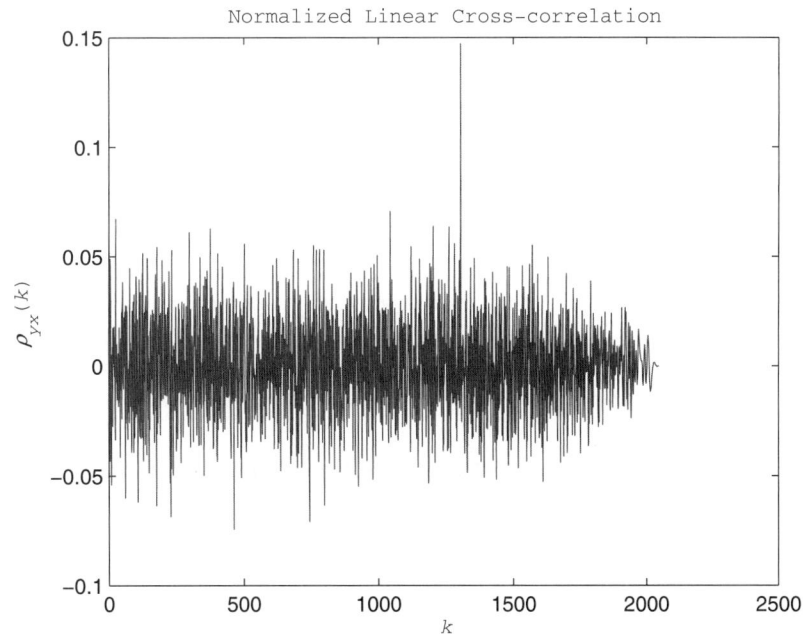

Normalized Linear Cross-correlation

2.11 Chapter Summary

Signals and Systems

Signal classification

This chapter focused on linear time-invariant discrete-time systems in the time domain. Discrete-time signals can be classified into a number of useful categories, including finite and infinite signals, causal and noncausal signals, periodic and aperiodic signals, bounded and unbounded signals, and energy and power signals. Some common signals include the unit impulse $\delta(k)$, the unit step $\mu(k)$, causal exponentials, and periodic power signals. A discrete-time system S processes a discrete-time input signal $x(k)$ to produce a discrete-time output signal $y(k)$. Discrete-time systems can be classified into linear and nonlinear systems, time-invariant and time-varying systems, causal and noncausal systems, stable and unstable systems, and passive and active systems.

System classification

Difference equations

A finite-dimensional linear time-invariant (LTI) discrete-time system S can be represented in the time domain by a constant-coefficient difference equation.

$$y(k) = \sum_{i=0}^{m} b_i x(k-i) - \sum_{i+1}^{n} a_i y(k-i), \quad y_0 \in R^n \tag{2.11.1}$$

Initial condition

The output of the system depends on the causal input $x(k)$ and the *initial condition* $y_0 = [y(-1), y(-2), \ldots, y(-n)]^T$. The output or response of S can be written as the sum of the zero-input response plus the zero-state response.

$$y(k) = y_{zi}(k) + y_{zs}(k) \tag{2.11.2}$$

Characteristic polynomial

The zero-input response $y_{zi}(k)$ is the part of the response that comes from the initial condition y_0 when the input is $x(k) = 0$. The zero-state response $y_{zs}(k)$ is the part of the response that comes from the input $x(k)$ when the initial condition is $y_0 = 0$. The key to the zero-input response is the *characteristic polynomial* of S.

$$a(z) = z^n + a_1 z^{n-1} + \cdots + a_n \tag{2.11.3}$$

Natural mode

Each root p_i of the characteristic polynomial generates a *natural mode* term of the form $c_i(k)(p_i)^k$ in the zero-input response. If the root p_i is a simple root that occurs only once, then the coefficient $c_i(k)$ is a constant. Otherwise, $c_i(k)$ is a polynomial in k of degree one less than the multiplicity of the root. The coefficients of the natural mode terms are computed by solving the equations that arise from applying the initial condition $y_{zi}(-i) = y(-i)$ for $1 \le i \le n$. The zero-state response will contain natural mode terms that are excited by the input $x(k)$ plus additional terms whose form depends on the type of input. The zero-state response to any input can be computed from the impulse response using convolution.

The Impulse Response and Convolution

Impulse response

The *impulse response* is the zero-state response of the system to the unit impulse input $\delta(k)$. If a system has an impulse response that is zero after a finite number of samples, then it is called an *FIR system*. Otherwise, it is an *IIR system*. The impulse response of an FIR system can be obtained directly from inspection of the input coefficients of the difference equation. A system is *BIBO stable* if and only if every bounded input is guaranteed to produced a bounded output. Otherwise, the system is *unstable*. The system S is stable if and only if the impulse response $h(k)$ is absolutely summable. All FIR systems are stable, but IIR systems may or may not be stable.

BIBO stable

Linear convolution

Given the impulse response $h(k)$, the zero-state response of the system S to any input can be obtained from it using linear convolution. If $h(k)$ and $x(k)$ are two causal signals, the *linear convolution* of $h(k)$ with $x(k)$ is defined

$$h(k) \star x(k) = \sum_{i=0}^{k} h(i)x(k-i), \quad k \ge 0 \tag{2.11.4}$$

When $h(k)$ is the impulse response of a linear discrete-time system and $x(k)$ is the system input, the zero-state response of the system is the linear convolution of $h(k)$ with $x(k)$.

$$y_{zs}(k) = h(k) \star x(k) \tag{2.11.5}$$

The convolution operation is commutative, so $h(k) \star x(k) = x(k) \star h(k)$. In addition, if $x(0) \neq 0$, then $h(k)$ can be recovered from $x(k)$ and $y(k)$ using a process known as *Deconvolution* *deconvolution*. If we set $h(0) = y(0)/x(0)$, the remaining samples of the impulse response are obtained recursively using

$$h(k) = \frac{1}{x(0)} \left\{ y(k) - \sum_{i=0}^{k-1} h(i)x(k-i) \right\}, \quad k \geq 1 \tag{2.11.6}$$

In terms of computation, convolution corresponds to the process of polynomial multiplication, while deconvolution corresponds to polynomial division. If $h(k)$ and $x(k)$ are both signals of length N, and $x(k)$ is replaced by its periodic extension $x_p(k)$, this results in an operation *Circular convolution* called the *circular convolution* of $h(k)$ with $x(k)$.

$$h(k) \circ x(k) = \sum_{i=0}^{N-1} h(i)x_p(k-i), \quad 0 \leq k < N \tag{2.11.7}$$

Circular convolution is a more efficient form of convolution that can be represented with matrix multiplication. Linear convolution of two finite signals can be implemented with circular convolution by zero-padding the signals.

$$h(k) \star x(k) = h_z(k) \circ x_z(k) \tag{2.11.8}$$

Correlation

Linear cross-correlation An operation that is closely related to convolution is correlation. The *linear cross-correlation* of an L-point signal $y(k)$ with an M-point signal $x(k)$ is defined as follows, where it is assumed that $M \leq L$.

$$r_{yx}(k) = \frac{1}{L} \sum_{i=0}^{L-1} y(i)x(i-k), \quad 0 \leq k < L \tag{2.11.9}$$

Here k is referred to as the *lag* variable because it represents the amount by which the second signal is delayed before the sum of products is computed. Linear cross-correlation can be used to measure the degree to which the shape of signal $x(k)$ is similar to the shape of signal $y(k)$. In particular, if $y(k)$ contains a scaled version of $x(k)$ delayed by d samples, then $r_{yx}(k)$ will exhibit a peak at $k = d$. The following normalized version of linear cross-correlation takes on values in the interval $[-1, 1]$.

$$\rho_{yx}(k) = \frac{r_{yx}(k)}{\sqrt{(M/L)r_{xx}(0)r_{yy}(0)}}, \quad 0 \leq k < L \tag{2.11.10}$$

Just as there is a circular version of convolution, there is also a circular version of cross-correlation. Suppose $x(k)$ and $y(k)$ are both of length N. If $x(k)$ is replaced by its periodic *Circular cross-correlation* extension $x_p(k)$, then this results in the following formulation of cross-correlation called *circular cross-correlation*.

$$c_{yx}(k) = \frac{1}{N} \sum_{i=0}^{N-1} y(i)x_p(i-k), \quad 0 \leq k < N \tag{2.11.11}$$

Num.	Learning Outcome	Sec.
1	Be able to classify discrete time signals into a variety of useful categories	2.2
2	Become familiar with common discrete-time signals and their characteristics	2.2
3	Know how to find the zero-input response of a difference equation for an arbitrary initial condition.	2.3
4	Know how to find the zero-state response of a difference equation for a nonzero input.	2.3
5	Know how to find the complete response of a difference equation both analytically and numerically.	2.3
6	Be able to represent discrete-time linear time-invariant systems graphically using block diagrams	2.4
7	Know how to compute the impulse response and classify systems based on the duration of the impulse response	2.5
8	Know how to perform linear and circular convolutions and compute the zero-state response using convolution	2.6
9	Understand what it means for a system to be stable and be able to determine stability from the impulse response.	2.7
10	Know how to use the GUI modules *g_systime* and *g_correlate* to interactively explore the input-output behavior of discrete-time systems	2.8

The following normalized version of circular cross-correlation takes on values in the interval $[-1, 1]$.

$$\sigma_{yx}(k) = \frac{c_{yx}(k)}{\sqrt{c_{xx}(0)c_{yy}(0)}}, \quad 0 \le k < N \tag{2.11.12}$$

There is a simple relationship between circular correlation and circular convolution. Circular cross-correlation of $y(k)$ with $x(k)$ is just a scaled version of circular convolution of $y(k)$ with $x(-k)$.

$$c_{yx}(k) = \frac{y(k) \circ x(-k)}{N}, \quad 0 \le k < N \tag{2.11.13}$$

GUI Modules

The FDSP toolbox includes GUI modules called *g_systime* and *g_correlate* that allow the user to interactively explore the input-output behavior of a discrete-time system in the time domain, and perform convolutions and correlations without any need for programming. Several common input signals are included, plus signals recorded from a PC microphone and user-defined signals saved in MAT files.

Learning Outcomes

This chapter was designed to provide the student with an opportunity to achieve the learning outcomes summarized in Table 2.5.

● ● ● ● ● ● ● ● ● ● ● ● ● ● ● ● ● ● ●

2.12 Problems

The problems are divided into Analysis and Design problems that can be solved by hand or with a calculator, GUI Simulation problems that are solved using GUI modules *g_systime* and *g_correlate*, and MATLAB Computation problems that require a user program.

2.12.1 Analysis and Design

2.1 Find the average power of the following signals.

(a) $x(k) = 10$
(b) $x(k) = 20\mu(k)$
(c) $x(k) = \text{mod}(k, 5)$
(d) $x(k) = a\cos(\pi k/8) + b\sin(\pi k/8)$
(e) $x(k) = 100[u(k + 10) - u(k - 10)]$
(f) $x(k) = j^k$

2.2 Classify each of the following systems as linear or nonlinear.

(a) $y(k) = 4[y(k - 1) + 1]x(k)$
(b) $y(k) = 6kx(k)$
(c) $y(k) = -y(k - 2) + 10x(k + 3)$
(d) $y(k) = .5y(k) - 2y(k - 1)$
(e) $y(k) = .2y(k - 1) + x^2(k)$
(f) $y(k) = -y(k - 1)x(k - 1)/10$

2.3 Classify each of the following systems as time-invariant or time-varying.

(a) $y(k) = [x(k) - 2y(k - 1)]^2$
(b) $y(k) = \sin[\pi y(k - 1)] + 3x(k - 2)$
(c) $y(k) = (k + 1)y(k - 1) + \cos[.1\pi x(k)]$
(d) $y(k) = .5y(k - 1) + \exp(-k/5)\mu(k)$
(e) $y(k) = \log[1 + x^2(k - 2)]$
(f) $y(k) = kx(k - 1)$

2.4 Classify each of the following systems as causal or noncausal.

(a) $y(k) = [3x(k) - y(k - 1)]^3$
(b) $y(k) = \sin[\pi y(k - 1)] + 3x(k + 1)$
(c) $y(k) = (k + 1)y(k - 1) + \cos[.1\pi x(k^2)]$
(d) $y(k) = .5y(k - 1) + \exp(-k/5)\mu(k)$
(e) $y(k) = \log[1 + y^2(k - 1)x^2(k + 2)]$
(f) $h(k) = \mu(k + 3) - \mu(k - 3)$

2.5 Consider the following system that consists of a gain of A and a delay of d samples.

$$y(k) = Ax(k - d)$$

(a) Find the impulse response $h(k)$ of this system.
(b) Classify this system as FIR or IIR.
(c) Is this system BIBO stable? If so, find $\|h\|_1$.
(d) For what values of A and d is this a passive system?
(e) For what values of A and d is this an active system?
(f) For what values of A and d is this a lossless system?

2.6 Consider the following sum of causal exponentials.

$$x(k) = [c_1(p_1)^k + c_2(p_2)^k]\mu(k)$$

(a) Using the inequalities in Appendix 2, show that

$$|x(k)| \le |c_1| \cdot |p_1|^k + |c_2| \cdot |p_2|^k$$

(b) Show that $x(k)$ is absolutely summable if $|p_1| < 1$ and $|p_2| < 1$. Find an upper bound on $\|x\|_1$

(c) Suppose $|p_1| < 1$ and $|p_2| < 1$. Find an upper bound on the energy E_x.

2.7 For each of the following signals, determine whether or not it is bounded. For the bounded signals, find a bound B_x.

(a) $x(k) = [1 + \sin(5\pi k)]\mu(k)$

(b) $x(k) = k(.5)^k \mu(k)$

(c) $x(k) = \left[\dfrac{(1+k)\sin(10k)}{1 + (.5)^k} \right] \mu(k)$

(d) $x(k) = [1 + (-1)^k]\cos(10k)\mu(k)$

2.8 Classify each of the following signals as bounded or unbounded.

(a) $x(k) = k\cos(.1\pi k)/(1 + k^2)$

(b) $x(k) = \sin(.1k)\cos(.2k)\delta(k - 3)$

(c) $x(k) = \cos(\pi k^2)$

(d) $x(k) = \tan(.1\pi k)[\mu(k) - \mu(k - 10]$

(e) $x(k) = k^2/(1 + k^2)$

(f) $x(k) = k\exp(-k)\mu(k)$

2.9 Classify each of the following signals as periodic or aperiodic. For the periodic signals, find the period M.

(a) $x(k) = \cos(.02\pi k)$

(b) $x(k) = \sin(.1k)\cos(.2k)$

(c) $x(k) = \cos(\sqrt{3}k)$

(d) $x(k) = \exp(j\pi/8)$

(e) $x(k) = \mathrm{mod}(k, 10)$

(f) $x(k) = \sin^2(.1\pi k)\mu(k)$

(g) $x(k) = j^{2k}$

2.10 Classify each of the following signals as causal or noncausal.

(a) $x(k) = \max\{k, 0\}$

(b) $x(k) = \sin(.2\pi k)\mu(-k)$

(c) $x(k) = 1 - \exp(-k)$

(d) $x(k) = \mathrm{mod}(k, 10)$

(e) $x(k) = \tan(\sqrt{2}\pi k)[\mu(k) + \mu(k - 100)]$

(f) $x(k) = \cos(\pi k) + (-1)^k$

(g) $x(k) = \sin(.5\pi k)/(1 + k^2)$

2.11 Classify each of the following signals as finite or infinite. For the finite signals, find the smallest integer N such that $x(k) = 0$ for $|k| > N$.

(a) $x(k) = \mu(k + 5) - \mu(k - 5)$

(b) $x(k) = \sin(.2\pi k)\mu(k)$

(c) $x(k) = \min(k^2 - 9, 0)\mu(k)$

(d) $x(k) = \mu(k)\mu(-k)/(1 + k^2)$

(e) $x(k) = \tan(\sqrt{2}\pi k)[\mu(k) - \mu(k - 100)]$

(f) $x(k) = \delta(k) + \cos(\pi k) - (-1)^k$

(g) $x(k) = k^{-k}\sin(.5\pi k)$

2.12 Consider the block diagram shown in Figure 2.32.

(a) Write a single difference equation description of this system.

(b) Write a system of difference equations for this system for $u_i(k)$ and $y(k)$.

FIGURE 2.32: A Block Diagram of the System in Problem 2.12

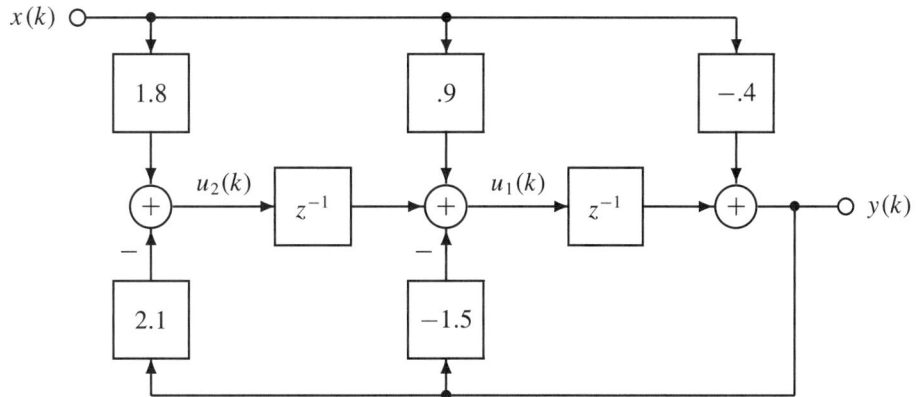

2.13 Consider the following linear time-invariant discrete-time system S.

$$y(k) = 1.8y(k-1) - .81y(k-2) - 3x(k-1)$$

(a) Find the characteristic polynomial $a(z)$ and express it in factored form.
(b) Write down the general form of the zero-input response $y_{zi}(k)$.
(c) Find the zero-input response when $y(-1) = 2$ and $y(-2) = 2$.

2.14 Consider the following linear time-invariant discrete-time system S.

$$y(k) = y(k-1) - .21y(k-2) + 3x(k) + 2x(k-2)$$

(a) Find the characteristic polynomial $a(z)$ and the input polynomial $b(z)$. Express $a(z)$ in factored form.
(b) Write down the general form of the zero-input response $y_{zi}(k)$.
(c) Find the zero-input response when the initial condition is $y(-1) = 1$ and $y(-2) = -1$.
(d) Write down the general form of the zero-state response when the input is $x(k) = 2(.5)^{k-1}\mu(k)$.
(e) Find the zero-state response using the input in (d).
(f) Find the complete response using the initial condition in (c) and the input in (d).

2.15 Consider the following linear time-invariant discrete-time system S. Sketch a block diagram of this IIR system.

$$y(k) = 3y(k-1) - 2y(k-2) + 4x(k) + 5x(k-1)$$

2.16 Consider the following linear time-invariant discrete-time system S.

$$y(k) = -.64y(k-2) + x(k) - x(k-2)$$

(a) Find the characteristic polynomial $a(z)$ and express it in factored form.
(b) Write down the general form of the zero-input response $y_{zi}(k)$, expressing it as a real signal.
(c) Find the zero-input response when $y(-1) = 3$ and $y(-2) = 1$.

2.17 Consider the following linear time-invariant discrete-time system S.

$$y(k) - .9y(k-1) = 2x(k) + x(k-1)$$

(a) Find the characteristic polynomial $a(z)$ and the input polynomial $b(z)$.

(b) Write down the general form of the zero-state response $y_{zs}(k)$ when the input is $x(k) = 3(.4)^k \mu(k)$.

(c) Find the zero-state response.

2.18 Consider the following linear time-invariant discrete-time system S. Sketch a block diagram of this FIR system.

$$y(k) = x(k) - 2x(k-1) + 3x(k-2) - 4x(k-4)$$

2.19 Consider the following linear time-invariant discrete-time system S, called an *auto-regressive* system. Sketch a block diagram of this system.

$$y(k) = x(k) - .8y(k-1) + .6y(k-2) - .4y(k-3)$$

2.20 Consider the following linear time-invariant discrete-time system S.

$$y(k) - y(k-2) = 2x(k)$$

(a) Find the characteristic polynomial of S and express it in factored form.

(b) Write down the general form of the zero-input response $y_{zi}(k)$.

(c) Find the zero-input response when $y(-1) = 4$ and $y(-2) = -1$.

2.21 Consider the following linear time-invariant discrete-time system S.

$$y(k) - 2y(k-1) + 1.48y(k-2) - .416y(k-3) = 5x(k)$$

(a) Find the characteristic polynomial $a(z)$. Using the MATLAB function *roots*, express $a(z)$ in factored form.

(b) Write down the general form of the zero-input response $y_{zi}(k)$.

(c) Write the equations for the unknown coefficient vector $c \in R^3$ as $Ac = y_0$, where $y_0 = [y(-1), y(-2), y(-3)]^T$ is the initial condition vector.

2.22 Consider the following linear time-invariant discrete-time system S.

$$y(k) = y(k-1) - .24y(k-2) + 3x(k) - 2x(k-1)$$

(a) Find the characteristic polynomial $a(z)$ and the input polynomial $b(z)$.

(b) Suppose the input is the unit step $x(k) = \mu(k)$. Write down the general form of the zero-state response $y_{zs}(k)$.

(c) Find the zero-state response to the unit step input.

2.23 Suppose $h(k)$ and $x(k)$ are the following signals of length L and M, respectively.

$$h = [3, 6, -1]^T$$
$$x = [2, 0, -4, 5]^T$$

(a) Let h_z and x_z be zero-padded versions of $h(k)$ and $x(k)$, respectively, of length $N = L + M - 1$. Construct h_z and x_z.

(b) Let $y_c(k) = h_z(k) \circ x_z(k)$. Find the circular convolution matrix $C(x_z)$ such that $y_c = C(x_z)h_z$.

(c) Use $C(x_z)$ to find $y_c(k)$.

(d) Use $y_c(k)$ to find the linear convolution $y(k) = h(k) \star x(k)$ for $0 \le k < N$.

2.24 Consider the following linear time-invariant discrete-time system S. Compute and sketch the impulse response of this FIR system.

$$y(k) = u(k-1) + 2u(k-2) + 3u(k-3) + 2u(k-4) + u(k-5)$$

2.25 Using the definition of linear convolution, show that for any signal $h(k)$

$$h(k) \star \delta(k) = h(k)$$

2.26 Consider a linear discrete-time system S with input x and output y. Suppose S is driven by an input $x(k)$ for $0 \le k < L$ to produce a zero-state output $y(k)$. Use deconvolution to find the impulse response $h(k)$ for $0 \le k < L$ if $x(k)$ and $y(k)$ are as follows.

$$x = [2, 0, -1, 4]^T$$
$$y = [6, 1, -4, 3]^T$$

2.27 Suppose $x(k)$ and $y(k)$ are the following finite signals.

$$x = [5, 0, -4]^T$$
$$y = [10, -5, 7, 4, -12]^T$$

(a) Write the polynomials $x(z)$ and $y(z)$ whose coefficient vectors are x and y, respectively. The leading coefficient corresponds to the highest power of z.
(b) Using long division, compute the quotient polynomial $q(z) = y(z)/x(z)$.
(c) Deconvolve $y(k) = h(k) \star x(k)$ to find $h(k)$ using (2.7.15) and (2.7.18). Compare the result with $q(z)$ from part (b).

2.28 Consider the following linear time-invariant discrete-time system S.

$$y(k) = -.25y(k-2) + x(k-1)$$

(a) Find the characteristic polynomial and the input polynomial.
(b) Write down the form of the impulse response $h(k)$.
(c) Find the impulse response. Use the identities in Appendix 2 to express $h(k)$ in real form.

2.29 Use Definition 2.3 to show that the linear convolution operator is distributive.

$$f(k) \star [g(k) + h(k)] = f(k) \star g(k) + f(k) \star h(k)$$

2.30 Use Definition 2.3 and the commutative property to show that the linear convolution operator is associative.

$$f(k) \star [g(k) \star h(k)] = [f(k) \star g(k)] \star h(k)$$

2.31 Consider the following linear time-invariant discrete-time system S. Suppose $0 < m \le n$ and the characteristic polynomial $a(z)$ has simple nonzero roots.

$$y(k) = \sum_{i=0}^{m} b_i x(k-i) - \sum_{i=1}^{n} a_i y(k-i)$$

(a) Find the characteristic polynomial $a(z)$ and the input polynomial $b(z)$.
(b) Find a constraint on $b(z)$ that ensures that the impulse response $h(k)$ does not contain an impulse term.

2.32 Suppose $h(k)$ and $x(k)$ are defined as follows.

$$h = [2, -1, 0, 4]^T$$
$$x = [5, 3, -7, 6]^T$$

(a) Let $y_c(k) = h(k) \circ x(k)$. Find the circular convolution matrix $C(x)$ such that $y_c = C(x)h$.
(b) Use $C(x)$ to find $y_c(k)$.

2.33 Consider the following linear time-invariant discrete-time system S.

$$y(k) = .6y(k-1) + x(k) - .7x(k-1)$$

(a) Find the characteristic polynomial and the input polynomial.
(b) Write down the form of the impulse response $h(k)$.
(c) Find the impulse response.

2.34 This problem establishes the normalized circular cross-correlation inequality $|\sigma_{yx}(k)| \le 1$. Let $x(k)$ and $y(k)$ be sequences of length N where $x_p(k)$ is the periodic extension of $x(k)$.
(a) Consider the signal $u(i, k) = ay(i) + x_p(i - k)$ where a is arbitrary. Show that

$$\frac{1}{N} \sum_{i=0}^{N-1} [ay(i) + x_p(i - k)]^2 = a^2 c_{yy}(0) + 2ac_{yx}(k) + c_{xx}(0) \ge 0$$

(b) Show that the inequality in part (a) can be written in matrix form as

$$[a, 1] \begin{bmatrix} c_{yy}(0) & c_{yx}(k) \\ c_{yx}(k) & c_{xx}(0) \end{bmatrix} \begin{bmatrix} a \\ 1 \end{bmatrix} \ge 0$$

(c) Since the inequality in part (b) holds for any a, the 2×2 coefficient matrix $C(k)$ is positive semi-definite which means that $\det[C(k)] \ge 0$. Use this fact to show that

$$c_{yx}^2(k) \le c_{xx}(0)c_{yy}(0), \quad 0 \le k < N$$

(d) Use the results from part (c) and the definition of normalized cross-correlation to show that

$$-1 \le \sigma_{yx}(k) \le 1, \quad 0 \le k < N$$

2.35 Let $x(k)$ be an N-point signal with average power P_x.
(a) Show that $r_{xx}(0) = c_{xx}(0) = P_x$
(b) Show that $\rho_{xx}(0) = \sigma_{xx}(0) = 1$

2.36 Suppose $y(k)$ is as follows.

$$y = [5, 7, -2, 4, 8, 6, 1]^T$$

(a) Construct a 3-point signal $x(k)$ such that $r_{yx}(k)$ reaches its peak positive value at $k = 3$ and $|x(0)| = 1$.
(b) Construct a 4-point signal $x(k)$ such that $r_{yx}(k)$ reaches its peak negative value at $k = 2$ and $|x(0)| = 1$.

2.37 Consider the following FIR system.

$$y(k) = \sum_{i=0}^{5} (1 + i)^2 x(k - i)$$

Let $x(k)$ be a bounded input with bound B_x. Show that $y(k)$ is bounded with bound $B_y = cB_x$. Find the minimum scale factor c.

2.38 Some books use the following alternative way to define the linear cross-correlation of an L point signal $y(k)$ with and M-point signal $x(k)$. Using a change of variable, show that this is equivalent to Definition 2.5

$$r_{yx}(k) = \frac{1}{L} \sum_{n=0}^{L-1-k} y(n + k)x(n)$$

2.39 From Proposition 2.1, a linear time-invariant discrete-time system S is BIBO stable if and only if the impulse response $h(k)$ is absolutely summable, that is, $\|h\|_1 < \infty$. Show that $\|h\|_1 < \infty$ is necessary for stability. That is, suppose that S is stable but $h(k)$ is not absolutely summable. Consider the following input where $h^*(k)$ denotes the complex conjugate of $h(k)$. (Proakis and Manolakis, 1992)

$$x(k) = \begin{cases} \dfrac{h^*(k)}{|h(k)|}, & h(k) \neq 0 \\ 0, & h(k) = 0 \end{cases}$$

(a) Show that $x(k)$ is bounded by finding a bound B_x.
(b) Show that S is not is BIBO stable, by showing that $y(k)$ is unbounded at $k = 0$.

2.40 Suppose $y(k)$ is as follows.

$$y = [8, 2, -3, 4, 5, 7]^T$$

(a) Construct a 6-point signal $x(k)$ such that $\sigma_{yx}(2) = 1$ and $|x(0)| = 6$.
(b) Construct a 6-point signal $x(k)$ such that $\sigma_{yx}(3) = -1$ and $|x(0)| = 12$.

2.41 Suppose $x(k)$ and $y(k)$ are defined as follows.

$$x = [4, 0, -12, 8]^T$$
$$y = [2, 3, 1, -1]^T$$

(a) Find the circular cross-correlation matrix $E(x)$ such that $c_{yx} = E(x)y$.
(b) Use $E(x)$ to find the circular cross-correlation $c_{yx}(k)$.
(c) Find the normalized circular cross-correlation $\sigma_{yx}(k)$.

2.42 Suppose $x(k)$ and $y(k)$ are defined as follows.

$$x = [5, 0, -10]^T$$
$$y = [1, 0, -2, 4, 3]^T$$

(a) Find the linear cross-correlation matrix $D(x)$ such that $r_{yx} = D(x)y$.
(b) Use $D(x)$ to find the linear cross-correlation $r_{yx}(k)$.
(c) Find the normalized linear cross-correlation $\rho_{yx}(k)$.

2.43 Consider a linear time-invariant discrete-time system S with the following impulse response. Find conditions on A and p that guarantee that S is BIBO stable.

$$h(k) = A(p)^k \mu(k)$$

2.12.2 GUI Simulation

2.44 Consider the following discrete-time system which is a narrow band *resonator* filter with sampling frequency of $f_s = 800$ Hz.

$$y(k) = .704y(k-1) - .723y(k-2) + .141x(k) - .141x(k-2)$$

Use GUI module *g_systime* to find the zero-input response for the following initial conditions. In each case, plot $N = 50$ points.
(a) $y_0 = [10, -3]^T$
(b) $y_0 = [-5, -8]^T$

2.45 Consider a discrete-time system with the following characteristic and input polynomials. Use GUI module *g_systime* to plot the step response using $N = 100$ points. The MATLAB *poly* function can be used to specify coefficient vectors a and b in terms of their roots, as discussed in Section 2.9.

$$a(z) = (z + .5 \pm j.6)(z - .9)(z + .75)$$

$$b(z) = 3z^2(z - .5)^2$$

2.46 Consider the following linear discrete-time system.

$$y(k) = 1.7y(k - 2) - .72y(k - 4) + 5x(k - 2) + 4.5x(k - 4)$$

Use GUI module *g_systime* to plot the following damped cosine input and the zero-state response to it using $N = 300$. To determine F_0, set $2\pi F_0 kT = .3\pi k$ and solve for F_0/f_s, where $T = 1/f_s$.

$$x(k) = .97^k \cos(.3\pi k)$$

2.47 Consider the following linear discrete-time system.

$$y(k) = -.4y(k - 1) + .19y(k - 2) - .104y(k - 3) + 6x(k) - 7.7x(k - 1) + 2.5x(k - 2)$$

Create a MAT-file called *prob2_47* that contains $fs = 100$, the appropriate coefficient vectors a and b, and the following input samples where $v(k)$ is white noise uniformly distributed over $[-.2, .2]$. Uniform white noise can be generated with the MATLAB function *rand*.

$$x(k) = k \exp(-k/50) + v(k), \quad 0 \le k < 500$$

(a) Print the MATLAB program used to create *prob2_47.mat*.
(b) Use GUI module *g_systime* and the User-defined option to plot the roots of the characteristic polynomial and the input polynomial.
(c) Plot the zero-state response on the input $x(k)$.

2.48 Consider the following discrete-time system. Use GUI module *g_systime* to simulate this system. Hint: You can enter the b vector in the edit box by using two statements on one line: i = 0:8; b = cos(pi*i/4)

$$y(k) = \sum_{i=0}^{8} \cos(\pi i /4)x(k - i)$$

(a) Plot the polynomial roots
(b) Plot and the impulse response using $N = 40$.

2.49 Consider the following two polynomials. Use *g_systime* to compute, plot, and save, in a data file, the coefficients of the quotient polynomial $q(z)$ and the remainder polynomial $r(z)$ where $b(z) = q(z)a(z) + r(z)$. Then *load* the saved file and display the coefficients of the quotient and remainder polynomials.

$$a(z) = z^2 + 3z - 4$$

$$b(z) = 4z^4 - z^2 - 8$$

2.50 Consider the following discrete-time system which is a *notch* filter with sampling interval $T = 1/360$ sec.

$$y(k) = .956y(k - 1) - .914y(k - 2) + x(k) - x(k - 1) + x(k - 2)$$

Use GUI module *g_systime* to find the output corresponding to the sinusoidal input $x(k) = \cos(2\pi F_0 kT)\mu(k)$. Do the following cases. Use the caliper option to estimate the steady state amplitude in each case.

(a) Plot the output when $F_0 = 10$ Hz.

(b) Plot the output when $F_0 = 60$ Hz.

2.51 Consider the following two polynomials. Use *g_systime* to compute, plot, and save, in a data file, the coefficients of the product polynomial $c(z) = a(z)b(z)$. Then *load* the saved file and display the coefficients of the product polynomial.

$$a(z) = z^2 - 2z + 3$$
$$b(z) = 4z^3 + 5z^2 - 6z + 7$$

2.52 Use the GUI module *g_correlate* to record the sequence of vowels "A","E","I", "O","U" in y. Play y to make sure you have a good recording of all five vowels. Then record the vowel "O" in x. Play x back to make sure you have a good recording of "O" that sounds similar to the "O" in y. Save these data in a MAT-file named *my_vowels*.

(a) Plot the inputs x and y showing the vowels.

(b) Plot the normalized cross-correlation of y with x using the *Caliper* option to mark the peak which should show the location of x in y.

(c) Based on the plots in (a), estimate the lag d_1 that would be required to get the "O" in x to align with the "O" in y. Compare this with the peak location d_2 in (b). Find the percent error relative to the estimated lag d_1. There will be some error due to the overlap of x with adjacent vowels and co-articulation effects in creating y.

2.53 The file *prob2_53.mat* contains two signals, x and y, and their sampling frequency fs. Use the GUI module *g_correlate* to load x, y, and fs.

(a) Plot $x(k)$ and $y(k)$.

(b) Plot the normalized linear cross-correlation $\rho_{yx}(k)$. Does $y(k)$ contain any scaled and shifted versions of $x(k)$? Determine how many, and use the Caliper option to estimate the locations of $x(k)$ within $y(k)$.

2.12.3 MATLAB Computation

2.54 Consider the following discrete-time system.

$$a(z) = z^4 - .3z^3 - .57z^2 + .115z + .0168$$
$$b(z) = 10(z + .5)^3$$

This system has four simple nonzero roots. Therefore the zero-input response consists of a sum of the following four natural mode terms.

$$y_{zi}(k) = c_1(p_1)^k + c_2(p_2)^k + c_3(p_3)^k + c_4(p_4)^k$$

The coefficients can be determined from the initial condition

$$y_0 = [y(-1), y(-2), y(-3), y(-4)]^T$$

Setting $y_{zi}(-k) = y(-k)$ for $1 \le k \le 4$ yields the following linear algebraic system in the coefficient vector $c = [c_1, c_2, c_3, c_4]^T$.

$$\begin{bmatrix} p_1^{-1} & p_2^{-1} & p_3^{-1} & p_4^{-1} \\ p_1^{-2} & p_2^{-2} & p_3^{-2} & p_4^{-2} \\ p_1^{-3} & p_2^{-3} & p_3^{-3} & p_4^{-3} \\ p_1^{-4} & p_2^{-4} & p_3^{-4} & p_4^{-4} \end{bmatrix} \begin{bmatrix} c_1 \\ c_2 \\ c_3 \\ c_4 \end{bmatrix} = y_0$$

Write a MATLAB program that uses *roots* to find the roots of the characteristic polynomial and then solves this linear algebraic system for the coefficient vector c using the MATLAB left division or \ operator when the initial condition is y_0. Print the roots and the coefficient vector c. Use *stem* to plot the zero-input response $y_{zi}(k)$ for $0 \le k \le 40$.

2.55 Consider the following pair of signals.

$$h = [1, 2, 3, 4, 5, 4, 3, 2, 1]^T$$
$$x = [2, -1, 3, 4, -5, 0, 7, 9, -6]^T$$

Verify that linear convolution and circular convolution produce different results by writing a MATLAB program that uses the FDSP function f_conv to compute the linear convolution $y(k) = h(k) \star x(k)$ and the circular convolution $y_c(k) = h(k) \circ x(k)$. Plot $y(k)$ and $y_c(k)$ below one another on the same screen.

2.56 Consider the discrete-time system in Problem 2.54. Write a MATLAB program that uses the FDSP function $f\_filter0$ to compute the zero-input response to the following initial condition. Use *stem* to plot the zero-input response $y_{zi}(k)$ for $-4 \le k \le 40$.

$$y_0 = [y(-1), y(-2), y(-3), y(-4)]^T$$

2.57 Consider the following polynomials

$$a(z) = z^4 + 4z^3 + 2z^2 - z + 3$$
$$b(z) = z^3 - 3z^2 + 4z - 1$$
$$c(z) = a(z)b(z)$$

Let $a \in R^5$, $b \in R^4$, and $c \in R^8$ be the coefficient vectors of $a(z)$, $b(z)$, and $c(z)$, respectively.
(a) Find the coefficient vector of $c(z)$ by direct multiplication by hand.
(b) Write a MATLAB program that uses *conv* to find the coefficient vector of $c(z)$ by computing c as the linear convolution of a with b.
(c) In the program, show that a can be recovered from b and c by using the MATLAB function *deconv* to perform deconvolution.

2.58 Consider the following discrete-time system.

$$y(k) = .95y(k-1) + .035y(k-2) - .462y(k-3) + .351y(k-4)$$
$$+ .5x(k) - .75x(k-1) - 1.2x(k-2) + .4x(k-3 - 1.2x)(k-4)$$

Write a MATLAB program that uses *filter* and *plot* to compute and plot the zero-state response of this system to the following input. Plot both the input and the output on the same graph.

$$x(k) = (k+1)^2(.8)^k \mu(k), \quad 0 \le k \le 100$$

2.59 Consider the following FIR filter. Write a MATLAB program that performs the following tasks.

$$y(k) = \sum_{i=0}^{20} \frac{(-1)^i x(k-i)}{10 + i^2}$$

(a) Use the function *filter* to compute and plot the impulse response $h(k)$ for $0 \le k < N$ where $N = 50$.
(b) Compute and plot the following periodic input.

$$x(k) = \sin(0.1\pi k) - 2\cos(0.2\pi k) + 3\sin(0.3\pi k), \quad 0 \le k < N$$

(c) Use *conv* to compute the zero-state response to the input $x(k)$ using convolution. Also compute the zero-state response to $x(k)$ using *filter*. Plot both responses on the same graph using a legend.

2.60 Consider the following running average filter. Write a MATLAB program that performs the following tasks.

$$y(k) = \frac{1}{10} \sum_{i=0}^{9} x(k-i), \quad 0 \le k \le 100$$

(a) Use *filter* and *plot* to compute and plot the zero-state response to the following input where $v(k)$ is a random white noise uniformly distributed over $[-.1, .1]$. Plot $x(k)$ and $y(k)$ below one another. Uniform white noise can be generated using the MATLAB function *rand*.

$$x(k) = \exp(-k/20)\cos(\pi k/10)\mu(k) + v(k)$$

(b) Add a third curve to the graph in part (a) by computing and plotting the zero-state response using *conv* to perform convolution.

2.61 Consider the following pair of signals.

$$h = [1, 2, 4, 8, 16, 8, 4, 2, 1]^T$$
$$x = [2, -1, -4, -4, -1, 2]^T$$

Verify that linear convolution can be achieved by zero padding and circular convolution by writing a MATLAB program that pads these signals with an appropriate number of zeros, and uses the FDSP toolbox function *f_conv* to compare the linear convolution $y(k) = h(k) \star x(k)$ with the circular convolution $y_{zc}(k) = h_z(k) \circ x_z(k)$. Plot the following.
(a) The zero-padded signals $h_z(k)$ and $x_z(k)$ on the same graph using a legend.
(b) The linear convolution $y(k) = h(k) \star x(k)$.
(c) The zero-padded circular convolution $y_{zc}(k) = h_z(k) \circ x_z(k)$.

2.62 Consider the following pair of signals of length $N = 8$.

$$x = [2, -4, 7, 3, 8, -6, 5, 1]^T$$
$$y = [3, 1, -5, 2, 4, 9, 7, 0]^T$$

Write a MATLAB program that performs the following tasks.
(a) Use the FDSP toolbox function *f_corr* to compute and plot the circular cross-correlation $c_{yx}(k)$.
(b) Compute and print $u(k) = x(-k)$ using the periodic extension $x_p(k)$.
(c) Verify that $c_{yx}(k) = [y(k) \circ x(-k)]/N$ by using the FDSP toolbox function *f_conv* to compute and plot the scaled circular convolution $w(k) = [u(k) \circ x(k)]/N$. Plot $c_{yx}(k)$ and $w(k)$ below one another on the same screen.

2.63 Consider the following pair of signals.

$$x = [2, -1, -4, -4, -1, 2]^T$$
$$y = [1, 2, 4, 8, 16, 8, 4, 2, 1]^T$$

Verify that linear cross-correlation can be achieved by zero-padding and circular cross-correlation by writing a MATLAB program that pads these signals with an appropriate number of zeros, and uses the FDSP toolbox function *f_corr* to compute the linear cross-correlation $r_{yx}(k)$ and the circular cross-correlation $c_{y_z x_z}(k)$. Plot the following.
(a) The zero-padded signals $x_z(k)$ and $y_z(k)$ on the same graph using a legend.
(b) The linear cross-correlation $r_{yx}(k)$ and the scaled zero-padded circular cross-correlation $(N/L)c_{y_z x_z}(k)$ on the same graph using a legend.

2.64 Consider the following pair of signals.

$$x = [2, -4, 3, 7, 6, 1, 9, 4, -3, 2, 7, 8]^T$$
$$y = [3, 2, 1, 0, -1, -2, -3, -2, -1, 0, 1, 2]^T$$

Verify that linear cross-correlation and circular cross-correlation produce different results by writing a MATLAB program that uses the FDSP function f_corr to compute the linear cross-correlation $r_{yx}(k)$ and the circular cross-correlation $c_{yx}(k)$. Plot $r_{yx}(k)$ and $c_{yx}(k)$ below one another on the same screen.

Discrete-time Systems in the Frequency Domain

• • • • • • • • • • • • • • • • •

Chapter Topics

3.1 Motivation

3.2 Z-transform Pairs

3.3 Z-transform Properties

3.4 Inverse Z-transform

3.5 Transfer Functions

3.6 Signal Flow Graphs

3.7 Stability in the Frequency Domain

3.8 Frequency Response

3.9 System Identification

3.10 GUI Software and Case Studies

3.11 Chapter Summary

3.12 Problems

• • • • • • • • • • • • • • •

3.1 Motivation

Recall that a discrete-time system is a system that processes a discrete-time input signal $x(k)$ to produce a discrete-time output signal $y(k)$. If the signals $x(k)$ and $y(k)$ are discrete in amplitude, as well as in time, then they are digital signals and the associated system is a digital signal processor or a digital filter. In this chapter we focus our attention on analyzing the input-output behavior of linear time-invariant discrete-time systems in the frequency domain. Together with Chapter 2, this material lays a mathematical foundation for subsequent chapters where we design digital filters and develop digital signal processing (DSP) algorithms.

An essential tool for the analysis of discrete-time systems is the Z-transform, a transformation that maps or transforms a discrete-time signal $x(k)$ into a function $X(z)$ of a complex variable z.

$$X(z) = Z\{x(k)\}$$

With the help of the Z-transform, the difference-equation description of a discrete-time system in Chapter 2 can be converted to a simple algebraic equation which is readily solved for the Z-transform of the output $Y(z)$. Applying the inverse Z-transform to $Y(z)$ then produces the zero-state response $y_{zs}(k)$. Important qualitative features of discrete-time systems also can be obtained with the help of the Z-transform. Recall that a discrete-time system is *stable* if and only if every bounded input signal is guaranteed to produce a bounded output signal. Stability is an essential characteristic of practical digital filters, and the easiest way to establish stability is with the Z-transform.

We begin this chapter by examining a number of practical problems that can be modeled using discrete-time systems and solved using the Z-transform. Next, the Z-transform is defined, its region of convergence is analyzed, and a basic table of Z-transform pairs is developed. The size of this table is then expanded by introducing a number of important Z-transform properties. The problem of inverting the Z-transform is then addressed using the synthetic division, partial fraction expansion, and residue methods. Next, two new representations of a discrete-time system are introduced. The first is the transfer function representation which includes a discussion of poles, zeros, and modes. The second is a compact graphical representation called the signal flow graph. A simple frequency domain criterion to determine when a system is BIBO stable is presented and the Jury stability test is introduced. The frequency response of a discrete-time systems is then introduced, and interpretations of it are provided in terms of the transfer function and the steady-state response to periodic inputs. Finally, a GUI module called *g_sysfreq* is presented that allows the user the explore the input-output behavior of discrete-time systems in the frequency domain without any need for programming. The chapter concludes with some case study examples, and a summary of discrete-time system analysis techniques.

3.1.1 Satellite Attitude Control

As an example of a discrete-time system, consider the problem of finding a discrete-equivalent of a sampled-data system. Recall from Chapter 1 that a *sampled-data* system is a system that contains both continuous-time signals and discrete-time signals. Consequently, any system that has a digital-to-analog converter (DAC) or an analog-to-digital converter (ADC) is a sampled-data system. For example, consider the feedback control system shown in Figure 3.1 that contains both.

The task of this feedback system is to control the angular position of a satellite that is spinning about one axis in space, as shown in Figure 3.2. Here $r(k)$ is the desired angular position of the satellite at discrete time k, and $y(k)$ is the actual angular position. Since there is no friction, the motion of the satellite can be modeled using Newton's second law as follows.

$$J\frac{d^2 y_a(t)}{dt^2} = \tau_a(t) \tag{3.1.1}$$

From Figure 3.1, the input $\tau_a(t)$ is the torque generated by the thrusters, and J is the moment of inertia about the axis of rotation. The digital controller acts on the *error* signal input

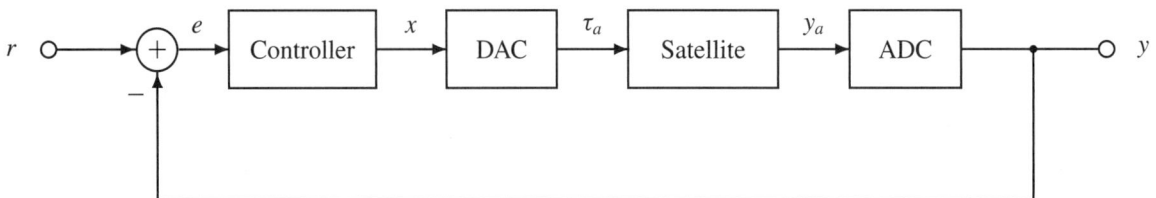

FIGURE 3.1: Single-axis Satellite Attitude Control System

FIGURE 3.2: A Spinning Satellite

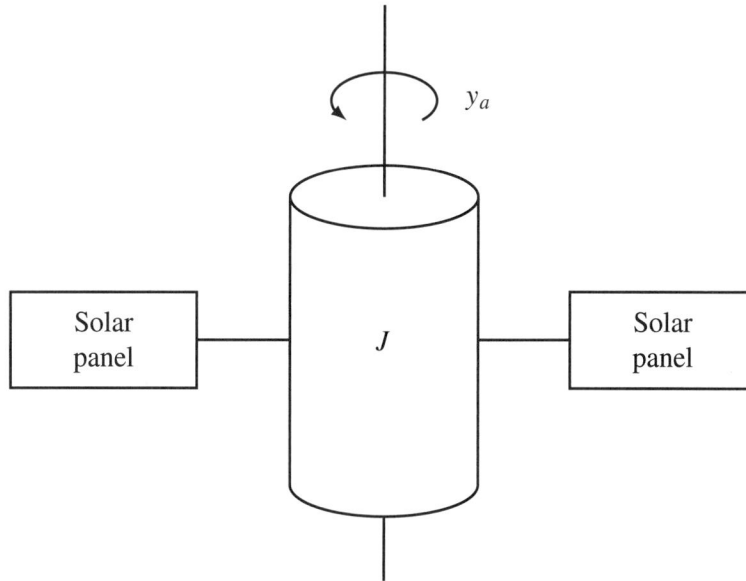

Controller

$e(k) = r(k) - y(k)$ to produce a control signal $x(k)$. For example, consider the following difference equation used to implement a simple controller.

$$x(k) = c[e(k) - e(k - 1)] \qquad (3.1.2)$$

Here the *controller gain c* is an engineering design parameter whose value can be chosen to satisfy some performance specification. The controller in (3.1.2) is an example of a finite impulse response or FIR discrete-time system.

Satellite model

Next suppose the DAC is modeled as a zero-order hold, as in Chapter 1. Then the discrete-equivalent of the DAC and the satellite can be modeled with the following hold-equivalent discrete-time system (Franklin et al., 1990), where $y(k) = y_a(kT)$.

$$y(k) = 2y(k - 1) - y(k - 2) + \left(\frac{T^2}{2J}\right) [x(k - 1) + x(k - 2)] \qquad (3.1.3)$$

Closed-loop system

If we substitute (3.1.2) into (3.1.3), and recall that $e(k) = r(k) - y(k)$, the entire closed-loop feedback system then can be modeled by the following discrete-equivalent system whose input is the desired angle $r(k)$ and output is the actual angle $y(k)$.

$$y(k) = (d - 1)y(k - 1) + dy(k - 2) + d[r(k - 1) + r(k - 2)] \qquad (3.1.4a)$$

$$d = \frac{cT^2}{2J} \qquad (3.1.4b)$$

Thus the sampled-data control system consists of three separate discrete-time systems, the controller in (3.1.2), the hold-equivalent of the DAC and the satellite in (3.1.3), and the overall closed-loop discrete-equivalent system in (3.1.4). Later in this chapter we will see that

Stable range

this control system is stable, and therefore operates successfully, for the following values of the controller gain.

$$0 < c < \frac{2J}{T^2} \qquad (3.1.5)$$

FIGURE 3.3: Digital Speech Synthesis

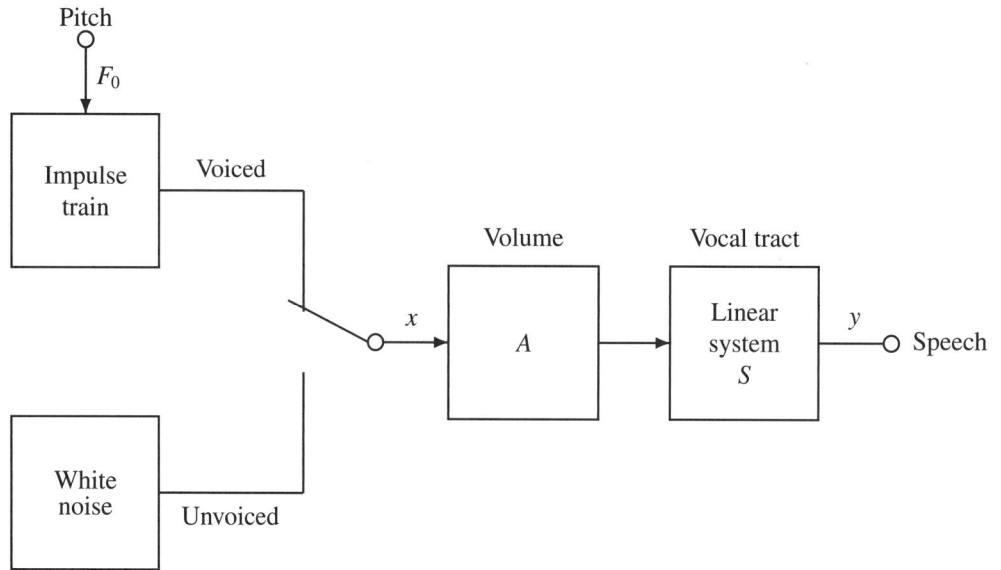

3.1.2 Modeling the Vocal Tract

One of the exciting application areas of DSP is the analysis and synthesis of human speech. Both speech recognition and speech synthesis are becoming increasingly commonplace as a user interface between humans and machines. An effective technique for generating synthetic speech is to use a discrete-time signal to represent the output from the vocal chords, and a slowly varying digital filter to model the vocal tract as illustrated by the block diagram in Figure 3.3 (Rabiner and Schafer, 1978; Markel and Gray, 1976).

Phoneme

Speech sounds can be decomposed into fundamental units called *phonemes* that are either voiced or unvoiced. Unvoiced phonemes are associated with turbulence in the vocal tract and are therefore modeled using filtered white noise. Unvoiced phonemes include the fricatives such as the s, sh, and f sounds, and the terminal sounds p, t, and k. Voiced phonemes are associated with the periodic excitation of the vocal chords. They include the vowels, nasal sounds, and transient terminal sounds such as b, d, and g. Voiced phonemes can be modeled as the response of a digital filter to a periodic impulse train with period M.

$$x(k) = \sum_{i=0}^{\infty} \delta(k - iM) \tag{3.1.6}$$

Pitch

If T is the sampling interval, then the period of the impulse train in seconds is $T_0 = MT$. The fundamental frequency or *pitch* of the speaker is then

$$F_0 = \frac{1}{MT} \tag{3.1.7}$$

Speaker pitch typically ranges from about 50 Hz to 400 Hz, with male speakers having a lower pitch, on the average, than female speakers. An illustration of a short segment of the vowel "O" is shown in Figure 3.4. To estimate the pitch of this speaker, we examine the distance between peaks in Figure 3.4. By using the FDSP toolbox function *f_caliper*, the period shown with plus marks in Figure 3.4 is about $T_0 = 6.9$ msec, which corresponds to a pitch of

$$F_0 \approx 145 \text{ Hz} \tag{3.1.8}$$

FIGURE 3.4:
Segment of
Recorded Vowel
"O"

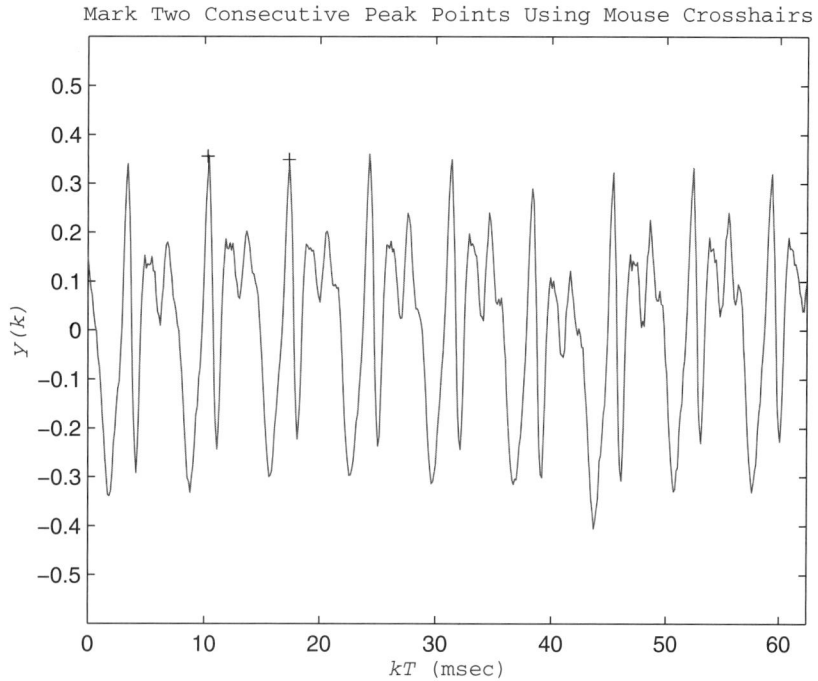

The linear system S in Figure 3.3 models the vocal tract cavity, including the throat, mouth,
Auto-regressive and lips. An effective model for most sounds is an nth order *auto-regressive system* where the
system coefficient vector of the input polynomial is simply $b = 1$.

$$y(k) = x(k) - \sum_{i=1}^{n} a_i y(k-i) \tag{3.1.9}$$

Finding a suitable parameter vector $a \in R^{n+1}$ for the vocal tract model, given the input $x(k)$,
System identification and output $y(k)$, is an example of *system identification*, a topic that is examined in Section 3.9.

●●●●●●●●●●●●●●●●

3.2 Z-transform Pairs

The Z-transform is a powerful tool that is useful for analyzing and solving linear discrete-time
systems.

DEFINITION

3.1: Z-transform

The *Z-transform* of a discrete-time signal $x(k)$ is a function $X(z)$ of a complex variable z
defined

$$X(z) \triangleq \sum_{k=-\infty}^{\infty} x(k)z^{-k}$$

From Definition 3.1 we see that the Z-transform is a power series in the variable z^{-1}. The
operation of performing a Z-transform can also be represented with the Z-transform *operator*

Z as follows.

$$Z\{x(k)\} = X(z) \tag{3.2.1}$$

note

Note that, by convention, the corresponding upper-case letter is used to denote the Z-transform of a time signal. For most practical signals, the Z-transform can be expressed in factored form as a ratio of two polynomials.

$$X(z) = \frac{b_0(z - z_1)(z - z_2) \cdots (z - z_m)}{(z - p_1)(z - p_2) \cdots (z - p_n)} \tag{3.2.2}$$

Here the roots of the numerator polynomial are called the *zeros* of $X(z)$, and the roots of the denominator polynomial are called the *poles* of $X(z)$. To compute the Z-transform directly from Definition 3.1, the following *generalized geometric series* is helpful where z is any real or complex-valued quantity (see Problem 3.17).

Generalized geometric series

$$\sum_{k=m}^{\infty} z^k = \frac{z^m}{1 - z}, \quad m \geq 0 \text{ and } |z| < 1 \tag{3.2.3}$$

3.2.1 Region of Convergence

For a given value of z, the power series in Definition 3.1 may or may not converge. To explore the region of convergence in the complex plane C, first note that every signal can be decomposed into the sum of two parts.

$$x(k) = x_c(k) + x_a(k) \tag{3.2.4}$$

Here the *causal part* $x_c(k)$ and the *anti-causal* part $x_a(k)$ are

$$x_c(k) \overset{\Delta}{=} x(k)\mu(k) \tag{3.2.5}$$
$$x_a(k) \overset{\Delta}{=} x(k)\mu(-k - 1) \tag{3.2.6}$$

For a general signal $x(k)$, the causal part $x_c(k)$ will have one region of convergence and the anti-causal part $x_a(k)$ will have another region of convergence. The overall region of convergence Ω_{ROC} includes the intersection of the two regions.

Example 3.1   **Region of Convergence**

To illustrate a Z-transform and its region of convergence, consider the following two-sided exponential signal.

$$x(k) = \begin{cases} a^k, & k \geq 0 \\ b^k, & k < 0 \end{cases}$$

Note that this can be written as a sum of causal and anti-causal parts as follows.

$$x(k) = \underbrace{a^k \mu(k)}_{x_c(k)} + \underbrace{b^k \mu(-k - 1)}_{x_a(k)}$$

Since the Z-transform is a linear operation, the transforms for the two parts can be computed separately and then added. Using the geometric series in (3.2.3)

$$X_c(z) = \sum_{k=0}^{\infty} a^k z^{-k}$$

$$= \sum_{k=0}^{\infty} (a/z)^k$$

$$= \frac{1}{1 - a/z}, \quad |a/z| < 1$$

$$= \frac{z}{z - a}, \quad |z| > |a|$$

Thus the region of convergence of the causal part is outside the circle of radius $|a|$ centered at the origin. The anti-causal Z-transform can be computed in a similar manner using a change of variable.

$$X_a(z) = \sum_{k=-\infty}^{-1} b^k z^{-k}$$

$$= \sum_{i=\infty}^{1} b^{-i} z^i, \quad i = -k$$

$$= \sum_{i=1}^{\infty} (z/b)^i$$

$$= \frac{z/b}{1 - z/b}, \quad |z/b| < 1$$

$$= \frac{-z}{z - b}, \quad |z| < |b|$$

In this case the region of convergence of the anti-causal part is inside the circle of radius $|b|$ centered at the origin. The complete Z-transform is then

$$X(z) = X_c(z) + X_a(z)$$

$$= \frac{z}{z - a} - \frac{z}{z - b}$$

$$= \frac{z(z - b) - z(z - a)}{(z - a)(z - b)}$$

$$= \frac{(a - b)z}{(z - a)(z - b)}, \quad |a| < |z| < |b|$$

Hence the region of convergence of $X(z)$ is an *annual ring* with inner radius $|a|$ and outer radius $|b|$ centered at the origin of the complex plane. Of course, if $|a| \geq |b|$, then the ring evaporates and the Z-transform does not exist because the region of convergence is the empty set. This occurs, for example, when $b = a$.

The results of Example 3.1 can be generalized in the following way. Suppose $\{p_1, \ldots, p_n\}$ are the poles of the causal part $x_c(k)$, and $\{q_1, \ldots, q_r\}$ are the poles of the anti-causal part $x_a(k)$, that survive any pole-zero cancellation in $X_c(z) + X_a(z)$. Define the radius of the innermost

FIGURE 3.5: Region of Convergence is Shaded for (a) General Signals, (b) Causal Signals. Here R_m is the radius of the innermost pole of the anti-causal part, and R_M is the radius of the outermost pole of the causal part

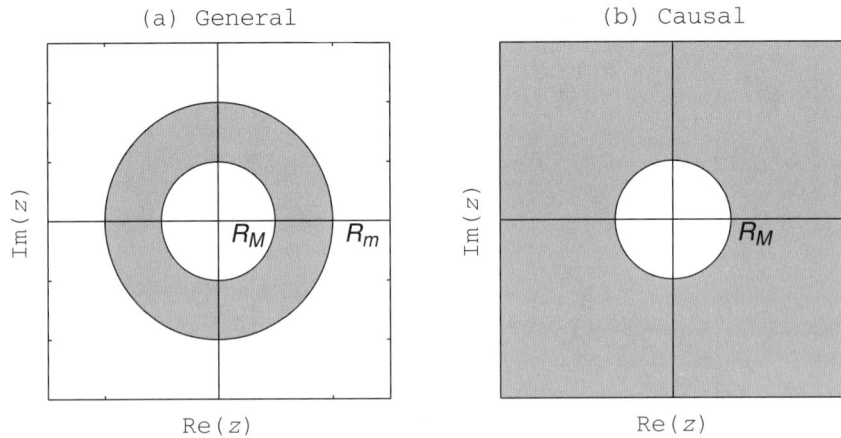

anti-causal pole and the outermost causal pole as follows.

$$R_m \triangleq \overset{r}{\underset{i=1}{\min}}\{|q_i|\} \tag{3.2.7}$$

$$R_M \triangleq \overset{n}{\underset{i=1}{\max}}\{|p_i|\} \tag{3.2.8}$$

The region of convergence of the causal part is $|z| > R_M$, and the region of convergence of the anti-causal part is $|z| < R_m$. That is, the causal part converges outside its outermost pole, and the anti-causal part converges inside its innermost pole. If $x(k)$ is causal, then the region of convergence is $|z| > R_M$, and if $x(k)$ is anti-causal, the region of convergence is $|z| < R_m$.

General signals   More generally, the region of convergence is

$$\Omega_{ROC} = \{z \in C : R_M < |z| < R_m\} \tag{3.2.9}$$

Plots of the regions of convergence for two important cases are shown in Figure 3.5. An important special case occurs when $x(k)$ is a finite signal. For a finite causal signal of length m, the Z-transform is

$$X(z) = x(0) + x(1)z^{-1} + \cdots + x(m-1)z^{1-m}$$
$$= \frac{x(0)z^{m-1} + x(1)z^{m-2} + \cdots + x(m-1)}{z^{m-1}} \tag{3.2.10}$$

Thus a finite causal signal of length $m > 1$ has $m - 1$ poles at $z = 0$, which means its region of convergence is $|z| > 0$. If $m = 1$ as in $x(k) = \delta(k)$, then the region of convergence is the entire complex plane. For a finite anticausal signal of length r, the Z-transform is

$$X(z) = x(-r)z^r + x(-r+1)z^{r-1} + \cdots + x(-1) \tag{3.2.11}$$

Thus a finite anti-causal signal of length r has no poles, which means the region of convergence is the entire complex plane $z \in C$. Finally, a general finite signal that contains both positive and negative samples has a combined region of convergence of $|z| > 0$. The most common *Causal signals*   cases are summarized in Table 3.1. Almost all of the signals that come up in practice are causal,

TABLE 3.1: ►
Regions of
Convergence

Signal	Length	Ω_{ROC}
Causal	Infinite	$|z| > R_M$
Anti-causal	Infinite	$|z| < R_m$
General	Infinite	$R_M < |z| < R_m$
Causal	Finite, $m > 1$	$|z| > 0$

either finite or infinite. For these signals the region of convergence is simply

$$\Omega_{ROC} = \{z \in C \ : \ |z| > R_M\} \tag{3.2.12}$$

3.2.2 Common Z-transform Pairs

The following examples illustrate the use of the geometric series to compute the Z-transform of the common discrete-time signals introduced in Section 2.2.

Example 3.2

Unit Impulse

Recall that the *unit impulse*, denoted $\delta(k)$, is a signal that is zero everywhere except at $k = 0$, where it takes on the value one. Using Definition 3.1 we find that the only nonzero term in the series is the constant term whose coefficient is unity. Consequently

$$Z\{\delta(k)\} = 1$$

Since the Z-transform of a unit impulse is constant, it has no poles, which means that its region of convergence is the entire complex plane $z \in C$.

Next recall that when the unit impulse $\delta(k)$ is summed from $-\infty$ to k, this produces the unit step $\mu(k)$.

Example 3.3

Unit Step

The *unit step*, denoted $\mu(k)$, is a causal infinite signal that takes on the value one for $k \geq 0$. Applying Definition 3.1, and using the geometric series in (3.2.3), we have

$$U(z) = \sum_{k=0}^{\infty} z^{-k}$$

$$= \sum_{k=0}^{\infty} (1/z)^k$$

$$= \frac{1}{1 - 1/z}, \quad |1/z| < 1$$

Negative powers of z can be cleared by multiplying the numerator and the denominator by z, which then yields

$$Z\{\mu(k)\} = \frac{z}{z - 1}, \quad |z| > 1$$

Note that the Z-transform has a zero at $z = 0$ and a pole at $z = 1$. Plots of the signal $\mu(k)$ and the pole-zero pattern of its Z-transform are shown in Figure 3.6.

FIGURE 3.6: Time Plot and Pole-zero Pattern of a Unit Step Signal $\mu(k)$

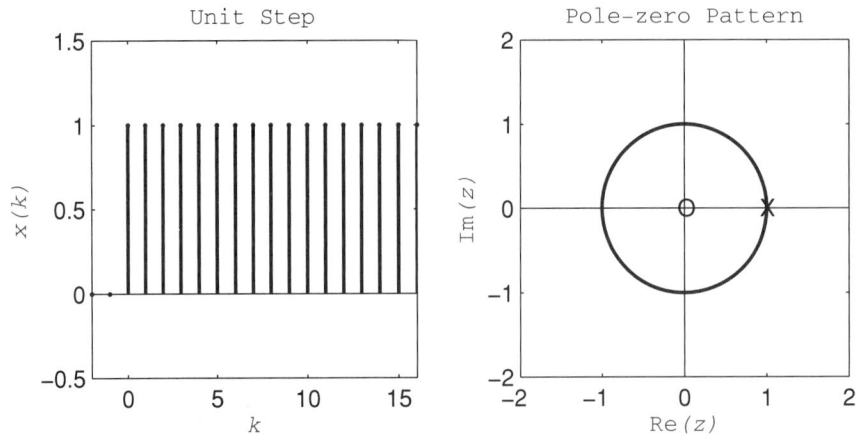

The unit step function $\mu(k)$ is useful because if we multiply another signal by $\mu(k)$, the result is a causal signal. By generalizing the unit step function slightly, we generate the most important Z-transform pair.

Example 3.4

Causal Exponential

Let c be real or complex and consider the following *causal exponential* signal generated by taking powers of c.

$$x(k) = c^k \mu(k)$$

Using Definition 3.1 and the geometric series in (3.2.3), we have

$$X(z) = \sum_{k=0}^{\infty} c^k z^{-k}$$

$$= \sum_{k=0}^{\infty} (c/z)^k$$

$$= \frac{1}{1 - c/z}, \quad |c/z| < 1$$

If we clear $1/z$ by multiplying the numerator and denominator by z, the Z-transform of the causal exponential is

$$Z\{c^k \mu(k)\} = \frac{z}{z - c}, \quad |z| > |c|$$

Plots of the causal exponential signal and its pole-zero pattern are shown in Figure 3.7. The first case, $c = 0.8$, corresponds to a damped exponential, and the second, $c = 1.1$, to a growing exponential. If the pole at $z = c$ is negative, then $x(k)$ oscillates between positive and negative values as it decays or grows. When $c = 1$, the causal exponential reduces to the unit step in Example 3.3.

FIGURE 3.7: Time Plots and Pole-zero Patterns of Causal Exponentials for Example 3.2.3 with $c = 0.8$ (Bounded) and $c = 1.1$ (Unbounded)

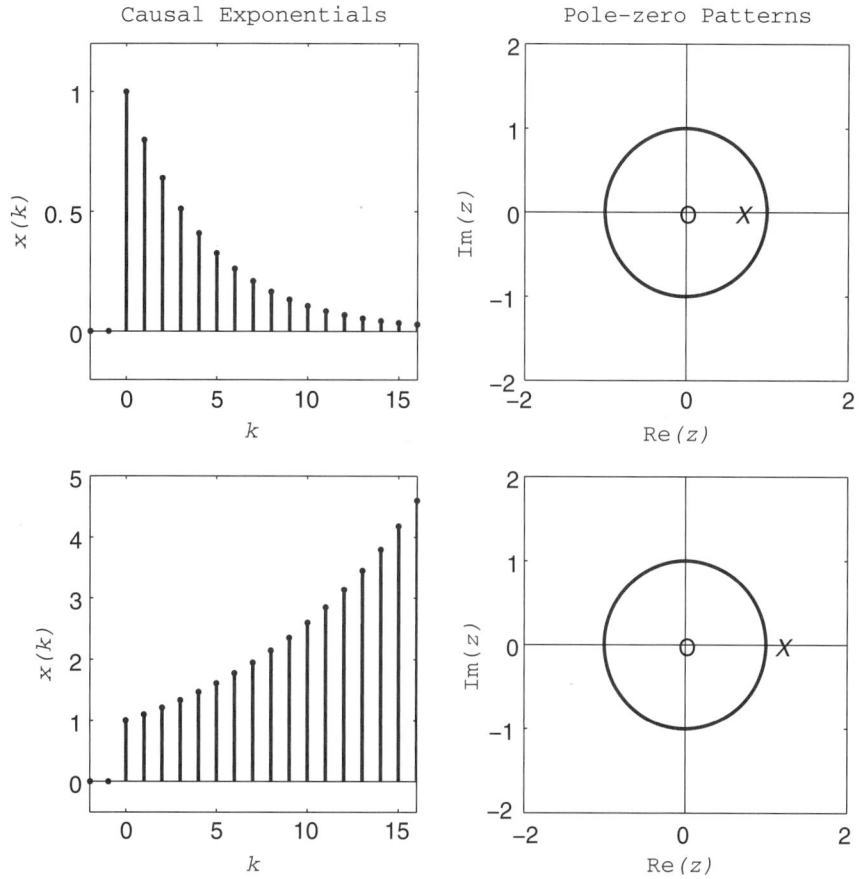

The causal exponential has a single real pole. Another important case corresponds to a complex conjugate pair of poles. The associated signal is referred to as an exponentially damped sinusoid.

Example 3.5

Exponentially damped Sine

Let c and d be real with $c > 0$, and consider the following causal exponentially-damped sine wave.

$$x(k) = c^k \sin(dk)\mu(k)$$

Using Euler's identity in Appendix 2, we can express $\sin(dk)$ as

$$\sin(dk) = \frac{\exp(jdk) - \exp(-jdk)}{j2}$$

Thus the signal $x(k)$ can be represented as

$$x(k) = \frac{c^k[\exp(jdk) - \exp(-jdk)]\mu(k)}{j2}$$

$$= \frac{[c\exp(jd)]^k\mu(k) - [c\exp(-jd)]^k\mu(k)}{j2}$$

FIGURE 3.8: Time Plot and Pole-zero Pattern of Exponentially damped Sine Wave for Example 3.5 with $c = .8$ and $d = .6$

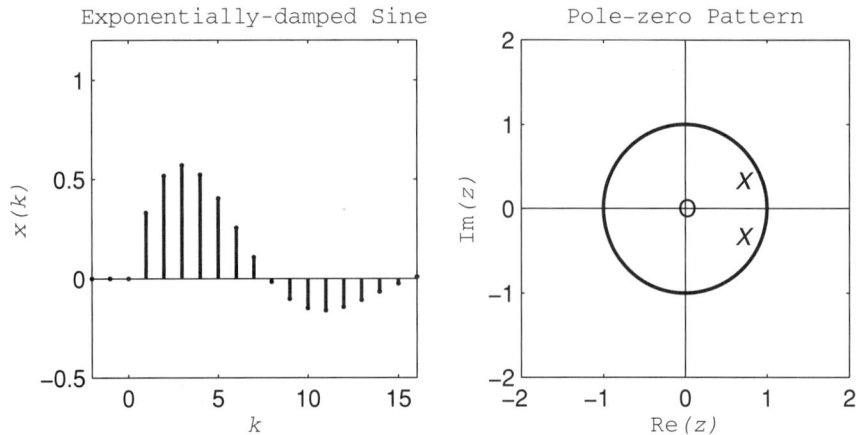

It is clear from Definition 3.1 that the Z-transform operator is a *linear* operator. That is, the transform of the sum of two signals is the sum of the transforms, and the transform of a scaled signal is just the scaled transform. Using this property, Euler's identity, and the results of Example 3.4, we have

$$X(z) = Z \left\{ \frac{[c \exp(jd)]^k \mu(k) - [c \exp(-jd)]^k \mu(k)}{j2} \right\}$$

$$= \frac{Z\{[c \exp(jd)]^k \mu(k)\}}{j2} - \frac{Z\{[c \exp(-jd)]^k \mu(k)\}}{j2}$$

$$= \frac{1}{j2} \left\{ \frac{z}{z - c \exp(jd)} - \frac{z}{z - c \exp(-jd)} \right\}, \quad |z| > c$$

$$= \frac{cz[\exp(jd) - \exp(-jd)]}{j2[z - c \exp(jd)][z - c \exp(-jd)]}, \quad |z| > c$$

$$= \frac{c \sin(d)z}{(z^2 - c[\exp(jd) + \exp(-jd)] + c^2)}, \quad |z| > c$$

Note that $X(z)$ has poles with a magnitude of c and a phase angle of $\pm d$. Finally, if we apply Euler's identity to the denominator, the Z-transform of the causal exponentially damped sine wave is

$$Z\{c^k \sin(dk)\mu(k)\} = \frac{c \sin(d)z}{z^2 - 2c \cos(d)z + c^2}, \quad |z| > c$$

Plots of the exponentially damped sine wave and its pole-zero pattern are shown in Figure 3.8. Note that the *magnitude* of the poles c determines whether, and how fast, the signal decays to zero, while the *angle* of the poles d determines the frequency of the oscillation.

A brief summary of the most important Z-transform pairs is shown in Table 3.2. Note that the unit step function is a special case of the causal exponential function with $c = 1$. Similarly, transforms of the sine and cosine are obtained from the damped sine and cosine by setting the damping factor to $c = 1$. A more complete table of Z-transform pairs can be found in Appendix 1.

Signal	Z-transform	Poles	Ω_{ROC}
$\delta(k)$	1	none	$z \in C$
$\mu(k)$	$\dfrac{z}{z-1}$	$z=1$	$|z|>1$
$k\mu(k)$	$\dfrac{z}{(z-1)^2}$	$z=1$	$|z|>1$
$c^k\mu(k)$	$\dfrac{z}{z-c}$	$z=c$	$|z|>|c|$
$k(c)^k\mu(k)$	$\dfrac{cz}{(z-c)^2}$	$z=c$	$|z|>|c|$
$\sin(dk)\mu(k)$	$\dfrac{\sin(d)z}{z^2-2\cos(d)z+1}$	$z=\exp(\pm jd)$	$|z|>1$
$\cos(dk)\mu(k)$	$\dfrac{[z-\cos(d)]z}{z^2-2\cos(d)z+1}$	$z=\exp(\pm jd)$	$|z|>1$
$c^k\sin(dk)\mu(k)$	$\dfrac{c\sin(d)z}{z^2-2c\cos(d)z+c^2}$	$z=c\exp(\pm jd)$	$|z|>c$
$c^k\cos(dk)\mu(k)$	$\dfrac{[z-c\cos(d)]z}{z^2-2c\cos(d)z+c^2}$	$z=c\exp(\pm jd)$	$|z|>c$

TABLE 3.2: ▶ Basic Z-transform Pairs

3.3 Z-transform Properties

3.3.1 General Properties

The effective size of Table 3.2 can be increased significantly by judiciously applying a number of the properties of the Z-transform.

Linearity Property

The most fundamental property of the Z-transform is the notion that the Z-transform operator is a *linear* operator. In particular, if $x(k)$ and $y(k)$ are two signals and a and b are two arbitrary constants, then from Definition 3.1 we have

$$Z\{ax(k)+by(k)\} = \sum_{k=-\infty}^{\infty}[ax(k)+by(k)]z^{-k}$$

$$= a\sum_{k=-\infty}^{\infty}x(k)z^{-k}+b\sum_{k=-\infty}^{\infty}y(k)z^{-k} \qquad (3.3.1)$$

Thus the Z-transform of the sum of two signals is just the sum of the Z-transforms of the signals. Similarly, the Z-transform of a scaled signal is just the scaled Z-transform of the signal.

Linearity property

$$Z\{ax(k)+by(k)\}=aX(z)+bY(z) \qquad (3.3.2)$$

Recall that Example 3.5 was an example that made use of the linearity property to compute a Z-transform. The region of convergence of the Z-transform of $ax(k)+by(k)$ includes the intersection of the regions of convergence of $X(z)$ and $Y(z)$.

FIGURE 3.9: The
r-sample Delay
Operator

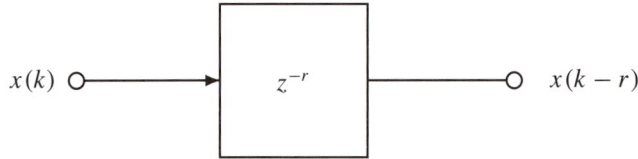

FIGURE 3.9: The *r*-sample Delay Operator

Delay Property

Perhaps the most widely used property, particularly for the analysis of linear difference equations, is the time shift or *delay* property. Let $r \geq 0$ denote the number of samples by which a causal discrete-time signal $x(k)$ is delayed. Applying Definition 3.1, and using the change of variable $i = k - r$, we have

$$Z\{x(k-r)\} = \sum_{k=0}^{\infty} x(k-r)z^{-k}$$

$$= \sum_{i=-r}^{\infty} x(i)z^{-(i+r)}, \quad i = k - r$$

$$= z^{-r} \sum_{i=0}^{\infty} x(i)z^{-i} \qquad (3.3.3)$$

Delay property   Consequently, delaying a signal by r samples is equivalent to multiplying its Z-transform by z^{-r}.

$$Z\{x(k-r)\} = z^{-r}X(z) \qquad (3.3.4)$$

In view of the delay property, we can regard z^{-1} as a *unit delay* operator and z^{-r} as a delay of r samples, as shown in Figure 3.9.

Example 3.6   **Pulse**

As a simple illustration of the linearity and delay properties, consider the problem of finding the Z-transform of a pulse of height b and duration r starting at $k = 0$. This signal can be written in terms of steps: as a step up of amplitude b at time $k = 0$ followed by a step down of amplitude b at time $k = r$.

$$x(k) = b[\mu(k) - \mu(k-r)]$$

Applying the linearity and the delay properties and using Table 3.2, we have

$$X(z) = b(1 - z^{-r})Z\{\mu(k)\}$$

$$= \frac{b(1-z^{-r})z}{z-1}$$

$$= \frac{b(z^r - 1)}{z^{r-1}(z-1)}, \quad |z| > 1$$

Note that when $b = 1$ and $r = 1$, this reduces to $X(z) = 1$, which is the Z-transform of the unit impulse.

Z-scale Property

Another property that can be used to produce new Z-transform pairs is obtained by multiplying a discrete-time signal by an exponential. In particular, let c be constant and consider the signal $c^k x(k)$.

$$
\begin{aligned}
Z\{c^k x(k)\} &= \sum_{k=-\infty}^{\infty} c^k x(k) z^{-k} \\
&= \sum_{k=-\infty}^{\infty} x(k) \left(\frac{z}{c}\right)^{-k}
\end{aligned}
\tag{3.3.5}
$$

Z-scale property   Consequently, multiplying a time signal by c^k is equivalent to scaling the Z-transform variable z by $1/c$.

$$
Z\{c^k x(k)\} = X\left(\frac{z}{c}\right)
\tag{3.3.6}
$$

A simple illustration of the *z-scale* property can be found in Table 3.2. Notice that the Z-transform of the causal exponential signal can be obtained from the Z-transform of the unit step by simply replacing z with z/c. To determine the region of convergence of the Z-transform of $c^k x(k)$, we replace z by z/c in the region of convergence of $X(z)$.

Time Multiplication Property

We can add an important family of entries to Table 3.2 by considering what happens when we take the derivative of the Z-transform of a signal.

$$
\begin{aligned}
\frac{dX(z)}{dz} &= \frac{d}{dz} \sum_{k=-\infty}^{\infty} x(k) z^{-k} \\
&= -\sum_{k=-\infty}^{\infty} k x(k) z^{-(k+1)} \\
&= -z^{-1} \sum_{k=-\infty}^{\infty} k x(k) z^{-k}
\end{aligned}
\tag{3.3.7}
$$

Time-multiplication property   Since the sum in (3.3.7) is just the Z-transform of $kx(k)$, we have

$$
Z\{kx(k)\} = -z\frac{dX(z)}{dz}
\tag{3.3.8}
$$

Thus multiplying a discrete-time signal by the time variable k is equivalent to taking the derivative of the Z-transform and scaling by $-z$.

Example 3.7

Unit Ramp

As an illustration of the time multiplication and Z-scale properties, consider the following signal which is referred to as a *unit ramp*.

$$x(k) \triangleq k\mu(k)$$

Applying the time multiplication property, we have

$$X(z) = -z\frac{d}{dz}Z\{\mu(k)\}$$

$$= -z\frac{d}{dz}\left\{\frac{z}{z-1}\right\}$$

$$= \frac{z}{(z-1)^2}$$

Thus the Z-transform of a unit ramp is

$$Z\{k\mu(k)\} = \frac{z}{(z-1)^2}, \quad |z| > 1$$

The time multiplication property can be applied repeatedly to obtain the Z-transform of additional signals from the family $k^m\mu(k)$ for $m \geq 0$. The results for $0 \leq m \leq 2$ are summarized in Appendix 1.

Given the Z-transform of the unit ramp, the Z-scale property can be applied to generalize the signal further. Consider the following causal exponential with a linear coefficient.

$$x(k) = k(c)^k\mu(k)$$

Using the Z-scale property and the Z-transform of the unit ramp,

$$X(z) = Z\{k\mu(k)\}|_{z=z/c}$$

$$= \left.\frac{z}{(z-1)^2}\right|_{z=z/c}$$

$$= \frac{z/c}{(z/c-1)^2}$$

$$= \frac{cz}{(z-c)^2}$$

Thus the Z-transform of a causal exponential with a linear coefficient is

$$Z\{k(c)^k\mu(k)\} = \frac{cz}{(z-c)^2}, \quad |z| > |c|$$

Note that when $c = 1$, this reduces to the Z-transform of the unit ramp.

Convolution Property

Perhaps the most important property of the Z-transform, when it comes to solving difference equations, is the convolution property. Suppose $h(k)$ and $x(k)$ are zero extended, as needed, so they are defined for all k. By interchanging the order of the summations and using a change of variable, one can show that the Z-transform of the convolution of two signals is just the *product* of the Z-transforms of the signals. Using Definitions 2.3 and 3.1

$$Z\{h(k) \star x(k)\} = \sum_{k=-\infty}^{\infty} [h(k) \star x(k)]z^{-k}$$

$$= \sum_{k=-\infty}^{\infty} \left[\sum_{i=-\infty}^{\infty} h(i)x(k-i) \right] z^{-k}$$

$$= \sum_{i=-\infty}^{\infty} h(i) \sum_{k=-\infty}^{\infty} x(k-i)z^{-k}$$

$$= \sum_{i=-\infty}^{\infty} h(i) \sum_{m=-\infty}^{\infty} x(m)z^{-(m+i)}, \quad m = k - i$$

$$= \sum_{i=-\infty}^{\infty} h(i)z^{-i} \sum_{m=-\infty}^{\infty} x(m)z^{-m} \tag{3.3.9}$$

Thus *convolution* in the time domain maps into *multiplication* in the Z-transform domain. The Z-transform domain is also referred to as the complex *frequency domain*.

Frequency domain convolution

$$Z\{h(k) \star x(k)\} = H(z)X(z) \tag{3.3.10}$$

Time Reversal Property

Another property that effectively expands the table of Z-transform pairs arises when we reverse time by replacing $x(k)$ by $x(-k)$.

$$Z\{x(-k)\} = \sum_{k=-\infty}^{\infty} x(-k)z^{-k}$$

$$= \sum_{i=\infty}^{-\infty} x(i)z^{i}, \quad i = -k$$

$$= \sum_{i=-\infty}^{\infty} x(i)(1/z)^{-i} \tag{3.3.11}$$

Thus reversing time is equivalent to taking the reciprocal of z.

$$Z\{x(-k)\} = X(1/z) \tag{3.3.12}$$

The time reversal property can be used, for example, to convert the Z-transform of a causal signal into the Z-transform of a noncausal signal. Note that to determine the region of convergence of the Z-transform of $x(-k)$, we replace z by $1/z$ in the region of converge of $X(z)$.

Correlation Property

Convolution in the time domain maps into multiplication in the Z-transform or frequency domain. Since cross-correlation and convolution are generally similar operations, it is not surprising that the correlation operation simplifies in the frequency domain as well. To see this, let two finite signals $y(k)$ and $x(k)$ of length L and M, respectively, be zero-extended as needed so they are defined for all k. From Definition 2.5, the cross-correlation of $y(k)$ with

$x(k)$ can be expressed as follows.

$$r_{yx}(k) = \frac{1}{L} \sum_{i=-\infty}^{\infty} y(i)x(i-k)$$

$$= \frac{1}{L} \sum_{i=-\infty}^{\infty} y(k)x[-(k-i)]$$

$$= \frac{y(k) \star x(-k)}{L} \qquad (3.3.13)$$

Thus cross-correlation of $y(k)$ with $x(k)$ is just a scaled version of convolution of $y(k)$ with $x(-k)$. Using the convolution property, we then have

$$Z\{r_{yz}(k)\} = \frac{Y(z)Z\{x(-k)\}}{L} \qquad (3.3.14)$$

But $x(-k)$ is the time reversed version of $x(k)$. Applying the time reversal property then yields the correlation property.

$$Z\{r_{yz}(k)\} = \frac{Y(z)X(1/z)}{L} \qquad (3.3.15)$$

Note that this is essentially the convolution property, but with scaling by $1/L$ and $X(z)$ replaced by $X(1/z)$.

3.3.2 Causal Properties

The Z-transform properties considered thus far all expand the effective size of the table of Z-transform pairs. There are also some properties that can serve as partial checks on the validity of a computed Z-transform pair. Suppose the signal $x(k)$ is causal. It then follows from Definition 3.1 that as $z \to \infty$, all of the terms go to zero except for the constant term that represents the initial value of a causal signal. This is referred to as the *initial value* theorem.

Initial value theorem

$$x(0) = \lim_{z \to \infty} X(z) \qquad (3.3.16)$$

The initial value $x(0)$ can be thought of as the value of a causal signal $x(k)$ at one end of the time range. The value of $x(k)$ at the other end of the range as $k \to \infty$ also can be obtained from $X(z)$ if we assume it is finite. Consider the following function of z.

$$X_1(z) = (z-1)X(z) \qquad (3.3.17)$$

Suppose that the poles of $X_1(z)$ are all strictly inside the unit circle. Therefore $X(z)$ has its poles inside the unit circle except for the possibility of a simple pole at $z = 1$. The terms in $x(k)$ associated with the poles inside the unit circle, even if they are multiple poles, will all decay to zero as $k \to \infty$. This will be shown in Section 3.4. Thus the steady-state solution as $k \to \infty$ consists of either zero or a term associated with a pole at $z = 1$. In Section 3.4 it is shown that the part of $x(k)$ associated with a pole at $z = 1$ is simply $X_1(1)$. Consequently, if $(z-1)X(z)$ has all its poles strictly inside the unit circle, then from the *final value theorem*

Final value theorem

$$x(\infty) = \lim_{z \to 1}(z-1)X(z) \qquad (3.3.18)$$

It is important to emphasize that the final value theorem is applicable *only* if the region of convergence of $(z-1)X(z)$ includes the unit circle.

Example 3.8

Initial and Final Value Theorems

Consider the Z-transform of a pulse of amplitude a and duration M that was previously computed in Example 3.6.

$$x(k) = b[\mu(k) - \mu(k-M)]$$

$$X(z) = \frac{b(z^M - 1)}{z^{M-1}(z-1)}$$

From (3.3.11), the initial value of $x(k)$ is

$$x(0) = \lim_{z \to \infty} X(z)$$
$$= \lim_{z \to \infty} \frac{b(z^M - 1)}{z^{M-1}(z-1)}$$
$$= b$$

Similarly, using (3.3.13), and noting that all of the poles of $(z-1)X(z)$ are inside the unit circle, we find that the final value of $x(k)$ is

$$x(\infty) = \lim_{z \to 1}(z-1)X(z)$$
$$= \lim_{z \to 1} \frac{b(z^M - 1)}{z^{M-1}}$$
$$= 0$$

These values are consistent with $x(k)$ being a pulse of amplitude b and duration M.

A summary of the basic properties of the Z-transform can be found in Table 3.3. Verification of the complex conjugate property is left as an exercise (see Problem 3.7).

TABLE 3.3: ► Z-transform Properties

Property	Description
Linearity	$Z\{ax(k) + by(k)\} = aX(z) + bY(z)$
Delay	$Z\{x(k-r)\} = z^{-r}X(z)$
Time multiplication	$Z\{kx(k)\} = -z\dfrac{dX(z)}{dz}$
Time reversal	$Z\{x(-k)\} = X(1/z)$
Z-scale	$Z\{a^k x(k)\} = X(z/a)$
Complex conjugate	$Z\{x^*(k)\} = X^*(z^*)$
Convolution	$Z\{h(k) \star x(k)\} = H(z)X(z)$
Correlation	$Z\{r_{yx}(k)\} = \dfrac{Y(z)X(1/z)}{L}$
Initial value	$x(0) = \lim_{z \to \infty} X(z)$
Final value	$x(\infty) = \lim_{z \to 1}(z-1)X(z),$ stable

● ● ● ● ● ● ● ● ● ● ● ● ● ● ● ● ● ● ●

3.4 Inverse Z-transform

Inverse transform

There are a number of application areas where it is relatively easy to find the Z-transform of the solution to a problem. To find the actual solution $x(k)$, we must return to the time domain by computing the *inverse* Z-transform of $X(z)$.

$$x(k) = Z^{-1}\{X(z)\} \tag{3.4.1}$$

3.4.1 Noncausal Signals

Normalized denominator

Z-transforms that are associated with linear time-invariant systems take the form of a ratio of two polynomials in z. That is, they are *rational polynomials* in z. By convention, the denominator polynomial $a(z)$ is always *normalized* to make its leading coefficient one.

$$X(z) = \frac{b_0 z^m + b_1 z^{m-1} + \cdots + b_m}{z^n + a_1 z^{n-1} + \cdots + a_n} = \frac{b(z)}{a(z)} \tag{3.4.2}$$

Proper rational polynomial

If $x(k)$ is a causal signal, then $X(z)$ will be a *proper rational polynomial* where $m \leq n$. That is, the number of zeros of $X(z)$ will be less than or equal to the number of poles. If $m > n$, $X(z)$ can be preprocessed by dividing the numerator polynomial $b(z)$ by the denominator polynomial $a(z)$ using long division. Recall from Section 2.7 that this produces a quotient polynomial $Q(z)$ and a remainder polynomial $R(z)$.

$$X(z) = Q(z) + \frac{R(z)}{a(z)} \tag{3.4.3}$$

The quotient polynomial will be of degree $m - n$ and can be expressed

$$Q(z) = \sum_{i=0}^{m-n} q_i z^i \tag{3.4.4}$$

For the special case when $m = n$, the quotient polynomial is simply the constant, b_0. Comparing (3.4.4) with the power series in Definition 3.1, the inverse Z-transform of the quotient polynomial is

$$q(k) = \sum_{i=0}^{m-n} q_i \delta(k + i) \tag{3.4.5}$$

Thus $Q(z)$ represents the anti-causal part of the signal, plus the constant term. The inverse Z-transform of the remaining part of $X(z)$ in (3.4.3), which represents the part of the signal for $k > 0$, can be obtained using the following techniques. For the remainder of this section, we assume that $X(z)$ represents a causal signal, or the causal part of a signal, which means that $X(z)$ is given in (3.4.2) with $m \leq n$.

3.4.2 Synthetic Division

Synthetic Division Method

If only a finite number of samples of $x(k)$ are required, then the synthetic division method can be used to invert $X(z)$. First we rewrite (3.4.2) in terms of negative powers of z. Recalling that $m \leq n$, one can multiply the numerator and denominator by z^{-n} which yields

$$X(z) = \frac{z^{-r}(b_0 + b_1 z^{-1} + \cdots b_m z^{-m})}{1 + a_1 z^{-1} + \cdots + a_n z^{-n}} \tag{3.4.6}$$

In this case $r = n - m$ is the difference between the number of poles and the number of zeros. The basic idea behind the synthetic division method is to perform long division of the numerator polynomial by the denominator polynomial in (3.4.6) to produce an infinitely long quotient polynomial. Then $X(z)$ can be written as

$$X(z) = z^{-r}[q_0 + q_1 z^{-1} + q_2 z^{-2} + \cdots] \tag{3.4.7}$$

If $q(k)$ is the signal whose kth sample is q_k, then comparing (3.4.7) with Definition 3.1 it is clear that $x(k)$ is $q(k)$ delayed by r samples. Therefore the inverse Z-transform of $X(z)$ using synthetic division is

$$x(k) = q(k - r)\mu(k - r) \tag{3.4.8}$$

Example 3.9 | **Synthetic Division**

As an illustration of the synthetic division method, consider the following Z-transform.

$$X(z) = \frac{z + 1}{z^2 - 2z + 3}$$

There are $n = 2$ poles and $m = 1$ zeros. Multiplying the top and bottom by z^{-2} yields

$$X(z) = \frac{z^{-1}(1 + z^{-1})}{1 - 2z^{-1} + 3z^{-2}}$$

Here $r = n - m = 1$. Performing long division yields

$$
\begin{array}{r}
1 + 3z^{-1} + 3z^{-2} - 3z^{-3} - 15z^{-4} + \cdots \\
\hline
1 - 2z^{-1} + 3z^{-2} \;\big|\; 1 + z^{-1}
\end{array}
$$

$$
\begin{aligned}
&\underline{1 - 2z^{-1} + 3z^{-2}} \\
&\quad 3z^{-1} - 3z^{-2} \\
&\quad \underline{3z^{-1} - 6z^{-2} + 9z^{-3}} \\
&\qquad\quad 3z^{-2} - 9z^{-3} \\
&\qquad\quad \underline{3z^{-1} - 6z^{-2} + 9z^{-3}} \\
&\qquad\qquad\quad -3z^{-3} - 9z^{-3} \\
&\qquad\qquad\quad \underline{-3z^{-3} + 6z^{-4} - 9z^{-4}} \\
&\qquad\qquad\qquad\quad -15z^{-4} + 9z^{-5}
\end{aligned}
$$

Applying (3.4.8) with $r = 1$, the inverse Z-transform of $X(z)$ is

$$x(k) = [0, 1, 3, 3, -3, -15, \cdots], \quad 0 \le k \le 5$$

Impulse Response Method

MATLAB functions In Section 3.7 it will be shown that there is a very easy numerical way to find a finite number of samples of the inverse Z-transform of $X(z)$. It is based on an interpretation of $x(k)$ as the impulse response of a discrete-time system with characteristic polynomial $a(z)$ and input polynomial

$b(z)$. The impulse response is computed using the MATLAB function *filter* as follows.

```
N = 100;                         % number of points
delta = [1 ; zeros(N-1,1)];      % unit impulse input
x = filter(b,a,delta);           % impulse response
stem(x)                          % plot x in discrete time
```

3.4.3 Partial Fractions

The synthetic division and impulse response methods are useful, but they have the drawback that they do not produce a *closed-form expression* for $x(k)$ as a function of k. Instead, they generate numerical values for a finite number of samples of $x(k)$. If it is important to find a closed-form expression for the inverse Z-transform of $X(z)$, then the method of partial fraction expansion can be used.

Simple Poles

The easiest way to find the inverse Z-transform of $X(z)$ is to locate $X(z)$ in a table of Z-transform pairs and read across the row to find the corresponding $x(k)$. Unfortunately, for most problems of interest, $X(z)$ does not appear in the table. The basic idea behind partial fractions is to write $X(z)$ as a sum of terms, each of which is in a Z-transform table. First, consider the case when the n poles of $X(z)$ are nonzero and simple. Then $X(z)$ in (3.4.2) can be written in factored form as

$$X(z) = \frac{b(z)}{(z - p_1)(z - p_2) \cdots (z - p_n)} \tag{3.4.9}$$

Next express $X(z)/z$ using partial fractions.

$$\frac{X(z)}{z} = \sum_{i=0}^{n} \frac{R_i}{z - p_i} \tag{3.4.10}$$

Residue   In this case $X(z)/z$ has $n + 1$ simple poles. Coefficient R_i is called the *residue* of $X(z)/z$ at pole p_i, where $p_0 = 0$ is the pole added by dividing $X(z)$ by z. A general approach to find the residues is to put the $n + 1$ partial fraction terms over a common denominator similar to (3.4.9) and then equate numerators. By equating the coefficients of like powers of z, this results in $n + 1$ equations in the $n + 1$ unknown residues.

Simple poles   Although this general approach will work, it is usually simpler to compute each residue directly. Multiplying both sides of (3.4.10) by $(z - p_k)$ and evaluating the result at $z = p_k$, we find that

$$R_k = \left. \frac{(z - p_k)X(z)}{z} \right|_{z=p_k}, \quad 0 \le k \le n \tag{3.4.11}$$

The term residue comes from the fact that R_i is what *remains* of $X(z)/z$ at $z = p_i$, after the pole at $z = p_i$ has been removed. Once the partial fraction representation in (3.4.10) is obtained, it is a simple matter to find the inverse Z-transform. First, multiply both sides of (3.4.10) by z.

$$X(z) = R_0 + \sum_{i=1}^{n} \frac{R_i z}{z - p_i} \tag{3.4.12}$$

Using the linearity property of the Z-transform and entries from Table 3.2, we find the inverse Z-transform of $X(z)$ is

$$x(k) = R_0\delta(k) + [R_1(p_1)^k + R_2(p_2)^k + \cdots + R_n(p_n)^k]\mu(k) \qquad (3.4.13)$$

Example 3.10

Simple Poles: Partial Fractions

As an illustration of the partial fraction expansion method with simple poles, consider the following Z-transform.

$$\frac{X(z)}{z} = \frac{10(z^2 + 4)}{z(z^2 - 2z - 3)}$$

$$= \frac{10(z^2 + 4)}{z(z + 1)(z - 3)}$$

Here $X(z)/z$ has simple real poles at $p_0 = 0$, $p_1 = -1$, and $p_2 = 3$. From (3.4.11), the residues are

$$R_0 = \left.\frac{10(z^2 + 4)}{(z + 1)(z - 3)}\right|_{z=0} = \frac{40}{-3}$$

$$R_1 = \left.\frac{10(z^2 + 4)}{z(z - 3)}\right|_{z=-1} = \frac{50}{4}$$

$$R_2 = \left.\frac{10(z^2 + 4)}{z(z + 1)}\right|_{z=3} = \frac{130}{12}$$

It then follows from (3.4.13) that a closed-form expression for the inverse Z-transform of $X(z)$ is

$$x(k) = \frac{-40}{3}\delta(k) + \left[\frac{25(-1)^k}{2} + \frac{65(3)^k}{6}\right]\mu(k)$$

Multiple Poles

Multiple poles

When the poles of $X(z)/z$ include multiple poles, poles that occur more than once, the partial fraction expansion takes on a different form. Suppose p is a pole of multiplicity $m \geq 1$. Recall from Chapter 2 that this would generate a natural mode term in the zero-input response of the form $c(k)p^k\mu(k)$, where $c(k)$ is a polynomial of degree $m - 1$. This type of term also appears in $x(k)$. For example, suppose $X(z)$ consists of a single nonzero pole at $z = p_1$ repeated n times. Then the partial fraction expansion is

$$\frac{X(z)}{z} = \frac{R_0}{z - p_0} + \frac{c_1}{z - p_1} + \frac{c_2}{(z - p_1)^2} + \cdots + \frac{c_n}{(z - p_1)^n} \qquad (3.4.14)$$

Residue R_0 can be found as before by using (3.4.11). To find the coefficient vector $c \in R^n$, one can put the terms in (3.4.14) over a common denominator and equate numerators. Equating the coefficients of like powers of z, results in $n + 1$ equations in the $n + 1$ variables $\{R_0, c_1, \ldots, c_n\}$.

Example 3.11

Multiple Poles: Partial Fractions

As an illustration of the partial fraction expansion method with multiple poles, consider the following Z-transform.

$$\frac{X(z)}{z} = \frac{2(z+3)}{z(z^2 - 4z + 4)}$$

$$= \frac{2(z+3)}{z(z-2)^2}$$

Here $X(z)/z$ has a simple pole at $p_0 = 0$ and a multiple pole of multiplicity $m_1 = 2$ at $p_1 = 2$. From (3.4.14), the form of the partial fraction expansion is

$$\frac{X(z)}{z} = \frac{R_0}{z} + \frac{c_1}{z-2} + \frac{c_2}{(z-2)^2}$$

Using (3.4.11), we find the residue of the pole at $p_0 = 0$ is

$$R_0 = \frac{2(z+3)}{(z-2)^2}\bigg|_{z=0} = \frac{6}{4} = 1.5$$

Next, substituting $R_0 = 1.5$ and putting the terms of the partial fraction expansion over a common denominator yields

$$\frac{X(z)}{z} = \frac{1.5(z-2)^2 + c_1 z(z-2) + c_2 z}{z(z-2)^2}$$

$$= \frac{1.5(z^2 - 4z + 4) + c_1(z^2 - 2z) + c_2 z}{z(z-2)^2}$$

$$= \frac{(1.5 + c_1)z^2 + (c_2 - 2c_1 - 6)z + 6}{z(z-2)^2}$$

The numerator of $X(z)/z$ is $b(z) = 2z + 6$. Equating coefficients of like powers of z yields $6 = 6$ and

$$1.5 + c_1 = 0$$
$$c_2 - 2c_1 + 6 = 2$$

From the first equation $c_1 = -1.5$. It then follows from the second equation that $c_2 = -7$. If we multiply both sides $X(z)/z$ by z, the partial fraction expansion in this case is

$$X(z) = 1.5 + \frac{-1.5z}{z-2} + \frac{-7z}{(z-2)^2}$$

$$= 1.5 - \frac{1.5z}{z-2} - \frac{(3.5)2z}{(z-2)^2}$$

From Table 3.2, the inverse Z-transform of $X(z)$ is then

$$x(k) = 1.5\delta(k) - [1.5(2)^k + 3.5k(2)^k]\mu(k)$$

$$= 1.5\delta(k) - (1.5 + 3.5k)2^k\mu(k)$$

Note that in this case the polynomial coefficient of the double pole at $z = 2$ is $c(z) = -(1.5 + 3.5z)$.

In general, $X(z)$ may have both simple poles and multiple poles. In this case the partial fraction expansion of $X(z)/z$ will be a combination of terms like those in (3.4.10) for the simple poles and those in (3.4.14) for the multiple poles with $n+1$ total terms. The residues of the simple poles can be computed directly using (3.4.11). To compute the remaining coefficients, one can put all of the terms over a common denominator, and then equate numerators.

Complex Poles

Complex conjugate pair

The poles of $X(z)$ can be real or complex. If the signal $x(k)$ is real, then complex poles appear in *conjugate pairs*. The techniques for simple and multiple poles can be applied to complex poles. When the poles appear as complex conjugate pairs, their residues will also be complex conjugates. The pair of terms in $x(k)$ associated with a complex conjugate pair of poles can be combined using Euler's identity, and the result will be a real damped sinusoidal term. Although this approach is certainly feasible, there is an alternative that avoids the use of complex arithmetic. When complex poles occur, it is often easier to work directly with real second-order terms. To illustrate, suppose $X(z)$ consists of the following strictly proper $(m < n)$ rational polynomial with poles at $z = c\exp(\pm jd)$.

$$X(z) = \frac{b_0 z + b_1}{[z - c\exp(jd)][z - c\exp(-jd)]} \qquad (3.4.15)$$

Note that if $X(z)$ starts out with a numerator of degree $m = 2$, one can always perform a long division step to produce a constant quotient plus a remainder term of the form of (3.4.15).

Delay property

The basic idea is to write $X(z)$ as a linear combination of the damped sine and cosine terms from Table 3.2. Since the numerators of these terms do not contain any constant terms, this form can be achieved by multiplying and dividing $X(z)$ by z.

$$X(z) = \frac{1}{z}\left\{\frac{b_0 z^2 + b_1 z}{[z - c\exp(jd)][z - c\exp(-jd)]}\right\} \qquad (3.4.16)$$

Note that the $1/z$ factor can be regarded as a delay of one sample. Next consider the numerator in (3.4.16). To make the numerator of $X(z)$ equal to a linear combination of the numerators of the damped sine and cosine terms in Table 3.2, we need to find f_1 and f_2 such that

$$\begin{aligned} b_0 z^2 + b_1 z &= f_1[c\sin(d)z] + f_2[z - c\cos(d)]z \\ &= f_2 z^2 + c[f_1\sin(d) - f_2\cos(d)]z \end{aligned} \qquad (3.4.17)$$

Equating coefficients of like powers of z yields $f_2 = b_0$ and $f_1 = [b_1/c + f_2\cos(d)]/\sin(d)$. Thus the weighting coefficients for the linear combination of sine and cosine terms are

$$f_1 = \frac{b_1 + cb_0\cos(d)}{\sin(d)} \qquad (3.4.18a)$$

$$f_2 = b_0 \qquad (3.4.18b)$$

Recalling the delay in (3.4.16), and using Table 3.2, we find the inverse Z-transform of the quadratic term is

$$x(k) = c^{k-1}\{f_1\sin[d(k-1)] + f_2\cos[d(k-1)]\}\mu(k-1) \qquad (3.4.19)$$

Notice from (3.4.19) that $x(0) = 0$. This is consistent with the result obtained by applying the initial value theorem to $X(z)$ in (3.4.15).

Example 3.12   **Complex Poles: Table**

To illustrate the case of complex poles, consider the following Z-transform.

$$X(z) = \frac{3z + 5}{z^2 - 4z + 13}$$

The poles of $X(z)$ are at

$$
\begin{aligned}
p_{1,2} &= \frac{4 \pm \sqrt{16 - 4(13)}}{2} \\
&= 2 \pm j3 \\
&= c \exp(\pm jd)
\end{aligned}
$$

In polar coordinates, the magnitude c and phase angle d are

$$
\begin{aligned}
c &= \sqrt{4 + 9} = 3.61 \\
d &= \arctan(3/2) = .983
\end{aligned}
$$

From (3.4.20), the weighting coefficients for the damped sine and cosine terms are

$$
\begin{aligned}
f_1 &= \frac{5 + 3.61(3)\cos(.983)}{\sin(.983)} = 13.2 \\
f_2 &= 3
\end{aligned}
$$

Finally, from (3.4.19), the inverse Z-transform is

$$x(k) = 3.61^{k-1}\{13.2\sin[.983(k-1)] + 3\cos[.983(k-1)]\}\mu(k-1)$$

3.4.4 Residue Method

The partial fraction method is effective for simple poles, but it can become cumbersome for multiple poles and furthermore it requires the use of a Z-transform table. There is an alternative method that does not require a table, and can be less work for multiple poles. It is based on the following elegant formulation of the inverse Z-transform from the theory of functions of a complex variable.

$$Z^{-1}\{X(z)\} = \frac{1}{j2\pi} \oint_C X(z)z^{k-1}dz \qquad (3.4.20)$$

Here (3.4.20) is a contour integral where C is a counterclockwise contour in the region of convergence of $X(z)$ that encloses all of the poles. The inverse Z-transform formulation in (3.4.20) is based on the Cauchy integral theorem (see Problem 3.32). Practical evaluation of $x(k)$ in (3.4.20) is achieved by using the *Cauchy residue theorem*.

Cauchy residue theorem

$$x(k) = \sum_{i=1}^{q} \text{Res}(p_i, k) \qquad (3.4.21)$$

Here p_i for $1 \leq i \leq q$ are the q distinct poles of $X(z)z^{k-1}$. The notation $\text{Res}(p_i, k)$ denotes the *residue* of $X(z)z^{k-1}$ at the pole $z = p_i$. To compute the residues, it is helpful to first factor the denominator polynomial as follows.

$$X(z) = \frac{b(z)}{(z - p_1)^{m_1}(z - p_2)^{m_2} \cdots (z - p_q)^{m_q}} \tag{3.4.22}$$

Simple residue

Here p_i is a pole of *multiplicity m_i* for $1 \leq i \leq q$. There are two cases. If p_i is a simple pole with multiplicity $m_i = 1$, then the residue is what *remains* of $X(z)z^{k-1}$ at the pole after the pole has been removed.

$$\text{Res}(p_i, k) = (z - p_i)X(z)z^{k-1}\Big|_{z=p_i} \quad \text{if} \quad m_i = 1 \tag{3.4.23}$$

If we compare (3.4.23) with (3.4.11), it is clear that for simple poles the residue method is the same amount of work as the partial fraction method where $R_i = \text{Res}(p_i, 1)$. However, there is no need to use a table in this case because $\text{Res}(p_i, k)$ includes the dependence on k.

Multiple residue

The second case is for multiple poles. When p_i is a pole of multiplicity $m_i > 1$, the expression for the residue in (3.4.23) has to be generalized as follows.

$$\text{Res}(p_i, k) = \frac{1}{(m_i - 1)!} \frac{d^{m_i - 1}}{dz^{m_i - 1}} \left\{ (z - p_i)^{m_i} X(z)z^{k-1} \right\}\Big|_{z=p_i} \quad \text{for } m_i > 1 \tag{3.4.24}$$

Thus the pole is again removed, but before the result is evaluated at the pole, it is differentiated $m_i - 1$ times and scaled by $1/(m_i - 1)!$. Note that the expression for a multiple-pole residue in (3.4.24) reduces to the simpler expression for a simple-pole residue in (3.4.23) when $m_i = 1$.

The residue method requires the same amount of computational effort as the partial fraction method for the case of simple poles, and it requires less computational effort for multiple poles because there is only a single term associated with a multiple pole, not m_i terms. In addition, the residue method does not require the use of a table.

The residue method does suffer from a drawback. Because $X(z)z^{k-1}$ depends on k, there can be some values of k that cause a pole at $z = 0$ to appear; when this happens, its residue must be included in (3.4.21) as well. For example, if $X(0) \neq 0$, then $X(z)z^{k-1}$ will have a pole at $z = 0$ when $k = 0$, but the pole disappears for $k > 0$. Fortunately, this problem can be easily resolved by separating $x(k)$ into two cases, $k = 0$ and $k > 0$. The initial value theorem can be used to compute $x(0)$. Letting $z \to \infty$, and recalling that $m \leq n$ yields

Initial value

$$x(0) = \begin{cases} b_0, & m = n \\ 0, & m < n \end{cases} \tag{3.4.25}$$

Note that the initial value can be written compactly as $x(0) = b_0\delta(n - m)$. The residue method is summarized in Algorithm 3.1.

1. Factor the denominator polynomial of $X(z)$ as in (3.4.22).
2. Set $x(0) = b_0\delta(n - m)$.
3. For $i = 1$ to q do
{

    If $m_i = 1$ then p_i is a simple pole and

$$\text{Res}(p_i, k) = (z - p_i)X(z)z^{k-1}\big|_{z=p_i}$$

    else p_i is a multiple pole and

$$\text{Res}(p_i, k) = \frac{1}{(m_i - 1)!}\frac{d^{m_i-1}}{dz^{m_i-1}}\left\{(z - p_i)^{m_i}X(z)z^{k-1}\right\}\bigg|_{z=p_i}$$

}
4. Set

$$x(k) = x(0)\delta(k) + \left[\sum_{i=1}^{q}\text{Res}(p_i, k)\right]\mu(k - 1)$$

Example 3.13 | **Simple Poles: Residue Method**

As a simple initial example, consider a Z-transform with two simple nonzero poles.

$$X(z) = \frac{z^2}{(z - a)(z - b)}$$

The initial value of $x(k)$ is $x(0) = 1$. The two residues are

$$\text{Res}(a, k) = \frac{z^{k+1}}{z - b}\bigg|_{z=a} = \frac{a^{k+1}}{a - b}$$

$$\text{Res}(b, k) = \frac{z^{k+1}}{z - a}\bigg|_{z=b} = \frac{b^{k+1}}{b - a}$$

Thus

$$x(k) = x(0)\delta(k) + [\text{Res}(a, k) + \text{Res}(b, k)]\mu(k - 1)$$
$$= \delta(k) + \left(\frac{a^{k+1} - b^{k+1}}{a - b}\right)\mu(k - 1)$$
$$= \left(\frac{a^{k+1} - b^{k+1}}{a - b}\right)\mu(k)$$

Example 3.14 | **Mixed Poles: Residue Method**

Next, consider a mixed case that has both simple and multiple poles.

$$X(z) = \frac{1}{(z - a)^2(z - b)}$$

The initial value of $x(k)$ is $x(0) = 0$. The residue of the multiple pole at $z = a$ is

$$
\begin{aligned}
\operatorname{Res}(a, k) &= \frac{d}{dz} \left\{ \frac{z^{k-1}}{z-b} \right\} \bigg|_{z=a} \\
&= \frac{(z-b)(k-1)z^{k-2} - z^{k-1}}{(z-b)^2} \bigg|_{z=a} \\
&= \frac{(a-b)(k-1)a^{k-2} - a^{k-1}}{(a-b)^2} \\
&= \frac{[(a-b)(k-1) - a]a^{k-2}}{(a-b)^2}
\end{aligned}
$$

The residue of the simple pole at $z = b$ is

$$
\begin{aligned}
\operatorname{Res}(b, k) &= \frac{z^{k-1}}{(z-a)^2} \bigg|_{z=b} \\
&= \frac{b^{k-1}}{(b-a)^2}
\end{aligned}
$$

Thus the inverse Z-transform of $X(z)$ is

$$
\begin{aligned}
x(k) &= x(0)\delta(k) + [\operatorname{Res}(a, k) + \operatorname{Res}(b, k)]\mu(k-1) \\
&= \left[\frac{[(a-b)(k-1) - a]a^{k-2}}{(a-b)^2} + \frac{b^{k-1}}{(b-a)^2} \right] \mu(k-1) \\
&= \left[\frac{[(a-b)(k-1) - a]a^{k-2} - b^{k-1}}{(a-b)^2} \right] \mu(k-1)
\end{aligned}
$$

MATLAB Function

There is a built-in MATLAB function available called *residue* that can be used to compute the residue terms for partial fraction expansions.

```
% RESIDUE: Compute residues and poles
%
% Usage:
%          [r,p,q] = residue (b,a);
% Pre:
%          b = vector of length m+1 containing numerator coefficients (m <= n)
%          a = vector of length n+1 containing denominator coefficients
% Post:
%          r = vector of length n containing the residues R_i in (3.4.13)
%          p = vector of length n containing the poles
%          q = residue R_0.
% Note:
%          If X(z) contains multiple poles, then the corresponding elements
%          of r are the coefficients c_i in (3.4.14)
```

● ● ● ● ● ● ● ● ● ● ● ● ● ● ● ● ● ●

3.5   Transfer Functions

3.5.1 The Transfer Function

The examples of linear discrete-time systems introduced thus far are all special cases of the following generic linear time-invariant difference equation which we refer to as the system S.

$$y(k) + \sum_{i=1}^{n} a_i y(k - i) = \sum_{i=0}^{m} b_i x(k - i) \tag{3.5.1}$$

By convention, the coefficient of the current output $y(k)$ has been normalized to one by dividing both sides of the equation, if needed, by a_0. Recall from Chapter 2 that the complete solution to (3.5.1) depends on both the causal input $x(k)$ and the initial condition $y_0 = [y(-1), y(-2), \ldots, y(-n)]^T$. That is, in general, the output $y(k)$ can be decomposed into the sum of two parts.

$$y(k) = y_{zi}(k) + y_{zs}(k) \tag{3.5.2}$$

Zero-input response   The first term, $y_{zi}(k)$, is the *zero-input response*. It is the part of the output that is generated by the initial condition. When the input is zero, $y(k) = y_{si}(k)$. For a system with stable natural modes

$$y_{zi}(k) \to 0 \quad \text{as} \quad k \to \infty \tag{3.5.3}$$

Zero-state response   The term $y_{zs}(k)$ is the *zero-state response*. The zero-state response is the part of the output that is generated by the input $x(k)$. When the initial condition is zero, $y(k) = y_{zs}(k)$. In view of (3.5.3), one can *measure* the zero-state response of a stable system by first waiting for $y_{zi}(k)$ to die out, and then exciting the system with the input and measuring the output.

The difference equation in (3.5.1) is merely one way to represent a discrete-time system. The following alternative representation, based on the Z-transform, is a more compact algebraic characterization.

DEFINITION
....
3.2: Transfer Function

Let $x(k)$ be a nonzero input to the system S, and let $y(k)$ be the output assuming the initial condition is zero. Then the *transfer function* of the system is defined

$$H(z) \triangleq \frac{Y(z)}{X(z)}$$

The transfer function is simply the Z-transform of the zero-state response divided by the Z-transform of the input. The transfer function $H(z)$ is a concise algebraic representation of the system. By multiplying both sides of the expression for $H(z)$ by $X(z)$, we get the *Frequency domain representation*   *frequency-domain* input-output representation

$$Y(z) = H(z)X(z) \tag{3.5.4}$$

A block diagram which shows the relationship between the input, the output, and the transfer function of a discrete-time system is shown in Figure 3.10.

FIGURE 3.10:
Transfer Function
Representation

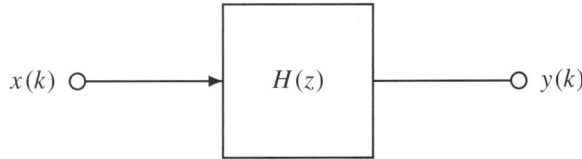

The transfer function of the system S can be determined using the time shift property of the Z-transform. Taking the Z-transform of both sides of (3.5.1) yields

$$Y(z) + \sum_{i=1}^{n} a_i z^{-i} Y(z) = \sum_{i=0}^{m} b_i z^{-i} X(i) \tag{3.5.5}$$

Factoring $Y(z)$ from the left-hand side and $X(z)$ from the right-hand side then yields

$$\left(1 + \sum_{i=1}^{n} a_i z^{-i}\right) Y(z) = \left(\sum_{i=0}^{m} b_i z^{-i}\right) X(z) \tag{3.5.6}$$

Finally, solving for $Y(z)/X(z)$ produces the following transfer function for the discrete-time system S.

$$H(z) = \frac{b_0 + b_1 z^{-1} + \cdots + b_m z^{-m}}{1 + a_1 z^{-1} + \cdots + a_n z^{-n}} \tag{3.5.7}$$

If we compare (3.5.7) with (3.5.1), it is apparent that the transfer function can be written down directly from *inspection* of the difference equation. The transfer function representation in (3.5.7) is in terms of z^{-1}. To convert it to positive powers of z, we multiply the top and bottom by z^n.

$$H(z) = \frac{z^{n-m}(b_0 z^m + b_1 z^{m-1} + \cdots + b_m)}{z^n + a_1 z^{n-1} + \cdots + a_n} \tag{3.5.8}$$

Note that when $m \le n$, the transfer function $H(z)$ can have $n - m$ zeros at $z = 0$, and when $m > n$, it can have $m - n$ poles at $z = 0$.

Example 3.15 **Transfer Function**

As a simple illustration of computing the transfer function, consider the following discrete-time system.

$$y(k) = 1.2y(k-1) - .32y(k-2) + 10x(k-1) + 6x(k-2)$$

Using (3.5.7), we see from inspection that

$$H(z) = \frac{10z^{-1} + 6z^{-2}}{1 - 1.2z^{-1} + .32z^{-2}}$$

Note that when the terms corresponding to delayed outputs are on the right-hand side of the difference equation, the *signs* of the coefficients must be reversed in the denominator of $H(z)$. To reformulate $H(z)$ in terms of positive powers of z, we multiply the numerator and the denominator by z^2, which yields the positive-power form of the transfer function.

$$H(z) = \frac{10z + 6}{z^2 - 1.2z + .32}$$

3.5.2 Zero-State Response

Zero-state response

One of the useful features of the transfer function representation is that it provides us with a simple means of computing the *zero-state response* of a discrete-time system for an arbitrary input $x(k)$. Taking the inverse Z-transform of both sides of (3.5.4) yields the following formulation of the zero-state response.

$$y_{zs}(k) = Z^{-1}\{H(z)X(z)\} \qquad (3.5.9)$$

Consequently, if the initial condition is zero, the output of the system is just the inverse Z-transform of the product of the transfer function and the Z-transform of the input. The following example illustrates this technique for computing the output.

Example 3.16   **Zero-state Response**

Consider the discrete-time system in Example 3.15. Suppose the input is the unit step $x(k) = \mu(k)$ and the initial condition is zero. Then the Z-transform of the output is

$$
\begin{aligned}
Y(z) &= H(z)X(z) \\
&= \left(\frac{10z + 6}{z^2 - 1.2z + .32} \right) \frac{z}{z - 1} \\
&= \frac{(10z + 6)z}{(z^2 - 1.2z + .32)(z - 1)}
\end{aligned}
$$

To find $y(k)$ we invert $Y(z)$ using the residue method in Algorithm 3.1. The denominator of $Y(z)$ is already partially factored. Applying the quadratic formula, we find that the remaining two roots are

$$
\begin{aligned}
p_{1,2} &= \frac{1.2 \pm \sqrt{1.44 - 1.28}}{2} \\
&= \frac{1.2 \pm .4}{2} \\
&= \{.8, .4\}
\end{aligned}
$$

The initial value of the output is

$$y(0) = b_0 \delta(m - n) = 0$$

The residues of the three poles are

$$
\text{Res}(.8, k) = \left. \frac{(10z + 6)z^k}{(z - .4)(z - 1)} \right|_{z=.8} = \frac{14(.8)^k}{.4(-.2)} = -175(.8)^k
$$

$$
\text{Res}(.4, k) = \left. \frac{(10z + 6)z^k}{(z - .8)(z - 1)} \right|_{z=.4} = \frac{10(.4)^k}{-.4(-.6)} = 41.7(.4)^k
$$

$$
\text{Res}(1, k) = \left. \frac{(10z + 6)z^k}{(z - .8)(z - .4)} \right|_{z=1} = \frac{16}{.2(.6)} = 133.3
$$

FIGURE 3.11: Zero-state Output for Example 3.16

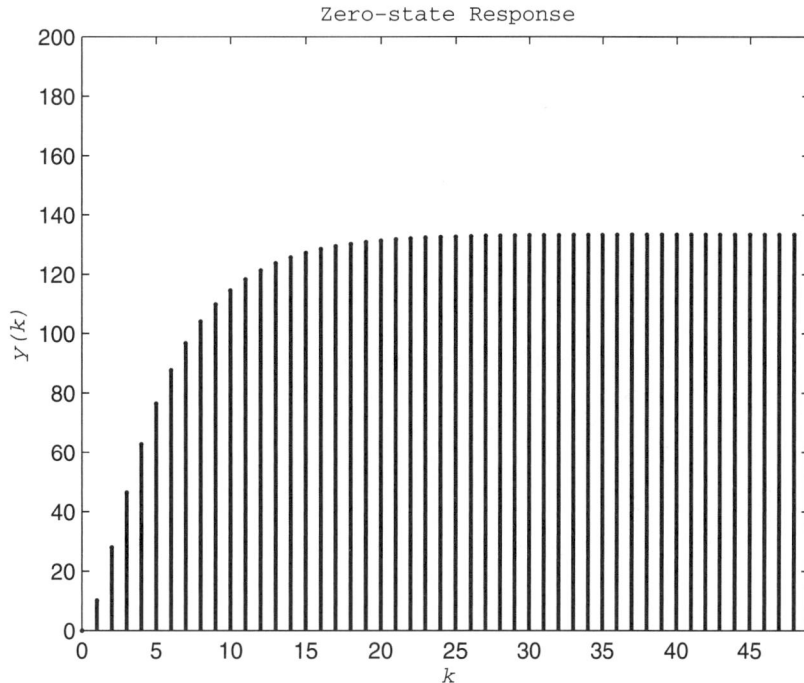

Finally, the zero-state response is

$$y(k) = y(0)\delta(k) + [\text{Res}(.8, k) + \text{Res}(.6, k) + \text{Res}(1, k)]\mu(k - 1)$$

$$= [133.3 - 175(.8)^k + 41.7(.4)^k]\mu(k - 1)$$

$$= [133.3 - 175(.8)^k + 41.7(.4)^k]\mu(k)$$

A plot of the step response, generated by running *exam3_16*, is shown in Figure 3.11.

3.5.3 Poles, Zeros, and Modes

Recall that the poles and zeros of a signal were defined earlier in Section 3.2. This concept also can be applied to discrete-time systems using the transfer function. Using the positive-power version of $H(z)$ in (3.5.8) and factoring the numerator and denominator results in the following *factored form* of the transfer function.

$$H(z) = \frac{b_0 z^{n-m}(z - z_1)(z - z_2) \cdots (z - z_m)}{(z - p_1)(z - p_2) \cdots (z - p_n)} \tag{3.5.10}$$

Poles, zeros   The roots of the numerator polynomial are called the *zeros* of the discrete-time system, and the roots of the denominator polynomial are called the *poles* of the system. Recall that the *multiplicity* of a pole or zero corresponds to the number of times it appears.

Example 3.17   **Poles and Zeros**

Consider the home mortgage system discussed in Case Study 2.2. The difference equation for this system is

$$y(k) = y(k - 1) + \left(\frac{r}{12}\right) y(k - 1) - x(k)$$

Thus the transfer function of this system is

$$H(z) = \frac{-1}{1 - (1 + r/12)z^{-1}} = \frac{-z}{z - (1 + r/12)}$$

The home mortgage system has one zero at $z = 0$ and one pole at $z = 1 + r/12$ where r is the annual interest rate, expressed as a fraction.

Poles and zeros have simple interpretations in the time domain as well. Suppose $Y(z) = H(z)X(z)$. Then the output $y(k)$ can be decomposed into the sum of two types of terms called *modes*.

$$y(k) = \text{natural modes} + \text{forced modes} \tag{3.5.11}$$

Multiple mode

Each *natural mode* term is generated by a pole of $H(z)$, while each *forced mode* term is generated by a pole of the input or *forcing* function $X(z)$. For a simple pole at p, the corresponding mode is of the form $c(p)^k$ for some constant c. More generally, if p is a pole of multiplicity r, then the mode is of the form

$$\text{multiple mode} = (c_0 + c_1 k + \cdots + c_{r-1} k^{r-1}) p^k \mu(k), \quad r \geq 1 \tag{3.5.12}$$

Thus a multiple pole is similar to a simple pole, except that the coefficient is a polynomial in k of degree $r - 1$. For a simple pole, $r = 1$ and the coefficient reduces to a polynomial of degree zero, a constant. Interpretation of poles of $Y(z)$ as modes of $y(k)$ is useful because it allows us to write down the *form* of $y(k)$, showing the dependence on k, directly from inspection of the factored form of $Y(z)$. For example, if $H(z)$ has poles at p_i for $1 \leq i \leq n$, and $X(z)$ has poles at q_i for $1 \leq i \leq r$, and if all the poles are simple, then the form of $y(k)$ is

$$y(k) = \sum_{i=1}^{n} c_i (p_i)^k \mu(k) + \sum_{i=1}^{r} d_i (q_i)^k \mu(k) \tag{3.5.13}$$

The first sum in (3.5.13) represents the natural modes and the second sum represents the forced modes. Of course, if there are multiple poles, including those caused by poles of $X(z)$ matching those of $H(z)$, then there will be fewer terms in (3.5.13), but some of the terms will have polynomial coefficients.

Pole-zero cancellation

The zeros of $H(z)$ and $X(z)$ also have simple interpretations in terms of $y(k)$. Since $Y(z) = H(z)X(z)$, there is a possibility of *pole-zero cancellation* between $H(z)$ and $X(z)$. For example, if $H(z)$ has a zero at $z = q$ and $X(z)$ has a pole at $z = q$, then the pole at $z = q$ will not appear in $Y(z)$ which means that there is no corresponding forced-mode term in $y(k)$. That is, the zeros of $H(z)$ can *suppress* certain forced-mode terms and prevent them from appearing in $y(k)$.

Similarly, with a judicious choice for $x(k)$, one can suppress natural-mode terms in $y(k)$. In particular, if $H(z)$ has a pole at $z = p$ and $X(z)$ has a zero at $z = p$, then the natural-mode term generated by the pole of $H(z)$ will not appear in the output $y(k)$. Finally, if a pole of $X(z)$ matches a pole of $H(z)$, then this effectively increases the multiplicity of the pole in $Y(z)$, a phenomenon known as harmonic forcing. We illustrate these ideas with the following example.

Example 3.18   **Cancelled Modes**

Consider the discrete-time system discussed in Example 3.16. The factored form of its transfer function is

$$H(z) = \frac{10(z + .6)}{(z - .8)(z - .4)}$$

Thus there will be up to two natural mode terms in $y(k)$. Next consider the following input signal.

$$x(k) = 10(-.6)^k \mu(k) - 4(-.6)^{k-1} \mu(k-1)$$

Using the properties of the Z-transform and Table 3.2, the Z-transform is

$$X(z) = 10 \left(\frac{z}{z+.6} \right) - 4z^{-1} \left(\frac{z}{z+.6} \right)$$

$$= \frac{10z - 4}{z + .6}$$

$$= \frac{10(z - .4)}{z + .6}$$

Consequently, there is potentially one forced-mode term in $y(k)$. From (3.5.9), the zero-state response is

$$y(k) = Z^{-1}\{H(z)X(z)\}$$

$$= Z^{-1}\left\{ \frac{100}{z - .8} \right\}$$

$$= y(0)\delta(k) + \text{Res}(.8, k)\mu(k-1)$$

$$= 100(.8)^{k-1}\mu(k-1)$$

For this particular input, neither the forced mode generated by the pole of $X(z)$ at $z = -.6$, nor the natural mode generated by the pole of $H(z)$ at $z = .4$, appears in the zero-state output $y(k)$ due to pole-zero cancellation.

Note that if we applied an input of the form $x(k) = .8^k \mu(k)$, then this would create a double pole in $Y(z)$ at $z = .8$. When a pole of $X(z)$ matches a pole of $H(z)$, this is referred to as *harmonic forcing*.

Harmonic forcing

Stable Modes

It is instructive to look at what happens to a multiple mode term in the limit as $k \to \infty$. A multiple mode term can be written as $c(k)p^k \mu(k)$ where $c(k)$ is a polynomial of degree $r - 1$ with r being the multiplicity of the pole. For $r > 1$, the polynomial factor will satisfy $|c(k)| \to \infty$ as $k \to \infty$. However, if $|p| < 1$, the exponential factor will satisfy $|p^k| \to 0$ as $k \to \infty$. To determine the value of the product $c(k)p^k$ in the limit as $k \to \infty$, first note that

$$\lim_{k \to \infty} |c(k)p^k| = \lim_{k \to \infty} \frac{|c(k)|}{|p|^{-k}}$$

$$= \lim_{k \to \infty} \frac{|c(k)|}{\exp[-k \log(|p|)]} \tag{3.5.14}$$

To evaluate the limit, L'Hospital's rule must be applied. The $(r-1)$th derivative of the polynomial $c(k)$ in (3.5.12) is the constant $(r-1)! \, c_{r-1}$. The $(r-1)$th derivative of the exponential is $[-\log(|p|)]^{r-1} \exp[-k \log(|p|)]$. Since $\log(|p|) < 0$, by L'Hospital's rule the limit is

$$\lim_{k \to \infty} c(k)p^k = 0 \quad \Leftrightarrow \quad |p| < 1 \tag{3.5.15}$$

Consequently, the exponential factor p^k goes to zero faster than the polynomial factor $c(z)$ goes to infinity if and only if $|p| < 1$. Note that (3.5.15) is the justification for the claim about *Stable modes* $y_{zi}(k)$ in (3.5.3). In summary, a *stable mode* is a mode that is associated with a pole that lies inside the unit circle.

3.5.4 DC Gain

When the poles of the system all lie strictly inside the unit circle, it is clear from (3.5.15) that all of the natural mode terms decay to zero with increasing time. In this case, it is *Step response* possible to determine the steady-state response of the system to a step input directly from $H(z)$. Suppose the input is a step of amplitude c. Then using (3.5.4) and the final value theorem

$$\lim_{k\to\infty} y(k) = \lim_{z\to 1}(z-1)Y(z)$$

$$= \lim_{z\to 1}(z-1)H(z)Z\{c\mu(k)\}$$

$$= \lim_{z\to 1}(z-1)H(z)\left(\frac{cz}{z-1}\right)$$

$$= H(1)c \tag{3.5.16}$$

Thus the steady-state response to a step of amplitude c is the constant $H(1)c$. A unit step input can be regarded as a cosine input $x(k) = \cos(2\pi F_0 k)\mu(k)$, whose frequency happens to be $F_0 = 0$ Hz. That is, a step or constant input is actually a DC input. The amount by which the amplitude of the DC input is scaled as it passes through the system to produce the steady-state *DC gain* output is called the *DC gain* of the system. From (3.5.16), it is evident that

$$\text{DC gain} = H(1) \tag{3.5.17}$$

Consequently, if the poles of the transfer function $H(z)$ are all strictly inside the unit circle, the DC gain of the system is simply $H(1)$. In Section 3.8, we will see that the gain of the system at other frequencies also can be obtained by evaluating $H(z)$ at appropriate values of z.

Example 3.19 **Comb Filter**

As an illustration of poles, zeros, and DC gain, consider the following transfer function where $0 < r < 1$.

$$H(z) = \frac{z^n}{z^n - r^n}$$

This is the transfer function of a *comb filter*, a system that is discussed in detail in Chapter 7. Note that $H(z)$ has n poles and n zeros. The zeros are all at the origin, while the n poles are spaced equally around a circle of radius r. A plot of $|H(z)|$ for the case $N = 6$ and $r = 0.9$ is shown in Figure 3.12. Note the placement of the six poles (the plot is clipped at $|H(z)| \le 2$) and the multiple zero at the origin. The DC gain of the comb filter is

$$\text{DC gain} = H(1) = \frac{1}{1 - r^n} = 2.134$$

FIGURE 3.12:
Magnitude of
Comb Filter
Transfer Function
with $n = 6$ and
$r = .9$

Magnitude of Transfer Function

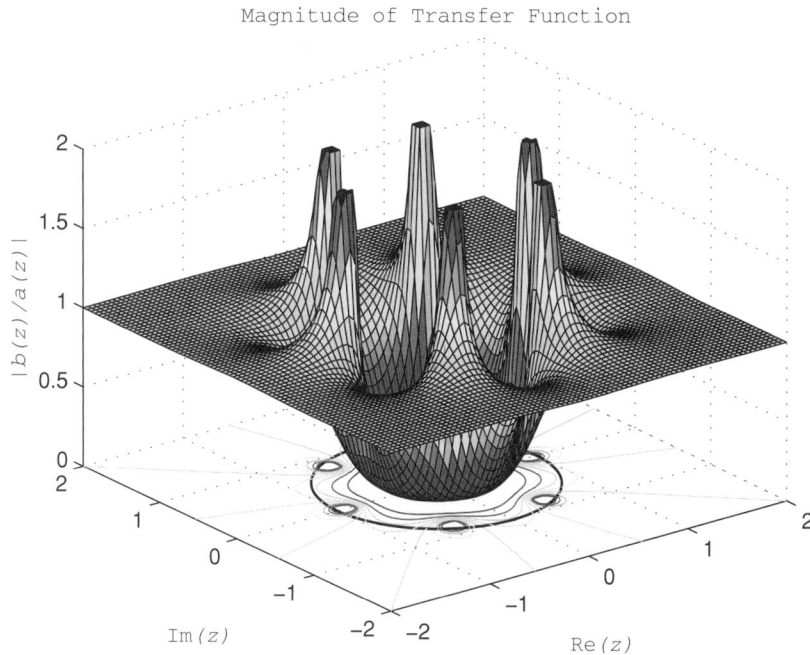

3.6 Signal Flow Graphs

Engineers often use diagrams to visualize relationships between variables and subsystems. In Chapter 2, this type of visualization was represented using block diagrams. Discrete-time systems also can be represented efficiently using a special type of diagram called a *signal flow graph*. A signal flow graph is a collection of *nodes* interconnected by *arcs*. Each node is represented with a dot, and each arc is represented with a directed line segment. An arc is an input arc if it enters a node and an output arc if it leaves a node. Each arc transmits or carries a signal in the direction indicated. When two or more input arcs enter a node, their signals are *added* together. That is, a node can be thought of as a summing junction with respect to its inputs. An output arc carries the value of the signal leaving the node. When an arc is labeled, the labeling indicates what type of operation is performed on the signal as it traverses the arc. For example, a signal might be scaled by a constant or, more generally, acted on by a transfer function. As a simple illustration, consider the signal flow graph shown in Figure 3.13 that consists of four nodes and three arcs.

Node, arc

Label

Observe that the node labeled x has two input arcs and one output arc. Thus the value of the signal leaving the node is $x = au + bv$. When an arc is not labeled, the default gain is

FIGURE 3.13: A
Simple Signal Flow
Graph

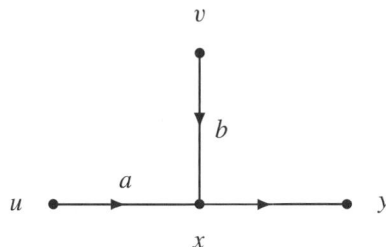

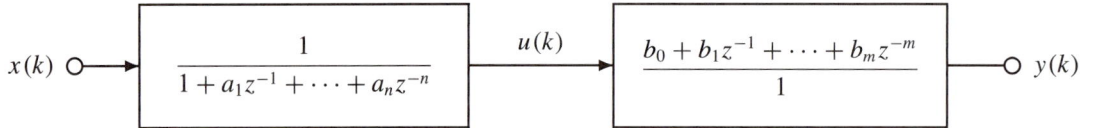

FIGURE 3.14: Decomposition of Transfer Function $H(z)$

assumed to be one. Therefore, the value of the signal at the output node is

$$y = x$$
$$= au + bv \tag{3.6.1}$$

In order to develop a signal flow graph for the discrete-time system S, we first factor the transfer function $H(z)$ into a product of two transfer functions as follows.

$$H(z) = \frac{b_0 + b_1 z^{-1} + \cdots b_m z^{-m}}{1 + a_1 z^{-1} + \cdots + a_n z^{-n}}$$
$$= \left(\frac{b_0 + b_1 z^{-1} + \cdots b_m z^{-m}}{1}\right)\left(\frac{1}{1 + a_1 z^{-1} + \cdots + a_n z^{-n}}\right)$$
$$= H_b(z)H_a(z) \tag{3.6.2}$$

Here $H(z)$ consists of two subsystems in series, one associated with the numerator polynomial, and the other with the denominator polynomial, as shown in Figure 3.14.

The intermediate variable $U(z) = H_a(z)X(z)$ is the output from the first subsystem and the input to the second subsystem.

Subsystem $H_a(z)$ processes input $x(k)$ to produce an intermediate variable $u(k)$, and then subsystem $H_b(z)$ acts on $u(k)$ to produce the output $y(k)$. The decomposition in Figure 3.14 can be represented by the following pair of difference equations.

$$u(k) = x(k) - \sum_{i=1}^{n} a_i u(k - i) \tag{3.6.3a}$$

$$y(k) = \sum_{i=0}^{m} b_i u(k - i) \tag{3.6.3b}$$

Given the decomposition in (3.6.3), the entire system can be represented with a signal flow graph, as shown in Figure 3.15 which illustrates the case $m = n = 3$.

FIGURE 3.15: Signal Flow Graph of a Discrete-time System, $m = n = 3$

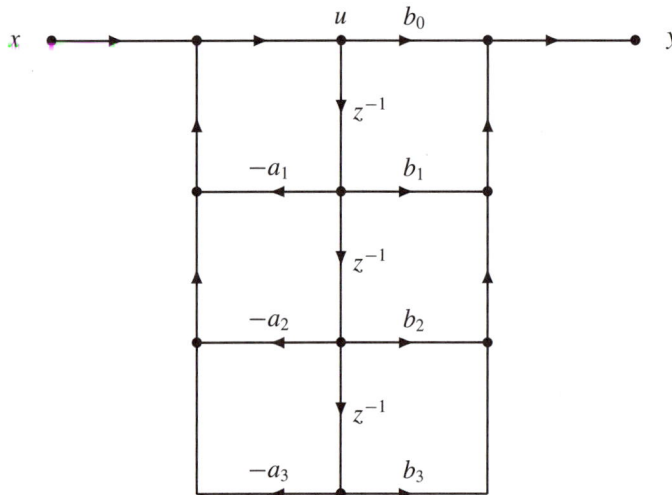

FIGURE 3.16: Signal Flow Graph of System in Example 3.20

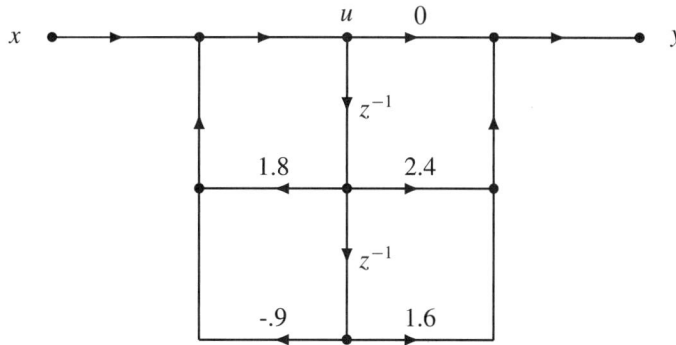

The output of the top center node is $u(k)$ and the nodes below it produce delayed versions of $u(k)$. The left-hand side of the signal flow graph ladder is a feedback system that implements (3.6.3a) to produce $u(k)$, while the right-hand side is a feed forward system that implements (3.6.3b) to produce $y(k)$. In the general case, the order of the signal flow graph will be $M = \max\{m, n\}$. If $m \neq n$, then some of the arcs will be missing or labeled with gains of zero.

Example 3.20

Signal Flow Graph

Consider a discrete-time system with the following transfer function.

$$H(z) = \frac{2.4z^{-1} + 1.6z^{-2}}{1 - 1.8z^{-1} + .9z^{-2}}$$

By inspection of Figure 3.15, it is evident that the signal flow graph of this system is as shown in Figure 3.16. Note how the gain of one of the arcs on the right-hand side is zero due to a missing term in the numerator. Given $H(z)$, the difference equation of this system is

$$y(k) = 2.4x(k-1) + 1.6x(k-2) + 1.8y(k-1) - .9y(k-2)$$

Note that we now have three distinct ways to represent a system. In the time domain there is the difference equation, in the frequency domain there is the transfer function, and in the graphical domain there is the signal flow graph, as shown in Figure 3.17. With some care, it should be possible to go directly from any one of these representations to another by inspection.

ARMA Models

AR model

The transfer function of the first subsystem in Figure 3.14 is a special case of the following structure.

$$H_{\text{AR}}(z) = \frac{b_0}{1 + a_1 z^{-1} + \cdots + a_n z^{-n}} \tag{3.6.4}$$

The system $H_{\text{AR}}(z)$, whose numerator is constant, is called an *auto-regressive* or AR model.

MA model

Similarly, the second subsystem in Figure 3.14 has the following structure.

$$H_{\text{MA}}(z) = b_0 + b_1 z^{-1} + \cdots + b_m z^{-m} \tag{3.6.5}$$

FIGURE 3.17: Alternative Ways to Represent a Discrete-time System

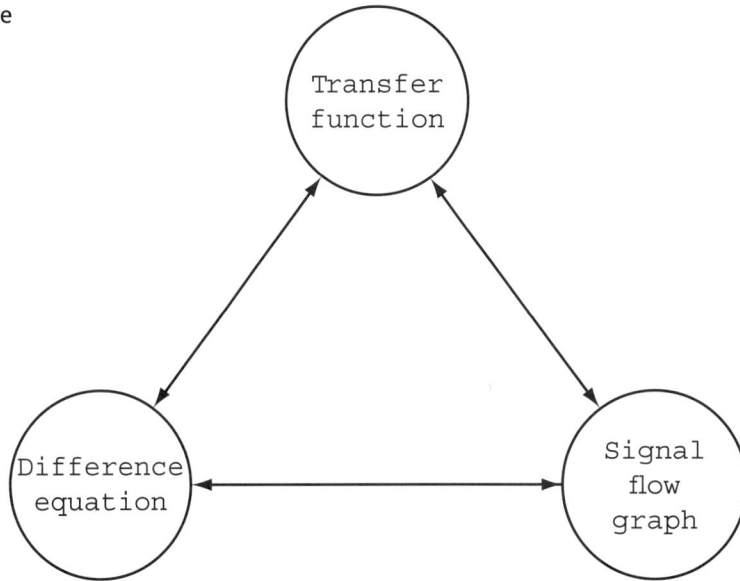

System $H_{\mathrm{MA}}(z)$, whose denominator is one, is called a *moving average* or MA model. The name "moving average" arises from the fact that if $b_k = 1/(m + 1)$, then the output is a running average of the last $m + 1$ samples. More generally, if the b_k are not equal, then $H_{\mathrm{MA}}(z)$ represents a weighted moving average of the last $m + 1$ samples.

ARMA model   The general case of a transfer function $H(z)$ for the system S in (3.6.2) is called an *autoregressive moving average* or ARMA model. From Figure 3.14, a general ARMA model can always be factored into a product of an AR model and a MA model.

$$H_{\mathrm{ARMA}}(z) = H_{\mathrm{AR}}(z)H_{\mathrm{MA}}(z) \tag{3.6.6}$$

For the signal flow graph representation in Figure 3.15, the left side of the ladder is the AR part and the right side of the ladder is the MA part. Note that the MA model in (3.6.5) is the FIR system discussed in Chapter 2. Both the AR model in (3.6.4) and the ARMA model in (3.6.2) are examples of IIR systems, with the AR model being a special case. An application where an AR model arises is the model of the human vocal tract introduced in Section 3.1.

• • • • • • • • • • • • • • • • • •

3.7   Stability in the Frequency Domain

3.7.1 Input-output Representations

Recall from Chapter 2 that the zero-state response of a system can be represented in the time domain as the convolution of the impulse response $h(k)$ with the input $x(k)$.

$$y(k) = h(k) \star x(k) \tag{3.7.1}$$

Input-output   This is sometimes referred to as the *input-output representation* of the system S in the time
representation   domain. An important property of the Z-transform is that convolution in the time domain maps into multiplication in the frequency domain. Taking the Z-transform of both sides of (3.7.1) leads to the input-output representation in the frequency domain.

$$Y(z) = H(z)X(z) \tag{3.7.2}$$

FIGURE 3.18: Two Input-output Representations of the Linear System S

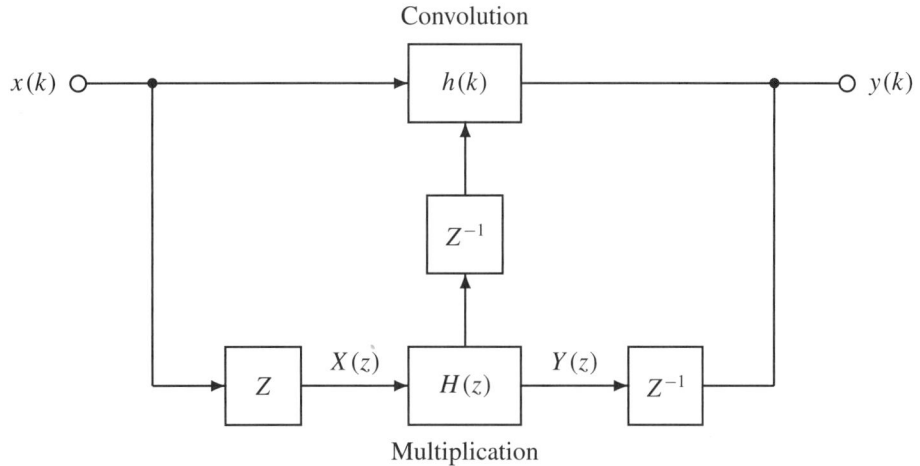

Suppose the input is the unit impulse $x(k) = \delta(k)$. Then in the time domain, the resulting output is the impulse response $y(k) = h(k)$. But in the frequency domain, $X(z) = 1$, so the resulting output in the frequency domain is $Y(z) = H(z)$. It follows that the transfer function is just the Z-transform of the impulse response or, putting it another way,

$$h(k) = Z^{-1}\{H(z)\} \tag{3.7.3}$$

That is, the Z-transform of the impulse response is the transfer function, and the inverse Z-transform of the transfer function is the impulse response. In effect, $h(k)$ and $H(z)$ are equivalent representations of the system S, one in the time domain and the other in the frequency domain. The relationship between $h(k)$ and $H(z)$ is summarized in Figure 3.18.

In view of (3.7.3), one way to find the inverse Z-transform of say $X(z)$ is to regard $X(z)$ as a transfer function of a linear time-invariant system. The inverse transform $x(k)$ is then just *Impulse response* the impulse response of this system. The impulse response can be computed numerically using *method* the MATLAB *filter* function. This is the basis of the impulse response method of numerically inverting the Z-transform presented in Section 3.4.

3.7.2 BIBO Stability

Practical discrete-time systems, particularly digital filters, tend to have one qualitative feature in common: they are stable. Recall from Chapter 2 that a system S is BIBO stable if and only if every bounded input is guaranteed to produce a bounded output. Proposition 2.1 provided a simple time-domain criterion for stability. A system is BIBO stable if and only if the impulse response is absolutely summable. There is an equivalent and even simpler test for stability in the frequency domain. Consider an input signal $x(k)$ and its Z-transform.

$$X(z) = \frac{d(z)}{(z - q_1)^{n_1}(z - q_2)^{n_2} \cdots (z - q_r)^{n_r}} \tag{3.7.4}$$

To determine under what conditions this signal is bounded, first note that each pole q_i of $X(z)$ generates a mode or term in $x(k)$ of form $c_i(k)q_i^k$ where $c_i(k)$ is a polynomial whose degree is one less that the multiplicity n_i of the pole. From (3.5.15), these modes go to zero as $k \to \infty$ if and only if the poles are strictly inside the unit circle. Therefore the terms of $x(k)$ associated with poles inside the unit circle are bounded. Any pole strictly outside the unit circle generates a mode that grows and therefore an unbounded $x(k)$.

Next consider poles on the unit circle. For simple poles on the unit circle the coefficient polynomial $c_i(k)$ is constant and the magnitude of the pole is $|q_i| = 1$, so the modes associated with these poles are also bounded although they do not go to zero. Finally, multiple poles on the unit circle have polynomial coefficients that grow with time whereas the exponential factor q_i^k does not go to zero, so these modes generate an unbounded $x(k)$. In summary, a signal is bounded if and only if its Z-transform has poles inside or on the unit circle with the poles on the unit circle being simple poles.

Next consider the output of the system. If the transfer function $H(z)$ has poles at p_i for $1 \le i \le n$, then $Y(z)$ can be expressed

$$Y(z) = \frac{b(z)d(z)}{(z - p_1)^{m_1} \cdots (z - p_s)^{m_s}(z - q_1)^{n_1} \cdots (z - q_r)^{n_r}} \tag{3.7.5}$$

Suppose $x(k)$ is bounded. Then $y(k)$ will be bounded if the poles of $H(z)$ are all strictly inside the unit circle. However, $y(k)$ will not be bounded for an arbitrary bounded $x(k)$ if $H(z)$ has a pole outside the unit circle or on the unit circle. The latter case can cause an unbounded $y(k)$ because even a simple pole of $H(z)$ on the unit circle could be matched by a simple pole of $X(z)$ at the same location, thus creating a double pole and an unstable mode. These observations lead to the following fundamental frequency-domain stability criterion.

PROPOSITION

3.1: BIBO Stability: Frequency Domain

A linear time-invariant system S with transfer function $H(z)$ is BIBO stable if and only if the poles of $H(z)$ satisfy

$$|p_i| < 1, \quad 1 \le i \le n$$

The poles of a stable system, whether simple or multiple, must all lie strictly inside the unit circle, as shown in Figure 3.19. Recall from (3.5.15) that this is equivalent to saying that all of the natural modes of the system must decay to zero.

FIGURE 3.19: Region of Stable Poles

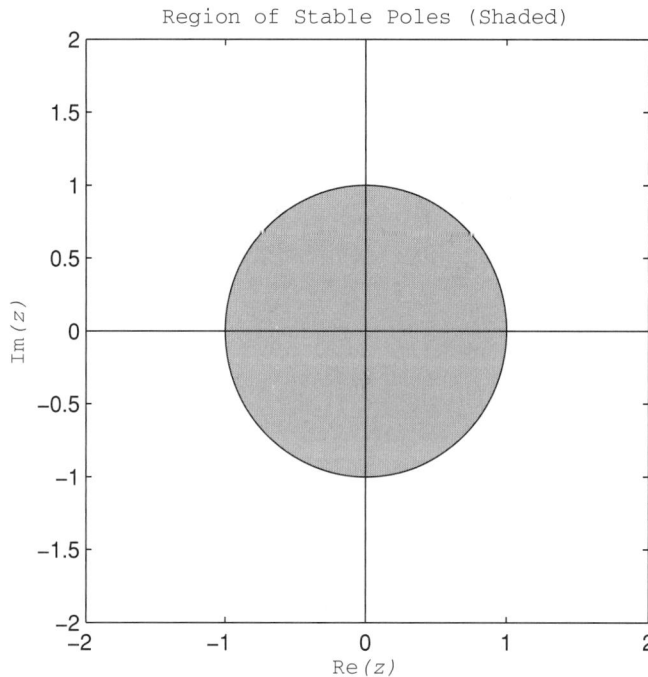

Region of Stable Poles (Shaded)

FIR Systems

From the time-domain analysis in Chapter 2 it was determined that there was one class of systems that was always stable: FIR systems.

$$y(k) = \sum_{i=0}^{m} b_i x(k - i) \tag{3.7.6}$$

When the unit impulse input is applied, this results in the following FIR impulse response.

$$h(k) = \sum_{i=0}^{m} b_i \delta(k - i) \tag{3.7.7}$$

FIR transfer function

From Proposition 2.1, a system is BIBO stable if and only if the impulse response $h(k)$ is absolutely summable. Since the impulse response of an FIR system is, by definition, finite, it is always absolutely summable. Therefore all FIR systems are BIBO stable. Applying the Z-transform to both sides of (3.7.7) and using the time shift property, we find the FIR transfer function is

$$
\begin{aligned}
H(z) &= \sum_{i=0}^{m} b_i z^{-i} \\
&= \frac{1}{z^m} \sum_{i=0}^{m} b_i z^{m-i}
\end{aligned} \tag{3.7.8}
$$

Here all the poles of $H(z)$ are well within the unit circle, namely, at $z = 0$. It follows from Proposition 3.1 that FIR systems are always BIBO stable.

IIR Systems

The stability of IIR systems, whether of the AR model type or the more general ARMA model type, may or may not be stable depending on the coefficients of the denominator polynomial.

Example 3.21

Unstable Transfer Function

Consider a discrete-time system with the following transfer function.

$$H(z) = \frac{10}{z + 1}$$

Marginally unstable

This system has a single pole on the unit circle at $z = -1$. Since it is not strictly inside, from Proposition 3.1 this is an example of an unstable system. When a discrete-time system has one or more simple poles on the unit circle, and the rest of its poles are inside the unit circle, it is sometimes referred to as a *marginally unstable* system. For a marginally unstable system there are natural modes that neither grow without bound nor decay to zero. From Algorithm 3.1, the impulse response of the system $H(z)$ is

$$
\begin{aligned}
h(k) &= h(0)\delta(k) + \text{Res}(-1, k)\mu(k - 1) \\
&= 10(-1)^{k-1}\mu(k - 1)
\end{aligned}
$$

Thus the pole at $z = -1$ produces a term or mode that oscillates between ± 1. Since this system is unstable, there must be at least one bounded input that produces an unbounded output. Consider the following input that is bounded by $B_x = 1$.

$$x(k) = (-1)^k \mu(k)$$

The Z-transform of the zero-state response is

$$Y(z) = H(z)X(z)$$
$$= \frac{10z}{(z+1)^2}$$

Applying Algorithm 3.1, we find the zero-state response is

$$y(k) = Z^{-1}\{Y(z)\}$$
$$= y(0)\delta(k) + \text{Res}(-1, k)\mu(k-1)$$
$$= 10k(-1)^{k-1}\mu(k-1)$$
$$= 10k(-1)^{k-1}\mu(k)$$

Harmonic forcing

Clearly, $y(k)$ is not bounded. Note that the forced mode of $X(z)$ at $z = -1$ exactly matches the marginally unstable natural mode of $H(z)$ at $z = -1$. This causes $Y(z)$ to have a double pole at $z = -1$, thereby generating a mode that grows. This phenomenon of driving a system with one of its natural modes is known has *harmonic forcing*. It tends to elicit a large response because the input reinforces the natural motion of the system.

3.7.3 The Jury Test

MATLAB functions

The stability criterion in Proposition 3.1 provides us with a simple and easy way to test for stability. In particular, if $a = [1, a_1, a_2, \ldots, a_n]^T$ denotes the $(n+1) \times 1$ coefficient vector of the denominator polynomial of $H(z)$, then the following MATLAB command returns the radius of the largest pole.

```
r = max(abs(roots(a)));              % radius of largest pole
```

In this case the system is stable if and only if $r < 1$, that is, if and only if the region of convergence of $H(z)$ includes the unit circle. There are instances when a system has one or more unspecified design parameters, and we want to determine a *range* of values for the parameters over which the system is stable. To see how this can be done, let $a(z)$ denote the denominator polynomial of the transfer function $H(z)$ where it is assumed, for convenience, that $a_0 > 0$.

$$a(z) = a_0 z^n + a_1 z^{n-1} + \cdots + a_n \tag{3.7.9}$$

Necessary stability conditions

The objective is to determine if $a(z)$ is a *stable* polynomial, a polynomial whose roots lie strictly inside the unit circle. There are two simple necessary conditions that a stable polynomial must satisfy, namely

$$a(1) > 0 \tag{3.7.10a}$$
$$(-1)^n a(-1) > 0 \tag{3.7.10b}$$

If the polynomial in question violates either (3.7.10a) or (3.7.10b), then there is no need to test further because it is known to be *unstable*.

Jury table

Suppose $a(z)$ passes the necessary conditions in (3.7.10), which means that it is a viable candidate for a stable polynomial. We then construct a table from the coefficients of $a(z)$ known as a *Jury table*. The case $n = 3$ is shown in Table 3.4.

Observe that the first two rows of the Jury table are obtained directly from inspection of the coefficients of $a(z)$. The rows appear in pairs, with the even rows being reversed versions of the odd rows immediately above them. Starting with row three, the odd rows are constructed

TABLE 3.4: ▶
The Jury Table,
$n = 3$

Row	Coefficients			
1	a_0	a_1	a_2	a_3
2	a_3	a_2	a_1	a_0
3	b_0	b_1	b_2	
4	b_2	b_1	b_0	
5	c_0	c_1		
6	c_1	c_0		
7	d_0			
8	d_0			

using 2×2 determinants as follows.

$$b_0 = \frac{1}{a_0} \begin{vmatrix} a_0 & a_3 \\ a_3 & a_0 \end{vmatrix}, \quad b_1 = \frac{1}{a_0} \begin{vmatrix} a_0 & a_2 \\ a_3 & a_1 \end{vmatrix}, \quad b_2 = \frac{1}{a_0} \begin{vmatrix} a_0 & a_1 \\ a_3 & a_2 \end{vmatrix} \tag{3.7.11a}$$

$$c_0 = \frac{1}{b_0} \begin{vmatrix} b_0 & b_2 \\ b_2 & b_0 \end{vmatrix}, \quad c_1 = \frac{1}{b_0} \begin{vmatrix} b_0 & b_1 \\ b_2 & b_1 \end{vmatrix} \tag{3.7.11b}$$

$$d_0 = \frac{1}{c_0} \begin{vmatrix} c_0 & c_1 \\ c_1 & c_0 \end{vmatrix} \tag{3.7.11c}$$

As a matter of convenience, any odd row can be scaled by a positive value. Once the Jury table is constructed, it is a simple matter to determine if the polynomial is stable. The polynomial *Jury test* $a(z)$ is stable if the first entry in each *odd* row of the Jury table is positive.

$$a_0 > 0, \quad b_0 > 0, \quad c_0 > 0, \quad \cdots \tag{3.7.12}$$

Example 3.22   **Jury Test**

Consider the following general second-order discrete-time system with design parameters a_1 and a_2. Suppose there is no pole-zero cancellation between $a(z)$ and $b(z)$.

$$H(z) = \frac{b(z)}{z^2 + a_1 z + a_2}$$

For this system

$$a(z) = z^2 + a_1 z + a_2$$

The first two rows of the Jury table are

$$J_2 = \begin{bmatrix} 1 & a_1 & a_2 \\ a_2 & a_1 & 1 \end{bmatrix}$$

If we apply (3.7.12), the elements in the third row are

$$b_0 = \begin{vmatrix} 1 & a_2 \\ a_2 & 1 \end{vmatrix} = 1 - a_2^2$$

$$b_1 = \begin{vmatrix} 1 & a_1 \\ a_2 & a_1 \end{vmatrix} = a_1(1 - a_2)$$

From the stability condition $b_0 > 0$, we have

$$|a_2| < 1$$

Given b_0 and b_1, the first four rows of the Jury table are

$$J_4 = \begin{bmatrix} 1 & a_1 & a_2 \\ a_2 & a_1 & 1 \\ 1 - a_2^2 & a_1(1 - a_2) & \\ a_1(1 - a_2) & a - a_2^2 & \end{bmatrix}$$

From (3.7.12), the element in the fifth row of the Jury table is

$$c_0 = \frac{1}{1 - a_2^2} \begin{vmatrix} 1 - a_2^2 & a_1(1 - a_2) \\ a_1(1 - a_2) & 1 - a_2^2 \end{vmatrix}$$

$$= \frac{(1 - a_2^2)^2 - a_1^2(1 - a_2)^2}{1 - a_2^2}$$

$$= \frac{(1 - a_2)^2[(1 + a_2)^2 - a_1^2]}{1 - a_2^2}$$

Since $|a_2| < 1$, the denominator and the first factor in the numerator are both positive. Thus the condition $c_0 > 0$ reduces to

$$(1 + a_2)^2 > a_1^2$$

We can break this inequality into two cases. If $a_1 \geq 0$, then $(1 + a_2) > a_1$ or

$$a_2 > a_1 - 1, \quad a_1 \geq 0$$

Here in the a_2 versus a_1 plane, a_2 must be above a line of slope 1 and intercept -1 when $a_1 \geq 0$. If $a_1 < 0$, then $1 + a_2 > -a_1$ or

$$a_2 > -a_1 - 1, \quad a_1 < 0$$

Here a_2 must be above a line of slope -1 and intercept -1 when $a_1 < 0$. Adding the constraint $|a_2| < 1$, this generates the stable region in parameter space shown in Figure 3.20. As long as the two coefficients lie in the shaded region known as the *stability triangle*, the second-order system will be BIBO stable.

Stability triangle

FIGURE 3.20: Stable Parameter Region for a Second-order System

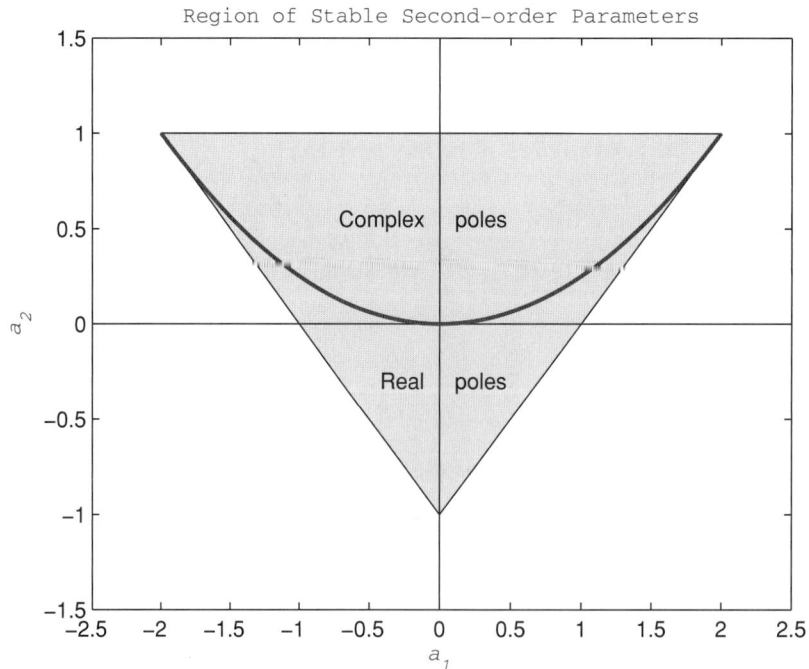

Region of Stable Second-order Parameters

The parabolic line in Figure 3.20 that separates real poles from complex poles is obtained by factoring the second-order denominator polynomial $a(z) = z^2 + a_1 z + a_2$, which yields poles at

$$p_{1,2} = \frac{-a_1 \pm \sqrt{a_1^2 - 4a_2}}{2} \qquad (3.7.13)$$

Complex poles   When $4a_2 > a_1^2$, the poles go from real to complex. Thus the parabola $a_2 = a_1^2/4$ separates the two regions in Figure 3.20. Recall that stable real poles generate natural modes in the form of damped exponentials. A stable complex conjugate pair of poles generates a natural mode that is an exponentially damped sinusoid. Although the stability result in Figure 3.20 applies only to second-order systems, it is quite useful because a practical way to implement a higher-order filter with transfer function $H(z)$ is as a product of second-order factors $H_i(z)$ and, if needed, one first-order factor.

$$H(z) = H_1(z)H_2(z) \cdots H_r(z) \qquad (3.7.14)$$

Cascade form   This is called the *cascade form*, and it is discussed in Chapter 7. This type of implementation has practical value because as the filter order increases, the locations of the poles can become highly sensitive to small errors in the polynomial coefficients. As a result, a direct form implementation of a high-order IIR filter can sometimes become unstable due to finite precision effects, whereas a cascade form implementation may remain stable.

3.8 Frequency Response

3.8.1 Frequency Response

The spectral characteristics or frequency content of a signal $x(k)$ can be modified in a desired manner by passing $x(k)$ through a linear discrete-time system to produce a second signal $y(k)$, as shown in Figure 3.21. In this case we refer to the system that processes $x(k)$ to produce $y(k)$ as a digital filter.

Typically, signal $x(k)$ will contain significant power at certain frequencies and less power or no power at other frequencies. The distribution of power across frequencies is called the power density spectrum. A digital filter is designed to *reshape* the spectrum of the input by removing certain spectral components and enhancing others. The manner in which the spectrum of $x(k)$ is reshaped is specified by the frequency response of the filter.

DEFINITION
3.3: Frequency Response

Let $H(z)$ be the transfer function of a stable linear system S, and let T be the sampling interval. The *frequency response* of the system is denoted $H(f)$ and defined as

$$H(f) \triangleq H(z)|_{z=\exp(j2\pi fT)}, \qquad 0 \le |f| \le f_s/2$$

FIGURE 3.21: Digital Filter with Transfer Function $H(z)$

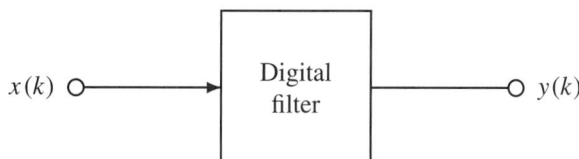

$x(k)$ ○——→ | Digital filter | ——○ $y(k)$

FIGURE 3.22:
Evaluation of the
Frequency
Response along
$|z| = 1$. The Exterior
Labels Denote
Fractions of f/f_s

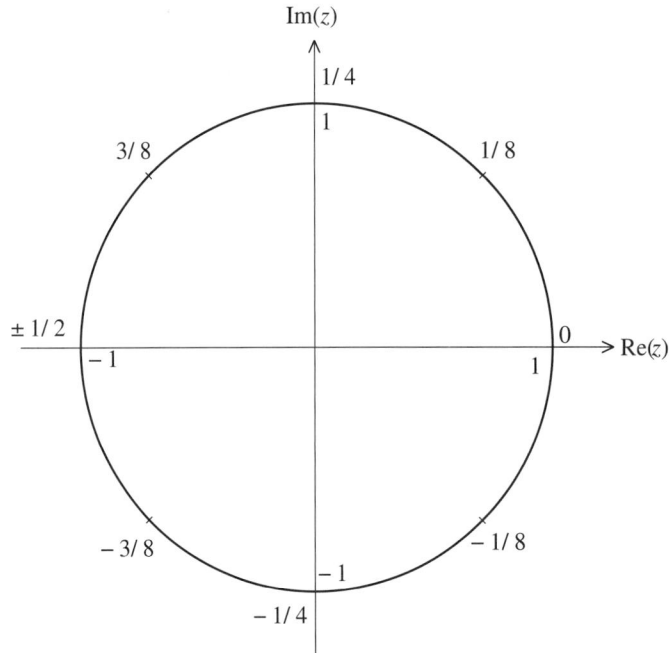

As the *frequency* f ranges over the interval $[-f_s/2,\ f_s/2]$, the argument $z = \exp(j2\pi f T)$ traces out the unit circle in a counter clockwise sense, as shown in Figure 3.22. In other words, the frequency response is just the transfer function $H(z)$ evaluated along the unit circle. Recall that the upper frequency limit $f_s/2$ represents a bound on frequencies that can be present in $x_a(t)$ without aliasing. In Chapter 1 we referred to $f_s/2$ as the folding frequency. For $H(f)$ to be well defined, the region of convergence of $H(z)$ must include the unit circle. This will be the case as long as the system S is BIBO stable.

A comment about notation is in order. Note that the same base symbol H is being used to denote both the transfer function $H(z)$ and the frequency response $H(f)$, two quantities that are distinct but related. An alternative approach would be to introduce separate symbols for each. However, the need for separate symbols will arise repeatedly, and using distinct symbols in each instance quickly leads to a proliferation of symbols that can become confusing in its own right. Instead, we will adopt the convention that the *argument type*, a complex z or a real f, will be used to dictate the meaning of H. Some authors use the notation $H(e^{j\omega})$ for the frequency response.

The frequency response is defined over the range $[-f_s/2,\ f_s/2]$, where negative frequencies correspond to the bottom half of the unit circle, and positive frequencies the top half. However, if $H(z)$ is generated by a difference equation with real coefficients, then all of the information about $H(f)$ is contained in the positive frequency range $[0,\ f_s/2]$. More specifically, if the coefficients of $H(z)$ are real, then the frequency response satisfies the following *symmetry* property where $H^*(f)$ denotes the *complex-conjugate* of $H(f)$.

Symmetry property

$$H(-f) = H^*(f), \quad 0 \le |f| \le f_s/2 \tag{3.8.1}$$

Magnitude response Since $H(f)$ is complex, it can be written in polar form as $H(f) = A(f)\exp[\phi(f)]$. The magnitude of $H(f)$ is called the *magnitude response* of the filter.

$$A(f) \overset{\Delta}{=} \sqrt{\text{Re}^2[H(f)] + \text{Im}^2[H(f)]}, \quad 0 \le |f| \le f_s/2 \tag{3.8.2}$$

Applying the symmetry property in (3.8.1) yields

$$\begin{aligned} A(-f) &= |H(-f)| \\ &= |H^*(f)| \\ &= |H(f)| \\ &= A(f) \end{aligned} \tag{3.8.3}$$

Phase response Consequently, the magnitude response of a real filter is an *even* function of f with $A(-f) = A(f)$. The phase angle $\phi(f)$ is called the *phase response* of the filter.

$$\phi(f) \overset{\Delta}{=} \tan^{-1}\left\{ \frac{\text{Im}[H(f)]}{\text{Re}[H(f)]} \right\}, \quad 0 \le |f| \le f_s/2 \tag{3.8.4}$$

In this case, applying the symmetry property in (3.8.1) yields

$$\begin{aligned} \phi(-f) &= \angle H(-f) \\ &= \angle H^*(f) \\ &= -\angle H(f) \\ &= -\phi(f) \end{aligned} \tag{3.8.5}$$

Consequently, the phase response of a real filter is an *odd* function of f with $\phi(-f) = -\phi(f)$.

3.8.2 Sinusoidal Inputs

There is a simple physical interpretation of the frequency response in terms of inputs and outputs. To see this it is helpful to first examine the steady-state response of the system S to the following complex sinusoidal input.

$$\begin{aligned} x(k) &= [\cos(2\pi fkT) + j\sin(2\pi kfT)]\mu(k) \\ &= \exp(j2\pi ktT)\mu(k) \\ &= [\exp(j2\pi fT]^k\mu(k) \\ &= p^k\mu(k) \end{aligned} \tag{3.8.6}$$

Thus the complex sinusoid is really a causal exponential input with a complex exponential factor $p = \exp(j2\pi fT)$. If the system S is stable, then all of the natural modes will go to zero as $k \to \infty$. Hence the *steady-state* response will consist of the forced mode associated with the pole of $X(z)$ at $p = \exp(j2\pi fT)$. If we apply Algorithm 3.1, the steady-state response is

$$y_{ss}(k) = \text{Res}(p, k) \tag{3.8.7}$$

The Z-transform of $x(k)$ is $X(z) = z/(z - p)$. Since $H(z)$ is stable and $|p| = 1$, the pole at $z = p$ is a simple pole and the residue is

$$\begin{aligned} \text{Res}(p, k) &= (z - p)Y(z)z^{k-1}\big|_{z=p} \\ &= (z - p)H(z)X(z)z^{k-1}\big|_{z=p} \\ &= H(z)z^k\big|_{z=p} \\ &= H(p)p^k \end{aligned} \tag{3.8.8}$$

Substituting (3.8.8) into (3.8.7) with $p = \exp(2\pi fT)$, and using the polar form of $H(f)$, we have

$$
\begin{aligned}
y_{ss}(k) &= H(p)p^k \\
&= H(f)\exp(j2\pi kfT) \\
&= A(f)\exp[j\phi(f)]\exp(j2\pi kfT) \\
&= A(f)\exp[j2\pi kfT + \phi(f)]
\end{aligned}
\tag{3.8.9}
$$

Consequently, when a complex sinusoidal input of frequency f is applied to a stable system S, the steady-state response is also a complex sinusoidal input of frequency f, but its amplitude has been scaled by $A(f)$ and its phase angle has been shifted by $\phi(f)$. The steady-state response to a real sinusoidal input behaves in a similar manner. Suppose

$$
\begin{aligned}
x(k) &= \cos(2\pi kfT) \\
&= .5[\exp(j2\pi kfT) + \exp(-j2\pi fT)]
\end{aligned}
\tag{3.8.10}
$$

Since the system S is linear, if we use the identities in Appendix 2, the steady-state response to $x(k)$ is

$$
\begin{aligned}
y_{ss}(k) &= .5[H(f)\exp(j2\pi kfT) + H(-f)\exp(-j2\pi kfT)] \\
&= .5\{H(f)\exp(j2\pi kfT) + [H(f)\exp(j2\pi fT)]^*\} \\
&= \mathrm{Re}\{H(f)\exp(j2\pi kfT)\} \\
&= \mathrm{Re}\{A(f)\exp[j2\pi kfT + \phi(f)]\} \\
&= A(f)\cos[2\pi kfT + \phi(f)]
\end{aligned}
\tag{3.8.11}
$$

Thus a real sinusoidal input behaves in the same way as a complex sinusoidal input when it is processed by the system S.

PROPOSITION
....
3.2: Frequency Response

Let $H(f) = A(f)\exp[j\phi(f)]$ be the frequency response of a linear system S, and let $x(k) = \cos(2\pi kfT)$ be a sinusoidal input where T is the sampling interval and $0 \le |f| \le f_s/2$. The steady state response to $x(k)$ is

$$
y_{ss}(k) = A(f)\cos[2\pi fkT + \phi(f)]
$$

In view of Proposition 3.2, the magnitude response and the phase response have simple interpretations which allow them to be measured directly. The magnitude response $A(f)$

Gain

Phase shift

specifies the *gain* of the system at frequency f, the amount by which the sinusoidal input $x(k)$ is amplified or attenuated. The phase response $\phi(f)$ specifies the *phase shift* of the system at frequency f, the number of radians by which the sinusoidal input $x(k)$ is advanced if positive or delayed if negative.

By designing a filter with a prescribed magnitude response, certain frequencies can be removed, and others enhanced. The design of digital FIR filters is the focus of Chapter 6, and the design of digital IIR filters is the focus of Chapter 7.

Example 3.23   **Frequency Response**

As an example of computing the frequency response of a discrete-time system, consider a second-order digital filter with the following transfer function.

$$
H(z) = \frac{z + 1}{z^2 + .64}
$$

Let $\theta = 2\pi f T$. Then from Definition 3.3 and Euler's identity, the frequency response is

$$H(f) = \left(\frac{z+1}{z^2 - .64}\right)\Bigg|_{z=\exp(j\theta)}$$

$$= \frac{\exp(j\theta) + 1}{\exp(j2\theta) - .64}$$

$$= \frac{\cos(\theta) + 1 + j\sin(\theta)}{\cos(2\theta) - .64 + j\sin(2\theta)}$$

The magnitude response of the filter is

$$A(f) = |H(f)|$$

$$= \frac{\sqrt{[\cos(2\pi f T) + 1]^2 + \sin^2(2\pi f T)}}{\sqrt{[\cos(4\pi f T) - .64]^2 + \sin^2(4\pi f T)}}$$

The phase response of the filter is

$$\phi(f) = \angle H(f)$$

$$= \tan^{-1}\left[\frac{\sin(2\pi f T)}{\cos(2\pi f T) + 1}\right] - \tan^{-1}\left[\frac{\sin(4\pi f T)}{\cos(4\pi f T) - .64}\right]$$

Plots of the magnitude response and the phase response for the case $f_s = 2000$ Hz are shown in Figure 3.23. Only the positive frequencies are plotted because S is a real system. Notice that frequencies near $f = 500$ Hz are enhanced by the filter, whereas the frequency component at $f = 1000$ Hz is eliminated.

FIGURE 3.23:
Frequency
Response of Filter

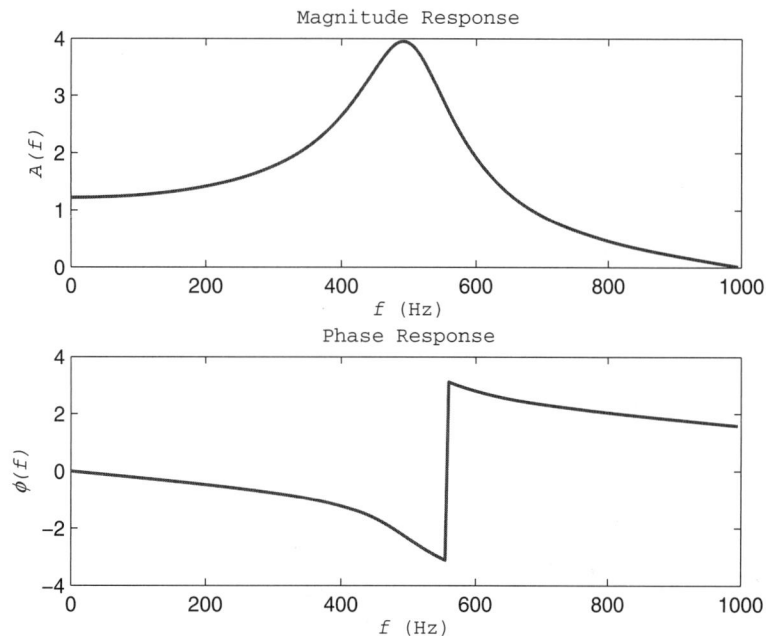

3.8.3 Periodic Inputs

The expression for the steady-state output in Proposition 3.2 can be generalized to periodic inputs. Suppose a continuous-time signal $x_a(t)$ is periodic with period T_0 and fundamental frequency $F_0 = 1/T_0$. The signal $x_a(t)$ can be approximated with a truncated Fourier series as follows.

$$x_a(t) \approx \frac{d_0}{2} + \sum_{i=1}^{M} d_i \cos(2\pi i F_0 t + \theta_i) \tag{3.8.12}$$

Let c_i denote the ith complex Fourier coefficient of $x_a(t)$. That is,

$$c_i = \frac{1}{T_0} \int_{-T_0/2}^{T_0/2} x_a(t) \exp(-j2\pi i F_0 t) dt, \quad i \geq 0 \tag{3.8.13}$$

From Appendix 1, the magnitude d_i and phase angle θ_i of the ith harmonic of $x_a(t)$ can then be obtained from c_i as $d_i = 2|c_i|$ and $\theta_i = \angle c_i$, respectively. Next let the samples $x(k) = x_a(kT)$ be a discrete-time input to a stable linear system with transfer function $H(z)$.

$$x(k) = \frac{d_0}{2} + \sum_{i=1}^{M} d_i \cos(2\pi i F_0 kT + \theta_i) \tag{3.8.14}$$

Since the system $H(z)$ is linear, it follows that the steady-state response to a sum of inputs is just the sum of the steady-state responses to the individual inputs. Setting $f = i F_0$ in Proposition 3.2, we find that the steady-state response to $\cos(2\pi i F_0 kT + \theta_i)$ is scaled by $A(i F_0)$ and shifted in phase by $\phi(i F_0)$. Thus the steady-state response to the sampled periodic input $x(k)$ is

$$y_{ss}(k) = \frac{A(0)d_0}{2} + \sum_{i=1}^{M} A(i F_0)d_i \cos[2\pi i F_0 kT + \theta_i + \phi(i F_0)] \tag{3.8.15}$$

It should be pointed out that there is a practical upper limit on the number of harmonics M. The sampling process will introduce aliasing if there are harmonics located at or above the folding frequency $f_s/2$. Therefore, for (3.8.15) to be valid, the number of harmonics must satisfy

$$M < \frac{f_s}{2F_0} \tag{3.8.16}$$

Example 3.24

Steady-state Response

As an illustration of using (3.8.15) to compute the steady-state response, consider the following stable first-order filter.

$$H(z) = \frac{.2z}{z - .8}$$

Let $\theta = 2\pi f T$. Then from Definition 3.3 and Euler's identity, the frequency response of this filter is

$$H(f) = \frac{.2z}{z - .8}\bigg|_{z=\exp(j\theta)}$$

$$= \frac{.2\exp(j\theta)}{\exp(j\theta) - .8}$$

$$= \frac{.2\exp(j\theta)}{\cos(\theta) - .8 + j\sin(\theta)}$$

The filter magnitude response is

$$A(f) = |H(f)|$$

$$= \frac{.2}{\sqrt{[\cos(2\pi f T) - .8]^2 + \sin^2(2\pi f T)}}$$

The filter phase response is

$$\phi(f) = \angle H(f)$$

$$= 2\pi f T - \tan^{-1}\left[\frac{\sin(2\pi f T)}{\cos(2\pi f T) - .8}\right]$$

Next suppose the input $x_a(t)$ is an even periodic pulse train of period T_0 where the pulses are of unit amplitude and radius $0 \le a \le T_0/2$. From the Fourier series table in Appendix 1, the truncated Fourier series of $x_a(t)$ is

$$x_a(t) \approx \frac{2a}{T_0} + \frac{4a}{T_0}\sum_{i=1}^{M}\text{sinc}(2i\,F_0 a)\cos(2\pi i\,F_0 t)$$

Recall from Chapter 1 that $\text{sinc}(x) = \sin(\pi x)/(\pi x)$. Suppose the period is $T_0 = .01$ sec and $a = T_0/4$, which corresponds to a square wave with fundamental frequency $F_0 = 100$ Hz. If we sample at $f_s = 2000$ Hz, then to avoid aliasing, the M harmonics must be below 1000 Hz. Thus from (3.8.16), the maximum number of harmonics used to approximate $x_a(t)$ should be $M = 9$. The sampled version of $x_a(t)$ is then

$$x(k) = \frac{1}{2} + \sum_{i=1}^{9}\text{sinc}\left(\frac{i}{2}\right)\cos(.1\pi i k)$$

Finally, from (3.8.13), the steady-state response to the periodic input $x(k)$ is

$$y_{ss}(k) = \frac{1}{2} + \sum_{i=1}^{9} A(100i)\text{sinc}\left(\frac{i}{2}\right)\cos[.1\pi i k + \phi(100i)]$$

Plots of $x(k)$ and $y_{ss}(k)$, obtained by running *exam3_25* from the driver program *f_dsp*, are shown in Figure 3.24. The oscillations or ringing in the approximation to the square wave are caused by the fact that only a finite number of harmonics are used (Gibb's phenomenon). The steady-state output is much smoother than $x(k)$ due to the low-pass nature of the filter $H(z)$.

FIGURE 3.24:
Steady-state
Response to
Periodic Pulse Train

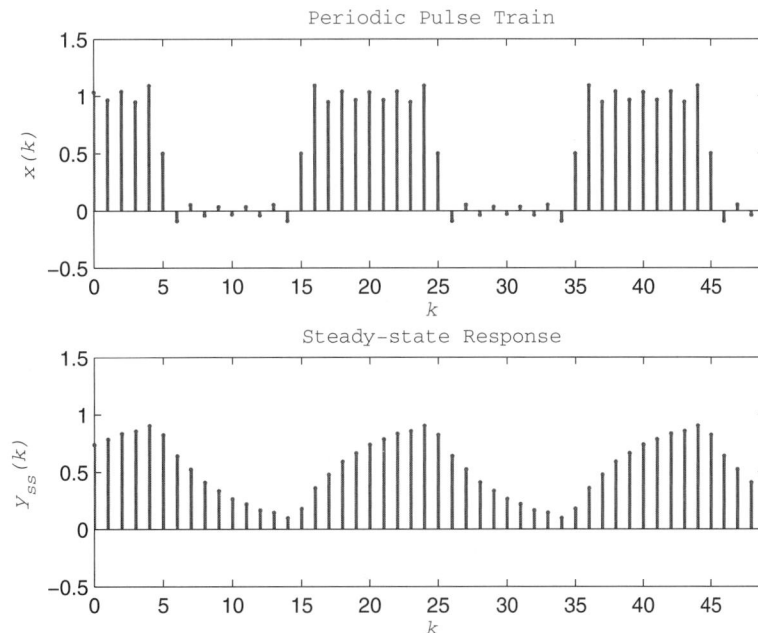

FDSP Functions

The FDSP toolbox contains the following function for finding the frequency response of a stable linear discrete-time system.

```
%F_FREQZ: Compute the frequency response of a discrete-time system
%
% Usage:
%        [H,f] =f_freqz (b,a,N,fs);
% Pre:
%        b       = numerator polynomial coefficient vector
%        a       = denominator polynomial coefficient vector
%        N       = frequency precision factor: df = fs/N
%        fs      = sampling frequency (default = 1)
% Post:
%        H = 1 by N+1 complex vector containing discrete
%            frequency response
%        f = 1 by N+1 vector containing discrete
%            frequencies at which H is evaluated
% Notes:
%        1. The frequency response is evaluated  along the
%           top half of the unit circle.  Thus f ranges
%           from 0 to fs/2.
%
%        2. H(z) must be stable.  Thus the roots of a(z) must
%           lie inside the unit circle.
```

Note that f_freqz is the discrete-time version of the continuous-time function f_freqs introduced in Chapter 1. To compute and plot the magnitude response and phase response, the following standard MATLAB functions can be used.

```
A = abs(H);                              % magnitude response
subplot(2,1,1)                           % place above
plot (f,A)                               % magnitude response plot
phi = angle(H);                          % phase response
subplot(2,1,2)                           % place below
plot (f,phi)                             % phase response plot
```

● ● ● ● ● ● ● ● ● ● ● ● ● ● ● ● ⋯

3.9   System Identification

Sometimes the desired input-output behavior of a system is known, but the parameters of the transfer function, difference equation, or signal flow graph are not. For example, a segment of speech can be measured as in Section 3.1, but the model of the system that produced that speech segment is unknown. The problem of identifying a transfer function $H(z)$ from an input

System identification   signal $x(k)$ and an output signal $y(k)$ is referred to as *system identification*. Here the system S can be thought of as a *black box*, as shown in Figure 3.25. The box is *black* in the sense that it is not possible to see, or know with certainty, the details of what is inside.

An effective way to identify a suitable transfer function is to first assume a particular form or structure for the black box model. For example, one could use an MA model, an AR model,

FIGURE 3.25: A Black Box System to be Identified

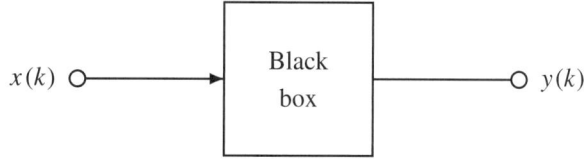

$x(k)$ ——→ Black box ——○ $y(k)$

or an ARMA model. Once a model is selected, the dimension or order of the model must be determined. If the selected structure is sufficiently general for the problem at hand and the dimension is sufficiently large, then it should be possible to *fit* the model to the data.

3.9.1 Least-squares Fit

For this introduction to system identification, we will employ an auto-regressive or AR structure because this is sufficiently general to allow for an infinite impulse response; yet it is mathematically simpler than the ARMA model when it comes to finding optimal parameter values. In Chapter 9, the question of system identification is revisited, this time in the context of adaptive MA models. An AR model has a transfer function that can be expressed as follows.

$$H(z) = \frac{1}{a_0 + a_1 z^{-1} + \cdots + a_n z^{-n}} \tag{3.9.1}$$

Notice that in this version of an AR transfer function, the numerator is unity, rather than b_0, and the denominator has not been normalized to $a_0 = 1$. This is done because it simplifies the problem of identifying the coefficient vector $a \in R^{n+1}$. Of course, if the numerator and denominator of (3.9.1) are divided by a_0, then this results in the standardized AR form with $b_0 = 1/a_0$. The difference equation associated with $H(z)$ is

$$\sum_{i=0}^{n} a_i y(k-i) = x(k) \tag{3.9.2}$$

System identification is based on sets of input-output measurements. Usually these measurements are real, but in some cases they can represent the desired behavior of a fictitious system. Suppose the following input-output data are available.

$$D = \{[x(k), y(k)] \in R^2 \mid 0 \le k < N\} \tag{3.9.3}$$

Given the data in D, the AR model in (3.9.2) will precisely fit the data if

$$\sum_{i=0}^{n} a_i y(k-i) = x(k), \quad 0 \le k < N \tag{3.9.4}$$

Let $p = n + 1$ denote the number of parameters. The N equations in (3.9.4) can be written in matrix form. Define an $N \times p$ coefficient matrix Y, and an $N \times 1$ column vector x as follows.

$$Y = \begin{bmatrix} y(0) & y(-1) & \cdots & y(-n) \\ y(1) & y(0) & \cdots & y(1-n) \\ \vdots & \vdots & & \vdots \\ y(N-1) & y(N-2) & \cdots & y(N-1-n) \end{bmatrix}, \quad x = \begin{bmatrix} x(0) \\ x(1) \\ \vdots \\ x(N-1) \end{bmatrix} \tag{3.9.5}$$

Notice that the ith column of Y consists of N samples of y starting with sample $y(-i)$. Elements of Y that correspond to negative time are associated with initial conditions of the system, and they can be taken to be zero. Given the definitions of x and Y, the N input-output constraints in (3.9.4) can be expressed in compact matrix form as

$$Ya = x \tag{3.9.6}$$

When the number of measurements matches the AR order $N = p$, the matrix Y is square. If the square matrix Y is nonsingular, then the coefficient vector a is unique and can be solved for as $a = Y^{-1}x$. Although this will cause the AR model to fit the input-output data exactly, it is typically not a practical solution because it is valid only for a limited number of samples. Experience shows that it is unlikely that the fit will be acceptable for $k > N$.

A better approach is to use a long recording of input-output data ($N \gg p$) that is more representative of the overall behavior of the unknown system. When $N > p$, the system of N equations in the p unknowns becomes an *over-determined* linear algebraic system, a system that does not in general have a solution. In this case, there will be a nonzero *residual error* vector that represents the difference between Ya and x.

Residual error

$$r(a) \triangleq Ya - x \tag{3.9.7}$$

The residual error vector $r(a) = 0$ if and only if a provides an exact fit to the input-output data D. Since an over-determined system does not in general have an exact solution because there are more constraints than unknowns, the next best thing is to find an $a \in R^p$ that will minimize the size of the residual error vector, that is, the square of the Euclidean norm. This leads to the *least-squares* error criterion.

Least squares

$$E_N = \sum_{i=0}^{N-1} r_i^2(a) \tag{3.9.8}$$

To minimize the error E_n, recall that the $p \times N$ matrix Y^T denotes the transpose of Y. Multiplying both sides of (3.9.6) on the left by Y^T yields

$$Y^T Ya = Y^T x \tag{3.9.9}$$

Here $Y^T Y$ is a smaller square matrix of dimension p. Suppose the output samples in D are such that the p columns of the coefficient matrix Y are linearly independent, which means that Y is of *full rank*. Then the square matrix $Y^T Y$ will be nonsingular. Multiplying both sides of (3.9.9) on the left by the inverse then produces the least-squares parameter vector.

Full rank

$$a = (Y^T Y)^{-1} Y^T x \tag{3.9.10}$$

Pseudo-inverse

The matrix $Y^+ = (Y^T Y)^{-1} Y^T$ is called the *pseudo-inverse* or Moore-Penrose inverse of Y (Noble, 1969). From linear algebra, the parameter vector a obtained by multiplying by the pseudo-inverse is a least-squares solution of (3.9.5). That is, it is a solution that minimizes E_N.

Although (3.9.10) is a convenient way to express the optimal parameter vector, to compute a it is not necessary to form $Y^T Y$, compute the inverse, and multiply. Instead, the following

MATLAB function

MATLAB command using the left division or backslash operator can be used to find the least-squares solution to an over-determined system.

```
a = Y \ x;                    % Least-squares solution of Ya = x
```

Example 3.25 **System Identification**

As an illustration of least-squares system identification with an AR model, suppose the input-output data D is generated by using $N = 100$ samples from the following "black box" system.

$$H_{data}(z) = \frac{2z^2 - .8z + .64}{z^2 - 1.2728z + .81}$$

Suppose the input is white noise uniformly distributed over the interval $[-1, 1]$. Since the order of the underlying black box system is normally not known, it can be useful to perform a system identification using different values for the AR model order n. A plot of the error E_N for $1 \le n \le 10$ obtained by running *exam3_26* is shown in Figure 3.26.

FIGURE 3.26: Least-squares Error E_N for the System Using $N = 100$ Points

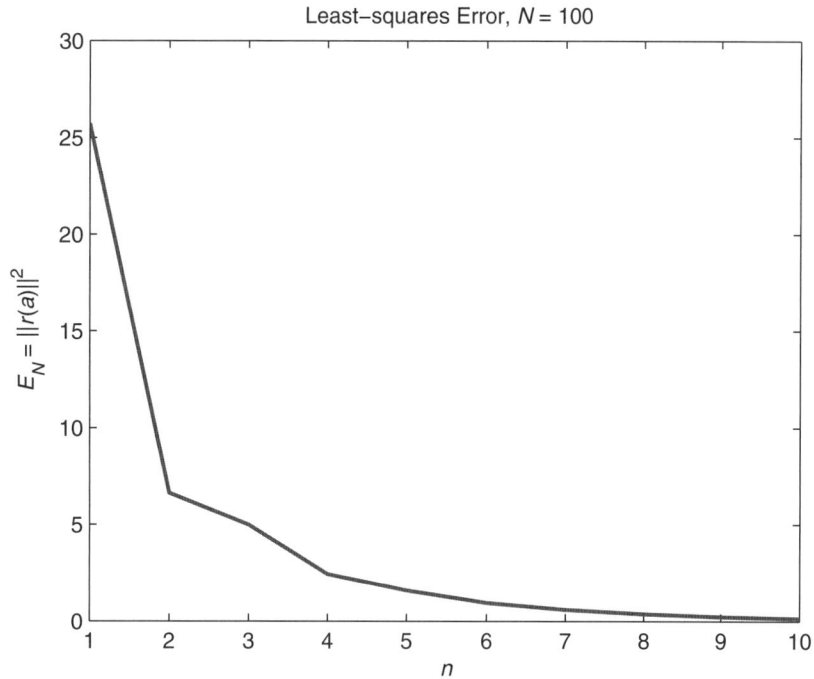

Least–squares Error, $N = 100$

FIGURE 3.27: Comparison of the Magnitude and Phase Responses of the Black Box System and the AR Model or Order $n = 10$

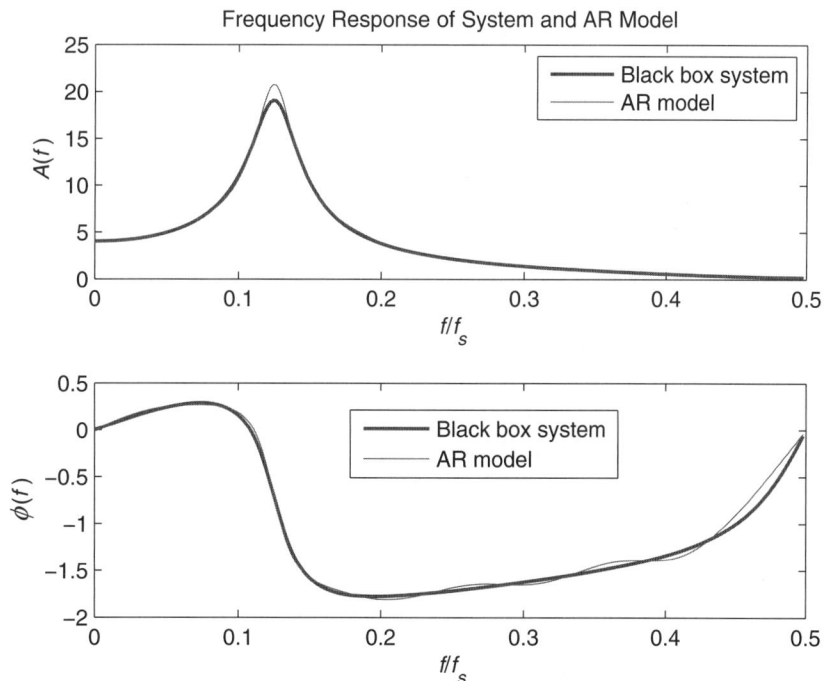

Frequency Response of System and AR Model

Notice how the error decreases with increasing n, showing an improved fit between the AR model and the black box system. The number of parameters in the black box system is five, while there are $p = n + 1$ parameters in the AR model, so they have the same number of parameters when $n = 4$. Of course, the structure of the black box system is more general than the AR structure, so one would not expect a perfect fit for $n = 4$. However, the least-square error is quite small for $n = 10$. The frequency responses of the system and the model when

$n = 10$ are compared in Figure 3.27. Although the fit is not exact, the overall qualitative features of both the magnitude response and the phase response agree quite well.

3.9.2 Persistently Exciting Inputs

When first encountered, the adjectives "persistent" and "exciting" might seem a bit strange when applied to an input signal, but both have important practical meanings. They are associated with the idea that the input signal generates outputs that are sufficiently "rich" to cause the coefficient matrix Y in (3.9.5) to have full rank. To see how a problem might arise, suppose the black box system S is stable and the initial condition is zero. If the input is $x(k) = 0$, then $y(k) = 0$ and $Y = 0$, so clearly Y does not have full rank in this case. More generally, suppose the input $x(k)$ has the following Z-transform.

$$X(z) = \frac{d(z)}{(z - q_1)(z - q_2) \cdots (z - q_n)} \tag{3.9.11}$$

If the poles of $X(z)$ are all strictly inside the unit circle, then, since S is stable, the output $y(k)$ will decay to zero.

$$y(k) \to 0 \quad \text{as} \quad k \to \infty \tag{3.9.12}$$

Persistence

In this case the bottom rows of Y begin to approach rows of zeros, so increasing the number of samples N does not contribute to improving the estimate of a. This is because the forced modes in $y(k)$ generated by the input $x(k)$ do not *persist* in time; they die out. To generate forced modes that do persist but not grow, the poles of $X(z)$ must be simple poles on the unit circle that will generate sinusoidal terms in $y(k)$. For example, poles at $q_{1,2} = \exp(j2\pi fT)$ produce a steady-state sinusoidal term in $y(k)$ of frequency f. Notice that this suggests that the input $x(k)$ should be a power signal rather than an energy signal.

Exciting

In order to get an accurate value for a, the other thing that is needed is to *excite* all of the natural modes of the system with the input. This can be achieved by driving the system with an input that contains many frequencies spread over the range $[0, f_s/2]$. Although this can be done with an input consisting of several sinusoids, a particularly effective input is a white noise input. As we shall see in Chapter 4, a white noise input is a broadband signal whose power is spread evenly over all frequencies. This is why a white noise input was used in Example 3.26.

Online identification

When an expensive practical system needs to be identified, for example, one that is being used in a manufacturing process, it may not be feasible to take the system offline and excite it with a test input $x(k)$ to generate the input-output data D. In instances like these, the system can remain in operation and be identified *online* by superimposing a small white noise signal on top of the nominal input signal. Thus if $u(k)$ is the input required for normal operation and $v(k)$ is a small white noise signal, then

$$x(k) = u(k) + v(k) \tag{3.9.13}$$

The discussion of persistently exciting inputs presented here is a brief informal introduction intended to make the reader aware that care must be taken in selecting the input. A more formal and detailed presentation of persistent excitation can be found, for example, in Haykin (2002).

FDSP Functions

The FDSP toolbox contains a function called *f_idar* for performing system identification using an AR model.

```
% F_IDAR: Identify an AR systems using input-output data
%
```

```
% Usage:
%          [a,E] = f_idar(x,y,n);
% Pre:
%          x = array of length N containing the input samples
%          y = array of length N containing the output samples
%          n = the order of the AR model (n < N)
% Post:
%          a = 1 by (n+1) coefficient vector of the AR system
%          E = the least square error
% Notes:
%          1. For a good fit, use N >> n.
%          2. The input x must be persistently exciting such
%                as white noise or a broadband input
```

● ● ● ● ● ● ● ● ● ● ● ● ● ● ●

3.10 GUI Software and Case Studies

This section focuses on applications of discrete-time systems. A graphical user interface module called *g_sysfreq* is introduced that allows the user to explore the input-output behavior of linear discrete-time systems in the frequency domain without any need for programming. Case study examples are then presented and solved using MATLAB.

3.10.1 g_sysfreq: Discrete-time System Analysis in the Frequency Domain

The graphical user interface module *g_sysfreq* allows the user to investigate the input-output behavior of linear discrete-time systems in the frequency domain. GUI module *g_sysfreq* features a display screen with tiled windows, as shown in Figure 3.28. The *Block Diagram* window in the upper-left corner of the screen contains a color-coded block diagram of the system under investigation. Below the block diagram are a number of edit boxes whose contents can be modified by the user. The edit boxes for *a* and *b* allow the user to select the coefficients of the numerator polynomial and the denominator polynomial of the following transfer function.

$$H(z) = \frac{b_0 + b_1 z^{-1} + \cdots + b_m z^{-m}}{1 + a_1 z^{-1} + \cdots + a_n^{-n}} \tag{3.10.1}$$

The numerator and denominator coefficient vectors can be edited directly by clicking on the shaded area and entering in new values. Any MATLAB statement or statements defining *a* and *b* can be entered. The Enter key is used to activate a change to a parameter. Additional scalar parameters that appear in edit boxes are associated with the damped cosine input.

$$x(k) = c^k \cos(2\pi F_0 kT)\mu(k) \tag{3.10.2}$$

They include the input frequency $0 \leq F_0 \leq f_s/2$, the exponential damping factor c, which is constrained to lie the interval $[-1, 1]$, and the sampling frequency f_s. The *Parameters* window also contains two push button controls. The push button controls play the signals $x(k)$ and $y(k)$ on the PC speaker using the current sampling rate f_s. This option is active on any PC with a sound card. It allows the user to hear the filtering effects of $H(z)$ on various types of inputs.

The *Type* and *View* windows in the upper-right corner of the screen allow the user to select both the type of input signal, and the viewing mode. The inputs include white noise uniformly

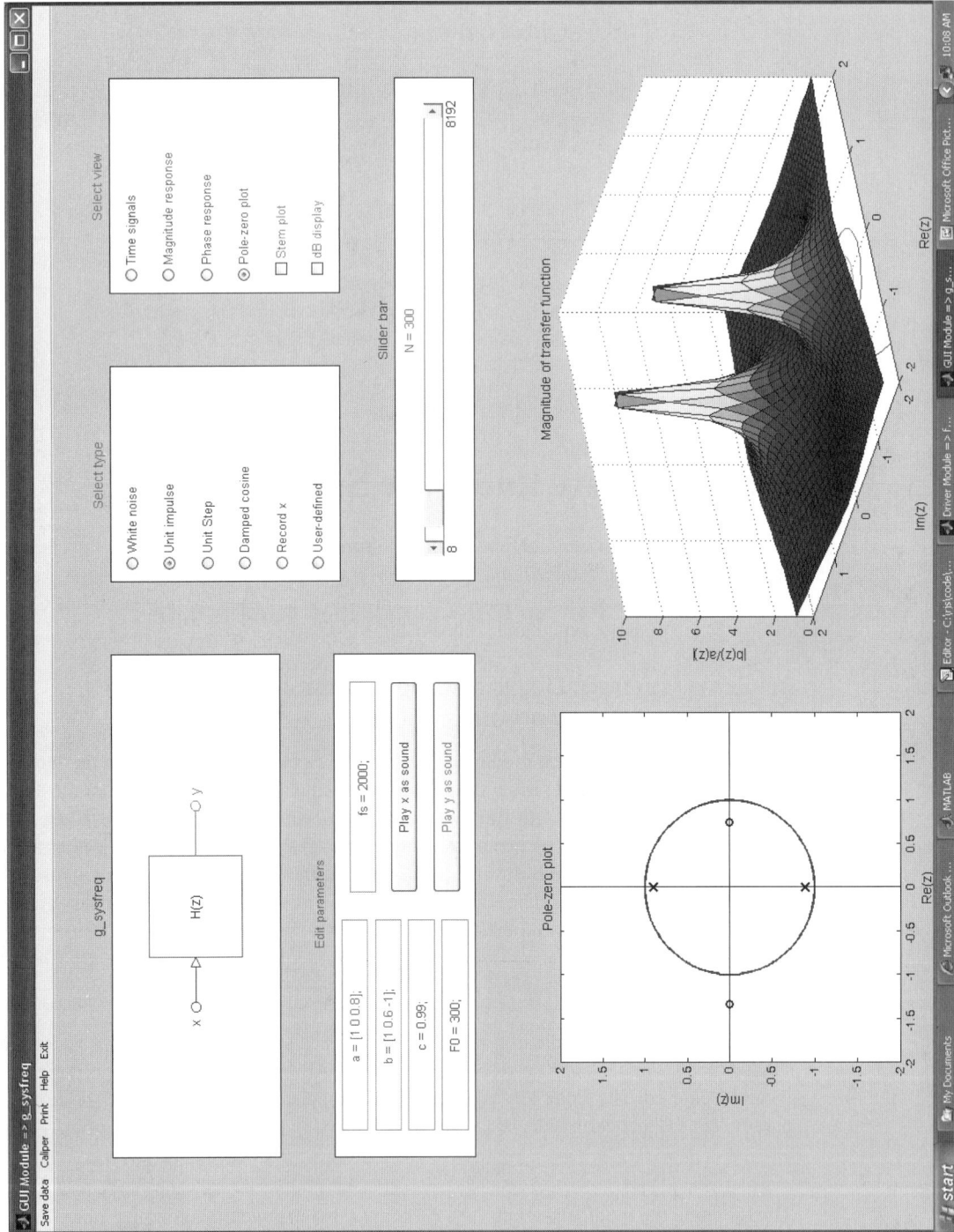

FIGURE 3.28: Display Screen of Chapter GUI Module *g_sysfreq*

distributed over $[-1, 1]$, a unit impulse input, a unit step input, the damped cosine input in (3.10.2), recorded sounds from a PC microphone, and user-defined inputs from a MAT file. The Recorded sound option can be used to record up to one second of sound at a sampling rate of $f_s = 8192$ Hz. For the User-defined option, a MAT file containing the input vector x, the sampling frequency f_s, and the coefficient vectors a, and b must be supplied by the user.

The *View* options include plots of the input $x(k)$, the output $y(k)$, the magnitude response $A(f)$, the phase response $\phi(f)$, and a pole-zero sketch. The magnitude response is either linear or logarithmic, depending on the status of the dB check box control. Similarly, the plots of the input and the output use continuous time or discrete time, depending on the status of the Stem plot check box control. The poles-zero sketch also includes a plot of the transfer function surface $|H(z)|$. The *Plot* window on the bottom half of the screen shows the selected view. The curves are color-coded to match the block diagram labels. The slider bar below the *Type* and *View* window allows the user to change the number of samples N.

The *Menu* bar at the top of the screen includes several menu options. The *Caliper* option allows the user to measure any point on the current plot by moving the mouse cross hairs to that point and clicking. The *Save data* option is used to save the current x, y, f_s, a, and b in a user-specified MAT file for future use. The User-defined input option can be used to reload this data. The *Print* option prints the contents of the plot window. Finally, the *Help* option provides the user with some helpful suggestions on how to effectively use module $g\_sysfreq$.

CASE STUDY 3.1 | ## Satellite Attitude Control

In Section 3.1, the following discrete-time model was introduced for a single-axis satellite attitude control system.

$$y(k) = (1 - d)y(k - 1) - dy(k - 2) + d[r(k - 1) + r(k - 2)]$$
$$d = \frac{cT^2}{2J}$$

Here $r(k)$ is the desired angular position of the satellite at the kth sampling time, and $y(k)$ is the actual angular position. The constants c, T, and J denote the controller gain, the sampling interval, and the satellite moment of inertia, respectively. By inspection of the difference equation, the transfer function of this control system is

$$H(z) = \frac{Y(z)}{R(z)}$$
$$= \frac{d(z^{-1} - z^{-2})}{1 - (1 - d)z^{-1} + dz^{-2}}$$
$$= \frac{d(z + 1)}{z^2 + (d - 1)z + d}$$

The controller gain c appearing in the expression for d is an engineering design parameter that must be chosen to satisfy some performance specification. The most fundamental performance constraint is that the control system be stable. The denominator polynomial of the transfer function is

$$a(z) = z^2 + (d - 1)z + d$$

We can apply the Jury test to determine a stable range for c. The first two rows of the Jury table are

$$J_2 = \begin{bmatrix} 1 & d - 1 & d \\ d & d - 1 & 1 \end{bmatrix}$$

From (3.7.12), the elements of the third row are

$$b_0 = \begin{vmatrix} 1 & d \\ d & 1 \end{vmatrix} = 1 - d^2$$

$$b_1 = \begin{vmatrix} 1 & d-1 \\ d & d-1 \end{vmatrix} = -(1-d)^2$$

From the stability condition $b_0 > 0$, we have

$$|d| < 1$$

The first four rows of the Jury table are

$$J_4 = \begin{bmatrix} 1 & d-1 & d \\ d & d-1 & 1 \\ 1-d^2 & -(1-d)^2 & \\ -(1-d)^2 & 1-d^2 & \end{bmatrix}$$

From (3.7.12), the element in the fifth row of the Jury table is

$$\begin{aligned} c_0 &= \frac{1}{1-d^2} \begin{vmatrix} 1-d^2 & -(1-d)^2 \\ -(1-d)^2 & 1-d^2 \end{vmatrix} \\ &= \frac{(1-d^2)^2 - (1-d)^4}{1-d^2} \\ &= \frac{(1-d)^2[(1+d)^2 - (1-d)^2]}{1-d^2} \\ &= \frac{4d(1-d)^2}{1-d^2} \end{aligned}$$

Since $|d| < 1$, the denominator and the second factor in the numerator are positive. Thus the stability condition $c_0 > 0$ reduces to $d > 0$. Together with $|d| < 1$, this yields $0 < d < 1$. From the definition of d in the original difference equation, we conclude that the control system is BIBO stable for the following range of controller gains.

$$0 < c < \frac{2J}{T^2}$$

To test the effectiveness of the control system, suppose $T = .1$ sec and $J = 5$ N-m-sec^2. Then $0 < c < 1000$ is the stable range. Suppose a ground station issues a command to rotate the satellite one quarter of a turn. Then the desired angular position signal is

$$r(k) = \left(\frac{\pi}{2}\right) \mu(k)$$

CASE STUDY 3.1   MATLAB function *case3_1* computes the resulting zero-state response for three different values of the controller gain c. It can be executed directly from the FDSP driver program *f_dsp*.

```
function case3_1

% CASE STUDY 3.1: Satellite attitude control

clc
f_header('Case Study 3.1: Satellite attitude control')
n = 21;
T = .1;                              % sampling interval
```

```
J = 5;                                % moment of inertia
c = [.1 3-sqrt(8) .5]*(2*J/T^2)       % controller gains
d = (T^2/(2*J))*c;
m = length(c);

% Compute step response

r = (pi/2)*ones(n,1);
for i = 1 : m
    a = [1 d(i)-1 d(i)];
    b = [0 d(i) d(i)];
    pole = roots(a)
    y(:,i) = (180/pi)*filter(b,a,r);
end

% Plot curves

figure
k = [0 : n-1];
for i = 1 : 3
    subplot (3,1,i)
    hp = stem (k,y(:,i),'filled','.');
    set (hp,'LineWidth',1.5)
    axis ([k(1) k(n) 0 150])
    box on
    switch i
    case 1,
        title ('Satellite step response')
        text (10,120,'Overdamped','HorizontalAlignment','center')
    case 2,
        ylabel ('{y(k)} (deg)')
        text (10,120,'Critically damped','HorizontalAlignment','center')
    case 3,
        xlabel ('{k}')
        text (10,120,'Underdamped','HorizontalAlignment','center')
    end
    hold on
    plot(k,(180/pi)*r,'r')
end
f_wait
```

When *case3_1* is executed, it produces the plot shown in Figure 3.29. Note that for all three controller gains, the satellite turns to the desired 90 degree orientation. The control system pole locations for the three controller gains are summarized in Table 3.5. When $c = 100$, there are two distinct real poles and a sluggish *overdamped* response results. When $c = 171.6$, there is a double real pole which results in a *critically damped* response. This is the fastest possible step response among those that do not overshoot the final position. When $c = 500$, there is a pair of complex conjugate poles which generate an oscillatory *underdamped* response.

FIGURE 3.29: Response of Satellite to a 90 Degree Step Input Using Different Controller Gains

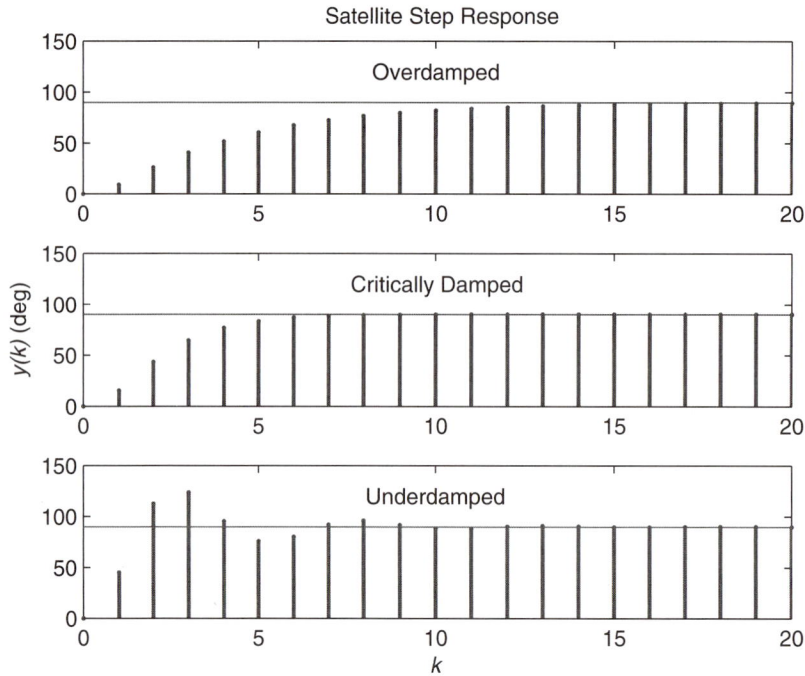

Satellite Step Response

TABLE 3.5: ▶ Control System Pole Locations for Different Controller Gains

c	Poles	Case
100	$p_{1,2} = .770, .130$	Overdamped
171.6	$p_{1,2} = .414, .414$	Critically damped
500	$p_{1,2} = .25 \pm .661j$	Underdamped

CASE STUDY 3.2

Speech Compression

Phoneme

Recall from Section 3.1 that speech production can be modeled using an AR discrete-time system, as shown in Figure 3.3. Fundamental speech components or *phonemes* are either voiced or unvoiced sounds. For the unvoiced phonemes such as the fricatives s, sh, and f, and the plosives p, t, and k, the AR model is driven by a random white noise input. For voiced sounds which include the vowels, nasal sounds, and transient terminal sounds such as b, d, and g, the input to the AR model is a periodic impulse train with period M.

$$x(k) = \sum_{i=0}^{\infty} \delta(k - iM)$$

If T is the sampling interval, then the period of the impulse train in seconds is $T_0 = MT$. The fundamental frequency or *pitch* of the speaker is then

Pitch

$$F_0 = \frac{1}{MT}$$

Speaker pitch typically ranges from about 50 Hz to 400 Hz. An illustration of a short segment of the vowel "O" was shown previously in Figure 3.4 where the pitch was estimated to be $F_0 \approx 113.6$ Hz. For the linear system S in Figure 3.3 that models air flow through the vocal tract including the throat, mouth, and lips, the following AR model can be used.

$$y(k) = b_0 x(k) - \sum_{i=1}^{n} a_i y(k - i)$$

Finding appropriate values for b_0 and a from the recorded $y(k)$ is a system identification problem that can be solved using the techniques discussed in Section 3.9. For an AR model to be valid, only a short segment of speech can be used. For example, by using segments of length $\tau = 20$ msec, the statistical characteristics of the speech remain effectively constant. That is, the speech signal can be assumed to be a *stationary signal* over an interval of this length. Typically speech is sampled at a sampling rate of $f_s = 8000$ Hz. Hence a segment of duration τ consists of N samples where

Stationary signal

$$N = f_s \tau$$
$$= 160$$

A direct brute force way to transmit speech over a communication channel is to send the samples themselves, $Y = \{y(k) \mid 0 \leq k < N\}$. The utility of using an AR model is that the coefficients of the model can be sent instead, and then the speech can be reconstructed at the receiver end (Rabiner and Schafer, 1978). Typically, a *frame* of speech transmitted in this manner includes the following information.

Frame

$$\text{frame} = [f_s, F_0, v, b_0, a_1, \dots, a_n]$$

Here f_s is the sampling frequency, F_0 is the pitch, v is a voiced/unvoiced switch, b_0 is the volume, and $a \in R^n$ specifies the remaining coefficients of the AR filter. Thus the total length of the frame is $p = n + 4$. If an effective model can be achieved for $p < N$, then fewer bits of data have to be sent over the communication channel. Consequently, a less expensive lower-bandwidth channel can be used. Typically an AR filter of order $n = 10$ is sufficient. In this case a savings of $146/160$ or 91.3 percent can be achieved using this data compression technique.

CASE STUDY 3.2   MATLAB function *case3_2* tests the idea of sending the model rather than the data. The user can try out different order AR models, listen to the original and the reconstructed sound segments, and view the compression achieved.

```
function case3_2

% CASE STUDY 3.2: Speech Compression

clc
f_header('Case Study 3.2: Speech Compression')
load case3_2          % audio data
tau = .02;            % segment duration
N = round(fs*tau);    % samples/segment

% Find model for voiced sound
n = f_prompt('Enter model order n',1,12,10);
M = f_prompt('Enter pitch period M',1,120,59);
x = zeros(N,1);
for i = 1 : N
    if mod(i-1,M) == 0
        x(i) = 1;
    end
end
q = round(length(y)/2);
yseg = y(q+1:q+N);
[a,E] = f_idar(x,yseg,n);
y = filter(1,a,x);
```

```
pole_radius = abs(roots(a))
if max(pole_radius) >= 1
    fprintf ('The AR model is unstable.\n')
end

% Play original and reconstructed sounds

P = 25;
f_wait
fprintf ('Repeated sound segment ...\n')
Yseg = repmat(yseg,P,1);
wavplay(Yseg)
f_wait
fprintf ('Repeated AR model ...\n')
Y = repmat(y,P,1);
wavplay(Y)
compress = 100*(N-n-4)/N;
fprintf ('\nCompression = %.1f percent\n\n',compress)
```

When *case3_2* is executed, the user is prompted for the AR model order n, and the pitch period M. Suggested default values are provided as a starting point, but the user is encouraged to see what happens as other values are used. One of the problems that can occur with an AR filter is that it can become unstable. This becomes more likely as the filter order increases and the pitch period varies.

CASE STUDY 3.3

Fibonacci Sequence and the Golden Ratio

There is a simple discrete-time system that can be used to produce a well-known sequence of numbers called the *Fibonacci sequence*.

$$y(k) = y(k - 1) + y(k - 2) + x(k)$$

The impulse response of this system is the Fibonacci sequence. Note that with $x(k) = \delta(k)$ and a zero initial condition, we have $y(0) = 1$. For $k > 0$, the next number in the sequence is just the sum of the previous two. This yields the following impulse response.

$$h(k) = [1, 1, 2, 3, 5, 8, 13, 21, 34, 55, 89, \ldots]$$

Fibonacci introduced this model in 1202 to describe how fast rabbits could breed under ideal circumstances (Cook, 1979). He starts with one male-female pair and assumes that at the end of each month they breed. One month later the female produces another male-female pair and the process continues. The number of pairs at the end of each month then follows the Fibonacci pattern. The Fibonacci numbers occur in a surprising number of places in nature. For example, the number of petals in flowers is often a Fibonacci number as can be seen in Table 3.6.

The system used to generate the Fibonacci sequence is an unstable system with $h(k)$ growing without bound as $k \to \infty$. However, it is of interest to investigate the *ratio* of successive samples of the impulse response. This ratio converges to a special number called the *golden ratio*.

$$\gamma \triangleq \lim_{k \to \infty} \left\{ \frac{h(k)}{h(k - 1)} \right\} \approx 1.618$$

TABLE 3.6: ▶
Fibonacci Numbers
and Flowers

Flower	Number of Petals
Iris	3
Wild rose	5
Delphinium	8
Corn marigold	13
Aster	21
Pyrethrum	34
Michelmas daisy	55

The golden ratio is noteworthy in that it has been used in Greek architecture as far back as 430 BC in the construction of the Parthenon, a temple to the goddess Athena. For example, the ratio of the width to the height of the front of the temple is γ.

Consider the problem of finding the exact value of the golden ratio. From the difference equation, the transfer function of the Fibonacci system is

$$H(z) = \frac{1}{1 - z^{-1} - z^{-2}}$$

$$= \frac{z^2}{z^2 - z - 1}$$

Factoring the denominator, we find that this system has poles at

$$p_{1,2} = \frac{1 \pm \sqrt{5}}{2}$$

From Algorithm 3.1, the initial value is $h(0) = 1$, and the residues at the two poles are

$$\text{Res}(p_1, k) = \frac{p_1^{k+1}}{p_1 - p_2}$$

$$\text{Res}(p_2, k) = \frac{p_2^{k+1}}{p_2 - p_1}$$

Thus the impulse response is

$$h(k) = h(0)\delta(k) + [\text{Res}(p_1, k) + \text{Res}(p_2, k)]\mu(k - 1)$$

$$= \delta(k) + \left(\frac{p_1^{k+1} - p_2^{k+1}}{p_1 - p_2} \right) \mu(k - 1)$$

$$= \left(\frac{p_1^{k+1} - p_2^{k+1}}{p_1 - p_2} \right) \mu(k)$$

Note that $|p_1| > 1$ and $|p_2| < 1$. Therefore, $p_2^{k+1} \to 0$ as $k \to \infty$. Thus we have

$$\gamma = \lim_{k \to \infty} \left\{ \frac{p_1^{k+1}}{p_1^k} \right\} = p_1$$

Consequently, the golden ratio is

$$\gamma = \frac{1 + \sqrt{5}}{2} = 1.6180339 \cdots$$

CASE STUDY 3.3   MATLAB function *case3_3* computes the impulse response of the Fibonacci system. It also computes the golden ratio, both directly and as a limit of the ratio $g(k) = h(k)/h(k - 1)$.

```
function case3_3

% CASE STUDY 3.3: Fibonacci sequence and the golden ratio
```

```
clc
f_header('Case Study 3.3: Fibonacci Sequence and the Golden Ratio')
N = 21;
gamma = (1 + sqrt(5))/2          % golden ratio
g = zeros(N,1);                  % estimates of gamma
a = [1 -1 -1];                   % denominator coefficients
b = 1;                           % numerator coefficients


% Estimate golden ratio with pulse response

[h,k] = f_impulse (b,a,N);
h
for i = 2 : N
   g(i) = h(i)/h(i-1);
end
figure
hp = stem (k(2:N),g(2:N),'filled','.');
set (hp,'LineWidth',1.5)
f_labels ('The golden ratio','{k}','{h(k)/h(k-1)}')
axis ([k(1) k(N) 0 3])
hold on
plot (k,gamma*ones(N),'r')
golden = sprintf ('\\gamma = %.6f',gamma);
text (10,2.4,golden,'HorizontalAlignment','center')
box on
f_wait
```

When *case3_3* is executed, it produces the plot shown in Figure 3.30 which graphs $g(k) = h(k)/h(k-1)$. It is evident that $g(k)$ rapidly converges to the golden ratio γ.

FIGURE 3.30:
Numerical
Approximation of
the Golden Ratio

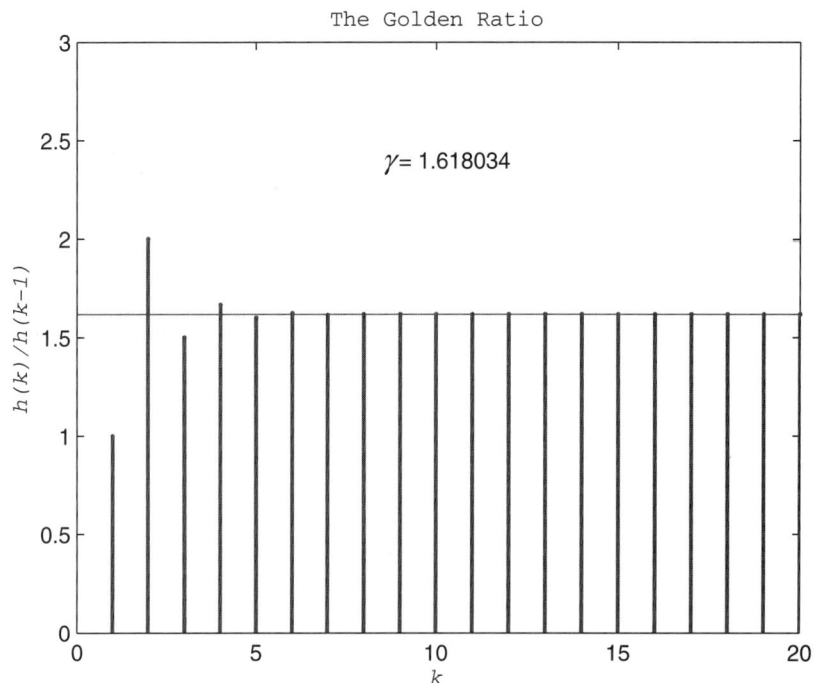

● ● ● ● ● ● ● ● ● ● ● ● ● ● ● ● ● ●

3.11 Chapter Summary

Z-transform

This chapter focused on the analysis of linear time-invariant discrete-time systems in the frequency domain using the Z-transform. The Z-transform is an essential analytical tool for digital signal processing. It is a transformation that maps a discrete-time signal $x(k)$ into a algebraic function $X(z)$ of a complex variable z.

$$X(z) = \sum_{k=-\infty}^{\infty} x(k)z^{-k}, \quad z \in \Omega_{ROC} \tag{3.11.1}$$

Poles, zeros
Region of
convergence

Typically $X(z)$ is a ratio of two polynomials in z. The roots of the numerator polynomial are called the *zeros* of $X(z)$, and the roots of the denominator polynomial are called the *poles* of $X(z)$. For general noncausal signals, the *region of convergence* Ω_{ROC} is an annular region centered at the origin of the complex plane. For anti-causal signals that are nonzero for $k < 0$, $X(z)$ converges inside the innermost pole, and for causal signals that are nonzero for $k \geq 0$, $X(z)$ converges outside the outermost pole.

Inverse transform

The Z-transform is usually found by consulting a table of Z-transform pairs. The size of the table is effectively enlarged by using the Z-transform properties. The initial and final value theorems allow us to recover the initial and final values of the signal $x(k)$ directly from its Z-transform. More generally, a finite number of samples of $x(k)$ can be obtained by inverting the Z-transform using the synthetic division method or the impulse response method. If a closed-form expression for $x(k)$ is desired, then the *inverse Z-transform* should be computed using either the partial fraction method with a table or the residue method.

$$x(k) = \frac{1}{j2\pi} \int_C X(z)z^{k-1} dz \tag{3.11.2}$$

Cauchy's residue method is the method of choice for most cases (real poles, single or multiple) because it can require less computational effort than the partial fraction method. Furthermore, unlike the partial fraction method, the residue method does not require the use of a table of Z-transform pairs.

Transfer Function

The *transfer function* of a discrete-time system is a compact algebraic representation of the system defined as the Z-transform of the output divided by the Z-transform of the input, assuming the initial condition is zero.

$$H(z) = \frac{Y(z)}{X(z)} \tag{3.11.3}$$

By using the linearity and time shift properties, the transfer function can be obtained directly from inspection of the difference equation representation of a discrete time system. A third representation of a discrete-time system is the signal flow graph which is a concise graphical description. It it possible to go directly from any of the three representations to another by inspection.

BIBO Stability

Impulse response

The time-domain equivalent of the transfer function is the impulse response. The *impulse response* is the zero-state response of the system when the input is the unit impulse $\delta(k)$. The easiest way to compute the impulse response $h(k)$ is to take the inverse Z-transform of the transfer function $H(z)$.

$$h(k) = Z^{-1}\{H(z)\} \tag{3.11.4}$$

Input-output representation

The zero-state response of the system to any input can be obtained from the impulse response using convolution. Convolution, in the time domain, maps into multiplication in the Z-transform domain. This leads to the *input-output representation* in the frequency domain.

$$Y(z) = H(z)X(z) \tag{3.11.5}$$

Stability test

A system is *BIBO stable* if and only if every bounded input is guaranteed to produced a bounded output. Otherwise, the system is *unstable*. Each pole of $H(z)$ generates a *natural mode* term in $y(k)$. For stable systems, the natural modes all decay to zero. A system is BIBO stable if and only if all of the poles of $H(z)$ lie strictly inside the unit circle of the complex plane. The Jury test is a tabular *stability test* that can be used to determine ranges for the system parameters over which a system is stable. All FIR systems are stable because their poles are all at the origin, but IIR systems may or may not be stable.

Frequency Response

The frequency response of a stable discrete-time system is the transfer function evaluated along the unit circle. If f_s is the sampling frequency and $T = 1/f_s$ is the sampling interval, then the frequency response is

$$H(f) = H(z)|_{z=\exp(j2\pi fT)}, \quad 0 \le |f| \le f_s/2 \tag{3.11.6}$$

Magnitude, phase response

A digital filter is a discrete-time system that is designed to have a prescribed frequency response. The magnitude $A(f) = |H(f)|$ is called the *magnitude response* of the filter, and the phase angle $\phi(f) = \angle H(f)$ is called the *phase response* of the filter. When a stable discrete-time system is driven by a cosine input with frequency $0 \le f \le f_s/2$, the steady-state output is a sinusoid of frequency f whose amplitude is scaled by $A(f)$ and whose phase is shifted by $\phi(f)$.

$$y_{ss}(k) = A(f)\cos[2\pi fkT + \phi(f)] \tag{3.11.7}$$

By designing a digital filter with a prescribed magnitude response, certain frequencies can be removed from the input signal, and other frequencies can be enhanced.

System Identification

System identification is the process of finding the parameters of a discrete-time model of a system using only input and output measurements. The different structures then can be used for the model, including an auto-regressive (AR) model, a moving-average (MA) model, and an auto-regressive moving-average (ARMA) model.

$$H(z) = \frac{b_0 + b_1 z^{-1} + \cdots + b_m z^{-m}}{1 + a_1 z^{-1} + \cdots + a_n z^{-n}} \tag{3.11.8}$$

A least-squares error criterion can be used to find the optimal parameter values as long as the number of input-output samples is at least as large as the number of parameters.

GUI Module

The FDSP toolbox includes a GUI module called *g_sysfreq* that allows the user to interactively investigate the input-output behavior of a discrete-time system in the frequency domain without any need for programming. Several common input signals are included, plus signals recorded

TABLE 3.7: ▶
Learning Outcomes
for Chapter 3

Num.	Learning Outcome	Sec.
1	Know how to compute the Z-transform using the geometric series	3.2
2	Understand how to use the Z-transform properties to expand the size of a table of Z-transform pairs	3.3
3	Be able to invert the Z transform using synthetic divisions, partial fractions, and the residue method	3.4
4	Be able to go back and forth between difference equations, transfer functions, impulse responses, and signal flow graphs	3.5–3.7
5	Understand the relationship between poles, zeros, and natural and forced modes of a linear system	3.5
6	Know how to find the impulse response and use it to compute the zero-state response to any input	3.7
7	Understand the differences between FIR systems and IIR systems, and know which system is always stable	3.7
8	Appreciate the significance of the unit circle in the Z-plane in terms of stability, and know how to evaluate stability using the Jury test	3.8
9	Be able to compute the frequency response from the transfer function	3.9
10	Know how to compute the steady-state output of a stable discrete-time system corresponding to a periodic input	3.9
11	Know how to use the GUI module *g_system* to investigate the input-output behavior of a discrete-time system	3.10

from a PC microphone and user-defined signals saved in MAT files. Viewing options include the time signals, the magnitude spectrum, the phase spectrum, and the pole-zero plot.

Learning Outcomes

This chapter was designed to provide the student with an opportunity to achieve the learning outcomes summarized in Table 3.7.

● ● ● ● ● ● ● ● ● ● ● ● ● ● ● ● ●

3.12   Problems

The problems are divided into Analysis and Design problems that can be solved by hand or with a calculator, GUI Simulation problems that are solved using GUI module *g_sysfreq*, and MATLAB Computation problems that require a user program.

3.12.1 Analysis and Design

3.1 Consider the following noncausal signal. Show that $X(z)$ does not exist for any scalar c. That is, show that the region of convergence of $X(z)$ is the empty set.

$$x(k) = c^k$$

3.2 Consider the following discrete-time signal.

$$x(k) = a^k \sin(bk + \theta)\mu(k)$$

(a) Use Table 3.1 and the trigonometric identities in Appendix 2 to find $X(z)$.
(b) Verify that $X(z)$ reduces to an entry in Table 3.2 when $\theta = 0$. Which one?
(c) Verify that $X(z)$ reduces to another entry in Table 3.2 when $\theta = \pi/2$. Which one?

3.3 Consider the following signal.

$$x(k) = \begin{cases} 10, & 0 \leq k < 4 \\ -2, & 4 \leq k < \infty \end{cases}$$

(a) Write $x(k)$ as a difference of two step signals.
(b) Use the time shift property to find $X(z)$. Express your final answer as a ratio of two polynomials in z.
(c) Find the region of convergence of $X(z)$.

3.4 Consider the following signal.

$$x(k) = \begin{cases} 2k, & 0 \leq k < 9 \\ 18, & 9 \leq k < \infty \end{cases}$$

(a) Write $x(k)$ as a difference of two ramp signals.
(b) Use the time shift property to find $X(z)$. Express your final answer as a ratio of two polynomials in z.
(c) Find the region of convergence of $X(z)$.

3.5 Use Appendix 1 and the properties of the Z-transform to find the Z-transform of the following cubic exponential signal. Simplify your final answer as much as you can.

$$x(k) = k^3 (c)^k \mu(k)$$

3.6 Suppose $X(z)$ converges on $\Omega_x = \{z/|z| > R_x\}$ and $Y(z)$ converges on $\Omega_y = \{z/|z| < R_y\}$.
(a) Classify $x(k)$ and $y(k)$ as to their type: causal, anti-causal, noncausal.
(b) Find a subset of the region of convergence of $ax(k) + by(k)$.
(c) Find the region of convergence of $c^k x(k)$.
(d) Find the region of convergence of $y(-k)$.

3.7 Let $x^*(k)$ denote the complex conjugate of $x(k)$. Show that the Z-transform of $x^*(k)$ can be expressed in terms of the Z-transform of $x(k)$ as follows. This is called the *complex conjugate* property.

Complex conjugate property

$$Z\{x^*(k)\} = X^*(z^*)$$

3.8 Consider the following finite anti-causal signal where $x(-1) = 4$.

$$x = [\cdots, 0, 0, 3, -7, 2, 9, 4]$$

(a) Find the Z-transform $X(z)$, and express it as a ratio of two polynomials in z.
(b) What is the region of convergence of $X(z)$?

3.9 Consider the following finite causal signal where $x(0) = 8$.

$$x = [8, -6, 4, -2, 0, 0, \cdots]$$

(a) Find the Z-transform $X(z)$, and express it as a ratio of two polynomials in z.
(b) What is the region of convergence of $X(z)$?

3.10 Consider the following causal signal.

$$x(k) = 2(.8)^{k-1} \mu(k)$$

(a) Find the Z-transform $X(z)$, and express it as a ratio of two polynomials in z.
(b) What is the region of convergence of $X(z)$?

3.11 Consider the following anti-causal signal.

$$x(k) = 5(-.7)^{k+1}\mu(-k-1)$$

(a) Find the Z-transform $X(z)$, and express it as a ratio of two polynomials in z.
(b) What is the region of convergence of $X(z)$?

3.12 Consider the following noncausal signal.

$$x(k) = 10(.6)^k\mu(k+2)$$

(a) Find the Z-transform $X(z)$, and express it as a ratio of two polynomials in z.
(b) What is the region of convergence of $X(z)$?

3.13 Consider the following finite noncausal signal where $x(0) = 3$.

$$x = [\cdots, 0, 0, 1, 2, 3, 2, 1, 0, 0, \cdots]$$

(a) Find the Z-transform $X(z)$, and express it as a ratio of two polynomials in z.
(b) What is the region of convergence of $X(z)$?

3.14 Consider the following pair of finite causal signals, each starting with sample $k = 0$.

$$x(k) = [1, 2, 3]$$
$$y(k) = [7, 2, 4, 6, 1]$$

(a) Find $X(z)$ as a ratio of polynomials in z, and find the region of convergence.
(b) Find $Y(z)$ as a ratio of polynomials in z, and find the region of convergence.
(c) Consider the cross-correlation of $y(k)$ with $x(k)$.

$$r_{yx}(k) = \frac{1}{5}\sum_{i=0}^{4} y(i)x(i-k), \quad 0 \le k < 5$$

Using the correlation property, find the Z-transform of the cross-correlation $r_{yx}(k)$ as a ratio of polynomials in z, and find the region of convergence.

3.15 Consider the following Z-transform.

$$X(z) = \frac{10(z-2)^2(z+1)^3}{(z-.8)^2(z-1)(z-.2)^2}$$

(a) Find $x(0)$ without inverting $X(z)$.
(b) Find $x(\infty)$ without inverting $X(z)$.
(c) Write down the *form* of $x(k)$ from inspection of $X(z)$. You can leave the coefficients of each term of $X(z)$ unspecified.

3.16 A student attempts to apply the final value theorem to the following Z-transform and gets the steady-state value $x(\infty) = -5$. Is this correct? If not, what is the value of $x(k)$ as $k \to \infty$? Explain your answer.

$$X(z) = \frac{10z^3}{(z^2 - z - 2)(z - 1)}, \quad |z| > 2$$

3.17 The basic *geometric series* in (2.2.14) is often used to compute Z-transforms. It can be generalized in a number of ways.

(a) Prove that the geometric series in (2.2.14) converges to $1/(1-z)$ for $|z| < 1$ by showing that

$$\lim_{N \to \infty} (1-z) \sum_{k=0}^{N} z^k = 1 \quad \Longleftrightarrow \quad |z| < 1$$

(b) Use (2.2.14) to establish (3.2.3). That is, show that

$$\sum_{k=m}^{\infty} z^k = \frac{z^m}{1-z}, \quad m \geq 0, \ |z| < 1$$

(c) Use the results of part (b) to show the following. Hint: Write the sum as a difference of two series.

$$\sum_{k=m}^{n} z^k = \frac{z^m - z^{n+1}}{1-z}, \quad n \geq m \geq 0, \ |z| < 1$$

(d) Show that the result in part (c) holds for *all* complex z by multiplying both sides by $1-z$ and simplifying the left-hand side.

3.18 Consider the following Z-transform.

$$X(z) = \frac{z^4 + 2z^3 + 3z^2 + 2z + 1}{z^4}, \quad |z| > 0$$

(a) Rewrite $X(z)$ in terms negative powers of z.
(b) Use Definition 3.1 to find $x(k)$.
(c) Verify that $x(k)$ is consistent with the initial value theorem.
(d) Verify that $x(k)$ is consistent with the final value theorem.

3.19 Let $h(k)$ and $x(k)$ be the following pair of signals.

$$h(k) = [1 - (.9)^k]\mu(k)$$
$$x(k) = (-1)^k \mu(k)$$

(a) Find $H(z)$ as a ratio of polynomials in z and its region of convergence.
(b) Find $X(z)$ as a ratio of polynomials in z and its region of convergence.
(c) Use the convolution property to find the Z-transform of $h(k) \star x(k)$ as a ratio of polynomials in z and its region of convergence.

3.20 In Problem 3.19 the region of convergence of the Z-transform of $h(k) \star x(k)$ is $\Omega_{ROC} = \Omega_H \cap \Omega_X$, where Ω_H is the region of convergence of $H(z)$, and Ω_X is the region of convergence of $X(z)$. Is this true in general? If not, find an example of an $H(z)$ and an $X(z)$ where Ω_{ROC} is larger than $\Omega_H \cap \Omega_X$.

3.21 Consider the following Z-transform.

$$X(z) = \frac{2z}{z^2 - 1}, \quad |z| > 1$$

(a) Find $x(k)$ for $0 \leq k \leq 5$ using the synthetic division method.
(b) Find $x(k)$ using the partial fraction method.
(c) Find $x(k)$ using the residue method.

3.22 Consider the following Z-transform. Find $x(k)$ using the time shift property and the residue method.

$$X(z) = \frac{100}{z^2(z - .5)^3}, \quad |z| > .5$$

3.23 Consider the following Z-transform.

$$X(z) = \frac{z^4 + 1}{z^2 - 3z + 2}, \quad |z| > 2$$

(a) Find the causal part of $x(k)$
(b) Find the anti-casual part of $x(k)$

3.24 Consider the following noncausal signal

$$x(k) = c^k \mu(-k)$$

(a) Using Definition 3.1 and the geometric series, find $X(z)$ as a ratio of two polynomials in z and its region of convergence.
(b) Verify the results of part (a) by instead finding $X(z)$ using Table 3.2 and the time reversal property.

3.25 Consider the following Z-transform. Use Algorithm 3.1 to find $x(k)$. Express your final answer as a real signal.

$$X(z) = \frac{1}{z^2 + 1}, \quad |z| > 1$$

3.26 Repeat Problem 3.25, but use Table 3.2 and the Z-transform properties.

3.27 Consider the following Z-transform. Find $x(k)$.

$$X(z) = \frac{5z^3}{(z^2 - z + .25)(z + 1)}, \quad |z| > 1$$

3.28 Consider a discrete-time system described by the following difference equation.

$$y(k) = y(k - 1) - .24y(k - 2) + 2x(k - 1) - 1.6x(k - 2)$$

(a) Find the transfer function $H(z)$.
(b) Write down the form of the natural mode terms of this system.
(c) Find the zero-state response to the step input $x(k) = 10\mu(k)$.
(d) Find the zero-state response to the causal exponential input $x(k) = .8^k \mu(k)$. Does a forced mode term appear in $y(k)$? If not, why not?
(e) Find the zero state response to the causal exponential input $x(k) = .4^k \mu(k)$. Is this an example of harmonic forcing? Why or why not?

3.29 Consider a running average filter of order $M - 1$.

$$y(k) = \frac{1}{M} \sum_{i=0}^{M-1} x(k - i)$$

(a) Find the transfer function $H(z)$. Express it as a ratio of two polynomials in z.

(b) Use the geometric series in (3.2.3) to show that an alternative form of the transfer function is as follows. *Hint*: Express $y(k)$ as a difference of two sums.

$$H(z) = \frac{z^M - 1}{M(z-1)z^{M-1}}$$

(c) Convert the transfer function in part (b) to a difference equation.

3.30 Consider a discrete-time system described by the following transfer function.

$$H(z) = \frac{3(z - .4)}{z + .8}$$

(a) Suppose the zero-state response to an input $x(k)$ is $y(k) = \mu(k)$. Find $X(z)$.
(b) Find $x(k)$.

3.31 Consider the following Z-transform.

$$X(z) = \frac{z}{z - 1}$$

(a) Find $x(k)$ if the region of convergence is $|z| > 1$.
(b) Find $x(k)$ if the region of convergence is $|z| < 1$.

3.32 The formulation of the inverse Z-transform using the contour integral in (3.4.20) is based on the *Cauchy integral theorem*. This theorem states if C is any counterclockwise contour that encircles the origin, then

Cauchy integral formulation

$$\frac{1}{j2\pi} \oint_C z^{k-1-i} dz = \begin{cases} 1, & i = k \\ 0, & i \neq k \end{cases}$$

Use Definition 3.1 and the Cauchy integral theorem to show that the Z-transform can be inverted as in (3.4.20). That is, show that

$$x(k) = \frac{1}{j2\pi} \oint_C X(z) z^{k-1} dz$$

3.33 Consider a discrete-time system described by the following transfer function.

$$H(z) = \frac{z + .5}{z - .7}$$

(a) Find an input $x(k)$ that creates a forced mode of the form $c(.3)^k$ and causes the natural mode term to disappear in the zero-state response.
(b) Find an input $x(k)$ that has no zeros and creates a forced mode of the form $(c_1 k + c_2)(.7)^k$ in the zero-state response.

3.34 When two signals are multiplied, this corresponds to one signal amplitude modulating the other signal. The following property of the Z-transform is called the *modulation property*.

Modulation property

$$Z\{h(k)x(k)\} = \frac{1}{j2\pi} \oint_C H(u)X\left(\frac{z}{u}\right) u^{-1} du$$

Use the Cauchy integral representation of a time signal in Problem 3.32 to verify the modulation property.

3.35 Consider a discrete-time system described by the following transfer function.

$$H(z) = \frac{4z^2 + 1}{z^2 - 1.8z + .81}$$

(a) Find the difference equation.
(b) Find the impulse response $h(k)$.
(c) Sketch the signal flow graph.

3.36 Consider a discrete-time system described by the signal flow graph shown in Figure 3.31.
(a) Find the transfer function $H(z)$.
(b) Find the impulse response $h(k)$.
(c) Find the difference equation.

FIGURE 3.31: Signal Flow Graph of System in Problem 3.36

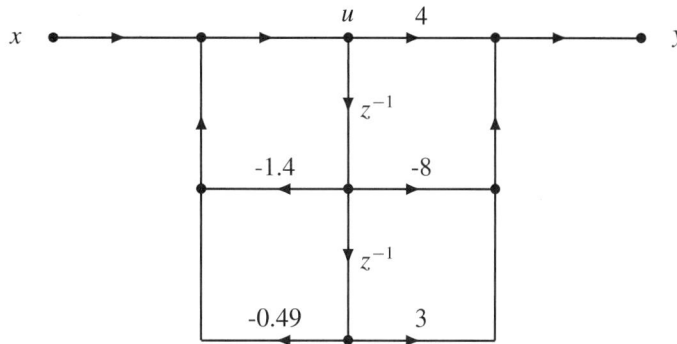

3.37 Find the overall transfer function $H(z) = Y(z)/X(z)$ of the system whose signal flow graph is shown in Figure 3.32. This is called a *parallel* configuration of $H_1(z)$ and $H_2(z)$.

FIGURE 3.32: Signal Flow Graph of a Parallel Configuration

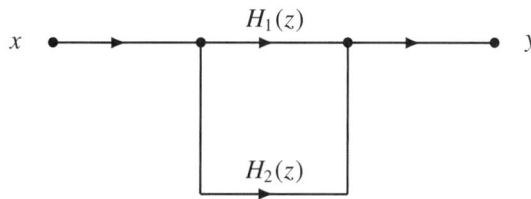

3.38 A discrete time system has poles at $z = \pm.5$ and zeros at $z = \pm j2$. The system has a DC gain of 20.
(a) Find the transfer function $H(z)$.
(b) Find the impulse response $h(k)$.
(c) Find the difference equation.
(d) Sketch the signal flow graph.

3.39 Consider a discrete-time system described by the signal flow graph shown in Figure 3.33 on page 222.
(a) Find the transfer function $H(z)$.
(b) Write the difference equations as a system of two equations.
(c) Write the difference equation as a single equation.

3.40 Consider a discrete-time system described by the following impulse response.

$$h(k) = [2 - .5^k + .2^{k-1}]\mu(k)$$

(a) Find the transfer function $H(z)$.
(b) Find the difference equation.
(c) Sketch the signal flow graph.

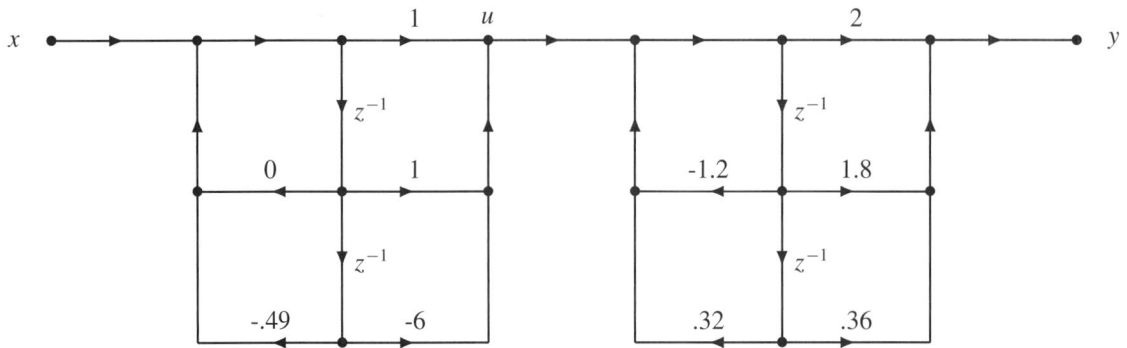

FIGURE 3.33: Signal Flow Graph of System in Problem 3.39

3.41 Find the overall transfer function $H(z) = Y(z)/X(z)$ of the system whose signal flow graph is shown in Figure 3.34. This is called a *feedback* configuration of $H_1(z)$ and $H_2(z)$.

FIGURE 3.34: Signal Flow Graph of a Feedback Configuration

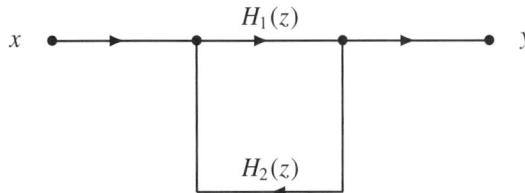

3.42 Consider a discrete-time system described by the following difference equation.

$$y(k) = .6y(k-1) + .16y(k-2) + 10x(k-1) + 5x(k-2)$$

(a) Find the transfer function $H(z)$.
(b) Find the impulse response $h(k)$.
(c) Sketch the signal flow graph.

3.43 Find the transfer function $H(z) = Y(z)/X(z)$ of the system whose signal flow graph is shown in Figure 3.35. This is called a *cascade* configuration of $H_1(z)$ and $H_2(z)$.

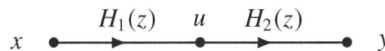

FIGURE 3.35: Signal Flow Graph of a Cascade Configuration

3.44 Consider a system with the following impulse response.

$$h(k) = 10(.5)^k \mu(k)$$

(a) Find the transfer function.
(b) Find the magnitude and phase responses.
(c) Find the fundamental frequency F_0, expressed as a fraction of f_s, of the following periodic input.

$$x(k) = \sum_{i=0}^{9} \frac{1}{1+i} \cos(.1\pi i k)$$

(d) Find the steady-state response $y_{ss}(k)$ to the periodic input in part (c). Express your final answer in terms of F_0.

3.45 For the system in Problem 3.44, consider the following following complex sinusoidal input.

$$x(k) = \cos(\pi k/3) + j \sin(\pi k/3)$$

(a) Find the frequency F_0 of $x(k)$, expressed as a fraction of f_s.
(b) Find the steady-state output $y_{ss}(k)$.

3.46 Consider the following first-order FIR system which is called a backwards Euler differentiator.

$$H(z) = \frac{z - 1}{Tz}$$

(a) Find the frequency response $H(f)$.
(b) Find and sketch the magnitude response $A(f)$.
(c) Find and sketch the phase response $\phi(f)$.
(d) Find the steady state response to the following periodic input.

$$x(k) = 2 \cos(.8\pi k) - \sin(.5\pi k)$$

3.47 Consider the following first-order IIR.

$$H(z) = \frac{z + .5}{z - .5}$$

(a) Find the frequency response $H(f)$.
(b) Find and sketch the magnitude response $A(f)$.
(c) Find and sketch the phase response $\phi(f)$.

3.48 Is the following system BIBO stable? Show your work.

$$H(z) = \frac{5z^2(z + 1)}{(z - .8)(z^2 + .2z - .8)}$$

3.49 Consider the following transfer function with parameter α.

$$H(z) = \frac{z^2}{(z - .8)(z^2 - z + \alpha)}$$

(a) Sketch the stability triangle in Figure 3.20; use it to find a range of values for the parameter α over which $H(z)$ is BIBO stable.
(b) For the stability limits in part (a), find the poles of $H(z)$.

3.50 Consider a system with the following impulse response.

$$h(k) = (-1)^k \mu(k)$$

(a) Find the transfer function $H(z)$.
(b) Find a bounded input $x(k)$ such that the zero-state response is unbounded.
(c) Find a bound for $x(k)$.
(d) Show that the zero-state response $y(k)$ is unbounded.

3.51 Consider the following discrete-time system.

$$H(z) = \frac{10z}{z^2 - 1.5z + .5}$$

(a) Find the poles and zeros of $H(z)$.
(b) Show that this system is BIBO unstable.

(c) Find a bounded input $x(k)$ that produces an unbounded output. Show that $x(k)$ is bounded. Hint: Use harmonic forcing.

(d) Find the zero-state response produced by the input in part (c) and show that it is unbounded.

3.52 Consider the following system that consists of a gain of A and a delay of d samples

$$y(k) = Ax(k - d)$$

(a) Find the transfer function, the poles, the zeros, and the DC gain.
(b) Is this system BIBO stable?. Why or why not?
(c) Find the impulse response of this system.
(d) Find the frequency response of this system.
(e) Find the magnitude response,.
(f) Find the phase response.

3.53 Consider the following second-order system.

$$H(z) = \frac{3(z + 1)}{z^2 - .81}$$

(a) Find the frequency response $H(f)$.
(b) Find and sketch the magnitude response $A(f)$.
(c) Find and sketch the phase response $\phi(f)$.
(d) Find the steady state response to the following periodic input.

$$x(k) = 10\cos(.6\pi k)$$

3.54 Consider the following transfer function with parameter β.

$$H(z) = \frac{z^2}{(z + .7)(z^2 + \beta z + .5)}$$

(a) Sketch the stability triangle in Figure 3.20; use it to find a range of values for the parameter β over which $H(z)$ is BIBO stable.
(b) For the stability limits in part (a), find the poles of $H(z)$.

3.55 Consider a discrete-time system described by the following transfer function.

$$H(z) = \frac{1}{z^2 + 1}$$

(a) Show that this system is BIBO unstable.
(b) Find a bounded $x(k)$ with bound $B_x = 1$ that produces an unbounded zero-state output.
(c) Find the Z-transform of the zero-state output when the input in part (b) is applied.

3.56 An alternative to using an AR model for system identification is to use an MA model. One important advantage of an MA model is that it is always stable.

$$H(z) = \sum_{i=0}^{m} b_i z^{-i}$$

(a) Let D be the input-output data in (3.9.3). Suppose $y = [y(0), \ldots, y(N - 1)]^T$, and let $b \in R^{m+1}$ be the parameter vector. Find an $N \times (m + 1)$ coefficient matrix U, analogous to Y in (3.9.5), such that the MA model agrees with the data D when $N = m + 1$ and

$$Ub = y$$

(b) Find an expression for the optimal least-squares b when $N > m + 1$.

3.12.2 GUI Simulation

3.57 Consider the system in Problem 3.53. Use GUI module $g\_sysfreq$ to plot the step response. Estimate the DC gain from the step response using the Caliper option.

3.58 Consider the system in Problem 3.53. Use GUI module $g\_sysfreq$ to perform the following tasks.
(a) Plot the pole-zero pattern. Is this system BIBO stable?
(b) Plot the response to white noise; use Caliper to mark the minimun point.

3.59 Consider the following linear discrete-time system.

$$H(z) = \frac{5z^{-2} + 4.5z^{-4}}{1 - 1.8z^{-2} + .81z^{-4}}$$

Use GUI module $g\_sysfreq$ to plot the following damped cosine input and the zero-state response to it.

$$x(k) = .96^k \cos(.4\pi k)$$

3.60 Consider the running average filter in Problem 3.29. Suppose $M = 10$. Use GUI module $g\_sysfreq$ to perform the following tasks.
(a) Plot the impulse response using $N = 100$ and stem plots.
(b) Plot the magnitude response using the linear scale.
(c) Plot the magnitude response using the dB scale.
(d) Plot the phase response.

3.61 Consider the following linear discrete-time system. Suppose the sampling frequency is $f_s = 1000$ Hz. Use GUI module $g\_sysfreq$ to plot the magnitude response using the linear scale and the phase response.

$$H(z) = \frac{10(z^2 + .8)}{(z^2 + .9)(z^2 + .7)}$$

3.62 Consider the following linear discrete-time system. Use GUI module $g\_sysfreq$ to plot the magnitude response and the phase response. Use $f_s = 100$ Hz, and use the dB scale for the magnitude response.

$$H(z) = \frac{5(z^2 + .9)}{(z^2 - .9)^2}$$

3.63 Consider the following linear discrete-time system.

$$H(z) = \frac{6 - 7.7z^{-1} + 2.5z^{-2}}{1 - 1.7z^{-1} + .8z^{-2} - .1z^{-3}}$$

Create a MAT-file called $prob3\_59$ that contains $fs = 100$, the appropriate coefficient vectors a and b, and the following input samples where $v(k)$ is white noise uniformly distributed over $[-.5, .5]$.

$$x(k) = k\exp(-k/50) + v(k), \quad 0 \le k < 500$$

Use GUI module $g\_sysfreq$ and the User-defined option to plot this input and the zero-state response to this input.

3.12.3 MATLAB Computation

3.64 Consider the following discrete-time system. Write a MATLAB program that performs the following tasks.

$$H(z) = \frac{2z^5 + .25z^4 - .8z^3 - 1.4z^2 + .6z - .9}{z^5 + .055z^4 - .85z^3 - .04z^2 + .49z - .32}$$

(a) Compute and display the poles, zeros, and DC gain. Is this system stable?
(b) Plot the poles and zeros using the FDSP toolbox function *f_pzplot*.
(c) Plot the transfer function surface using *f_pzsurf*.

3.65 Consider the following discrete-time system.

$$H(z) = \frac{10z^3}{z^4 - .81}$$

Write a MATLAB program that performs the following tasks.

(a) Use *f_freqz* to compute the magnitude response and the phase response at $M = 500$ points, assuming $f_s = 200$ Hz. Plot them as a 2 by 1 array of plots
(b) Use *filter* to compute the zero-state response to the following periodic input with $F_0 = 10$ Hz. Compute the steady state response $y_{ss}(k)$ to $x(k)$ using the magnitude and phase responses evaluated at $f = F_0$. Plot the zero-state response and the steady-state response on the same graph using a legend.

$$x(k) = 3\cos(2\pi F_0 kT)\mu(k), \quad 0 \le k \le 100$$

3.66 Consider the following discrete-time system.

$$H(z) = \frac{1.5z^4 - .4z^3 - .8z^2 + 1.1z - .9}{z^4 - .95z^3 - .035z^2 + .462z - .351}$$

Write a MATLAB program that uses *filter* and *plot* to compute and plot the zero-state response of this system to the following input. Plot both the input and the output on the same graph.

$$x(k) = (k+1)(.9)^k\mu(k), \quad 0 \le k \le 100$$

3.67 The MAT file *prob3_67* contains an input signal x, an output signal y, and a sampling frequency fs. Write a MATLAB program that performs system identification with these data by performing the following tasks.

(a) Load x, y, and fs from *prob3_67* and use *f_idar* to compute an AR model of order $n = 8$. Print the coefficient vector a
(b) Plot the first 100 samples of the data $y(k)$ and the AR model output $Y(k)$ on the same graph using a legend.

3.68 System identification can be performed using an MA model instead of the AR model discussed in Section 3.9. Recall that this was the focus of Problem 3.56.

(a) Write a function called *f_idma*, similar to the FDSP function *f_idar*, that performs system identification using an MA model. The calling sequence should be as follows.

```
% F_IDMA: MA system identification
%
% Usage:
%        [b,E] = f_idma (x,y,m);
% Pre:
%        x = vector of length N containing the input samples
%        y = vector of length N containing the output samples
```

```
%          m = the order of the MA model (m < N)
% Post:
%          b = vector of length m+1 containing the least-squares
%                 coefficients
%          E = least squares error
```

(b) Test your *f_idma* function by solving Problem 3.67, but using *f_idma* in place of *f_idar*. Use an MA model of order *m* = 20.

(c) Print your user documentation for *f_idma* using the command help f_idma.

CHAPTER

4

Fourier Transforms and Spectral Analysis

Chapter Topics

4.1 Motivation

4.2 Discrete-time Fourier Transform (DTFT)

4.3 Discrete Fourier Transform (DFT)

4.4 Fast Fourier Transform (FFT)

4.5 Fast Convolution and Correlation

4.6 White Noise

4.7 Auto-correlation

4.8 Zero-padding and Spectral Resolution

4.9 Spectrogram

4.10 Power Density Spectrum Estimation

4.11 GUI Software and Case Studies

4.12 Chapter Summary

4.13 Problems

4.1 Motivation

Recall that the Z-transform starts with a discrete-time signal $x(k)$ and transforms it into a function $X(z)$ of a complex variable z. In this chapter we focus on important special cases of the Z-transform that apply when the region of convergence includes the unit circle. The first case involves evaluating the Z-transform along the unit circle itself. This leads to the

DTFT   *discrete-time Fourier transform* or DTFT.

$$X(f) = \sum_{k=-\infty}^{\infty} x(k) \exp(-j2\pi k f T), \quad -f_s/2 \le f \le f_s/2$$

Signal spectrum The DTFT, denoted $X(f) = \mathrm{DTFT}\{x(k)\}$, is referred to as the *spectrum* of the signal $x(k)$. The spectrum reveals how the average power of $x(k)$ is distributed over the frequencies in the interval $[-f_s/2, f_s/2]$. The spectrum of the impulse response of a linear system is the frequency response of the system.

DFT

The second important special case involves applying the Z-transform to an N-point signal, and evaluating it at N points uniformly spaced around the unit circle. This leads to the *discrete Fourier transform* or DFT.

$$X(i) = \sum_{i=0}^{N-1} x(k) \exp(j2\pi ikT), \quad 0 \le i < N$$

The DFT, denoted $X(i) = \text{DFT}\{x(k)\}$, is a sampled version of the DTFT. The DFT provides the same type of spectral information as the DTFT, but with lower resolution. In particular, $|X(i)|^2/N$ is the average power of $x(k)$ at frequency $f_i = if_s/N$ for $0 \le i < N$. For computational purposes, the DFT is more efficient than the DTFT because it consists of only N terms, and it needs to be evaluated only at N discrete frequencies. There is a highly efficient

FFT

implementation of the DFT called the *fast Fourier transform* or FFT. The computational effort of the DFT grows as the square of the length of the signal N, whereas the computational effort of the FFT grows at the much slow rate of $N \log_2(N)$. The FFT can be used to develop fast versions of both convolution and correlation.

We begin this chapter by introducing a number of examples where spectral analysis can be put to use. Next the DTFT and its inverse are introduced. The DTFT is used to compute the spectra of discrete-time signals of infinite duration. Many of the properties of the DTFT are inherited directly from the Z-transform. The DFT and its inverse are then defined, and some useful properties of the DFT based on symmetry are developed. A simple graphical relationship between the Z-transform, the DTFT, and the DFT is presented. The highly efficient FFT implementation of the DFT is then derived using the decimation in time approach. This is followed by a comparison of the relative computational effort required by the FFT and the DFT in terms of the number of floating point operations or FLOPs.

The use of the DFT to compute the magnitude, phase, and power density spectra of finite signals is then presented. This is followed by an introduction to white noise. White noise is an important type of random signal that is useful for signal modeling and system identification. It can be characterized in an elegant way in terms of its auto-correlation and its spectrum. The use of the DFT to approximate the frequency response of a discrete-time system is then examined as is the use of zero padding to interpolate between discrete frequencies. Next the spectrogram is introduced to characterize signals whose spectral characteristics evolve with time. This is followed by an examination of techniques that are used to numerically estimate a continuous power density spectrum. Finally, a GUI module called *g_spectra* is introduced that allows the user to interactively explore the spectral characteristics of a variety of discrete-time signals without any need for programming. The chapter concludes with some case study examples, and a summary of Fourier transforms and spectral analysis.

4.1.1 Fourier Series

Consider a periodic continuous-time signal $x_a(t)$ with period T_0. For example, $x_a(t)$ might represent the hum or whine of a rotating machine where the fundamental frequency $F_0 = 1/T_0$ changes with the speed of rotation. Since $x_a(t)$ is periodic, it can be approximated by a truncated

Fourier series

Fourier series using M harmonics.

$$x_a(t) \approx \sum_{i=-(M-1)}^{M-1} c_i \exp(ji2\pi F_0 t) \tag{4.1.1}$$

This is the complex form of the Fourier series with the ith Fourier coefficient being

$$c_i = \frac{1}{T_0} \int_0^{T_0} x_a(t) \exp(-ji2\pi F_0 t) dt \tag{4.1.2}$$

Fourier series coefficients of common periodic waveforms can be found in Appendix 1. Next suppose $x_a(t)$ is converted to a discrete-time signal $x(k)$ by sampling it at $N = 2M$ points using a sampling frequency of $f_s = N F_0$. Thus the sampling interval is $T = T_0/N$ and

$$x(k) = x_a(kT), \quad 0 \le k < N \tag{4.1.3}$$

Note that the N samples of $x_a(t)$ span one period. To compute the ith Fourier coefficient of $x_a(t)$, we approximate the integral in (4.1.2) with a sum. Recalling that $F_0 = f_s/N$ and $T = T_0/N$, this yields

$$
\begin{aligned}
c_i &= \frac{1}{T_0} \int_0^{T_0} x_a(t) \exp(-ji2\pi F_0 t) \\
&\approx \frac{1}{T_0} \sum_{k=0}^{N-1} x_a(kT) \exp(-ji2\pi F_0 kT) T, \quad N \gg 1 \\
&= \frac{T}{T_0} \sum_{k=0}^{N-1} x(k) \exp(-ji2\pi f_s kT/N) \\
&= \frac{1}{N} \sum_{k=0}^{N-1} x(k) \exp(-jik2\pi/N), \quad 0 < |i| < N/2
\end{aligned}
\tag{4.1.4}
$$

Thus the Fourier coefficients of the periodic signal $x_a(t)$ can be approximated using the samples $x(k)$. Interestingly enough, the summation part of (4.1.4) is the *discrete Fourier transform* or DFT of the samples. Let $X(i) = \text{DFT}\{x(k)\}$. Then the Fourier coefficients of $x_a(t)$ can be approximated as follows.

DFT

$$c_i \approx \frac{X(i)}{N}, \quad 0 \le i < N/2 \tag{4.1.5}$$

The DFT produces $M = N/2$ complex Fourier coefficients $\{c_0, c_1, \ldots, c_{M-1}\}$. To obtain the remaining coefficients in (4.1.1), observe from (4.1.2) that for real $x_a(t)$ the coefficients corresponding to negative values of i are the *complex conjugates* of the coefficients corresponding to positive values of i. Thus

$$c_{-i} \approx \frac{X^*(i)}{N}, \quad 0 \le i < N/2 \tag{4.1.6}$$

4.1.2 DC Wall Transformer

Many electronic items receive power from batteries or from a DC wall transformer. A DC wall transformer is an inexpensive power supply that approximates the constant voltage produced by a battery. A block diagram of a typical DC wall transformer is shown in Figure 4.1.

FIGURE 4.1: DC Wall Transformer

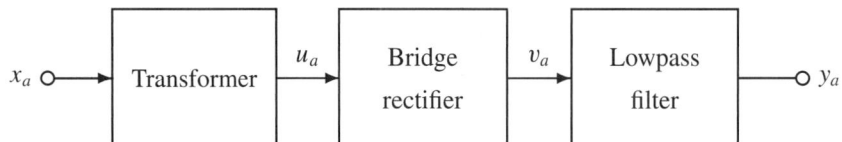

x_a O⟶ [Transformer] ⟶u_a [Bridge rectifier] ⟶v_a [Lowpass filter] ⟶O y_a

The transformer block steps the 120 volt 60 Hz sinusoidal AC input signal x_a down to a lower voltage AC signal u_a. The full wave bridge rectifier consists of four diodes arranged in a configuration that takes the absolute value of the AC signal u_a. Thus the signals $x_a(t)$, $u_a(t)$, and $v_a(t)$ can be modeled as follows where $0 < \alpha < 1$ depends on the desired DC voltage.

$$x_a(t) = 120\sqrt{2}\sin(120\pi t) \tag{4.1.7a}$$

$$u_a(t) = \alpha x_a(t) \tag{4.1.7b}$$

$$v_a(t) = |u_a(t)| \tag{4.1.7c}$$

The full wave bridge rectifier output v_a has a DC component or average value $d_0/2$, plus a periodic component that has a fundamental frequency of $F_0 = 120$ Hz. To produce a pure DC output similar to that of a battery, the nonconstant part of v_a must be filtered out with the lowpass RC filter section whose transfer function is

$$H_a(s) = \frac{1}{RCs + 1} \tag{4.1.8}$$

Ripple voltage

Of course, the lowpass filter is not an ideal filter, so some of the nonconstant part of v_a survives in the wall transformer output y_a in the form of a small AC *ripple* voltage. The end result is that a DC wall transformer output can be modeled as a DC component $d_0/2$, plus a small periodic ripple component with a fundamental frequency of 120 Hz.

$$y_a(t) = \frac{d_0}{2} + \sum_{i=1}^{M-1} d_i \cos(240i\pi t + \theta_i) \tag{4.1.9}$$

Total harmonic distortion

For an ideal wall transformer, there is no AC ripple, so $d_i = 0$ for $i > 0$. Thus we can measure the quality of the wall transformer output (or any other DC power supply output, for that matter) by using the *total harmonic distortion* THD, as follows.

$$P_y = \frac{d_0^2}{4} + \frac{1}{2}\sum_{i=1}^{M-1} d_i^2 \tag{4.1.10a}$$

$$\text{THD} = \frac{100(P_y - d_0^2/4)}{P_y} \ \% \tag{4.1.10b}$$

Notice that this definition of total harmonic distortion is similar to that used in Chapter 1 for an ideal amplifier except that in (4.1.10b), the term $d_0^2/4$ is removed from the numerator, whereas in (1.1.5), the term $d_1^2/2$ was removed from the numerator. This is because for an ideal DC power supply the output should be the zeroth harmonic, while for an ideal amplifier the output should be the first harmonic. The term $d_0^2/4$ represents the power of the DC term or zeroth harmonic, while $d_i^2/2$ represents the power of the ith harmonic for $i > 0$.

From Appendix 1, the coefficients d_i and θ_i of the cosine series in (4.1.9) can be obtained from the complex Fourier series coefficients in (4.1.2) as follows.

$$d_i = 2|c_i|, \quad 0 \le i < M \tag{4.1.11a}$$

$$\theta_i = \tan^{-1}\left(\frac{-\text{Im}\{c_i\}}{\text{Re}\{c_i\}}\right), \quad 0 \le i < M \tag{4.1.11b}$$

To measure the harmonic distortion THD we sample the output at $N = 2M$ points using a sampling frequency of $f_s = NF_0$. If $Y(i) = \text{DFT}\{y_a(kT)\}$, then from (4.1.5) and (4.1.11) we have

$$d_i = \frac{2|Y(i)|}{N}, \quad 0 \le i < M \tag{4.1.12a}$$

$$\theta_i = \tan^{-1}\left(\frac{-\text{Im}\{Y(i)\}}{\text{Re}\{Y(i)\}}\right), \quad 0 \le i < M \tag{4.1.12b}$$

FIGURE 4.2: A
Linear Discrete-time
System with
Transfer Function
$H(z)$

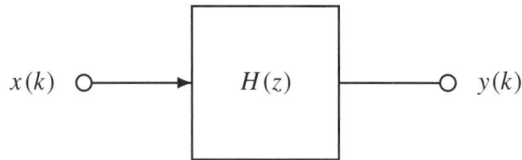

$$x(k) \; \circ \!\!\!\longrightarrow \boxed{\;H(z)\;} \!\!\!\longrightarrow \!\circ \; y(k)$$

4.1.3 Frequency Response

The DFT also can be used to characterize a linear discrete-time system or digital filter. Consider the discrete-time system shown in Figure 4.2. Recall from Chapter 3 that the system transfer function is the Z-transform of the zero-state response divided by the Z-transform of the input.

$$H(z) = \frac{Y(z)}{X(z)} \tag{4.1.13}$$

Frequency response

Also recall from Chapter 3 that if the system is stable and we evaluate the transfer function along the unit circle using $z = \exp(j2\pi f T)$, the resulting function of f is called the *frequency response*. The samples of the frequency response can be easily approximated using the DFT. In particular, if $X(i) = \text{DFT}\{x(k)\}$ and $Y(i) = \text{DFT}\{y(k)\}$, then the approximate frequency response evaluated at discrete frequency $f_i = i f_s / N$ Hz is as follows, where the accuracy of the approximation improves as N increases.

$$H(i) = \frac{Y(i)}{X(i)}, \quad 0 \le i < N \tag{4.1.14}$$

The frequency response specifies how much each sinusoidal input gets scaled in magnitude and shifted in phase as it passes through the system. For the DFT method of evaluating the frequency response in (4.1.14), the input signal $x(k)$ should be chosen such that it has power at all frequencies of interest so as to avoid division by zero. For example, we can let $x(k)$ be a unit impulse or a random white noise signal. Since $H(k)$ is complex, it can be expressed in polar form in terms of its magnitude and phase angle.

$$H(i) = A(i) \exp[j\phi(i)], \quad 0 \le i < N \tag{4.1.15}$$

Once $H(i)$ is known, the steady-state response of the system to a sinusoidal input at a discrete frequency can be obtained from inspection. For example, suppose $x(k) = c \sin(2\pi f_n kT)$ for some $0 \le n < N$. Then the steady-state output generated by this input is scaled in amplitude by $A(n)$ and shifted in phase by $\phi(n)$.

$$y_{ss}(k) = A(n) c \sin[2\pi f_n kT + \phi(n)] \tag{4.1.16}$$

As an illustration, consider a stable second-order discrete-time system characterized by the following transfer function.

$$H(z) = \frac{10 + 20z^{-1} - 5z^{-2}}{1 - .2z^{-1} - .63z^{-2}} \tag{4.1.17}$$

This system is stable with poles at $z = .5$ and $z = -.9$. Suppose the sampling frequency is $f_s = 100$ Hz and the DFT is evaluated using $N = 256$ points. The resulting magnitude response $A(i)$ and phase response $\phi(i)$ for $0 \le i \le N/2$ are shown in Figure 4.3. Note that only the positive frequencies are plotted. Values of i in the range $N/2 < i < N$ correspond to the values of $z = \exp(j2\pi f_i T)$ that are in the lower half of the unit circle and therefore represent negative frequencies.

FIGURE 4.3:
Magnitude
Response and Phase
Response of System
in (4.1.17) Using
$f_s = 100$ Hz and
$N = 256$

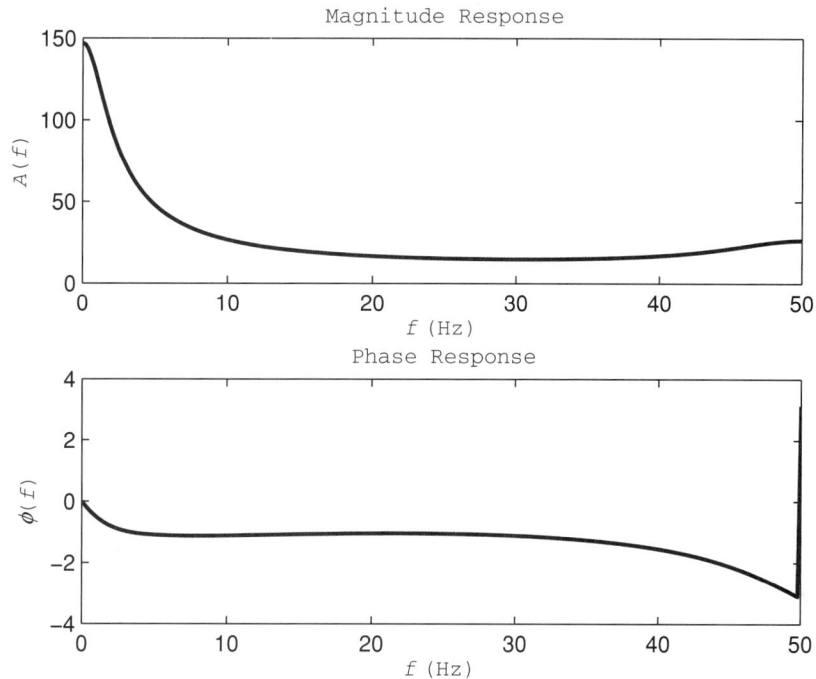

4.2 Discrete-time Fourier Transform (DTFT)

In Chapters 2 and 3 we focused on digital signal processing from the point of view of the system that acts on the input $x(k)$ to produce the output $y(k)$. In this chapter we look more carefully at the characteristics of the signals themselves.

4.2.1 DTFT

Each discrete-time signal can be characterized in terms of how its average power is distributed over frequencies in the interval $[-f_s/2, f_s/2]$. A particularly useful tool for this purpose is the discrete-time Fourier transform or DTFT. It is a transformation that maps a discrete *time* signal into a continuous *frequency* signal as follows.

DEFINITION
....
4.1: DTFT

The *discrete-time Fourier transform* or DTFT of $x(k)$ is denoted $X(f) = \text{DTFT}\{x(k)\}$ and defined as

$$X(f) \triangleq \sum_{k=-\infty}^{\infty} x(k) \exp(-jk2\pi fT), \quad 0 \leq |f| \leq f_s/2$$

Note that the DTFT is really just the Z-transform $X(z)$ evaluated along the unit circle using $z = \exp(j2\pi fT)$. That is,

$$X(f) = X(z)|_{z=\exp(j2\pi fT)}, \quad 0 \leq |f| \leq f_s/2 \qquad (4.2.1)$$

As a consequence, $X(f)$ is well defined if and only if the region of convergence of $X(z)$ includes the unit circle. Given the analysis in Chapters 2 and 3, this means that for a causal signal, the DTFT $X(f)$ exists if $x(k)$ is absolutely summable or if the poles of $X(z)$ all lie strictly inside the unit circle.

Signal spectrum   The DTFT of $x(k)$ is referred to as the *spectrum* of the signal $x(k)$. Since the spectrum $X(f)$ is complex, it can be represented in polar form as $X(f) = A_x(f) \exp[j\phi_x(f)]$ where

$$A_x(f) = |X(f)| \tag{4.2.2a}$$
$$\phi_x(f) = \angle X(f) \tag{4.2.2b}$$

In this case, $A_x(f)$ is called the *magnitude spectrum* of $x(k)$, and $\phi_x(f)$ is called the *phase spectrum* of $x(k)$.

The time signal $x(k)$ can be recovered from its spectrum $X(f)$ using the inverse transform. To isolate sample $x(i)$, multiply $X(f)$ by the complex conjugate exponential $\exp(ji2\pi fT)$, and integrate over one period.

$$
\begin{aligned}
\int_{-f_s/2}^{f_s/2} X(f) \exp(ji2\pi fT) df &= \int_{-f_s/2}^{f_s/2} \sum_{k=0}^{\infty} x(k) \exp(jk2\pi fT) \exp(ji2\pi fT) df \\
&= \sum_{k=0}^{\infty} x(k) \int_{-f_s/2}^{f_s/2} \exp[j(i-k)2\pi fT] df \\
&= \sum_{k=0}^{\infty} x(k) f_s \delta(i-k) \\
&= x(i) f_s
\end{aligned}
\tag{4.2.3}
$$

Here we have used the fact that the complex exponential is periodic with period f_s when $i \neq k$. It is valid to interchange the order of the integral and the sum because $x(k)$ is absolutely summable. Solving (4.2.3) for $x(i)$ then yields the *inverse* DTFT or IDTFT.

$$x(k) = \frac{1}{f_s} \int_{-f_s/2}^{f_s/2} X(f) \exp(jk2\pi fT) df, \quad |k| \geq 0 \tag{4.2.4}$$

The DTFT in Definition 4.1 is sometimes called the *analysis* equation because it decomposes a signal into its spectral components. The IDTFT in (4.2.4) is then called the *synthesis* equation because it reconstructs or synthesizes the signal from its spectral components.

Periodic property   From Euler's identity, $\exp(j2\pi fT)$ is *periodic* with period f_s. It then follows from Definition 4.1 that the spectrum of $x(k)$ is also periodic with period f_s.

$$X(f + f_s) = X(f) \tag{4.2.5}$$

The periodic nature of $X(f)$ also follows from the observation that $X(f)$ is $X(z)$ evaluated along the unit circle, with each period corresponding to trip around the circle. Since $X(f)$ is periodic with period f_s, the frequency is typically restricted to the interval $[-f_s/2, f_s/2]$.

Symmetry property   If the time signal $x(k)$ is real, then the frequency interval can be restricted still further. This is a consequence of the *symmetry* property of the DTFT. When $x(k)$ is real, it follows from Definition 4.1 that

$$X^*(f) = X(-f) \tag{4.2.6}$$

Property	Equation	$x(k)$
Periodic	$X(f + f_s) = X(f)$	General
Symmetry	$X^*(f) = X(-f)$	Real
Even magnitude	$A_x(-f) = A_x(f)$	Real
Odd phase	$\phi_x(-f) = -\phi(f)$	Real

Thus the spectrum of $x(k)$ at negative frequencies is just the complex conjugate of the spectrum of $x(k)$ at positive frequencies. Just as was the case with the frequency response in Chapter 3, the symmetry property can be used to show that the magnitude spectrum of a real signal is an *even* function of f, and the phase spectrum of a real signal is an *odd* function of f.

Real signals

$$A_x(-f) = A_x(f) \tag{4.2.7a}$$

$$\phi_x(-f) = -\phi_x(f) \tag{4.2.7b}$$

Consequently, for real signals all of the essential information about the spectrum is contained in the nonnegative frequency range $[0, f_s/2]$. A summary of the symmetry properties of the DTFT is shown in Table 4.1

Example 4.1

Spectrum of Causal Exponential

As an illustration of using the DTFT to analyze a discrete-time signal, consider the following causal exponential.

$$x(k) = c^k \mu(k)$$

From Table 3.1, the Z-transform of this signal is

$$X(z) = \frac{z}{z - c}$$

$$= \frac{1}{1 - cz^{-1}}$$

Here $X(z)$ has a pole at $z = c$. Thus for the DTFT to converge, it is necessary that $|c| < 1$. From (4.2.1), the spectrum of $x(k)$ is

$$X(f) = \frac{1}{1 - c\exp(-j2\pi fT)}$$

Using Euler's identity, we find that the magnitude spectrum of the causal exponential is

$$A_x(f) = |X(f)|$$

$$= \frac{1}{\sqrt{[1 - c\cos(2\pi fT)]^2 + c^2\sin^2(2\pi fT)}}$$

$$= \frac{1}{\sqrt{1 - 2c\cos(2\pi fT) + c^2}}$$

Similarly, the phase spectrum is

$$\phi_x(f) = \angle X(f)$$

$$= -\tan^{-1}\left[\frac{c\sin(2\pi fT)}{1 - c\cos(2\pi fT)}\right]$$

FIGURE 4.4: Magnitude and Phase Spectra of Causal Exponential with $c = 0.8$

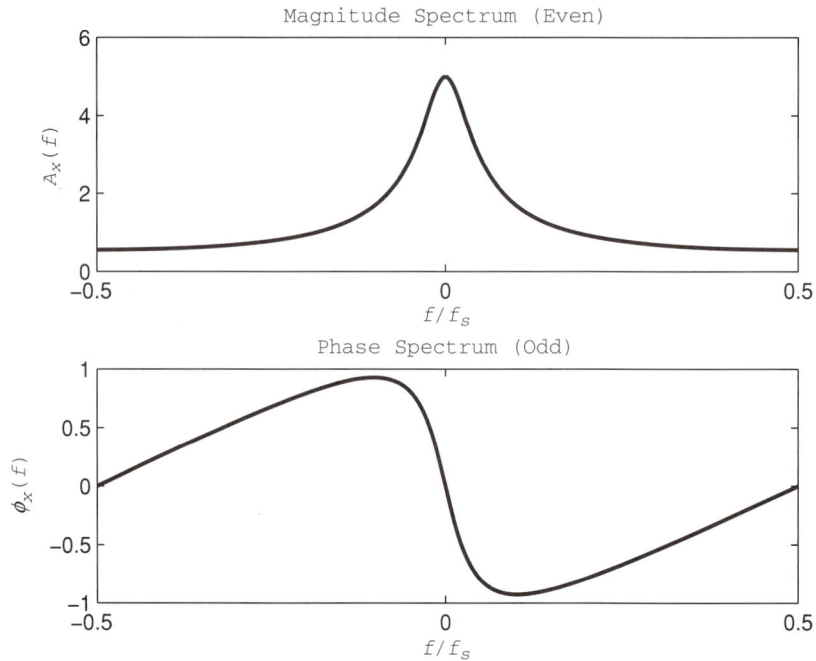

Plots of the magnitude spectra and the phase spectra, obtained by running *exam4_1*, are shown in Figure 4.4. Note how the magnitude spectrum is even and the phase spectrum is odd. Also observe that *normalized frequency* is plotted along the abscissa because $x(k)$ may or may not have been obtained by sampling an underlying continuous-time signal. Thus the independent variable is

Normalized frequency

$$\hat{f} \triangleq \frac{f}{f_s}$$

Based on normalized frequency, $\hat{f}$ is equivalent to setting the sampling interval to $T = 1$ sec.

Signals can be categorized or classified based on the part of the overall spectrum or frequency range that they occupy (Proakis and Manolakis, 1992). Table 4.2 summarizes some practical signals that include examples taken from biomedical, geological, and communications applications. Note the impressive range of frequencies (26 orders of magnitude!) going from circadian rhythms that oscillate at approximately one cycle per day to lethal gamma rays that are faster than a billion billion Hz.

4.2.2 Properties of the DTFT

The DTFT has several properties that it inherits directly from the Z-transform. There are a number of additional properties that are specific to the DTFT itself.

Time Shift Property

An example of a property that comes directly from the Z-transform is the time shift property. Recall that multiplying the Z-transform by z^{-r} is equivalent to delaying the signal $x(k)$ by r

TABLE 4.2: ▶
Frequency Ranges
of Some Practical
Signals

Signal Type	Frequency Range (Hz)
Circadian rhythm	$1.1 \times 10^{-5} - 1.2 \times 10^{-5}$
Earthquake	$10^{-2} - 10^{1}$
Electrocardiogram (ECG)	$0 - 10^{2}$
Electroencephalogram (EEG)	$0 - 10^{2}$
AC power	$5 \times 10^{2} - 6 \times 10^{2}$
Wind	$10^{2} - 10^{3}$
Speech	$10^{2} - 4 \times 10^{3}$
Audio	$2 \times 10^{1} - 2 \times 10^{4}$
AM radio	$5.4 \times 10^{5} - 1.6 \times 10^{6}$
FM radio	$8.8 \times 10^{7} - 1.08 \times 10^{8}$
Cell phone	$8.1 \times 10^{8} - 9 \times 10^{8}$
TV	$3 \times 10^{8} - 9.7 \times 10^{8}$
GPS	$1.52 \times 10^{9} - 1.66 \times 10^{9}$
Shortwave radio	$3 \times 10^{6} - 3 \times 10^{9}$
Radar, microwave	$3 \times 10^{9} - 3 \times 10^{12}$
Infrared light	$3 \times 10^{12} - 4.3 \times 10^{14}$
Visible light	$4.3 \times 10^{14} - 7.5 \times 10^{14}$
Ultraviolet light	$7.5 \times 10^{14} - 3 \times 10^{17}$
X-ray	$3 \times 10^{17} - 3 \times 10^{19}$
Gamma ray	$5 \times 10^{19} - 10^{21}$

samples. Since $X(f)$ is $X(z)$ evaluated along the unit circle, the factor z^{-r} is $\exp(-j2\pi rfT)$. Thus the DTFT version of the *time shift* property is

Time shift property

$$\text{DTFT}\{x(k-r)\} = \exp(-j2\pi rfT)X(f) \tag{4.2.8}$$

Frequency Shift Property

Frequency shift property

There is a dual to the time shift property called the frequency shift property. Using the IDFT in (4.2.4) and a change of variable

$$\text{IDFT}\{X(f-F_0)\} = \frac{1}{f_s}\int_{-f_s/2}^{f_s/2} X(f-F_0)\exp(jk2\pi fT)df$$

$$= \frac{1}{f_s}\int_{-f_s/2+F_0}^{f_s/2+F_0} X(F)\exp[jk2\pi(F+F_0)T]dF, \quad F = f - F_0$$

$$= \frac{\exp(jk2\pi F_0 T)}{f_s}\int_{-f_s/2}^{f_s/2} X(F)\exp(jk2\pi FT)dF$$

$$= \exp(jk2\pi F_0 T)x(k) \tag{4.2.9}$$

Thus the *frequency shift* property of the DTFT is

$$\text{DTFT}\{\exp(jk2\pi F_0 T)x(k)\} = X(f - F_0) \qquad (4.2.10)$$

Parseval's Identity

Recall from Chapter 2 that absolutely summable signals are square summable and therefore energy signals. For energy signals, there is a simple relationship between the time signal and its spectrum. Using Definition 4.1, we have

$$\int_{-f_s/2}^{f_s/2} X(f)Y^*(f)df = \int_{-f_s/2}^{f_s/2} \sum_{k=-\infty}^{\infty} x(k)\exp(-j2\pi kfT) \sum_{i=-\infty}^{\infty} y^*(i)\exp(j2\pi ifT)df$$

$$= \sum_{k=-\infty}^{\infty}\sum_{i=-\infty}^{\infty} x(k)y^*(i)\int_{-f_s/2}^{f_s/2}\exp[-j2\pi(k-i)fT]df$$

$$= \sum_{k=-\infty}^{\infty}\sum_{i=-\infty}^{\infty} x(k)y^*(i)f_s\delta(k-i)$$

$$= f_s\sum_{k=-\infty}^{\infty} x(k)y^*(k) \qquad (4.2.11)$$

This leads to the DTFT version of the following result known as *Parseval's identity*.

PROPOSITION
....
4.1: Parseval's Identity:
DTFT

Let $x(k)$ and $y(k)$ be absolutely summable with discrete-time Fourier transforms $X(f)$ and $Y(f)$, respectively. Then

$$\sum_{k=-\infty}^{\infty} x(k)y^*(k) = \frac{1}{f_s}\int_{-f_s/2}^{f_s/2} X(f)Y^*(f)df$$

Note the when $y(k) = x(k)$ in Proposition 4.1, the left-hand side reduces to the energy of the signal $x(k)$.

$$\sum_{k=-\infty}^{\infty} |x(k)|^2 = \frac{1}{f_s}\int_{-f_s/2}^{f_s/2} |X(f)|^2 df \qquad (4.2.12)$$

Energy density

Thus Parseval's identity provides us with a different way to compute the energy E_x using the spectrum $X(f)$. With this in mind, the *energy density* of $x(k)$ is defined as follows.

$$S_x(f) \triangleq |X(f)|^2 \qquad (4.2.13)$$

Note that the total energy of a signal is just the integral of the energy density. More generally, for a real signal where the energy density is an even function, the amount of energy in the nonnegative frequency band $[f_1, f_2]$ is

$$E(f_1, f_2) = 2\int_{f_1}^{f_2} S_x(f)df \qquad (4.2.14)$$

The factor two in (4.2.14) accounts for the negative frequencies. The total energy is $E_x = E(0, f_s/2)$. Recall that the auto-correlation of $x(k)$ evaluated at a lag of $k = 0$ is another way to express the total energy $E_x = r_{xx}(0)$.

Wiener-Khintchine Theorem

One of the properties that is inherited from the Z-transform is the correlation property. Recall that when $z = \exp(j2\pi fT)$, replacing z by $1/z$ is equivalent to replacing f by $-f$. Therefore from Table 3.3, the DTFT of the cross-correlation $r_{yx}(k)$ of finite signals of length L is

$$R_{yx}(f) = \frac{Y(f)X(-f)}{L} \qquad (4.2.15)$$

Some authors define cross-correlation using infinite signals and therefore do not divide by the signal length L. The definition of finite cross-correlation introduced in Chapter 2 is used here because it is more consistent with the generalization of cross-correlation to random signals that is used in Chapter 9.

It is of interest to consider the case of auto-correlation when $y(k) = x(k)$. In this case $X(f)X(-f) = |X(f)|^2 = S_x(f)$. This leads to the *Wiener-Khintchine theorem* which says that the DTFT of the auto-correlation of $x(k)$ is a scaled version of the energy density spectrum of $x(k)$.

$$R_{xx}(f) = \frac{S_x(f)}{L} \qquad (4.2.16)$$

The properties of the DTFT are summarized in Table 4.3. Most of these properties have direct analogs with the Z-transform properties in Table 3.3, and are obtained by substituting $z = \exp(j2\pi fT)$.

TABLE 4.3: ▶
DTFT Properties

Property	Time Signal	DTFT
Linearity	$ax(k) + by(k)$	$aX(f) + bY(f)$
Time shift	$x(k - r)$	$\exp(-j2\pi r fT)X(f)$
Frequency shift	$\exp(jk2\pi F_0 T)x(k)$	$X(f - F_0)$
Time reversal	$x(-k)$	$X(-f)$
Complex conjugate	$x^*(k)$	$X^*(-f)$
Convolution	$h(k) \star x(k)$	$H(f)X(f)$
Correlation	$r_{yx}(k)$	$\dfrac{Y(f)X(-f)}{L}$
Wiener-Khintchine	$r_{xx}(k)$	$\dfrac{S_x(f)}{L}$
Parseval	$\displaystyle\sum_{k=-\infty}^{\infty} x(k)y^*(k)$	$\dfrac{1}{f_s}\displaystyle\int_{-f_s/2}^{f_s/2} X(f)Y^*(f)df$
	$\displaystyle\sum_{k=-\infty}^{\infty} |x(k)|^2$	$\dfrac{1}{f_s}\displaystyle\int_{-f_s/2}^{f_s/2} |X(f)|^2 df$

Example 4.2

IDTFT of Ideal Lowpass Characteristic

An ideal lowpass filter with a cutoff frequency of $0 < F_c < f_s/2$ has a phase response of $\phi(f) = 0$ and a magnitude response consisting of a rectangular window of radius F_c.

$$H_{\text{low}}(f) = \mu(f + F_c) - \mu(f - F_c), \quad 0 \le |f| \le f_s/2$$

Here the rectangle window is expressed using a step up at $f = -F_c$, followed by a step down at $f = F_c$. Consider the problem of finding the impulse response $h_{\text{low}}(k)$. Using the expression for the IDTFT in (4.2.4) and the identities from Appendix 2

$$
\begin{aligned}
h_{\text{low}}(k) &= \frac{1}{f_s} \int_{-f_s/2}^{f_s/2} H_{\text{low}}(f) \exp(j2\pi kfT)df \\
&= \frac{1}{f_s} \int_{-F_c)}^{F_c} \exp(j2\pi kfT)df \\
&= \frac{1}{f_s} \left. \frac{\exp(j2\pi kfT)}{j2\pi kT} \right|_{-F_c}^{F_c} \\
&= \frac{\exp(j2\pi kF_cT) - \exp(-j2\pi kF_cT)}{j2\pi k} \\
&= \frac{\sin(2\pi kF_cT)}{\pi k} \\
&= \frac{2F_cT \sin(2\pi kF_cT)}{2\pi kF_cT}
\end{aligned}
$$

Recall from Section 1.2 that $\text{sinc}(x) = \sin(\pi x)/(\pi x)$. Thus the ideal lowpass impulse response can be written in terms of the sinc function as

$$h_{\text{low}}(k) = 2F_cT \, \text{sinc}(2kF_cT)$$

Plots of $h_{\text{low}}(k)$ and $H_{\text{low}}(f)$ for the case when $F_c = f_s/4$ are shown in Figure 4.5. Notice that the impulse response is noncausal. Therefore, an ideal lowpass characteristic cannot be achieved with a physically realizable filter. In Chapters 6 and 7 we examine a variety of ways to approximate the ideal lowpass characteristic with FIR and IIR digital filters.

FIGURE 4.5: Impulse Response (a) and Frequency Response (b) of an Ideal Lowpass Filter with Cutoff Frequency $F_0 = f_s/4$.

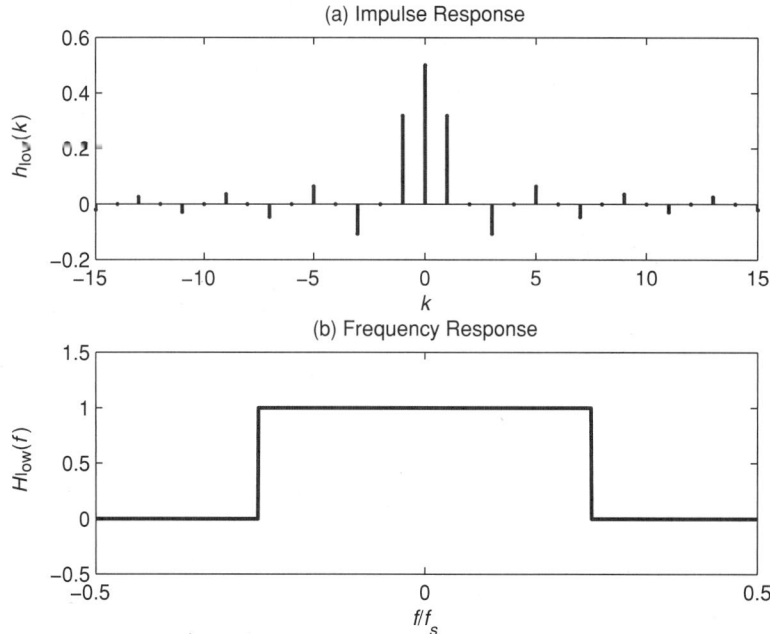

TABLE 4.4: ▶
Basic DTFT Pairs

$x(k)$	$X(f)$	Parameters
$\delta(k)$	1	—
$c^k \mu(k)$	$\dfrac{\exp(j2\pi fT)}{\exp(j2\pi fT) - c}$	$|c| < 1$
$k(c)^k \mu(k)$	$\dfrac{c\exp(j2\pi fT)}{[\exp(j2\pi fT) - c]^2}$	$|c| < 1$
$2F_c T \operatorname{sinc}(2kF_c T)$	$\mu(f + F_c) - \mu(f - F_c)$	$0 < F_c < f_s/2$
$\mu(k+r) - \mu(k - r - 1)$	$\dfrac{\sin[\pi(2r+1)f]}{\sin(\pi f)}$	—

A list of basic DTFT pairs is shown in Table 4.4. This was constructed by starting with the stable Z-transform pairs in Table 3.2 and adding some additional pairs.

● ● ● ● ● ● ● ● ● ● ● ● ● ● ● ● ● ●

4.3 Discrete Fourier Transform (DFT)

4.3.1 DFT

In this section we focus on causal finite signals. The discrete Fourier transform or DFT can be regarded as a special case of the discrete-time Fourier transform. Recall from Definition 4.1 that the DTFT of a causal signal $x(k)$ is

$$X(f) = \sum_{k=0}^{\infty} x(k) \exp(-jk2\pi fT), \quad 0 \le |f| \le f_s/2 \qquad (4.3.1)$$

Although the DTFT is an important tool for analyzing discrete-time signals, it suffers from certain practical limitations when used as a computational tool. One drawback is that a direct evaluation of $X(f)$ using the definition requires an infinite number of floating point operations or FLOPs. This is compounded by a second computational drawback, namely, that the transform itself must be evaluated at an infinite number frequencies f. The first limitation can be removed by focusing on signals of finite duration. Since $X(f)$ is just $X(z)$ evaluated along the unit circle, $X(f)$ converges for signals that are absolutely summable. But if $x(k)$ is absolutely summable, then $|x(k)| \to 0$ as $k \to \infty$. Consequently, for sufficiently large values of N, $X(f)$ can be approximated by the following finite sum.

$$X(f) \approx \sum_{k=0}^{N-1} x(k) \exp(-jk2\pi fT) \qquad (4.3.2)$$

To address the second limitation, we sample $X(f)$ by evaluating it at N equally spaced values of f. In particular, consider the following *discrete frequencies* equally spaced over one period of $X(f)$.

$$f_i = \frac{if_s}{N}, \quad 0 \le i < N \qquad (4.3.3)$$

Bin frequencies   The discrete frequencies f_i are sometimes referred to as *bin frequencies*. Let z_i be the point in the complex plane corresponding to frequency f_i.

$$z_i = \exp(ji2\pi/N) \qquad (4.3.4)$$

Notice that $|z_i| = 1$. Thus the N evaluation points in (4.3.4) are equally spaced around the unit circle, as shown in Figure 4.6 for the case $N = 8$. Observe that the unit circle is traversed

FIGURE 4.6: Evaluation
Points of DFT

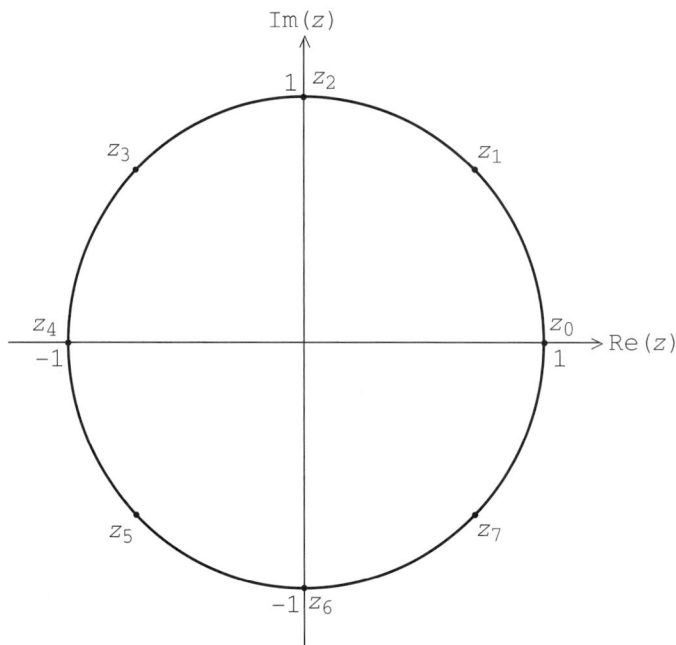

in the counterclockwise direction with $z_0 = 1$, $z_{N/4} = j$, $z_{N/2} = -1$, and $z_{3N/4} = -j$. The formulation of the discrete Fourier transform can be simplified by introducing the following factor which corresponds to z_i for $i = -1$.

$$W_N \triangleq \exp(-j2\pi/N) \qquad (4.3.5)$$

Nth root of unity
Orthogonal property

If we use Euler's identity, it is clear that $W_N^N = 1$. Consequently, the factor W_N can be thought of as the *N*th *root of unity*. Note that W_N^{ik} can be regarded as a function of i or k. This function has a number of interesting symmetry properties including the following *orthogonal property*, whose proof is left as an exercise (see Problem 4.19).

$$\sum_{i=0}^{N-1} W_N^{ik} = N\delta(k), \quad 0 \le k < N \qquad (4.3.6)$$

The discrete values of z in (4.3.4) can be reformulated in terms of the *N*th root of unity as $z_i = W_N^{-i}$. When this value for z_i is substituted into the truncated expression for the DTFT in (4.3.2), the resulting transformation from discrete-time $x(k)$ to discrete-frequency $X(i)$ is called the discrete Fourier transform or DFT.

DEFINITION
....

4.2: DFT

Let $x(k)$ be a causal *N*-point signal, and let $W_N = \exp(-j2\pi/N)$. The *discrete Fourier transform* of $x(k)$, denoted $X(i) = \text{DFT}\{x(k)\}$, is defined

$$X(i) \triangleq \sum_{k=0}^{N-1} x(k)W_N^{ik}, \quad 0 \le i < N$$

In terms of notation, it should pointed out that the same base symbol is being used to denote the Z-transform $X(z)$ the DTFT $X(f)$, and the DFT, $X(i)$. This is consistent with the convention adopted earlier where the argument type, a complex z, a real f, or an integer i, is used to distinguish between the different cases and dictate the meaning of X. This approach is used in order to avoid a proliferation of different, but related, symbols.

The continuous-time Fourier transform $X_a(f) = F\{x_a(t)\}$ has an inverse whose form is almost identical to the original transform. The same is true of the discrete Fourier transform.

IDFT   The *inverse of the DFT*, which is denoted $x(k) = \text{IDFT}\{X(i)\}$, is computed as follows.

$$x(k) = \frac{1}{N} \sum_{i=0}^{N-1} X(i) W_N^{-ki}, \quad 0 \le k < N \tag{4.3.7}$$

To verify that (4.3.7) does indeed represent the IDFT, we use (4.3.6) and Definition 4.2.

$$\sum_{i=0}^{N-1} X(i) W_N^{-ki} = \sum_{i=0}^{N-1} \left[\sum_{m=0}^{N-1} x(m) W_N^{im} \right] W_N^{-ki}$$

$$= \sum_{m=0}^{N-1} x(m) \sum_{i=0}^{N-1} W_N^{(m-k)i}$$

$$= \sum_{m=0}^{N-1} x(m) N \delta(m - k)$$

$$= N x(k) \tag{4.3.8}$$

Thus the IDFT is identical to the DFT except that W_N has been replaced by its complex conjugate $W_N^* = W_N^{-1}$, and the final result is normalized by N. In practical terms this means that any algorithm devised to compute the DFT can also be used, with only minor modification, to compute the IDFT as well.

4.3.2 Matrix Formulation

The DFT can be interpreted as a transformation or mapping from a vector of input samples x to a vector of output samples X.

$$x = [x(0), x(1), \ldots, x(N-1)]^T \tag{4.3.9a}$$
$$X = [X(0), X(1), \ldots, X(N-1)]^T \tag{4.3.9b}$$

Here x and X are $N \times 1$ column vectors, written as transposed rows to conserve space. Since the DFT is a linear transformation from x to X, it can be represented by an $N \times N$ matrix. Consider, in particular, the matrix $W_{ik} = W_N^{ik}$ where the row index i and column index k are assumed to start from zero. For example, for the case $N = 5$

$$W = \begin{bmatrix} W_N^0 & W_N^0 & W_N^0 & W_N^0 & W_N^0 \\ W_N^0 & W_N^1 & W_N^2 & W_N^3 & W_N^4 \\ W_N^0 & W_N^2 & W_N^4 & W_N^6 & W_N^8 \\ W_N^0 & W_N^3 & W_N^6 & W_N^9 & W_N^{12} \\ W_N^0 & W_N^4 & W_N^8 & W_N^{12} & W_N^{16} \end{bmatrix} \tag{4.3.10}$$

Matrix DFT

Note that W is symmetric. Comparison of (4.3.9) and (4.3.10) with Definition 4.2 reveals that the DFT can be expressed in vector form as

$$X = Wx \qquad (4.3.11)$$

For small values of N, (4.3.11) can be used to compute the DFT. As we shall see, for moderate to large values of N, a much more efficient FFT implementation of the DFT is the method of choice.

The vector form of the DFT also can be used to compute the IDFT. Multiplying both sides of (4.3.11) on the left by W^{-1} yields $x = W^{-1}X$. If we compare this equation with (4.3.7), we find that there is no need to explicitly invert the matrix W because $W^{-1} = W^*/N$. Thus the vector form for the IDFT is

Matrix IDFT

$$x = \frac{W^* X}{N} \qquad (4.3.12)$$

The following examples illustrate computation of the DFT and the IDFT for small values of N.

Example 4.3

DFT

As an example of computing a DFT using the vector form, suppose the input samples are as follows.

$$x = [3, -1, 0, 2]^T$$

Thus $N = 4$ and from (4.3.5)

$$W_4 = \cos\left(\frac{2\pi}{4}\right) - j\sin\left(\frac{2\pi}{4}\right)$$
$$= -j$$

Next from (4.3.10) and (4.3.11), the DFT of x is

$$X = Wx$$

$$= \begin{bmatrix} W_4^0 & W_4^0 & W_4^0 & W_4^0 \\ W_4^0 & W_4^1 & W_4^2 & W_4^3 \\ W_4^0 & W_4^2 & W_4^4 & W_4^6 \\ W_4^0 & W_4^3 & W_4^6 & W_4^9 \end{bmatrix} x$$

$$= \begin{bmatrix} 1 & 1 & 1 & 1 \\ 1 & -j & -1 & j \\ 1 & -1 & 1 & -1 \\ 1 & j & -1 & -j \end{bmatrix} \begin{bmatrix} 3 \\ -1 \\ 0 \\ 2 \end{bmatrix}$$

$$= \begin{bmatrix} 4 \\ 3 + j3 \\ 2 \\ 3 - j3 \end{bmatrix}$$

Note that even though the signal $x(k)$ is real, its DFT $X(i)$ is complex.

Example 4.4

IDFT

As a numerical check, suppose we compute the IDFT of the result from Example 4.3.

$$X = [4, 3 + j3, 2, 3 - j3]^T$$

Using $N = 4$, the matrix W from Example 4.3, and (4.3.12), we have

$$x = \frac{W^* X}{N}$$

$$= \frac{1}{4} \begin{bmatrix} 1 & 1 & 1 & 1 \\ 1 & -j & -1 & j \\ 1 & -1 & 1 & -1 \\ 1 & j & -1 & -j \end{bmatrix}^* \begin{bmatrix} 4 \\ 3 + j3 \\ 2 \\ 3 - j3 \end{bmatrix}$$

$$= \frac{1}{4} \begin{bmatrix} 1 & 1 & 1 & 1 \\ 1 & j & -1 & -j \\ 1 & -1 & 1 & -1 \\ 1 & -j & -1 & j \end{bmatrix} \begin{bmatrix} 4 \\ 3 + j3 \\ 2 \\ 3 - j3 \end{bmatrix}$$

$$= \frac{1}{4} \begin{bmatrix} 12 \\ -4 \\ 0 \\ 8 \end{bmatrix} = \begin{bmatrix} 3 \\ -1 \\ 0 \\ 2 \end{bmatrix} \checkmark$$

4.3.3 Fourier Series and Discrete Spectra

Periodic signals have a special form of spectrum called a discrete spectrum. Suppose $x_a(t)$ is a periodic continuous-time signal with period T_0 and fundamental frequency $F_0 = 1/T_0$. Then from Appendix 1, $x_a(t)$ can be expanded into a complex Fourier series

$$x_a(t) = \sum_{i=-\infty}^{\infty} c_i \exp(ji2\pi F_0 t) \tag{4.3.13}$$

where the ith complex Fourier coefficient is

$$c_i = \frac{1}{T_0} \int_0^{T_0} x_a(t) \exp(-ji2\pi F_0 t)dt, \quad 0 \le |i| < \infty \tag{4.3.14}$$

Power signals   Let $x(k) = x_a(kT)$ be the kth sample of $x_a(t)$ using a sampling interval of T. Since $x(k)$ is a *power signal*, to examine the spectrum of $x(k)$, we need to generalize $X(f)$ to include the possibility of impulse terms such as $\delta_a(f - F_0)$. By starting with $X(f) = \delta_a(f - F_0)$, applying the inverse DTFT, and using the sifting property of the unit impulse, we can show that the DTFT of a complex exponential in the time domain is a shifted impulse in the frequency domain.

$$\text{DTFT}\{f_s \exp(j2\pi F_0 kT)\} = \delta_a(f - F_0) \tag{4.3.15}$$

From (4.3.15) and the linearity of the DTFT, the spectrum of the periodic signal $x(k)$ is then

$$X(f) = \frac{1}{f_s} \sum_{i=-\infty}^{\infty} c_i \delta_a(f - iF_0) \tag{4.3.16}$$

Discrete spectrum

Notice that $X(f)$ in (4.3.16) is zero everywhere except at the harmonic frequencies $i F_0$, where it contains impulses of strength c_i/f_s. Since $X(f)$ is zero except at integer multiples of F_0, it is referred to as a discrete-frequency spectrum or simply a *discrete spectrum*. For a periodic signal with a discrete spectrum, all of the power is concentrated at the fundamental frequency F_0 and its harmonics. This includes the zeroth harmonic if the DC component or average value of $x(k)$ is nonzero.

Fourier Coefficients

The periodic continuous-time signal $x_a(t)$ in (4.3.13) can be approximated by truncating the Fourier series to M harmonics.

$$x_a(t) \approx \sum_{i=-(M-1)}^{M-1} c_i \exp(ji2\pi F_0 t) \tag{4.3.17}$$

Suppose that $x_a(t)$ is sampled at $N = 2M$ points using a sampling rate of $f_s = NF_0$. In this case the N samples cover one period of $x_a(t)$ with $T_0 = NT$. If the number of samples per period is sufficiently large, then over the ith sampling interval the integrand in (4.3.14) can be approximated by its initial value. This leads to the following approximation for the ith Fourier coefficient.

$$
\begin{aligned}
c_i &\approx \frac{1}{NT} \sum_{k=0}^{N-1} x_a(kT) \exp(-ji2\pi F_0 kT)T \\
&= \frac{1}{N} \sum_{k=0}^{N-1} x(k) \exp(-jik2\pi/N) \\
&= \frac{1}{N} \sum_{k=0}^{N-1} x(k) W_N^{ik} \\
&= \frac{X(i)}{N}, \quad 0 \le i < M
\end{aligned}
\tag{4.3.18}
$$

Fourier coefficients

Hence the *Fourier coefficients* of $x_a(t)$ can be obtained from the DFT of the samples of $x_a(t)$. For a real signal $x_a(t)$, it follows from (4.3.14) that $c_{-i} = c_i^*$. Thus the complete set of Fourier coefficients is

$$c_i = \begin{cases} \dfrac{X(i)}{N}, & 0 \le i < M \\ \dfrac{X^*(i)}{N}, & -M < i < 0 \end{cases} \tag{4.3.19}$$

The Fourier series is often expressed in real form using either sines and cosines or cosines with phase shift.

$$x_a(t) = \frac{d_0}{2} + \sum_{i=1}^{\infty} d_i \cos(2\pi i F_0 t + \theta_i) \tag{4.3.20}$$

The magnitude d_i and phase angle θ_i of the ith harmonic also can be obtained using the DFT. From Appendix 1 and (4.3.19), we have

$$d_i \approx \frac{2|X(i)|}{N} \tag{4.3.21a}$$
$$\theta_i \approx \angle X(i) \tag{4.3.21b}$$

In general, DFT sample $X(i)$ specifies the magnitude and phase angle of the ith spectral component of $x(k)$, where $x(k)$ can be thought of as one cycle of a longer periodic signal

Signal spectra $x_p(k)$. The DFT can be written in polar form as $X(i) = A_x(i) \exp[j\phi_x(i)]$. In this case the *magnitude spectrum* $A_x(i)$ and *phase spectrum* $\phi_x(i)$ are defined as follows for $0 \le i < N$.

$$A_x(i) \triangleq |X(i)| \tag{4.3.22a}$$
$$\phi_x(i) \triangleq \angle X(i) \tag{4.3.22b}$$

The average power of $x(k)$ at discrete frequency $f_i = i f_s / N$ can be determined from the *power density spectrum* which is just the square of the magnitude spectrum, normalized by N.

$$S_x(i) \triangleq \frac{|X(i)|^2}{N}, \quad 0 \le i < N \tag{4.3.23}$$

Example 4.5

Spectra

As a very simple example of a discrete time signal and its spectra, consider the unit impulse $x(k) = \delta(k)$. Using Definition 4.2

$$X(i) = \sum_{k=0}^{N-1} x(k) W_N^{ik}$$
$$= \sum_{k=0}^{N-1} \delta(k) W_N^{ik}$$
$$= 1, \quad 0 \le i < N$$

It then follows from (4.3.22) and (4.3.23) that the magnitude, phase, and power density spectra of the unit impulse for $0 \le i < N$ are

$$A_x(i) = 1$$
$$\phi_x(i) = 0$$
$$S_x(i) = \frac{1}{N}$$

The fact that $S_x(i)$ is a constant nonzero value for all i means that the unit impulse has its power distributed evenly over all N discrete frequencies.

FDSP Functions

The FDSP toolbox contains the following function for evaluating the magnitude, phase, and power density spectra of a finite discrete-time signal using the DFT.

```
% F_SPEC: Compute magnitude, phase, and power density spectra
%
% Usage:
%        [A,phi,S,f] = f_spec (x,N,fs);
```

```
% Pre:
%          x  = vector of length M containing signal samples
%          N  = optional integer specifying number of points
%               in spectra (default M). If N > M, then N-M
%               zeros are padded to x.
%          fs = optional sampling frequency in Hz (default 1)
% Post:
%          A   = vector of length N containing magnitude spectrum
%          phi = vector of length N containing phase spectrum (radians)
%          S   = vector of length N containing power density spectrum
%          f   = vector of length N containing the discrete evaluation
%                frequencies: 0 <= f(i) <= (N-1)*fs/N.
%
```

For continuous-time signals, the Fourier transform $X_a(f) = F\{x_a(t)\}$ is used to compute the spectrum. Together with the DTFT and the DFT this makes three Fourier transforms for computing signal spectra. For comparison they are summarized in Table 4.5. They differ from one another in the types of independent variables, continuous or discrete.

The DTFT is a special case of the Z-transform evaluated along the unit circle. For causal finite signals, the DFT is a sampled version of the DTFT. For the DTFT and the DFT to exist, the region of convergence of the Z-transform must include the unit circle. If $x(k)$ is causal, then $x(k)$ must be absolutely summable or, equivalently, the poles of $X(z)$ must lie strictly inside the unit circle. The relationships between $X(z)$, $X(f)$, and $X(i)$ are summarized in Figure 4.7.

4.3.4 DFT Properties

Like the Z-transform and the DTFT, the DFT has a number of important properties. To begin with, the Nth root of unity W_N satisfies some useful symmetry properties. For convenient reference, the symmetry properties of W_N are summarized in Table 4.6. Each entry can be verified using (4.3.5). Note that the first four properties verify that W_N^k moves clockwise around the unit circle as k ranges from 0 to N. The remaining properties are useful for developing important properties of the DFT and the fast Fourier transform or FFT.

Periodic Property

Note from Definition 4.2 that the dependence of $X(i)$ on i occurs only in the exponent of W_N. As a result, $X(i)$ can be regarded as a function that is defined for *all* integer values of i, not just for $0 \le i < N$. It is of interest to examine what happens when we go outside the range

TABLE 4.5: ▶
Comparison of
Fourier Transforms

Transform	Symbol	Time	Frequency
Fourier transform	$X_a(f)$	Continuous	Continuous
Discrete-time Fourier transform	$X(f)$	Discrete	Continuous
Discrete Fourier transform	$X(i)$	Discrete	Discrete

FIGURE 4.7:
Relationships
between the
Z-transform, the
DTFT, and the DFT
of Causal Signals

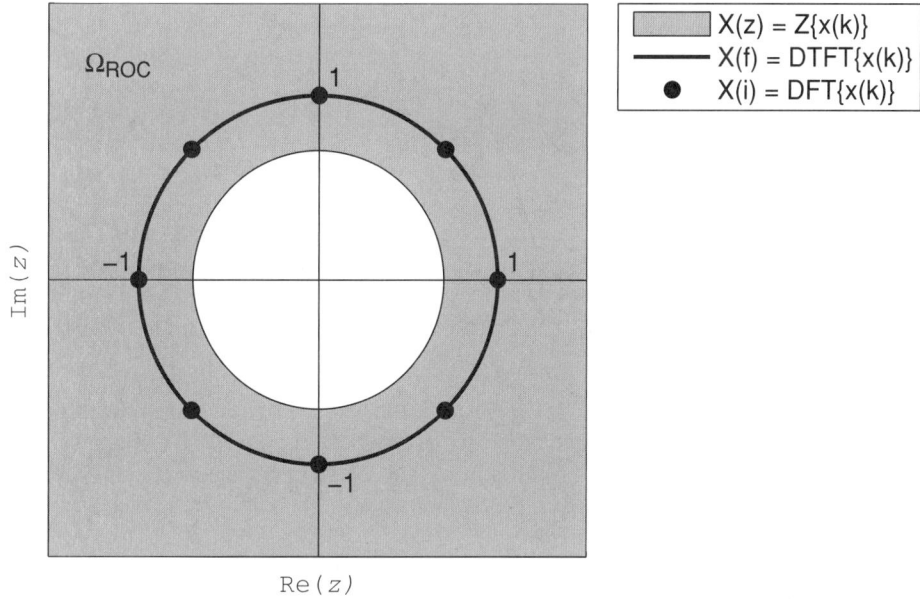

from 0 to $N - 1$. First, note from entry 4 of Table 4.6 that $W_N^{Nk} = 1$ for every integer k. We use this result to demonstrate that $X(i)$ is periodic.

$$X(i + N) = \sum_{k=0}^{N-1} x(k) W_N^{(i+N)k}$$

$$= \sum_{k=0}^{N-1} x(k) W_N^{ik}$$

$$= X(i) \qquad (4.3.24)$$

Periodic property   Thus the DFT is periodic with a period of N. This is analogous to $X(f)$ being periodic with period f_s.

Midpoint Symmetry Property

If the signal $x(k)$ is *real*, then all of the information needed to reconstruct the N real points of $x(k)$ is contained in the first $N/2$ complex points of $X(i)$. To see this, we compute the DFT, starting at the end and working backwards. In particular from Definition 4.2, Table 4.6, and

TABLE 4.6: ▶
Symmetry
Properties of W_N

Property	Description	Property	Description
1	$W_N^{N/4} = -j$	5	$W_N^{(i+N)k} = W_N^{ik}$
2	$W_N^{N/2} = -1$	6	$W_N^{i+N/2} = -W_N^{i}$
3	$W_N^{3N/4} = j$	7	$W_N^{2i} = W_{N/2}^{i}$
4	$W_N^{N} = 1$	8	$W_N^{*} = W_N^{-1}$

(4.3.24), we have

$$
\begin{aligned}
X(N-i) &= \sum_{k=0}^{N-1} x(k) W_N^{(N-i)k} \\
&= \sum_{k=0}^{N-1} x(k) W_N^{-ik} W_N^{Nk} \\
&= \sum_{k=0}^{N-1} x(k) W_N^{-ik} \\
&= X^*(i)
\end{aligned}
\tag{4.3.25}
$$

The most important consequence of (4.3.25) lies in the observation that for a real signal the DFT contains *redundant* information. Recall that $X(i)$ can be expressed in polar form as $X(i) = A_x(i) \exp[j\phi_x(i)]$, where $A_x(i)$ is the magnitude spectrum and $\phi_x(i)$ is the phase spectrum. In many cases of practical interest, N is a power of two and therefore even. When N is even, it follows from (4.3.25) that the magnitude and phase spectra exhibit the following symmetry about the *midpoint* $X(N/2)$.

Midpoint symmetry property

$$
A_x(N/2+i) = A_x(N/2-i), \quad 0 \le i < N/2 \tag{4.3.26a}
$$
$$
\phi_x(N/2+i) = -\phi_x(N/2-i), \quad 0 \le i < N/2 \tag{4.3.26b}
$$

Thus the magnitude spectrum $A_x(i)$ exhibits *even* symmetry about the midpoint, while the phase spectrum $\phi_x(i)$ exhibits *odd* symmetry about the midpoint. The following numerical example illustrates this important observation.

Example 4.6   **DFT Midpoint Symmetry**

Consider the following real signal of length $N = 256$.

$$
x(k) = [.8^k - (-.9)^k]\mu(k), \quad 0 \le k < 256
$$

The DFT of this signal can be found by running the script *exam4_6*. Plots of the resulting magnitude spectrum $A_x(i)$ and phase spectrum $\phi_x(i)$ for $0 \le i < N$ are shown in Figure 4.8. Observe the even symmetry of $A_x(i)$ about sample $N/2 = 128$ and the odd symmetry of $\phi_x(i)$ about $N/2$. In this case, all of the information about the 256 real points in $x(k)$ is encoded in the first 128 complex points of $X(i)$.

A summary of the symmetry properties of the DFT is shown in Table 4.7. These are analogous to the symmetry properties of the infinite-dimensional DTFT in Table 4.3.

Linearity Property

The DFT is a *linear* transformation from one N-point sequence into another. That is, if $x(k)$ and $y(k)$ are signals and a and b are constants, then from Definition 4.2.

$$
\begin{aligned}
\text{DFT}\{ax(k)+by(k)\} &= \sum_{k=0}^{N-1} [ax(k)+by(k)]W_N^{ik} \\
&= a\sum_{k=0}^{N-1} x(k)W_N^{ik} + b\sum_{k=0}^{N-1} y(k)W_N^{ik} \\
&= aX(i)+bY(i)
\end{aligned}
\tag{4.3.27}
$$

Linearity property

Thus the DFT of the sum of two signals is just the sum of the DFTs of the signals. Similarly, the DFT of a scaled signal is just the scaled DFT of the signal.

FIGURE 4.8: Midpoint Symmetry of Magnitude and Phase Spectra of Signal

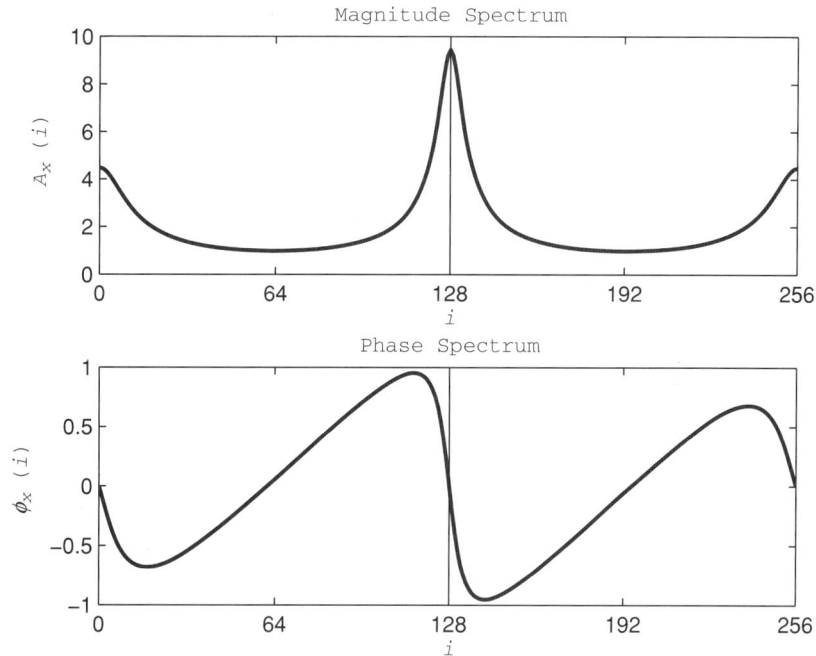

Time Reversal Property

Periodic extension

The next property requires that we work with the periodic extension of $x(k)$. Recall from the discussion of circular convolution in Section 2.7 that the periodic extension of $x(k)$ is denoted $x_p(k)$ and defined in terms of the MATLAB *mod* function as follows.

$$x_p(k) \triangleq x[\text{mod}(k, N)] \tag{4.3.28}$$

Therefore, $x_p(k) = x(k)$ for $0 \leq k < N$ and $x_p(k)$ extends $x(k)$ periodically in both positive and negative directions. Given that $x_p(k)$ is defined for all k, consider the replacement of k by $-k$. This is called time reversal or reflection. Using $x_p(k + N) = x_p(k)$, Table 4.6, and a change of variable

$$
\begin{aligned}
\text{DFT}\{x_p(-k)\} &= \sum_{k=0}^{N-1} x_p(-k) W_N^{ik} \\
&= \sum_{k=0}^{N-1} x_p(N - k) W_N^{ik}, \quad m = N - k \\
&= \sum_{m=N}^{1} x_p(m) W_N^{i(N-m)} \\
&= \sum_{m=1}^{N} x_p(m) W_N^{-im}
\end{aligned}
\tag{4.3.29}
$$

TABLE 4.7: ▶ Symmetry Properties of the DFT

Property	Equation	$x(k)$
Periodic	$X(i + N) = X(i)$	General
Symmetry	$X^*(i) = X(N - i)$	Real
Even magnitude	$A_x(N/2 + i) = A_x(N/2 - i)$	Real, N even
Odd phase	$\phi_x(N/2 + i) = -\phi(N/2 - i)$	Real, N even

Here we have made use of the identity $W_N^N = 1$. Next note that $x_p(m) = x(m)$ for $1 \leq m < N$. Furthermore, $x_p(N) = x(0)$ and $W_N^{iN} = W_N^{i0}$. Thus for *real* $x(k)$, the DFT can be rewritten as

$$
\begin{aligned}
\text{DFT}\{x_p(-k)\} &= \sum_{m=0}^{N-1} x(m) W_N^{-im} \\
&= X^*(i)
\end{aligned}
\tag{4.3.30}
$$

Time reversal property   Consequently, for real signals, reversing time is equivalent to taking the complex conjugate of the DFT.

Circular Shift Property

Another important property that makes use of the periodic extension $x_p(k)$ is the circular shift property. Suppose we shift $x_p(k)$ by r samples. Then, using a change of variables, we have

$$
\begin{aligned}
\text{DFT}\{x_p(k-r)\} &= \sum_{k=0}^{N-1} x_p(k-r) W_N^{ik} \\
&= \sum_{q=-r}^{N-1-r} x_p(q) W_N^{i(q+r)}, \quad q = k-r \\
&= W_N^{ir} \sum_{q=-r}^{N-1-r} x_p(q) W_N^{iq}
\end{aligned}
\tag{4.3.31}
$$

Just as $x_p(q)$ is a periodic function of q with period N, from Table 4.6, the factor W_N^{iq} is also a periodic function of q of period N. Consequently, the term $x_p(q) W_N^{iq}$ in (4.3.26) is periodic with period N which means that the summation over one period can start at $q = 0$ rather than $q = -r$ without affecting the result. Thus

$$
\begin{aligned}
\text{DFT}\{x_p(k-r)\} &= W_N^{ir} \sum_{q=0}^{N-1} x_p(q) W_N^{iq} \\
&= W_N^{ir} X(i)
\end{aligned}
\tag{4.3.32}
$$

Circular shift property   The time shift by r samples is referred to as a circular shift because $x_p(k)$ is a periodic extension of $x(k)$. One can think of the N samples of $x(k)$ as being *wrapped around* the unit circle of the complex plane shown previously in Figure 4.6. The signal $x_p(k-r)$ then represents a counterclockwise shift of $x(k)$ by r samples.

Circular Convolution Property

Recall from Definition 2.4 in Section 2.7 that the circular convolution of $h(k)$ with $x(k)$ is defined in terms of the periodic extension $x_p(k)$ as follows.

$$
h(k) \circ x(k) = \sum_{i=0}^{N-1} h(i) x_p(k-i)
\tag{4.3.33}
$$

Just as convolution maps into multiplication under both the Z-transform and the DTFT, the same is true of the DFT in terms of circular convolution. Using the circular shift property and Definition 4.2

$$
\begin{aligned}
\text{DFT}\{h(k) \circ x(k)\} &= \text{DFT}\left[\sum_{m=0}^{N-1} h(m)x_p(k-m)\right] \\
&= \sum_{m=0}^{N-1} h(m)\text{DFT}\{x_p(k-i)\} \\
&= \sum_{m=0}^{N-1} h(m)W_N^{im}X(i) \\
&= H(i)X(i)
\end{aligned}
\tag{4.3.34}
$$

Circular convolution property

Thus circular convolution maps into multiplication under the DFT. This is analogous to the linear convolution property of the Z-transform and the DTFT.

Circular Correlation Property

Closely related to the operation of circular convolution is circular correlation. Recall from Definition 2.6 in Section 2.8 that the circular convolution of $y(k)$ with $x(k)$ is defined in terms of the periodic extension $x_p(k)$ as follows.

$$
c_{yx}(k) = \frac{1}{N}\sum_{i=0}^{N-1} y(i)x_p(i-k)
\tag{4.3.35}
$$

One of the properties of circular cross-correlation from Table 2.4 in Chapter 2 was that circular cross-correlation can be computed using circular convolution as follows.

$$
c_{yx}(k) = \frac{y(k) \circ x(-k)}{N}
\tag{4.3.36}
$$

Circular correlation property

Applying the circular convolution property and the time reversal property to real signals then leads to the *circular correlation property*.

$$
\text{DFT}\{c_{yx}(k)\} = \frac{X(i)Y^*(i)}{N}
\tag{4.3.37}
$$

Parseval's Identity

Parseval's identity is a simple and elegant relationship between a time signal and its transform. There are several versions of Parseval's identity, depending on the independent variable (continuous or discrete) and the signal duration (finite or infinite). For example, the DTFT version of Parseval's identity is given in Proposition 4.1. The DFT version of Parseval's identity is very similar. From (4.3.6)

$$\sum_{i=0}^{N-1} X(i)Y^*(i) = \sum_{i=0}^{N-1} \left[\sum_{k=0}^{N-1} x(k)W_N^{ki} \sum_{m=0}^{N-1} y^*(m) W_N^{-mi} \right]$$

$$= \sum_{k=0}^{N-1} \sum_{m=0}^{N-1} x(k)y^*(m) \sum_{i=0}^{N-1} W_N^{(k-m)i}$$

$$= \sum_{k=0}^{N-1} \sum_{m=0}^{N-1} x(k)y^*(m) N\delta(k-m)$$

$$= N \sum_{k=0}^{N-1} x(k)y^*(k) \tag{4.3.38}$$

The end result is the DFT version of Parseval's identity.

PROPOSITION

4.2: Parseval's Identity: DFT

Let $x(k)$ and $y(k)$ be two N-point time signals with discrete Fourier transforms $X(i)$ and $Y(i)$, respectively. Then

$$\sum_{k=0}^{N-1} x(k)y^*(k) = \frac{1}{N} \sum_{i=0}^{N-1} X(i)Y^*(i)$$

Parseval's identity

Note the when $y(k) = x(k)$ in Proposition 4.2, the left-hand side reduces to the energy of the signal $x(k)$.

$$\sum_{k=0}^{N-1} |x(k)|^2 = \frac{1}{N} \sum_{i=0}^{N-1} |X(i)|^2 \tag{4.3.39}$$

As an illustration of how we can apply *Parseval's identity*, recall that the average power of an N-point signal is defined

$$P_x \triangleq \frac{1}{N} \sum_{k=0}^{N-1} |x(k)|^2 \tag{4.3.40}$$

Power density spectrum

Comparing (4.3.40) with (4.3.39), we see that Parseval's identity provides us with an alternative frequency-domain version of average power. In particular, recalling that $S_x(i) = |X(i)|^2/N$ is the *power density spectrum*, it follows that

$$P_x = \frac{1}{N} \sum_{i=0}^{N-1} S_x(i) \tag{4.3.41}$$

Thus the average power is just the average of the power density spectrum of $x(k)$, hence the name power density spectrum.

Example 4.7

Parseval's Identity

To verify Parseval's identity, consider the signal $x(k)$ from Example 4.3.

$$x = [3, -1, 0, 2]^T$$

In this case $N = 4$. A direct time-domain computation of the average power using (4.3.40) yields

$$P_x = \frac{1}{N}\sum_{k=0}^{N-1}|x(k)|^2$$
$$= .25(9+1+0+4)$$
$$= 3.5$$

From Example 4.3, the DFT of $x(k)$ is

$$X = [4, 3+j3, 2, 3-j3]^T$$

Thus from (4.3.23), the power density spectrum is

$$S_x = .25[16, 18, 4, 18]^T$$
$$= [4, 4.5, 1, 4.5]^T$$

Finally, from (4.3.41), the frequency-domain method of determining the average power is

$$P_x = \frac{1}{N}\sum_{i=0}^{N-1}S_x(i)$$
$$= .25(4+4.5+1+4.5)$$
$$= 3.5 \checkmark$$

A summary of the properties of the DFT can be found in Table 4.8. The last property, called the Wiener-Khintchine theorem, is derived in Section 4.7.

TABLE 4.8: ▶
DFT Properties

Property	Time Signal	DFT	Comments
Linearity	$ax(k)+by(k)$	$aX(i)+bY(i)$	General
Time reversal	$x_p(-k)$	$X^*(i)$	Real x
Circular shift	$x_p(k-r)$	$W_N^{ir}X(i)$	General
Circular convolution	$x(k)\circ y(k)$	$X(i)Y(i)$	General
Circular correlation	$c_{yx}(k)$	$\dfrac{Y(i)X^*(i)}{N}$	Real x
Wiener-Khintchine	$c_{xx}(k)$	$S_x(i)$	General
Parseval	$\sum_{k=0}^{N-1}x(k)y^*(k)$	$\frac{1}{N}\sum_{i=0}^{N-1}X(i)Y^*(k)$	
	$\sum_{k=0}^{N-1}|x(k)|^2$	$\frac{1}{N}\sum_{i=0}^{N-1}|X(i)|^2$	

• • • • • • • • • • • • • • • • • •

4.4 Fast Fourier Transform (FFT)

The discrete Fourier transform or DFT is an important computational tool that is widely used in DSP. As with any computational method, it can be rated in terms of how the computational effort grows as the size of the problem increases. Recall that the DFT of an N-point signal $x(k)$ is

$$X(i) = \sum_{k=0}^{N-1} x(k) W_N^{ik}, \quad 0 \le i < N \tag{4.4.1}$$

Complex FLOP

The N^2 values of W_N^{ik} do not depend on $x(k)$, so they can be precomputed once and stored in an $N \times N$ matrix W. Each point $X(i)$ then requires N complex multiplications. Since there are N values of i, the total number of complex *floating point operations* or FLOPs required to compute the entire DFT is

$$n_{\text{DFT}} = N^2 \text{ FLOPs} \tag{4.4.2}$$

Algorithm order

Thus the computational effort, measured in complex multiplications, grows as the square of the size of the problem N. In this case we say that the DFT computation is of order $O(N^2)$. More generally, a computational algorithm is of *order* $O(N^p)$ if and only if for some constant c, the number of computations n satisfies

$$n \approx cN^p, \quad N \gg 1 \tag{4.4.3}$$

4.4.1 Decimation in Time FFT

The popularity of the DFT arises, in part, from the fact that there is an implementation of it that dramatically decreases the computational time, particularly for large values of N. To see how the improvement in speed is achieved, suppose that the number of points N is a power of two. That is, $N = 2^r$ for some integer r where

$$r = \log_2(N) \tag{4.4.4}$$

Time decimation

Next suppose we *decimate* the N-point signal $x(k)$ into two $N/2$-point signals $x_e(k)$ and $x_o(k)$, corresponding to the even and odd indices or subscripts of x, respectively.

$$x_e \overset{\Delta}{=} [x(0), x(2), \ldots, x(N-2)]^T \tag{4.4.5a}$$
$$x_o \overset{\Delta}{=} [x(1), x(3), \ldots, x(N-1)]^T \tag{4.4.5b}$$

The DFT in (4.4.1) then can be recast as two sums, one corresponding to the even values of k, and the other corresponding to the odd values of k. Recalling from Table 4.6 that $W_N^{2k} = W_{N/2}^k$, we have

$$
\begin{aligned}
X(i) &= \sum_{k=0}^{N/2-1} x(2k) W_N^{2ki} + \sum_{k=0}^{N/2-1} x(2k+1) W_N^{(2k+1)i} \\
&= \sum_{k=0}^{N/2-1} x_e(k) W_N^{2ki} + W_N^i \sum_{k=0}^{N/2-1} x_o(k) W_N^{2ki} \\
&= \sum_{k=0}^{N/2-1} x_e(k) W_{N/2}^{ki} + W_N^i \sum_{k=0}^{N/2-1} x_o(k) W_{N/2}^{ki} \\
&= X_e(i) + W_N^i X_o(i), \quad 0 \le i < N
\end{aligned}
\tag{4.4.6}
$$

Here $X_e(i) = \mathrm{DFT}\{x_e(k)\}$ and $X_o(i) = \mathrm{DFT}\{x_o(k)\}$ are $N/2$-point transforms of the even and odd parts of $x(k)$, respectively. Thus (4.4.6) decomposes the original N-point problem into two $N/2$-point subproblems with N complex multiplications and N complex additions required to merge the solutions. Each $N/2$-point DFT requires $N^2/4$ FLOPs. Hence for large values of N, the number of floating point operations required to implement the even-odd decomposition in (4.4.6) is

$$
n_{\mathrm{eo}} \approx \frac{N^2}{2} \quad \text{FLOPs}, \quad N \gg 1
\tag{4.4.7}
$$

Comparing (4.4.7) with (4.4.2), we see that the computational effort has been reduced by a factor of two for large values of N. It is helpful to break the *merging formula* in (4.4.6) into two cases where $0 \le i < N/2$.

$$
X(i) = X_e(i) + W_N^i X_o(i)
\tag{4.4.8a}
$$

$$
X(i + N/2) = X_e(i + N/2) + W_N^{i+N/2} X_o(i + N/2)
\tag{4.4.8b}
$$

Since $X_e(i)$ and $X_o(i)$ are $N/2$-point transforms, we know from the periodic property in Table 4.7 that $X_e(i + N/2) = X_e(i)$ and similarly for $X_o(i)$. Furthermore, from Table 4.6 we have $W_N^{i+N/2} = -W_N^i$. Thus (4.4.8) can be simplified as follows where $0 \le i < N/2$.

$$
Y(i) = W_N^i X_o(i)
\tag{4.4.9a}
$$

$$
X(i) = X_e(i) + Y(i)
\tag{4.4.9b}
$$

$$
X(i + N/2) = X_e(i) - Y(i)
\tag{4.4.9c}
$$

Computational butterfly

The merging formula is written as three equations using a temporary variable $Y(i)$. This increases storage by one complex scalar, but it reduces the number of complex multiplications from two to one. The computations in (4.4.9) are referred to as an ith order *butterfly*. The name arises from the flow graph representation of (4.4.9) shown in Figure 4.9. With a little imagination, the "wings" of the butterfly are apparent.

FIGURE 4.9: Signal Flow Graph of ith Order Butterfly Computation

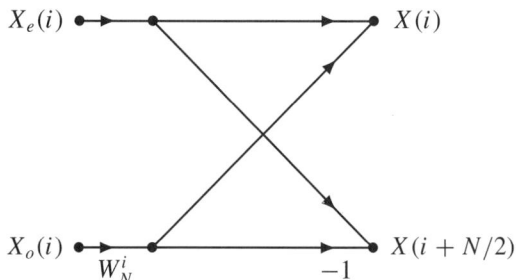

FIGURE 4.10:
Even/Odd
Decomposition of
DFT with $N = 8$

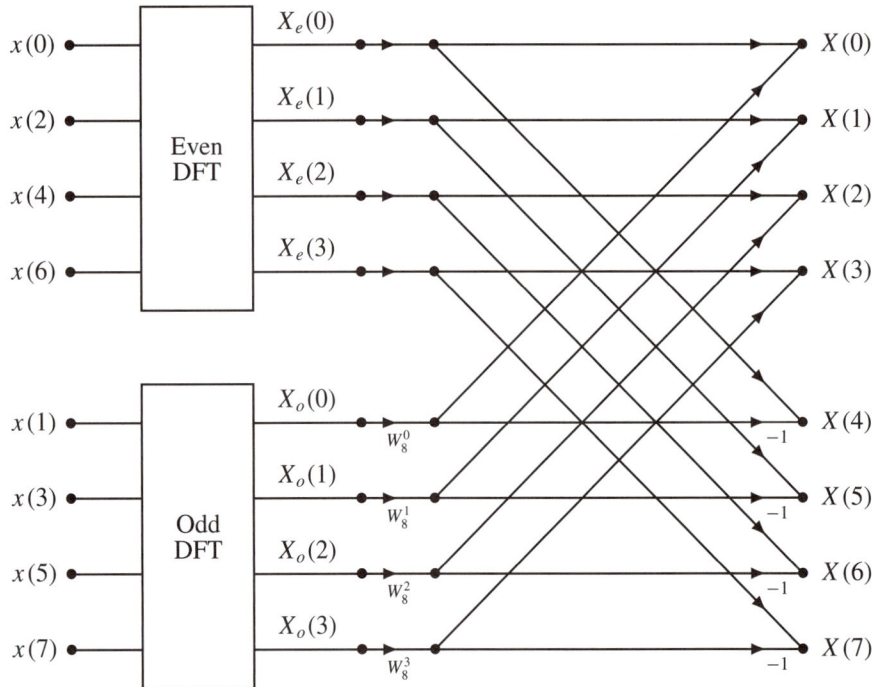

The even-odd decomposition of the DFT in (4.4.9) consists of two $N/2$-point DFTs plus $N/2$ interleaved butterfly computations. A block diagram that illustrates the case $N = 8$ is shown in Figure 4.10.

The beauty of the even-odd decomposition technique in Figure 4.10 is that there is nothing to prevent us from applying it again! For example, the even samples $x_e(k)$ and the odd samples $x_o(k)$ can each be decimated into even and odd parts. In this way we can decompose each $N/2$-point transform into a pair of $N/4$-point transforms. Since $N = 2^r$, this process can be continued a total of r times. In the end we are left with a collection of elementary two-point DFTs. Interestingly enough, the two-point DFT is simply a zeroth-order butterfly. The resulting algorithm is called the decimation in time *fast Fourier transform* or FFT (Cooley and Tukey, 1965). The signal flow graph for the case $N = 8$ is shown in Figure 4.11. Note how there are r iterations, and each iteration consists of $N/2$ butterfly computations.

FFT

Observe from Figure 4.11 that the input to the first iteration has been *scrambled* due to repeated decimation into even and odd subsequences. As it turns out, there is a simple numerical relationship between the original DFT order and the scrambled FFT order. This becomes apparent when we look at the binary representations of the indices, as shown in Table 4.9 for the case $N = 8$. Note how the scrambled FFT indices are obtained by taking the normal DFT indices, converting to binary, reversing the bits, and then converting back to decimal.

The following MATLAB-based algorithm is designed to interchange the elements of the input vector x using the bit reversal process shown in Table 4.9.

ALGORITHM

4.1: Bit Reversal

```
1. b = dec2bin(x,N);        % convert to binary string
2. c = b(N:-1:1);           % reverse bits
3. x = bin2dec(c);          % convert back to decimal value
```

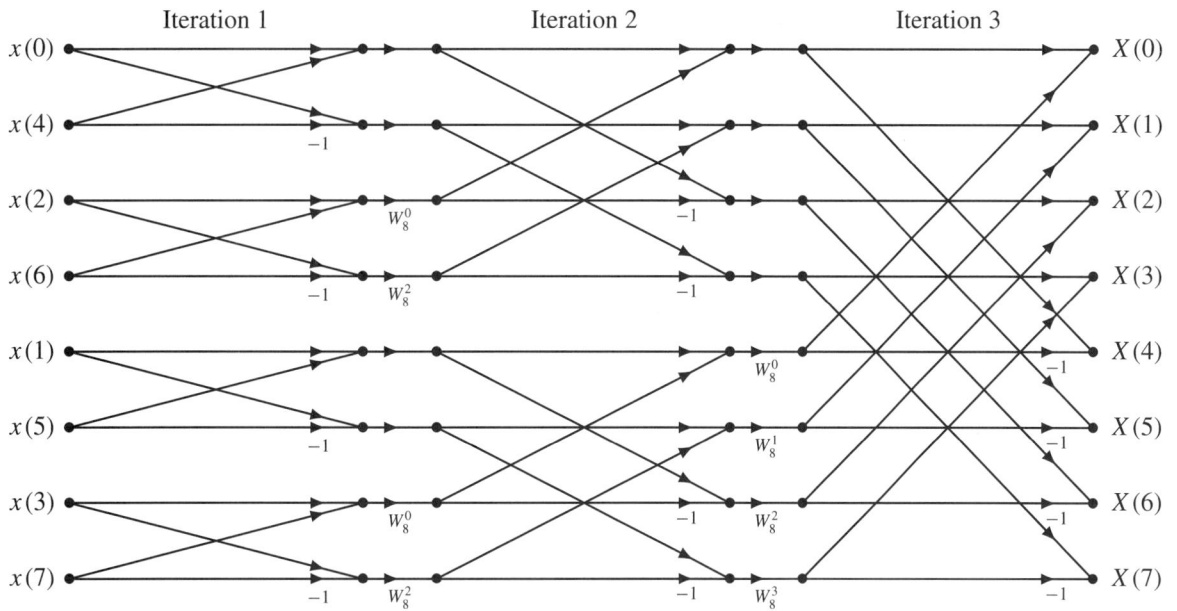

FIGURE 4.11: Signal Flow Graph of Decimation in Time FFT with $N = 8$

The MATLAB function *dec2bin* converts the decimal value x into a binary string of zeros and ones. Step 2 reverses the string elements, and then *bin2dec* in step 3 converts the bit-reverses string back to a decimal value. If Algorithm 4.1 is called twice, it restores the bits of x to the original order. That is, Algorithm 4.1 is its own inverse.

The overall decimation in time FFT is summarized in Algorithm 4.2. The input vector is scrambled in step 1 with a call to Algorithm 4.1. The $rN/2$ butterfly computations are performed in step 2, and the final result is copied from x to X in step 3. To analyze step 2 it is helpful to refer to Figure 4.11. Step 2 consists of three loops. The outer loop goes through the r iterations moving from left to right in Figure 4.11. Notice that for each iteration there are groups of butterflies. In step 2, s is the spacing between groups, g is the number of groups, and b is the number of butterflies per group. The parameter b is also the wingspan of the butterfly. The second loop goes through the g groups associated with the current iteration, and the third and innermost loop goes through the b butterflies of each group. The first three equations in

TABLE 4.9: ▶
Scrambled FFT
Order Using Bit
Reversal, $N = 8$

DFT Order	Forward Binary	Reverse Binary	FFT Order
0	000	000	0
1	001	100	4
2	010	010	2
3	011	110	6
4	100	001	1
5	101	101	5
6	110	011	3
7	111	111	7

the innermost loop compute the weight w and the butterfly location n, while the last three equations compute the butterfly outputs as in (4.4.9).

ALGORITHM

4.2: FFT

1. Call Algorithm 4.1 to scramble the input vector x.
2. For $i = 1$ to r do   % iteration
   {
   (a) Compute

$$s = 2^i \qquad \text{% group spacing}$$
$$g = N/s \qquad \text{% number of groups}$$
$$b = s/2 \qquad \text{% butterflies/group}$$

   (b) For $k = 0$ to $g - 1$ do
      {
      For $m = 0$ to $b - 1$ do
         {

$$\theta = -2\pi m g / N$$
$$w = \cos(\theta) + j\sin(\theta)$$
$$n = ks + m$$
$$y = wx(n + b)$$
$$x(n) = x(n) + y$$
$$x(n + b) = x(n) - 2y$$

         }
      }
   }
3. For $k = 0$ to $N - 1$ set $X(k) = x(k)$.

4.4.2 FFT Computational Effort

Although the derivation of the FFT requires some attention to detail, the payoff is worthwhile because the end result is an algorithm that is dramatically faster than the DFT for practical values of N. To see how much faster, note from Figure 4.11 that there are r iterations and $N/2$ butterflies per iteration. Given the weight W_N^i, the butterfly computation in (4.4.9) requires one complex multiplication. Thus there are $rN/2$ complex multiplications required for the FFT. It then follows from (4.4.4) that the computational effort of the FFT is

FFT speed

$$n_{\text{FFT}} = \frac{N \log_2(N)}{2} \text{ FLOPs} \qquad (4.4.10)$$

Thus the FFT algorithm is of order $O[N \log_2(N)]$ while the DFT is of order $O(N^2)$. A graphical comparison of the number of FLOPs required by the DFT and the FFT is shown in

FIGURE 4.12:
Computational
Effort of DFT and
FFT for $1 \leq N \leq 256$

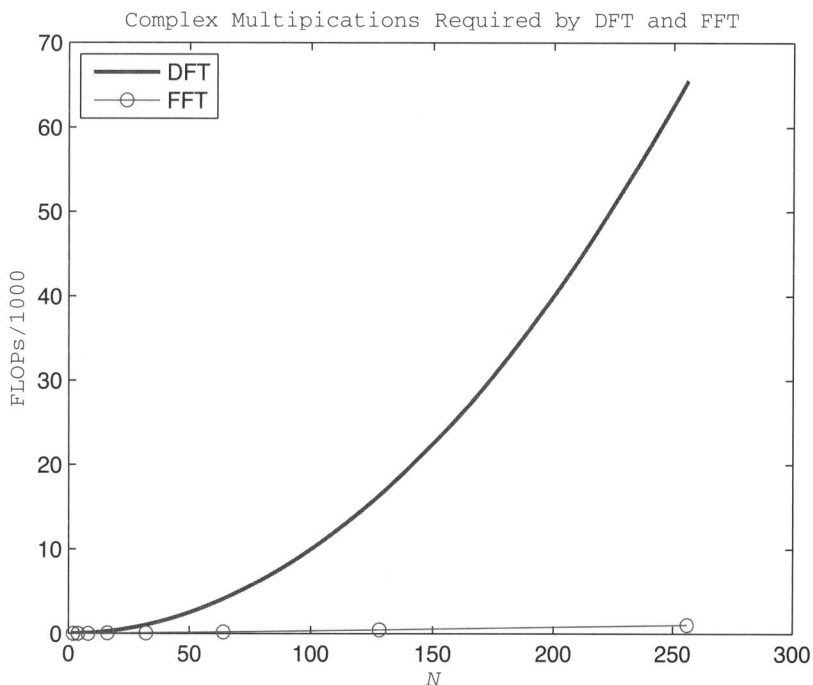

Complex Multipications Required by DFT and FFT

Figure 4.12 for $1 \leq N \leq 256$. Even over this modest range of values for N, the improvement in the *speed* of the FFT in comparison with the DFT is quite dramatic. For many practical problems, larger values of N in the range 1024 to 8192 are often used. For the case $N = 1024$, the DFT requires 1.049×10^6 FLOPs, while the FFT requires only 5.12×10^3 FLOPs. In this instance a speed improvement by a factor of 204.8, or more than two orders of magnitude, is achieved.

Given the highly efficient FFT, there are a number of DSP applications where the best way to solve a time domain problem is to use the following steps.

ALGORITHM

4.3: Problem domain

1. Transform to the frequency domain using the FFT.
2. Solve the problem in the frequency domain.
3. Transform back to the time domain using the IFFT.

Recall that the formulation for the inverse discrete Fourier transform or IDFT is very similar to that of the DFT, namely,

$$x(k) = \frac{1}{N} \sum_{i=0}^{N-1} X(i) W_N^{-ki} \tag{4.4.11}$$

One way to devise an algorithm for the IDFT is to modify Algorithm 4.2 by adding an input parameter which specifies which direction we want to transform, forward or reverse. Still another approach is to use Algorithm 4.2 itself, without modification. Recall from Table 4.6 that the complex conjugate of W_N is just $W_N^* = W_N^{-1}$. This being the case, we can reformulate (4.4.11) in terms of the FFT as follows.

$$x(k) = \frac{1}{N} \sum_{i=0}^{N-1} X(i) W_N^{-ki}$$

$$= \frac{1}{N} \sum_{i=0}^{N-1} X(i)(W_N^{ki})^*$$

$$= \frac{1}{N} \left(\sum_{i=0}^{N-1} X^*(i) W_N^{ki} \right)^*$$

$$= \left(\frac{1}{N} \right) \text{FFT}^* \{X^*(i)\} \qquad (4.4.12)$$

Thus we can implement an IFFT of $X(i)$ by taking the complex conjugate of the FFT of $X^*(i)$ and then normalizing the result by N. These steps are summarized in the following algorithm. Note that this approach takes advantage of the fact that the FFT can be applied to either a real or a complex signal.

ALGORITHM

4.4: IFFT

1. For $k = 0$ to $N - 1$ set $x(k) = X^*(k)$.
2. Call Algorithm 4.2 to compute $X(i) = \text{FFT}\{x(k)\}$.
3. For $k = 0$ to $N - 1$ set $x(k) = X^*(k)/N$.

4.4.3 Alternative FFT Implementations

There is one drawback to the FFT, as implemented, that is evident from Figure 4.11. For the FFT the number of points N must be a power of two, while the DFT is defined for all $N \geq 1$. Often this is not a serious limitation because the user may be at liberty to choose a value for N, say, in collecting data samples for an experiment. In still other cases, it may be possible to *pad* the signal $x(k)$ with a sufficient number of zeros to make N a power of two. Because N is a power of two, the formulation of the FFT in Algorithm 4.2 is called a *radix-two* version of the FFT. It is also possible to factor N in other ways and achieve alternative versions of the FFT (Ingle and Proakis, 2000). In general the alternative versions will have computational speeds that lie between the DFT and the radix-two FFT shown in Figure 4.11.

An alternative to the decimation in time method summarized in Algorithm 4.2 is something called the decimation in frequency method (see, e.g., Schilling and Harris, 2000). The decimation in frequency method starts by decomposing $x(k)$ into a first half $0 \leq k < N/2$, and a second half $N/2 \leq k < N$. The merging formula for the decimation in frequency method is $X(2i) = \text{DFT}\{a(k)\}$ and $X(2i+1) = \text{DFT}\{b(k)\}$, where

$$a(k) = x(k) + x(k + N/2) \qquad (4.4.13a)$$
$$b(k) = [x(k) - x(k + N/2)]W_N^k \qquad (4.4.13b)$$

The process proceeds in a manner generally similar to the decimation in time approach and again leads to $rN/2$ butterfly computations. In this case it is the output vector $X(i)$ that ends up being scrambled, hence the name decimation in frequency.

MATLAB Functions

There are a number of MATLAB functions for computing the FFT and signal spectra that are very simple to use.

```
% FFT: Compute a fast Fourier transform
%
% Usage:
%         X = fft(x,N);
% Pre:
%         x = vector of length M containing samples to be transformed
%         N = optional number of samples to transform.  If N > M, x is
%             zero-padded with N-M samples.  The spacing between
%             discrete frequencies is Delta_F = f_s/N;
% Post:
%         X = complex vector containing the DFT of x.
```

There is also a function x = ifft(X) that is used to perform the inverse FFT. Once the FFT is obtained, the magnitude, phase, and power density spectra can be easily obtained. Recall that the FDSP function *f_spec* can also be used.

```
X   = fft(x,N);                          % Compute N-point FFT
A   = abs(X);                            % Magnitude spectrum
phi = angle(X);                          % Phase spectrum
S   = A.^2/N;                            % Power density spectrum
```

● ● ● ● ● ● ● ● ● ● ● ● ● ● ● ● ● ●

4.5   Fast Convolution and Correlation

In this section we again focus on causal finite signals. The FFT provides us with a very efficient way to implement two important operations, convolution and correlation. The basic idea is to transform the problem into the frequency domain with the FFT, perform the operation in the frequency domain, and then transform back to the time domain with the IFFT.

4.5.1 Fast Convolution

Suppose $h(k)$ is of length L and $x(k)$ is of length M. From (2.7.6), the linear convolution of $h(k)$ with $x(k)$ is

$$h(k) \star x(k) = \sum_{i=0}^{L} h(i)x(k-i), \quad 0 \le k < L + M - 1 \tag{4.5.1}$$

For computational purposes, the most important result from Chapter 2 was that the linear convolution of two finite signals can be implemented using circular convolution with zero padding.

$$h(k) \star x(k) = h_z(k) \circ x_z(k) \tag{4.5.2}$$

Zero padding

Here $h_z(k)$ is the zero-padded version of $h(k)$ obtained by appending $M + p$ zeros, and $x_z(k)$ is the zero-padded version of $x(k)$ using $L + p$ zeros where $p \geq -1$. Thus the common length of $h_z(k)$ and $x_z(k)$ is $N = L + M + p$. Next, recall from from the properties of the DFT in Table 4.8 that

$$\text{DFT}\{h_z(k) \circ x_z(k)\} = H_z(i)X_z(i) \tag{4.5.3}$$

That is, circular convolution in the time domain maps into multiplication in the frequency domain using the DFT. Consequently, an effective way to perform circular convolution is

$$h_z(k) \circ x_z(k) = \text{IDFT}\{H_z(i)X_z(i)\}, \quad 0 \leq k < N \tag{4.5.4}$$

In view of (4.5.4), we now have in place all of the tools needed to perform a practical, highly efficient, linear convolution of two finite signals. All that is needed is to make the signal length $N = L + M + p$ be a power of two. Since (4.5.2) holds for any $p \geq -1$, consider the following value for N.

$$N = \text{nextpow2}(L + M - 1) \tag{4.5.5}$$

Here the MATLAB function *nextpow2* finds the smallest power of two that is greater than or equal to its calling argument. Therefore, selecting N as in (4.5.5) ensures that N is the smallest power of two such that $N \geq L + M - 1$. For this value of N, the highly efficient radix-two FFT can be used to compute the DFTs of $h_z(k)$ and $x_z(k)$. This results in the following version of linear convolution called *fast convolution*

Fast convolution

$$h(k) \star x(k) = \text{IFFT}\{H_z(i)X_z(i)\}, \quad 0 \leq k < L + M - 1 \tag{4.5.6}$$

A block diagram of the fast convolution operation is shown in Figure 4.13. Note that this is an example of Algorithm 4.3 where the problem is transformed into the frequency domain where it can be solved more efficiently. Even though fast convolution involves several

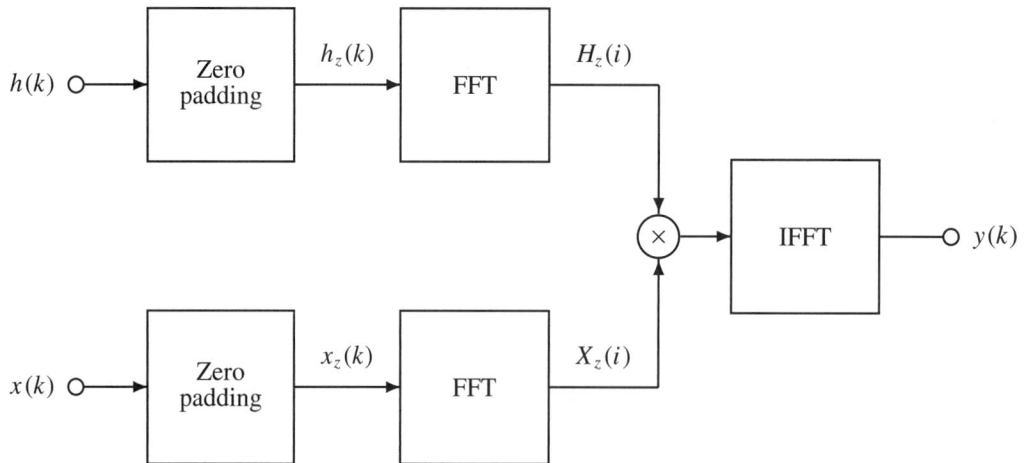

FIGURE 4.13: Fast Linear Convolution

steps, there is a value for N beyond which fast convolution is more efficient than the direct computation of linear convolution.

Computational Effort

To simplify the analysis of the computational effort, suppose the signals $h(k)$ and $x(k)$ are both of length L, where L is a power of two. In this case zero padding to length $N = 2L$ is sufficient. From (4.4.10), the two FFTs in Figure 4.13 each require $(N/2)\log_2(N)$ complex floating point operations or FLOPs, while the inverse FFT requires $(N/2)\log_2(N) + 1$ FLOPs. The multiplication of $H_z(i)$ times $X_z(i)$ for $0 \leq i < N$ requires an additional N FLOPs. Hence the total number of complex FLOPs is $(3N/2)\log_2(N) + N + 1$. The direct computation of linear convolution in (4.5.1) does not involve any complex arithmetic, so to provide a fair comparison, we should count the number of real multiplications. The product of two complex numbers can be expressed as follows.

$$(a + jb)(c + jd) = ac - bd + j(ad + bc) \tag{4.5.7}$$

Real FLOPs

Consequently, each complex multiplication requires four real multiplications. Recalling that $N = 2L$, the total number of real FLOPs required to perform a fast linear convolution of two L-point signals is then

$$n_{\text{fast}} = 12L\log_2(2L) + 8L + 4 \text{ FLOPs} \tag{4.5.8}$$

Next consider the number of real multiplications required to implement linear convolution directly. If we set $M = L$ in (4.5.1), the number of real multiplications or FLOPs is

$$n_{\text{dir}} = 2L^2 \text{ FLOPs} \tag{4.5.9}$$

Comparing (4.5.9) with (4.5.8), we see that for small values of L, a direct computation of linear convolution will be faster. However, the L^2 term grows faster than the $L\log_2(2L)$ term, so eventually the fast convolution will outperform direct convolution. A plot of the number of real FLOPs required by the two methods for signal lengths in the range $2 \leq L \leq 1024$ is shown in Figure 4.14. The two methods require roughly the same number of FLOPs for $L \leq 32$. However, fast linear convolution is superior to direct linear convolution for signal lengths in the range $L \geq 64$, and as L increases, it becomes significantly faster.

| Example 4.8 |

Fast Convolution

To illustrate the use of fast convolution, consider a linear discrete-time system with the following transfer function.

$$H(z) = \frac{.98\sin(\pi/24)z}{z^2 - 1.96\cos(\pi/24)z + .9604}$$

From the Z-transform pairs in Table 3.2, the impulse response of this system is

$$h(k) = .98^k \sin(\pi k/24)\mu(k)$$

Next suppose this system is driven with the following exponentially damped sinusoidal input.

$$x(k) = k^2(.99)^k \cos(\pi k/48)\mu(k)$$

Since both $h(k)$ and $x(k)$ decay to zero, we can approximate them as L-point signals when L is sufficiently large. The zero-state response of the system can be obtained by running *exam4_8*. It computes the linear convolution of $h(k)$ with $x(k)$ using fast convolution with $L = 512$.

FIGURE 4.14:
Comparison of
Computational
Effort Required for
Linear Convolution
of Two *L*-point
Signals

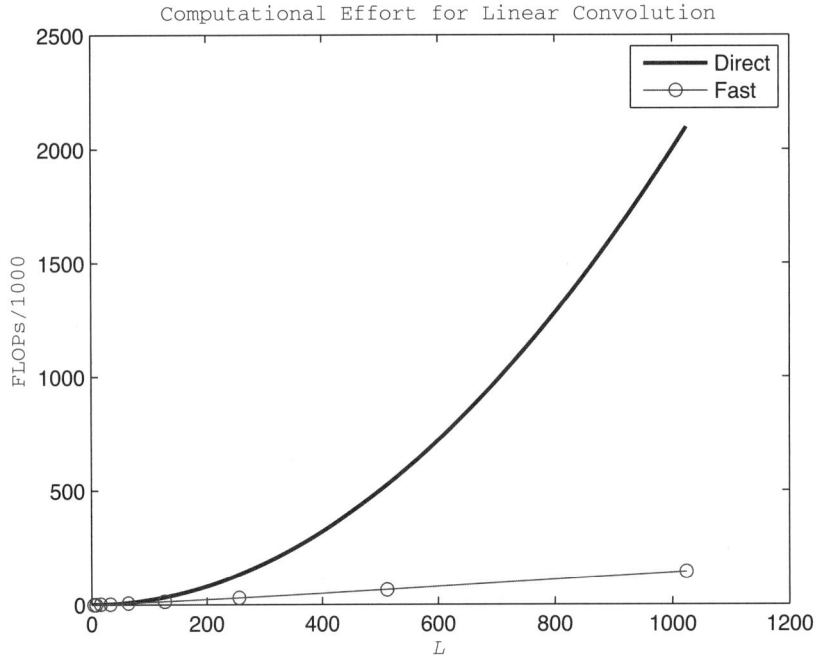

FIGURE 4.15: Zero-
state Response of
$x(k)$ Using Fast
Linear Convolution

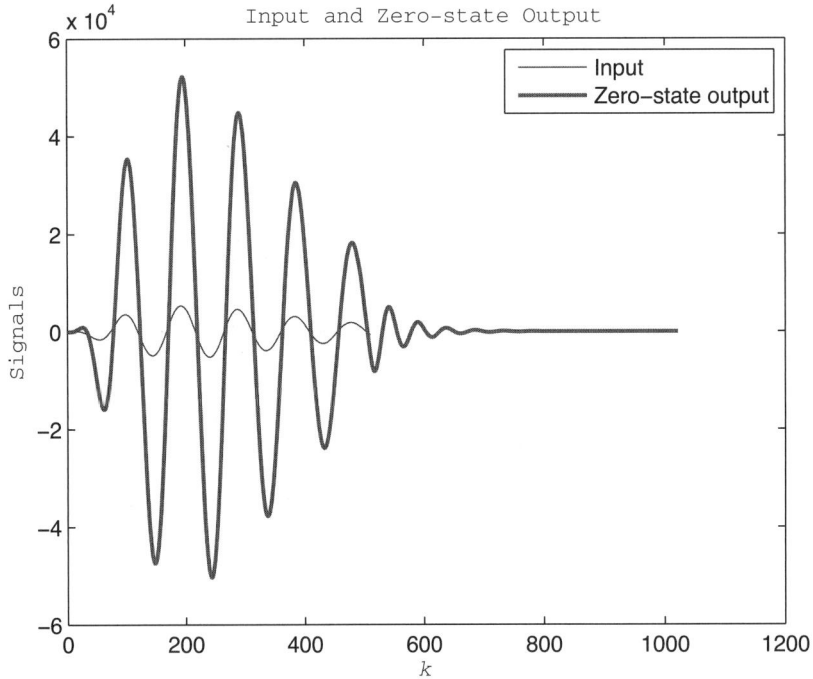

Plots of the input $x(k)$ and zero-state output $y(k)$ are shown in Figure 4.15. In this case the total number of real FLOPs was $n_{fast} = 4.67 \times 10^4$. This is in contrast to the direct method which would require $n_{dir} = 5.24 \times 10^5$ FLOPs. This corresponds to a saving of 91.1 %.

*4.5.2 **Fast Block Convolution**

The efficient formulation of convolution based on the FFT assumes that the two signals are of finite duration. There are some applications where the input $x(k)$ is available continuously and is of indefinite duration. For example, the input might represent a long speech signal obtained from a microphone. In cases like these, the number of input samples M will be very large, and computation of an FFT of length $N = L + M + p$ may not be practical. Another potential drawback of batch processing is that none of the samples of the filtered output are available until the entire N points have been processed.

These difficulties associated with very long inputs can be addressed by using a technique known as *block convolution*. Suppose the impulse response consists of L samples and the input contains M samples where $M \gg L$. The basic idea is to break up the input signal into blocks or sections of length L. Each of these blocks is convolved with the L-point impulse response $h(k)$. If the results are combined in the proper way, the original $(L + M - 1)$-point convolution can be recovered.

To see how this is achieved, first note that the $x(k)$ can be padded with up to $L - 1$ zeros, if needed, such that the length of the zero-padded input $x_z(k)$ is QL for some integer $Q > 1$. The zero-padded input then can be expressed as a sum of Q blocks of length L as follows.

$$x_z(k) = \sum_{i=0}^{P-1} x_i(k - iL), \quad 0 \le k < M \tag{4.5.10}$$

Here the Q blocks, or subsignals of length L, are extracted from the original signal $x(k)$ using a window of length L as follows.

$$x_i(k) \triangleq \begin{cases} x(k + iL), & 0 \le k < L \\ 0, & \text{otherwise} \end{cases} \tag{4.5.11}$$

Subsignal $x_i(k)$ is the segment of $x(k)$ starting at $k = Li$, but it has been shifted so that it starts at $k = 0$. From (4.5.10) and (4.5.1), the linear convolution of $h(k)$ with $x(k)$ is then

$$\begin{aligned} h(k) * x(k) &= \sum_{i=0}^{Q-1} h(k) * x_i(k - iL) \\ &= \sum_{i=0}^{Q-1} y_i(k - iL) \end{aligned} \tag{4.5.12}$$

Here $y_i(k)$ is the convolution of $h(k)$ with the ith subsignal $x_i(k)$. That is,

$$y_i(k) = h(k) * x_i(i), \quad 0 \le i < Q \tag{4.5.13}$$

The block convolutions in (4.5.13) are between two L-point signals. If these signals are padded with $L + p$ zeros where $p \ge -1$ and $2L + p$ is a power of two, then a radix-two FFT can be used. The results must then be shifted and added as in (4.5.12). The resulting procedure, known as the *overlap-add method* of block convolution, is summarized in the following algorithm,

ALGORITHM

4.5: Fast Block Convolution

1. Compute

$$M_{\text{save}} = M$$
$$r = L - \text{mod}(M, L)$$
$$M = M + r$$
$$x_z = [x(0), \ldots, x(M_{\text{save}} - 1), 0, \ldots, 0]^T \in R^M$$
$$Q = \frac{M}{L}$$
$$N = 2^{\text{ceil}[\log_2(2L-1)]}$$
$$h_z = [h(0), \ldots, h(L - 1), 0, \ldots, 0]^T \in R^N$$
$$H_z = \text{FFT}\{h_z(k)\}$$
$$y_0 = [0, \ldots, 0]^T \in R^{L(Q-1)+N}$$

2. For $i = 0$ to $Q - 1$ compute

$$x_i(k) = x_z(k + iL), \quad 0 \le k < L$$
$$x_{iz}(k) = [x_i(0), \ldots, x_i(L - 1), 0, \ldots, 0]^T \in R^N$$
$$X_{iz}(i) = \text{FFT}\{x_{iz}(k)\}$$
$$y_i(k) = \text{IFFT}\{H_z(i)X_{iz}(i)\}$$
$$y_0(k) = y_0(k) + y_i(k - Li), \quad Li \le k < Li + 2N - 1$$

3. Set

$$y(k) = y_0(k), \quad 0 \le k < L + M_{save} - 1$$

In step 1 of Algorithm 4.5, the original number of input samples is saved in M_{save}. If $\text{mod}(M, L) > 0$, then M is not an integer multiple of L. In this case r zeros are padded to the end of x so that the length of $x_z(k)$ is $M = QL$, where Q is an integer representing the number of blocks. Next N is computed so that the zero-padded length of h satisfies $N \ge 2L - 1$ and is a power of two. Since H_z has to be computed only once, it is also computed in step 1. The Q block convolutions are then performed in step 2 and the results are overlapped and added using y_0 for storage. Finally, the relevant part of $y_0(k)$ is extracted in step 3. There is a closely related alternative to the overlap-add method in Algorithm 4.5 called the overlap-save method (see, e.g., Oppenheim et al., 1999)

Example 4.9   **Fast Block Convolution**

As an illustration of the fast block convolution technique, suppose the impulse response is

$$h(k) = .8^k \sin\left(\frac{\pi k}{4}\right), \quad 0 \le k < 12$$

Thus $L = 12$. Next, let the input consist of white noise uniformly distributed over $[-1, 1]$.

$$x(k) = v(k), \quad 0 \le k < 70$$

In this case $M_{save} = 70$. The number of samples in the zero-padded version of x is

$$\begin{aligned} M &= M + [L - \text{mod}(M, L)] \\ &= 70 + 12 - \text{mod}(70, 12) \\ &= 82 - 10 \\ &= 72 \end{aligned}$$

Thus $Q = 72/12$ and there are exactly $Q = 6$ blocks of length 12 in $x_z(k)$. Since both h and x_i are of length L, the minimum length of the zero-padded versions of h and x_i will be $L + L + p$ for $p \ge -1$. We can choose $N = 2L + p$ to be a power of two as follows.

$$\begin{aligned} q &= \text{nextpow2}(2L - 1) \\ &= 32 \end{aligned}$$

Plots of the impulse response $h(k)$, the zero-padded input $x_z(k)$, and the output $y(k)$ are shown in Figure 4.16. These we generated by running *exam4_9*. The zero-padded version of the input is sectioned into Q blocks, each of length L. Script *exam4_9* also computes the convolution in the direct manner. Both outputs are plotted in Figure 4.16, where it can be seen that they are identical. For this example, a modest value for M was used so that the results are easier to visualize.

FIGURE 4.16: Block Convolution of $h(k)$ with $x(k)$ where $L = 12$, $M = 32$, $Q = 3$, and $N = 32$

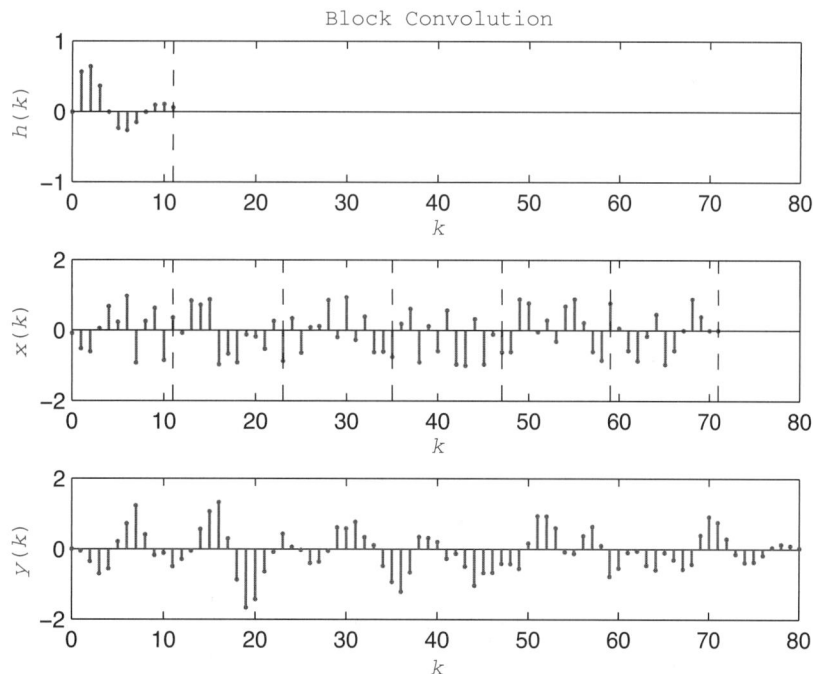

FDSP Functions

The FDSP toolbox contains the following functions for performing fast convolutions and fast block convolutions.

```
% F_CONV:      Fast linear or circular convolution
% F_BLOCKCONV: Fast linear block convolution
%
% Usage:
%          y = f_conv (h,x,circ);
%          y = f_blockconv (h,x);
% Pre:
%          h    = vector of length L containing pulse
%                 response signal
%          x    = vector of length M containing input signal
%          circ = optional convolution type code (default: 0)
%
%                 0 = linear convolution
%                 1 = circular convolution (requires M = L)
% Post:
%          y = vector of length L+M-1 containing the
%              convolution of h with x. If circ = 1,
%              y is of length L.
% Note:
%      If h is the impulse response of a discrete-time
%      linear system and x is the input, then y is the
%      zero-state response when circ = 0.
```

4.5.3 Fast Correlation

Practical cross-correlations often involve long signals, so it is important to develop a numerical implementation of linear cross-correlation that is more efficient than the direct method. Recall from Definition 2.5 that if $y(k)$ is a signal of length L and $x(k)$ is a signal of length $M \leq L$, then the linear cross-correlation of $y(k)$ with $x(k)$ is

$$r_{yx}(k) = \frac{1}{L} \sum_{i=0}^{L-1} y(i)x(i-k), \quad 0 \leq k < L \tag{4.5.14}$$

Just as was the case with convolution, linear cross-correlation can be achieved using circular cross-correlation with zero padding. In particular, from Table 2.4

$$r_{yx}(k) = \left(\frac{N}{L}\right) c_{y_z x_z}(k) \tag{4.5.15}$$

Here $y_z(k)$ is a zero-padded version of $y(k)$ with $M + p$ zeros appended. Similarly, $x_z(k)$ is a zero-padded version of $x(k)$ with $L + p$ zeros where $p \geq -1$. Therefore x_z and y_z are both signals of length $N = L + M + p$. Next recall from the properties of the DFT in Table 4.8 that

$$\text{DFT}\{c_{y_z x_z}(k)\} = \frac{Y_z(i)X_z^*(i)}{N} \tag{4.5.16}$$

Therefore, an effective way to perform circular cross-correlation is

$$c_{y_z x_z}(k) = \frac{\text{IDFT}\{Y_z(i)X_z^*(i)\}}{N} \tag{4.5.17}$$

To convert (4.5.17) to a more efficient implementation based on the FFT, we need to make the signal length $N = L + M + p$ be a power of two. Since (4.5.15) holds for any $p \geq -1$, consider the following value for N.

$$N = \text{nextpow2}(L + M - 1) \tag{4.5.18}$$

For this value of N, a radix-two FFT can be used in place of the DFT. Using (4.5.15) and (4.5.17), we then arrive at the following highly efficient version of linear cross-correlation called *fast cross-correlation* based on Algorithm 4.3.

Fast correlation

$$r_{yx}(k) = \frac{\text{IFFT}\{Y_z(i)X_z^*(i)\}}{L}, \quad 0 \leq k < L \tag{4.5.19}$$

Note the strong similarity between fast convolution in (4.5.6) and the fast correlation in (4.5.19). The only differences are that $Y_z(i)$ is replaced by its complex conjugate $Y_z^*(i)$, the final result is scaled by $1/L$, and it is evaluated only for $0 \leq k < L$. A block diagram of the fast correlation operation is shown in Figure 4.17. Just as with fast convolution, there is a value for L beyond which fast correlation is more efficient than the direct computation of cross-correlation using (4.5.14).

Computational Effort

The analysis of the computational effort for fast cross-correlation is similar to that for fast convolution. For simplicity, suppose the signals $x(k)$ and $y(k)$ are both of length L where L is a power of two. Then from (4.5.18), zero padding to length $N = 2L$ is sufficient. If we proceed as was done with convolution, the number of real FLOPs required to perform a fast linear cross-correlation of two L-point signals is

$$n_{\text{fast}} = 12L \log_2(2L) + 8L + 6 \quad \text{FLOPs} \tag{4.5.20}$$

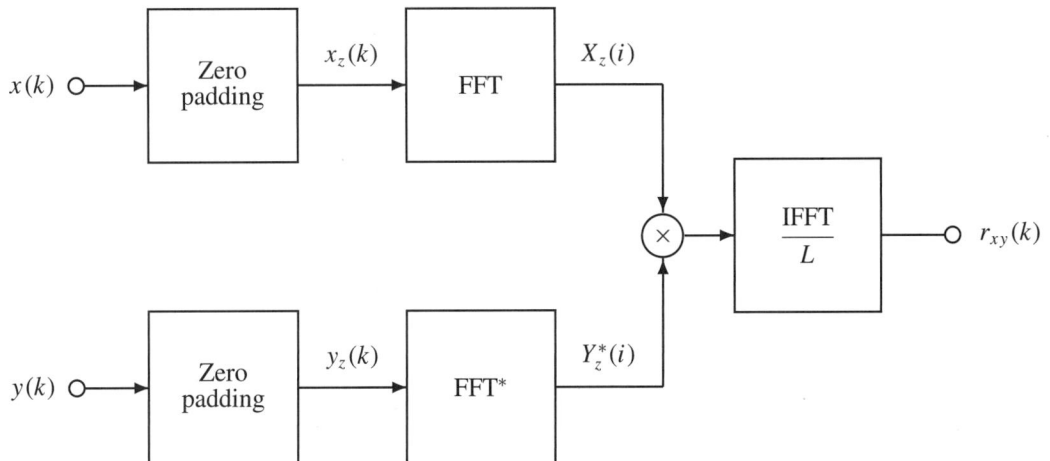

FIGURE 4.17: Fast Linear Cross-correlation

FIGURE 4.18:
Comparison of
Computational
Effort for Linear
Cross-correlation of
Two L-point Signals

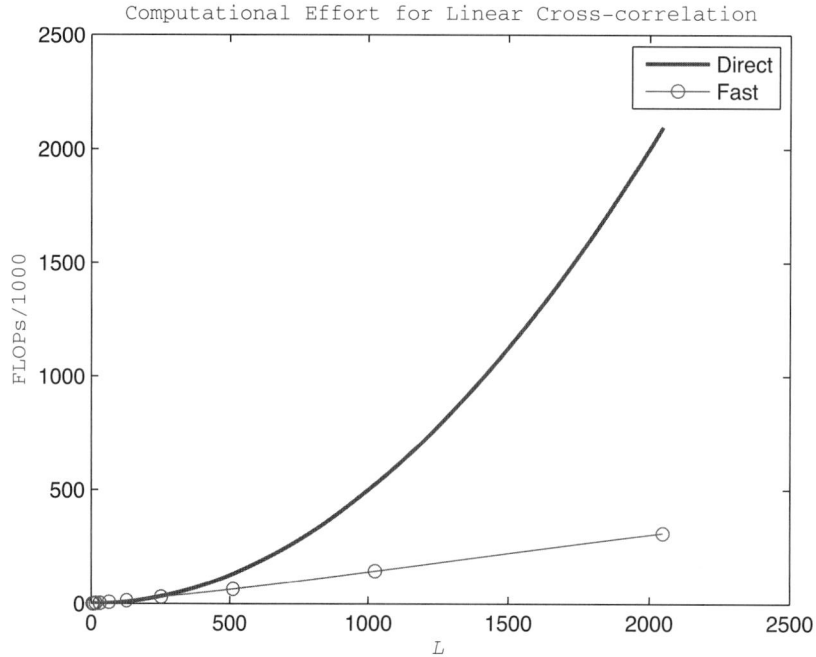

Next consider the number of real multiplications required to implement linear cross-correlation directly. A direct application of (4.5.14) yields $L^2 + L$ multiplications. However, the lower limit on the sum in (4.5.14) can be replaced by $i = k$ because $x(k)$ is causal. This reduces the number of real multiplications by approximately a factor of two.

$$n_{\text{dir}} = \frac{L^2}{2} + L \quad \text{FLOPs} \tag{4.5.21}$$

Comparing (4.5.21) with (4.5.20), we again see that for small values of L, a direct computation of linear cross-correlation will be faster. However, the L^2 term grows faster than the $L \log_2(2L)$ term, so eventually fast cross-correlation will outperform direct cross-correlation. A plot of the number of real FLOPs required by the two methods for signal lengths in the range $2 \leq L \leq 2048$ is shown in Figure 4.18. The two methods require roughly the same number of FLOPs for $L = 256$. However, fast linear cross-correlation is superior to direct linear cross-correlation for signal lengths in the range $L > 512$, and as L increases it becomes significantly faster.

Example 4.10

Fast Linear Correlation

To illustrate the use of fast correlation, let $L = 1024$ and $M = 512$ and consider the following pair of signals where $v(k)$ is white noise uniformly distributed over the interval $[-1, 1]$.

$$x(k) = \frac{3k}{M} \exp\left(\frac{-4k}{M}\right) \sin\left(\frac{5\pi k^2}{M}\right), \quad 0 \leq k < M$$

$$y(k) = x_z(k - p) + v(k), \quad 0 \leq k < L$$

We refer to $x(k)$ as a multi-frequency *chirp* signal because it contains a range of frequencies due to the k^2 factor in the sine term. Here $x_z(k)$ denotes the zero-padded extension of $x(k)$. Thus $x_z(k - p)$ is just $x(k)$ shifted to the right by p samples. For this example, $p = 279$. A plot of these two signals, obtained by running *exam4_10*, is shown in Figure 4.19. Also computed is the normalized linear cross-correlation of $y(k)$ with $x(k)$, as shown in Figure 4.20.

FIGURE 4.19: A Pair of Discrete-time Signals

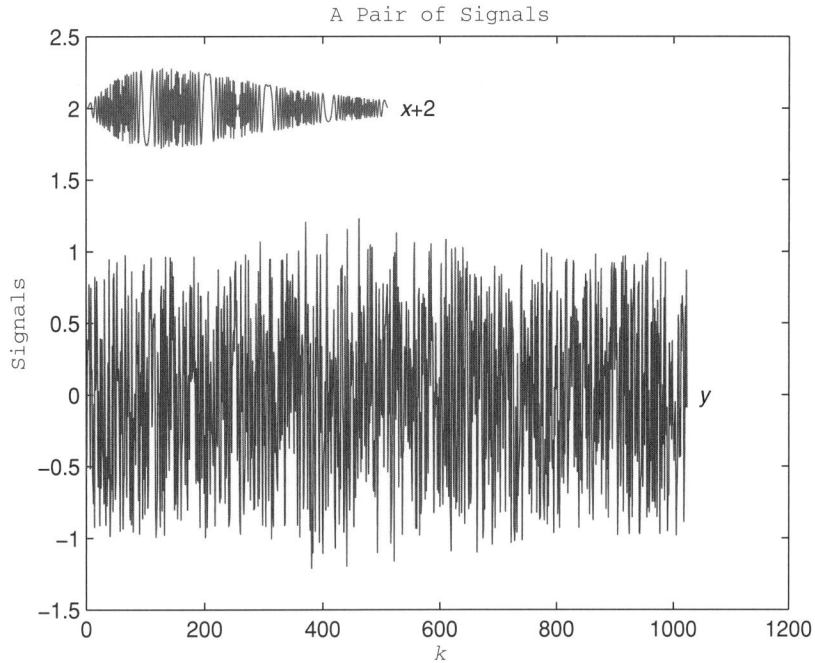

FIGURE 4.20: Normalized Cross-correlation of the Signals in Figure 4.19

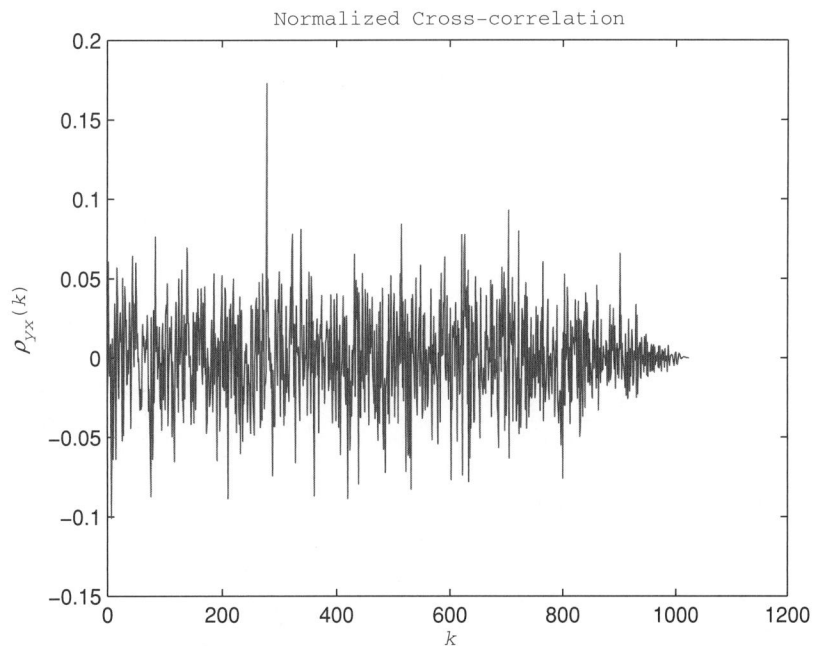

Observe that the peak correlation occurs at $\rho_{yx}(279) = .173$, as expected. This indicates that cross-correlation has succeeded in detecting and identifying the location of the chirp $x(k)$ within $y(k)$. The number of real FLOPs required by fast cross-correlation in this case was $n_{\text{fast}} = 1.02 \times 10^5$. This is in contrast the direct computation using (4.5.14), which requires $n_{\text{dir}} = 5.25 \times 10^5$ real FLOPs. This corresponds to a savings of 80.6 %.

FDSP Functions

The FDSP toolbox contains the following function for computing fast linear and circular cross-correlations.

```
% F_CORR: Fast cross-correlation of two discrete-time signals
%
% Usage:
%         r = f_corr (y,x,circ,norm)
% Pre:
%         y    = vector of length L containing first signal
%         x    = vector of length M <= L containing second signal
%         circ = optional correlation type code (default 0):
%
%                 0 = linear correlation
%                 1 = circular correlation
%
%         norm = optional normalization code (default 0):
%
%                 0 = no normalization
%                 1 = normalized cross-correlation
% Post:
%         r = vector of length L contained selected cross-
%             correlation of y with x.
% Notes:
%         To compute auto-correlation use x = y.
```

If the MATLAB Signal Processing Toolbox is available, there is a function called *xcorr* for performing cross-correlations. It computes $r_{yx}(k)$ for both positive and negative lags k, and it also provides for different types of normalization.

4.6   White Noise

In this section we examine the spectral characteristics of an important type of random signal called white noise. White noise is useful because it provides an effective way to model physical signals, signals that are typically corrupted with noise. For example, the quantization error associated with an analog-to-digital converter (ADC) can be modeled with white noise. Another important application area is in the identification of linear discrete-time systems. White noise input signals are particularly suitable as test signals for system identification because they contain power at all frequencies and therefore excite all of the natural modes of the system under investigation.

4.6.1 Uniform White Noise

Uniform probability density

Let x be a random variable that takes on values in the interval $[a, b]$. If each value of x is equally likely to occur, then we say that x is *uniformly distributed* over the interval $[a, b]$. A uniformly distributed random variable can be characterized by the following *probability density* function.

FIGURE 4.21:
Probability Density
Function
Corresponding to a
Uniform
Distribution over
$[a, b]$ where $a = -3$
and $b = 7$

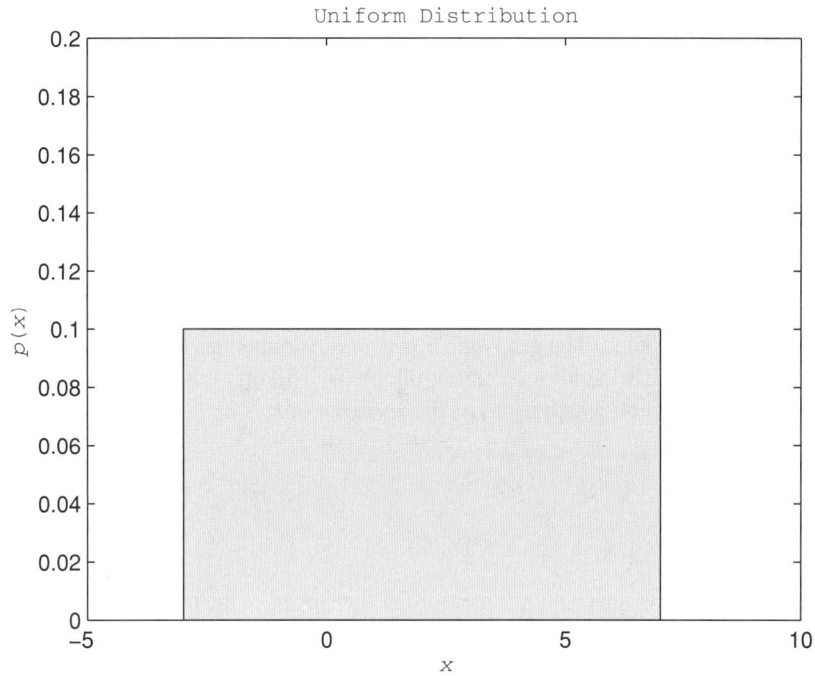

$$p(x) = \begin{cases} \dfrac{1}{b - a}, & a \leq x \leq b \\ 0, & \text{otherwise} \end{cases} \qquad (4.6.1)$$

A graph of the uniform probability density function is shown in Figure 4.21 for the case $[a, b] = [-3, 7]$. Note that the area under a probability density curve is always one. For any probability density function, the probability that a random variable x lies in an interval $[c, d]$ can be computed as

$$P_{[c,d]} = \int_c^d p(x)dx \qquad (4.6.2)$$

Random variables are characterized by their statistical properties. To define a variety of statistical characteristics the following operator is useful.

DEFINITION
....
4.3: Expected Value

Let x be a random variable with probability density $p(x)$. The *expected value* of $f(x)$ is denoted $E[f(x)]$ and defined

$$E[f(x)] \triangleq \int_{-\infty}^{\infty} f(x)p(x)dx$$

Moment

Note that for the case of uniform weighting, the expected value of $f(x)$ can be interpreted as the average value of $f(x)$ over the interval $[a, b]$. More generally, the expected value represents a *weighted average* with weighting $p(x)$. Given the expected value, the kth *moment* of x is defined as $E[x^k]$ for $k \geq 1$. Thus the kth moment is just the expected value of the polynomial x^k.

Mean   The most fundamental moment is the first moment, which is called the *mean* of x.

$$\mu \triangleq E[x] \qquad (4.6.3)$$

The mean μ is the average value about which the random variables are distributed. For random variables uniformly distributed over the interval $[a, b]$, a direct application of Definition 4.3 reveals that the mean is $\mu = (a + b)/2$.

Once the mean is determined, a second set of moments called the central moments can be *Central moment*   computed. The kth *central moment* is defined as $E[(x - \mu)^k]$. The central moments specify the distribution of x about the mean. The first nonzero central moment is the second central *Variance*   moment, which is called the *variance* of x.

$$\sigma^2 \triangleq E[(x - \mu)^2] \qquad (4.6.4)$$

The variance σ^2 is a measure of the spread of the random variable about the mean. The square root of the variance σ is called the *standard deviation* of x.

MATLAB Functions

MATLAB, like most programming languages, provides a facility for generating uniformly distributed random numbers. To create an array of N random numbers v uniformly distributed over $[a, b]$, the following code fragment can be used.

```
v = a + (b-a)*rand(N,1);                    % Uniform random numbers
```

In general, rand(N, M) returns an $N \times M$ matrix of random numbers uniformly distributed over the interval $[0, 1]$. The offset by a and scaling by $b - a$ then maps $[0, 1]$ into $[a, b]$.

Uniform white noise   The signal v is referred to as *uniform white noise*. More specifically, it is white noise that is uniformly distributed over the interval $[a, b]$. The reason for the term *white* arises from the fact that just as white light contains all colors, a white noise signal contains power at all frequencies.

The average power of a random variable x is the second moment $E[x^2]$. For the uniformly distributed random variable v, the average power can be computed using Definition 4.3 and the probability density function in (4.6.1), as follows.

$$\begin{aligned}
P_v &= E[v^2] \\
&= \int_{-\infty}^{\infty} v^2 p(v) dv \\
&= \frac{1}{b-a} \int_{a}^{b} v^2 dv \\
&= \left. \frac{x^3}{3(b-a)} \right|_{a}^{b}
\end{aligned} \qquad (4.6.5)$$

Average power   Consequently, the average power of white noise uniformly distributed over the interval $[a, b]$ is

$$P_v = \frac{b^3 - a^3}{3(b - a)} \qquad (4.6.6)$$

For the special case $[a, b] = [-c, c]$, this reduces to $P_v = c^2/3$. These results are summarized in Appendix 2 for convenient reference. Recall from (4.3.18) that the power density spectrum $S_x(i) = |X(i)|^2/N$ specifies the amount of power at frequency $f_i = if_s/N$. For a white noise signal, the power density spectrum is flat as can be seen from the following numerical example.

Example 4.11

Uniform White Noise

Suppose $a = -5$, $b = 5$, and $N = 512$. Then from (4.6.6), the average power of the white noise signal $v(k)$ is

$$\begin{aligned}
P_v &= \frac{5^3 - (-5)^3}{3(5 - (-5))} \\
&= \frac{250}{30} = 8.333
\end{aligned}$$

Plots of a uniform white noise signal and its power density spectrum can be obtained by running *exam4_11*. The time signal $v(k)$ for $0 \le k < N$ is shown in Figure 4.22, and its power density spectrum $S_v(i)$ for $0 \le i < N/2$ is shown in Figure 4.23. Note how the power density spectrum is, at least roughly, flat and nonzero, indicating that there is power at all $N/2$ discrete frequencies. The uneven nature of the computed power density spectrum in Figure 4.22 is addressed later in the chapter, where techniques for estimating a smoother power density spectrum are introduced.

FIGURE 4.22: White Noise Uniformly Distributed over $[-5, 5]$ with $N = 512$

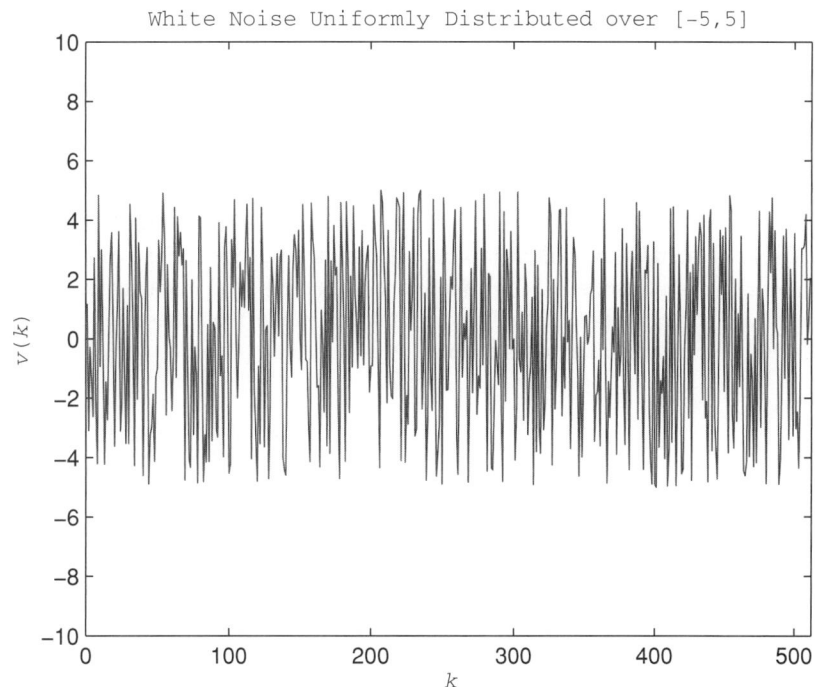

FIGURE 4.23: Power Density Spectrum of Uniformly Distributed White Noise

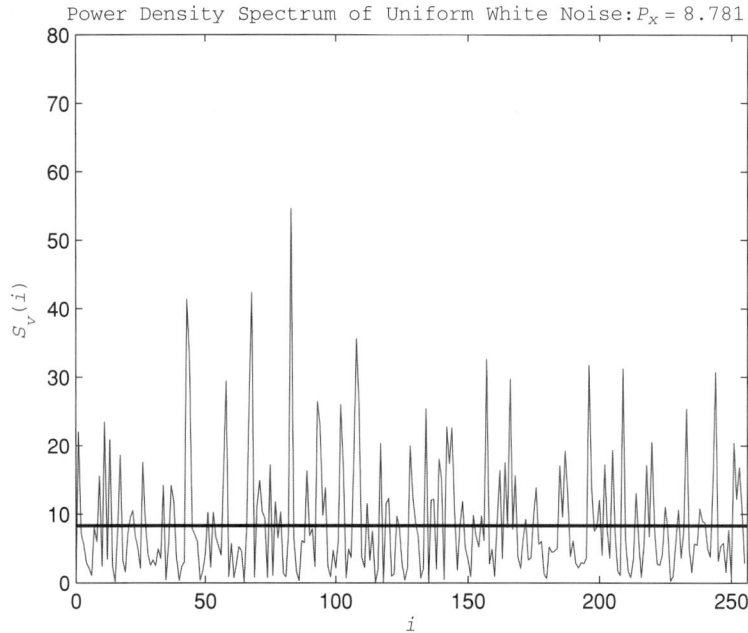

Power Density Spectrum of Uniform White Noise: $P_x = 8.781$

The horizontal line in Figure 4.23 specifies the theoretical average power P_v from (4.6.6). The actual average power of $v(k)$ can be computed directly for comparison. Applying the time-domain formula in (4.3.40) yields

$$P_V = \frac{1}{N} \sum_{i=0}^{N-1} v^2(k) = 8.781$$

As the signal length N increases, the difference between the actual average power P_V and the theoretical average power P_v decreases.

4.6.2 Gaussian White Noise

Uniformly distributed white noise is appropriate in those instances where the value of the signal is constrained to lie within a fixed interval. For example, the quantization noise of an n-bit bipolar ADC with reference voltage V_r lies in the interval $[-V_r, V_r]$. However, there are other cases where it is more natural to model the noise using a normal or *Gaussian* probability density function, as follows.

Gaussian probability density

$$p(x) = \frac{1}{\sigma\sqrt{2\pi}} \exp\left[\frac{-(x-\mu)^2}{2\sigma^2}\right] \qquad (4.6.7)$$

Here the two parameters are the mean μ and the standard deviation σ. The mean specifies where the random values are centered, while the standard deviation specifies the spread of the random values about the mean. A plot of the bell-shaped Gaussian probability density function is shown in Figure 4.24 for the case where $\mu = 0$ and $\sigma = 1$.

FIGURE 4.24: Gaussian Probability Density Function with $\mu = 0$ and $\sigma = 1$

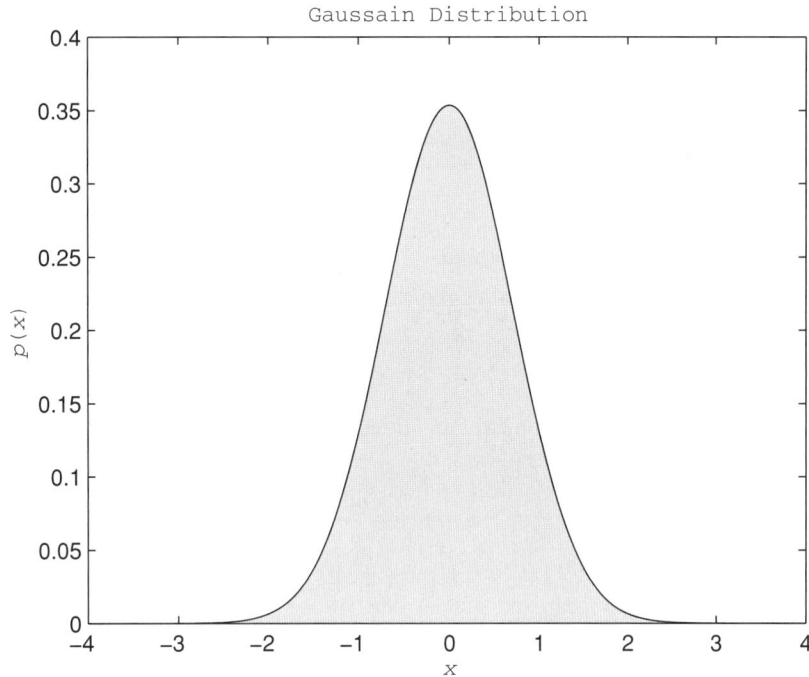

Gaussain Distribution

MATLAB Functions

Random numbers with a normal or Gaussian distribution can be generated using the MATLAB function *randn*. To create an array of N random numbers v with a Gaussian distribution of mean *mu* and standard deviation *sigma*, the following code fragment can be used.

```
v = mu + sigma*randn(N,1);                  % Gaussian random numbers
```

In general, randn(N, M) returns an $N \times M$ matrix of random numbers with a normal or Gaussian distribution with a mean of zero and a standard deviation of one. The offset by *mu* and scaling by *sigma* then maps this into a distribution with the desired mean and standard deviation.

Gaussian white noise

The random signal v is referred to as *Gaussian white noise* with mean μ and standard deviation σ. Again, it is called white noise because $v(k)$ contains power at all frequencies. Later, we will see how to filter it to produce *colored* noise.

For a Gaussian random variable v, the average power can be computed as $E[v^2]$ using the probability density function in (4.6.7). To simplify the final result, we restrict our attention to the important special case of zero-mean noise. Using a table of integrals (Dwight, 1961), we have

$$
\begin{aligned}
P_v &= E[v^2] \\
&= \int_{-\infty}^{\infty} v^2 p(v)\,dv \\
&= \frac{2}{\sigma\sqrt{2\pi}} \int_0^{\infty} v^2 \exp\left(\frac{-v^2}{2\sigma^2}\right) dv, \quad \mu = 0 \\
&= \frac{2}{\sigma\sqrt{2\pi}} \left(\frac{\sqrt{8\pi}\,\sigma^3}{4}\right)
\end{aligned} \tag{4.6.8}
$$

Consequently, the average power of zero-mean Gaussian white noise with standard deviation σ is just the variance σ^2.

$$P_v = \sigma^2 \tag{4.6.9}$$

Example 4.12

Gaussian White Noise

Consider the following continuous-time periodic signal produced by an AM mixer with frequencies F_1 and F_2.

$$x_a(t) = \sin(2\pi F_1 t)\cos(2\pi F_2 t)$$

From the trigonometric identities in Appendix 2, the product of two sinusoids produces sum and difference frequencies.

$$x_a(t) = \frac{\sin[2\pi(F_1 + F_2)t] + \sin[2\pi(F_1 - F_2)t]}{2}$$

Suppose $F_1 = 300$ Hz, $F_2 = 100$ Hz, and $x_a(t)$ is corrupted with Gaussian noise $v(k)$, with zero mean and standard deviation $\sigma = .8$. If the result is then sampled at $f_s = 1$ kHz using $N = 1024$ samples, the corresponding discrete-time signal is

$$x(k) = \sin(.3\pi k)\cos(.1\pi k) + v(k), \quad 0 \le k < N$$

From (4.6.9), the average power of the noise term is

$$P_v = .64$$

The signal $x(k)$ and its power density spectrum $S_x(i)$ can be obtained by running *exam4_12*. The first quarter of the noise-corrupted time-signal is shown in Figure 4.25, and its power

FIGURE 4.25:
Periodic Signal
Corrupted with
Zero-mean
Gaussian White
Noise

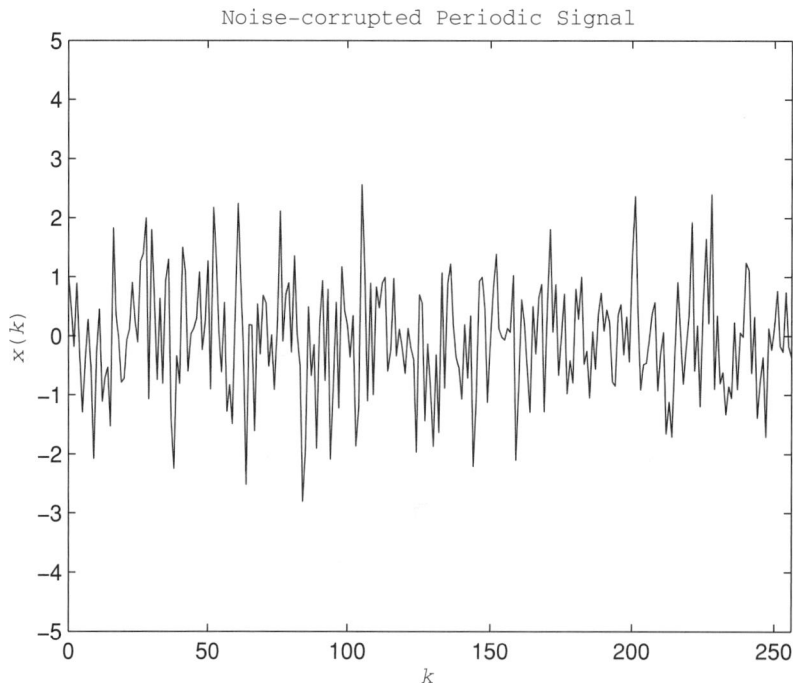

Noise-corrupted Periodic Signal

FIGURE 4.26: Power Density Spectrum of Noise-corrupted Periodic Signal

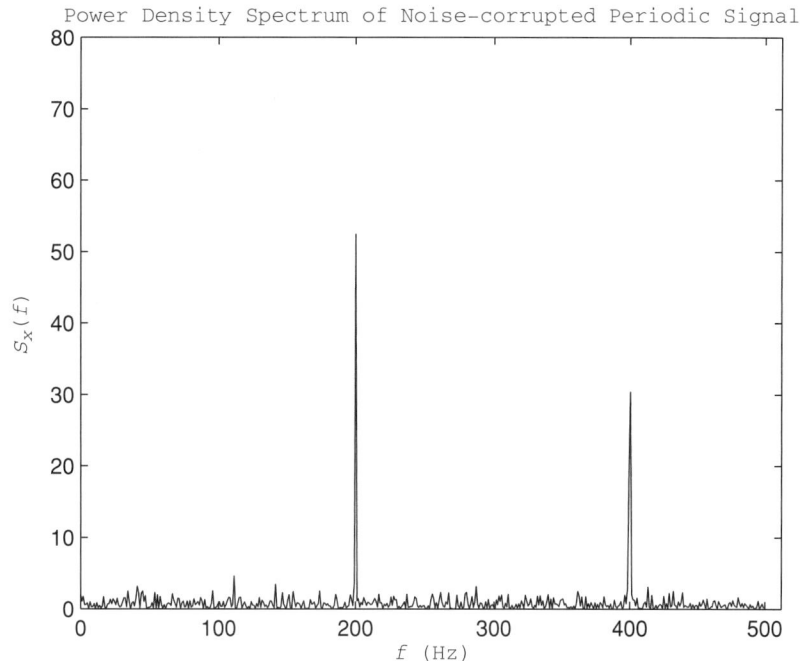

Power Density Spectrum of Noise-corrupted Periodic Signal

density spectrum is shown in Figure 4.26. Note how it is difficult to tell from the time plot in Figure 4.25 that $x(k)$ contains a periodic component due to the presence of the noise. However, the two spectral components at the sum frequency $F_1 + F_2 = 400$ Hz and the difference frequency $F_1 - F_2 = 200$ Hz are evident from the power density spectrum plot in Figure 4.26 where two distinct spikes are present. The Gaussian white noise also is apparent in Figure 4.26 as low level power that is distributed over all frequencies.

The power density spectrum plot in Figure 4.26 uses the independent variable $f = if_s/N$, rather than i, in order to facilitate interpretation of the frequency. Thus it is a plot of $S_x(f)$ versus f rather than $S_x(i)$ versus i.

FDSP Functions

As a convenience to the user, the FDSP toolbox contains the following functions for generating uniform and Gaussian white noise signals.

```
% F_RANDINIT: Initialize the random number generator
% F_RANDU: Generate uniform random numbers
% F_RANDG: Generate Gaussian random numbers
%
% Usage:
%            f_randinit (seed)
%        A = f_randu (m,n,a,b);
%        A = f_randg (m,n,mu,sigma);
% Pre:
%        seed  = nonnegative integer.  Each seed produces a
%                different pseudo-random sequence.
%        m     = number of rows
```

```
%       n      = number of columns
%       a      = lower limit for the uniform distribution
%       b      = upper limit for the uniform distribution
%       mu     = mean of the Gaussian distribution
%       sigma  = standard deviation of the Gaussian distribution
% Post:
%       A = m by n matrix of random numbers
```

The MATLAB function *rand* can be used to control the sequence of random numbers generated by successive calls to *f_randu* and *f_randg*.

```
rand ('state',s)                          % Initialize random number generator
```

When *rand* is called with 'state' as its first calling argument, the second calling argument is an integer *s* that represents the initial state of the random number generator. The default state is $s = 0$. Each $s \geq 0$ produces a different pseudo-random sequence.

4.7   Auto-correlation

In this section we use correlation techniques to process noise-corrupted signals. Recall that the cross-correlation of a signal with itself is called auto-correlation. Consider the case of circular auto-correlation.

DEFINITION

4.4: Circular Auto-correlation

Let $x(k)$ be an N-point signal and let $x_p(k)$ be its periodic extension. The *circular auto-correlation* of $x(k)$ is denoted $c_{xx}(k)$ and defined

$$c_{xx}(k) \stackrel{\Delta}{=} \frac{1}{N} \sum_{i=k}^{N-1} x(i)x_p(i-k), \quad 0 \leq k < N$$

As the notation suggests, auto-correlation is a special case of cross-correlation with $y = x$. Since $x_p(0) = x(0)$, it follows from Definition 4.4 that circular auto-correlation at a lag of $k = 0$ is simply the average power of $x(k)$. Thus the average power can be expressed in terms of circular auto-correlation as

$$P_x = c_{xx}(0) \tag{4.7.1}$$

Auto-correlation can be normalized just as cross-correlation. From (4.7.1), the *normalized circular auto-correlation*, denoted $\sigma_{xx}(k)$ is

$$\sigma_{xx}(k) = \frac{c_{xx}(k)}{P_x}, \quad 0 \leq k < N \tag{4.7.2}$$

In view of (4.7.1), it follows that $\sigma_{xx}(0) = 1$. Since $|\sigma_{xx}(k)| \leq 1$, this means that the normalized circular auto-correlation always reaches its peak value of one at a lag of $k = 0$.

4.7.1   Auto-correlation of White Noise

White noise has a particularly simple auto-correlation. To see this, let $v(k)$ be a random white noise signal of length N with a mean of $\mu = 0$. Recall that the mean of a random signal is *Ergodic signal*   the expected value $E[v(k)]$. If a random signal has the property that it is *ergodic*, then the expected value of $f\{v(k)\}$ for a function f can be approximated by replacing the ensemble average (which depends on the probability density) with the simpler time average.

$$E[f\{v(k)\}] \approx \frac{1}{N}\sum_{k=0}^{N-1} f\{v(k)\} \tag{4.7.3}$$

From Definition 4.4, the circular auto-correlation of $v(k)$ can be expressed in terms of expected values, as follows.

$$\begin{aligned}c_{vv}(k) &= \frac{1}{N}\sum_{i=0}^{N-1} v(i)v_p(i-k)\\ &\approx E[v(i)v_p(i-k)]\end{aligned} \tag{4.7.4}$$

The samples of a white noise signal are statistically independent of one another. For *statistically independent* random variables, the expected value of the product is equal to the product of the expected values. Since the signal $v(i)$ has zero mean, it follows that

Statistically independent signals

$$\begin{aligned}c_{vv}(k) &\approx E[v(i)v_p(i-k)]\\ &= E[v(i)]E[v_p(i-k)]\\ &= 0, \quad k \neq 0\end{aligned} \tag{4.7.5}$$

Uncorrelated signals

When two signals are statistically independent and one or both have zero mean, the expected value of the product is zero. In this case we say that the two signals are *uncorrelated*.

For the case $k = 0$, $c_{vv}(0) = P_v$ where P_v is the average power of $v(k)$. Combining the two cases, the circular auto-correlation of a zero-mean white noise signal with average power P_v can be expressed as

$$c_{vv}(k) \approx P_v\delta(k), \quad 0 \leq k < N \tag{4.7.6}$$

Hence the circular auto-correlation of zero-mean white noise is simply an impulse of strength P_v at $k = 0$. As the value of N increases, the approximation in (4.7.6) becomes more accurate. The same analysis that was used to develop the approximation in (4.7.6) can be applied to linear auto-correlation as well, and the result is identical. Thus for $N \gg 1$, the linear auto-correlation of zero-mean white noise can be approximated as

$$r_{vv}(k) \approx P_v\delta(k), \quad 0 \leq k < N \tag{4.7.7}$$

To numerically verify (4.7.7), suppose $v(k)$ is Gaussian white noise with mean $\mu = 0$ and standard deviation $\sigma = 1$. The normalized linear auto-correlation for the case $N = 1024$ is shown in Figure 4.27. Since the auto-correlation is normalized, the theoretical result should be $\rho_{vv}(k) = \delta(k)$. When circular auto-correlation is used, the only difference is that there is no narrowing of the tail of $\sigma_{vv}(k)$ for large values of k.

FIGURE 4.27:
Normalized Linear
Auto-correlation
of Zero-mean
Gaussian White
Noise

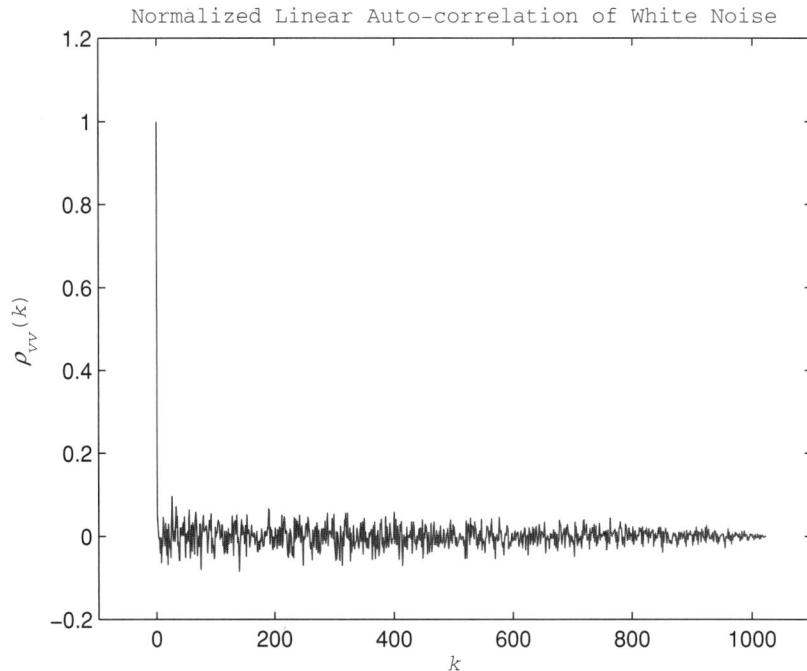

Normalized Linear Auto-correlation of White Noise

4.7.2 Power Density Spectrum

There is a simple and elegant relationship between the power density spectrum and circular auto-correlation. Recall that the power density spectrum specifies the distribution of power over the discrete frequencies. For an N-point signal $x(k)$, the power density spectrum is as follows where $X(i)$ is the DFT of $x(k)$.

$$S_x(i) \triangleq \frac{|X(i)|^2}{N}, \quad 0 \leq i < N \tag{4.7.8}$$

To determine the relationship between $S_x(i)$ and $c_{xx}(k)$, we use the cross-correlation property of the DFT from Table 4.8 with $y = x$. Applying the DFT to $c_{xx}(k)$ yields

$$\begin{aligned} C_{xx}(i) &= \text{DFT}\{c_{xx}(k)\} \\ &= \frac{X(i)X^*(i)}{N} \\ &= \frac{|X(i)|^2}{N}, \quad 0 \leq i < N \end{aligned} \tag{4.7.9}$$

Combining (4.7.8) and (4.7.9) then leads to the following alternative formulation of the power density spectrum.

$$S_x(i) = C_{xx}(i), \quad 0 \leq i < N \tag{4.7.10}$$

Thus the DFT of the circular auto-correlation of $x(k)$ is the power density spectrum. This is the DFT version of the *Wiener-Khintchine* theorem, and it is listed as one of the DFT properties in Table 4.8. A block diagram illustrating the Wiener-Khintchine theorem is shown in Figure 4.28.

FIGURE 4.28: Power Density Spectrum Using Circular Auto-correlation

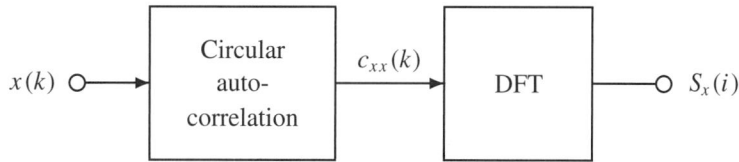

White noise

It is instructive to apply (4.7.10) to white noise. Suppose $v(k)$ is zero-mean white noise with average power P_v. Using (4.7.6) and (4.7.10), we have

$$
\begin{aligned}
S_v(i) &= C_{vv}(i) \\
&\approx \text{DFT}\{P_v\delta(k)\} \\
&= P_v
\end{aligned}
\tag{4.7.11}
$$

Hence zero-mean white noise with average power P_v has a power density spectrum that is *flat* and equal to P_v. It is for this reason that we refer to the noise as *white* because it contains power at all frequencies just as white light contains all colors.

Example 4.13

Power Density Spectrum

To illustrate the application of Figure 4.28, let $N = 512$ and consider a signal $x(k)$ that consists of a double pulse of width $M = 8$ centered at $k = N/2$.

$$
x(k) = \begin{cases}
0, & 0 \le k < N/2 - M \\
1, & N/2 - M \le k < N/2 \\
-1, & N/2 \le k < N/2 + M \\
0, & N/2 + M \le k < N
\end{cases}
$$

When *exam4_13* is run, it computes the power density spectrum $S_x(i)$ by finding the DFT of the circular auto-correlation, as in Figure 4.28. The resulting plots of $x(k)$ and its power density spectrum $S_x(i)$ are shown in Figure 4.29.

FIGURE 4.29: Power Density Spectrum of Double Pulse of Width $M = 8$ Using Circular Auto-correlation

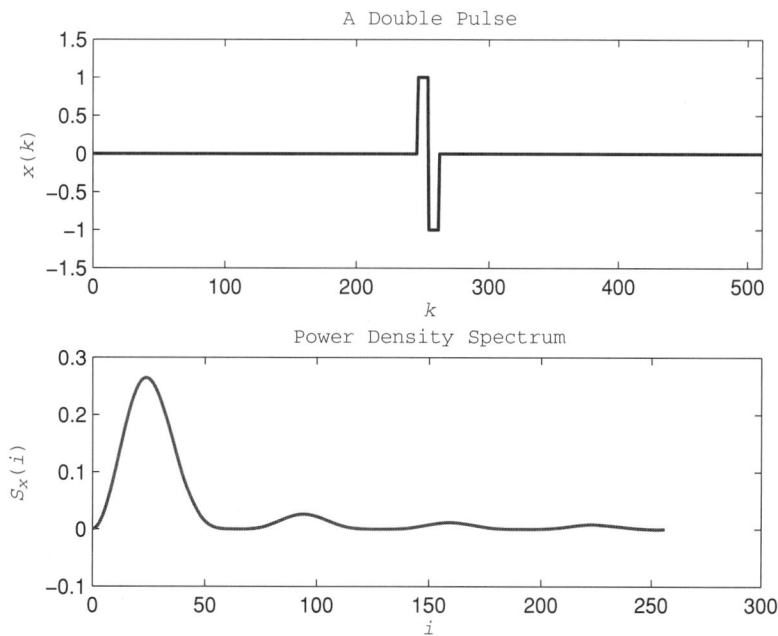

4.7.3 Extracting Periodic Signals from Noise

Practical signals are often corrupted with noise. Suppose $x(k)$ is a periodic signal with period M. We can model a noisy version of $x(k)$ as follows.

$$y(k) = x(k) + v(k), \quad 0 \le k < N \tag{4.7.12}$$

Here $v(k)$ represents additive zero-mean white noise that may arise, for example, from the measurement process or perhaps because $x(k)$ is transmitted over a noisy communication channel.

Period Estimation

Our initial objective is to estimate the period of $x(k)$ using $y(k)$. Since $v(k)$ contains power at all frequencies, completely removing $v(k)$ with a filtering operation is not an option. Instead, we can use correlation techniques. Consider the circular auto-correlation of the noisy signal $y(k)$. Let $y_p(k)$ and $v_p(k)$ be the periodic extensions of the N-point signals $y(k)$ and $v(k)$, respectively. Using (4.7.4), Definition 4.4, and the fact that the expected value operator is linear, we have

$$
\begin{aligned}
c_{yy}(k) &\approx E[y(i)y_p(i-k)] \\
&= E[\{x(i) + v(i)\}\{x_p(i-k) + v_p(i-k)\}] \\
&= E[x(i)x_p(i-k) + x(i)v_p(i-k) + v(i)x_p(i-k) + v(i)v_p(i-k)] \\
&= E[x(i)x_p(i-k)] + E[x(i)v_p(i-k)] + E[v(i)x_p(i-k)] + E[v(i)v_p(i-k)] \\
&\approx c_{xx}(k) + c_{xv}(k) + c_{vx}(k) + c_{vv}(k)
\end{aligned}
\tag{4.7.13}
$$

Typically, the noise $v(k)$ is statistically independent of the signal $x(k)$. Since $E[v(k)] = 0$, this means that the circular cross-correlation terms $c_{xv}(k)$ and $c_{vx}(k)$ are both zero. Then using (4.7.6) simplifies the circular auto-correlation of the noisy signal $y(k)$ to

$$c_{yy}(k) \approx c_{xx}(k) + P_v\delta(k) \tag{4.7.14}$$

Consequently, $c_{yy}(0) = P_x + P_v$, where P_x is the average power of the signal, and P_v is the average power of the noise. For $k > 0$ we have $c_{yy}(k) \approx c_{xx}(k)$. That is, the effect of auto-correlation is to average out or *reduce* the noise, so the circular auto-correlation of $y(k)$ is less noisy than $y(k)$ itself. Not only does the circular auto-correlation reduce noise, but it is also *Periodic signal* periodic with the same period as $x(k)$. In particular, using $x(k + M) = x(k)$, we have

$$
\begin{aligned}
c_{xx}(k + M) &= \frac{1}{N}\sum_{i=0}^{N-1} x(i)x_p(i - k - M) \\
&= \frac{1}{N}\sum_{i=0}^{N-1} x(i)x_p(i - k) \\
&= c_{xx}(k)
\end{aligned}
\tag{4.7.15}
$$

Thus the circular auto-correlation of a periodic signal is itself periodic with the same period, but it is less noisy.

$$c_{xx}(k + M) = c_{xx}(k), \quad 0 \le k < N - M \tag{4.7.16}$$

Since $c_{xx}(k)$ is periodic with period M, and $c_{yy}(k) \approx c_{xx}(k)$ for $k > 0$, it follows that the auto-correlation of $y(k)$ will be periodic with period M. We can estimate the period of $x(k)$ using a distinctive reference point such as a peak in $c_{yy}(k)$.

FIGURE 4.30: A Noisy Periodic Signal

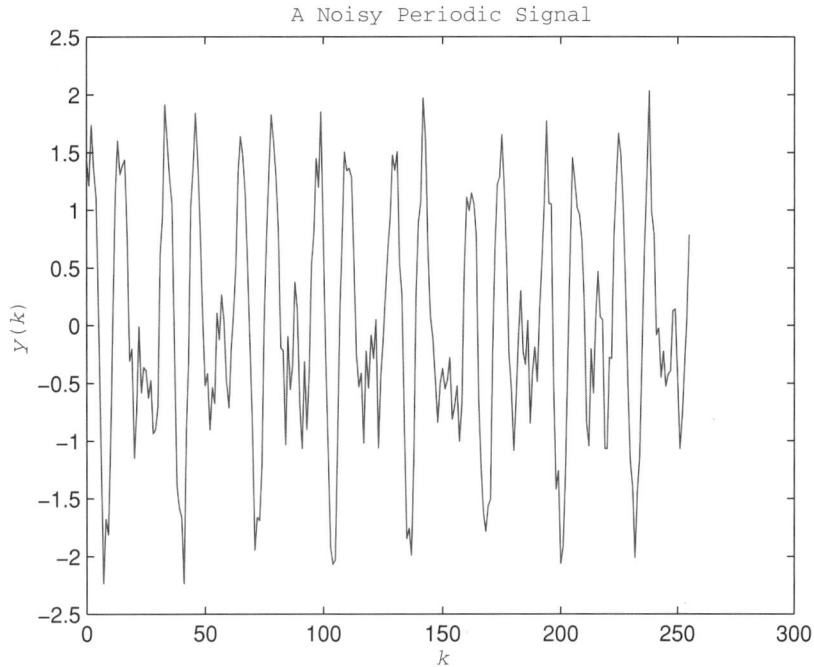
A Noisy Periodic Signal

Example 4.14

Period Estimation

Suppose $N = 256$. Consider the following periodic signal which includes two sinusoidal components.

$$x(k) = \cos\left(\frac{32\pi k}{N}\right) + \sin\left(\frac{48\pi k}{N}\right)$$

Note that the cosine term has period $N/16$ and the sine term has period $N/24$. Thus the period of $x(k)$ is the common period

$$M = \frac{N}{8}$$
$$= 32$$

Suppose $y(k)$ is a noisy version of $x(k)$, as in (4.7.12) where $v(k)$ is white noise uniformly distributed over the interval $[-.5, .5]$. A plot of $y(k)$, obtained by running *exam4_14*, is shown in Figure 4.30. Note that the periodic nature of the underlying signal $x(k)$ is apparent, but it is difficult to precisely estimate the period due the presence of the noise. By contrast, the circular auto-correlation of $y(k)$ is much less noisy, as can be seen from the plot in Figure 4.31. Using the FDSP function *f_caliper* and rounding, the period of $c_{yy}(k)$ is $M = 32$.

Signal Estimation

Once the period of a noise-corrupted periodic signal has been determined, we can use this information to extract the signal itself from the noise. Suppose $x(k)$ is an N-point signal that

FIGURE 4.31:
Circular Auto-correlation of Noisy Periodic Signal in Figure 4.30

Mark Two Consecutive Peak Points with Mouse Cross Hairs!

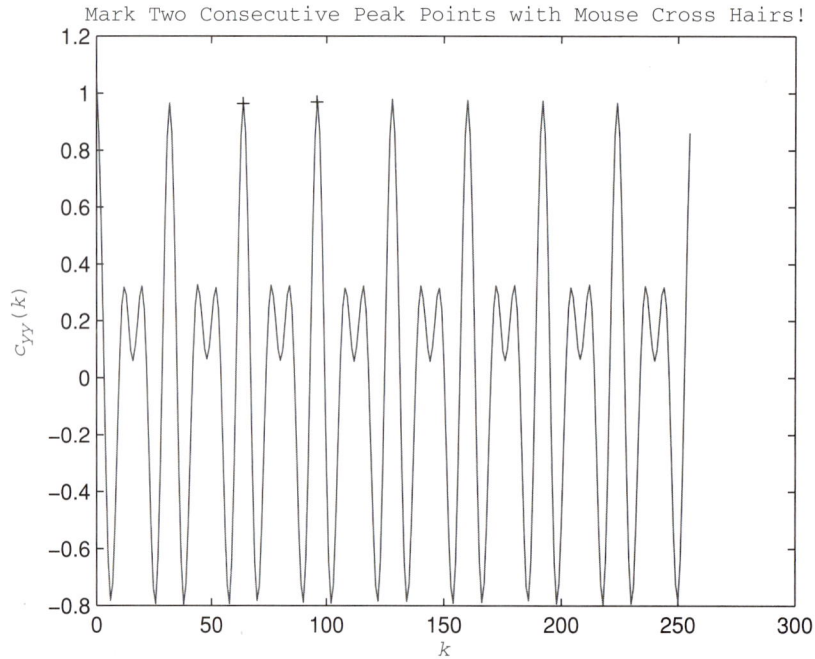

is periodic with period $M \ll N$. Then the number of complete cycles of $x(k)$ in $y(k)$ is

$$L = \text{floor}\left(\frac{N}{M}\right) \qquad (4.7.17)$$

Next let $\delta_M(k)$ be an N-point periodic impulse train with period M. One can represent $\delta_M(k)$ as follows.

$$\delta_M(k) = \sum_{i=0}^{L-1} \delta(k - iM), \quad 0 \le k < N \qquad (4.7.18)$$

Suppose $y(k)$ is a noisy version of $x(k)$ that has been corrupted by zero-mean white noise $v(k)$, as in (4.7.12). The underlying periodic signal $x(k)$ can be extracted from the noisy signal $y(k)$ by cross-correlating with $\delta_M(k)$. To see this, let $y_p(k)$ be the periodic extension of $y(k)$. Recalling the time-reversal property of circular cross-correlation in Table 2.4 and using Definition 4.4, we find that the circular cross-correlation of $y(k)$ with $\delta_M(k)$ is

$$
\begin{aligned}
c_{y\delta_M}(k) &= c_{\delta_M y}(-k) \\
&= \frac{1}{N} \sum_{i=0}^{N-1} \delta_M(i) y_p(i+k) \\
&= \frac{1}{N} \sum_{i=0}^{N-1} \left[\sum_{q=0}^{L-1} \delta(i - qM) \right] y_p(i+k) \\
&= \frac{1}{N} \sum_{q=0}^{L-1} y_p(qM + k) \qquad (4.7.19)
\end{aligned}
$$

FIGURE 4.32:
Extracting a
Periodic Signal of
Period *M* from
Noise Using Circular
Cross-correlation

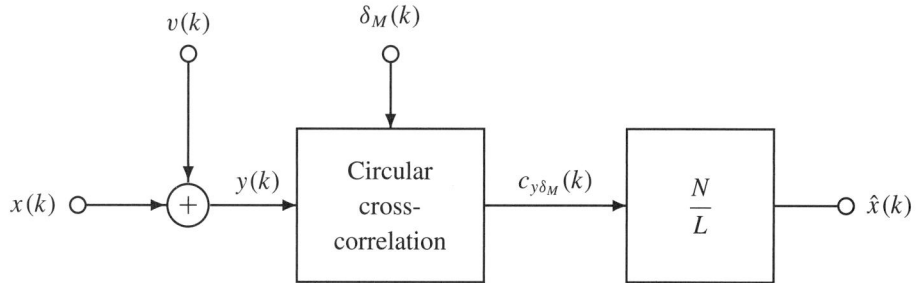

Next the expression for $y(k)$ in (4.7.12) can be substituted in (4.7.19). Recalling that $x(k)$ is periodic with period M, we then have

$$c_{y\delta_M}(k) = \frac{1}{N}\sum_{q=0}^{L-1} x_p(qM+k) + \frac{1}{N}\sum_{q=0}^{L-1} v_p(qM+k)$$

$$= \frac{1}{N}\sum_{q=0}^{L-1} x_p(k) + \frac{1}{N}\sum_{q=0}^{L-1} v_p(qM+k)$$

$$= \frac{Lx(k)}{N} + \frac{1}{N}\sum_{q=0}^{L-1} v_p(qM+k), \quad 0 \le k < N \qquad (4.7.20)$$

For each k, the last term in (4.7.20) represents a sum of L statistically independent noise terms. Since $v(k)$ is zero-mean white noise, this means that for $L \gg 1$ the last term is approximately zero. Thus for $N \gg M$ we have the following method of *extracting a periodic signal* from noise. The underlying periodic signal $x(k)$ can be approximated using $\hat{x}(k)$ where

Periodic signal extraction

$$\hat{x}(k) = \left(\frac{N}{L}\right) c_{y\delta_M}(k), \quad 0 \le k < N \qquad (4.7.21)$$

A block diagram summarizing the steps required to extract a periodic signal from noise using circular cross-correlation is shown in Figure 4.32.

Example 4.15

Extracting a Periodic Signal from Noise

To illustrate the application of Figure 4.32 to extract a periodic signal from noise, let $N = 256$ and consider the following noisy periodic signal.

$$x(k) = \cos\left(\frac{32\pi k}{N}\right) + \sin\left(\frac{48\pi k}{N}\right)$$
$$y(k) = x(k) + v(k)$$

Suppose $v(k)$ is white noise uniformly distributed over $[-.5, .5]$. The signal $y(k)$ was considered previously in Example 4.14 and is shown in Figure 4.30. The analysis of the auto-correlation of $y(k)$ in Figure 4.31 revealed the period of $x(k)$ to be $M = 32$. Therefore the number of complete cycles of $x(k)$ in $y(k)$ is

$$L = \text{floor}\left(\frac{N}{M}\right)$$
$$= 8$$

FIGURE 4.33:
Comparison of
Periodic Signal $x(k)$
and Estimate $\hat{x}(k)$
Extracted from $y(k)$
Using Circular
Cross-correlation

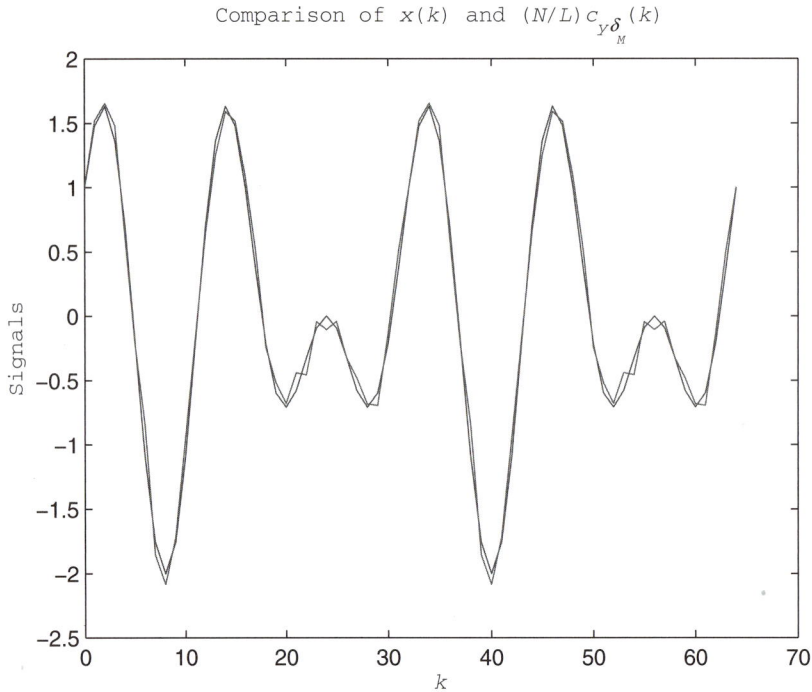

Comparison of $x(k)$ and $(N/L)c_{y\delta_M}(k)$

The extraction of $x(k)$ from the noise-corrupted $y(k)$ using the technique in Figure 4.32 is performed by running *exam4_15*. The estimate of $x(k)$ using circular cross-correlation is

$$\hat{x}(k) = \left(\frac{N}{L}\right) c_{y\delta_M}(k)$$

A plot comparing the first two periods of $x(k)$ with the estimate $\hat{x}(k)$ is shown in Figure 4.33. The reconstruction, in this case, is quite reasonable but it is not exact, given that the noise term in (4.7.20) is only approximately zero. Note that the estimate $\hat{x}(k)$ should improve as N/M increases.

For the periodic signal extraction in Example 4.15, the period of the periodic component of $y(k)$ consists of an integer number of samples M. It is also possible for the periodic component of $y(k)$ to have a period T_0 that is not an integer multiple of T. For this more general case, a modified version of the technique in Figure 4.32 should be applied. When $T_0 \neq MT$ for an integer M, then the periodic impulse train $\delta_M(k)$ should be generalized in such a way that the lone unit pulse at multiples of M samples is replaced by a pair of adjacent smaller pulses whose amplitudes sum to unity. In particular, let

$$M = \text{floor}\left(\frac{T_0}{T}\right) \tag{4.7.22}$$

Then a fraction α of the unit pulse will occur at multiples of sample M, and the remaining fraction $1 - \alpha$ will occur at multiples of sample $M + 1$ where

$$\alpha = 1 - \frac{T_0}{T} + M \tag{4.7.23}$$

Notice that if $T_0 = MT$, then $\alpha = 1$ and all of the unit pulse is applied at multiples of sample M. Furthermore, as $T_0 \to (M + 1)T$, $\alpha \to 0$ and $(1 - \alpha) \to 1$.

4.8 Zero Padding and Spectral Resolution

4.8.1 Frequency Response Using the DFT

Recall that the frequency response $H(f)$ of a stable linear system was defined as the transfer function $H(z)$ evaluated along the unit circle. Alternatively, $H(f)$ is the DTFT of the impulse response $h(k)$. For a causal system,

$$H(f) = \sum_{k=0}^{\infty} h(k) \exp(-jk2\pi fT) \tag{4.8.1}$$

Put another way, the frequency response is the spectrum of the impulse response. The magnitude spectrum $A(f) = |H(f)|$ is the magnitude response of the system, and the phase spectrum $\phi(f) = \angle H(f)$ is the phase response of the system. There is a simple way to approximate the frequency response using the DFT, and in certain instances this approximation is exact. Let $H(i)$ denote the DFT of the first N samples of $h(k)$.

$$H(i) = \text{DFT}\{h(k)\}, \quad 0 \le i < N \tag{4.8.2}$$

Next, suppose we evaluate the frequency response at the ith discrete frequency $f_i = if_s/N$. Recalling that $W_N = \exp(-j2\pi/N)$ and using (4.8.1), we find

$$
\begin{aligned}
H(f_i) &= \sum_{k=0}^{\infty} h(k) \exp(-jki2\pi/N) \\
&= \sum_{k=0}^{\infty} h(k) W_N^{ik} \\
&= \sum_{k=0}^{N-1} h(k) W_N^{ik} + \sum_{k=N}^{\infty} h(k) W_N^{ik} \\
&= H(i) + \sum_{k=N}^{\infty} h(k) W_N^{ik}
\end{aligned}
\tag{4.8.3}
$$

Thus the difference between $H(f_i)$ and $H(i)$ is represented by the *tail* of the DTFT of the impulse response $h(k)$. The magnitude of this difference can be bounded in the following way.

$$
\begin{aligned}
|H(f_i) - H(i)| &= \left| \sum_{k=N}^{\infty} h(k) W_N^{ik} \right| \\
&\le \sum_{k=N}^{\infty} |h(k) W_N^{ik}| \\
&= \sum_{k=N}^{\infty} |h(k)| \cdot |W_N^{ik}| \\
&= \sum_{k=N}^{\infty} |h(k)|
\end{aligned}
\tag{4.8.4}
$$

Notice that the upper bound in (4.8.4) no longer depends on i. For a stable filter, the impulse response is absolutely summable, so the right-hand side is finite and goes to zero as $N \to \infty$.

Consequently, the DFT of the impulse response can be used to approximate the discrete-time frequency response as long as the number of samples N is sufficiently large.

$$H(i) \approx H(f_i), \quad 0 \leq i \leq \frac{N}{2} \tag{4.8.5}$$

In (4.8.5) we restrict i to the interval $0 \leq i \leq N/2$ because, for a real system, $H(i)$ exhibits symmetry about the midpoint $i = N/2$, as in Table 4.7. In effect, all of the essential information is contained in the subinterval $0 \leq i \leq N/2$, which corresponds to positive frequencies.

For an important class of digital filters, the approximation in (4.8.5) is exact. To see this, recall that if $H(z)$ is an FIR filter of order m, then $h(k) = 0$ for $k > m$. It then follows from (4.8.4) that the upper bound on the error is zero for $N > m$. That is, if $H(z)$ is an FIR filter of order m and $N > m$, then $H(i)$ is an exact representation of the N samples of the frequency response. The frequency response $H(i)$ can be expressed in polar form as $H(i) = A(i) \exp[j\phi(i)]$ where the $A(i)$ is the *magnitude response*, and $\phi(i)$ is the *phase response*.

$$A(i) \overset{\Delta}{=} |H(i)| \tag{4.8.6a}$$

$$\phi(i) \overset{\Delta}{=} \angle H(i) \tag{4.8.6b}$$

Example 4.16	**Discrete-time Frequency Response**

As an example of using (4.8.2) to find the frequency response, consider a running average filter of order $M - 1$.

$$y(k) = \frac{1}{M} \sum_{i=0}^{M-1} x(k - i)$$

Running average filter — The impulse response of this FIR filter is a pulse of amplitude $1/M$ and duration M samples starting at $k = 0$.

$$h(k) = \frac{1}{M} \sum_{i=0}^{M-1} \delta(k - i)$$

The frequency response $H(i) = \text{DFT}\{h(k)\}$ for the case $M = 8$, $N = 1024$, and $f_s = 200$ Hz can be obtained by running *exam4_16*. Note that since $N = 2^{10}$, the FFT implementation can be used. The resulting magnitude response $A(f)$ and phase response $\phi(f)$ are shown in Figure 4.34. There are $M/2$ lobes in the magnitude response, and the phase response has $M/2$ jump discontinuities between the lobes but is otherwise linear. The independent variable used in Figure 4.34 is $f = i f_s / N$.

The relatively large side lobes in the magnitude response reveal that a running average is not particularly effective as a *lowpass* filter because the stopband gain outside the main lobe is relatively large. If a *weighted average* of the last M samples is used instead, the size of the side lobes can be reduced. For example, consider the following weighted average that uses weighting based on something called the Hanning window.

$$y(k) = \frac{1}{M} \sum_{i=0}^{M-1} \left[1 - \cos\left(\frac{2\pi k}{M-1}\right) \right] x(k - i)$$

The magnitude response for this weighted average filter is shown in Figure 4.35. Note how that amplitudes of the side lobes have been significantly reduced as a result of the weighting, although the main lobe is now wider.

FIGURE 4.34:
Frequency
Response of
Running Average
Filter with $M = 8$,
$N = 1024$, and
$f_s = 200$ Hz

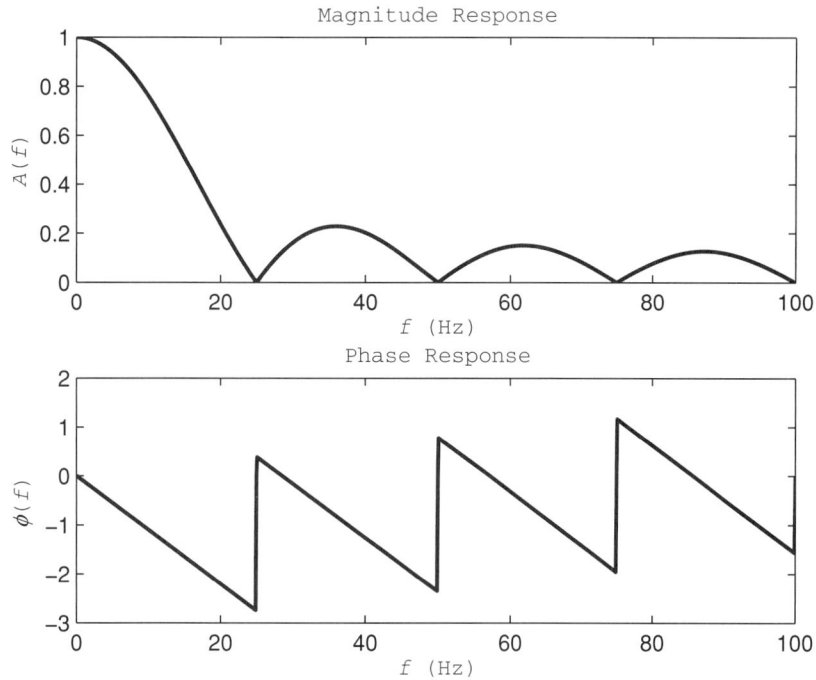

FIGURE 4.35:
Magnitude
Response of
Weighted Average
Filter with Hanning
Window
Weighting, $m = 8$,
$N = 1024$, and
$f_s = 200$ Hz

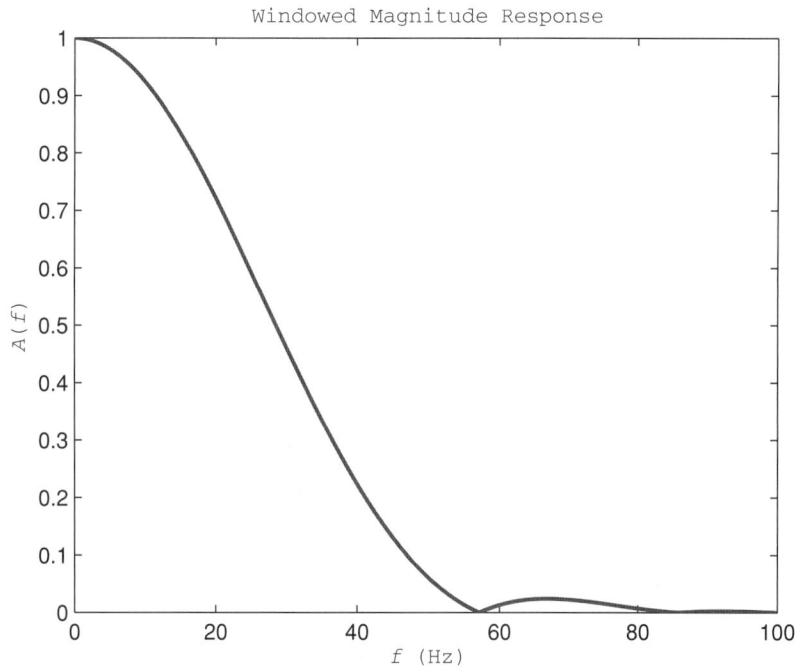

Decibel Scale (dB)

The plot of the magnitude response in Figure 4.34 is referred to as a *linear* plot because both axes use linear scales. If there is a large frequency range or a large range in the values of the magnitude response, then logarithmic units are often used. When plotting the magnitude response of a filter, or the magnitude spectrum of a signal, the logarithmic unit that is typically

FIGURE 4.36:
Magnitude
Response of
Running Average
Filter in
Example 4.16 Using
the Logarithmic
Decibel Scale

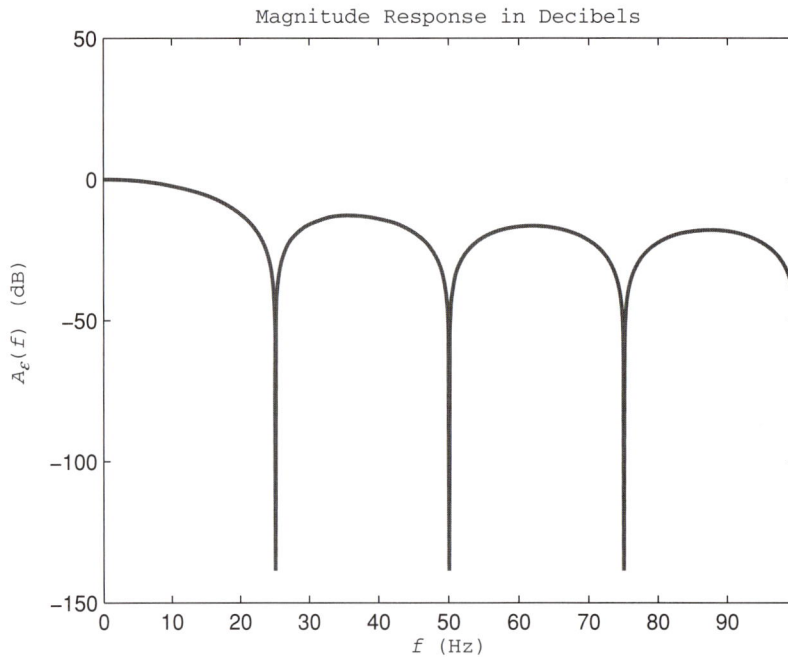

Magnitude Response in Decibels

Decibel   used is the decibel which is abbreviated dB. A *decibel* is defined as 10 times the logarithm (base 10) of the signal power. Since power corresponds to $|H(i)|^2$, this translates to 20 times $\log_{10}$ of $|H(i)|$.

$$A(i) \triangleq 20 \log_{10}(|H(i)|) \ \ \text{dB} \tag{4.8.7}$$

One of the useful characteristics of the dB scale is that it allows us to better quantify how close to zero the magnitude response gets. Notice from (4.8.7) that a gain of unity corresponds to zero dB, and that every reduction in $|H(i)|$ by a factor of 10 corresponds to a change of -20 dB. For an ideal filter, the signal attenuation in the stopband is complete, so the magnitude response is zero. However, in practice, Wiener and Paley (1934) showed that the magnitude response of a causal system can not be identically zero over a continuum of frequencies; it can only be zero at isolated frequencies. By using decibels, we can see how close to zero the magnitude response gets. However, there is one problem that arises with the use of decibels. If the magnitude response does go to zero at certain isolated frequencies, as in Figure 4.34, then from (4.8.7) this corresponds to a decibel value of minus infinity. To accommodate this situation we usually replace $|H(i)|$ in (4.8.7) as follows

$$A_\epsilon(i) = 20 \log_{10}(\max\{|H(i)|, \epsilon\}) \ \ \text{dB} \tag{4.8.8}$$

Here $\epsilon > 0$ is taken to be a very small number. For example, the single precision machine epsilon $\epsilon_M = 1.19 \times 10^{-7}$ might be used. A plot of the magnitude response of the filter in Example 4.16 using $\epsilon = \epsilon_M$ and the logarithmic dB scale is shown in Figure 4.36.

FDSP Functions

Recall from Section 3.8 that the FDSP function *f_freqz* can be used to compute the discrete-time frequency response. Function *f_freqz* is the discrete equivalent of the continuous-time frequency response function *f_freqs* introduced in Chapter 1. If the Signal Processing Toolbox is available, there is a function called *freqz* that is available for computing the discrete frequency response.

4.8.2 Zero Padding

Zero padding

There is a simple way to interpolate signal spectra without increasing the sampling rate. The length of the signal can be increased from N to M by padding $x(k)$ with $M-N$ zeros as follows.

$$x_z(k) \triangleq \begin{cases} x(k), & 0 \le k < N \\ 0, & N \le k < M \end{cases} \tag{4.8.9}$$

Here $x_z(k)$ denotes the *zero-padded* version for $x(k)$. Suppose $S_z(i)$ is the power density spectrum of $x_z(k)$, but scaled by M/N because no new power was added by the zero padding. That is,

$$S_z(i) \triangleq \frac{|X_z(i)|^2}{N}, \quad 0 \le i < M \tag{4.8.10}$$

Frequency precision

where $X_z(i) = \text{DFT}\{x_z(k)\}$. Then the *frequency precision*, or space between adjacent discrete frequencies, of $S_z(i)$ is

$$\Delta f_z = \frac{f_s}{M} \tag{4.8.11}$$

The effect of zero padding is to *interpolate* between the original points of the spectrum. In particular, if $M = qN$ for some integer q, there will be $q-1$ *new* points between each of the original N points. The original points remain unchanged.

$$X_z(qi) = X(i), \quad 0 \le i < N \tag{4.8.12}$$

By decreasing Δf_z through zero padding, the location of an individual sinusoidal spectral component can be more precisely determined as can be seen from the following example.

Example 4.17

Zero Padding

As an illustration of how zero padding can be put to effective use, suppose $f_s = 1024$ Hz and $N = 256$. Consider a noise-free signal with a single sinusoidal spectral component at $F_0 = 330.5$ Hz.

$$x(k) = \cos(2\pi F_0 kT), \quad 0 \le k < N$$

If we zero-pad $x(k)$ by a factor of eight, then $M = 8N = 2048$. Thus the frequency precision before and after zero padding is

$$\Delta f_x = \frac{f_s}{N} = 4 \text{ Hz}$$

$$\Delta f_z = \frac{f_s}{M} = .5 \text{ Hz}$$

The corresponding sinusoidal frequency indices for the two cases are

$$i_x = \frac{F_0}{\Delta f_x} = 82.625$$

$$i_z = \frac{F_0}{\Delta f_z} = 661$$

Notice that i_x is not an integer, so the rounded value, 83, must be used. It follows that the peaks in the power density spectra will occur at

$$f_x = 83\Delta f_x = 332 \text{ Hz}$$
$$f_z = 661\Delta f_z = 330.5 \text{ Hz}$$

FIGURE 4.37: Power Density Spectra Using Zero Padding of $N = 256$ Samples of $x(k)$ to Create $M = 2048$ Samples of $x_z(k)$

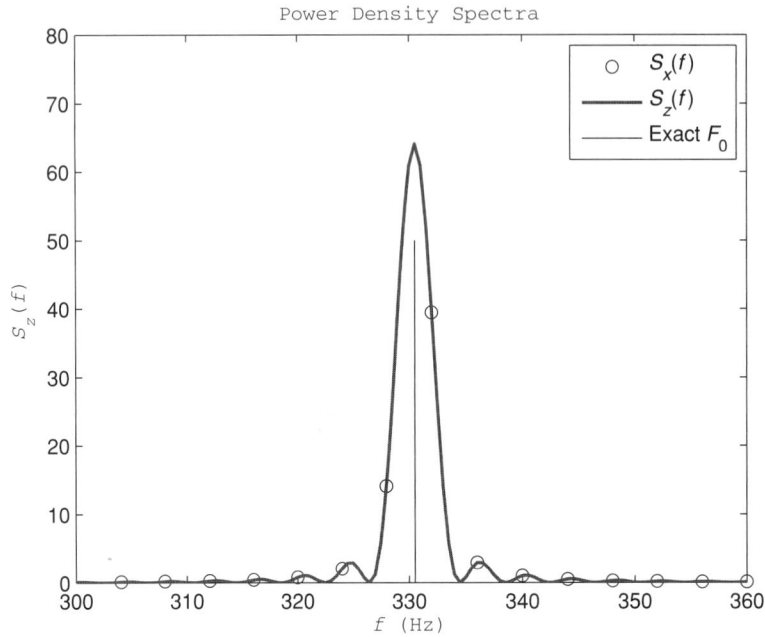

Consequently, with a frequency precision of $\Delta f_z = .5$ Hz, the zero-padded power density spectrum should be able to locate the sinusoidal component exactly. The two power density spectra can be computed by running *exam4_17*. The resulting plot is shown in Figure 4.37. To clarify the display, only the frequency range $300 \leq f \leq 360$ Hz is plotted. The low-precision power density spectrum $S_x(f)$ is plotted with isolated points, while the high-precision power density spectrum $S_z(f)$ is plotted as points connected by straight lines. Note how $S_z(f)$ interpolates between the points of $S_x(f)$. In this case, $M/N = 8$, so there are seven points of $S_z(f)$ between each pair of points in $S_x(f)$.

Ringing   The ringing in $S_z(f)$ arises because we are multiplying an infinitely long $x(k)$ by a *rectangular window* to obtain $x_z(k)$. As we shall see later in this chapter, we can reduce the side lobes, at the expense of broadening the main lobe, by multiplying by a different window.

4.8.3 Spectral Resolution

Although zero padding allows us to identify the precise location of an *isolated* sinusoidal frequency component, it is less helpful when it comes to trying to *resolve* the difference between two closely spaced sinusoidal components. When sinusoidal components are too close together, their peaks in the power density spectrum tend to merge into a single broader peak. The basic problem with zero padding is that it does not add any new information to the signal. The frequency resolution, or smallest frequency *difference* that one can detect, is limited by the sampling frequency and the number of nonzero samples. The following ratio, called the *Rayleigh limit*, represents the value of the frequency resolution.

Rayleigh limit

$$\Delta F = \frac{f_s}{N} \tag{4.8.13}$$

Frequency resolution

Thus there is a difference between the frequency precision, which is the spacing between discrete frequencies of the DFT, and the *frequency resolution*, which is the smallest difference in frequencies that can be reliably detected. Zero padding can be helpful in distinguishing between two spectral components, but we cannot go below the Rayleigh limit in (4.8.13). Observe that since $f_s = 1/T$, the Rayleigh limit is really just the reciprocal of the length of the original signal $\tau = NT$. Thus for a given sampling frequency, to improve the frequency resolution, we must add more samples.

Example 4.18

Frequency Resolution

As an illustration of the detection of two closely spaced sinusoidal components, suppose $f_s = 1024$ Hz and $N = 1024$. Consider the following noise-free signal with spectral components at $F_0 = 330$ Hz and $F_1 = 331$ Hz.

$$x(k) = \sin(2\pi F_0 kT) + \cos(2\pi F_1 kT), \quad 0 \le k < N$$

If we zero-pad $x(k)$ by a factor of two, then $M = 2N = 2048$. Thus the frequency precision before and after zero padding is

$$\Delta f_x = \frac{f_s}{N} = 1 \text{ Hz}$$

$$\Delta f_z = \frac{f_s}{M} = .5 \text{ Hz}$$

The two power density spectra can be computed by running *exam4_18*. The resulting plots are shown in Figures 4.38 and 4.39, respectively. To clarify the display, only the frequency range of $320 \le f \le 340$ Hz is plotted in each case. The low-precision power density spectrum in Figure 4.38 shows a single broad peak centered at the midpoint,

$$F_m = \frac{F_0 + F_1}{2} = 330.5 \text{ Hz}$$

FIGURE 4.38: Power Density Spectrum of Two Closely Spaced Sinusoidal Components with $f_s = 1024$ Hz and $N = 1024$

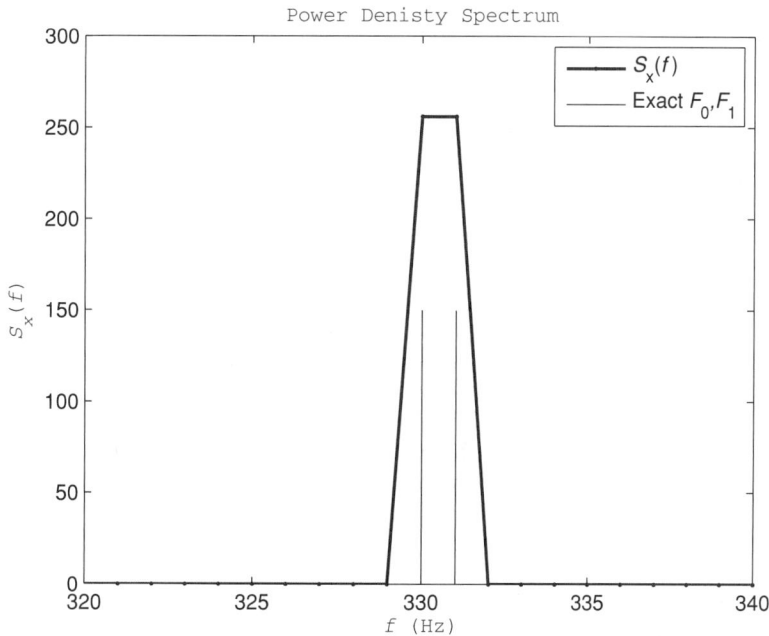

FIGURE 4.39:
Spectral Resolution
of Two Closely
Spaced Sinusoidal
Components Using
Zero Padding with
$f_s = 1024$ Hz,
$N = 1024$, and
$M = 2048$

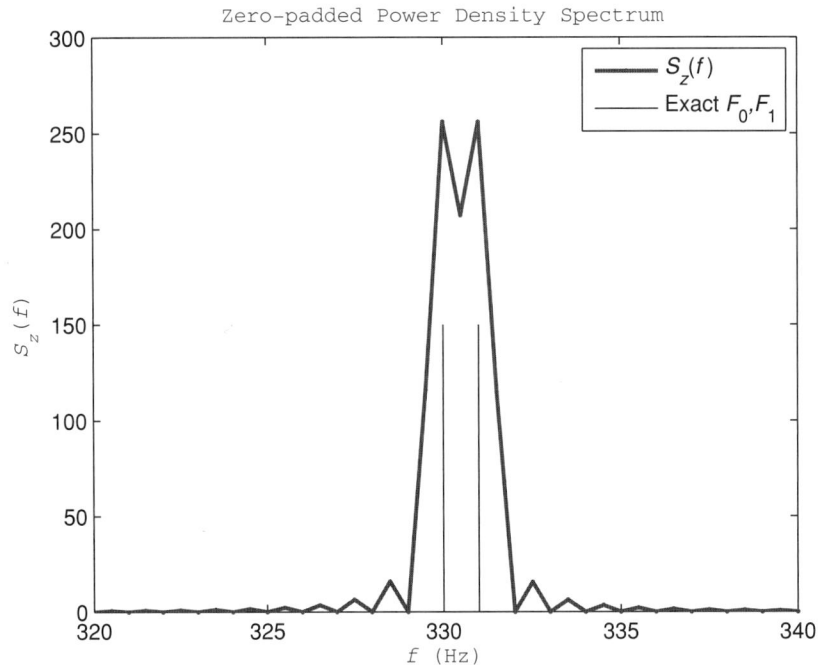

By zero padding we can decrease the frequency increment from 1 Hz to .5 Hz. This results in the plot shown in Figure 4.39 where a pair of peaks begins to emerge, but with some overlap. The power density spectrum between the peaks decreases sufficiently to make the two peaks discernible, but it does not go to zero between the peaks. From (4.8.13), the Rayleigh limit is

$$\Delta F = \frac{f_s}{N} = 1 \text{ Hz}$$

Interestingly enough, increasing M beyond 2048 does *not* make the two peaks more discernible because the separation between F_0 and F_1 is the Rayleigh limit.

$$|F_1 - F_0| = \Delta F$$

Of course, we can decrease the Rayleigh limit by adding more samples. If the number of samples is doubled to $N = 2048$, then the resulting power density spectrum, without any zero padding, is shown in Figure 4.40. Here it is clear that the two peaks are now separated.

MATLAB Functions

The concept of zero padding is useful even when the motivation is not to improve the frequency precision. If the number of samples N is not a power of two, then a radix-two FFT cannot be used. Instead, a DFT or perhaps a less efficient alternative form of the FFT is needed. This problem can be circumvented by simply padding $x(k)$ with enough zeros to make the new length M the next power of two. The following code fragment can be used.

```
M = nextpow2(length(x));        % M = next power of 2
X = fft(x,M);                   % Zero-pad to M samples
```

FIGURE 4.40: Spectral Resolution of Two Closely Spaced Sinusoidal Components Using Improved Frequency Resolution with $f_s = 1024$ Hz and $N = 2048$

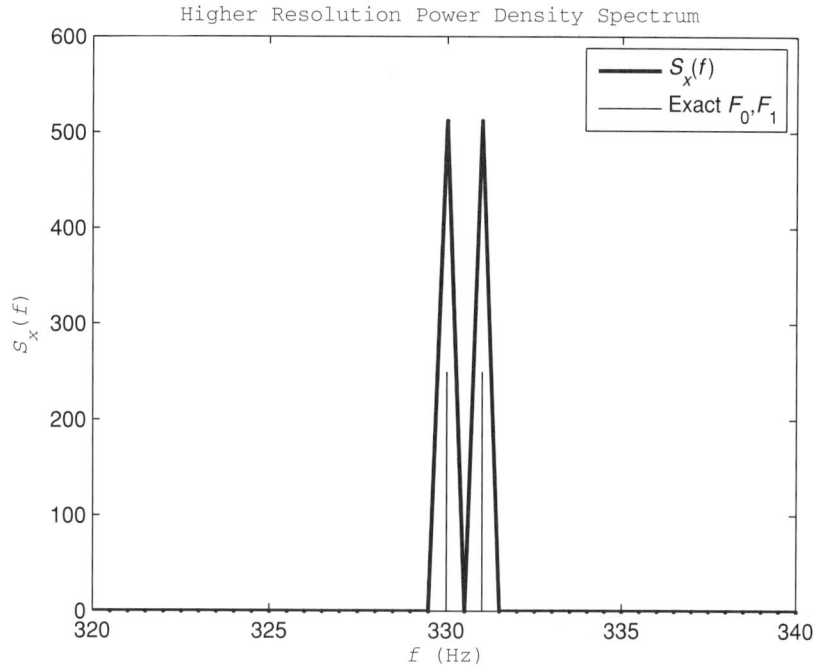

For example, if $N = 1000$, then padding $x(k)$ with 24 zeros will increase the number of samples to $M = 2^{10} = 1024$. This increases storage requirements by 2.4 percent, but in exchange it significantly decreases the computational time. Recall that a 1000-point DFT requires $N^2 = 10^6$ complex FLOPs. However, a 1024-point FFT requires only $(M/2)\log_2(M) = 5120$ complex FLOPs. Thus at a cost of 2.4 percent in storage space, we have purchased an increase in speed by a factor of 99.5, a real bargain!

• • • • • • • • • • • • • •

4.9 Spectrogram

4.9.1 Data Windows

Many signals of practical interest are sufficiently long that their spectral characteristics can be thought of as changing with time. For example, a segment of recorded speech can be broken down into more basic units such as words, syllables, or phonemes. Each basic unit will have its own distinct spectral characteristics. One way to capture the fact that the spectral characteristics are changing with time is to compute a sequence of short overlapping DFTs, sometimes called a *short term Fourier transform* or STFT. For example, suppose $x(k)$ is a long N-point signal where $N = LM$ for integers L and M. Then $x(k)$ can be decomposed into $2M - 1$ overlapping subsignals of length L, as shown Figure 4.41 for the case $M = 5$.

If the subsignals are denoted $x_m(k)$ for $0 \leq m < 2M - 1$, then the mth subsignal can be extracted from the original signal $x(k)$ as

$$x_m(k) \triangleq x(mL/2 + k), \quad 0 \leq m < 2M - 1, \ 0 \leq k < L \tag{4.9.1}$$

FIGURE 4.41:
Decomposition of
$x(k)$ into $2M - 1$
Overlapping
Subsignals of
Length L where
$M = 5$

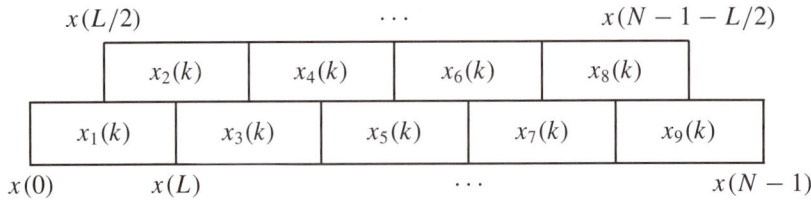

Subsignal

A useful way to look at subsignal $x_m(k)$ is as a product of the original signal $x(k)$ times a window function $w_R(k)$.

$$x_m(k) = w_R(k - mL/2)x(k) \qquad (4.9.2)$$

Rectangular window

If we compare (4.9.2) with (4.9.1), the window function is a *rectangular window* of unit amplitude that is zero for all k except $0 \le k < L$. In terms of step functions

$$w_R(k) \stackrel{\Delta}{=} \mu(k) - \mu(k - L) \qquad (4.9.3)$$

Spectral leakage

When a signal is multiplied by a rectangular window in the time domain, this creates something called *spectral leakage* in the frequency domain. An example of this phenomenon can be seen in the frequency response of the running average filter in Example 4.16. There the impulse response was a rectangular pulse of height $1/M$ and width M. The magnitude response in Figure 4.34 included several side lobes in addition to the larger main lobe. These side lobes are a manifestation of spectral leakage that occurs when there is an abrupt transition or vertical edge in the time domain.

The side lobes can be reduced, at the expense of making the main lobe wider, by using a "softer" or less abrupt window, as was done in Figure 4.35 using the Hanning window. Nonrectangular data windows go to zero more gradually than the rectangular window, and for these windows, the frequency domain leakage outside the main lobe is reduced. Some popular window functions for generating data windows of width L are summarized in Table 4.10.

Plots of the rectangular, Hanning, Hamming, and Blackman windows in Table 4.10 for the case $L = 256$ are shown in Figure 4.42. Note how all of the windows except the rectangular window $w_R(k)$ decrease gradually. With the exception of the Hamming window, they all taper to zero by the time the edge of the window is reached. The spectral characteristics of data windows are examined in more detail in Chapter 6 where they are used in the design of digital FIR filters.

TABLE 4.10: ▶
Window Functions
for Data Windows
of Width L

Number	Name	$w(k)$
0	Rectangular	$w_R(k)$
1	Hanning	$\left[.5 - .5\cos\left(\dfrac{2\pi k}{L}\right)\right] w_R(k)$
2	Hamming	$\left[.54 - .46\cos\left(\dfrac{2\pi k}{L}\right)\right] w_R(k)$
3	Blackman	$\left[.42 - .5\cos\left(\dfrac{2\pi k}{L}\right) + .08\sin\left(\dfrac{4\pi k}{L}\right)\right] w_R(k)$

FIGURE 4.42: Window Functions for
$L = 250$:
0 = Rectangular,
1 = Hanning,
2 = Hamming, and
3 = Blackman

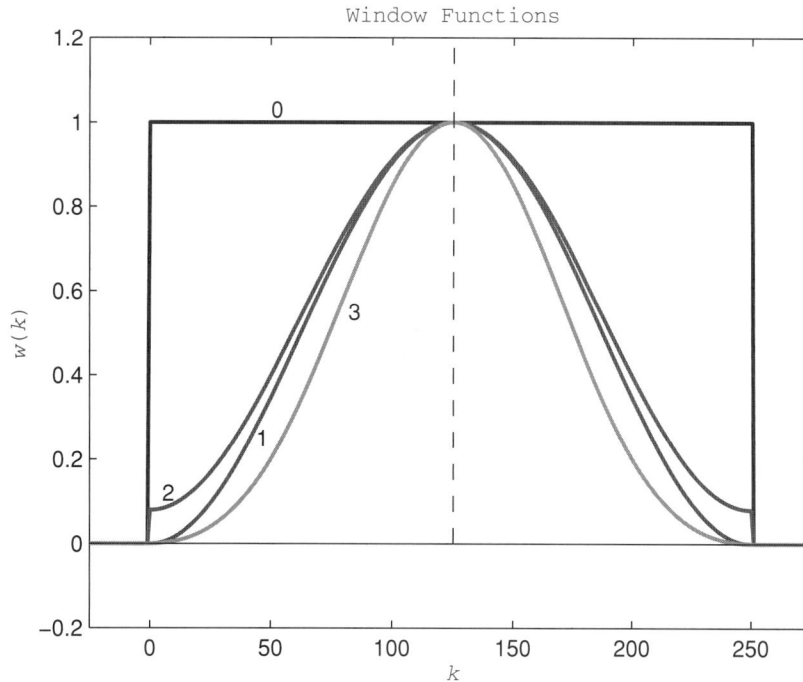

Window Functions

4.9.2 Spectrogram

The $2M - 1$ subsignals of length L in (4.9.1) overlap one another by $L/2$ samples. Suppose we take the DFTs of windowed versions of $x_m(k)$ for $0 \leq m < 2M - 1$ and arrange them as the rows of a matrix. This matrix, called a spectrogram, reveals how the spectral characteristics of the signal $x(k)$ change with time.

DEFINITION

4.5: Spectrogram

Let $x(k)$ be an N-point signal decomposed into $2M - 1$ overlapping subsignals $x_m(k)$ of length L, as in (4.9,1), and let $w(k)$ be a window function from Table 4.10. The *spectrogram* of $x(k)$ is a $(2M - 1) \times L$ matrix G defined as

$$G(m, i) \triangleq |\text{DFT}\{w(k)x_m(k)\}|, \quad 0 \leq m < 2M - 1, \quad 0 \leq i < L$$

The spectrogram $G(m, i)$ is a collection of short DFTs parameterized by the starting time. The first independent variable m specifies the starting time, in increments of $L/2$, while the second variable i specifies the frequency, in increments of f_s/L. The purpose of the window function $w(k)$ is to reduce frequency domain leakage. When the basic rectangular window is used, the values of $G(m, i)$ tend to get smeared or spread out along the frequency dimension due to leakage.

The spectrogram is typically displayed as a two-dimensional contour plot. Note that $G(m, i) \geq 0$. The two-dimensional display uses different colors or shades of a color to denote the values of $G(m, i)$ within certain bands or levels. Thus a spectrogram is a contour plot of slices or level surfaces of the windowed magnitude spectra. Note that if $x(k)$ is real, then $G(m, i)$ will exhibit even symmetry about the line $i = L/2$. Consequently, only the $(2M - 1) \times L/2$ submatrix on the left needs to be plotted for real $x(k)$.

FIGURE 4.43:
Recording of
Vowels at
$f_s = 8000$ Hz

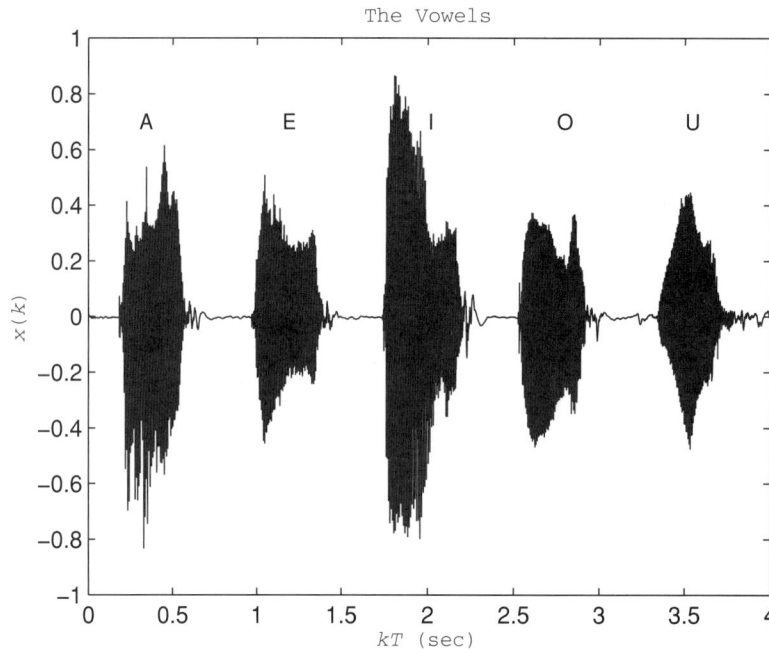

The Vowels

Example 4.19

Spectrogram

Consider the signal $x(k)$ shown in Figure 4.43, which consists of a four-second recording of the vowels {A, E, I, O, U} spoken in succession. In this case the sampling rate was $f_s = 8000$ Hz and the number of samples was $N = 32000$. If we choose $L = 256$, then $M = 125$ and G is a 250×256 matrix. Letting $T = 1/f_s$, the time and frequency increment are as follows.

$$\Delta t = \frac{LT}{2} = 16 \text{ msec}$$

$$\Delta f = \frac{f_s}{L} = 31.25 \text{ Hz}$$

When *exam4_19* is run, it computes $G(m, i)$. The resulting spectrogram (just the first half) using $p = 12$ levels and a rectangular window is shown in Figure 4.44. Note how each vowel has its own distinct set of contour lines. For example, the vowel "I" has significant power between zero and 1500 Hz, while the vowel "A" has power between zero and about 600 Hz, with some additional power centered around 2500 Hz. The use of a rectangular window (no tapering) tends to smear the spectrogram features parallel to the frequency axis due to the leakage phenomenon. A somewhat cleaner spectrogram can be obtained by using the Hamming window, as shown in Figure 4.45. Here the islands associated with peaks in the magnitude response are more isolated from one another.

Wavelets

Subsignal overlaps in Figure 4.41 other than $L/2$ can be used to compute the spectrogram. In addition, a more flexible approach is to decompose the signal $x(k)$ into low frequency and high frequency parts using the *discrete wavelet transform* or DWT. For a discussion of wavelets and their use in time-frequency analysis, the interested reader is referred, for example, to (Burrus and Guo, 1997).

FIGURE 4.44:
Spectrogram of
Vowels Using a
Rectangular
Window and 12
Levels

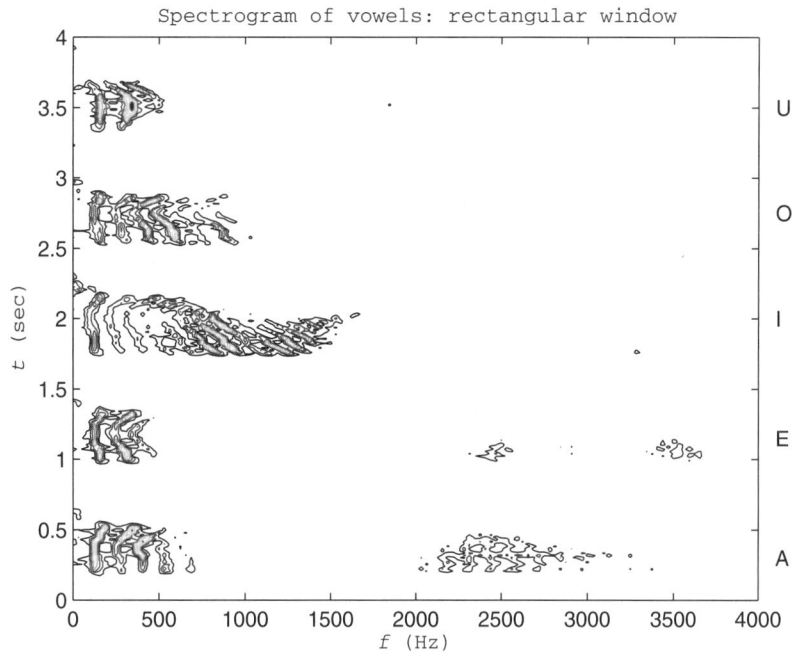

Spectrogram of vowels: rectangular window

FIGURE 4.45:
Spectrogram of
Vowels Using a
Hamming Window
and 12 Levels

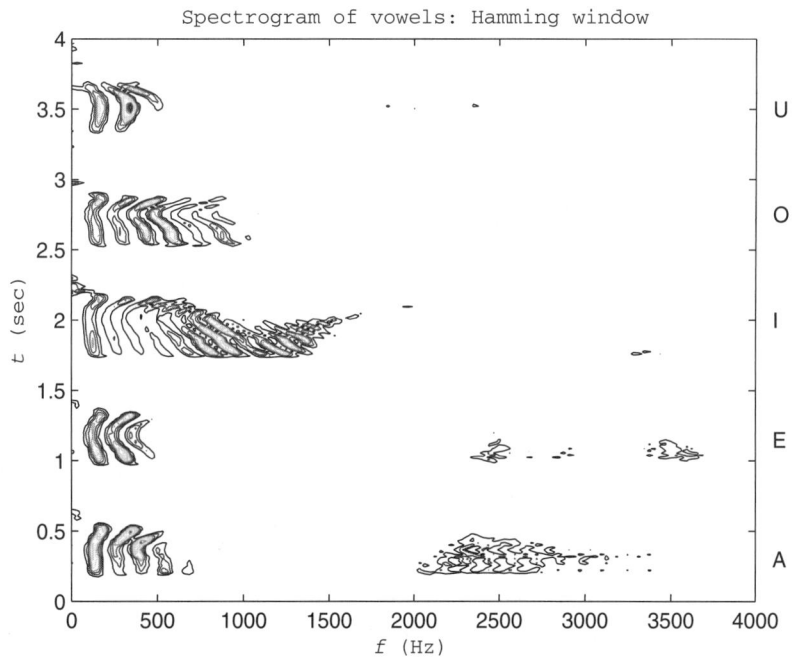

Spectrogram of vowels: Hamming window

FDSP Functions

The FDSP toolbox contains the following function for computing the spectrogram of a discrete-time signal.

```
% F_SPECGRAM: Compute spectrogram of a signal
%
```

```
% Usage:
%         [G,f,t] = f_specgram (x,L,fs,win)
% Pre:
%         x   = vector of length N samples
%         L   = subsignal length
%         fs  = sampling frequency
%         win = window type
%
%                   0 = rectangular
%                   1 = Hanning
%                   2 = Hamming
%                   3 = Blackman
% Post:
%         G = (2M-1) by L matrix containing spectrogram where M = N/L.
%         f = vector of length L/2 containing frequency values
%         t = vector of length 2M-1 containing time values
```

● ● ● ● ● ● ● ● ● ● ● ● ● ● ● ● ● ●

4.10   Power Density Spectrum Estimation

In this section we develop techniques for estimating the underlying continuous-time power density spectrum (PDS) of a signal when the length of the data sequence N is large. In addition, we focus on the practical problem of detecting the presence and location of one or more sinusoidal components buried in a signal that is corrupted with noise. The problem is made more challenging when the frequencies of the unknown sinusoidal components do not correspond *Bin frequency* to any of the discrete frequencies $f_i = i f_s / N$ which are sometimes called the *bin frequencies*.

The topic of power density spectrum estimation is one that has been studied in some depth (Proakis and Manolakis, 1992; Ifeachor and Jervis, 2002; Prat, 1997). The treatment presented here includes a basic presentation of the classical nonparametric estimation methods. Recall that the power density spectrum of an N-point signal is given by

$$S_x(i) = \frac{|X(i)|^2}{N}, \quad 0 \le i < N \tag{4.10.1}$$

Here $S_x(i)$ specifies the average power contained in the ith harmonic of the periodic extension $x_p(k)$ of the N-point signal $x(k)$. The average power of $x(k)$ is the average of the power density spectrum.

$$P_x = \frac{1}{N} \sum_{i=0}^{N-1} S_x(i) \tag{4.10.2}$$

4.10.1   Bartlett's Method

Periodogram The formulation of the power density spectrum in (4.10.1) is sometimes called a *periodogram*. If the sequence $x(k)$ is long, then a more reliable way to estimate the power density spectrum is

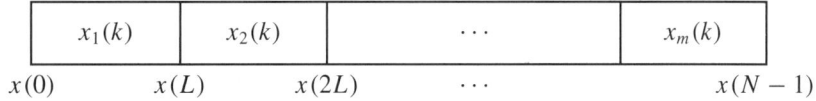

FIGURE 4.46: Partition of $x(k)$ into M Subsignals of Length L

to partition $x(k)$ into a number of *subsignals*. Suppose $N = LM$ for a pair of positive integers *Subsignal* L and M. Then $x(k)$ can be partitioned into M subsignals, each of length L, as shown in Figure 4.46.

If the subsignals are denoted $x_m(k)$ for $0 \le m < M$, then the mth subsignal can be extracted from the original signal $x(k)$ as

$$x_m(k) \stackrel{\Delta}{=} x(mL + k), \quad 0 \le k < L \tag{4.10.3}$$

Next let $X_m(i) = \text{DFT}\{x_m(k)\}$ for $0 \le m < M$. The estimated power density spectrum of $x(k)$ is obtained by taking the average of the M individual power density spectra.

$$S_B(i) = \frac{1}{LM} \sum_{m=0}^{M-1} |X_m(i)|^2, \quad 0 \le i < L \tag{4.10.4}$$

Average periodogram   The power density spectrum estimate $S_B(i)$ is called an *average periodogram* or Bartlett's method (Bartlett, 1948). Note that one of the consequences of Bartlett's method is that it changes the frequency precision of the power density spectrum. If f_s is the sampling frequency, then the frequency precision, or space between adjacent discrete frequencies, is

$$\Delta f = \frac{f_s}{L} \tag{4.10.5}$$

This is in contrast to the periodogram method in (4.10.1) where the frequency precision is $\Delta f = f_s/N$. Since $N = LM$, the frequency precision of Bartlett's method is larger by a factor of M. This loss of precision is offset by the observation that the variance in the estimated power density spectrum is reduced by the same factor. We can approximate the variance of the average periodogram as follows.

$$\sigma_B^2 = \frac{1}{L} \sum_{i=0}^{L-1} [S_B(i) - P_x]^2 \tag{4.10.6}$$

Ideally, the power density spectrum of white noise should be a flat line at $S_x(i) = P_x$, as in (4.7.11). By reducing the variance of the computed power density spectrum, Bartlett's method comes closer to this ideal, as can be seen in the following example.

Example 4.20   **Bartlett's Method: White Noise**

As an illustration of how the average periodogram can improve the estimate of the power density spectrum, let $x(k)$ be white noise uniformly distributed over the interval $[-5, 5]$. Then from (4.4.6) the average power of the white noise signal is

$$P_v = \frac{5^3 - (-5)^3}{3[5 - (-5)]}$$
$$= \frac{250}{30}$$
$$= 8.333$$

FIGURE 4.47:
Periodograms of
White Noise
Uniformly
Distributed over
$[-5, 5]$ with
$N = 4096$,
(a) Average
Periodogram with
$L = 512$, $M = 8$,
$\sigma_B^2 = 9.940$,
(b) Single
Periodogram with
$L = 4096$, $M = 1$,
$\sigma_x^2 = 76.225$

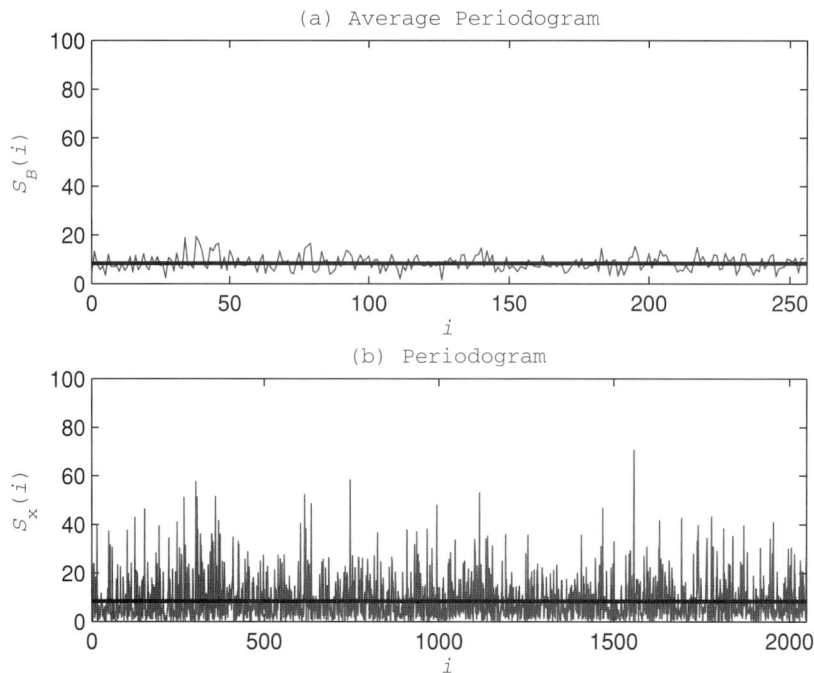

(a) Average Periodogram

(b) Periodogram

Suppose the length of $x(k)$ is $N = 2048$. One way to factor N is $L = 256$ and $M = N/L = 8$. In this case we compute eight periodograms, each of length 256. The resulting average periodogram can be obtained by running *exam4_20*. A plot of the estimated power density spectrum $S_B(i)$ is shown in Figure 4.47. This uniform white noise signal was previously analyzed in Example 4.11. Comparing Figure 4.47(a) with Figure 4.23, we see that the average periodogram estimate is noticeably flatter and closer to the ideal characteristic, $P_v = 8.333$. The actual average power is also closer to the theoretical value of P_v as a result of using the longer signal of length $N = 2048$. In this case we have

$$P_x = \frac{1}{N} \sum_{k=0}^{N-1} x^2(k)$$
$$= 8.581$$

The reduction in the variance of the estimate is a consequence of the averaging of M periodograms. From (4.10.6), the variance of the estimate in this case is

$$\sigma_B^2 = \frac{1}{L} \sum_{i=0}^{L-1} [S_B(i) - P_x]^2$$
$$= 8.979$$

The curious reader might wonder if perhaps the same improvement in the estimate of the power density spectrum might be achieved with a single periodogram by simply increasing N from 512 in Example 4.11 to the 2048 samples used in Example 4.20. This is an intuitively appealing conjecture, but unfortunately it does not hold, as can be seen in Figure 4.47(b) which uses $L = 2048$ and $M = 1$. Although the average power $P_x = 8.581$ does come closer to P_v as N increases, it is apparent that the variance in the power density spectrum curve does *not* decrease in this case. In particular, by using (4.10.6), the variance using a single periodogram of length $N = 2048$ is

$$\sigma_x^2 = 73.621$$

If we take the ratio of the two variances, we find that $\sigma_x^2/\sigma_B^2 = 8.200$. Thus the improvement in the variance achieved by Bartlett's average periodogram method was by a factor of approximately $M = 8$.

Next consider the practical problem of detecting the presence, and precise location, of unknown sinusoidal components in a signal $x(k)$.

Example 4.21   **Bartlett's Method: Periodic Input**

Suppose the sampling rate is $f_s = 1024$ Hz and the number of samples is $N = 512$. If Bartlett's method is used with $L = 128$ and $M = 4$, then from (4.10.5), the frequency precision is

$$\Delta f = \frac{f_s}{L}$$
$$= 8 \text{ Hz}$$

Consider a signal $x(k)$ that contains two sinusoidal components, one at frequency $F_0 = 200$ Hz, and the other at frequency $F_1 = 331$ Hz.

$$x(k) = \sin(2\pi F_0 kT) - \sqrt{2}\cos(2\pi F_1 kT)$$

Observe that F_0 is an integer multiple of Δf, so F_0 corresponds to discrete frequency i_0 with

$$i_0 = \frac{F_0}{\Delta f}$$
$$= 25$$

However, F_1 is not an integer multiple of the frequency precision Δf. Therefore, F_1 falls between two discrete frequencies at

$$i_1 = \frac{F_1}{\Delta f}$$
$$= 41.375$$

The average periodogram estimate of the power density spectrum of $x(k)$ can be obtained by running *exam4_21*. The resulting plot of $S_B(f)$ is shown in Figure 4.48. Here we have used the independent variable $f = if_s/L$ to facilitate interpretation of the frequencies. Note that there are two spikes indicating the presence of two sinusoidal components.

The first spike is centered at $F_0 = 200$ Hz, as expected. This spike is symmetric about $f = F_0$, and it is relatively narrow, with a width of ± 8 Hz which corresponds to the frequency precision Δf. Next consider the spectral spike associated with F_1. Unfortunately, this spike is not as narrow as the low-frequency spike, particularly near the bottom. Furthermore, the high-frequency spike is also *not centered* at $F_1 = 331$ Hz. Instead the peak of the second spike occurs below F_1 at discrete frequency

$$f_{41} = \frac{41 f_s}{L}$$
$$= 328 \text{ Hz}$$

The estimate of the location of the second spectral component is low by .91 percent. Thus the estimated power density spectrum does an effective job in detecting sinusoidal components

FIGURE 4.48: Average Periodogram of Signal with $N = 512$, $L = 128$, and $M = 4$

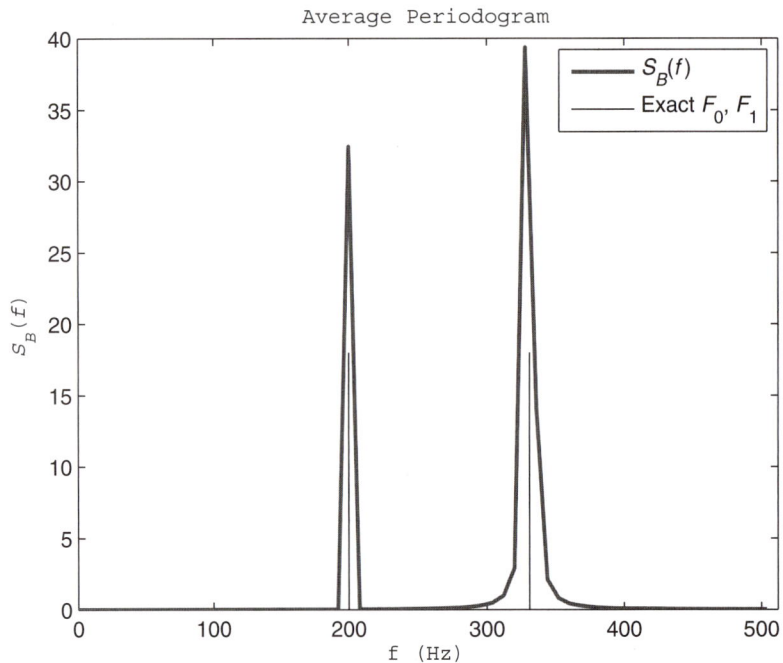

Average Periodogram

located at the discrete frequencies, but the performance is less impressive for frequencies located between the discrete frequencies.

4.10.2 Welch's Method

The broad spike in Figure 4.48 associated with a spectral component that is not aligned with one of the discrete frequencies, $f_i = i f_s / L$, is a manifestation of the leakage phenomenon.

Leakage Basically, the spectral power that should be concentrated at a single frequency has spread out or *leaked* into adjacent frequencies. To reduce the leakage, Welch (1967) proposed modifying the average periodogram approach in two ways. Rather than partition the signal into M distinct subsignals, as in Figure 4.46, we instead decompose $x(k)$ into a number of overlapping subsignals. If $N = LM$, then a total of $2M - 1$ subsignals of length L can be constructed using an overlap of $L/2$, as shown previously in Figure 4.41. If the subsignals are denoted $x_m(k)$ for $0 \leq m < 2M - 1$, then mth subsignal can be extracted from the original signal $x(k)$ as

$$x_m(k) \stackrel{\Delta}{=} x(mL/2 + k), \quad 0 \leq k < L \tag{4.10.7}$$

Recall that the extraction of $x_0(k)$ from $x(k)$ can be thought of as multiplication of $x(k)$

Rectangular window by the following *rectangular window* of width L. The other subsignals can be extracted in a similar manner with a shifted rectangular window.

$$w_R(k) = \mu(k) - \mu(k - L) \tag{4.10.8}$$

When the DFT of subsignal $x_m(k)$ is computed, it can be thought of as finding the Fourier coefficients of the periodic extension of $x_m(k)$. However, if subsignal $x_m(k)$ does not represent an integer number of cycles of a given sinusoidal component, then the periodic extension of $x_m(k)$ will have a jump discontinuity. This discontinuity leads to the well-known Gibb's phenomenon of the Fourier series, a phenomenon that causes ringing or oscillations near the

FIGURE 4.49:
Modified Average
Periodogram of
Noisy Periodic
Signal

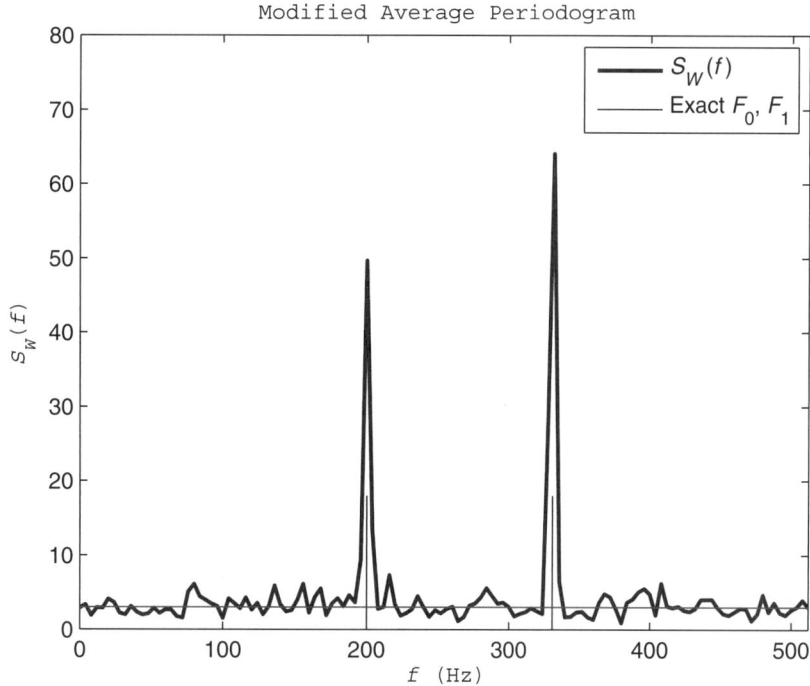

discontinuity when the signal is reconstructed. This ringing manifests itself in the frequency domain as the leakage observed in the second spectral spike in Figure 4.49.

If T is the sampling interval, then the length of subsignal $x_m(k)$ is $\tau = LT$. Thus for a sinusoidal component at frequency F_0, the number of cycles per subsignal is

$$
\begin{aligned}
i_0 &= F_0 \tau \\
&= \frac{L F_0}{f_s}
\end{aligned}
\tag{4.10.9}
$$

Since $f_i = i f_s / L$, it follows that discrete frequency f_i contains exactly i cycles per subsignal. For the first spectral component in Example 4.21 at $F_0 = 200$ Hz, the number of cycles per subsignal was $i_0 = 25$. However, for the troublesome spectral component at $F_1 = 331$ Hz, the number of cycles per subsignal was $i_1 = F_1 \tau = 41.375$. Thus there was a jump discontinuity associated with the second spectral component.

In order to reduce the effects of jump discontinuities in the periodic extensions of the subsignals, Welch proposed that the subsignals be multiplied by a window other than the rectangular window. If the window goes to zero gradually, rather than abruptly, as in (4.10.8), then the jump discontinuities in the periodic extensions of the subsignals can be eliminated. A number of popular candidates for data window functions were summarized in Table 4.10, and plots of these window functions were shown previously in Figure 4.42. Given a window $w(k)$ of width L, we can compute the DFT of the mth windowed subsequence of $x(k)$. An estimate of the power density spectrum is then obtained by averaging the $2M - 1$ overlapping windowed periodograms.

$$
S_W(i) = \frac{1}{L(2M - 1)} \sum_{m=0}^{2M-2} |\text{DFT}\{w(k)x_m(k)\}|^2, \quad 0 \le i < L \tag{4.10.10}
$$

Modified average periodogram

Power density spectrum estimate $S_W(i)$ is called a *modified average periodogram* or Welch's method. The overlap between subsignals could be increased or decreased. However, the 50% overlap has been shown to improve certain statistical properties of the estimate (Welch, 1967). In the case of the Hanning window, the 50% overlap means that each sample is equally weighted (see Problem 4.40).

Example 4.22

Welch's Method: Noisy Periodic Input

As an illustration of Welch's method, consider a noise-corrupted version of the signal from Example 4.21. That is, suppose $F_0 = 200$ Hz, $F_1 = 331$ Hz, and $v(k)$ is white noise uniformly distributed over the interval $[-3, 3]$.

$$x(k) = \sin(2\pi F_0 kT) - \sqrt{2}\cos(2\pi F_1 kT) + v(k), \quad 0 \le k < N$$

Suppose $N = 1024$ and we factor N as $L = 256$, $M = N/L = 4$. Thus Welch's method uses $2M - 1 = 7$ subsignals, each of length $L = 256$. From (4.10.5), the frequency precision is

$$\Delta f = \frac{f_s}{L}$$
$$= 4 \text{ Hz}$$

In this case the discrete frequency indices of F_0 and F_1 are

$$i_0 = \frac{F_0}{\Delta f} = 50$$
$$i_1 = \frac{F_1}{\Delta f} = 82.75$$

An estimate of the power density spectrum of $x(k)$ using Welch's method with the Hanning window can be obtained by running script *exam4_22*. The resulting plot of $S_W(f)$ is shown in Figure 4.49. Here we have used the independent variable $f = if_s/L$ to facilitate interpretation of the frequencies. Note that there are two spikes indicating the presence of two sinusoidal components. In addition there is power uniformly distributed over all frequencies which represents the white noise term $v(k)$. The average power of the white noise is $P_v = 3$, which is indicated by the horizontal line. The two peaks occur at

$$f_{50} = 50\Delta f = 200 \text{ Hz}$$
$$f_{83} = 83\Delta f = 332 \text{ Hz}$$

In this case there is improvement in the estimate of F_1 in comparison with Example 4.21, but only because L has been doubled, which improves the frequency precision from 8 Hz to 4 Hz. The important difference between Figure 4.49 and Figure 4.48 lies in the observation that the width of the second peak is now comparable to the width of the first peak. This reduction in width is a result of the windowing which reduces the leakage into nearby frequencies. The peaks in Figure 4.49 are actually somewhat wider than the frequency precision of ± 4 Hz. This is the price that is paid for the leakage reduction through windowing. The spectral characteristics of data windows are investigated in more detail in Chapter 6 where they are used to improve the performance of digital FIR filters.

FDSP Functions

The FDSP toolbox contains the following function for estimating the power density spectrum using both Bartlett's method and Welch's method.

```
% F_PDS: Compute estimated power density spectrum
%
% Usage:
%        [S,f,Px] = f_pds (x,N,L,fs,win,meth);
% Pre:
%        x    = vector of length n containing input samples
%        N    = total number of samples.  If N > n, then x is padded
%               with N - n zeros.
%        L    = length of subsequence to use.  L must be an integer factor
%               of N.  That is N = LM for a pair of integers L and M.
%        fs   = sampling frequency
%        win  = window type to be used
%
%               0 = rectangular
%               1 = Hanning
%               2 = Hamming
%               3 = Blackman
%
%        meth = an integer method selector.
%
%               0 = Bartlett's average periodogram
%               1 = Welch's modified average periodogram
%
% Post:
%        S  = 1 by L vector containing estimate of power density spectrum
%        f  = 1 by L vector containing frequencies at which S is
%             evaluated (0 to (L-1)fs/L).
%        Px = average power of x
```

If the Signal Processing Toolbox is available, the object *spectrum* and the method *psd* can be used to compute the power density spectrum.

● ● ● ● ● ● ● ● ● ● ● ● ● ● ● ●

4.11 GUI Software and Case Studies

This section focuses on applications of Fourier transforms and the spectral analysis of discrete-time signals. A graphical user interface module called *g_spectra* is introduced that allows the user to interactively explore the magnitude, phase, and power density spectra of discrete-time signals without any need for programming. Case study examples are then presented and solved using MATLAB.

g_spectra: Spectral Analysis of Discrete-time Signals

The graphical user interface module *g_spectra* allows the user to investigate the spectral characteristics of a variety of discrete-time signals. GUI module *g_spectra* features a display screen with tiled windows, as shown in Figure 4.50. The *Block Diagram* window in the upper-left corner of the screen contains a color-coded block diagram of the operation being performed. This module computes the DFT of an N-point signal $x(k)$.

$$X(i) = \text{DFT}\{x(k)\}, \quad 0 \leq i < N \tag{4.11.1}$$

Below the block diagram are a number of *edit boxes* containing parameters that can be modified by the user. The parameters include the sampling frequency fs, the frequency of a cosine input $F0$, a bound b on zero-mean uniform white noise, and a clipping threshold c. Changes to the parameters are activated with the Enter key. To the right of the edit boxes are pushbutton and check box controls. The Play x as sound pushbutton plays the signal $x(k)$ as sound on the PC speaker. The dB Display check box control toggles the graphical display between a linear scale and a logarithmic dB scale. The Add noise check box adds white noise uniformly distributed over $[-b, b]$ to the input $x(k)$. Finally, the Clip check box activates clipping of the input $x(k)$ to the interval $[-c, c]$.

The *Type* and *View* windows in the upper-right corner of the screen allow the user to select both the type of input signal $x(k)$ and the viewing mode. The inputs include several common signals that can optionally include white noise depending on the status of the Add noise check box. There are two inputs that the user can customize. The Record sound input allows the user to record one second of sound from the PC microphone at a sampling rate of $fs = 8192$ Hz. Proper recording can be verified by using the Play x as sound pushbutton. The User-defined input prompts for the name of a user-supplied MAT file that contains up to 8192 samples of $x(k)$ plus the sampling frequency fs. If no fs is present in the MAT file, then the current value for the sampling frequency is used. Near the bottom of the *Type* and *View* windows is a slider bar that allows the user to control the number of samples N. When N is increased, the currently selected input is padded with zeros. If the input is then reselected, the signal is recomputed and extended to N nonzero samples.

The viewing options include the time signal $x(k)$, the magnitude spectrum $A(f)$, the phase spectrum $\phi(f)$, and the estimated power density spectrum $S_W(f)$ where

$$f_i = \frac{i f_s}{N}, \quad 0 \leq i \leq N/2 \tag{4.11.2}$$

The power density spectrum estimate uses Welch's modified average periodogram method with $L = N/4$. There are also viewing options for displaying the data window and the spectrogram of x, using overlapping subsignals of length $L = N/8$. The *Plot* window on the bottom half of the screen shows the selected view. The curves are color-coded to match the block diagram in the upper left corner of the screen. The magnitude response and power density spectrum plots can be displayed using either the linear scale or the logarithmic dB scale depending on the status of the dB Display check box.

The *Menu* bar at the top of the screen includes four options. The *Data window* option allows the user to choose a data window to be used for the power density spectrum estimate and the spectrogram. The *Caliper* option allows the user to measure any point on the current plot by moving the mouse crosshairs to that point and clicking. The *Save data* option is used to save the current x and fs in a user-specified MAT file for future use. Files created in this manner can be reloaded with the User-defined input option. The *Print* option prints the contents of the plot window. Finally, the *Help* option provides the user with some helpful suggestions on how to effectively use module *g_spectra*.

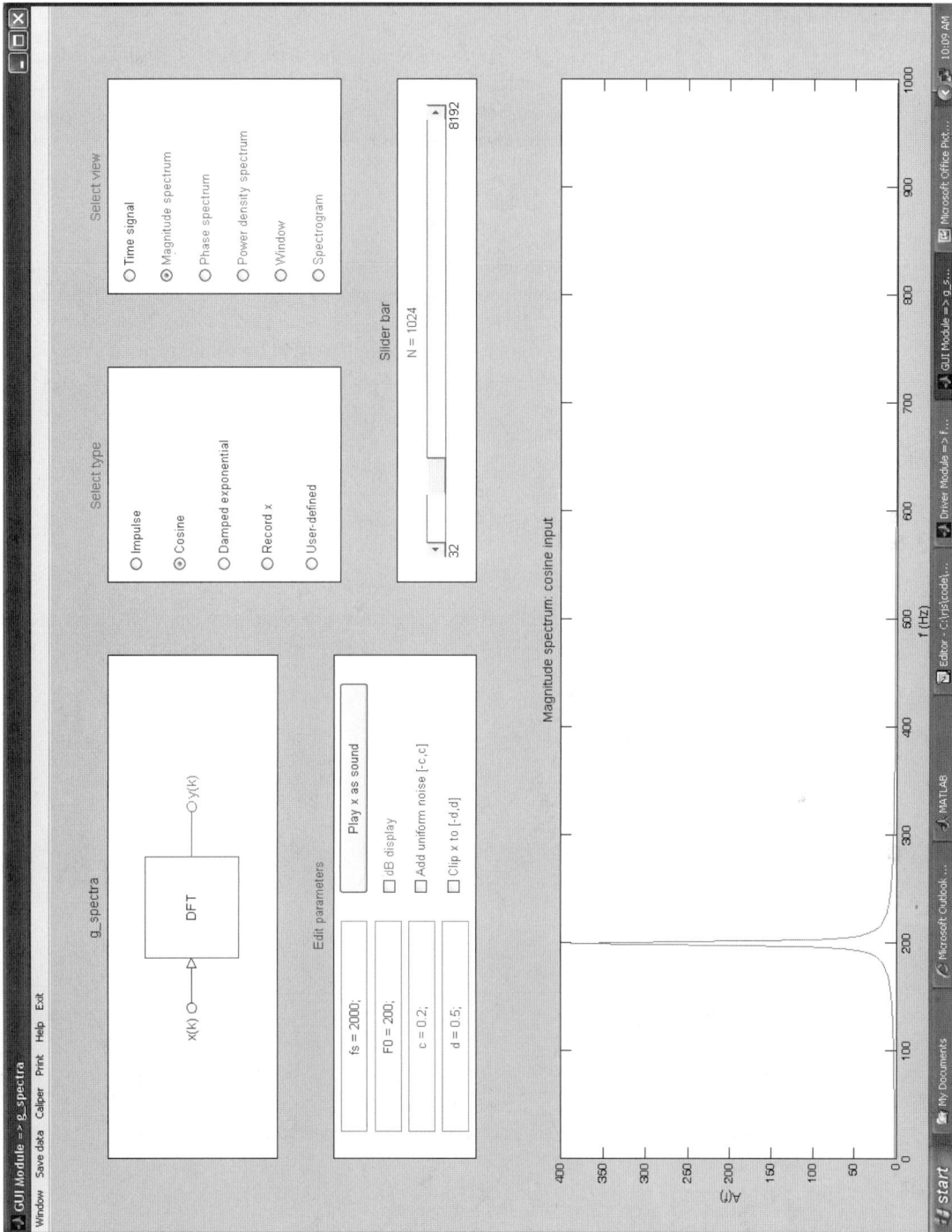

FIGURE 4.50: Display Screen of Chapter GUI Module *g_spectra*

<table>
<tr><td>CASE STUDY 4.1</td></tr>
</table>

Signal Detection

One of the application areas of spectral analysis is the detection of signals buried in noise. Suppose the sampling rate is f_s and the signal to be analyzed includes M sinusoidal components with frequencies $F_1, F_2, \ldots, F_M$, where $0 \leq F_i \leq f_s/2$ are *unknown* and must be determined from a spectral analysis of the following signal

$$x(k) = \sum_{i=1}^{M} \sin(2\pi F_i kT) + v(k), \quad 0 \leq k < N$$

Here $v(k)$ is additive white noise uniformly distributed over the interval $[-b, b]$. For example, $v(k)$ might represent measurement noise or noise picked up in a communication channel over which $x(k)$ is transmitted.

CASE STUDY 4.1

The unknown frequencies can be detected and identified by running *case4_1* which uses $N = 1024$, $f_s = 2000$ Hz, and $b = 2$.

```
function case4_1

% CASE STUDY 4.1: Signal detection

f_header('Case Study 4.1: Signal detection')

% Prompt for simulation parameters

seed = f_prompt ('Enter initial state of random number generator',0,10000,3000);
M = f_prompt ('Enter number of random sinusoidal terms',0,10,3);
rand ('state',seed);
N = 1024;
k = 0 : N-1;
b = 2;
x = -b + 2*b*rand(1,N);
fs = 2000;
T = 1/fs;
F = (fs/2)*rand(1,M);
for i = 1 : M
    x = x + sin(2*pi*F(i)*k*T);
end

% Plot portion of signal

figure
t = k*T;
plot (t(1:N/8),x(1:N/8))
axis([t(1),t(N/8),-5,5])
f_labels ('Noisy time signal with sinusoidal components','{\itt} (sec)','\it{x(t)}')
f_wait

% Compute and plot power density spectrum

figure
A = abs(fft(x));
S_x = A.^2/N;
f = linspace (0,(N-1)*fs/N,N);
plot (f(1:N/2),S_x(1:N/2))
set(gca,'Xlim',[0,fs/2])
```

```
f_labels ('Power density spectrum','{\itf} (Hz)','\it{S_x(f)}')

% Find frequencies with user-supplied threshold

S_max = max(S_x);
s = f_prompt ('Enter threshold for locating peaks',0,S_max,.7*S_max);
ipeak = S_x > s;
for i = 1 : N/2
   if ipeak(i) == 1
       fprintf ('f = %.0f Hz\n',f(i))
   end
end
f_wait
```

The first part of the *case4_1* prompts the user for an initial state for the random number generator and for the number of unknown frequencies M. Each state generates a different set of M random frequencies. A plot of the first $N/8$ samples of the signal $x(k)$ generated using the default responses is shown in Figure 4.51. Note that because of the additive white noise, it is not clear from direct inspection of $x(k)$ whether it has *any* sinusoidal components, much less how many there are and where they are located. However, the existence of sinusoidal spectral components is apparent when we examine the power density spectrum plot $S_x(f)$ shown in Figure 4.52.

In this case there are three sinusoidal components corresponding to the three distinct peaks. To determine the locations of these peaks, the user is prompted for a threshold value s, so that all f for which $S_x(f) > s$ can be displayed. Using the default threshold value, the three

FIGURE 4.51: A Noise-corrupted Signal with Unknown Sinusoidal Components, $N = 1024$, and $f_s = 2000$ Hz

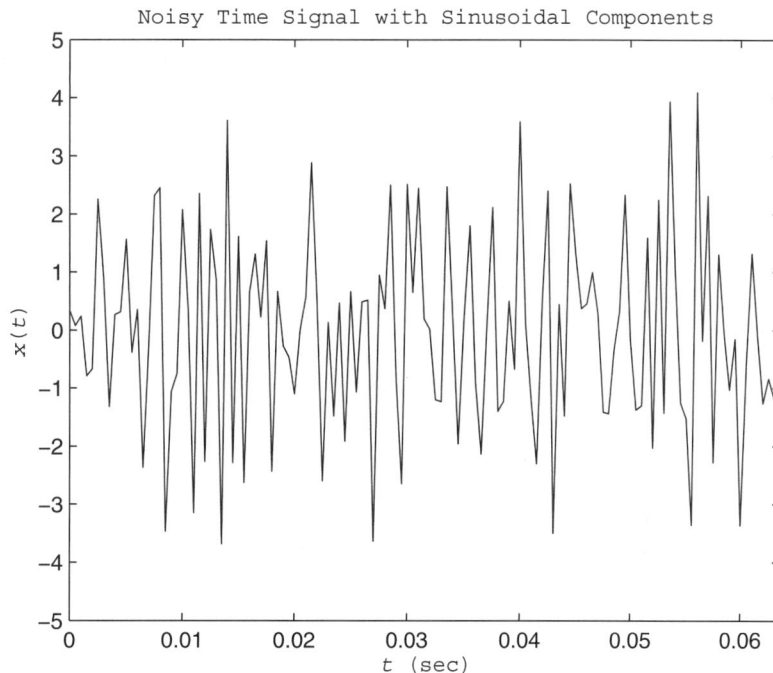

FIGURE 4.52: Power Density Spectrum of the Signal in Figure 4.51

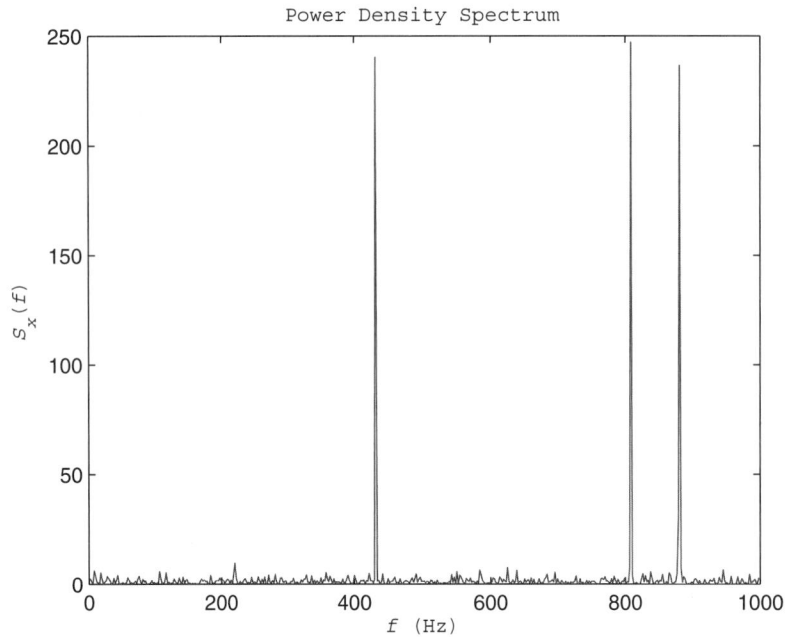

unknown frequencies are identified to be

$$F_1 = 432 \text{ Hz}$$
$$F_2 = 809 \text{ Hz}$$
$$F_3 = 881 \text{ Hz}$$

CASE STUDY 4.2

Distortion Due to Clipping

Many a parent has had the experience of having a child turn the music up so loud that the sound begins to distort. The distortion is caused by the fact that the amplifier or the speakers are being overdriven to the point that they are no longer operating in their linear range. This type of distortion can be modeled with a saturation nonlinearity where the output is *clipped* at lower and upper limits.

$$y = \text{clip}(x, a, b) \triangleq \begin{cases} a, & -\infty < x < a \\ x, & a \leq x \leq b \\ b, & b < x < \infty \end{cases}$$

When the input x is in the interval $[a, b]$, there is a simple linear relationship, $y = x$. Outside this range, y saturates to a for $x < a$ or b for $x > b$. A graph of a clipped cosine for the case $[a, b] = [-.7, .7]$ is shown in Figure 4.53.

In a sound system, clipping typically occurs because the magnitude of the amplified output signal exceeds the DC power supply level of the amplifier. To illustrate this phenomenon, consider a cosine input of unit amplitude and frequency $F_0 = 625$ Hz. Using a sampling frequency of $f_s = 20$ kHz and $N = 32$ yields one period of $x(k)$.

$$x(k) = \cos(1300\pi kT), \quad 0 \leq k < N$$

FIGURE 4.53:
Saturation of *x* due to Clipping to the Interval [−.7, .7]

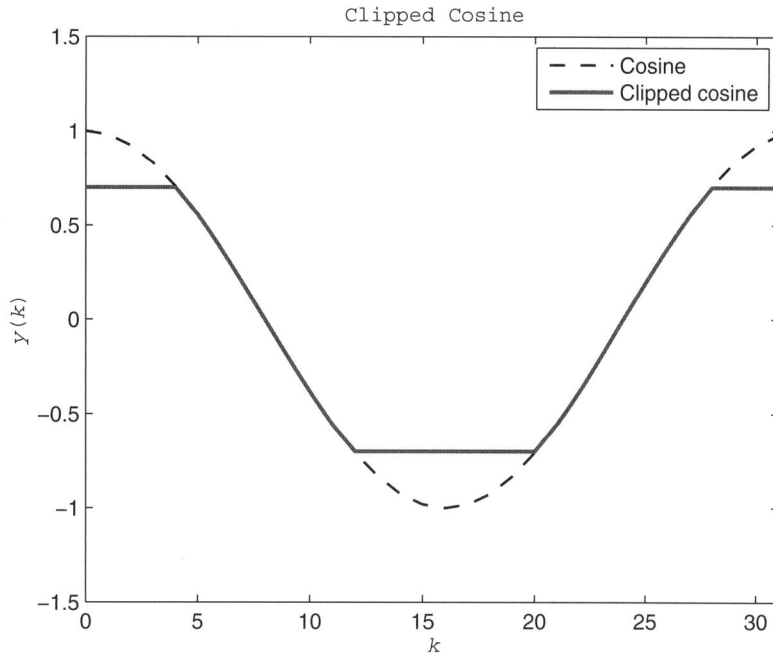

Suppose this signal is sent through a saturation nonlinearity with a clipping interval, $[-c, c]$,

$$y(k) = \text{clip}[x(k), -c, c], \quad 0 \le k < N$$

A plot of the resulting input and output for the case $c = 0.7$ is shown in Figure 4.53. We can quantify the amount of distortion caused by clipping by computing the total harmonic distortion THD. Output $y(k)$ is a sampled version of an underlying periodic signal $y_a(t)$ that can be approximated by the following truncated Fourier series.

$$y_a(t) = \frac{d_0}{2} + \sum_{i=1}^{N/2-1} d_i \cos(2\pi F_0 t + \theta_i)$$

Recall from (4.1.12) that the Fourier series coefficients can be obtained directly from the DFT of $y(k)$. In particular, if $A(i)$ is the magnitude spectrum of $y(k)$, then

$$d_i = \frac{2A(i)}{N}, \quad 0 \le i < N/2$$

Each term of the Fourier series has an average power associated with it. From (4.1.10a), the total average power of $y_a(t)$ is

$$P_y = \frac{d_0^2}{4} + \frac{1}{2} \sum_{i=1}^{N/2} d_i^2$$

The total harmonic distortion THD is the average power of the unwanted harmonics expressed as a percentage of the total average power. Thus from (4.1.10b) we have

$$\text{THD} = \frac{100(P_y - d_1^2/2)}{P_y} \%$$

CASE
STUDY
4.2 The total harmonic distortion can be obtained by running *case4_2*.

```
function case4_2

% CASE STUDY 4.2: Distortion due to clipping

f_header('Case Study 4.2: Distortion due to clipping')

% Construct input and output

N = 32;
k = 0 : N-1;
fs = 20000;
T = 1/fs;
F_0 = fs/N;
c = 0.70;
x = cos(2*pi*F_0*k*T);
y = f_clip(x,-c,c);

% Plot clipped signal

figure
h = plot (k,x,'--k',k,y);
set (h(2),'LineWidth',1.5)
set (h(1),'LineWidth',1.0)
axis([k(1),k(N),-1.5,1.5])
legend ('Cosine','Clipped Cosine')
f_labels ('Clipped cosine','\it{k}','\it{y(k)}')
f_wait

% Compute total harmonic distortion

A = abs(fft(y));
d = 2*A/N;
Delta_f = fs/N;
i = round(F_0/Delta_f) + 1;
P_y = d(1)^2/4 + (1/2)*sum(d(2:N/2).^2)
D = 100*(P_y - (d(i)^2)/2)/P_y

% Compute and plot magnitude spectrum

figure
i = 1 : N/2;
hp = stem (i-1,A(i),'filled','.');
set (hp,'LineWidth',1.5);
f_labels('Magnitude spectrum','\it{i}','\it{A(i)}')
set(gca,'Xlim',[0,N/2])
box on
f_wait
```

FIGURE 4.54:
Magnitude
Spectrum of
Clipped Signal
Cosine,
THD = 1.93 %

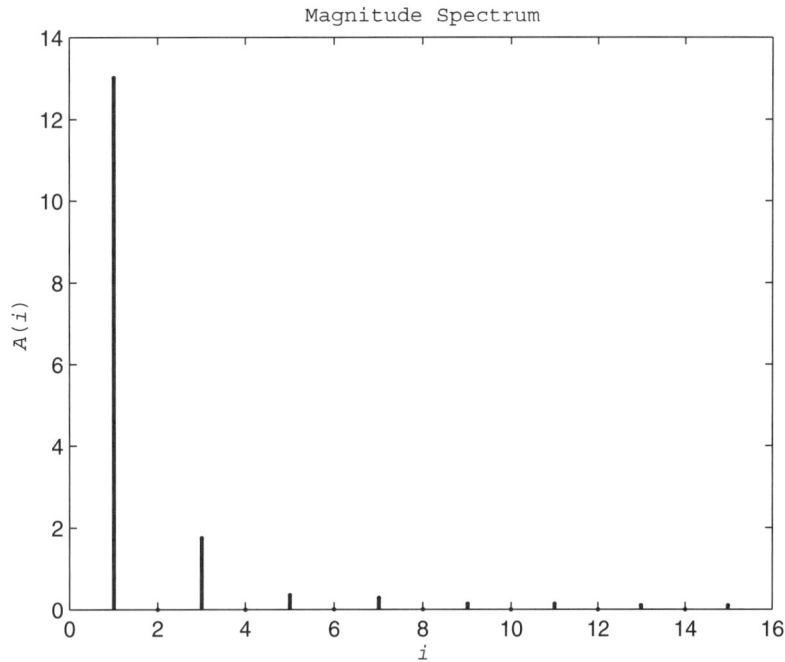

When *case_2* is run, it generates the clipping plot in Figure 4.53 and the magnitude spectrum plot shown in Figure 4.54. Note the presence of power at the third, fifth, and other odd harmonics due to the clipping operation. If we use a clipping threshold of $c = .7$, the total harmonic distortion in this case is found to be

$$\text{THD} = 1.93\ \%$$

For audio signals, the THD of a pure tone is one measure of the quality of the sound.

● ● ● ● ● ● ● ● ● ● ● ● ● ● ● ● ● ●

4.12   Chapter Summary

Discrete-time Fourier Transform (DTFT)

This chapter focused on Fourier transforms and the spectral analysis of discrete-time signals. For absolutely summable signals, the region of convergence of the Z-transform includes the unit circle. The *discrete-time Fourier transform* is obtained by evaluating the Z-transform $X(z)$ along the unit circle using $z = \exp(j2\pi fT)$. This produces a mapping from discrete time to *DTFT* continuous frequency called the DTFT.

$$X(f) = \sum_{k=-\infty}^{\infty} x(k)\exp(j2\pi kfT), \quad -f_s/2 \le f \le f_s/2 \qquad (4.12.1)$$

Spectrum   The function of $X(f)$ is called the *spectrum* of $x(k)$. The magnitude $A_x(f) = |X(f)|$ is the *magnitude spectrum*, and the phase angle $\phi_x(f) = \angle X(f)$ is the *phase spectrum*. The spectrum $X(f)$ is periodic with period f_s. For real signals, the spectrum satisfies the symmetry condition

$$X(-f) = X^*(f) \qquad (4.12.2)$$

This implies that real signals have an even magnitude spectrum and an odd phase spectrum. For real signals, all of the essential information is contained in the range $0 \leq f \leq f_s/2$ that corresponds to the positive frequencies along the top half of the unit circle. The DTFT inherits properties from the Z-transform, and it has additional properties including Parseval's identity.

Discrete Fourier Transform (DFT)

When the DTFT is applied to an N-point signal, and $X(f)$ is evaluated at N discrete frequencies equally spaced around the unit circle, the resulting transform is called the *discrete-Fourier transform* or DFT. Let W_N denote the Nth root of unity.

$$W_N = \exp(-j2\pi/N) \tag{4.12.3}$$

DFT   Then the DFT, which is denoted $X(i) = \text{DFT}\{x(k)\}$, is defined in terms of W_N as

$$X(i) = \sum_{k=0}^{N-1} x(k)W_N^{ik}, \quad 0 \leq i < N \tag{4.12.4}$$

Since W_N is complex, $X(i)$ will be complex. We can express the DFT in polar form as $X(i) = A_x(i)\exp[j\phi_x(i)]$, where

$$A_x(i) = |X(i)| \tag{4.12.5a}$$
$$\phi_x(i) = \angle X(i) \tag{4.12.5b}$$

Power density spectrum   The magnitude $A_x(i)$ is called the *magnitude spectrum* of $x(k)$, and the phase angle $\phi_x(i)$ is called the *phase spectrum* of $x(k)$. The *power density spectrum* of $x(k)$ is defined as

$$S_x(i) = \frac{|X(i)|^2}{N} \tag{4.12.6}$$

For a signal $x(k)$ of length N, the DFT is periodic with period N. In addition, if $x(k)$ is real, the spectrum satisfies a midpoint symmetry condition

$$X^*(i) = X(N-i) \tag{4.12.7}$$

The magnitude and power density spectra exhibit even symmetry about the midpoint $i = N/2$, and the phase spectrum exhibits odd symmetry about the midpoint. Consequently, for real *Bin frequencies*   signals, all of the essential information is contained in the range $0 \leq i \leq N/2$ that corresponds to the positive discrete frequencies, also called *bin frequencies*.

$$f_i = \frac{if_s}{N}, \quad 0 \leq i \leq \frac{N}{2} \tag{4.12.8}$$

Here $f_s = 1/T$ is the sampling frequency, and the highest discrete frequency $f_{N/2}$ is the folding frequency $f_s/2$.

Just as light can be decomposed into different colors, signals have power that is distributed over different frequencies. If $x_p(k)$ denotes the periodic extension of $x(k)$, then the average power of $x_p(k)$ at discrete frequency f_i is given by $P_i = S_x(i)$. The total average power of $x(k)$ is

$$P_x = \frac{1}{N}\sum_{i=0}^{N-1} S_x(i) \tag{4.12.9}$$

White noise $v(k)$ with average power P_v is a random signal whose power density spectrum is flat and equal to P_v. From the Wiener-Khintchine theorem, the DFT of the circular auto-correlation of a signal is equal to the power density spectrum of the signal. Thus the circular auto-correlation of white noise is

$$c_{vv}(k) = P_v \delta(k) \tag{4.12.10}$$

Fast Fourier Transform (FFT)

FLOPs

FFT

When the number of samples N is an integer power of two, a highly efficient implementation of the DFT called the *fast Fourier transform* or FFT is available. For large values of N, the number of complex floating point operations or FLOPs required to perform a DFT is approximately N^2, while the number of complex FLOPs required for an FFT is only $(N/2)\log_2(N)$. Thus the FFT is P times faster than the DFT, where

$$P = \frac{N}{2\log_2(N)} \tag{4.12.11}$$

For $N = 1024$ the FFT is about 200 times faster, and for $N = 8192$ it is more that 1200 times faster than the DFT.

Spectral Analysis of Signals

The frequency response of a stable linear discrete-time system with transfer function $H(z)$ can be approximated at discrete frequencies $f_i = if_s/N$ using $H(i) = \mathrm{DFT}\{h(k)\}$, where $h(k)$ is the system impulse response. For an IIR system the approximation becomes increasingly accurate as the number of samples N increases, and for an FIR system of order m, the approximation is exact for $N > m$.

Frequency precision

Zero-padding

The spacing between discrete frequencies, $\Delta f = f_s/N$, is called the *frequency precision*. The frequency precision of an N-point signal $x(k)$ can be improved by appending zeros to the end of $x(k)$. If $M - N$ zeros are added, this results in an M-point *zero-padded* signal $x_z(k)$, whose spectrum is the same as $x(k)$, but with a finer frequency precision of

$$\Delta f = \frac{f_s}{M} \tag{4.12.12}$$

Frequency resolution

Using zero padding, isolated sinusoidal spectral components of $x(k)$ can be more accurately detected and located. Power density spectrum peaks of closely-spaced sinusoids tend to merge into one broad peak due to *spectral leakage*. The smallest frequency difference that can be reliably detected is called the *frequency resolution*. The frequency resolution ΔF is the reciprocal of the duration of $x(k)$ and is referred to as the *Rayleigh limit*.

$$\Delta F = \frac{f_s}{N} \tag{4.12.13}$$

Spectrogram

Spectrogram

Some practical signals, such as speech and music, are sufficiently long that their spectral characteristics can be thought of as changing with time. A long signal $x(k)$ can be partitioned into $2M - 1$ overlapping subsignals of length L, and these subsignals can be windowed and then transformed with L-point DFTs. When the resulting magnitude spectra are arranged as the rows of a $(2M - 1) \times L$ matrix, it is called the *spectrogram* of $x(k)$.

$$G(m, i) = |\mathrm{DFT}\{w(k)x(mL/2 + k)\}| \tag{4.12.14}$$

The spectrogram shows how the spectrum changes with time. The first independent variable m specifies the starting time in increments of $L/2$ samples, while the second independent variable i specifies the discrete frequency in increments of f_s/L.

Power Density Spectrum Estimation

A number of techniques have been proposed for obtaining improved estimates of the underlying continuous-time power density spectrum of a signal. The basic definition of $S_x(i)$ in (4.12.6) is called a *periodogram*. Improved estimates with a smaller variance can be obtained using Bartlett's *average periodogram* method, and Welch's *modified average periodogram* method.

Periodogram

Average periodograms are computed by partitioning a long signal $x(k)$ into subsignals of length L. For Bartlett's method, the subsignals do not overlap, while for Welch's method they overlap

Data windows

by $L/2$ samples. Welch's method also multiplies the subsignals by *data windows* which taper gradually to zero at each end. The use of data windows tends to reduce the effects of the *spectral leakage*, a computational artifact that leads to overly broad spikes in the estimated power density spectrum.

GUI Module

The FDSP toolbox includes a GUI module called *g_spectra* that allows the user to perform spectral analysis of discrete-time signals without any need for programming. Several common signals are included, plus signals recorded from a PC microphone and user-defined signals saved in MAT files. The signals can be noise-corrupted or noise-free and clipped or unclipped. Viewing options include the magnitude spectrum, the phase spectrum, the estimated power density spectrum, and the spectrogram.

Learning Outcomes

This chapter was designed to provide the student with an opportunity to achieve the learning outcomes summarized in Table 4.11.

TABLE 4.11: ▶
Learning Outcomes for Chapter 4

Num.	TCHLearning Outcome	Sec.
1	Know how to compute the spectra of discrete-time signals using the discrete-time Fourier transform or DTFT	4.2
2	Know how to use the DFT to find the magnitude, phase, and power density spectra of finite signals	4.3
3	Know how to apply, and use the properties of, the discrete Fourier transform or DFT	4.3–4.4
4	Know how to compute the fast Fourier transform (FFT) using decimation in time	4.5
5	Be able to compare the computational complexity of the DFT and the FFT in terms of the number of FLOPs	4.5
6	Understand how to characterize white noise, and why this random signal is useful for signal modeling and system testing	4.6
7	Know how to compute the frequency response of a stable linear discrete-time system using the DFT	4.7
8	Understand how zero padding can be used to interpolate between discrete frequencies and improve the frequency precision	4.8
9	Understand how to estimate the power density spectrum of a signal using Bartlett's and Welch's methods	4.9
10	Understand what a spectrogram is and how it is used to characterize signals whose spectral characteristics change with time	4.10
11	Know how to use GUI module *g_spectra* to perform spectral analysis of discrete-time signals and systems	4.11

• • • • • • • • • • • • • • • •

4.13   Problems

The problems are divided into Analysis and Design problems that can be solved by hand or with a calculator, GUI Simulation problems that are solved using GUI modules *g_correlate* and *g_spectra*, and MATLAB Computation problems that require a user program.

4.13.1 Analysis and Design

4.1   Verify the following values of $W_N^k = \exp(-j2\pi k/N)$ appearing in Table 4.6.

$$W_N^k = \begin{cases} -j, & k = N/4 \\ -1, & k = N/2 \\ j, & k = 3N/4 \\ 1, & k = N \end{cases}$$

4.2   Using the results of Problem 4.1, verify the following properties of $W_N = \exp(-j2\pi/N)$ appearing in Table 4.6.

(a) $W_N^{(i+N)k} = W_N^{ik}$

(b) $W_N^{i+N/2} = -W_N^i$

(c) $W_N^{2i} = W_{N/2}^i$

(d) $W_N^* = W_N^{-1}$

4.3   The following scalar, c, is real. Find its value. *Hint*: Use Euler's identity.

$$c = j^j$$

4.4   Suppose a signal $x(k)$ has the following magnitude spectrum.

$$A_x(f) = \cos(\pi f T), \quad 0 \le |f| \le f_s/2$$

(a) Find the energy density spectrum $S_x(f)$.

(b) Find the total energy E_x.

(c) Find the energy is contained in the range $0 \le |f| \le \alpha f_s$ where $0 \le \alpha \le .5$.

4.5   Consider the following causal finite signal with $x(0) = 1$.

$$x(k) = [1, 2, 1]^T$$

(a) Find the spectrum $X(f)$.

(b) Find the magnitude spectrum $A_x(f)$.

(c) Find the phase spectrum $\phi_x(f)$.

4.6   Let $x_a(t)$ be periodic with period T_0, and let $x(k)$ be a sampled version of $x_a(t)$ using sampling interval T.

(a) For what values of T is $x(k)$ periodic? Provide an example.

(b) For what values of T is $x(k)$ not periodic? Provide an example.

4.7   If one allows for the possibility that $X(f)$ can contain impulses of the form $\delta_a(f)$, then the table of DTFT pairs can be expanded. Using the inverse DTFT of an impulse, find the DTFT of $x(k)$ where c is an arbitrary constant.

$$x(k) = c$$

4.8 Using Euler's identity, find the inverse DTFT of the following signals.

(a) $X_1(f) = \dfrac{\delta_a(f - F_0) + \delta_a(f + F_0)}{2}$

(b) $X_2(f) = \dfrac{\delta_a(f - F_0) - \delta(f + F_0)}{j2}$

4.9 Consider the following discrete-time signal.

$$x(k) = c\cos(2\pi F_0 kT + \theta)$$

(a) Find a and b such that $x(k) = a\cos(2\pi F_0 kT) + b\sin(2\pi F_0 kT)$
(b) Use part (a) and Problem 4.8 to find $X(f)$.

4.10 Consider the following discrete-time signal.

$$x = [2, -1, 3]^T$$

(a) Find the third root of unity W_3.
(b) Find the 3×3 DFT transformation matrix W.
(c) Use W to find the DFT of x.
(d) Find the inverse DFT transformation matrix W^{-1}.
(e) Find the discrete-time signal x whose DFT is given by

$$X = [3, -j, j]^T$$

4.11 Find the DTFT of the following signals where $|c| < 1$.
(a) $x(k) = c^k \cos(2\pi F_0 kT)\mu(k)$
(b) $x(k) = c^k \sin(2\pi F_0 kT)\mu(k)$

4.12 Consider the following signal where $|c| < 1$.

$$x(k) = k^2 c^k \mu(k)$$

(a) Using Appendix 1, find the spectrum $X(f)$.
(b) Find the magnitude spectrum $A_x(f)$.
(c) Find the phase spectrum $\phi_x(f)$.

Frequency differentiation property

4.13 Show that the DTFT satisfies the following property called the *frequency differentiation property*.

$$\text{DTFT}\{kTx(k)\} = \left(\frac{j}{2\pi}\right)\frac{dX(f)}{df}$$

Modulation property

4.14 Recall from Problem 3.34 that the Z-transform satisfies the following *modulation property*.

$$Z\{h(k)x(k)\} = \frac{1}{j2\pi}\oint_C H(u)X\left(\frac{z}{u}\right)u^{-1}du$$

Use this result and the relationship between the Z-transform and the DTFT to show an equivalent *modulation property* of the DTFT. Here multiplication in the time domain maps into convolution in the frequency domain.

$$\text{DTFT}\{h(k)x(k)\} = \frac{1}{f_s}\int_{-f_s/2}^{f_s/2} H(\lambda)X(f - \lambda)d\lambda \qquad (4.12.1)$$

4.15 Consider an N-point signal $x(k)$. Find the smallest integer N such that a radix-two FFT of $x(k)$ is at least 100 times as fast as the DFT of $x(k)$ when speed is measured in complex FLOPs.

4.16 Consider the following discrete-time signal where $|c| < 1$.

$$x(k) = c^k, \quad 0 \le k < N$$

(a) Find $X(i)$
(b) Use the geometric series to simplify $X(i)$ as much as possible.

4.17 Consider the following discrete-time signal.

$$x = [1, 2, 1, 0]^T$$

(a) Find $X(i) = \text{DFT}\{x(k)\}$.
(b) Compute the magnitude spectrum $A_x(i)$.
(c) Compute the phase spectrum $\phi_x(i)$.
(d) Compute the power density spectrum $S_x(i)$.

4.18 Suppose $x(k)$ is real with $X(i) = \text{DFT}\{x(k)\}$.
(a) Show that $X(0)$ is real.
(b) Show that when N is even, $X(N/2)$ is real.

Orthogonal property 4.19 The following *orthogonal property* of W_N was used to derive the IDFT.

$$\sum_{i=0}^{N-1} W_N^{ik} = N\delta(k), \quad 0 \le k < N$$

The finite geometric series in Problem 3.17d is valid for any complex z. Use this to verify the orthogonality property of W_N.

4.20 Recall that the DFT of an N-point signal is periodic with period N. One of the properties of the DFT is the *conjugate property*

$$\text{DFT}\{x^*(k)\} = X^*(-i)$$

This property can be used to compute two real DFTs of length N using a single complex DFT of length N. Let $a(k)$ and $b(k)$ be real and consider the complex signal

$$c(k) = a(k) + jb(k), \quad 0 \le k < N$$

Using the identities in Appendix 2, and the conjugate property, show that

$$A(i) = \frac{C(i) + C^*(-i)}{2}$$

$$B(i) = \frac{C(i) - C^*(-i)}{j2}$$

4.21 Consider the following discrete-time signal.

$$x = [-1, 2, 2, 1]^T$$

(a) Find the average power P_x.
(b) Find the DFT of x.
(c) Verify Parseval's identity this case.

4.22 Compute the following DFT pairs for N-point signals.
(a) If $x(k) = \delta(k)$, find $X(i)$.
(b) If $X(i) = \delta(i)$, find $x(k)$.

4.23 Suppose $x(k)$ is a real N-point signal. Show that the spectrum of $x(k)$ satisfies the following symmetry properties.
(a) $\text{Re}\{X(i)\} = \text{Re}\{X(N-i)\}$.
(b) $\text{Im}\{X(i)\} = -\text{Im}\{X(N-i)\}$.

4.24 Let $x(k)$ be an N-point signal. Starting with the definition of average power in (4.3.4), use Parseval's identity to show that the average power is the average of the power density spectrum.

4.25 Consider the following discrete-time signal.

$$x = [12, 4, -8, 16]^T$$

(a) Starting with (2.8.2), but replacing x with x_p, find the circular auto-correlation matrix $E(x)$ such that $c_{xx} = E(x)x$.
(b) Use $E(x)$ to find the circular auto-correlation $c_{xx}(k)$.
(c) Find the normalized circular auto-correlation $\sigma_{xx}(k)$.

4.26 A white noise signal $v(k)$ is uniformly distributed over the interval $[-a, a]$. Suppose $v(k)$ has the following circular auto-correlation.

$$c_{vv}(k) = 8\delta(k), \quad 0 \leq k < 1024$$

(a) Find the interval bound a.
(b) Sketch the power density spectrum of $v(k)$.

4.27 Let $v(k)$ be an N-point white noise signal with mean μ_v and variance σ_v^2. Show that the average power, the mean, and the variance are related as follows.

$$P_v \approx \mu_v^2 + \sigma_v^2$$

4.28 Let $v(k)$ be an N-point white noise signal with mean μ_v and variance σ_v^2. Show that the circular auto-correlation of $v(k)$ is

$$c_{vv}(k) \approx \mu_v^2 + \sigma_v^2\delta(k)$$

4.29 Let $v(k)$ be an N-point white noise signal with mean μ_v and variance σ_v^2. Using the results of Problem 4.28, show that the power density spectrum of $v(k)$ is

$$S_v(i) \approx \sigma_v^2 + N\mu_v^2\delta(i)$$

4.30 Let v be a random variable that is uniformly distributed over the interval $[a, b]$.
(a) Find the mth statistical moment $E[v^m]$ for $m \geq 0$.
(b) Verify that $E[v^m] = P_v$ in (4.6.6) when $m = 2$.

4.31 Let x be a random variable whose probability density function is given in Figure 4.55 on page 327.
(a) What is the probability that $-.5 \leq x \leq .5$?
(b) Find $E[x^2]$.

4.32 Consider the following discrete-time signal.

$$x = [10, -5, 20, 0, 15]^T$$

(a) Using (2.8.2), find a linear auto-correlation matrix $D(x)$ such that $r_{xx} = D(x)x$.
(b) Use $D(x)$ to find the linear auto-correlation $r_{xx}(k)$.
(c) Using Definition 2.5, find the normalized linear auto-correlation $\rho_{xx}(k)$.
(d) Find the average power P_x.

FIGURE 4.55: Probability
Density Function
for Problem 4.31

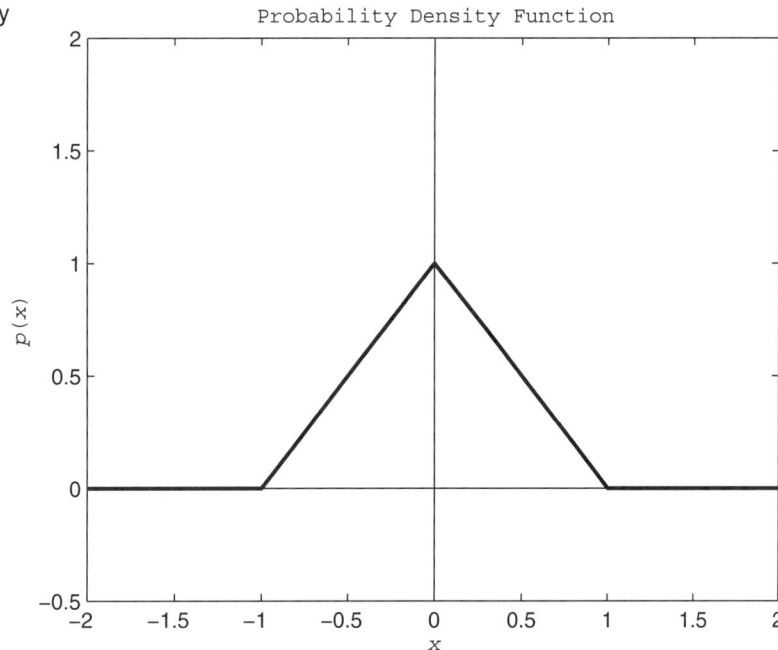

Probability Density Function

4.33 Suppose $h(k)$ and $x(k)$ are both of length $L = 2048$.
  (a) Find the number of real FLOPs for a fast linear convolution of $h(k)$ with $x(k)$.
  (b) Find the number of real FLOPs for a direct linear convolution of $h(k)$ with $x(k)$.
  (c) Express the answer to (a) as a percentage of the answer to (b).

4.34 Suppose $h(k)$ is of length L, and $x(k)$ is of length M. Let L and M be powers of two with $M \geq L$.
  (a) Find the number of real FLOPs for a fast linear convolution of $h(k)$ with $x(k)$. Does your answer agree with (4.5.8) when $M = L$?
  (b) Find the number of real FLOPs for a direct linear convolution of $h(k)$ with $x(k)$. Does your answer agree with (4.5.9) when $M = L$?

4.35 Suppose L is a power of two and $M = QL$ for some positive integer Q. Let n_{block} be the number of real FLOPs needed to compute a fast block convolution of an L-point signal $h(k)$ with an M-point signal $x(k)$. Find n_{block}.

4.36 Use the DFT to solve the following.
  (a) Recover $x(k)$ from $c_{yx}(k)$ and $y(k)$.
  (b) Recover $y(k)$ from $c_{yx}(k)$ and $x(k)$.

4.37 Suppose $x(k)$ and $y(k)$ are both of length $L = 4096$.
  (a) Find the number of real FLOPs for a fast linear cross-correlation of $y(k)$ with $x(k)$.
  (b) Find the number of real FLOPs for a direct linear cross-correlation of $y(k)$ with $x(k)$.
  (c) Express the answer to (a) as a percentage of the answer to (b).

4.38 Suppose $y(k)$ is of length L and $x(k)$ is of length $M \leq L$.
  (a) Find the number of real FLOPs for a fast linear cross-correlation of $y(k)$ with $x(k)$. Does your answer agree with (4.5.20) when $M = L$?

(b) Find the number of real FLOPs for a direct linear cross-correlation of $y(k)$ with $x(k)$. Does your answer agree with (4.5.21) when $M = L$?

4.39 Consider the following digital filter where $|a| < 1$.

$$H(z) = \frac{1}{1 - az^{-1}}$$

(a) Find the impulse response $h(k)$.
(b) Find the frequency response $H(f)$.
(c) Let $H(i)$ be the N-point DFT of $h(k)$, and let $f_i = if_s/N$. Given an arbitrary $\epsilon > 0$, use (4.8.4) to find a lower bound n such that for $N \geq n$,

$$|H(i) - H(f_i)| \leq \epsilon \quad \text{for} \quad 0 \leq i < N$$

4.40 One of the problems with using data windows to reduce the Gibb's phenomenon in the periodic extension of an N-point signal $x(k)$ is that the samples are no longer weighted equally when computing an estimate of the power density spectrum. This is particularly the case when no overlap of subsignals is used.
(a) Use the trigonometric identities in Appendix 2 to show that the Hanning window in Table 4.10 can be expressed as

$$w(k) = .5 + .5 \cos \left[\frac{2\pi(k - L/2)}{L} \right], \quad 0 \leq k < L$$

(b) If a 50% overlap of subsignals is used for the power density spectrum estimate, then each overlapped sample gets counted twice, once with weight $w(k)$ and once with weight $w(k+L/2)$. Show that if the Hanning window is used, the overlapped samples are weighted equally. Find the total weight for each overlapped sample.
(c) Are there any other windows in Table 4.10 for which the total weighting of the overlapped samples is uniform when a 50% overlap is used? If so, which ones?

4.41 Consider the spectrogram in Definition 4.5. Suppose the signal $x(k)$ is real.
(a) Find the number of complex FLOPs needed if the DFT is used.
(b) Find the number of complex FLOPs needed if the FFT is used.

4.42 A signal $x_a(t)$ is sampled at $N = 300$ points using a sampling rate of $f_s = 1600$ Hz. Let $x_z(k)$ be a zero-padded version of $x(k)$ using $M - N$ zeros. Suppose a radix-two FFT is used to find $X_z(i)$
(a) Find a lower bound on M that ensures that the frequency precision of $X_z(i)$ is no larger than 2 Hz.
(b) How much faster or slower is the FFT of $x_z(k)$ in comparison with the DFT of $x(k)$? Express your answer as a ratio of the computational effort of the FFT to the computational effort of the DFT.

4.43 Consider the spectrogram in Definition 4.5.
(a) Modify the spectrogram definition using zero padding so the frequency precision is improved by a factor of two.
(b) Compute the percent increase in computational effort for the modified spectrogram in comparison with the original spectrogram assuming the FFT is used. Use complex FLOPs to measure the computational effort and assume $x(k)$ is real.
(c) Does the modified spectrogram have improved frequency resolution? If not, how can the frequency resolution be improved and what is the tradeoff?

4.13.2 **GUI Simulation**

4.44 Use the GUI module *g_spectra* to plot the power density spectrum of a noise-free cosine input using the default parameter values. Use the dB scale and do the following cases.
(a) Rectangular window
(b) Hanning window
(c) Hamming window
(d) Blackman window

4.45 Using the GUI module *g_correlate*, select the impulse train input. This sets $y(k)$ to a periodic input, and $x(k)$ to an impulse train whose period matches the period of $y(k)$. Set $L = 4096$ and $M = 4096$.
(a) Plot the noise-corrupted periodic input $y(k)$ and the periodic impulse train $x(k)$.
(b) Plot the normalized circular auto-correlation of $y(k)$.
(c) Plot the normalized circular cross-correlation $\sigma_{yx}(k)$. This should be proportional to $y(k)$, but with the noise reduced.

4.46 Use the GUI module *g_spectra* to plot the following characteristics of a noise-corrupted damped exponential input using the default parameter values. Use the linear scale.
(a) Time signal
(b) Magnitude spectrum
(c) Power density spectrum (Blackman window)
(d) Blackman window

4.47 Use the GUI module *g_spectra* to plot the spectrogram of the following signals. Use $f_s = 3000$ Hz and $N = 2048$ samples for each.
(a) Cosine of unit amplitude and frequency $F_0 = 400$ Hz
(b) Cosine of unit amplitude and frequency $F_0 = 400$ Hz, clipped to $[-.5, .5]$
(c) Cosine of unit amplitude and frequency $F_0 = 400$ Hz, plus white noise uniformly distributed over $[-1.5, 1.5]$

4.48 Using the GUI module *g_spectra*, record the word HELLO. Play it back to make sure it is recorded properly. Save it in a MAT-file called *hello*. Then reload it as a User-defined input. Plot the following spectral characteristics.
(a) Magnitude spectrum
(b) Power density spectrum (Hamming window)
(c) Spectrogram

4.49 Consider the signal shown in Figure 4.56 on page 330 which contains one or more sinusoidal components corrupted with white noise. The complete signal $x(k)$ and the sampling frequency f_s are stored in the file *prob4_49.mat*. Use the GUI module *g_spectra* to plot the following spectral characteristics.
(a) The power density spectrum (Hamming window). Use the Caliper option to estimate the frequencies of the sinusoidal components.
(b) The spectrogram (Hamming window).

4.50 Using the GUI module *g_correlate*, select the white noise input. Set the scale factor to $c = 0$.
(a) Plot $x(k)$ and $y(k)$. What is the range of values over which the uniform white noise is distributed?
(b) Verify that $r_{yy}(k) \approx P_y\delta(k)$ by plotting the auto-correlation of $y(k)$.
(c) Use the *Caliper* option to estimate P_y.
(d) Verify that this estimate of P_y is consistent with the theoretical value in (4.6.6).

FIGURE 4.56: Noise-corrupted Signal with Unknown Sinusoidal Components (Samples 0 to $N/8$)

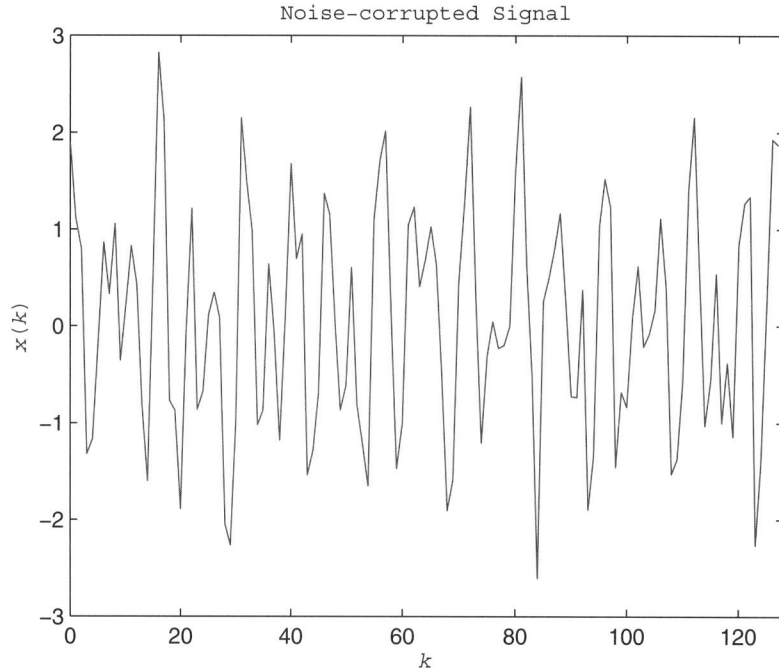
Noise-corrupted Signal

4.51 Using the GUI module *g_correlate*, select the periodic input.
   (a) Plot $x(k)$ and $y(k)$.
   (b) Plot the normalized circular auto-correlation $\sigma_{yy}(k)$. Notice how the noise has been reduced.
   (c) Estimate the period of $y(k)$ in seconds by estimating the period of σ_{yy}.

4.52 Consider the following noise-corrupted periodic signal with a sampling frequency of $f_s = 1600$ Hz and $N = 1024$. Here $v(k)$ is white noise uniformly distributed over $[-1, 1]$.

$$x(k) = \sin(600\pi kT)\cos^2(200\pi kT) + v(k), \quad 0 \le k < N$$

Create a MAT-file called *prob4_52* containing x and f_s. Then use *g_spectra* to plot the following.
   (a) Magnitude spectrum
   (b) Power density spectrum using Welch's method (rectangular window)
   (c) Power density spectrum using Welch's method (Blackman window)

4.53 Use the GUI module *g_spectra* to perform the following analysis of the vowels. Play back the sound in each case to make sure you have a good recording.
   (a) Record one second of the vowel "A", save it, and plot the time signal.
   (b) Record one second of the vowel "E", save it, and plot the time signal.
   (c) Record one second of the vowel "I", save it, and plot the time signal.
   (d) Record one second of the vowel "O", save it, and plot the time signal.
   (e) Record one second of the vowel "U", save it, and plot the time signal.

4.54 A signal stored in *prob4_54.mat* contains white noise plus a single sinusoidal component whose frequency does not correspond to any of the discrete frequencies. Use GUI module *g_spectra* to plot the following spectral characteristics.
   (a) The magnitude spectrum of $x(k)$ using the linear scale.

(b) The power density spectrum of $x(k)$ using the Blackman window. Use the Caliper option to estimate the frequency of the sinusoidal component.

4.13.3 MATLAB Computation

4.55 Consider the following digital filter of order n where $n = 11$ and $r = .98$.

$$H(z) = \frac{(1+r^n)(1-z^{-n})}{2(1-r^n z^{-n})}$$

Suppose $f_s = 2200$ Hz. Write a program that uses *filter* to do the following.
(a) Compute and plot the impulse response $h(k)$ for $0 \le k < N$ where $N = 1001$.
(b) Compute and plot the magnitude response $A(f)$ for $0 \le f \le f_s/2$.
(c) What type of filter is this, FIR or IIR? Which frequencies get rejected by this filter?

Dead zone

4.56 In addition to saturation due to clipping, another common type of nonlinearity is the *dead-zone* nonlinearity shown in Figure 4.57. The algebraic representation of a dead zone of radius a is as follows.

$$F(x, a) \triangleq \begin{cases} 0, & 0 \le |x| \le a \\ x, & a < |x| < \infty \end{cases}$$

Suppose $f_s = 2000$ Hz, and $N = 100$. Consider the following input signal where $0 \le k < N$ corresponds to one cycle.

$$x(k) = \cos(40\pi kT), \quad 0 \le k < N$$

Let the dead-zone radius be $a = .25$. Write a MATLAB program that does the following.
(a) Compute and plot $y(k) = F[x(k), a]$ versus k.
(b) Compute and plot the magnitude spectrum of $y(k)$.

FIGURE 4.57: Dead-zone Nonlinearity of Radius a

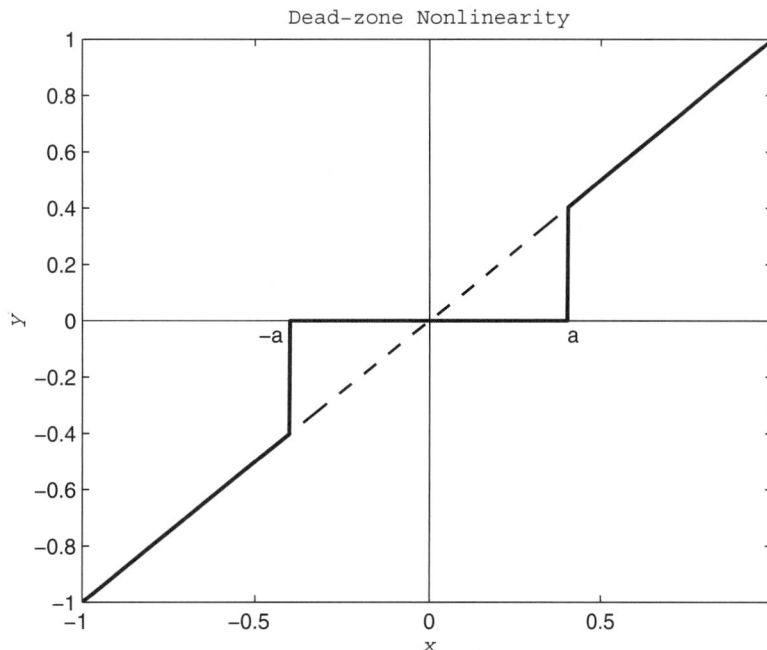

Dead-zone Nonlinearity

(c) Using the DFT, compute and print the total harmonic distortion of $y(k)$ caused by the dead zone. Here, if d_i and θ_i for $0 \le i < M$ are the cosine form Fourier coefficients of $y(k)$ with $M = N/2$, then

$$\text{THD} = \frac{100(P_y - d_1^2/2)}{P_y}\%$$

4.57   Consider the following N-point periodic signal of period M. Suppose $M = 128$ and $N = 1024$.

$$x(k) = 1 + 3\cos\left(\frac{2\pi k}{M}\right) - 2\sin\left(\frac{4\pi k}{M}\right), \quad 0 \le k < N$$

Let $y(k)$ be a noise-corrupted version of $x(k)$ where $v(k)$ is white noise uniformly distributed over $[-1, 1]$.

$$y(k) = x(k) + v(k), \quad 0 \le k < N$$

The objective of this problem is to study how *sensitive* the periodic signal extraction technique is to the estimate of the period M.

$$\hat{x}_m(k) = \left(\frac{N}{L}\right) c_{y\delta_m}(k)$$

Write a program which performs the following tasks.
(a) Compute and plot the noise-corrupted periodic signal $y(k)$.
(b) Compute and plot on the same graph $x(k)$ and $\hat{x}_m(k)$ for $m = M - 5$ using a legend.
(c) Compute and plot on the same graph $x(k)$ and $\hat{x}_m(k)$ for $m = M$ using a legend.
(d) Compute and plot on the same graph $x(k)$ and $\hat{x}_m(k)$ for $m = M + 5$ using a legend.

4.58   Let $h(k)$ and $x(k)$ be two N-point white noise signals uniformly distributed over $[-1, 1]$. Recall that the MATLAB function *conv* can be used to compute linear convolution. Write a MATLAB program which uses *tic* and *toc* to compute the computational time t_{dir} of *conv* and the computational time t_{fast} of the FDSP toolbox function f_conv for the cases $N = 4096$, $N = 8192$, and $N = 16384$.
(a) Print the two computational times t_{dir} and t_{fast} for $N = 4096, 8192$, and 16384.
(b) Plot t_{dir} versus $N/1024$ and t_{tast} versus $N/1024$ on the same graph and include a legend.

4.59   Write a program which creates a 1×2048 vector x of white noise uniformly distributed over $[-.5, .5]$. The program should then compute and display the following.
(a) The average power P_x, the predicted average power P_v, and the percent error in P_x.
(b) Plot the estimated power density spectrum using Bartlett's method with $L = 512$. Use a y-axis range of $[0, 1]$. In the plot title, print L and the estimated variance σ_B^2 of the power density spectrum.
(c) Repeat part (b), but use $L = 32$.

4.60   Let $x(k)$ be an N-point white noise signal uniformly distributed over $[-1, 1]$ where $N = 4096$. Write a program that performs the following tasks.
(a) Create $x(k)$ and then compute and plot the normalized circular auto-correlation $\sigma_{xx}(k)$.
(b) Compute $c_{xx}(k)$, and use the result to compute and plot the power density spectrum of $x(k)$.
(c) Compute and print the average power P_x.

4.61   Consider the following linear discrete-time system. Write a MATLAB program that performs the following tasks.

$$H(z) = \frac{z}{z^2 - 1.4z + .98}$$

(a) Compute and plot the impulse response $h(k)$ for $0 \le k < L - 1$ where $L = 500$.
(b) Construct an M-point white noise input $x(k)$ that is distributed uniformly over $[-5, 5]$ where $M = 10000$. Use the FDSP toolbox function $f\_blockconv$ to compute the zero-state response $y(k)$ to the input $x(k)$ using block convolution. Plot $y(k)$ for $9500 \le k < 10000$.
(c) Print the number of FFTs and the lengths of the FFTs used to perform the block convolution.

4.62 Consider the following digital filter of order $m = 2p$ where $p = 20$.

$$H(z) = \sum_{i=0}^{2p} b_i z^{-i}$$
$$b_p = .5$$
$$b_i = \frac{[.54 - .46\cos(\pi i/p)]\{\sin[.75\pi(i - p)] - \sin[.25\pi(i - p)]\}}{\pi(i - p)}, \quad i \ne p$$

Suppose $f_s = 200$ Hz. Write a program that uses *filter* to do the following.
(a) Compute and plot the impulse response $h(k)$ for $0 \le k < N$ where $N = 64$.
(b) Compute and plot the magnitude response $A(f)$ for $0 \le f \le f_s/2$.
(c) What type of filter is this, FIR or IIR? What range of frequencies gets passed by this filter?

4.63 Let $x_a(t)$ be a periodic pulse train of period T_0. Suppose the pulse amplitude is $a = 10$, and the pulse duration is $\tau = T_0/5$, as shown in Figure 4.58 for the case $T_0 = 1$. This signal can be represented by the following cosine form Fourier series.

$$x_a(t) = \frac{d_0}{2} + \sum_{i=1}^{\infty} d_i \cos\left(\frac{2\pi it}{T_0} + \theta_i\right)$$

Write a MATLAB program that uses the DFT to compute coefficients d_0 and (d_i, θ_i) for $1 \le i < 16$. Plot d_i and θ_i using a 2×1 array of plots and the MATLAB function *stem*.

FIGURE 4.58: Periodic Pulse Train with $a = 10$ and $T_0 = 1$

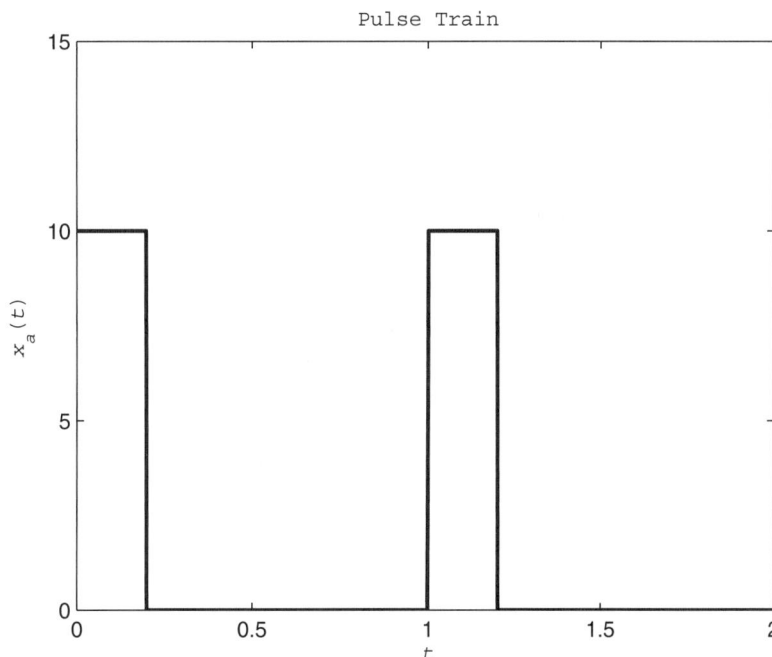

4.64 Consider the following noise-corrupted periodic signal with a sampling frequency of $f_s = 1600$ Hz and $N = 1024$.

$$x(k) = \sin^2(400\pi kT)\cos^2(300\pi kT) + v(k), \quad 0 \le k < N$$

Here $v(k)$ is zero-mean Gaussian white noise with a standard deviation of $\sigma = 1/\sqrt{2}$. Write a program that performs the following tasks.
(a) Compute and plot the power density spectrum $S_x(f)$ for $0 \le f \le f_s/2$.
(b) Compute and print the average power of $x(k)$ and the average power of $v(k)$.

4.65 Repeat Problem 4.56, but using $f_s = 1000$ Hz, $N = 50$ samples, and the cubic nonlinearity

$$F(x) = x^3$$

PART II

Digital Filter Design

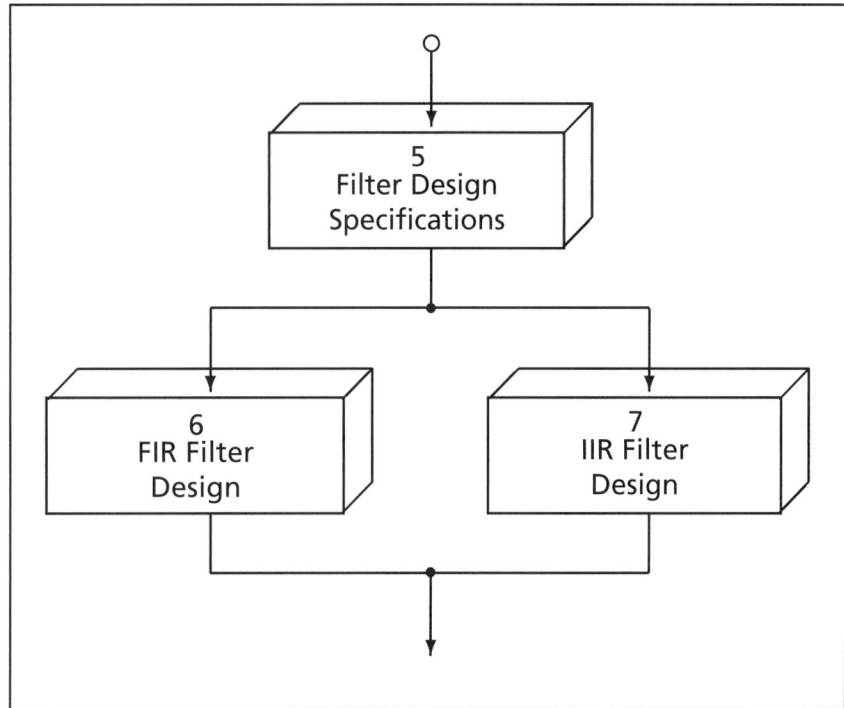

```
                    ○
                    │
                    ▼
          ┌─────────────────┐
          │        5        │
          │  Filter Design  │
          │ Specifications  │
          └─────────────────┘
                    │
          ┌─────────┴─────────┐
          ▼                   ▼
   ┌───────────┐       ┌───────────┐
   │     6     │       │     7     │
   │ FIR Filter│       │ IIR Filter│
   │  Design   │       │  Design   │
   └───────────┘       └───────────┘
          │                   │
          └─────────┬─────────┘
                    │
                    ▼
```

CHAPTER 5

Filter Design Specifications

Chapter Topics

5.1 Motivation

5.2 Frequency-selective Filters √

5.3 Linear-phase and Zero-phase Filters √

5.4 Minimum-phase and Allpass Filters √

5.5 Quadrature Filters

5.6 Notch Filters and Resonators

5.7 Narrowband Filters and Filter Banks

5.8 Adaptive Filters

5.9 GUI Software and Case Study

5.10 Chapter Summary

5.11 Problems

5.1 Motivation

The remaining chapters focus largely on the design and application of various types of digital filters. To lay a foundation for filter design, it is helpful to first consider certain fundamental characteristics that filters have in common. One characteristic of frequency-selective filters is that they are constructed to meet certain *design specifications*. For example, the design specifications dictate which frequencies or spectral components of the input are passed by the filter, which are rejected, and the degree to which rejected frequencies are blocked by the filter. These characteristics are specified using a desired magnitude response $A(f)$.

Design specifications

$$H(f) = A(f) \exp[j\phi(f)]$$

Often the phase response $\phi(f)$ is left unspecified, but in certain cases the design specifications call for a certain type of phase response as well, such as a linear phase response, $\phi(f) = -2\pi\tau f$, that corresponds to a delay of τ, or a zero phase response $\phi(f) = 0$, which can be realized only with a noncausal filter.

Each filter can be realized physically in hardware, or mathematically in software, using any one of several filter structures. For example, both FIR and IIR transfer functions can be realized with the following cascade structure where the L blocks are second-order subsystems.

$$H(z) = b_0 H_1(z) H_2(z) \cdots H_L(z)$$

When infinite-precision arithmetic is used, all of the different filter structures are equivalent in terms of their input-output characteristics. However, when finite-precision arithmetic is used to implement the filter, some of the structures are superior to others in terms of their sensitivity to detrimental finite word length effects.

We begin this chapter by introducing examples of filter specifications and structures. The frequency-selective filter design problem is then formulated by presenting a set of filter design specifications, both linear and logarithmic, for the desired magnitude response $A(f)$. Next the notion of a linear-phase filter is introduced, and four types of FIR linear-phase filters are presented. This is followed by a decomposition of a general filter into a minimum-phase part whose phase lag is as small as possible, and an allpass part whose magnitude response is constant. Next a Hilbert transformer filter is introduced that produces a sinusoidal output that is delayed by a quarter of a cycle, so that the input and output are in phase quadrature. This is followed by a discussion of filters that reject or pass isolated frequency components using notch filters and resonators. The use of multirate techniques to implement narrowband filters and filter banks for frequency-division multiplexing is then presented. This is followed by an introduction to adaptive transversal filters and their use in solving problems whose characteristics evolve with time. Finally, a GUI module called *g_filters* is introduced that allows the user to construct filters from design specifications and examine finite precision effects such as coefficient quantization, all without any need for programming. The chapter concludes with a case study example, and a summary of filter design specifications and filter types.

5.1.1 Filter Design Specifications

Perhaps the most common type of digital filter is a lowpass filter. A digital lowpass filter is a filter that removes the higher frequencies but passes the lower frequencies. An example of a

Lowpass filter   magnitude response of a digital *lowpass filter* is shown in Figure 5.1. This is a fourth-order

FIGURE 5.1: Magnitude Response of a Lowpass Chebyshev-I Filter of Order $n = 4$

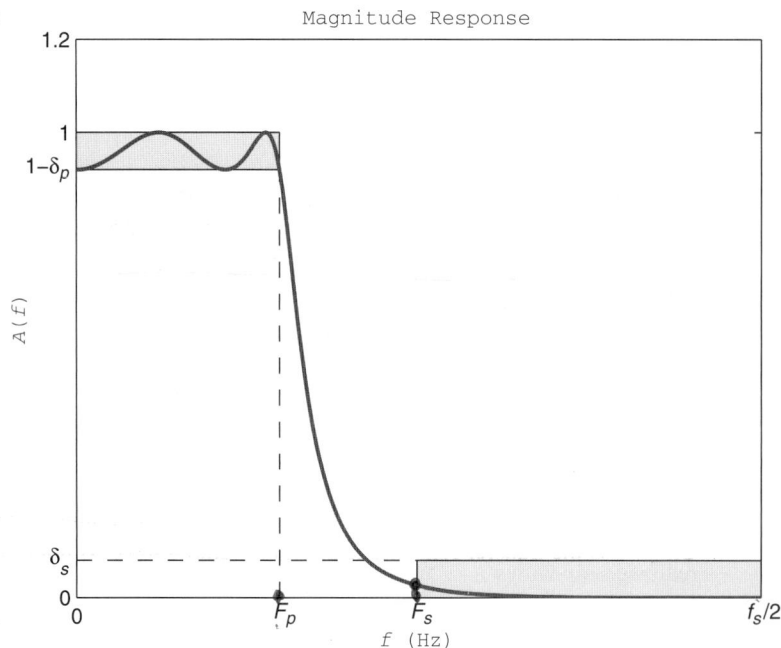

Chebyshev-I filter. The design of FIR filters is discussed in Chapter 6, and the design of IIR filters, including Chebyshev filters, is discussed in Chapter 7.

Passband

Passband

The shaded areas in Figure 5.1 represent the design specifications. The shaded region in the upper-left corner represents the filter *passband,* and the shaded region in the lower-right corner represents the filter stopband. Notice that the passband has width F_p and height δ_p. That is, the desired magnitude response must meet, or exceed, the following *passband specification*.

$$1 - \delta_p \le A(f) \le 1, \quad 0 \le f \le F_p \tag{5.1.1}$$

Passband specification

Here $0 < F_p < f_s/2$ is the *passband cutoff frequency*, and $\delta_p > 0$ is the *passband ripple*. The passband ripple can be made small, but must be positive for a physically realizable filter. It is called a ripple factor because the magnitude response sometimes oscillates within the passband, as shown in Figure 5.1. However, for some filters such as Butterworth filters and Chebyshev-II filters, the magnitude response decreases monotonically within the passband. For the filter in Figure 5.1, the passband cutoff frequency is $F_p/f_s = .15$, and the passband ripple is $\delta_p = .08$.

Stopband

Stopband

Similar to the passband, the shaded *stopband* region in the lower-right corner of Figure 5.1 has width $f_s/2 - F_s$ and height δ_s. Thus the desired magnitude response must meet, or exceed, the following *stopband specification*.

$$0 \le A(f) \le \delta_s, \quad F_s \le f \le f_s/2 \tag{5.1.2}$$

It is evident from Figure 5.1 that the passband specification is met exactly, whereas the stopband specification is exceeded in this case. Here $F_p < F_s < f_s/2$ is the *stopband cutoff frequency*, and $\delta_s > 0$ is the *stopband attenuation*. Again the stopband attenuation can be made small, but must be positive for a physically realizable filter. For the filter in Figure 5.1, the stopband cutoff frequency is $F_s/f_s = .25$ and the stopband attenuation is $\delta_s = .08$.

Stopband specification

Transition Band

Transition band

Notice that there is a significant part of the spectrum that is left unspecified. The frequency band $[F_p, F_s]$ between the passband and the stopband is called the *transition band*. The width of the transition band can be made small but it must be positive for a physically realizable filter. Indeed, as the passband ripple, the stopband attenuation, and the transition bandwidth all approach zero, the required order of the filter approaches infinity. The limiting special case of the filter with $\delta_p = 0$, $\delta_s = 0$, and $F_s = F_p$ is an ideal lowpass filter.

5.1.2 Filter Realization Structures

Each digital filter has a number of alternative realizations depending on which filter structure is used. Filter realization structures are considered in detail at the end of Chapter 6 (for FIR filters) and Chapter 7 (for IIR filters). The different filter realizations are equivalent to one another as long as infinite-precision arithmetic is used. To illustrate some filter realization structures, consider the fourth-order lowpass filter in Figure 5.1. Using design techniques covered in Chapter 7, the transfer function of this Chebyshev-I lowpass filter is

$$H(z) = \frac{.0095 + .0379z^{-1} + .0569z^{-2} + .0379z^{-3} + .0095z^{-4}}{1 - 2.2870z^{-1} + 2.5479z^{-2} - 1.4656z^{-3} + .3696z^{-4}} \tag{5.1.3}$$

FIGURE 5.2: Signal
Flow Graph of a
Direct Form II
Realization of the
Fourth-order
Chebyshev-I Filter

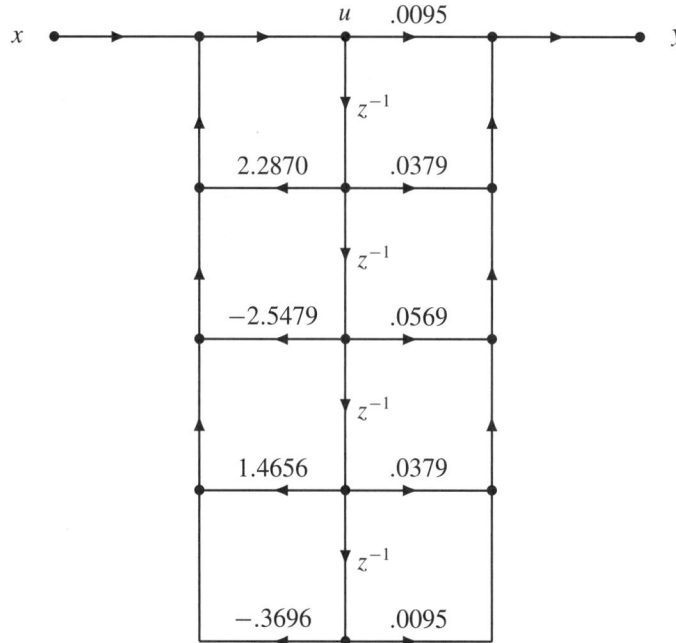

Direct Form II

Filter realization structures can be represented graphically using the signal flow graphs intro-duced in Section 3.6. For example, a direct form II realization of $H(z)$ is shown in Figure 5.2. Recall that the nodes are summing nodes, and arcs without labels have a default gain of unity. Notice that the gains of the branches correspond directly to the coefficients of the numerator and denominator polynomials of $H(z)$. This is a characteristic of direct form realizations that sets them apart from the indirect forms.

Filter realizations can also be represented mathematically using the difference equations associated with the signal flow graph. The direct form II filter realization shown in Figure 5.2 can be implemented with the following pair of difference equations where $u(k)$ is an intermediate variable.

$$u(k) = x(k) + 2.2870x(k-1) - 2.5479x(k-2)$$
$$+ 1.4656x(k-3) - .3696x(k-4) \tag{5.1.4a}$$
$$y(k) = .0095u(k) + .0379u(k-1) + .0569u(k-2)$$
$$+ .0379u(k-3) + .0095u(k-4) \tag{5.1.4b}$$

Cascade Form

To develop an alternative to the direct form II realization in Figure 5.2, we first recast the transfer function in (5.5.3) in terms of positive powers of z which yields

$$H(z) = \frac{.0095z^4 + .0379z^3 + .0569z^2 + .0379z + .0095}{z^4 - 2.2870z^3 + 2.5479z^2 - 1.4656z + .3696} \tag{5.1.5}$$

The .0095 can be factored from the numerator. The resulting numerator polynomial and de-nominator polynomial then can be factored into zeros and poles. Suppose complex-conjugate pairs of zeros are grouped together and similarly for complex-conjugate pairs of poles. The transfer function then can be written as a product of two second-order transfer functions, each

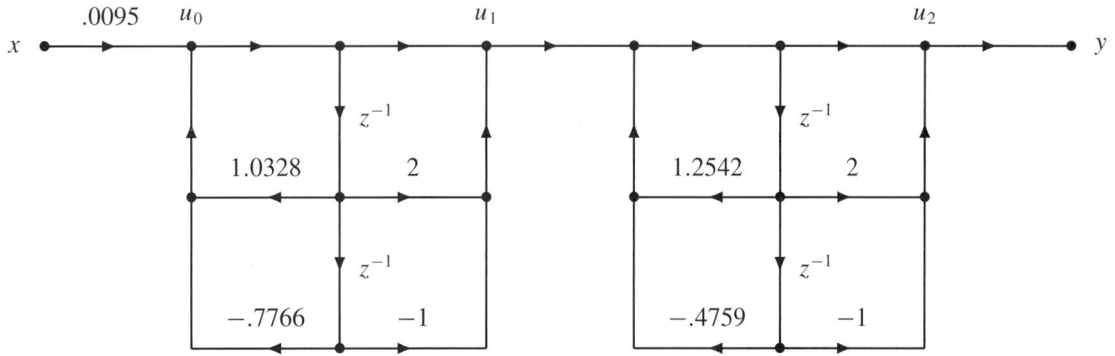

FIGURE 5.3: Signal-flow Graph of a Cascade-form Realization of the Fourth-order Chebyshev-I Filter Using Direct Form II Realizations for the Second-order Blocks

Cascade realization   with real coefficients. This is called a *cascade* form *realization*.

$$H(z) = .0095 H_1(z) H_2(z) \qquad (5.1.6)$$

There are several possible formulations of the two second-order blocks depending on how the zeroes and poles are ordered and grouped together. One such ordering from Chapter 7 is

$$H_1(z) = \frac{1 + 2z^{-1} + z^{-2}}{1 - 1.0328z^{-1} + .7766z^{-2}} \qquad (5.1.7)$$

$$H_2(z) = \frac{1 + 2z^{-1} + z^{-2}}{1 - 1.2542z^{-1} + .4759z^{-2}} \qquad (5.1.8)$$

Each of the second-order blocks can be realized using one of the direct forms. For example, a signal-flow graph which uses direct form II realizations from Section 3.6 for the two blocks is shown in Figure 5.3.

As with the higher-order direct-form realization, the signal flow graph of the cascade realization can be implemented with a system of difference equations as follows.

$$u_0(k) = .0095x(k) \qquad (5.1.9a)$$

$$u_1(k) = u_0(k) + 2u_0(k-1) - u_0(k-2)$$
$$+ 1.0328u_1(k-1) - .7766u_1(k-2) \qquad (5.1.9b)$$

$$u_2(k) = u_1(k) + 2u_1(k-1) - u_1(k-2)$$
$$+ 1.2542u_2(k-1) - .4749u_2(k-2) \qquad (5.1.9c)$$

$$y(k) = u_2(k) \qquad (5.1.9d)$$

Quantization Error

When a filter is implemented in software using MATLAB, double-precision floating point arithmetic is used for all calculations. This typically involves 64 bits of precision which corresponds to about 16 decimal digits for the mantissa or fractional part, and the remaining bits used to represent the exponent. For convenience of display, only four decimal places are shown in Figures 5.2 and 5.3. In most instances, double-precision arithmetic is a good approximation to infinite-precision arithmetic, so no significant finite word length effects are apparent. However, if a filter is implemented on specialized DSP hardware, or if storage space or speed requirements dictate the need to use single-precision floating point arithmetic or integer fixed-point arithmetic, then finite word length effects can begin to manifest themselves.

FIGURE 5.4:
Magnitude
Responses of
Fourth-order
Chebyshev-I Lowpass
Filter Using *N* Bits of
Precision to
Represent the
Coefficients

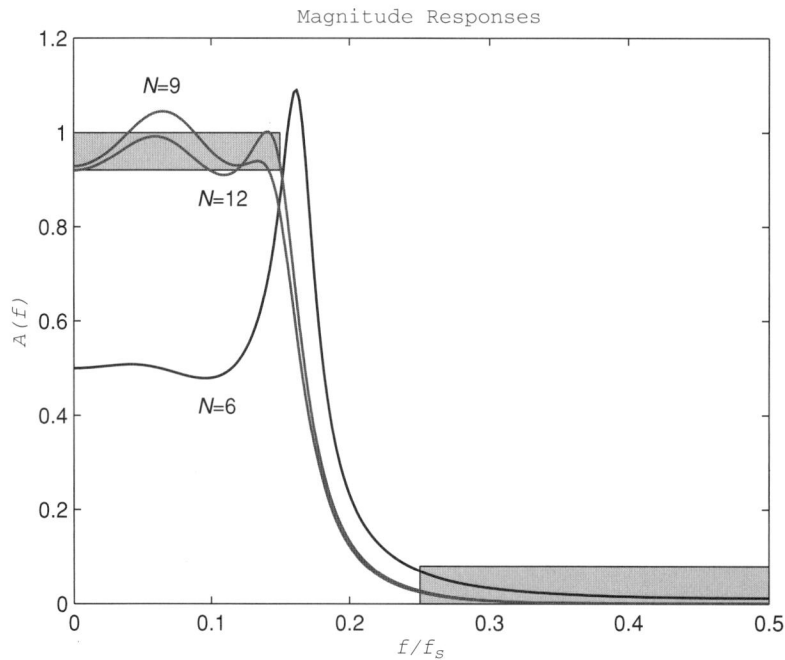

FIGURE 5.4: Magnitude Responses of Fourth-order Chebyshev-I Lowpass Filter Using *N* Bits of Precision to Represent the Coefficients

To illustrate the detrimental effects that limited precision can have, suppose the coefficients of the fourth-order Chebyshev-I lowpass filter in Figure 5.1 are represented using N bits. The resulting magnitude responses for three cases are shown in Figure 5.4. Comparing Figure 5.4 with Figure 5.1, we see that the case using $N = 12$ bits is essentially correct, but the lower-precision cases, $N = 6$ and $N = 9$, have magnitude responses that differ significantly from the double-precision version shown in Figure 5.1. Interestingly enough, if the precision is lowered still further to $N = 4$ bits, the coefficient quantization error becomes so large that the poles of the filter migrate outside the unit circle at which point the implementation becomes unstable!

● ● ● ● ● ● ● ● ● ● ● ● ● ● ● ●

5.2 Frequency-selective Filters

A digital filter is a discrete-time system that reshapes the spectrum of the input signal to produce desired spectral characteristics in the output signal. Recall from Definition 3.3 that a stable system with transfer function $H(z)$ has the following *frequency response* where f_s is the sampling rate.

$$H(f) \triangleq H(z)|_{z=\exp(j2\pi fT)}, \quad 0 \le |f| \le \frac{f_s}{2} \tag{5.2.1}$$

Thus the frequency response is just the transfer function evaluated along the unit circle. The complex-valued function $H(f)$ can be expressed in polar form as $H(f) = A(f)\exp[j\phi(f)]$, where $A(f)$ denotes the *magnitude response* and $\phi(f)$ denotes the *phase response* of the filter.

$$A(f) \triangleq |H(f)|, \quad 0 \le |f| \le \frac{f_s}{2} \tag{5.2.2a}$$

$$\phi(f) \triangleq \angle H(f), \quad 0 \le |f| \le \frac{f_s}{2} \tag{5.2.2b}$$

In Proposition 3.2, it was shown that for a stable system $H(z)$ the steady-state response to the sinusoidal input $x(k) = \sin(2\pi F_0 kT)$ is

$$y(k) = A(F_0)\sin[2\pi F_0 kT + \phi(F_0)] \tag{5.2.3}$$

Gain

Thus the magnitude response $A(F_0)$ can be interpreted as the *gain of the filter* at frequency F_0. It specifies the amount by which a sinusoidal signal of frequency F_0 is scaled as it passes

Phase shift

through the filter. Similarly, the phase response $\phi(F_0)$ can be interpreted as the *phase shift* of the filter at frequency F_0. It specifies the number of radians by which a sinusoidal signal of frequency F_0 gets advanced as it passes through the filter.

By designing a filter with a specified $A(f)$ or $\phi(f)$ we can control the spectral characteristics of the output signal $y(k)$. Most digital filters are designed to produce a desired magnitude response $A(f)$. However, there are specialized filters, such as allpass filters, that are designed

Linear phase

to produce a desired phase response. A particularly useful phase response is a *linear* phase response of the form

$$\phi(f) = -2\pi\tau f \tag{5.2.4}$$

Linear-phase filters have the property that each spectral component gets delayed by the same amount, namely, τ seconds. Consequently, the spectral components of the input signal that survive at the filter output are not otherwise distorted. Although it is possible to approximate linear-phase filters in the passband with IIR filters (e.g., with Bessel filters), it turns out that it is much simpler to use FIR filters to design linear-phase filters. Linear-phase FIR filters are discussed in detail in Section 5.3.

5.2.1 Linear Design Specifications

There are many specialized frequency-selective filters that one can consider. However, the most common filters fall into four basic categories: lowpass, highpass, bandpass and bandstop. The magnitude responses of *ideal* versions of the four basic filter types are shown in Figure 5.5.

FIGURE 5.5: Ideal Magnitude Responses of the Four Basic Filter Types

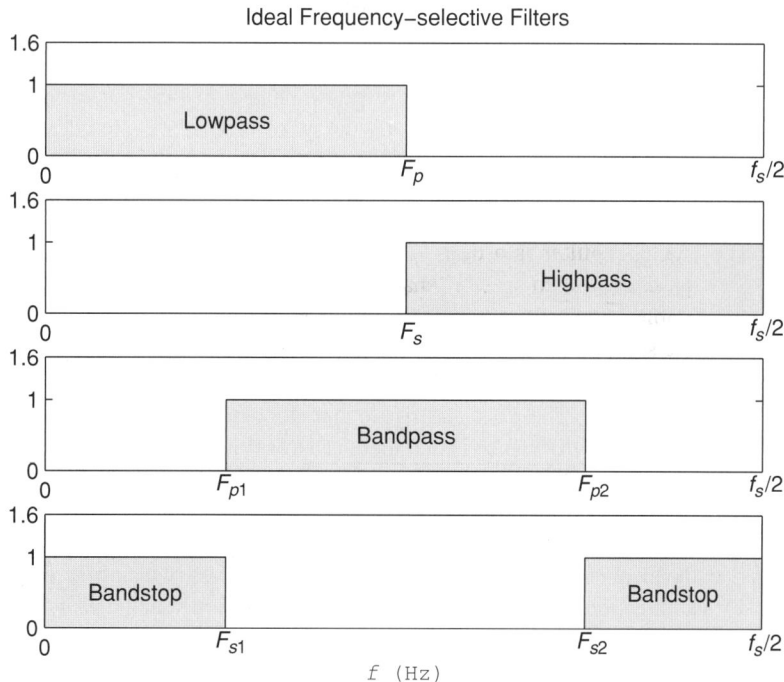

Passband

Stopband

Note that in each case the upper frequency limit is the folding frequency $f_s/2$, because this is the highest frequency that the digital filter can process. Recall from Chapter 1 that analog signals at higher frequencies get aliased back into the range $[0, f_s/2]$ during the sampling process. The range of frequencies over which $A(f) = 1$ is called the *passband*, and the range of frequencies over which $A(f) = 0$ is called the *stopband*. One of the advantages of digital filters is that the passband gain can be set to a value greater than one, if desired, in which case the signal is amplified in the passband. Analog filters can also have passbands with gains greater than one, but they must be implemented as active filters, rather than passive filters.

There is a fundamental result that limits what kind of filters can be used to achieve the idealized frequency response characteristics in Figure 5.5. It is called the Paley-Wiener theorem (1934).

PROPOSITION
....

5.1: Paley-Wiener

Theorem

Let $H(f)$ be the frequency response of a stable causal filter with $A(f) = |H(f)|$. Then

$$\int_{-f_s/2}^{f_s/2} |\log[A(f)]| df < \infty$$

All of the ideal frequency-selective filters in Figure 5.5 have the property that they exhibit complete attenuation of the signal ($\delta_s = 0$) over a stopband of nonzero length. Since $\log(0) = -\infty$, it follows from the Paley-Wiener theorem that none of the ideal filters can be causal. That is, the magnitude response of a causal filter can go to zero only at isolated frequencies such as the $m/2$ frequencies of the running average filter in Example 4.16, not over a nonzero range of frequencies. Note that this is consistent with the analysis of an ideal lowpass filter found in Example 4.2. There the impulse response for a cutoff frequency of F_c was found to be

$$h_{\text{low}}(k) = 2F_c T \text{sinc}(2\pi k F_c T) \tag{5.2.5}$$

Since $h_{\text{low}}(k) \neq 0$ for $k < 0$, this filter is not causal and therefore does not have a real-time physical realization. In spite of Proposition 5.1, causal filters can be designed that closely *approximate* ideal frequency response characteristics. For example, the impulse response in (5.2.5) can be multiplied by a window of radius m centered at $k = 0$ and then delayed by m samples to make it causal. This general approach is the basis for one of the FIR design techniques presented in Chapter 6.

In addition to the constraint on the stopband, a practical filter must also contain a transition band separating the passband from the stopband. A more realistic design specification for the magnitude response of a lowpass filter is shown in Figure 5.6. There are two differences worth noting. First, both the passband and the stopband are specified by a *range* of acceptable values for the desired magnitude response. The parameter δ_p is called the *passband ripple* because the magnitude response often oscillates within the passband. Similarly, δ_s is called the *stopband attenuation*. The passband ripple and stopband attenuation can be made small, but not zero. The second difference is that there is a transition band of width $F_s - F_p$ between the passband and the stopband. Again, the width of the transition band can be made small (at the expense of the filter order), but not zero.

The desired magnitude response must fall within the shaded area in Figure 5.6. Note how the ideal cutoff F_p in Figure 5.5 has been split into two cutoff frequencies, F_p and F_s, to create a transition band in Figure 5.6. A practical design specification for the magnitude response of a highpass filter is shown in Figure 5.7. Again, a transition band has been created by splitting the ideal cutoff frequency F_s into two cutoff frequencies, F_s and F_p.

FIGURE 5.6: Linear Magnitude Response Specifications for a Lowpass Filter

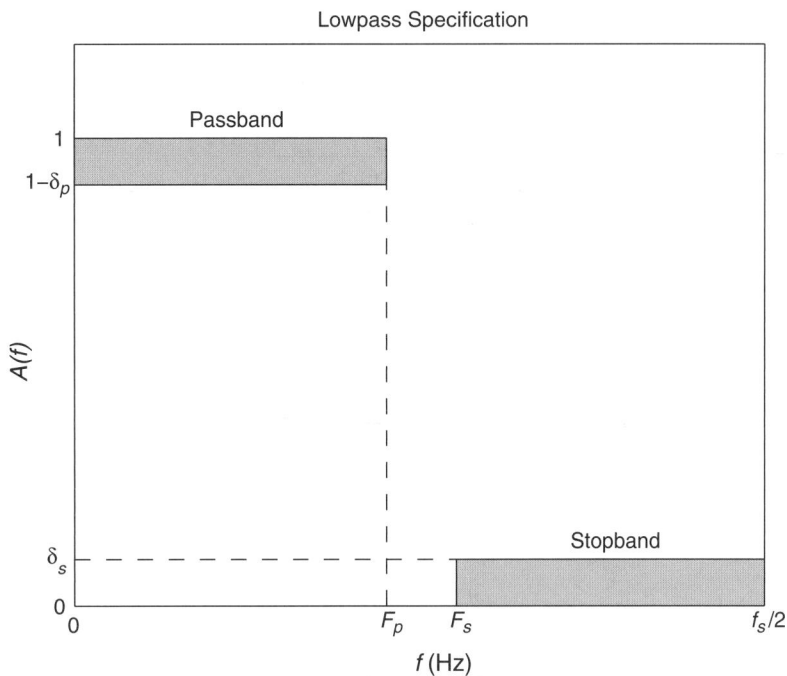

FIGURE 5.7: Linear Magnitude Response Specifications for a Highpass Filter

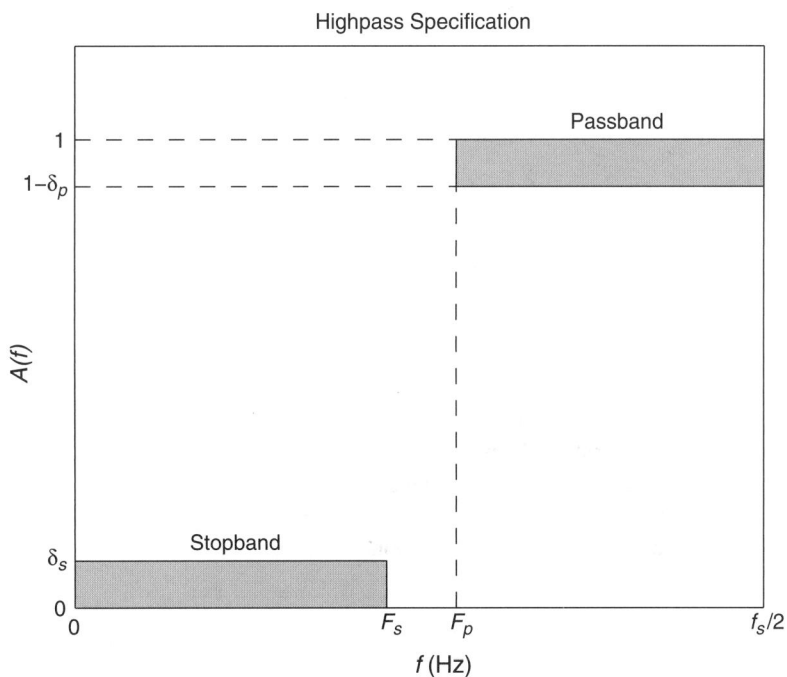

The design specification for the magnitude response of a bandpass filter is a bit more involved because there are two transition bands bracketing the passband, as can be seen in Figure 5.8. Thus there are four cutoff frequencies plus a passband ripple δ_p and a stopband attenuation δ_s. The design specification for the magnitude response of a stopband filter also has two transition bands, this time bracketing the stopband, as can be seen in Figure 5.9.

FIGURE 5.8: Linear Magnitude Response Specifications for a Bandpass Filter

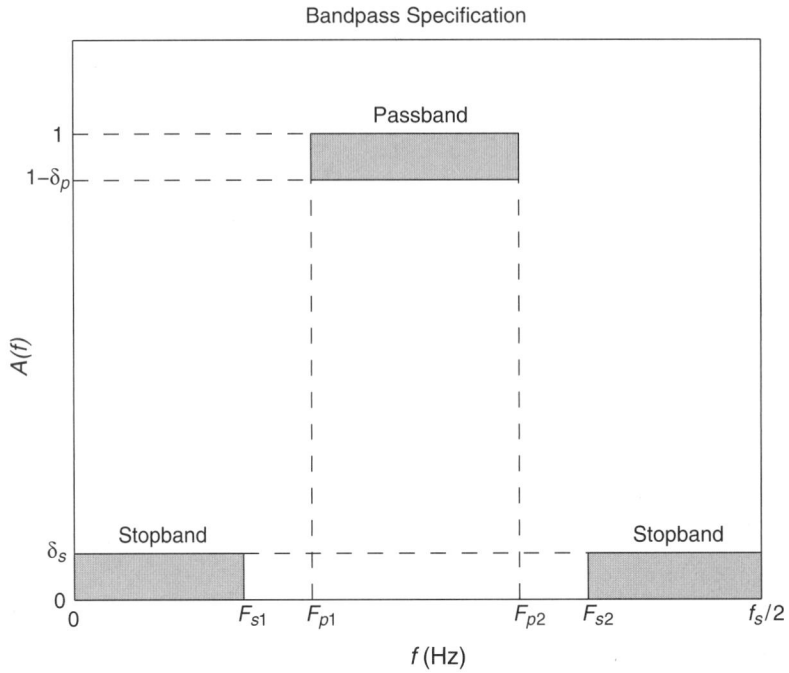

FIGURE 5.9: Linear Magnitude Response Specifications for a Bandstop Filter

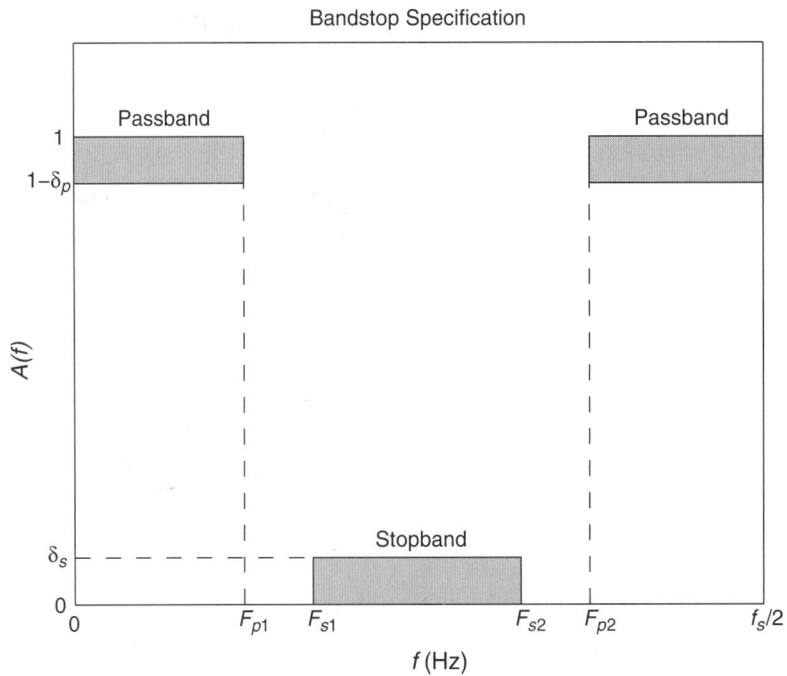

Example 5.1   **Linear Design Specifications**

As a simple illustration of design specifications for a desired magnitude response, consider the following first-order IIR filter.

$$H(z) = \frac{.5(1-c)(1+z^{-1})}{1-cz^{-1}}$$

Recasting $H(z)$ in terms of positive powers of z, we have

$$H(z) = \frac{.5(1-c)(z+1)}{z-c}$$

Thus $H(z)$ has a zero at $z = -1$ and a pole at $z = c$. For the filter to be stable, it is necessary that the pole satisfy $|c| < 1$. Before we compute the complete frequency response, we can evaluate the filter gain at the two ends of the spectrum. Setting $f = 0$ in $z = \exp(2\pi f T)$ yields $z = 1$. Thus the low-frequency or DC gain of the filter is

$$A(0) = |H(z)|_{z=1}$$
$$= \frac{.5(1-c)2}{1-c}$$
$$= 1$$

Next, setting $f = f_s/2$ in $z = \exp(2\pi f T)$ yields $z = -1$. Thus the high-frequency gain of the filter is

$$A(f_s/2) = |H(z)|_{z=-1}$$
$$= \frac{.5(1-c)0}{-1-c}$$
$$= 0$$

It follows that $H(z)$ is a lowpass filter with a passband gain of one. To make the example specific, suppose $c = .5$. Then from (5.2.1) the frequency response of this IIR filter is

$$H(f) = H(z)|_{z=\exp(j2\pi f T)}$$
$$= \frac{.25[\exp(j2\pi f T) + 1]}{\exp(j2\pi f T) - .5}$$
$$= \frac{.25[(\cos(2\pi f T) + 1) + j\sin(2\pi f T)]}{(\cos(2\pi f T) - .5) + j\sin(2\pi f T)}$$

The magnitude response of the lowpass filter is then

$$A(f) = |H(f)|$$
$$= \frac{.25\sqrt{[\cos(2\pi f T) + 1]^2 + \sin^2(2\pi f T)}}{\sqrt{[\cos(2\pi f T) - .5]^2 + \sin^2(2\pi f T)}}$$

For this simple first-order filter, suppose the cutoff frequencies for the transition band are taken to be

$$F_p = .1 f_s$$
$$F_s = .4 f_s$$

The magnitude response decreases monotonically in this case. Consequently, the passband ripple satisfies $1 - \delta_p = A(F_p)$ or

$$\delta_p = 1 - A(F_p)$$
$$= 1 - \frac{.25\sqrt{[\cos(.2\pi) + 1]^2 + \sin^2(.2\pi)}}{\sqrt{[\cos(.2\pi) - .5]^2 + \sin^2(.2\pi)}}$$
$$= .2839$$

FIGURE 5.10:
Magnitude Response
of First-order IIR
Lowpass Filter

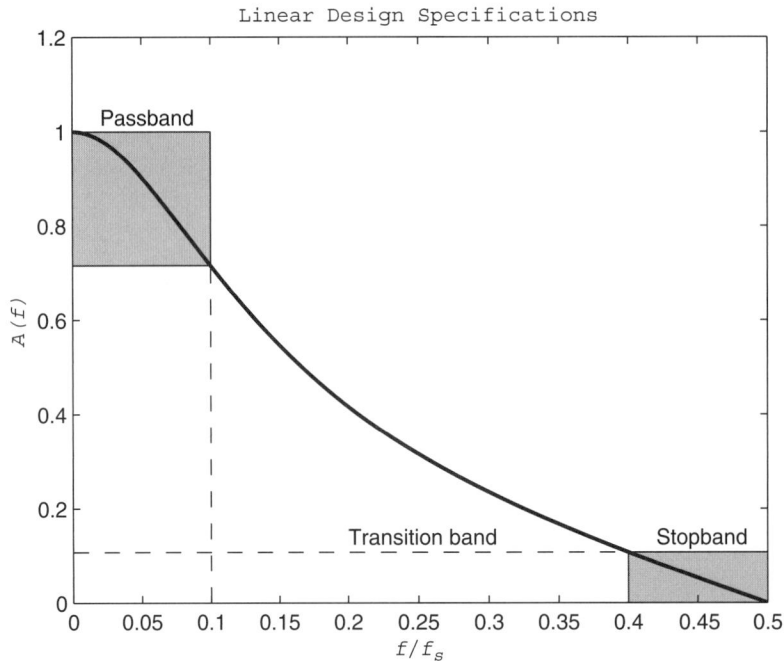

Linear Design Specifications

Similarly, from Figure 5.6, the stopband attenuation satisfies

$$\delta_s = A(F_s)$$

$$= \frac{.25\sqrt{[\cos(.8\pi) + 1]^2 + \sin^2(.8\pi)}}{\sqrt{[\cos(.8\pi) - .5]^2 + \sin^2(.8\pi)}}$$

$$= .1077$$

Normalized
frequency
A plot of the magnitude response of the first-order filter is shown in Figure 5.10. For convenience, *normalized frequency* $\hat{f} = f/f_s$ is used as the independent variable in this case. For this filter, the passband ripple, the stopband attenuation, and the transition bandwidth are all relatively large because this is the lowest-order IIR filter possible.

5.2.2 Logarithmic Design Specifications (dB)

The filter specifications in Figure 5.6 through Figure 5.9 are referred to as *linear* specifications because they are applied to the actual value of $A(f)$. It is also common to use a logarithmic *Decibel* specification that represents the value of the magnitude response using the *decibel* or dB scale.

$$A(f) \triangleq 10\log_{10}\{|H(f)|^2\} \text{ dB} \qquad (5.2.6)$$

A logarithmic design specification for the magnitude response of a lowpass filter is shown in Figure 5.11. Note that the passband ripple in dB is A_p, and stopband attenuation in dB is A_s. The dB scale is useful to show the degree of attenuation in the stopband. The lowpass specifications in Figure 5.6 and Figure 5.11 are equivalent. Using (5.2.6), we find the logarithmic

FIGURE 5.11:
Logarithmic
Magnitude Response
Specifications for a
Lowpass Filter

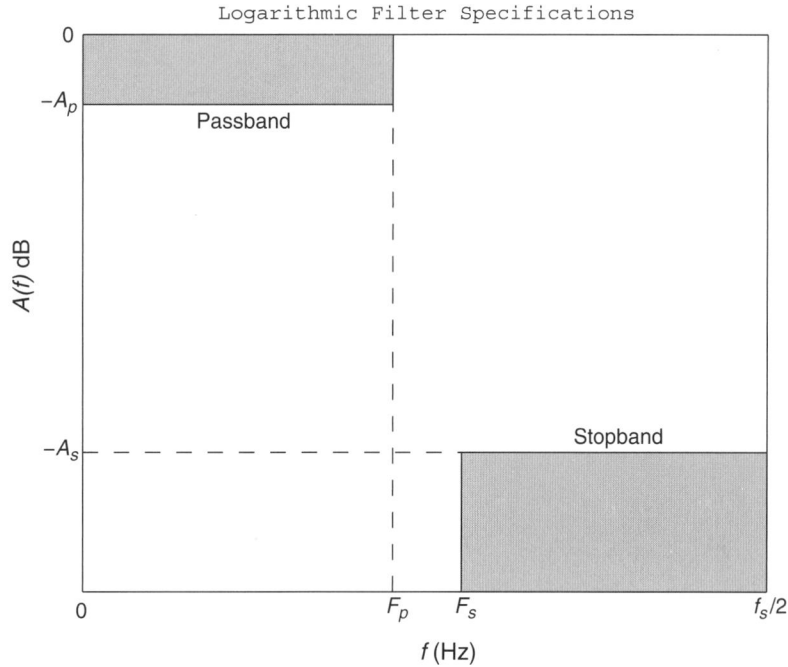

Logarithmic Filter Specifications

Linear to logarithmic specifications

specifications can be expressed in terms of the linear specifications as

$$A_p = -20\log_{10}(1 - \delta_p) \text{ dB} \qquad (5.2.7a)$$
$$A_s = -20\log_{10}(\delta_s) \text{ dB} \qquad (5.2.7b)$$

Logarithmic to linear specifications

Similarly, if we solve (5.2.7) for δ_p and δ_s, the linear specifications can be expressed in terms of the logarithmic specifications as

$$\delta_p = 1 - 10^{-A_p/20} \qquad (5.2.8a)$$
$$\delta_s = 10^{-A_s/20} \qquad (5.2.8b)$$

Example 5.2   **Logarithmic Design Specifications**

To facilitate a comparison of the two types of filter design specifications, consider the following first-order IIR system.

$$H(z) = \frac{.25(1 + z^{-1})}{1 - .5z^{-1}}$$

This filter, which was considered in Example 5.1, has a passband cutoff frequency of $F_p = .1f_s$ and a stopband cutoff frequency of $F_s = .4f_s$. Using (5.2.7a), and the linear passband ripple from Example 5.1, we arrive at the following equivalent passband ripple in dB.

$$A_p = -20\log_{10}(1 - .2839)$$
$$= 2.901 \text{ dB}$$

FIGURE 5.12:
Magnitude Response
of First-order IIR
Lowpass Filter Using
the dB Scale

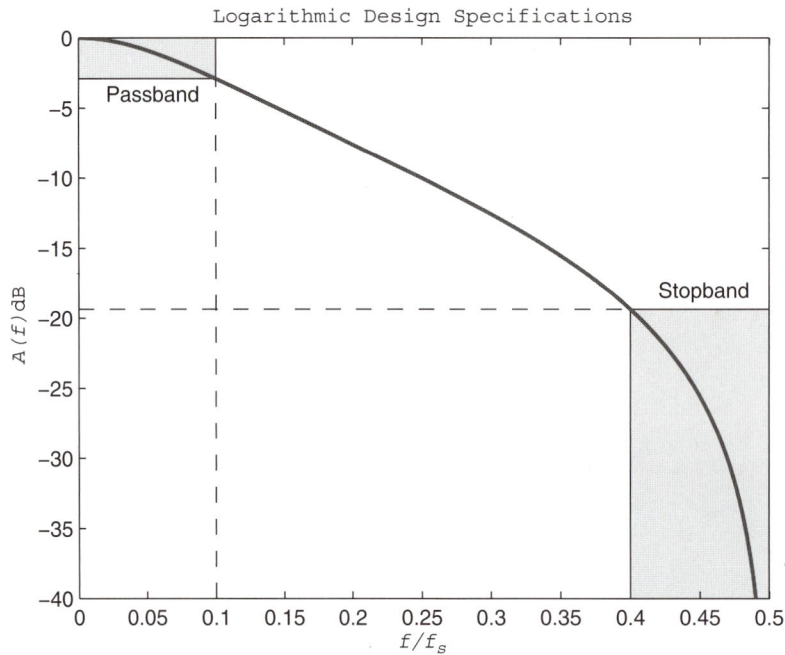

Next, if we use (5.2.7b) and the linear stopband attenuation from Example 5.1, the equivalent logarithmic stopband attenuation is

$$A_s = -20\log_{10}(.1077)$$
$$= 9.358 \text{ dB}$$

A plot of the magnitude response in dB can be obtained by running *exam5_2* with the results shown in Figure 5.12. Note that the display range had to be clipped (at -40 dB), because $A(f_s/2) = 0$ which means that $A(f_s/2) = -\infty$ dB.

5.3   Linear-phase and Zero-phase Filters

5.3.1   Linear Phase

The filter design specifications discussed thus far are specifications on the desired magnitude response of the filter. It is also possible to design filters with prescribed phase responses. To illustrate the type of phase responses that are desirable and achievable, consider the following analog system.

$$H_a(s) = \exp(-\tau s) \qquad (5.3.1)$$

Delay line Recall from the properties of the Laplace transform (Appendix 1) that $H_a(s)$ represents an ideal *delay line* of delay τ. Thus the input is delayed by τ but is not otherwise distorted by the system. The frequency response of the delay line is $H_a(f) = \exp(-j2\pi\tau f)$. Consequently, if we regard the delay line as a filter, it is an allpass filter with $A(f) = 1$ and with a phase response of

$$\phi_a(f) = -2\pi\tau f \qquad (5.3.2)$$

The linear phase response in (5.3.2) represents a pure delay. To interpret the meaning of a nonlinear phase response, it is helpful to introduce the concept of group delay.

DEFINITION

5.1: Group Delay

Let $\phi(f)$ be the phase response of a linear system. The *group delay* of the system is denoted $D(f)$ and defined

$$D(f) \triangleq \left(\frac{-1}{2\pi}\right) \frac{d\phi(f)}{df}$$

The group delay is the negative of the slope of the phase response, scaled by $1/(2\pi)$. Observe from (5.3.2), that the group delay of a delay line is simply $D(f) = \tau$. That is, for a delay line the group delay specifies the amount by which the signal is delayed as it is processed by the filter. For most filters, the group delay $D(f)$ is not constant. In these cases, we can interpret $D(f)$ as the amount by which the spectral component at frequency f gets delayed as it is processed by the filter. Given the notion of group delay, we are now in a position to define what is meant by a generalized linear-phase digital filter.

DEFINITION

5.2: Linear-phase Filter

Let F_z denote the set of frequencies at which the magnitude response is $A(f) = 0$. A digital filter $H(z)$ is a *linear- phase filter* if and only if there exists a constant τ such that

$$D(f) = \tau, \quad f \notin F_z$$

A digital filter is a linear-phase filter if the group delay is constant except possibly at frequencies at which the magnitude response is zero. These spectral components do not appear in the filter output, so the group delay is not meaningful for $f \in F_z$. Typically F_z is a finite isolated set of frequencies and often F_z is the empty set. Note that Definition 5.1 and Definition 5.2 imply the following general form for a linear phase response.

$$\phi(f) = \alpha + \beta(f) - 2\pi\tau f \tag{5.3.3}$$

Here α is a constant and $\beta(f)$ is piecewise constant with jump discontinuities permitted at the frequencies F_z at which $A(f) = 0$.

Amplitude Response

An alternative way to characterize a linear-phase filter is in terms of the frequency response which can be written in the following general form.

$$H(f) = A_r(f) \exp[j(\alpha - 2\pi\tau f)] \tag{5.3.4}$$

Amplitude response

Here the factor $A_r(f)$ is real, but it can be both *positive and negative*. It is referred to as the *amplitude response* of $H(z)$. This is to distinguish it from the magnitude response $A(f)$, which is never negative. Taking the magnitudes of both sides of (5.3.4) we see that the amplitude response and the magnitude response are related to one another as follows.

$$|A_r(f)| = A(f) \tag{5.3.5}$$

The points at which $A_r(f) = 0$ are the points F_z, where the phase can abruptly change by π as $A_r(f)$ changes sign. Thus the piecewise-constant function $\beta(f)$ in (5.3.3) jumps between zero and π each time the amplitude response $A_r(f)$ changes sign.

Bessel filter

A linear phase characteristic in the passband can be achieved in the analog domain by an IIR Bessel filter. However, the linear-phase feature does not survive the analog-to-digital transformation. Consequently, it is better to start with a digital FIR filter as follows.

$$H(z) = \sum_{i=0}^{m} b_i z^{-i} \tag{5.3.6}$$

For an FIR filter, there is a simple symmetry condition on the coefficients that guarantees a linear phase response. First we illustrate this with a special case.

Example 5.3

Even Symmetry

Consider an FIR filter of order $m = 4$ having the following transfer function.

$$H(z) = c_0 + c_1 z^{-1} + c_2 z^{-2} + c_1 z^{-3} + c_0 z^{-4}$$

Recall that for an FIR filter, $h(k) = b_k$ for $0 \le k \le m$. Thus the impulse response $h = [c_0, c_1, c_2, c_1, c_0]^T$, in this case $h(k)$ exhibits even symmetry about the midpoint $k = m/2$. The frequency response of this filter, in terms of $\theta = 2\pi f T$, is

$$
\begin{aligned}
H(f) &= H(z)|_{z=\exp(j\theta)} \\
&= c_0 + c_1 \exp(-j\theta) + c_2 \exp(-j2\theta) + c_1 \exp(-j3\theta) + c_0 \exp(-j4\theta) \\
&= \exp(-j2\theta)[c_0 \exp(j2\theta) + c_1 \exp(j\theta) + c_2 + c_1 \exp(-j\theta) + c_0 \exp(-2j\theta)]
\end{aligned}
$$

Combining terms with identical coefficients, and using Euler's identity, we have

$$
\begin{aligned}
H(f) &= \exp(-j2\theta)\{c_0[\exp(j2\theta) + \exp(-j2\theta)] + c_1[\exp(j\theta) + \exp(-j\theta)] + c_2\} \\
&= \exp(-j2\theta)\{[2c_0 \cos(2\theta) + 2c_1 \cos(\theta) + c_2]\} \\
&= \exp(-j4\pi f T)A_r(f)
\end{aligned}
$$

Comparing with (5.3.4), we see that this is a linear-phase system with phase offset $\alpha = 0$ and delay $\tau = 2T$. In this instance, the amplitude response $A_r(f)$ is the following *even* function that may be positive or negative.

$$A_r(f) = 2c_0 \cos(4\pi f T) + 2c_1 \cos(2\pi f T) + c_2$$

The even symmetry of $h(k)$ about the midpoint $k = m/2$ is one way to obtain a linear-phase filter. Another approach is to use odd symmetry of $h(k)$ about $k = m/2$, as can be seen from the following example.

Example 5.4

Odd Symmetry

Consider an FIR filter of order $m = 4$ having the following transfer function.

$$H(z) = c_0 + c_1 z^{-1} - c_1 z^{-3} - c_0 z^{-4}$$

Recalling that $h(k) = b_k$, we see that for this filter the impulse response $h = [c_0, c_1, 0, -c_1, -c_0]^T$, exhibits odd symmetry about the midpoint $k = m/2$. The frequency response of this filter, in terms of $\theta = 2\pi f T$, is

$$
\begin{aligned}
H(f) &= H(z)|_{z=\exp(j\theta)} \\
&= c_0 + c_1 \exp(-j\theta) - c_1 \exp(-j3\theta) - c_0 \exp(-j4\theta) \\
&= \exp(-j2\theta)[c_0 \exp(j2\theta) + c_1 \exp(j\theta) - c_1 \exp(-j\theta) - c_0 \exp(-2j\theta)]
\end{aligned}
$$

Combining terms with identical coefficients, and using Euler's identity, we have

$$
\begin{aligned}
H(f) &= \exp(-j2\theta)\{c_0[\exp(j2\theta) - \exp(-j2\theta)] + c_1[\exp(j\theta) - \exp(-j\theta)]\} \\
&= j \exp(-j2\theta)[2c_0 \sin(2\theta) + 2c_1 \sin(\theta)] \\
&= \exp[j(\pi/2 - 4\pi f T)]A_r(f)
\end{aligned}
$$

Comparing with (5.3.4), we see that this is a linear-phase system with phase offset $\alpha = \pi/2$ and delay $\tau = 2T$. In this instance, the amplitude response $A_r(f)$, is the following *odd* function that may be positive or negative.

$$A_r(f) = 2c_0 \sin(4\pi f T) + 2c_1 \sin(2\pi f T)$$

The results of Examples 5.1 and 5.2 can be generalized. In particular, it is possible to show by examining additional cases that an FIR filter of order m is a linear-phase filter when it satisfies the following symmetry condition.

Let $H(z)$ be an FIR filter of order m. Then $H(z)$ is a linear-phase filter with group delay $D(f) = mT/2$ if and only if the impulse response $h(k)$ satisfies the symmetry condition

$$h(k) = \pm h(m-k), \quad 0 \le k \le m$$

The symmetry constraint in Proposition 5.2 says that the impulse response must exhibit *even symmetry* about the midpoint $k = m/2$ when the plus sign is used, or *odd symmetry* about the midpoint when the minus sign is used. We can further decompose the symmetry condition into cases where the filter order m is even or odd, and this results in a total of four linear-phase filter types as summarized in Table 5.1.

Observe that when the filter order m is even, there is a middle sample $h(m/2)$, about which the impulse response exhibits either even or odd symmetry. When m is odd, the symmetry is about a point that is midway between a pair of samples. The impulse responses of the four linear-phase filter types are shown in Figure 5.13 for the case $h(k) = k^2$ for $k \le \text{floor}(m/2)$. Note that the middle sample of the type 3 filter must satisfy $h(m/2) = -h(m/2)$, which means that $h(m/2) = 0$ in this case. For filters with even symmetry, the *phase offset* is $\alpha = 0$, and the amplitude response in (5.3.4) is an even function. For filters with odd symmetry, $\alpha = \pi/2$, and the amplitude response is an odd function. For type 1 and type 2 filters, the impulse responses are the numerical equivalent of *palindromes*, words that are the same whether they are spelled forwards or backwards.

The symmetry condition that guarantees a linear phase response also imposes certain constraints on the zeros of an FIR filter. To see this we start with (5.3.6) and apply Proposition 5.2 with a change of variable.

$$H(z) = \sum_{i=0}^{m} h(i)z^{-i}$$

$$= \pm \sum_{i=0}^{m} h(m-i)z^{-i}$$

$$= \pm \sum_{k=m}^{0} h(k)z^{-(m-k)}, \quad k = m-i$$

$$= \pm z^{-m} \sum_{k=0}^{m} h(k)z^{k} \tag{5.3.7}$$

TABLE 5.1: ▶ Linear-phase FIR Filters, $D(f) = mT/2$

Type	Midpoint Symmetry $h(k)$	Filter Order m	Phase Offset α	Amplitude Response $A_r(f)$	End-point Zeros	Passband
1	Even	Even	0	Even	None	All
2	Even	Odd	0	Even	$H(-1)=0$	Lowpass
3	Odd	Even	$\pi/2$	Odd	$H(\pm 1)=0$	Bandpass
4	Odd	Odd	$\pi/2$	Odd	$H(1)=0$	Highpass

FIGURE 5.13: Impulse Responses of Four Types of Generalized Linear-phase Filters

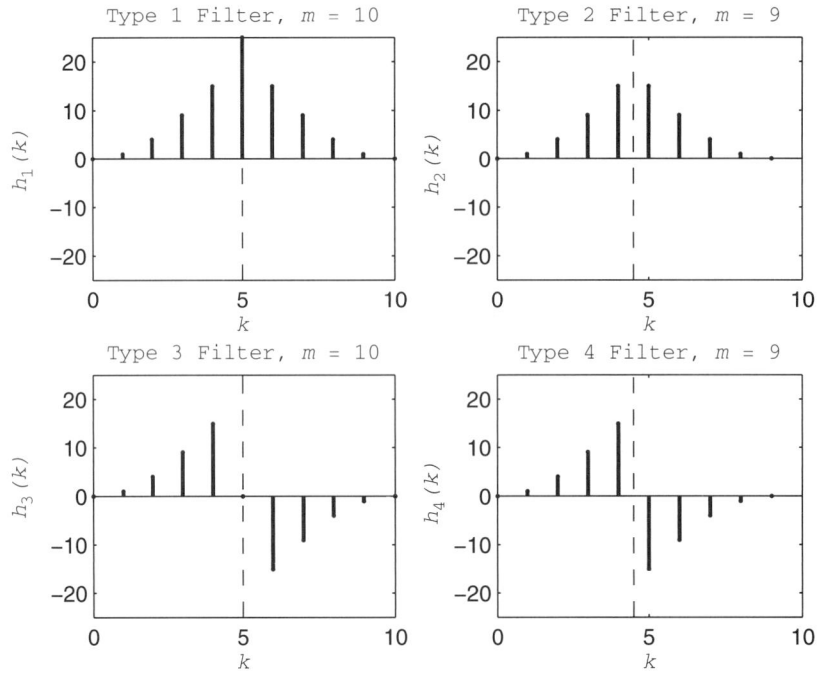

Symmetry condition

The final summation on the right-hand side of (5.3.7) is simply $H(z^{-1})$. Thus the linear-phase symmetry condition, expressed in terms of the transfer function, is

$$H(z) = \pm z^{-m} H(z^{-1}) \tag{5.3.8}$$

Here the plus sign in (5.3.8) applies to filter types 1 and 2 that exhibit even symmetry about the midpoint $m/2$, and the minus sign applies to filter types 3 and 4 that exhibit odd symmetry. The frequency domain symmetry condition (5.3.8) places constraints on the locations of the zeros of $H(z)$. For example, consider a type 2 filter with even symmetry and odd order. Using the plus sign and evaluating (5.3.8) at $z = -1$ yields $H(-1) = -H(-1)$. Therefore every type 2 linear-phase filter has a zero at $z = -1$. Consequently, the passband of the a type 2 filter is lowpass or possibly bandpass.

Next consider a type 3 filter with odd symmetry and even order. Using the minus sign and evaluating (5.3.8) at $z = -1$ yields $H(-1) = -H(-1)$, which means that $z = -1$ is also a zero of the type 3 filter. Evaluating (5.3.8) with the minus sign at $z = 1$ yields $H(1) = -H(1)$. Hence a type 3 filter has zeros at $z = \pm 1$. It follows that the passband of a type 3 filter is bandpass.

Finally, consider a type 4 filter where both the symmetry and the order are odd. Using the minus sign and evaluating (5.3.8) at $z = 1$ yields $H(1) = -H(1)$, which means that $z = 1$ is a zero of $H(z)$. Thus a type 4 filter has a passband that is highpass or possibly bandpass. The zeros and passbands of the four linear-phase filter types are summarized in the last two columns of Table 5.1. Note that the type 1 filter (even symmetry, even order) is the most general in that it does not contain zeros at either end of the frequency range.

In addition to the zeros at the two ends of the frequency range, the constraint on $H(z)$ in (5.3.8) also implies that complex zeros must appear in certain patterns. Let $z = r \exp(j\phi)$ be a complex zero with $r > 0$. Then from (5.3.8), $z^{-1} = r^{-1} \exp(-j\phi)$ must also be a zero of $H(z)$.

Linear-phase zeros

Furthermore, if the coefficient vector b is real, then zeros must occur in complex-conjugate pairs. Hence for $r \neq 0$, the zeros will appear in groups of four and satisfy the following

FIGURE 5.14: Poles and Zeros of the Type 1 FIR Linear-phase Filter of Order $m = 6$

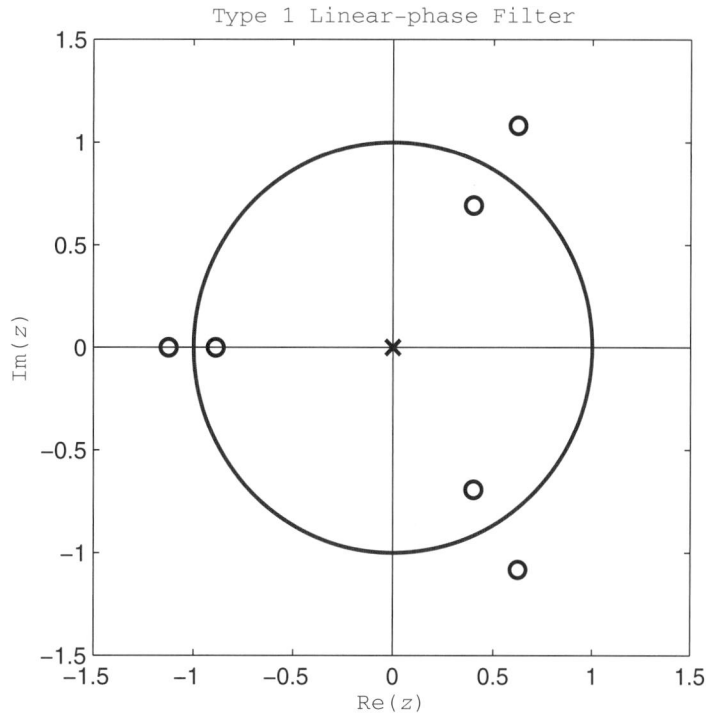

Type 1 Linear-phase Filter

reciprocal symmetry.

$$Q = \{r \exp(\pm j\phi), r^{-1} \exp(\mp j\phi)\} \tag{5.3.9}$$

When the zeros are real, this yields $\phi = 0$ or $\phi = \pi$, and the set Q in (5.3.9) reduces to a pair of reciprocal real zeros. A pole-zero plot for a type 1 FIR linear-phase filter showing a typical arrangement of both real and complex zeros is shown in Figure 5.14. Since it is an FIR filter, the poles are all located at the origin.

The importance of linear-phase filters is that they do not distort the signals that are processed by the filter. This is significant in applications, such as speech or music, where the effects of phase distortion can be noticeable. To verify the effect of a linear-phase filter on an individual spectral component, suppose $x(k) = \cos(2\pi f k T)$. For a filter with group delay $D(f) = mT/2$, the phase response is $\phi(f) = -\pi m T f$. Thus from Proposition 3.2, the steady-state output is

$$
\begin{aligned}
y_{ss}(k) &= A(f)\cos[2\pi f k T + \phi(f)] \\
&= A(f)\cos(2\pi f k T - \pi m T f) \\
&= A(f)\cos[2\pi f(k - m/2)T] \\
&= A(f)x(k - m/2)
\end{aligned}
\tag{5.3.10}
$$

Consequently, the spectral component at frequency f is scaled by $A(f)$ and delayed by $D(f) = mT/2$, but is not otherwise affected by the filter. A summary of the steady-state input-output characteristics of linear-phase filters is shown in Figure 5.15.

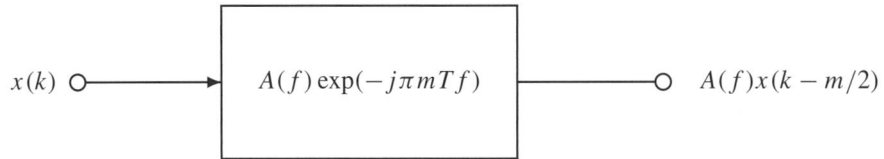

FIGURE 5.15: Steady-state Output of a Linear-phase FIR Filter with Group Delay $D(f) = mT/2$ when $x(k) = \cos(2\pi fkT)$

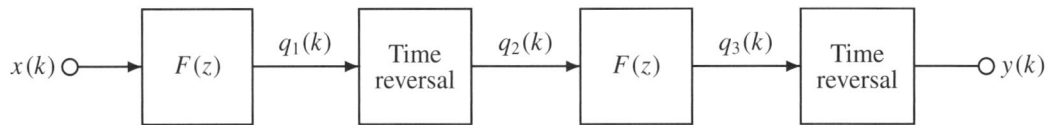

FIGURE 5.16: Noncausal Zero-phase FIR Filter of Order $2m$

5.3.2 Zero-phase Filters

Zero-phase filter

Suppose the frequency response $H(f)$ is real and non-negative. Since the imaginary part is identically zero, this means that $H(z)$ is a *zero-phase* filter.

$$\phi(f) = 0, \quad 0 \le |f| \le f_s/2 \tag{5.3.11}$$

A zero-phase filter is a linear-phase filter with a group delay of zero. Just as with an ideal frequency-selective filter, a zero-phase filter can not be realized with a causal system. However, if we relax the causality constraint and focus in input signals of finite length, then there is a relatively simple procedure for implementing a zero-phase filter. The key is the time-reversal property of the DFT. Recall from Table 4.8 that for a real N-point signal $x(k)$ with periodic extension $x_p(k)$,

Time reversal property

$$\text{DFT}\{x_p(-k)\} = X^*(i) \tag{5.3.12}$$

For $0 \le k < N$, the time-reversed periodic extension of $x(k)$ is just $x(N-k)$, where $x(N) = x(0)$. Suppose $F(z)$ is a linear-phase FIR filter of order m. Then $F(z)$ will have frequency response of $F(i)$ and a group delay of $D(f) = mT/2$. Let the input to the filter be $x(k)$ and the output be $q_1(k)$, as shown in Figure 5.16. Then in the frequency domain

$$Q_1(i) = F(i)X(i) \tag{5.3.13}$$

Next suppose the filter output $q_1(k)$ is time reversed to produce a new signal $q_2(k) = q_1(N-k)$. Using (5.3.13) and the time-reversal property in (5.3.12)

$$Q_2(i) = F^*(i)X^*(i) \tag{5.3.14}$$

The time-reversed output is then used as in input to a second copy of $F(z)$, as shown in Figure 5.16. From (5.3.14), the resulting output is

$$Q_3(i) = F(i)F^*(i)X^*(i) \tag{5.3.15}$$

Note that the first use of $F(z)$ caused a delay $mT/2$ in $q_1(k)$. By reversing $q_1(k)$ to produce $q_2(k)$ and then processing $q_2(k)$ with $F(z)$ again, this has the effect of causing a time advance of $mT/2$, thus cancelling the time delay. The resulting output $q_3(k)$ is then time-reversed using $y(k) = x_3(N-k)$ to cancel the original time reversal. From (5.3.15) and the time-reversal property in (5.3.12), this yields

$$Y(i) = F^*(i)F(i)X(i) \qquad (5.3.16)$$

Noncausal filter Finally, $F^*(i)F(i) = |F(i)|^2$. Thus the noncausal system in Figure 5.16 has the following real non-negative frequency response whose phase response is $\phi(i) = 0$.

$$H(i) = |F(i)|^2 \qquad (5.3.17)$$

Since the FIR filter $F(z)$ processes the signal twice, the noncausal filter $H(z)$ is of order $2m$. To design and implement a noncausal FIR filter with zero phase and a prescribed magnitude response, the following algorithm can be used.

ALGORITHM

5.1: Zero-phase filter

1. Pick a desired magnitude response $A_d(i) \geq 0$ for $0 \leq i < N$.
2. Using techniques from Chapter 6, design a linear-phase filter $F(z)$ of order m with magnitude response $A(i) = \sqrt{A_d(i)}$.
3. Implement the noncausal filter $H(z)$ as follows for $0 \leq k < N$ where $q_i(N) = q_i(0)$ for $1 \leq i \leq 3$.

$$q_1(k) = \sum_{i=0}^{m} f_i x(k-i)$$

$$q_2(k) = q_1(N-k)$$

$$q_3(k) = \sum_{i=0}^{m} f_i q_2(k-i)$$

$$y(k) = q_3(N-k)$$

Example 5.5

Zero-phase Filter

To illustrate a filter with zero phase shift, suppose $f_s = 200$ Hz and consider an input that includes two sinusoidal components with frequencies $F_0 = 20$ Hz and $F_1 = 60$ Hz.

$$x_1(k) = 2\cos(2\pi F_0 kT)$$
$$x_2(k) = -3\sin(2\pi F_1 kT)$$
$$x(k) = x_1(k) + x_2(k)$$

Using design techniques presented in Chapter 6, suppose $F(z)$ consists of a lowpass FIR filter of order $m = 30$ with cutoff frequency

$$F_c = \frac{F_0 + F_1}{2}$$

The noncausal filter $H(z)$ should then block component $x_2(k)$ while it passes component $x_1(k)$ undistorted and with no phase shift. Plots of $x(k)$ and the filter output $y(k)$ generated by *exam5-5* are shown in Figure 5.17.

Notice that after a brief startup transient, the output $y(k)$ closely matches component $x_1(k)$ as expected. However, in addition to the startup transient there is also a transient of length approximately $m/2$ at the end of the time signal. This is a characteristic zero-phase filters. The ending transient is actually the startup transient of the second stage in Figure 5.16 where the signal is run backward through $F(z)$ to cancel the time delay. If a suitable initial condition is selected for stage 2, the second transient can be reduced. Alternatively, if the signal length

FIGURE 5.17: Input and Output of a Noncausal Zero-phase FIR Filter of Order $2m = 60$

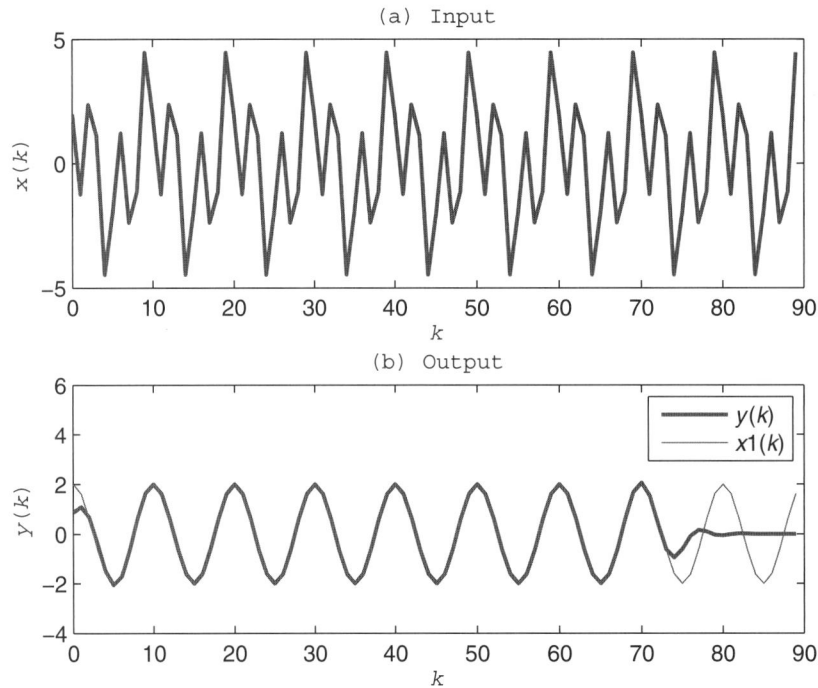

(a) Input

(b) Output

N is large in comparison with the filter order m, then the transients will occupy only a small fraction of the filter output signal.

FDSP Functions

The FDSP toolbox contains the following function that implements Algorithm 5.1 to perform zero-phase filtering. If the MATLAB Signal Processing Toolbox is available, then the function *filtfilt* can be used to perform zero-phase filtering.

```
% F_ZEROPHASE Compute the output of noncausal zero-phase filter
%
% Usage:
%        y = f_zerophase(b,x);
% Pre:
%        b = array of length m+1 containing the FIR filter coefficients
%        x = array of length N containing the input samples
% Post:
%        y = array of length N containing the output samples
```

5.4   Minimum-phase and Allpass Filters

Every digital filter with a rational transfer function can be expressed as a product of two specialized filters. The first one is called a minimum-phase filter, and the second is an allpass filter.

5.4.1 Minimum-phase Filters

The magnitude response, by itself, does not provide enough information to completely specify a filter. Indeed, among IIR filters having m zeros, there are up to 2^m distinct filters, each having an identical magnitude response $A(f)$. To see this, recall that a rational IIR transfer function can be written as a ratio of two polynomials in z.

$$H(z) = \frac{b(z)}{a(z)} \tag{5.4.1}$$

Since the polynomial $b(z)$ has real coefficients, the complex conjugate $b^*(z)$ can be obtained by replacing z by z^*. On the unit circle, $z = \exp(j2\pi fT)$ which means that $z^* = z^{-1}$. Thus the square of the magnitude response can be expressed as follows.

$$
\begin{aligned}
A^2(f) &= \left.\frac{|b(z)|^2}{|a(z)|^2}\right|_{z=\exp(j2\pi fT)} \\
&= \left.\frac{b(z)b^*(z)}{|a(z)|^2}\right|_{z=\exp(j2\pi fT)} \\
&= \left.\frac{b(z)b(z^{-1})}{|a(z)|^2}\right|_{z=\exp(j2\pi fT)}
\end{aligned} \tag{5.4.2}
$$

Next suppose $H(z)$ has a zero at $z = c$ with $c \neq 0$. Then $b(z)$ and $b(z^{-1})$ can be written in partially factored form as

$$b(z) = (z - c)b_0(z) \tag{5.4.3a}$$
$$b(z^{-1}) = (z^{-1} - c)b_0(z^{-1}) \tag{5.4.3b}$$

If the factor $z - c$ in $b(z)$ is interchanged with the corresponding factor $z^{-1} - c$ in $b(z^{-1})$, the product $b(z)b(z^{-1})$ does not change which means $A^2(f)$ in (5.4.2) does not change. Replacing the factor $z - c$ with the factor $z^{-1} - c$ is equivalent to replacing the zero at $z = c$ with a zero at its reciprocal $z = c^{-1}$, and scaling by a constant. To determine the constant, first note that

$$z^{-1} - c = z^{-1}(1 - cz) = -cz^{-1}(z - c^{-1}) \tag{5.4.4}$$

When the magnitude response of $H(z)$ is computed, we evaluate $H(z)$ along the unit circle, which means $|z^{-1}| = 1$. Thus a new numerator polynomial which does not change the magnitude response of $H(z)$ is as follows

$$B(z) = \frac{-c(z - c^{-1})b(z)}{(z - c)} \tag{5.4.5}$$

Since this can be done with any of the m zeros of $b(z)$; there are up to 2^m distinct combinations of zeros of $H(z)$ that all yield filters with identical magnitude responses. The differences between these filters lie in their phase responses $\phi(f)$.

DEFINITION

5.3: Minimum-phase Filter

A digital filter $H(z)$ is a *minimum-phase* filter if and only if all of its zeros lie inside or on the unit circle. Otherwise, it is a *nonminimum-phase* filter.

Every IIR filter $H(z)$ can be converted to a minimum-phase filter with the same magnitude response by replacing the zeros outside the unit circle with their reciprocals. The term *minimum phase* arises from the fact that the net phase change of a minimum-phase filter, over the frequency range $[0, f_s/2]$, is

$$\phi(f_s/2) - \phi(0) = 0 \tag{5.4.6}$$

Minimum phase

Nonminimum-phase filters have at least one zero outside the unit circle. It can be shown (Proakis and Manolakis, 1988) that if $H(z)$ has p zeros outside the unit circle, then the net phase change is $\phi(f_s/2) - \phi(0) = -p\pi$. Consequently, among filters that have the same magnitude response, the minimum-phase filter is the filter that has the smallest amount of phase lag.

Example 5.6

Minimum-phase Filter

To illustrate how different filters can have identical magnitude responses, consider the following second-order IIR filter.

$$H_{00}(z) = \frac{2(z+.5)(z-.5)}{(z-.5)^2+.25}$$

This is a stable IIR filter with poles at $z = .5 \pm j.5$. It is also a minimum-phase filter because the zeros at $z = \pm.5$ are both inside the unit circle. There are three other IIR filters that have an identical magnitude response, which we obtain by using the reciprocal of the first zero, the second zero, or both and multiplying by the negative of the original zero, as in (5.4.5). Thus the transfer functions of other three filters are.

$$H_{10}(z) = \frac{(z+2)(z-.5)}{(z-.5)^2+.25}$$
$$H_{01}(z) = \frac{-(z+.5)(z-2)}{(z-.5)^2+.25}$$
$$H_{11}(z) = \frac{-.5(z+2)(z-2)}{(z-.5)^2+.25}$$

Pole-zero plots of the four equivalent filters are shown in Figure 5.18. Only filter $H_{00}(z)$ is a minimum-phase filter. Since filter $H_{11}(z)$ has all of its zeros outside the unit circle, it is called a *maximum-phase* filter, while $H_{10}(z)$ and $H_{01}(z)$ are called *mixed-phase* filters.

FIGURE 5.18: Pole-zero Plots of Four Filters Having the Same $A(f)$

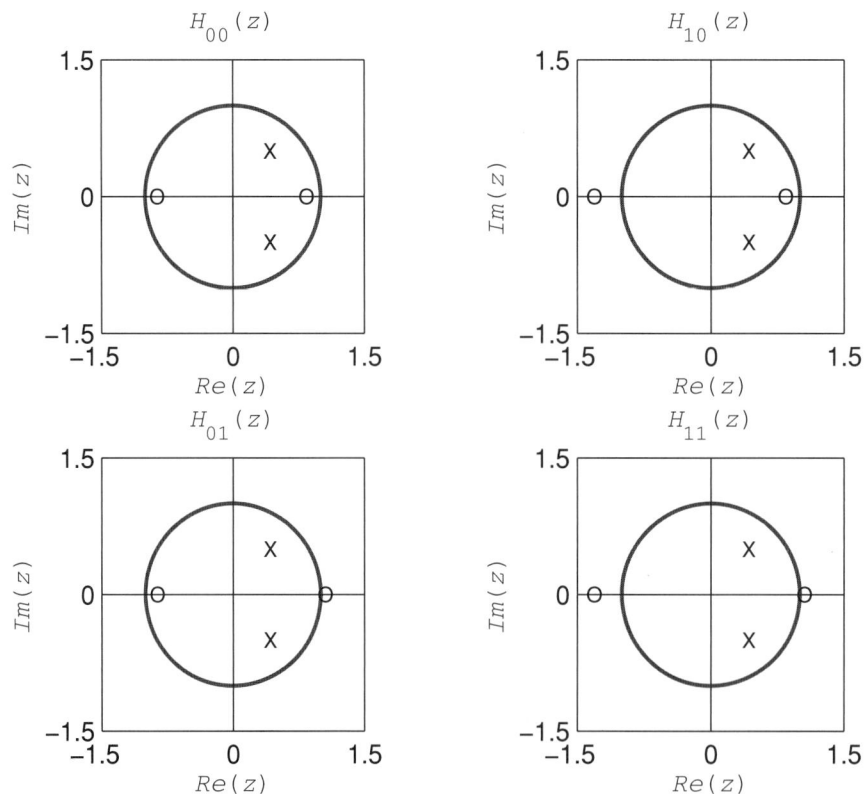

FIGURE 5.19: Identical Magnitude Responses of the Four Filters

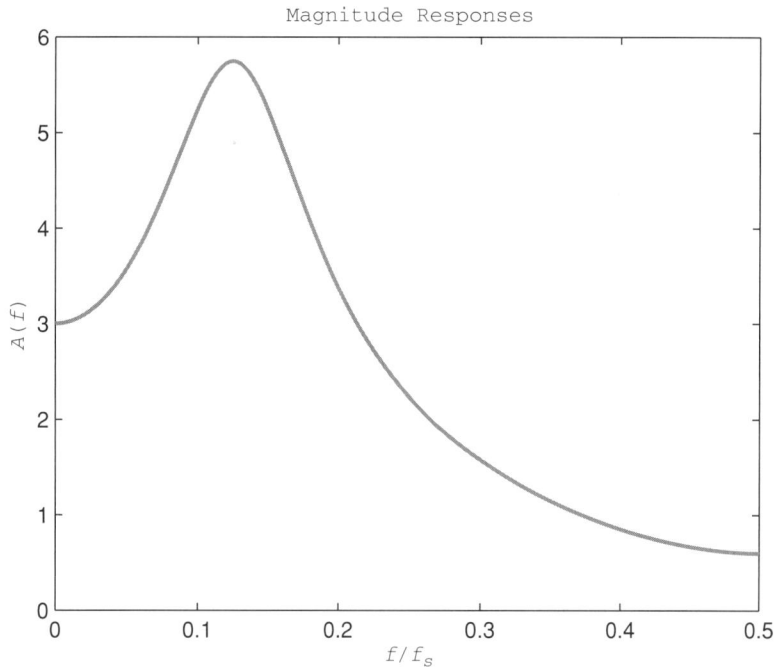

Magnitude Responses

FIGURE 5.20: The Phase Responses of the Four Filters

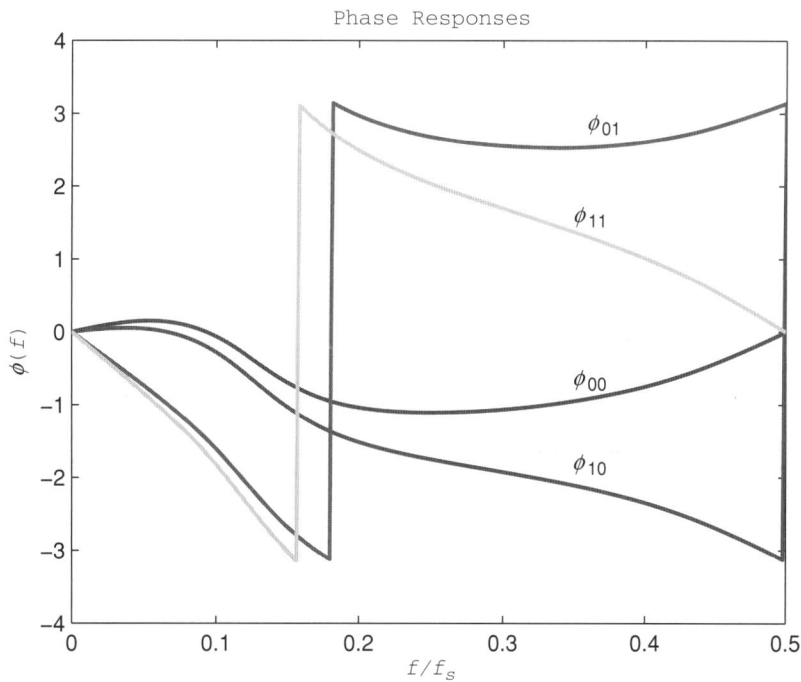

Phase Responses

Plots of the four magnitude responses are shown in Figure 5.19 where it is evident that they are all identical. However, the four phase responses are distinct from one another, as can be seen in the plot in Figure 5.20. Note that the net phase change for the minimum-phase filter is zero. For the two mixed-phase filters, $H_{10}(z)$ and $H_{01}(z)$, the net phase change is $-\pi$. Finally, the maximum-phase filter, $H_{11}(z)$, has a net phase change of -2π.

Every rational IIR filter can be converted to minimum-phase form by replacing each zero outside the unit circle by its reciprocal and scaling by the negative of the zero. If the original filter has a pair of complex conjugate zeros at $z = r \exp(\pm j\theta)$ with $r > 1$, then both zeros must be replaced by zeros at $z = r^{-1} \exp(\mp j\theta)$ in order for the coefficients of the new filter to remain real.

5.4.2 Allpass Filters

Another important class of IIR filters that can be used to provide phase compensation is the allpass filter. As the name implies, an *allpass* filter is a filter that passes all spectral components equally because it has a flat magnitude response.

DEFINITION

5.4: Allpass Filter

A digital filter $H(z)$ is an *allpass* filter if and only if it has the following magnitude response.

$$A(f) = 1, \quad 0 \le |f| \le f_s/2$$

Reflective structure

There are no special constraints on the phase response of an allpass filter. Allpass filters have transfer function coefficients with the following *reflective structure*.

$$H_{\text{all}}(z) = \frac{a_n + a_{n-1}z^{-1} + \cdots + z^{-n}}{1 + a_1 z^{-1} + \cdots + a_n z^{-n}} = \frac{z^{-n}a(z)}{a(z^{-1})} \tag{5.4.7}$$

Notice that the numerator polynomial is just the denominator polynomial $a(z^{-1})$, but with the coefficients reversed. To see how this gives a flat magnitude response, let $A(f)$ denote the magnitude response of the FIR filter $H(z) = a(z^{-1})$ corresponding to the denominator in (5.4.7). Since the magnitude response is even and $|z| = 1$ on the unit circle, the magnitude response of the filter in (5.4.7) is

$$A_{\text{all}}(f) = |H_{\text{all}}(z)|_{z=\exp(j2\pi fT)}$$

$$= \left| \frac{z^{-n}a(z)}{a(z^{-1})} \right|_{z=\exp(j2\pi fT)}$$

$$= \frac{|z^{-n}| \cdot |a(z)|}{|a(z^{-1})|} \bigg|_{z=\exp(j2\pi fT)}$$

$$= \frac{A(-f)}{A(f)}$$

$$= 1 \tag{5.4.8}$$

The process of converting a filter $H(z)$ to minimum-phase form can be thought of as multiplication of $H(z)$ by a transfer function $F(z)$. To illustrate, suppose $H(z)$ has a single zero at $z = c$ where c lies outside the unit circle. Then replacing this zero with one at $z = c^{-1}$ and multiplying by $-c$ is equivalent to multiplying $H(z)$ by

$$F(z) = \frac{-c(z - c^{-1})}{z - c} \tag{5.4.9}$$

If $z = c$ is the only zero of $H(z)$ outside the unit circle, then the minimum-phase version of $H(z)$ can be expressed as

$$H_{\min}(z) = F(z)H(z) \tag{5.4.10}$$

Next consider the characteristics of the transfer function $F(z)$ used to convert $H(z)$ to minimum-phase form. Note from (5.4.9) that

$$
\begin{aligned}
F(z) &= \frac{-c(z - c^{-1})}{z - c} \\
&= \frac{-cz + 1}{z - c} \\
&= \frac{-c + z^{-1}}{1 - cz^{-1}}
\end{aligned}
\tag{5.4.11}
$$

Comparing (5.4.11) with (5.4.7), we see that $F(z)$ is an allpass filter with $a = [1, -c]^T$. Although $F(z)$ was developed using only one zero outside the unit circle, the process can be repeated for any number of zeros, with the resulting allpass filter having factors similar to (5.4.9).

The magnitude response of the inverse of a filter is just the inverse of the magnitude response of the filter. Hence if $H_{all}(z) = F^{-1}(z)$, then $H_{all}(z)$ is an allpass filter. Furthermore, multiplying both sides of (5.3.10) on the left by $H_{all}(z)$, we conclude that every rational IIR transfer function $H(z)$ can be decomposed into the product of an allpass filter $H_{all}(z)$ times a minimum-phase filter $H_{min}(z)$.

PROPOSITION
....

5.3: Minimum-phase
Allpass Decomposition

Let $H(z)$ be a rational IIR transfer function, and let $H_{min}(z)$ be the minimum-phase form of $H(z)$. Then there exists a stable allpass filter H_{all} such that

$$
H(z) = H_{all}(z) H_{min}(z)
\tag{5.4.12}
$$

A block diagram of the decomposition into allpass and minimum-phase parts is shown in Figure 5.21. The minimum-phase part is the minimum-phase form of $H(z)$. Therefore the magnitude response of $H_{min}(z)$ is identical to the magnitude response of $H(z)$. The allpass part is the inverse of the system that transforms $H(z)$ into its minimum-phase form. Consequently, the allpass part $H_{all}(z)$ is always stable.

Another way to characterize an allpass filter is in terms of its poles and zeros. For each pole of $H_{all}(z)$ at $z = c$, there is a matching zero at its reciprocal $z = c^{-1}$. Thus allpass filters always have the same number of poles and zeros, with the poles and zeros forming reciprocal pairs. The following algorithm summarizes the steps needed to decompose a general IIR filter into its minimum-phase and allpass parts.

FIGURE 5.21:
Decomposition of
IIR Filter into
Allpass and
Minimum-phase
Parts

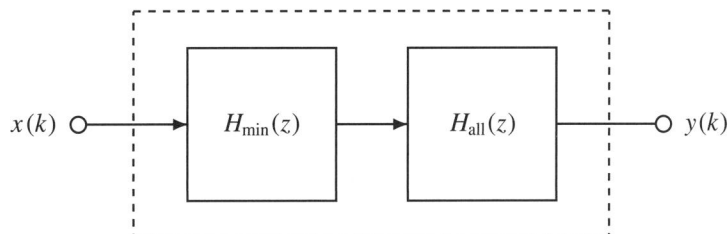

ALGORITHM

5.2: Minimum-phase
Allpass Decomposition

1. Set $H_{\min}(z) = H(z)$, and $H_{\text{all}}(z) = 1$. Factor the numerator polynomial of $H(z)$ as follows.

$$b(z) = b_0(z - z_1)(z - z_2) \cdots (z - z_m)$$

2. For $i = 1$ to m, do
{
    If $|z_i| > 1$ then compute

$$F(z) = \frac{-z_i z + 1}{z - z_i}$$
$$H_{\min}(z) = F(z)H_{\min}(z)$$
$$H_{\text{all}}(z) = F^{-1}(z)H_{\text{all}}(z)$$

}

Example 5.7

Minimum-phase Allpass Decomposition

As an illustration of the decomposition of an IIR transfer function into allpass and minimum-phase parts, consider the following digital filter.

$$H(z) = \frac{.2[(z + .5)^2 + 1.5^2]}{z^2 - .64}$$

This is a stable IIR filter with real poles at $p_{1,2} = \pm .8$ and complex-conjugate zeros at $z_{1,2} = -.5 \pm j1.5$, as shown in Figure 5.22. Since the zeros are both outside the unit circle, this is a maximum-phase filter. The minimum-phase form is obtained by replacing the zeros

FIGURE 5.22: Pole-zero Plot of System

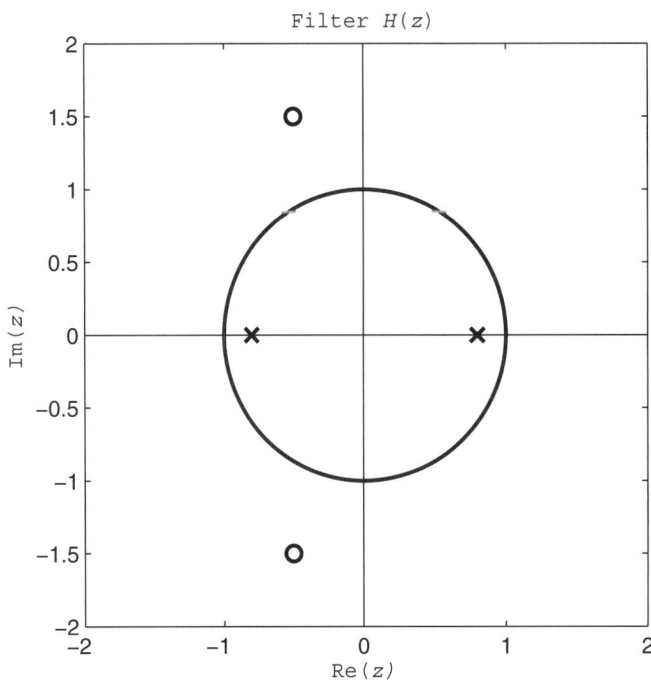

by their reciprocals and multiplying by the negatives of the zeros. The new zeros are

$$z_{3,4} = \frac{1}{z_{1,2}}$$

$$= \frac{1}{-.5 \pm j1.5}$$

$$= \frac{-.5 \mp j1.5}{.25 + 2.25}$$

$$= -.2 \mp j.6$$

The product $z_1 z_2 = |z_1|^2$. Thus the minimum-phase form of $H(z)$ is

$$H_{\min}(z) = \frac{|z_1|^2 .2(z - z_3)(z - z_4)}{z^2 - .64}$$

$$= \frac{(.25 + 2.25).2[(z + .2)^2 + .6^2]}{z^2 - .64}$$

$$= \frac{.5[(z + .2)^2 + .6^2]}{z^2 - .64}$$

Since both zeros were replaced, the allpass part is just the original numerator divided by the numerator of $H_{\min}(z)$.

$$H_{\text{all}}(z) = \frac{.2[(z + .5)^2 + 1.5^2]}{.5[(z + .2)^2 + .6^2]}$$

$$= \frac{.4[(z + .5)^2 + 1.5^2]}{(z + .2)^2 + .6^2}$$

A plot of the original and decomposed magnitude responses and the original and decomposed phase responses is shown in Figure 5.23. Note that $A_{\min}(f) = A(f)$ and $A_{\text{all}}(f) = 1$, as expected. It is also evident that $\phi_{\min}(f)$ has less phase lag than $\phi(f)$.

FIGURE 5.23:
Magnitude and Phase Plots of $H(z)$, $H_{\text{all}}(z)$, and $H_{\min}(z)$

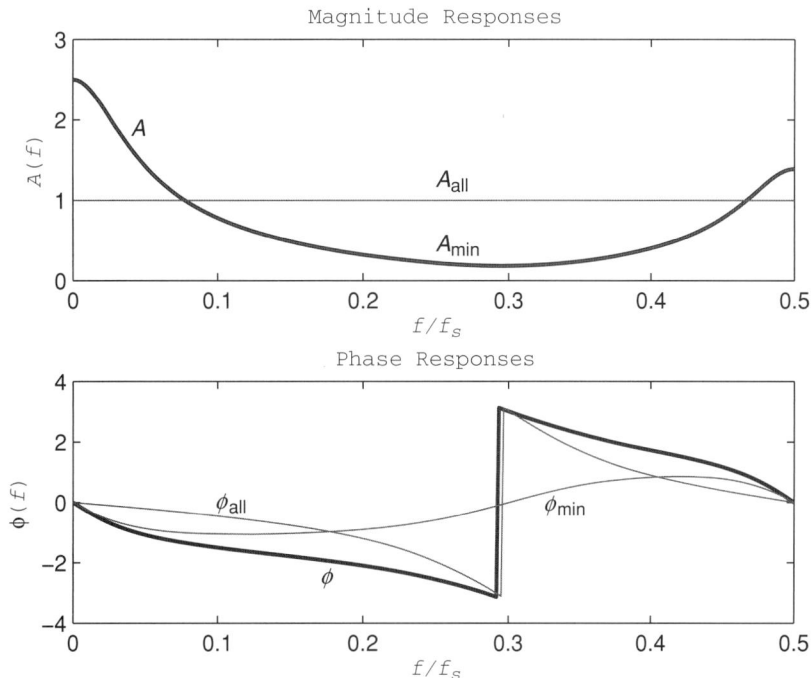

5.4.3 **Inverse Systems and Equalization**

One of the application areas of minimum-phase systems is the construction of inverse systems. Suppose $H(z) = b(z)/a(z)$ is a stable system with frequency response $H(f)$. The inverse of $H(z)$ is the system whose transfer function is the reciprocal, $H^{-1}(z) = a(z)/b(z)$. If $H^{-1}(z)$ is stable, then it has a frequency response $H^{-1}(f) = 1/H(f)$. Consequently, the frequency response of a cascade or series configuration of the two systems is $H^{-1}(f)H(f) = 1$. In this case we say that $H^{-1}(z)$ is an *equalizer* for the system $H(z)$ in the sense that it cancels out the effects of $H(z)$ with

$$H^{-1}(z)H(z) = 1 \qquad (5.4.13)$$

The problem arises when $H^{-1}(z)$ is not stable. This will occur if $H(z)$ has zeros on or outside the unit circle because they become poles of the reciprocal system. Suppose the design objective for an equalizer system is more limited. If the objective is to cancel only the magnitude response characteristics of $H(z)$, then the minimum-phase form of $H(z)$ can be used. In particular, suppose $A(f) > 0$, which means that $H(z)$ has no zeros on the unit circle. Then $H_{\min}(z)$ has no zeros on or outside the unit circle, which means that $H_{\min}^{-1}(z)$ is stable. Consider the following equalized system.

$$H_{\text{equal}}(z) = H_{\min}^{-1}(z)H(z) \qquad (5.4.14)$$

To examine the magnitude response of the equalized system, first note from the minimum-phase allpass decomposition in (5.4.12) that

$$H_{\text{equal}}(z) = H_{\text{all}}(z) \qquad (5.4.15)$$

Recall from Algorithm 5.2 that the poles of $H_{\text{all}}(z)$ are the reciprocals of the zeros of $H(z)$ that lie outside the unit circle. Therefore $H_{\text{all}}(z)$ is stable. Since $H_{\text{all}}(z)$ is an allpass filter, it follows that $H_{\min}^{-1}(z)$ serves as a *magnitude equalizer* with the magnitude response of the equalized system being

Magnitude equalizer

$$A_{\text{equal}}(f) = 1 \qquad (5.4.16)$$

Therefore $H_{min}^{-1}(z)$ cancels or equalizes the magnitude response of $H(z)$, but not the phase response. In Chapter 6, a two-stage quadrature filter design technique is presented that allows us to design a filter that equalizes both the magnitude and the phase response of a stable $H(z)$, but it introduces a delay in $H_{\text{equal}}(z)$. Equalizers with delay also can be designed using adaptive filters, as discussed in Chapter 9.

FDSP Functions

The FDSP toolbox contains the following function for decomposing a transfer function into minimum-phase and allpass parts.

```
% F_MINALL: Factor a filter into minimum-phase and allpass parts
%
% Usage:
%          [B_min,A_min,B_all,A_all] = f_minall (b,a)
% Pre:
%          b   = vector of length m+1 containing coefficients
%                of numerator polynomial.
```

```
%        a   = vector of length n+1 containing coefficients
%              of denominator polynomial (n >= m).
% Post:
%        B_min = (q+1) by 1 vector containing numerator
%                coefficients of minimum-phase part
%        A_min = (r+1) by 1 vector containing denominator
%                coefficients of minimum-phase part
%        B_all = (s+1) by 1 vector containing numerator
%                coefficients of allpass part
%        A_all = (s+1) by 1 vector containing denominator
%                coefficients of allpass part
```

5.5 Quadrature Filters

5.5.1 Differentiator

Phase quadrature

A pair of periodic signals is said to be in *phase quadrature* if one signal leads or lags the other by a quarter of a cycle. A common example of a quadrature pair is the sine and cosine. Certain specialized filters have steady-state outputs that are in phase quadrature with the input. Perhaps the simplest example is a differentiator. A continuous-time differentiator has transfer function $H_a(s) = s$ and frequency response $H_a(f) = j2\pi f$. In order to approximate the differentiator with a causal mth order linear-phase filter, consider a discrete-time frequency response with a delay of $mT/2$.

$$H(f) = j2\pi fT \exp(j\pi mfT) \tag{5.5.1}$$

Backward Euler

In real-time applications such as digital feedback control, a PID controller is often implemented by approximating the derivative term with a first-order *backward Euler* difference, as follows.

$$y(k) \approx \frac{x(k) - x(k-1)}{T} \tag{5.5.2}$$

This numerical approximation replaces $dx(t)/dt$ at $t = kT$ with the slope of the straight line connecting sample $x(k-1)$ to sample $x(k)$. The FIR transfer function of the backward Euler differentiator is

$$\begin{aligned} H(z) &= \frac{1 - z^{-1}}{T} \\ &= \frac{z-1}{Tz} \end{aligned} \tag{5.5.3}$$

Thus $H(z)$ has a pole at $z = 0$ and a zero at $z = 1$. Using Euler's identity, the frequency response in this case is

$$\begin{aligned} H(f) &= \frac{1 - \exp(-j2\pi fT)}{T} \\ &= j2\exp(-j\pi fT)\left[\frac{\exp(j\pi fT) - \exp(-j\pi fT)}{j2T}\right] \\ &= \left[\frac{j2\sin(\pi fT)}{T}\right]\exp(-j\pi fT) \end{aligned} \tag{5.5.4}$$

FIGURE 5.24:
Magnitude
Response of the
Backward Euler
Approximation to a
Differentiator

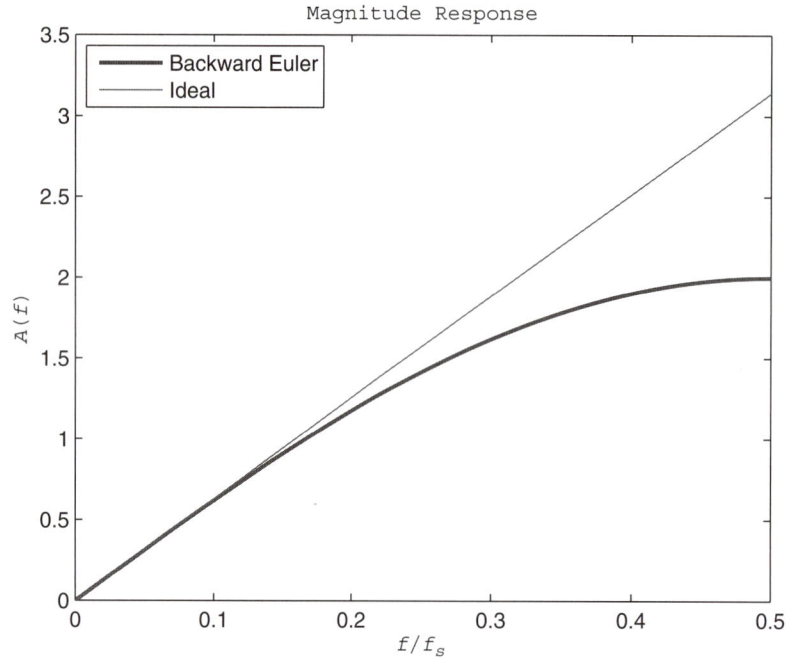

A plot of the magnitude response of the backward Euler differentiator is shown in Figure 5.24 along with the ideal differentiator magnitude response. For normalized frequencies in the range $[0, .1]$ the magnitude response approximation is close, but then it deteriorates increasingly for higher frequencies.

Since the backward Euler approximation is a linear-phase FIR filter of order $m = 1$, it includes a delay of $T/2$ or half a sample. If we ignore the effects of this delay, the remaining phase response matches that of the differentiator exactly. This is a consequence of the fact that the backward Euler impulse response is

$$h = \frac{1}{T}[1, -1] \tag{5.5.5}$$

Consequently, $h(k)$ exhibits odd symmetry about the midpoint $k = m/2$. Since $H(z)$ is of odd order, it follows from Table 5.1 that the backward Euler differentiator is a type 4 linear-phase FIR filter with a phase offset of $\alpha = \pi/2$, which is exactly what is needed for a differentiator.

To obtain an approximation that has a more accurate magnitude response, we can employ a higher-order type 4 linear-phase FIR filter. Using design techniques presented in Chapter 6, the following FIR filter of order $m = 5$ is obtained.

$$
\begin{aligned}
y(k) = {} & .0509x(k) - .1415x(k-1) + 1.2732x(k-2) \\
& -1.2732x(k-3) + .1415x(k-4) - .0509x(k-5)
\end{aligned} \tag{5.5.6}
$$

Notice the odd symmetry in the coefficient vector $b \in R^6$, which is the impulse response. A plot of the magnitude response of this fifth-order FIR filter is shown in Figure 5.25. Although it is not exact, it does show a marked improvement over the backward Euler differentiator, particularly at the higher frequencies. Taking the delay of 2.5 samples into account, the phase response with an phase offset of $\alpha = \pi/2$ is exact.

FIGURE 5.25:
Magnitude
Response of a Type
4 Linear-phase
Approximation to a
Differentiator
Using a Filter of
Order $m = 5$

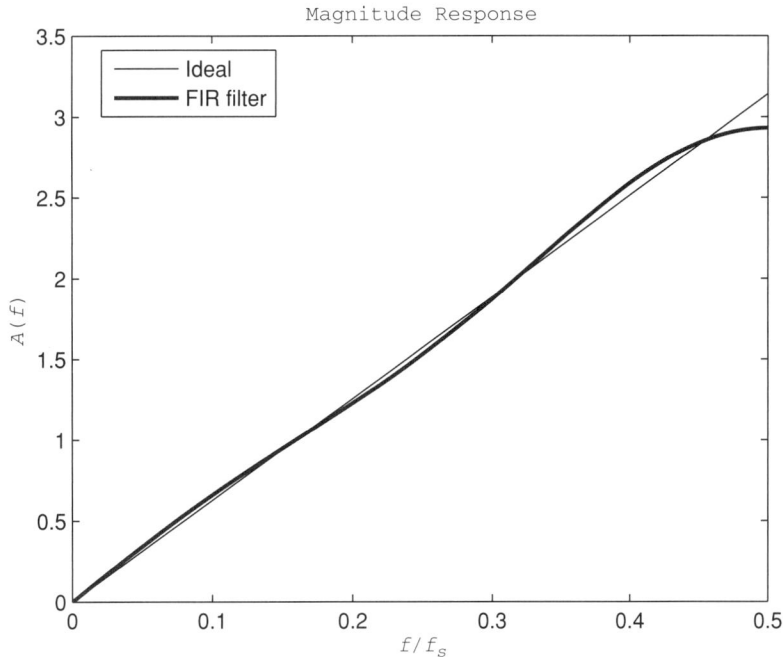

Magnitude Response

5.5.2 Hilbert Transformer

Hilbert transformer

Another filter that produces a steady state output that is in phase quadrature with a sinusoidal input is called a phase shifter or a *Hilbert transformer*. It is an allpass filter that has the following idealized desired frequency response.

$$H_d(f) = -j\,\mathrm{sgn}(f), \quad 0 \le |f| \le f_s/2 \tag{5.5.7}$$

Here *sgn* denotes the *signum* or sign function defined as $\mathrm{sgn}(f) = f/|f|$ for $f \ne 0$ and $\mathrm{sgn}(0) = 0$. Thus

$$\mathrm{sgn}(f) \triangleq \begin{cases} 1, & f > 0 \\ 0, & f = 0 \\ -1, & f < 0 \end{cases} \tag{5.5.8}$$

Notice that the magnitude response is $|H_d(f)| = 1$ and there is a constant phase shift of $\angle H_d(f) = -\pi/2$ for $0 \le f < f_s/2$. Like the ideal frequency selective filters, the Hilbert transformer cannot be realized with a causal system. This can be seen by applying the inverse DTFT (see Problem 5.25) which yields the following impulse response.

$$h_d(k) = \begin{cases} \dfrac{1 - \cos(k\pi)}{k\pi}, & k \ne 0 \\ 0, & k = 0 \end{cases} \tag{5.5.9}$$

To obtain a causal linear-phase approximation using an FIR filter of order m, we include a delay of $mT/2$ in the desired frequency response.

$$H_h(f) = -j\,\mathrm{sgn}(f)\exp(-j\pi m f T), \quad 0 \le |f| \le f_s/2 \tag{5.5.10}$$

Since the factor j in (5.5.10) inserts a phase shift of $-\pi/2$, this suggests the use of a linear-phase filter from Table 5.1 that has a phase offset of $\alpha = \pi/2$ and an amplitude response of $A_r(f) = -1$. Either a type 3 or a type 4 linear-phase filter could be used. As an illustration, suppose a type 3 linear-phase filter of order $m = 40$ is designed using techniques from

FIGURE 5.26:
Magnitude
Response of a Type
3 Linear-phase
Approximation to a
Hilbert Transformer
Using a Filter of
Order $m = 40$

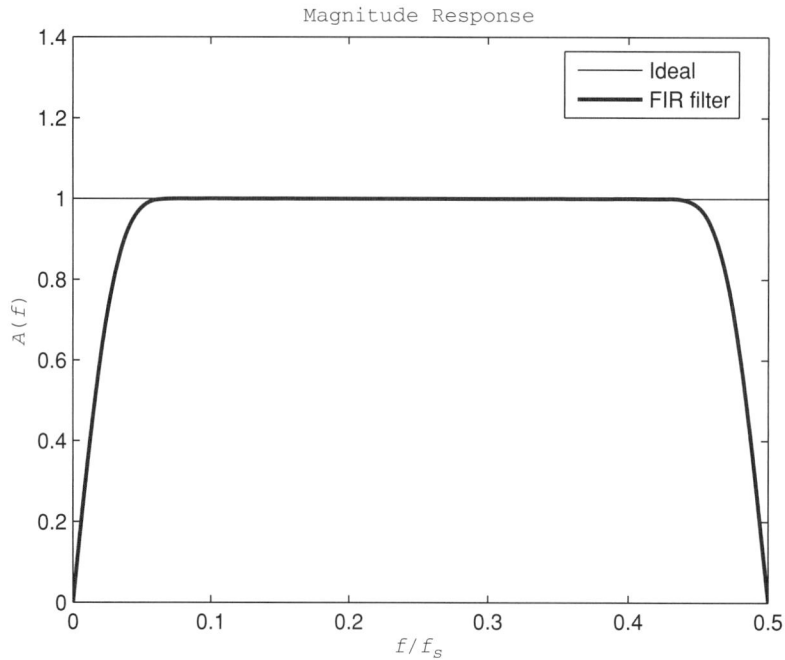

Chapter 6. The resulting magnitude response is shown in Figure 5.26. It is evident that this approximation to the Hilbert transformer is actually a wideband bandpass filter rather than the ideal allpass filter. This is because a type 3 linear-phase FIR filter has zeros at $H(\pm 1) = 0$ which correspond to the two ends of the frequency range. As the filter order m increases, the passband can be made wider, but the gain will always be zero at DC and at $f_s/2$.

In order to produce a pair of signals $x_1(k)$ and $x_2(k)$ that are in phase quadrature, suppose the input consists of a cosine of frequency f.

$$x(k) = \cos(2\pi f kT) \tag{5.5.11}$$

The Hilbert transformer in (5.5.10) will shift the phase of $x(k)$ by $-\pi/2$, but it will also delay the signal by $mT/2$. Thus the total phase shift of the Hilbert transformer is

$$\phi(f) = -\frac{\pi}{2} - \pi m f T \tag{5.5.12}$$

Suppose $x_2(k)$ is produced by passing $x(k)$ through a Hilbert transformer $H_h(z)$, as in Figure 5.27. To ensure that $x_1(k)$ and $x_2(k)$ are exactly $\pi/2$ radius apart, $x_1(k)$ must be the

FIGURE 5.27:
Generation of
Phase-Quadrature
Signals and a
Half-band Output
Using a Hilbert
Transformer

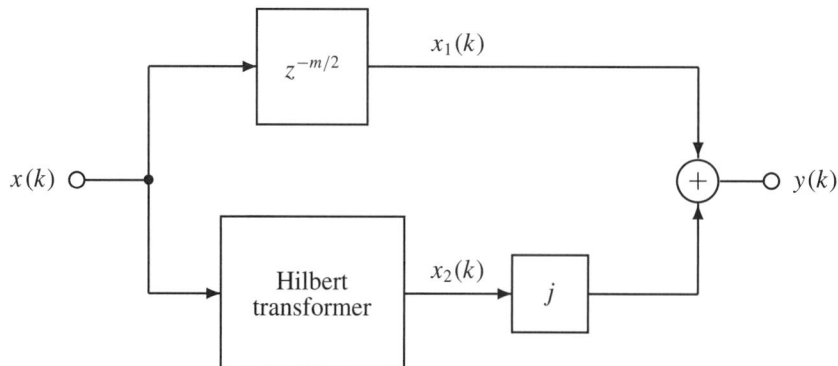

output of an allpass filter with an identical delay of $mT/2$ so that the delays of $x_1(k)$ and $x_2(k)$ match. Recall that m is even, so this can be achieved with an integer delay of $m/2$ samples using a filter with transfer function

$$F(z) = z^{-m/2} \tag{5.5.13}$$

For the cosine input in (5.5.11), the steady-state values of $x_1(k)$ and $x_2(k)$ in Figure 5.27 will then be delayed versions of a cosine and a sine as follows.

$$x_1(k) = \cos[2\pi f(k - m/2)T] \tag{5.5.14a}$$
$$x_2(k) = A_h(f)\sin[2\pi f(k - m/2)T] \tag{5.5.14b}$$

Note that the Hilbert transformer output $x_2(k)$ has a frequency-dependent amplitude $A_h(f)$. Thus $x_2(k)$ is a phase-shifted version of $x_1(k)$ so long as $A_h(f)$ approximates the allpass characteristic $A_h(f) \approx 1$.

In Chapter 6 a general two-stage quadrature filter design technique will be introduced where stage 1 generates $x_1(k)$ and $x_2(k)$, as in Figure 5.17. Stage 2 will then process $x_1(k)$ and $x_2(k)$ and combine the results to produce an overall FIR filter from *both* magnitude and phase specifications.

Hilbert transformers are used in a number of applications in communications and speech processing. For example, the Hilbert transformer can be used, as shown in the second half of *Complex signal* Figure 5.27, to create a *complex signal* of the following form.

$$y(k) = x_1(k) + jx_2(k) \tag{5.5.15}$$

Recall that if a signal is real, then its magnitude spectrum is an even function, and its phase spectrum is an odd function. When we generalize to complex signals, this symmetry constraint no longer applies. Indeed, when the Hilbert transformer amplitude response is ideal with $A_h(f) = 1$, the spectrum of the complex output signal $y(k)$ is

$$
\begin{aligned}
Y(f) &= X_1(f) + jX_2(f) \\
&= [\exp(-j\pi mfT) + jH_h(f)]X(f) \tag{5.5.16}
\end{aligned}
$$

From (5.5.10), $jH_h(f) = \pm\exp(-j\pi mfT)$, depending on the sign of f. Thus the spectrum of $y(k)$ is

$$
Y(f) = \begin{cases} 2\exp(-j\pi mfT)X(f), & 0 \le f < f_s/2 \\ 0, & -f_s/2 < f < 0 \end{cases} \tag{5.5.17}
$$

Notice that the spectrum of the complex output $y(k)$ is *zero* over the negative frequency range. Put another way, $Y(z) = 0$ along the bottom half of the unit circle. For this reason, $y(k)$ in *Half-band signal* Figure 5.27 is referred to as a *half-band* signal.

The complex signal $y(k)$ is the discrete-equivalent of an analytic signal because its spectrum is zero for $-f_s/2 < f < 0$. From (5.5.17) it is evident that $y(k)$ contains all of the information needed to reconstruct the original signal $x(k)$, but it occupies only half of the bandwidth. As a result, $y(k)$ can be transmitted more efficiently than $x(k)$. Recall from the frequency shift property of the DTFT in Table 4.3 that if a signal is modulated by a complex exponential, this shifts its spectrum.

$$q(k) = \exp(j2\pi F_0 kT)y(k) \tag{5.5.18}$$
$$Q(f) = Y(f - F_0) \tag{5.5.19}$$

Frequency-division multiplexing Using this technique, several half-band signals can be translated to different regions of the spectrum and then transmitted simultaneously. This technique, which is referred to as *frequency-division multiplexing*, is discussed in Chapter 8.

5.5.3 Digital Oscillator

A discrete-time system that generates a phase quadrature pair at a fixed frequency F_0 is called a *digital oscillator*. A sinusoidal oscillator of frequency F_0 will produce the following quadrature pair.

$$x_1(k) = \cos(2\pi F_0 kT) \tag{5.5.20a}$$
$$x_2(k) = \sin(2\pi F_0 kT) \tag{5.5.20b}$$

To derive a linear system whose output is $x_1(k)$, we express $x_1(k)$ as a linear combination of $x_1(k-1)$ and $x_2(k-1)$. For convenience, let $\alpha = 2\pi F_0 T$. Using the cosine of the sum trigonometric identity from Appendix 2,

$$\begin{aligned} x_1(k) &= \cos[2\pi F_0(k-1)T + \alpha] \\ &= \cos[2\pi F_0(k-1)T]\cos(\alpha) - \sin[2\pi F_0(k-1)T]\sin(\alpha) \\ &= \cos(\alpha)x_1(k-1) - \sin(\alpha)x_2(k-1) \end{aligned} \tag{5.5.21}$$

Applying a similar analysis to $x_2(k)$ using the sine of the sum trigonometric identity yields

$$x_2(k) = \sin(\alpha)x_1(k-1) + \cos(\alpha)x_2(k-1) \tag{5.5.22}$$

Recalling that $\alpha = 2\pi F_0 T$, we find that the quadrature pair in (5.5.20) satisfies the following vector difference equation.

$$\begin{bmatrix} x_1(k) \\ x_2(k) \end{bmatrix} = \underbrace{\begin{bmatrix} \cos(2\pi F_0 T) & -\sin(2\pi F_0 T) \\ \sin(2\pi F_0 T) & \cos(2\pi F_0 T) \end{bmatrix}}_{C(F_0)} \begin{bmatrix} x_1(k-1) \\ x_2(k-2) \end{bmatrix} \tag{5.5.23}$$

Note that the two-dimensional system in (5.5.23) has no input. Instead, it is a nonzero initial condition that starts the oscillation. If we evaluate (5.5.20) at $k = 0$, the required initial condition is $x_1(0) = 1$ and $x_2(0) = 0$. The coefficient matrix $C(F_0)$ in (5.5.22) has a simple interpretation. Let $X(k) \in R^2$ be a 2×1 column vector with elements $x_1(k)$ and $x_2(k)$. Then (5.5.23) and the initial condition can be written in vector form as

$$X(k) = C(F_0)X(k-1), \quad X(0) = [1, 0]^T \tag{5.5.24}$$

Rotation matrix The 2×2 coefficient matrix $C(F_0)$ is a *rotation matrix*. When the vector $X(k-1)$ is multiplied by $C(F_0)$, it rotates $X(k-1)$ counterclockwise about the origin by angle $\alpha = 2\pi F_0 T$ to produce $X(k)$. In this way, the solution to (5.5.24) starts at $X(0) = [1, 0]^T$ and traverses the unit circle counterclockwise in the x_2 versus x_1 plane.

Chebyshev polynomials, first kind Once a quadrature pair is generated by a digital oscillator, it can be used to synthesize a more general periodic signal by post processing $x_1(k)$ and $x_2(k)$ to generate harmonics. To see how, consider the following classical family of orthogonal polynomials called the *Chebyshev polynomials of the first kind*. The first two Chebyshev polynomials of the first kind are $T_0(x) = 1$ and $T_1(k) = x$. Subsequent polynomials are generated recursively as follows.

$$T_i(x) = 2xT_{i-1}(x) - T_{i-2}(x), \quad i \geq 2 \tag{5.5.25}$$

The Chebyshev polynomials have many interesting properties but the one of interest here is their use in generating even harmonics.

$$T_i[\cos(\theta)] = \cos(i\theta), \quad i \geq 0 \tag{5.5.26}$$

Since $x_1(k)$ is a cosine, this means that any even periodic function can be generated from $x_1(k)$ using a suitable linear combination of the Chebyshev polynomials of the first kind.

Chebyshev polynomials, second kind

To generate an odd periodic function, we can use the *Chebyshev polynomials of the second kind*. The first two Chebyshev polynomials of the second kind are $U_0(x) = 1$ and $U_1(x) = 2x$. The recursive relationship for generating the remaining Chebyshev polynomials is similar to (5.5.25), namely,

$$U_i(x) = 2xU_{i-1}(x) - U_{i-2}(x), \quad i \geq 0 \tag{5.5.27}$$

To generate odd harmonics using the Chebyshev polynomials of the second kind, both the cosine and the sine must be available as a quadrature pair.

$$\sin(\theta)U_{i-1}[\cos(\theta)] = \sin(i\theta), \quad i \geq 1 \tag{5.5.28}$$

Suppose the frequency $F_0 \ll f_s$. To avoid aliasing, the ith harmonic must be below the folding frequency $f_s/2$. Thus the number of harmonics that can be generated without aliasing is

$$r < \frac{f_s}{2F_0} \tag{5.5.29}$$

A general periodic signal $y(k)$ with period $T_0 = 1/F_0$ can be approximated with r harmonics using the following truncated Fourier series.

$$y(k) = a_0 + \sum_{i=1}^{r} a_i \cos(2\pi i F_0 T) + b_i \sin(2\pi i F_0 T) \tag{5.5.30}$$

Given the quadrature pair in (5.5.20) and the harmonic generation properties in (5.5.26) and (5.5.28), the periodic signal $y(k)$ can be expressed as follows where $f_e(x)$ and $f_o(x)$ are polynomials.

$$y(k) = f_e[x_1(k)] + x_2(k)f_o[x_1(k)] \tag{5.5.31}$$

A block diagram which shows the digital oscillator and the post processing needed to form a periodic output $y(k)$ is shown in Figure 5.28. The polynomials $f_e(x)$ and $f_o(x)$ represent the even and odd parts of $y(k)$, respectively. From (5.5.30), they are formed from the Chebyshev polynomials as follows.

$$f_e(x) = \sum_{i=0}^{r} a_i T_i(x) \tag{5.5.32}$$

$$f_o(x) = \sum_{i=1}^{r} b_i U_{i-1}(x) \tag{5.5.33}$$

As an illustration of the use of a digital oscillator to generate an arbitrary periodic waveform, suppose $f_s = 1000$ Hz and $F_0 = 25$ Hz. Up to $r = 19$ harmonics can be generated without aliasing. Suppose the coefficient vectors are $a = [0, 1, .5, .25]^T$ and $b = [1, -.5, .25]^T$ and $N = 200$ points are computed. A plot of the resulting output $y(k)$ versus the first quadrature signal $x_1(k)$ is shown in Figure 5.29. Also plotted is $x_2(k)$ versus $x_1(k)$ which generates a circle, as expected. The closed curve corresponding to $y(k)$ versus $x_1(k)$ shows that $y(k)$ is indeed periodic with period $F_0 = 25$ Hz.

FIGURE 5.28:
Generation of a
Periodic Output
from a Quadrature
Pair

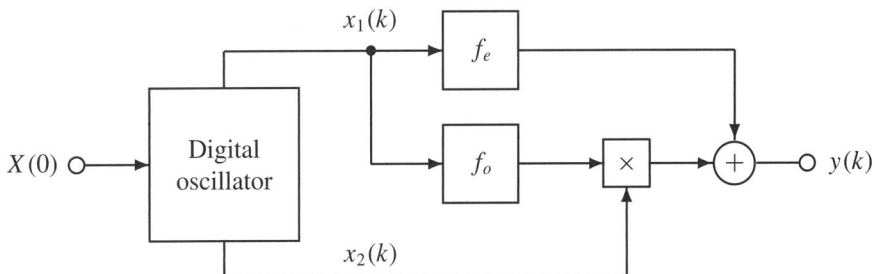

FIGURE 5.29: A Numerical Example of the Quadrature Pair and a Periodic Output Generated from Them

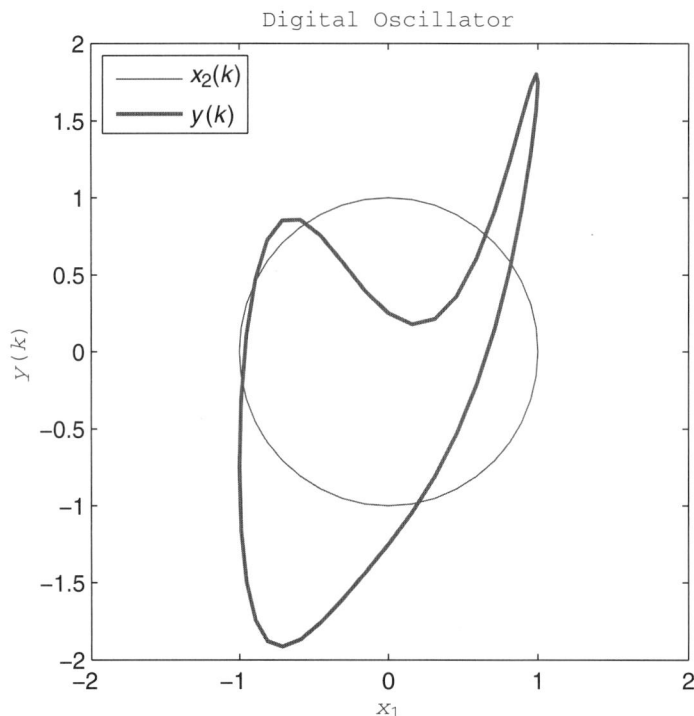

● ● ● ● ● ● ● ● ● ● ● ● ● ● ● ● ● ●

5.6   Notch Filters and Resonators

5.6.1 Notch Filters

Notch filter

The frequency selective filters in Section 5.2 all have the property that the stopbands are of nonzero length. There is a class of filters called *notch filters* where the stopband consists of a single frequency. To define a design specification for a notch filter, it is helpful to use the following continuous-frequency version of the unit impulse called a *unit pulse*.

$$\delta_1(f) \triangleq \begin{cases} 1, & f = 0 \\ 0, & f \neq 0 \end{cases} \tag{5.6.1}$$

Note that $\delta_1(f)$ is similar to the unit impulse $\delta_a(f)$, except that it takes on a value of one at $f = 0$. Given the unit pulse function in (5.6.1), a notch filter with a notch at F_0 is a filter whose magnitude response is

$$A_{\text{notch}}(f) \triangleq 1 - \delta_1(f - F_0), \quad 0 \leq |f| \leq f_s/2 \tag{5.6.2}$$

Hence a notch filter is a filter designed to pass all frequencies except F_0 where F_0 is completely blocked by the filter. Note that no explicit constraint is placed on the phase response. Notch filters can be approximated with low-order IIR filters using techniques discussed in Chapter 7. Perhaps the simplest example of a notch filter is a filter designed to block DC, that is, a filter with a notch at $F_0 = 0$ Hz. If a noncausal filter is permitted, then an exact DC-blocking filter is easily implemented. All one has to do is set the output $y(k)$ equal to the input $x(k)$ minus its mean.

For a causal realization of a DC notch filter, consider a first-order IIR filter with the following structure.

$$H_{DC}(z) = \frac{c(z-1)}{z-r}, \quad 0 \ll r < 1 \tag{5.6.3}$$

Recall that along the unit circle, $z = \exp(j2\pi fT)$, so DC corresponds to $z = 1$. Thus the zero of $H_{DC}(z)$ at $z = 1$ ensures that the DC component of $x(k)$ will be blocked. To ensure a passband gain of one, we can set $H_{DC}(-1) = 1$ where $z = -1$ corresponds to the folding frequency. Setting $H_{DC}(-1) = 1$ and solving for c yields

$$c = \frac{1+r}{2} \tag{5.6.4}$$

Notice that since $0 \ll r < 1$, this implies that $c \approx 1$. For an effective notch filter, it is also required that $A_{notch}(f) \approx 1$ for $f \neq F_0$. This can be achieved by placing the pole at $z = r$ very close to the zero at $z = 1$. Let F_c be the 3 dB cutoff frequency of the notch filter. That is, F_c is the frequency at which

$$A_{notch}^2(F_c) = .5 \tag{5.6.5}$$

For a DC notch filter, $F_c \ll f_s/2$. Thus $2\pi F_c T \ll \pi$, which means $\cos(2\pi F_c T) \approx 1$ and $\sin(2\pi F_c T) \approx 2\pi F_c T$. Recall also that $r \approx 1$. With these approximations and (5.6.3), the square of the magnitude response at $f = F_c$ is

$$
\begin{aligned}
A_{notch}^2(F_c) &= \frac{|c[\exp(j2\pi F_c T) - 1]|^2}{|\exp(j2\pi F_c T) - r|} \\
&= \frac{c^2\{[\cos(2\pi F_c T) - 1]^2 + \sin^2(2\pi F_c T)\}}{[\cos(2\pi F_c T) - r]^2 + \sin^2(2\pi F_c T)} \\
&\approx \frac{(2\pi F_c T)^2}{(1-r)^2 + (2\pi F_c)^2}
\end{aligned}
\tag{5.6.6}
$$

Setting $A_{notch}^2(F_c) = .5$ as in (5.6.5) then yields

$$(2\pi F_c T)^2 \approx (1-r)^2 \tag{5.6.7}$$

Bandwidth   The *bandwidth* of the stopband or notch is $\Delta F = 2F_c$. Solving (5.6.7) for r and expressing the final result in terms of the notch bandwidth yields the following design formula for r.

$$r \approx 1 - \frac{\pi \Delta F}{f_s} \tag{5.6.8}$$

Thus to design a simple IIR filter that blocks DC with a notch of width ΔF, we use (5.6.8) to determine the pole radius r, and then (5.6.4) to determine the gain c.

Example 5.8   **DC Notch Filter**

As an illustration of a notch filter that blocks DC, suppose $f_s = 100$ Hz, and consider the sequence of three filters with notch bandwidths of $\Delta F = [.4, .2, .1]$ Hz. From (5.6.8) and (5.6.4), the corresponding poles and gains are

$$r = [.9969, .9937, .9874]$$

$$c = [.9984, .9969, .9937]$$

Plots of the magnitude response are shown in Figure 5.30. Note that only 4% of the frequency range is plotted in order to more clearly see the notch. Although the notch filter was not designed

FIGURE 5.30:
Magnitude
Responses of Notch
Filters Blocking DC

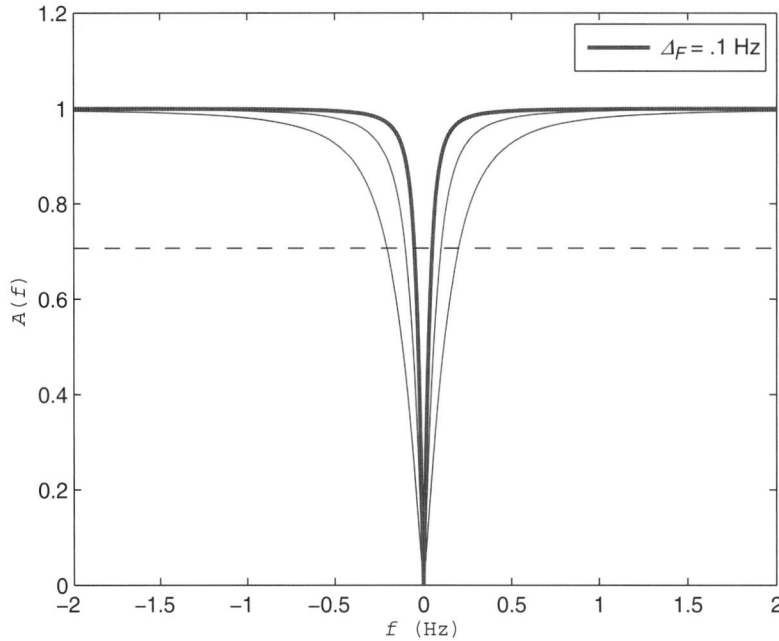

to satisfy a phase response specification, because the pole and zero are so close together, their phase contributions almost cancel except near DC. Consequently, the phase response of $H_{DC}(z)$ will be close to zero within the passband.

For most notch filters, the notch frequency is not $F_0 = 0$ but is instead somewhere in the range $0 < F_0 \leq f_s/2$. For a notch at frequency F_0, there must be a zero on the unit circle at $z = \exp(j2\pi F_0 T)$. Hence the radius of the zero is one and the angle is

$$\theta_0 = 2\pi F_0 T \tag{5.6.9}$$

Since $z = \exp(j\theta_0)$ is complex, there will be a complex-conjugate pair of zeros at $z = \exp(\pm j\theta_0)$ and a matching complex-conjugate pair of poles at $z = r\exp(\pm j\theta_0)$ where $0 \ll r < 1$. Thus the general form of a notch filter is

$$H_{notch}(z) = \frac{c[z - \exp(j\theta_0)][z - \exp(-j\theta_0)]}{[z - r\exp(j\theta_0)][z - r\exp(-j\theta_0)]} \tag{5.6.10}$$

The poles and zeros occur in conjugate pairs, so the coefficients of H_{notch} are real. Techniques to determine c and r are presented in Chapter 7 along with design examples. Given the notch filter magnitude response specification in (5.6.2), notch filters with more than one notch frequency can be realized by using a *series* configuration of individual notch filters, also called a cascade configuration. An implementation of a double-notch filter with notches at F_0 and F_1 using a series configuration of two notch filters is shown in Figure 5.31. If multiple notch frequencies are equally spaced to include a fundamental F_0 and its harmonics, then a special structure *Inverse comb filter*   called an *inverse comb filter* can be used. The design of comb filters is discussed in Chapter 7.

5.6.2 Resonators

Resonators and notch filers form a complementary pair. Whereas an ideal notch filter is designed to stop a single frequency and pass all others, an ideal resonator is design to pass single

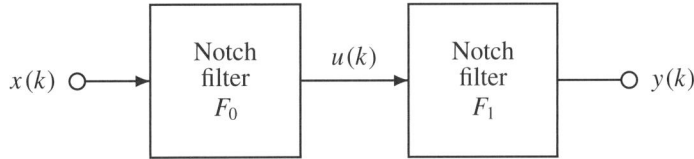

FIGURE 5.31: A Double-notch Filter with Notches at F_0 and F_1 Using a Series Configuration of Two Notch Filters

frequency and stop all others. That is, it is designed to "resonate" at a specific frequency. Consequently, the magnitude response specification of a resonator with resonant frequency F_0 is

$$A_{\text{res}}(f) \overset{\Delta}{=} \delta_1(f - F_0), \quad 0 \leq |f| \leq f_s/2 \tag{5.6.11}$$

If $A_{\text{notch}}(f)$ is the magnitude response of a notch filter with notch frequency F_0, and $A_{\text{res}}(f)$ is the magnitude response of a resonator with resonant frequency F_0, then

$$A_{\text{notch}}^2(f) + A_{\text{res}}^2(f) = 1, \quad 0 \leq |f| \leq f_s/2 \tag{5.6.12}$$

Power-complementary pair

In this case we say that $H_{\text{notch}}(z)$ and $H_{\text{res}}(z)$ form a *power-complementary* pair. The complementary pair relationship in (5.6.12) suggests that one way to design a resonator is

$$H_{\text{res}}(z) = 1 - H_{\text{notch}}(z) \tag{5.6.13}$$

If this approach is applied to the DC notch filter in (5.6.3), then after simplification using (5.6.4) this results in the following DC resonator.

$$H_{\text{dc}}(z) = \frac{.5(1 - r)(z + 1)}{z - r} \tag{5.6.14}$$

Notice that the coefficients of the numerator can be very small. Direct substitution reveals that $H_{\text{dc}}(1) = 1$ and $H_{\text{dc}}(-1) = 0$.

An alternative approach to a general resonator with resonant frequency $0 < F_0 \leq f_s/2$ is to put poles at $z = r\exp(\pm j\theta_0)$ to create the resonance, and zeros at $z = \pm 1$ to ensure a bandpass characteristic.

$$H_{\text{res}}(z) = \frac{c(z^2 - 1)}{[z - r\exp(j\theta_0)][z - r\exp(-j\theta_0)]} \tag{5.6.15}$$

Techniques to determine c and r are presented in Chapter 7 together with examples. Given the resonator magnitude response specification in (5.6.11), resonators with more than one resonant frequency can be realized by using a *parallel* configuration of individual resonators. An implementation of a double-resonator with resonance frequencies at F_0 and F_1 using a parallel configuration of two resonators is shown in Figure 5.32. If multiple resonator frequencies are equally spaced to include a fundamental F_0 and its harmonics, then a special structure called a *comb filter* can be used as described in Chapter 7.

Comb filter

Example 5.9

Power-complementary pair

As an illustration of a power-complementary pair, consider a DC resonator designed using (5.6.14). To more clearly see the shape of the resonance, suppose a fairly wide resonance bandwidth of $\Delta F = f_s/20$ is used which represents 5 % of the frequency range. From (5.6.8), this results in a pole radius of

$$r = .8429$$

FIGURE 5.32: A Double-resonator with Resonance Frequencies at F_0 and F_1 Using a Parallel Configuration of Two Resonators

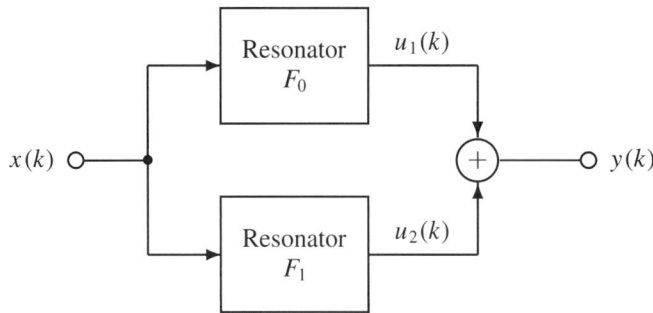

FIGURE 5.33: Frequency Responses of a Power-Complementary Pair, a DC Resonator and a DC Notch Filter with $\Delta F = f_s/20$

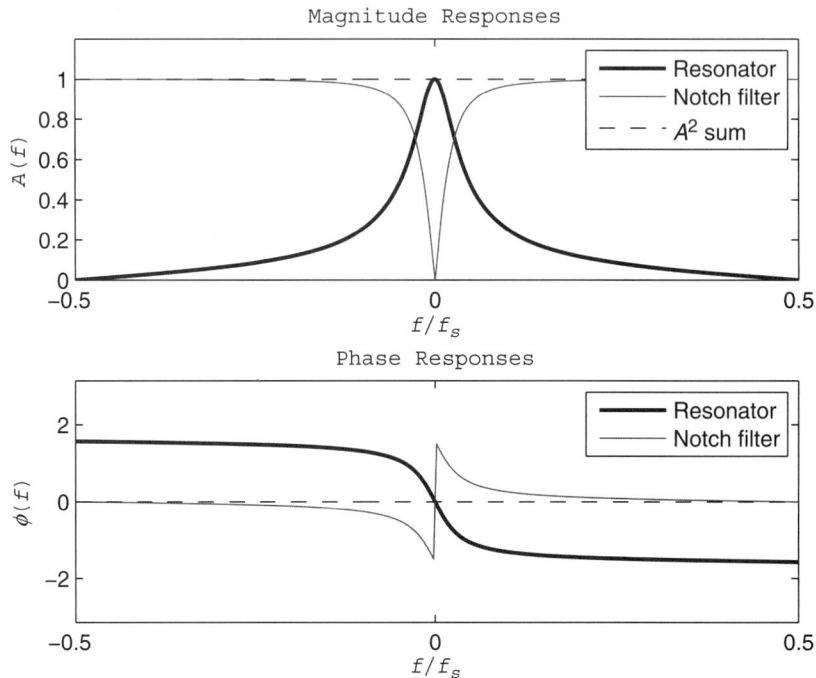

Plots of the magnitude response and the phase response of the resulting DC resonator $H_{dc}(z)$ are shown in Figure 5.33. Also shown are the magnitude and phase responses of the complementary DC notch filter $H_{\mathrm{DC}}(z)$. Note how the magnitude responses form a power-complementary pair even though these filters are not ideal. It is also clear that the phase responses are flat for frequencies well away from the resonance and the notch.

5.7 Narrowband Filters and Filter Banks

5.7.1 Narrowband Filters

Narrowband filter

Suppose a lowpass or a highpass filter has a cutoff frequency of F_c. As the sampling frequency f_s is increased, the normalized cutoff frequency F_c/f_s will approach zero. Filters with very small normalized cutoff frequencies are called *narrowband filters*. Narrowband filters have practical applications, but they can be challenging to design. An effective way to implement a narrowband

FIGURE 5.34: Analog Multirate Design of a Narrowband Filter

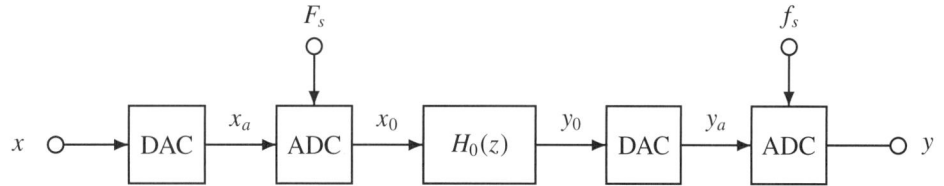

filter is to convert it into a filter whose normalized cutoff frequency lies somewhere near the middle of the Nyquist range with $F_c/f_s \approx .25$. Since the cutoff frequency F_c is typically fixed by the application, this necessitates changing the sampling frequency f_s. Suppose, however, that the sampled input $x(k)$ is already available and must be used.

A brute force analog approach to lowering the sampling rate is summarized in Figure 5.34. Here the input $x(k)$ is first converted back to analog form $x_a(t)$ using a DAC. Next $x_a(t)$ is *resampled* with an ADC at a lower sampling rate F_s. This results in a new discrete-time signal $x_0(k)$ that can be filtered by $H_0(z)$ to produce $y_0(k)$. The process is then reversed to restore the original sampling rate. First, $y_0(k)$ is converted to $y_a(t)$ using a DAC, and then $y(k)$ is obtained from $y_a(t)$ by sampling at the original sampling rate f_s using an ADC. This analog approach does have the advantage that the new sampling rate F_s, seen by $H_0(z)$, can be arbitrary. However, it is an expensive, hardware-intensive, multi-stage approach that requires four converters, and each converter introduces quantization noise.

Decimator, Interpolator

Interestingly enough, the sampling rate f_s can be changed with a purely discrete-time system, a system that is linear but time-varying. When the sampling rate is reduced by an integer factor M, this is called to as a *decimator*. If the sampling rate is increased by an integer factor L, this is called an *interpolator*. More generally, the sampling rate can be changed by a factor L/M which is called a *rational* sampling rate converter. The design and application of discrete sampling rate converters is presented in Chapter 8. There are a number of applications that arise when the sampling rate is allowed to change. This area of investigation, which is the focus of Chapter 8, is referred to as *multirate signal processing*.

Multirate signal processing

The multirate application of interest here is the design of a narrowband filter. Suppose an integer decimator is used to reduce the sampling rate to $F_s = f_s/M$. In order to move the cutoff frequency F_c to the middle of the Nyquist range, it is necessary that $F_c/F_s \approx .25$. Thus the sampling rate reduction factor M can be computed as follows.

$$M = \text{round}\left(\frac{f_s}{4F_c}\right) \tag{5.7.1}$$

The multirate narrowband filter design technique is summarized in Figure 5.35. First the input $x(k)$ is down-sampled by a factor of M with a decimator. This is indicated in Figure 5.35 by the symbol $\downarrow M$. Next the down-sampled signal $x_0(k)$ is filtered by a filter $H_0(z)$ with cutoff frequency $F_c \approx F_s/4$. Finally, the original sampling rate is restored by up-sampling $y_0(k)$ by the factor M using an interpolator, denoted $\uparrow M$, to produce $y(k)$.

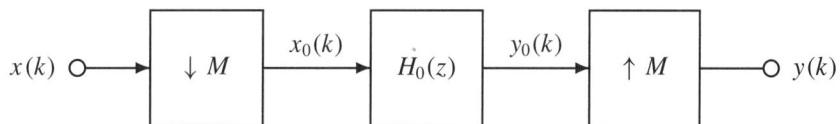

FIGURE 5.35: Multirate Design of a Narrowband Filter Using a Decimator and an Interpolator

FIGURE 5.36:
Magnitude
Response of a
Multirate
Narrowband
Lowpass Filter of
Order $m = 80$ Using
Rate Conversion
Factor $M = 10$

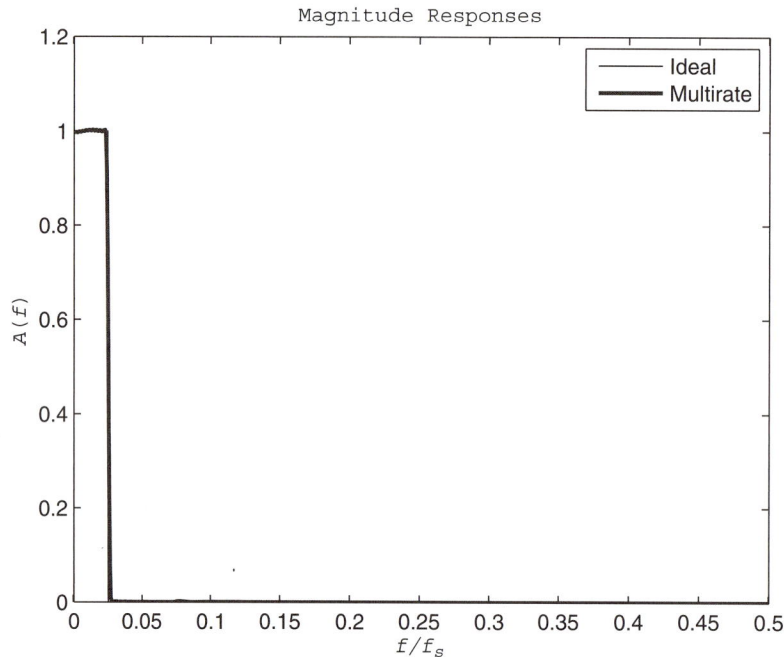

Magnitude Responses

Example 5.10

Narrowband Filter

As an illustration of the design of a narrowband filter, suppose the original sampling rate is $f_s = 1000$ Hz, and it is desired to design a lowpass filter with a cutoff frequency of $F_c = 25$ Hz which is 1/20 the folding frequency. From (5.7.1), the rate conversion factor for the decimator and the interpolator is

$$M = \text{round}\left[\frac{1000}{4(25)}\right] = 10$$

Using techniques presented in Chapters 6 and 8, a linear-phase FIR filter $H_0(z)$ of order $m = 80$ can be designed using the Hamming window. The resulting magnitude response of the multirate narrowband lowpass filter is shown in Figure 5.36. It is evident that a close approximation to the desired response is obtained.

In Chapter 8 it is shown that sampling rate converters with a rate conversion factor of M require a lowpass digital anti-aliasing or anti-imaging filter with a cutoff frequency of $f_s/(2M)$. Since this is itself a narrowband filter when $M \gg 1$, it would seem that we are still confronted with the problem of implementing a narrowband filter. However, if $M \gg 1$, then the overall rate conversion factor M can be factored into a product of smaller integer factors.

$$M = M_1 \cdot M_2 \cdots M_p \tag{5.7.2}$$

The sampling rate converter then can be implemented using a *series* connection of p smaller sampling rate converters with rate conversion factors $M_1, M_2, \ldots M_p$. In this way a narrowband filter with $M \gg 1$ can be realized using a series of less narrow filters. This multistage approach can be used, for example, when $M > 10$.

FIGURE 5.37:
Magnitude
Responses of a
Four-Filter Filter
Bank

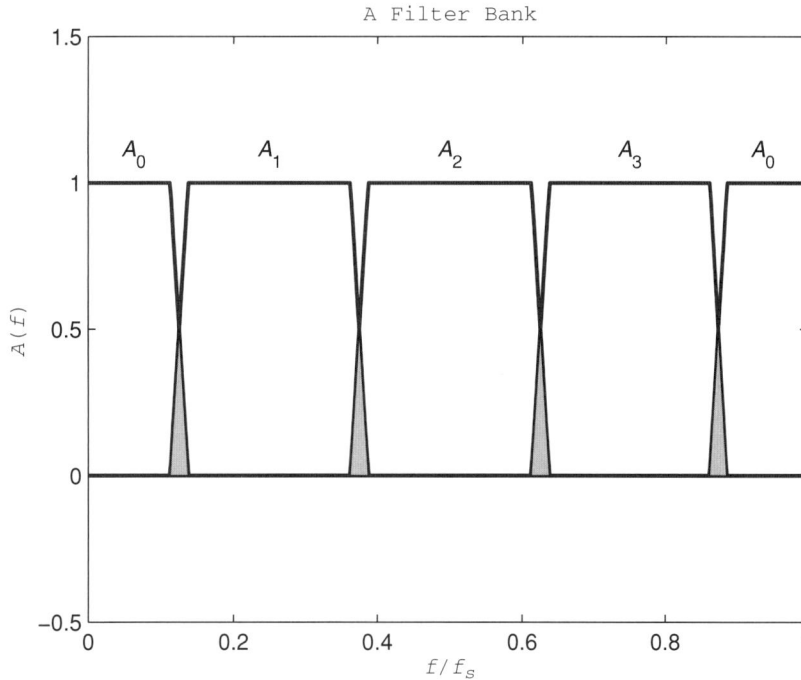

5.7.2 Filter Banks

In addition to narrowband filters with small cutoff frequencies, there are applications that make use of bandpass filters with narrow passbands. A *filter bank* is a parallel connection of filters with nonoverlapping passbands where the collective passband occupies the entire Nyquist range.

$$\sum_{i=0}^{N-1} A_i(f) = 1 \tag{5.7.3}$$

As an illustration, the magnitude responses of a filter bank consisting of four filters are shown in Figure 5.37. A parallel configuration of N filters in a filter bank can have either a common input or a common output. Consider the case when the N filters are driven by the same input $x(k)$, as shown in Figure 5.38. Note that output $y_i(k)$ represents the part of the

FIGURE 5.38: A
Four-filter Analysis
Filter Bank

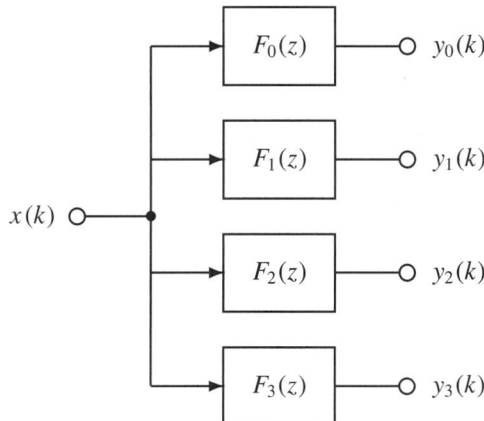

input $x(k)$ whose spectrum lies in the ith *subband*. The configuration in Figure 5.38 is referred to as an *analysis filter bank* because it decomposes the input into subsignals.

An important application of analysis filter banks arises when one attempts to send several low bandwidth signals simultaneously over a single high bandwidth communication channel. Suppose $x_i(k)$ is a bandlimited signal of bandwidth B that represents information to be transmitted. Subsignal $x_i(k)$ occupies the part of the spectrum represented by $-B \leq f \leq B$ where it is assumed that $B \ll f_s$. Next suppose $F_0 \geq 2B$. Consider a new complex subsignal $u_i(k)$ that is constructed by modulating $x_i(k)$ with a complex exponential of frequency iF_0.

$$u_i(k) = \exp(j2\pi iF_0 kT) \cdot x_i(k), \quad 0 \leq i < N \tag{5.7.4}$$

Recall from the frequency shift property of the DTFT that if $x_i(k)$ is modulated by the complex exponential, this shifts the spectrum of $x_i(k)$ to the right by iF_0 Hz.

$$U_i(f) = X_i(f - iF_0), \quad 0 \leq i < N \tag{5.7.5}$$

Since $F_0 \geq 2B$, the shifted magnitude spectra $A_i = |U_i|$ of the subsignals will not overlap. In particular, the spectrum of $U_i(f)$ will be centered at iF_0 with a radius of B, as shown in Figure 5.39 for the case $N = 4$. A composite input $x(k)$ is then constructed by summing the modulated subsignals as follows.

$$x(k) = \sum_{i=0}^{N-1} u_i(k) \tag{5.7.6}$$

Suppose the ith filter in an analysis filter bank has a magnitude response centered at iF_0 for $0 \leq i < N$. When this filter bank is driven with the composite input $x(k)$, the output $y_i(k)$ will be $x_i(k)$, but with its spectrum shifted to the right by iF_0. Consequently, the original subsignal $x_i(k)$ then can be recovered from $y_i(k)$ by modulating it with the complex conjugate of the

FIGURE 5.39: Subsignal Magnitude Spectra, $A_i = |U_i|$, Occupying Different Parts of the Nyquist Spectrum

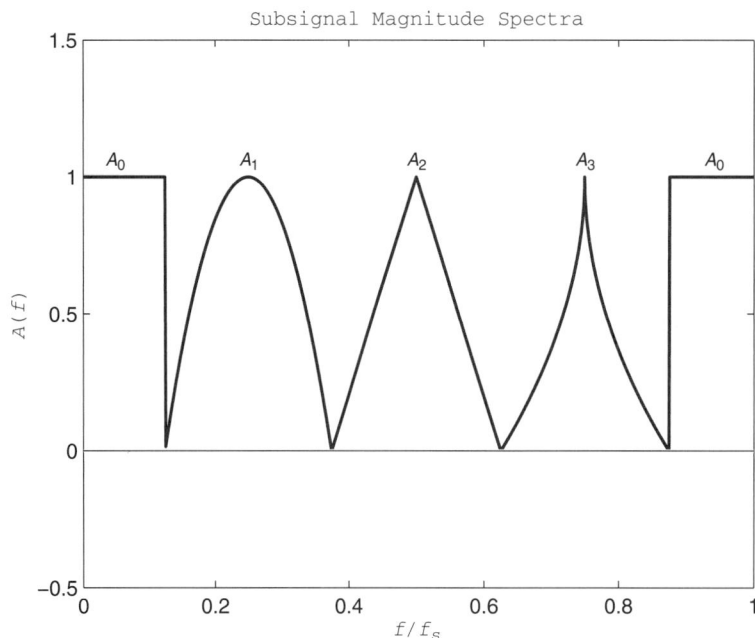

FIGURE 5.40: A Five-filter Synthesis Filter Bank

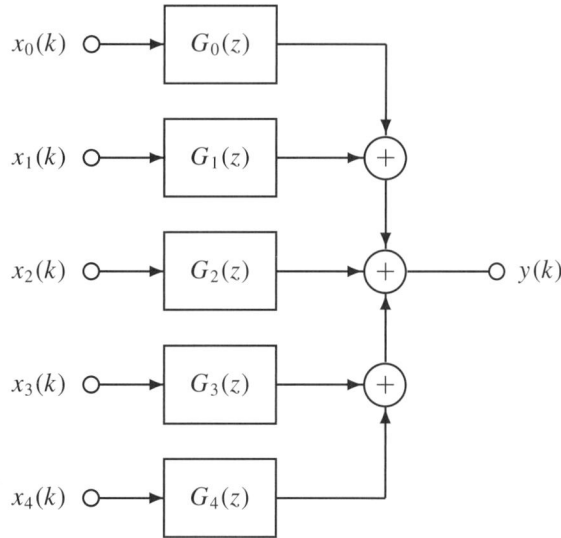

complex exponential.

$$x_i(k) \approx \exp(-j2\pi i F_0 kT) \cdot y_i(k), \quad 0 \le i < N \tag{5.7.7}$$

The extraction of $x_i(k)$ with an analysis filter bank using (5.7.6) is only approximate because the filters in the filter bank are not ideal. The technique of simultaneously transmitting several low bandwidth subsignals over a single high bandwidth channel using different parts of the spectrum for each subsignal is called *frequency-division multiplexing*. It is also possible to interleave the time samples of subsignals when they are transmitted, and this is referred to as *time-division multiplexing*.

Frequency-division multiplexing

Another way to configure the parallel filters in a filter bank is with a common summed output, as shown in Figure 5.40. Since the subsignals are combined to form a single output $y(k)$, this is called a *synthesis filter bank*. Systems which employ both analysis and synthesis filter banks are discussed in Chapter 8.

Synthesis filter bank

5.8 Adaptive Filters

The final type of digital filter that we consider is an adaptive filter. Filter design specifications for adaptive filters are distinct from the others because they are not formulated in the frequency domain using magnitude and phase responses. Instead, the design specifications of an adaptive filter are expressed in the time domain. The most common structure for a adaptive filter is a time-varying FIR filter, also called a *transversal* filter.

Transversal filter

$$y(k) = \sum_{i=0}^{m} w(i)x(k-i) \tag{5.8.1}$$

The $(m+1) \times 1$ coefficient vector $w(k)$ is a called the *weight vector*. Although the transversal FIR structure is not as general as an IIR structure, it does have one important redeeming feature. As long as $w(k)$ remains bounded, the transversal filter will be bounded-input bounded-output (BIBO) stable.

FIGURE 5.41: An
Adaptive Filter

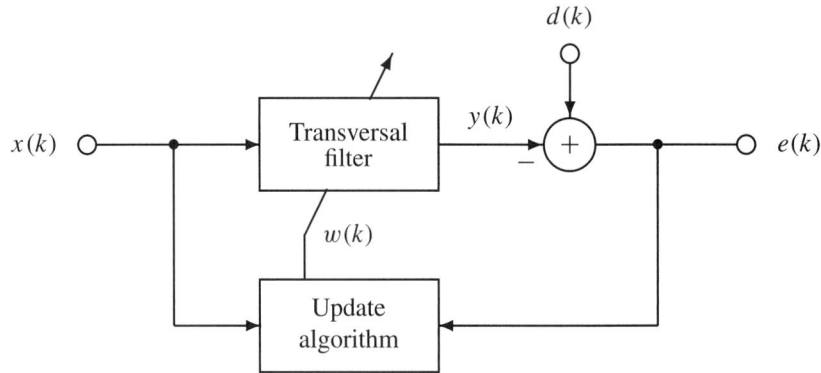

The adaptive filter design problem is formulated by provided the designer with a pair of time signals, the input $x(k)$ and a *desired output*, $d(k)$. There are several application areas for adaptive filters, including system identification, channel equalization, signal prediction, and noise cancellation, all described in Chapter 9. They differ from one another mainly in the way the input and the desired output are formulated. A block diagram of an adaptive filter is shown in Figure 5.41.

The unusual notation with a diagonal arrow through the transversal filter block is symbolic of "dial" where turning the dial corresponds to adjusting the parameters of the transversal filter. The output of a transversal filter can be formulated in a compact way by introducing the following vector of past inputs called a *state vector*.

State vector

$$u(k) \triangleq [x(k), x(k-1), \ldots, x(k-m)]^T \tag{5.8.2}$$

Given the weight vector $w(k)$ and the state vector $u(k)$, the filter output is simply the dot product of $w(k)$ with $u(k)$.

$$y(k) = w^T(k)u(k) \tag{5.8.3}$$

To adjust the parameter vector $w(k)$, an optimization procedure is used. Note from Figure 5.41 that the *error signal $e(k)$* is the difference between the desired output $d(k)$ and the filter output $y(k)$.

$$e(k) = d(k) - y(k) \tag{5.8.4}$$

Since the objective is to make the filter output closely follow the desired output, this is equivalent to minimizing the size of the error. In particular, the time-domain design specification for an adaptive filter is to minimize the expected value of the square of the error, also called the *mean square error*.

Mean square error

$$\epsilon(w) \triangleq E[e^2(k)] \tag{5.8.5}$$

To find a weight vector that minimizes the mean square error $\epsilon(w)$, we use an iterative optimization procedure is used. Suppose $w(0) \in R^{m+1}$ is an initial guess. The method of steepest descent is used to search for an optimal w.

$$w(k) = w(k-1) - \mu \nabla \epsilon[w(k-1)], \quad k \geq 1 \tag{5.8.6}$$

Here $\mu > 0$ is the step size which must be sufficiently small to ensure convergence. The gradient vector $\nabla \epsilon(w) = \partial \epsilon(w)/\partial w$ represents the direction of increasing mean square error. For the purposes of computing the gradient, the instantaneous value of the squared error $e^2(k)$

LMS method

can be used. This simplification leads to the following update formula for the weights called the *least mean square* or LMS method (Widrow and Sterns, 1985)

$$w(k) = w(k-1) + 2\mu e(k)u(k), \quad k \geq 1 \tag{5.8.7}$$

A detailed analysis of the LMS method including a convergence analysis can be found in Chapter 9. For example, in Chapter 9 it is shown that there is a tradeoff in selecting μ between the speed of convergence and the amount of residual noise in the steady-state solution. Techniques where μ is allowed to vary with k are also considered.

Example 5.11 Adaptive Filter

To illustrate the use of an adaptive filter, consider the problem of identifying a linear discrete-time system from input-output measurements. Suppose the input $x(k)$ consists of white noise uniformly distributed over the interval $[-1, 1]$, and the desired output $d(k)$ is produced by applying the input to the following IIR system with poles at $z = .4 \pm j.6$, zeros at $z = \pm .9$, and a gain factor of two.

$$y(k) = 2x(k) - 1.62x(k-2) + .8y(k-1) - .52y(k-2)$$

Next suppose an adaptive transversal filter of order $m = 12$ is used. If the step size is $\mu = .01$ and $N = 1000$ iterations are performed starting with an initial guess of $w(0) = 0$, this results in an FIR model whose magnitude and phase response is shown in Figure 5.42. Note that although there is some error, the fit is quite close for both the magnitude and the phase in spite of the fact that the desired response was generated by a more general IIR system. The fit can be improved by increasing m, but at the expense of using a higher order model.

FIGURE 5.42:
Magnitude and Phase Responses of the Adaptive Filter of Order $m = 12$

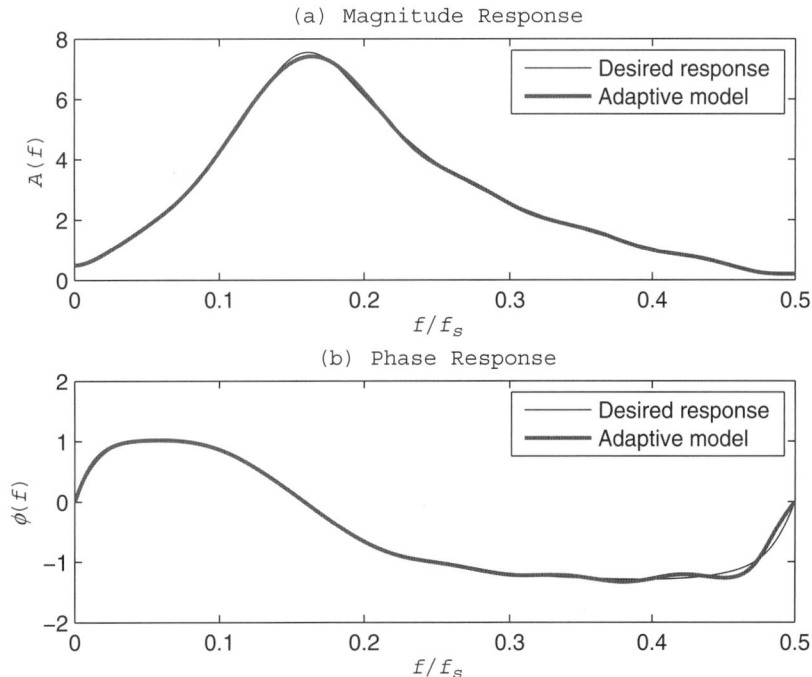

Pseudo-filters

Although the design specifications of an adaptive filter are specified in the time domain in the form of minimizing the mean square error, they can be recast in a more familiar form by using the notion of a pseudo-filter. Let $F = \{f_0, f_1, \ldots, f_{N-1}\}$ be a set of frequencies in the range $[0, f_s/2]$. Consider a composite input $x(k)$ that consists of sinusoids at these frequencies.

$$x(k) = \sum_{i=0}^{N-1} \cos(2\pi f_i k T) \tag{5.8.8}$$

Next suppose that at frequency f_i the desired gain is A_i and the desired phase shift is ϕ_i for $0 \le i < N$. Then the desired steady-state response to the input $x(k)$ is

$$d(k) = \sum_{i=0}^{N} C_i A_i \cos(2\pi f_i k T + \phi_i) \tag{5.8.9}$$

Here the constants $C_i > 0$ are *relative weights* which can be used by the designer to indicate that one frequency is more important than another. Often uniform weighting is used with $C_i = 1$ for $0 \le i < N$. However, if the fit to a desired magnitude or phase response at, say, f_j is difficult to achieve, then frequency f_j can be given increased weight in the optimization procedure by increasing C_j. Since the filter that produced the desired response in (5.8.9) does not necessarily exist, it is referred to as a *pseudo-filter*.

Pseudo-filter

● ● ● ● ● ● ● ● ● ● ● ● ● ● ● ● ● ● ●

5.9   GUI Software and Case Study

This section focuses on design specifications for various types of digital filters. A graphical user interface module called *g_filters* is introduced that allows the user to interactively explore the magnitude and phase characteristics, the poles and zeros, and the impulse response of digital filters without any need for programming. A case study example is presented and solved using MATLAB.

5.9.1   g_filters: Evaluation of Digital Filter Characteristics

The graphical user interface module *g_filters* allows the user to construct filters from design specifications and also evaluate coefficient quantization effects, all without any need for programming. GUI module *g_filters* features a display screen with tiled windows, as shown in Figure 5.43.

The *Block diagram* window in the upper left-hand corner contains a block diagram of the filter under investigation which can be an FIR filter, an IIR filter, or a user-defined filter. The FIR filters are designed with the windowing method discussed in Chapter 6, while the IIR filters are Butterworth filters created with the bilinear transformation discussed in Chapter 7. The following transfer function is used where $a = 1$ for the FIR filters.

$$H(z) = \frac{b_0 + b_1 z^{-1} + \cdots b_m z^{-m}}{1 + a_1 z^{-1} + \cdots + a_n z^{-n}} \tag{5.9.1}$$

The *Parameters* window below the block diagram displays edit boxes containing the filter parameters. The contents of each edit box can be directly modified by the user with the changes activated with the Enter key. Parameters *F0, F1, B,* and *fs* are the lower cutoff frequency, upper cutoff frequency, transition bandwidth, and sampling frequency, respectively.

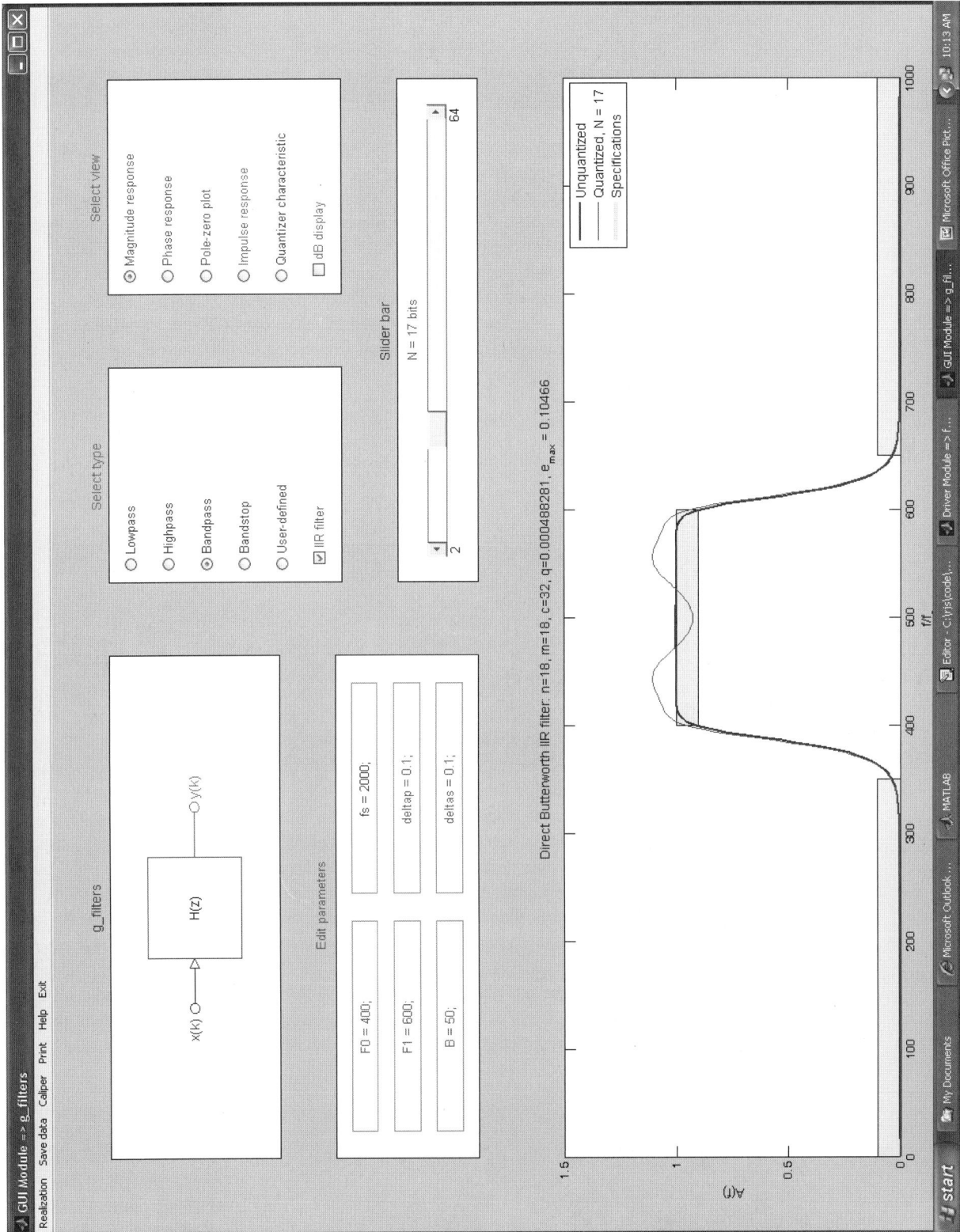

FIGURE 5.43: Display Screen of Chapter GUI Module *g_filters*

The lowpass filter uses cutoff frequency *F0*, the highpass filter uses cutoff frequency *F1*, and the bandpass and bandstop filters use both *F0* and *F1*. The parameters *deltap* and *deltas* specify the passband ripple and stopband attenuation, respectively.

The *Type* and *View* windows in the upper-right corner of the screen allow the user to select both the type of filter and the viewing mode. There are two categories of *filter types*. First, the user must select either an FIR filter or an IIR filter. Within each of these types, the user can then select the basic frequency-selective filter type: lowpass, highpass, bandpass, or bandstop. There is also a user-defined filter whose parameters are defined in a user-supplied MAT file that contains *a*, *b*, and *fs*. The *View options* include magnitude responses, phase responses, pole-zero plots, impulse responses, and the coefficient quantizer input-output characteristic. The dB check box toggles between logarithmic (dB) and linear displays of the magnitude response. When dB is checked, the passband ripple and stopband attenuation factors in the *Parameter* window also change to their logarithmic equivalents, Ap and As.

Just below the *Type* and *View* windows is a horizontal slider bar that controls N, the number of bits of precision used for coefficient quantization. To determine the quantization level, first $c_{\max}$ is computed such that

$$c_{\max} = \max\{|a_1|, \ldots, |a_n|, |b_0|, \ldots, |b_m|\} \qquad (5.9.2)$$

The scale factor c is then set to the next higher power as follows.

$$c = \text{nextpow2}(c_{\max}) \qquad (5.9.3)$$

Coefficient quantization

This way, a fixed-point representation of the filter coefficients *a* and *b* uses $M = \log_2(c)$ bits for the integer part, and $N - M$ bits for the fraction part. This corresponds to the following *coefficient quantization* level.

$$q = \frac{c}{2^{N-1}} \qquad (5.9.4)$$

As the number of bits of precision used to represent *a* and *b* decreases, the filter performance begins to deteriorate. Finite precision effects for FIR filters and IIR filters are discussed at the ends of Chapters 6 and 7, respectively.

All of the view options, except the quantizer characteristic, display two cases for comparison. The first is a double precision floating point filter which approximates the *unquantized* case, and the second is a fixed-point N-bit *quantized* filter using coefficient quantization. Higher-order IIR filters can become unstable for small values of N. When this happens, the migration of the quantized poles outside the unit circle can be viewed directly using the pole-zero plot option. The *Plot* window along the bottom half of the screen shows the selected view.

The *Menu* bar at the top of the screen includes several *menu options*. The *Caliper* option allows the user to measure any point on the current plot by moving the mouse crosshairs to that point and clicking. The *Save data* option is used to save the current *a*, *b*, *fs*, *x*, and *y* in a user-specified MAT file for future use. Files created in this manner can be loaded with the User-defined input option. The *Realization* option allows the user to choose between direct and cascade form realization structures. The cascade realization implements the *n*th order filter using a series of $L = \text{ceil}\{N/2\}$ second-order filters.

$$H(z) = b_0 H_1(z) \cdots H_L(z) \qquad (5.9.5)$$

The cascade realization has superior numerical properties compared to the direct realization, and this can be verified by comparing the two while the number of bits of precision is allowed to vary. The *Print* option prints the contents of the plot window. Finally, the *Help* option provides the user with some helpful suggestions on how to use module *g_filters* effectively.

CASE STUDY 5.1	**Highpass Elliptic Filter**

To illustrate of the use of filter design specifications and the effects of coefficient quantization, suppose $f_s = 200$ Hz and consider the problem of constructing a highpass digital filter to meet the following design specifications.

$$F_s = 40 \text{ Hz} \tag{5.9.6a}$$
$$F_p = 42 \text{ Hz} \tag{5.9.6b}$$
$$\delta_p = .05 \tag{5.9.6c}$$
$$\delta_s = .05 \tag{5.9.6d}$$

There are many filters that can meet or exceed these specifications. Design techniques for FIR filters are discussed in Chapter 6 and for IIR filters are discussed in Chapter 7. Notice that the width of the transition band is relatively small at

$$B = |F_p - F_s|$$
$$= 2 \text{ Hz} \tag{5.9.7}$$

CASE
STUDY 5.1

As B, δ_p, and δ_s are decreased, the required filter order increases. As we shall see in Chapter 7, the filter with the smallest order that still meets the specifications is an elliptic filter. An elliptic filter is an IIR filter that is optimal in the sense that the magnitude response contains ripples of equal amplitude in both the passband and the stopband. The coefficients of an elliptic filter can be obtained by using the FDSP toolbox function *f_ellipticz* presented in Chapter 7. The design of elliptic filters is discussed in detail in Chapter 7 where the theory behind *f_ellipticz* is introduced.

```
function case5_1

% EXAMPLE 5.12: Highpass elliptic filter

f_header('Example 5.12: Highpass elliptic filter')
fs = 200;
F_s = f_prompt ('Enter stopband cutoff frequency',0,fs/2,40);
F_p = f_prompt ('Enter passband cutoff frequency',F_s,fs/2,42);
delta_p = f_prompt ('Enter passband ripple factor',0,.5,.05);
delta_s = f_prompt ('Enter stopband attenuation factor',0,.5,.05);
N = f_prompt ('Enter number of bits of precision',2,64,10);

% Compute filter coefficients

f_type = 1;
[b,a] = f_ellipticz (F_p,F_s,delta_p,delta_s,f_type,fs);
n = length(a) - 1

% Compute quantized filter coefficients

c = max(abs([a(:) ; b(:)]));
c = 2^ceil(log(c)/log(2))
q = c/2^(N-1)
a_q = f_quant (a,q,0);
b_q = f_quant (b,q,0);
```

```
% Compute and plot magnitude responses

p = 250;
[H,f] = f_freqz (b,a,p,fs);
A = abs(H);
[H_q,f] = f_freqz (b_q,a_q,p,fs);
A_q = abs(H_q);
figure
h = plot (f,A,f,A_q);
set (h(1),'LineWidth',1.5)
f_labels ('Magnitude responses','\it{f}','\it{A(f)}')
q_str = sprintf ('Quantized, {\\itN} = %d',N);
legend ('Unquantized',q_str)
hold on
fill ([0 F_s F_s 0],[0 0 delta_s delta_s],'c')
fill ([F_p fs/2 fs/2 F_p],[1-delta_p 1-delta_p 1 1],'c')
h = plot (f,A,f,A_q);
set (h(1),'LineWidth',1.5)
f_wait


% Pole-zeros plots

figure
subplot (2,2,1)
f_pzplot(b,a,'Unquantized filter');
axis ([-1.5 1.5 -1.5 1.5])
for N = [10 8 6]
    q = c/(2^(N-1));
    a_q = f_quant (a,q,0);
    b_q = f_quant (b,q,0);
    switch N
        case 10, subplot (2,2,2);
        case 8, subplot (2,2,3);
        case 6, subplot (2,2,4);
    end
    caption = sprintf ('Quantized filter, {\\itN} = %d',N);
    f_pzplot (b_q,a_q,caption);
    axis ([-1.5 1.5 -1.5 1.5])
end
f_wait
```

When *case5_1* is run, it produces an elliptic filter of order $n = 6$. It then computes and plots two magnitude responses, as shown in Figure 5.44. The second magnitude response is that of a quantized elliptic highpass filter using coefficient quantization with a scale factor of $c = 4$ and $N = 10$ bits. Note that whereas the unquantized (double precision floating point) filter meets the design specifications, the quantized filter clearly does not.

The loss in fidelity of the magnitude response can be attributed to the fact that the poles and zeros of the quantized filter tend to move from their optimal positions as the quantization level increases. *Exam5_12* generates four plots of the poles and zeros corresponding to different levels of quantization, as shown in Figure 5.45. Inspection reveals that both the poles and

FIGURE 5.44:
Magnitude
Responses of
Unquantized and
Quantized Elliptic
Highpass Filters of
Order $n = 6$

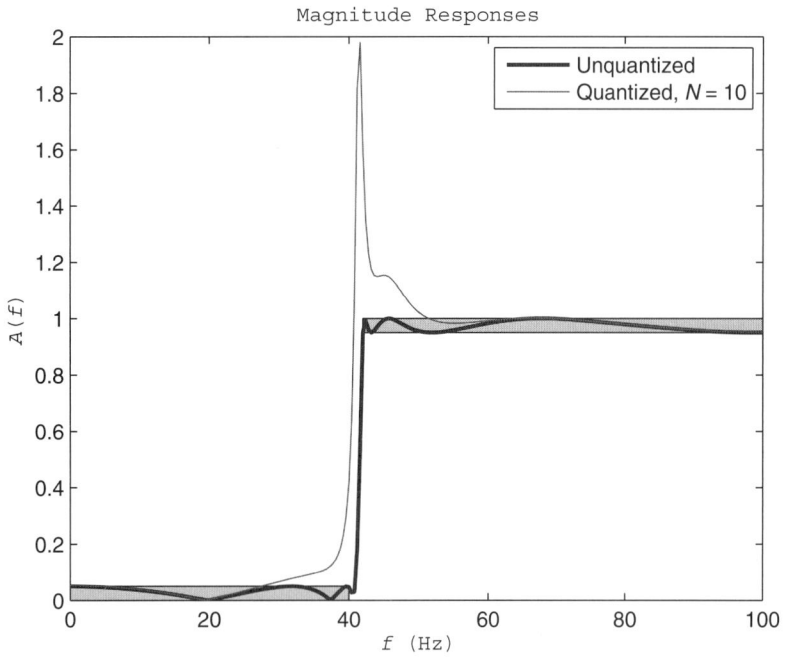

FIGURE 5.45: Pole-zero Plots of Elliptic Highpass Filters of Order $n = 6$ Using Quantized and Unquantized Coefficients

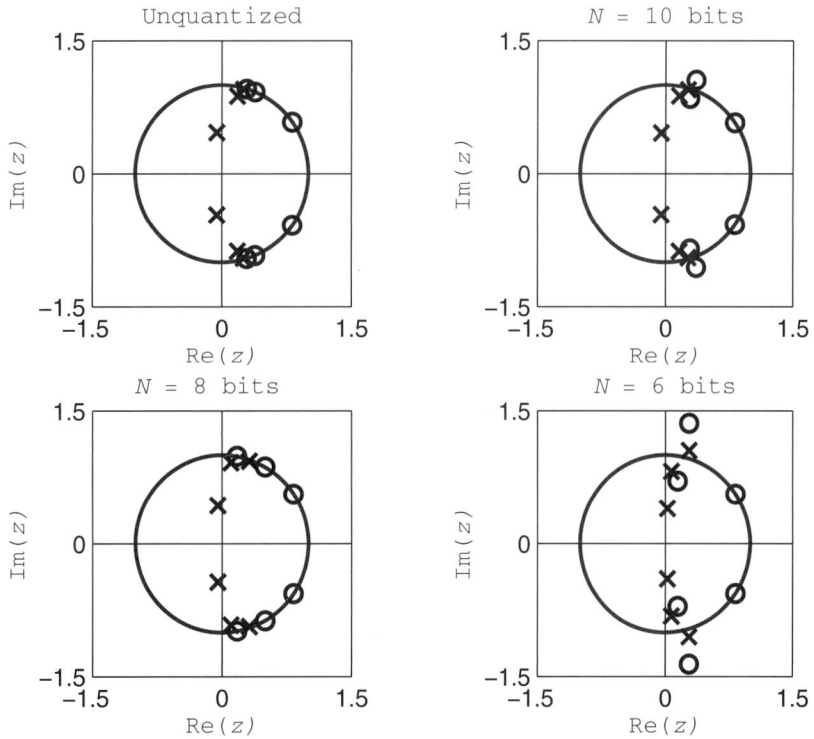

the zeros move significantly as the number of bits for coefficient quantization decreases from $N = 10$ to $N = 6$. This can be attributed to the fact that the roots of a polynomial are very sensitive to the values of the polynomial coefficients. For the final case of $N = 6$, the poles have migrated outside the unit circle, which means that this implementation does not even *have* a frequency response because the filter has become unstable.

5.10 Chapter Summary

Magnitude Response Specifications

This chapter focused on filter design specifications, and a survey of digital filter types. Both FIR and IIR filters were examined. The following IIR transfer function reduces to an FIR transfer function when $a_i = 0$ for $1 \le i \le n$.

$$H(z) = \frac{b_0 + b_1 z^{-1} + \cdots + b_m z^{-m}}{1 + a_1 z^{-1} + \cdots + a_n z^{-n}} \tag{5.10.1}$$

Lowpass specification

Filter design specifications are usually formulated in terms of a desired magnitude response $A(f) = |H(f)|$. The basic types of frequency-selective filters are lowpass, highpass, bandpass, and bandstop filters. For a *lowpass filter*, the design specifications are as follows.

$$1 - \delta_p \le A(f) \le 1, \quad 0 \le f \le F_p \tag{5.10.2a}$$

$$0 \le A(f) \le \delta_s, \quad F_s \le f \le f_s/2 \tag{5.10.2b}$$

Here (5.10.2a) is the passband specification, and (5.10.2b) is the stopband specification. For the passband, $0 < F_p < f_s/2$ is the passband cutoff frequency and $\delta_p > 0$ is the *passband ripple* factor. For the stopband, $F_p < F_s < f_s/2$ is the stopband cutoff frequency, and $\delta_s > 0$ is the *stopband attenuation* factor. Left unspecified is the band of frequencies between the

Transition band

passband and the stopband, which is called the *transition band*. The width of the transition band is

$$B = |F_s - F_p| \tag{5.10.3}$$

As the transition bandwidth B, passband ripple δ_p, and stopband attenuation δ_s approach zero, the order of the filter required to meet the specifications approaches infinity. The limiting special case of $B = 0$, $\delta_p = 0$, and $\delta_s = 0$ is an ideal lowpass filter. Bandpass filters have two stopbands bracketing a passband, and bandstop filters have two passbands bracketing

Logarithmic scale

a stopband. The filter magnitude response is often represented using a logarithmic scale of decibels (dB) as follows.

$$A(f) = 20 \log_{10}\{|H(f)|\} \text{ dB} \tag{5.10.4}$$

The ripple and attenuation factors δ_p and δ_s have logarithmic equivalents A_p and A_s, which are expressed in units of dB. The logarithmic scale can be useful for showing the degree of attenuation in the stopband.

Phase Response Specifications

Linear-phase filters

Often the desired phase response of a filter is left unspecified. One noteworthy exception is the design of *linear-phase* filters. Linear-phase filters with a phase response of $\phi(f) = -\tau f$

delay all spectral components of the input by the constant τ, and therefore do not distort the spectral components that pass through the filter. Although a linear phase characteristic can be approximated in the passband with an IIR Bessel filter, the simplest way to design an exact linear-phase filter is to use an FIR filter with an impulse response that satisfies the following *Symmetry constraint* linear-phase *symmetry constraint*.

$$h(k) = \pm h(m - k), \quad 0 \le k \le m \tag{5.10.5}$$

The symmetry constraint in (5.10.5) yields a filter with a constant group delay of $\tau = mT/2$. *FIR impulse* It is a direct constraint on the FIR coefficients because for an FIR filter, the nonzero part of the *response impulse response* is specified by the numerator coefficients.

$$h(k) = \begin{cases} b_k, & 0 \le k \le m \\ 0, & m < k < \infty \end{cases} \tag{5.10.6}$$

If the plus sign is used in (5.10.6), the impulse response $h(k)$ is a palindrome that exhibits even symmetry about the midpoint $k = m/2$; otherwise it exhibits odd symmetry. There are four types of linear-phase FIR filters, depending on whether the symmetry is even or odd and the filter order m is even or odd. The most general linear-phase filter is a type 1 filter with even symmetry and even order. The other three filter types have zeros at one or both ends of the frequency range, and are sometimes used for specialized applications. The zeros of a linear-phase FIR filter that are not on the unit circle occur in groups of four because for every zero at $z = r \exp(j\phi)$, there is a reciprocal zero at $z = r^{-1} \exp(j\phi)$.

Minimum-phase filter Every IIR transfer function $H(z)$ has a *minimum-phase* form $H_{\min}(z)$, whose magnitude response is the same as that of $H(z)$, but whose phase response has the least amount of phase lag possible. The minimum-phase form of $H(z)$ can be obtained as follows.

$$H_{\min}(z) = H_{\text{all}}^{-1}(z) H(z) \tag{5.10.7}$$

Allpass filter Here $H_{\text{all}}(z)$ is a *allpass filter* that is constructed from the zeros of $H(z)$ that lie outside the unit circle. An allpass filter passes all spectral components equally with a magnitude response of $A_{\text{all}}(f) = 1$.

Filter Types

In addition to the four basic frequency-selective filters, there are a wide variety of specialized filters. Filters that have a steady-state output that is shifted by a quarter of a cycle when a sinu- *Quadrature filter* soidal input is applied are *quadrature filters*. For example, a differentiator is a quadrature filter. *Hilbert transformer* Another important example of a quadrature filter is a *Hilbert transformer*, whose frequency response is

$$H_h(f) = \begin{cases} -j, & 0 \le f \le f_s/2 \\ j, & -f_s/2 < f < f \end{cases} \tag{5.10.8}$$

With the help of a Hilbert transformer, a complex signal whose spectrum occupies only half the bandwidth of the original signal can be generated.

Notch filter Another useful class of filters includes notch filters and resonators. A *notch filter* is a filter that is designed to block a single isolated frequency F_0 and pass all others.

$$H_{\text{notch}}(f) = 1 - \delta_1(f - F_0) \tag{5.10.9}$$

Resonator A *resonator*, by contrast, is the dual of a notch filter. It passes a single frequency F_0 and blocks all others.

$$H_{\text{res}}(f) = \delta_1(f - F_0) \tag{5.10.10}$$

Notch filters and resonators can be implemented using low-order IIR filters. Notch filters with multiple notches can be realized using a series or cascade configuration of individual notch filters. Likewise, resonators with multiple resonant frequency can be realized using a parallel configuration of individual resonators. Resonators and notch filters with a complete set of equally spaced frequencies are called *comb filters*.

Narrowband filter

A *narrowband filter* is a lowpass or a highpass filter whose normalized cutoff frequency satisfies $F_c/f_s \ll 1$. Narrowband filters can be effectively implemented by using sampling rate converters that lower the sampling rate to $F_s = f_s/M$ where $F_c/F_s \approx .25$. A sampling rate converter that lowers the sampling rate is a *decimator*, and one that raises the sampling rate is an *interpolator*.

Filter bank

Bandpass filters with narrow passbands are used in parallel *filter banks* where the entire Nyquist spectrum is partitioned into a set of nonoverlapping magnitude responses.

$$\sum_{i=0}^{N-1} A_i(f) = 1 \tag{5.10.11}$$

Filter banks with a common input are called *analysis banks* because they decompose a signal into its parts, and filter banks that produce a common output are called *synthesis banks* because they reconstruct a signal from it parts. By modulating bandlimited subsignals with complex exponentials and using analysis filter banks, several low bandwidth signals can be transmitted simultaneously over a single high bandwidth channel. This process is known as *frequency-division multiplexing*.

Frequency-division multiplexing

Adaptive filters

Adaptive filters are distinct from the other filters in a number of ways. They are time-varying discrete-time systems, and the design specifications for an *adaptive filter* are expressed in the time domain rather than the frequency domain. An adaptive transversal filter is an FIR filter whose weights are updated iteratively to minimize the mean square error between the filter output $y(k)$ and a desired output $d(k)$. Adaptive filters enjoy widespread application in areas such as system identification, inverse filtering or equalization, signal prediction, and noise cancellation.

Design, analysis, and applications of the different filter types are considered in detail in the remaining chapters of this text. A summary of the filter types and where they are investigated is shown in Table 5.2.

The FDSP toolbox includes a GUI module called *g_filters* that allows the user to construct filters from design specifications and examine the detrimental effects of coefficient quantization without any need for programming. The filters include FIR and IIR filters, lowpass, highpass, bandpass, and bandstop filters, plus user-defined filters.

Learning Outcomes

This chapter was designed to provide the student with an opportunity to achieve the learning outcomes summarized in Table 5.3

TABLE 5.2: ▶
Filter Types

Type	Chapters
Frequency-selective filters	6, 7
Linear-phase and zero-phase filters	6
Quadrature filters	6
Resonators	7
Notch filters	7
Narrowband filters	8
Filter banks	8
Adaptive filters	9

TABLE 5.3: ▶
Learning Outcomes
for Chapter 5

Num.	Learning Outcome	Sec.
1	Know how to specify the design characteristics of frequency-selective filters	5.2
2	Be able to go back and forth between linear and logarithmic design specifications	5.2
3	Understand what a linear-phase filter is and what kind of FIR impulse response is required	5.3
4	Know how to decompose a general transfer function into its minimum-phase and all-pass factors	5.4
5	Understand what it means for signals to be in phase quadrature and how to generate them with a digital oscillator	5.5
6	Know what functions notch filters, resonators, and comb filters are designed to perform	5.6
7	Understand what narrowband filter are and how sampling rate converters are used to design them	5.7
8	Know how analysis and synthesis filter bands are constructed and applied	5.7
9	Understand the differences between adaptive and fixed filters	5.8
10	Be able to use the GUI module *g_filters* to explore filter design specifications and filter types	5.9

● ● ● ● ● ● ● ● ● ● ● ● ● ● ● ● ● ●

5.11 Problems

The problems are divided into Analysis and Design problems that can be solved by hand or with a calculator, GUI Simulation problems that are solved using GUI module *g_filters*, and MATLAB Computation problems.

5.11.1 Analysis and Design

5.1 Consider a type 1 linear-phase FIR filter of order $m = 2$ with coefficient vector $b = [1, 1, 1]^T$.
(a) Find the transfer function $H(z)$.
(b) Find the amplitude response $A_r(f)$.
(c) Find the zeros of $H(z)$.

5.2 A bandpass filter has a sampling frequency of $f_s = 2000$ Hz and satisfies the following design specifications.

$$[F_{s1}, F_{p1}, F_{p2}, F_{s2}, \delta_p, \delta_s] = [200, 300, 600, 700, .15, .05]$$

(a) Find the logarithmic passband ripple A_p.
(b) Find the logarithmic stopband attenuation A_s.
(c) Using a logarithmic scale, sketch the shaded passband and stopband regions that $A(f)$ must lie within.

5.3 Consider a type 3 linear-phase FIR filter of order $m = 2$ with coefficient vector $b = [1, 0, -1]^T$.
(a) Find the transfer function $H(z)$.
(b) Find the amplitude response $A_r(f)$.
(c) Find the zeros of $H(z)$.

5.4 Consider a type 2 linear-phase FIR filter of order $m = 1$ with coefficient vector $b = [1, 1]^T$.
   (a) Find the transfer function $H(z)$.
   (b) Find the amplitude response $A_r(f)$.
   (c) Find the zeros of $H(z)$.

5.5 Consider the following FIR filter.

$$H(z) = 1 + 2z^{-1} + 3z^{-2} - 3z^{-3} - 2z^{-4} - z^{-5}$$

   (a) Is this a linear-phase filter? If so, what is the type?
   (b) Sketch a signal flow graph showing a direct-form II realization of $H(z)$ as in Section 5.1.

5.6 Consider a type 4 linear-phase FIR filter of order $m = 1$ with coefficient vector $b = [1, -1]^T$.
   (a) Find the transfer function $H(z)$.
   (b) Find the amplitude response $A_r(f)$.
   (c) Find the zeros of $H(z)$.

5.7 Let $H(z)$ be an arbitrary FIR transfer function of order m. Show that $H(z)$ can be written as a sum of two linear-phase transfer functions $H_e(z)$ and $H_o(z)$, where $h_e(k)$ exhibits even symmetry about $k = m/2$ and $h_o(k)$ exhibits odd symmetry about $k = m/2$. Hint: Add and subtract $h(m - k)$.

$$H(z) = H_e(z) + H_o(z)$$

5.8 Consider the following FIR filter.

$$H(z) = 1 + z^{-1} - 5z^{-2} + z^{-3} - 6z^{-4}$$

   (a) Is this a linear-phase filter? If so, what is the type?
   (b) Sketch a signal flow graph showing a direct-form II realization of $H(z)$ as in Section 5.1.
   (c) Using the MATLAB function *roots*, find the zeros of $H(z)$. Then sketch a signal flow graph showing a cascade form realization of $H(z)$.

5.9 This question focuses on the concept of the amplitude response of a filter.
   (a) Show how to compute the magnitude response from the amplitude response.
   (b) Suppose the magnitude response equals the amplitude response for $0 \leq f \leq F_0$, but for $f > F_0$ they differ. What happens to the phase response at $f = F_0$?

5.10 Recall from Table 5.1 that linear-phase FIR filters of types 2–4 all have zeros at $z = -1$ or $z = 1$ or both. A type 1 linear-phase FIR filter is more general.
   (a) Show that for a type 1 linear-phase FIR filter, symmetry constraint (5.3.8) does not imply that $H(z)$ has a zero at $z = -1$.
   (b) Show that for a type 1 linear-phase FIR filter, symmetry constraint (5.3.8) does not imply that $H(z)$ has a zero at $z = 1$.

5.11 Suppose $H(z)$ is a type 4 linear-phase FIR filter.

$$H(z) = c_0 + c_1 z^{-1} + c_2 z^{-2} + c_3 z^{-3}$$

   (a) Find the amplitude response of this filter.
   (b) Find the phase offset α and group delay $D(f)$ of this filter.

5.12 Suppose $H(z)$ is a type 2 linear-phase FIR filter.

$$H(z) = c_0 + c_1 z^{-1} + c_2 z^{-2} + c_3 z^{-3}$$

   (a) Find the phase offset α and group delay $D(f)$ of this filter.
   (b) Find the amplitude response of this filter.

5.13 Consider the following running average filter.

$$H(z) = \frac{1}{10} \sum_{i=0}^{9} z^{-i}$$

(a) Write down the difference equation for this filter.
(b) Convert this filter to a noncausal zero-phase filter. That is, write down the difference equations for the zero-phase version of the running average filter. You can use $f_i = \sqrt{(1/10)}$ in Algorithm 5.1.

5.14 Suppose the impulse response of an FIR filter of order $m = 5$ is as follows where the X terms are to be determined.

$$h = [2, 4, 2, X, X, X]$$

(a) Assuming $H(z)$ is a linear-phase filter, sketch the complete impulse response. If there are multiple solutions, sketch each of them.
(b) For each solution in part (a), indicate the linear-phase FIR filter type.
(c) For each solution in part (a), find the phase offset α and the group delay $D(f)$.

5.15 Consider the following FIR filter of order m known as a *running average* filter.

$$H(z) = \frac{1 + z^{-1} + \cdots + z^{-(M-1)}}{M}$$

(a) Find the impulse response of this filter.
(b) Is this a linear-phase filter? If so, what type?
(c) Find the group delay of this filter.

5.16 Suppose $H(z)$ is a stable filter with $A(f) = 0$ for $.1 \le f/f_s \le .2$. Show that $H(z)$ is not causal.

5.17 A bandstop filter has a sampling frequency of $f_s = 200$ Hz and satisfies the following design specifications.

$$[F_{p1}, F_{s1}, F_{s2}, F_{p2}, A_p, A_s] = [30, 40, 60, 80, 2, 30]$$

(a) Find the linear passband ripple δ_p.
(b) Find the linear stopband attenuation δ_s.
(c) Using a linear scale, sketch the shaded passband and stopband regions that $A(f)$ must lie within.

5.18 A linear-phase FIR filter $H(z)$ of order $m = 8$ has zeros at $z = \pm j.5$ and $z = \pm.8$.
(a) Find the remaining zeros of $H(z)$ and sketch the poles and zeros in the complex plane.
(b) The DC gain of the filter is 2. Find the filter transfer function $H(z)$.
(c) Suppose the input signal gets delayed by 20 msec as it passes through this filter. What is the sampling frequency f_s?

5.19 Consider the following first-order IIR filter.

$$H(z) = \frac{.4(1 - z^{-1})}{1 + .2z^{-1}}$$

(a) Compute and sketch the magnitude response $A(f)$.
(b) What type of filter is this (lowpass, highpass, bandpass, bandstop)?
(c) Suppose $F_p = .4 f_s$. Find the passband ripple δ_p.
(d) Suppose $F_s = .2 f_s$. Find the stopband attenuation δ_s.

5.20 Let $X(k) = [x_1(k), x_2(k)]^T$. A digital oscillator that produces two sinusoidal outputs $x_1(k)$ and $x_2(k)$ that are in phase quadrature can be obtained using a first-order two-dimensional system of the following form.

$$X(k) = AX(k-1), \quad X(0) = c$$

(a) Find a coefficient matrix A which produces an oscillator with frequency $F_0 = .3f_s$.
(b) Find an initial condition vector c that produces the solution

$$X(k) = \begin{bmatrix} \cos(.6\pi k) \\ \sin(.6\pi k) \end{bmatrix}$$

(c) Find a coefficient matrix A and an initial condition vector c that produces the solution

$$X(k) = \begin{bmatrix} d\cos(2\pi F_0 kT + \psi) \\ d\sin(2\pi F_0 kT + \psi) \end{bmatrix}$$

5.21 Suppose the following quadrature pair of sinusoidal signals with frequency F_0 and unit amplitude is available.

$$X(k) = \begin{bmatrix} \cos(2\pi F_0 kT) \\ \sin(2\pi F_0 kT) \end{bmatrix}$$

(a) Find the Chebyshev polynomials of the first kind $T_i(x)$ for $0 \le i \le 3$.
(b) Find the Chebyshev polynomials of the second kind $U_i(x)$ for $0 \le i \le 3$.
(c) Let $X(k) = [x_1(k), x_2(k)]^T$. Find polynomials f and g such that

$$f[x_1(k)] + x_2(k)g[x_1(k)] = \cos^3(2\pi F_0 kT) + 2\sin^2(2\pi F_0 kT)$$

5.22 Consider the following IIR filter.

$$H(z) = \frac{10(z^2 - 4)(z^2 + .25)}{(z^2 + .64)(z^2 - .16)}$$

(a) Find $H_{\min}(z)$, the minimum-phase version of $H(z)$.
(b) Sketch the poles and zeros of $H_{\min}(z)$.
(c) Find an allpass filter $H_{all}(z)$ such that $H(z) = H_{all}(z)H_{\min}(z)$.
(d) Sketch the poles and zeros of $H_{all}(z)$.

5.23 Suppose $H(z)$ is an allpass filter with input $x(k)$ and output $y(k)$ whose magnitude response satisfies the following constraint.

$$A(f) = 1, \quad |f| \le f_s/2$$

(a) Show that $|Y(f)| = |X(f)|$.
(b) Use Parseval's identity to show that $H(z)$ is a *lossless* system. That is, show that the energy of $y(k)$ is equal to the energy of $x(k)$.

$$\sum_{k=-\infty}^{\infty} y^2(k) = \sum_{k=-\infty}^{\infty} x^2(k)$$

5.24 Consider the following IIR filter.

$$H(z) = \frac{2z^2 + 5z + 2}{z^2 - 1}$$

   (a) Find the minimum-phase form of $H(z)$.
   (b) Find a magnitude equalizer $G(z)$ such that $G(z)H(z)$ is an allpass filter with magnitude response $A(f) = 1$.

5.25 An ideal Hilbert transformer has the following frequency response.

$$H_d(f) = -j\,\text{sgn}(f), \quad 0 \le |f| < f_s/2$$

   (a) Show that a Hilbert transformer is an allpass filter.
   (b) Using the inverse DTFT, show that the impulse response of an ideal Hilbert transformer is

$$h_d(k) = \begin{cases} \dfrac{1 - \cos(k\pi)}{k\pi}, & k \neq 0 \\ 0, & k = 0 \end{cases}$$

5.26 Suppose $H(z)$ is a filter with input $x(k)$ and output $y(k)$ whose magnitude response satisfies the following constraint.

$$A(f) \le 1, \quad |f| \le f_s/2$$

   (a) Show that $|Y(f)| \le |X(f)|$.
   (b) Use Parseval's identity to show that $H(z)$ is a *passive* system. That is, show that the energy of $y(k)$ is less than or equal to the energy of $x(k)$.

$$\sum_{i=0}^{\infty} y^2(k) \le \sum_{i=0}^{\infty} x^2(k)$$

5.27 Let $H(z)$ be a nonzero linear-phase FIR filter of order $m = 2$.
   (a) Is it possible for $H(z)$ to be a minimum-phase filter? If so, construct an example. If not, why not?
   (b) Is it possible for $H(z)$ to be an allpass filter? If so, construct an example. If not, why not?

5.28 Consider the following IIR filter.

$$H(z) = \frac{2(z + 1.25)(z^2 + .25)}{z(z^2 - .81)}$$

   (a) Find the minimum-phase version of this system, and sketch its poles and zeros.
   (b) Find the maximum-phase version of this system, and sketch its poles and zeros.
   (c) How many transfer functions with real coefficients have the same magnitude response as $H(z)$?

5.29 The following IIR filter has two parameters α and β. For what values of these parameters is this an allpass filter?

$$H(z) = \frac{1 + 3z^{-1} + (\alpha + \beta)z^{-2} + 2z^{-3}}{2 + (\alpha - \beta)z^{-1} + 3z^{-2} + z^{-3}}$$

5.30 The general form for a notch filter with a notch at $F_0 \neq 0$ is given in (5.6.10) where $\theta_0 = 2\pi F_0 T$.

$$H_{\text{notch}}(z) = \frac{c[z - \exp(j\theta_0)][z - \exp(-j\theta_0)]}{[z - r\exp(j\theta_0)][z - r\exp(-j\theta_0)]}$$

(a) Rewrite $H_{\text{notch}}(z)$ as a ratio of two polynomials with real coefficients.
(b) Find a value for the gain factor c such that $H_{\text{notch}}(f) = 1$ at $f = 0$.

5.31 Consider the mean square error performance criterion used for the adaptive filter in Figure 5.41.

$$\epsilon(w) = E[e^2(k)]$$

(a) Suppose the mean square error is approximated as $\epsilon(w) = e^2(k)$. Find the gradient vector $\nabla\epsilon(w) = \partial\epsilon(w)/\partial w$. Express your final answer in terms of the state vector of past inputs u.
(b) Using your expression for $\nabla\epsilon(w)$ from part (a) and the step size μ, show that the steepest descent method for approximating w in (5.8.6) reduces to the LMS method.

5.32 Suppose the following two filters are resonators with resonant frequencies at F_0 and F_1, respectively. Write the difference equation of a double-resonator with resonant frequencies at F_0 and F_1.

$$H_0(z) = \frac{b_0z^2 + b_1z + b_2}{z^2 + a_1z + a_2}$$

$$H_1(z) = \frac{B_0z^2 + B_1z + B_2}{z^2 + A_1z + A_2}$$

5.33 Consider the DC resonator in (5.6.14).

$$H_{\text{dc}}(z) = \frac{.5(1 - r)(z + 1)}{z - r}$$

(a) Find the impulse response $h(k)$.
(b) Find the difference equation.

5.34 Suppose the following two filters are notch filters with notches at F_0 and F_1, respectively. Write the difference equation of a double-notch filter with notches at F_0 and F_1.

$$H_0(z) = \frac{b_0z^2 + b_1z + b_2}{z^2 + a_1z + a_2}$$

$$H_1(z) = \frac{B_0z^2 + B_1z + B_2}{z^2 + A_1z + A_2}$$

5.35 Consider the DC notch filter in (5.6.3).

$$H_{\text{DC}}(z) = \frac{.5(1 + r)(z - 1)}{z - r}$$

(a) Find the impulse response $h(k)$.
(b) Find the difference equation.

5.36 The general form for a resonator with a resonant frequency of F_0 is given in (5.6.15) where $\theta_0 = 2\pi F_0 T$.

$$H_{\text{res}}(z) = \frac{c(z^2 - 1)}{[z - r\exp(j\theta_0)][z - r\exp(-j\theta_0)]}$$

(a) Rewrite $H_{\text{res}}(z)$ as a ratio of two polynomials with real coefficients.
(b) Find a value for the gain factor c such that $|H_{\text{res}}(f)| = 1$ at $f = F_0$.

FIGURE 5.46: Equalizer Design Using an Adaptive Filter

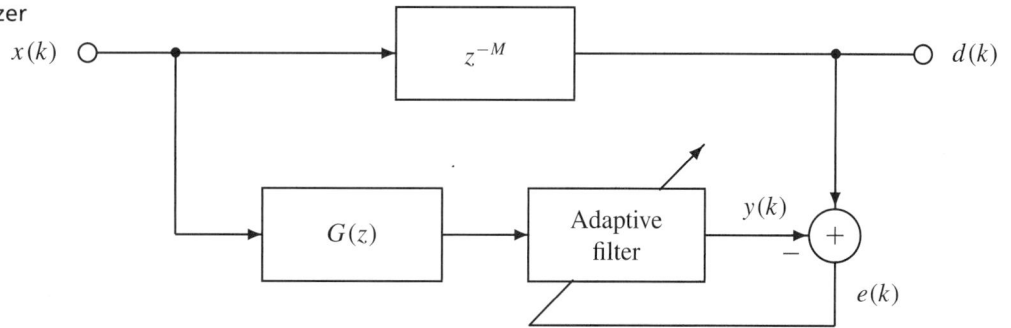

5.37 Using the results from Problem 5.30 and (5.6.8), design a notch filter $H_{\text{notch}}(z)$ that has a notch at $F_0 = .1 f_s$ and a notch bandwidth of $\Delta F = .01 f_s$.

5.38 Consider the adaptive filter shown in Figure 5.46. This configuration can be used to design an equalizer with delay. Here $G(z)$ is a stable IIR filter. Suppose the adaptive filter converges to an FIR filter $H(z)$ with error $e(k) = 0$. Let

$$G_{\text{equal}}(z) = G(z) H(z)$$

(a) Show that $G_{\text{equal}}(z)$ is an allpass filter with $A_{\text{equal}}(f) = 1$.
(b) Show that $G_{\text{equal}}(z)$ is a linear-phase filter with $\phi_{\text{equal}}(f) = -2\pi M T f$.

5.39 Using the results from Problem 5.36 and (5.6.8), design a resonator $H_{\text{res}}(z)$ that has a resonant frequency at $F_0 = .4 f_s$ and a bandwidth of $\Delta F = .02 f_s$.

5.40 Consider the filter bank with $m = 2$ filters shown in Figure 5.47.
(a) Find an expression for the magnitude response $A_0(f)$ in the transition band $[.22 f_s, .28 f_s]$.
(b) Find the 3 dB cutoff frequency F_0 of the first filter.

FIGURE 5.47: A Power-complementary Pair of Filters

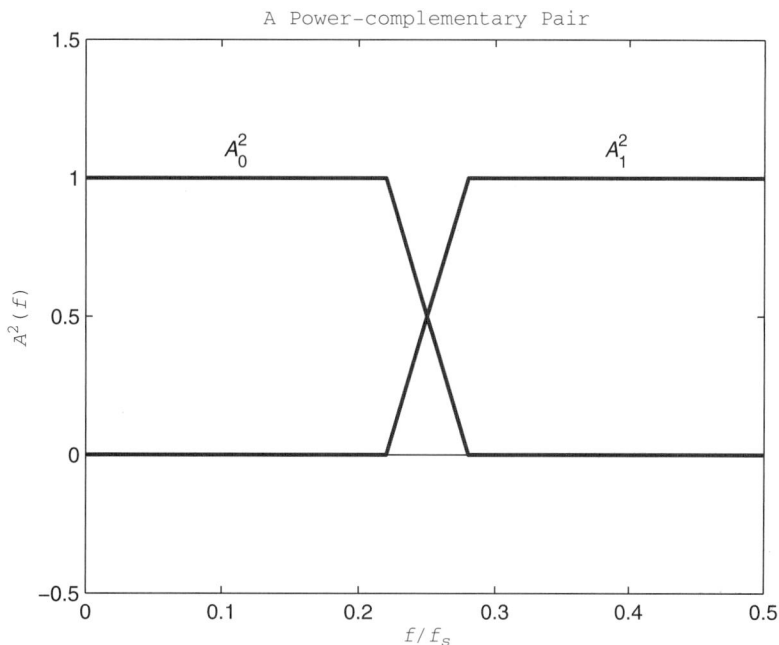

(c) Find an expression for the magnitude response $A_1(f)$ in the transition band $[.22f_s, .28f_s]$.

(d) Find the 3 dB cutoff frequency F_1 of the second filter.

(e) Show that the two filters form a *power-complementary pair* with

$$A_0^2(f) + A_1^2(f) = 1$$

5.41 Consider the filter bank with $m = 2$ filters shown in Figure 5.48.

(a) Find an expression for the magnitude response $A_0(f)$ in the transition band $[.2f_s, .3f_s]$.

(b) Find the 3 dB cutoff frequency F_0 of the first filter.

(c) Find an expression for the magnitude response $A_1(f)$ in the transition band $[.2f_s, .3f_s]$.

(d) Find the 3 dB cutoff frequency F_1 of the second filter.

(e) Show that the two filters form a *magnitude-complementary pair* with

$$A_0(f) + A_1(f) = 1$$

FIGURE 5.48: A Magnitude-complementary Pair of Filters

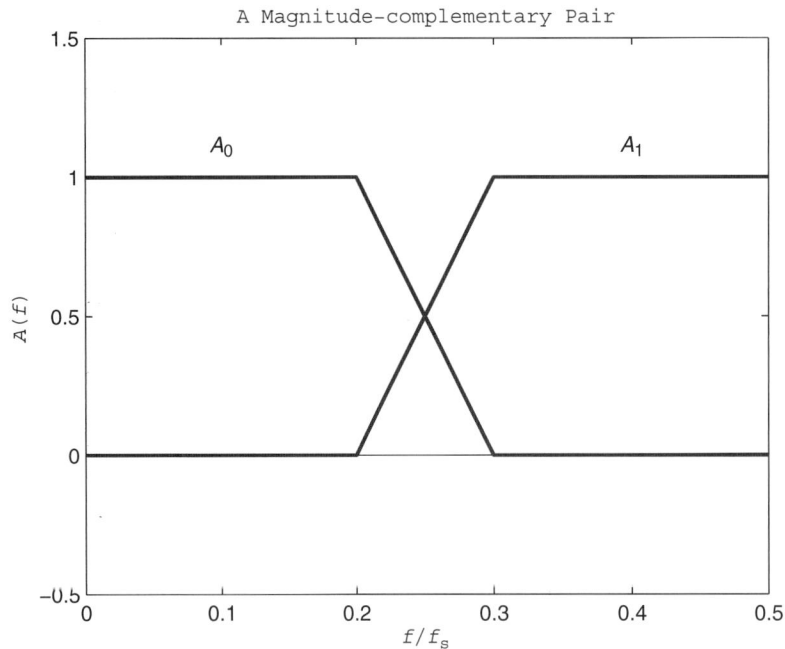

5.42 Consider the problem of designing a lowpass narrowband filter. Suppose the sampling frequency is $f_s = 20$ kHz.

(a) The desired lowpass cutoff frequency is $F_c = 50$ Hz. Find the sampling rate reduction factor M such that if $F_s = f_s/M$; then the new normalized cutoff frequency will be $F_c = .25F_s$.

(b) A cascade configuration of three rate converters with rate conversion factors M_1, M_2, M_3 can be used to implement a multistage sampling rate converter. Factor M from part (a) as follows, where the maximum of $\{M_1, M_2, M_3\}$ is as small as possible.

$$M = M_1 M_2 M_3$$

5.11.2 GUI Simulation

5.43 Use the GUI module *g_filters* and select an FIR lowpass filter. Adjust the parameter values to $f_s = 100$ Hz, $F_0 = 30$ Hz, and $B = 10$ Hz.
(a) Plot the magnitude response using the dB scale.
(b) Plot the phase response.
(c) Plot the impulse response. Is this a linear-phase filter? If so, what type?

5.44 Use the GUI module *g_filters* and select an IIR highpass filter. Adjust the number of bits of precision N to highest value that still makes the quantized filter go unstable.
(a) Plot the unstable pole-zero plot
(b) Increase N by one so the quantized filter becomes stable. Then plot the impulse response.

5.45 Use the GUI module *g_filters* and select the User-defined filter option. Load the filter in MAT-file *prob5_45. mat*. Set the number of bits for coefficient quantization to $N = 10$.
(a) Plot the magnitude response using a direct form realization.
(b) Plot the phase response using a direct form realization.
(c) Plot the magnitude response using a cascade form realization.
(d) Plot the phase response using a cascade form realization.

5.46 Use the GUI module *g_filters* and select an FIR bandstop filter. Adjust the number of bits of precision N until the quantization level q is larger than .005.
(a) Plot the magnitude response.
(b) Plot the pole-zero plot

5.47 A notch filter is a filter that is designed to remove a single frequency. Consider the following transfer function for a notch filter.

$$H(z) = \frac{.9766(1 + z^{-1} + z^{-2})}{1 + .9764z^{-1} + .9534z^{-2}}$$

Create a MAT-file called that contains $fs = 1000$, and the a and b for this filter. Then use the User-defined option of GUI module *g_filters* to load this filter. Set $N = 6$ bits.
(a) Plot the magnitude response. Use the Caliper option to estimate the notch frequency.
(b) Plot the phase response.
(c) Plot the pole-zero pattern.

5.48 Consider the following *running average filter*. Create a MAT-file called *prob5_48.mat* that contains $fs = 300$, a, and b for this filter.

$$y(k) = \frac{1}{10} \sum_{i=0}^{9} x(k - i)$$

Use the GUI module *g_filters* with the User-defined option to load this filter.
(a) Plot the magnitude response.
(b) Plot the phase response.
(c) Plot the pole-zero plot.
(d) Plot the impulse response. Is this a linear-phase filter? If so, what type?

5.49 The derivative of an analog signal $x_a(t)$ can be approximated numerically by taking differences between the samples of the signal using the following first-order *backwards Euler differentiator*.

$$y(k) = \frac{x(k) - x(k - 1)}{T}$$

Create a MAT-file called *prob5_49.mat* that contains $f_s = 10$, a, and b for this filter. Then use GUI module *g_filters* with the User-defined option to load this filter.
(a) Plot the magnitude response.
(b) Plot the phase response.
(c) Plot the impulse response. Is this a linear-phase filter? If so, what type?

5.50 Use the GUI module *g_filters* to analyze an IIR bandpass filter. Adjust the number of bits of precision N until the quantized filter first goes unstable. Then increase N by 1.
(a) Plot the magnitude response
(b) Plot the pole-zero pattern

5.11.3 MATLAB Computation

5.51 An *inverse comb filter* (see Chapter 7) is a filter that eliminates set of isolated equally-spaced frequencies from a signal. Consider the following inverse comb filter that has *n teeth*.

$$H(z) = \frac{b_0(1 - z^{-n})}{1 - r^n z^{-n}}$$

Here the filter gain is $b_0 = (1 + r^n)/2$. Suppose $n = 8$, $r = .96$, and $f_s = 300$ Hz. Write a MATLAB program that uses *f_freqz* to compute the frequency response. Compute both the unquantized frequency response (set bits $= 64$), and the frequency response with coefficient quantization using $N = 8$ bits. Plot both magnitude responses on a single plot using the linear scale and a legend.

5.52 A *comb filter* (see Chapter 7) is a filter that extracts a set of isolated equally spaced frequencies from a signal. Consider the following comb filter that has *n teeth*.

$$H(z) = \frac{b_0}{1 - r^n z^{-n}}$$

Here the filter gain is $b_0 = 1 - r^n$. Suppose $n = 10$, $r = .98$, and $f_s = 300$ Hz. Write a MATLAB program that uses *f_freqz* to compute the frequency response. Compute both the unquantized frequency response (set bits $= 64$), and the frequency response with coefficient quantization using $N = 4$ bits. Plot both magnitude responses on a single plot using the linear scale and a legend.

5.53 Consider the following FIR system.

$$G(z) = 3 - 4z^{-1} + 2z^{-2} + 7z^{-3} + 4z^{-4} + 9z^{-5}$$

Suppose $G(z)$ is driven by $N = 500$ samples of white noise $x(k)$ uniformly distributed over $[-10, 10]$. Let $D(z) = G(z)X(z)$ represent the desired output. Write a MATLAB program that performs the following tasks.
(a) Compute the optimal weight w for an adaptive transversal filter of order m using the LMS method. Start from an initial guess of $w(0) = 0$ and choose a step size μ that ensures convergence. Compute and display the final w for three cases: $m = 3$, $m = 5$, and $m = 7$. Also display the coefficient vector b of $G(z)$.
(b) Let $H(z)$ be the transfer function of the transversal filter using the final weights when $m = 7$. Create a 2×1 array of plots. Plot the magnitude responses of $G(z)$ and $H(z)$ on the first plot with a legend. Plot the phase responses of $G(z)$ and $H(z)$ on the second plot with a legend.

5.54 Consider the following IIR filter.

$$H(z) = \frac{1 + 1.75z^{-2} - .5z^{-4}}{1 + .4096z^{-4}}$$

(a) Write a MATLAB program that uses f_minall to compute and print the coefficients of the minimum-phase and allpass parts of $H(z)$.

(b) Use the MATLAB $subplot$ command to plot the magnitude responses $A(f)$, $A_{min}(f)$ and $A_{all}(f)$ on a single screen using three separate plots.

(c) Repeat part (b), but for the phase responses.

(d) Use f_pzplot to plot the poles and zeros of $H(z)$, $H_{min}(z)$ and $H_{all}(z)$ on one screen using three separate square plots.

CHAPTER 6

FIR Filter Design

Chapter Topics

6.1 Motivation

6.2 Windowing Method ✓

6.3 Frequency-sampling Method

6.4 Least-squares Method ✓

6.5 Equiripple Filters

6.6 Differentiators and Hilbert Transformers

6.7 Quadrature Filters

6.8 Filter Realization Structures

*6.9 Finite Word Length Effects

6.10 GUI Software and Case Study

6.11 Chapter Summary

6.12 Problems

6.1 Motivation

A digital filter is a discrete-time system that is designed to reshape the spectrum of the input signal in order to produce desired spectral characteristics in the output signal. In this chapter we focus on a specific type of digital filter, the finite impulse response or FIR filter. An mth-order FIR filter is a discrete-time system with the following generic transfer function.

$$H(z) = b_0 + b_1 z^{-1} + \cdots + b_m z^{-m}$$

FIR filters offer a number of important advantages in comparison with IIR filters. The nonzero part of the impulse response of an FIR filter is simply $h(k) = b_k$ for $0 \leq k \leq m$. Consequently, an FIR impulse response can be obtained directly from inspection of the transfer function or the difference equation. Since the poles of an FIR filter are all located at the origin, FIR filters are always stable. FIR filters tend to be less sensitive to finite word length effects. Unlike IIR filters, quantized FIR filters cannot become unstable or exhibit limit cycle oscillations. FIR filters can be designed to closely approximate arbitrary magnitude responses if the order of the filter is allowed to be sufficiently large. Furthermore, if symmetry is used; the phase response of an FIR filter can be made to be linear. This is an important characteristic because it means that different spectral components of the input signal are delayed by the same amount as they are processed by the filter. A linear-phase filter does not distort a signal within the passband; it only delays it. With a quadrature filter approach, FIR filters can be designed to meet both magnitude and phase specifications as long as sufficient delay is included.

Although impressive control of the frequency response can be achieved with FIR filters, these filters do suffer from some limitations in comparison with IIR filters. One fundamental drawback occurs when we attempt to design a sharp frequency-selective filter with a narrow transition band, similar to an IIR elliptic filter. To meet the same design specifications with an FIR filter, a much higher-order filter is required. The increased order means larger storage requirements and a longer computational time. Computational time is particularly important in real-time applications where inter-sample signal processing must be performed before each new ADC sample arrives. High-order FIR filters are more suitable for offline batch processing where the entire input signal is available ahead of time.

We begin this chapter by introducing an example of an application of FIR filters. Next, several methods for FIR filter design are presented. Most of the design methods produce linear-phase FIR filters with a propagation delay of $\tau = mT/2$, where m is the filter order and T is the sampling interval. The first design method is the windowing method, a simple and effective technique that is based on a soft or gradual truncation of the delayed impulse response. Next, a frequency sampling method is presented that uses the inverse DFT to compute the filter coefficients. The frequency sampling method can be optimized with the inclusion of transition band samples. This is followed by the least-squares method. This is an optimization method that uses an arbitrary set of discrete frequencies and a user-selectable weighting function. Another optimization method based on the Parks-McClellan algorithm is used to design a sharp equiripple frequency-selective FIR filter that is analogous to the IIR elliptic filter. All of these methods can be used to construct a filter with a prescribed magnitude response and a linear phase response. Next, the focus turns to specialized linear-phase FIR filters such as the Hilbert transformer that produce signals that are in phase quadrature. These signals form the basis for a general two-stage quadrature filter that can be designed to meet both magnitude response and phase response specifications as long as sufficient delay is included. Finally, a GUI module called *g_fir* is introduced that allows the user to design and evaluate a variety of FIR filters without any need for programming. The chapter concludes with a case study example, and a summary of FIR filter design techniques.

6.1.1 Numerical Differentiators

In engineering applications there are instances where it is useful to obtain the derivative of an analog signal from the samples of the signal. For example, an estimate of velocity might be obtained from position measurements, or an acceleration estimate might be obtained from velocity measurements. Numerical differentiation is a challenging practical problem because the process is highly sensitive to the presence of noise. As an introduction to the topic, we examine the use of simple low-order FIR filters to numerically approximate the differentiation process. The objective is to design a digital equivalent to the following analog system.

$$H_a(s) = s \qquad (6.1.1)$$

Suppose the analog signal to be differentiated, $x_a(t)$, is approximated with a linear polynomial in the neighborhood of $t = kT$.

$$x_a(t) \approx x(k) + c(t - kT) \tag{6.1.2}$$

Recall that $x(k) = x_a(kT)$. Let $\dot{x}_a(t) = dx_a(t)/dt$ denote the derivative of $x_a(t)$. Then from (6.1.2) we have $\dot{x}_a(kT) = c$. Evaluating (6.1.2) at $t = (k-1)T$, and solving for c, this yields the following first-order approximation for the derivative.

$$y_1(k) = \frac{x(k) - x(k-1)}{T} \tag{6.1.3}$$

Backward Euler differentiator The formulation in (6.1.3) is called a *backward Euler* numerical approximation to the derivative. Note that it is an FIR filter of order $m = 1$ with coefficient vector $b = [1/T, -1/T]$. A block diagram of the first-order backward Euler differentiator is shown in Figure 6.1.

The FIR filter model of the differentiation process can be improved if we approximate $x_a(t)$ with a quadratic polynomial as follows.

$$x_a(t) \approx x(k) + c(t - kT) + d(t - kT)^2 \tag{6.1.4}$$

Again, $\dot{x}_a(kT) = c$. To determine the parameter c, we evaluate (6.1.4) at $t = (k-1)T$ and at $t = (k-2)T$, which yields

$$x(k-1) = x(k) - Tc + T^2 d \tag{6.1.5a}$$
$$x(k-2) = x(k) - 2Tc + 4T^2 d \tag{6.1.5b}$$

If (6.1.5b) is subtracted from four times (6.1.5a), the dependence on d drops out. The resulting equation can then be solved for c, which yields the following second-order differentiator.

$$y_2(k) = \frac{3x(k) - 4x(k-1) + x(k-2)}{2T} \tag{6.1.6}$$

Second-order differentiator This FIR filter is called a *second-order backward differentiator*. It is backward, because it makes use of current and past samples, not future samples. This is important if the differentiator is to be used in real-time applications such as feedback control. A block diagram of the second-order backward differentiator is shown in Figure 6.2.

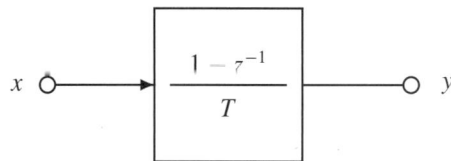

FIGURE 6.1: Block Diagram of a First-order Backward Euler Differentiator

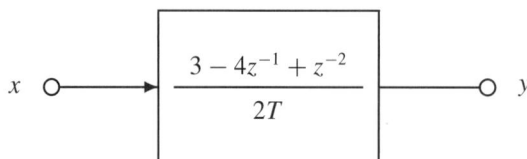

FIGURE 6.2: Block Diagram of a Second-order Backward Differentiator

It would be tempting to continue this process using higher-order polynomial approxima-tions of $x_a(t)$. Unfortunately, one quickly reaches a point of diminishing return in terms of accuracy. This is because higher-degree polynomials are not effective in modeling the under-ling trend of signals. Even though they can be made to go through all of the samples, they tend to do so by oscillating wildly between the samples as the polynomial degree increases. Later, we will examine a different design approach that can be used to approximate a delayed version of a differentiator. To examine the effectiveness of the differentiator in Figure 6.2, consider the following noise-free input signal.

$$x_a(t) = \sin(\pi t) \tag{6.1.7}$$

Plots of $x_a(t)$, $\dot{x}_a(t)$, and $y_2(k)$ are shown in Figure 6.3. Here a sampling frequency of $f_s = 20\,\text{Hz}$ is used. After a short start-up transient, the approximation is effective in this case.

6.1.2 Signal-to-noise Ratio

Although the approximation to the derivative in Figure 6.3 is effective for the noise-free signal in (6.1.7), when noise is added to $x_a(t)$, the numerical approximations to the derivative rapidly deteriorate. To illustrate, consider the noise-corrupted signal

$$y(k) = x(k) + v(k) \tag{6.1.8}$$

Here $v(k)$ is white noise uniformly distributed over the interval $[-c, c]$. To specify the size of the noise relative to the size of the signal, the following concept be used.

FIGURE 6.3: Numerical Approximation to the Derivative of a Sinusoidal Input Using an FIR Filter

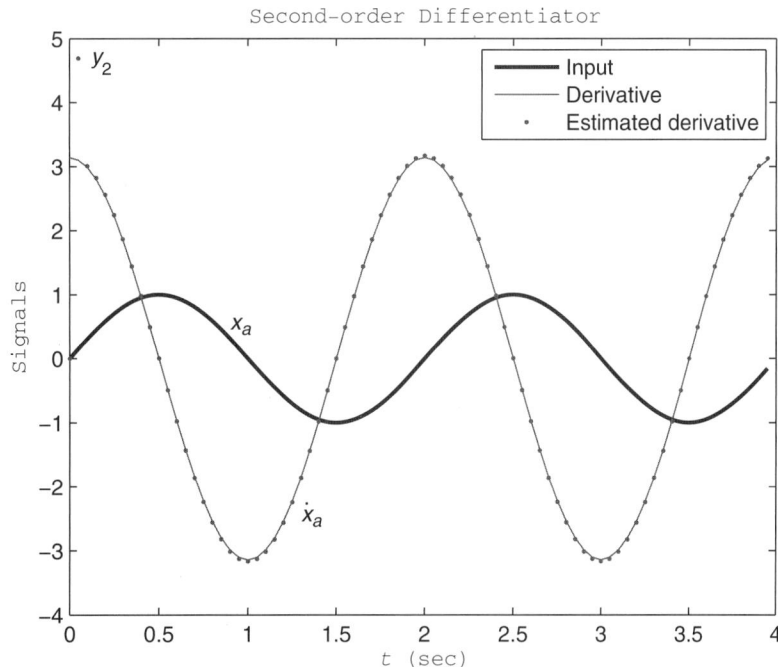

DEFINITION

6.1: Signal-to-noise
Ratio

Let $y(k) = x(k) + v(k)$ represent a signal $x(k)$ that is corrupted with noise $v(k)$. If P_x is the average power of the signal, and P_v is the average power of the noise, then the *signal-to-noise ratio* is defined as

$$\text{SNR}(y) \triangleq 10 \log_{10}\left(\frac{P_x}{P_v}\right) \text{ dB}$$

The average power is the mean or expected value of the square of the signal, $P_x = E[x^2(k)]$. A signal that is all noise has a signal-to-noise ratio of minus infinity, while a noise-free signal has a signal-to-noise ratio of plus infinity. When the signal-to-noise ratio is zero dB, the power of the signal is equal to the power of the noise.

Using the trigonometric identities from Appendix 2, one can show that the average power of a sinusoid of amplitude A is $A^2/2$. From (4.6.6) or Appendix 2, the average power of white noise uniformly distributed over $[-c, c]$ is $P_v = c^2/3$. Suppose $c = .01$, which is only a modest amount of noise. Then from Definition 6.1, the signal-to-noise ratio of the noise corrupted sinusoid in (6.1.8) is

$$\text{SNR}(y) = 10 \log_{10}\left[\frac{1/2}{(.01)^2/3}\right]$$
$$= 10 \log_{10}\left(\frac{10^4}{6}\right)$$
$$= 32.22 \text{ dB} \tag{6.1.9}$$

When the noise-corrupted sinusoid in (6.1.8) is processed with the second-order differentiator, the results are as shown in Figure 6.4. Note that $x_a(t)$ itself is not significantly distorted because the signal-to-noise ratio is relatively large. However, the numerical approximation to $\dot{x}_a(t)$ does not fare well in comparison with the noise-free case in Figure 6.3. This is

FIGURE 6.4: Numerical Approximations to the Derivative of a Noisy Sinusoidal Input Using a Second-order FIR Filter

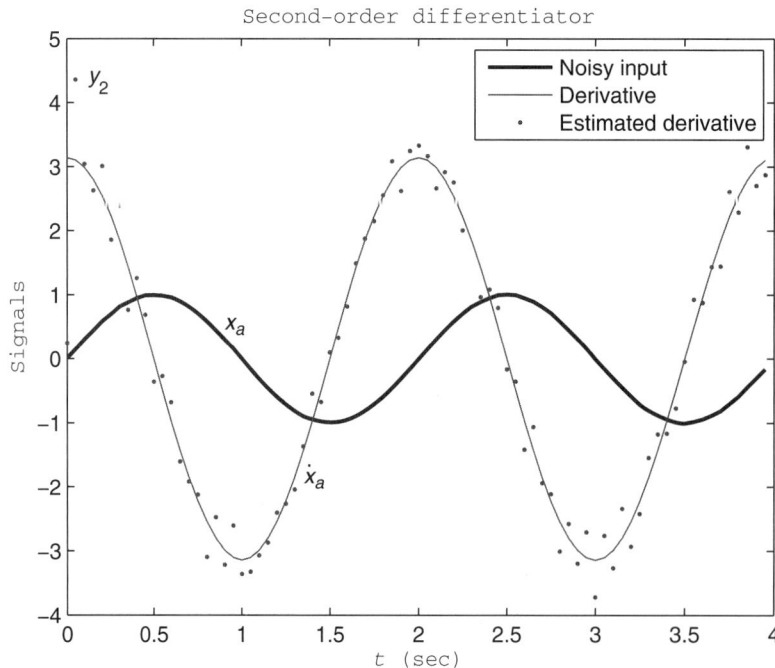

because the differentiation process *amplifies* the high-frequency part of the noise. In this particular instance, most of the noise could have been removed by first preprocessing $y(k)$ with a narrowband resonator filter with resonant frequency $F_0 = 1$ Hz. Resonator filters are discussed in Chapter 7. However, for a more general broadband signal, $x_a(t)$, this is not a viable option because the filtering would remove important spectral components of $x_a(t)$ itself.

Statistically independent signals

Typically, additive white noise $v(k)$ has zero mean and is *statistically independent* of the signal $x(k)$, which means that $E[x(k)v(k)] = E[x(k)]E[v(k)]$. In this case, there is a simple relationship between the average power of the noise-corrupted signal and the average power of the noise-free signal. Here

$$
\begin{aligned}
P_y &= E[y^2(k)] \\
&= E[\{x(k) + v(k)\}^2] \\
&= E[x^2(k) + 2x(k)v(k) + v^2(k)] \\
&= E[x^2(k)] + E[2x(k)v(k)] + E[v^2(k)] \\
&= P_x + 2E[x(k)]E[v(k)] + P_v
\end{aligned}
\tag{6.1.10}
$$

For zero-mean noise, $E[v(k)] = 0$, in which case (6.1.10) reduces to

$$
P_y = P_x + P_v
\tag{6.1.11}
$$

Zero-mean white noise

Thus for a signal $x(k)$ that is corrupted with *zero-mean white noise* $v(k)$, the average power of the noise-corrupted signal $y(k)$ is just the sum of the average power of the signal plus the average power of the noise. Typically, $y(k)$ is known or can be measured. If the average power of either the signal $x(k)$ or the noise $v(k)$ are known, or can be computed, then the average power of the other can be determined using (6.1.11). The signal-to-noise ratio of $y(k)$ then can be determined using Definition 6.1.

● ● ● ● ● ● ● ● ● ● ● ● ● ● ● ● ● ●

6.2 Windowing Method

In this section we examine a simple technique for designing a linear-phase FIR filter with a pre-scribed magnitude response. First, we briefly revisit the filter design specifications introduced in Chapter 5. For mth-order linear-phase FIR filters, the design specifications are formulated in terms of the desired *amplitude response*, $A_r(f)$, first introduced in Section 5.3.2.

Amplitude response

$$
H(f) = A_r(f) \exp[j(\alpha - \pi m f T)]
\tag{6.2.1}
$$

Recall that the amplitude response is real but can be positive or negative. An FIR lowpass design specification based on the desired amplitude response is shown in Figure 6.5. Note that the passband ripple parameter, δ_p, now represents the radius of a region centered about $A_r(f) = 1$, within which the amplitude response must lie. The same is true for the stopband attenuation. These specifications based on the amplitude response also carry over to highpass, bandpass, and bandstop filters.

FIGURE 6.5: Amplitude Response Specification for an FIR Lowpass Filter with $A(f) = |A_r(f)|$

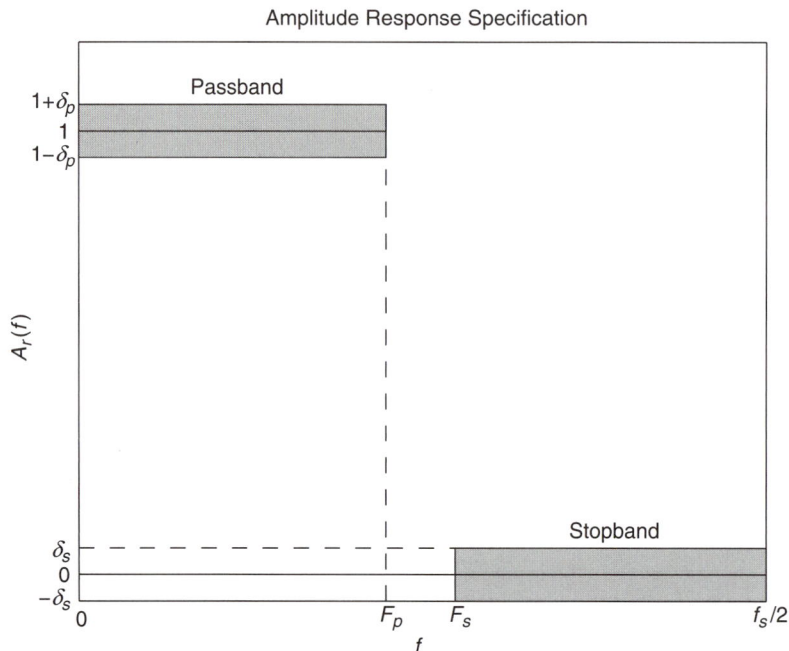

Amplitude Response Specification

6.2.1 Truncated Impulse Response

The basic idea behind the windowing method is to first truncate the desired impulse response to a finite number of samples and delay the impulse response, as needed, so that it is causal. A filter is then designed with an impulse response that matches the delayed truncated impulse response. Except for the zero-phase filters, the filters that we have considered thus far have all been causal filters, with $h(k) = 0$ for $k < 0$. In order to develop a general design technique based on idealized filters, we consider noncausal filters as well. When $h(k)$ is noncausal, the DTFT of $h(k)$ gives rise to the following frequency response.

$$H(f) = \sum_{k=-\infty}^{\infty} h(k) \exp(-j2\pi k f T) \tag{6.2.2}$$

Using Euler's identity in (6.2.2), we see that $H(f)$ is periodic with period f_s. Consequently, one can interpret (6.2.2) as a complex Fourier series of the periodic function $H(f)$. The kth coefficient of the Fourier series is

$$h(k) = \frac{1}{f_s} \int_{-f_s/2}^{f_s/2} H(f) \exp(j2\pi k f T) df, \quad -\infty < k < \infty \tag{6.2.3}$$

Using (6.2.3), we can recover the impulse response $h(k)$ from a desired frequency response $H(f)$. Since we are interested in designing a linear-phase filter that does not distort the input signal, the frequency response must satisfy the linear-phase symmetry constraint introduced in Chapter 5.

Type 1 and Type 2 Filters

Let m be the filter order, and suppose $h(k)$ exhibits even symmetry about $k = m/2$. Then from (6.2.1) and Table 5.1, the desired frequency response for a type 1 or a type 2 causal linear-phase filter with a group delay of $\tau = mT/2$ is

$$H(f) = A_r(f) \exp(-j\pi mfT) \ \} \text{type 1 and 2} \qquad (6.2.4)$$

For a type 1 or type 2 linear-phase filter, the amplitude response, $A_r(f)$, is a real even function that is specified by the filter designer. The desired magnitude response is $A(f) = |A_r(f)|$. Using (6.2.4), Euler's identity, and the fact that $A_r(f)$ is even, we find that the expression for $h(k)$ in (6.2.3) becomes

$$
\begin{aligned}
h(k) &= \frac{1}{f_s} \int_{-f_s/2}^{f_s/2} A_r(f) \exp(-j\pi mfT) \exp[j2\pi kfT] df \\
&= T \int_{-f_s/2}^{f_s/2} A_r(f) \exp[j2\pi(k - .5m)fT] df \\
&= T \int_{-f_s/2}^{f_s/2} A_r(f)\{\cos[2\pi(k - .5m)fT] + j\sin[2\pi(k - .5m)fT]\} df \\
&= T \int_{-f_s/2}^{f_s/2} A_r(f) \cos[2\pi(k - .5m)fT] df \qquad (6.2.5)
\end{aligned}
$$

Since the integrand in (6.2.5) is even, the integral can be performed over the positive frequencies and doubled. This results in the following impulse response for a type 1 or type 2 linear-phase filter of order m.

$$h(k) = 2T \int_0^{f_s/2} A_r(f) \cos[2\pi(k - .5m)fT] df, \quad 0 \le k \le m \quad (6.2.6)$$

Type 3 and Type 4 Filters

Next, suppose the impulse response exhibits odd symmetry about $k = m/2$. Then, from (6.2.1) and Table 5.1, the desired frequency response for a causal type 3 or a type 4 filter with a group delay of $\tau = mT/2$ is

$$H(f) = j A_r(f) \exp(-j\pi mfT) \ \} \text{type 3 and 4} \qquad (6.2.7)$$

Note that we have used the fact that $\exp(j\pi/2) = j$. For a type 3 or type 4 linear-phase filter, the amplitude response, $A_r(f)$, is a real odd function that is specified by the designer to get the desired magnitude response, $A(f) = |A_r(f)|$. Using (6.2.7), Euler's identity, and the fact that $A_r(f)$ is odd, we find that the expression for the impulse response in (6.2.3) becomes

$$
\begin{aligned}
h(k) &= \frac{j}{f_s} \int_{-f_s/2}^{f_s/2} A_r(f) \exp[j2\pi(k - .5m)fT] df \\
&= jT \int_{-f_s/2}^{f_s/2} A_r(f)\{\cos[2\pi(k - .5m)fT] + j\sin[2\pi(k - .5m)fT]\} df \\
&= -T \int_{-f_s/2}^{f_s/2} A_r(f) \sin[2\pi(k - .5m)fT] df \qquad (6.2.8)
\end{aligned}
$$

TABLE 6.1: ▶
Impulse Responses
of Ideal Frequency-
selective Linear-
phase Type 1 Filters
of Order $m = 2p$

Filter	$h(p)$	$h(k), 0 \le k \le m, k \ne p$
Lowpass	$2F_0 T$	$\dfrac{\sin[2\pi(k-p)F_0 T]}{\pi(k-p)}$
Highpass	$1 - 2F_0 T$	$\dfrac{-\sin[2\pi(k-p)F_0 T]}{\pi(k-p)}$
Bandpass	$2(F_1 - F_0)T$	$\dfrac{\sin[2\pi(k-p)F_1 T] - \sin[2\pi(k-p)F_0 T]}{\pi(k-p)}$
Bandstop	$1 - 2(F_1 - F_0)T$	$\dfrac{\sin[2\pi(k-p)F_0 T] - \sin[2\pi(k-p)F_1 T]}{\pi(k-p)}$

Here we have used the fact that $\exp(j\pi/2) = j$. Again the integrand in (6.2.8) is even, so the integral can be performed over the positive frequencies and doubled. This yields the following impulse response for a type 3 or type 4 linear-phase filter of order m.

$$h(k) = -2T \int_0^{f_s/2} A_r(f) \sin[2\pi(k - .5m)fT]df, \quad 0 \le k \le m \quad (6.2.9)$$

Recall that the type 1 linear-phase filter (even symmetry, even order) is the most general in the sense that there are no zeros at $f = 0$ or $f = f_s/2$. Using (6.2.6) with order $m = 2p$, we summarize the impulse responses for the four ideal frequency-selective filters in Table 6.1.

Example 6.1 **Truncated Impulse-response Filter**

As an illustration of an FIR filter designed by the truncated impulse response approach, suppose $f_s = 100$ Hz, and consider the problem of designing a lowpass filter with cutoff frequency $F_0 = f_s/4$. Suppose $m = 40$ is used to approximate $H(f)$. Then $p = 20$ and from Table 6.1, the filter coefficients are $h(p) = .5$ and

$$h(k) = \frac{\sin[.5\pi(k - p)]}{\pi(k - p)}$$
$$= .5 \operatorname{sinc}[.5(k - p)], \quad 0 \le k \le m$$

The delay in this case is

$$\tau = pT$$
$$= .2 \text{ sec}$$

The impulse response of this filter can be obtained by running *exam6_1* using *f_dsp*. From the resulting plot, shown in Figure 6.6, we see that this is indeed a type 1 linear-phase FIR filter with a palindrome-type impulse response, $h(k)$, centered about $k = 20$. Notice that the ideal sinc function impulse response is truncated to a radius of $p = m/2$ samples, and then delayed by p samples to make it causal.

The magnitude response and phase response generated by *exam6_1* are shown in Figure 6.7. Although the magnitude response is an effective approximation to the ideal lowpass characteristic, it is evident that there is *ringing* or oscillation, particularly in the neighborhood of the cutoff frequency, $F_0 = 25$ Hz. The approximation to an ideal lowpass characteristic can be improved by increasing the filter order. However, even for large values of p, significant ringing persists near $f = F_0$. This is an inherent characteristic of the truncated impulse response that is present whenever there is a jump discontinuity in the function being approximated. The *Gibb's phenomenon* oscillation near a jump discontinuity is referred to as *Gibb's phenomenon*.

FIGURE 6.6:
Palindrome Impulse
Response of Type 1
Linear-phase FIR
Filter with $m = 40$

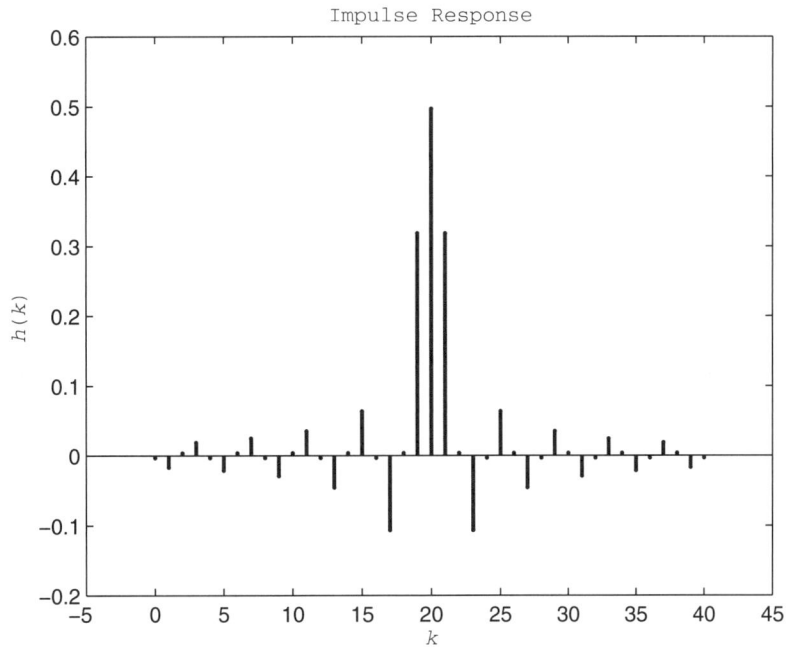

FIGURE 6.7: Frequency
Response of
Truncated Impulse
Response FIR Filter
with $m = 40$

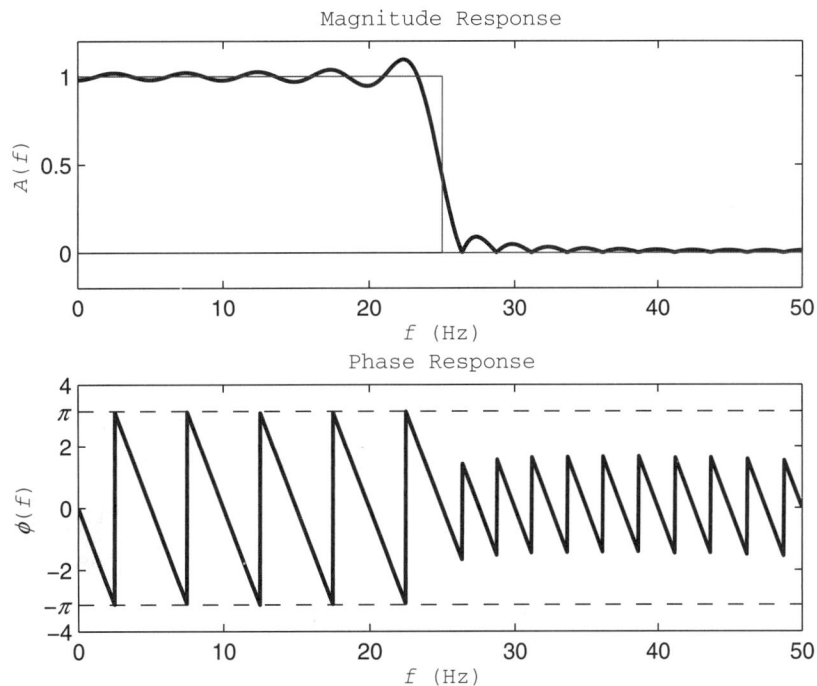

It is of interest to observe the phase response carefully. The jump discontinuities in the passband occurring at $\phi = -\pi$ are an artifact of the fact that the phase is computed modulo 2π, so at $-\pi$ it wraps around to π. However, the jump discontinuities of amplitude π in the stopband are discontinuities that occur because of a sign change in the amplitude response, $A_r(f)$. Notice that all of these jumps occur at points in the set F_z where $A(f) = 0$.

6.2.2 Windowing

The beauty of the truncated impulse response method is that it allows us to design a magnitude response of prescribed shape. However, if the desired magnitude response contains one or more jump discontinuities, then the oscillations caused by Gibb's phenomenon effectively prohibit the design of filters having a very small passband ripple or stopband attenuation. Fortunately, there is a trade-off we can make that reduces the ripples at the expense of increasing the width of the transition band. To see how this can be done, first notice that a filter transfer function can be rewritten as follows.

$$H(z) = \sum_{i=-\infty}^{\infty} w_R(i)h(i)z^{-i} \tag{6.2.10}$$

Rectangular window

This corresponds to a causal filter of order m when $w_R(i)$ is the following *rectangular window* of order m.

$$w_R(i) \triangleq \begin{cases} 1, & 0 \leq i \leq m \\ 0, & \text{otherwise} \end{cases} \tag{6.2.11}$$

Thus truncation of the impulse response to $0 \leq i \leq m$ is equivalent to multiplication of the impulse response by a rectangular window. It is the abrupt truncation of the impulse response that causes the oscillations associated with Gibb's phenomenon. The amplitude of the oscillations can be decreased by tapering the impulse response to zero more gradually.

Windowed coefficients

Recall that for an FIR filter the numerator coefficients are $b_i = h(i)$. Therefore, if $w(i)$ represents a window of order m, then the tapered filter coefficients are

$$b_i = w(i)h(i), \quad 0 \leq i \leq m \tag{6.2.12}$$

There are many windows that have been proposed. Some of the more popular fixed windows are summarized in Table 6.2. They include the rectangular window (also called the boxcar window), the Hanning window, the Hamming window, and the Blackman window. Recall from Chapter 4 that these are the same data windows that were used with the spectrogram and with Welch's method of estimating the power density spectrum of a signal.

Plots of the windows are shown in Figure 6.8 for the case $m = 60$. Notice that they all are symmetric about $i = m/2$ and attain a peak value of $w(i) = 1$ at the midpoint. With the exception of the rectangular window, they all gradually taper to zero at the end points $i = 0$ and $i = m$, except for the Hamming window which is near zero. Multiplication by a tapered window can be thought of as a form of *soft* truncation as opposed to the hard or abrupt truncation of the rectangular window. Since the windows are all palindromes with $w(i) = w(m - i)$, filters using the windowed coefficients in (6.2.12) continue to be linear-phase filters.

TABLE 6.2: ▶ Windows of Order m

Type	Name	$w(i)$, $0 \leq i \leq m$
0	Rectangular	1
1	Hanning	$.5 - .5\cos\left(\dfrac{\pi i}{.5m}\right)$
2	Hamming	$.54 - .46\cos\left(\dfrac{\pi i}{.5m}\right)$
3	Blackman	$.42 - .5\cos\left(\dfrac{\pi i}{.5m}\right) + .08\cos\left(\dfrac{2\pi i}{.5m}\right)$

FIGURE 6.8: Windows Used to Taper Truncated Impulse Response: 0 = rectangular, 1 = Hanning, 2 = Hamming, and 3 = Blackman

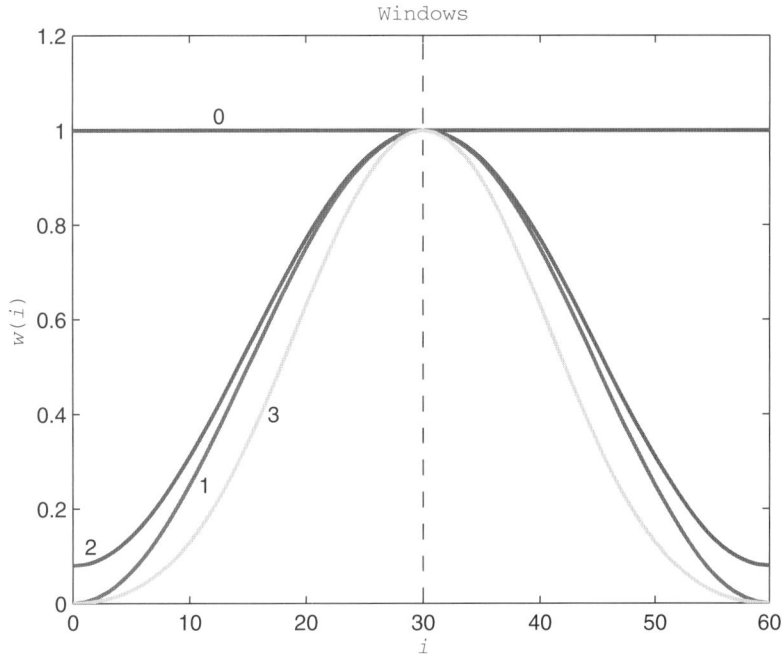

Example 6.2

Windowed Lowpass Filter

To illustrate the effects of the different windows, consider the design of a lowpass filter with cutoff frequency $F_0 = f_s/4$. Suppose $m = 40$ and $p = m/2$. Then from Table 6.1, Table 6.2, and (6.2.12), the filter coefficients using the rectangular window are

$$b_i = .5\,\mathrm{sinc}[.5(i - p)], \quad 0 \le i \le m$$

A plot of the rectangular magnitude response obtained by running *exam6_2*, using normalized frequency f/f_s, is shown in Figure 6.9. Here the logarithmic dB scale is used for the magnitude response because it better illustrates the amount of attenuation in the stopband which is 21 dB or roughly a factor of 10 in this case.

Next consider the same filter, but using the Hanning window. From Table 6.2, the filter coefficients are

$$b_i = .25[1 + \cos(\pi i/p)]\,\mathrm{sinc}[.5(i - p)], \quad 0 \le i \le m$$

A plot of the Hanning magnitude response is shown in Figure 6.10. Notice that the attenuation in the stopband is now 44 dB, and furthermore, the response is also more flat in the passband in comparison with the rectangular window in Figure 6.9. However, this improvement in passband ripple and stopband attenuation is achieved at the expense of a what is clearly a wider transition band.

Still better stopband attenuation can be obtained using the Hamming window. From Table 6.2, the filter coefficients are

$$b_i = .5[.54 + .46\cos(\pi i/p)]\,\mathrm{sinc}[.5(i - p)], \quad 0 \le i \le m$$

A plot of the Hamming magnitude response is shown in Figure 6.11. In this case, the stopband attenuation is 53 dB and the lobes are roughly constant throughout most of the stopband.

FIGURE 6.9: Lowpass
Magnitude
Response Using a
Rectangular
Window, $m = 40$.

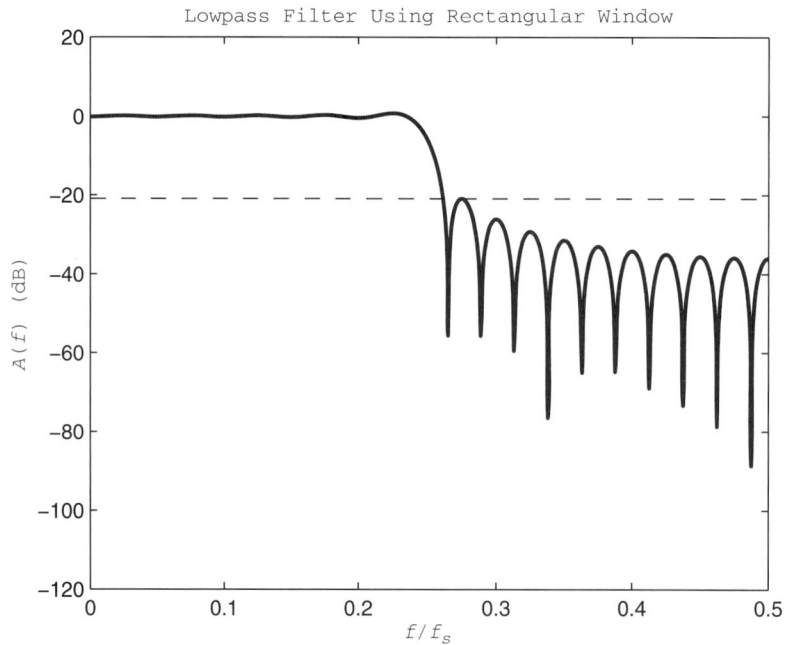

FIGURE 6.9: Lowpass Magnitude Response Using a Rectangular Window, $m = 40$.

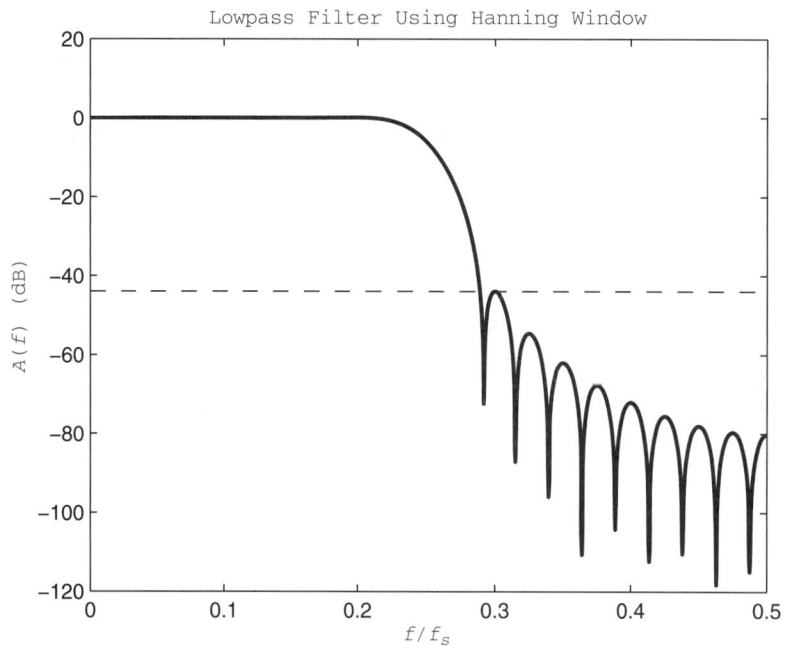

FIGURE 6.10: Lowpass Magnitude Response Using a Hanning Window, $m = 40$

Finally, the maximum stopband attenuation can be obtained using the Blackman window. From Table 6.2, the filter coefficients are

$$b_i = .5[.42 + .5\cos(\pi i/p) + .08\cos(2\pi i/p)]\operatorname{sinc}[.5(i - p)], \quad 0 \leq i \leq m$$

FIGURE 6.11: Lowpass Magnitude Response Using a Hamming Window, $m = 40$

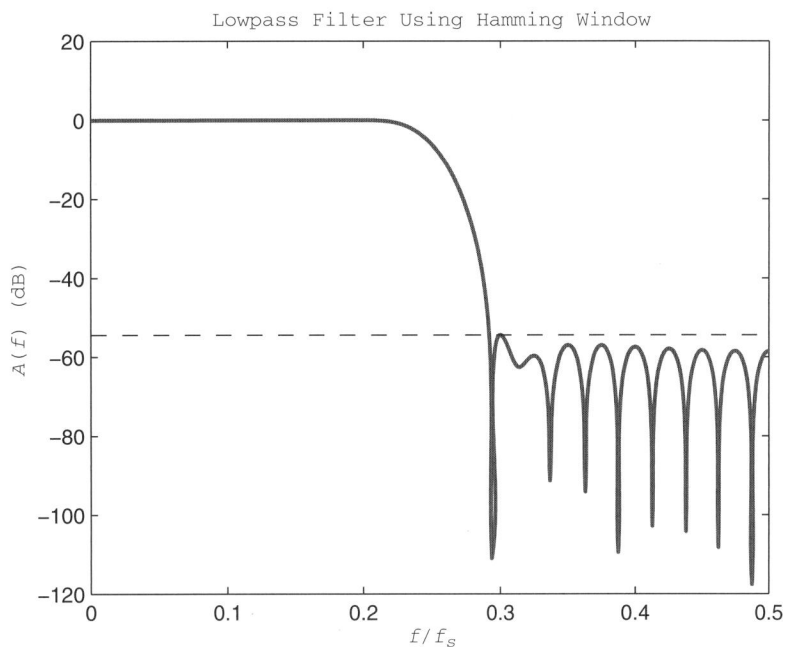

Lowpass Filter Using Hamming Window

FIGURE 6.12: Lowpass Magnitude Response Using a Blackman Window, $m = 40$

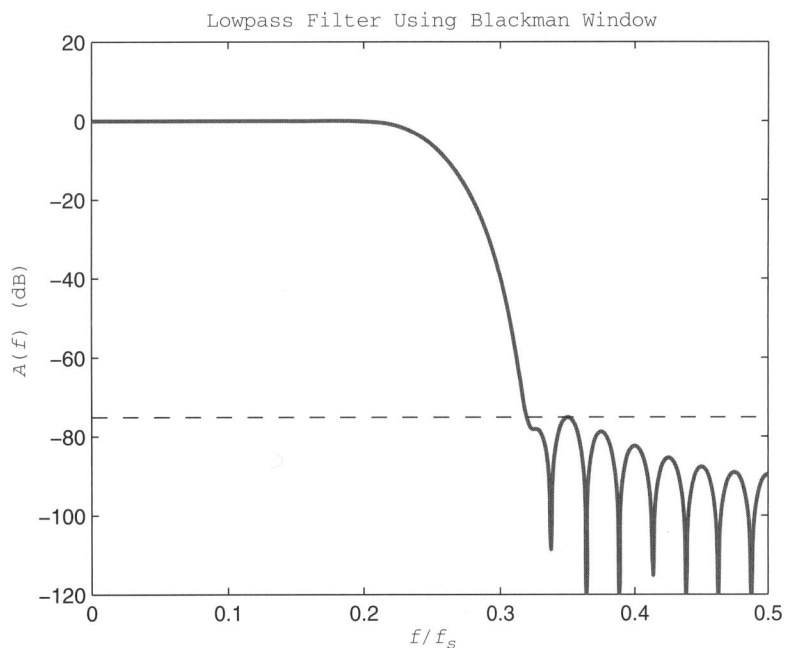

Lowpass Filter Using Blackman Window

A plot of the Blackman magnitude response is shown in Figure 6.12. Here the stop band attenuation is 75 dB, which is 54 dB better than with the rectangular window. However, the choice of the Blackman window is not clear-cut because the transition band is largest in this case.

TABLE 6.3: ▶
Design
Characteristics
of Windows

Window Type	$\hat{B} = \|F_s - F_p\|/f_s$	δ_p	δ_s	A_p (dB)	A_s (dB)
Rectangular	$\dfrac{.9}{m}$	.0819	.0819	.742	21
Hanning	$\dfrac{3.1}{m}$	.0063	.0063	.055	44
Hamming	$\dfrac{3.3}{m}$	.0022	.0022	.019	53
Blackman	$\dfrac{5.5}{m}$	.00017	.00017	.0015	75.4

Transition bandwidth

The improvements in passband ripple and stopband attenuation achieved by using windowing come at the cost of a wider *transition band*. However, the width of the transition band for each window can be controlled by the filter order m. A summary of the filter design characteristics using the different windows can be found in Table 6.3.

Kaiser Windows

Additional windows have been proposed, including the Bartlett, Lanczos, Tukey, Dolph-Chebyshev, and Kaiser windows. The Kaiser window, $w_K(i)$, is a near-optimal window based on the use of zeroth-order modified Bessel functions of the first kind (Kaiser, 1966).

$$w_K(i) = \frac{I_0\{\beta(1 - [(i - p)/p]^2)^{1/2}\}}{I_0(\beta)}, \quad 0 \le i \le m \tag{6.2.13}$$

Here I_0 is a zeroth-order modified Bessel function of the first kind. It is obtained by solving a certain family of differential equations. To evaluate $I_0(x)$, the MATLAB function besseli$(0, x)$ can be used. Unlike the fixed windows in Table 6.2, the Kaiser window has an adjustable shape parameter, $\beta \ge 0$, that allows the user to control the trade-off between the main lobe width and the side lobe amplitudes. When $\beta = 0$, the Kaiser window reduces to the rectangular window. For a given window size m, increasing β leads to an increase in the stopband attenuation A_s. However, this also causes the main lobe to get wider, thereby increasing the width of the transition band. The width of the main lobe can be decreased by increasing m. Kaiser (1974) developed the following approximations for determining suitable values for β and m, given a desired stopband attenuation A_s and a desired normalized transition band $\hat{B} = |F_s - F_p|/f_s$.

$$\beta \approx \begin{cases} .1102(A_s - 8.7), & A_s > 50 \\ .5842(A_s - 21)^{.4} + .07886(A - 21), & 21 \le A_s \le 50 \\ 0, & A_s < 21 \end{cases} \tag{6.2.14}$$

$$m \approx \frac{A_s - 8}{4.568\pi\,\hat{B}} \tag{6.2.15}$$

Recall that $A_s = 21$ dB corresponds to a rectangular window. We summarize the steps of the windowed FIR filter design procedure with the following algorithm.

ALGORITHM
....
6.1: Windowed FIR Filter

1. Pick $m > 0$ and a window w.

2. For $i = 0$ to m compute

$$b_i = w(i)2T \int_0^{f_s/2} A_r(f) \cos[2\pi(i - .5m)fT]df$$

3. Set

$$H(z) = \sum_{i=0}^{m} b_i z^{-i}$$

The FIR filter produced by Algorithm 6.1 is a type 1 or a type 2 linear-phase filter with delay $\tau = mT/2$ and an even amplitude response, $A_r(f)$. To design a type 3 or type 4 filter, the desired amplitude response should be odd and the expression for b_i in step 2 should be based on (6.2.9) instead of (6.2.6).

Example 6.3

Windowed Bandpass Filter

Consider the problem of designing a bandpass filter with cutoff frequencies $F_0 = f_s/8$ and $F_1 = 3f_s/8$. Suppose the Blackman window is used. Using Table 6.1, Table 6.2, and Algorithm 6.1, the filter coefficients are $b_p = .5$ and for $0 \le i \le m, i \ne p$

$$b_i = \frac{[.42 + .5\cos(\pi i/p) + .08\cos(2\pi i/p)]\{\sin[.75\pi(i - p)] - \sin[.25\pi(i - p)]\}}{\pi(i - p)}$$

Suppose $m = 80$. A plot of the magnitude response, obtained by running *exam6_3*, is shown in Figure 6.13. It is clear that the passband and stopband ripples have been effectively reduced

FIGURE 6.13: Magnitude Response of Bandpass Filter Using a Hamming Window, $m = 80$

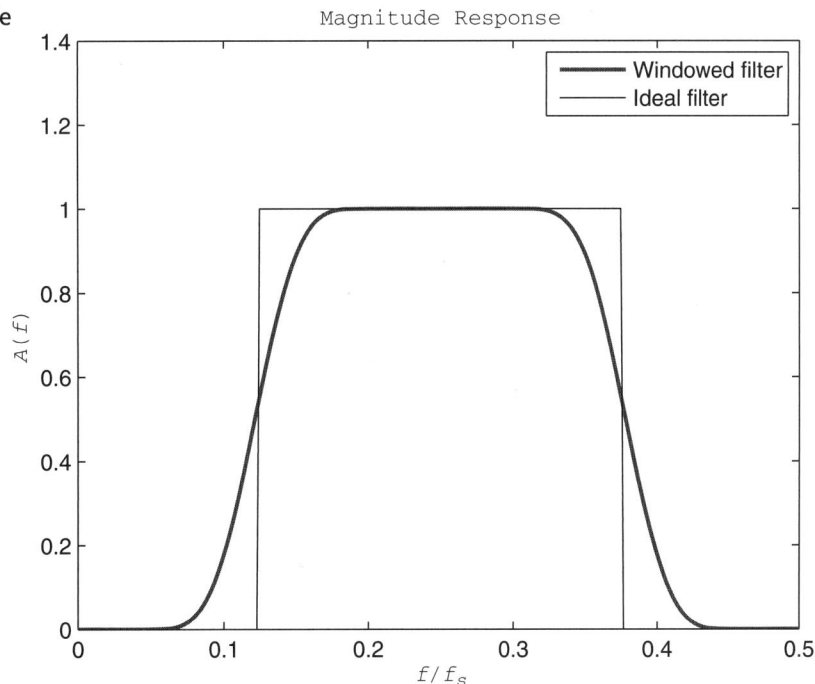

in comparison with Figure 6.7. From Table 6.3, the normalized width of the transition band in this case is

$$\frac{\Delta F}{f_s} = \frac{5.5}{80}$$
$$= .069$$

FDSP Functions

The FDSP toolbox contains two functions for designing a linear-phase FIR filter using the windowing method. The first function, *f_firideal*, uses design specifications of ideal frequency-selective filters.

```
% F_FIRIDEAL Design an ideal linear-phase frequency-selective windowed FIR filter
%
% Usage:
%        b = f_firideal (f_type,F,m,fs,win)
% Pre:
%        f_type = integer selecting the frequency-selective filter type
%
%                 0 = lowpass
%                 1 = highpass
%                 2 = bandpass
%                 3 = bandstop
%
%        F     = scalar or vector of length two containing the
%                cutoff frequency or frequencies.
%        m     = filter order (even)
%        fs    = sampling frequency
%        win   = the window type to be used:
%
%                 0 = rectangular
%                 1 = Hanning
%                 2 = Hamming
%                 3 = Blackman
% Post:
%        b    = 1 by m+1 vector of filter coefficients.
```

A general linear-phase FIR filter with an arbitrary amplitude response can be designed with the windowing method using the function *f_firwin*. The first calling argument, *fun*, is the name of a user-supplied function that specifies the desired amplitude response.

```
% F_FIRWIN: Design a general windowed FIR filter
%
% Usage:
%        b = f_firwin (@fun,m,fs,win,sym,p)
```

Continued on p. 423

Continued from p. 422

```
% Pre:
%        fun = name of user-supplied function that
%              specifies the desired amplitude
%              response of the filter. Usage:
%
%              [A, theta] = fun(f,fs,p)
%
%              Here f is the frequency, fs is the sampling
%              frequency, and p is an option parameter
%              vector containing things like cutoff
%              frequencies,etc. Output A is a the
%              desired amplitude response. Optional output theta
%              is included for compatability with f_firquad.
%              Set theta = zeros(size(f)) for a
%              linear-phase filter.
%
%        m   = the filter order
%        fs  = sampling frequency
%        win = the window type to be used:
%
%              0 = rectangular
%              1 = Hanning
%              2 = Hamming
%              3 = Blackman
%
%        sym = symmetry of impulse response.
%
%              0 = even symmetry of h(k) about k = m/2
%              1 = odd symmetry of h(k) about k = m/2
%
%        p   = an optional vector of length contained
%              design parameters to be passed to fun.
%              For example p might contain cutoff
%               frequencies or gains.
% Post:
%        b = 1 by m+1 vector of filter coefficients.
```

● ● ● ● ● ● ● ● ● ● ● ● ● ● ● ⋯

6.3 Frequency-sampling Method

An alternative technique for designing a linear-phase FIR filter with a prescribed magnitude response is the frequency-sampling method. As the name implies, this method is based on using samples of the desired frequency response.

6.3.1 **Frequency Sampling**

Suppose there are N frequency samples uniformly distributed over the range $0 \leq f < f_s$ with the ith discrete frequency being

$$f_i = \frac{i f_s}{N}, \quad 0 \leq i < N$$

Recall from Section 4.8.1 that the samples of the frequency response of an FIR filter can be obtained directly from the DFT of the impulse response. In particular, for an FIR filter of order $m = N - 1$, we have $H(f_i) = H(i)$ for $0 \leq i < N$ where $H(i) = \text{DFT}\{h(k)\}$. Taking the inverse DFT we then arrive at the following expression for the impulse response of the desired FIR filter.

$$h(k) = \text{IDFT}\{H(f_i)\}, \quad 0 \leq k < N \qquad (6.3.1)$$

Interpolated response

The filter in (6.3.2) has a frequency response $H(f)$ that *interpolates*, or passes through, the N samples. Next, consider the problem of placing constraints on $H(f)$ that ensure that the filter is a linear-phase filter. Suppose $h(k)$ is a linear-phase impulse response exhibiting even symmetry about the midpoint $k = m/2$. Then from (6.2.4) the frequency response of this type 1 or type 2 filter ($\alpha = 0$) is as follows

$$H(f) = A_r(f) \exp(-j\pi m f T) \qquad (6.3.2)$$

Here the amplitude response, $A_r(f)$, is a real even function that specifies the desired magnitude response, $A(f) = |A_r(f)|$. Recalling the expression for the IDFT in (4.3.7), and using (6.3.1) through (6.3.3), the kth sample of the impulse response can be written as

$$
\begin{aligned}
h(k) &= \frac{1}{N} \sum_{i=0}^{N-1} H(f_i) \exp(j2\pi i k/N) \\
&= \frac{1}{N} \sum_{i=0}^{N-1} A_r(f_i) \exp(-j\pi m f_i T) \exp(j2\pi i k/N) \\
&= \frac{1}{N} \sum_{i=0}^{N-1} A_r(f_i) \exp(-j\pi m i/N) \exp(j2\pi i k/N) \\
&= \frac{1}{N} \sum_{i=0}^{N-1} A_r(f_i) \exp[j2\pi i (k - .5m)/N] \\
&= \frac{1}{N} \sum_{i=0}^{N-1} A_r(f_i)\{\cos[2\pi i (k - .5m)/N] + j \sin[2\pi i (k - .5m)/N]\} \qquad (6.3.3)
\end{aligned}
$$

For a real $h(k)$, the sine terms in (6.3.4) cancel one another. The $i = 0$ term can be treated separately, in which case the expression for $h(k)$ in (6.3.4) reduces to

$$h(k) = \frac{A_r(f_0)}{N} + \frac{1}{N} \sum_{i=1}^{N-1} A_r(f_i) \cos[2\pi i (k - .5m)/N] \qquad (6.3.4)$$

Using the symmetry properties of the DFT in Table 4.7, one can show that the contributions of the i term and the $N - i$ term are identical which means they can be combined. Recalling that $b_k = h(k)$ for an FIR filter, we arrive at the following expression for the coefficients of a linear-phase *frequency-sampled filter* of order m where $m = N - 1$.

Frequency-sampled filter

$$b_k = \frac{A_r(0)}{m+1} + \frac{2}{m+1} \sum_{i=1}^{\text{floor}(m/2)} A_r(f_i) \cos\left[\frac{2\pi i(k - .5m)}{m+1}\right],$$
$$0 \le k \le m \tag{6.3.5}$$

Note that for a type 1 filter the order m is even, in which case $\text{floor}(m/2) = m/2$. For a type 2 filter m is odd.

Example 6.4

Frequency-sampled Lowpass Filter

To illustrate the frequency sampling method, consider the problem of designing a lowpass filter with cutoff frequency $F_0 = f_s/4$. Suppose a filter of order $m = 20$ is used. In this case the samples of the desired amplitude response are

$$A_r(f_i) = \begin{cases} 1, & 0 \le i \le 5 \\ 0, & 6 \le i \le 10 \end{cases}$$

Next, from (6.3.6) the filter coefficients are

$$b_k = \frac{1}{21} + \frac{2}{21} \sum_{i=1}^{5} \cos\left[\frac{2\pi i(k - 10)}{21}\right], \quad 0 \le k \le 20$$

A plot of the magnitude response, obtained by running *exam6_4*, is shown in Figure 6.14. Note the significant ripples in the magnitude response between the samples. The stopband attenuation is more easily seen in the logarithmic plot which reveals that $A_s = 15.6$ dB.

6.3.2 Transition-band Optimization

The ringing or ripple in the magnitude response evident in Figure 6.14 is caused by the abrupt transition from passband to stopband in the desired magnitude response. One way to reduce these oscillations is to taper the filter coefficients using a data window. Recall that the trade-off involved in using a window to reduce ripple is a wider transition band. With the frequency sampling method, one can forego the use of a window and instead explicitly specify $A(f)$ in the transition band by including one or more *transition-band samples*. This increase in the width of the transition band has the effect of improving the passband ripple and stopband attenuation, as can be seen from the following example.

Transition-band samples

FIGURE 6.14: Magnitude
Response of a
Frequency-sampled
Lowpass Filter with
$m = 20$

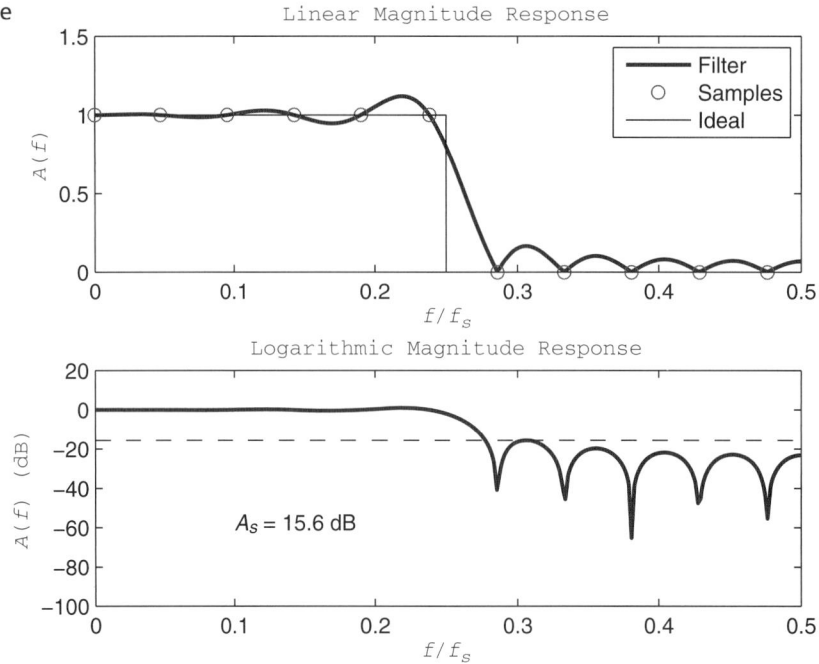

Example 6.5   **Filter with Transition-band Sample**

Again consider the problem of designing a lowpass filter with cutoff frequency $F_0 = f_s/4$.
Suppose a filter of order $m = 20$ is used as in Example 6.4. However, in this case we insert a
single transition band sample as follows.

$$A_r(f_i) = \begin{cases} 1, & 0 \le i \le 5 \\ .5, & i = 6 \\ 0, & 7 \le i \le 10 \end{cases}$$

Using (6.3.6), we find that the filter coefficients are

$$b_k = \frac{1}{21} + \frac{2}{21}\left\{ \sum_{i=1}^{5} \cos\left[\frac{2\pi i(k-10)}{21}\right] + .5\cos\left[\frac{2\pi 6(k-10)}{21}\right]\right\}, \quad 0 \le k \le 20$$

A plot of the logarithmic magnitude response, obtained by running *exam6_5*, is shown in
Figure 6.15. Comparing the results with Figure 6.14, it is clear that the stopband attenuation
has increased from 15.6 dB to 29.5 dB. The passband ripple has also been reduced, but at the
expense of a wider transition band.

The transition band sample that was inserted in the desired magnitude response in
Example 6.5 was $A_r(f_6) = .5$, which corresponds to a straight line interpolation between
the end of the ideal passband and the start of the ideal stopband. Clearly, other choices for the
value of the transition band sample are also possible. One can use this extra degree of freedom
to shape the magnitude response in the transition band in order to maximize the stopband
attenuation, as can be seen in the following example.

FIGURE 6.15: Magnitude
Response of a
Lowpass Filter with
$m = 20$ and One
Transition-band
Sample, $A_r(f_6) = .5$

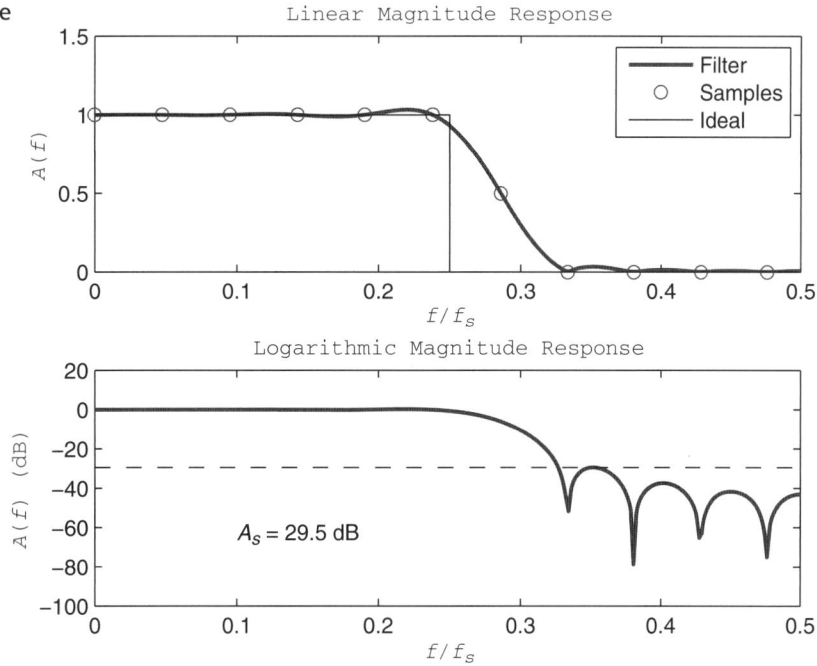
Linear Magnitude Response

Logarithmic Magnitude Response

$A_s = 29.5$ dB

Example 6.6

Filter with Optimal Transition-band Sample

Again consider the problem of designing a lowpass filter of order $m = 20$ with cutoff frequency $F_0 = f_s/4$. In this case we insert a general transition band sample as follows.

$$A_r(f_i) = \begin{cases} 1, & 0 \le i \le 5 \\ x, & i = 6 \\ 0, & 7 \le i \le 10 \end{cases}$$

Using (6.3.6), we find that the filter coefficients are

$$b_k(x) = \frac{1}{21} + \frac{2}{21}\left\{\sum_{i=1}^{5}\cos\left[\frac{2\pi i(k-10)}{21}\right] + x\cos\left[\frac{2\pi 6(k-10)}{21}\right]\right\}, \quad 0 \le k \le 20$$

The problem is to find a value for x that maximizes the stopband attenuation A_s. A simple way to achieve this is to compute the stopband attenuation, $A_s(x)$, for three distinct values of x in the range $0 < x < 1$. For example, suppose the values $x = [.25, .5, .75]^T$ are used. Consider a quadratic polynomial passed through these three data points.

$$A_s(x) = c_1 + c_2 x + c_3 x^2$$

The coefficient vector, $c = [c_1, c_2, c_3]^T$, of the polynomial that interpolates the data points must satisfy the following system of three linear algebraic equations.

$$\begin{bmatrix} 1 & x_1 & x_1^2 \\ 1 & x_2 & x_2^2 \\ 1 & x_3 & x_3^2 \end{bmatrix}\begin{bmatrix} c_1 \\ c_2 \\ c_3 \end{bmatrix} = \begin{bmatrix} A_s(x_1) \\ A_s(x_2) \\ A_s(x_3) \end{bmatrix}$$

Solving this system for c then yields a polynomial model of the stopband attenuation as a function of the transition band sample. The stopband attenuation is maximized by differentiating

FIGURE 6.16: Magnitude Response of a Lowpass Filter with $m = 20$ and an Optimal Transition-band Sample, $A_r(f_6) = .388$

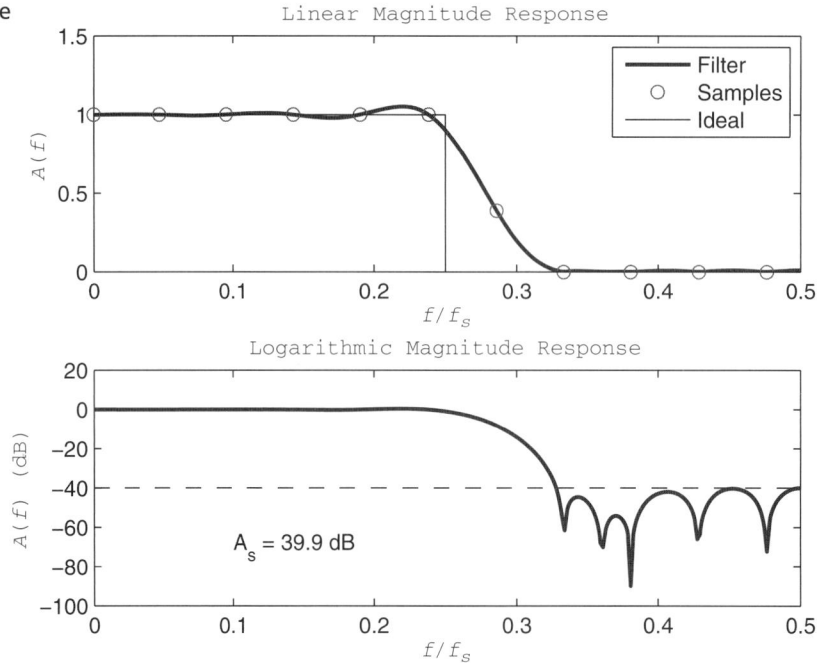

$A_s(x)$, setting the result to zero, and solving for x. This yields the following optimal transition sample value.

$$x_{max} = \frac{-c_2}{2c_3}$$

This optimization procedure is implemented in *exam6_6*. Running *exam6_6* from *f_dsp* produces a coefficient vector of $c = [18.35, 62.74, -80.83]^T$. Thus the optimal transition band sample is

$$A_r(f_6) = x_{max}$$
$$= \frac{-62.74}{2(-80.83)}$$
$$= .388$$

A plot of the optimum magnitude response is shown in Figure 6.16. By using an optimal value for the transition sample, we find that the stopband attenuation has increased to $A_s = 39.9$ dB.

The notion of using samples in the transition band can be extended to more than one sample with a corresponding improvement in the stopband attenuation. Rabiner et al. (1970) have developed tables of optimal transition band samples for FIR filters of different lengths and different numbers of transition band samples. For example, when a filter of order $m = 15$ with five passband samples and two transition band samples is used, a stopband attenuation in excess of 100 dB can be achieved.

The frequency-sampling method also can be used to design linear-phase FIR filters whose impulse responses exhibit odd symmetry about $k = m/2$. From (6.2.7), this corresponds to a filter with the following type of frequency response.

$$H(f) = jA_r(f)\exp(-j\pi mfT) \tag{6.3.6}$$

Using (6.3.1) through (6.3.3), we can write the kth sample of the impulse response as follows, where the number of samples is $N = m + 1$.

$$h(k) = \frac{1}{N} \sum_{i=0}^{N-1} H(f_i) \exp(j2\pi ik/N)$$

$$= \frac{j}{N} \sum_{i=0}^{N-1} A_r(f_i) \exp(-j\pi m f_i T) \exp\{j2\pi i(k/N)\}$$

$$= \frac{j}{N} \sum_{i=0}^{N-1} A_r(f_i) \exp\{j[2\pi i(k - .5m)/N]\}$$

$$= \frac{j}{N} \sum_{i=0}^{N-1} A_r(f_i)\{\cos[2\pi i(k - .5m)/N] + j \sin[2\pi i(k - .5m)/N]\} \qquad (6.3.7)$$

For a real $h(k)$ the cosine terms in (6.3.8) cancel one another and we have

$$h(k) = \frac{-1}{N} \sum_{i=0}^{N-1} A_r(f_i) \sin[2\pi i(k - .5m)/N] \qquad (6.3.8)$$

Since $A_r(f)$ is an odd function, the $i = 0$ terms drops out because $A_r(0) = 0$. Using the symmetry properties of the DFT in Table 4.7, one can show that the contributions of the i term and the $N - i$ term are identical. Thus we can sum half of the terms and double the result. Recalling that $b_k = h(k)$ for an FIR filter, we arrive at the following expression for the

Frequency-sampled filter

coefficients of a linear-phase *frequency-sampled filter* of order m where $m = N - 1$.

$$b_k = \frac{-2}{m + 1} \sum_{i=1}^{\text{floor}(m/2)} A_r(f_i) \sin\left[\frac{2\pi i(k - .5m)}{m + 1}\right], \qquad 0 \le k \le m \qquad (6.3.9)$$

Note that for a type 3 filter the order m is even, in which case $\text{floor}(m/2) = m/2$. For a type 4 filter m is odd.

FDSP Functions

The FDSP toolbox contains the following function for designing a linear-phase FIR filter using the frequency-sampling method.

```
%F_FIRSAMP: Design a frequency-sampled FIR filter
%
% Usage:
%          b = f_firsamp (A,m,fs,sym)
```

Continued on p. 430

Continued from p. 429

```
% Pre:
%          A   = 1 by floor(m/2)+1 array containing the
%                samples of the desired amplitude response.
%
%                |A(i)| = |H(f_i)|
%
%                Here the ith discrete frequency is
%
%                f_i = (i-1)fs/(m+1)
%
%                where fs is the sampling frequency.
%          m   = order of filter
%          fs  = the sampling frequency in Hz
%          sym = symmetry of pulse response.
%
%                0 = even symmetry of h(k) about k = m/2
%                1 = odd symmetry of h(k) about k = m/2
% Post:
%          b   = 1 by m+1 vector of filter coefficients.
```

• • • • • • • • • • • • • • • •

6.4 Least-squares Method

Least squares

The frequency response of a digital filter is periodic with period f_s. Because the windowing method uses a truncated Fourier series expansion of the desired amplitude response, $A_d(f)$, this method produces a filter that is optimal in the sense that it minimizes the following objective.

$$J = \int_0^{f_s/2} [A_d(f) - A_r(f)]^2 df \tag{6.4.1}$$

The actual amplitude response $A_r(f)$ is real but can be positive and negative. For a type 1 or a type 2 linear-phase filter it is even, and for a type 3 or a type 4 linear-phase filter it is odd. An alternative approach to filter design is to use a discrete version of the objective J. Let $\{F_0, F_1, \ldots, F_p\}$ be a set of $p + 1$ distinct frequencies with $F_0 = 0$, $F_p = f_s/2$, and

$$F_0 < F_1 < \cdots < F_p \tag{6.4.2}$$

Discrete least squares

The spacing between the discrete frequencies is often uniform, as in $F_i = if_s/(2p)$, but this is not required. Next, let $w(i) > 0$ be a *weighting function* were $w(i)$ specifies the relative importance of discrete frequency F_i. The special case, $w(i) = 1$ for $0 \leq i \leq p$ is referred to as *uniform weighting*. A weighted discrete version of the objective function in (6.4.1) can be formulated as follows.

$$J_p = \sum_{i=0}^p w^2(i)[A_r(F_i) - A_d(F_i)]^2 \tag{6.4.3}$$

A filter design technique that minimizes J_p is called a *least-squares method*. The filter amplitude response depends on the filter coefficient vector b, but the exact form of this dependence depends on the type of linear-phase filter used. To illustrate the least-squares method, consider the most general linear-phase filter, a type 1 filter of order m where $m \le 2p$.

$$H(z) = \sum_{i=0}^{m} b_i z^{-i} \qquad (6.4.4)$$

Recall from Table 5.1 that for a type 1 filter, m is even and the impulse response satisfies the even symmetry condition $h(m - k) = h(k)$. For FIR filters in general, $b_i = h(i)$, which means $b_{m-i} = b_i$ for $0 \le i \le m$. For convenience, let $\theta = 2\pi f T$ and let $m = 2r$. One then can write the frequency response of $H(z)$ as follows.

$$H(f) = \sum_{i=0}^{m} b_i \exp(-ji\theta)$$

$$= \exp(-jr\theta) \sum_{i=0}^{m} b_i \exp[-j(i - r)\theta] \qquad (6.4.5)$$

Since m is even, the middle or rth term can be separated out from the sum. The remaining terms then can be combined in pairs using the symmetry condition, $b_{m-i} = b_i$, and Euler's identity.

$$H(f) = \exp(-jr\theta)\{b_r + \sum_{i=0}^{r-1} b_i \exp[-j(i - r)\theta] + b_{m-i} \exp[-j(m - i - r)\theta]\}$$

$$= \exp(-jr\theta)\{b_r + \sum_{i=0}^{r-1} b_i[\exp[-j(i - r)\theta] + \exp[j(i - r - m + 2r)\theta]]\}$$

$$= \exp(-jr\theta)\{b_r + \sum_{i=0}^{r-1} b_i[\exp[-j(i - r)\theta] + \exp[j(i - r)\theta]]\}$$

$$= \exp(-jr\theta)\{b_r + 2\sum_{i=0}^{r-1} b_i \cos[(i - r)\theta]\}$$

$$= A_r(f) \exp(-jr\theta) \qquad (6.4.6)$$

For convenience, define $c_i = b_i$ for $i \ne r$ and $c_r = b_r/2$. Recalling that $\theta = 2\pi f T$, we then arrive at the following amplitude response for a type 1 linear-phase FIR filter of order $2r$.

$$A_r(f) = 2\sum_{i=0}^{r} c_i \cos[2\pi(i - r)fT] \qquad (6.4.7)$$

Now that we have an expression that shows the dependence of $A_r(f)$ on c, we can proceed to find an optimal value for c and therefore b. From (6.4.3), the objective function is

$$J_p(c) = \sum_{i=0}^{p} w^2(i) \left[2\sum_{k=0}^{r} c_k \cos[2\pi(k - r)F_i T] - A_d(F_i)\right]^2 \qquad (6.4.8)$$

To find an optimal value for the coefficient vector c, set $\partial J_p(c)/\partial c = 0$ and solve for c. Rather than take explicit partial derivatives, it is helpful to reformulate (6.4.8) in terms of vector notation. Let the $(p+1) \times (r+1)$ matrix G and the $(p+1) \times 1$ column vector d be defined as follows.

$$G_{ik} \overset{\Delta}{=} 2w(i)\cos[2\pi(k-r)F_iT], \quad 0 \le i \le p, \ 0 \le k \le r \tag{6.4.9a}$$

$$d_i \overset{\Delta}{=} w(i)A_d(F_i), \qquad\qquad 0 \le i \le p \tag{6.4.9b}$$

If $c = [c_0, c_1, \ldots, c_r]^T$ is the unknown coefficient vector, then the objective function in (6.4.8) can be written in compact vector form as

$$J_p(c) = (Gc - d)^T(Gc - d) \tag{6.4.10}$$

Since $p \ge r$, the linear algebraic system $Gc = d$ is *overdetermined* with more equations than unknowns, so that in general there is no c such that $Gc = d$. To find the coefficient vector c that minimizes $J_p(c)$, multiply $Gc = d$ on the left by G^T. This yields the *normal equations*.

Normal equations

$$G^T G c = G^T d \tag{6.4.11}$$

The solution to (6.6.11) is the coefficient vector of the least-squares filter, the filter that minimizes $J_p(c)$. Note that $G^T G$ is a square $(r+1) \times (r+1)$ matrix of full rank. Consequently, the optimal coefficient vector can be expressed as

$$c = (G^T G)^{-1} G^T d \tag{6.4.12}$$

Pseudo-inverse
The matrix $G^+ = (G^T G)^{-1} G^T$ is the call *pseudo-inverse* of G. Normally, one does not solve for c using the pseudo-inverse. Instead, the normal equations in (6.6.11) are solved directly because this takes only about one-third as many FLOPs for large r. The normal equations can become ill-condition for large values of r. Once the $(r+1) \times 1$ vector c is determined, the original $(m+1) \times 1$ coefficient vector b is then obtained as follows.

$$b_i = \begin{cases} c_i, & 0 \le i < r \\ 2c_r, & i = r \\ c_{2r-i}, & r < i \le 2r \end{cases} \tag{6.4.13}$$

Example 6.7   **Least-squares Bandpass Filter**

To illustrate the least-squares method, consider the problem of designing a bandpass filter with a piecewise-linear magnitude response that features a passband of $3f_s/16 \le |f| \le 5f_s/16$ and transition bands of width $f_s/32$. Suppose $p = 40$ and uniformly spaced discrete frequencies are used.

$$F_i = \frac{if_s}{2p}, \quad 0 \le i \le p$$

Two cases are considered. The first one uses uniform weighting, and the second weights the passband samples by $w(i) = 10$. In both cases, the order of the FIR filter is $m = 40$. The results, obtained by running *exam6_6*, are shown in Figure 6.17. Note how by weighting the passband samples more heavily, the passband ripple is reduced, but at the expense of less attenuation in the stopband.

FIGURE 6.17: Magnitude Responses of a Least-squares Bandpass Filter of Order $m = 64$ Using $p + 1 = 129$ Discrete Frequencies and Both Uniform Weighting and Increased Passband Weighting

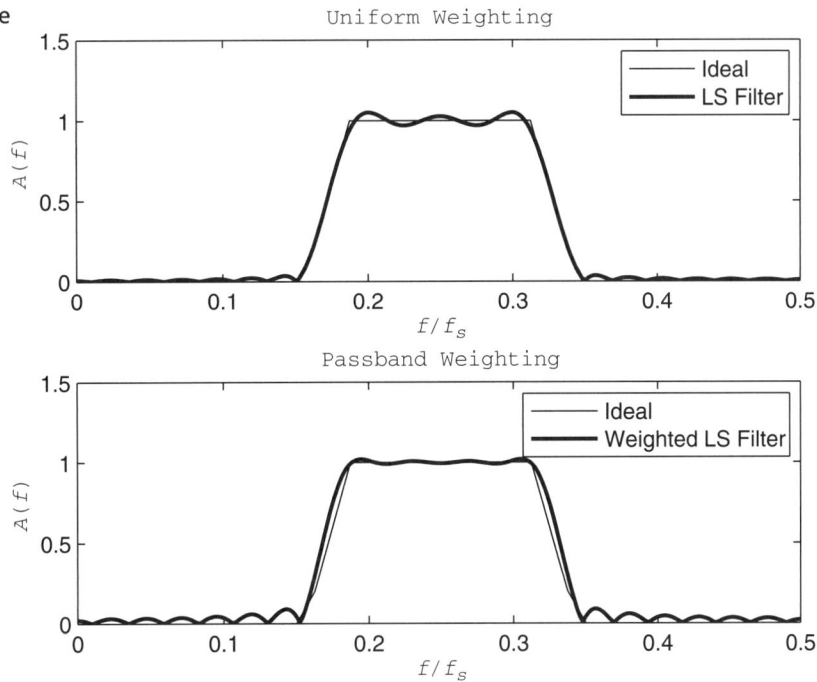

FDSP Functions

The FDSP toolbox contains the following function for designing a type 1 or type 2 linear-phase FIR filter using the least-squares method.

```
% F_FIRLS: Design a least-squares FIR filter
%
% Usage:
%          b   = f_firls (F,A,m,fs,w)
% Pre:
%          F   = 1 by (p+1) vector containing discrete
%                frequencies with F(1) = 0 and F(p+1)=fs/2.
%          A   = 1 by (p+1) vector containing samples of
%                desired amplitude response at the discrete
%                frequencies.
%
%                |A(i)| = |H_d(F_i)|
%
%          m   = order of filter (1 <= m <= 2*p)
%          fs  = the sampling frequency in Hz
%          w   = 1 by (p+1) vector containing weighting
%                factors.  If w is not present, then
%                uniform weighting is used.
% Post:
%          b   = 1 by m+1 vector of filter coefficients.
```

6.5   Equiripple Filters

The FIR filter design methods discussed thus far all involve optimization. The windowing method (with a rectangular window) minimizes the mean-squared error in (6.4.1). The frequency-sampling method uses optimal placement of transition band samples to maximize the stopband attenuation. The least-squares method minimizes a weighted sum of squares of errors at discrete frequencies as in (6.4.3). In this section we examine an optimization technique that is based on minimizing the maximum of the absolute value of the error within the passband and the stopband.

6.5.1   Minimax Error Criterion

To simplify the development, we focus our attention on the most general type of linear-phase FIR filter, a type 1 filter. Consequently, the filter order is even with $m = 2p$, and the impulse response $h(k)$ exhibits even symmetry about the midpoint $k = p$. From (6.2.4), this means that the frequency response can be represented as follows.

$$H(f) = A_r(f) \exp(-j2\pi p f T) \tag{6.5.1}$$

Here the filter amplitude response, $A_r(f)$, is a real even function of f. Like the frequency response, the amplitude response is periodic with period f_s. Therefore, the amplitude response can be approximated with the following truncated trigonometric Fourier series.

$$A_r(f) = \sum_{i=0}^{p} d_i \cos(2\pi i f T) \tag{6.5.2}$$

Notice that there are no sine terms because $A_r(f)$ is an even function. For a windowed filter with the rectangular window, the coefficients d_i are the Fourier coefficients. In this case the resulting filter minimizes the mean-squared error in (6.4.1). However, there is another way to compute the d_i that minimizes an alternative objective. To see this, it is helpful to reformulate the truncated cosine series in (6.5.2) using Chebyshev polynomials. Recall from Section 5.5 that the first two Chebyshev polynomials of the first kind are $T_0(x) = 1$ and $T_1(x) = x$. The

Chebyshev polynomials

remaining *Chebyshev polynomials* can be defined recursively as follows.

$$T_k(x) = 2x T_{k-1}(x) - T_{k-2}(x), \quad k \geq 2 \tag{6.5.3}$$

Thus $T_k(x)$ is a polynomial of degree k. The Chebyshev polynomials are a classic family of orthogonal polynomials that have many useful properties (Schilling and Lee, 1988). In particular, recall from (5.5.25) that

$$T_k[\cos(\theta)] = \cos(k\theta), \quad k \geq 0 \tag{6.5.4}$$

In view of this harmonic generating property, the truncated cosine series in (6.5.2) can be recast in terms of Chebyshev polynomials as

$$A_r(f) = \sum_{k=0}^{p} d_k T_k[\cos(2\pi f T)] \tag{6.5.5}$$

Thus $A_r(f)$ can be thought of as a trigonometric polynomial, a polynomial in $x = \cos(2\pi f T)$. To formulate an alternative error criterion, let $A_d(f)$ be the desired amplitude response, and let $w(f) > 0$ be a weighting function. Then the weighted *error* at frequency f is defined as

$$E(f) \triangleq w(f)[A_d(f) - A_r(f)] \tag{6.5.6}$$

TABLE 6.4: ▶
Frequency Bands
for the Four Basic
Frequency-selective
Filters

Filter Type	F
Lowpass	$[0, F_p] \cup [F_s, f_s/2]$
Highpass	$[0, F_s] \cup [F_p, f_s/2]$
Bandpass	$[0, F_{s1}] \cup [F_{p1}, F_{p2}] \cup [F_{s2}, f_s/2]$
Bandstop	$[0, F_{p1}] \cup [F_{s1}, F_{s2}] \cup [F_{p2}, f_s/2]$

A logical way to select the weighting function is to set $w(f) = 1/\delta_p$ in the passband, and $w(f) = 1/\delta_s$ in the stopband. This way, if one of the specifications is more stringent than the other, it will be given a higher weight because it is more difficult to satisfy. The weighting function in (6.5.6) can be scaled by a positive constant without changing the nature of the optimization problem. This leads to the following normalized weighting function.

$$w(f) = \begin{cases} \delta_s/\delta_p, & f \in \text{passband} \\ 1, & f \in \text{stopband} \end{cases} \tag{6.5.7}$$

Frequency-selective filter specifications place constraints on the amplitude response in the passband and the stopband, but not in the transition band. Let F denote the set of frequencies over which the amplitude response is specified. Then F is the following subset of $[0, f_s/2]$, where $\cup$ denotes the union of the two sets.

$$F \stackrel{\Delta}{=} \text{passband} \cup \text{stopband} \tag{6.5.8}$$

The frequency bands for the four basic frequency-selective filter types are summarized in Table 6.4.

Given the filter specification frequencies and the weighting function, the design objective is then to find a Chebyshev coefficient vector $d \in R^{p+1}$ that solves the following optimization problem.

Minimax criterion

$$\min_{d \in R^{p+1}} [\max_{f \in F} \{|E(f)|\}] \tag{6.5.9}$$

The performance criterion in (6.5.9) is called the *minimax* criterion because it minimizes the maximum of the absolute value of the error over the passband and the stopband.

Next, consider the problem of determining the filter impulse response once an optimal Chebyshev coefficient vector d has been found. This can be achieved using the DTFT. Since $H(f) = \text{DTFT}\{h(k)\}$ is the frequency response, the impulse response can be recovered from $H(f)$ using the inverse DTFT.

$$h(k) = \frac{1}{f_s} \int_{-f_s/2}^{f_s/2} H(f) \exp(jk2\pi fT) df \tag{6.5.10}$$

If the amplitude response is as in (6.5.2), then from (6.5.1) we have

$$\begin{aligned} h(k) &= \frac{1}{f_s} \int_{-f_s/2}^{f_s/2} A_r(f) \exp(-jp2\pi fT) \exp(jk2\pi fT) df \\ &= \frac{1}{f_s} \int_{-f_s/2}^{f_s/2} A_r(f) \exp[j(k-p)2\pi fT] df \\ &= \frac{1}{f_s} \int_{-f_s/2}^{f_s/2} \sum_{i=0}^{p} d_i \cos(i2\pi fT) \exp[j(k-p)2\pi fT] df \\ &= \frac{1}{f_s} \sum_{i=0}^{p} d_i \int_{-f_s/2}^{f_s/2} \cos(i2\pi fT) \exp[j(k-p)2\pi fT] df \end{aligned} \tag{6.5.11}$$

First, consider the case $k = p$. Here the exponent in (6.5.11) is zero, so the exponential factor disappears and the impulse response simplifies to

$$h(p) = \frac{1}{f_s} \sum_{i=0}^{p} d_i \int_{-f_s/2}^{f_s/2} \cos(i2\pi fT)df$$
$$= d_0 \tag{6.5.12}$$

Next, consider the case $k \neq p$. When Euler's identity is used, the terms in (6.5.11) involving odd functions of f disappear because the range of integration is symmetric about $f = 0$. Using the product of cosines, trigonometric identity from Appendix 2 then yields the following.

$$h(k) = \frac{1}{f_s} \sum_{i=0}^{p} d_i \int_{-f_s/2}^{f_s/2} \cos(i2\pi fT)\{\cos[(k-p)2\pi fT] + j\sin[(k-p)2\pi fT]\}df$$
$$= \frac{1}{f_s} \sum_{i=0}^{p} d_i \int_{-f_s/2}^{f_s/2} \cos(i2\pi fT)\cos[(k-p)2\pi fT]df$$
$$= \frac{1}{2f_s} \sum_{i=0}^{p} d_i \int_{-f_s/2}^{f_s/2} \{\cos[(i+[k-p])2\pi fT) + \cos[(i-[k-p])2\pi fT]\}df$$
$$= \frac{d_{p-k}}{2}, \quad 0 \leq k < p \tag{6.5.13}$$

Similarly, for the range $p < k \leq 2p$, the $i = k - p$ term survives and we have $h(k) = d_{k-p}/2$. To summarize, the impulse response of a type 1 linear-phase filter can be recovered from the Chebyshev coefficient vector d as follows.

$$h(k) = \begin{cases} \dfrac{d_{p-k}}{2}, & 0 \leq k < p \\ d_p, & k = p \\ \dfrac{d_{k-p}}{2}, & p < k \leq 2p \end{cases} \tag{6.5.14}$$

6.5.2 Parks-McClellan Algorithm

The formulation of the amplitude response $A_r(f)$ as a polynomial in $x = \cos(2\pi fT)$ effectively transforms the filter design problem into a polynomial approximation problem over the set F. Let δ denote the optimal value of the minimax performance criterion.

$$\delta = \min_{d \in R^{p+1}} [\max_{f \in F} \{|E(f)|\}] \tag{6.5.15}$$

Parks and McClellan (1972a,b) applied the *alternation theorem* from the theory of polynomial approximation to solve for d. This theorem is due to Remez (1957).

PROPOSITION
....
6.1: Alternation
Theorem

The function $A_r(f)$ in (6.5.2) solves the minimax optimization problem in (6.5.15) if and only if there exists at least $p + 2$ extremal frequencies $F_0 < F_1 < \cdots < F_{p+1}$ in F such that $E(F_{i+1}) = -E(F_i)$ and

$$|E(F_i)| = \delta, \quad 0 \leq i < p + 2$$

Extremal frequencies are frequencies at which the magnitude of the error achieves its extreme or maximum value within the passband or the stopband. Extremal frequencies include local minima, local maxima, and band edge frequencies. The name, alternation theorem, arises from the fact that the sign of the error alternates as one traverses the extremal frequencies. An

FIGURE 6.18: Optimal Amplitude Response Using the Minimax Optimization Criterion with $m = 12$

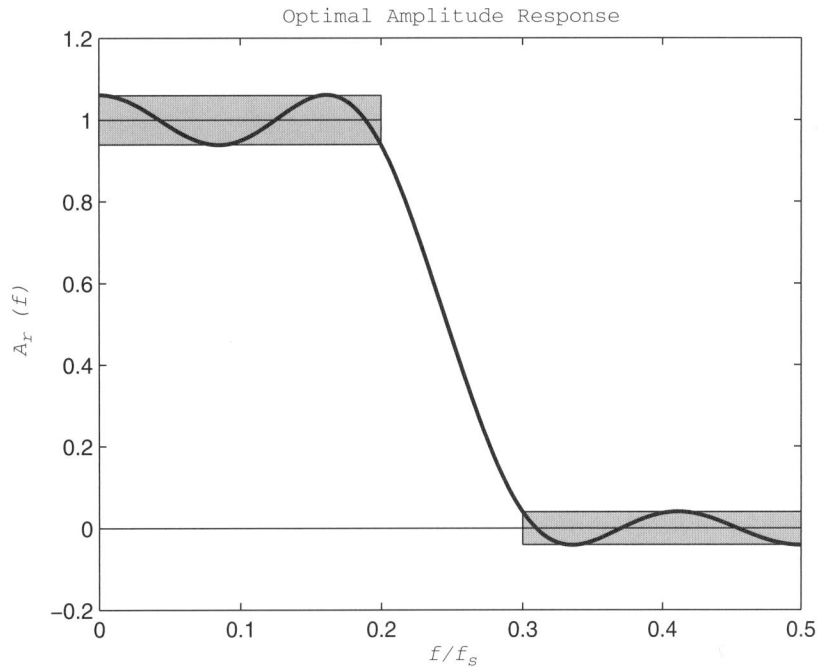

Optimal Amplitude Response

example of an optimal amplitude response for a lowpass filter is shown in Figure 6.18. Notice that the local extrema are associated with ripples in the amplitude response and, within each band, these ripples are of the same size. It is for this reason that an optimal minimax filter is called an *equiripple* filter.

Equiripple filter

For the equiripple filter shown in Figure 6.18, the passband ripple is $\delta_p = 0.06$, the stopband attenuation is $\delta_s = 0.04$, and the filter order is $m = 12$. Observe that there are four extrema frequencies in the passband and another four in the stopband. Thus the number of extrema frequencies is $p + 2 = 8$, so from Proposition 6.1 the amplitude response in Figure 6.18 is optimal. The need for at least $p + 2$ extremal frequencies arises from the following observations. Since $A_r(f)$ is a polynomial of degree p, there can be up to $p - 1$ local minima and local maxima where the slope of $A_r(f)$ is zero. For a lowpass or highpass filter, the optimal amplitude response also goes through the interior passband and stopband edge points.

$$A_r(F_p) = 1 - \delta_p \tag{6.5.16a}$$
$$A_r(F_s) = \delta_s \tag{6.5.16b}$$

That makes $p + 1$ extremal frequencies. In addition at least one, and perhaps both, of the endpoint frequencies, $F = 0$ and $F = f_s/2$, are also extremal frequencies. Therefore, for a lowpass or a highpass filter the number of extremal frequencies in the optimal amplitude response will be either $p + 2$ or $p + 3$. Bandpass and bandstop filters can have up to $p + 5$ extremal frequencies because there are two additional band edge frequencies. From Proposition 6.1, an optimal equiripple amplitude response must have at least $p + 2$ extremal frequencies.

In order to determine the optimal Chebyshev coefficient vector $d \in R^{p+1}$, one starts with the alternation theorem. Using the definition of $E(f)$ in (6.5.6), we have

$$w(F_i)[A_d(F_i) - A_r(F_i)] = (-1)^i \delta, \quad 0 \le i < p + 2 \tag{6.5.17}$$

These equations can be recast as

$$A_r(F_i) + \frac{(-1)^i \delta}{w(F_i)} = A_d(F_i), \quad 0 \le i < p + 2 \tag{6.5.18}$$

Let $\theta_i = 2\pi F_i T$ for $0 \le i < p+2$. Then substituting for $A_r(F_i)$, using (6.5.2) yields

$$\sum_{k=0}^{p} d_k \cos(k\theta_i) + \frac{(-1)^i \delta}{w(F_i)} = A_d(F_i), \quad 0 \le i < p+2 \tag{6.5.19}$$

Next let $c = [d_0, \cdots, d_p, \delta]^T$ represent a vector of unknown parameters. Then the $p+2$ equations in (6.5.19) can be rewritten as the following vector equation.

$$\begin{bmatrix} 1 & \cos(\theta_0) & \cdots & \cos(p\theta_0) & \frac{1}{w(F_0)} \\ 1 & \cos(\theta_1) & \cdots & \cos(p\theta_1) & \frac{-1}{w(F_1)} \\ & & \vdots & & \\ 1 & \cos(\theta_p) & \cdots & \cos(p\theta_p) & \frac{(-1)^p}{w(F_p)} \\ 1 & \cos(\theta_{p+1}) & \cdots & \cos(p\theta_{p+1}) & \frac{(-1)^{p+1}}{w(F_{p+1})} \end{bmatrix} \underbrace{\begin{bmatrix} d_0 \\ d_1 \\ \vdots \\ d_p \\ \delta \end{bmatrix}}_{c} = \begin{bmatrix} A_d(F_0) \\ A_d(F_1) \\ \vdots \\ A_d(F_p) \\ A_d(F_{p+1}) \end{bmatrix} \tag{6.5.20}$$

Given a set of extremal frequencies, one could solve (6.5.20) for the Chebyshev coefficient vector d and the parameter δ. Unfortunately, we do not know the extremal frequencies, so we have to estimate them using an iterative process known as the Remez exchange algorithm (Remez, 1957). To locate the $p+2$ extremal frequencies, we start by choosing an initial guess. For example, they might be equally spaced over the set F. Next, (6.5.20) is solved for d and δ. Once d is known, the error function $E(f)$ in (6.5.6) can be evaluated. Typically, $E(f)$ is evaluated on a dense grid of at least $16m$ points in F. If $|E(f)| < \delta + \epsilon$ for some small tolerance ϵ, then convergence has been achieved. Otherwise, a new set of extremal frequencies is determined and the process is repeated. This iterative process is summarized as follows.

ALGORITHM

6.2: Equiripple FIR Filter

1. Pick a filter order $m = 2p$, $N > 0$, and $\epsilon > 0$. Set $k = 1$, and pick $M \ge 16m$. Compute the initial extrema frequencies F_i for $0 \le i < p+2$ to be equally spaced over the set F.
2. Do
   {
   (a) Solve (6.5.20) for d and δ.
   (b) Evaluate $E_k = E(F_k)$ for $0 \le k < M$ where the dense F_k are equally spaced over F. Compute the peak error.

   $$\|E\| = \max_{k=0}^{M-1}\{|E_k|\}$$

   (d) If $\|E\| \ge \delta + \epsilon$ then
      (1) Find the extrema frequencies (local minima and maxima) in the dense set E_k.
      (2) Eliminate the excess extremal frequencies.

Continued on p. 439

Continued from p. 438

(e) Set $k = k + 1$.
}
3. While $\|E\| \geq \delta + \epsilon$ and $k < N$
4. Compute h using (6.5.14), and set

$$H(z) = \sum_{i=0}^{m} h(i) z^{-i}$$

If Algorithm 6.2 terminates with $k = N$, then either ϵ or N should be increased. The computed minimax value δ may or may not satisfy the stopband specification $\delta \leq \delta_s$ depending on the filter order. If the ripples in the resulting amplitude response exceed the filter specifications, then the filter order m should be increased. Kaiser has proposed the following estimate for the *equiripple filter order* needed to meet a given design specification (Rabiner et al., 1975).

Equiripple filter order

$$m \approx \text{ceil} \left\{ \frac{-[10 \log_{10}(\delta_p \delta_s) + 13]}{14.6 \hat{B}} + 1 \right\} \tag{6.5.21}$$

Here $\hat{B} = |F_s - F_p|/f_s$ is the normalized width of the transition band. This value can be used as a starting point for choosing m. If the filter specifications are not met, then m can be gradually increased until they are met or exceeded.

Much of the computational effort in Algorithm 6.2 occurs in steps 2a-b, but these steps can be made more efficient. To see this, let α_i be defined as follows, where $\prod$ denotes the product.

$$\alpha_i = \frac{(-1)^i}{\displaystyle\prod_{k=0, k \neq i}^{p+1} [\cos(\theta_i) - \cos(\theta_k)]} \tag{6.5.22}$$

Parks and McClellan (1972a,b) have shown that the parameter δ can be computed separately as follows.

$$\delta = \frac{\displaystyle\sum_{i=0}^{p+1} \alpha_i A_d(F_i)}{\displaystyle\sum_{i=0}^{p+1} \alpha_i / w(F_i)} \tag{6.5.23}$$

Given δ, the terms in (6.5.19) involving δ can be brought over to the right-hand side. The new augmented right-hand side vector is then

$$h_i = A_d(F_i) - \frac{(-1)^i \delta}{w(F_i)}, \quad 0 \leq i < p + 1 \tag{6.5.24}$$

It then follows that (6.5.19) can be rewritten as

$$A_r(F_i) = h_i, \quad 0 \leq i < p + 1 \tag{6.5.25}$$

Thus the value of $A_r(f)$ is known at $p + 1$ of the extremal frequencies. Since $A_r(f)$ is known to be a polynomial of degree p, there is no need to solve (6.5.20) for the coefficient vector d at each iteration. This is a potentially time-consuming step when p is large. Instead, the points

$A_r(F_i)$ can be used to construct a Lagrange interpolating polynomial (Parks and McClellan, 1972a,b). From $\theta_i = 2\pi F_i T$ and $\theta = 2\pi f T$, the kth Lagrange interpolating polynomial is

$$L_i(f) = \frac{\prod_{k=0,k\neq i}^{p} [\cos(\theta) - \cos(\theta_k)]}{\prod_{k=0,k\neq i}^{p} [\cos(\theta_i) - \cos(\theta_k)]} \tag{6.5.26}$$

Note that $L_k(F_i) = \delta(i - k)$ where $\delta(i)$ is the unit impulse. Using this orthogonality property and (6.5.25), the amplitude response can be represented as follows.

$$A_r(f) = \sum_{i=0}^{p+1} h_i L_i(f) \tag{6.5.27}$$

This removes the need to solve the linear algebraic system in step 2b, a process that takes about $(p+1)^3/3$ FLOPs. Once Algorithm 6.2 has converged, (6.5.20) can be solved once for the final Chebyshev coefficient vector d, and then (6.5.14) can be used to find the impulse response $h(k)$.

Example 6.8   **Equiripple Filter**

As an illustration of the equiripple filter design technique, consider the problem of designing a bandpass filter with $f_s = 200$ Hz and the following design specifications.

$$(F_{s1}, F_{p1}, F_{p2}, F_{s2}) = (36, 40, 60, 64) \text{ Hz}$$
$$(\delta_p, \delta_s) = (0.02, 0.02)$$

The transition band is fairly narrow in this case, with a normalized width of

$$\hat{B} = \frac{F_{p1} - F_{s1}}{f_s}$$
$$= 0.04$$

From (6.5.21) an initial estimate of the filter order required to meet these specifications is

$$m \approx \text{ceil}\left\{ \frac{-[10\log_{10}(0.0004) + 13]}{14.6(0.04)} + 1 \right\}$$
$$= 73$$

The coefficients of the equiripple filter can be computed by running script *exam6_13*. The estimate of $m \approx 73$ is a bit low and the resulting filter does not quite meet the specifications. The magnitude response for the case $m = 84$ is shown in Figure 6.19 where the equiripple nature of the magnitude response is apparent.

It is somewhat difficult to visualize the degree to which the specifications are satisfied when we view Figure 6.19. Instead, the stopband attenuation is easier to see when the magnitude response is plotted using the logarithmic dB scale. From (5.2.7), the passband ripple and stopband attenuation in dB are

$$A_p = 0.1755 \text{ dB}$$
$$A_s = 33.98 \text{ dB}$$

A plot of the magnitude response in units of dB is shown in Figure 6.20. Here it is clear that for a filter of order $m = 84$, the specified stopband attenuation of $A_s = 33.98$ dB is exceeded. Therefore, the filter order might be reduced somewhat and still meet the specification.

FIGURE 6.19: Linear Magnitude Response of the Optimal Equiripple Bandpass Filter, $m = 84$

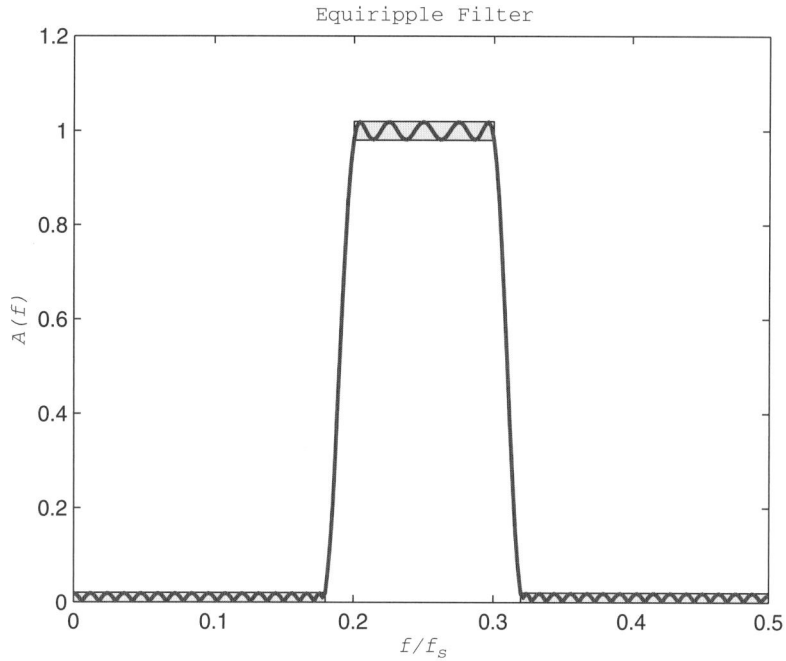

FIGURE 6.20: Logarithmic Magnitude Response of the Optimal Equiripple Bandpass Filter in dB, $m = 84$

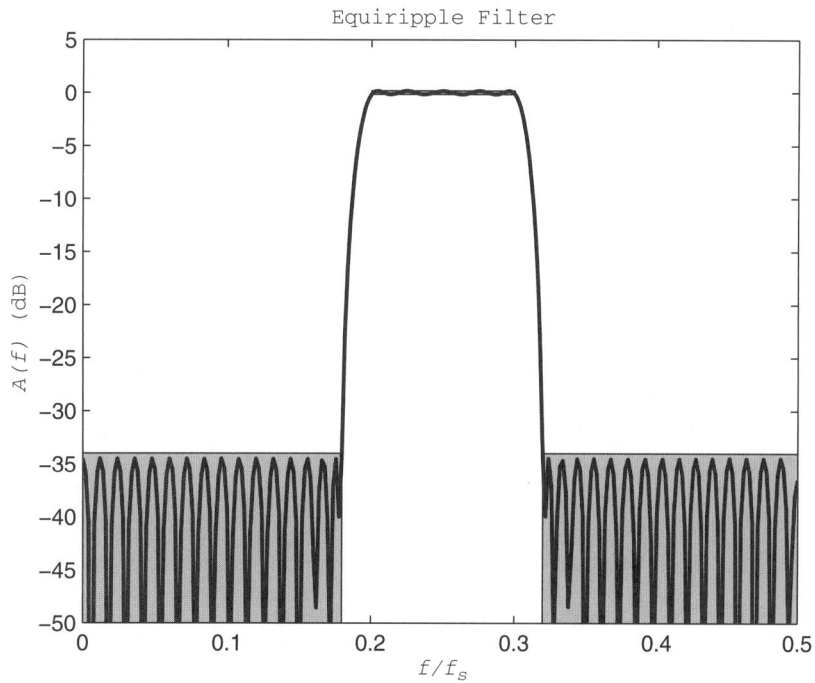

FDSP Functions

The FDSP toolbox contains the following function for designing an optimal equiripple FIR filter using the Parks-McLellan algorithm.

```
% F_FIRPARKS: Design a Parks-McClellan equiripple FIR filter
%
% Usage:
%        b = f_firparks (m,Fp,Fs,deltap,deltas,ftype,fs)
% Pre:
%        m      = filter order (m >= 2)
%        Fp     = passband cutoff frequency or frequencies
%        Fs     = stopband cutoff frequency or frequencies
%        deltap = passband ripple
%        deltas = stopband attenuation
%        ftype  = filter type
%
%                     0 = lowpass
%                     1 = highpass
%                     2 = bandpass (Fp and Fs are vectors)
%                     3 = bandstop (Fp and Fs are vectors)
%
%        fs     = sampling frequency in Hz
% Post:
%        b = 1 by (m+1) coefficient vector of numerator
%            polynomial
%        n = optional estimate of filter order needed to
%            meet specs.
%
% Note: This function uses an equiripple FIR filter
%       implementation based on a free C version developed
%       by Jake Janovetz (janovetz@uiuc.edu).
```

6.6 Differentiators and Hilbert Transformers

In this section we investigate some specialized linear-phase FIR filters that have impulse responses and amplitude responses that exhibit odd symmetry. Filters with odd symmetry have a phase offset of $\alpha = \pi/2$. Consequently, after the effects of the group delay have been removed, the steady-state output of the filter is in *phase quadrature* with a sinusoidal input.

Phase quadrature

6.6.1 Differentiators

Recall that in Section 6.1 we considered the problem of designing low-order differentiators based on numerical approximations of the slope of the input signal. An alternative approach is to design a filter whose frequency response corresponds to that of a differentiator. Recall that

Differentiator   an analog *differentiator* is a system with the following frequency response

$$H_a(f) = j2\pi f \tag{6.6.1}$$

Thus a differentiator has a constant phase response of $\phi_a(f) = \pi/2$, and a magnitude response that increases linearly with f. If a causal linear-phase FIR filter is used to approximate a differentiator, then we must allow for a group delay of $\tau = mT/2$. Thus the problem is to design a digital filter with the following frequency response.

$$H_d(f) = j2\pi f T \exp(-j\pi m f T) \tag{6.6.2}$$

Note that a T is included in the amplitude response $A_r(f) = 2\pi f T$, because a digital differentiator processes frequencies only in the range $0 \le f \le f_s/2$. The group delay in this case is $\tau = mT/2$.

In order to choose between a type 3 linear-phase FIR filter and a type 4 filter, it is helpful to look at the zeros of $H(f)$. Recall from Table 5.1 that a type 3 filter with even order m has a zero at $f = 0$ and a zero at $f = f_s/2$. This is in contrast to a type 4 filter with an odd order m that has a zero only at $f = 0$. If $A_r(f) = 2\pi f T$, then $A_r(0) = 0$ and $A_r(f_s/2) = \pi$. Thus the type 4 linear-phase filter appears to be the filter of choice. The effectiveness of the type 4 filter in comparison with the type 3 filter is illustrated with the following example.

Example 6.9 | **Differentiator**

Consider a linear-phase FIR filter of order m with an impulse response $h(k)$ that exhibits odd symmetry about the midpoint $k = m/2$. Suppose the windowing method is used to design $H_d(z)$. Since $H(f)$ does not contain any jump discontinuities, a rectangular window should suffice. From (6.2.9) the filter coefficients are

$$b_i = -2T \int_0^{f_s/2} A_r(f) \sin[2\pi(i - .5m)fT]df$$

$$= -2T \int_0^{f_s/2} 2\pi f T \sin[2\pi(i - .5m)fT]df$$

To simplify the integral, let $\theta = 2\pi f T$. Then the expression for b_i becomes

$$b_i = \frac{-1}{\pi} \int_0^{\pi} \theta \sin[(i - .5m)\theta]d\theta$$

$$= \frac{-1}{\pi} \left\{ \frac{\sin[(i - .5m)\theta]}{(i - .5m)^2} - \frac{\theta \cos[(i - .5m)\theta]}{i - .5m} \right\} \Big|_0^{\pi}$$

Thus the filter coefficients for the differentiator are

$$b_i = \frac{\cos[\pi(i - .5m)]}{i - .5m} - \frac{\sin[\pi(i - .5m)]}{\pi(i - .5m)^2}, \quad 0 \le i \le m$$

Note that, for a type 3 filter, m is even and $(i - .5m)\pi$ is a multiple of π. In this case, when $i \ne .5m$, the sine terms drop out. If we use L'Hospital's rule and put both terms over a common denominator, the midpoint case $i = m/2$ yields $b_{m/2} = 0$. Hence for a type 3 filter

$$b_i = \frac{\cos[\pi(i - .5m)]}{i - .5m}, \quad 0 \le i \le m, \ i \ne .5m$$

FIGURE 6.21: A Differentiator Using a Type 3 Linear-phase FIR Filter of Order $m = 12$ and a Rectangular Window

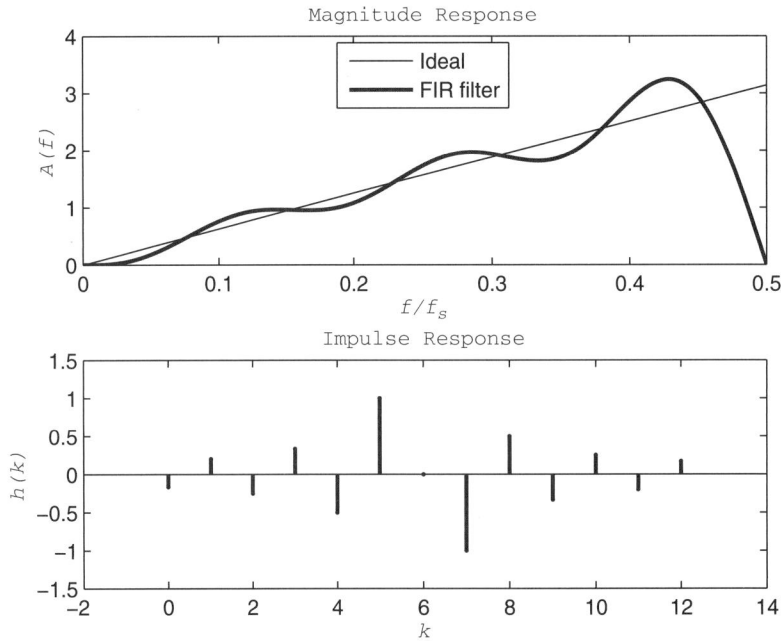

For a type 4 filer, m is odd and the factor $i - .5m$ never goes to zero. In this case $(i - .5m)\pi$ is an odd multiple of $\pi/2$ and the cosine terms drop out. Hence for a type 4 filter

$$b_i = -\frac{\sin[\pi(i - .5m)]}{\pi(i - .5m)^2}, \quad 0 \le i \le m$$

Plots of the magnitude and impulse responses for a type 3 filter with $m = 12$ are shown in Figure 6.21. These were generated by running *exam6_9*. Notice that the zero at $f = f_s/2$ causes significant ringing in the magnitude response, particularly at higher frequencies. In this case, the impulse response goes to zero relatively slowly, suggesting a poor fit.

When a type 4 filter is used, the results are more effective. Plots of the magnitude and impulse responses for the case $m = 11$ are shown in Figure 6.22. Even though this is a lower-order filter than the type 3 filter, it is clearly a better fit, with some error evident at higher frequencies. Notice how the impulse response goes to zero relatively quickly.

Noncausal Differentiator

Offline noncausal differentiator

It should be pointed out that the phase response of the FIR differentiator is $\phi(f) = \pi/2 - \pi mfT$, which corresponds to $\phi(0) = \pi/2$ and a group delay of $\tau = mT/2$. A constant phase response of $\phi(f) = \pi/2$, corresponding to the pure differentiator in (6.6.1), can be achieved if a non-causal implementation is used. A noncausal filter is obtained by multiplying $H_d(z)$ by z^p where $p = \text{floor}(m/2)$. Thus the transfer function of the noncausal filter is

$$H_D(z) = \sum_{i=0}^{m} b_i z^{p-i} \tag{6.6.3}$$

FIGURE 6.22: A Differentiator Using a Type 4 Linear-phase FIR Filter of Order $m = 11$ and a Rectangular Window

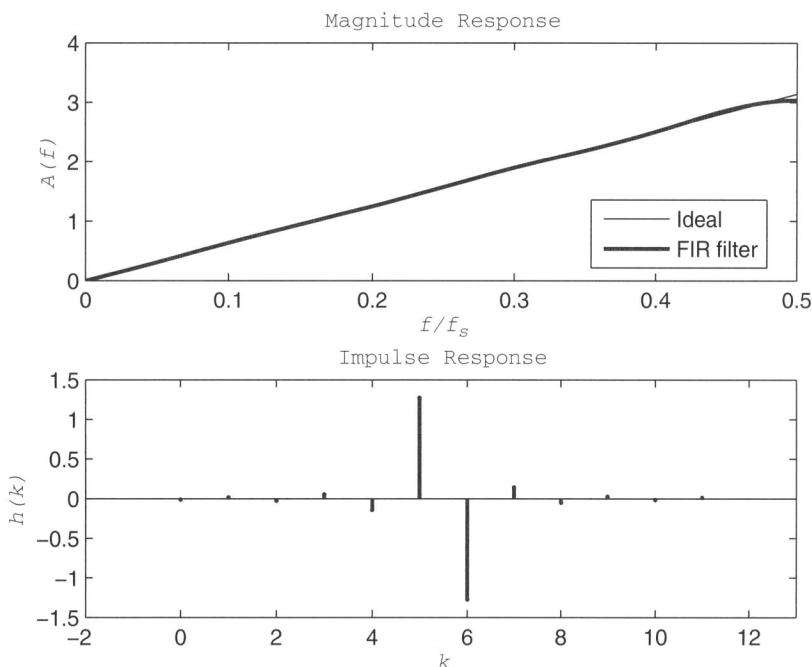

A noncausal filter can be used when offline batch processing is used with the entire input signal available ahead of time. If the differentiator requires a real-time implementation, then a causal filter must be used.

6.6.2 Hilbert Transformers

Hilbert transformer

Another example of a quadrature filter that can be effectively realized with a type 3 or type 4 linear-phase FIR filter is the Hilbert transformer. The analog version of a *Hilbert transformer* is a system with the following frequency response.

$$H_a(f) = -j\text{sgn}(f) \tag{6.6.4}$$

Recall from (5.5.8) that *sgn* denotes the *signum* or sign function defined $\text{sgn}(f) = f/|f|$ for $f \neq 0$ and $\text{sgn}(0) = 0$. Thus a Hilbert transformer imparts a constant phase shift of $-\pi/2$ for $f > 0$ and $\pi/2$ for $f < 0$. Hilbert transformers are used in a number of applications in communications and speech processing. Recall from Section 5.5.2 that Hilbert transformers can be used to generate complex *half band* signals whose spectrum is zero for $f < 0$.

To implement a Hilbert transformer with a causal linear-phase FIR filter, we insert a delay of $\tau = mT/2$, where m is the filter order. From (6.6.4), this yields the following frequency response for a digital Hilbert transformer.

$$H_h(f) = -j\text{sgn}(f) \exp(-j\pi m f T), \quad -f_s/2 < f < f_s/2 \tag{6.6.5}$$

From Table 5.1, this corresponds to the frequency response of a linear-phase FIR filter with a phase offset of $\alpha = \pi/2$ and an amplitude response of $A_r(f) = -\text{sgn}(f)$. Thus a type 3 or a type 4 filter can be used.

Example 6.10   **Hilbert Transformer**

Consider a linear-phase FIR filter of order m with an impulse response $h(k)$ that exhibits odd symmetry about $k = m/2$. Suppose the windowing method is used to design $H(z)$. From (6.2.9), the filter coefficients, using $b_i = h(i)$, are

$$b_i = -2T \int_0^{f_s/2} A_r(f) \sin[2\pi(i - .5m)fT]df$$

$$= 2T \int_0^{f_s/2} \sin[2\pi(i - .5m)fT]df$$

$$= \left\{ \frac{-2T \cos[2\pi(i - .5m)fT]}{2\pi(i - .5m)T} \right\}\Big|_0^{f_s/2}$$

Thus the filter coefficients for the Hilbert transformer using a rectangular window are

$$b_i = \frac{1 - \cos[\pi(i - .5m)]}{\pi(i - .5m)}, \quad 0 \le i \le m$$

When m is odd as in a type 4 filter, the factor $i - .5m$ never goes to zero. In this case $(i - .5m)\pi$ is an odd multiple of $\pi/2$ and the cosine term drops out. Thus for a type 4 filter

$$b_i = \frac{1}{\pi(i - .5m)}, \quad 0 \le i \le m$$

For a type 3 filter with m even, L'Hospital's rule applied to the case $i = m/2$ yields $b_{m/2} = 0$. Plots of the magnitude and impulse responses for the type 3 case with $m = 30$ are shown in Figure 6.23 where a rectangular window is used. These were generated by running *exam6_10*. Notice that this approximates a Hilbert transformer, but only over a limited range of frequencies. Often this is sufficient depending on the application. The ripples can be reduced by tapering

FIGURE 6.23: A Hilbert Transformer Using a Type 3 Linear-phase FIR Filter of Order $m = 30$ and a Rectangular Window

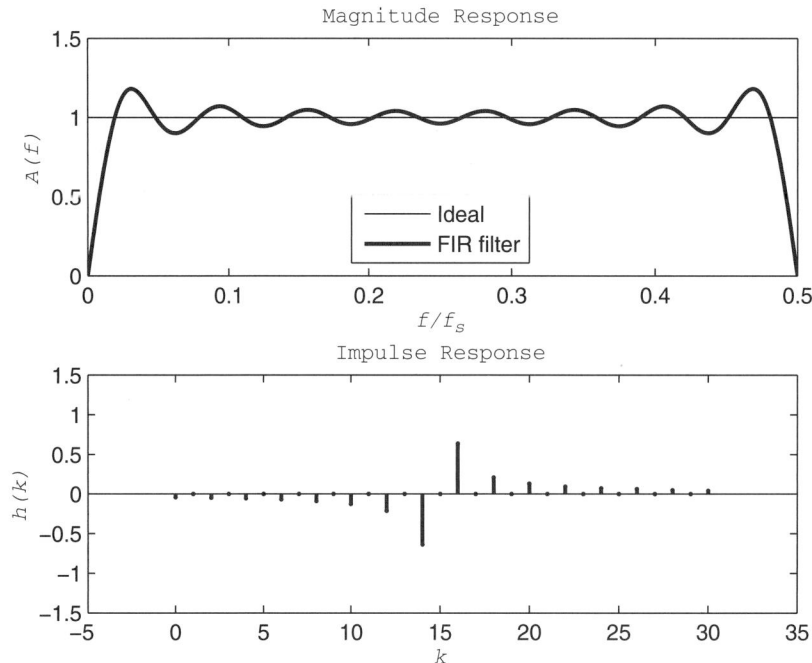

FIGURE 6.24: A
Hilbert Transformer
Using a Type 3
Linear-phase FIR
Filter of Order
$m = 30$ and a
Hamming Window

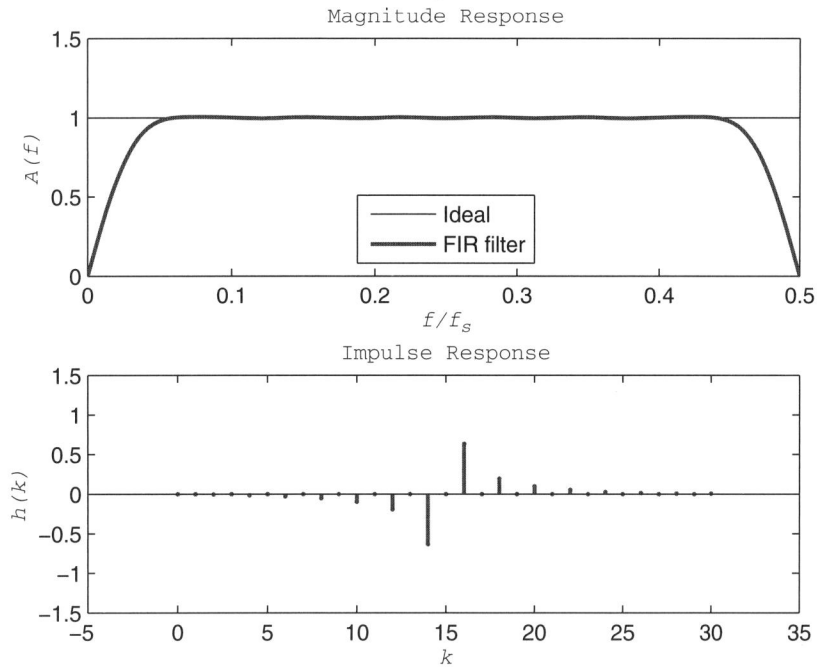

FIGURE 6.24: A Hilbert Transformer Using a Type 3 Linear-phase FIR Filter of Order $m = 30$ and a Hamming Window

the coefficients with a data window. Plots of the magnitude and impulse responses for a type 3 filter with $m = 30$ and the Hamming window are shown in Figure 6.24. It is evident that the magnitude response is smoother in the passband, but the width of the passband has been reduced somewhat in this case.

FDSP Functions

The FDSP toolbox contains the following functions for designing a differentiator filter and a Hilbert transformer filter.

```
% F_DIFFERENTIATOR: Design a differentiator linear-phase FIR filter
% F_HILBERT: Design a Hilbert transformer using a linear-phase FIR filter
%
% Usage:
%          b = f_differentiator(m,win);
%          b = f_hilbert(m,win);
% Pre:
%          m   = the filter order
%          win = the window type to be used:
%
%                0 = rectangular
%                1 = Hanning
%                2 = Hamming
%                3 = Blackman
% Post:
%          b = 1 by (m+1) array of FIR filter coefficients
```

6.7   Quadrature Filters

In this section a general two-stage quadrature filter is developed that is designed to meet both amplitude response and phase response specifications, but with a group delay. Stage one makes use of a Hilbert transformer to produce a pair of signals in phase quadrature. Stage two applies linear-phase FIR filters to these signals to generate the real and imaginary parts of the desired frequency response.

6.7.1   Generation of a Quadrature Pair

The overall structure of the two-stage quadrature filter is shown in Figure 6.25. It features two parallel branches, with each branch consisting of a cascaded pair of linear-phase FIR filters.

The purpose of stage one is to generate a quadrature pair that is synchronized with a sinusoidal input. Suppose the input is a cosine of frequency f.

$$x(k) = \cos(2\pi f k T) \tag{6.7.1}$$

From Figure 6.25, the FIR filter in the upper left corner is a pure delay of $m/2$ samples. It is assumed that m is even, so $m/2$ is an integer. Thus the magnitude response is $A(f) = 1$, and the phase response is $\phi(f) = -\pi m f T$. It follows that the steady-state output of the upper branch of stage one is simply $x(k)$ delayed by $m/2$ samples.

$$x_1(k) = \cos[2\pi f(k - .5m)T] \tag{6.7.2}$$

The FIR filter in the lower left corner of Figure 6.25 is a Hilbert transformer. If the order of the FIR filter is m, then this corresponds to a type 3 linear-phase filter with frequency response

$$H_h(f) = A_h(f)\exp[-j(\pi m f T + \pi/2)], \quad 0 < f < f_s/2 \tag{6.7.3}$$

For m sufficiently large, the magnitude response is $A_h(f) \approx 1$. Note that the phase is linear, with a group delay of $mT/2$ and an offset of $-\pi/2$. Thus the steady-state output of the lower

FIGURE 6.25: A Two-stage Quadrature Filter

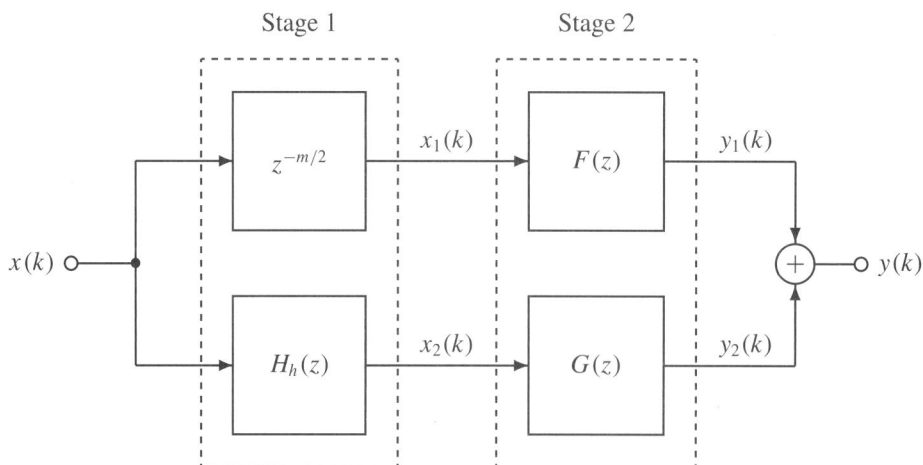

branch of stage one is

$$x_2(k) = A_h(f)\cos[2\pi f(k - .5m)T - \pi/2]$$
$$= A_h(f)\sin[2\pi f(k - .5m)T]$$
$$\approx \sin[2\pi f(k - .5m)T], \quad 0 < f < f_s/2 \tag{6.7.4}$$

It follows from (6.7.2) and (6.7.4) that the stage one outputs, $x_1(k)$ and $x_2(k)$, form a quadrature pair. The relative phase relationship of $\pi/2$ is exact, and the amplitude of $x_1(k)$ is exact at $A_1(f) = 1$. However, the amplitude of $x_2(k)$ is $A_2(f) \approx 1$ over an interval within the range $0 < f < f_s/2$. Since $H_h(z)$ is a type 3 FIR linear-phase filter, $A_h(0) = A_h(f_s/2) = 0$.

Example 6.11	**Quadrature Pair**

To illustrate the effectiveness of stage one in generating a quadrature pair, consider the case of a linear-phase filter of order $m = 30$ with a Hamming window. The magnitude response of this Hilbert transformer was shown previously in Figure 6.25. To directly view the amplitude and phase relationships between $x_1(k)$ and $x_2(k)$, suppose the filter $H_h(z)$ is driven by the cosine input in (6.7.1). Steady-state plots of $x_2(k)$ versus $x_1(k)$ for $N = 100$ values of f are shown in Figure 6.26 where

$$f_i = \frac{if_s}{2N}, \quad 0 \le i < N$$

The elliptical patterns correspond to frequencies near the two ends of the Nyquist range. The darker circle of unit radius corresponds to the range of frequencies over which the quadrature signals both have unit amplitude.

FIGURE 6.26: Phase Plane Plots Corresponding to Different Frequencies Generated by Stage One Using a Hilbert Transformer of Order $m = 60$ with a Hamming Window

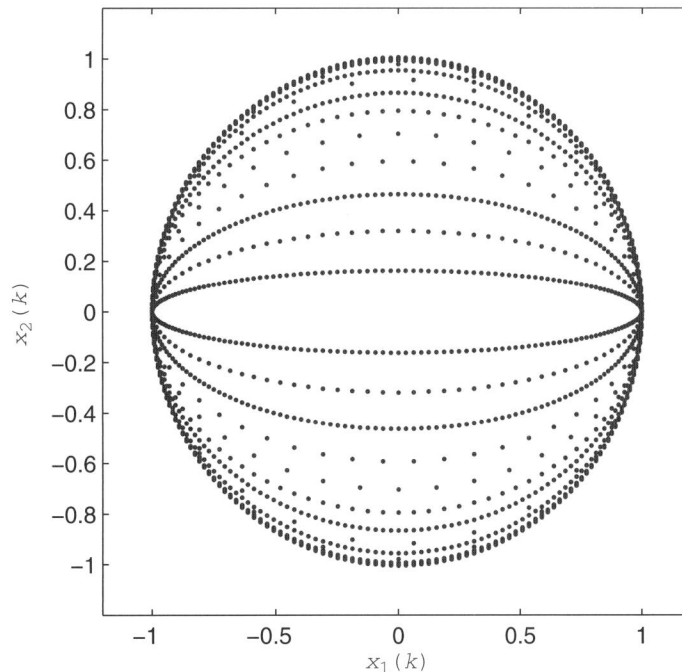

6.7.2 Quadrature Filter

Next, consider the problem of designing an FIR filter with a desired magnitude response $A_d(f)$ and a desired phase response $\theta_d(f)$, subject to the constraint that the total phase response $\phi_d(f)$ also includes a linear-phase term representing a constant group delay of $\tau_q = mT$. Thus the total phase is

$$\phi_d(f) = \theta_d(f) - 2\pi mfT \qquad (6.7.5)$$

Residual phase

Here $\theta_d(f)$ will be referred to as the *residual phase*, the phase that remains after the effects of the group delay have been removed. Suppose the FIR filters $F(z)$ and $G(z)$ in stage two of Figure 6.25 are type 1 linear-phase filters of order m. Let $A_f(f)$ and $A_g(f)$ denote the amplitude responses of $F(z)$ and $G(z)$, respectively, and let input $x(k)$ be a cosine of frequency f, as in (6.7.1). Since $F(z)$ is linear-phase, from (6.7.2) the steady-state output of $F(z)$ is

$$y_1(k) = A_f(f) \cos[2\pi f(k - m)T] \qquad (6.7.6)$$

Next, suppose the Hilbert transformer bandwidth is sufficiently large that $A_h(f) \approx 1$ for the frequency of interest. Since $G(z)$ is linear-phase, from (6.7.4) the steady state output of $G(z)$ is

$$y_2(k) \approx A_g(f) \sin[2\pi f(k - m)T] \qquad (6.7.7)$$

If we make use of the cosine of the sum trigonometric identity from Appendix 2, the desired steady-state output of the quadrature filter with magnitude response $A_d(f)$ and phase response $\phi_d(f) = \theta_d(f) - 2\pi mfT$ is

$$\begin{aligned} y(k) &= A_d(f) \cos[2\pi f(k - m)T + \theta_d(f)] \\ &= A_d(f) \cos[\theta_d(f)] \cos[2\pi f(k - m)T] \\ &\quad - A_d(f) \sin[\theta_d(f)] \sin[2\pi f(k - m)T] \end{aligned} \qquad (6.7.8)$$

From (6.7.6), (6.7.7), and Figure 6.24, the quadrature filter output is

$$\begin{aligned} y(k) &= y_1(k) + y_2(k) \\ &\approx A_f(f) \cos[2\pi f(k - m)T] + A_g(f) \sin[2\pi f(k - m)T] \end{aligned} \qquad (6.7.9)$$

Amplitude responses

Equating the expressions for $y(k)$ in (6.7.8) and (6.7.9), we find that the desired *amplitude responses* for the stage two linear-phase filters are

$$A_f(f) = A_d(f) \cos[\theta_d(f)] \qquad (6.7.10)$$
$$A_g(f) = -A_d(f) \sin[\theta_d(f)] \qquad (6.7.11)$$

Quadrature filter

Here $F(z)$ and $G(z)$ are designed to approximate $A_f(f)$ and $A_g(f)$, respectively. The *quadrature filter* transfer function from Figure 6.25 is

$$H_q(z) = z^{-m/2} F(z) + H_h(z)G(z) \qquad (6.7.12)$$

Since $z^{-m/2}$, $H_h(z)$, $F(z)$, and $G(z)$ are all linear-phase FIR filters of order m, the two-stage quadrature filter $H_q(z)$ is an FIR filter of order $2m$. It will have a group delay of $\tau_q = mT$, a magnitude response that approximates $A_d(f)$, and a residual phase response that approximates $\theta_d(f)$.

To interpret the contributions of $F(z)$ and $G(z)$, note that the desired frequency response $H_d(f)$, minus the delay, can be written in terms of its real and imaginary parts as follows.

$$\hat{H}_d(f) = A_d(f)\cos[\theta_d(f)] + jA_d(f)\sin[\theta_d(f)] \tag{6.7.13}$$

Comparing (6.7.13) with (6.7.10) and (6.7.11), we see that filter $F(z)$ is used to design the real part of the desired frequency response, and filter $G(z)$ is used to design the imaginary part. In summary, the four linear-phase FIR filters that make up the quadrature filter in Figure 6.25 perform the following functions. The FIR filter in the upper left provides a delay of $mT/2$ for synchronization, Hilbert transformer $H_h(z)$ provides a phase shift of $\pi/2$ to produce a quadrature pair, $F(z)$ approximates the real part of the desired frequency response, and $G(z)$ approximates the imaginary part.

Since the Hilbert transformer is a type 3 linear-phase filter, its frequency response must satisfy the end-point zero constraints, $A_h(0) = A_h(f_s/2) = 0$. From Figure 6.25 this means that in the steady state $y_2(k) = 0$ when $f = 0$ and when $f = f_s/2$. This is not as serious a limitation as, at first, it might appear. Recall that the two frequency limits $f = 0$ and $f = f_s/2$ correspond to $z = \pm 1$. However, for a real system with desired transfer function $H_d(z)$, the points $H_d(\pm 1)$ will be real. Thus for a real system, the desired phase response will satisfy

$$\text{mod}[\phi_d(f), \pi] = 0, \quad f \in \{0, f_s/2\} \tag{6.7.14}$$

Since the desired phase angle for a real filter is a multiple of π at the two end points, it follows that the imaginary part of the desired frequency response will be zero at $f = 0$ and $f = f_s/2$.

Example 6.12	**Quadrature Filter**

As an illustration of a quadrature filter, suppose the desired magnitude response consists of three sections as follows, where fT represents normalized frequency.

$$A_d(f) = \begin{cases} .5(1 + 8fT), & 0 \le fT < .125 \\ 64(fT - .125)^2, & .125 \le fT \le .375 \\ 1 + .5\sin[16\pi(fT - .375)], & .375 < fT \le .5 \end{cases}$$

Next, let the residual phase response be the following.

$$\theta_d(f) = \pi(1 - .25fT) + 2\sin(8\pi fT)$$

Suppose the linear-phase filters are of order $m = 160$ with a Hamming window. This produces an overall group delay of $\tau = 160T$. Plots of the magnitude response and the residual phase response of the quadrature filter, obtained by running *exam6_12*, are shown in Figure 6.27. Although the fit is not exact, it is an effective fit of *both* the desired magnitude response and the desired residual phase response.

Quadrature filters include linear-phase filters as a special case. Observe that when the residual phase response is set to $\theta_d(f) = 0$, this corresponds to linear phase. From (6.7.1), this implies $A_g(f) = 0$ and $A_f(f) = A_d(f)$, which means that $F(z)$ is an mth-order linear-phase filter and $G(z) = 0$. Given the delay of $m/2$ samples in stage one, the quadrature filter will have the accuracy of an mth-order linear-phase filter, but with a group delay of $\tau_q = mT$.

The desired magnitude response in Figure 6.27 goes to zero only at a single point, $f = .25f_s$. Recall from the Paley-Wiener theorem in Proposition 5.1 that if the magnitude response is identically zero over an interval of nonzero length, then an exact realization is not possible with a causal filter. Examples include the ideal lowpass, highpass, bandpass, and bandstop frequency-selective filters. It is of interest to see what happens when the quadrature design method is applied to such a filter. From (6.7.10) and (6.7.11), the magnitude response and the

FIGURE 6.27: Magnitude
and Residual Phase
Responses of the
Two-stage
Quadrature Filter
for Example 6.12
with $m = 120$ and a
Hamming Window

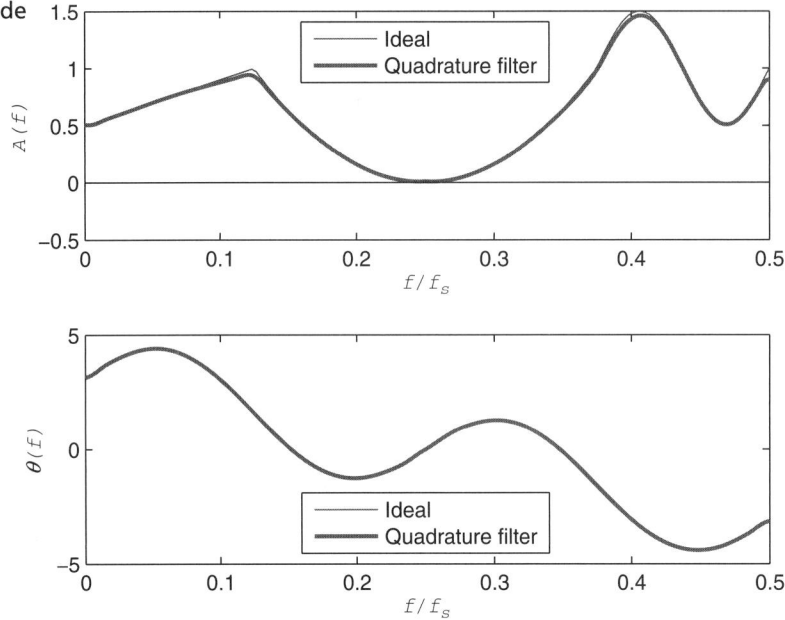

residual phase response of the quadrature filter can be expressed in terms of the amplitude responses of the second stage filters as follows.

$$A_q(f) = \sqrt{A_f^2(f) + A_g^2(f)} \tag{6.7.15}$$

$$\theta_q(f) = \arctan\left[\frac{-A_g(f)}{A_f(f)}\right] \tag{6.7.16}$$

The expression for $A_q(f)$ is well-behaved and should yield a good approximation except possibly near the ends of the Nyquist range where the magnitude response of the Hilbert transformer begins to decrease. However, the expression for $\theta_q(f)$ is *not* well-behaved when $A_d(f) = 0$. Indeed, if $A_d(f) = 0$ over a stopband interval, then from (6.7.10) and (6.7.11) it follows that both $A_f(f) = 0$ and $A_g(f) = 0$. But from (6.7.15), this implies that the residual phase response over the idealized stopband is $\arctan(0/0)$, which is numerically not well-defined. Put another way, the phase angle is not meaningful when the magnitude is zero. As a simple numerical remedy, let $\delta_s > 0$ be a small stopband attenuation factor. One can then replace the desired magnitude response in (6.7.10) and (6.7.11) as follows, which effectively bounds $A_d(f)$ away from zero.

$$A_f(f) = \max\{A_d(f), \delta_s\}\cos[\theta(f)] \tag{6.7.17}$$

$$A_g(f) = -\max\{A_d(f), \delta_s\}\sin[\theta(f)] \tag{6.7.18}$$

Example 6.13   Frequency-selective Filter

To illustrate the application of the quadrature method to an idealized frequency-selective filter, consider the following highpass filter.

$$A(f) = \mu(fT - .25)$$

As a desired residual phase response, consider the following sinusoid.

$$\theta(f) = \pi\sin[6\pi(fT)^2];$$

Suppose the linear-phase filters are of order $m = 150$ with a Blackman window. Let the stopband attenuation factor be $\delta_s = .01$. Plots of the resulting magnitude response and residual phase response, obtained by running *exam6_12*, are shown in Figure 6.28. Again, the fit is not

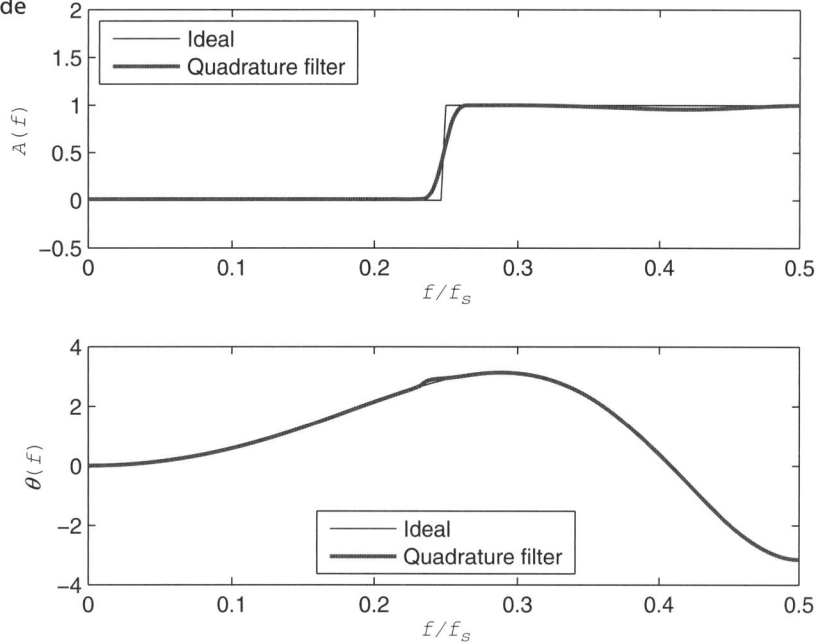

FIGURE 6.28: Magnitude and Residual Phase Responses of the Two-stage Quadrature Filter for Example 6.12 with $m = 150$, a Blackman window, and $\delta_s = .01$

exact, but it is a reasonable fit of both the desired magnitude response and the desired residual phase response, including over the stopband.

Generalized least-squares filter

An alternative way to design a filter with both a prescribed magnitude response and a prescribed residual phase response is to *generalize* the least-squares design method of Section 6.4. In particular, the integrand of the objective function J in (6.4.1) can be modified by adding a second term consisting of the square of the difference between the desired residual phase response and the actual residual phase response.

6.7.3 Equalizer Design

A particularly nice application of quadrature filters arises in the design of equalizer filters. Recall from Proposition 5.3 that every rational transfer function $H(z)$ can be decomposed into the product of an allpass part, $H_{\text{all}}(z)$, and a minimum-phase part, $H_{\text{min}}(z)$. In Section 5.4.3 it

Magnitude equalization

was shown that the inverse of the minimum-phase part serves as a *magnitude equalizer* for the original system.

$$H_{\text{equal}}(z) = H_{\text{min}}^{-1}(z) H(z) \tag{6.7.19}$$

From Proposition 5.3, $H_{\text{equal}}(z) = H_{\text{all}}(z)$ which means that $A_{\text{equal}}(f) = 1$. That is, post-processing by the inverse of the minimum-phase part of $H(z)$ eliminates distortion in the magnitude response, but not the phase response.

Optimal Delay

A more complete equalizer is one that eliminates both magnitude distortion and phase distortion. To achieve this more complete form of equalization, suppose the original system has a magnitude response of $A_d(f)$, and a phase response of $\phi_d(f)$.

$$H(f) = A_d(f) \exp[j\phi_d(f)] \tag{6.7.20}$$

First, we decompose the total phase response $\phi_d(f)$ into a linear-phase part, $-2\pi f \tau_d$, and a residual phase part, $\theta_d(f)$. To find an optimal value for the linear-phase delay τ_d, consider

$N + 1$ values of f equally spaced over the Nyquist range.

$$f_i = \frac{i f_s}{2N}, \quad 0 \le i \le N \tag{6.7.21}$$

Let $f \in R^{N+1}$ be a column vector whose ith element is f_i, and let $\phi \in R^{N+1}$ be the column vector of corresponding values of the phase.

$$\phi_i = \phi_d(f_i), \quad 0 \le i \le N \tag{6.7.22}$$

Suppose a delay τ is chosen to minimize a sum of the squares of the difference between $\phi_d(f)$ and $-2\pi f \tau$.

$$E(\tau) = \sum_{i=0}^{N} (\phi_i + 2\pi f_i \tau)^2 \tag{6.7.23}$$

Optimal delay   Setting $dE(\tau)/d\tau = 0$ and solving for τ yields the following least-squares value for the delay τ where the sums are expressed as dot products of the column vectors f and ϕ.

$$\tau = \left(\frac{-1}{2\pi}\right) \frac{\phi^T f}{f^T f} \tag{6.7.24}$$

If $\phi(f)$ decreases with f, which is typical of a many physical systems, then $\phi^T f < 0$, which means that the *optimal delay* τ will be positive. The τ in (6.7.24) ensures that the linear phase term $-2\pi f \tau$ accounts for as much of the total phase $\phi_d(f)$ as possible in the sense of minimizing $E(\tau)$. The value of τ in general will not be an integer multiple of T. Let τ_d be the nearest integer multiple of the sampling interval T. That is

$$M = \text{round}\left(\frac{\tau}{T}\right) \tag{6.7.25}$$

$$\tau_d = MT \tag{6.7.26}$$

The residual phase $\theta_d(f)$ is then the phase remaining after the effects of a delay of τ_d have been removed.

$$\theta_d(f) = \phi_d(f) + 2\pi f \tau_d \tag{6.7.27}$$

The optimal delay term extracted from $\phi_d(f)$ to produce a residual phase term $\theta_d(f)$ is useful for quadrature filter design in general, because it tends to minimize the magnitude of the residual phase term. When $|\theta_d(f)| \le \pi$, there will be no jump discontinuities in $\theta_d(f)$, and this makes $\theta_d(f)$ easier to approximate with the mth-order stage two filters, $F(z)$ and $G(z)$.

Equalizer Design

To design an equalizer using a quadrature filter, it is necessary to choose a desired magnitude response $A_D(f)$ and residual phase response $\theta_D(f)$. Suppose the original magnitude response $A_d(f)$ is positive. To invert $A_d(f)$ and $\theta_d(f)$ we use

$$A_D(f) = \frac{1}{A_d(f)} \tag{6.7.28}$$

$$\theta_D(f) = -\theta_d(f) \tag{6.7.29}$$

Equalized system   This should produce in a quadrature filter $H_q(z)$ that can be used to equalize the original system $H(z)$.

$$H_{\text{equal}}(z) = H_q(z) H(z) \tag{6.7.30}$$

Equalized output   To the extent that a quadrature filter of order $2m$ does achieve a magnitude response that matches $A_D(f)$ and a residual phase response that matches $\theta_D(f)$, the equalized system will be allpass system with unit magnitude response, zero residual phase response, and a group delay of $\tau_q = (M + m)T$. Hence, if $x(k)$ is the input to the equalized system, the steady state output will be

$$y(k) \approx x[k - (M + m)] \qquad (6.7.31)$$

In this way, the amplitude and phase distortion created by passing $x(k)$ through $H(z)$ can be cancelled. The original signal $x(k)$ is *restored*, but with a delay of $M + m$ samples.

Example 6.14

Equalizer Filter

As an illustration of the use of a quadrature filter to design an equalizer or magnitude and phase compensator, consider a system with the following magnitude and phase responses.

$$A_d(f) = \frac{1.1}{.1 + (fT - .2)^2/.09}$$

$$\phi_d(f) = -20\pi fT + \pi \sin^2(2\pi kT)$$

Suppose the sampling frequency is $f_s = 100 \, \text{Hz}$. From (6.7.24) and with $N = 300$, the optimal least-squares delay associated with the phase response is

$$\tau = .0925$$

From (6.7.25), a delay of $M = 9$ samples is used to compute the desired residual phase response $\theta_d(f)$ in (6.7.27). Suppose a quadrature filter of order $2m$ with $m = 100$ and a Hamming window is used. Running *exam6_14* results in the three magnitude response plots shown in Figure 6.29. Here (a) shows the original magnitude response $A_d(f)$, (b) shows the magnitude response $A_q(f)$ of the equalizer, and (c) shows the equalized magnitude response which is close to unity except near the ends of the Nyquist range where the Hilbert transformer approximation is less accurate. The three residual phase responses are shown in Figure 6.30. Again (a) shows the original residual phase response $\theta_d(f)$, (b) shows the residual phase response $\theta_q(f)$ of the equalizer, and (c) shows the equalized residual phase response which is close to zero throughout indicating effective cancellation of the phase distortion.

FIGURE 6.29: Magnitude Responses of (a) System $H(z)$, (b) Equalizer Filter $H_q(z)$, and (c) Equalized System $H_{\text{equal}}(z) = H_q(z)H(z)$

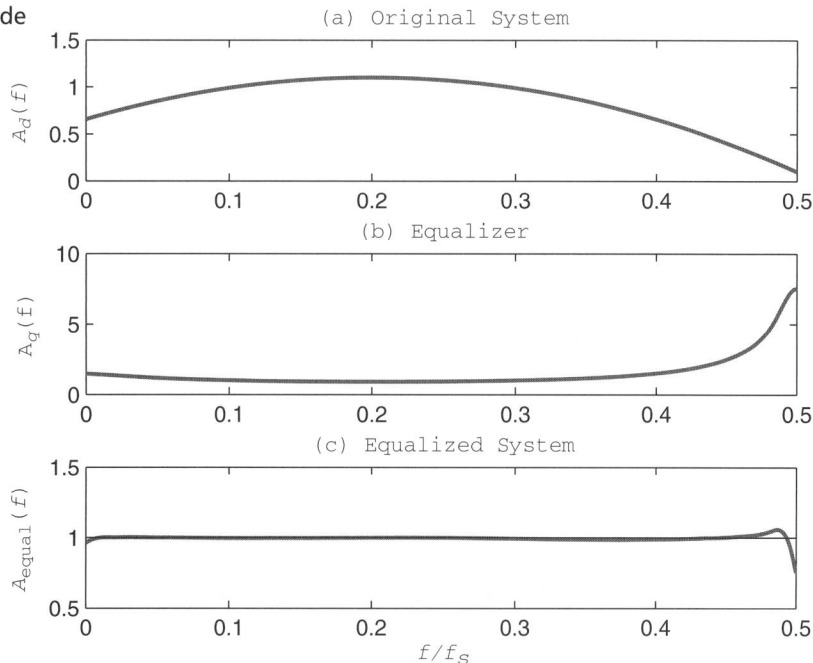

FIGURE 6.30: Residual Phase Responses of (a) System $H(z)$, (b) Equalizer Filter $H_q(z)$, and (c) Equalized System $H_{equal}(z) = H_q(z)H(z)$

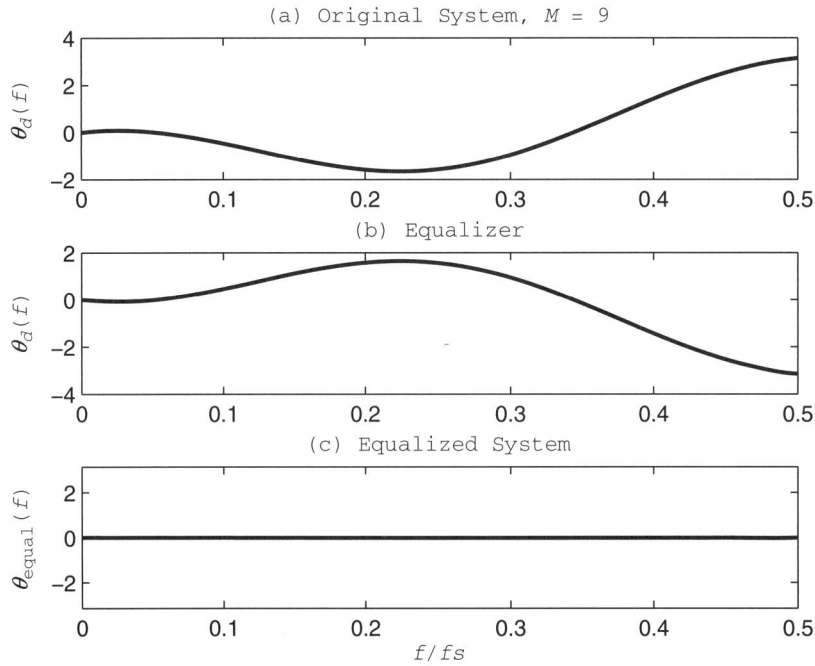

(a) Original System, $M = 9$

(b) Equalizer

(c) Equalized System

FDSP Functions

The FDSP toolbox contains the following function for designing a quadrature filter.

```
% F_FIRQUAD: Design a general two-stage quadrature filter
%
% Usage:
%          b = f_firquad (fsys,m,fs,win,deltas,q)
% Pre:
%          fsys    = handle of function that specifies the desired
%                    magnitude response and residual phase response.
%                    Usage:
%
%                    [A,theta] = fsys(f,fs,q);
%
%                    Here f is the frequency, fs is the sampling
%                    frequency, and q is an optional parameter
%                    vector containing things like cutoff
%                    frequencies,etc. Output A is a the desired
%                    desired magnitude response and output theta
%                    is the desired residual phase response at f. Since
%                    there is a group delay of tau = mT where
%                    T = 1/fs, the total phase response will be:
%
%                    phi(f) = theta(f) - 2*pi*f*m*T
```

Continued on p. 457

Continued from p. 456

```
%         m       = order each subfilter (must be even)
%         fs      = sampling frequency
%         win     = the window type to be used:
%
%                    0 = rectangular
%                    1 = Hanning
%                    2 = Hamming
%                    3 = Blackman
%
%         deltas = stopband attenuation factor (0 < deltas << 1)
%         q      = an optional vector of length contained
%                    design parameters to be passed to fun.
%                    For example q might contain cutoff
%                     frequencies or gains.
% Post:
%         b = 1 by 2m+1 array of filter coefficients.
```

● ● ● ● ● ● ● ● ● ● ● ● ● ● ● ● ● ● ●

6.8   Filter Realization Structures

In this section we investigate alternative configurations that can be used to realize or implement FIR filters with signal flow graphs. These filter realization structures differ from one another with respect to memory requirements, computational time, and sensitivity to finite word length effects. An mth-order FIR filter has the following transfer function.

$$H(z) = b_0 + b_1 z^{-1} + \cdots + b_m z^{-m} \tag{6.8.1}$$

6.8.1 Direct Forms

Tapped Delay Line

By taking the inverse Z-transform of $Y(z) = H(z)X(z)$, using (6.8.1) and the delay property, we find that this results in the following time domain representation of an mth-order FIR filter.

$$y(k) = \sum_{i=0}^{m} b_i x(k - i) \tag{6.8.2}$$

The representation in (6.8.2) is a *direct* representation because the coefficients of the difference equations are obtained directly from inspection of the transfer function. A signal flow graph of a direct form realization, for the case $m = 3$, is shown in Figure 6.31. This structure is called a *tapped delay line* or a *transversal filter*.

FIGURE 6.31:
Tapped-delay-line
Realization of FIR
Filter, $m = 3$

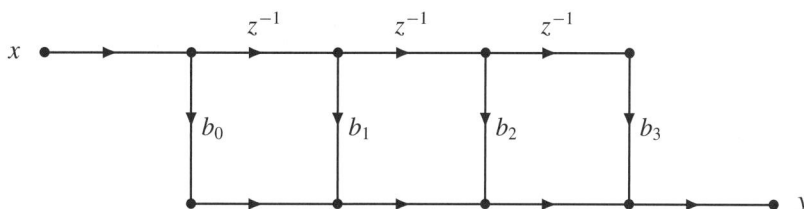

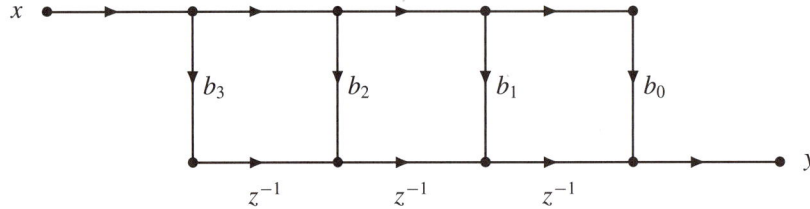

Transposed Tapped Delay Line

Every signal flow graph satisfies something called the flow graph reversal theorem.

PROPOSITION

6.2: Flow graph reversal theorem

If each arc of a signal flow graph is reversed and the input and output labels are interchanged, then the resulting signal flow graph is equivalent.

Applying the flow graph reversal theorem to the direct form graph in Figure 6.31, and drawing the final result with the input on the left rather than the right, we arrive at the FIR filter realization shown in Figure 6.32, which is called a *transposed tapped delay line* filter realization.

Linear-phase Form

The tapped delay line realizations in Figures 6.31 and 6.32 are general realizations that are valid for any FIR filter. However, many FIR filters are designed to be linear-phase FIR filters. Recall that $b_i = h(i)$ for an FIR filter. If the linear-phase symmetry constraint in Proposition 5.2 is recast in terms of the FIR filter coefficients, this yields

$$b_k = \pm b_{m-k}, \quad 0 \leq k \leq m \tag{6.8.3}$$

By exploiting the symmetry constraint we can develop a filter realization that requires only about half as many floating point multiplications (FLOPs). To illustrate the method, consider the most general linear-phase filter, the type 1 filter with even symmetry and even order. If we let $p = m/2$, it is possible to rewrite (6.8.2) as

$$y(k) = b_p x(k - p) + \sum_{i=0}^{p-1} b_i [x(k - i) + x(k - m + i)] \tag{6.8.4}$$

Thus the number of multiplications has been reduced from m to $m/2+1$. Similar expressions for $y(k)$ can be developed for the other three types of linear-phase FIR filters (see Problem 6.18). A signal flow graph realization of a type 1 linear-phase FIR filter is shown in Figure 6.33 for the case $m = 6$.

Observe from Figure 6.33 that there are $m/2+1$ floating point multiplications, one for each distinct coefficient. However, the number of delay elements is m, so there is no reduction in the memory requirements. A comparison of the direct-form FIR filter realizations is summarized in Table 6.5, where it can be seen that the three realizations are identical in terms of storage requirements. In each case the number of floating-point operations or FLOPs grows linearly with the order of the filter. For large values of m, the linear-phase form has approximately .5 times as many multiplications, but about 1.5 times as many additions.

FIGURE 6.33: Direct-form Realization of a Type 1 Linear-phase FIR Filter, $m = 6$

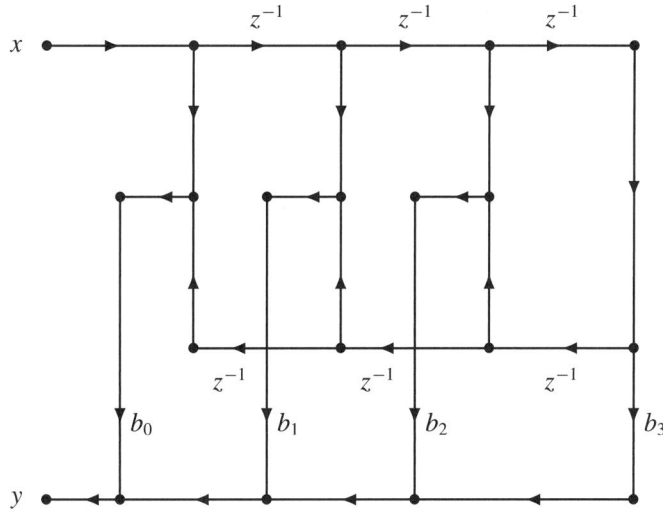

TABLE 6.5: ▶ Comparison of Direct-form Realizations of FIR Filter of Order m

Direct Form	Storage Elements	Additions	Multiplications
Tapped delay line	m	$m + 1$	$m + 1$
Transposed tapped delay line	m	$m + 1$	$m + 1$
Linear phase	m	$3n/2 + 1$	$m/2 + 1$

6.8.2 Cascade Form

The direct forms have the virtue that they can be obtained easily from direct inspection of the transfer function. However, the direct forms also suffer from a practical drawback. As the order of the filter, m, increases, the direct form filters become increasingly sensitive to the finite word length effects that are discussed in Section 6.9. For example, the roots of a polynomial can be very sensitive to small changes in the coefficients of the polynomial, particularly as the degree of the polynomial increases. To develop a realization that will be less sensitive to the effects of finite precision, it is helpful to first recast the transfer function in (6.8.1) in terms of positive powers of z. If the numerator is then factored, this yields the factored form

$$H(z) = \frac{b_0(z - z_1)(z - z_2) \cdots (z - z_m)}{z^m} \tag{6.8.5}$$

Since $H(z)$ has m poles at $z = 0$, an FIR filter is always stable. The coefficients of $H(z)$ are assumed to be real, so complex zeros occur in conjugate pairs. The representation in (6.8.5) can be recast as a product of M second-order subsystems as follows, where $M = \text{floor}[(m + 1)/2]$.

$$H(z) = b_0 H_1(z) \cdots H_M(z) \tag{6.8.6}$$

This is called a *cascade form* realization, and a block diagram for the case $M = 2$ is shown in Figure 6.34. A second-order block $H_i(z)$ is constructed from two poles at $z = 0$ and either

FIGURE 6.34:
Cascade-form Block
Diagram, $M = 2$

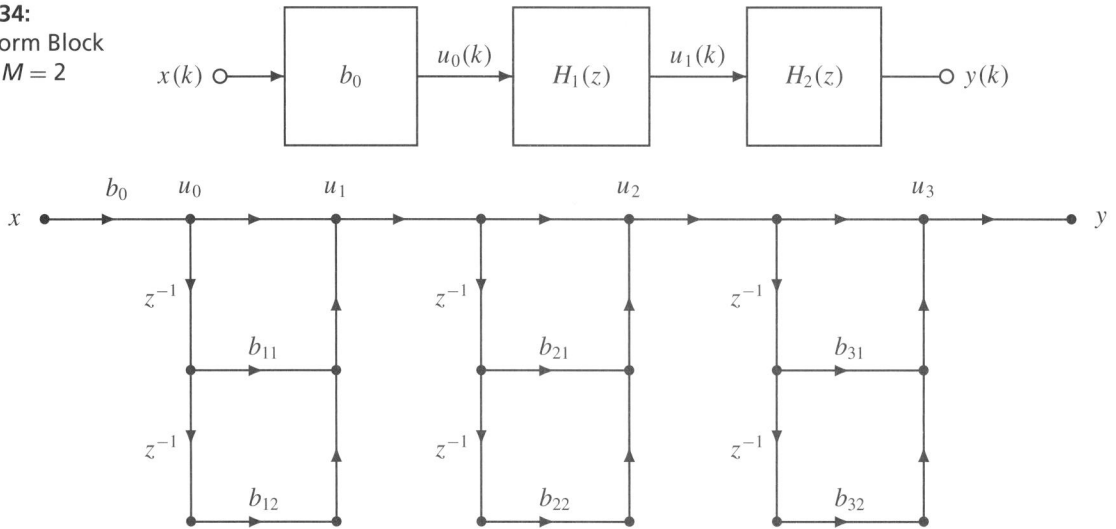

FIGURE 6.35: Cascade-form Realization of FIR Filter, $M = 3$

two real zeros or a complex conjugate pair of zeros. This way, the coefficients of $H_i(z)$ are guaranteed to be real.

$$H_i(z) = 1 + b_{i1}z^{-1} + b_{i2}z^{-2}, \quad 1 \le i \le M \qquad (6.8.7)$$

If $H_i(z)$ is constructed from zeros z_i and z_j, then the coefficients can be computed using sums and products as follows for $1 \le i \le M$.

$$b_{i1} = -(z_i + z_j) \qquad (6.8.8a)$$
$$b_{i2} = z_i z_j \qquad (6.8.8b)$$

Let u_i denote the output of the ith second-order block. Then from (6.8.6) and (6.8.7), a cascade form realization is characterized by the following time domain equations.

$$u_0(k) = b_0 x(k) \qquad (6.8.9a)$$
$$u_i(k) = u_{i-1}(k) + b_{i1}u_{i-1}(k - 1) + b_{i2}u_{i-1}(k - 2), \quad 1 \le i \le M \qquad (6.8.9b)$$
$$y(k) = r_M(k) \qquad (6.8.9c)$$

If m is even, there will be M subsystems, each of order two, and if m is odd, there will be $M - 1$ second-order subsystems plus one first-order subsystem. The coefficients of a first-order subsystem are obtained from (6.8.8) by setting $z_j = 0$.

Either of the tapped-delay-line forms can be used to realize the second-order blocks in (6.8.7). A block diagram of the overall structure of a cascade form realization, for the case $M = 3$, is shown in Figure 6.35. Since the cascade form coefficients must be computed using (6.8.8), rather than obtained directly from inspection of $H(z)$, the cascade form realization is example of an *indirect* form.

Example 6.15   **FIR Cascade Form**

As an illustration of a cascade form realization of an FIR filter, consider the following fifth-order transfer function.

$$H(z) = \frac{3(z - .6)(z + .3)[(z - .4)^2 + .25](z + .9)}{z^5}$$

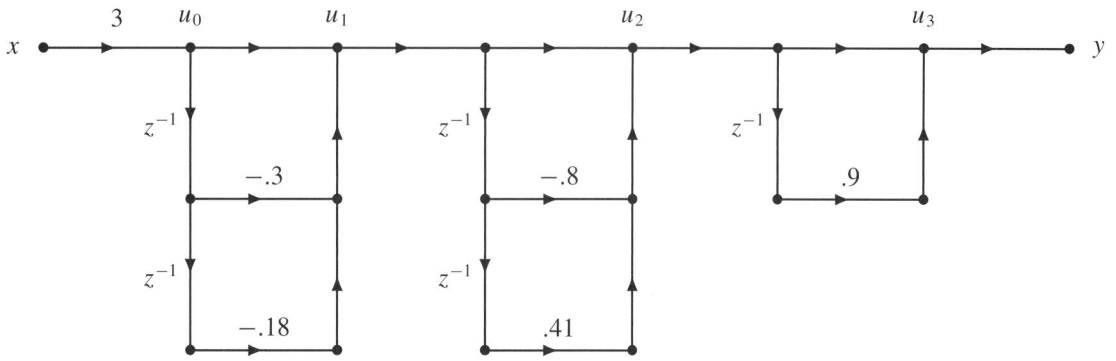

FIGURE 6.36: Cascade-form Realization of Filter

Inspection of $H(z)$ reveals that the zeros are

$$z_1 = .6$$
$$z_2 = -.3$$
$$z_{3,4} = .4 \pm j.5$$
$$z_5 = -.9$$

Suppose $H_1(z)$ is a block associated with the real zeros, $z = .6$ and $z = -.3$, $H_2(z)$ is associated with the complex conjugate pair of zeros, and $H_3(z)$ is a first-order block associated with the zero, $z = -.9$. Running *exam6_14* we get the following subsystems.

$$b_0 = 3$$
$$H_1(z) = 1 - .3z^{-1} - .18z^{-2}$$
$$H_2(z) = 1 - .8z^{-1} + .41z^{-2}$$
$$H_3(z) = 1 + .9z^{-1}$$

A signal flow graph of the resulting cascade form realization of the fifth-order FIR filter is as shown in Figure 6.36.

6.8.3 Lattice Form

Another indirect form that finds applications in speech processing and adaptive systems is the lattice form realization shown in Figure 6.37 for the case $m = 2$. The time domain equations for a lattice form realization of order m are expressed in terms of the intermediate variables u_i and v_i. From Figure 6.37, we have

$$u_0(k) = b_0 x(k) \tag{6.8.10a}$$

$$v_0(k) = u_0(k) \tag{6.8.10b}$$

$$u_i(k) = u_{i-1}(k) + K_i v_{i-1}(k-1), \quad 1 \le i \le m \tag{6.8.10c}$$

$$v_i(k) = K_i u_{i-1}(k) + v_{i-1}(k-1), \quad 1 \le i \le m \tag{6.8.10d}$$

$$y(k) = u_m(k) \tag{6.8.10e}$$

Thus, an mth-order lattice-form realization has m stages. Coefficient K_i of the ith stage is called a *reflection coefficient*. To determine the vector of m reflection coefficients, it is useful to introduce the following operation which is applicable to an mth-order FIR filter.

Reflection coefficient

$$z^{-m} H(z^{-1}) = \sum_{i=0}^{m} b_{m-i} z^{-i} \tag{6.8.11}$$

FIGURE 6.37: Lattice-form Realization of FIR Filter, $m = 2$

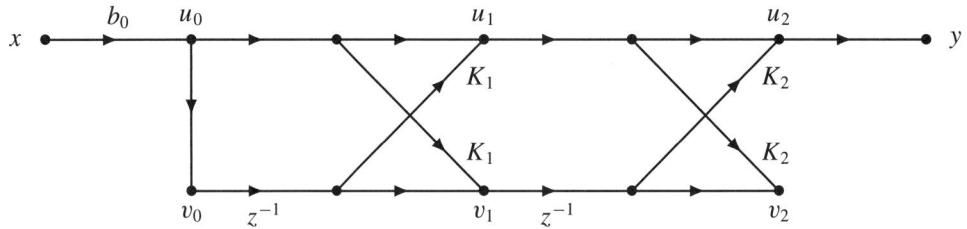

Comparing (6.8.11) with (6.8.1), we see that $z^{-m}H(z^{-1})$ is simply the polynomial obtained by *reversing* the coefficients of $H(z)$. The following algorithm can be used to compute the reflection coefficients.

ALGORITHM

6.3: Lattice-form Realization

1. Factor $H(z)$ into $H(z) = b_0 A_m(z)$ and compute

$$B_m(z) = z^{-m} A_m(z^{-1})$$

$$K_m = \lim_{z \to \infty} B_m(z)$$

2. For $i = m$ down to 2 compute
{

$$A_{i-1}(z) = \frac{A_i(z) - K_i B_i(z)}{1 - K_i^2}$$

$$B_{i-1}(z) = z^{-(i-1)} A_{i-1}(z^{-1})$$

$$K_{i-1} = \lim_{z \to \infty} B_{i-1}(z)$$

}

Algorithm 6.3 produces b_0 and an $m \times 1$ vector of reflection coefficients, K, as long as $|K_i| \neq 1$ for $1 \leq i \leq m$. If $|K_i| = 1$, then $A_{i-i}(z)$ has a zero on the unit circle. In this case, this zero can be factored out and the algorithm applied to the reduced-order polynomial.

Example 6.16

FIR Lattice Form

As an illustration of a lattice form realization of an FIR filter, consider the following second-order transfer function.

$$H(z) = 2 + 6z^{-1} - 4z^{-2}$$

Applying step 1 of Algorithm 6.3 yields $b_0 = 2$ and

$$A_2(z) = 1 + 3z^{-1} - 2z^{-2}$$

$$B_2(z) = -2 + 3z^{-1} + z^{-2}$$

$$K_2 = -2$$

FIGURE 6.38: Lattice-form Realization of FIR Filter

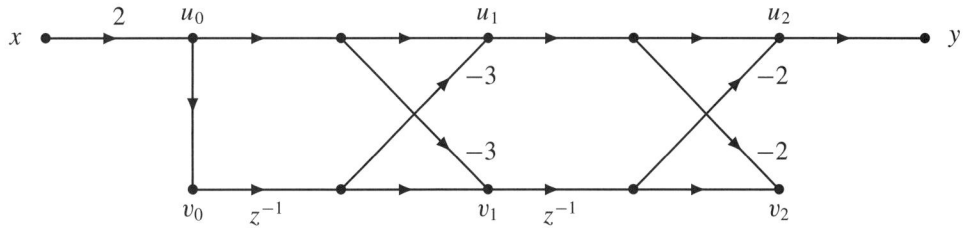

Next, applying step 2 with $i = 2$

$$A_1(z) = \frac{1 + 3z^{-1} - 2z^{-2} + 2(-2 + 3z^{-1} + z^{-2})}{1 - 4}$$

$$= \frac{-3 + 9z^{-1}}{-3}$$

$$= 1 - 3z^{-1}$$

$$B_1(z) = -3 + z^{-1}$$

$$K_1 = -3$$

Thus $b_0 = 2$, and the reflection coefficient vector is $K = [-3, -2]^T$. A signal flow graph of the lattice form realization is shown in Figure 6.38.

There are additional indirect forms that have been proposed and are used. Included among these are the frequency-sampling realization, and parallel form realizations (see, e.g., Proakis and Manolakis, 1992).

MATLAB Toolbox

The FDSP toolbox contains the following functions for computing indirect form realizations of a FIR transfer functions.

```
% F_CASCADE: Find cascade form digital filter realization
% F_LATTICE: Find lattice form FIR filter realization
%
% Usage:
%          [B,A,b_0] = f_cascade (b)
%          [K,b_0]   = f_lattice (b)
% Pre:
%          b = vector of length m+1 containing coefficients
%              of numerator polynomial.
```

Continued on p. 464

Continued from p. 463

```
% Post:
%      B    = N by 3 matrix containing coefficients of
%             numerators of second-order blocks.
%      A    = N by 3 matrix containing coefficients of
%             denominators of second-order blocks.
%      b_0 = numerator gain
%      K    = 1 by m vector containing reflection
%             coefficients
% Notes:
%      1. It is required that b(1)<>0.  Otherwise factor
%         out a z^-1 and then find the cascade form
%      2. This algorithm assumes that $|K(i)| ~= 1$
%         for 1 <= i <= m.
```

To evaluate the cascade and lattice form filters, the outputs from the calls to *f_cascade* and *f_lattice* are used as inputs to the following filter evaluation functions.

```
% F_FILTCAS: Compute output of cascade form filter realization
% F_FILTLAT: Compute output of lattice form filter realization
%
% Usage:
%      y = f_filtcas (B,A,b_0,x)
%      y = f_filtlas (K,b_0,x)
% Pre:
%      B    = N by 2 matrix containing numerator
%             coefficients of second-order blocks.
%      A    = N by 3 matrix containing denominator
%             coefficients of second-order blocks.
%      b_0 = numerator gain factor
%      x    = vector of length p containing samples of
%             input signal.
%      K    = 1 by m vector containing reflection
%             coefficients
% Post::
%      y = vector of length p containing samples of
%          output signal assuming zero initial
%          conditions.
```

*6.9 Finite Word Length Effects

When a filter is implemented, finite precision must be used to represent the values of the signals and the filter coefficients. The resulting reductions in filter performance caused by going from infinite precision to finite precision are called *finite word length effects*. If a filter is implemented in software using MATLAB, then double-precision floating-point arithmetic is used. This corresponds to 64 bits of precision or about 16 significant decimal digits. In

MATLAB precision

most instances, this is a sufficiently good approximation to infinite-precision arithmetic that no significant finite word length effects are apparent. However, if a filter is implemented on specialized DSP hardware (Kuo and Gan, 2005), or storage or speed requirements dictate the need to use single-precision floating-point arithmetic or fixed-point arithmetic, then finite word length effects begin to manifest themselves.

6.9.1 Binary Number Representation

Finite word length effects depend on the method used to represent numbers. If a filter is implemented in software on a PC, then typically a binary *floating-point representation* of numbers is used where some of the bits are used to represent the mantissa, or fractional part, and the remaining bits are reserved to represent the exponent. For example, MATLAB uses $N = 64$ bits with 53 bits reserved for the mantissa and 11 bits for the exponent. Floating-point representations have the advantage that they can represent very large and very small numbers; hence issues of overflow and scaling typically do not come into play. However, the spacing between adjacent floating point values is not constant, instead it is proportional to the magnitude of the number.

An alternative to the floating-point representation is the *fixed-point representation* which does not have a field of bits reserved to represent the exponent. Both floating-point and fixed-point representations can be found on specialized DSP hardware. The fixed-point representation is faster and more efficient than the floating-point representation. Furthermore, the precision, or spacing between adjacent values, is uniform throughout the range of numerical values represented. Unfortunately, this range typically is much smaller than for floating-point numbers, so one has to be concerned with the possibility of overflow. Although finite word length effects appear in both floating-point and fixed-point arithmetic, they are less severe for floating-point representations, particularly double-precision representations such as those used with MAT-LAB and C++.

In this section we will focus our attention on the use of fixed-point numerical representations. An N-bit binary fixed-point number b, consisting of N *bi*nary dig*its*, or *bits*, can be expressed as follows.

Bit

$$b = b_0 b_1 \cdots b_{N-1}, \quad 0 \le b_i \le 1 \tag{6.9.1}$$

Typically the number b is normalized with the binary point appearing between bits b_0 and b_1. In this case, the decimal equivalent of a positive number is

$$x = \sum_{i=1}^{N-1} b_i 2^{-i} \tag{6.9.2}$$

Bit b_0 is reserved to hold the sign of x, with $b_0 = 0$ indicating a positive number and $b_0 = 1$ a negative number. There are several schemes for encoding negative numbers, including sign magnitude, one's complement, offset binary, and two's complement (Gerald and Wheatley, 1989). The most popular method is the two's complement representation. The two's complement representation of a negative number is obtained by complementing each bit, and then adding one to the least significant bit. Any carry past the sign bit is ignored. One of the virtues of two's complement arithmetic occurs when several numbers are added. If the sum fits within N bits, then the total will be correct even if the intermediate results or partial sums overflow!

The range of values that can be represented in (6.9.2) is $-1 \le x < 1$. Of course, many values of interest may fall outside this normalized range. Larger values can be accommodated by using a fixed scale factor c. Notice from (6.9.2) that scaling by $c = 2^M$ effectively moves the binary point M places to the right. This increases the range to $-c \le x < c$, but causes a

FIGURE 6.39:
Fixed-point
Representation of
N-bit Number
Using Scale Factor
$c = 2^M$

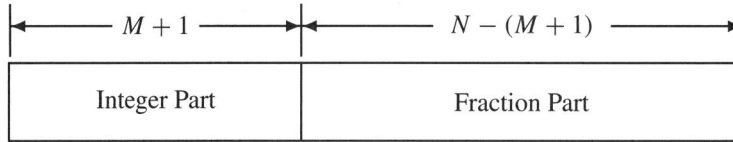

Quantization level

loss of precision. The precision, or spacing between adjacent values, is constant and is called the *quantization level*.

$$q = \frac{c}{2^{N-1}} \tag{6.9.3}$$

When a scale factor of $c = 2^M$ is used, this corresponds to reserving $M + 1$ bits for the integer part, including the sign, and the remaining $N - (M + 1)$ bits for the fractional part, as shown in Figure 6.39.

6.9.2 Input Quantization Error

Recall from Chapter 1 that the value of the input signal $x(k)$ is *quantized* to a finite number of bits as a result of analog-to-digital (ADC) conversion. To model quantization, it is helpful to introduce the following operator.

DEFINITION

6.2: Quantization
Operator

Let N be the number of bits used to represent a real value x. The *quantized* version of x is denoted $Q_N(x)$ and defined as

$$Q_N(x) \triangleq q \text{ floor} \left(\frac{x + q/2}{q} \right)$$

The form of quantization in Definition 6.2 uses *rounding* which can be seen from the $q/2$ term in the numerator. If this term is removed, then the resulting quantization operation uses *truncation*. For convenience, we will assume that quantization by rounding to N bits is used. A graph of the nonlinear input-output characteristic of the quantization operator for the case $N = 4$ is shown in Figure 6.40.

If an ADC has a precision of N bits, then the quantized output from the ADC, denoted $x_q(k)$, is computed as follows.

$$x_q(k) = Q_N[x(k)] \tag{6.9.4}$$

Although the deterministic nonlinear model in (6.9.4) is accurate, it is more useful for analysis purposes to replace it by an equivalent linear *statistical* model. The quantized signal can be regarded as an infinite-precision signal $x(k)$ plus a *quantization error* term, $\Delta x(k)$, as shown in Figure 6.41. Thus the quantized signal is

Quantization error

$$x_q(k) = x(k) + \Delta x(k) \tag{6.9.5}$$

Recall from Chapter 1 that if $|x_a(t)| \leq c$, then $|\Delta x(k)| \leq q/2$, where q in (6.9.3) is the ADC quantization level. Consequently, when rounding is used, the quantization error can

FIGURE 6.40:
Input-output
Characteristic of
Quantization
Operator for $N = 4$
Using Rounding

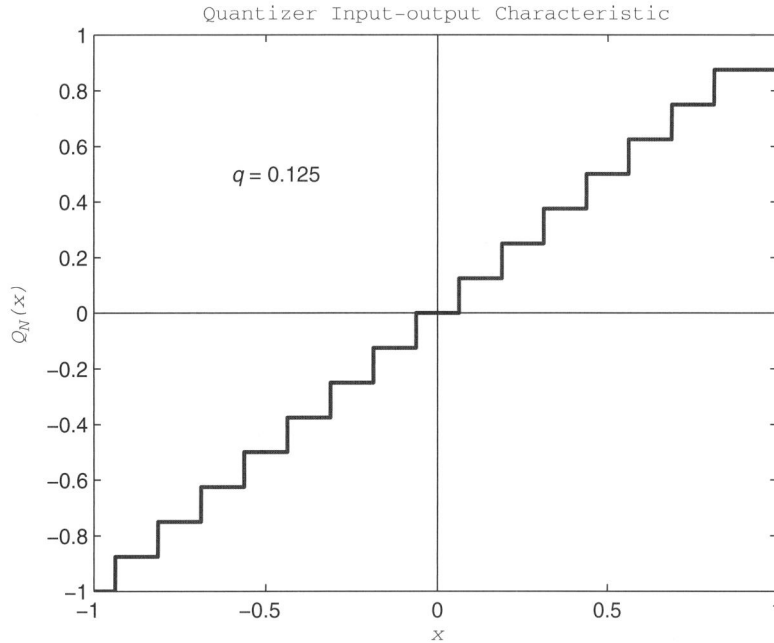

Quantizer Input-output Characteristic

$q = 0.125$

FIGURE 6.41: Linear
Statistical Model of
Input Quantization

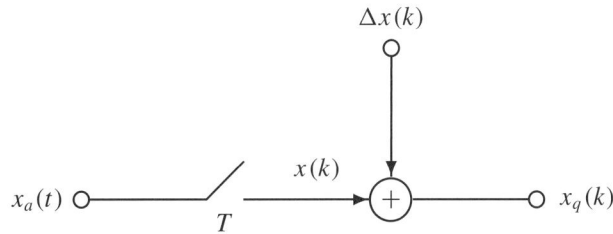

be modeled as *white noise* that is uniformly distributed over the interval $[-q/2, q/2]$. The probability density of ADC quantization noise is as shown in Figure 6.42.

A convenient measure of the size of the quantization noise is the average power. For zero-mean noise, the average power is the variance $\sigma_x^2 = E[\Delta x^2]$. Using Definition 4.3 and the probability density in Figure 6.42, we find that the average power of the ADC quantization noise is

$$
\begin{aligned}
\sigma_x^2 &= \int_{-\infty}^{\infty} x^2 p(x) dx \\
&= \frac{1}{q} \int_{-q/2}^{q/2} x^2 dx \\
&= \frac{q^2}{12}
\end{aligned}
\tag{6.9.6}
$$

The quantization noise associated with the input signal is filtered by the system $H(z)$ and appears in the system output as

$$
y_q(k) = y(k) + \Delta y(k)
\tag{6.9.7}
$$

To determine the average power of the output noise, first note that for a linear system the response to the input noise $\Delta x(k)$ is the output noise $\Delta y(k)$. That is, if $h(k)$ is the

FIGURE 6.42: Probability
Density of ADC
Quantization Noise
Using Rounding

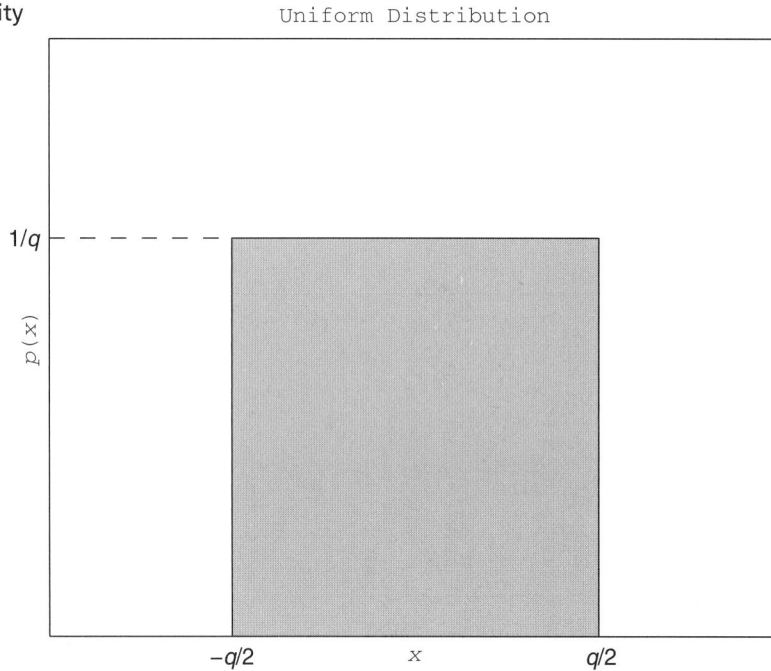

FIGURE 6.42: Probability Density of ADC Quantization Noise Using Rounding

impulse response of the system, then subtracting the noise-free output from the complete output yields

$$\Delta y(k) = \sum_{i=0}^{m} h(i) \Delta x(k - i) \qquad (6.9.8)$$

The input quantization noise $\Delta x(k)$ is zero-mean white noise. Consequently, the average power of the output noise is

$$
\begin{aligned}
E[\Delta y^2(k)] &= E\left[\sum_{i=0}^{m} h(i) \Delta x(k - i) \sum_{p=0}^{m} h(p) \Delta x(k - p) \right] \\
&= \sum_{i=0}^{m} \sum_{p=0}^{m} h(i) h(p) E[\Delta x(k - i) \Delta x(k - p)] \\
&= \sum_{i=0}^{m} h^2(i) E[\Delta x^2(k - i)] \\
&= \left[\sum_{i=0}^{m} h^2(i) \right] E[\Delta x^2(k)]
\end{aligned} \qquad (6.9.9)
$$

Thus the average power of the output noise is proportional to the average power of the input quantization noise as follows.

$$\sigma_y^2 = \Gamma \sigma_x^2 \qquad (6.9.10)$$

Power gain The constant of proportionality Γ is called the *power gain*. From (6.9.9), the power gain is computed from the impulse response $h(k)$ as follows.

$$\Gamma \triangleq \sum_{k=0}^{m} h^2(k) \qquad (6.9.11)$$

Example 6.17

Input Quantization Noise

As an illustration of ADC quantization effects, suppose an ADC with a precision of $N = 8$ bits is used to sample an input signal with range $|x_a(t)| \leq 10$. From (6.9.3) the ADC quantization level is

$$q = \frac{10}{2^7}$$
$$= .0781$$

It then follows from (6.9.6) that the average power of the ADC quantization noise at the input is

$$\sigma_x^2 = \frac{.0781^2}{12}$$
$$= 5.0863 \times 10^{-4}$$

Next, suppose the quantization noise is passed through the following FIR filter.

$$H(z) = 10 \sum_{i=0}^{30} .9^i z^{-i}$$

The impulse response of this FIR filter is

$$h(k) = 10(.9)^k [\mu(k) - \mu(k - 31)]$$

Using the geometric series and (6.9.11), we find that the power gain of the filter is

$$\Gamma = \sum_{k=0}^{30} [10(.9)^k]^2$$
$$= 100 \left[\sum_{k=0}^{\infty} .81^k - \sum_{k=31}^{\infty} .81^k \right]$$
$$= 100 \left[\frac{1}{1 - .81} - \frac{.81^{31}}{1 - .81} \right]$$
$$= 525.37$$

Finally, with (6.9.10), the average power of the ADC quantization noise appearing at the filter output is

$$\sigma_y^2 = \Gamma \sigma_x^2$$
$$= 526.37(5.0863 \times 10^{-4})$$
$$= .2672$$

Observe that even though the noise power at the input is quite small, the noise power at the output can be large.

6.9.3 Coefficient Quantization Error

FIR filter coefficients are also quantized when they are stored in fixed length memory locations. Let the unquantized or infinite-precision version of the transfer function be

$$H(z) = \sum_{i=0}^{m} b_i z^{-i} \tag{6.9.12}$$

Suppose the elements of the coefficient vector are quantized to N bits to yield $b_q = Q_N(b)$, where Q_N is the N-bit quantization operator introduced in Definition 6.2. The quantized coefficient vector b_q can be expressed as follows.

$$b_q = b + \Delta b \tag{6.9.13}$$

Coefficient quantization error — Here Δb is the *coefficient quantization error*. If $|b_i| \leq c$, then the elements of the vector Δb can be modeled as random numbers uniformly distributed over $[-q/2, q/2]$, where q is the quantization level given in (6.9.3). For an FIR filter the effects of coefficient quantization are relatively easy to model. Let $H_q(z)$ denote the transfer function using quantized coefficients. From (6.9.12) and (6.9.13)

$$\begin{aligned} H_q(z) &= \sum_{i=0}^{m} (b_i + \Delta b_i) z^{-i} \\ &= \sum_{i=0}^{m} b_i z^{-i} + \sum_{i=0}^{m} \Delta b_i z^{-i} \\ &= H(z) + \Delta H(z) \end{aligned} \tag{6.9.14}$$

Here the subsystem $\Delta H(z)$ is the coefficient quantization error transfer function

$$\Delta H(z) = \sum_{i=0}^{m} \Delta b_i z^{-i} \tag{6.9.15}$$

From (6.9.14) it is clear that quantization of the FIR coefficients is equivalent to introducing the error system $\Delta H(z)$ in parallel with the unquantized system, as shown in Figure 6.43.

One can place a simple upper bound on the effects of coefficient quantization on the frequency response $H(f)$. If we use $|\Delta b_i| \leq q/2$, the error in the system magnitude response can be bounded as follows.

$$\begin{aligned} \Delta A(f) &= |\sum_{i=0}^{m} \Delta b_i \exp(-ji2\pi f T)| \\ &\leq \sum_{i=0}^{m} |\Delta b_i \exp(-ji2\pi f T)| \\ &= \sum_{i=0}^{m} |\Delta b_i| \leq \frac{(m+1)q}{2} \end{aligned} \tag{6.9.16}$$

FIGURE 6.43: The Effects of Coefficient Quantization on an FIR Filter $H(z)$

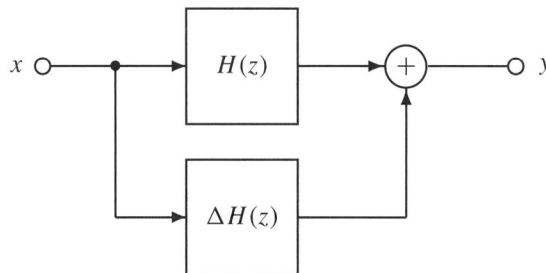

Magnitude error

It then follows from (6.9.3) that for an mth-order FIR filter with coefficients $|b_i| \le c$ that are quantized to N bits, the error in the magnitude of the frequency response satisfies

$$\Delta A(f) \le \frac{(m+1)c}{2^N} \qquad (6.9.17)$$

The upper bound in (6.9.17) is a *conservative* one that assumes a worst case in which each of the coefficient quantization errors has the same sign and is the maximum value possible.

Example 6.18

FIR Coefficient Quantization Error

As an illustration of FIR coefficient quantization noise, consider a bandpass filter of order $m = 64$ designed using the least-squares method. Suppose the coefficients are quantized using $N = 8$ bits. Evaluation of the $(m + 1)$ coefficients reveals that they lie in the range $|b_i| \le c$ where $c = 1$. With (6.9.17), the error in the magnitude response is bounded as follows.

$$\Delta A(f) \le \frac{64 + 1}{2^8}$$
$$= .2539$$

When *exam6_17* is run, it produces the two magnitude response plots shown in Figure 6.44. The first plot approximates the unquantized case using double-precision floating-point arithmetic. The second plot uses a tapped delay line direct form realization with quantized coefficients. It is apparent from inspection that quantizing to $N = 8$ bits introduces error, particularly in the stopband. The signal attenuation for the quantized case is not nearly as good as it is for the unquantized case.

FIGURE 6.44: Magnitude Responses of a Least-Squares Bandpass Filter of Order $m = 64$ for (a) Unquantized Coefficients, and (b) Coefficients Quantized to $N = 8$ Bits

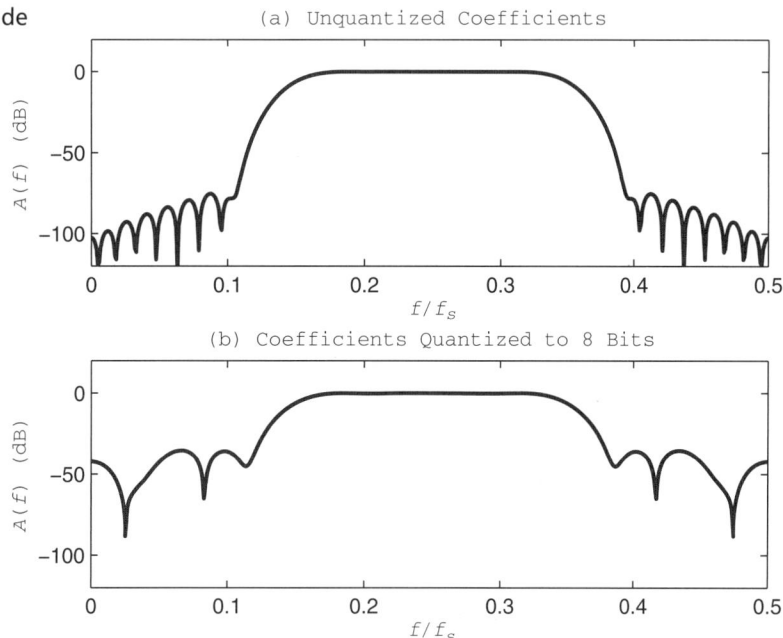

Unit Circle Zeros

Another way to evaluate the effects of coefficient quantization is to look at the locations of the zeros. The roots of a polynomial can be very sensitive to small changes in the coefficients of the polynomial, particularly for higher-degree polynomials. As a consequence, higher-order direct form realizations of $H(z)$ can be sensitive to coefficient quantization error. For FIR filters, the most important case corresponds to zeros on the unit circle. Zeros on the unit circle are qualitatively important because they produce complete attenuation of the input signal at specific frequencies. For example, suppose it is desired to remove the frequency F_0. This is achieved by placing zeros at $z = \pm \exp(j\theta_0)$ where $\theta_0 = 2\pi F_0 T$. A second-order FIR filter that completely attenuates frequency F_0 is then

$$
\begin{aligned}
H(z) &= \frac{[z - \exp(j\theta_0)][z - \exp(-j\theta_0)]}{z^2} \\
&= \frac{z^2 - [\exp(j\theta_0) + \exp(-j\theta_0)]z + 1}{z^2} \\
&= \frac{z^2 - 2\cos(\theta_0)z + 1}{z^2} \\
&= 1 - 2\cos(\theta_0)z^{-1} + z^{-2}
\end{aligned}
\tag{6.9.18}
$$

Note that $H(z)$ has a coefficient vector of $b = [1, -2\cos(\theta_0), 1]^T$. If b is quantized, this will result in a small change in θ_0, but no change b_0 or b_2 because they can be represented exactly. Consequently, the quantized zero will *remain* on the unit circle, although the angle (i.e., the frequency) may change. It follows that if $H(z)$ is realized using a cascade of quantized second-order blocks, then zeros on the unit circle will be *preserved* although their frequencies will change slightly.

Linear-phase Block

Most FIR filters are linear-phase filters, so it is useful to examine what effect coefficient quantization has on this important property. Recall that the most general type 1 linear-phase FIR filter of order m can be realized with the direct form shown in Figure 6.33. From (6.8.4), the difference equation of this direct form linear-phase realization is as follows where $p = m/2$.

$$
y(k) = b_p x(k - p) + \sum_{i=0}^{p-1} b_i[x(k - i) + x(k - m + i)]
\tag{6.9.19}
$$

Linear phase   To preserve the *linear phase* response of a type 1 filter, it is necessary that $b_i = b_{m-i}$ for $0 \leq i \leq m$. But from the structure in (6.9.19), it is apparent that b_i and b_{m-i} are implemented with the *same* coefficient. Consequently, if the $p + 1$ coefficients in (6.9.19) are quantized, the resulting quantized filter will still be a linear-phase filter; only its magnitude response will be affected. That is, the linear-phase feature of the direct form realization in (6.9.19) is *unaffected* by coefficient quantization.

As it turns out, a linear-phase response can be preserved even when a cascade-form realization is used. Recall that linear-phase filters have the property that if $z = r\exp(j\phi)$ is a zero with $r \neq 1$, then so is its reciprocal, $z = r^{-1}\exp(-j\phi)$. For a filter with real coefficients, complex zeros appear in conjugate pairs. Consequently, complex zeros that are not on the unit circle appear in groups of four. To preserve this grouping, one can use the following fourth-order

FIGURE 6.45: Cascade-form Realization of a Fourth-order Linear-phase Block

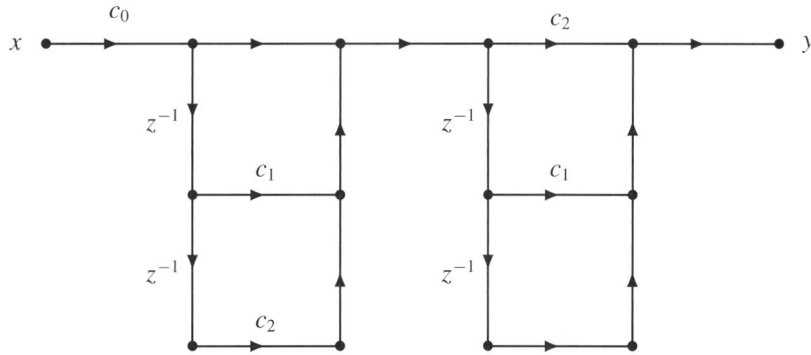

block in a cascade-form realization.

$$
\begin{aligned}
H(z) &= \frac{[z - r\exp(j\phi)][z - r\exp(-j\phi)][z - r^{-1}\exp(-j\phi)][z - r^{-1}\exp(j\phi)]}{z^4} \\
&= \frac{[z^2 - 2r\cos(\phi)z + r^2][z^2 - 2r^{-1}\cos(\phi)z + r^{-2}]}{z^4} \\
&= \frac{[z^2 - 2r\cos(\phi)z + r^2][r^2z^2 - 2r\cos(\phi)z + 1]}{r^2z^4} \\
&= c_0(1 + c_1 z^{-1} + c_2)(c_2 + c_1 z^{-1} + 1)
\end{aligned}
\tag{6.9.20}
$$

Note that this factored fourth-order linear-phase block can be realized as a cascade of two second-order blocks where the distinct coefficients are $c = [r^{-2}, -2r\cos(\phi), r^2]^T$. The coefficient order is reversed for the two blocks. If c is quantized, this reciprocal zero relationship is preserved. Hence the quantized fourth-order block is still a linear-phase system. A block diagram of a cascade-form realization of a fourth-order linear-phase block is shown in Figure 6.45.

6.9.4 Roundoff Error, Overflow, and Scaling

The arithmetic used to compute the filter output must also be performed using finite precision. The time domain representation of an FIR filter using a tapped delay line direct form realization is

$$
y(k) = \sum_{i=0}^{m} b_i x(k - i)
\tag{6.9.21}
$$

Roundoff error

If the coefficients are quantized to N bits and the signals are quantized to N bits, then the product terms in (6.9.21) will each be of length $2N$ bits. When the products are then rounded to N bits, the resulting error is called *roundoff error*. It can be modeled as white noise with a separate white noise source for each product. Assuming the roundoff noise sources are statistically independent of one another, the noise sources associated with the products can be combined into a single error term as follows, where Q_N denotes the N-bit quantization operator.

$$
e(k) = \sum_{i=0}^{m} Q_N[b_i x(k - i)] - b_i x(k - i)
\tag{6.9.22}
$$

For a tapped delay line direct form realization, this results in the equivalent linear model of roundoff error shown in Figure 6.46 for the case $m = 2$.

FIGURE 6.46: Linear Model of Product Roundoff Error in an FIR Filter, $m = 2$

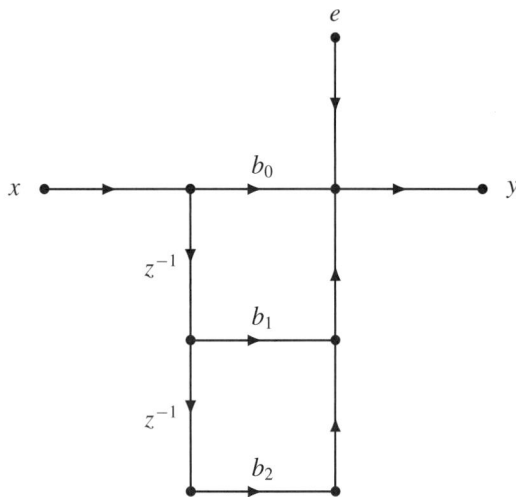

Suppose both the coefficient b_i and the input $x(k - i)$ lie in the range $[-c, c]$. Then the quantization level is q, as given in (6.9.3). Each roundoff error noise is uniformly distributed over $[-q/2, q/2]$ and has average power $\sigma_x^2 = a^2/12$. Since the $m + 1$ sources are assumed to be statistically independent, their contributions can be added which means that the average power of the noise appearing at the filter output is

$$\sigma_y^2 = \frac{(m + 1)q^2}{12} \qquad (6.9.23)$$

The roundoff noise can be reduced further if a special hardware architecture is used. Some DSP processors use a $2N$-bit double-length accumulator to store the results of the multiplications in (6.9.21). When this hardware configuration is used, it is only the final sum, $y(k)$, that is quantized to N bits. As a consequence, the noise term in (6.9.22) simplifies to

$$e(k) = Q_N \left[\sum_{i=0}^{n} b_i x(k - i) \right] - y(k) \qquad (6.9.24)$$

In this case there is only one roundoff noise source instead of $m + 1$ as in (6.9.22). The end result is that the average power of the roundoff error output noise in (6.9.23) is reduced by a factor of $m + 1$. This makes the use of a double-length accumulator an attractive hardware option for implementing a direct form FIR filter.

Overflow

Overflow error

Another source of error occurs as a result of the summing operation in (6.9.21). The sum of several N-bit numbers will not necessarily fit within N bits. When the sum is too large to fit, this results in *overflow error*. Overflow errors can cause a significant change in the filter output. This type of error can eliminated, or significantly reduced, by scaling either the input or the filter coefficients. If $|x(k)| \leq c$, the FIR filter output in (6.9.21) is also bounded as follows.

$$
\begin{aligned}
|y(k)| = |\sum_{i=0}^{m} b_i x(k-i)| \\
\leq \sum_{i=0}^{m} |b_i x(k-i)| \\
= \sum_{i=0}^{m} |b_i| \cdot |x(k-i)| \\
\leq c \sum_{i=0}^{m} |b_i|
\end{aligned}
\tag{6.9.25}
$$

Thus $|y(k)| \leq c\|b\|_1$, where $\|b\|_1$ is the L_1 *norm* of the coefficient vector b. That is,

$$
\|b\|_1 \overset{\Delta}{=} \sum_{i=0}^{m} |b_i|
\tag{6.9.26}
$$

From (6.9.25) it is apparent that addition overflow at the output is eliminated (i.e., $|y(k)| \leq c$) when the input signal $x(k)$ is scaled by s_1 using scale factor $s_1 = 1/\|b\|_1$. A signal flow graph of a third-order direct form FIR filter realization that uses scaling to prevent overflow is shown in Figure 6.47.

Scaling using the L_1 norm is effective in preventing overflow, but it does suffer from a practical drawback. Roundoff noise and ADC quantization noise are not affected significantly by scaling. As a result, when the input is scaled by s_1, the resulting reduction in signal strength can cause a corresponding reduction in the signal-to-noise ratio. Less severe forms of scaling can be used that eliminate most, but not all, overflow. For example, if the input signal is a pure sinusoid, then overflow from this type of periodic input can be eliminated by using scaling that is based on the filter magnitude response.

$$
\|b\|_\infty \overset{\Delta}{=} \max_{0 \leq f \leq f_s/2} \{A(f)\}
\tag{6.9.27}
$$

FIGURE 6.47: Scaling to Prevent Addition Overflow in an FIR Filter, $m = 3$

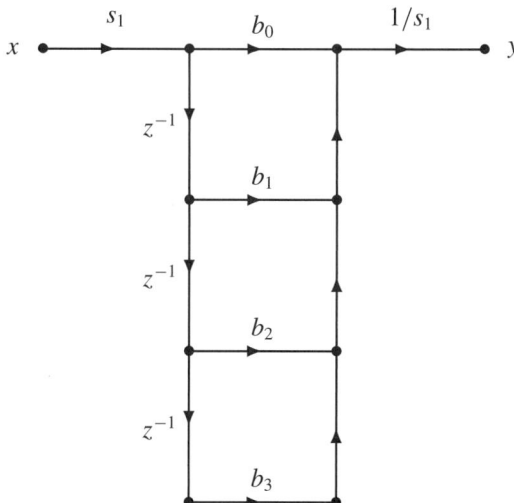

Overflow from a pure sinusoidal input is prevented if the input signal $x(k)$ is scaled by $s_\infty = 1/\|b\|_\infty$. Another common form of scaling uses the L_2 or Euclidean norm.

$$\|b\|_2 \triangleq \left(\sum_{i=0}^{m} |b_i|^2 \right)^{1/2} \tag{6.9.28}$$

Again the scale factor is $s_2 = 1/\|b\|_2$. One advantage of the L_2 norm is that, like the L_1 norm, it is easy to compute. The three norms can be shown to satisfy the following relationship.

$$\|b\|_2 \le \|b\|_\infty \le \|b\|_1 \tag{6.9.29}$$

Example 6.19

FIR Overflow and Scaling

As an illustration of the prevention of overflow by scaling, suppose $|x(k)| \le 5$, and consider the following FIR filter.

$$H(z) = \sum_{i=0}^{20} \frac{z^{-i}}{1+i}$$

Here $c = 5$. Running *exam6_19* produces the following values for the scale factors for this filter of order $m = 20$.

$$s_1 = 3.6454$$
$$s_2 = 1.2643$$
$$s_\infty = 3.6454$$

In this instance it turns out that the L_1 and L_∞ scale factors are identical. For this FIR system $s_\infty = A(0)$ because the magnitude response achieves its peak at $f = 0$, as can be seen from the plot shown in Figure 6.48.

FIGURE 6.48: Magnitude Responses of FIR Filter

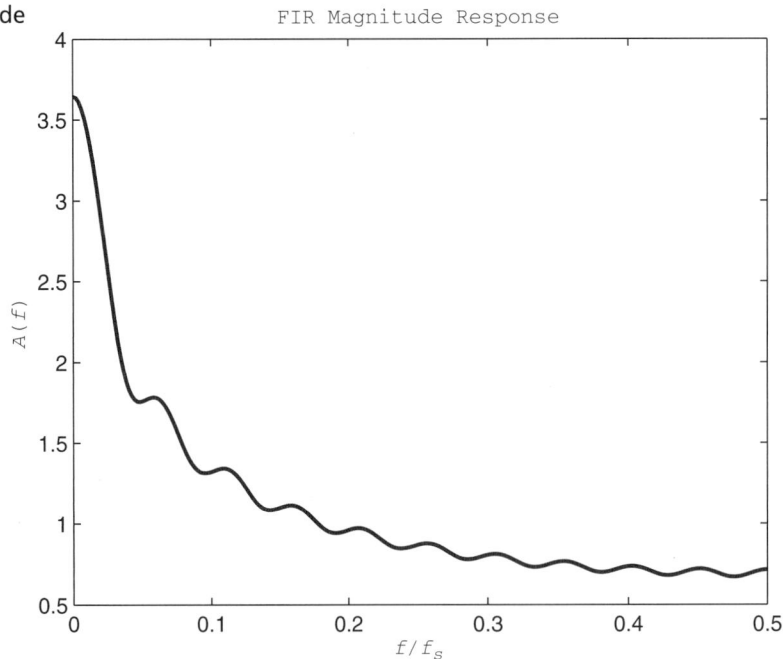

● ● ● ● ● ● ● ● ● ● ● ● ● ● ● ● ⋯⋯

6.10 GUI Software and Case Study

This section focuses on the design and realization of FIR filters. A graphical user interface module called *g_fir* is introduced that allows the user to design and implement FIR filters, all without any need for programming. A case study example is presented and solved using MATLAB.

g_fir: Design and Implement FIR Filters

The FDSP toolbox includes a graphical user interface module called *g_fir* that allows the user to design a variety of FIR filters. GUI module *g_fir* features a display screen with tiled windows, as shown in Figure 6.49. The upper left-hand *Block diagram* window contains a block diagram of the FIR filter under investigation. It is an *m*th-order filter with the following transfer function.

$$H(z) = \sum_{i=0}^{m} b_i z^{-i} \tag{6.10.1}$$

Edit boxes   The *Parameters* window below the block diagram displays *edit boxes* containing the filter parameters. The contents of each edit box can be directly modified by the user, with the Enter key used to activate changes. The parameters $F0$, $F1$, B, and fs are the lower cutoff frequency, upper cutoff frequency, transition bandwidth, and sampling frequency, respectively. The parameters *deltap* and *deltas* are the passband ripple factor and the stopband attenuation factor, respectively. For a lowpass filter the passband cutoff is $F0$, and for a highpass filter the passband cutoff is $F1$. Bandpass filters have a passband of $[F0, F1]$, and bandstop filters have a stopband of $[F0, F1]$. In all cases B is the width of the transition band.

Type options   The *Type* and *View* windows in the upper-right corner of the screen allow the user to select both the type of filter and the viewing mode. The filter types include lowpass, highpass, bandpass, and bandstop filters. Also included is a user-defined filter. The desired magnitude response and residual phase response of the user-defined filter are specified in a user-supplied M-file that has the following calling sequence where u_sys is a user-supplied name.

```
[A,theta]=u_sys(f,fs);     %Frequency response
```

When u_sys is called with frequency f and sampling frequency fs, it must evaluate the desired magnitude response and residual phase response at the vector f, and return the results in the vectors A and *theta*, respectively. The output *theta* is needed only for the quadrature filter type. For all other filters types, the residual phase response is theta = zeros(size(f)).

View options   The *View* options include the magnitude response, the phase response, the impulse response, a pole-zero plot, and the window used with the windowed design method. The dB check box toggles the magnitude response display between linear and logarithmic scales. When it is checked, the passband ripple and stopband attenuation in the *Parameters* window change to their logarithmic equivalents, Ap and As, respectively. The *Plot* window along the bottom half of the screen shows the selected view. Below the *Type* and *View* windows is a horizontal slider bar that allows the user to directly control the filter order m.

Menu options   The *Menu* bar at the top of the screen includes several menu options. The *Caliper* option allows the user to measure any point on the current plot by moving the mouse cross hairs to that point and clicking. The *Save data* option is used to save the current a, b, fs, x, and y in a

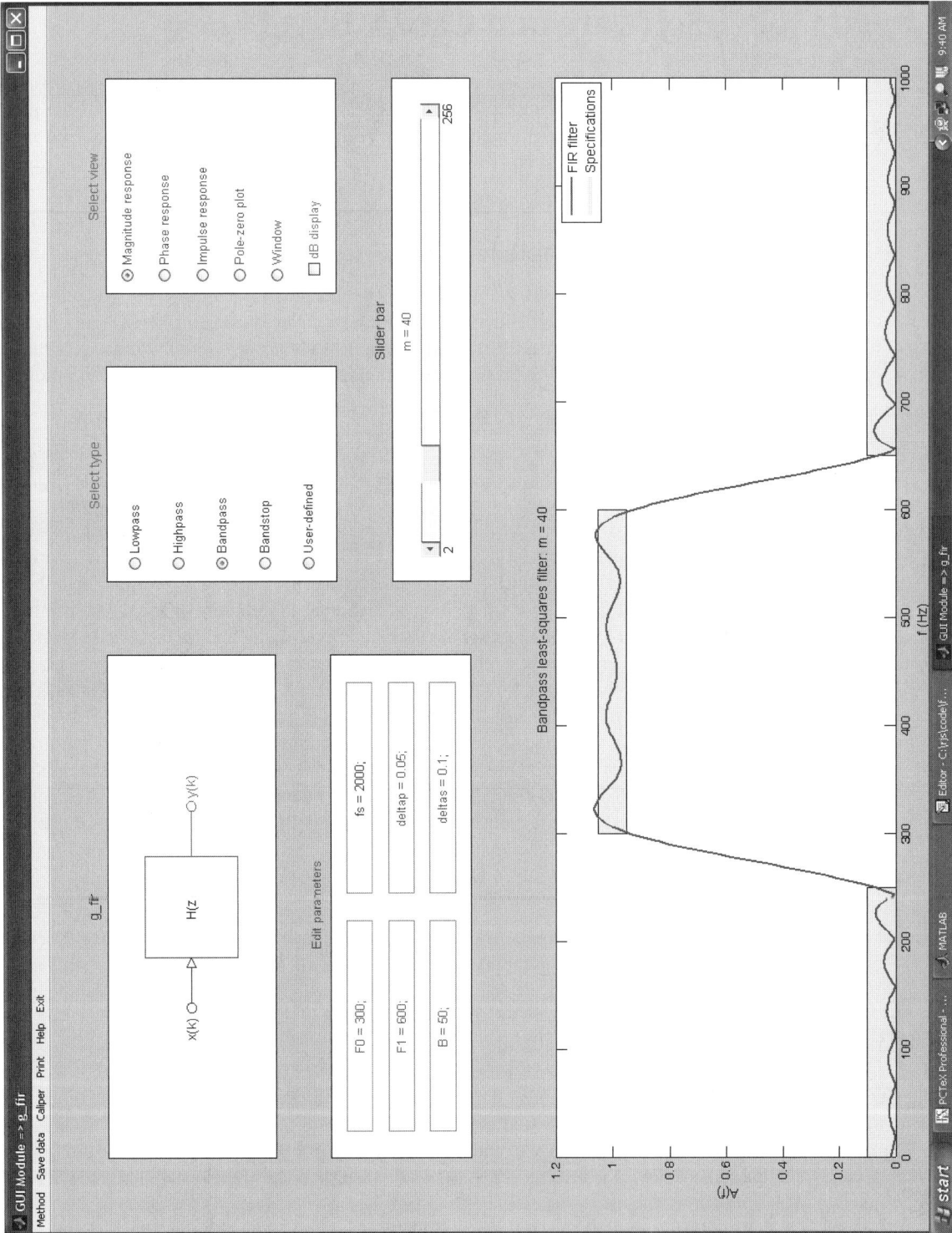

FIGURE 6.49: Display Screen of Chapter GUI Module *g_fir*

user-specified MAT file for future use. Files created in this manner can be loaded into other GUI modules such as *g_filters*. The *Method* option allows the user to select the filter design method from the windowed, frequency-sampled, least-squares, equiripple, and quadrature methods. The *Print* option prints the contents of the plot window. Finally, the *Help* option provides the user with some helpful suggestions on how to effectively use module *g_fir*.

Recall that GUI module *g_filters* was discussed previously in Section 5.9. If *g_filters* are used, different filter realization structures can be investigated, and the effects of coefficient quantization error can be explored. To analyze an FIR filter designed with *g_fir* in this manner, use the Save option from *g_fir*, and then the User-defined option in *g_filters*. All of the discrete-time GUI modules can export and import filter parameters in this manner.

CASE STUDY 6.1

Bandstop Filter Design: A Comparison

In order to illustrate the different design methods for constructing an FIR filter, suppose $f_s = 2000$ Hz, and consider the problem of designing a bandstop filter to meet the following specifications.

$$(F_{p1}, F_{s1}, F_{s2}, F_{p2}) = (200, 300, 700, 800) \text{ Hz} \tag{6.10.2a}$$

$$(\delta_p, \delta_s) = (0.04, 0.02) \tag{6.10.2b}$$

From (5.2.7), the corresponding passband ripple and stopband attenuation in dB are

$$A_p = 0.36 \text{ dB} \tag{6.10.3a}$$

$$A_s = 33.98 \text{ dB} \tag{6.10.3b}$$

To facilitate a comparison of the five design methods covered in this chapter, a common filter order of $m = 80$ is used for all cases. Plots of the resulting magnitude responses can be obtained by running the *exam6_20* from the driver program *f_dsp*.

CASE
STUDY
6.1

```
function case6_1

% CASE STUDY 6.1: FIR bandstop filter design

f_header('Case Study 6.1: FIR bandstop filter design\n\n')
deltap = 0.04
deltas = 0.02
Ap = -20*log10(1 - deltap)
As = -20*log10(deltas)
fs = 2000;
T = 1/fs;
Fp = [200 800]
Fs = [300 700]
sym = 0;
filt = 3;
win = 1;
m = f_prompt ('Enter filter order',0,120,80);

% Compute windowed filter

p = [Fp(1), Fs(1), Fs(2), Fp(2)];
b = f_firwin (@bandstop,m,fs,win,sym,p);
```

Continued on p. 480

Continued from p. 479

```
        show_filter(b,m,fs,Fp,Fs,Ap,As,'Windowed Filter')

        % Compute frequency-sampled filter

        M = floor(m/2) + 1;
        F = linspace (0,fs/2,M);
        A = bandstop (F,fs,p);
        b = f_firsamp (A,m,fs,sym);
        show_filter(b,m,fs,Fp,Fs,Ap,As,'Frequency-sampled Filter')

        % Compute least-squares filter

        w = (deltas/deltap)*ones(size(F));
        istop = (F >= Fs(1)) & (F <= Fs(2));
        w(istop) = 1;
        b = f_firls (F,A,m,fs,w);
        show_filter(b,m,fs,Fp,Fs,Ap,As,'Least-squares Filter')

        % Compute equiripple filter

        b = f_firparks (m,Fp,Fs,deltap,deltas,filt,fs);
        show_filter(b,m,fs,Fp,Fs,Ap,As,'Equiripple Filter')

        % Compute quadrature filter

        win = 3;
        deltaS = .0001;
        b = f_firquad (@bandstop,m,fs,win,deltaS,p);
        show_filter(b,m,fs,Fp,Fs,Ap,As,'Quadrature Filter')

        function [A,theta] = bandstop (f,fs,p)

        % BANDSTOP: Amplitude response of bandstop filter
        %
        %           p(1) = Fp1
        %           p(2) = Fs1
        %           p(3) = Fs2
        %           p(4) = Fp2

        A = zeros(size(f));
        for i = 1 : length(f)
            if (f(i) <= p(1) | f(i) >= p(4))
                A(i) = 1;
            elseif (f(i) > p(1) & f(i) < p(2))
                A(i) = 1 - (f(i) - p(1))/(p(2)-p(1));
            elseif (f(i) > p(3) & f(i) < p(4))
                A(i) = (f(i) - p(3))/(p(4) - p(3));
            end
        end
```

Continued on p. 481

Continued from p. 480

```
    theta=zeros(size(f));

    function show_filter(b,m,fs,Fp,Fs,Ap,As,caption)

    % SHOW_FILTER: Display the magnitude response in dB

    figure N = 250;
    Amin = 80;
    Amax = 20;
    [H,f] = f_freqz (b,1,N,fs);
    AdB = 20*log10(max(abs(H),eps));
    istop = (f >= Fs(1)) & (f <= Fs(2));
    Astop = -max(AdB(istop))
    cap2 = sprintf ([caption ' {m} = %d, {As} = %.1f dB'],m,Astop);
    f_labels (cap2,'{f} (Hz)','{A(f)} (dB)')
    axis ([0 fs/2 -Amin Amax])
    hold on
    box on
    fill ([0 Fs(1) Fs(1) 0],[-Ap -Ap Ap Ap],'c')
    fill ([Fs(1) Fs(2) Fs(2) Fs(1)],[-Amin -Amin -As -As],'c')
    fill ([Fs(2) fs/2 fs/2 Fs(2)],[-Ap -Ap Ap Ap],'c')
    plot (f,AdB,'LineWidth',1.5)
    f_wait
```

The first plot generated by *case6_1* corresponds to a windowed filter and is shown in Figure 6.50. In this case a Hanning window was used. It is apparent that the windowed filter does not meet the stopband specification because the transition band is too wide. The stopband attenuation in this case is $A_{\text{stop}} = 22.7$ dB.

The second plot generated by *case6_1*, which corresponds to the frequency-sampled method, is shown in Figure 6.51. Here the spacing between the frequency samples is

$$\Delta F = \frac{f_s}{m}$$
$$= 25 \text{ Hz} \tag{6.10.4}$$

From (6.10.2), the width of the transition band is $B = 100$ Hz. Consequently, there are three samples in the interior of each transition band. On the surface it appears that the frequency-sampled magnitude response may have met the stopband specification. However, a close inspection of Figure 6.51 reveals that near $f = F_{s2}$, the magnitude response is larger than the A_s specification, so the stopband attenuation ends up being only $A_{\text{stop}} = 24.5$ dB in this case.

The third plot generated by *case6_1*, corresponding to the least-squares method, is as shown in Figure 6.52. To facilitate comparison with the equiripple method, the weighting vector for the least-squares method was constructed using (6.7.7). From (6.10.2) we have $\delta_s/\delta_p = .5$, so this resulted in the following weighting vector.

$$w(i) = \begin{cases} .5, & f_i \in \text{passband} \\ 1, & f_i \in \text{stopband} \end{cases} \tag{6.10.5}$$

The least-squares method also requires that the weights be specified in the transition band, and in this case they were set to the passband value. The resulting magnitude response in

FIGURE 6.50: Magnitude Response of a Windowed Bandstop Filter Using a Hanning Window, $m = 80$

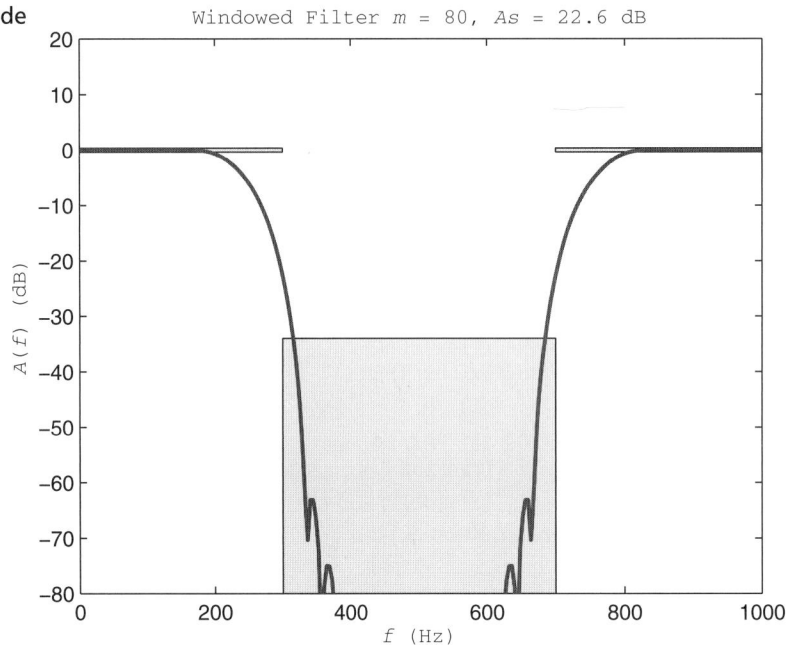
Windowed Filter $m = 80$, $As = 22.6$ dB

FIGURE 6.51: Magnitude Response of a Frequency-sampled Bandstop Filter, $m = 80$

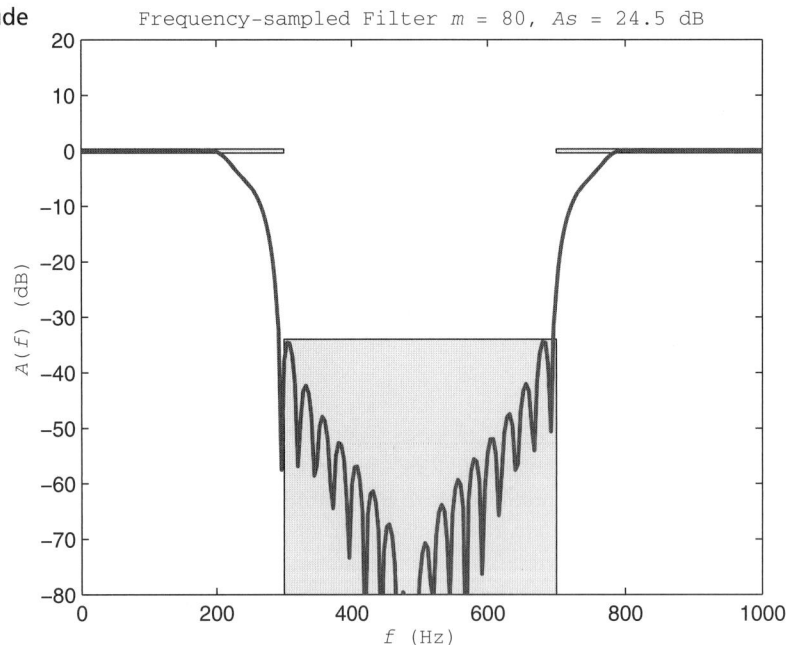
Frequency-sampled Filter $m = 80$, $As = 24.5$ dB

Figure 6.52 does meet the design specifications using a filter of order $m = 80$ with a stopband attenuation of $A_{\text{stop}} = 34.5$ dB in comparison with the specification of $A_s = 33.98$ dB.

The fourth plot generated by *case6_1* corresponds to the optimal equiripple design method and is as shown in Figure 6.53. It is apparent from Figure 6.53 that the equiripple filter easily meets, and in fact exceeds, the stopband specification. The stopband attenuation achieved in this case is $A_{\text{stop}} = 72.8$ dB. Recall that each 20 dB corresponds to a reduction in gain by a factor of 10. Therefore the stopband gain is somewhere between 10^{-4} and 10^{-3} for this filter.

FIGURE 6.52: Magnitude Response of a Least-squares Bandstop Filter, $m = 80$

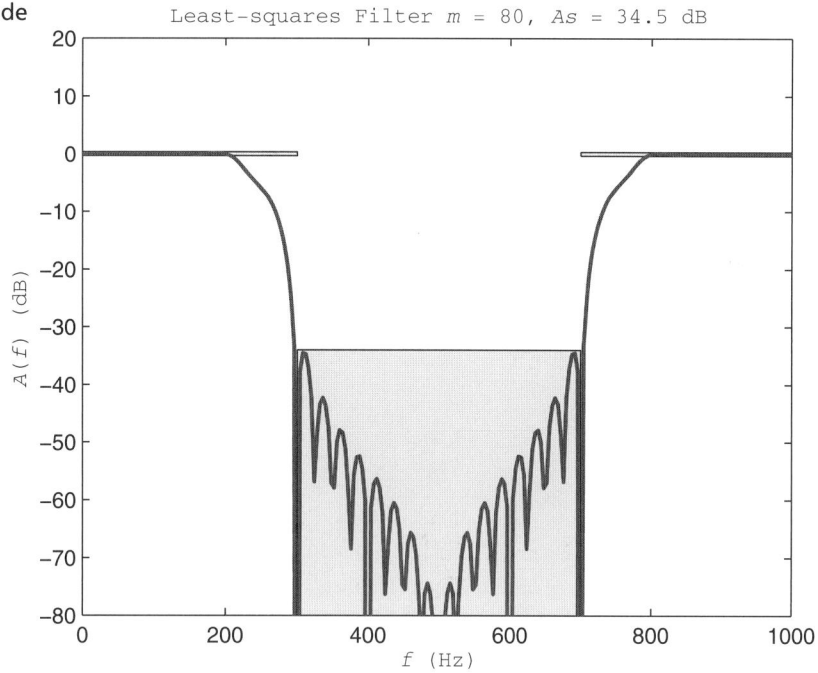

Least-squares Filter $m = 80$, $As = 34.5$ dB

FIGURE 6.53: Magnitude Response of an Equiripple Bandstop Filter, $m = 80$

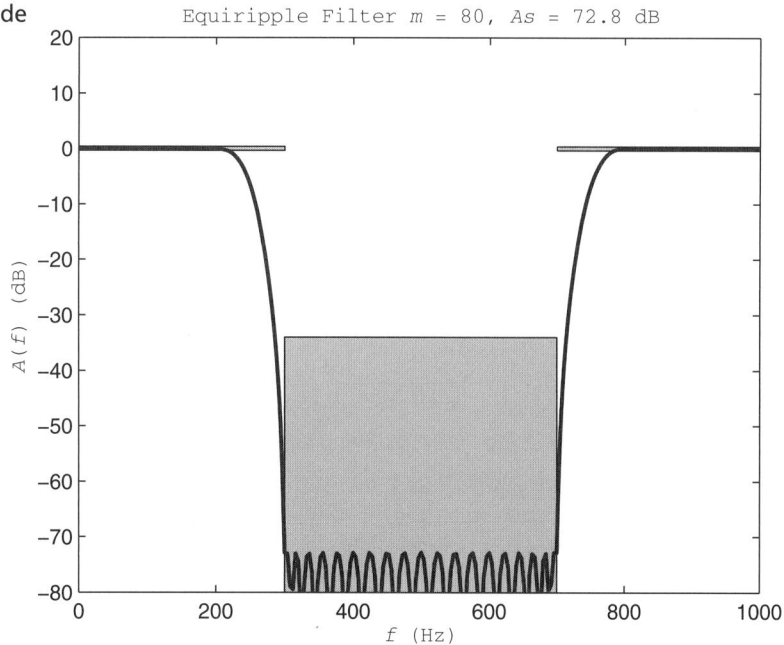

Equiripple Filter $m = 80$, $As = 72.8$ dB

The last plot generated by *case6_1*, corresponding to the quadrature method, is shown in Figure 6.54. The quadrature method is more general than the other methods because the residual phase response can be specified. To facilitate a comparison, the residual phase response was set to $\theta_d(f) = 0$. The resulting magnitude response is shown in Figure 6.54. The filter order was $m = 80$ with a Blackman window. To demonstrate that small values of stopband

FIGURE 6.54: Magnitude Response of a Quadrature Bandstop Filter with a Blackman Window and $\delta_s = .0001$, $m = 80$

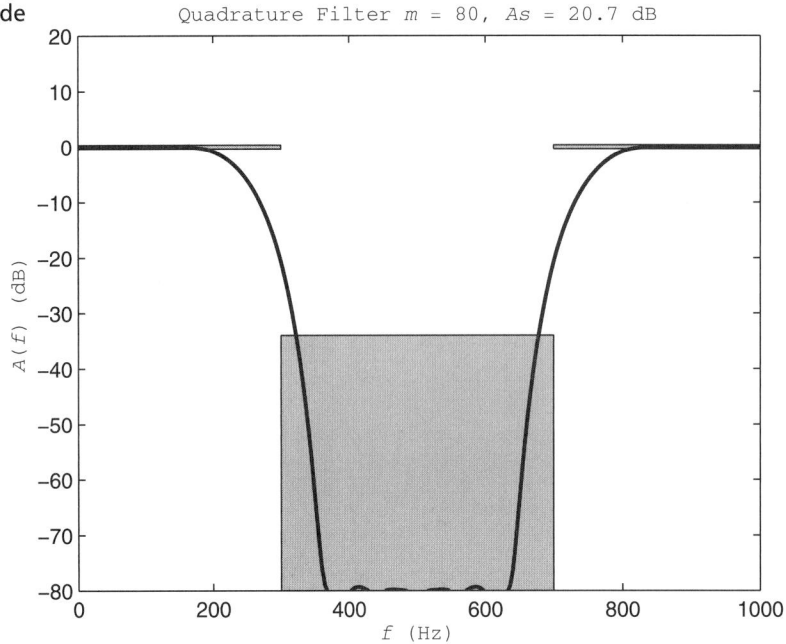

Quadrature Filter $m = 80$, $As = 20.7$ dB

attenuation can be used, it was set to $\delta_s = .0001$. Note that the stopband attenuation appears to closely track the specification throughout most of the stopband. However, it does not satisfy it at the edges of the stopband where the stopband attenuation is $A_{\text{stop}} = 20.7$ dB.

6.11   Chapter Summary

Finite Impulse Response Filters

This chapter focused on the design of finite impulse response or FIR digital filters having the following transfer function.

$$H(z) = \sum_{i=0}^{m} b_i z^{-i} \tag{6.11.1}$$

FIR filters offer a number of important advantages in comparison with IIR filters. FIR filters are always stable, regardless of the values of the filter coefficients. FIR filters are also less sensitive to finite word length effects. The impulse response of an FIR filter can be obtained directly from inspection of the transfer function or the difference equation as follows.

$$h(k) = \begin{cases} b_k, & 0 \le k \le m \\ 0, & m < k < \infty \end{cases} \tag{6.11.2}$$

Techniques are available for designing FIR filters that closely approximate arbitrary magnitude responses if the order of the filter is allowed to be sufficiently large. One drawback of FIR filters is that they require higher-order filters than IIR filters that satisfy the same design specifications. This implies larger storage requirements and longer computational times, a consideration that may be important for real-time signal processing applications.

Linear-phase Filters

The phase response of FIR filters can be made to be linear. A linear-phase response is an important characteristic because it means that different spectral components of the input signal are delayed by the same amount as they are processed by the filter. A linear-phase filter does not distort a signal within the passband; it only delays it by $\tau = mT/2$. The symmetry constraint on the impulse response that ensures that an FIR filter has linear phase is

$$h(k) = \pm h(m - k), \quad 0 \le k \le m \tag{6.11.3}$$

If the plus sign is used, the impulse response $h(k)$ is a palindrome that exhibits even symmetry about the midpoint $k = m/2$; otherwise it exhibits odd symmetry. There are four types of linear-phase FIR filters depending on whether the symmetry is even or odd and the filter order is even or odd. The most general linear-phase filter is a type 1 filter with even symmetry and even order. The other three filter types have zeros at one or both ends of the frequency range, and they are sometimes used for specialized applications such as the design of differentiators and Hilbert transformers.

Filter Design Methods

Four techniques were presented for designing an mth-order linear-phase FIR filter, and one for a $2m$-th-order FIR filter. The first four all introduce a constant group delay corresponding to half the filter length to make the impulse response causal. The windowing method is a truncated impulse response technique that tapers the coefficients with a data window to reduce ringing in the amplitude response caused by the Gibb's phenomenon. There is a trade-off between the reduction in ringing and the width of the transition band. Popular windows include the rectangular, Hanning, Hamming, and Blackman windows. When the rectangular window is used, the resulting filter minimizes the following mean-square error where $A_d(r)$ is the desired amplitude response and $A_r(f)$ is the actual amplitude response.

$$J = \int_0^{f_s/2} |A_d(f) - A_r(f)|^2 df \tag{6.11.4}$$

The frequency sampling method uses m equally spaced frequencies and the IDFT to compute the filter coefficients. Oscillations in the resulting magnitude response can be reduced by including one or more frequency samples in a transition band. The values of the transition band samples can be optimized to maximize the stopband attenuation.

The least-squares method is a direct optimization method that uses an arbitrary set of distinct discrete frequencies. It minimizes a weighted sum of squares of the error between the desired and the actual amplitude responses. Finding the coefficients requires solving a linear algebraic system of order $p + 1$, where $p = m/2$. If significant error exists at certain frequencies, these frequencies can be given additional weight to redistribute the error. The windowed, frequency-sampled, and least-squares methods can be used to design general linear-phase FIR filters with prescribed amplitude responses.

The fourth method is the optimal equiripple method, a technique that minimizes the maximum of the absolute value of the error in the passband and the stopband. Equiripple filters have amplitude responses that have ripples of equal magnitude in the passband and in the stopband. For a given set of frequency-selective filter design specifications, equiripple filters tend to be of lower order than filters designed with the windowed, frequency-sampled, and least-squares methods. To minimize the maximum of the absolute value of the error in the passband and the stopband, the amplitude response of an optimal equiripple filter must satisfy the following equations.

$$A_r(F_i) + \frac{(-1)^i \delta}{w(F_i)} = A_d(F_i), \quad 0 \le i < p + 2 \tag{6.11.5}$$

Here $w(f) > 0$ is a weighting function, and the F_i are extremal frequencies in the passband and the stopband where the magnitude of the error achieves its maximum value of δ. The FIR amplitude response $A_r(f)$ can be shown to be a polynomial in $x = \cos(2\pi f T)$ of degree $p = \text{floor}(m/2)$. The following normalized weighting function is used to design an equiripple frequency-selective filter.

$$w(f) = \begin{cases} \delta_s/\delta_p, & f \in \text{passband} \\ 1, & f \in \text{stopband} \end{cases} \qquad (6.11.6)$$

The last filter design method is a quadrature filter of order $2m$. Quadrature filters differ from the others in that they are designed to meet both magnitude response and residual phase response specifications. The residual phase response is the phase response remaining after the phase associated with a constant group delay has been removed. A quadrature filter of order $2m$ has the following transfer function where $p = m/2$.

$$H(z) = z^{-p} F(z) + H_h(z) G(z) \qquad (6.11.7)$$

Quadrature filters use a delay of $m/2$ samples and a Hilbert transformer, $H_h(z)$, in the first stage to create a pair of signals that are in phase quadrature. Linear-phase FIR filters, $F(z)$ and $G(z)$, are then applied to these signals to approximate the real and imaginary parts of the desired frequency response, respectively. The desired magnitude response $A_d(f)$ must have a stopband attenuation, $\delta_s > 0$, for the phase response to be well defined.

Filter Realization Structures

There are a number of alternative signal flow graph realizations of FIR filters. Direct form realizations have the property that the gains in the signal flow graphs are obtained directly from inspection of the transfer function. For FIR filters these include the tapped delay line, the transposed tapped delay line, and a direct form realization for linear-phase filters that requires only about half as many floating point multiplications. There are also a number of indirect realizations whose parameters must be computed from the original transfer function. The indirect forms decompose the original transfer function into lower-order blocks by combining complex conjugate pairs of zeros. For example, FIR filters can be realized with the following cascade-form realization, which is based on factoring $H(z)$.

$$H(z) = b_0 H_1(z) \cdots H_M(z) \qquad (6.11.8)$$

Here $M = \text{floor}[(m + 1)/2]$ and the $H_i(z)$ are second-order blocks with real coefficients except for $H_M(z)$, which is a first-order block when the filter order m is odd. Another FIR filter realization is the lattice form realization that consists of m blocks and a signal flow graph that resembles a lattice ladder structure. All of the filter realizations are equivalent to one another in terms of their overall input-output behavior when infinite precision arithmetic is used.

Finite Word Length Effects

Finite word length effects arise when a filter is implemented in either hardware or software. They are caused by the fact that both the filter parameters and the filter signals must be represented using a finite number of bits of precision. Both floating point and fixed point numerical representations can be used. MATLAB uses a 64-bit double-precision floating point representation that minimizes finite word length effects. When an N-bit fixed point representation is used for values in the range $[-c, c]$, the quantization level, or spacing between adjacent values, is

$$q = \frac{c}{2^{N-1}} \qquad (6.11.9)$$

Typically, the scale factor is $c = 2^M$ for some integer $M \geq 0$. This way, $M + 1$ bits are used to represent the integer part including the sign, and the remaining $N - (M + 1)$ bits are used for the fraction part.

Quantization error can arise from ADC quantization, input quantization, coefficient quantization, and product roundoff quantization. It is modeled as additive white noise uniformly distributed over $[-q/2, q/2]$. Another source of error is overflow error that can occur when several finite precision numbers are added. Overflow error can be eliminated by proper scaling of the input. The roots of a polynomial are very sensitive to the changes in the polynomial coefficients, particularly for higher-degree polynomials. Quantized FIR filters do not become unstable because all of the poles are at the origin. More generally, FIR filters are less sensitive to finite word length effects than IIR filters. Indirect form realizations tend to be less sensitive to finite word length effects because the block transfer functions are only of second order.

GUI Modules

The FDSP toolbox includes a GUI module called *g_fir* that allows the user to design and implement FIR filters without any need for programming. The filters include lowpass, highpass, bandpass, and bandstop filters plus a user-defined filter whose amplitude response and residual phase response are specified in an M file. Different design methods can be compared with each other and against the design specifications as the user varies the filter order m. The design methods include the windowed method, the frequency-sampled method, the least-squares method, the optimal equiripple method, and the quadrature method. The FDSP toolbox also includes a GUI module called *g_filters*, described in Section 5.9, that allows the user to investigate different filter realization structures and the effects of coefficient quantization error.

Learning Outcomes

This chapter was designed to provide the student with an opportunity to achieve the learning outcomes summarized in Table 6.6.

TABLE 6.6: ▶ Learning Outcomes for Chapter 6

Num.	Learning Outcome	Sec.
1	Understand the relative advantages and disadvantages of FIR filters in comparison with IIR filters	6.1
2	Know how to measure the signal-to-noise ratio	6.1
3	Be able to design a linear-phase FIR filter with a prescribed magnitude response using the windowing method	6.2
4	Understand why windows are used and what the trade-offs are	6.2
5	Be able to design a linear-phase FIR filter with a prescribed magnitude response using the frequency sampling method	6.3
6	Know how to insert frequency samples in the transition band to control filter performance	6.3
7	Be able to design a linear-phase FIR filter with a prescribed magnitude response using the least-squares method	6.4
8	Be able to design a linear-phase equiripple frequency-selective filter using the Parks-McClellan algorithm	6.5
9	Know how to design FIR differentiators and Hilbert transformers	6.6
10	Be able to design a FIR filter with prescribed magnitude and residual phase responses using the quadrature method	6.7
11	Understand the benefits of different filter realization structures	6.8
12	Be aware of detrimental finite word length effects and know how to minimize them	6.9
13	Know how to use the GUI module *g_fir* to design and analyze digital FIR filters without any programming	6.10

6.12   Problems

The problems are divided into Analysis and Design problems that can be solved by hand or with a calculator, GUI Simulation problems that are solved using GUI module g_fir, and MATLAB Computation problems that require a user program.

6.12.1   Analysis and Design

6.1 Consider the problem of designing an ideal linear-phase bandstop FIR filter with the windowing method using the Blackman window. Find the coefficients of a filter of order $m = 40$ using the following cutoff frequencies.

$$(f_s, F_{s1}, F_{s2}) = (10, 2, 4)\ \text{kHz}$$

6.2 Consider the problem of designing a type 1 linear-phase windowed FIR filter with the following desired amplitude response.

$$A_r(f) = \cos(\pi f T), \quad 0 \le |f| \le f_s/2$$

Suppose the filter order is even with $m = 2p$. Find the impulse response $h(k)$ using a rectangular window. Simplify the expression for $h(k)$ as much as possible.

6.3 Suppose the windowing method is used to design an mth-order lowpass FIR filter. The candidate windows include rectangular, Hanning, Hamming, and Blackman.
(a) Which window has the smallest transition band?
(b) Which window has the smallest passband ripple A_p?
(c) Which window has the largest stopband attenuation A_s?

6.4 Consider the problem of using the windowing method to design a lowpass filter to meet the following specifications.

$$(f_s, F_p, F_s) = (200, 30, 50)\ \text{Hz}$$
$$(A_p, A_s) = (.02, 50)\ \text{dB}$$

(a) Which types of windows can be used to satisfy these design specifications?
(b) For each of the windows in part (a), find the minimum order of filter m that will satisfy the design specifications.
(c) Assuming an ideal piecewise-constant amplitude response is used, find an appropriate value for the cutoff frequency F_c.

6.5 Suppose a lowpass filter of order $m = 10$ is designed using the windowing method with the Hanning window and $f_s = 2000$ Hz.
(a) Estimate the width of the transition band.
(b) Estimate the linear passband ripple and stopband attenuation.
(c) Estimate the logarithmic passband ripple and stopband attenuation.

6.6 Consider the problem of designing an mth-order type 3 linear-phase FIR filter having the following amplitude response.

$$A_r(f) = \sin(2\pi f T), \quad 0 \leq f \leq f_s/2$$

(a) Assuming $m = 2p$ for some integer p, find the coefficients using the windowing method with the rectangular window.
(b) Find the filter coefficients using the windowing method with the Hamming window.

6.7 Consider the following noise-corrupted periodic signal. Here $v(k)$ is white noise uniformly distributed over $[-.5, .5]$.

$$x(k) = 3 + 2\cos(.2\pi k)$$
$$y(k) = x(k) + v(k)$$

(a) Find the average power of the noise-free signal $x(k)$.
(b) Find the signal-to-noise ratio of $y(k)$.
(c) Suppose $y(k)$ is sent through an ideal lowpass filter with cutoff frequency $F_0 = .15 f_s$ to produce $z(k)$. Is the signal $x(k)$ affected by this filter? Find the signal-to-noise ratio of $z(k)$.

6.8 A linear-phase FIR filter is designed with the windowing method using the Hanning window. The filter meets its transition bandwidth specification of 200 Hz exactly with a filter of order $m = 30$.
(a) What is the sampling rate f_s?
(b) Find the filter order needed to achieve the same transition bandwidth using the Hamming window.
(c) Find the filter order needed to achieve the same transition bandwidth using the Blackman window.

6.9 Consider the problem of designing a filter to approximate a differentiator. Use the frequency sampling method to design a type 3 linear-phase filter of order $m = 40$ that approximates a differentiator, but with a delay $m/2$ samples. That is, find simplified expressions for the coefficients of a filter with the following desired amplitude response.

$$A_r(f) = 2\pi f T$$

6.10 Consider a type 3 linear-phase FIR filter of order $m = 2p$. Find a simplified expression for the amplitude response $A_r(f)$ similar to (6.4.7), but for a type 3 linear-phase FIR filter.

6.11 Consider the problem of constructing an equiripple bandstop filter of order $m = 40$. Suppose the design specifications are as follows.

$$(f_s, F_{p1}, F_{s1}, F_{s2}, F_{p2}) = (200, 20, 30, 50, 60) \text{ Hz}$$
$$(\delta_p, \delta_s) = (.05, .03)$$

(a) Let r be the number of extremal frequencies in the optimal amplitude response. Find a range for r.
(b) Find the set of specification frequencies, F.
(c) Find the weighting function $w(f)$.
(d) Find the desired amplitude response $A_d(f)$.

(e) The amplitude response $A_r(f)$ is a polynomial in x. Find x in terms of f, and find the polynomial degree.

6.12 Use the results of Problem 6.10 to derive the normal equations for the coefficients of a least-squares type 3 linear-phase filter. Specifically, find expressions for the coefficient matrix G and the right-hand side vector d, and show how to obtain the filter coefficients from the solution of the normal equations.

6.13 Suppose the equiripple design method is used to construct a highpass filter to meet the following specifications. Estimate the required filter order.

$$(f_s, F_s, F_p) = (100, 20, 30) \text{ kHz}$$
$$(A_p, A_s) = (.2, 32) \text{ dB}$$

6.14 Consider the following FIR filter. Find a cascade form realization of this filter and sketch the signal flow graph.

$$H(z) = \frac{10(z^2 - .6z - .16)[(z - .4)^2 + .25]}{z^4}$$

6.15 Consider the problem of designing a type 1 linear-phase bandpass FIR filter using the frequency sampling method. Suppose the filter order is $m = 60$. Find a simplified expression for the filter coefficients using the following ideal design specifications.

$$(f_s, F_{p1}, F_{p2}) = (1000, 100, 300) \text{ Hz}$$

6.16 Consider the problem of constructing an equiripple lowpass filter of order $m = 4$ satisfying the following design specifications.

$$(f_s, F_p, F_s) = (10, 2, 3) \text{ Hz}$$
$$(\delta_p, \delta_s) = (.05, .1)$$

Suppose the initial guess for the extremal frequencies is as follows.

$$(F_0, F_1, F_2, F_3) = (0, F_p, F_s, f_s/2)$$

(a) Find the weights $w(F_i)$ for $0 \le i \le 3$.
(b) Find the desired amplitude response values $A_d(F_i)$ for $0 \le i \le 3$.
(c) Find the extremal angles $\theta_i = 2\pi F_i T$ for $0 \le i \le 3$.
(d) Write down the vector equation that must be solved to find the Chebyshev coefficient vector d and the parameter δ. You do not have to solve the equation; just formulate it.

6.17 Suppose $F(z)$ and $G(z)$ are the following FIR filters.

$$F(z) = 1 + 2z^{-1} + z^{-2}$$
$$G(z) = 2 + z^{-1} + 2z^{-2}$$

(a) Show that $F(z)$ and $G(z)$ are type 1 linear-phase FIR filters.
(b) Find the amplitude responses $A_f(f)$ and $A_g(f)$.
(c) Assuming $F(z)$ and $G(z)$ are used to construct a quadrature filter using an ideal Hilbert transformer, find the magnitude response $A_q(f)$ and the residual phase response $\theta_q(f)$.

6.18 Find an efficient direct form realization for a linear-phase filter of order $m = 2p$ similar to (6.8.4), but applicable to a type 3 filter. Sketch the signal flow graph for the case $m = 4$.

6.19 Consider the following FIR filter.

$$H(z) = 3 + 4z^{-1} + 6z^{-2} + 4z^{-3} + 3z^{-4}$$

Suppose the input signal lies in the range $|x(k)| \le 10$. Find the scale factor for the input that ensures that the filter output will not overflow the range $|y(k)| \le 10$.

6.20 Consider the following FIR filter.

$$H(z) = \frac{(z^2 + 25)(z^2 + .04)}{z^4}$$

(a) Show that this is a type 1 linear-phase filter.
(b) Sketch a signal flow graph realization of $H(z)$ that is still a linear-phase system even when the coefficients are quantized.

6.21 A high-order FIR filter is realized as a cascade of second-order blocks.
(a) Suppose the filter has a sampling rate of $f_s = 300$ Hz and a zero at $z_0 = \exp(j\pi/3)$. Find a nonzero periodic signal $x(k)$ that gets completely attenuated by the filter.
(b) If a zero of a second-order block starts out on the unit circle, will the radius of the zero change as a result of the coefficient quantization? That is, will the zero still be on the unit circle?
(c) If a zero of a second-order block starts out on the unit circle, will the angle of the zero change as a result of the coefficient quantization? That is, will the frequency of the zero change?

6.22 Suppose the coefficients of an FIR filter of order $m = 30$ all lie within the range $|b_i| \le 4$. Assuming they are quantized to $N = 12$ bits, find an upper bound on the error in the magnitude of the frequency response caused by coefficient quantization.

6.23 Consider the following FIR filter. Find a lattice form realization of this filter and sketch the signal flow graph.

$$H(z) = 1 + 2z^{-1} + 3z^{-2} + 4z^{-3}$$

6.24 Consider the system shown in Figure 6.55. The ADC has a precision of 10 bits and an input range of $|x_a(t)| \le 10$. The transfer function of the digital filter is

$$H(z) = \frac{3z^2 - 2z}{z^2 - 1.2z + .32}$$

(a) Find the quantization level of the ADC.
(b) Find the average power of the quantization noise at the input x.
(c) Find the power gain of $H(z)$.
(d) Find the average power of the quantization noise at the output y.

FIGURE 6.55: ADC Quantization Noise

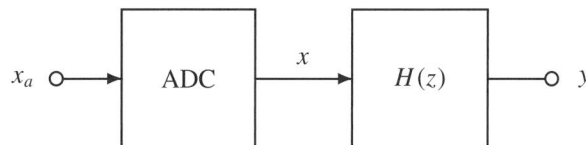

6.25 Suppose a 12-bit fixed point representation is used to represent values in the range $-10 \le x < 10$.
(a) How many distinct values of x can be represented?
(b) What is the quantization level, or spacing between adjacent values?

6.26 Consider the problem of designing a quadrature filter with the following frequency response. To simplify the final answer, you can assume that the Hilbert transformer component of the quadrature filter is ideal.

$$H(f) = \begin{cases} 5j \exp(-j\pi 20 fT), & 0 < f < f_s/2 \\ 0, & f = 0, \pm f_s/2 \\ -5j \exp(-j\pi 20 fT), & -f_s/2 < f < 0 \end{cases}$$

(a) Find the magnitude response $A(f)$ and the residual phase response $\theta(f)$.
(b) Suppose windowed filters with a Hamming window are used. Find $F(z)$ and $G(z)$

6.27 Suppose a 16-bit fixed point representation is used for values in the range $|x| \le 8$.
(a) How many distinct values of x can be represented?
(b) What is the quantization level, or spacing between adjacent values?
(c) How many bits are used to represent the integer part (including the sign)?
(d) How many bits are used to represent the fraction part?

6.12.2 **GUI Simulation**

6.28 Use the GUI module *g_fir* to construct a windowed highpass filter using the Hamming window.
(a) Plot the linear magnitude response and use the Caliper option to measure the actual width of the transition band.
(b) Plot the phase response.
(c) Plot the impulse response.

6.29 Use the GUI module *g_fir* to design a windowed lowpass filter. Set the width of the transition band to $B = 150$ Hz. For each of the following cases, find the lowest value for the filter order m that meets the specifications. Plot the linear magnitude response in each case.
(a) Rectangular window
(b) Hanning window
(c) Hamming window
(d) Blackman window

6.30 Use the GUI module *g_fir* to design a windowed bandstop filter with the Hanning window to meet the following specifications. Adjust the filter order to the lowest value that meets the design specifications.

$$(f_s, F_{p1}, F_{s1}, F_{s2}, F_{p2}) = (100, 20, 25, 35, 40) \text{ Hz}$$
$$(\delta_p, \delta_s) = (.05, .05)$$

(a) Plot the magnitude response using the linear scale.
(b) Save filter parameters $a, b,$ and *fs* in *prob6_30.mat*. Then use GUI module *g_filters* to load these as a user-defined filter. Adjust the number of bits used for coefficient quantization to $N = 6$. Plot the linear magnitude responses.

6.31 Use the GUI module *g_fir* to design a frequency-sampled bandstop filter to meet the following specifications. Adjust the filter order to the lowest value that meets the design specifications.

$$(f_s, F_{p1}, F_{s1}, F_{s2}, F_{p2}) = (100, 20, 25, 35, 40) \text{ Hz}$$
$$(\delta_p, \delta_s) = (.05, .05)$$

(a) Plot the magnitude response using the linear scale.
(b) Save filter parameters $a, b,$ and fs in *prob6_31.mat*. Then use GUI module *g_filters* to load these as a user-defined filter. Adjust the number of bits used for coefficient quantization to $N = 6$. Plot the linear magnitude responses.

6.32 Write an amplitude response function, called *prob6_32.m*, for the following user-defined filter (see *u_fir1* for an example).

$$A_r(f) = \frac{\cos(\pi f^2/100)}{1 + f^2}, \quad 0 \le f \le 10 \text{ Hz}$$

Using GUI module *g_fir*, set $fs = 20$ Hz and select a frequency-sampled filter. Then use the User-defined option to load this filter. Plot the following cases.
(a) Magnitude response, $m = 10$
(b) Magnitude response, $m = 20$
(c) Magnitude response, $m = 40$
(d) Impulse response, $m = 40$

6.33 Use the GUI module *g_fir* to design a least-squares bandpass filter to meet the following specifications. Adjust the filter order to the lowest value that meets the design specifications.

$$(f_s, F_{s1}, F_{p1}, F_{p2}, F_{s2}) = (2000, 300, 400, 600, 700) \text{ Hz}$$
$$(A_p, A_s) = (.4, 30) \text{ dB}$$

(a) Plot the magnitude response using the dB scale.
(b) Save filter parameters $a, b,$ and fs in *prob6_33.mat*. Then use GUI module *g_filters* to load these as a user-defined filter. Adjust the number of bits used for coefficient quantization to $N = 6$. Plot the linear magnitude responses.

6.34 Use the GUI module *g_fir* and the User-defined option to load the filter in file *u_fir1*. Adjust the filter order to $m = 90$. Plot the linear magnitude response for each of the following cases.
(a) Windowed filter with Blackman window
(b) Least-squares filter

6.35 Write an amplitude response function called *prob6_35.m* for the following user-defined filter (see *u_fir1* for an example).

$$A_r(f) = 2\left|\cos\left(\frac{2\pi f}{f_s}\right)\right|$$

Then use the User-defined option of GUI module *g_fir* to load this filter. Select a least-squares filter. Plot the linear magnitude response for the following three cases.
(a) $m = 10$
(b) $m = 20$
(c) $m = 40$

6.36 Use the GUI module *g_fir* to design an optimal equiripple bandpass filter to meet the following specifications. Adjust the filter order to the lowest value that meets the design specifications.

$$(f_s, F_{s1}, F_{p1}, F_{p2}, F_{s2}) = (2000, 300, 400, 600, 700) \text{ Hz}$$
$$(A_p, A_s) = (.4, 30) \text{ dB}$$

(a) Plot the magnitude response using the dB scale.

(b) Save filter parameters in *prob6_36.mat*. Then use GUI module *g_filters* to load these as a user-defined filter. Adjust the number of bits used for coefficient quantization to $N = 6$. Plot the linear magnitude responses.

6.37 Write an amplitude response and residual phase response function called *prob6_37.m* for the following user-defined filter (see *u_fir1* for an example).

$$A_r(f) = 10f/fs$$
$$\theta(f) = \pi \sin(20f//fs)$$

Set the filter order to $m = 150$, and select the quadrature filter. Then use the User-defined option of GUI module *g_fir* to load this filter.

(a) Print your amplitude response and residual phase response functions.

(b) Plot the linear magnitude response

(c) Plot the phase response

6.38 Write an amplitude response and residual phase response function called *prob6_38.m* for the following user-defined filter (see *u_fir1* for an example).

$$A_r(f) = .5\{1 + \text{sgn}[\sin(8\pi fT)]\}$$
$$\theta(f) = 0$$

Set $\delta_s = .001$ and $m = 120$. Then use the User-defined option of GUI module *g_fir* to load this filter. Plot the following.

(a) The linear magnitude response of a least-squares filter.

(b) The pole-zero pattern of a least-squares filter.

(c) The linear magnitude response of a quadrature filter.

(d) The pole-zero pattern of a quadrature filter.

6.12.3 MATLAB Computation

6.39 Write a MATLAB program that uses *f_firwin* to design a linear-phase highpass FIR filter of order $m = 60$ with stopband cutoff frequency $F_s = 20$ Hz, passband cutoff frequency $F_p = 30$, and sampling frequency $f_s = 100$ Hz. Use a Blackman window, and make the desired amplitude response piecewise-constant with cutoff $F_c = (F_s + F_p)/2$.

(a) Use *f_freqz* to compute and plot the magnitude response using the dB scale.

(b) Use Table 6.3, the *hold on* command, and the *fill* function to add a shaded area showing the predicted stopband attenuation A_s.

6.40 Write a MATLAB program that uses *f_firideal* to design a linear-phase lowpass FIR filter of order $m = 40$ with passband cutoff frequency $F_p = f_s/5$ and stopband cutoff frequency $F_s = f_s/4$, where the sampling frequency is $f_s = 100$ Hz. Use a rectangular window, and set the ideal cutoff frequency to the middle of the transition band. Use *f_freqz* to compute and plot the magnitude response using the linear scale. Then use Table 6.3, the *hold on* command, and the *fill* function to add the following items to your magnitude response plot.

(a) A shaded area showing the passband ripple δ_p.

(b) A shaded area showing the stopband attenuation δ_s.

6.41 Write a MATLAB program that constructs the following signal where $f_s = 200$ Hz. Here $v(k)$ is white noise uniformly distributed over $[-1, 1]$, $F_1 = 10$ Hz, $F_2 = 30$ Hz, and $N = 4096$. Use a random number generator seed of 100 to produce $v(k)$.

$$x(k) = 4\sin(2\pi F_1 kT)\cos(2\pi F_2 kT), \quad 0 \le k < N$$
$$y(k) = x(k) + v(k), \qquad\qquad\qquad 0 \le k < N$$

(a) Compute P_x and P_v directly from the samples. Use Definition 6.1.1 to compute and print the signal-to-noise ratio of $y(k)$.
(b) Compute P_y directly from the samples. Use P_v, (6.1.1), and Definition 6.1.1 to compute and print the signal-to-noise ratio of $y(k)$.
(c) Compute and print the percent error of the estimate of the SNR found in part (b) relative to the SNR found in part (a).
(d) Plot the magnitude spectrum of $y(k)$ showing the signal and the noise.

6.42 Write a MATLAB program that uses f_firwin to design a type 1 linear-phase FIR filter of order $m = 80$ using $f_s = 1000$ Hz and the Hamming window to approximate the following amplitude response. Use f_freqz to compute the magnitude response.

$$A_r(f) = \begin{cases} \left(\dfrac{f}{250}\right)^2, & 0 \le |f| < 250 \\[4mm] .5\cos\left[\dfrac{\pi(f-250)}{500}\right], & 250 \le |f| < 500 \end{cases}$$

(a) Plot the linear magnitude response.
(b) On the same graph, add the desired magnitude response and a legend.

6.43 Write a MATLAB program that uses $f\_firideal$ to design a linear-phase highpass FIR filter of order $m = 30$ with stopband cutoff frequency $F_s = 20$ Hz, passband cutoff frequency $F_p = 30$, and sampling frequency $f_s = 100$ Hz. Use a Hanning window, and set the ideal cutoff frequency to the middle of the transition band.
(a) Use f_freqz to compute and plot the magnitude response using the dB scale.
(b) Use Table 6.3, the *hold on* command, and the *fill* function to add a shaded area showing the predicted stopband attenuation A_s.

6.44 Write a MATLAB program that uses $f\_firideal$ to design a linear-phase highpass FIR filter of order $m = 40$ with stopband cutoff frequency $F_s = 20$ Hz, passband cutoff frequency $F_p = 30$, and sampling frequency $f_s = 100$ Hz. Use a Hamming window, and set the ideal cutoff frequency to the middle of the transition band.
(a) Use f_freqz to compute and plot the magnitude response using the dB scale.
(b) Use Table 6.3, the *hold on* command, and the *fill* function to add a shaded area showing the predicted stopband attenuation A_s.

6.45 Write a MATLAB program that uses function $f\_firsamp$ to design a linear-phase bandpass FIR filter of order $m = 40$ using the frequency sampling method. Use a sampling frequency of $f_s = 200$ Hz, and a passband of $F_p = [20, 60]$ Hz. Use f_freqz to compute and plot the linear magnitude response. Add the frequency samples using a separate plot symbol and a legend. Do the following cases.
(a) No transition band samples (ideal amplitude response)
(b) One transition band sample of amplitude .5 on each side of the passband.

6.46 The Chebyshev polynomials have several interesting properties. Write a MATLAB program that uses the FDSP toolbox function $f\_chebpoly$ and the *subplot* command to construct a 2×2 array of plots of the Chebyshev polynomials $T_k(x)$ for $1 \leq k \leq 4$. Use the plot range $-1 \leq x \leq 1$. Using induction and your observations of the plots, list as many general properties of $T_k(x)$ as you can. Use the help command for instructions on how to use $f\_chebpoly$.

6.47 Write a MATLAB function called $u\_firorder$ which estimates the order of an equiripple filter required to meet given design specifications using (6.5.21). The calling sequence for $u\_firorder$ should be as follows.

```
% U_FIRORDER: Estimate required order for FIR equiripple filter
%
% Usage:
%        m = u_firorder (deltap,deltas,Bhat);
% Pre:
%        deltap = passband ripple
%        deltas = stopband attenuation
%        Bhat = normalized transition bandwidth
% Post:
%        m = estimated FIR equiripple order
```

Test your function by plotting a family of curves on one graph. For the kth curve use $deltap = deltas = \delta$ where $\delta = .03k$ for $1 \leq k \leq 3$. Plot m versus $Bhat$ for $.01 \leq Bhat \leq .1$ and include a legend.

6.48 Write a MATLAB program that uses the function $f\_firparks$ to design an equiripple lowpass filter to meet the following design specifications where $f_s = 4000$ Hz. Find the lowest-order filter that meets the specifications.

$$(F_p, F_s) = (1200, 1400) \text{ Hz}$$
$$(\delta_p, \delta_s) = (.03, .04)$$

(a) Print the minimum filter order and the estimated order based on (6.5.21).
(b) Plot the linear magnitude response.
(c) Use *fill* to add shaded areas to the plot showing the design specifications.

6.49 Write a MATLAB program that uses function f_firls to design a least-squares linear-phase FIR filter of order $m = 30$ with sampling frequency $f_s = 400$ and the following amplitude response.

$$A_r(f) = \begin{cases} \dfrac{f}{100}, & 0 \leq |f| < 100 \\ \dfrac{200 - f}{100}, & 100 \leq |f| \leq 200 \end{cases}$$

Select $2m$ equally spaced discrete frequencies, and use uniform weighting. Use f_freqz to compute and plot both magnitude response (ideal and actual) on the same graph.

6.50 Write a MATLAB program that uses function $f\_firsamp$ to design a linear-phase bandstop FIR filter of order $m = 60$ using the frequency sampling method. Use a sampling frequency of $f_s = 20$ kHz, and a stopband of $F_s = [3, 8]$ kHz. Use f_freqz to compute and plot the linear

magnitude response. Add the frequency samples using a separate plot symbol and a legend. Do the following cases.
(a) No transition band samples (ideal amplitude response)
(b) One transition band sample of amplitude .5 on each side of the stopband.

6.51 Consider the following FIR impulse response. Suppose the filter order is $m = 30$.

$$h(k) = \frac{k+1}{m}, \quad 0 \le k \le m$$

(a) Write a MATLAB program that uses $f\_cascade$ to compute a cascade form realization of this filter. Print the gain b_0 and the block coefficients, B and A.
(b) Suppose the sampling frequency is $f_s = 400$ Hz. Use f_freqz to compute the frequency response using a cascade form realization. Compute both the unquantized frequency response (set bits = 64), and the frequency response with coefficient quantization using 8 bits. Plot both magnitude responses on a single plot using the dB scale and a legend.

6.52 Using function $f\_firquad$ and Example 6.14 as a starting point, write a MATLAB program that designs an equalizer for a system $H(z)$ with the following magnitude and phase responses. Use filters of order $m = 160$, $\delta_s = .001$, and the Hamming window.

$$A_d(f) = \exp[-(fT - .25)^2/.01]$$
$$\phi_d(f) = -10\pi(fT)^2 + \sin(5\pi fT)$$

(a) Print the optimal delay τ and the total delay τ_q of the equalizer.
(b) Print a 3×1 array of plots showing the magnitude responses of the original system, the equalizer, and the equalized system similar to Figure 6.29.
(c) Print a 3×1 array of plots showing the residual phase responses of the original system, the equalizer, and the equalized system similar to Figure 6.30.
(d) Plot the impulse response of the equalizer filter.

6.53 Write a MATLAB program that uses the function $f\_firparks$ to design an equiripple highpass filter to meet the following design specifications where $f_s = 300$ Hz. Find the lowest-order filter that meets the specifications.

$$(F_s, F_p) = (90, 110) \ Hz$$
$$(\delta_p, \delta_s) = (.02, .03)$$

(a) Print the minimum filter order and the estimated order based on (6.5.21).
(b) Plot the linear magnitude response.
(c) Use $fill$ to add shaded areas to the plot showing the design specifications.

6.54 Consider the following FIR transfer function.

$$H(z) = \sum_{i=0}^{20} \frac{z^{-i}}{1+i}$$

(a) Write a MATLAB program that uses $f\_lattice$ to compute a lattice form realization of this filter. Print the gain and the reflection coefficients of the blocks.
(b) Suppose the sampling frequency is $f_s = 600$ Hz. Use f_freqz to compute the frequency response using a lattice form realization. Compute both the unquantized frequency response (e.g., 64 bits), and the frequency response with coefficient quantization using $N = 8$ bits. Plot both magnitude responses on a single plot using the dB scale and a legend.

6.55 Write a MATLAB program that uses the function *f_hilbert* to compute a Hilbert transformer filter using a Blackman window. Do the following cases.

(a) Use *f_freqz* to compute and plot the magnitude responses for $m = 40$ and $m = 80$ on the same graph. Also show the ideal magnitude response and add a legend. Specify the filter type in the title.

(b) Use *f_freqz* to compute and plot the magnitude responses for $m = 41$ and $m = 81$ on the same graph. Also show the ideal magnitude response and add a legend. Specify the filter type in the title.

IIR Filter Design

Chapter Topics

7.1   Motivation
7.2   Filter Design by Pole-zero Placement
7.3   Filter Design Parameters
7.4   Classical Analog Filters
7.5   Bilinear transformation Method
7.6   Frequency Transformations
7.7   Filter Realization Structures
*7.8 Finite Word Length Effects
7.9   GUI Software and Case Study
7.10 Chapter Summary
7.11 Problems

7.1 Motivation

Just as mechanical filters are used in pipes to block the flow of certain particles, discrete-time systems can be used as digital filters to block the flow of certain signals. A digital filter is a discrete-time system that is designed to reshape the spectrum of the input signal in order to produce desired spectral characteristics in the output signal. Thus a digital filter is a frequency-selective filter that modifies the magnitude spectrum and the phase spectrum by selecting certain spectral components and inhibiting others. In this chapter we investigate the design of digital infinite impulse response (IIR) filters having the following generic transfer function.

$$H(z) = \frac{b_0 + b_1 z^{-1} + \cdots + b_m z^{-m}}{1 + a_1 z^{-1} + \cdots + a_n z^{-n}}$$

Recall that $H(z)$ is an IIR filter if $a_i \neq 0$ for some $i \geq 1$. Otherwise, $H(z)$ is an FIR filter. Simple specialized IIR filters can be designed directly using a careful placement of poles and zeros. A more general technique is to start with a normalized lowpass analog filter and put it through a series of transformations to convert it to a desired digital filter. First, a frequency transformation is used to map the normalized lowpass analog filter into a specified frequency-selective filter. This is followed by a bilinear transformation that converts the analog filter into an equivalent digital filter.

We begin this chapter by introducing some practical examples of applications of IIR filters. Next, simple techniques based on pole-zero placement and gain matching are introduced for direct design of specialized IIR filters such as resonators, notch filters, and comb filters. This is followed by a presentation of the classical analog lowpass filters including Butterworth, Chebyshev, and elliptic filters. An analog-to-digital filter transformation technique called the bilinear transformation method is then presented. Next, frequency transformations are introduced that map normalized lowpass filters into lowpass, highpass, bandpass, and bandstop filters. Finally, a GUI module called g_iir is introduced that allows the user to design and evaluate a variety of digital IIR filters without any need for programming. The chapter concludes with a case study example, and a summary of IIR filter design techniques.

7.1.1 Tunable Plucked-string Filter

Computer-generated music is a natural application area for IIR filter design (Steiglitz, 1996). A simple, yet highly effective, building block for the synthesis of musical sounds is the tuned plucked-string filter shown in Figure 7.1. The output from this type of filter can be used, for example, to synthesize the sound from a stringed instrument such as a guitar.

The design parameters of the plucked-string filter in Figure 7.1 are the sampling frequency f_s, the pitch parameter $0 < c < 1$, the feedback delay L, and the feedback attenuation factor $0 < r < 1$. Later, the components of this type of filter are examined in detail. For now, consider the question of developing an expression for the overall transfer function $H(z) = Y(z)/X(z)$ of the plucked-string filter. The block with input $e(k)$ and output $w(k)$ is a first-order lowpass filter with transfer function

$$F(z) \triangleq \frac{W(z)}{E(z)}$$
$$= \frac{1 + z^{-1}}{2} \tag{7.1.1}$$

Recall from (5.4.7) that the block with input $w(k)$ and output $y(k)$ is a first-order allpass filter. Allpass filters pass all frequencies equally because they have flat magnitude responses. The purpose of an allpass filter is to change the phase of the input and thereby introduce some delay. The transfer function of the allpass filter in Figure 7.1 is as follows, where $0 < c < 1$ is

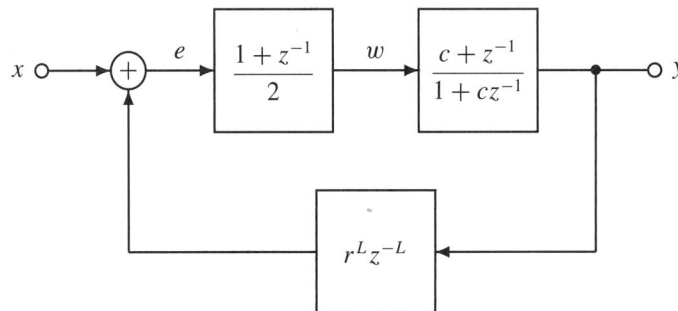

FIGURE 7.1: A Tunable Plucked-string Filter

the pitch parameter.

$$G(z) \triangleq \frac{Y(z)}{W(z)}$$

$$= \frac{c + z^{-1}}{1 + cz^{-1}} \tag{7.1.2}$$

The key to finding the overall transfer function is to compute the Z-transform of the summing junction output $E(z)$. From (7.1.1), (7.1.2), and Figure 7.1

$$
\begin{aligned}
E(z) &= X(z) + r^L z^{-L} Y(z) \\
&= X(z) + r^L z^{-L} G(z) W(z) \\
&= X(z) + r^L z^{-L} G(z) F(z) E(z)
\end{aligned}
\tag{7.1.3}
$$

Solving (7.1.3) for $E(z)$ yields

$$E(z) = \frac{X(z)}{1 - r^L z^{-L} G(z) F(z)} \tag{7.1.4}$$

From (7.1.1) and (7.1.2), the Z-transform of the output can be expressed as

$$
\begin{aligned}
Y(z) &= G(z) W(z) \\
&= G(z) F(z) E(z) \\
&= \frac{G(z) F(z) X(z)}{1 - r^L z^{-L} G(z) F(z)}
\end{aligned}
\tag{7.1.5}
$$

It follows that the overall transfer function of the tunable plucked-string filter is

$$
\begin{aligned}
H(z) &= \frac{Y(z)}{X(z)} \\
&= \frac{G(z) F(z)}{1 - r^L z^{-L} G(z) F(z)}
\end{aligned}
\tag{7.1.6}
$$

To verify that this is an IIR filter, substitute the expressions for $F(z)$ and $G(z)$, from (7.1.1) and (7.1.2), respectively, into (7.1.6), which yields

$$H(z) = \frac{\left(\dfrac{c + z^{-1}}{1 + cz^{-1}}\right)\left(\dfrac{1 + z^{-1}}{2}\right)}{1 - r^L z^{-L}\left(\dfrac{c + z^{-1}}{1 + cz^{-1}}\right)\left(\dfrac{1 + z^{-1}}{2}\right)} \tag{7.1.7}$$

Multiplying the top and bottom of (7.1.7) by $(1 + cz^{-1})$ and combining terms then leads to the following simplified transfer function for the tunable plucked-string filter.

$$H(z) = \frac{.5[c + (1 + c)z^{-1} + z^{-2}]}{1 + cz^{-1} - .5r^L[cz^{-L} + (1 + c)z^{-(L+1)} + z^{-(L+2)}]} \tag{7.1.8}$$

Since the denominator polynomial satisfies $a(z) \neq 1$, this is indeed an IIR filter. The plucked-string sound is generated by the filter output when the input is an impulse or a short burst of white noise.

The frequency response of the plucked-string filter consists of a series of peaks or *resonances* that decay gradually, depending on the value of the feedback attenuation parameter r. To tune the first resonant frequency, the parameters L and c are used. Suppose the sampling *Pitch* frequency is f_s and the desired value for the first resonance, or *pitch*, is F_0. Then L and c can be

FIGURE 7.2:
Magnitude
Response of
Plucked-string
Filter: $L = 59$,
$c = .8272$, $r = .999$

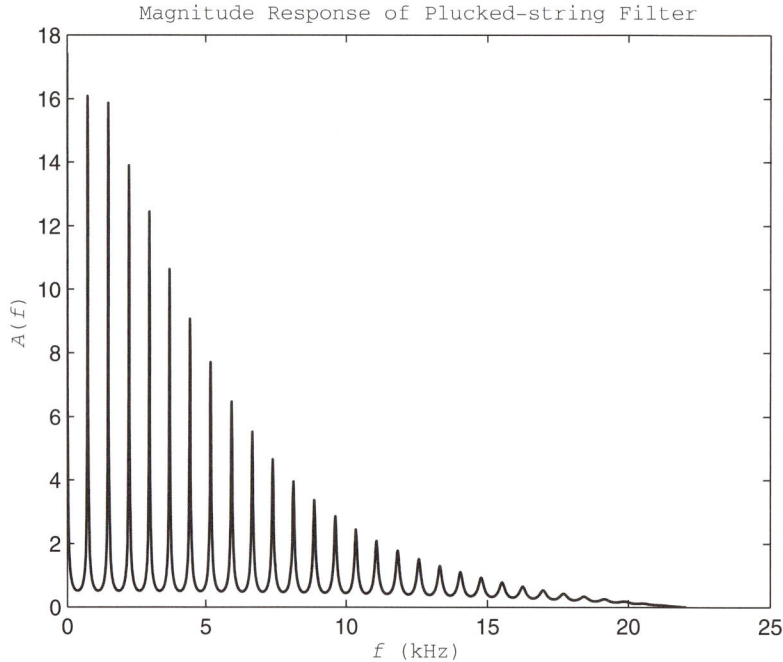

Magnitude Response of Plucked-string Filter

computed as follows (Jaffe and Smith, 1983).

$$L = \text{floor}\left(\frac{f_s - .5F_0}{F_0}\right) \tag{7.1.9a}$$

$$\delta = \frac{f_s - (L + .5)F_0}{F_0} \tag{7.1.9b}$$

$$c = \frac{1 - \delta}{1 + \delta} \tag{7.1.9c}$$

To make the discussion specific, suppose the sampling frequency is $f_s = 44.1$ kHz, a value commonly used in digital recording. Next, suppose the desired location of the first resonance is $F_0 = 740$ Hz. Applying (7.1.9) then yields $L = 59$ and $c = .8272$. If the feedback attenuation factor is set to $r = .999$, then this results in the plucked-string filter magnitude response shown in Figure 7.2. Interestingly enough, the resonant frequencies are almost, but not quite, harmonically related (see Steiglitz, 1996). When Figure 7.2 is generated by running *fig7_2* from *f_dsp*, the sound made by the filter output is also played on the PC speakers. Give it a try, and let your ears be the judge!

7.1.2 Colored Noise

As a second example of an application of IIR filters, consider the problem of creating a test signal with desired spectral characteristics. One of the most popular test signals is white noise because it contains power at all frequencies. In particular, an N-point white noise signal $v(k)$ with average power P_v has the following power-density spectrum.

$$S_v(f) \approx P_v, \quad 0 \le f < \frac{f_s}{2} \tag{7.1.4}$$

The term *white* arises from the fact that $x(k)$ contains power at all frequencies just as white light is composed of all colors. If the natural frequencies of a linear system are confined to an

FIGURE 7.3:
Generation of
Colored Noise $x(k)$
from White Noise
$v(k)$

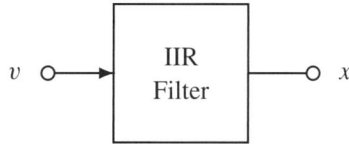

v ○──→ [IIR Filter] ──○ x

interval $[F_0, F_1]$, then it is more efficient to excite these natural modes with a signal that has its power restricted to the frequency range $[F_0, F_1]$. Since only a subset of the entire range of frequencies is represented, this type of signal is sometimes referred to as *colored* noise. For example, low-frequency noise might be thought of as *red* and high-frequency noise as *blue*. The desired power density spectrum for colored noise in the interval $[F_0, F_1]$ is

Colored noise

$$S_x(f) = \begin{cases} 0, & 0 \le f < F_0 \\ P_x, & F_0 \le f \le F_1 \\ 0, & F_1 \le f < f_s/2 \end{cases} \tag{7.1.5}$$

A simple scheme for generating colored noise with a desired power density spectrum is shown in Figure 7.3. The basic approach is to start with white noise, which is easily constructed, and then pass it through a digital filter that removes the undesired spectral components. The digital filter can be either an IIR filter or an FIR filter. However, because a linear phase response is not crucial, the colored noise can be generated more efficiently using an IIR filter.

The ideal filter is a bandpass filter with a low-frequency cutoff of F_0, a high-frequency cutoff of F_1, and a passband gain of one, as shown in Figure 7.4. The removal of signal power below F_0 Hz and above F_1 Hz means that the average power of the colored noise $x(k)$ will be the following fraction of the average power of the white noise $v(k)$.

$$P_x = \frac{2(F_1 - F_0)P_v}{f_s} \tag{7.1.6}$$

FIGURE 7.4: Magnitude
Response of an
Ideal Bandpass
Filter

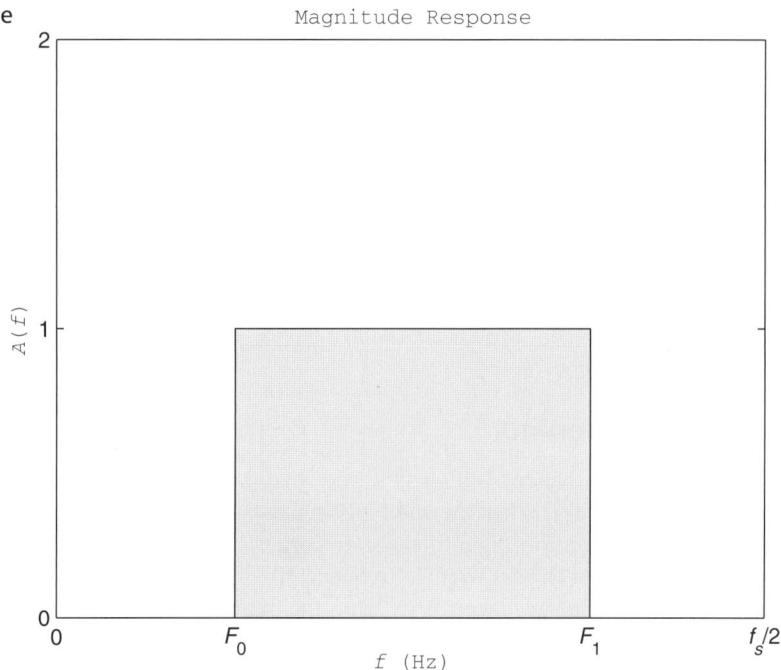

FIGURE 7.5: Power-density Spectrum of Colored Noise Created by Sending White Noise through a Bandpass Filter with $[F_0, F_1] = [150, 300]$ Hz

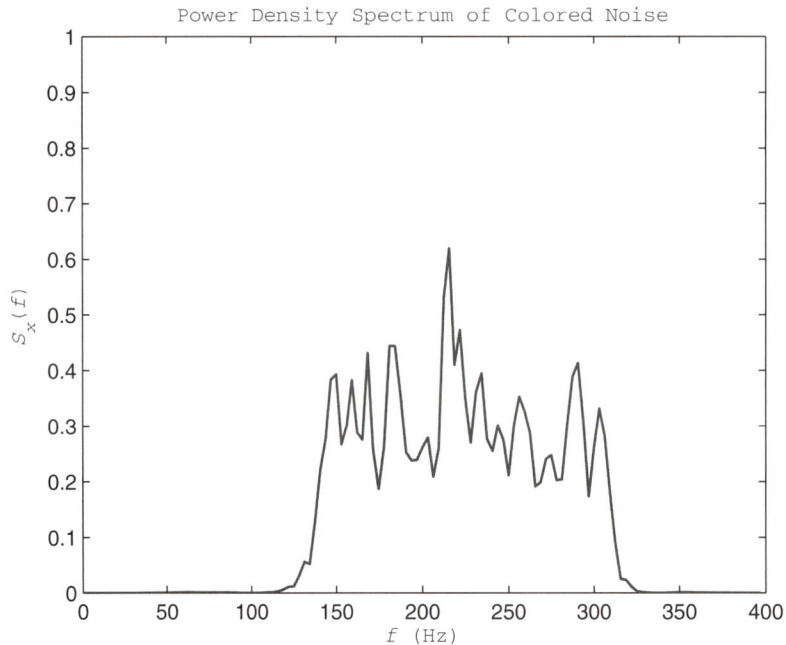

To make the example specific, suppose the sampling frequency is $f_s = 800$ Hz, and the desired frequency band for the colored noise is $[F_0, F_1] = [150, 300]$ Hz. Let $v(k)$ consist of $N = 2048$ samples of white noise uniformly distributed over $[-1, 1]$. Using a design technique covered later in the chapter, a tenth-order elliptic bandpass filter can be constructed. When the white noise is passed through this IIR filter, this results in colored noise with the power-density spectrum shown in Figure 7.5. It is evident that there is still a small amount of power outside the desired range [150, 300] Hz. This is due to the fact that practical filters, unlike ideal filters, have a transition band between the passband and the stopband. The width of the transition band can be reduced by going to a higher-order filter, but it can not be eliminated completely.

7.2 Filter Design by Pole-zero Placement

In addition to the basic frequency-selective filters, there are a number of specialized filters that arise in applications. A direct design procedure is available for these filters based on gain matching and a careful placement of poles and zeros.

7.2.1 Resonator

Recall that a bandpass filter is a filter that passes signals whose frequencies lie within an interval $[F_0, F_1]$. When the width of the passband is small in comparison with f_s, we say that the filter is a *narrowband* filter. An important limiting special case of a narrowband filter is a filter designed to pass a single frequency $0 < F_0 < f_s/2$. Such a filter is called a *resonator* with a *resonant frequency* of F_0. Thus the frequency response of an ideal resonator is

Resonant frequency

$$H_{\text{res}}(f) = \delta_1(f - F_0), \quad 0 \le f \le f_s/2 \tag{7.2.1}$$

Here $\delta_1(f)$ denotes the unit pulse $\delta(k)$ but with the integer argument k replaced by a real argument f. That is,

$$\delta_1(f) \triangleq \begin{cases} 1, & f = 0 \\ 0, & f \neq 0 \end{cases} \tag{7.2.2}$$

Angle of F_0

A resonator can be used to extract a single frequency component, or a very narrow range of frequencies, from a signal. A simple way to design a resonator is to place a pole near the point on the unit circle that corresponds to the resonant frequency F_0. Recall that as the angles of the points along the top half of the unit circle go from 0 to π, the frequency f ranges from 0 to $f_s/2$. Thus the angle corresponding to frequency F_0 is

$$\theta_0 = \frac{2\pi F_0}{f_s} \tag{7.2.3}$$

The radius of the pole must be less than one for the filter transfer function $H_{res}(z)$ to be stable. Furthermore, if the coefficients of the denominator of $H_{res}(z)$ are to be real, then complex poles must occur in conjugate pairs. One can ensure that the resonator completely attenuates the two end frequencies, $f = 0$ and $f = f_s/2$, by placing zeros at $z = 1$ and $z = -1$, respectively. These constraints yield a resonator transfer function with the following factored form.

$$H_{res}(z) = \frac{b_0(z-1)(z+1)}{[z - r\exp(j\theta_0)][z - r\exp(-j\theta_0)]} \tag{7.2.4}$$

Using Euler's identity from Appendix 2, we find that the resonator transfer function can be simplified and expressed as a ratio of two polynomials.

$$\begin{aligned} H_{res}(z) &= \frac{b_0(z^2-1)}{z^2 - r[\exp(j\theta_0) + \exp(-j\theta_0)]z + r^2} \\ &= \frac{b_0(z^2-1)}{z^2 - r2\text{Re}\{\exp(j\theta_0)\}z + r^2} \\ &= \frac{b_0(z^2-1)}{z^2 - 2r\cos(\theta_0)z + r^2} \end{aligned} \tag{7.2.5}$$

There are two design parameters that remain to be determined, the pole radius r, and the gain factor b_0. To achieve a sharp filter that is highly selective, one needs $r \approx 1$, but for stability, it is essential that $r < 1$. Suppose ΔF denotes the *radius* of the 3 dB passband of the filter. Thus $|H_{res}(f)| \geq 1/\sqrt{2}$ for f in the range $[F_0 - \Delta F, F_0 + \Delta F]$. For a narrowband filter $\Delta F \ll f_s$. In this case the following approximation, derived in Section 5.6, can be used to estimate the *pole radius*.

Pole radius

$$r \approx 1 - \frac{\Delta F \pi}{f_s} \tag{7.2.6}$$

The *gain* factor b_0 in the numerator of $H_{res}(z)$ is inserted to ensure that the passband gain is one. The value of z corresponding to the nominal center of the passband is $z_0 = \exp(j2\pi F_0 T)$.

Gain   Setting $|H(z_0)| = 1$ in (7.2.5) and solving for b_0 yields the following expression for the *gain factor*.

$$b_0 = \frac{|\exp(j2\theta_0) - 2r\cos(\theta_0)\exp(j\theta_0) + r^2|}{|\exp(j2\theta_0) - 1|} \tag{7.2.7}$$

The resonator transfer function, in terms of negative powers of z, is then

$$H_{\text{res}}(z) = \frac{b_0(1 - z^{-2})}{1 - 2r\cos(\theta_0)z^{-1} + r^2 z^{-2}} \tag{7.2.8}$$

Example 7.1   **Resonator Filter**

Suppose the sampling frequency is $f_s = 1200$ Hz, and it is desired to design a resonator to meet the following specifications.

$$F_0 = 200 \text{ Hz}$$
$$\Delta F = 6 \text{ Hz}$$

From (7.2.3), the pole angle is

$$\theta_0 = \frac{2\pi(200)}{1200}$$
$$= \frac{\pi}{3}$$

Clearly, $\Delta F \ll f_s$ in this case. Thus, from (7.2.6), the pole radius is

$$r = 1 - \frac{6\pi}{1200}$$
$$= .9843$$

Next, from (7.2.7), the scalar multiplier b_0 required for a passband gain of one is

$$b_0 = \frac{|\exp(j2\pi/3) - 2(.9843)\cos(\pi/3)\exp(j\pi/3) + (.9843)^2|}{|\exp(j2\pi/3) - 1|}$$
$$= .0156$$

Finally, from (7.2.8), the transfer function of the resonator is

$$H_{\text{res}}(z) = \frac{.0156(1 - z^{-2})}{1 - 2(.9843)\cos(\pi/3)z^{-1} + (.9843)^2 z^{-2}}$$
$$= \frac{.0156(1 - z^{-2})}{1 - .9843z^{-1} + .9688z^{-2}}$$

A plot of the poles and zeros of the resonator, obtained by running *exam7_1*, is shown in Figure 7.6. The complex conjugate pair of poles just inside the unit circle causes the magnitude response of the resonator to peak near $f = F_0$. This is evident from the plot of the resonator magnitude response, as shown in Figure 7.7.

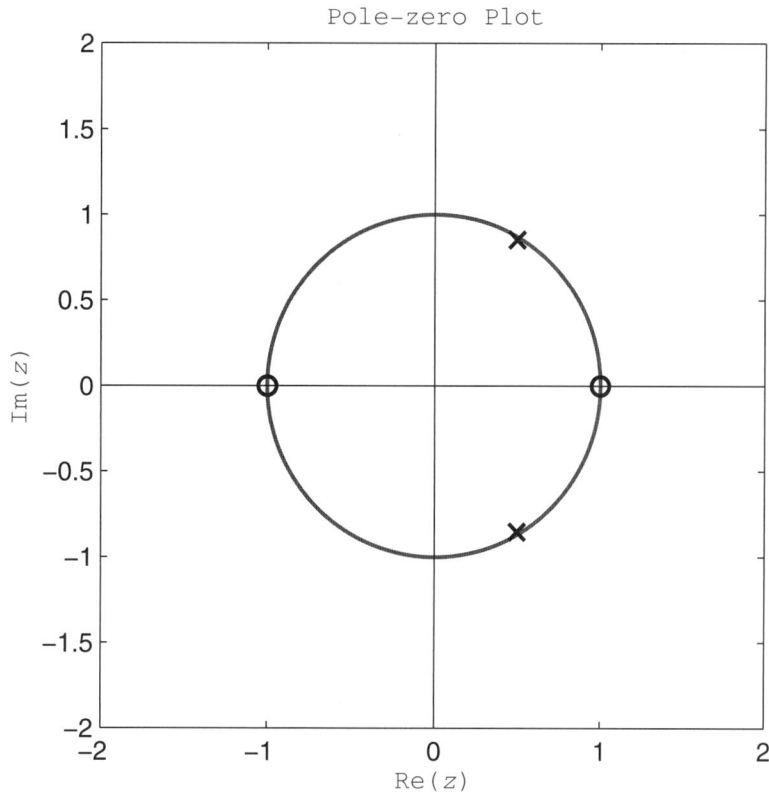

FIGURE 7.6: Poles and Zeros of a Resonator

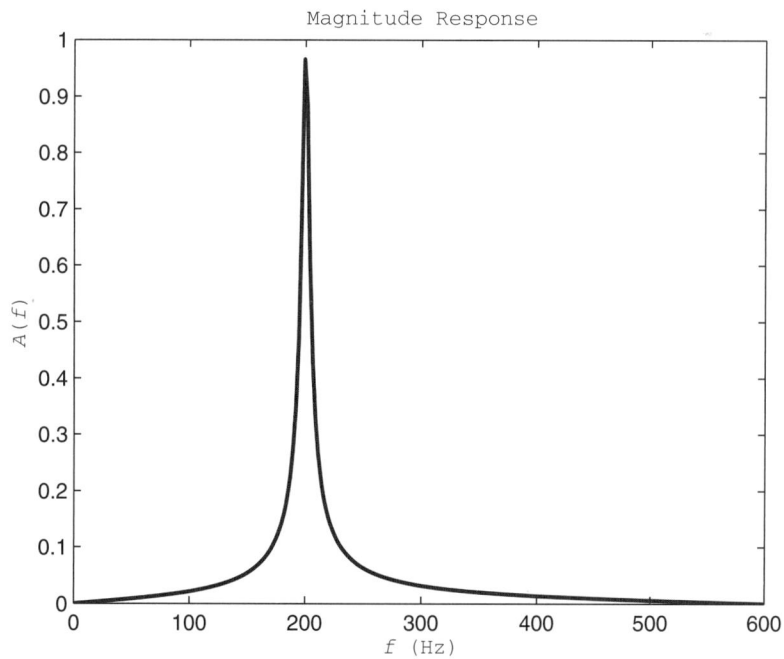

FIGURE 7.7: Magnitude Response of a Resonator with $F_0 = 200$ Hz

7.2.2 Notch Filter

Another specialized filter that occurs in applications is the *notch filter*. A notch filter can be thought of as a limiting special case of a bandstop filter where the width of the stopband goes to zero. That is, a notch filter is a filter designed to remove a single frequency F_0 called the *notch frequency*. The frequency response of an ideal notch filter is

Notch frequency

$$H_{\text{notch}}(f) = 1 - \delta_1(f - F_0), \quad 0 \le f \le f_s/2 \tag{7.2.9}$$

To design a notch filter we start by placing a zero at the point on the unit circle that corresponds to the notch frequency F_0. The angle θ_0, associated with frequency F_0, was given previously in (7.2.3). Placing a zero at $z_0 = \exp(j2\pi F_0 T)$ ensures that $H_{\text{notch}}(F_0) = 0$ as desired. However, it does not leave us with a design parameter to control the 3 dB bandwidth of the stopband. This is achieved by placing a pole at the same angle θ_0, but just inside the unit circle with $r < 1$. Since the poles and zeros must occur in complex-conjugate pairs (for real coefficients) this yields a transfer function for a notch filter with the following factored form.

$$H_{\text{notch}}(z) = \frac{b_0[z - \exp(j\theta_0)][z - \exp(-j\theta_0)]}{[z - r\exp(j\theta_0)][z - r\exp(-j\theta_0)]} \tag{7.2.10}$$

Using Euler's identity, we find that the transfer function in (7.2.10) again can be simplified and expressed as a ratio of two polynomials with the final result being

$$H_{\text{notch}}(z) = \frac{b_0(z^2 - 2\cos(\theta_0)z + 1)}{z^2 - 2r\cos(\theta_0)z + r^2} \tag{7.2.11}$$

There are two design parameters to be specified, the pole radius r, and the gain factor b_0. Just as with the resonator, to achieve a sharp filter that is highly selective requires $r \approx 1$, but for stability and to avoid canceling the zero, it essential that $r < 1$. If ΔF denotes the radius of the 3 dB stopband of the filter, and if $\Delta F \le f_s$, then the approximation for r in (7.2.6) can be used.

Gain

The *gain* factor b_0 in the numerator of $H_{\text{notch}}(z)$ is inserted to ensure that the passband gain is one. The passband includes both $f = 0$ and $f = f_s/2$. To set the DC gain to one, set $|H_{\text{notch}}(1)| = 1$ in (7.2.11) and solve for b_0, which yields

$$b_0 = \frac{|1 - 2r\cos(\theta_0) + r^2|}{2|1 - \cos(\theta_0)|} \tag{7.2.12}$$

Alternatively, the high-frequency gain can be set to one using $|H_{\text{notch}}(-1)| = 1$. The notch filter transfer function, in terms of negative powers of z, is then

$$H_{\text{notch}}(z) = \frac{b_0[1 - 2\cos(\theta_0)z^{-1} + z^{-2}]}{1 - 2r\cos(\theta_0)z^{-1} + r^2 z^{-2}} \tag{7.2.13}$$

Example 7.2   **Notch Filter**

Suppose the sampling frequency is $f_s = 2400$ Hz, and it is desired to design a notch filter to meet the following specifications.

$$F_0 = 800 \text{ Hz}$$
$$\Delta F = 18 \text{ Hz}$$

From (7.2.3), the angle of the zero is

$$\theta_0 = \frac{2\pi(800)}{2400}$$

$$= \frac{2\pi}{3}$$

In this case $\Delta F \ll f_s$. Thus from (7.2.6), the pole radius is

$$r = 1 - \frac{18\pi}{2400}$$
$$= .9764$$

Next, from (7.2.12), the scalar multiplier b_0 required for a DC passband gain of one is

$$b_0 = \frac{|1 - 2(.9764)\cos(2\pi/3) + (.9764)^2|}{2|1 - \cos(2\pi/3)|}$$
$$= .9766$$

Finally, from (7.2.13), the transfer function of the notch filter is

$$H_{\text{notch}}(z) = \frac{.9766[1 - 2\cos(2\pi/3)z^{-1} + z^{-2}]}{1 - 2(.9764)\cos(2\pi/3)z^{-1} + (.9764)^2 z^{-2}}$$

$$= \frac{.9766(1 + z^{-1} + z^{-2})}{1 + .9764z^{-1} + .9534z^{-2}}$$

A plot of the poles and zeros of the notch filter, obtained by running *exam7_2*, is shown in Figure 7.8. Note how the poles "almost" cancel the zeros. The complex conjugate pair of zeros

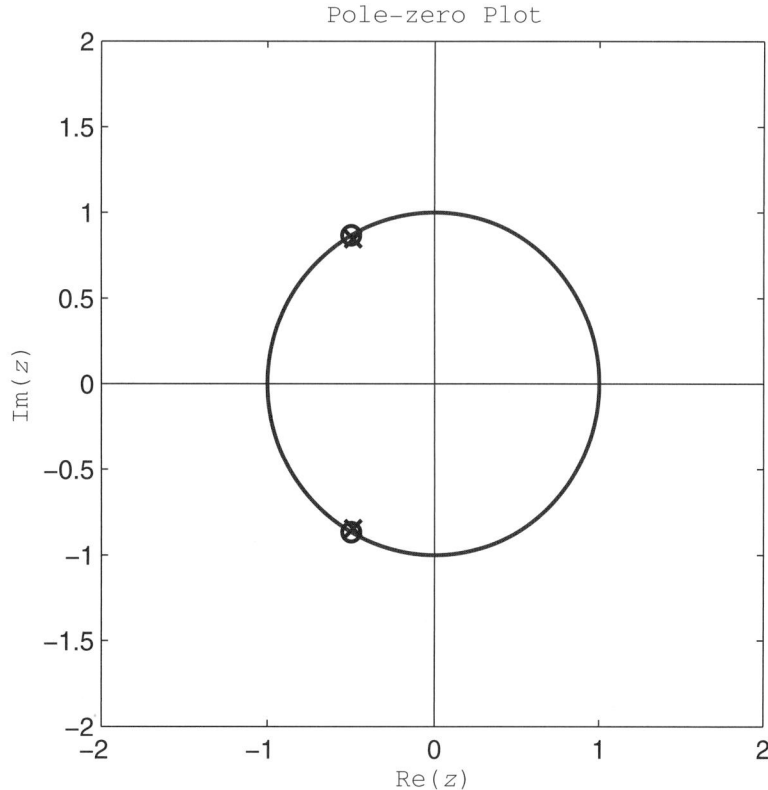

FIGURE 7.8: Poles and Zeros of a Notch Filter

FIGURE 7.9: Magnitude Response of a Notch Filter with $F_0 = 800$ Hz

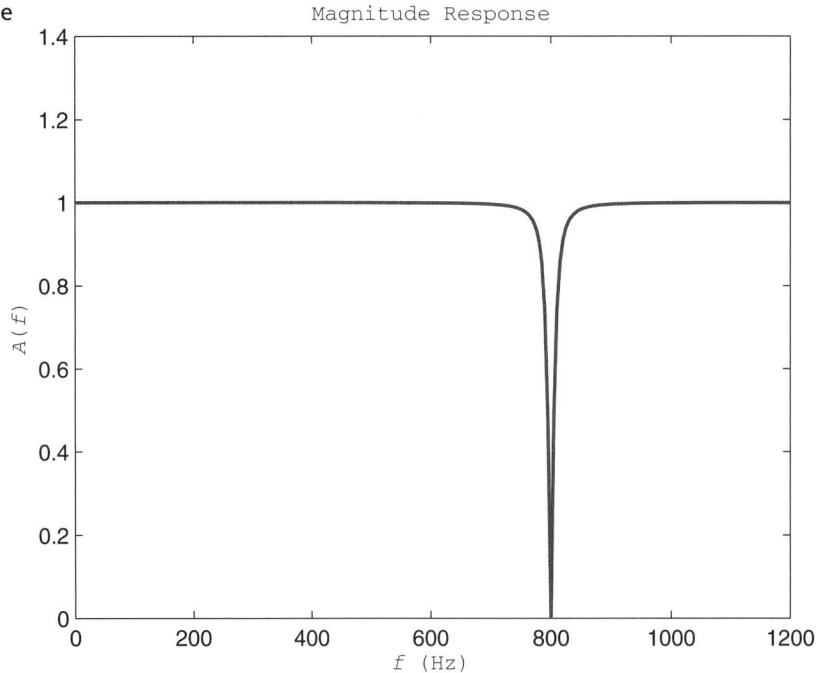

Magnitude Response

on the unit circle cause the magnitude response of the notch filter to go to zero at $f = F_0$. This is apparent from the plot of the notch filter magnitude response shown in Figure 7.9.

7.2.3 Comb Filters

Another specialized filter, one that includes a resonator as a special case, is the *comb filter*. A comb filter is a narrowband filter that has several equally spaced passbands starting at $f = 0$. In the limit as the widths of the passbands go to zero, the comb filter is a filter that passes DC, a fundamental frequency F_0, and several of its harmonics. Thus an ideal comb filter of order n has the following frequency response where $F_0 = f_s/n$.

$$H_{\text{comb}}(f) - \sum_{i=0}^{\text{floor}(n/2)} \delta_1(f - i F_0), \quad 0 \leq f \leq f_s/2 \tag{7.2.14}$$

Note that if n is even, then floor$(n/2) = n/2$. Consequently, for a comb filter of even order there are $n/2 + 1$ resonant frequencies in the range $[0, f_s/2]$, and for a comb filter of odd order, there are only $(n-1)/2 + 1$ resonant frequencies. Odd order comb filters do not have a resonant frequency at $f = f_s/2$. Since the resonant frequencies are equally spaced, a comb filter of order n has a very simple transfer function, namely,

$$H_{\text{comb}}(z) = \frac{b_0}{1 - r^n z^{-n}} \tag{7.2.15}$$

Thus $H_{\text{comb}}(z)$ has n zeros at the origin, and the poles of $H_{\text{comb}}(z)$ correspond to the n roots of r^n. That is, the poles are equally spaced around a circle of radius $r < 1$. The distribution of poles for the even case $n = 10$ and $r = .9843$ is shown in Figure 7.10.

FIGURE 7.10: Poles
and Zeros of a
Comb Filter of
Order $n = 10$

Pole-zero Plot

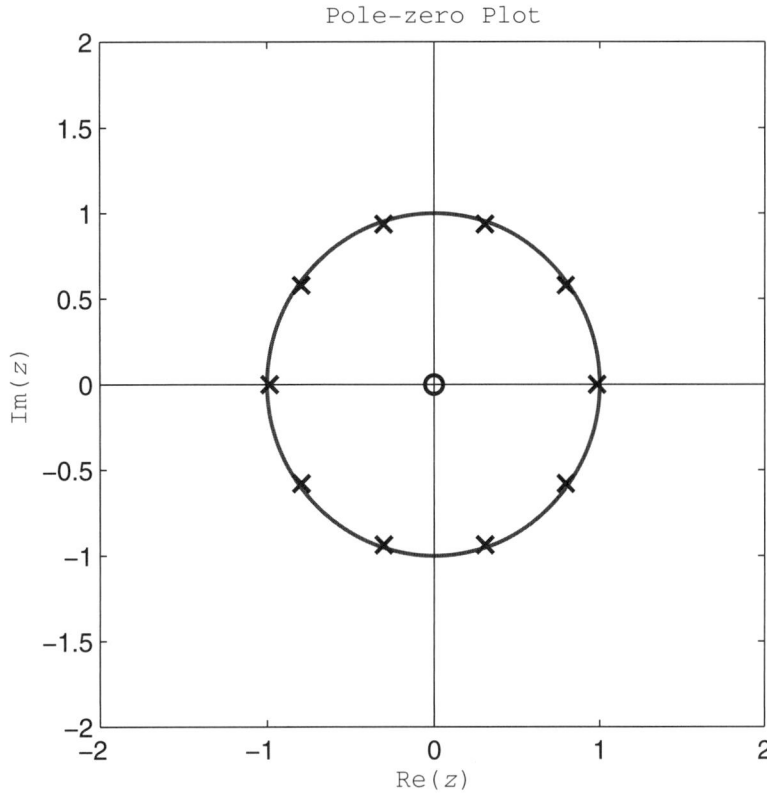

To achieve a highly selective comb filter, one needs $r \approx 1$, but for stability it is necessary that $r < 1$. The gain constant can be selected, such that the passband gain at DC is one. Setting $|H_{\text{comb}}(1)| = 1$ in (7.2.15) and solving for b_0 yields

Gain

$$b_0 = 1 - r^n \tag{7.2.16}$$

A plot of the magnitude response of the comb filter in Figure 7.10, corresponding to the case $n = 10$, is shown in Figure 7.11. Here the sampling frequency was selected to be $f_s = 200$ Hz, and the 3 dB radius was $\Delta F = 1$ Hz.

Just as the comb filter is a generalization of the resonator, there is a generalization of the notch filter called an *inverse comb filter* that eliminates DC, the fundamental notch frequency F_0, and several of its harmonics. An ideal inverse comb filter of order n has the following frequency response where $F_0 = f_s/n$.

$$H_{\text{inv}}(f) = 1 - \sum_{i=0}^{\text{floor}(n/2)} \delta_1(f - iF_0), \quad 0 \le f \le f_s/2 \tag{7.2.17}$$

In addition to having zeros equally spaced around the unit circle, the inverse comb filter also has equally spaced poles just inside the unit circle. Thus the transfer function of an inverse

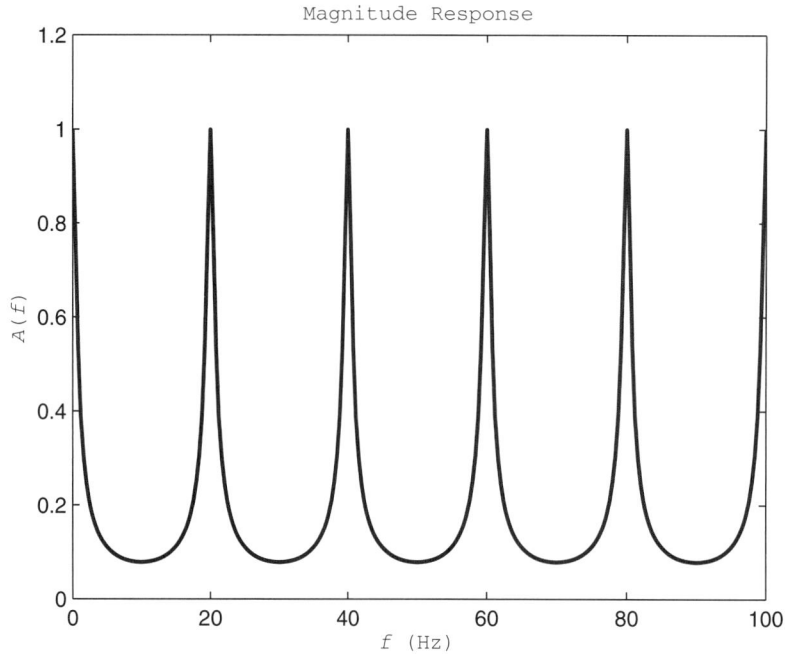

comb filter of order n has the following form.

$$H_{\text{inv}}(z) = \frac{b_0(1 - z^{-n})}{1 - r^n z^{-n}} \qquad (7.2.18)$$

The distribution of poles and zeros for the odd-order case, $n = 11$ and $r = .9857$, is shown in Figure 7.12. Note how there are no poles or zeros at $z = -1$ because n is odd.

Gain   To pass the frequencies between the harmonics of F_0, one needs $r \approx 1$, but for stability and to avoid pole-zero cancellation it is necessary that $r < 1$. The gain constant can be selected such that the passband gain at $f = F_0/2$ is one where $F_0 = f_s/n$. The point on the unit circle corresponding to the middle of the first passband is $z_1 = \exp(j\pi/n)$. Setting $|H_{\text{inv}}(z_1)| = 1$ in (7.2.18) and solving for b_0 yields

$$b_0 = \frac{1 + r^n}{2} \qquad (7.2.19)$$

A plot of the magnitude response of the inverse comb filter in Figure 7.12, corresponding to the case $n = 11$, is shown in Figure 7.13. Here the sampling frequency was selected to be $f_s = 2200$, and the 3 dB radius was $\Delta F = 10$ Hz.

There are a number of applications of comb filters and inverse comb filters. For example, suppose the input signal is a noise-corrupted periodic signal with a known fundamental frequency of F_0. Let $H_{\text{comb}}(z)$ be a comb filter of order n, and suppose the sampling frequency satisfies

$$f_s = nF_0 \qquad (7.2.20)$$

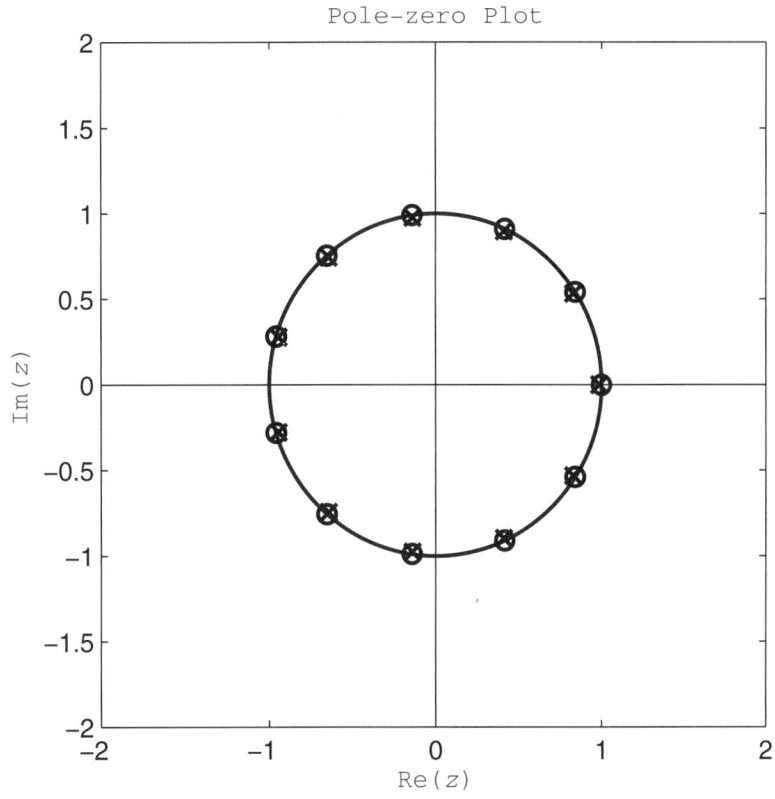

FIGURE 7.12: Poles and Zeros of an Inverse Comb Filter of Order $n = 11$

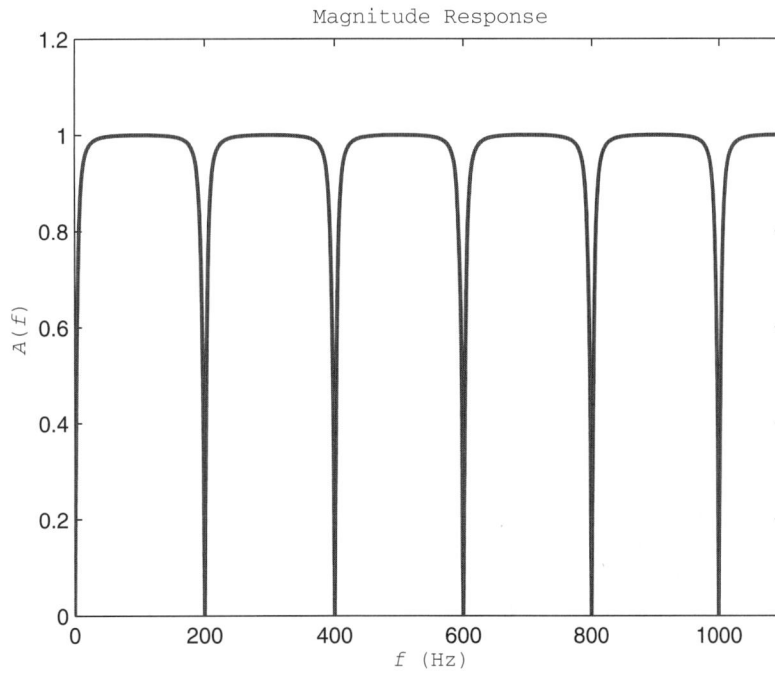

FIGURE 7.13: Magnitude Response of an Inverse Comb Filter with $n = 11$, $f_s = 2200$ Hz, and $\triangle F = 10$ Hz

The first resonant frequency of the comb filter is then F_0. Consequently, the comb filter can be used to extract the first $n/2$ harmonics of the periodic component of the input signal in this case.

Another application arises when an input signal includes a periodic noise component that needs to be removed. For example, a sensitive acoustic or a biomedical measurement may be corrupted by the 60 Hz "hum" of overhead fluorescent lights. In this case an inverse comb filter can be used to remove this periodic noise.

FDSP Functions

The FDSP toolbox contains the following functions for designing IIR filters by gain matching and pole-zero placement.

```
% F_IIRRES:   Design an IIR resonator filter
% F_IIRNOTCH: Design an IIR notch filter
% F_IIRCOMB:  Design an IIR comb filter
% F_IIRINV:   Design an IIR inverse comb filter
%
% Usage:
%       [b,a] = f_iirres   (F0,DeltaF,fs)
%       [b,a] = f_iirnotch (F0,DeltaF,fs)
%       [b,a] = f_iircomb  (n,DeltaF,fs)
%       [b,a] = f_iirinv   (n,DeltaF,fs)
% Pre:
%       F0     = resonant or notch frequency
%       DeltaF = 3dB radius
%       fs     = sampling frequency
%       n      = filter order
% Post:
%       b = 1 by (n+1) numerator coefficient vector
%       a = 1 by (n+1) denominator coefficient vector
```

● ● ● ● ● ● ● ● ● ● ● ● ● ● ● ⋯

7.3 Filter Design Parameters

Prototype filter

The most widely used design procedure for digital IIR filters starts with a normalized lowpass analog filter called a *prototype* filter. One then transforms the prototype filter into a desired frequency-selective digital filter. There are four classical families of analog lowpass filters that are typically used as prototype filters, and each family is optimal in some sense. Before examining them, it is helpful to introduce two filter design parameters that are derived from the filter design specifications. For convenience, the lowpass filter design specifications first introduced in Chapter 5 are displayed in Figure 7.14. Recall that F_p is the passband cutoff frequency, F_s is the stopband cutoff frequency, δ_p is the passband ripple, and δ_s is the stopband attenuation. Left unspecified is the magnitude response in the transition band whose width is $B = F_s - F_p$. Let $A_a(f)$ denote the desired analog magnitude response. Then the passband and stopband specifications for an analog filter can be represented by separate inequalities as

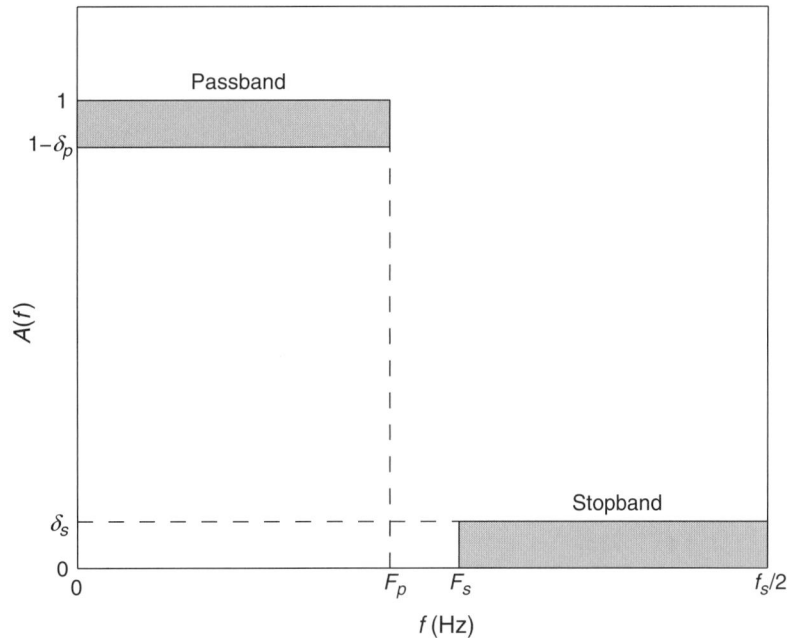

FIGURE 7.14: Design Specifications of an IIR Lowpass Filter

follows.

$$1 - \delta_p \leq A_a(f) \leq 1 , \quad 0 \leq |f| \leq F_p \tag{7.3.1a}$$

$$0 \leq A_a(f) \leq \delta_s , \quad F_s \leq |f| < \infty \tag{7.3.1b}$$

The design specifications in (7.3.1) are linear design specifications. In order to better reveal the amount of attenuation in the stopband, the magnitude response is sometimes plotted in units of dB using the logarithmic scale, $A(f) = 20 \log_{10}\{|H(f)|\}$. The logarithmic equivalents of the passband ripple and stopband attenuation are

$$A_p = -20 \log_{10}(1 - \delta_p) \tag{7.3.2a}$$

$$A_s = -20 \log_{10}(\delta_s) \tag{7.3.2b}$$

For example, a stop attenuation of $\delta_s = .01$ corresponds to $A_s = 40$ dB, and each reduction by a factor of ten generates an increase of 20 dB. The design procedures for the classical analog filters are based on linear specifications. However, if the user instead starts out with logarithmic specifications, they can be converted to equivalent linear specifications as follows.

$$\delta_p = 1 - 10^{-A_p/20} \tag{7.3.3a}$$

$$\delta_s = 10^{-A_s/20} \tag{7.3.3b}$$

The development of design formulas for the classical analog filters can be streamlined by introducing the following two filter design parameters that are obtained from the filter design

TABLE 7.1: ▶
Filter Design
Parameters

Filter Type	Selectivity Factor	Discrimination Factor
Ideal	$r = 1$	$d = 0$
Practical	$r < 1$	$d > 0$

specifications (Porat, 1997).

$$r \triangleq \frac{F_p}{F_s} \tag{7.3.4a}$$

$$d \triangleq \left[\frac{(1 - \delta_p)^{-2} - 1}{\delta_s^{-2} - 1} \right]^{1/2} \tag{7.3.4b}$$

Selectivity factor

The first parameter $0 < r < 1$ is called the *selectivity factor*. Note that for an ideal filter there is no transition band, so $F_s = F_p$. Consequently, for an ideal filter, the selectivity factor is $r = 1$, whereas for a practical filter, $r < 1$.

Discrimination factor

The second parameter $d > 0$ is called the *discrimination factor*. Observe that when $\delta_p = 0$, the numerator in (7.4.4b) goes to zero, so $d = 0$. Similarly when $\delta_s = 0$, the denominator in (7.4.4b) goes to infinity, so again $d = 0$. Hence for an ideal filter the discrimination factor is $d = 0$, whereas as for a practical filter $d > 0$. See Table 7.1 for a summary of the filter design parameter characteristics.

Example 7.3   **Filter Design Parameters**

As a simple illustration of filter design parameters, consider the problem of designing a lowpass analog filter to meet the following logarithmic design specifications.

$$(F_p, F_s) = (400, 500) \text{ Hz}$$
$$(A_p, A_s) = (.5, 35) \text{ dB}$$

First, convert from logarithmic to linear specifications. From (7.3.3a), the required passband ripple is

$$\delta_p = 1 - 10^{-.5/20}$$
$$= .0559$$

Similarly, from (7.3.3b), the required stopband attenuation is

$$\delta_s = 10^{-35/20}$$
$$= .0178$$

For this filter, the width of the transition band is $B = 100$ Hz. Thus the selectivity factor r is less than one, and from (7.4.4a), we have

$$r = .8$$

Finally, both the passband ripple and the stopband attenuation are positive. Thus the discrimination factor d will also be positive. From (7.4.4b)

$$d = \left[\frac{(1 - .0559)^{-2} - 1}{(.0178)^{-2} - 1} \right]^{1/2}$$
$$= .006213$$

Classical analog filters can be designed by starting with a desired magnitude response and working backwards to determine the poles, zeros, and gain. To apply this reverse procedure, it is necessary to first develop a relationship between an analog transfer function $H_a(s)$ and the square of its magnitude response. Recall that the frequency response of an analog filter is defined as follows.

$$H_a(f) = H_a(s)|_{s=j2\pi f} \tag{7.3.5}$$

The frequency response $H_a(f)$ can be expressed in polar form as $H_a(f) = A_a(f)\exp[j\phi_a(f)]$, where $A_a(f)$ is the magnitude response, and $\phi_a(f)$ is the phase response. Since the coefficients of $H_a(s)$ are real, the square of the magnitude response can be expressed as follows, where $H_a^*(s)$ denotes the complex conjugate of $H_a(s)$.

$$
\begin{aligned}
A_a^2(f) &= |H_a(f)|^2 \\
&= |H_a(s)|_{s=j2\pi f}^2 \\
&= \{H_a(s)H_a^*(s)\}|_{s=j2\pi f} \\
&= \{H_a(s)H_a(-s)\}|_{s=j2\pi f} \tag{7.3.6}
\end{aligned}
$$

Instead of replacing s by $j2\pi f$ on the right-hand side of (7.3.6), one can replace f by $s/(j2\pi)$ on the left-hand side. This yields the following fundamental relationship between the transfer function and its *squared magnitude response*.

Squared magnitude response

$$H_a(s)H_a(-s) = A_a^2\left(\frac{s}{j2\pi}\right) \tag{7.3.7}$$

Each of the classical analog filters can be characterized by its squared magnitude response. The relationship in (7.3.7) then can be employed to synthesize the filter transfer function.

● ● ● ● ● ● ● ● ● ● ● ● ● ● ● ●

7.4   Classical Analog Filters

The most popular design procedure for digital IIR filters is to start with a normalized lowpass analog filter and then transform it into an equivalent frequency-selective digital filter.

7.4.1   Butterworth Filters

The first family of analog filters that we consider are the Butterworth filters. A Butterworth filter of order n is a lowpass analog filter with the following squared magnitude response.

$$A_a^2(f) = \frac{1}{1 + (f/F_c)^{2n}} \tag{7.4.1}$$

3 dB cutoff

Notice from (7.4.1) that $A_a^2(F_c) = .5$. Frequency F_c is called the *3 dB cutoff* frequency because

$$20\log_{10}\{A_a(F_c)\} \approx -3 \text{ dB} \tag{7.4.2}$$

FIGURE 7.15:
Squared Magnitude
Response of a
Lowpass
Butterworth Filter
of Order $n = 4$ with
$F_c = 1$ Hz

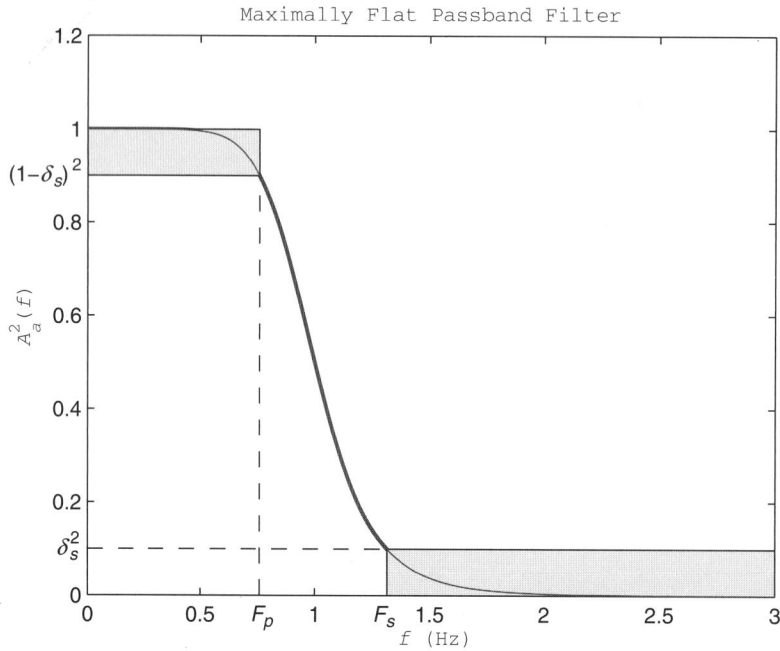

Maximally Flat Passband Filter

A plot of the squared magnitude response for a Butterworth filter of order $n = 4$ with a 3 dB cutoff frequency of $F_c = 1$ Hz is shown in Figure 7.15.

The poles of $H_a(s)$ can be recovered from the squared magnitude response. From (7.4.1) and the relationship in (7.3.7),

$$H_a(s) H_a(-s) = A_a^2 \left(\frac{s}{j2\pi} \right)$$

$$= \frac{1}{1 + [s/(j2\pi F_c)]^{2n}}$$

$$= \frac{(j2\pi F_c)^{2n}}{s^{2n} + (j2\pi F_c)^{2n}}$$

$$= \frac{(-1)^n (2\pi F_c)^{2n}}{s^{2n} + (-1)^n (2\pi F_c)^{2n}} \tag{7.4.3}$$

Thus the poles p_k of $H_a(s) H_a(-s)$ lie on a circle of radius $2\pi F_c$ at angles θ_k, where

$$\theta_k = \frac{(2k + 1 + n)\pi}{2n}, \quad 0 \le k < 2n \tag{7.4.4a}$$

$$p_k = 2\pi F_c \exp(j\theta_k), \quad 0 \le k < 2n \tag{7.4.4b}$$

Normalized filter   A *normalized* lowpass filter is a lowpass filter whose cutoff frequency is $F_c = 1/(2\pi)$ Hz, which corresponds to a radian cutoff frequency of $\Omega_c = 1$ rad/sec. For a normalized lowpass Butterworth filter, the poles are equally spaced around the unit circle with a separation of π/n radians. Two cases are illustrated in Figure 7.16, corresponding to an odd order ($n = 5$) and an even order ($n = 6$). Note that in either case the first n poles all lie in the left half of the complex plane. One associates the left-half plane poles $\{p_0, p_1, \ldots, p_{n-1}\}$ with $H_a(s)$

FIGURE 7.16: Poles of Normalized Lowpass Butterworth Filters

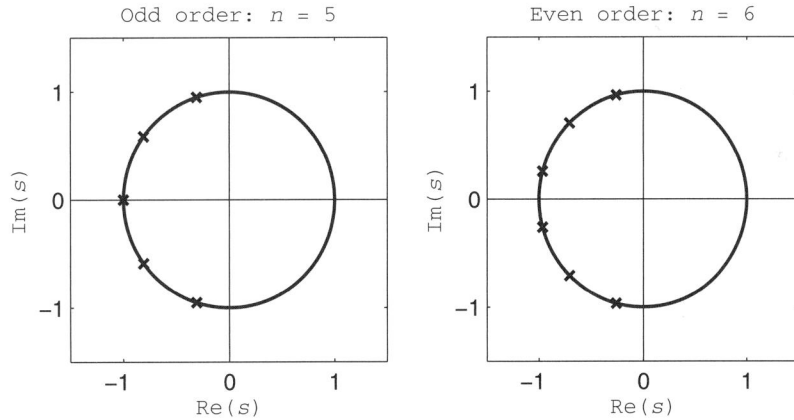

in (7.4.3), and the right-half plane poles $\{p_n, p_{n+1}, \ldots, p_{2n-1}\}$ with $H_a(-s)$. This way, filter $H_a(s)$ is guaranteed to be stable. The transfer function of an nth-order lowpass Butterworth filter with cutoff frequency F_c is then

$$H_a(s) = \frac{(2\pi F_c)^n}{(s - p_0)(s - p_1) \cdots (s - p_{n-1})} \tag{7.4.5}$$

Butterworth filters have a number of useful qualitative properties. For example, notice that the magnitude response decreases *monotonically* starting from $A_a(0) = 1$. For high frequencies, the asymptotic attenuation of an nth-order filter is $20n$ dB per decade. That is

$$20 \log_{10}\{A_a(10f)\} \approx 20 \log_{10}\{A_a(f)\} - 20n \text{ dB}, \quad f \gg F_c \tag{7.4.6}$$

Perhaps the most noteworthy property of Butterworth filters is that the first $2n - 1$ derivatives of the squared magnitude response are zero at $f = 0$. Consequently, among filters of order n, the Butterworth filter magnitude response is as flat as possible at $f = 0$. For this reason, *Maximally flat* Butterworth filters are called *maximally flat* filters.

The two design parameters available with Butterworth filters are the filter order n and the cutoff frequency F_c. Suppose it is desired to develop a lowpass Butterworth filter satisfying the linear design specification in Figure 7.15.

Then from (7.4.1), the passband and stopband specification constraints are

$$\frac{1}{1 + (F_p/F_c)^{2n}} = (1 - \delta_p)^2 \tag{7.4.7a}$$

$$\frac{1}{1 + (F_s/F_c)^{2n}} = \delta_s^2 \tag{7.4.7b}$$

The passband constraint in (7.4.7a) and the stopband constraint in (7.4.7b) can each be solved for F_c^{2n}. By setting these two expressions for F_c^{2n} equal, one can eliminate the cutoff frequency parameter F_c. Solving the resulting equation for the filter order n, and using the design parameters in (7.3.4), then yields

$$n = \text{ceil}\left[\frac{\ln(d)}{\ln(r)}\right] \tag{7.4.8}$$

Thus the required filter order can be expressed directly in terms of the discrimination factor d and the selectivity factor r defined in (7.3.4). The function *ceil* in (7.4.8) is used because the expression in the square brackets may not be an integer. The *ceil* function rounds up to the next integer value. Since n is rounded up, typically this means that the passband and the stopband specifications will be exceeded (more than met). To meet the passband constraint exactly, one can solve (7.4.7a) for F_c, which yields

$$F_{cp} = \frac{F_p}{[(1 - \delta_p)^{-2} - 1]^{1/(2n)}} \tag{7.4.9}$$

In this case, the stopband constraint is exceeded. Similarly, to meet the stopband constraint exactly, one solves (7.4.7b) for F_c, which yields the slightly simpler expression

$$F_{cs} = \frac{F_s}{(\delta_s^{-2} - 1)^{1/(2n)}} \tag{7.4.10}$$

In this case the passband constraint is exceeded. Finally, both constraints can be exceeded (assuming the expression for n is not already an integer) when the cutoff frequency is set to the average.

$$F_c = \frac{F_{cp} + F_{cs}}{2} \tag{7.4.11}$$

The design formulas in (7.4.8) through (7.4.11) are all based on the linear design specifications in Figure 7.15. If the logarithmic design specifications are used instead, then (7.3.3) should be applied first to convert A_p and A_s into δ_p and δ_s, respectively.

Example 7.4   **Butterworth Filter**

As an illustration of the use of the design formulas, consider the problem of designing a lowpass Butterworth filter to meet the following linear design specifications.

$$F_p = 1000 \text{ Hz}$$
$$F_s = 2000 \text{ Hz}$$
$$\delta_p = .05$$
$$\delta_s = .05$$

From (7.3.4), the selectivity and discrimination factors are

$$r = .5$$
$$d = \left(\frac{.95^{-2} - 1}{.05^{-2} - 1}\right)^{1/2}$$
$$= .0165$$

Thus, from (7.4.8), the minimum filter order is

$$n = \text{ceil}\left[\frac{\ln(.0165)}{\ln(.5)}\right]$$
$$= \text{ceil}(5.9253)$$
$$= 6$$

Next, from (7.4.9), the cutoff frequency for which the passband specification is met exactly is

$$F_{cp} = \frac{1000}{(.95^{-2} - 1)^{1/12}}$$

$$= 1203.8 \text{ Hz}$$

Similarly, from (7.4.10), the cutoff frequency for which the stopband specification is met exactly is

$$F_{cs} = \frac{2000}{(.05^{-2} - 1)^{1/12}}$$

$$= 1209.0 \text{ Hz}$$

Any cutoff frequency in the range $F_{cp} \leq F_c \leq F_{cs}$ will meet or exceed both specifications. For example, $F_c = 1206$ Hz will suffice. From (7.3.2), the equivalent logarithmic passband and stopband specifications are $A_p = .4455$ dB, and $A_s = 26.02$ dB.

Butterworth filter transfer functions can be designed directly using (7.4.4) and (7.4.5). There is also an alternative table-based approach that works well for low-order filters. It starts with a normalized lowpass Butterworth filter and then makes use of a simple *frequency transformation*. Let $H_n(s)$ denote the transfer function of a normalized nth-order Butterworth lowpass filter, a filter with a 3 dB radian cutoff frequency of $\Omega_c = 1$ rad/sec.

$$H_n(s) = \frac{a_n}{s^n + a_1 s^{n-1} + \cdots + a_n} \tag{7.4.12}$$

The coefficients of the denominator polynomials for the first few normalized Butterworth lowpass filters are summarized in Table 7.2.

Next, let F_c denote the desired 3 dB cutoff frequency in Hz. The transfer function $H_a(s)$ can be obtained by replacing s in (7.4.12) with s/Ω_c where $\Omega_c = 2\pi F_c$. Thus if $a(s)$ is as

Lowpass Butterworth

given in Table 7.2, then an nth-order *Butterworth lowpass* filter with radian cutoff frequency Ω_c has the following transfer function.

$$H_a(s) = \frac{\Omega_c^n a_n}{s^n + \Omega_c a_1 s^{n-1} + \cdots + \Omega_c^n a_n} \tag{7.4.13}$$

The replacement of s with s/Ω_c is an example of a frequency transformation that maps a normalized lowpass filter into a general lowpass filter. Later, other examples of frequency transformations are presented that convert normalized lowpass filters into highpass, bandpass, and bandstop filters.

TABLE 7.2: ▶
Denominators of
Normalized
Lowpass
Butterworth Filters
with $a_0 = 1$

n	a_1	a_2	a_3	a_4	a_5	a_6	a_7	a_8
1	1	0	0	0	0	0	0	0
2	1.414214	1	0	0	0	0	0	0
3	2	2	1	0	0	0	0	0
4	2.613126	3.414214	2.613126	1	0	0	0	0
5	3.236068	5.236068	5.236068	3.236068	1	0	0	0
6	3.863703	7.464102	9.14162	7.464102	3.863703	1	0	0
7	4.493959	10.09783	14.59179	14.59179	10.09783	4.493959	1	0
8	5.125831	13.13707	21.84615	25.68836	21.84615	13.13707	5.125831	1

| **Example 7.5** | **Butterworth Transfer Function** |

As an illustration of the frequency-transformation method using Table 7.2, consider the problem of designing a transfer function for a third-order lowpass Butterworth filter with a cutoff frequency of $F_c = 10$ Hz. In this case $\Omega_c = 20\pi$, and from Table 7.2

$$H_a(s) = \frac{2.481 \times 10^5}{s^3 + 125.7s^2 + 7896s + 1.481 \times 10^5}$$

7.4.2 Chebyshev-I Filters

The magnitude responses of Butterworth filters are smooth and flat because of the maximally flat property. However, a drawback of the maximally flat property is that the transition band of a Butterworth filter is not as narrow as it could be. An effective way to decrease the width of the transition band is to allow ripples or oscillations in the passband or the stopband. The following Chebyshev-I filter of order n is designed to allow n ripples within the passband.

$$A_a^2(f) = \frac{1}{1 + \epsilon^2 T_n^2(f/F_p)} \tag{7.4.14}$$

Here n is the filter order, F_p is the passband frequency, $\epsilon > 0$ is a *ripple factor* parameter, and $T_n(x)$ is a polynomial of degree n called a *Chebyshev polynomial* of the first kind. Recall from Section 5.5.3 that the Chebyshev polynomials can be generated recursively. The first two Chebyshev polynomials are $T_0(x) = 1$ and $T_1(x) = x$. The remaining polynomials are then computed from the previous two according to the recurrence relation

$$T_{k+1}(x) = 2x T_k(x) + T_{k-1}(x), \quad k \geq 1 \tag{7.4.15}$$

Therefore $T_2(x) = 2x^2 - 1$ and so on. The first few Chebyshev polynomials are summarized in Table 7.3.

The Chebyshev polynomials have many interesting properties. For example, $T_n(x)$ is an odd function when n is odd, and an even function when n is even. Furthermore, $T_n(1) = 1$ for all n. For the purpose of filter design, the most important property is that $T_n(x)$ oscillates in the interval $[-1, 1]$ when $|x| \leq 1$, and $T_n(x)$ is monotonic when $|x| > 1$. This oscillation causes the square of the magnitude response of a Chebyshev-I filter to have ripples of equal size in the passband and be monotonically decreasing outside of the passband. A plot of the squared magnitude response is shown in Figure 7.17 for the case $n = 4$ with $F_p = 1$ Hz. Note

TABLE 7.3: ▶
Chebyshev Polynomials of the First Kind

n	$T_n(x)$
0	1
1	x
2	$2x^2 - 1$
3	$4x^3 - 3x$
4	$8x^4 - 8x^2 + 1$
5	$16x^5 - 20x^3 + 5x$
6	$32x^6 - 48x^4 + 18x^2 - 1$
7	$64x^7 - 112x^5 + 56x^3 - 7x$
8	$128x^8 - 256x^6 + 160x^4 - 32x^2 + 1$

that because $T_n(1) = 1$, it follows from (7.4.14) that, at the edge of the passband,

$$A_a^2(F_p) = \frac{1}{1 + \epsilon^2} \quad (7.4.16)$$

Therefore the ripple factor parameter ϵ specifies the size of the passband ripple of the filter. By setting $1/(1 + \epsilon^2) = (1 - \delta_p)^2$ and solving for ϵ, it follows that a desired passband ripple, δ_p, can be achieved by setting the ripple factor parameter to

$$\epsilon = \left[(1 - \delta_p)^{-2} - 1\right]^{1/2} \quad (7.4.17)$$

Notice from Figure 7.17 that not only are the n ripples in $A_a^2(f)$ confined to the passband, but they are all of the same amplitude, δ_p. Because of this characteristic, Chebyshev filters are called *equiripple* filters. More specifically, Chebyshev-I filters are optimal in the sense that they are equiripple in the passband. At the start of the passband the squared magnitude response is either one or $1/(1 + \epsilon^2)$ depending on whether n is odd or even, respectively.

Equiripple filter

$$A_a^2(0) = \begin{cases} 1, & n \text{ odd} \\ \dfrac{1}{1 + \epsilon^2}, & n \text{ even} \end{cases} \quad (7.4.18)$$

Unlike the poles of a Butterworth filter that are on a circle, the poles of a Chebyshev-I filter are on an ellipse. The minor and major axes of the ellipse are computed as follows where $F_0 = F_p$.

$$\alpha = \epsilon^{-1} + \sqrt{\epsilon^{-2} + 1} \quad (7.4.19a)$$
$$r_1 = \pi F_0(\alpha^{1/n} - \alpha^{-1/n}) \quad (7.4.19b)$$
$$r_2 = \pi F_0(\alpha^{1/n} + \alpha^{-1/n}) \quad (7.4.19c)$$

FIGURE 7.17: Squared Magnitude Responses of a Chebyshev-I Lowpass Filter of Order $n = 4$ with $F_p = 1$ Hz

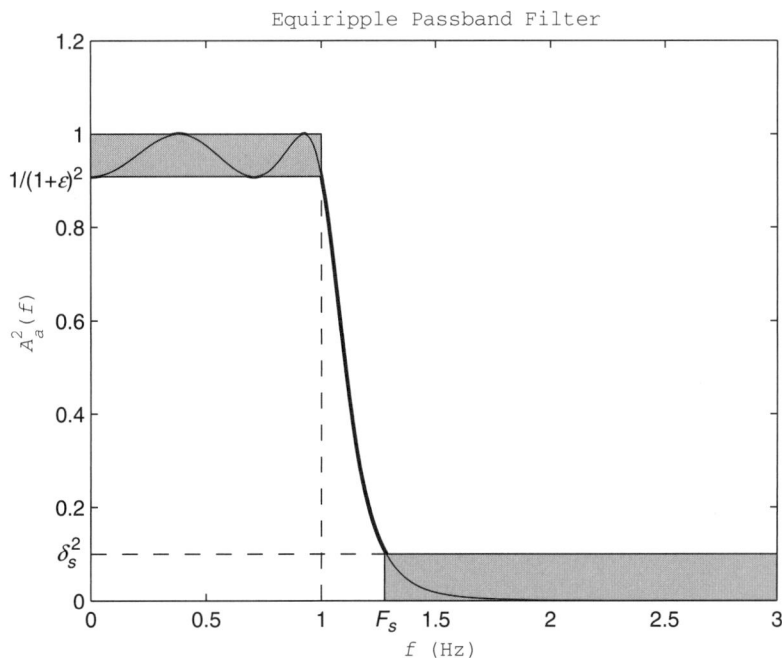

The angles at which the poles are located are the same as those for a Butterworth filter, namely

$$\theta_k = \frac{(2k + 1 + n)\pi}{2n}, \quad 0 \le k < n \tag{7.4.20}$$

If the poles are expressed in rectangular form as $p_k = \sigma_k + j\omega_k$, then the real and imaginary parts of the poles are

$$\sigma_k = r_1 \cos(\theta_k), \quad 0 \le k < n \tag{7.4.21a}$$

$$\omega_k = r_2 \sin(\theta_k), \quad 0 \le k < n \tag{7.4.21b}$$

The DC gain of the Chebyshev-I filter is $A_a(0)$, as given in (7.4.18). Let $\beta = (-1)^n p_0 p_1 \cdots p_{n-1}$. Then the transfer function of an nth-order Chebyshev-I filter is as follows.

$$H_a(s) = \frac{\beta A_a(0)}{(s - p_0)(s - p_1) \cdots (s - p_{n-1})} \tag{7.4.22}$$

The only design parameter that remains to be determined for a Chebyshev-I filter is the filter order n. The minimal filter order depends on the filter design specifications. Using the selectivity and discrimination factors in (7.3.4)

$$n = \text{ceil} \left[\frac{\ln(d^{-1} + \sqrt{d^{-2} - 1})}{\ln(r^{-1} + \sqrt{r^{-2} - 1})} \right] \tag{7.4.23}$$

Unlike a Butterworth filter, a Chebyshev-I filter always meets the passband specification exactly as long as the ripple factor ϵ is chosen as in (7.4.17). The stopband specification will be exceeded when the expression inside the square brackets in (7.4.23) is less than the integer filter order n.

Example 7.6   **Chebyshev-I Filter**

As an illustration of the use of the Chebyshev design formulas, consider the problem of designing a lowpass Chebyshev-I filter to meet the same design specifications as in Example 7.4. From Example 7.4, the selectivity and discrimination factors are

$$r = .5$$
$$d = .0165$$

Then from (7.4.23), the minimum filter order is

$$n = \text{ceil} \left[\frac{\log[(.0165)^{-1} + \sqrt{(.0165)^{-2} - 1}]}{\log[(.5)^{-1} + \sqrt{(.5)^{-2} - 1}]} \right]$$

$$= \text{ceil}(3.6449)$$

$$= 4$$

Note that by permitting ripples in the passband, the design specification can be met with a Chebyshev-I filter of order $n = 4$. This is in contrast to the maximally flat Butterworth filter response in Example 7.4 that required a filter of order $n = 6$ for the same specifications.

7.4.3 **Chebyshev-II Filters**

The use of the term Chebyshev-I filter suggests that there must also be a Chebyshev-II filter, and this is indeed the case. A Chebyshev-II filter is an equiripple filter that has the ripples in the stopband rather than the passband. This is achieved by having the following squared magnitude response.

$$A_a^2(f) = \frac{\epsilon^2 T_n^2(F_s/f)}{1 + \epsilon^2 T_n^2(F_s/f)} \qquad (7.4.24)$$

The design parameter n for the Chebyshev-II filter is the same as that for the Chebyshev-I filter. However, in this case the magnitude response oscillates in the stopband and is monotonically decreasing outside the stopband. A plot of the squared magnitude response is shown in Figure 7.18 for the case $n = 4$ with $F_s = 1$ Hz. Recalling that $T_n(1) = 1$, it follows from (7.4.24) that at the edge of the stopband

$$A_a^2(F_s) = \frac{\epsilon^2}{1 + \epsilon^2} \qquad (7.4.25)$$

In this case the ripple factor parameter ϵ specifies the size of the stopband attenuation of the filter. By setting $\epsilon^2/(1 + \epsilon^2) = \delta_s^2$ and solving for ϵ, it follows that a desired stopband attenuation δ_s can be achieved by setting the ripple factor parameter to

$$\epsilon = \delta_s(1 - \delta_s^2)^{-1/2} \qquad (7.4.26)$$

Again notice from Figure 7.18 that there are n ripples in $A_a^2(f)$ confined to the stopband, and they are all of the same amplitude, δ_s. Consequently, Chebyshev-II filters are optimal in

FIGURE 7.18:
Squared Magnitude Responses of a Chebyshev-II Lowpass Filter of Order $n = 4$ with $F_s = 1$ Hz

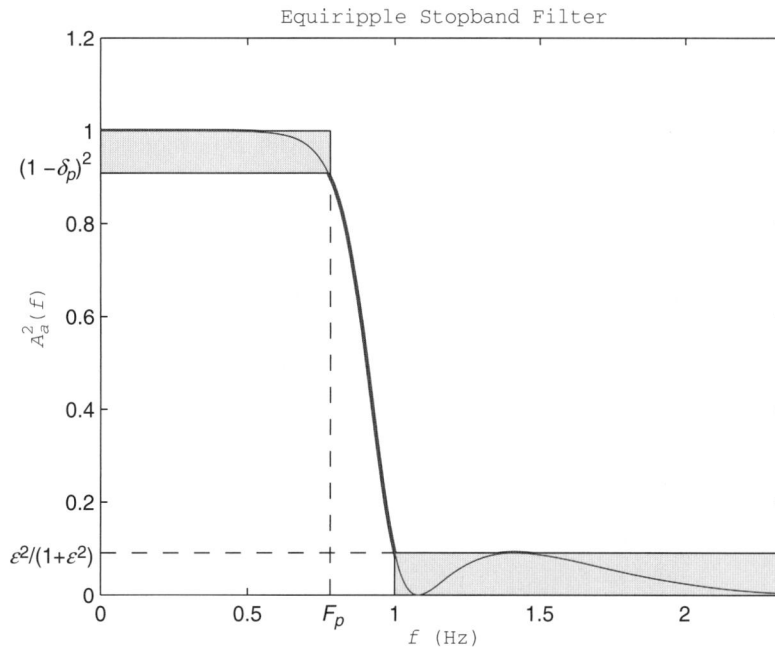

the sense that they are *equiripple* in the stopband. At the end of the stopband the squared magnitude response is either zero or $\epsilon^2/(1 + \epsilon^2)$, depending on whether n is odd or even, respectively. That is,

$$\lim_{f \to \infty} A_a^2(f) = \begin{cases} 0, & n \text{ odd} \\ \dfrac{\epsilon^2}{1 + \epsilon^2}, & n \text{ even} \end{cases} \tag{7.4.27}$$

Because f/F_p in the Chebyshev-I magnitude response has been replaced by F_s/f in the Chebyshev-II magnitude response, the poles of the Chebyshev-II filter are located at the reciprocals of the poles of the Chebyshev-I filter. That is, if $p_k = \sigma_k + j\omega_k$ are the poles defined in (7.4.21), but with $F_0 = F_s$, then the Chebyshev-II poles are

$$q_k = \frac{(2\pi F_s)^2}{p_k}, \quad 0 \le k < n \tag{7.4.28}$$

Note from (7.4.24) that the numerator of $A_a^2(f)$ is not constant. This means that a Chebyshev-II filter also has either n or $n - 1$ finite zeros. They are located along the imaginary axis at

$$r_k = \frac{j2\pi F_s}{\sin(\theta_k)}, \quad 0 \le k < n \tag{7.4.29}$$

When n is even, there are n finite zeros, as indicated in (7.4.29). However, when n is odd, there are only $n - 1$ finite zeros. Indeed, when n is odd, observe from (7.4.20) that $\theta_{(n-1)/2} = \pi$. Thus $r_{(n-1)/2}$ is an infinite zero in this case.

For every Chebyshev-II filter the DC gain is $A_a(0) = 1$. Let $\beta = q_0 q_1 \cdots q_{n-1}/(r_0 r_1 \cdots r_{n-1})$ where $r_{(n-1)/2}$ is left out if n is odd. Then the transfer function of an nth-order Chebyshev-II filter is

$$H_a(s) = \frac{\beta(s - r_0)(s - r_1) \cdots (s - r_{n-1})}{(s - q_0)(s - q_1) \cdots (s - q_{n-1})} \tag{7.4.30}$$

Again, when n is odd, the numerator factor $(s - r_{(n-1)/2})$ is left out of (7.4.30). The minimum order for a Chebyshev-II filter is the same as the minimum order for a Chebyshev-I filter and is given in (7.4.23). Thus the Chebyshev-II filter has a smaller transition band than the Butterworth filter, but like the Butterworth filter, it is monotonic in the passband. A Chebyshev-II filter will always meet the stopband specification exactly as long as the ripple factor ϵ is chosen as in (7.4.26). The passband specification will be exceeded when the expression inside the square brackets in (7.4.23) is less than the integer filter order n.

7.4.4 Elliptic Filters

The last classical lowpass analog filter that we consider is the elliptic or Cauer filter. Elliptic filters are filters that are equiripple in both the passband and the stopband. In that sense, elliptic IIR filters are similar to the equiripple FIR filters constructed in Chapter 6 using the Parks-McLellan algorithm. The squared magnitude response of an nth-order elliptic filter is as follows.

$$A_a^2(f) = \frac{1}{1 + \epsilon^2 U_n^2(f/F_p)} \tag{7.4.31}$$

FIGURE 7.19:
Squared Magnitude
Responses of an
Elliptic Lowpass
Filter of Order $n = 4$
with $F_p = 1$ Hz

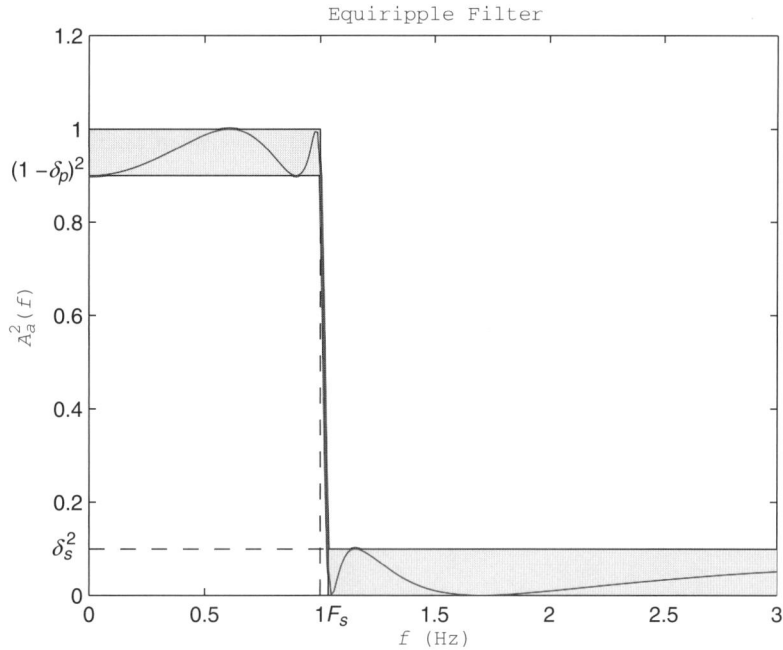

Here U_n is an nth-order Jacobian elliptic function, also called a Chebyshev rational function (Porat, 1997). By permitting ripples in both the passband and the stopband, elliptic filters achieve very narrow transition bands. A plot of the squared magnitude response is shown in Figure 7.19 for the case $n = 4$ with $F_p = 1$ Hz.

The design parameter ϵ for an elliptic filter is identical to that of the Chebyshev-I filter. Since $U_n(1) = 1$ for all n, it follows from (7.4.31) that (7.4.16) holds. This in turn means that the passband specification can be met exactly if the ripple factor ϵ is set to satisfy (7.4.17). Elliptic filters are considerably more complex to analyze and design than Butterworth and Chebyshev filters. Finding the zeros and poles of an elliptic filter involves the iterative solution of nonlinear algebraic equations, equations whose terms include integrals (Parks and Burrus, 1987). Let us focus, instead, on the remaining design parameter, the minimal filter order. Let $g(x)$ denote the following function which is called a *complete elliptic integral* of the first kind.

$$g(x) = \int_0^{\pi/2} \frac{d\theta}{\sqrt{1 - x^2 \sin^2(\theta)}} \tag{7.4.32}$$

Recalling the definitions of the selectivity and discrimination factors in (7.3.4), we find that the required order for an elliptic filter to meet the design specifications is

$$n = \text{ceil}\left[\frac{g(r^2)g(\sqrt{1 - d^2})}{g(\sqrt{1 - r^2})g(d^2)}\right] \tag{7.4.33}$$

Elliptic filters always meet the passband specification exactly as long as the ripple factor ϵ is chosen as in (7.4.17). The stopband specification will be exceeded when the expression in the square brackets in (7.4.33) is smaller than the filter order n.

TABLE 7.4: ▶
Summary of
Classical Analog
Filters

Analog Filter	Passband	Stopband	Transition Band	Exact Specification
Butterworth	Monotonic	Monotonic	Broad	Either
Chebyshev-I	Equiripple	Monotonic	Narrow	Passband
Chebyshev-II	Monotonic	Equiripple	Narrow	Stopband
Elliptic	Equiripple	Equiripple	Very narrow	Passband

Example 7.7

Elliptic Filter

For comparison, consider an elliptic lowpass filter that meets the same specifications that were used in Examples 7.4 and 7.6. From Example 7.4, the selectivity and discrimination factors are

$$r = .5$$
$$d = .0165$$

The elliptic integral function in (7.4.32) can be evaluated numerically using the MATLAB function *ellipke*. If we run *exam7_7*, the filter order is

$$n = \text{ceil}\left[\frac{g(.25)g(\sqrt{1-(.0165)^2})}{g(.75)g[(.0165)^2]}\right]$$
$$= \text{ceil}(2.9061)$$
$$= 3$$

In this case, note that by permitting ripples in both the passband and the stopband, we see that the design specification can be met with an elliptic filter of order $n = 3$. This is in contrast to the Chebyshev filters that required $n = 4$, and the Butterworth filter that required $n = 6$.

Although the elliptic filter is the filter of choice if the only criterion is to minimize the order, the other classical analog filters are often used as well because they tend to have better (more linear) phase response characteristics. A summary of the essential characteristics of the classical analog filters is shown in Table 7.4.

FDSP Functions

The FDSP toolbox contains four functions for computing the coefficients of the classical lowpass analog filters, and a function for computing the analog frequency response. Recall from Chapter 1, that the FDSP function *f_freqs* can be used to compute the frequency response of an analog filter.

```
% F_BUTTERS:   Design lowpass analog Butterworth filter
% F_CHEBY1S:   Design Chebyshev-I lowpass analog filter
% F_CHEBY2S:   Design a Chebyshev-II lowpass analog filter
% F_ELLIPTICS: Design elliptic lowpass analog filter
%
% Usage:
%       [b,a] = f_butters   (Fp,Fs,deltap,deltas,n)
%       [b,a] = f_cheby1s   (Fp,Fs,deltap,deltas,n)
%       [b,a] = f_cheby2s   (Fp,Fs,deltap,deltas,n)
%       [b,a] = f_elliptics (Fp,Fs,deltap,deltas,n)
```

Continued on p. 529

Continued on p. 528

```
% Pre:
%        Fp      = passband cutoff frequency in Hz
%        Fs      = stopband cutoff frequency in Hz
%                  (Fs > Fp)
%      deltap = passband ripple
%      deltas = stopband attenuation
%        n       = an optional integer specifying the filter
%                  order.  If n is not present, the smallest
%                  order which meets the specifications is
%                  used.
% Post:
%        b = coefficient vector of numerator polynomial
%        a = 1 by (n+1) coefficient vector of denominator
%              polynomial
```

● ● ● ● ● ● ● ● ● ● ● ● ● ● ● ● ●

7.5 **Bilinear transformation Method**

Now that a collection of analog prototype filters is in place, the next task is to transform an analog filter into an equivalent digital filter. Although a number of approaches are available, they all must satisfy the fundamental qualitative constraint that a stable analog filter $H_a(s)$ transform into a stable digital filter $H(z)$. The classical analog nth-order filters discussed in Section 7.4 each have n distinct poles p_k. Therefore, $H_a(s)$ can be written in partially factored form as

$$H_a(s) = \frac{b(s)}{(s - p_0)(s - p_1) \cdots (s - p_{n-1})} \tag{7.5.1}$$

A design technique that is highly effective is based on replacing integration with a discrete-time numerical approximation. Recall that an integrator has the continuous-time transfer function $H_0(s) = 1/s$. The time-domain input-output representation of an integrator is

$$y_a(t) = \int_0^t x_a(\tau) d\tau \tag{7.5.2}$$

To approximate the area under the curve $x_a(t)$ numerically, let $x(k) = x_a(kT)$ where T is the sampling interval. Consider the trapezoids formed by connecting the samples with straight lines, as shown in Figure 7.20. This is equivalent to using a piecewise-linear approximation to $x_a(t)$. Suppose $y(k)$ denotes the approximation to the integral at time $t = kT$. The approximation at time kT is the approximation at time $(k-1)T$ plus the area of the kth trapezoid. From Figure 7.20, the kth trapezoid has width T and average height $[x(k-1) + x(k)]/2$. Thus,

$$y(k) = y(k-1) + \left(\frac{T}{2}\right)[x(k-1) + x(k)] \tag{7.5.3}$$

Trapezoid integrator   The approximation in (7.5.3) is called a *trapezoid rule integrator*. Taking the Z-transform of both sides of (7.5.3) and using the delay property yields $(1 - z^{-1})Y(z) = (T/2)(1 + z^{-1})X(z)$.

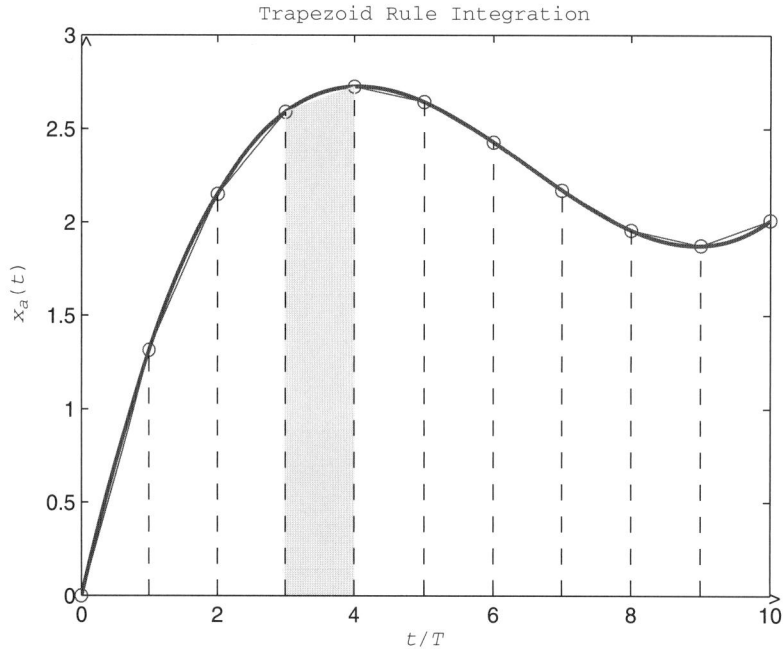

Thus the transfer function of a trapezoid rule integrator is

$$H_0(z) = \frac{T}{2} \left(\frac{1 + z^{-1}}{1 - z^{-1}} \right) \tag{7.5.4}$$

Note that replacing $1/s$ by $H_0(z)$ is equivalent to replacing s by $1/H_0(z)$. Therefore one can approximate the integration process with a trapezoid rule integrator by making the following substitution for s in the analog filter transfer function $H_a(s)$.

$$H(z) = H_a(s)|_{s=g(z)} \tag{7.5.5}$$

Here $g(z) = 1/H_0(z)$. That is, the substitution $s = g(z)$ in (7.5.5) uses

$$g(z) = \frac{2}{T} \left(\frac{z - 1}{z + 1} \right) \tag{7.5.6}$$

Bilinear transformation

The transformation from $H_a(s)$ to $H(z)$ in (7.5.5) is called a bilinear transformation, and filter designs based on it use the *bilinear transformation* method. Before designing a filter using the bilinear transformation, we find it is helpful to examine the relationship between z and $s = g(z)$ in more detail. First, note that the transformation can be inverted. That is, one can solve (7.5.6) for z which yields

$$z = \frac{2 + sT}{2 - sT} \tag{7.5.7}$$

Next, suppose s is expressed in rectangular form as $s = \sigma + j\omega$. Substituting this into (7.5.7) and taking the magnitude of both sides yields

$$|z| = \frac{\sqrt{(2 + \sigma T)^2 + (\omega T)^2}}{\sqrt{(2 - \sigma T)^2 + (\omega T)^2}} \tag{7.5.8}$$

FIGURE 7.21:
Bilinear
Transformation
from the *s* Plane
onto the *z* Plane

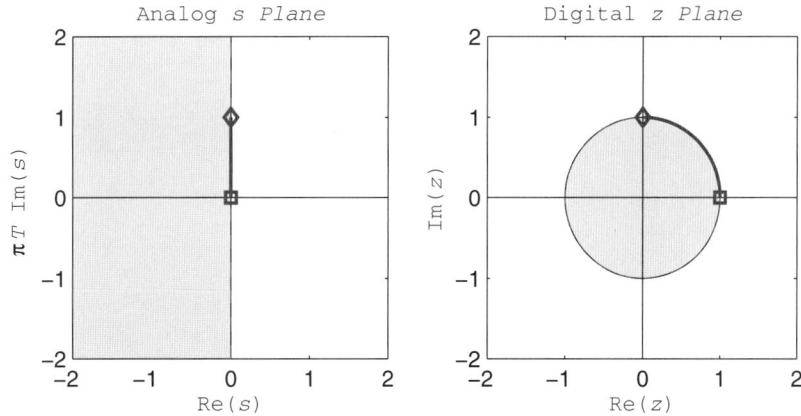

Notice that if $\sigma = 0$, then $|z| = 1$. Thus the bilinear transformation maps the imaginary axis of the s plane into the unit circle of the z plane. Furthermore, if $\sigma < 0$, then $|z| < 1$, which means that the left half of the s plane is mapped into the interior of the unit circle in the z plane. Similarly, when $\sigma > 0$ the right half of the s plane is mapped into the exterior of the unit circle in the z plane. Therefore the bilinear transformation satisfies the fundamental property that it is guaranteed to transform a stable analog filter $H_a(s)$ into a stable digital filter $H(z)$. A graphical summary of the bilinear transformation is shown in Figure 7.21.

The bilinear transformation in Figure 7.21 maps the entire imaginary axis of the s plane onto the unit circle of the z plane as in (7.5.7). In so doing, the infinite *analog frequencies* range, $0 \leq F < \infty$, gets compressed or warped into the finite *digital frequency* range, $0 \leq f < f_s/2$. To develop the relationship between F and f, let $s = j2\pi F$ denote a point on the imaginary axis of the s plane, and let $z = \exp(j2\pi f T)$ denote the corresponding point on the unit circle of the z plane. Setting $s = g(z)$ in (7.5.6) and using Euler's identity yields

$$j2\pi F = \frac{2}{T}\left[\frac{\exp(j2\pi f T) - 1}{\exp(j2\pi f T) + 1}\right]$$

$$= \frac{2}{T}\left(\frac{\exp(j\pi f T)[\exp(j\pi f T) - \exp(-j\pi f T)]}{\exp(j\pi f T)[\exp(j\pi f T) + \exp(-j\pi f T)]}\right)$$

$$= \frac{2}{T}\left[\frac{j2\sin(\pi f T)}{2\cos(\pi f T)}\right]$$

$$= \frac{j2\tan(\pi f T)}{T} \tag{7.5.9}$$

Solving (7.5.9) for F then results in the following relationship between the digital filter frequency f and the analog filter frequency F.

$$F = \frac{\tan(\pi f T)}{\pi T} \tag{7.5.10}$$

Frequency warping

The transformation from f to F in (7.5.10) is called *frequency warping* because it represents an expansion of the finite digital frequency range $0 \leq f < f_s/2$ into the infinite analog frequency range $0 \leq F < \infty$. The nonlinear frequency warping curve is shown in Figure 7.22. Note that there is an asymptote at the folding frequency $f_d = f_s/2$.

FIGURE 7.22:
Frequency Warping
Caused by the
Bilinear
Transformation

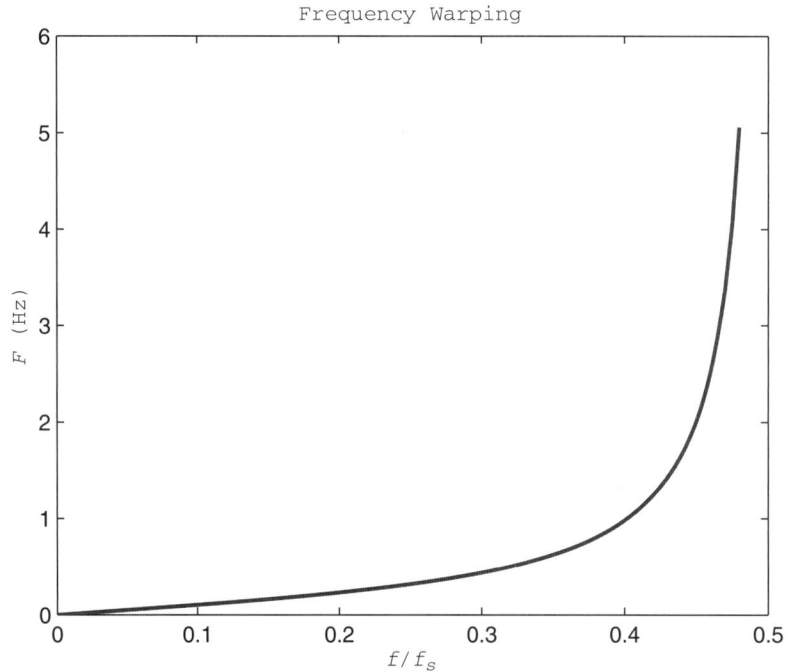
Frequency Warping

The mapping from digital frequency f to analog frequency F in (7.5.10) can be inverted by solving (7.5.10) for f. Multiplying (7.5.10) by πT, taking the arctangent of both sides, and dividing by πT yields

$$f = \frac{\tan^{-1}(\pi F T)}{\pi T} \qquad (7.5.11)$$

When the bilinear transformation from s to z in (7.5.7) is performed, the analog frequencies in the range $0 \le F < \infty$ get compressed into digital frequencies in the range $0 \le f < f_s/2$, as indicated in (7.5.11). One can take this nonlinear compression into account in filter design by first *prewarping* each desired digital cutoff frequency f_c into a corresponding analog cutoff frequency F_c, using (7.5.10). When the bilinear transformation is then performed, these prewarped cutoff frequencies get warped back into the original desired digital cutoff frequencies as in (7.5.11). The overall design procedure for the bilinear-transformation method is summarized in the following algorithm where it is assumed that $m \le n$.

ALGORITHM

7.1: Bilinear Transformation Method

1. Prewarp all digital cutoff frequencies, f_i, into corresponding analog cutoff frequencies, F_i, using (7.5.10).
2. Construct an analog prototype filter $H_a(s)$ using the prewarped cutoff frequencies.
3. If $H_a(s)$ is low order, compute $H(z) = H_a[g(z)]$ using (7.5.6). For a higher-order $H_a(s)$, the following steps can be used.
   (a) Factor the numerator and denominator of $H_a(s)$ as follows.

$$H_a(s) = \frac{\beta_0(z - u_1) \cdots (z - u_m)}{(z - v_1) \cdots (z - v_n)}$$

Continued on p. 533

Continued on p. 532

(b) Compute the digital zeros and poles as follows using (7.5.7).

$$z_i = \frac{2 + u_i T}{2 - u_i T}, \quad 1 \le i \le m$$

$$p_i = \frac{2 + v_i T}{2 - v_i T}, \quad 1 \le i \le n$$

(c) Compute the digital filter gain using

$$b_0 = \frac{\beta_0 T^{n-m}(2 - u_1 T) \cdots (2 - u_m T)}{(2 - v_1 T) \cdots (2 - v_n T)}$$

(d) Construct the factored form of the digital filter as follows.

$$H(z) = \frac{b_0(z + 1)^{n-m}(z - z_1) \cdots (z - z_m)}{(z - p_1) \cdots (z - p_n)}$$

4. Express $H(z)$ as a ratio of two polynomials in z^{-1}.

Algorithm 7.1 assumes that $H_a(s)$ is a proper rational polynomial which means that $m \le n$. If $m < n$, then $H_a(s)$ will have $n - m$ zeros at $s = \infty$. Note from step 3d that these $n - m$ high-frequency zeros get mapped into zeros at $z = -1$, the highest digital frequency that $H(z)$ can process.

Example 7.8

Bilinear transformation Method

As an illustration of using Algorithm 7.1 to design a digital lowpass filter, suppose the sampling frequency is $f_s = 20$ Hz, and consider the following lowpass design specifications.

$$(f_0, f_1) = (2.5, 7.5) \text{ Hz}$$

$$(\delta_p, \delta_s) = (.1, .1)$$

Here f_0 and f_1 denote the desired passband and stopband frequencies, respectively. From step 1 of Algorithm 7.5.1, the prewarped passband and stopband frequencies are

$$F_0 = \frac{\tan(2.5\pi/20)}{\pi/20}$$

$$= 2.637 \text{ Hz}$$

$$F_1 = \frac{\tan(7.5\pi/20)}{\pi/20}$$

$$= 15.37 \text{ Hz}$$

Suppose the analog prototype filter used is a lowpass Butterworth filter. From (7.4.8), the minimum order for the filter is

$$n = \text{ceil} \left\{ \frac{\log \left[\dfrac{.9^{-2} - 1}{.1^{-2} - 1} \right]}{2 \log \left(\dfrac{2.637}{15.37} \right)} \right\}$$

$$= \text{ceil}(1.715)$$

$$= 2$$

Next, suppose the cutoff frequency F_c is selected to meet the passband specification exactly. Then from (7.4.9), the required cutoff frequency is

$$F_c = \frac{2.637}{(.9^{-2} - 1)^{1/4}}$$
$$= 3.789 \text{ Hz}$$

Thus the radian cutoff frequency is $\Omega_c = 2\pi F_c = 23.81$ rad/sec. From Table 7.2, and (7.4.13), the transfer function of a prewarped Butterworth filter of order $n = 2$ is

$$H_a(s) = \frac{\Omega_c^2}{s^2 + \sqrt{2}\Omega_c s + \Omega_c^2}$$

Since $H_a(s)$ is a low-order filter, one can apply step 3 of Algorithm 7.1 using direct substitution. Thus from (7.5.6), the discrete-equivalent transfer function $H(z)$ is

$$H(z) = H_a[g(z)]$$
$$= \frac{\Omega_c^2}{g^2(z) + \sqrt{2}\Omega_c g(z) + \Omega_c^2}$$
$$= \frac{\Omega_c^2}{\left[\dfrac{2(z-1)}{T(z+1)}\right]^2 + \sqrt{2}\Omega_c\left[\dfrac{2(z-1)}{T(z+1)}\right] + \Omega_c^2}$$
$$= \frac{(T\Omega_c)^2(z+1)^2}{4(z-1)^2 + 2\sqrt{2}T\Omega_c(z-1)(z+1) + (T\Omega_c)^2(z+1)^2}$$
$$= \frac{(T\Omega_c)^2(z+1)^2}{4(z^2 - 2z + 1) + 2\sqrt{2}T\Omega_c(z^2 - 1) + (T\Omega_c)^2(z^2 + 2z + 1)}$$

FIGURE 7.23:
Magnitude
Response of a
Digital IIR Filter
Obtained by a
Bilinear
Transformation of
the Prewarped
Analog
Butterworth Filter:
$n = 2$ and
$F_c = 3.789$ Hz

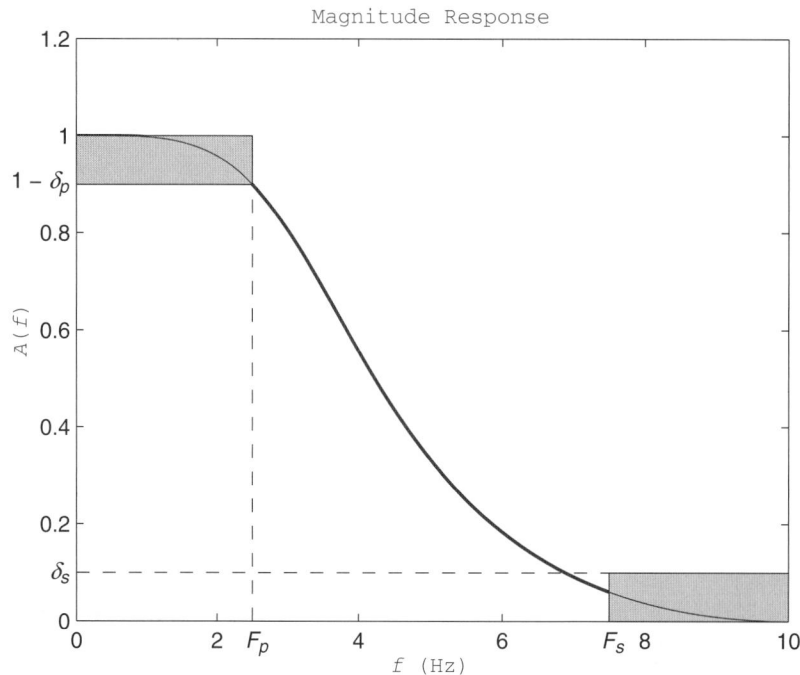

Next, the terms in the denominator are combined, the denominator is normalized, and the numerator and denominator are multiplied by z^{-2}. The final result after substitution of $T = 1/20$ and $\Omega_c = 23.81$ is then

$$H(z) = \frac{.1613(1 + 2z^{-1} + z^{-2})}{1 - .5881z^{-1} + .2334z^{-2}}$$

A plot of the digital filter magnitude response is shown in Figure 7.23. Note how the passband specification $1 - \delta_p \le A(f) \le 1$ for $0 \le f \le 2.5$ Hz is met exactly, and the stopband specification $0 \le A(f) \le \delta_s$ for $7.5 \le f \le 10$ Hz is exceeded.

FDSP Functions

The FDSP toolbox contains the following function for performing a digital-to-analog filter transformation using the bilinear-transformation method.

```
% F_BILIN: Bilinear analog to digital filter transformation
%
% Usage:
%        [B,A] = f_bilin (b,a,fs)
% Pre:
%        b  = vector of length m+1 containing coefficients
%             of analog numerator polynomial.
%        a  = vector of length n+1 containing coefficients
%             of analog denominator polynomial (n >= m).
%        fs = sampling frequency in Hz
% Post:
%        B = (n+1) by 1 vector containing coefficients of
%            digital numerator polynomial.
%        A = (n+1) by 1 vector containing coefficients of
%            digital denominator polynomial.
% Notes:
%        The critical frequencies of H(s) must first be
%        prewarped using:
%
%            F = tan(pi*f*T)/(pi*T)
```

7.6 Frequency Transformations

At this point, the tools are in place to design digital lowpass filters. One starts with a normalized classical analog lowpass filter $H_a(s)$. Next, the cutoff frequency is prewarped using (7.5.10). Finally, the bilinear transformation $s = g(z)$ in (7.5.6) is applied to produce a digital equivalent filter $H(z)$. In this section we use frequency transformations to extend this technique so it is also applicable to the other frequency-selective filters such as highpass, bandpass, and bandstop filters.

$H_a(s)$	$D(s)$
Lowpass with cutoff Ω_0	$\dfrac{s}{\Omega_0}$
Highpass with cutoff Ω_0	$\dfrac{\Omega_0}{s}$
Bandpass with cutoffs Ω_0, Ω_1	$\dfrac{s^2 + \Omega_0\Omega_1}{(\Omega_1 - \Omega_0)s}$
Bandstop with cutoffs Ω_0, Ω_1	$\dfrac{(\Omega_1 - \Omega_0)s}{s^2 + \Omega_0\Omega_1}$

7.6.1 Analog Frequency Transformations

Recall that one way to design a lowpass Butterworth filter with a radian cutoff frequency of Ω_0 is to start with a normalized lowpass Butterworth filter and replace s with s/Ω_0. This is an example of a frequency transformation. Using this same basic approach, one can transform a normalized lowpass filter into other types of frequency-selective filters such as highpass, bandpass, and bandstop filters. To illustrate the procedure, consider the following normalized lowpass Butterworth filter of order $n = 1$ taken from (7.4.12) and Table 7.2.

$$H_{\text{norm}}(s) = \frac{1}{s + 1} \tag{7.6.1}$$

Suppose s is replaced, not by s/Ω_0, but by Ω_0/s. The resulting transfer function is then

$$
\begin{aligned}
H_a(s) &= H_{\text{norm}}(\Omega_0/s) \\
&= \frac{1}{\Omega_0/s + 1} \\
&= \frac{s}{s + \Omega_0}
\end{aligned}
\tag{7.6.2}
$$

Notice that $A_a(0) = 0$ and $A_a(\infty) = 1$, which means that $H_a(s)$ is a highpass filter. Furthermore, $A_a(\Omega_0) = 1/\sqrt{2}$. Thus the frequency transformation $D(s) = \Omega_0/s$ maps a normalized lowpass filter into a highpass filter with a 3 dB cutoff frequency of Ω_0 rad./sec.

It is also possible to transform a normalized lowpass filter into a bandpass filter. Recall that a bandpass filter has a low-frequency cutoff, Ω_0, and a high-frequency cutoff, Ω_1. Since a bandpass filter has two cutoff frequencies, the complex frequency variable s must be replaced by a quadratic polynomial of s in order to double the order of the transfer function. In particular, if s is replaced by $D(s) = (s^2 + \Omega_0\Omega_1)/[(\Omega_1 - \Omega_0)s]$, then the resulting filter is a bandpass filter with the desired cutoff frequencies.

Just as the highpass transformation is the reciprocal of the lowpass transformation, the bandstop transformation is the reciprocal of the bandpass transformation. A summary of the four basic frequency transformations can be found in Table 7.5. If we use these transformations, a normalized lowpass transfer function $H_{\text{norm}}(s)$ with a cutoff frequency of $\Omega_c = 1$ rad/sec can be converted into an arbitrary lowpass, highpass, bandpass, or bandstop transfer function $H_a(s)$.

Example 7.9   **Lowpass to Bandpass**

As an illustration of the frequency-transformation method, consider the problem of designing an analog bandpass filter. Suppose the desired cutoff frequencies are $F_0 = 5$ Hz and $F_1 = 15$ Hz. Thus the corresponding radian cutoff frequencies are $\Omega_0 = 10\pi$ rad/sec and $\Omega_1 = 30\pi$ rad/sec. As a starting point, consider the first-order lowpass Butterworth filter in (7.6.1). If we use the third entry of Table 7.5, the bandpass filter transfer function is

$$H_a(s) = H_{\text{norm}}[D(s)]$$

$$= \cfrac{1}{\cfrac{s^2 + \Omega_0\Omega_1}{(\Omega_1 - \Omega_0)s} + 1}$$

$$= \frac{(\Omega_1 - \Omega_0)s}{s^2 + (\Omega_1 - \Omega_0)s + \Omega_0\Omega_1}$$

$$= \frac{20\pi s}{s^2 + 20\pi s + 300\pi^2}$$

Given the classical analog lowpass filters in Section 7.4, and the frequency transformations in Table 7.5, it is possible to design a variety of analog frequency-selective filters. The bilinear analog-to-digital filter transformation in Section 7.5 then can be applied to convert these analog filters to equivalent digital filters. Before the bilinear transformation is applied, all cutoff frequencies must be prewarped using (7.5.10). The following example illustrates the use of the bilinear transformation method to construct a bandpass filter.

Example 7.10 — Digital Bandpass Filter

Consider the second-order analog bandpass filter from Example 7.9. That is, suppose the desired cutoff frequencies are $F_0 = 5$ Hz, and $F_1 = 15$ Hz, and the sampling rate is $f_s = 50$ Hz. If we apply (7.5.10), the prewarped radian cutoff frequencies are

$$\Omega_0 = \frac{2\pi \tan(\pi F_0 T)}{\pi T}$$
$$= 100 \tan(.2\pi)$$
$$= 32.49$$

$$\Omega_1 = \frac{2\pi \tan(\pi F_1 T)}{\pi T}$$
$$= 100 \tan(.6\pi)$$
$$= 137.64$$

From Example 7.9, the transfer function of a second-order Butterworth bandpass filter with cutoff frequencies of Ω_0 and Ω_1 is

$$H_a(s) = \frac{(\Omega_1 - \Omega_0)s}{s^2 + (\Omega_1 - \Omega_0)s + \Omega_0\Omega_1}$$

$$= \frac{105.15s}{s^2 + 105.15s + 4472.1}$$

Next, apply the bilinear transformation to convert $H_a(s)$ into an equivalent digital filter. Using (7.5.6) yields

$$H(z) = H_a[g(z)]$$

$$= \frac{105.15g(z)}{g^2(z) + 105.15g(z) + 4472.1}$$

$$= \frac{105.15\left[\dfrac{2(z-1)}{T(z+1)}\right]}{\left[\dfrac{2(z-1)}{T(z+1)}\right]^2 + 105.15\left[\dfrac{2(z-1)}{T(z+1)}\right] + 4472.1}$$

FIGURE 7.24:
Frequency
Response of a
Second-order
Digital Bandpass
Filter

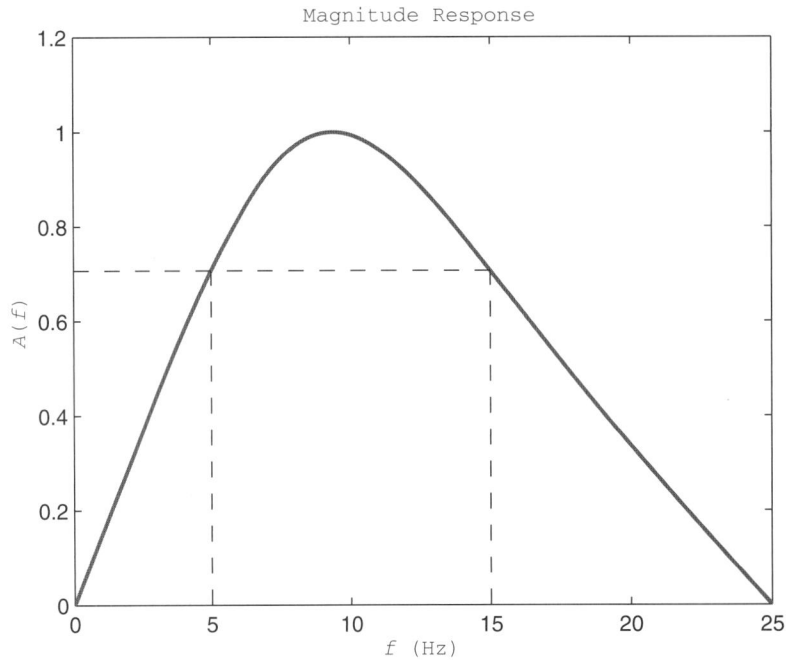

$$= \frac{1.0515(z-1)(z+1)}{(z-1)^2 + 1.0515(z-1)(z+1) + .44721(z+1)^2}$$

$$= \frac{1.0515(z^2-1)}{(z^2 - 2z + 1) + 1.0515(z^2 - 1) + .44721(z^2 + 2z + 1)}$$

Combining like terms, normalizing the denominator, and multiplying top and bottom by z^{-2} then results in the following digital bandpass filter using the bilinear transformation method

$$H(z) = \frac{.4208(1 - z^{-2})}{1 - .4425z^{-1} + .1584z^{-2}}$$

A plot of the magnitude response of $H(z)$, obtained by running *exam7_10*, is shown in Figure 7.24. Although this filter is far from ideal because of its low order, it does have the correct 3 dB cutoff frequencies.

The design technique based on an analog frequency transformation of a lowpass prototype filter is summarized in Figure 7.25.

FIGURE 7.25: Digital
Filter Design Using
an Analog
Frequency
Transformation

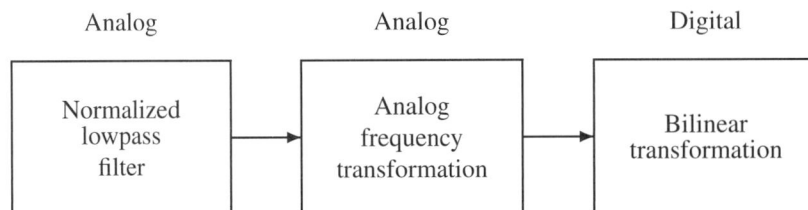

7.6.2 Digital Frequency Transformations

Frequency transformations from lowpass filters to other frequency-selective filters also can be done in the discrete-time or digital domain. In this case the bilinear analog-to-digital filter transformation is applied to a normalized lowpass prototype filter. The resulting lowpass digital filter is then transformed to a lowpass, highpass, bandpass, or bandstop filter using a digital frequency transformation.

Let $H_{\text{low}}(z)$ be a digital lowpass filter with a cutoff frequency of F_c. This is converted to another frequency-selective filter by replacing z with a frequency transformation $D(z)$ as follows.

$$H(z) = H_{\text{low}}[D(z)] \tag{7.6.3}$$

The transformation $D(z)$, must satisfy certain qualitative properties. First, it must map a rational polynomial $H_{\text{low}}(z)$ into a rational polynomial $H(z)$. This means that $D(z)$ itself must be a ratio of polynomials. Because $D(z)$ is a frequency response transformation, it should map the unit circle into the unit circle. Evaluating $D(z)$ along the unit circle yields the frequency response $D(f)$. Thus the magnitude response must satisfy

$$|D(f)| = 1, \quad 0 \le f < f_s \tag{7.6.4}$$

Notice that the magnitude response constraint in (7.6.4) is an allpass characteristic. Hence $D(z)$ must be an allpass filter, as in (5.4.7). To maintain stability, the transformation $D(z)$ must also map the interior of the unit circle into the interior of the unit circle. Constantinides (1970) has developed four basic digital frequency transformations that are summarized in Table 7.6. The source filter is a lowpass filter with a cutoff frequency of F_c and the destination filter is the filter listed in column one.

The design technique based on a digital frequency transformation of a lowpass filter is summarized in Figure 7.26.

TABLE 7.6: ▶
Digital Frequency Transformations, $H(z) = H_{\text{low}}[D(z)]$

$H(z)$	$D(z)$	Coefficients
Lowpass with cutoff F_0	$\dfrac{-(z - a_0)}{a_0 z - 1}$	$a_0 = \dfrac{\sin[\pi(F_c - F_0)]}{\sin[\pi(F_c + F_0)]}$
Highpass with cutoff F_0	$\dfrac{z - a_0}{a_0 z - 1}$	$a_0 = \dfrac{\cos[\pi(F_c + F_0)]}{\cos[\pi(F_c - F_0)]}$
Bandpass with cutoffs F_0, F_1	$\dfrac{-(z^2 + a_0 z + a_1)}{a_1 z^2 + a_0 z + 1}$	$\alpha = \dfrac{\cos[\pi(F_1 + F_0)]}{\cos[\pi(F_1 - F_0)]}$
		$\beta = \tan(\pi F_c) \cot[\pi(F_1 - F_0)]$
		$a_0 = \dfrac{-2\alpha\beta}{\beta + 1}$
		$a_1 = \dfrac{\beta - 1}{\beta + 1}$
Bandstop with cutoffs F_0, F_1	$\dfrac{z^2 + a_0 z + a_1}{a_1 z^2 + a_0 z + 1}$	$\alpha = \dfrac{\cos[\pi(F_1 + F_0)]}{\cos[\pi(F_1 - F_0)]}$
		$\beta = \tan(\pi F_c) \tan[\pi(F_1 - F_0)]$
		$a_0 = \dfrac{-2\alpha}{\beta + 1}$
		$a_1 = \dfrac{1 - \beta}{1 + \beta}$

FIGURE 7.26: Digital Filter Design Using a Digital Frequency Transformation

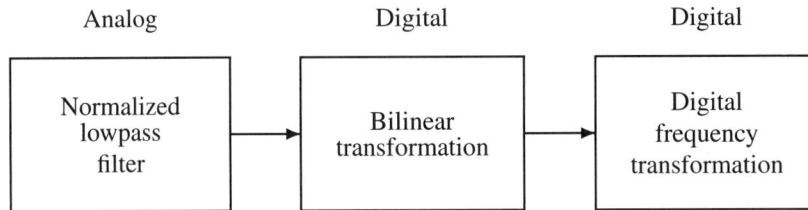

Analog

Digital

Digital

Normalized lowpass filter

Bilinear transformation

Digital frequency transformation

FDSP Functions

The FDSP toolbox contains the following functions for designing classical frequency-selective IIR digital filters using the method summarized in Figure 7.25. Recall that the function *f_freqz* can be used to compute the frequency response of a digital filter.

```
% F_BUTTERZ:   Design a Butterworth IIR digital filter
% F_CHEBY1Z:   Design a Chebyshev-I IIR digital filter
% F_CHEBY2Z:   Design a Chebyshev-II IIR digital filter
% F_ELLIPTICZ: Design elliptic IIR digital filter
%
% Usage:
%        [b,a] = f_butterz   (Fp,Fs,deltap,deltas,ftype,fs,n)
%        [b,a] = f_cheby1z   (Fp,Fs,deltap,deltas,ftype,fs,n)
%        [b,a] = f_cheby2z   (Fp,Fs,deltap,deltas,ftype,fs,n)
%        [b,a] = f_ellipticz (Fp,Fs,deltap,deltas,ftype,fs,n)
% Pre:
%        Fp     = passband cutoff frequency or frequencies
%        Fs     = stopband cutoff frequency or frequencies
%        deltap = passband ripple
%        deltas = stopband attenuation
%        ftype  = filter type
%
%                 0 = lowpass
%                 1 = highpass
%                 2 = bandpass (Fp and Fs are vectors)
%                 3 = bandstop (Fp and Fs are vectors)
%
%        fs     = sampling frequency in Hz
%        n      = an optional integer specifying the
%                 filter order.  If n is not present, an
%                 estimate of order required to meet the
%                 specifications is used.
% Post:
%        b = 1 by (n+1) coefficient vector of numerator
%            polynomial
%        a = 1 by (n+1) coefficient vector of denominator
%            polynomial
```

● ● ● ● ● ● ● ● ● ● ● ● ● ● ●

7.7 Filter Realization Structures

In this section we investigate alternative configurations that can be used to realize IIR filters with signal flow graphs. These filter realizations differ from one another with respect to memory requirements, computational time, and sensitivity to finite word length effects.

7.7.1 Direct Forms

An nth-order IIR filter has a transfer function $H(z)$ that can be expressed as a ratio of two polynomials.

$$H(z) = \frac{b_0 + b_1 z^{-1} + \cdots + b_n z^{-n}}{1 + a_1 z^{-1} + \cdots + a_n z^{-n}} \tag{7.7.1}$$

For convenience, we have assumed that the degree of the numerator is equal to the degree of the denominator because this simplifies and streamlines the treatment. This is not a limiting restriction because one can always pad the numerator or denominator coefficient vector with zeros, as needed, to make them the same length.

Direct Form I

The simplest realization of $H(z)$ is based on factoring the transfer function into its autoregressive and moving average parts as follows.

$$H(z) = \underbrace{\left(\frac{1}{1 + a_1 z^{-1} + \cdots + a_n z^{-n}}\right)}_{H_{\text{ar}}(z)} \underbrace{\left(\frac{b_0 + b_1 z^{-1} + \cdots + b_n z^{-n}}{1}\right)}_{H_{\text{ma}}(z)} \tag{7.7.2}$$

If $U(z)$ denotes the output of the moving average subsystem $H_{\text{ma}}(z)$, then the input-output description of the IIR filter can be written in terms of the intermediate variable $U(z)$ as follows.

$$U(z) = H_{\text{ma}}(z)X(z) \tag{7.7.3a}$$
$$Y(z) = H_{\text{ar}}(z)U(z) \tag{7.7.3b}$$

If we take the inverse Z-transform of (7.7.3), using (7.7.2) and the delay property, an IIR filter can be represented in the time domain by the following pair of difference equations.

$$u(k) = \sum_{i=0}^{n} b_i x(k-i) \tag{7.7.4a}$$

$$y(k) = u(k) - \sum_{i=1}^{n} a_i y(k-i) \tag{7.7.4b}$$

Direct form I   The representation in (7.7.4) is called a *direct form I* realization. It is a *direct* representation because the coefficients of the difference equations can be obtained directly from inspection of the transfer function. A signal flow graph of a direct form I realization, for the case $n = 3$, is shown in Figure 7.27. Note how the left side of the signal flow graph implements the moving average part in (7.7.4a), and the right side implements the autoregressive part in (7.7.4b).

Direct Form II

There is a very simple change that can be made to (7.7.2) to generate an alternative direct form realization that has some advantages over the direct form I structure. Suppose the order of the

FIGURE 7.27: Direct
Form I Realization,
$n = 3$

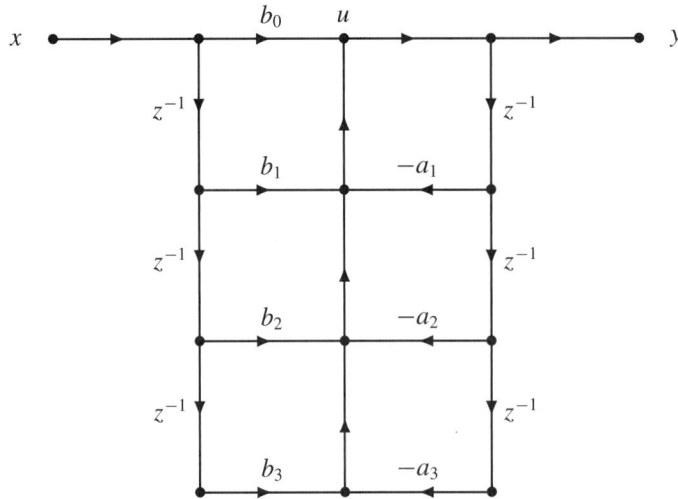

autoregressive and moving average subsystems is interchanged. Clearly, this does not affect the overall transfer function.

$$H(z) = \underbrace{\left(\frac{b_0 + b_1 z^{-1} + \cdots + b_n z^{-n}}{1}\right)}_{H_{\mathrm{ma}}(z)} \underbrace{\left(\frac{1}{1 + a_1 z^{-1} + \cdots + a_n z^{-n}}\right)}_{H_{\mathrm{ar}}(z)} \tag{7.7.5}$$

Next, let $U(z)$ denote the output of the autoregressive subsystem, $H_{\mathrm{ar}}(z)$. Then the input-output description of the overall filter in terms of the intermediate variable $U(z)$ is as follows.

$$U(z) = H_{\mathrm{ar}}(z)X(z) \tag{7.7.6a}$$
$$Y(z) = H_{\mathrm{ma}}(z)U(z) \tag{7.7.6b}$$

Taking the inverse Z-transform of (7.7.6), using (7.7.5) and the delay property, we can represent, an IIR filter in the time domain by the following difference equations.

$$u(k) = x(k) - \sum_{i=1}^{n} a_i x(k - i) \tag{7.7.7a}$$

$$y(k) = \sum_{i=0}^{n} b_i u(k - i) \tag{7.7.7b}$$

Direct form II   This alternative representation of an IIR filter is called a *direct form II* realization. A signal flow graph of a direct form II realization, for the case $n = 3$, is shown in Figure 7.28. In this case, the left side of the signal flow graph implements the autoregressive part in (7.7.7a), and the right side implements the moving average part in (7.7.7b).

It is of interest to compare the signal flow graphs in Figure 7.27 and Figure 7.28. Each arc associated with a delay element requires one memory or storage element to implement. Consequently, direct form I requires a total of $2n$ storage elements, whereas direct form II requires only n storage elements. Since the minimum number of storage elements required for an nth-order filter is n, direct form II is an example of a *canonic* representation.

Transposed Direct Form II

Just as was the case with FIR filters, one can apply the flow graph reversal theorem in Proposition 6.2 to generate a transposed direct form realization by reversing the directions of all arcs

FIGURE 7.28: Direct Form II Realization, $n = 3$

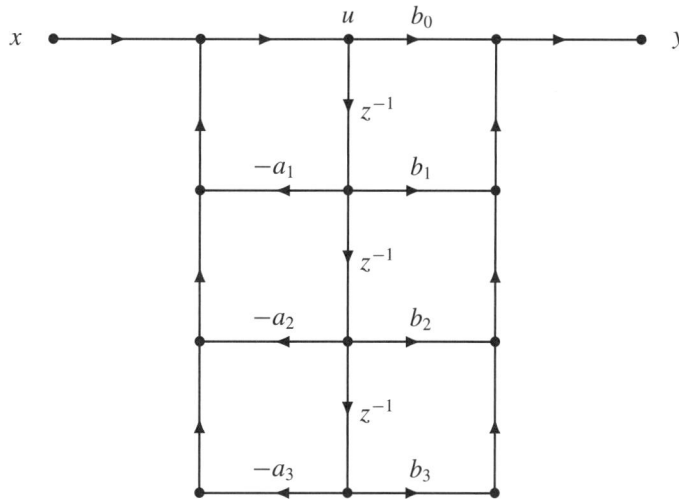

and interchanging the labels of the input and output. Redrawing the signal flow graph so the input is on the left yields the *transposed direct form II* realization shown in Figure 7.29 for the case $m = 3$.

The difference equations describing the transposed direct form II realization can be obtained directly from inspection of Figure 7.29. Here a vector of intermediate variables $u = [u_1, u_2, \ldots, u_n]^T$ is used. The intermediate variables are defined recursively starting at the bottom of the signal flow graph and moving up the center. The last equation is then the output equation.

$$u_1(k) = b_n x(k) - a_n y(k) \tag{7.7.8a}$$

$$u_i(k) = b_i x(k) - a_i y(k) + u_{i-1}(k), \quad 2 \le i \le n \tag{7.7.8b}$$

$$y(k) = b_0 x(k) + u_n(k - 1) \tag{7.7.8c}$$

FIGURE 7.29: Transposed Direct Form II Realization, $n = 3$

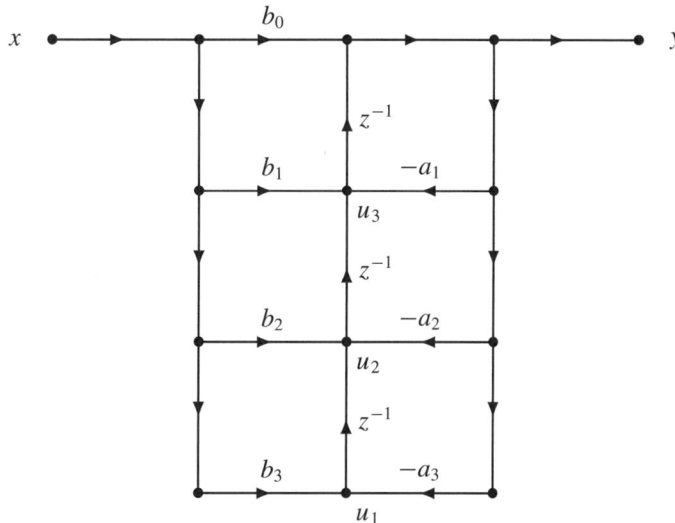

TABLE 7.7: ►
Comparison of
Direct-form
Realizations of IIR
Filter

Direct Form	Storage Elements	Additions	Multiplications
I	$2n$	$2n$	$2n-1$
II	n	$2n$	$2n-1$
Transposed II	n	$2n$	$2n-1$

The direct form realizations also can be compared in terms of the required computational effort. Each summing junction node with m inputs requires $m-1$ floating-point additions, and each arc with a constant nonunity gain requires one floating-point multiplication. The results of the comparison are summarized in Table 7.7, where it can be seen that the three direct form realizations are identical in terms of the computational time, measured in floating-point operations or FLOPs. In each case the number of FLOPs grows linearly with the order of the filter. However, the direct form II realizations have the advantage that they require only half as much memory.

7.7.2 Parallel Form

Just as with FIR filters, there are a number of IIR indirect forms whose coefficients are derived from the original coefficient vectors, a and b. To develop the indirect form realization structures, consider the following partially factored form of the transfer function.

$$H(z) = \frac{b(z)}{(z-p_1)(z-p_2)\cdots(z-p_n)} \tag{7.7.9}$$

Here p_k is the kth pole of the filter. If $H(z)$ has poles at $z=0$, these poles represent pure delays that can be removed and treated separately. Consequently, it is assumed that $p_k \neq 0$ for $1 \leq k \leq n$. For most practical filters, the nonzero poles are distinct from one another. Suppose $H(z)/z$ is expressed in partial fraction form. Multiplying both sides by z then results in the following representation of $H(z)$, if we assume the poles are distinct.

$$H(z) = R_0 + \sum_{i=1}^{n} \frac{R_i z}{z - p_i} \tag{7.7.10}$$

Here R_i is the residue of $H(z)/z$ at the ith pole with $p_0 = 0$. From (7.7.10)

$$R_0 = H(0) \tag{7.7.11a}$$

$$R_i = \left. \frac{(z-p_i)H(z)}{z} \right|_{z=p_i}, \quad 1 \leq i \leq n \tag{7.7.11b}$$

The problem with using (7.7.10) directly for a signal flow graph realization is that the poles and residues are often complex. If $H(z)$ has real coefficients, then the complex poles and residues will appear in conjugate pairs. Consequently, one can rewrite $H(z)$ as a sum of N second-order subsystems with real coefficients as follows where $N = \text{floor}\{(n+1)/2\}$.

$$H(z) = R_0 + \sum_{i=1}^{N} H_i(z) \tag{7.7.12}$$

Parallel realization   This is called a *parallel form* realization. The ith second-order subsystem is constructed by combining pairs of terms in (7.7.10) associated with either real poles or complex conjugate pairs of poles. This way, the second-order coefficients will be real. Combining the terms associated with poles p_i and p_j, simplifying, and expressing the final result in terms of negative powers of z yields

$$H_i(z) = \frac{b_{i0} + b_{i1}z^{-1}}{1 + a_{i1}z^{-1} + a_{i2}z^{-2}}, \quad 1 \leq i \leq m \tag{7.7.13}$$

Note that $b_{i2} = 0$. The real coefficients of the second-order block can be expressed in terms of the poles and residues as follows for $1 \leq i \leq N$.

$$b_{i0} = R_i + R_j \tag{7.7.14a}$$
$$b_{i1} = -(R_i p_j + R_j p_i) \tag{7.7.14b}$$
$$a_{i1} = -(p_i + p_j) \tag{7.7.14c}$$
$$a_{i2} = p_i p_j \tag{7.7.14d}$$

Let u_i denote the output of the ith second-order block. Then from (7.7.12) and (7.7.13), a parallel form realization is characterized in the time domain by the following difference equations.

$$u_i(k) = b_{i0}x(k) + b_{i1}x(k-1) - a_{i1}u_i(k-1)$$
$$-a_{i2}u_i(k-2), \quad 1 \leq i \leq N \tag{7.7.15a}$$
$$y(k) = R_0 x(k) + \sum_{i=1}^{N} u_i(k) \tag{7.7.15b}$$

Note that if n is even, then there will be N subsystems, each of order two, whereas if n is odd, there will be $N - 1$ second-order subsystems and one first-order subsystem. The coefficients of the first-order subsystem are obtained from (7.7.14) by setting $R_j = 0$ and $p_j = 0$.

Any of the direct forms can be used to realize the second-order blocks in (7.7.13). A block diagram showing the overall structure of a parallel form realization, for the case $N = 2$, is shown in Figure 7.30. Since the parallel form coefficients must be computed using (7.7.14), rather than obtained directly from inspection of $H(z)$, the parallel form realization is an example of an *indirect* form.

FIGURE 7.30:
Parallel-form Block
Diagram, $N = 2$

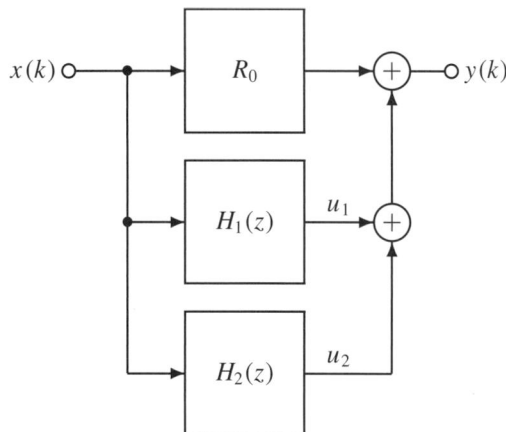

Example 7.11 **IIR Parallel Form**

As an illustration of a parallel form realization of an IIR filter, consider the following fourth-order transfer function.

$$H(z) = \frac{2z(z^3 + 1)}{[(z + .3)^2 + .16](z - .8)(z + .7)}$$

Inspection of $H(z)$ reveals that the poles are

$$p_{1,2} = -.3 \pm j.4$$
$$p_3 = .8$$
$$p_4 = -.7$$

Suppose $H_1(z)$ is a block associated with the complex conjugate pair of poles, and $H_2(z)$ is associated with the real poles. Running *exam7_11* produces the following three subsystems.

$$R_0 = 0$$
$$H_1(z) = \frac{3.266 - 2.134z^{-1}}{1 + .6z^{-1} + .25z^{-2}}$$
$$H_2(z) = \frac{-1.266 + 3.220z^{-1}}{1 - .1z^{-1} - .56z^{-2}}$$

Suppose a direct form II realization is used for the second-order blocks. The resulting signal flow graph of a parallel form realization of the fourth-order filter is as shown in Figure 7.31.

FIGURE 7.31: Parallel-form Realization

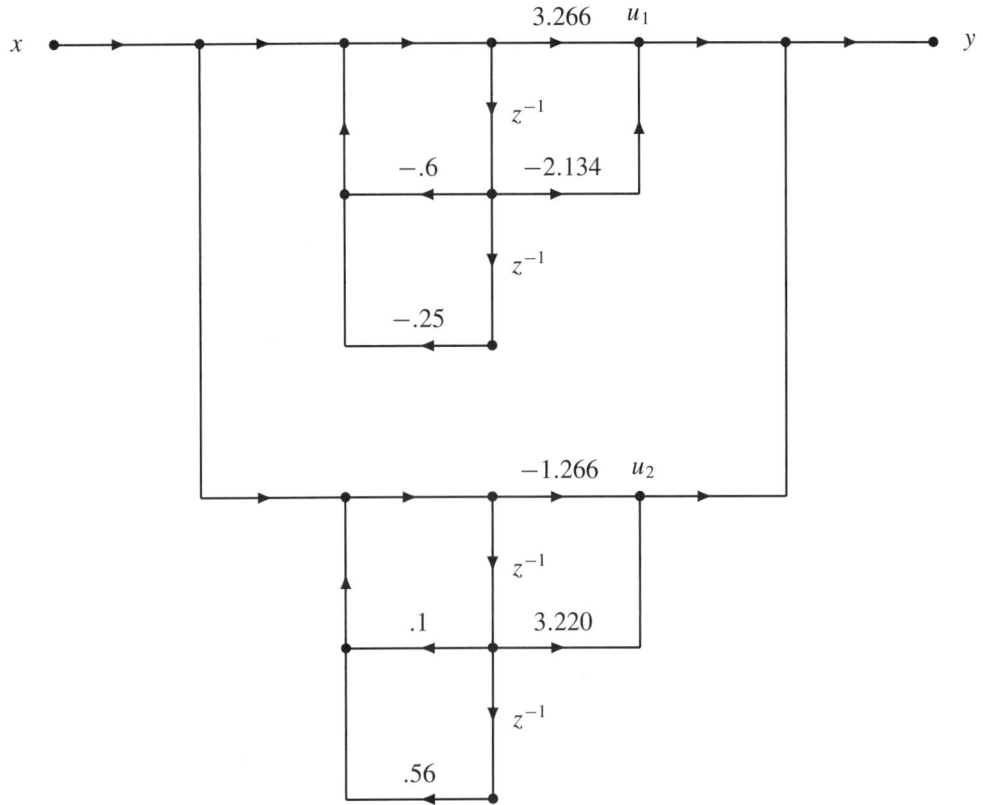

7.7.3 Cascade Form

An even simpler way to decompose $H(z)$ into lower-order subsystems is to factor both the numerator and the denominator of $H(z)$ as follows.

$$H(z) = \frac{b_0(z - z_1)(z - z_2) \cdots (z - z_n)}{(z - p_1)(z - p_2) \cdots (z - p_n)} \tag{7.7.16}$$

Note that if the degree of the numerator in (7.7.1) is $m < n$, then the factored representation in (7.7.16) will have $n - m$ zeros at $z_k = 0$. Since the coefficients of $H(z)$ are assumed to be real, complex poles and zeros will occur in conjugate pairs. The representation in (7.7.16) can be recast as a product of N second-order subsystems as follows where $N = \text{floor}[(n + 1)/2]$.

$$H(z) = b_0 H_1(z) \cdots H_N(z) \tag{7.7.17}$$

Cascade realization
This is called a *cascade form* realization. Second-order block $H_i(z)$ is constructed from either two real zeros or a complex-conjugate pair of zeros (and similarly for the poles). This way, the coefficients of $H_i(z)$ are guaranteed to be real.

$$H_i(z) = \frac{1 + b_{i1}z^{-1} + b_{i2}z^{-2}}{1 + a_{i1}z^{-1} + a_{i2}z^{-2}}, \quad 1 \le i \le N \tag{7.7.18}$$

If $H_i(z)$ is constructed from zeros z_i and z_j and poles p_q and p_r, then the four coefficients can be computed using sums and products as follows for $1 \le i \le N$.

$$b_{i1} = -(z_i + z_j) \tag{7.7.19a}$$
$$b_{i2} = z_i z_j \tag{7.7.19b}$$
$$a_{i1} = -(p_q + p_r) \tag{7.7.19c}$$
$$a_{i2} = p_q p_r \tag{7.7.19d}$$

Let u_i denote the output of the ith second-order block. Then from (7.7.17) and (7.7.18), a cascade form realization is characterized by the following time domain equations.

$$u_0(k) = b_0 x(k) \tag{7.7.20a}$$
$$u_i(k) = u_{i-1}(k) + b_{i1}u_{i-1}(k - 1) + b_{i2}u_{i-1}(k - 2)$$
$$\quad -a_{i1}u_i(k - 1) - a_{i2}u_i(k - 2), \quad 1 \le i \le N \tag{7.7.20b}$$
$$y(k) = u_N(k) \tag{7.7.20c}$$

Just as with the parallel form, if n is even, there will be N subsystems, each of order two, while if n is odd, there will be $N - 1$ second-order subsystems plus one first-order subsystem. The coefficients of a first-order subsystem are obtained from (7.7.19) by setting $z_j = 0$ and $p_r = 0$.

Any of the direct forms can be used to realize the second-order blocks in (7.7.18). A block diagram of the overall structure of a cascade form realization, for the case $N = 2$, is shown in Figure 7.32. Since the cascade form coefficients must be computed using (7.7.19), rather than

FIGURE 7.32: Cascade-form Block Diagram, $N = 2$

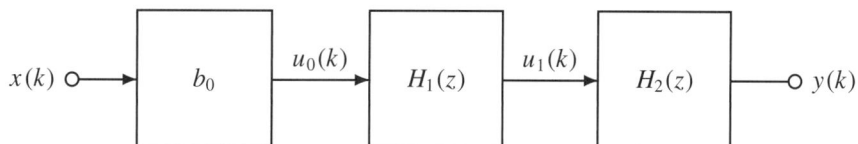

obtained directly from inspection of $H(z)$, the cascade form realization is another example of an *indirect* form.

Example 7.12

IIR Cascade Form

To compare the cascade form realization with the parallel form realization, consider the fourth-order transfer function introduced earlier in Example 7.11.

$$H(z) = \frac{2z(z^3 + 1)}{[(z + .3)^2 + .16](z - .8)(z + .7)}$$

The poles are listed in Example 7.11. There is a single zero at $z = 0$, and the remaining zeros are the three roots of -1, which are equally spaced around the unit circle.

$$z_1 = -1$$
$$z_{2,3} = \cos(\pi/3) \pm j\sin(\pi/3)$$
$$z_4 = 0$$

Suppose $H_1(z)$ is a block associated with the complex-conjugate pairs of zeros $\{z_2, z_3\}$ and poles $\{p_1, p_2\}$, and $H_2(z)$ is associated with the real zeros and poles. Running *exam7_12* produces the following three subsystems.

$$b_0 = 2$$
$$H_1(z) = \frac{1 - z^{-1} + z^{-2}}{1 + .6z^{-1} + .25z^{-2}}$$
$$H_2(z) = \frac{1 + z^{-1}}{1 - .1z^{-1} - .56z^{-2}}$$
$$H_2(z) = \frac{1 + z^{-1}}{1 - .1z^{-1} - .56z^{-2}}$$

If a direct form II realization is used for the second-order blocks, then the resulting signal flow graph of a cascade form realization of the fourth-order filter is as shown in Figure 7.33.

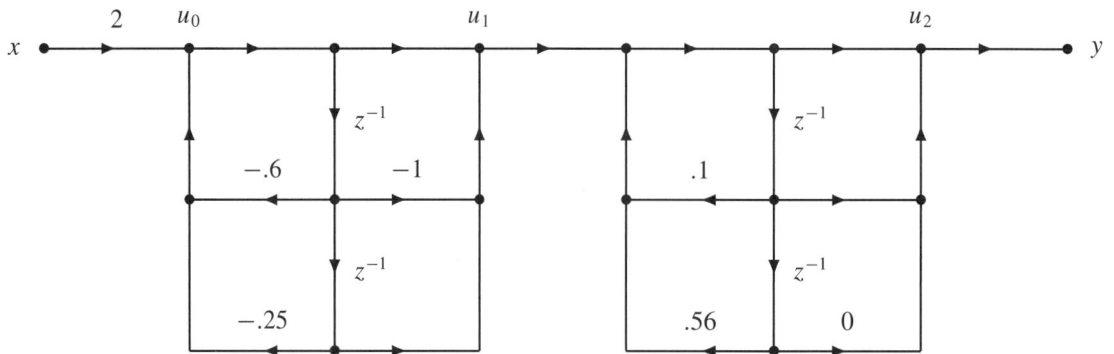

FIGURE 7.33: Cascade-form Realization

One advantage that the cascade form has over the parallel form is that there is considerable flexibility in the way the zeros and poles can be grouped together to form the second-order blocks. Let $b_i(z)$ be the numerator associated with the ith pair of zeros, and let $a_j(z)$ be the denominator associated with the jth pair of poles. Since i and j range from 1 to N, there are a total of $N!$ possible orderings of the numerators (and similarly for the denominators). Hence the total number of ways the numerators and denominators can be combined to form second-order blocks is $p = (N!)^2$. All of the orderings are equivalent if infinite precision arithmetic is used. For finite-precision filters, it is recommended that pairs of zeros and poles that are close to one another be grouped together in order to reduce occurrence of block outputs that are very large or very small. First, the pole closest to the unit circle is paired with the nearest zero. This process is repeated until all of the poles and zeros are paired. Finally, it is recommended that the blocks be ordered either in terms of increasing pole distance from the unit circle or in terms of decreasing pole distance (Jackson, 1986).

FDSP Functions

The FDSP toolbox contains the following functions for computing indirect form realizations of an IIR transfer function.

```
% F_PARALLEL: Find parallel form filter realization
% F_CASCADE:  Find cascade form digital filter realization
%
% Usage:
%        [B,A,R0] = f_parallel (b,a)
%        [B,A,b0] = f_cascade  (b,a)
% Pre:
%        b = vector of length n+1 containing coefficients
%            of numerator polynomial.
%        a = vector of length n+1 containing coefficients
%            of denominator polynomial.
% Post:
%        B   = N by 3 matrix containing coefficients of
%              numerators of second-order blocks.
%        A   = N by 3 matrix containing coefficients of
%              denominators of second-order blocks.
%        R0 = constant term of parallel form realization.
%        b0 = numerator gain
% Notes:
%        1. It is required that length(b) = length(a). If
%           needed, pad b with trailing zeros.
%        2. It is required that b(1)<>0.  Otherwise factor
%           out a z^-1 and then find the parallel or
%           cascade form
```

Once the parameters of the indirect forms are obtained by calls to *f_parallel* or *f_cascade*, the indirect forms can be evaluated using the following FDSP functions.

```
% F_FILTPAR: Compute output of parallel form filter realization
% F_FILTCAS: Compute output of cascade form filter realization
%
% Usage:
%        y = f_filtpar (B,A,R0,x)
%        y = f_filtcas (B,A,b0,x)
```

Continued on p. 550

Continued on p. 549

```
% Pre:
%         B  = N by 2 matrix containing numerator
%              coefficients of second-order blocks.
%         A  = N by 3 matrix containing denominator
%               coefficients of second-order blocks.
%         R0 = direct term of parallel form realization
%         x  = vector of length p containing samples
%               of input signal.
%         b0 = numerator gain factor
% Post:
%         y = vector of length p containing samples of
%              output signal assuming zero initial
%              conditions.
%
% Note: The arguments B, A, 0, and b0 are obtained by
%       calling f_parallel or f_cascade.
```

● ● ● ● ● ● ● ● ● ● ● ● ● ● ● ● ● ●

*7.8   Finite Word Length Effects

IIR filters are more general than FIR filters and, because of this, there are certain additional finite word length effects that are peculiar to IIR filters. We begin by examining finite word length effects that the two types of filters have in common. For example, input quantization error for an IIR filter is the same as that for an FIR filter. The only difference is that when the power gain Γ in (6.9.11) is used to compute the power of the quantization noise at the output, there are an infinite number of terms to sum for an IIR filter instead of the finite number for an FIR filter.

7.8.1   Coefficient Quantization Error

The effects of coefficient quantization error are somewhat more involved for an IIR filter because one has to consider both the poles and the zeros, not just the zeros as with an FIR filter. Recall that if the coefficients range over the interval $[-c, c]$ and N bits are used to represent

Quantization level   the coefficients, then the *quantization level* is

$$q = \frac{c}{2^{N-1}} \tag{7.8.1}$$

If $c = 2^M$, then for a fixed-point representation, $M + 1$ bits are used for the integer part (including the sign), and $N - (M + 1)$ bits are used for the fractional part. Consider an IIR filter with the following transfer function.

$$H(z) = \frac{b_0 + b_1 z^{-1} + \cdots + b_m z^{-m}}{1 + a_1 z^{-1} + \cdots + a_n z^{-n}} \tag{7.8.2}$$

The filter parameters a and b must be quantized because they are stored in fixed length memory locations. Assuming the coefficients of $H(z)$ are quantized to N bits, this results in the following quantized transfer function where $Q_N(x)$ is the staircase-like quantization operator in Definition 6.2.

$$H_q(z) = \frac{Q_N(b_0) + Q_N(b_1)z^{-1} + \cdots + Q_N(b_m)z^{-m}}{1 + Q_N(a_1)z^{-1} + \cdots + Q_N(a_n)z^{-n}} \tag{7.8.3}$$

Pole Locations

One way to evaluate the effects of coefficient quantization is to look at the locations of the poles and zeros. Recall that the roots of a polynomial can be very sensitive to small changes in the coefficients of the polynomial, particularly for higher-degree polynomials. As a consequence, high-order direct form realizations of $H(z)$ can be very sensitive to coefficient quantization error. For example, if $H(z)$ is a narrowband filter with poles clustered just inside the unit circle, then some of those poles may migrate across the unit circle and render a direct form realization unstable. Even if the poles do not cross the unit circle, movement of a pole or a zero near the unit circle can cause a significant change in the magnitude response.

Given the sensitivity of the poles and zeros to coefficient quantization, the preferred realizations are the indirect cascade and parallel forms based on second-order blocks. For both of these realizations, the poles are decoupled from one another with each pair of poles associated with its own second-degree polynomial. For the cascade realization, this is also true for the zeros. However, for the parallel form realization, the zeros are more sensitive to coefficient quantization because the residues in (7.7.11) depend on all of the coefficients of $H(z)$.

It is of interest to examine a typical second-order block in more detail. Suppose a complex conjugate pair of poles is located at $p = r \exp(\pm j\phi)$. Then the transfer function of this block can be written as follows.

$$\begin{aligned} H(z) &= \frac{b(z)}{[z - r\exp(j\phi)][z - r\exp(-j\phi)]} \\ &= \frac{b(z)}{z^2 - r[\exp(j\phi) + \exp(-j\phi)]z + r^2} \\ &= \frac{b(z)}{z^2 - 2r\cos(\phi)z + r^2} \end{aligned} \tag{7.8.4}$$

The coefficient vector of the denominator of a second-order block is $a = [1, -2r\cos(\phi), r^2]^T$. The denominator coefficient vector can be recast in terms of the pole p as follows.

$$a = [1, -2\operatorname{Re}(p), |p|^2]^T \tag{7.8.5}$$

Note that the real part of the pole is proportional to a_1, but the radius of the pole is proportional to $\sqrt{a_2}$. This nonlinear dependence of the pole radius on coefficient a_2 means that achievable pole locations using quantized versions of a will not be equally spaced. From the stability triangle in Figure 3.20, for stable poles coefficient a_1 must be in the range $(-c, c)$ where $c = 2$. The distribution of possible pole locations of stable poles for the case $N = 5$ is shown in Figure 7.34. Note that not only is the grid of possible pole locations not uniform, it is also sparse in the vicinity of the real axis.

The problem of nonuniform placement of poles can be circumvented by using a coupled-form realization of the denominators of the second-order blocks (Rabiner et al, 1970). This realization features parameters that correspond directly to the real and imaginary parts of the poles. As a result, coefficient quantization produces a uniform grid of realizable pole locations. A disadvantage of the coupled-form realization is that the number of multiplications is increased in comparison with a direct form realization.

FIGURE 7.34: Realizable Stable Pole Locations for a Quantizing Second-order Block with c = 2 and N = 5

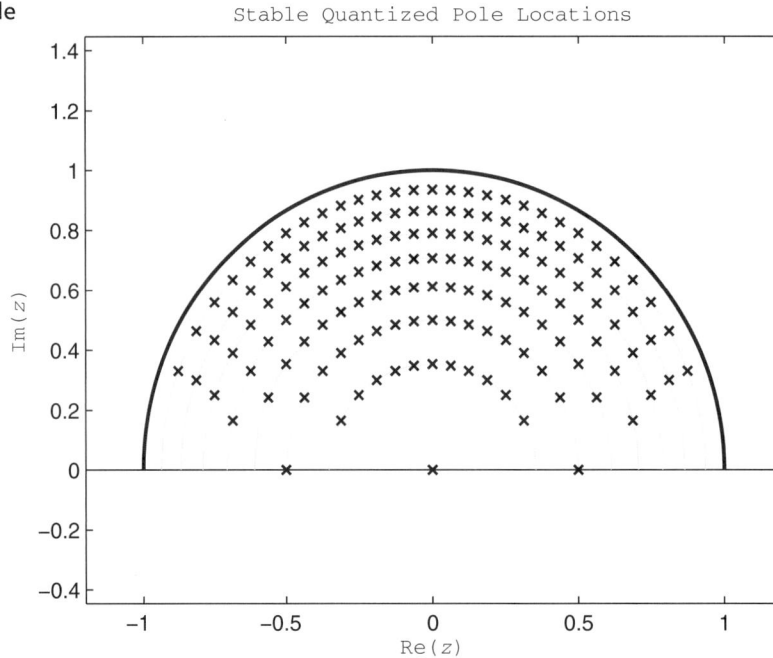

Zero Placement

The placement of zeros of a second-order block is also constrained as in Figure 7.34. For some filters of interest the zeros are on the unit circle. In these cases $r = 1$ and the second-order block transfer function simplifies to

$$H(z) = \frac{b_0(z^2 - 2\cos(\phi)z + 1)}{a(z)} \tag{7.8.6}$$

Zeros on unit circle

Since $b_2 = b_0$, the zeros of the quantized transfer function *remain* on the unit circle; only their angles (frequencies) change. Thus zeros of cascade form filters that are on the unit circle are relatively insensitive to coefficient quantization. Examples of filters with *zeros on the unit circle* include notch filters and inverse comb filters.

Example 7.13   **IIR Coefficient Quantization**

As an illustration of the detrimental effects of coefficient quantization error, consider a comb filter designed to extract a finite number of isolated equally spaced frequencies. From (7.2.15) and (7.2.16), for a filter of order $n = 9$ with a pole radius of $r = .98$, the comb filter transfer function is

$$H(z) = \frac{.1663}{1 - .8337z^{-9}}$$

This is a relatively benign example because all but two of the coefficients can be represented exactly; only b_0 and a_9 have quantization error. Suppose $c = 2$ and $N = 4$ bits are used to quantize the coefficients. Thus the coefficient values are in the range $[-2, 2]$, and the quantization level is $q = .25$. A comparison of the magnitude responses for the double-precision floating-point case (64 bits), and the N-bit fixed-point case is shown in Figure 7.35, where it is evident that there is a difference in the magnitude responses as would be expected given the low precision. The attenuation between the frequencies to be extracted is clearly

FIGURE 7.35: Magnitude Responses of a Comb Filter Using a Double-precision Floating-point Implementation and a Fixed-point Representation with $c = 2$ and $N = 4$

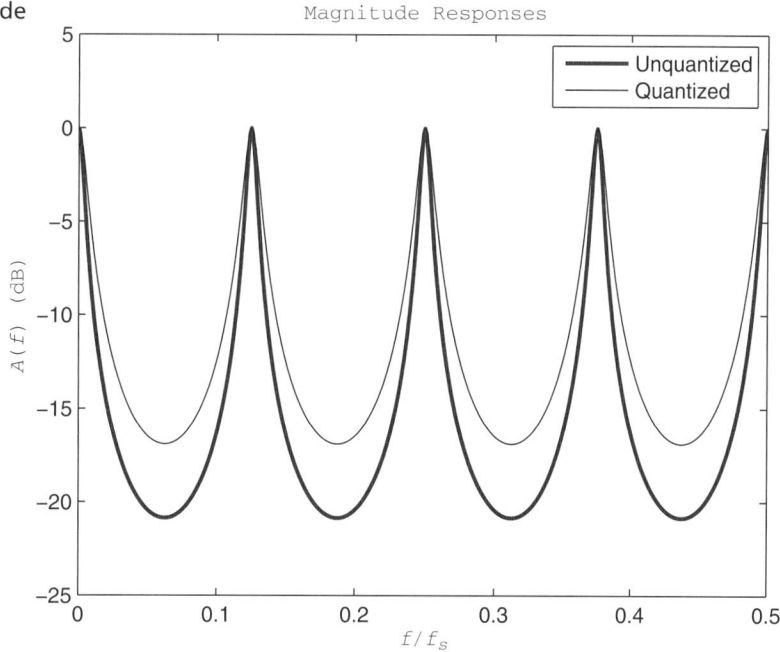

Magnitude Responses

inferior for the quantized filter. When $N \geq 12$, the two magnitude responses are more or less indistinguishable. For $N < 4$ the quantized system $H_q(z)$ becomes unstable.

7.8.2 Roundoff Error, Overflow, and Scaling

As with FIR filters, the arithmetic used to compute an IIR filter output must be performed with finite precision. For example, the output of a general IIR filter can be computed as follows using a direct form II realization.

$$u(k) = \sum_{i=1}^{n} a_i u(k - i) + x(k) \tag{7.8.7a}$$

$$y(k) = \sum_{i=0}^{m} b_i u(k - i) \tag{7.8.7b}$$

Roundoff error

If the coefficients are quantized to N bits, and the signals are quantized to N bits, then the product terms will each be of length $2N$ bits. When the products are rounded to N bits, the resulting *roundoff error* can be modeled as uniformly distributed white noise. In this instance there are several sources of noise. If we assume the roundoff noise sources are statistically independent of one another, the noise sources associated with the input terms can be combined, and similarly for the noise sources associated with the output terms.

$$e_a(k) \triangleq \sum_{i=1}^{n} Q_N[a_i u(k - i)] - a_i u(k - i) \tag{7.8.8a}$$

$$e_b(k) \triangleq \sum_{i=0}^{n} Q_N[b_i u(k - i)] - b_i u(k - i) \tag{7.8.8b}$$

For a second-order direct form II realization, this results in the equivalent linear model of roundoff error displayed in Figure 7.36 for the case $m = n = 2$.

FIGURE 7.36: Linear Model of Product Roundoff Error in a Direct Form II Realization

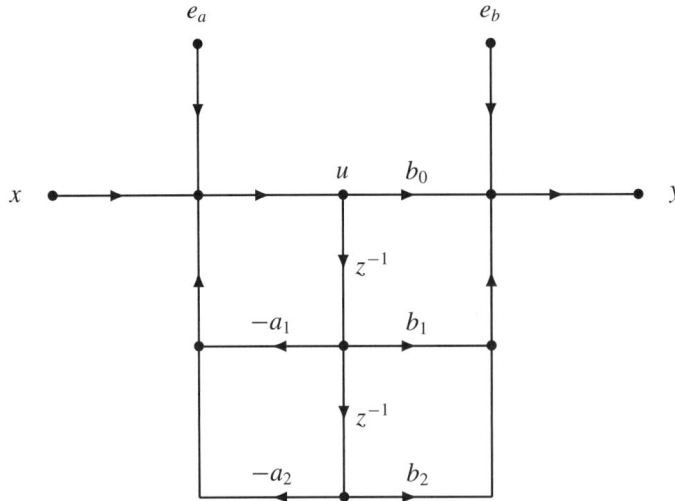

For an IIR filter, the roundoff noise appearing at the output will depend on the power gain Γ. If $h(k)$ is the impulse response of a stable IIR filter, then the power gain is

$$\Gamma \triangleq \sum_{k=0}^{\infty} h^2(k) \tag{7.8.9}$$

It can be shown (e.g., Oppenheim et al, 1999) that the average power of the roundoff noise appearing at the filter output is as follows, where q is the signal quantization level in (7.8.1) and Γ is the power gain.

$$\sigma_y^2 = \frac{(\Gamma n + m + 1)q^2}{12} \tag{7.8.10}$$

Overflow

Overflow error

Another source of error occurs as a result of the summing operations in (7.8.7). The sum of several N-bit numbers will not always fit within N bits. When the sum is too large to fit, this results in *overflow error*. A single overflow error can cause a significant change in filter performance. This is because when a two's-complement representation overflows, even by a small amount, it goes from a large positive number to a large negative number or conversely. This can be seen from the overflow characteristic shown in Figure 7.37, where it is assumed that the numbers are fractions in the interval $-1 \le x < 1$.

Clipping

There are a number of approaches to compensating for overflow. One way is to detect overflow and then *clip* the summing junction output to the maximum value that can be represented. This clipping or saturation characteristic is shown with the dashed line in Figure 7.37. *Clipping* tends to reduce, but not eliminate, the detrimental effects of overflow error.

Scaling

Another approach is to eliminate overflow from occurring at all by the use of *scaling*. Let $h_i(k)$ be the impulse response measured at the output of the ith summing node. For the second-order

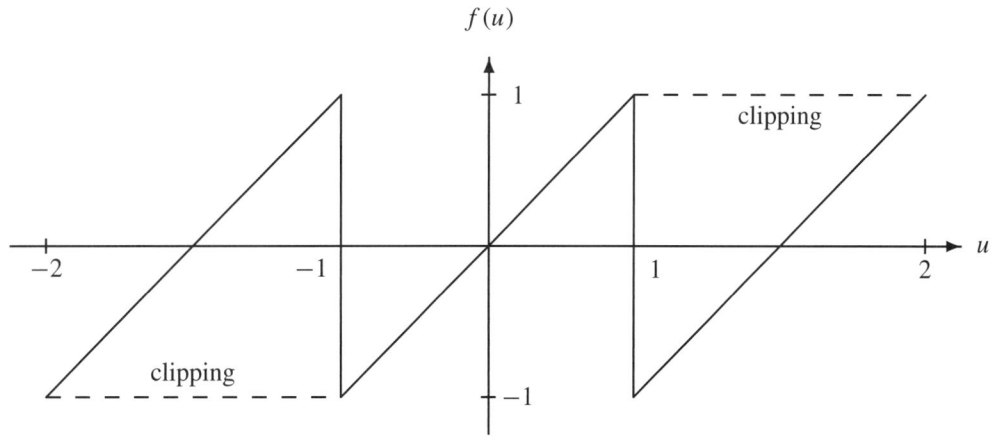

FIGURE 7.37: Overflow Characteristic of Two's Complement Addition

block shown in Figure 7.36, there are two summing nodes with

$$h_1(k) = Z^{-1} \left\{ \frac{1}{a(z)} \right\} \tag{7.8.11a}$$

$$h_2(k) = Z^{-1} \left\{ \frac{b(z)}{a(z)} \right\} \tag{7.8.11b}$$

Notice that $h_1(k)$ is the impulse response of the auto-regressive or all-pole part, and $h_2(k)$ is the complete impulse response of $H(z)$. If $y_i(k)$ is the output of the ith summing node, and $|x(k)| \le c$, then

$$
\begin{aligned}
|y_i(k)| &= |\sum_{p=0}^{\infty} h_i(p)x(k-p)| \\
&\le \sum_{p=0}^{\infty} |h_i(p)x(k-p)| \\
&= \sum_{p=0}^{\infty} |h_i(p)| \cdot |x(k-p)| \\
&\le c \sum_{p=0}^{\infty} |h_i(p)| \tag{7.8.12}
\end{aligned}
$$

Thus $|y_i(k)| \le c\|h_i\|_1$, where $\|h_i\|$ is the L_1 norm of h_i. That is,

$$\|h_i\|_1 \overset{\Delta}{=} \sum_{k=0}^{\infty} |h_i(k)| \tag{7.8.13}$$

Notice that the L_1 norm of h_i in (7.8.13) is an infinite series version of the L_1 norm of the coefficient vector b in (6.9.26). Recall from Section 2.9 that if the system $H_i(z)$ is BIBO stable, then the impulse response $h_i(k)$ is absolutely summable. Consequently, for stable filters the infinite series in (7.8.13) will converge.

Suppose there are a total of r summing nodes in the signal flow graph. Then addition overflow can be prevented if the input signal $x(k)$ is scaled by s_1, where *scale factor* s_p is

Scale factors

FIGURE 7.38: Scaling to Prevent Addition Overflow in a Second-order Direct Form II Block

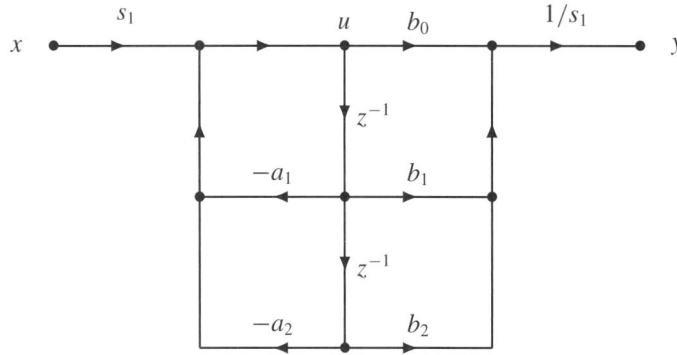

defined as follows.

$$s_p = \frac{1}{\max_{i=1}^{r}\{\|h_i\|_p\}}, \quad 1 \le p \le \infty \qquad (7.8.14)$$

A signal flow graph of a general second-order direct form II realization that uses scaling to prevent overflow is shown in Figure 7.38.

Although scaling using the L_1 norm is effective in preventing overflow, it does suffer from a practical drawback. Roundoff noise and input quantization noise are not significantly affected by scaling. As a consequence, when the input is scaled by s_1, the resulting reduction in signal strength can cause a corresponding reduction in the signal-to-noise ratio. A less severe form of scaling can be used that eliminates most, but not all, overflow. In those instances where overflow does occur, it can be compensated for using clipping. If the input signal is a pure sinusoid, then overflow from this type of periodic input can be eliminated by using scaling that is based on the filter magnitude response. Here the L_∞ norm of h_i is used where

$$\|h_i\|_\infty \overset{\Delta}{=} \max_{0 \le f \le f_s/2}\{A_i(f)\} \qquad (7.8.15)$$

Addition overflow from a pure sinusoidal input is prevented if the input signal $x(k)$ is scaled by s_∞, where s_∞ is computed as in (7.8.14). The most common form of scaling uses the L_2 or energy norm which is defined as follows.

$$\|h_i\|_2 \overset{\Delta}{=} \left(\sum_{k=0}^{\infty} |h_i(k)|^2\right)^{1/2} \qquad (7.8.16)$$

Again the scale factor s_2 is computed using (7.8.14). One advantage of the L_2 norm is that it is relatively easy to compute. In fact, for certain realizations, closed-form expressions for s_2 can be computed in terms of the filter coefficients (Ifeachor and Jervis, 2002). Note that the L_2 norm of h_i in (7.8.16) is a generalization of the Euclidean norm of the coefficient vector b in (6.9.28). Like their finite-dimensional counterparts, the L_1, L_2, and L_∞ norms can be shown to satisfy the following relationship.

$$\|h\|_2 \le \|h\|_\infty \le \|h\|_1 \qquad (7.8.17)$$

Example 7.14   **IIR Overflow and Scaling**

As an illustration of the prevention of overflow by scaling, suppose $|x(k)| \leq 5$, and consider the following IIR filter.

$$H(z) = \frac{4z^{-1}}{1 - .64z^{-2}}$$

Here $c = 5$, $b = [0, 4, 0]^T$, and $a = [1, 0, -.64]^T$. Expressing $H(z)$ in terms of positive powers of z and factoring the denominator yields

$$H(z) = \frac{4z}{(z - .8)(z + .8)}$$

From (7.8.11a), the impulse response of the auto-regressive part of $H(z)$ is

$$h_1(k) = Z^{-1} \left\{ \frac{1}{(z - .8)(z + .8)} \right\}$$

$$= .625[(.8)^{k-1} - (-.8)^{k-1}]\mu(k - 1)$$

$$= .78125[(.8)^k - (-.8)^k]\mu(k)$$

Similarly, from (7.8.11b), the impulse response of $H(z)$ is

$$h_2(k) = Z^{-1} \left\{ \frac{4z}{(z - .8)(z + .8)} \right\}$$

$$= 2.5[(.8)^k - (-.8)^k]\mu(k)$$

Notice that $h_2(k) \geq h_1(k)$ for $k \geq 0$. Therefore it is sufficient to compute the norm of $h(k) = h_2(k)$. From (7.8.13), the L_1 norm is

$$\|h\|_1 = 2.5 \sum_{k=0}^{\infty} |(.8)^k - (-.8)^k|$$

$$= 5 \sum_{i=0}^{\infty} (.8)^{2i}$$

$$= 5 \sum_{i=0}^{\infty} (.64)^i$$

$$= \frac{5}{1 - .64}$$

$$= 13.89$$

Thus a scale factor that will eliminate fixed-point overflow is $s_1 = 1/13.89$ or

$$s_1 = .072$$

A signal flow graph of a direct form II realization with scaling to avoid overflow is shown in Figure 7.39.

7.8.3 Limit Cycles

For IIR filters, there is an unusual finite word length effect that is sometimes observed when the input goes to zero. Recall that if a filter is stable and the input goes to zero, then the output

FIGURE 7.39: Scaling to Prevent Addition Overflow

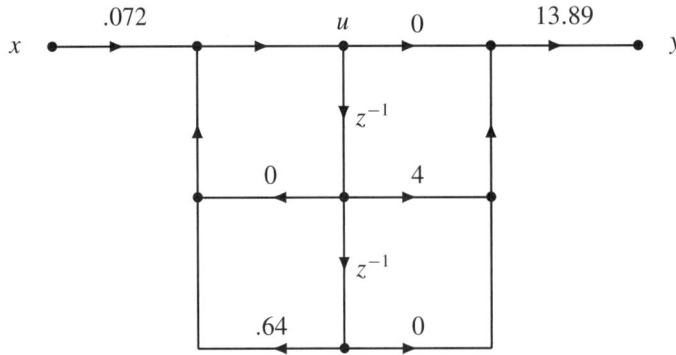

should approach zero as the natural mode terms die out. However, for finite-precision IIR filters, the output sometimes approaches a nonzero constant or it oscillates. These zero-input oscillations are a nonlinear phenomenon called *limit cycles*. There are two types of limit cycles. The first is a limit cycle that is caused by overflow error. Overflow limit cycles can be quite large in amplitude, but they can be eliminated if the outputs of the summing junctions are clipped. Of course, scaling by s_1 will also eliminate this type of limit cycle. The second type of limit cycle is a small limit cycle of amplitude q that can occur as a result of product roundoff error.

Limit cycle

Example 7.15

Limit Cycle

As an illustration of a limit cycle caused by product roundoff error, consider the following quantized first-order IIR system.

$$y(k) = Q_N[-.7y(k-1)] + 3x(k)$$

Suppose $N = 4$ bits are used with a scale factor of $c = 4$. In this case the quantization level is

$$q = \frac{4}{2^3} = .5$$

Next consider the impulse response of the quantized system, $h_q(k)$. The result, computed using *exam7_15*, is shown in Figure 7.40. Note that even though $H(z)$ clearly is stable with a pole at $p = -.7$, the steady-state response does not go to zero. Instead it oscillates with period two and amplitude q because of the product roundoff error. For comparison, the unquantized impulse response, $h(k)$, is also displayed.

In IIR filters, limit cycle solutions can be small limit cycles associated with roundoff error as in Figure 7.40, or large amplitude limit cycles associated with overflow. For FIR filters, limit cycles are *not* possible because there are no feedback paths to sustain an oscillation. Indeed, in comparison with IIR filters, FIR filters are less sensitive to finite word length effects in general. Along with their guaranteed stability, this is one of the advantages of FIR filters that accounts for their popularity.

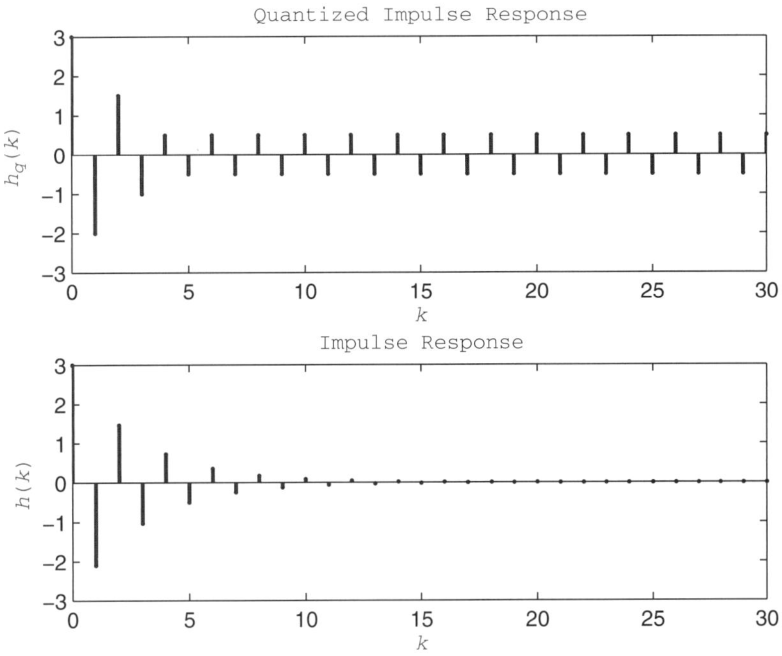

FIGURE 7.40: Limit Cycle in Impulse Response Caused by Product Roundoff Error using $N = 4$ Bits and a Scale Factor of $c = 4$

Quantized Impulse Response

Impulse Response

FDSP Functions

The FDSP toolbox contains the following functions for evaluating finite word length effects using different filter realization structures.

```
% F_FILTER1:  Compute the quantized zero-state response of an IIR filter
% F_IMPULSE:  Compute the quantized impulse response of an IIR filter
% F_FREQZ:    Compute frequency response of an IIR filter using the DFT
%
% Usage:
%         y    = f_filter1 (b,a,x,bits,realize);
%       [h,k]  = f_impulse (b,a,N,bits,realize);
%       [H,f]  = f_freqz (b,a,N,fs,bits,realize);
% Pre:
%         b      = coefficient vector of numerator polynomial
%         a      = coefficient vector of denominator polynomial
%         x      = a vector of length N containing the input
%         bits   = optional integer specifying the number of
%                    fixed-point bits used for coefficient quantization.
%                    The default is double precision floating-point
%         realize = optional integer specifying the realization
%                    structure to use.  The default is to use the
%                    direct form of MATLAB function filter.
%
```

Continued on p. 560

Continued on p. 559

```
%                      0 = direct form
%                      1 = cascade form
%                      2 = lattice form (FIR) or parallel form (IIR)
%
%           N       = number of samples
%           fs      = sampling frequency (default = 1)
% Post:
%           y = N by 1 vector containing the zero-state response
%           h = N by 1 vector containing the impulse response
%           k = N by 1 vector containing discrete times
%           H = 1 by N+1 complex vector containing the frequency response
%           f = 1 by N+1 vector containing discrete frequencies (0 to fs/2)
% Note:
%           For the parallel form, the poles of H(z) must be distinct.
```

7.9   GUI Software and Case Stud1

This section focuses on the design and realization of IIR filters. A graphical user interface module called *g_iir* is introduced that allows the user to design and implement IIR filters, without any need for programming. A case study example is presented and solved using MATLAB.

g_iir: Design and implement IIR filters

GUI Module

The FDSP toolbox includes a graphical user interface module called *g_iir* that allows the user to design a variety of IIR filters. *GUI module g_iir* features a display screen with tiled windows, as shown in Figure 7.41. The upper left-hand *Block diagram* window contains a block diagram of the IIR filter under investigation. It is an nth-order filter with the following transfer function.

$$H(z) = \frac{b_0 + b_1 z^{-1} + \cdots + b_n z^{-m}}{1 + a_1 z^{-1} + \cdots + a_n z^{-n}} \tag{7.9.1}$$

Edit boxes

The *Parameters* window below the block diagram displays edit boxes containing the filter parameters. The contents of each edit box can be directly modified by the user, with the changes activated with the Enter key. Parameters $F0$, $F1$, B, and fs are the lower cutoff frequency, upper cutoff frequency, transition bandwidth, and sampling frequency, respectively. The lowpass filter uses cutoff frequency $F0$, the highpass filter uses cutoff frequency $F1$, and the bandpass and bandstop filters use both $F0$ and $F1$. For resonators and notch filters, $F0$ is used. The parameters *deltap* and *deltas* specify the passband ripple and stopband attenuation, respectively.

Type options

The *Type* and *View* windows in the upper-right corner of the screen allow the user to select both the type of frequency-selective filter and the viewing mode. The filter types include a resonator filter; a notch filter; lowpass, highpass, bandpass, and bandstop filters; and a user-defined filter whose coefficients are defined in a user-supplied MAT file that contains a, b, and

View options

fs. The *View* options include the magnitude response, the phase response, and a pole-zero plot

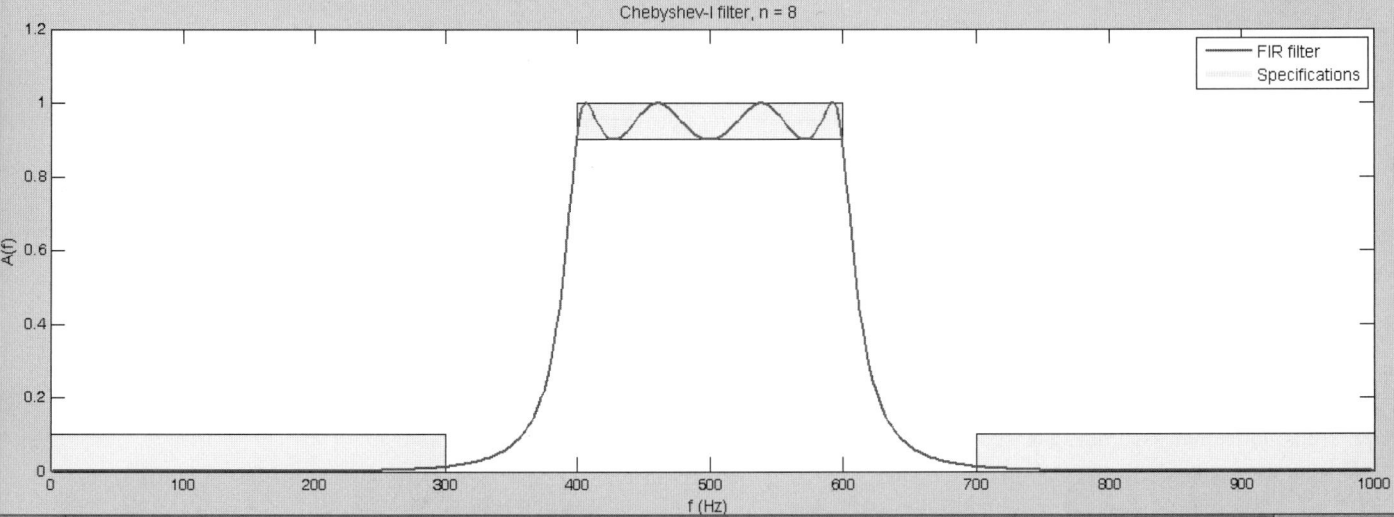

that also contains the impulse response. The *Plot* window along the bottom half of the screen shows the selected view.

Just below the view options is a checkbox control that toggles the magnitude response display between linear and logarithmic scales. When it is checked, the passband ripple and stopband attenuation in the *Parameters* window change to their logarithmic equivalents, A_p and A_s, respectively. Below the *Type* and *View* windows is a horizontal slider bar that allows the user to directly control the filter order n. Note that the filter order may or may not meet all of the design specifications depending on the value of n. Furthermore, for some filters, when n is set too high, the filter implementation can become unstable due to finite word length effects.

Menu options   The *Menu* bar at the top of the screen includes several menu options. The *Caliper* option allows the user to measure any point on the current plot by moving the mouse crosshairs to that point and clicking. The *Save data* option is used to save a, b, fs, x, and y in a user-specified MAT file for future use. Files created in this manner subsequently can be loaded with the User-defined filter option. The *Prototype* option allows the user to select the analog prototype filter (Butterworth, Chebyshev-I, Chebyshev-II, or elliptic) for use with the basic frequency-selective filter types. The *Print* option prints the contents of the *plot* window to a printer or a file. Finally, the *Help* option provides the user with some helpful suggestions on how to effectively use module *g_iir*.

CASE STUDY 7.1   Reverb Filter

Music generated in a concert hall sounds rich and full because it arrives at the listener along multiple paths, both direct and through a series of reflections. This reverberation effect can be simulated by processing the sound signal with a *reverb* filter (Steiglitz, 1996). The speed of sound in air at room temperature is about $v = 345$ m/sec. Suppose the distance from the music source to the listener is d m. If f_s denotes the sampling rate, then the distance in samples is

$$L = \text{floor}\left(\frac{f_s d}{v}\right) \tag{7.9.2}$$

Sound is attenuated as it travels through air and becomes dispersed. Let $0 < r < 1$ be the factor by which sound is attenuated as it propagates from the source to the listener. To roughly approximate a reverberation effect, suppose the echoes from one or more reflections arrive as multiples of L samples and are attenuated by powers of r. If there are n echoes, or paths of increasing length, then this effect can be modeled by the following difference equation.

$$y(k) = \sum_{i=1}^{n} r^i x(k - Li) \tag{7.9.3}$$

In the limit as the number of echoes, n, approaches infinity we arrive at the following geometric series transfer function for multiple echoes.

$$F(z) = \sum_{i=1}^{\infty} r^i z^{-Li}$$
$$= \frac{rz^{-L}}{1 - rz^{-L}} \tag{7.9.4}$$

Observe that $F(z)$ is essentially a comb filter with poles equally spaced around a circle of radius r. A block diagram of the comb filter, that explicitly shows the feedback path taken by the echoes, is shown in Figure 7.42.

To develop a more refined model of the reverberation effect, one should take into consideration the observation that high-frequency sounds tend to get absorbed more than low-frequency

FIGURE 7.42: Basic Comb Filter Used to Model Multiple Echoes

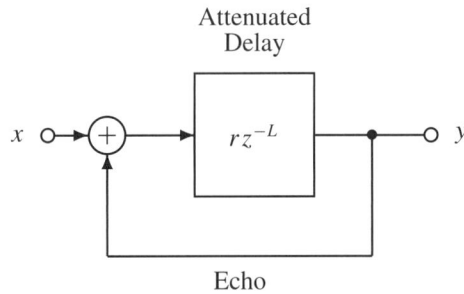

FIGURE 7.43: Lowpass Comb Filter that Includes Effects of Frequency-dependent Sound Absorption

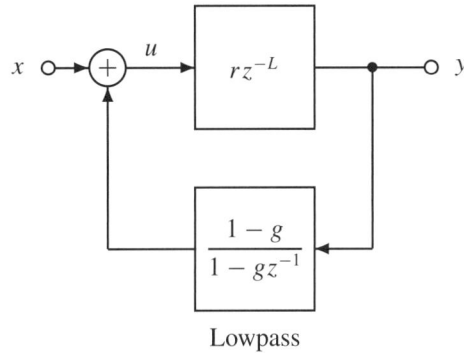

sounds. This effect can be included by inserting a first-order lowpass filter, $D(z)$, into the feedback path of the basic comb filter (Moorer, 1979).

$$D(z) = \frac{1-g}{1-gz^{-1}} \qquad (7.9.5)$$

Here the real pole $0 < g < 1$ controls the cutoff frequency of the lowpass characteristic. A block diagram showing this more refined lowpass comb filter is shown in Figure 7.43. The transfer function of the lowpass comb filter can be obtained from Figure 7.43 by solving for the summing junction output signal $U(z)$. Working backwards around the loop, we have

$$\begin{aligned} U(z) &= X(z) + D(z)Y(z) \\ &= X(z) + D(z)rz^{-L}U(z) \end{aligned} \qquad (7.9.6)$$

Solving (7.9.6) for $U(z)$ then yields

$$U(z) = \frac{X(z)}{1 - rz^{-L}D(z)} \qquad (7.9.7)$$

Finally, from Figure 7.43, (7.9.7) and (7.9.5), the output of the lowpass comb filter is

$$\begin{aligned} Y(z) &= rz^{-L}U(z) \\ &= \frac{rz^{-L}X(z)}{1 - rz^{-L}D(z)} \\ &= \frac{rz^{-L}X(z)}{1 - rz^{-L}(1-g)/(1-gz^{-1})} \\ &= \frac{rz^{-L}(1-gz^{-1})X(z)}{1 - gz^{-1} - r(1-g)z^{-L}} \end{aligned} \qquad (7.9.8)$$

Multiplying both sides of (7.9.8) by z^{L+1}, we find that the overall transfer function of the lowpass comb filter, in terms of positive powers of z, is

$$C(z) = \frac{r(z-g)}{z[z^L - gz^{L-1} - r(1-g)]} \qquad (7.9.9)$$

The reverberation effect can be made much fuller and richer when multiple lowpass comb filters with different delays and cutoff frequencies are used. Moorer (1979) has proposed using multiple lowpass comb filters in parallel followed by an allpass filter. The allpass filter inserts a frequency-dependent phase shift, or delay, but does not change the magnitude response. Recall from (5.4.7) that an allpass filter is a filter whose numerator and denominator exhibit reverse symmetry. For example, the following Mth-order allpass filter can delay signals up to M samples.

$$G(z) = \frac{c + z^{-M}}{1 + cz^{-M}} \qquad (7.9.10)$$

The overall configuration for a reverb filter, featuring P lowpass comb sections, is shown in Figure 7.44. The composite transfer function of the reverb filter is as follows.

$$H(z) = G(z) \sum_{i=1}^{P} C_i(z) \qquad (7.9.11)$$

To implement the reverb filter in Figure 7.44, the following parameters can be used (Moorer, 1979). For the allpass filter, set

$$c = .7 \qquad (7.9.12a)$$
$$M = \text{floor}(.006 f_s) \qquad (7.9.12b)$$

FIGURE 7.44: A Reverb Filter

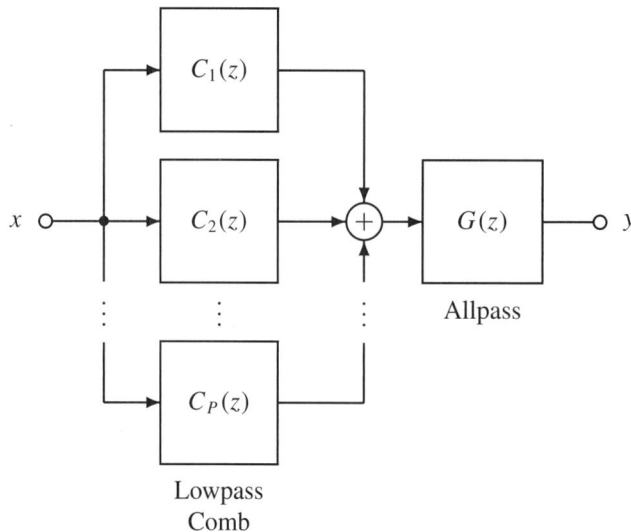

This corresponds to a delay of .006 sec. Next, use $P = 6$ lowpass comb filters with the following attenuation factor, delays, and poles.

$$r = .83 \tag{7.9.13a}$$
$$L = \text{floor}\{[.050, .056, .061, .068, .072, .078]f_s\} \tag{7.9.13b}$$
$$g = [.24, .26, .28, .29, .30, .32] \tag{7.9.13c}$$

CASE
STUDY 7.1 The FDSP toolbox contains a function called f_reverb that computes the reverb filter output using the parameters in (7.9.12) and (7.9.13). The reverb filter can be tested by running $case7\_1$ from f_dsp.

```
function case7_1

% CASE STUDY 7.1: Reverb Filter'

f_header ('Case study 7.1: Reverb filter')

% Plot impulse response

fs = 8000;
N = 8192;
x = [1,zeros(1,N-1)];
y = x;
[h,n] = f_reverb(x,fs);
n
figure
stem ([1:N-1],h(2:N),'filled','.')
f_labels ('Impulse response','{k}','{h(k)}')
axis ([0 9000 -0.4 0.6])
box on
f_wait

% Plot magnitude response

H = fft(h);
A = abs(H);
f = linspace (0,(N-1)*fs/N,N);
figure
plot (f(1:N/2),A(1:N/2))
f_labels ('Magnitude response','{f} (Hz)','{A(f)}')
f_wait

% Get sound and put it through reverb filter

tau = 3.0;
choice = 0;
p = floor(8000/tau);
z = zeros(p,1);
while choice ~= 4
    choice = menu ('Please select one','record sound','play back (normal)',...
                'play back (reverb)','exit');
```

Continued on p. 566

Continued on p. 566

```
    switch (choice)
    case 1,
        [z,cancel] = f_getsound (z,tau,fs);
        if ~cancel
            y = f_reverb (z,fs);
        end
    case 2,
        soundsc (z,fs)
    case 3,
        soundsc (y,fs);
    end
end
```

When *case7_1* is run, it first computes the impulse response shown in Figure 7.45. Unlike many of the discrete-time systems we have investigated, the impulse response of this system takes a very long time to die out because of the significant delays representing echos in the system.

The second part of *case7_1* computes the magnitude response shown in Figure 7.46. The interactions of the multiple lowpass comb filters provide a magnitude response that is broadband, yet exhibits detailed variation. This is due, in part, to the very high order of the reverb filter. From (7.9.9), (7.9.10), and Figure 7.44, the total filter order is as follows.

$$n = M + P + L_1 + \cdots + L_P \tag{7.9.14}$$

Recall from (7.9.12) and (7.9.13) that the allpass order M and the delays L_i are proportional to the sampling frequency which is set to $f_s = 8000$ Hz in *case7_1*. From (7.9.12) through (7.9.14), this results in an IIR reverb filter order n where

$$
\begin{aligned}
n &= \text{floor}(.006 f_s) + 6 + \text{floor}\{[.050 + .056 + .061 + .068 + .072 + .078] f_s\} \\
&= 3134 \tag{7.9.15}
\end{aligned}
$$

FIGURE 7.45: Impulse Response of a Reverb Filter

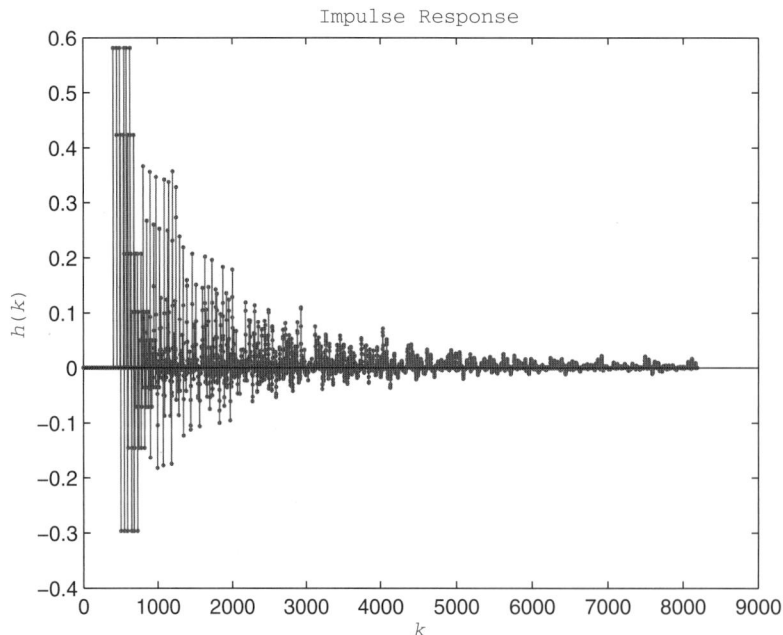

FIGURE 7.46:
Magnitude
Response of a
Reverb Filter

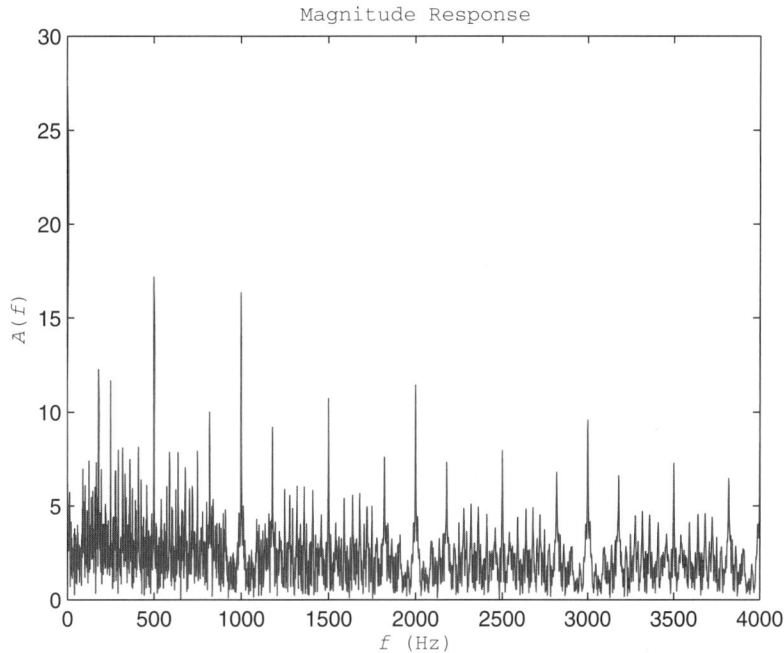

Magnitude Response

Since the reverb filter is stable, this means that all 3134 poles must be inside the unit circle! The final segment of *case7_1* displays a menu that allows the user to record up to four seconds of sound from the PC microphone, and then play it back on the PC speaker both with and without reverb filtering. The differences in the sound are distinct, and quite striking. Give it a try!

7.10   Chapter Summary

Infinite Impulse Response Filters

This chapter focused on the design of infinite impulse response or IIR digital filters with the following transfer function.

$$H(z) = \frac{b_0 + b_1 z^{-1} + \cdots + b_m z^{-m}}{1 + a_1 z^{-1} + \cdots + a_n z^{-n}} \tag{7.10.1}$$

The four basic frequency-selective filter types are lowpass, highpass, bandpass, and bandstop filters. For an ideal filter, the passband gain is $A(f) = 1$, the stopband is $A(f) = 0$, and there is no transition band. However, for a practical filter of finite order there must be a transition band separating the passband and the stopband. Furthermore, for a physically realizable filter the passband gain is not constant but lies instead within an interval $[1 - \delta_p, 1]$ where $\delta_p > 0$ is the *passband ripple*. Similarly, the stopband gain lies within an interval $[0, \delta_s]$ where $\delta_s > 0$ is the stopband attenuation. The filter magnitude response is often represented using a logarithmic scale of decibels (dB) as follows.

Passband ripple

$$A(f) = 20 \log_{10}\{|H(f)|\} \text{ dB} \tag{7.10.2}$$

Stopband attenuation

The ripple and attenuation specifications, δ_p and δ_s, have logarithmic equivalents, A_p and A_s, that are expressed in units of dB. The logarithmic scale is useful for showing the amount of *attenuation* in the *stopband*.

Resonator, notch filter

There are a number of specialized IIR filters that can be designed using pole-zero placement and gain matching. These include a *resonator*, which is designed to pass a single frequency, and a *notch filter*, which is designed to reject a single frequency. Generalizations of these two basic filters include the comb filter, which is designed to pass several harmonically related frequencies, and the inverse comb filter, which is designed to reject several harmonically related frequencies. Comb filters and inverse comb filters can be used to pass or reject a periodic input that is corrupted with noise if the first resonant frequency is set to match the fundamental frequency of the periodic input.

Classical Analog Prototype Filters

Prototype filters

A highly effective and widely used approach for designing frequency-selective IIR filters starts with an analog *prototype filter* and then transforms it to an equivalent digital filter. There are four classical families of analog filters that are used as prototype filters, and each is optimal in some sense. Butterworth filters have the property that their magnitude responses are as flat as possible in the passband. Butterworth filters are easy to design, but they have a relatively wide transition band. The transition band can be made more narrow by allowing ripples in the magnitude response. Chebyshev-I filters have ripples of equal amplitude in the passband and meet the passband specifications exactly, whereas Chebyshev-II filters have ripples of equal amplitude in the stopband and meet the stopband specification exactly. Elliptic filters have a narrow transition band that is achieved by allowing ripples of equal size in both the passband and the stopband. Thus elliptic IIR filters are similar to the equiripple FIR filters designed with the Parks-McLellan algorithm.

Design parameters

For the classical filters, the filter order required to meet a given design specification can be computed using two *design parameters* called the selectivity factor, r, and the discrimination factor, d. For an ideal filter, $r = 1$ and $d = 0$, whereas for a practical filter, $r < 1$ and $d > 0$. The classical analog lowpass filters $H_a(s)$ can be designed by starting with the magnitude response $A_a(f)$ and working backwards to determine the poles, zeros, and gain using the following fundamental relationship.

$$H_a(s)H_a(-s) = A_a^2\left(\frac{s}{j2\pi}\right) \tag{7.10.3}$$

Bilinear Analog-to-digital Filter Transformation

Normalized filter

Analog frequency-selective prototype filters of various types can be obtained by applying frequency transformations to a *normalized* lowpass filter, a filter whose radian cutoff frequency is $\Omega_0 = 1$ rad/sec. For bandpass and bandstop filters, these frequency transformations double the order of the filter. It is also possible to perform frequency transformations on digital lowpass

Bilinear transformation

filters. The most commonly used analog-to-digital filter transformation technique is the *bilinear transformation* method. This technique maps a stable analog filter $H_a(s)$ into a stable digital filter $H(z)$ using the following change of variables in $H_a(s)$.

$$s = \frac{2}{T}\left(\frac{z-1}{z+1}\right) \tag{7.10.4}$$

Frequency warping

The bilinear transformation method maps the imaginary axis of the s plane onto the unit circle of the z plane. The resulting compression of frequencies is called *frequency warping*, and it must be taken into account when the critical frequencies of the filter are specified. That is, in the construction of the analog prototype filter, each of the desired cutoff frequencies should first be prewarped using

$$F = \frac{\tan(\pi f T)}{\pi T} \qquad (7.10.5)$$

Music synthesis   IIR filters can be used for music synthesis and to create special sound effects. For example, the impulse response of the tunable plucked string filter can be adjusted to emulate the sound produced by a variety of stringed instruments. The rich full sound of concert halls can be reproduced by introducing special sound effects using a reverb filter. A reverb filter is a high-order IIR filter that is implemented as a parallel configuration of comb filters, each having a lowpass filter in its feedback path, followed by an allpass filter.

Filter Realization Structures

Direct forms   There are a number of alternative signal flow graph realizations of IIR filters. *Direct form* realizations have the property that the gains in the signal flow graphs are obtained directly from inspection of the transfer function. For IIR filters, direct form realizations include the direct form I, direct form II, and transposed direct form II structures. The direct form II realizations are canonic in the sense that they require the minimum number of memory locations to store past signal samples.

There are also a number of indirect realizations whose parameters must be computed from the original transfer function. The indirect forms decompose the original transfer function into lower-order blocks by combining complex conjugate pairs of poles and zeros. For example, IIR *Cascade form*   filters can be realized with the following *cascade-form* realization which is based on factoring $H(z)$.

$$H(z) = b_0 H_1(z) \cdots H_N(z) \qquad (7.10.6)$$

Here $N = \text{floor}[(n+1)/2]$ and the $H_i(z)$ are second-order blocks with real coefficients except for $H_N(z)$, which is a first-order block when the filter order n is odd. For IIR filters, *Parallel form*   an additional indirect filter structure is the *parallel-form* realization that is based on a partial fraction expansion of $H(z)$.

$$H(z) = R_0 + \sum_{i=1}^{N} H_i(z) \qquad (7.10.7)$$

Again, $N = \text{floor}[(n+1)/2]$ and each $H_i(z)$ is a second-order block with real coefficients except $H_N(z)$, which is a first-order block when n is odd. All of the filter realizations are equivalent to one another in terms of their overall input-output behavior if infinite precision arithmetic is used.

Finite Word Length Effects

Finite word length effects arise when a filter is implemented in either hardware or software. They are caused by the fact that both the filter parameters and the filter signals must be represented using a finite number of bits of precision. Both floating-point and fixed-point numerical representations can be used. MATLAB uses a double-precision 64-bit floating-point representation that tends to minimize finite word length effects. When an N-bit fixed-point

Quantization level   representation is used for values in the range $[-c, c]$, the *quantization level*, or spacing between adjacent values, is

$$q = \frac{c}{2^{N-1}} \qquad (7.10.8)$$

Typically, the scale factor is $c = 2^M$ for some integer $M \geq 0$. This way, $M + 1$ bits are used to represent the integer part including the sign, and the remaining $N - (M + 1)$ bits are used for the fraction part.

Quantization error can arise from input or ADC quantization, coefficient quantization, and product roundoff quantization. It is modeled using additive white noise uniformly distributed over $[-q/2, q/2]$. Another source of error is overflow error that can occur when several finite precision numbers are added. Overflow error can be reduced by clipping or eliminated by proper scaling of the input. The roots of a polynomial are very sensitive to the changes in the polynomial coefficients, particularly for high-degree polynomials. It is for this reason that IIR filter implementations can become unstable when their poles migrate across the unit circle as a result of quantization error. In addition, overflow error and roundoff error can cause steady-state oscillations in an IIR filter output after the input goes to zero. These oscillations are called *Limit cycle*   *limit cycles*. For IIR filters, the indirect form realizations tend to be less sensitive to finite word length effects because the block transfer functions are only of second order.

GUI Module

The FDSP toolbox includes a GUI module called *g_iir* that allows the user to design and compare IIR filters without any need for programming. The featured filters include resonators; notch filters; lowpass, highpass, bandpass, and bandstop filters; and user-defined filters specified in a MAT file. The filter order can be adjusted directly using a slider bar. The FDSP toolbox also includes a GUI module called *g_filters*, described in Section 5.9, that allows the user to import filters from *g_iir*. In this way, one can compare different realization structures and explore the effects of coefficient quantization error.

Learning Outcomes

This chapter was designed to provide the student with an opportunity to achieve the learning outcomes summarized in Table 7.8.

TABLE 7.8: ▶
Learning Outcomes
for Chapter 7

Num.	Learning Outcome	Sec.
1	Know how to design a tunable plucked string filter for music synthesis	7.1
2	Be able to design resonators, notch filters, and comb filters using pole-zero placement with gain matching	7.2
3	Understand the characteristics and relative advantages of classical lowpass analog Butterworth, Chebyshev, and elliptic filters	7.4
4	Know how to convert an analog filter into an equivalent digital filter using the bilinear transformation method	7.5
5	Be able to convert a lowpass filter into a lowpass, highpass bandpass, or bandstop filter using frequency transformation	7.6
6	Understand the benefits of different filter realization structures	7.7
7	Be aware of detrimental finite word length effects and know how to minimize them	7.8
8	Know how to use the GUI module *g_iir* to design and analyze digital IIR filters without any need for programming	7.9
9	Know how to add special effects to music and speech using reverb filters	7.9

7.11   Problems

The problems are divided into Analysis and Design problems that can be solved by hand or with a calculator, GUI Simulation problems that are solved using GUI module g_iir, and MATLAB Computation problems that require a user program.

7.11.1 Analysis and Design

7.1 Find the minimum order n of an analog elliptic filter that will meet the follow design specifications. You can use the MATLAB function $ellipke$ to evaluate an elliptic integral of the first kind.

$$[F_p, F_s, \delta_p, \delta_s] = [100, 200, .03, .05]$$

7.2 Sketch the poles and zeros of an analog lowpass Butterworth filter of order $n = 8$ that has a 3 dB cutoff frequency of $F_c = 1/\pi$ Hz.

7.3 Consider the problem of designing a notch filter that has two notch frequencies. Suppose the sampling frequency is $f_s = 360$ Hz.
(a) Design a notch filter $H_0(z)$ that has a notch frequency at $F_0 = 60$ Hz and a 3 dB stopband radius of 2 Hz.
(b) Design a notch filter $H_1(z)$ that has a notch frequency at $F_0 = 90$ Hz and a 3 dB stopband radius of 2 Hz.
(c) Combine $H_0(z)$ and $H_1(z)$ to produce a notch filter $H(z)$ that has notches at $F_0 = 60$ Hz and $F_1 = 90$ Hz. *Hint*: Use one of the indirect forms.
(d) Sketch the signal flow graph of $H(z)$ using direct form II realizations for the blocks $H_0(z)$ and $H_1(z)$.

7.4 Design a second-order analog lowpass Chebyshev-I filter, $H_a(s)$, using $F_p = 10$ Hz and $\delta_p = .1$.

7.5 Consider the problem of designing an analog lowpass Chebyshev-I filter to meet the following design specifications. Find the minimum order of the filter.

$$[F_p, F_s, \delta_p, \delta_s] = [100, 200, .03, .05]$$

7.6 Consider the problem of designing a notch filter that eliminates the frequency $F_0 = 60$ Hz.
(a) Suppose the notch filter pole is at the angle $\theta_0 = \pi/3$. Find the sampling frequency f_s.
(b) Design a notch filter $H_{\text{notch}}(z)$ that has a 3 dB stopband radius of $\Delta F = 1$ Hz.
(c) Sketch the signal flow graph using a transposed direct form II realization.

7.7 Consider the problem of designing a resonator that has two resonant frequencies. Suppose the sampling frequency is $f_s = 360$ Hz.
(a) Design a resonator $H_0(z)$ that has a resonant frequency at $F_0 = 90$ Hz and a 3 dB passband radius of 3 Hz.
(b) Design a resonator $H_1(z)$ that has a resonant frequency of $F_1 = 120$ Hz and a 3 dB passband radius of 4 Hz.
(c) Combine $H_0(z)$ and $H_1(z)$ to produce a resonator $H(z)$ that has resonant frequencies at $F_0 = 90$ Hz and $F_1 = 120$ Hz. *Hint*: Use one of the indirect forms.
(d) Sketch the signal flow graph of $H(z)$ using direct form II realizations for the blocks $H_0(z)$ and $H_1(z)$.

7.8 Consider the problem of designing a resonator that extracts the frequency $F_0 = 100$ Hz.
(a) Find a sampling frequency f_s that places the resonator pole at an angle of $\theta_0 = \pi/2$.
(b) Design a resonator $H_{res}(z)$ that has a 3 dB passband radius of $\Delta F = 2$ Hz.
(c) Sketch a signal flow graph using a direct form II realization.

7.9 Consider the problem of designing a filter whose impulse response emulates the sound from a stringed musical instrument. Suppose the sampling frequency is $f_s = 44.1$ kHz and the desired resonant frequency or pitch is $F_0 = 480$ Hz.
(a) Find the feedback parameter L and the pitch parameter c in Figure 7.1.
(b) Suppose the attenuation factor is $r = .998$. Find the tunable plucked-string filter transfer function $H(z)$.

7.10 Consider the problem of designing a lowpass analog Butterworth filter to meet the following specifications.

$$[F_p, F_s, \delta_p, \delta_s] = [300, 500, .1, .05]$$

(a) Find the minimum filter order n.
(b) For what cutoff frequency F_c is the passband specification exactly met?
(c) For what cutoff frequency F_c is the stopband specification exactly met?
(d) Find a cutoff frequency F_c for which $H_a(s)$ exceeds both the passband and the stopband specification.

7.11 Consider an input $y(k)$ that consists of a signal of interest, $x(k)$, plus a disturbance, $d(k)$.

$$y(k) = x(k) + d(k), \quad 0 \le k < N$$

Suppose that when the sampling rate is f_s, the disturbance $d(k)$ is periodic with a period of $L = 12$. Design an inverse comb filter $H_{inv}(z)$ that removes harmonics zero through $L/2$ of $d(k)$ from $y(k)$. Use a 3 dB passband radius of $\Delta F = f_s/200$.

7.12 Consider an input signal $y(k)$ that consists of a periodic component $x(k)$ plus a random white noise component $v(k)$.

$$y(k) = x(k) + v(k), \quad 0 \le k < 256$$

Suppose the sampling rate is f_s and this results in a signal $x(k)$ that is periodic with a period of $L = 16$. Design a comb filter $H_{comb}(z)$ that passes harmonics zero through $L/2$ of $x(k)$. Use a 3 dB passband radius of $\Delta F = f_s/100$.

7.13 Find the transfer function $H(s)$ of a third-order analog lowpass Butterworth filter that has a 3 dB cutoff frequency of $F_c = 4$ Hz.

7.14 Consider the problem of designing a lowpass analog filter $H_a(s)$ to meet the following specifications.

$$[F_p, F_s, \delta_p, \delta_s] = [1000, 1200, .05, .02]$$

(a) Find the passband ripple and stopband attenuation in units of dB.
(b) Find the selectivity factor r.
(c) Find the discrimination factor d.

7.15 Consider the following design specifications for a lowpass analog filter.

$$[F_p, F_s, \delta_p, \delta_s] = [50, 60, .05, .02]$$

Find the minimum-order filter needed to meet these specifications using the following classical analog filters.
(a) Butterworth filter
(b) Chebyshev-I filter
(c) Chebyshev-II filter

7.16 Consider the following first-order analog filter.

$$H_a(s) = \frac{s}{s + 4\pi}$$

(a) What type of frequency-selective filter is this (lowpass, highpass, bandpass, or bandstop)?
(b) What is the 3 dB cutoff frequency f_0 of this filter?
(c) Suppose $f_s = 10$ Hz. Find the prewarped cutoff frequency F_0.
(d) Design a digital equivalent filter $H(z)$ using the bilinear-transformation method.

7.17 Consider the following IIR system.

$$H(z) = \frac{z^3}{(z - .8)(z^2 - z + .24)}$$

(a) Expand $H(z)$ into partial fractions.
(b) Sketch a parallel form signal flow graph realization by combining the two poles that are closest to the unit circle into a second-order block.

7.18 Let $f_{clip}(x)$ be the following unit clipping nonlinearity.

$$f_{clip}(x) \triangleq \begin{cases} -1, & -\infty < x < -1 \\ x, & -1 \leq x \leq 1 \\ 1, & 1 < x < \infty \end{cases}$$

Show how f_{clip} can be used to eliminate limit cycles due to overflow error by sketching a modified direct form II signal flow graph of a second-order IIR block. You can assume all values are represented as fractions.

7.19 Consider the following IIR filter. Suppose 8-bit fixed-point arithmetic is used to implement this filter using a scale factor of $c = 4$.

$$H(z) = \frac{2z}{z + .7}$$

(a) Find the quantization level q.
(b) Find the power gain of this filter.
(c) Find the average power of the product round-off error.

7.20 Consider the following IIR filter.

$$H(z) = \frac{.5}{z + .9}$$

(a) Sketch a direct form II signal flow graph of $H(z)$.
(b) Suppose all filter variables are represented as fixed-point numbers, and the input is constrained to $|x(k)| \leq c$ where $c = z$. Find a scale factor s_1 that eliminates summing junction overflow error.

(c) Sketch a modified direct form II signal flow graph of $H(z)$ that implements scaling to eliminate summing junction overflow.

7.21 For the system in Problem 7.20, find a scale factor s_∞ that will eliminate summing junction overflow when the input is a pure sinusoid of amplitude $c \leq 5$.

7.22 The simplest digital equivalent filter is one that preserves the impulse response of $H_a(s)$. Let $h_a(t)$ denote the desired impulse response.

$$h_a(t) = L^{-1}\{H_a(s)\}$$

Next let T be the sampling interval. The objective is to design a digital filter $H(z)$ whose impulse response $h(k)$ satisfies

$$h(k) = h_a(kT), \quad k \geq 0$$

Impulse invariant method

Thus the impulse response of $H(z)$ consists of samples of the impulse response of $H_a(s)$. This design technique, which preserves the impulse response, is called the *impulse-invariant method*. Suppose $H_a(s)$ is a stable, strictly proper, rational polynomial with n distinct poles $\{p_1, p_2, \cdots, p_n\}$.
(a) Expand $H_a(s)/s$ into partial fractions.
(b) Find the impulse response $h_a(t)$.
(c) Sample $h_a(t)$ to find the impulse response $h(k)$.
(d) Find the transfer function $H(z)$.

7.23 Consider the following analog prototype filter of order $n = 2$.

$$H_a(s) = \frac{6}{s^2 + 5s + 6}$$

(a) Find the poles of $H_a(s)/s$.
(b) Find the residues of $H_a(s)/s$ at each pole.
(c) Find a digital equivalent transfer function using the impulse-invariant method in Problem 7.22. You can assume the sampling interval is $T = .5$ sec.

7.24 Find the transfer function $H(s)$ of a second-order highpass Butterworth filter that has a 3 dB cutoff frequency of $F_c = 5$ Hz.

7.25 Sketch a direct form II signal flow graph realization of the following difference equation.

$$y(k) = 10x(k) + 2x(k-1) - 4x(k-2) + 5x(k-3) - .7y(k-2) + .4y(k-3)$$

7.26 Sketch a direct form I signal flow graph realization of the following IIR transfer function.

$$H(z) = \frac{.8 - 1.2z^{-1} + .4z^{-3}}{1 - .9z^{-1} + .6z^{-2} + .3z^{-3}}$$

7.27 Find the transfer function $H(s)$ of a fourth-order bandpass Butterworth filter that has 3 dB cutoff frequencies of $F_0 = 2$ Hz and $F_1 = 4$ Hz.

7.28 Sketch a transposed direct form II signal flow graph realization of the following transfer function.

$$H(z) = \frac{1 - 2z^{-1} + 3z^{-2} - 4z^{-3}}{1 + .8z^{-1} + .6z^{-2} + .4z^{-3}}$$

7.29 Consider the following analog filter that has n poles and m zeros with $m \leq n$.

$$H_a(s) = \frac{\beta(s - z_1)(s - z_2) \cdots (s - z_m)}{(s - p_1)(s - p_2) \cdots (s - p_n)}$$

An alternative way to convert an analog filter into a digital filter is to map each pole and zero of $H_a(s)$ into a corresponding pole and zero of $H(z)$ using $z = \exp(sT)$. This yields

$$H(z) = \frac{b_0(z+1)^{n-m}[z - \exp(z_1 T)][z - \exp(z_2 T)] \cdots [z - \exp(z_m T)]}{[z - \exp(p_1 T)][z - \exp(p_2 T)] \cdots [z - \exp(p_n T)]}$$

Note that if $n > m$, then $H_a(s)$ has $n - m$ zeros at $s = \infty$. These zeros are mapped into the highest digital frequency, $z = -1$. The gain factor b_0 is selected such that the two filters have the same passband gain. For example, if $H_a(s)$ is a lowpass filter, then $H_a(0) = H(1)$. This method, which is analogous to Alg. 7.1 but using a different transformation, is called the *Matched Z-transform method*. *matched Z-transform* method. Use the matched Z-transform method to find a digital equivalent of the following analog filter. You can assume $T = .2$. Match the gains at DC.

$$H_a(s) = \frac{10s + 1}{s^2 + 3s + 2}$$

7.30 Consider the following IIR system.

$$H(z) = \frac{2(z^2 + .64)(z^2 - z + .24)}{(z^2 + 1.2z + .27)(z^2 + .81)}$$

(a) Sketch the poles and zeros of $H(z)$.
(b) Sketch a cascade form signal flow graph realization by grouping the complex zeros with the complex poles. Use a direct form II realization for each block.

7.11.2 GUI Simulation

7.31 Use the GUI module *g_iir* to design a lowpass Butterworth filter. Adjust the filter order to the lowest value that meets or exceeds the specifications. Plot the following.
(a) The linear magnitude response
(b) The phase response. Is this a linear-phase filter?
(c) The pole-zero plot

7.32 Use the GUI module *g_iir* to design a bandstop elliptic filter. Adjust the filter order to the lowest value that meets or exceeds the specifications. Plot the following.
(a) The linear magnitude response
(b) The phase response. Is this a linear-phase filter?
(c) The pole-zero plot

7.33 Create a MAT-file called *prob7_33.mat* that contains b, a, and fs for an inverse comb filter of order $n = 12$ using $f_s = 1000$ Hz and a 3 dB radius of $\Delta F = 2$ Hz. Then use the GUI module *g_iir* and the User-defined option to load this filter.
(a) Plot the linear magnitude response.
(b) Plot the phase response
(c) Plot the pole-zero pattern.

7.34 Use the GUI module *g_iir* to design a resonator filter with a resonant frequency of $F_0 = 300$ Hz.
(a) Plot the linear magnitude response. Use the *Caliper* option to mark the peak.
(b) Plot the phase response. Is this a linear-phase filter?
(c) Plot the pole-zero plot.

7.35 Use the GUI module *g_iir* to design a bandpass Chebyshev-II filter. Adjust the filter order to the lowest value that meets or exceeds the specifications. Plot the following.
(a) The linear magnitude response
(b) The phase response. Is this a linear-phase filter?
(c) The pole-zero plot

7.36 Use the GUI module *g_iir* to design a highpass Chebyshev-I filter. Adjust the filter order to the lowest value that meets or exceeds the specifications. Plot the following.
(a) The linear magnitude response
(b) The phase response. Is this a linear-phase filter?
(c) The pole-zero plot

7.37 Use the GUI module *g_iir* to construct a Chebyshev-I lowpass filter. Plot the linear magnitude response for the following cases.
(a) Adjust the filter order n to the highest value that does not meet the specifications.
(b) Adjust the filter order n to the lowest value that meets or exceeds the specifications.

7.38 Use the GUI module *g_iir* to design a notch filter with a notch frequency of $F_0 = 200$ Hz, and a sampling frequency of $f_s = 1200$ Hz.
(a) Plot the linear magnitude response.
(a) Plot the phase response. Is this a linear-phase filter?
(c) Plot the impulse response.

7.39 Use the GUI module *g_iir* to design a Butterworth bandpass filter. Find the smallest-order filter that meets or exceeds the following design specifications.

$$[f_s, F_{s1}, F_{p1}, F_{p2}, F_{s2}] = [2000, 300, 400, 600, 700] \text{ Hz}$$

$$[A_p, A_s] = [.6, 30] \text{ dB}$$

(a) Plot the magnitude response using the dB scale.
(b) Plot the pole-zero pattern.
(c) Save a, b, and *fs* in a MAT-file named *prob7_39*. Then use GUI module *g_filters* to load this as a user-defined filter. Adjust the number of bits used for coefficient quantization to $N = 12$. Plot the linear magnitude responses.

7.40 Use the GUI module *g_iir* to design a Chebyshev-I bandpass filter. Find the smallest-order filter that meets or exceeds the following design specifications.

$$[f_s, F_{s1}, F_{p1}, F_{p2}, F_{s2}] = [2000, 300, 400, 600, 700] \text{ Hz}$$
$$[\delta_p, \delta_s] = [.05, .03]$$

(a) Plot the linear magnitude response.
(b) Plot the pole-zero pattern.
(c) Save a, b, and *fs* in a MAT-file named *prob7_40*. Then use GUI module *g_filters* to load this as a user-defined filter. Adjust the number of bits used for coefficient quantization to $N = 10$. Plot the linear magnitude responses.

7.41 Use the GUI module *g_iir* to design an elliptic bandpass filter. Find the smallest-order filter that meets or exceeds the following design specifications.

$$[f_s, F_{s1}, F_{p1}, F_{p2}, F_{s2}] = [2000, 350, 400, 600, 650] \text{ Hz}$$
$$[\delta_p, \delta_s] = [.04, .02]$$

(a) Plot the linear magnitude response.
(b) Plot the pole-zero pattern.
(c) Save a, b, and fs in a MAT-file named *prob7_41*. Then use GUI module *g_filters* to load this as a user-defined filter. Adjust the number of bits used for coefficient quantization to $N = 9$. Plot the linear magnitude responses.

7.42 Use the GUI module *g_iir* to design a Butterworth bandpass filter. Find the smallest-order filter that meets or exceeds the following design specifications.

$$[f_s, F_{p1}, F_{s1}, F_{s2}, F_{p2}] = [100, 20, 25, 35, 40] \text{ Hz}$$
$$[\delta_p, \delta_s] = [.05, .02]$$

(a) Plot the magnitude response using the dB scale.
(b) Plot the pole-zero pattern.
(c) Save a, b, and fs in a MAT-file named *prob7_42*. Then use GUI module *g_filters* to load this as a user-defined filter. Adjust the number of bits used for coefficient quantization to $N = 16$. Plot the linear magnitude responses.

7.43 Use the GUI module *g_iir* to design a Chebyshev-II bandstop filter. Find the smallest-order filter that meets or exceeds the following design specifications.

$$[f_s, F_{p1}, F_{s1}, F_{s2}, F_{p2}] = [20000, 2500, 3000, 4000, 4500] \text{ Hz}$$
$$[\delta_p, \delta_s] = [.04, .03]$$

(a) Plot the linear magnitude response.
(b) Plot the pole-zero pattern.
(c) Save a, b, and fs in a MAT-file named *prob7_43*. Then use GUI module *g_filters* to load this as a user-defined filter. Adjust the number of bits used for coefficient quantization to $N = 17$. Plot the linear magnitude responses.

7.44 Use the GUI module *g_iir* to design an elliptic bandstop filter. Find the smallest-order filter that meets or exceeds the following design specifications.

$$[f_s, F_{p1}, F_{s1}, F_{s2}, F_{p2}] = [20, 6.5, 7, 8, 8.5] \text{ Hz}$$
$$[\delta_p, \delta_s] = [.02, .015]$$

(a) Plot the linear magnitude response.
(b) Plot the pole-zero pattern.
(c) Save a, b, and fs in a MAT-file named *prob7_44*. Then use GUI module *g_filters* to load this as a user-defined filter. Adjust the number of bits used for coefficient quantization to $N = 14$. Plot the linear magnitude responses.

7.45 Use the GUI module *g_iir* and the User-defined option to load the filter in MAT-file *u_iir1*.
(a) Plot the linear magnitude response. What type of filter is this?
(b) Plot the phase response
(c) Plot the impulse response.

7.11.3 MATLAB Computation

7.46 Write a MATLAB program that uses *f_ellipticz* to design a digital elliptic bandpass filter that meets the following design specifications.

$$[f_s, F_{s1}, F_{p1}, F_{p2}, F_{s2}, \delta_p, \delta_s] = [1600, 250, 350, 550, 650, .06, .04]$$

(a) Find the smallest filter order that meets the specifications. Print the order.

(b) Use *f_freqz* to compute and plot the magnitude response.

(c) Use *fill* to add shaded areas showing the design specifications.

7.47 Write a MATLAB program that uses *f_butters* and *f_low2highs* to design an analog Butter-worth highpass filter to meet the following design specifications.

$$[F_s, F_p, A_p, A_s] = [4, 6, .5, 24]$$

(a) Print the filter order, δ_p, and δ_s.

(b) Use *f_freqs* to compute and plot the magnitude response for $0 \leq f \leq 2F_p$ using the linear scale.

(c) Use *fill* to add shaded areas showing the design specifications on the magnitude response plot.

7.48 Write a MATLAB program that uses *f_cheby1s* and *f_low2bps* to design an analog Chebyshev-I bandpass filter to meet the following design specifications.

$$[F_{s1}, F_{p1}, F_{p2}, F_{s2}, A_p, A_s] = [35, 45, 60, 70, .4, 28]$$

(a) Print the filter order, δ_p, and δ_s.

(b) Use *f_freqs* to compute and plot the magnitude response for $0 \leq f \leq 2F_{s2}$ using the linear scale.

(c) Use *fill* to add shaded areas showing the design specifications on the magnitude response plot.

7.49 Write a MATLAB program that uses *f_elliptics* to design an analog elliptic lowpass filter to meet the following design specifications.

$$[F_p, F_s, \delta_p, \delta_s] = [10, 20, .04, .02]$$

(a) Print the filter order.

(b) Use *f_freqs* to compute and plot the magnitude response for $0 \leq f \leq 2F_s$.

(c) Use *fill* to add shaded areas showing the design specifications on the magnitude response plot.

7.50 Write a MATLAB program that uses *f_butters* and *f_bilin* to find the digital equivalent $H(z)$ of a sixth-order lowpass Butterworth filter using the bilinear transformation method. Suppose the sampling frequency is $f_s = 10$ Hz. Prewarp the analog cutoff frequency so that the digital cutoff frequency comes out to be $F_c = 1$ Hz.

(a) Plot the impulse response, $h(k)$.

(b) Use *f_pzplot* to plot the poles and zeros of $H(z)$.

(c) Use *f_freqz* to compute and plot the magnitude response, $A(f)$. Add the ideal magnitude response and a plot legend.

7.51 Write a MATLAB program that uses *f_cheby2z* to design a digital Chebyshev-II bandpass filter that meets the following design specifications.

$$[f_s, F_{s1}, F_{p1}, F_{p2}, F_{s2}, \delta_p, \delta_s] = [1600, 250, 350, 550, 650, .06, .04]$$

(a) Find the smallest filter order that meets the specifications. Print the order.

(b) Use *f_freqz* to compute and plot the magnitude response.

(c) Use *fill* to add shaded areas showing the design specifications.

7.52 Write a MATLAB program that uses *f_butters* to design an analog Butterworth lowpass filter to meet the following design specifications.

$$[F_p, F_s, \delta_p, \delta_s] = [10, 20, .04, .02]$$

(a) Print the filter order.
(b) Use *f_freqs* to compute and plot the magnitude response for $0 \leq f \leq 2F_s$.
(c) Use *fill* to add shaded areas showing the design specifications on the magnitude response plot.

7.53 Write a MATLAB program that uses *f_butterz* to design a digital Butterworth bandstop filter that meets the following design specifications.

$$[f_s, F_{p1}, F_{s1}, F_{s2}, F_{p2}, \delta_p, \delta_s] = [2000, 200, 300, 600, 700, .05, .03]$$

(a) Find the smallest filter order that meets the specifications. Print the order.
(b) Use *f_freqz* to compute and plot the magnitude response.
(c) Use *fill* to add shaded areas showing the design specifications.

7.54 Write a MATLAB program that uses *f_cheby1z* to design a digital Chebyshev-I bandstop filter that meets the following design specifications.

$$[f_s, F_{p1}, F_{s1}, F_{s2}, F_{p2}, \delta_p, \delta_s] = [2000, 200, 300, 600, 700, .05, .03]$$

(a) Find the smallest filter order that meets the specifications. Print the order.
(b) Use *f_freqz* to compute and plot the magnitude response.
(c) Use *fill* to add shaded areas showing the design specifications.

7.55 Write a MATLAB function called *f_filtnorm* that returns the L_p norm, $\|h\|_p$, of a digital filter. The function *f_filtnorm* should use the following calling sequence.

```
% F_FILTNORM: Return L_p norm of filter H(z) = b(z)/a(z)
%
% Usage:
%        d = f_filtnorm (b,a,p)
% Pre:
%        b = vector of length m+1 containing coefficients of
%            numerator polynomial.
%        a = vector of length n+1 containing coefficients of
%            denominator polynomial.
%        p = integer specifying norm type. Use p = Inf for
%            the infinity norm
% Post:
%        d = the L\_p norm, ||h||\_p
```

Test *f_filtnorm* by writing a MATLAB script that computes and prints the L_1, L_2, and L_∞ norms of the comb filter in Problem 5.46. Verify that (5.8.16) holds in this case.

7.56 Write a MATLAB program that uses *f_cheby2s* to design an analog Chebyshev-II lowpass filter to meet the following design specifications.

$$[F_p, F_s, \delta_p, \delta_s] = [10, 20, .04, .02]$$

(a) Print the filter order.

(b) Use f_freqs to compute and plot the magnitude response for $0 \le f \le 2F_s$.

(c) Use *fill* to add shaded areas showing the design specifications on the magnitude response plot.

7.57 Write a MATLAB program that uses $f\_cheby1s$ to design an analog Chebyshev-I lowpass filter to meet the following design specifications.

$$[F_p, F_s, \delta_p, \delta_s] = [10, 20, .04, .02]$$

(a) Print the filter order.

(b) Use f_freqs to compute and plot the magnitude response for $0 \le f \le 2F_s$.

(c) Use *fill* to add shaded areas showing the design specifications on the magnitude response plot.

PART III

Advanced Signal Processing

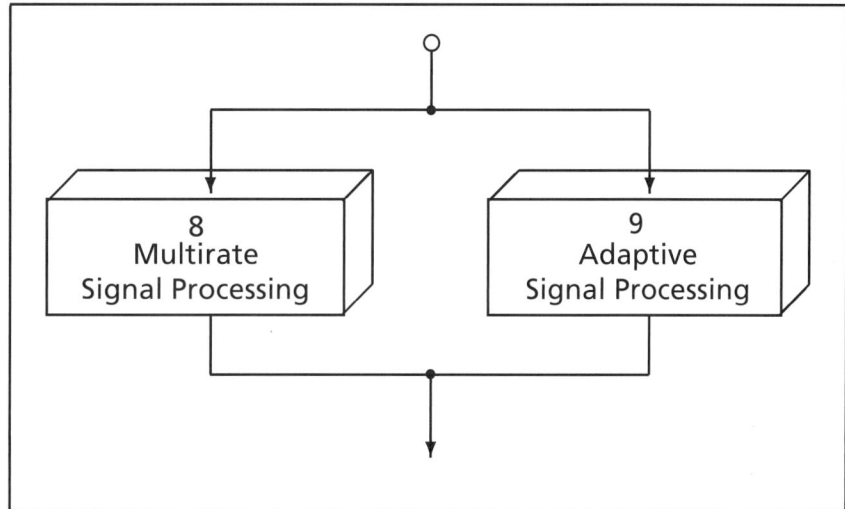

8 Multirate Signal Processing	9 Adaptive Signal Processing

CHAPTER 8

Multirate Signal Processing

Chapter Topics

8.1 Motivation

8.2 Integer Sampling Rate Converters

8.3 Rational Sampling Rate Converters

8.4 Multirate Filter Realization Structures

8.5 Narrowband Filters and Filter Banks

8.6 A Two-channel QMF Bank

8.7 Oversampling ADC

8.8 Oversampling DAC

8.9 GUI Software and Case Study

8.10 Chapter Summary

8.11 Problems

8.1 Motivation

All of the discrete-time systems encountered thus far have signals that are sampled at a single fixed sampling rate f_s. If this assumption is relaxed to allow some of the signals to be sampled at one rate while others are sampled at a higher or a lower rate, this leads to a *multirate* system. Multirate systems can offer important advantages over fixed-rate systems in terms of overall performance. One of the simplest examples of a multirate system is a sampling rate *decimator* that decreases the sampling rate of a discrete-time signal by an integer factor M.

Multirate system

$$y(k) = \sum_{i=0}^{m} b_i x(Mk - i)$$

Here output $y(k)$ is a filtered version of the input $x(k)$, but the input is evaluated only at every Mth sample. Extracting every Mth sample effectively reduces the sampling rate by M. The filtering operation is needed in order to preserve the spectral characteristics of the down-sampled signal. It is also possible to increase the sampling rate by an integer factor L using an *interpolator*. More generally, sampling rate converters can be designed where the ratio of the output sampling frequency to the input sampling frequency is an arbitrary rational number L/M. Modern high-performance DSP systems exploit the benefits of multirate systems. For example, multirate techniques can be used to design analog-to-digital converters (ADCs) and digital-to-analog converters (DACs) with improved noise immunity. Another important class of applications is the design of banks of narrowband filters such as those used in frequency-division multiplexing and demultiplexing.

We begin this chapter by introducing some examples of applications of multirate systems. Next, the design of integer sampling rate decimators and interpolators is presented. These rate converter building blocks are then used to construct rational sampling rate converters, both single stage and multistage. This is followed by an investigation of efficient realization structures for sampling rate converters based on polyphase filters. Next the discussion turns to multirate system applications starting with the design of narrowband filters and filter banks. This is followed by an analysis of a two-channel quadrature-mirror filter bank. The improved performance characteristics of oversampling ADCs are presented next. This is followed by a analogous presentation applied to oversampling DACs. Finally, a GUI module called *g_multirate* is introduced that allows the user to design and evaluate multirate DSP systems without any need for programming. The chapter concludes with a case study example, and a summary of multirate signal processing techniques.

8.1.1 Narrowband Filter Banks

Several signals can be transmitted simultaneously over a single communication channel by allocating a separate band of frequencies for each signal. This technique, known as frequency-division multiplexing or subband processing, requires the use of a bank of narrowband filters. In this way each filter can be used to extract a different signal. The magnitude responses of a bank of six narrowband filters are shown in Figure 8.1. Notice that to maximize the number

FIGURE 8.1:
Magnitude
Responses of a
Bank of Six
Narrowband Filters

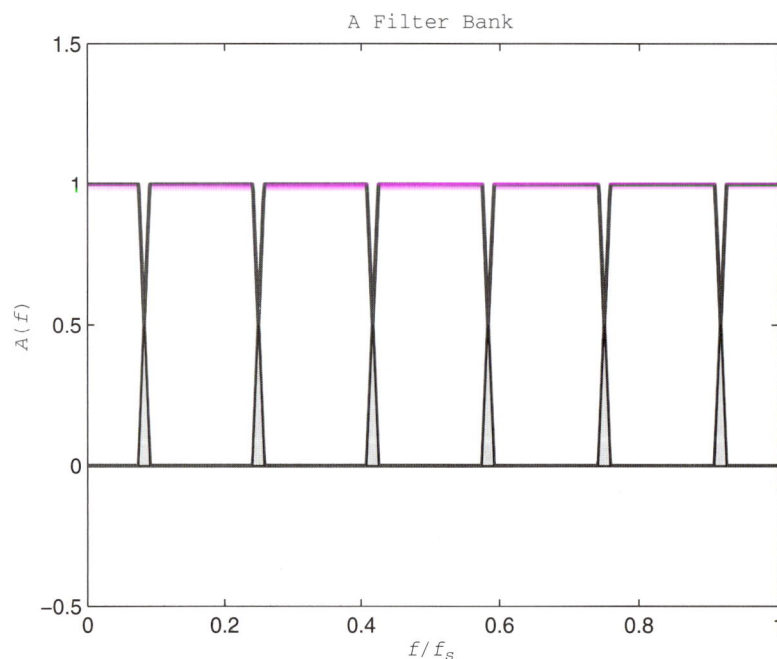

of filters in the bank, their transition bands overlap with one another, as shown in the shaded regions.

Narrowband filter   A filter is referred to as a *narrowband* filter when the width of its passband (or its stopband) is small in comparison with the sampling frequency f_s. For example, let $B_p = F_{p2} - F_{p1}$ denote the width of the passband of a bandpass filter. This filter is a narrowband filter if

$$B_p \ll f_s \tag{8.1.1}$$

Narrowband lowpass and highpass filters can be defined in an analogous way. The challenge in designing a bank of narrowband filters arises when one considers the required width of the transition band. Consider a bank of N narrowband filters. Since the discrete-time frequency response is periodic with period f_s, one can take the frequency range to be $[0, f_s]$ rather than $-f_s/2$ to $f_s/2$. Then the ith filter will be centered at $F_i = i f_s / N$ for $0 \le i \le N$ and will have a maximum passband width of

$$B_p \approx \frac{f_s}{N} \tag{8.1.2}$$

In order to maximize the use of the spectrum, the width of the transition band should be small in comparison with the width of the passband. Consequently, for a narrowband filter, the width of the transition band is very small in comparison with f_s. As an illustration, suppose a filter bank of $N = 10$ filters is to be designed, and suppose the width of the transition band is set to $B_t = B_p/20$. Then the normalized width of the transition band is

$$
\begin{aligned}
B &= \frac{B_t}{f_s} \\
&= \frac{B_p}{20 f_s} \\
&= .005
\end{aligned}
\tag{8.1.3}
$$

This design requirement is quite severe. To see what it implies, suppose the passband ripple and stopband attenuation are as follows for each filter in the bank.

$$(\delta_p, \delta_s) = (.01, .02) \tag{8.1.4}$$

If the equiripple FIR filter design method is used to design the filters, then from (6.5.21) the estimated order of the filter required to meet the design specification is

$$
\begin{aligned}
m &\approx \text{ceil} \left\{ \frac{-[10 \log_{10}(\delta_p \delta_s) + 13]}{14.6B} + 1 \right\} \\
&= \text{ceil} \left\{ \frac{-[10 \log_{10}(.0002) + 13]}{14.6(.005)} + 1 \right\} \\
&= 330
\end{aligned}
\tag{8.1.5}
$$

Clearly, a very high-order filter is needed to meet the narrowband design specification in this case. Recall that if an alternative FIR design method is used, such as a windowed, frequency-sampled, or least-squares filter, the required filter order will be even higher. Implementing a filter of such a high order brings with it a host of problems including significant memory requirements, lengthy processing time, and potentially debilitating finite word length effects. Fortunately, by using a multirate design with a multistage polyphase realization, these difficulties can be reduced significantly and the performance of the narrowband filter can be improved.

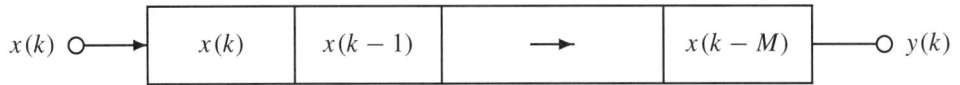

FIGURE 8.2: Delay of Discrete-time Signal Using an *M*-sample Shift Register

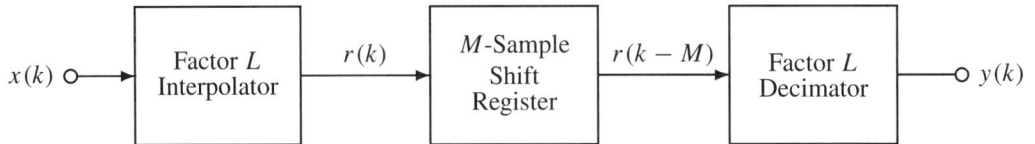

FIGURE 8.3: Intersample Delay of Discrete-time Signal Using a Multirate System

8.1.2 Fractional Delay Systems

A design task that occurs repeatedly in different applications is the problem of delaying a discrete-time signal without otherwise distorting it. If the desired delay is an integer multiple of the sampling interval T, then this is achieved easily. One can allocate a memory buffer in the form of a shift register of length M as shown in Figure 8.2. Here the signal shifted out the output end will be a delayed version of the input with a delay of $\tau = MT$.

Intersample delay   It is more challenging to design a system where the delay is not an integer multiple of the sampling interval, but instead involves an *intersample delay*. In effect, what is required is an allpass filter with a phase response of

$$\phi(f) = -2\pi f \tau \tag{8.1.6}$$

Recall from (5.4.7) that allpass IIR filters can be designed easily by enforcing a reflective symmetry constraint on the coefficients. Similarly, linear-phase FIR filters of order m can be designed with a group delay of $\tau = mT/2$. However, the design of fixed-rate allpass IIR filters with an arbitrary group delay presents a problem. Fortunately, by using multirate techniques this design problem becomes more manageable. The basic idea is to first increase the sampling rate by a factor L. One then delays the up-sampled signal by $0 < M < L$ samples using a shift register. This is followed by decreasing the sampling rate by a factor L to restore the original sampling frequency. The processing steps are summarized in Figure 8.3.

The factor L interpolator in Figure 8.3 increases the sampling rate by L so that the intermediate signal $r(k)$ is sampled at the rate $f_r = Lf_s$. A brute-force analog approach to changing the sampling rate is to convert $x(k)$ from digital to analog with a DAC, and then sample the result at the new rate with an ADC. A drawback of this analog approach is that it introduces additional quantization and aliasing errors. Fortunately, sampling rate converters that avoid these types of error can be designed by working strictly in the discrete-time domain. Once $x(k)$ has been up-sampled to produce $r(k)$, this intermediate signal is then delayed by an integer number of samples M using the shift register in Figure 8.2. If the length of the shift register is in the *Fractional delay*   range $0 < M < L$, this produces a *fractional* or intersample delay when viewed in terms of the original sampling rate f_s. In particular, the delay introduced by the shift register block in Figure 8.3 is

$$\tau = \left(\frac{M}{L}\right) T \tag{8.1.7}$$

Finally, the factor L decimator down-samples the delayed signal $v(k-M)$ by L, thereby restoring the original sampling frequency. It should be pointed out that interpolator and decimator blocks in Figure 8.3 include linear-phase FIR lowpass filters. Although these processing steps will also introduce delays, these delays are integer multiples of the original sampling interval.

● ● ● ● ● ● ● ● ● ● ● ● ● ● ● ● ● ●

8.2 Integer Sampling Rate Converters

Modern high-performance DSP systems often make use of *multirate* systems: systems where some of the signals are sampled at one frequency and others are sampled at another frequency. For example, the need for a sharp high-order analog anti-aliasing prefilter can be avoided if oversampling is used and the sampling rate is later reduced to the desired value.

A conceptually simple way to change the sampling rate is to convert a discrete-time signal from digital to analog with a DAC, and then resample the analog signal with an ADC using the desired sampling frequency. This brute force approach to sampling rate conversion has the advantage that the new sampling rate can be any value achievable by the ADC. However, a drawback is that the DAC and ADC introduce additional quantization noise and aliasing error. In this section, techniques are introduced that avoid these drawbacks by implementing sampling rate converters entirely in the discrete-time domain.

8.2.1 Sampling Rate Decimator

Decimator

Let us begin with a relatively simple problem, namely, a reduction in the sampling rate by an integer factor M. A sampling rate converter that reduces the sampling rate is called a *decimator* because one is removing samples. Let $x(k)$ be the discrete-time signal obtained by sampling an analog signal $x_a(t)$ at the rate f_s. If $T = 1/f_s$ is the sampling interval, then

$$x(k) = x_a(kT) \qquad (8.2.1)$$

The objective is to start with $x(k)$ and synthesize a new discrete-time signal $y(k)$ that corresponds to sampling $x_a(t)$ at the reduced rate, $f_M = f_s/M$, where M is a positive integer. Since M is an integer, it would appear that this can be accomplished by simply extracting every Mth sample of $x(k)$ as follows.

$$x_M(k) = x(Mk) \qquad (8.2.2)$$

The problem with this basic approach is that it does not take into account the frequency content of the two signals. If the original signal $x(k)$ is sampled in a manner that avoids aliasing, then from the sampling theorem, the analog signal $x_a(t)$ must be bandlimited to less than $f_s/2$ Hz. However, to avoid aliasing with the reduced-rate signal $x_M(k)$, the analog signal must be bandlimited to less than $f_s/(2M)$ Hz. Thus to eliminate aliasing in $x_M(k)$, one must first pass $x(k)$ through a lowpass filter with a cutoff frequency of $F_M = f_s/(2M)$.

$$H_M(f) \triangleq \begin{cases} 1, & 0 \leq |f| < F_M \\ 0, & F_M \leq |f| \leq f_s/2 \end{cases} \qquad (8.2.3)$$

Digital anti-aliasing filter

Unlike an analog anti-aliasing filter associated with an ADC, the filter in (8.2.3) is a *digital anti-aliasing* filter. Sampling rate decimation by an integer factor M is summarized in the block diagram shown in Figure 8.4. Note that it is standard practice to denote sampling rate reduction in (8.2.2), also called *down-sampling*, with the down-arrow notation ↓.

Since the anti-aliasing filter in Figure 8.4 is a digital filter, any of the linear-phase FIR filter design techniques discussed in Chapter 6 can be applied to design this lowpass filter. If nonlinear phase distortion is not of concern, then an IIR filter can be used for $H_M(z)$. When

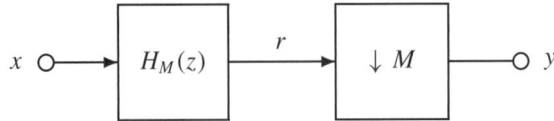

FIGURE 8.4: Sampling Rate Decimation by an Integer Factor, M

$H_M(z)$ is implemented as an FIR filter of order m, the output of the sampling rate decimator can be expressed in the time domain as follows.

$$y(k) = \sum_{i=0}^{m} b_i x(Mk - i) \tag{8.2.4}$$

The replacement of k on the right-hand side of (8.2.4) by Mk sets the down-sampler apart from a normal linear time-invariant FIR filter. The down-sampling operation in (8.2.4) continues to be a linear operation. However, if the input $x(i)$ is delayed by n samples, the output will not be delayed by n samples except when n is a multiple of M. Consequently, a decimator is a linear *time-varying* system.

Example 8.1

Integer Decimator

As an illustration of sampling rate decimation, consider the following analog input signal.

$$x_a(t) = \sin(2\pi t) - .5 \cos(4\pi t)$$

Let the sampling frequency be $f_s = 40$ Hz. Suppose the objective is to decimate the samples $x(k)$ by a factor of $M = 2$. From (8.2.3) the required lowpass filter has a gain of $H_M(0) = 1$ and a cutoff frequency of $F_M = 5$ Hz. Suppose a windowed linear-phase filter of order $m = 20$ with a Hanning window is used. Since the original signal was oversampled and is bandlimited to 2 Hz, the FIR filter does not have any appreciable effect in this instance. The decimator output is obtained by running *exam8_1*. Plots of the original samples and the decimated samples are shown in Figure 8.5. It is apparent from inspection that, after an initial start-up transient, the decimated samples faithfully reproduce the original signal. Note that there is a delay of $m/2 = 10$ of the original samples caused by the linear-phase filter.

8.2.2 Sampling Rate Interpolator

Next, consider the dual problem of designing a converter that increases the sampling rate by an integer factor L. A sampling rate converter that increases the sampling rate is called a *Interpolator* *interpolator* because one is inserting new samples that interpolate between the original samples. Here the objective is to synthesize a discrete-time signal $y(k)$ that corresponds to sampling $x_a(t)$ at the increased rate of $f_L = Lf_s$, where L is a positive integer. Since L is an integer, every Lth sample of the new signal $x_L(k)$ will correspond to a sample of the original signal $x(k)$. There are potentially many ways to interpolate between the original samples. The easiest is to simply insert $L - 1$ zero samples between each of the original samples as follows.

$$x_L(k) = \begin{cases} x(k/L), & |k| = 0, L, 2L, \cdots \\ 0, & \text{otherwise} \end{cases} \tag{8.2.5}$$

FIGURE 8.5: Sampling Rate Decimation by an Integer Factor $M = 2$ Using an FIR Filter of Order $m = 20$ with a Hamming Window

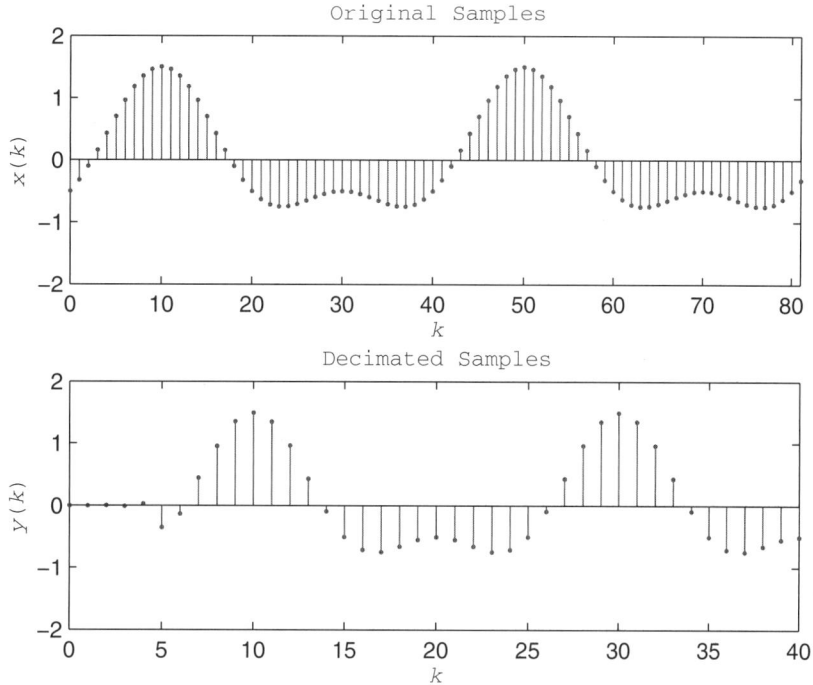

A helpful way to view $x_L(k)$ is in terms of the following periodic impulse train with period L.

$$\delta_L(k) \stackrel{\Delta}{=} \sum_{i=-\infty}^{\infty} \delta(k - Li) \tag{8.2.6}$$

The signal $x_L(k)$ is the signal $x(k/L)$ amplitude modulated by the periodic impulse train, $\delta_L(k)$. That is,

$$x_L(k) = x(k/L)\delta_L(k) \tag{8.2.7}$$

Note that $x(k/L)$ is not defined except when k is an integer multiple of L. However, the product in (8.2.7) is well defined for all k because $\delta_L(k) = 0$ when k is not a multiple of L. One can always replace k/L in (8.2.7) by *floor*(k/L) without changing the result.

The effect of using zero samples for interpolation can be seen by looking at the Z-transform of the interpolated signal. Using the change of variable $i = k/L$

$$
\begin{aligned}
X_L(z) &= \sum_{k=0}^{\infty} x(k/L)\delta_L(k)z^{-k} \\
&= \sum_{i=0}^{\infty} x(i)z^{-Li} \\
&= \sum_{i=0}^{\infty} x(i)(z^L)^{-i}
\end{aligned}
\tag{8.2.8}
$$

The Z-transform of the up-sampled signal can be expressed in terms of the Z-transform of the input as

$$X_L(z) = X(z^L) \tag{8.2.9}$$

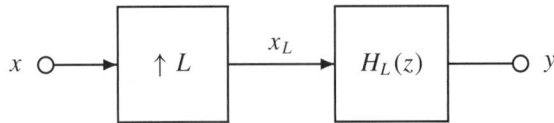

FIGURE 8.6: Sampling Rate Interpolation by an Integer Factor, *L*

Recall that the spectrum of a discrete-time signal can be obtained from the Z-transform by evaluating the Z-transform along the unit circle. If we replace z in (8.2.9) by $\exp(j2\pi fT)$, the spectrum of the interpolated signal is as follows.

$$X_L(f) = X(Lf), \quad 0 \le |f| \le f_s/2 \tag{8.2.10}$$

Thus the spectrum of the interpolated signal $x_L(k)$ is an L-fold replication of the spectrum of the original signal $x(k)$, with each replication centered at a multiple of f_s/L. These $L-1$ images of the original spectrum must be removed by passing $x_L(k)$ through a lowpass anti-imaging filter with a cutoff frequency of $F_L = .5f_s/L$.

$$H_L(f) \triangleq \begin{cases} L, & 0 \le |f| < F_L \\ 0, & F_L \le |f| \le .5f_s \end{cases} \tag{8.2.11}$$

Note that the passband gain of the anti-imaging filter has been set to $H_L(0) = L$. This is done to compensate for the fact that the average value of $x_L(k)$ is $1/L$ times the average value of $x(k)$ due to the presence of the zero samples. Unlike an analog anti-imaging filter associated *Digital anti-imaging* with a DAC, the filter in (8.2.11) is a *digital anti-imaging* filter. Sampling rate interpolation *filter* by an integer factor of L is summarized in the block diagram shown in Figure 8.6. Again it is standard practice to denote a sampling rate increase in (8.2.5), also called *up-sampling*, with the up-arrow notation ↑.

Since the anti-imaging filter in Figure 8.6 is a digital filter, any of the linear-phase FIR filter design techniques introduced in Chapter 6 can be used to design this lowpass filter. Alternatively, an IIR filter can be used if nonlinear phase distortion is not a concern. When $H_L(z)$ is implemented as an FIR filter of order m, the output of the sampling rate interpolator can be expressed in the time domain as follows.

$$y(k) = \sum_{i=0}^{m} b_i \delta_L(k-i)x\left(\frac{k}{L} - \frac{i}{L}\right) \tag{8.2.12}$$

Example 8.2   **Integer Interpolator**

As an illustration of sampling rate interpolation, consider the same analog input signal used in Example 8.1.

$$x_a(t) = \sin(2\pi t) - .5\cos(4\pi t)$$

Suppose the sampling frequency is $f_s = 20$ Hz. Consider the problem of interpolating the samples $x(k)$ by a factor of $L = 3$. From (8.2.11), the required lowpass filter has a gain of $H_L(0) = 3$ and a cutoff frequency of $F_L = 10/3$ Hz. Suppose a windowed linear-phase filter of order $m = 20$ with a Hanning window is used. The insertion of two zero samples between each of the original samples causes high-frequency images of the original spectrum to appear

FIGURE 8.7: Sampling Rate Interpolation by an Integer Factor $L = 3$, Using an FIR Filter of Order $m = 20$ with a Hanning Window

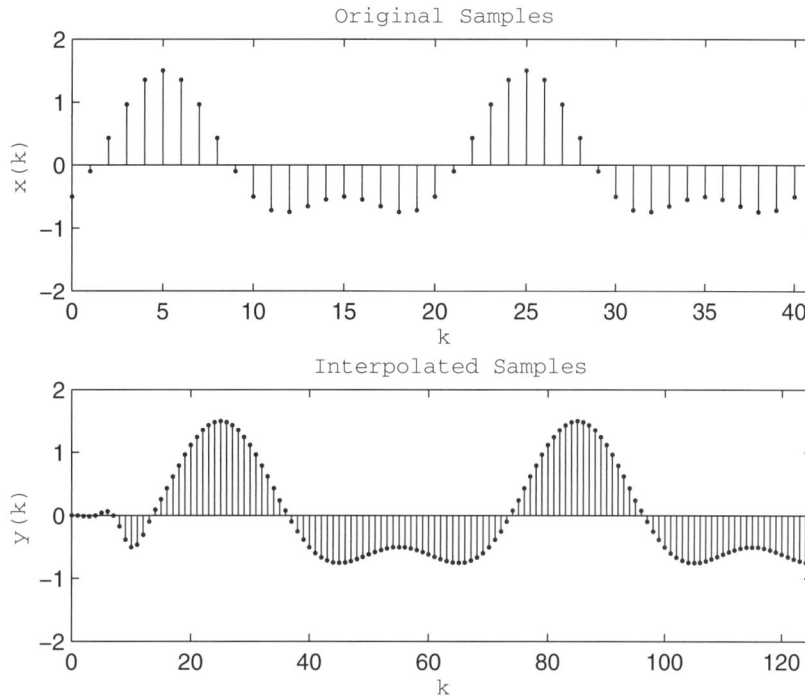

that must be removed by the anti-imaging filter. The interpolator output is obtained by running *exam8_2*. Plots of the original samples and the interpolated samples are shown in Figure 8.7. It is apparent from inspection that the interpolated samples have filled in between the original samples and preserved the wave shape in this case. It may seem counterintuitive that inserting a run of zero samples can interpolate between existing samples. It is the inclusion of the lowpass filter that effectively recovers the unique underlying bandlimited analog signal.

• • • • • • • • • • • • • • • • • •

8.3   Rational Sampling Rate Converters

8.3.1   Single-stage Converters

Sampling rate conversion by integer factors is useful, but can be too restrictive in some practical applications. For example, digital audio tape (DAT) used in sound recording studios has a sampling rate of $f_s = 48$ kHz, while a compact disc (CD) is recorded at a sampling rate of $f_s = 44.1$ kHz. In order to convert music from one format to the other, a noninteger change in the sampling rate is required. Fortunately, all the tools are in place to realize a much larger set of sampling rate converters. The basic approach is to first interpolate the signal by a factor of L, and then decimate the result by a factor of M. The net effect of this cascade configuration of an interpolator followed by a decimator is to change the sampling rate by a *rational* factor L/M. That is,

$$f_S = \left(\frac{L}{M}\right) f_s \qquad (8.3.1)$$

A block diagram of a rational sampling rate converter is shown in Figure 8.8. If $L/M < 1$, then the system in Figure 8.8 is a rational decimator, and if $L/M > 1$, it is a rational interpolator.

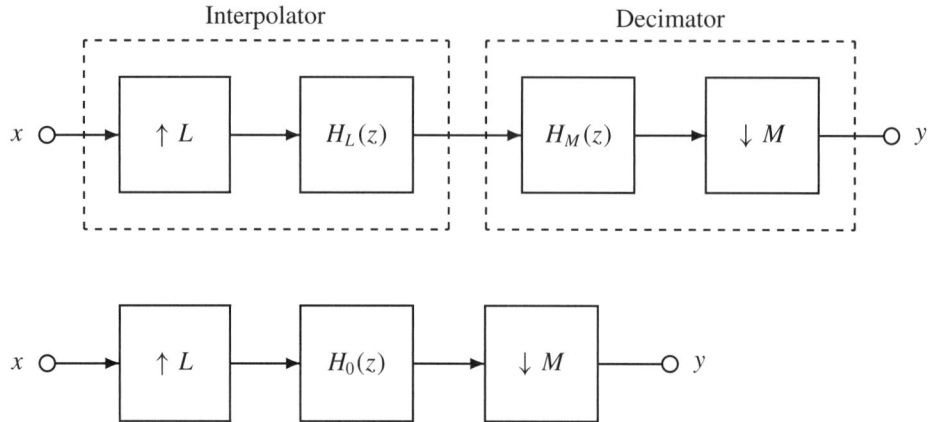

FIGURE 8.9: Simplified Rational Sampling Rate Converter with a Composite Anti-aliasing and Anti-imaging Filter and a Conversion Factor, L/M

The interpolation in Figure 8.8 is done first in order to work at the higher sampling rate, thereby preserving the original spectral characteristics of $x(k)$. Moreover, this ordering has an added benefit because the cascade configuration of the two lowpass filters can be combined into a single equivalent lowpass filter with a frequency response of $H_0(f) = H_L(f)H_M(f)$. The simplified configuration is shown in Figure 8.9. The passband gain of this composite anti-aliasing and anti-imaging filter is $H_0(0) = L$, and the cutoff frequency is F_0, where

$$F_0 = \min \left\{ \frac{f_s}{2L}, \frac{f_s}{2M} \right\} \tag{8.3.2}$$

Thus the frequency response of the composite digital anti-aliasing and anti-imaging filter is

$$H_0(f) = \begin{cases} L, & 0 \le |f| < F_0 \\ 0, & F_0 < |f| \le f_s/2 \end{cases} \tag{8.3.3}$$

If the composite filter is a linear-phase FIR filter of order m, then the output of a rational sampling rate converter can be expressed as follows in the time domain.

$$y(k) = \sum_{i=0}^{m} b_i \delta_L(Mk - i) x \left(\frac{Mk - i}{L} \right) \tag{8.3.4}$$

As a partial check, observe that (8.3.4) reduces to the decimator special case in (8.2.4) when $L = 1$ because $\delta_1(k) = 1$. Similarly, (8.3.4) reduces to the interpolator special case in (8.2.12) when $M = 1$.

Example 8.3

Rational Sampling Rate Converter

As an illustration of sampling rate conversion by a rational factor, consider the following analog input signal.

$$x_a(t) = \cos(2\pi t) + .8 \sin(4\pi t)$$

Suppose the sampling frequency is $f_s = 20$ Hz, and consider the problem of changing the sampling rate of $x(k)$ by a factor of $L/M = 3/2$. In this case the required lowpass filter has a

FIGURE 8.10: Sampling Rate Conversion by a Rational Factor $L/M = 3/2$ Using an FIR Filter of Order $m = 20$ with a Hamming Window

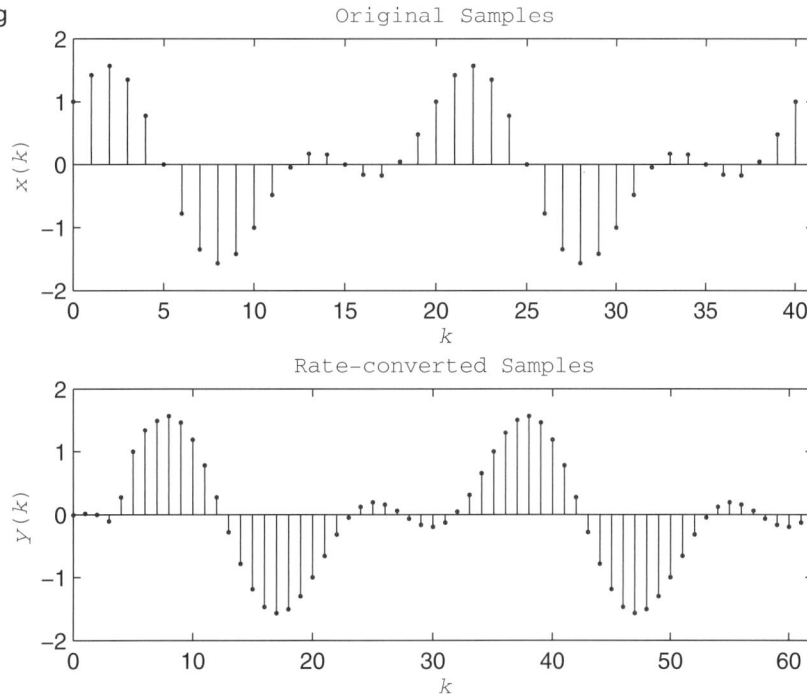

gain of $H_0(0) = 3$ and a cutoff frequency of

$$F_0 = \min\left\{\frac{20}{6}, \frac{20}{4}\right\}$$

$$= \frac{10}{3} \text{ Hz}$$

Suppose a windowed linear-phase filter of order $m = 20$ with a Hamming window is used. The converter output is obtained by running *exam8_3*. Plots of the original samples and the rate-converted samples are shown in Figure 8.10. It is apparent from inspection that the interpolated samples have filled in between the original samples with three new samples for each pair of original samples.

8.3.2 Multistage Converters

In some practical applications, the values for L or M can be relatively large. This presents some special challenges when it comes to implementation. If either L or M is large, the composite anti-aliasing and anti-imaging filter $H_0(z)$ will be a narrowband lowpass filter with a cutoff of $F_0 \ll f_s$. Narrowband linear-phase filter specifications are difficult to meet and can require very high-order FIR filters. This in turn can mean a large computational time and detrimental finite word length effects. The latter drawback can be mitigated by using a *multistage* sampling *Multistage converter* rate converter. The basic idea is to factor the desired conversion ratio into a product of ratios, each of which uses smaller values for L and M.

$$\frac{L}{M} = \left(\frac{L_1}{M_1}\right)\left(\frac{L_2}{M_2}\right)\cdots\left(\frac{L_r}{M_r}\right) \tag{8.3.5}$$

One can then implement r lower-order *stages* separately and configure them in a cascade, as shown in Figure 8.11 for the case $r = 2$. The optimal number of stages and the optimal

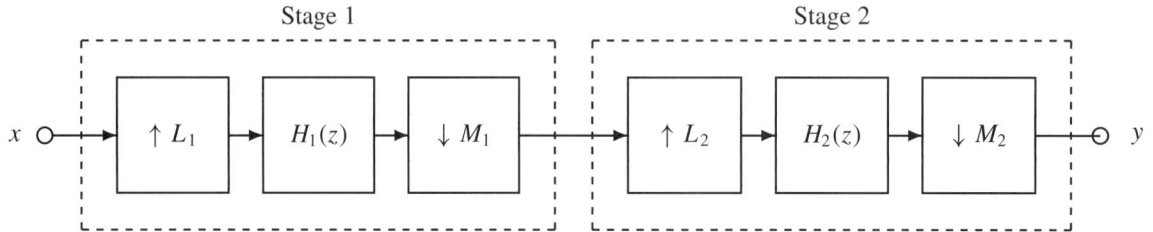

FIGURE 8.11: A Multistage Sampling Rate Converter with $r = 2$ Stages

factoring of L/M can be determined based on minimizing the computational time and the storage requirements (Crochiere and Rabiner, 1975, 1976).

Example 8.4 **DAT to CD**

Consider the problem of designing a sampling rate converter that will transform a signal $x(k)$ that was sampled using the standard digital audio tape (DAT) format to a signal $y(k)$ that is suitable playing on a compact disc (CD) drive. Since the CD sampling rate of 44.1 kHz is smaller than the DAT sampling rate of 48 kHz, this requires a rational decimator. The required frequency conversion ratio is

$$\frac{L}{M} = \frac{44.1}{48}$$
$$= \frac{441}{480}$$
$$= \frac{147}{160}$$

Consequently, this application requires a rational decimator with $L = 147$, and $M = 160$. From (8.3.2) and (8.3.3), a single-stage composite anti-aliasing and anti-imaging filter $H_0(f)$ must have a passband gain of $H_0(0) = 147$ and a cutoff frequency of

$$F_0 = \min\left\{\frac{24}{147}, \frac{24}{160}\right\} \text{ kHz}$$
$$= 150 \text{ Hz}$$

Thus the ideal frequency response for the composite anti-aliasing and anti-imaging filter is

$$H_0(f) = \begin{cases} 147, & 0 \le |f| < 150 \\ 0, & 150 \le |f| < 24000 \end{cases}$$

This is clearly a narrowband lowpass filter with a normalized cutoff frequency of $F_0/f_s = .003125$. A direct single-stage implementation would require a very high-order linear-phase FIR filter. This can be avoided if a multistage implementation is used. For example, the following three conversion ratios all have single-digit integer factors.

$$\frac{147}{160} = \left(\frac{7}{8}\right)\left(\frac{7}{5}\right)\left(\frac{3}{4}\right)$$

Using this multistage approach, a DAT-to-CD converter can be implemented using two decimators and one interpolator. To convert from CD to DAT format, the reciprocals (two interpolators and a decimator) can be used. A detailed design of a CD-to-DAT sampling rate converter is presented later.

FDSP Functions

The FDSP toolbox contains the following function for performing integer and rational sampling rate conversion. If the MATLAB Signal Processing Toolbox is available, then the function *decimate*, *interp*, and *resample* can be used to change the sampling rate.

```
% F_DECIMATE: Reduce sampling rate by factor M.
% F_INTERPOL: Increase sampling rate by factor L.
% F_RATECONV: Convert sampling rate by rational factor L/M.
%
% Usage:
%         [y,b] = f_decimate (x,fs,M,m,f_type,alpha)
%         [y,b] = f_interpol (x,fs,L,m,f_type,alpha)
%         [y,b] = f_rateconv (x,fs,L,M,m,f_type,alpha)
% Pre:
%         x     = a vector of length P containing the input
%                 samples
%         fs    = sampling frequency of x
%         M     = an integer specifying the conversion
%                 factor (M >= 1)
%         L     = an integer specifying the conversion
%                 factor (L >= 1).
%         m     = the order of the lowpass FIR anti-
%                 aliasing anti-imaging filter.
%         f_type = the FIR filer type to be used:
%
%                 0 = windowed (rectangular)
%                 1 = windowed (Hanning)
%                 2 = windowed (Hamming)
%                 3 = windowed (Blackman)
%                 4 = frequency-sampled
%                 5 = least-squares
%                 6 = equiripple
%
%         alpha = an optional scaling factor for the
%                 cutoff frequency of the FIR filter.
%                 Default: alpha = 1.  If present, the
%                 cutoff frequency used for the anti-
%                 aliasing filter H_0(z) is
%
%                 F_c = alpha*fs/(2M)
%                 F_c = alpha*fs/(2L)
% Post:
%         y = a 1 by N vector containing the output
%             samples. Here N = floor(P/M).
%         b = a 1 by (m+1) vector containing FIR filter
%             coefficients
```

Continued on p. 596

Continued from p. 595

```
% Notes:
%           If L or M are relatively large (e.g., greater
%           than 10), then it is the responsibility of the user
%           to perform the rate conversion in stages using
%           multiple calls.  Otherwise, the required value
%           for m can be very large.
```

● ● ● ● ● ● ● ● ● ● ● ● ● ● ●

8.4 Multirate Filter Realization Structures

Sampling rate converters have a considerable amount of built-in redundancy in terms of the required computational effort. In the case of a decimator with $M \gg 1$, all of the input samples are processed by the lowpass filter, but only every Mth sample of the filter output is used. An analogous observation holds for an interpolator with $L \gg 1$. Here most of the samples that are being processed by the lowpass filter are zero samples inserted between the original samples. Consequently, many of the floating-point operations are multiplications by zero.

8.4.1 Polyphase Decimator

To develop efficient realization structures for rate converters, first consider a decimator. Recall from Figure 8.4 that an integer factor of M decimator consists of a lowpass filter with cutoff frequency $F_M = f_s/(2M)$ followed by a down-sampler, $\downarrow M$. Suppose the lowpass anti-aliasing filter $H_M(z)$ is an FIR filter.

$$H_M(z) = \sum_{i=0}^{m} h(i)z^{-i} \tag{8.4.1}$$

Let $p = \text{floor}(m/M)$ be the number of segments of length M in the impulse response h. For convenience, pad h with $M-1$ zeros so that $h(i)$ is defined for $0 \le i < m + M - 1$. One can then define the following sequence of subfilters where $E_n(z)$ uses every Mth sample of $h(i)$ starting with sample n (Bellanger et al, 1976).

$$E_0(z) \triangleq \sum_{i=0}^{p} h(Mi)z^{-i}$$

$$E_1(z) \triangleq \sum_{i=0}^{p} h(Mi + 1)z^{-i}$$

$$\vdots$$

$$E_{M-1}(z) \triangleq \sum_{i=0}^{p} h(Mi + M - 1)z^{-i} \tag{8.4.2}$$

Here $z^{-n}E_n(z^M)$ processes every Mth sample of its input starting with sample n for $0 \le n < M$. Consequently, all of the samples can be processed using the following representation of $H_M(z)$.

$$H_M(z) = \sum_{i=0}^{M-1} z^{-i} E_i(z^M) \tag{8.4.3}$$

FIGURE 8.12: A Factor of M Decimator Using a Polyphase Decomposition of the Anti-aliasing Lowpass Filter with $M = 4$

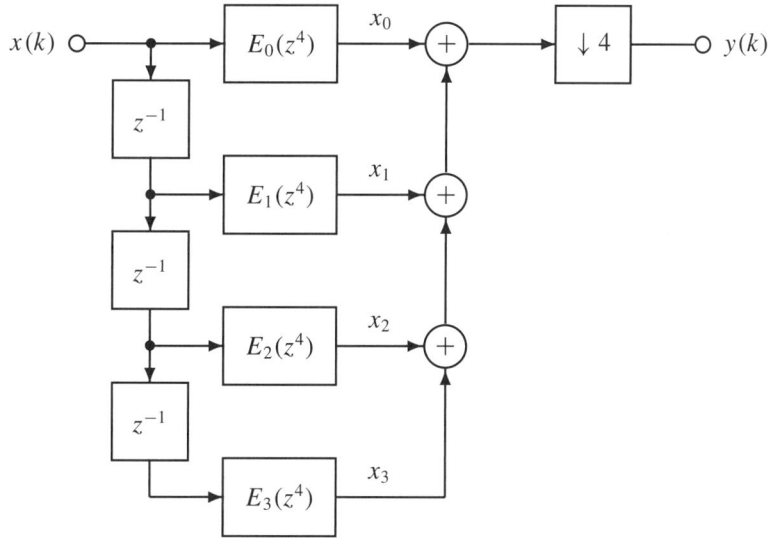

FIGURE 8.13: A More Efficient Realization of a Factor of M Decimator Using a Polyphase Decomposition of the Anti-aliasing Lowpass Filter with $M = 4$

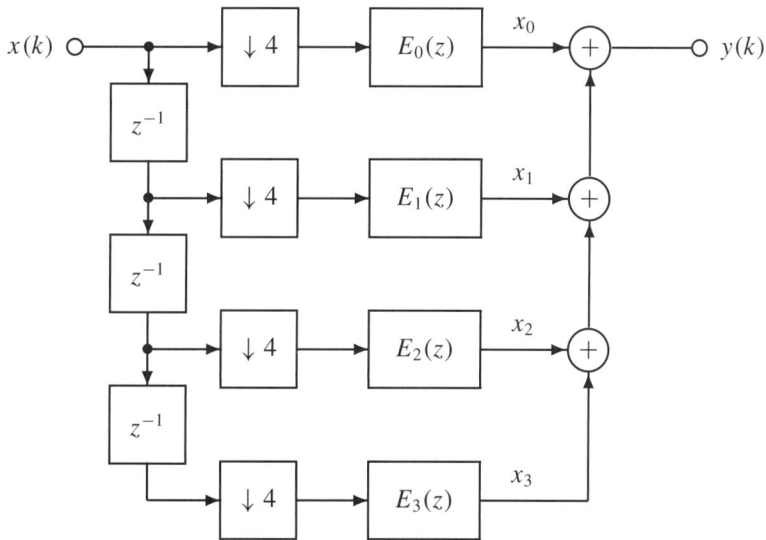

Polyphase decomposition

This is called an M-channel *polyphase decomposition* of $H_M(z)$. It is a parallel form realization where the ith branch operates on the ith phase of the input which includes samples $\{x(i), x(M+i), x(2M+i), \ldots\}$. A block diagram of factor of M decimator using a polyphase decomposition for the anti-aliasing filter is shown in Figure 8.12 for the case $M = 4$.

Notice that the sampling rate of the input signals processed by the subfilters $E_i(z^M)$ in Figure 8.12 is the original sampling rate f_s because the down-sampling occurs as the last operation. The down-sampler can be pushed backward through the summing junctions, in which case a copy of it appears in each of the parallel branches following the subfilters $E_i(z^M)$. Thus each subfilter is cascaded with its own down-sampler block. The order of the subfilter and down-sampler blocks can be interchanged. However, when this is done, to maintain the same input-output relationship, z^M must be replaced by z as an argument of $E_i(z)$. The resulting equivalent realization of the decimator is shown in Figure 8.13.

Although Figure 8.13 has more blocks than Figure 8.12, due to the replication of the down-samplers, the realization in Figure 8.13 is actually more efficient. Note that replacing z^M by z in $E_i(z)$ effectively reduces the *length* of the subfilter by a factor of M. Furthermore, the input signals that are now driving the subfilters in Figure 8.13 have been down-sampled by a factor of M, so the samples are arriving at a slower rate. For convenience, suppose the floating-point operations or FLOPs are measured using multiplications. The number of FLOPs/sec that are needed to compute subfilter output $x_i(k)$ is $\rho_i = [(m+1)/M]f_s/M$. Since there are M subfilters in Figure 8.13, the required computational rate for the polyphase output $y(k)$ is

$$\rho_M = \frac{(m+1)f_s}{M} \text{ FLOPs/sec} \tag{8.4.4}$$

This is in contrast to a direct realization of $H_M(z)$ in (8.4.1) that requires $(m+1)f_s$ FLOPs/sec. Thus the polyphase decimator realization is more efficient by a factor of M.

8.4.2 Polyphase Interpolator

Next, consider a polyphase realization of an interpolator. Recall from Figure 8.6 that an integer factor of L interpolator consists of an up-sampler $\uparrow L$ followed by a lowpass filter with cutoff frequency $F_L = f_s/(2L)$. Suppose the lowpass anti-imaging filter $H_L(z)$ is an FIR filter.

$$H_L(z) = \sum_{i=0}^{m} h(i)z^{-i} \tag{8.4.5}$$

As with the decimator, let $p = \text{floor}(m/L)$ be the number of segments of length L in the impulse response h. For convenience, pad h with $L-1$ zeros so that $h(i)$ is defined for $0 \le i < m+L-1$. The sequence of subfilters is then defined as before but in this case there are L phases of the signal and L subfilters to process them.

$$F_0(z) \triangleq \sum_{i=0}^{p} h(Li)z^{-i}$$

$$F_1(z) \triangleq \sum_{i=0}^{p} h(Li+1)z^{-i}$$

$$\vdots$$

$$F_{L-1}(z) \triangleq \sum_{i=0}^{p} h(Li+L-1)z^{-i} \tag{8.4.6}$$

Here $z^{-n}F_n(z^L)$ processes every Lth sample of its input starting with sample n for $0 \le n < L$. Hence all of the samples can be processed using the following representation of $H_L(z)$.

$$H_L(z) = \sum_{i=0}^{L-1} z^{-i}F_i(z^L) \tag{8.4.7}$$

Polyphase decomposition — This is an L-channel *polyphase* decomposition of $H_L(z)$. It is a parallel form realization where the ith branch operates on the ith phase of the input which includes samples $\{x(i), x(L+i), x(2L+i), \ldots\}$. A block diagram of factor of L decimator using a polyphase decomposition for the anti-imaging filter is shown in Figure 8.14 for the case $L = 3$.

Notice that the sampling rate of the input signals processed by the subfilters $F_i(z^L)$ in Figure 8.14 is Lf_s because the up-sampling occurs as the first operation. The up-sampler can be pushed forward through the pickoff points, in which case a copy of it appears in each parallel branch preceding the subfilters $F_i(z^L)$. Thus each subfilter is cascaded with its own up-sampler

FIGURE 8.14: A Factor of *L* Interpolator Using a Polyphase Decomposition of the Anti-imaging Lowpass Filter with $L = 3$

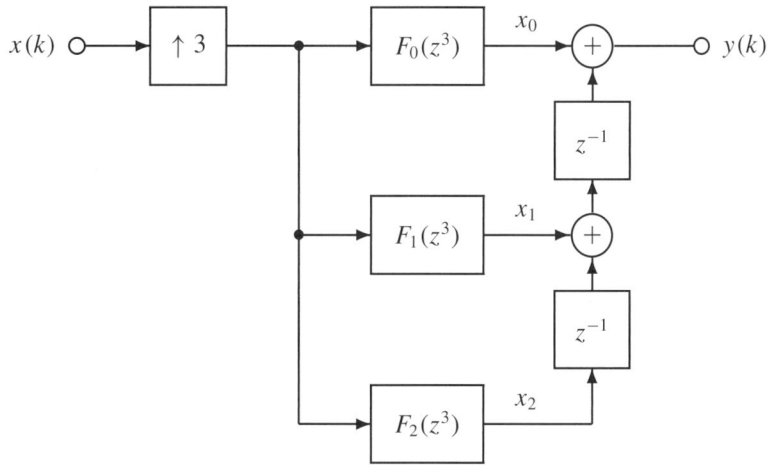

FIGURE 8.15: A More Efficient Realization of a Factor of *L* Interpolator Using a Polyphase Decomposition of the Anti-imaging Lowpass Filter with $L = 3$

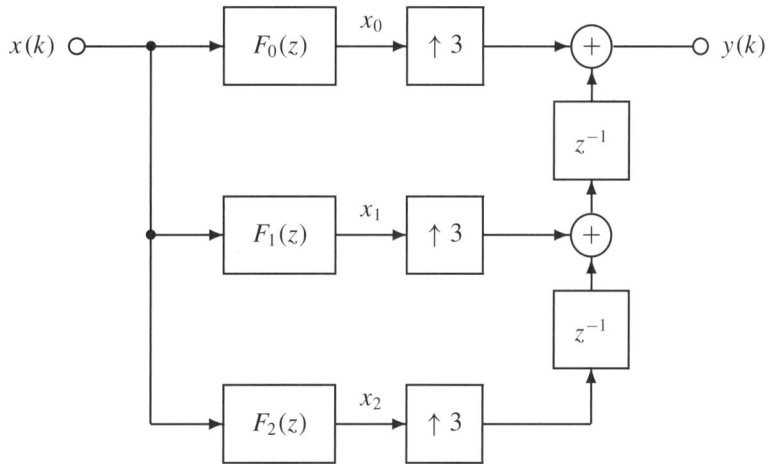

block. The order of the subfilter and up-sampler blocks can be interchanged. However, when this is done, to maintain the same input-output relationship, z^L must be replaced by z as an argument of $F_i(z)$. The resulting equivalent realization of the interpolator is shown in Figure 8.15.

Again, although Figure 8.15 has more blocks than Figure 8.14 due to the replication of the up-samplers, the realization in Figure 8.15 is actually more efficient. Note that replacing z^L by z in $F_i(z)$ effectively reduces the *length* of the subfilter by a factor of L. Furthermore, the input signals that are now driving the subfilters in Figure 8.15 have not been up-sampled by a factor of L, so the samples are arriving at a slower rate. The number of FLOPs/sec that are needed to compute subfilter output $x_i(k)$ is $\rho_i = [(m+1)/L]f_s/L$. Since there are L subfilters in Figure 8.15, the required computational rate for the polyphase output $y(k)$ is

$$\rho_L = \frac{(m+1)f_s}{L} \text{ FLOPs/sec} \qquad (8.4.8)$$

This is in contrast to a direct realization of $H_L(z)$ in (8.4.5) that requires $(m+1)f_s$ FLOPs/sec. Thus the polyphase interpolator realization is more efficient by a factor of L. Additional savings in computational effort can be achieved when linear-phase filters are used. Recall that linear-phase filters satisfy the symmetry constraint $h(m-k) = \pm h(k)$ for $0 \le k \le m$. This redundancy in the filter coefficients can be exploited using an approach analogous to Figure 6.31, but in the context of the polyphase structure.

8.5 Narrowband Filters and Filter Banks

Now that we have a means of changing the sampling rate of a discrete-time signal, this technique can be put to work in a number of practical ways.

8.5.1 Narrowband Filters

A narrowband filter is a sharp filter whose passband or stopband is small in comparison with the sampling frequency. To implement a narrowband filter, an IIR filter such as an elliptic filter might be used, but this introduces nonlinear phase distortion. Implementations of linear-phase narrowband filters typically require very high-order FIR filters. This implies increased memory requirements, longer computational times, and more significant finite word length effects. The latter problem can be reduced by using a multirate design of a narrowband filter. Suppose the ideal filter specification is to pass frequencies in the range $0 \leq |f| \leq F_0$ where $F_0 \ll f_s$.

$$H(f) = \begin{cases} 1, & 0 \leq |f| \leq F_0 \\ 0, & F_0 < |f| \leq f_s/2 \end{cases} \tag{8.5.1}$$

The first step of the multirate method is to reduce the sampling rate by an integer factor M. This has the effect of increasing the relative width of the passband by a factor of M. Setting *Decimation factor* $MF_0 \leq f_s/4$ yields the following upper bound on the *decimation factor M*.

$$M \leq \frac{f_s}{4F_0} \tag{8.5.2}$$

For the maximum value of M the new cutoff frequency is $MF_0 = .25 f_s$. Consequently, a reduction in the sampling rate transforms a narrowband filter, $H(z)$, into a wideband filter, $G(z)$, with a cutoff frequency that is up to one-fourth of the sampling rate.

$$G(f) = \begin{cases} 1, & 0 \leq |f| \leq MF_0 \\ 0, & MF_0 < |f| \leq f_s/2 \end{cases} \tag{8.5.3}$$

The wideband or regular filter $G(z)$ is easier to implement than a narrowband filter. To complete the process, the original sampling frequency must be restored using sampling rate interpolation by a factor of M. The resulting overall implementation of a multirate narrowband filter is shown in the block diagram in Figure 8.16. The following example compares a multirate narrowband design with a conventional fixed-rate design.

Example 8.5   **Multirate Narrowband Filter**

To illustrate of the multirate technique, consider the problem of designing an ideal lowpass filter with a cutoff frequency of $F_0 = f_s/32$. From (8.5.2), the decimation factor must satisfy

$$M \leq \frac{f_s}{4F_0}$$
$$= 8$$

$$x(k) \circ\!\!\longrightarrow \boxed{H_M(z)} \longrightarrow \boxed{\downarrow M} \longrightarrow \boxed{G(z)} \longrightarrow \boxed{\uparrow M} \longrightarrow \boxed{H_M(z)} \longrightarrow\!\!\circ y(k)$$

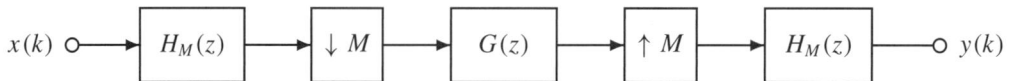

FIGURE 8.16: A Multirate Narrowband Filter Using a Rate Conversion Factor M

FIGURE 8.17:
Magnitude
Responses of
Narrowband
Lowpass Filters
Using a Fixed-rate
Design with
$m = 240$ and a
Multirate Design
with $m = 80$ Using
a Rate Conversion
Factor $M = 8$

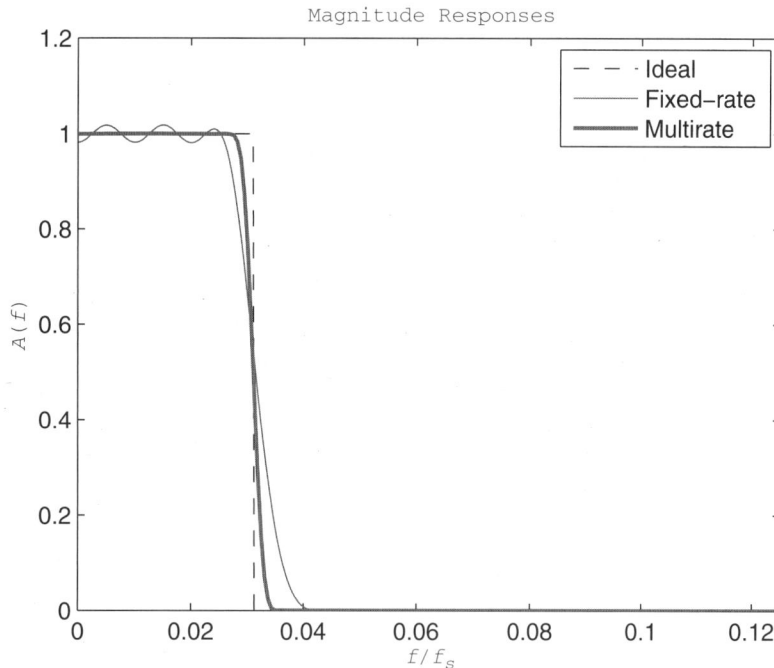

Suppose $M = 8$, and the windowing method is used to design both $H(z)$ and $G(z)$. When *exam8_5* is run, it generates the magnitude responses shown in Figure 8.17. To clarify the display, only the first quarter of the frequency range, $0 \leq f \leq f_s/8$, is shown. The fixed-rate magnitude response corresponds to a windowed FIR filter of order $m = 240$ using the Blackman window. For comparison, the multirate magnitude response uses a windowed FIR filter of order $m = 80$ with the Blackman window. The anti-aliasing and anti-imaging filters $H_M(z)$ are also FIR filters of order $m = 80$. Thus the two approaches are roughly comparable in terms of memory requirements and computational time. However, the multirate design is less sensitive to finite word length effects because it is a cascade of three filters of order $m = 80$, instead of one filter of order $m = 240$. It is evident from inspection of Figure 8.17 that the multirate design is superior to the fixed-rate design in terms of the width of the transition band. The passband ripple of the fixed-rate design can be reduced by decreasing m, but this is achieved at the expense of further increases in the width of the transition band.

8.5.2   Filter Banks

The narrowband lowpass filter designed in Example 8.5 could instead have been a narrowband bandpass or highpass filter. By using a combination of lowpass, bandpass, and highpass filters, the entire spectrum $[-f_s/2, f_s/2]$ can be covered with a bank of N *subband filters*. The magnitude responses for a filter bank consisting of $N = 4$ subband filters is shown in Figure 8.18. Since the discrete-time frequency response is periodic with period f_s, the spectrum in Figure 8.18 is plotted over the positive frequencies $[0, f_s]$ rather than $[-f_s/2, f_s/2]$. Notice that the transition bands of the subband filters have nonzero widths and that adjacent transition bands overlap. This way the entire spectrum is used and the overall filter bank is effectively an allpass filter. The ith frequency band is called the ith *channel*, and breaking the entire spectrum into N channels is called *frequency-division multiplexing*.

Frequency-division
multiplexing

FIGURE 8.18:
Magnitude
Responses of a
Bank of $N = 3$
Subband Filters

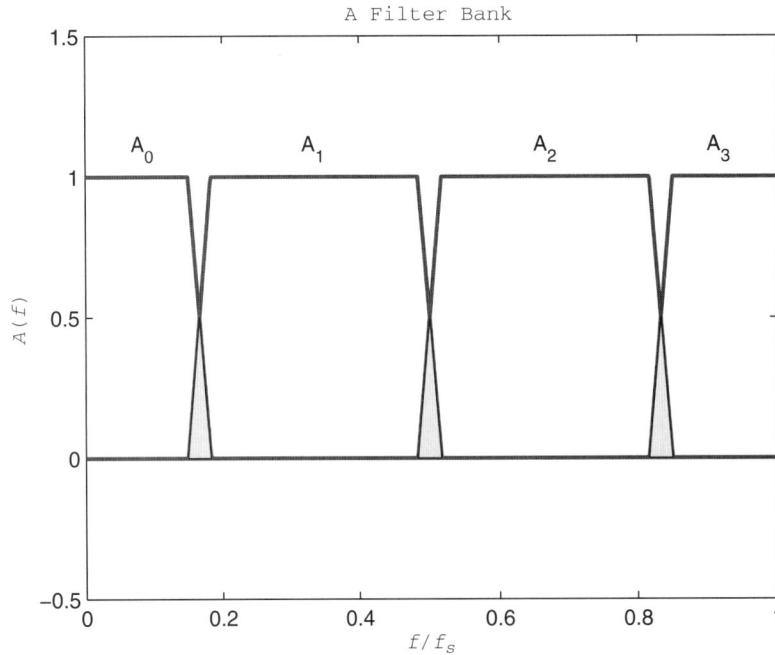

A Filter Bank

FIGURE 8.19: Analysis
and Synthesis Filter
Banks

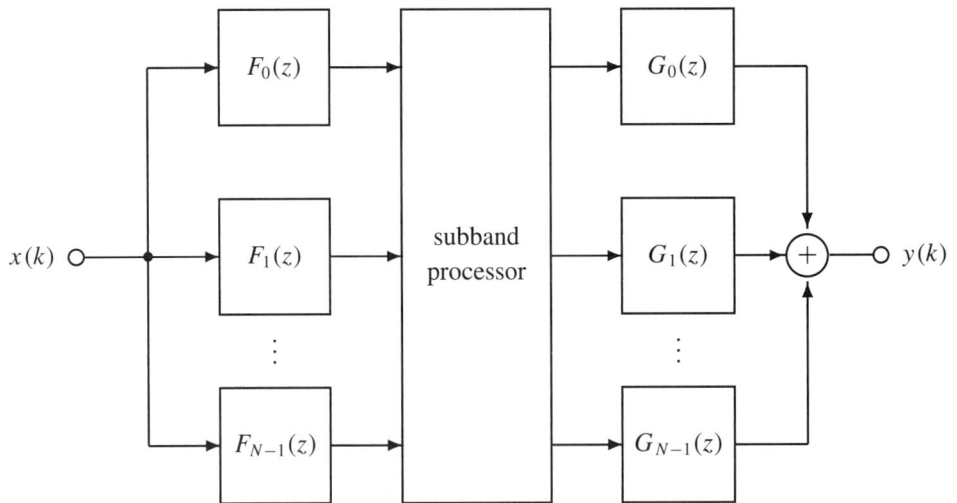

A filter bank is implemented as a parallel configuration of filters, as shown in Figure 8.19. The parallel configuration on the left side of Figure 8.19 is called an *analysis bank* because it decomposes the overall spectrum into N subbands. Each subband is processed separately. Depending on the application, there may also be a second parallel configuration of N filters as shown on the right side of Figure 8.19. This is called a *synthesis bank* because it recombines the subsignals into a single composite signal $y(k)$.

Analysis bank

Synthesis bank

For the bank of N filters shown in Figure 8.19, the width of each subband is $1/N$ times the width of the overall spectrum, $[0, f_s]$. Since each subband is of width f_s/N, the bandlimited subsignals can be down-sampled or decimated by a factor of N. This makes processing of

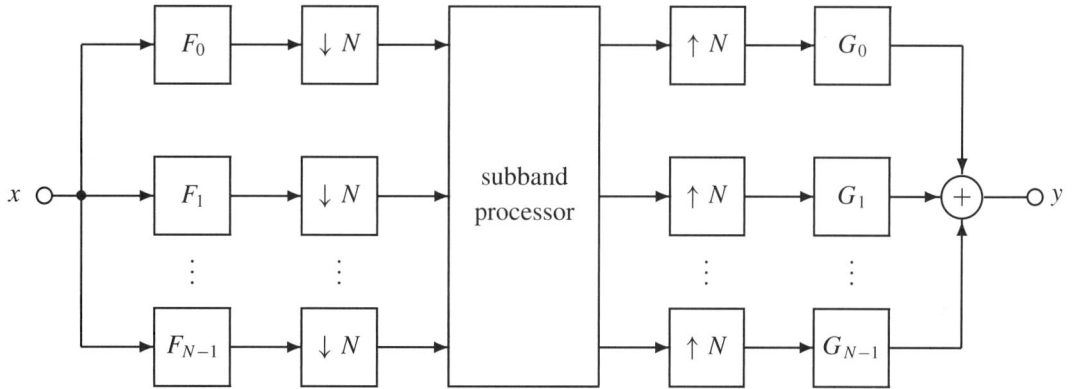

FIGURE 8.20: Decimated and Interpolated Filter Banks

the separate channels more efficient. Following the subband processing, the subsignals are up-sampled or interpolated by a factor of N to restore the original sampling rate. Finally, the up-sampled signals are recombined into a single signal in the synthesis filter bank on the right. The resulting configuration, called a decimated and interpolated filter bank, is shown in Figure 8.20.

Normally, time signals that are filtered are real-valued. When this assumption is relaxed to include complex-valued time signals, there is a simple way to synthesize a high-bandwidth composite signal $x(k)$ that contains several low-bandwidth subsignals. The essential step is to shift the spectrum of each subsignal so that it occupies a particular band in the overall spectrum. This can be achieved by using the *frequency shift* property of the DTFT from Table 4.3.

$$\text{DTFT}\{\exp(jk2\pi F_i T)x(k)\} = X(f - F_i) \qquad (8.5.4)$$

Note that if the kth sample of a signal $x(k)$ is scaled by the complex exponential, $\exp(jk2\pi F_i T)$, this shifts the spectrum of $x(k)$ to the right by F_i Hz. For example, suppose $x_i(k)$ is a bandlimited *subsignal* with bandwidth $B < f_s/N$ for $0 \le i < N$. Then the spectrum of $x_i(k)$ can be shifted to the right and centered at frequency $F_i = if_s/N$ by creating the following complex-valued signal.

$$y_i(k) = \exp(jk2\pi F_i T)x_i(k)$$

$$= \exp\left(\frac{jki2\pi}{N}\right)x_i(k)$$

$$= W_N^{-ki}x_i(k) \qquad (8.5.5)$$

Here the notation $W_N = \exp(-j2\pi/N)$ was used previously with the DFT in Chapter 4. Because $x_i(k)$ occupies only $1/N$ times the total bandwidth, one first up-samples $x_i(k)$ by a factor of N before modulating it as in (8.5.5). When the resulting subsignals are then combined, *Uniform DFT bank* this produces the synthesis filter bank shown in Figure 8.21. This is called a *uniform DFT* filter bank. Observe that the same prototype lowpass filter, $H_N(z)$, can be used to remove the images generated by the up-sampling. Modulation by W_N^{-ki} causes the spectrum of $x_i(k)$ to be shifted to the ith subband of $[0, f_s]$. Whereas the subsignal $x_i(k)$ may be real, the *composite* high-bandwidth signal $x(i)$ is complex.

FIGURE 8.21: Signal Synthesis Using a Uniform DFT Filter Bank

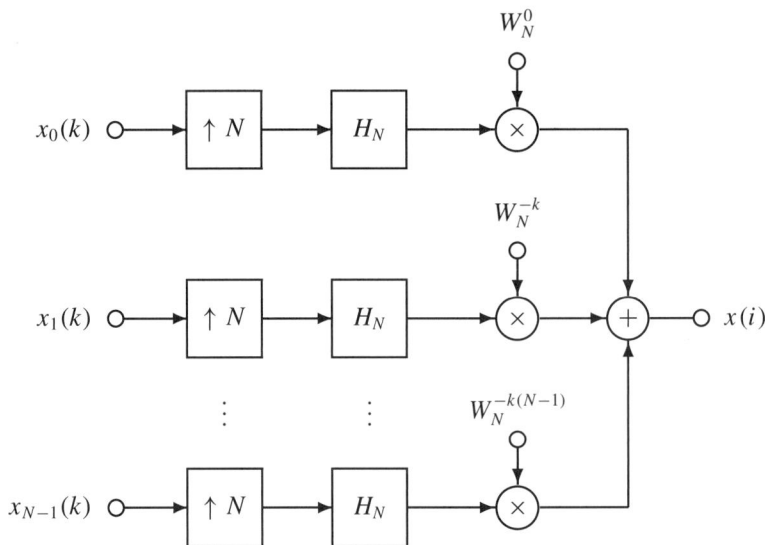

FIGURE 8.22: Signal Analysis Using a Uniform DFT Filter Bank

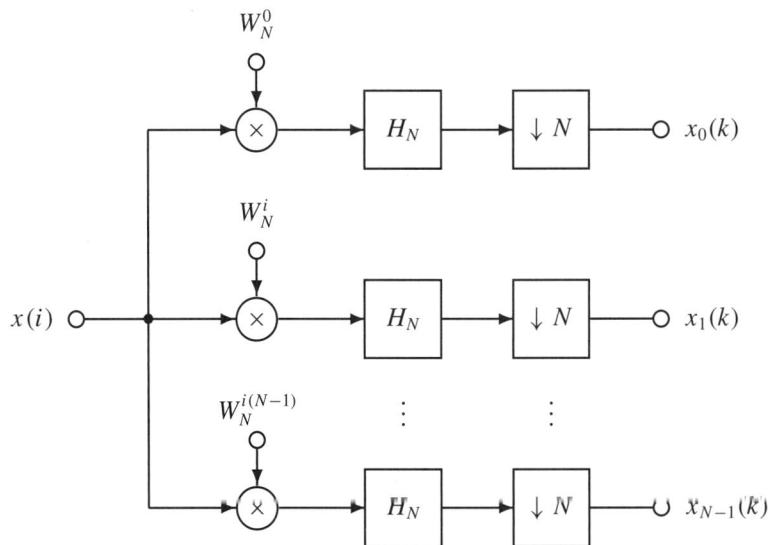

The signal synthesized by the filter bank in Figure 8.21 can be decomposed with an analysis filter bank. First, the signal $x(i)$ is modulated by W_N^{ik}. This has the effect of shifting the kth subband of $x(i)$ back to the origin. This subsignal is then down-sampled to cancel the effects of the up-sampling that was used in the synthesis bank. The end result is the uniform DFT analysis filter bank shown in Figure 8.22.

Example 8.6   **Signal Synthesis**

To illustrate the process of signal synthesis using a uniform DFT filter bank, let the number of filters in the bank be $N = 4$, and let the sampling frequency be $f_s = 10$ Hz. Suppose the subsignals are of length $p = 64$, and suppose all subsignals are bandlimited to $|f| \leq F_0$ where $F_0 = f_s/4$. The subsignals can be defined in terms of their spectra as follows,

FIGURE 8.23: Spectra of Four Bandlimited Subsignals with a Bandwidth of $B = f_s/4$

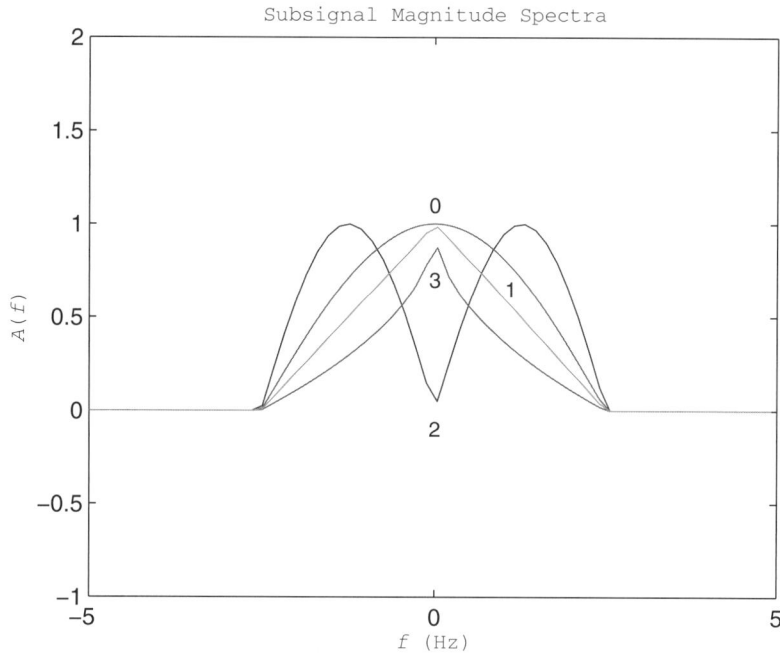

Subsignal Magnitude Spectra

where $f_i = i f_s/p$ for $0 \leq i < p$.

$$X_0(i) = \cos\left(\frac{\pi f_i}{2F_0}\right)$$

$$X_1(i) = 1 - |f_i|/F_0$$

$$X_2(i) = \left|\sin\left(\frac{\pi f_i}{F_0}\right)\right|$$

$$X_3(i) = 1 - (|f_i|/F_0)^2$$

In each case, the phase spectra are zero. Plots of the magnitude spectra are shown in Figure 8.23. After interchanging the first and second halves of the spectra using the MATLAB function *fftshift*, the time signals are then recovered from the spectra as follows using the inverse DFT.

$$x_q(k) = \text{IDFT}\{X_q(i)\}, \quad 0 \leq q < 4$$

Next, $x_q(k)$ is up-sampled by a factor of $N = 4$, as in Figure 8.21. The lowpass anti-imaging filter used was an windowed filter of order $m = 120$ using a Blackman window. For $N = 4$, the modulation factor is

$$W_4 = \exp(-j2\pi/4)$$
$$= \cos(\pi/2) - j\sin(\pi/2)$$
$$= j$$

Thus from Figure 8.21, the complex composite signal $x(k)$ is

$$x(k) = x_0(k) + W_4 x_1(k) + W_4^2 x_2(k) + W_4^3 x_3(k)$$
$$= x_0(k) + jx_1(k) - x_2(k) - jx_3(k)$$
$$= x_0(k) - x_2(k) + j[x_1(k) - x_3(k)]$$

FIGURE 8.24: Real and Imaginary Parts of the Composite High-bandwidth Signal $x(k)$ Containing Four Subsignals Using Frequency-division Multiplexing

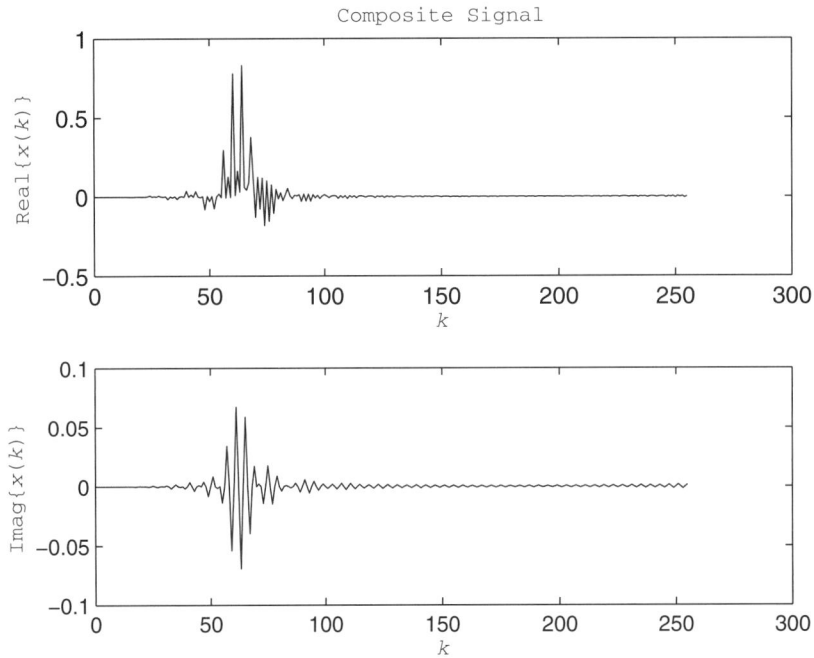

FIGURE 8.25: Magnitude Spectrum of the Composite High-bandwidth Signal Showing the Spectra of Subsignals in Each of Four Bands Using a Windowed Blackman Anti-imaging Filter

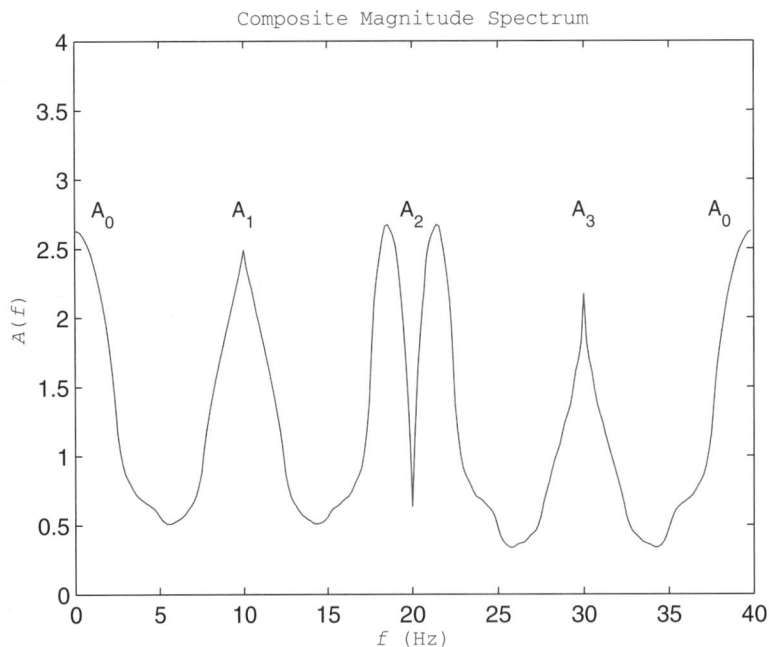

Plots of the real and imaginary parts of the composite signal $x(k)$, obtained by running *exam8_6*, are shown in Figure 8.24. Note that, due to the up-sampling, $x(k)$ is now of length $Np = 256$. Next, the magnitude spectrum of $x(k)$ is computed using

$$A(i) = |\text{FFT}\{x(k)\}|, \quad 0 \le i < Np$$

After interchanging the first and second halves of $A(i)$ using the MATLAB function *fftshift*, the resulting magnitude spectrum of $x(k)$ is shown in Figure 8.25. It is clear that the spectra

FIGURE 8.26:
Magnitude
Spectrum of the
Composite
High-bandwidth
Signal Showing
the Spectra of
Subsignals in Each
of Four Bands Using
a Windowed
Blackman
Anti-imaging Filter
with the Cutoff
Frequency Reduced
by Factor $\alpha = .5$

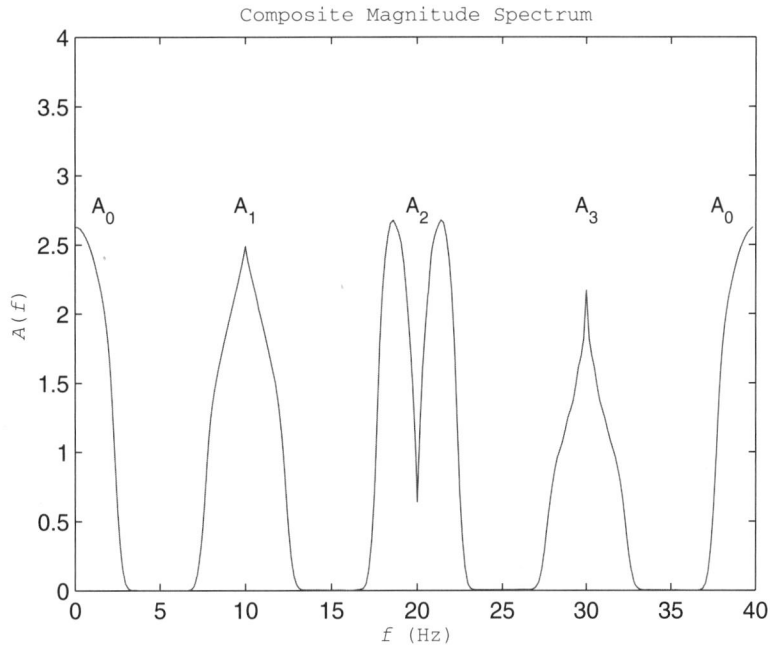

Composite Magnitude Spectrum

of $x_i(k)$ have been shifted and centered at $F_i = i f_s / N$ for $0 \leq i < N$. There is some overlap between subspectra, perhaps caused by the nonideal nature of the anti-imaging filter. Notice in Figure 8.23 that each of the subsignals occupies only half of the subband. Since this is the case, it should be possible to reduce the cutoff frequency of the of the anti-imaging filter $H_N(z)$ by a factor of $\alpha = .5$ so that

$$F_c = \frac{\alpha f_s}{2N}$$

The resulting magnitude spectrum of $x(k)$ using this lower cutoff frequency is shown in Figure 8.26. Notice that this has effectively eliminated the spectral overlap between the subbands. It should now be possible to use an analysis filter bank such as the uniform DFT filter bank in Figure 8.22 to extract the individual subsignals from $x(k)$.

8.6 A Two-channel QMF Bank

There are a number of applications where a signal $x(k)$ is decomposed into subsignals using an analysis filter bank (Mitra, 2001). Since the subsignals have a bandwidth that is small in comparison with f_s, they can be down-sampled. Separate low-frequency processing of the subsignals then can lead to efficiencies.

As an illustration of subband processing, consider the two-channel filter bank shown in Figure 8.27. Here the subsignals are encoded using an appropriate coding scheme (Esteban and Galand, 1977). Since each subband has its own spectral characteristics, some subsignals might be encoded using fewer bits than others. The encoded signals are either saved on a storage device or transmitted over a communications channel using *time-division multiplexing*. At the receiver end, the process is reversed. First, the signal is demultiplexed and decoded to resolve

Time-division multiplexing

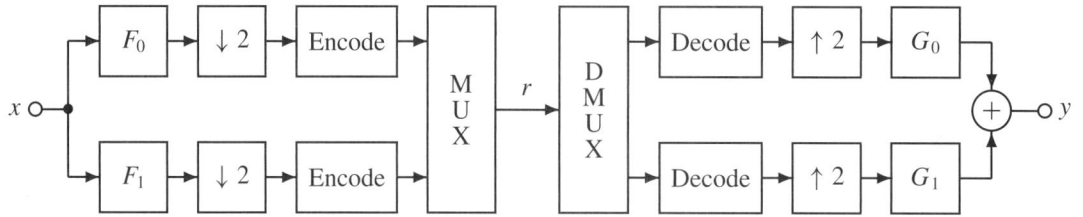

FIGURE 8.27: Efficient Transmission or Storage of a Signal $x(k)$ Using Subband Encoding and a Two-channel Filter Bank

FIGURE 8.28: A Two-channel Quadrature Mirror (QMF) Bank

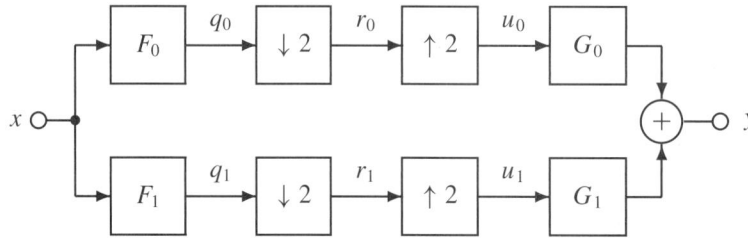

it into subsignals. The subsignals are up-sampled to restore the original sampling rate and then recombined into a single signal $y(k)$ using a synthesis filter bank.

If an efficient encoding scheme is used, the bandwidth of the channel used to transmit the time-multiplexed signal $r(k)$ will be less than the bandwidth needed to transmit the original signal $x(k)$ directly. Assuming that the encoding errors, the transmission errors, and the rate-conversion errors are sufficiently small, the output $y(k)$ will be a scaled and delayed version of the input $x(k)$. If the information loss due to encoding and channel distortion error are ignored, then the distortion caused by down-sampling and up-sampling can be analyzed using the simplified system shown in Figure 8.28. This system is referred to as a *two-channel quadrature mirror filter bank* or a QMF bank.

QMF bank

8.6.1 Rate Converters in the Frequency Domain

To model the effects of sampling rate conversion in the QMF bank, up-sampling and down-sampling must be formulated in the frequency domain. For a factor of L interpolator, recall from (8.2.9) that the Z-transform of the up-sampled signal $x_L(k)$ can be expressed in terms of the Z-transform of the input signal $x(k)$ as

$$X_L(z) = X(z^L) \tag{8.6.1}$$

Next, consider a factor of M decimator. Here every Mth sample of the input $x(k)$ appears in the down-sampled output $x_M(k)$.

$$x_M(k) = x(Mk) \tag{8.6.2}$$

Let $\hat{x}_M(i)$ be the signal that agrees with $x(i)$ at every Mth sample, but is otherwise zero.

$$\hat{x}_M(i) = \begin{cases} x(i), & |i| = 0, M, 2M, \ldots \\ 0, & \text{otherwise} \end{cases} \tag{8.6.3}$$

Recall from (8.2.6) that $\delta_M(k)$ is the notation used to represent a periodic impulse train with period M. The signal $\hat{x}_M(k)$ can be thought of as the periodic signal $\delta_M(k)$ amplitude modulated by $x(k)$.

$$\hat{x}_M(i) = \delta_M(i)x(i) \tag{8.6.4}$$

Since $\hat{x}_M(Mk) = x(Mk)$, x can be replaced with $\hat{x}_M$ in (8.6.2) without changing the result. Furthermore, because the intervening samples of $\hat{x}_M$ are zero, the change of variable $i = Mk$ can be used as follows in computing $X_M(z)$.

$$
\begin{aligned}
X_M(z) &= \sum_{k=-\infty}^{\infty} \hat{x}_M(Mk)z^{-k} \\
&= \sum_{i=-\infty}^{\infty} \hat{x}_M(i)z^{-i/M} \\
&= \sum_{i=-\infty}^{\infty} \hat{x}_M(i)(z^{1/M})^{-i} \tag{8.6.5}
\end{aligned}
$$

Thus the Z-transform of the down-sampled signal $x_M(k)$ can be expressed in terms of the Z-transform of $\hat{x}_M(k)$ as

$$X_M(z) = \hat{X}_M(z^{1/M}) \tag{8.6.6}$$

To represent $X_M(z)$ in terms of $X(z)$, one must examine $\hat{x}_M(k)$ in more detail. Recall from the discussion of the DFT in Chapter 4 that $W_M = \exp(-j2\pi/M)$. If we use the orthogonal property of W_M in (4.3.6), the periodic impulse train $\delta_M(k)$ can be expressed as the following sum of complex exponentials.

$$\delta_M(k) = \frac{1}{M} \sum_{i=0}^{M-1} W_M^{ik} \tag{8.6.7}$$

From (8.6.4) and (8.6.7), the Z-transform of $\hat{x}_M(k)$ is

$$
\begin{aligned}
\hat{X}_M(z) &= \sum_{k=-\infty}^{\infty} \delta_M(k)x(k)z^{-k} \\
&= \frac{1}{M} \sum_{i=0}^{M-1} \sum_{k=-\infty}^{\infty} W_M^{ik}x(k)z^{-k} \\
&= \frac{1}{M} \sum_{i=0}^{M-1} \sum_{k=-\infty}^{\infty} x(k)\left(W_M^{-i}z\right)^{-k} \\
&= \frac{1}{M} \sum_{i=0}^{M-1} X\left(W_M^{-i}z\right) \tag{8.6.8}
\end{aligned}
$$

From, combining (8.6.6) and (8.6.8), the Z-transform of the down-sampled signal $x_M(k)$ is

$$X_M(z) = \frac{1}{M} \sum_{i=0}^{M-1} X\left(W_M^{-i}z^{1/M}\right) \tag{8.6.9}$$

For the QMF bank in Figure 8.28, $M = 2$. If we use $W_2^0 = 1$ and $W_2^1 = -1$, the Z-transform of the output for the factor of two down-sampler simplifies to

$$X_2(z) = .5[X(z^{1/2}) + X(-z^{1/2})] \tag{8.6.10}$$

8.6.2 An Alias-free QMF Bank

Given the frequency domain representations of the up-sampled and down-sampled signals, one can now analyze the two-channel QMF bank in Figure 8.28. From (8.6.10), the outputs $r_i(k)$ from the down-samplers are

$$
\begin{aligned}
R_i(z) &= .5[Q_i(z^{1/2}) + Q_i(-z^{1/2})] \\
&= .5[F_i(z^{1/2})X(z^{1/2}) + F_i(-z^{1/2})X(-z^{1/2})], \quad 0 \leq i \leq 1
\end{aligned}
\tag{8.6.11}
$$

Next, from (8.6.11) and (8.6.1) with $L = 2$, the outputs from the up-samplers are

$$
\begin{aligned}
U_i(z) &= R_i(z^2) \\
&= .5[F_i(z)X(z) + F_i(-z)X(-z)], \quad 0 \leq i \leq 1
\end{aligned}
\tag{8.6.12}
$$

Finally, the output $y(k)$ of the QMF bank is

$$
\begin{aligned}
Y(z) &= G_0(z)U_0(z) + G_1(z)U_1(z) \\
&= .5[G_0(z)F_0(z) + G_1(z)F_1(z)]X(z) \\
&\quad + .5[G_0(z)F_0(-z) + G_1(z)F_1(-z)]X(-z)
\end{aligned}
\tag{8.6.13}
$$

Thus the overall input-output relationship of the QMF filter bank can be written in terms of two transfer functions.

$$
Y(z) = H(z)X(z) + D(z)X(-z)
\tag{8.6.14}
$$

Here $H(z)$ represents the transmission through the system if there were no rate conversion blocks, and $D(z)$ represents *aliasing* effects caused by the down-samplers and the up-samplers.

$$
H(z) = .5[F_0(z)G_0(z) + F_1(z)G_1(z)]
\tag{8.6.15}
$$

$$
D(z) = .5[F_0(-z)G_0(z) + F_1(-z)G_1(z)]
\tag{8.6.16}
$$

The objective in choosing the analysis and synthesis bank filters is to select filters such that $D(z) = 0$ and $H(z) = cz^{-m}$. That way, $y(k)$ will be a delayed and scaled replica of $x(k)$. There are a number of ways to make $D(z) = 0$. For example, one solution is

$$
G_0(z) = F_1(-z)
\tag{8.6.17}
$$

$$
G_1(z) = -F_0(-z)
\tag{8.6.18}
$$

Suppose $F_1(z)$ is selected such that

$$
F_1(z) = F_0(-z)
\tag{8.6.19}
$$

To interpret the frequency response characteristics of the two analysis bank filters, first note that $-1 = \exp(j2\pi f_s T/2)$. Therefore on the unit circle where $z = \exp(j2\pi f T)$,

$$
-z = \exp[j2\pi(f_s/2 + f)T]
\tag{8.6.20}
$$

Next substitute $z = \exp(j2\pi f T)$ into $F_1(z)$ to get the frequency response $F_1(f)$. For a real filter with symmetry about $f = f_s/2$, this yields

$$
\begin{aligned}
|F_1(f)| &= |F_0(f_s/2 + f)| \\
&= |F_0(f_s/2 - f)|
\end{aligned}
\tag{8.6.21}
$$

Consequently, if $F_0(z)$ is a lowpass filter, then $F_1(z)$ will be *mirror image* highpass filter, hence the term mirror image quadrature filter bank. Given the choice of $F_1(z)$ in (8.6.19), it follows from (8.6.17) that to eliminate aliasing

$$G_0(z) = F_0(z) \tag{8.6.22}$$

Notice from (8.5.18), (8.6.19), and (8.6.22) that all of the QMF filters can be expressed in terms of the lowpass anti-aliasing filter $F_0(z)$. From (8.6.15), the overall transfer function of the QMF bank is

$$H(z) = .5[F_0^2(z) - F_0^2(-z)] \tag{8.6.23}$$

Example 8.7	**Alias-free Two-channel QMF Bank**

As a simple example of a two-channel QMF bank of filters, suppose the first analysis filter is

$$F_0(z) = 1 + z^{-1}$$

It then follows from (8.6.19) that the other analysis bank filter is

$$\begin{aligned} F_1(z) &= F_0(-z) \\ &= 1 - z^{-1} \end{aligned}$$

Next, from (8.6.17) and (8.6.18), the two filters in the synthesis bank are

$$\begin{aligned} G_0(z) &= F_1(-z) \\ &= 1 + z^{-1} \\ G_1(z) &= -F_0(-z) \\ &= -1 + z^{-1} \end{aligned}$$

This guarantees that $D(z) = 0$. Finally, from (8.6.23), the overall QMF transfer function is

$$\begin{aligned} H(z) &= 0.5[F_0^2(z) - F_0^2(-z)] \\ &= .5[(1 + z^{-1})^2 - (1 - z^{-1})^2] \\ &= .5[1 + 2z^{-1} + z^{-2} - (1 - 2z^{-1} + z^{-2}] \\ &= 2z^{-1} \end{aligned}$$

Thus the output $y(k) = 2x(k - 1)$ is the reconstructed input $x(k)$, except that it is scaled by two and delayed by one sample.

For general FIR filters of order $m > 1$, the exact reconstruction in Example 8.7 is not possible, but it can be approximated. Suppose $F_0(z)$ is an mth-order type 1 linear-phase FIR lowpass filter. Thus m is even, and $b_{m-i} = b_i$ for $0 \le i \le m$.

$$F_0(z) = \sum_{i=0}^{m} b_i z^{-i} \tag{8.6.24}$$

The remaining analysis and synthesis bank filters in the QMF bank will also be linear-phase FIR filters of order m. Consequently, the QMF bank will be linear phase with a constant group delay of $\tau = mT$. For the magnitude response of the QMF bank to be one, it follows from (8.6.23) that the analysis bank filters should satisfy

$$|F_0(f)|^2 + |F_1(f)|^2 = 1, \quad 0 \le |f| \le f_s/2 \tag{8.6.25}$$

FIGURE 8.29:
Squared Magnitude
Responses of the
Analysis Filters and
the Overall QMF
Bank

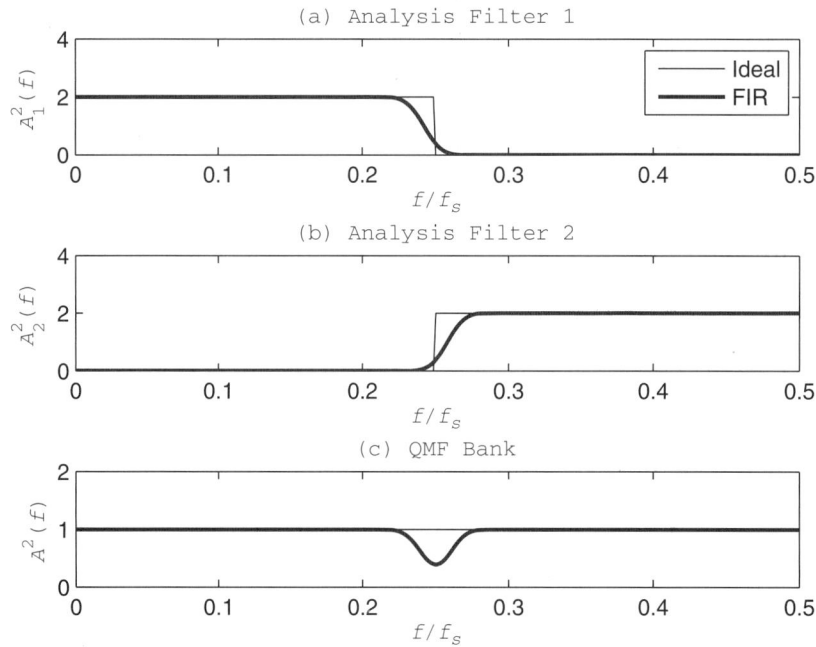

Thus $F_0(z)$ and $F_1(z)$ must form a power-complementary pair. If $F_0(z)$ and $F_1(z)$ could be found to satisfy (8.6.25), this would result in a perfect reconstruction QMF bank.

Example 8.8   **Two-channel QMF Bank**

As an approximation to a perfect reconstruction QMF bank, suppose $m = 80$ and $F_1(z)$ is designed as a lowpass filter with a cutoff frequency $F_0 = f_s/4$ using the windowing method with a Hamming window. The passband gain is set to $A = \sqrt{2}$ so that $A^2/2 = 1$. When *exam8_8* is run, it produces squared the magnitude response plots shown Figure 8.29. The analysis filters in (a) and (b) are lowpass and highpass, respectively, thus decomposing $[0, f_s/2]$ into two subbands. The overall squared magnitude response in (c) approximates (8.6.25) except near the cutoff frequency $F_0 = .f_s/4$, where the effects of the transition band are apparent.

8.7   Oversampling ADC

8.7.1   Anti-aliasing Filters

One of the practical difficulties associated with analog-to-digital conversion is the need for a lowpass analog anti-aliasing prefilter to bandlimit the signal to less than half of the sampling rate. Sharp high-order analog filters have characteristics that can drift with time and even cause them to become unstable. Using multirate techniques, some of the anti-aliasing task can be transferred to the digital domain and thereby allow for a simpler low-order analog filter.

Suppose the range of frequencies of interest for an analog signal $x_a(t)$ is $0 \le |F| \le F_a$. Normally, to avoid aliasing, one must bandlimit $x_a(t)$ to F_a Hz with a sharp analog lowpass

filter, and then sample at a rate f_s that is greater than twice the bandwidth. Instead, consider
oversampling $x_a(t)$ using the following increased sampling rate.

Oversampling

$$f_s = 2MF_a \tag{8.7.1}$$

Here M is an integer greater than one. Recall that this corresponds to oversampling by a factor
of $f_s/(2F_a) = M$. Oversampling by a factor of M significantly reduces the requirements for
the anti-aliasing filter. The analog anti-aliasing filter now has to satisfy the following frequency
response specification.

$$H_a(f) = \begin{cases} 1, & 0 \le |f| \le F_a \\ 0, & MF_a \le |f| < \infty \end{cases} \tag{8.7.2}$$

Even though $H_a(f)$ is still an ideal filter with no passband ripple and complete stopband
attenuation, the width of the transition band is no longer zero, but is instead $\Delta f = (M-1)F_a$.
Given a large transition band, $H_a(s)$ can be approximated with a simple inexpensive low-order
filter such as a first- or second-order Butterworth filter realizable by a single integrated circuit,
as discussed in Section 1.5. Recall from (7.4.1) that the magnitude response of an nth-order

Butterworth lowpass filter

analog *Butterworth lowpass filter* with a cutoff frequency of F_a is

$$A_n(f) = \frac{1}{\sqrt{1 + (f/F_a)^{2n}}} \tag{8.7.3}$$

The tradeoff of an increased sampling rate for a simpler analog anti-aliasing filter leads to a
discrete-time signal that is sampled at a rate that is significantly higher than twice the maximum
frequency of interest. Following the sampling operation, one can reduce the sampling rate to

Oversampling ADC

the minimum value needed using a decimator. The resulting structure, called an *oversampling
ADC*, is shown in the block diagram in Figure 8.30. Note that it has *two* anti-aliasing filters, a
low-order analog filter $H_a(s)$ with cutoff frequency F_a, and a high-order digital filter $H_M(z)$,
also with cutoff frequency F_a.

The use of a simpler analog anti-aliasing filter is the main benefit of the oversampling ADC,
but it is not the only one. Another advantage becomes apparent upon examination of the ADC
quantization noise. Recall from the discussion of quantization in (6.9.3) that if $|x_a(t)| \le c$,

ADC quantization level

then the *quantization level* or precision of an N-bit ADC is

$$q = \frac{c}{2^{N-1}} \tag{8.7.4}$$

The quantized output of the ADC can be represented as follows, where $x(k)$ is the exact value
and $v(k)$ is the quantization error.

$$x_q(k) = x(k) + v(k) \tag{8.7.5}$$

Assuming rounding is used, the quantization error $v(k)$ can be modeled as white noise uniformly
distributed over the interval $[-q/2, q/2]$. It was shown in (6.9.6) that the average power of the

Quantization noise power

quantization noise $\sigma_v^2 = E[v(k)^2]$ can be expressed in terms of q as

$$\sigma_v^2 = \frac{q^2}{12} \tag{8.7.6}$$

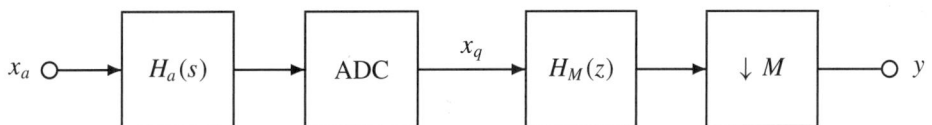

FIGURE 8.30: Oversampling ADC Using Sampling Rate Decimation by an Integer Factor *M*

In Figure 8.30, the quantized signal $x_q(k)$ is processed by a digital anti-aliasing filter $H_M(z)$, and then down-sampled. The down-sampler block does not affect the average power of $x_q(k)$. This is because even though only every Mth sample is extracted, thereby lowering the total energy, the resulting signal is also shorter by a factor of M. To examine the effects of the filter $H_M(z)$ on the quantization noise, recall from (6.9.10) that the average power of the noise at the filter output is

$$\sigma_y^2 = \Gamma \sigma_v^2 \tag{8.7.7}$$

Power gain Here Γ is the *power gain* of $H_M(z)$. If the digital anti-aliasing filter $H_M(z)$ is an FIR filter of order m, then from (6.9.11) the power gain can be expressed in terms of the impulse response $h_M(k)$ as

$$\Gamma = \sum_{k=0}^{m} h_M^2(k) \tag{8.7.8}$$

The expression for the power gain can be simplified further. Using Parseval's identity in Proposition 4.3, the power gain can be recast in terms of $H_M(f) = \text{DTFT}\{h_M(k)\}$ as

$$\begin{aligned}
\Gamma &= \frac{1}{f_s} \int_{-.5f_s}^{.5f_s} |H_M(f)|^2 df \\
&= \frac{1}{f_s} \int_{-.5f_s/M}^{.5f_s/M} df \\
&= \frac{1}{M}
\end{aligned} \tag{8.7.9}$$

If we combine (8.7.7) and (8.7.9), the average power of the quantization noise appearing at the output of the oversampling ADC is

$$\sigma_y^2 = \frac{\sigma_v^2}{M} \tag{8.7.10}$$

Consequently, oversampling by a factor of M has the beneficial effect of reducing the quantization noise power by a factor of M. This is achieved because oversampling by M effectively spreads the noise power out over the frequency range $[-f_s/2, f_s/2]$. The digital anti-aliasing filter $H_M(z)$, with cutoff frequency $F_0 = f_s/(2M)$, then removes most of the quantization noise.

The reduction in the quantization noise power in (8.7.10) can be interpreted as an increase in the number of *effective bits* of precision. For example, let B be the number of bits of precision using oversampling by a factor of M, and let N be the number of bits of precision without oversampling. Using (8.7.4), (8.7.6), and (8.7.10), and equating the quantization noise power for the two cases yields

$$\frac{c^2}{12M[2^{2(B-1)}]} = \frac{c^2}{12[2^{2(N-1)}]} \tag{8.7.11}$$

ADC effective precision Canceling the common terms, taking reciprocals, and then taking the base-2 logarithm of each side, one arrives at the following expression for the required precision when oversampling is used.

$$B = N - \log_2(M)/2 \tag{8.7.12}$$

Note that oversampling by a factor of $M = 4$ decreases the required precision of the ADC by 1 bit. That is, the signal-to-quantization noise ratio of an N-bit ADC without oversampling is the same as the signal-to-quantization noise ratio of an $(N - 1)$-bit ADC with oversampling by a factor of $M = 4$. Both systems have the same quantization error and therefore the same accuracy. For the more general case, oversampling by $M = 4^r$ reduces the required ADC precision by r bits.

| Example 8.9 | **Oversampling ADC** |

To illustrate the effectiveness of oversampling in the analog-to-digital conversion process, let the analog anti-aliasing filter be an nth-order Butterworth filter with the magnitude response in (8.7.3). The objective of this example is to use oversampling to ensure that the maximum aliasing the error is sufficiently small. The maximum of magnitude of the aliasing error occurs at the folding frequency, $f_d = f_s/2$. At this frequency the *aliasing error scale factor* is

Aliasing error factor

$$\epsilon_n \overset{\Delta}{=} A_n(f_d)$$

If oversampling by a factor of M is used, then $f_s = 2MF_a$ which means $f_d = MF_a$. Using (8.7.3) to achieve an aliasing error scale factor of ϵ requires

$$\frac{1}{\sqrt{1 + M^{2n}}} \leq \epsilon$$

Taking reciprocals and squaring both sides then yields

$$1 + M^{2n} \geq \epsilon^{-2}$$

Solving for M,

$$M \geq (\epsilon^{-2} - 1)^{.5/n}$$

For example, if a second-order Butterworth filter is used, then $n = 2$. Suppose the desired value aliasing error scale factor is $\epsilon = .01$. In this case the required sampling rate conversion factor is

$$M = \text{ceil}\{(10^4 - 1)^{1/4}\}$$
$$= \text{ceil}(9.9997)$$
$$= 10$$

The calculation of M can be verified graphically using Figure 8.31, which was generated by running *exam8_9*. The curves display the aliasing error scale factor ϵ_n versus the oversampling factor M for the first four Butterworth filters, $1 \leq n \leq 4$. For clarity of display, the ordinate is the aliasing error scale factor in units of dB. It is evident that very low aliasing error can be achieved by using a combination of oversampling and a sufficiently high-order Butterworth anti-aliasing filter.

8.7.2 Sigma-delta ADC

The power density spectrum of the quantization noise $v(k)$ is flat over the frequency range $0 \leq |f| \leq f_s/2$, and is equal to σ_v^2 in (8.7.6).

$$P_v(f) = \sigma_v^2 \qquad (8.7.13)$$

When the lowpass filter $H_M(z)$ in Figure 8.30 is applied to $v(k)$, it removes the fraction of the noise that occupies the stopband, thereby decreasing the average power of the noise by a

FIGURE 8.31:
Aliasing-error Scale
Factor ϵ_n in dB
versus the
Oversampling
Factor M Using
Butterworth
Anti-aliasing Filters
of Order n

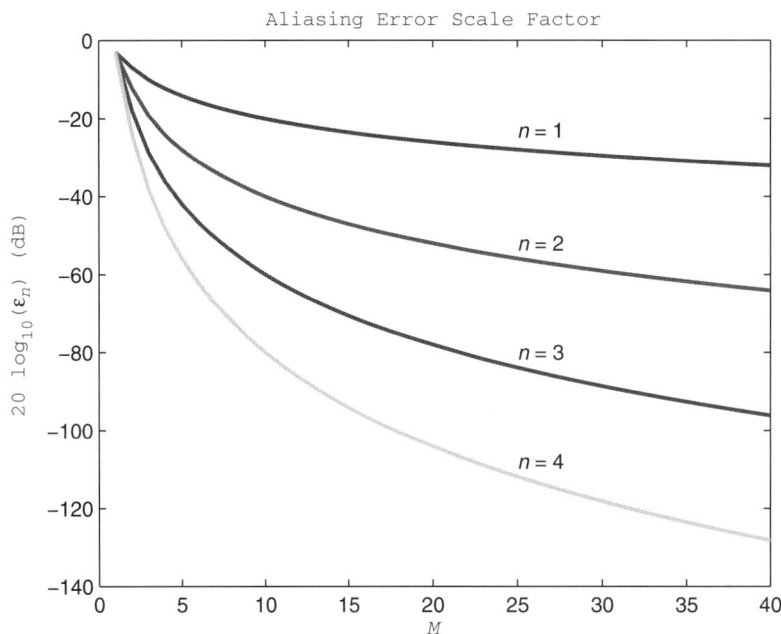

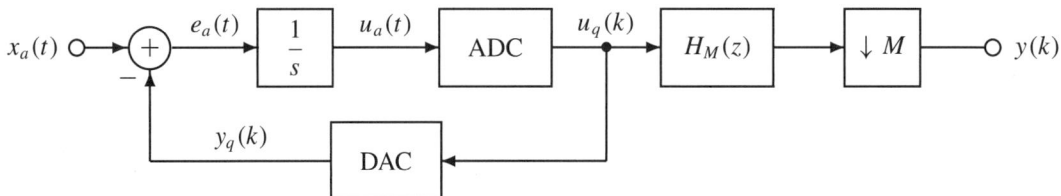

Aliasing Error Scale Factor

FIGURE 8.32: A Sigma-delta ADC Using Oversampling by a Factor of M

factor of M. The improvement in the signal-to-quantization noise ratio would be even more dramatic if the power density spectrum of the quantization noise were not flat but instead had a higher fraction of its power in the stopband $H_M(z)$.

By using a different quantization scheme called *sigma-delta modulation*, the spectrum of the quantization noise can be reshaped such that even more of the power lies outside the passband of $H_M(z)$ (Candy and Temes, 1992). A block diagram of a sigma-delta ADC is shown in Figure 8.32.

To analyze the input-output behavior of the sigma-delta ADC, it is helpful to first convert it to a discrete-equivalent form. First, consider the integrator block with input $e_a(t)$ and output $u_a(t)$. If a backward Euler approximation with a normalized sampling interval of $T = 1$ is used to represent integration, then the discrete-equivalent transfer function $H_I(z) = U(z)/E(z)$ of the integrator block is

$$H_I(z) = \frac{1}{1 - z^{-1}} \tag{8.7.14}$$

Next, the ADC can be modeled as an N-bit quantizer, $u_q(k) = Q_N[u(k)]$. To develop a linear model of the quantization process, the quantized ADC output $u_q(k)$ is represented as the unquantized signal $u(k)$ plus quantization error $v(k)$.

$$u_q(k) = u(k) + v(k) \tag{8.7.15}$$

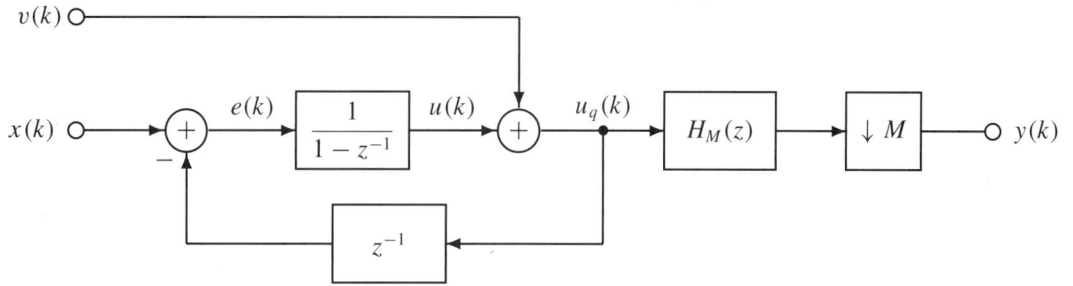

FIGURE 8.33: Linear Discrete-equivalent of a 1-bit Sigma-delta ADC

Here the quantization error $v(k)$ is white noise uniformly distributed over $[-q/2, q/2]$ with variance $\sigma_v^2 = q^2/12$ as in (8.7.6). Sigma-delta ADCs typically use a very high sampling rate f_s and a very low precision N. In fact, they often use the lowest number of bits possible, $N = 1$. In this case, the quantizer $u_q(k) = Q_1[u(k)]$ reduces to a simple comparator. Thus a 1-bit ADC has outputs of $\pm c$ depending on whether its input is positive or negative, respectively. The signal $u_q(k)$ can be thought of as a representation of $x_a(t)$ using pulse count modulation. The 1-bit DAC is modeled as z^{-1}, which accounts for a processing delay of one sample. Substituting these models for the subsystems in Figure 8.32, one arrives at the discrete-equivalent linear model of the sigma-delta ADC in Figure 8.33.

Example 8.10	**Sigma-delta Quantization**

As an illustration of the operation of the sigma-delta ADC in Figure 8.32, suppose the ADC input range is $c = 10$. Consider the following analog input.

$$x_a(t) = 8\cos(2\pi F_a t)$$

Suppose $F_a = 1/128$, and an oversampling factor of $M = 64$ is used. This results in a sampling rate of $f_s = 2MF_a = 1$ Hz. Let the total number of samples be $p = 8192$, which corresponds to M cycles of $x_a(t)$. When *exam8_10* is run, it produces the plots in Figure 8.34. Figure 8.34a shows one cycle of the sampled input. In Figure 8.34b, it is apparent that the positive pulses are denser for more positive values of $u_a(t)$ and the negative pulses are denser for more negative values of $x_a(t)$. The magnitude spectrum of $u_q(k)$ is shown in Figure 8.34c. Note the large spike at $f = F_a$ but many other discrete spectra at higher frequencies. Finally, the output $y(k)$ in Figure 8.34d is obtained by removing the spectral power of $X_q(f)$ outside of $F_M = f_s/(2M)$, and then reconstructing $y(k)$ using the inverse DFT. It is clear that $y(k)$ is a reasonable reproduction of $x_a(kT)$ even though a very crude 1-bit quantization is used.

The linear model of a sigma-delta ADC in Figure 8.33 is a two-input system with inputs $x(k)$ and $v(k)$. To find the Z-transform of the quantized output $u_q(k)$, note from Figure 8.33 that

$$\begin{aligned} U_q(z) &= U(z) + V(z) \\ &= H_I(z)E(z) + V(z) \\ &= H_I(z)[X(z) - z^{-1}U_q(z)] + V(z) \end{aligned} \qquad (8.7.16)$$

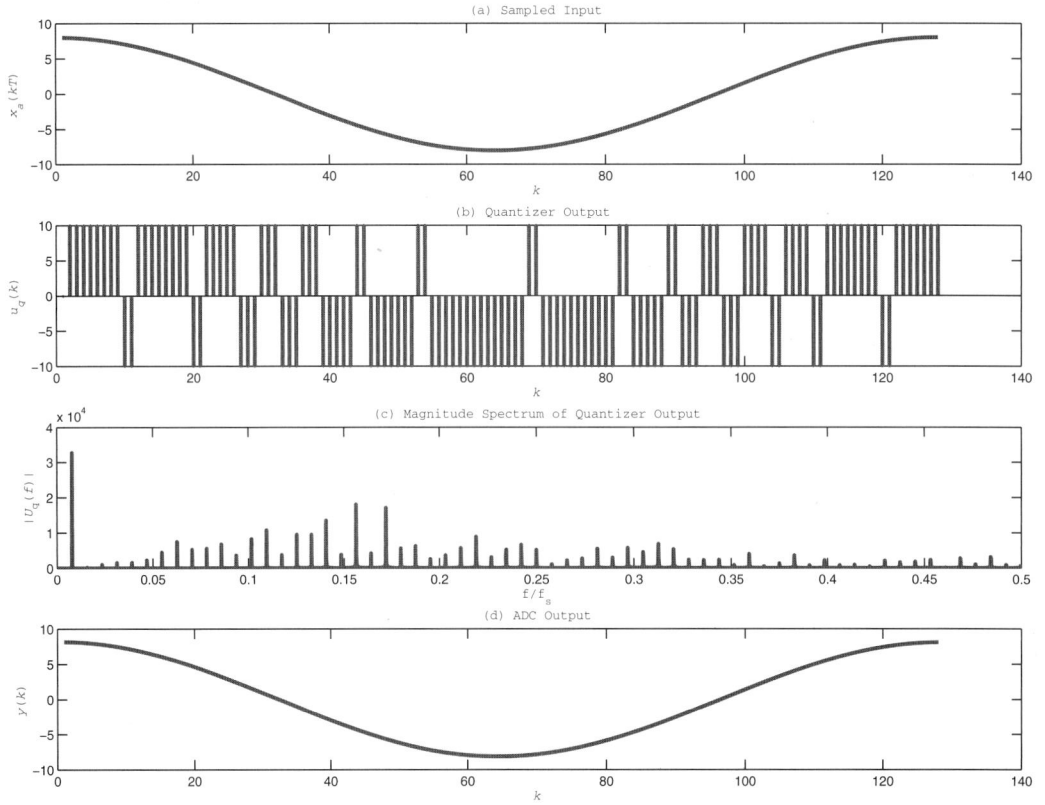

FIGURE 8.34: Output of the 1-bit Sigma-delta ADC

Solving (8.7.16) for $U_q(z)$ and then substituting for $H_I(z)$ using (8.7.14) yields

$$
\begin{aligned}
U_q(z) &= \left[\frac{1}{1 + z^{-1}H_I(z)} \right] [H_I(z)X(z) + V(z)] \\
&= (1 - z^{-1})[H_I(z)X(z) + V(z)] \\
&= X(z) + (1 - z^{-1})V(z)
\end{aligned}
\tag{8.7.17}
$$

The quantization noise appearing at the output will be a lowpass filtered version of $R(z) = (1 - z^{-1})V(z)$. It is the factor $1 - z^{-1}$ that changes or *reshapes* the spectral characteristics of the noise. In particular, using Euler's identity the power density spectrum is

$$
\begin{aligned}
P_r(f) &= |1 - \exp(-j2\pi fT)|^2 \sigma_v^2 \\
&= |\exp(-j\pi fT)(\exp(j\pi fT) - \exp(-j\pi fT)|^2 \sigma_v^2 \\
&= |\exp(-j\pi fT)|^2 |j2\sin(\pi fT)|^2 \sigma_v^2 \\
&= 4\sin^2(\pi fT)\sigma_v^2
\end{aligned}
\tag{8.7.18}
$$

The lowpass filter has a gain of one and a cutoff frequency of $f_s/(2M)$. Since the down-sampler does not change the average power, the average power of the quantization noise appearing at the output of the sigma-delta ADC is

$$P_v = \frac{1}{f_s} \int_{-.5f_s/M}^{.5f_s/M} P_r(f)\,df$$

$$= 4T\sigma_v^2 \int_{-.5f_s/M}^{.5f_s/M} \sin^2(\pi fT)\,df \qquad (8.7.19)$$

Because a sigma-delta ADC typically uses only a 1-bit quantizer, it compensates for this by oversampling by a factor $M \gg 1$. For $|f| \le f_s/(2M)$, $\sin(\pi fT) \approx \pi fT$. Using this approximation, the average power of the quantization noise at the output simplifies to

$$P_v \approx 4T\sigma_v^2 \int_{-.5f_s/M}^{.5f_s/M} (\pi Tf)^2 df$$

$$= \left(4\pi^2 T^3 \sigma_v^2\right)\left(\frac{f^3}{3}\right)\Bigg|_{-.5f_s/M}^{.5f_s/M}$$

$$= \left(4\pi^2 T^3 \sigma_v^2\right)\left(\frac{f_s^3}{12M^3}\right)$$

$$= \left(\frac{\pi^2}{3M^3}\right)\sigma_v^2 \qquad (8.7.20)$$

If a B-bit ADC is used in the sigma-delta ADC, then the quantization level from (8.7.4) is $q = c/2^{B-1}$. From (8.7.6) and (8.7.20), the average power of the sigma-delta ADC quantization noise in terms of the input amplitude c is

$$P_v \approx \frac{\pi^2 c^2}{36[2^{2(B-1)}]M^3} \qquad (8.7.21)$$

Due to the presence of the M^3 factor in the denominator, the quantization noise power decreases rapidly with M. As before, the reduction in the quantization noise power in (8.7.21) can be interpreted as an increase in the number of *effective bits* of precision. Let B be the number of bits of precision of a B-bit sigma-delta ADC using oversampling by a factor of M, and let N be the number of bits of precision of a regular ADC without oversampling. Using (8.7.4), (8.7.6), and (8.7.21), and equating the quantization noise power for the two cases yields

$$\frac{\pi^2 c^2}{36[2^{2(B-1)}]M^3} \approx \frac{c^2}{12[2^{2(N-1)}]} \qquad (8.7.22)$$

ADC effective precision

Canceling the common terms, taking reciprocals, taking the base-2 logarithm of each side, and simplifying the result, one arrives at the following expression for the required precision when a sigma-delta ADC is used.

$$B \approx N - \log_2(3/\pi^2)/2 - 3\log_2(M)/2 \qquad (8.7.23)$$

From (8.7.23) it is apparent that each increase in M by a factor of four leads to a decrease of three bits in the required precision for the ADC. A comparison in the reductions of the number of bits required to achieve a given quantization noise power P_v for oversampling with direct quantization and oversampling with sigma-delta quantization is shown in Table 8.1.

In column two of Table 8.1, note that if oversampling by a factor of $M = 64$ is used, an oversampling 6-bit ADC achieves the same accuracy as a 9-bit ADC without oversampling. From column three of Table 8.1, it is evident that a 1-bit sigma-delta ADC actually achieves

TABLE 8.1: ▶
ADC Bit Reduction
Achieved by
Oversampling

M	Oversampling ADC	Sigma-delta ADC
4	1	2.1
16	2	5.1
64	3	8.1
256	4	11.1

lower quantization noise and therefore better accuracy than a 9-bit ADC with no oversampling. It should be pointed out that oversampling by a large factor M means that the anti-aliasing filter $H_M(z)$ with cutoff frequency $F_M = f_s/(2M)$ is a narrowband filter that can be a challenge to implement. For example, for $M = 64$, the cutoff frequency is $f_M = .0078 f_s$. Increased savings in the number of bits can be achieved by higher-order sigma-delta ADCs, but the higher-order systems tend to be less stable (Oppenheim et al, 1999).

8.8 Oversampling DAC

8.8.1 Anti-imaging Filters

Just as oversampling can be used to ease the requirements on the analog anti-aliasing prefilter of an ADC, it can also be used to ease the requirements on the analog anti-imaging postfilter of a DAC. Recall that a DAC can be modeled as a zero-order hold with a transfer function of

$$H_0(s) = \frac{1 - \exp(-Ts)}{2} \qquad (8.8.1)$$

The DAC output is a piecewise-constant signal containing spectral images centered at multiples of the sampling frequency. Suppose the range of frequencies of interest for the analog output signal $y_a(t)$ is $0 \le |f| \le F_a$ where $f_s = 2F_a$. Normally, the piecewise-constant DAC output is passed through a sharp analog lowpass filter with a cutoff frequency of $f_s/2$ to remove the spectral images that are centered at multiples of f_s. Instead, consider an increase in the sampling rate to $f_s = 2LF_a$, where L is an integer greater than one, before performing the digital-to-analog conversion. This oversampling by a factor of L spreads the spectral images in the DAC output out so that now they are centered at multiples of Lf_s, thus making them easier to filter out. In particular, the analog anti-imaging filter now has to satisfy the following frequency response specification.

$$H_a(f) = \begin{cases} 1, & 0 \le |f| \le F_a \\ 0, & LF_a \le |f| < \infty \end{cases} \qquad (8.8.2)$$

Even though $H_a(f)$ is still an ideal filter with no passband ripple and complete stopband attenuation, the width of the transition band is no longer zero, but is instead $\Delta f = (L-1)F_a$. Given a large transition band, $H_a(s)$ can be approximated with a simple inexpensive low-order filter such as a first- or second-order analog Butterworth filter.

Oversampling DAC   The two-step process of sampling rate interpolation followed by a DAC is called *oversampling digital-to-analog conversion*. The overall structure of an oversampling DAC is shown in the block diagram in Figure 8.35. Note that it has *two* anti-imaging filters, a high-order digital filter $H_L(z)$ with cutoff frequency F_a, and a low-order analog filter $H_a(s)$, also with cutoff frequency F_a.

Just as with an ADC, the digital anti-imaging filter $H_L(z)$ effectively reduces the quantization noise power because it removes frequencies in the expanded stopband, $F_a \le |f| \le Lf_s/2$.

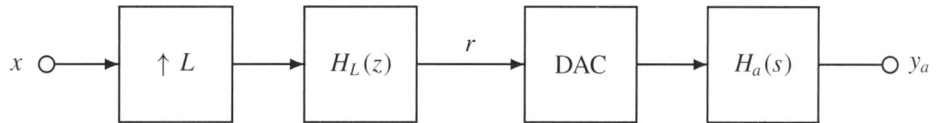

FIGURE 8.35: Oversampling DAC Using Sampling Rate Interpolation and Passband Equalization.

As a result, the average power of the quantization noise at the DAC output is as in (8.7.9), but with M replaced by L. That is,

$$\sigma_y^2 = \frac{\sigma_x^2}{L} \tag{8.8.3}$$

8.8.2 Passband Equalization

DAC magnitude response

The use of a sampling rate interpolator opens up additional opportunities in terms of improving system performance in the passband, $0 \leq |f| \leq F_a$. From (8.8.1) and Euler's identity, the *magnitude response* of the zero-order hold model of the DAC using a sampling frequency of Lf_s is

$$
\begin{aligned}
A_0(f) &= |H_0(s)|_{s=j2\pi f} \\
&= \left| \frac{1 - \exp(-j2\pi f T/L)}{j2\pi f} \right| \\
&= \left| \frac{\exp(-j\pi f T/L)[\exp(j\pi f T/L) - \exp(-j\pi f T/L)]}{j2\pi f} \right| \\
&= \left| \frac{\exp(-j\pi f T/L) \sin(\pi f T/L)}{\pi f} \right| \\
&= \left| \frac{\sin(\pi f T/L)}{\pi f} \right| \\
&= \frac{T|\mathrm{sinc}(\pi f T/L)|}{L}
\end{aligned}
\tag{8.8.4}
$$

A plot of the magnitude response of the DAC is shown in Figure 8.36. The side lobes are what cause the images of the baseband spectrum to appear at the output. These images must be removed with the analog anti-imaging filter $H_a(s)$.

Within the passband, the DAC shapes the magnitude spectrum of the signal using the scale factor $A_0(f)$. The effects of the DAC within the passband can be compensated for by using a more general version of the anti-imaging filter $H_L(z)$. In particular, one can *equalize* the effects of the DAC, within the frequency band $0 \leq |f| \leq F_a$, by using a digital anti-imaging filter with the following frequency response.

DAC magnitude equalizer

$$
H_L(f) = \begin{cases} \dfrac{L}{T \, |\mathrm{sinc}(\pi f T/L)|}, & 0 \leq |f| \leq F_a \\[2ex] 0, & F_a < |f| < f_s/2 \end{cases}
\tag{8.8.5}
$$

When $L \gg 1$, the spectral distortion in the passband due to the DAC is small because $\mathrm{sinc}(\pi f T/L) \approx 1$. However, if a low-order analog anti-imaging filter is used, then there can

FIGURE 8.36:
Magnitude
Response of a
Zero-order Hold
Model of a DAC

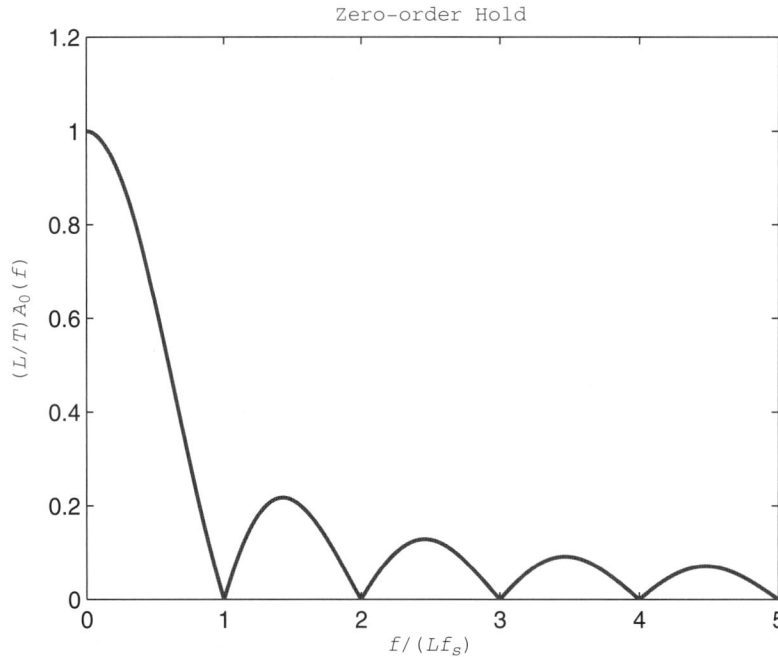

be a fairly significant ripple within the passband. This ripple can be compensated for as well with the digital filter $H_L(z)$. Suppose an nth-order Butterworth filter is used for the analog anti-imaging filter. Then from (8.7.2) the effects of both the DAC and the analog postfilter can be equalized, within the passband, by using a digital anti-imaging filter with the following frequency response. This way, the overall magnitude response is flat within the passband.

Magnitude equalizer

$$H_L(f) = \begin{cases} \dfrac{L\sqrt{1 + (f/F_a)^{2n}}}{T\,|\mathrm{sinc}(\pi f T/L)|}, & 0 \le |f| \le F_a \\ 0, & F_a < |f| < f_s/2 \end{cases} \qquad (8.8.6)$$

Example 8.11   **Oversampling DAC**

To illustrate the effectiveness of oversampling in the digital-to-analog conversion process, let the analog anti-imaging filter be an nth-order Butterworth filter with the magnitude response in (7.6.2). The objective of this example is to use oversampling to ensure that the magnitude spectra of the images are sufficiently small. Since the Butterworth magnitude response decreases monotonically, the spectral images are all scaled by a factor of at least $A_n(f_d)$, where $f_d = f_s/2$ is the folding frequency. Thus the scaling factor for the spectral images is

Aliasing error factor

$$\epsilon_n \overset{\Delta}{=} A_n(f_d)$$

If oversampling by a factor of L is used, then $f_s = 2LF_a$, which means $f_d = LF_a$. Using (8.7.2), to achieve a spectral image scaling factor of ϵ requires

$$\frac{1}{\sqrt{1 + L^{2n}}} \le \epsilon$$

FIGURE 8.37:
Magnitude
Response of
Equalized Digital
Anti-imaging
Filter $H_L(z)$,
Compensating for
the Effects of the
DAC and the
Analog
Anti-imaging Filter
when $n = 1$ and
$L = 20$

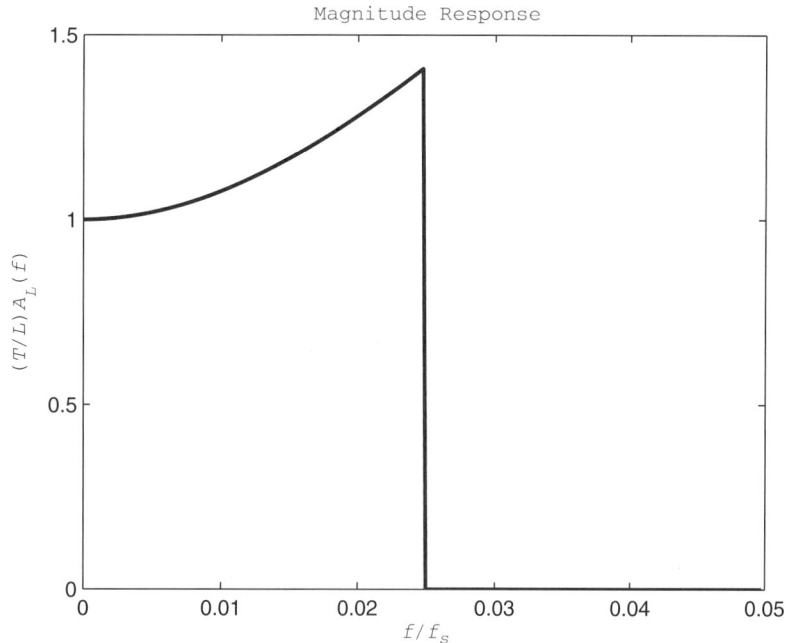

FIGURE 8.37:
Magnitude
Response of
Equalized Digital
Anti-imaging
Filter $H_L(z)$,
Compensating for
the Effects of the
DAC and the
Analog
Anti-imaging Filter
when $n = 1$ and
$L = 20$

Taking reciprocals, and squaring both sides then yields

$$1 + L^{2n} \geq \epsilon^{-2}$$

Solving for L,

$$L \geq (\epsilon^{-2} - 1)^{.5/n}$$

For example, if a first-order Butterworth filter is used, then $n = 1$. Suppose the desired spectral image scaling factor value is $\epsilon = .05$. In this case the sampling rate conversion factor required is

$$L = \text{ceil}\{(400 - 1)^{1/2}\}$$
$$= \text{ceil}(19.975)$$
$$= 20$$

The magnitude response of the equalized digital anti-imaging filter $H_L(z)$ is shown in Figure 8.37. It was generated by running *exam8_10*. For clarity of display, only the range $0 \leq |f| \leq f_s/L$ is shown. Since $L = 20$ in this case, almost all of the passband compensation is included to cancel the effects of the first-order Butterworth filter. The DAC (zero-order hold) magnitude response is essentially flat in the passband because $L \gg 1$.

● ● ● ● ● ● ● ● ● ● ● ● ● ● ● ●

8.9   GUI Software and Case Study

This section focuses on the design and realization of multirate systems. A graphical user interface module called *g_multirate* is introduced that allows the user to design rate converters and evaluate multirate systems, without any need for programming. A case study example is presented and solved using MATLAB.

g_multirate: Design and evaluate multirate systems

The FDSP toolbox includes a graphical user interface module called *g_multirate* that allows the user to perform rational rate conversion and evaluate multirate systems. GUI module *g_multirate* features a display screen with tiled windows, as shown in Figure 8.38. The upper left-hand *Block diagram* window contains a block diagram of the multirate system under investigation. It is a single-stage rational sampling rate converter with a rate conversion factor L/M described by the following time domain equation.

$$y(k) = \sum_{i=0}^{m} b_i \delta_L (Mk - i) x \left(\frac{Mk - i}{L} \right) \tag{8.9.1}$$

The *Parameters* window below the block diagram displays *edit boxes* containing simulation parameters. The contents of each edit box can be directly modified by the user with the Enter key used to activate the changes. Parameter fs is the sampling frequency, L is the interpolation factor, M is the decimation factor, and c is a damping factor used with the damped-cosine input. There are also two pushbutton controls that play the input signal $x(k)$ and the output signal $y(i)$, respectively, on the PC speaker.

The *Type* and *View* windows in the upper-right corner of the screen allow the user to select both the type of input and the viewing mode. The input types include uniformly distributed white noise, a damped cosine with a frequency of $f_s/10$ and a damping factor c^k, an amplitude-modulated (AM) sine wave, and a frequency-modulated (FM) sine wave. The Record sound option prompts the user to record one second of audio data using the PC microphone. Finally, the User-defined option prompts the user for the name of a MAT file containing a vector of samples x and the sampling frequency fs.

The *View* options include the time signals $x(k)$ and $y(i)$ and their magnitude spectra. Additional viewing options include the magnitude response, phase response, and impulse response of the combined anti-aliasing and anti-imaging filter.

$$H_0(z) = \sum_{i=0}^{m} b_i z^{-i} \tag{8.9.2}$$

The impulse response plot also plots the poles and zeros of the filter. There is also a dB check box control that toggles the magnitude spectra and magnitude response plots between linear and logarithmic scales. The filter order m is directly controllable with the horizontal slider bar below the *Type* and *View* windows. The *Plot* window along the bottom half of the screen shows the selected view.

The *Menu* bar at the top of the screen includes several menu options. The *Caliper* option allows the user to measure any point on the current plot by moving the mouse crosshairs to that point and clicking. The *Save data* option is used to save the current x, y, fs, a, and b in a user-specified MAT file for future use. Files created in this manner subsequently can be loaded with the User-defined input option. The *Filter* option allows the user to select the filter type for the combined anti-aliasing and anti-imaging filter $H_0(z)$. The choices include windowed (rectangular, Hanning, Hamming, or Blackman), frequency-sampled, least-squares, and equiripple filters. The *Print* option prints the contents of the *Plot* window. Finally, the *Help* option provides the user with some helpful suggestions on how to effectively use module *g_multirate*.

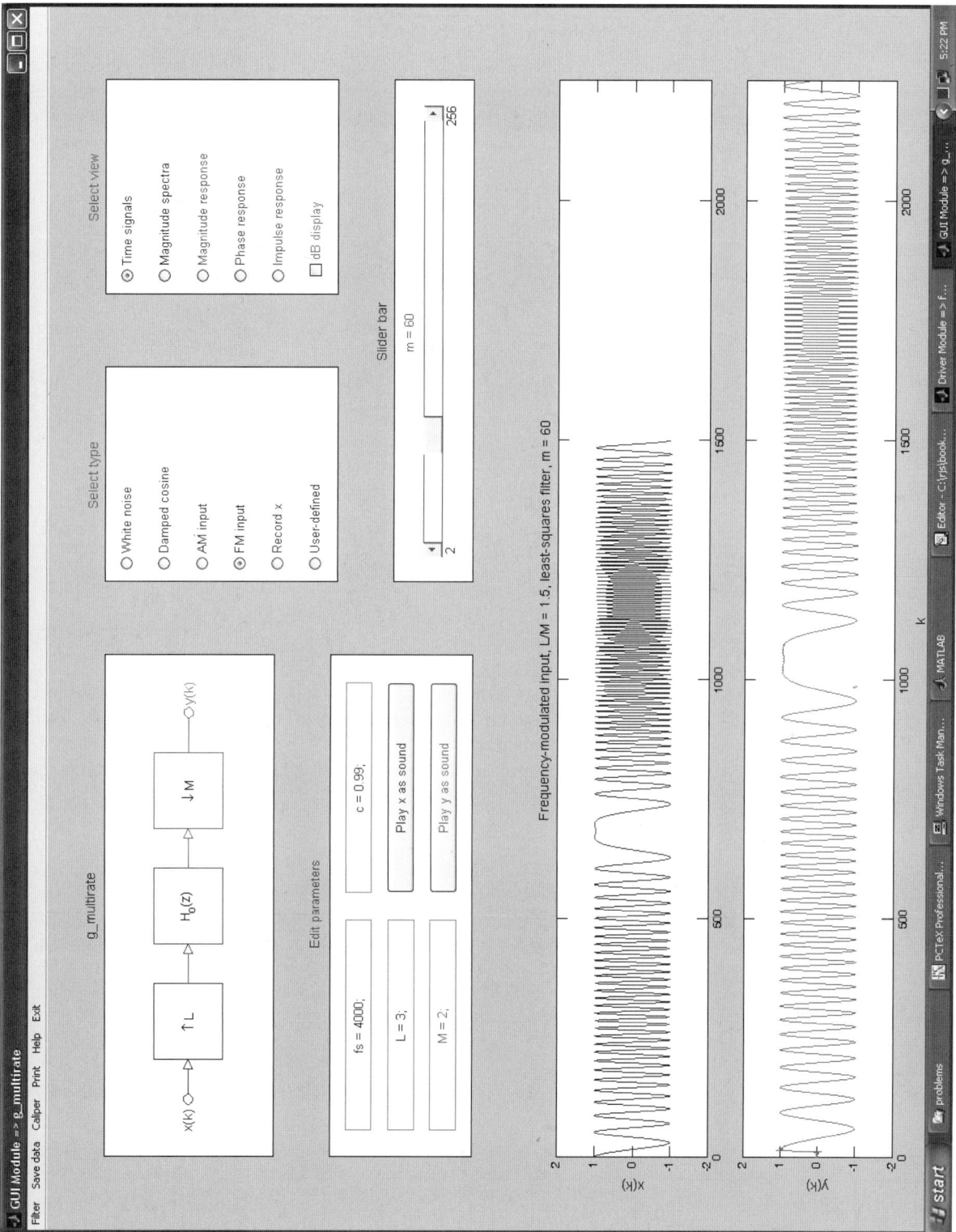

FIGURE 8.38: Display Screen of Chapter GUI Module *g_multirate*

Sampling Rate Converter (CD to DAT)

A multistage sampling rate converter is a good vehicle for illustrating the topics covered in this chapter. Consider the problem of converting music from compact disc (CD) format to digital audio tape (DAT) format. A CD is sampled at a rate of 44.1 kHz, while a DAT is sampled at a somewhat higher rate of 48 kHz. Thus the required frequency conversion ratio to go from CD format to DAT format is

$$\frac{L}{M} = \frac{48}{44.1}$$
$$= \frac{480}{441}$$
$$= \frac{160}{147} \tag{8.9.3}$$

First, suppose a single-stage sampling rate converter is to be used. From (8.3.2), the combined anti-aliasing and anti-imaging filter $H_0(f)$ must have a cutoff frequency of

$$F_0 = \min\left\{\frac{44100}{2(147)}, \frac{44100}{2(160)}\right\}$$
$$= 138.9 \text{ Hz} \tag{8.9.4}$$

The required passband gain is $H_0(0) = 160$. Hence, from (8.3.3), the desired frequency response for the ideal lowpass filter is

$$H_0(f) = \begin{cases} 160, & 0 \le |f| < 138.9 \\ 0, & 138.9 \le |f| < 22500 \end{cases} \tag{8.9.5}$$

Filter $H_0(z)$ is a *narrowband* filter with normalized cutoff frequency of $F_0/f_s = .003025$. Since a direct single-stage implementation would required a very high-order linear-phase FIR filter, instead consider a multistage implementation. Example 8.4 examined the case of converting from DAT to CD. Using the reciprocals of those conversion factors results in the following three-stage implementation.

$$\frac{160}{147} = \left(\frac{8}{7}\right)\left(\frac{5}{7}\right)\left(\frac{4}{3}\right) \tag{8.9.6}$$

Thus, a CD-to-DAT converter can be implemented using two rational interpolators and one rational decimator. A block diagram of the three-stage sampling rate converter is shown in Figure 8.39. Here $r_1(k)$ and $r_2(k)$ are intermediate output signals from stages one and two, respectively. The sampling rates of $r_1(k)$ and $r_2(k)$ are

$$f_1 = \left(\frac{8}{7}\right)44.1 = 50.4 \text{ kHz} \tag{8.9.7a}$$

$$f_2 = \left(\frac{5}{7}\right)50.4 = 36.0 \text{ kHz} \tag{8.9.7b}$$

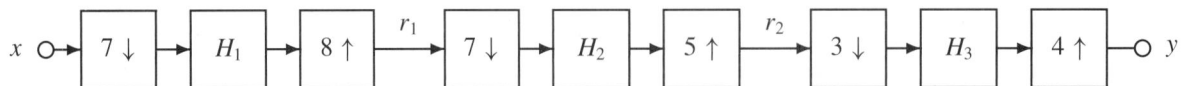

FIGURE 8.39: A Three-stage CD-to-DAT Sampling Rate Converter

From (8.3.2) and (8.9.6), the three cutoff frequencies of the combined anti-aliasing and anti-imaging filters are calculated as follows.

$$F_1 = \min\left\{\frac{44100}{2(8)}, \frac{44100}{2(7)}\right\} = 2756.3 \text{ Hz} \tag{8.9.8a}$$

$$F_2 = \min\left\{\frac{50400}{2(5)}, \frac{50400}{2(7)}\right\} = 3600 \text{ Hz} \tag{8.9.8b}$$

$$F_3 = \min\left\{\frac{436000}{2(4)}, \frac{36000}{2(3)}\right\} = 4500 \text{ Hz} \tag{8.9.8c}$$

It then follows from (8.3.3) that the ideal frequency responses of the three stage filters are

$$H_1(f) = \begin{cases} 8, & 0 \le |f| < 2756.3 \\ 0, & 2756.3 \le |f| < 22500 \end{cases} \tag{8.9.9a}$$

$$H_2(f) = \begin{cases} 5, & 0 \le |f| < 3600 \\ 0, & 3600 \le |f| < 25200 \end{cases} \tag{8.9.9b}$$

$$H_3(f) = \begin{cases} 4, & 0 \le |f| < 4500 \\ 0, & 4500 \le |f| < 18000 \end{cases} \tag{8.9.9c}$$

Suppose the stage filters are each implemented using a windowed linear-phase FIR filter of order $m = 60$ with the Hamming window. Plots of the magnitude responses of the three filters are shown in Figure 8.40.

FIGURE 8.40:
Magnitude Responses of the Three Stage Filters

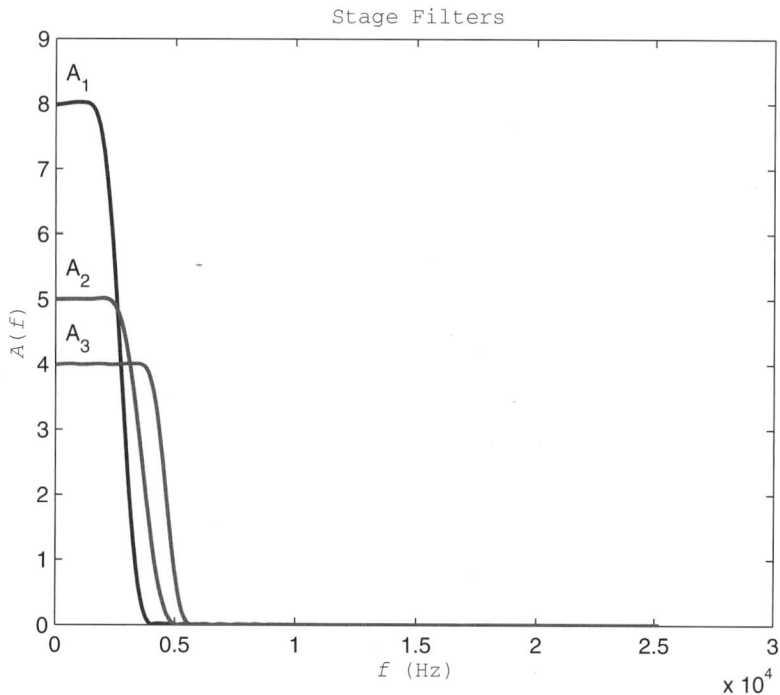

CASE
STUDY 8.1   Recall that the FDSP toolbox contains a function called *f_rateconv* for performing a single-stage rational sampling rate conversion by L/M. The three-stage conversion is implemented with multiple calls to *f_rateconv* by running the *case8_1* from *f_dsp*.

```
function case8_1

% Case Study 8.1: CD-to-DAT sampling rate converter

f_header('Case Study 8.1: CD-to-DAT sampling rate converter')
f_CD = 44100;
f_DAT = 48000;
L = [8 5 4];
M = [7 7 3];
m = 60;
win = 2;
sym = 0;
fs1 = L(1)*f_CD/M(1);
fs2 = L(2)*fs1/M(2);
fs = [f_CD,fs1,fs2];

% Compute stage filter magnitude responses

stg = f_prompt ('\nCompute stage filters separately (0=no,1=yes)',0,1,0);
if stg
   r = 250;
   for i = 1 : 3
      F(i) = (fs(i)/2)*min(1/L(i),1/M(i));
      p = [0 F(i) F(i) 0];
      b = L(i)*f_firwin ('f_firamp',m,fs(i),win,sym,p);
      [H,f(i,:)] = f_freqz (b,1,r,fs(i));
      A(i,:) = abs(H);
   end
   figure
   plot (f',A','LineWidth',1.5)
   f_labels ('Stage filters','{f} (Hz)','{A(f)}')
   x = 400;
   text (x,L(1)+.4,'{A_1}')
   text (x,L(2)+.4,'{A_2}')
   text (x,L(3)+.4,'{A_3}')
   f_wait
end

% Sample an input signal at CD rate

d = 2;
x = zeros(d*f_CD,1);
[x,cancel] = f_getsound (x,d,f_CD);
if cancel
   return
```

Continued on p. 629

Continued from p. 628

```
end
f_wait ('Press any key to play back sound at 44.1 kHz ...')
soundsc (x,f_CD)
f_wait ('Press any key to play back sound at 48 kHz ...')
soundsc (x,f_DAT)
figure
plot(x);
f_labels ('Use the mouse to select the start of a short speech segment ...','{k}','{x(k)}')
[k1,x1] = f_caliper(1);

% Convert segment of it to DAT rate

f_wait ('Press any key to rate convert the selected segment...')
p = floor(k1) : floor(k1)+400;
r1 = f_rateconv (x(p),fs(1),L(1),M(1),m,win);
r2 = f_rateconv (r1,fs(2),L(2),M(2),m,win);
y = f_rateconv (r2,fs(3),L(3),M(3),m,win);

% Plot segments

figure
subplot(4,1,1)
k = p - p(1);
plot(k,x(p),'LineWidth',1.5)
f_labels ('','','{x(k)}')
subplot(4,1,2)
plot(0:length(r1)-1,r1,'LineWidth',1.5)
f_labels ('','','{r_1(k)}')
subplot(4,1,3)
plot(0:length(r2)-1,r2,'LineWidth',1.5)
f_labels ('','','{r_2(k)}')
subplot(4,1,4)
plot(0:length(y)-1,y,'LineWidth',1.5)
f_labels ('','{k}','{y(k)}')
f_wait
```

When *case8_1* is run, it first computes the magnitude responses of the three stage filters as shown in Figure 8.40. It then prompts the user to speak into the microphone and records the response at the CD rate. The recorded speech is then played back at both the CD rate and the higher DAT rate for comparison by the ear. Next, the recorded speech is displayed in a plot, and crosshairs appear. The user should use the mouse to select the start of a speech segment. The selected segment is then converted from the CD rate to the DAT rate in three stages with the results as shown in Figure 8.41. Note that although all the segments appear to be the same shape, a close inspection of the scales along the horizontal axes shows that each is at a different sampling rate.

FIGURE 8.41:
Rate-converted
Segments of
Recorded Speech

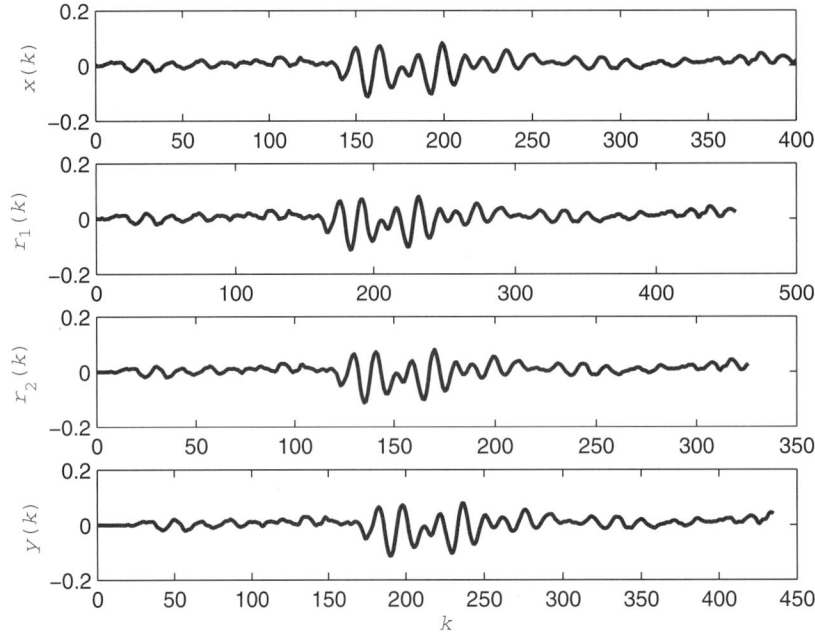

8.10 Chapter Summary

Rate Converters

This chapter focused on multirate signal processing techniques. Every multirate system contains at least one rate converter. A sampling rate converter is a linear time-varying system that changes the sampling rate of a discrete-time signal without converting the signal back to analog form. Instead, the resampling is done in the digital domain. The simplest sampling rate converter decreases the sampling frequency by an integer factor M. This is called a *decimator*, and it can be implemented with the following time-domain equation.

Decimator

$$y(k) - \sum_{i=0}^{m} b_i x(Mk - i) \tag{8.10.1}$$

The FIR filter with impulse response $h_M(k) = b_k$ for $0 \le k \le m$ is a lowpass filter with a passband gain of one and a cutoff frequency of $F_M = f_s/(2M)$. This filter is included to preserve the spectral characteristics of $x(k)$. Decreasing the sampling rate by a factor of M is called *down-sampling*, and it is represented in block diagrams with the symbol $\downarrow M$. The effect of down-sampling is to spread the spectrum out by a factor of M. Hence, to avoid aliasing, $H_M(z) = Z\{h_M(k)\}$ is inserted as a digital anti-aliasing filter.

Down-sampling

It is also possible to increase the sampling rate by an integer factor L. This is called an *interpolator*, and it can be implemented with the following time-domain equation.

Interpolator

$$y(k) = \sum_{i=0}^{m} b_i \delta_L(k - i) x\left(\frac{k - i}{L}\right) \tag{8.10.2}$$

Here $\delta_L(k)$ is a periodic train of impulses of period L starting at $k = 0$. The FIR filter with impulse response $h_L(k) = b_k$ for $0 \le k \le m$ is a lowpass filter with a passband gain of L

and a cutoff frequency of $F_L = f_s/(2L)$. Again, this filter is included to preserve the spectral characteristics of $x(k)$. Increasing the sampling rate by a factor of L is called *up-sampling*, and it is represented in block diagrams with the symbol $\uparrow L$. The effect of up-sampling is to compress the spectrum by a factor of L. Because the spectrum is periodic, this generates images of the original spectrum that must be removed by $H_L(z) = Z\{h_L(k)\}$, which is a digital anti-imaging filter.

Up-sampling

More general rate conversion by a rational factor L/M can be achieved by using a cascade configuration of an interpolator with rate conversion factor L followed by a decimator with rate conversion factor M. This is called a *rational rate* converter, and it can be implemented using the following time-domain equation.

Rational rate converter

$$y(k) = \sum_{i=0}^{m} b_i \delta_L(Mk - i)x\left(\frac{Mk - i}{L}\right) \tag{8.10.3}$$

Because the interpolator is followed by the decimator, the anti-imaging post-filter of the interpolator can be combined with the anti-aliasing pre-filter of the decimator. The cutoff frequency of the cascade combination of the two filters is $F_0 = \min\{F_L, F_M\}$ and the passband gain is L. Thus, the desired frequency response of the ideal combined anti-aliasing and anti-imaging filter is

Rational converter filter

$$H_0(f) = \begin{cases} L, & 0 \le |f| \le F_0 \\ 0, & F_0 < |f| < f_s/2 \end{cases} \tag{8.10.4}$$

Typically, anti-aliasing and anti-imaging filters are implemented with FIR filters because linear-phase characteristics can be easily included that delay the signal in the passband, but do not otherwise distort it. IIR filters, by contrast, can achieve sharper transition bands than FIR filters of the same order, but they do so at the expense of introducing nonlinear phase distortion.

To simplify the filter design, rational sampling rate converters with large values for L or M are implemented as a cascade configuration of lower-order rate converters. This is called a *multistage* rate converter, and it is based on the following factorization of the rate conversion factor.

Multistage rate converter

$$\frac{L}{M} = \left(\frac{L_1}{M_1}\right)\left(\frac{L_2}{M_2}\right) \cdots \left(\frac{L_p}{M_p}\right) \tag{8.10.5}$$

Polyphase filter

Both interpolators and decimators can be implemented efficiently using *polyphase filters*. If the rate conversion factor is N, then the ith *phase* of the input signal includes every N sample starting with sample i. A parallel combination of N polyphase filters is used, one for each phase. The output samples are obtained by summing the outputs of the parallel branches. Polyphase rate converter realizations reduce the number of floating-point multiplications or FLOPs per output sample by a factor of N.

Narrowband Filters and Filter Banks

There are many applications of multirate systems in modern DSP systems. For example, if it is necessary to delay a discrete-time signal by a fraction of a sample, the signal can be up-sampled by L, delayed by $0 < M < L$ using a shift register, and then down-sampled by L to restore the sampling rate. Using linear-phase FIR filters of order m achieves an intersample delay of

Intersample delay

$$\tau = \left(m + \frac{L}{M}\right)T \tag{8.10.6}$$

Another example is the design of a narrowband lowpass filter, a filter with a cutoff frequency satisfying $F_0 \ll f_s$. Here the signal is down-sampled to spread out the spectrum, filtered with an easier to implement wideband filter with a cutoff frequency $F_c \approx f_s/4$, and then up-sampled to restore the original sampling rate. Narrowband filters centered at submultiples of f_s can be configured in parallel to form filter banks. An *analysis bank* decomposes a signal $x(k)$ into subsignals, $x_i(k)$, occupying subbands of $[0, f_s]$. The subsignals are then down-sampled and processed separately. A *synthesis bank* takes the processed subband signals $y_i(k)$, up-samples them, and combines them into a single broadband signal $y(k)$. This technique, called *subband processing*, can be used to efficiently transmit several low-bandwidth signals over a single high-bandwidth channel using frequency-division multiplexing. An important special case of a filter bank system is the two-channel quadrature mirror filter bank or QMF bank. A two-channel QMF bank can achieve perfect reconstruction of the input (scaled and delayed) by using analysis filters that form a power complementary pair.

Oversampling ADCs and DACs

Oversampling ADC

Another class of applications of multirate systems arises in the implementation of high-performance analog-to-digital and digital-to-analog converters. An *oversampling* ADC uses oversampling by a factor of M followed by a sampling rate decimator. This technique reduces the requirements on the analog anti-aliasing filter by inserting a transition band of width $B = (M - 1)F_a$, where F_a is the bandwidth of the analog signal. This means that a high-order analog anti-aliasing filter with a sharp cutoff can be replaced with a less expensive low-order analog anti-aliasing filter. The oversampling ADC has the added benefit that the average power of the quantization noise appearing at the output is reduced by a factor of M. The quantization noise level can be reduced still further by modifying the spectrum of the quantization error using the sigma-delta quantization method. Sigma-delta ADCs based on 1-bit quantization have quantization error, and therefore accuracy, that is superior to regular ADCs as long as the oversampling factor M is made sufficiently large.

Oversampling DAC

Oversampling also can be used to implement a high-performance DAC. An *oversampling* DAC consists of a sampling rate interpolator with a conversion factor of L followed by a DAC. The effect of the up-sampling is to reduce the requirements on the analog anti-imaging filter by inserting a transition band of width $B = (L - 1)F_a$, where F_a is the bandwidth of the signal. Again this means that a high-order analog anti-imaging filter with a sharp cutoff can be replaced with a less expensive, low-order analog anti-imaging filter. Like the oversampling ADC, the oversampling DAC has the added benefit that the average power of the quantization noise appearing at the output is reduced by a factor of L.

For an oversampling DAC, the effects of both the zero-order hold and the analog anti-imaging filter $H_a(s)$ can be *equalized*, within the passband, by using a more general digital anti-imaging filter for the interpolator. For example, suppose the analog anti-imaging filter is an nth-order Butterworth filter. Then the overall magnitude response of the oversampling DAC

Passband Equalization

will be flat within the passband if the following digital anti-imaging filter is used.

$$H_L(f) = \begin{cases} \dfrac{L\sqrt{1 + (f/F_a)^{2n}}}{T\,|\mathrm{sinc}(\pi f T/L)|}, & 0 \le |f| \le F_a \\ 0, & F_a < |f| < f_s/2 \end{cases} \tag{8.10.7}$$

GUI Module

The FDSP toolbox includes a GUI module called *g_multirate* that allows the user to design and evaluate rational sampling rate converters. Rate conversion can be applied to a variety of inputs including recorded sound and user-defined inputs stored in MAT files. The choices for the anti-aliasing anti-imaging filters include windowed, frequency-sampled, least-squares, and equiripple linear-phase FIR filters.

Num.	Learning Outcome	Sec.
1	Know how to design and implement integer decimators and interpolators	8.2
2	Know how to design and implement rational sampling rate converters, both single stage and multistage	8.3
3	Be able to efficiently implement sampling rate converters using polyphase filter realizations	8.4
4	Be able to apply multirate techniques to design narrowband filters	8.5
5	Be able to apply multirate techniques to design filter banks	8.5
6	Understand how to design an oversampling ADC and know what benefits are achieved	8.6
7	Understand how to design an oversampling DAC and know what benefits are achieved	8.7
8	Know how to use the GUI module *g_multirate* to design and evaluate multirate DSP systems	8.8

TABLE 8.2: ▶
Learning Outcomes
for Chapter 8

Learning Outcomes

This chapter was designed to provide the student with an opportunity to achieve the learning outcomes summarized in Table 8.2.

● ● ● ● ● ● ● ● ● ● ● ● ● ● ● ● ● ●

8.11 Problems

The problems are divided into Analysis and Design problems that can be solved by hand or with a calculator, GUI Simulation problems that are solved using GUI module *g_multirate*, and MATLAB Computation problems that require a user program.

8.11.1 Analysis and Design

8.1 Suppose a multirate signal processing application requires a sampling rate conversion factor of $L/M = 3.15$.
   (a) Find the required frequency response of the ideal anti-aliasing and anti-imaging digital filter assuming a single-stage converter is used.
   (b) Factor L/M into a product of two rational numbers whose numerators and denominators are less than 10.
   (c) Sketch a block diagram of a multistage sampling rate converter based on your factoring of L/M from part (b).
   (d) Find the required frequency responses of the ideal combined anti-aliasing and anti-imaging digital filters for each of the stages in part (c).

8.2 Consider the problem of designing a rational sampling rate converter with a frequency conversion factor of $L/M = 2/3$.
   (a) Sketch a block diagram of the sampling rate converter.
   (b) Find the required frequency response of the ideal anti-aliasing and anti-imaging digital filter assuming f_s is the sampling rate of $x(k)$.
   (c) Use Tables 6.1 and 6.2 to design an anti-aliasing and anti-imaging filter of order $m = 50$ using the windowing method with the Blackman window.
   (d) Find the difference equation for the sampling rate converter.

8.3 Consider the problem of designing a sampling rate decimator with a down-sampling factor of $M = 8$.
  (a) Sketch a block diagram of the sampling rate decimator.
  (b) Find the required frequency response of the ideal anti-aliasing digital filter assuming f_s is the sampling rate of $x(k)$.
  (c) Using Tables 6.1 and 6.2, design an anti-aliasing filter of order $m = 40$ using the windowing method with a Hanning window.
  (d) Find the difference equation for the sampling rate decimator.

8.4 Suppose a signal is sampled at a rate of $f_s = 10$ Hz. Consider the problem of using the variable delay system in Figure 8.42 to implement an overall delay of $\tau = 2.38$ sec.
  (a) Find the smallest interpolation factor L that is needed.
  (b) Suppose the linear-phase FIR filters are of order $m = 50$. How much delay is introduced by the two lowpass filters?
  (c) What length of shift register M is needed to achieve the overall delay?

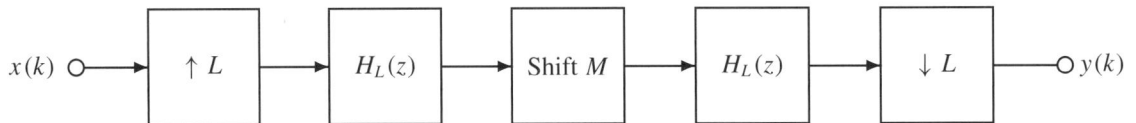

FIGURE 8.42: A Variable Delay System

8.5 Consider the problem of designing a rational sampling rate converter with a frequency conversion factor of $L/M = 5/4$.
  (a) Sketch a block diagram of the sampling rate converter.
  (b) Find the required frequency response of the ideal anti-aliasing and anti-imaging digital filter assuming f_s is the sampling rate of $x(k)$.
  (c) Use Tables 6.1 and 6.2 to design an anti-aliasing and anti-imaging filter of order $m = 50$ using the windowing method with the Blackman window.
  (d) Find the difference equation for the sampling rate converter.

8.6 Suppose a multirate signal processing application requires a sampling rate conversion factor of $L/M = .525$.
  (a) Find the required frequency response of the ideal anti-aliasing and anti-imaging digital filter assuming a single-stage converter is used.
  (b) Factor L/M into a product of two rational numbers whose numerators and denominators are less than 10.
  (c) Sketch a block diagram of a multistage sampling rate converter based on your factoring of L/M from part (b).
  (d) Find the required frequency responses of the ideal combined anti-aliasing and anti-imaging digital filters for each of the stages in part (c).

8.7 Consider the sampling rate interpolator shown previously in Figure 8.6. The input $x(k)$ is sampled at rate f_s and has a triangular magnitude spectrum $A_x(f)$, as shown in Figure 8.43. Suppose the up-sampling factor is $L = 3$.
  (a) Sketch the spectrum of the zero-interpolated signal $x_L(k)$ defined in (8.2.5) for $0 \le |f| \le f_s/2$.
  (b) Sketch the magnitude response of the ideal anti-imaging filter $H_L(z)$.
  (c) Sketch the magnitude spectrum of $y(k)$ for $0 \le |f| \le f_s/2$.

8.8 Consider the problem of designing a sampling rate interpolator with an up-sampling factor of

FIGURE 8.43:
Magnitude
Spectrum of $x(k)$
in Problem 8.7

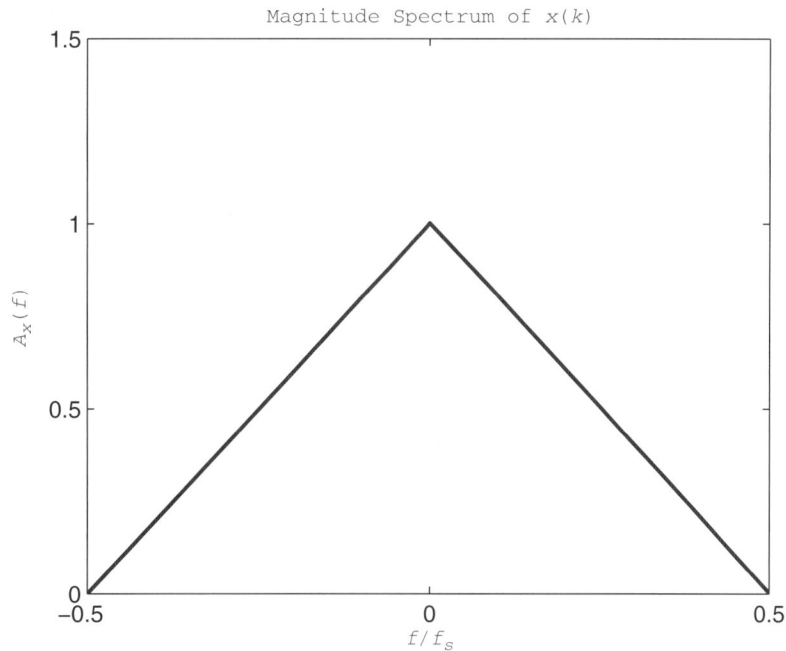

Magnitude Spectrum of $x(k)$

$L = 10$.

(a) Sketch a block diagram of the sampling rate interpolator.

(b) Find the required frequency response of the ideal anti-imaging digital filter assuming f_s is the sampling rate of $x(k)$.

(c) Using Tables 6.1 and 6.2, design an anti-imaging filter of order $m = 30$ using the windowing method with a Hamming window.

(d) Find the difference equation for the sampling rate interpolator.

8.9 Consider the variable delay system shown in Figure 8.42 where the sampling rate of $x(k)$ is f_s. Suppose $H_L(z)$ is a linear-phase FIR filter of order m. Find an expression for the total delay that takes into account the interpolator delay, the shift register delay, and the decimator delay.

8.10 Consider the problem of designing a multirate narrowband lowpass FIR filter as shown in Figure 8.44. Suppose the sampling frequency is $f_s = 8000$ Hz and the cutoff frequency is $F_0 = 200$ Hz.

(a) Find the largest integer frequency conversion factor M that can be used.

(b) Using (6.3.6), design an anti-aliasing filter $H_M(z)$ of order $m = 32$ using the frequency-sampled method. Do not use any transition band samples.

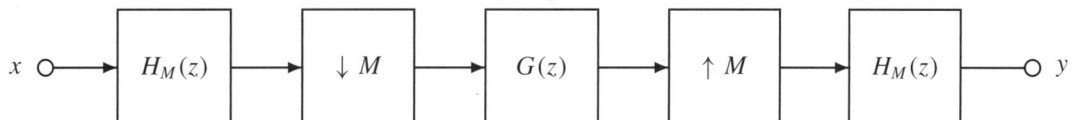

$$x \; \circ \longrightarrow \boxed{H_M(z)} \longrightarrow \boxed{\downarrow M} \longrightarrow \boxed{G(z)} \longrightarrow \boxed{\uparrow M} \longrightarrow \boxed{H_M(z)} \longrightarrow \circ \; y$$

FIGURE 8.44: A Multirate Narrowband FIR Filter

8.11 Consider a polyphase filter realization of a rational rate converter with rate conversion factor $L/M = 4/3$.

(a) Suppose the following FIR filter is used for the anti-aliasing filter of the decimator part.

Find the filters $E_i(z)$ for a polyphase realization of $H_M(z)$.

$$H_M(z) = \sum_{i=0}^{40} b_i z^{-i}$$

(b) Suppose the following FIR filter is used for the anti-imaging filter of the interpolator part. Find the filters $F_i(z)$ for a polyphase realization of $H_L(z)$.

$$H_L(z) = \sum_{i=0}^{40} c_i z^{-i}$$

(c) Sketch a block diagram of a polyphase realization of the rational rate converter using a cascade configuration of an interpolator followed by a decimator.

8.12 Consider the following FIR filter.

$$H(z) = 1 + 3z^{-1} + 5z^{-2} + \cdots + 23z^{-11}$$

(a) Find polyphase filters $E_i(z)$ such that

$$H(z) = \sum_{i=0}^{1} z^{-i} E_i(z^2)$$

(b) Find polyphase filters $E_i(z)$ such that

$$H(z) = \sum_{i=0}^{5} z^{-i} E_i(z^6)$$

(c) Which of the two polyphase realizations of $H(z)$ is faster in terms of the number of floating-point multiplications per output sample? How many times faster is it than a direct implementation of $H(z)$?

8.13 Consider the problem of designing an interpolator with $f_s = 12$ Hz and an up-sampling factor $L = 4$.
(a) What is the sampling rate of the output signal?
(b) Sketch the desired magnitude response of the ideal anti-imaging filter $H_L(z)$.
(c) Suppose the anti-imaging filter is a windowed filter of order $m = 20$ using the Hanning window. Use Tables 6.1 and 6.2 to find the impulse response, $h_L(k)$.
(d) Suppose a polyphase realization is used. Find the transfer functions $F_i(z)$ of the polyphase filters.
(e) Sketch a block diagram of a polyphase filter realization of the interpolator.

8.14 Consider an integer interpolator with an up-sampling factor of L and a linear-phase FIR anti-imaging filter of order m.
(a) Find n_L, the number of floating-point multiplications (FLOPs) needed to compute each sample of the output using a direct realization.
(b) Suppose a polyphase filter realization is used to implement the interpolator. Find N_L, the number of FLOPs needed to compute each sample of the output.
(c) Express N_L as a percentage of n_L.

8.15 Consider the following FIR filter.

$$H(z) = 2 + 4z^{-1} + 6z^{-2} + \cdots + 24^{-11}$$

(a) Find polyphase filters $E_i(z)$ such that

$$H(z) = \sum_{i=0}^{3} z^{-i} E_i(z^4)$$

(b) Find polyphase filters $E_i(z)$ such that

$$H(z) = \sum_{i=0}^{2} z^{-i} E_i(z^3)$$

(c) Which of the two polyphase realizations of $H(z)$ is faster in terms of the number of floating point-multiplications per output sample? How many times faster is it than a direct implementation of $H(z)$?

8.16 Consider a polyphase filter realization of a rational rate converter with rate conversion factor $L/M = 2/3$.

(a) Suppose the following FIR filter is used for the anti-aliasing filter of the decimator part. Find the filters $E_i(z)$ for a polyphase realization of $H_M(z)$.

$$H_M(z) = \sum_{i=0}^{30} b_i z^{-i}$$

(b) Suppose the following FIR filter is used for the anti-imaging filter of the interpolator part. Find the filters $F_i(z)$ for a polyphase realization of $H_L(z)$.

$$H_L(z) = \sum_{i=0}^{30} c_i z^{-i}$$

(c) Sketch a block diagram of a polyphase realization of the rational rate converter using a cascade configuration of an interpolator followed by a decimator.

8.17 Consider the problem of designing a decimator with $f_s = 60$ Hz and a down-sampling factor of $M = 3$.

(a) What is the sampling rate of the output signal?
(b) Sketch the desired magnitude response of the ideal anti-aliasing filter $H_M(z)$.
(c) Suppose the anti-aliasing filter is a windowed filter of order $m = 32$ using the Hamming window. Use Tables 6.1 and 6.2 to find the impulse response, $h_M(k)$.
(d) Suppose a polyphase realization is used. Find the transfer functions $E_i(z)$ of the polyphase filters.
(e) Sketch a block diagram of a polyphase filter realization of the decimator.

8.18 Consider the problem of designing a complex highpass filter with the following ideal magnitude response.

$$A(f) = \begin{cases} 0, & 0 \le f < F_0 \\ 1, & F_0 \le f < f_s \end{cases}$$

(a) Let $B = f_s - F_0$ be the width of the passband, and consider the problem of designing a lowpass filter $G(z)$ with cutoff frequency $F_c = B/2$. Using Tables 6.1 and 6.2, find the impulse response $g(k)$ for a filter of order $m = 50$ using the windowing method with the Blackman window.
(b) Using the frequency shift property in (8.5.4) and $g(k)$, find the impulse response $h(k)$ of the complex highpass filter with a cutoff frequency of F_0.
(c) Is the magnitude response of $H(z)$ an even function of f? Why or why not?
(d) Is the magnitude response of $H(z)$ a periodic function of f? If so, what is the period?

FIGURE 8.45: Two-band Magnitude Response of Problem 8.19

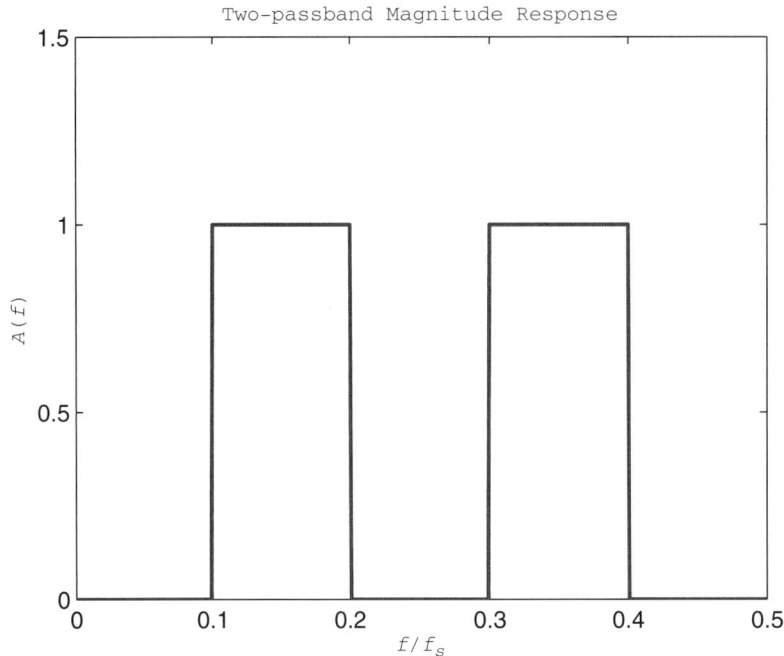

8.19 Consider the problem of designing a complex two-band filter with the magnitude response shown in Figure 8.45.

(a) Let $B = .1f_s$ be the width of each passband, and consider the problem of designing a lowpass filter $G(z)$ with cutoff frequency $F_c = B/2$. Using Tables 6.1 and 6.2, find the filter impulse response $g(k)$ for a filter of order $m = 80$ using the windowing method with the Hanning window.

(b) Using the frequency shift property in (8.5.4) and $g(k)$, find the impulse response $h_1(k)$ of the complex passband filter with cutoff frequencies $.1f_s$ and $.2f_s$.

(c) Using the frequency shift property in (8.5.4) and $g(k)$, find the impulse response $h_2(k)$ of the complex passband filter with cutoff frequencies $.3f_s$ and $.4f_s$.

(d) Using $h_1(k)$ and $h_2(k)$, find the impulse response $h(k)$ of a filter whose magnitude response approximates $A(f)$ in Figure 8.45.

(e) Sketch a block diagram of $H(z)$ using blocks $H_1(z)$ and $H_2(z)$.

8.20 Let $x_M(k) = x(Mk)$ be the output of a factor of M down-sampler with input $x(k)$. Recall from (8.6.9) that the Z-transform of $x_M(k)$ can be written in terms of the Z-transform of $x(k)$ as follows, where $W_M = \exp(-j2\pi/M)$

$$X_M(z) = \frac{1}{M} \sum_{i=0}^{M-1} X(W_M^{-i} z^{1/M})$$

Using the definitions of W_M and the DTFT, show that the spectrum of $x_M(k)$ is

$$X_M(f) = \frac{1}{M} \sum_{i=0}^{M-1} X\left(\frac{f + if_s}{M}\right)$$

8.21 Consider the design of a two-channel QMF bank. Suppose the first analysis bank filter has transfer function $F_0(z) = 2 - z^{-3}$.

(a) Find the remaining analysis and synthesis filter bank transfer functions that ensure an alias-free QMF bank.

(b) Find the overall QMF bank transfer function $H(z)$.

(c) Find the output $y(k)$ in terms of the input $x(k)$.

8.22 Consider the two-band filter whose magnitude response was shown previously in Figure 8.45. Find the power gain Γ of this filter.

8.23 Consider an ideal lowpass filter with a passband gain of $A \geq 1$ and a cutoff frequency of $F_c < f_s/2$. For what value of F_c is the power gain equal to one?

8.24 Consider the 10-bit oversampling ADC shown in Figure 8.46 with analog inputs in the range $|x_a(t)| \leq 5$.

(a) Find the average power of the quantization noise of the quantized input, $x_q(k)$.

(b) Suppose a second-order Butterworth filter is used for the analog anti-aliasing prefilter. The objective is to reduce the aliasing error by a factor of $\epsilon = .005$. Find the minimum required oversampling factor M.

(c) Find the average power of the quantization noise at the output $y(k)$ of the oversampling ADC.

(d) Suppose $f_s = 1000$ Hz. Sketch the ideal magnitude response of the digital anti-aliasing filter $H_M(f)$.

(e) Using Tables 6.1 and 6.2, design a linear-phase FIR filter of order $m = 80$ whose frequency response approximates $H_M(f)$ using the windowing method with a Hanning window.

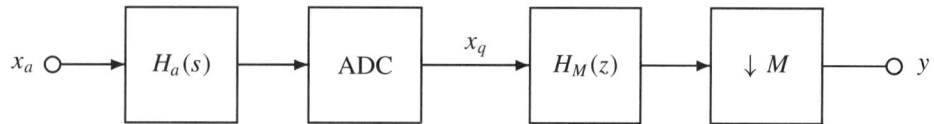

FIGURE 8.46: An Oversampling ADC with an Oversampling Factor M

8.25 A 12-bit oversampling ADC oversamples by a factor of $M = 64$. To achieve the same average power of the quantization noise at the output, but without using oversampling, how many bits are required?

8.26 Suppose an analog signal in the range $|x_a(t)| \leq 5$ is sampled with a 10-bit oversampling ADC with an oversampling factor of $M = 16$. The output of the ADC is passed through an FIR filter $H(z)$ as shown in Figure 8.47 where

$$H(z) = 1 - 2z^{-1} + 3z^{-2} - 2z^{-3} + z^{-4}$$

(a) Find the quantization level q.

(b) Find the power gain of the filter $H(z)$.

(c) Find the average power of the quantization noise at the system output, $y(k)$.

(d) To get the same quantization noise power, but without using oversampling, how many bits are required?

FIGURE 8.47: A Discrete-time Multirate System

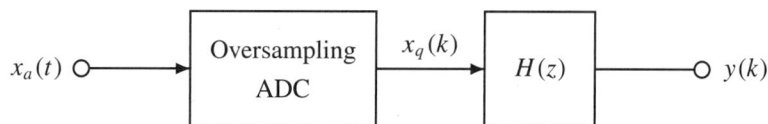

8.27 Consider the 10-bit oversampling DAC shown in Figure 8.48 with analog outputs in the range $|y_a(t)| \leq 10$.

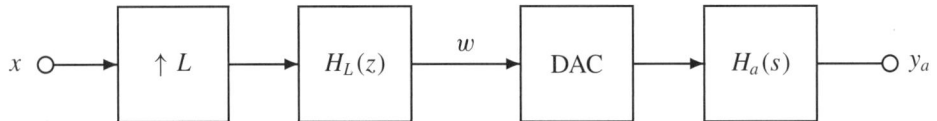

FIGURE 8.48: An Oversampling DAC with an Oversampling Factor L

(a) Suppose a first-order Butterworth filter is used for the analog anti-imaging postfilter. The objective is to reduce the imaging error by a factor of $\epsilon = .05$. Find the minimum required oversampling factor L.

(b) Find the average power of the quantization noise at the output of the DAC.

(c) Suppose $f_s = 2000$ Hz. Find the ideal frequency response of the digital anti-imaging filter $H_L(f)$. Include passband equalizer compensation for both the analog anti-imaging filter and the zero-order hold.

8.28 Consider an integer decimator with down-sampling factor M and a linear-phase FIR anti-aliasing filter of order m.

(a) Find n_M, the number of floating-point multiplications (FLOPs) needed to compute each sample of the output using a direct realization.

(b) Suppose a polyphase filter realization is used to implement the decimator. Find N_M, the number of FLOPs needed to compute each sample of the output.

(c) Express N_M as a percentage of n_M.

8.29 Consider the problem of designing a complex passband filter with the following ideal magnitude response.

$$A(f) = \begin{cases} 0, & 0 \leq f < F_0 \\ 1, & F_0 \leq f \leq F_1 \\ 0, & F_1 < f < f_s \end{cases}$$

(a) Let $B = F_1 - F_0$ be the width of the passband, and consider the problem of designing a lowpass filter $G(z)$ with cutoff frequency $F_c = B/2$. Using Tables 6.1 and 6.2, find the impulse response $g(k)$ for a filter of order $m = 60$ using the windowing method with the Hamming window.

(b) Using the frequency shift property in (8.5.4) and $g(k)$, find the impulse response $h(k)$ of the complex passband filter with cutoff frequencies F_0 and F_1.

(c) Is the magnitude response of $H(z)$ an even function of f? Why or why not?

(d) Is the magnitude response of $H(z)$ a periodic function of f? If so, what is the period?

8.11.2 GUI Simulation

8.30 Using the GUI module *g_multirate*, select the amplitude-modulated (AM) input. Reduce the sampling rate of the input using an integer decimator with a down-sampling factor of $M = 2$. Use a windowed filter with the Hanning window, and plot the following.

(a) The time signals.

(b) Their magnitude spectra.

(c) The filter magnitude response.

(d) The filter impulse response.

8.31 Using the GUI module *g_multirate*, adjust the filter order to $m = 80$. Print the magnitude responses of the following anti-aliasing and anti-imaging filters using the dB scale.

(a) Windowed filter with the Hanning window.

(b) Windowed filter with the Hamming window.
(c) Equiripple filter.

8.32 Using the GUI module *g_multirate*, select the frequency-modulated (FM) input. Increase the sampling frequency of the input using an interpolator with an up-sampling factor of $L = 3$. Use a windowed filter with the Hamming window, and plot the following.
(a) The time signals.
(b) Their magnitude spectra.
(c) The filter magnitude response.
(d) The filter phase response.

8.33 Using the GUI module *g_multirate*, select the damped cosine input. Set the damping factor to $c = .995$, the up-sampling factor to $L = 2$, and the down-sampling factor to $M = 3$. Plot the following.
(a) The time signals.
(b) The magnitude spectra.

8.34 Using the GUI module *g_multirate*, print the magnitude responses of the following anti-aliasing and anti-imaging filters using the linear scale.
(a) Windowed filter with the Blackman window.
(b) Frequency-sampled filter.
(c) Least-squares filter.

8.35 Using the GUI module *g_multirate*, record the word *hello* in x. Play it back to make sure it is a good recording. Save the recording in a MAT-file named *prob8_35.mat* using the Save option. Then reload it using the User-defined option. Play it back with and without rate conversion to hear the difference. Plot the following.
(a) The time signals.
(b) Their magnitude spectra.
(c) The filter magnitude response.
(d) The filter impulse response.

8.36 Use the GUI module *g_multirate* and the User-defined input option to load the MAT-file *u_multirate1*. Convert the sampling rate using $L = 4$ and $M = 3$ and a frequency-sampled filter. Plot the following.
(a) The time signals. What word is recorded?
(b) Their magnitude spectra.
(c) The filter impulse response.

8.11.3 MATLAB Computation

8.37 Write a MATLAB function called *u_narrowband* that uses the FDSP toolbox functions *f_firideal* and *f_rateconv* to compute the zero-state response of the multirate narrowband lowpass filter previously shown in Figure 8.44. The calling sequence for *u_narrowband* is as follows.

```
% U_NARROWBAND: Computer output of multirate narrowband lowpass filter
%
% Usage:
%        [y,M] = u_narrowband (x,F0,win,fs,m);
% Pre:
%        x   = array of length N containing input samples
%        F0  = lowpass cutoff freqeuncy (F0 <= fs/4)
```

```
%          win = window type
%
%                   0 = rectangular
%                   1 = Hanning
%                   2 = Hamming
%                   3 = Blackman
%
%          fs = sampling frequency
%          m  = filter order (even)
% Post:
%          y = array of length N containing output samples
%          M = frequency conversion factor used
```

Use the maximum frequency conversion factor possible. Test function *u_narrowband* by writing a script that uses it to design a lowpass filter with a cutoff frequency of $F0 = 10$ Hz, a sampling frequency of $f_s = 400$ Hz, and a filter order of $m = 50$. Plot the following.
(a) The narrowband filter impulse response.
(b) The narrowband filter magnitude response and the ideal magnitude response on the same graph with a legend.

8.38 Consider the following periodic analog signal with three harmonics.

$$x_a(t) = \sin(2\pi t) - 3\cos(4\pi t) + 2\sin(6\pi t)$$

Suppose this signal is sampled at $f_s = 24$ Hz using $N = 50$ samples to produce a discrete-time signal $x(k) = x_a(kT)$ for $0 \le k < N$. Write a MATLAB program that uses *f_interpol* to interpolate this signal by converting it to a sampling rate of $F_s = 72$ Hz. For the anti-imaging filter, use a least-squares filter of order $m = 50$. Use the *subplot* command and the *stem* function to plot the following discrete-time signals on the same screen.
(a) The original signal $x(k)$.
(b) The resampled signal $y(k)$ below it using a different color.

8.39 Consider the following periodic analog signal with three harmonics.

$$x_a(t) = \cos(2\pi t) - .8\sin(4\pi t) + .6\cos(6\pi t)$$

Suppose this signal is sampled at $f_s = 64$ Hz using $N = 120$ samples to produce a discrete-time signal $x(k) = x_a(kT)$ for $0 \le k < N$. Write a MATLAB program that uses *f_decimate* to decimate this signal by converting it to a sampling rate of $F_s = 32$ Hz. For the anti-aliasing filter, use a windowed filter of order $m = 40$ with the Hamming window. Use the *subplot* command and the *stem* function to plot the following discrete-time signals on one screen.
(a) The original signal $x(k)$.
(b) The resampled signal $y(k)$ below it using a different color.

8.40 Consider the following periodic analog signal with three harmonics.

$$x_a(t) = 2\cos(2\pi t) + 3\sin(4\pi t) - 3\sin(6\pi t)$$

Suppose this signal is sampled at $f_s = 30$ Hz using $N = 50$ samples to produce a discrete-time signal $x(k) = x_a(kT)$ for $0 \le k < N$. Write a MATLAB program that uses *f_rateconv* to convert it to a sampling rate of $F_s = 50$ Hz. For the anti-aliasing and anti-imaging filter, use a frequency-sampled filter of order $m = 60$. Use the *subplot* command and the *stem* function to plot the following discrete-time signals on the same screen.
(a) The original signal $x(k)$.
(b) The resampled signal $y(k)$ below it using a different color.

8.41 Write a function called *u_synbank* that synthesizes a composite signal $x(i)$ from N low-bandwidth subsignals $x_i(k)$ using a uniform DFT synthesis filter bank. The calling sequence for *u_synbank* is as follows.

```
% U_SYNBANK: Synthesize a complex composite signal from subsignals using a DFT
%
% Usage:
%          x = u_synbank (X,m,alpha,win,fs);
% Pre:
%          X      = p by N matrix containing subsignal i in column i
%          m      = order of anti-imaging filter
%          alpha  = relative cutoff frequency: F_0 = alpha*fs/(2N)
%          win    = an integer specifying the desired window type
%
%                  0 = rectangular
%                  1 = Hanning
%                  2 = Hamming
%                  3 = Blackman
%
%          fs = sampling frequency
% Post:
%          x = complex vector of length q = Np containing samples of composite
%              signal. x contains N frequency-multiplexed subsignals. The
%              bandwidth of x is N*fs/2 and the ith subsignal is in band i
```

Test function *u_synbank* by writing a program that uses the FDSP toolbox function *f_subsignals* to construct a 32 by 4 matrix X with the samples of the kth subsignal in column k. The function *f_subsignals* produces signals whose spectra are shown in Figure 8.23. Use $alpha = .5$, $fs = 200$ Hz, and a windowed filter of order $m = 90$ with a Hamming window. Save x and fs in a MAT-file named *prob8_41* and plot the following.

(a) The real and imaginary parts of the complex composite signal $x(i)$. Use *subplot* to construct a 2×1 array of plots on one screen.

(b) The magnitude spectrum $A(f) = |X(f)|$ for $0 \le f \le fs$.

8.42 Write a function called *u_analbank* that analyzes a composite signal $x(i)$ and decomposes it into N low-bandwidth subsignals $x_i(k)$ using a uniform DFT analysis filter bank. The calling sequence for *u_analbank* is as follows.

```
% U_ANALBANK: Analyze a complex composite signal into subsignals using a DFT
%
% Usage:
%          X = u_analbank (x,N,m,alpha,win,fs);
% Pre:
%          x      = complex vector of length q = Np containing samples of compos
%                   signal. x contains N frequency-multiplexed subsignals. The
%                   bandwidth of x is N*fs/2 and the ith subsignal is in band i
%          N      = number of subsignals in x
%          m      = order of anti-imaging filter
%          alpha  = relative cutoff frequency: F_0 = alpha*fs/(2N)
%          win    = an integer specifying the desired window type
%
```

```
%                         0 = rectangular
%                         1 = Hanning
%                         2 = Hamming
%                         3 = Blackman
%
%          fs      = sampling frequency
% Post:
%          X       = p by N matrix containing subsignal i in column i
```

Test function *u_analbank* by writing a script that analyzes the composite signal $x(i)$ obtained from the solution to Problem 8.41. That is, load MAT-file *prob8_41*. Use *alpha* $= .5$, and a windowed filter of order $m = 90$ with a Hamming window. Plot the following.

(a) The magnitude spectrum $A(f) = |X(f)|$ for $0 \le f \le fs$.

(b) The magnitude spectra of the subsignals extracted from X. Use *subplot* to construct a 2×2 array of plots on one screen.

CHAPTER 9

Adaptive Signal Processing

Chapter Topics

9.1 Motivation

9.2 Mean Square Error

9.3 The Least Mean Square (LMS) Method

9.4 Performance Analysis of LMS Method

9.5 Modified LMS Methods

9.6 Adaptive FIR Filter Design

9.7 The Recursive Least-Squares (RLS) Method

9.8 Active Noise Control

9.9 Nonlinear System Identification

9.10 GUI Software and Case Study

9.11 Chapter Summary

9.12 Problems

9.1 Motivation

The digital filters investigated in previous chapters have one fundamental characteristic in common. The coefficients of these filters are fixed; they do not evolve with time. When the filter parameters are allowed to vary with time, this leads to a powerful new family of digital filters called *adaptive* filters. This chapter focuses on an adaptive FIR type of filter called a *transversal* filter that has the following generic form.

Adaptive, transversal filter

$$y(k) = \sum_{i=0}^{m} w_i(k)x(k-i)$$

Notice that the constant FIR coefficient vector b has been replaced by a time-varying vector $w(k)$ of length $m + 1$ called the *weight* vector. The adaptive filter design problem consists of

645

developing an algorithm for updating the weight vector $w(k)$ to ensure that the filter satisfies some design criterion. For example, the objective might be to get the filter output $y(k)$ to track a desired output $d(k)$ as time increases. The transversal filter structure has an important qualitative advantage over an adaptive IIR filter structure. Once the weight vector has converged, the resulting filter is guaranteed to be stable.

We begin this chapter by introducing some examples of applications of adaptive filters. Four broad classes of applications are examined including system identification, channel equalization, signal prediction, and noise cancellation. Next the mean square error design criterion is formulated and a closed-form filter solution called the Weiner filter is developed. A simple and elegant method for updating the filter weights called the least mean square or LMS method is then developed. The performance characteristics of the LMS method are investigated, including bounds on the step size that ensure convergence, estimates of the convergence rate, and estimates of the steady-state error. A number of modifications to the basic LMS method are then presented, including the normalized LMS method, the correlation LMS method, and the leaky LMS method. An adaptive design technique for FIR filters using pseudo-filter input-output specifications is then introduced. Next, a recursive adaptive filter technique called the recursive least-squares or RLS method is presented. A more general LMS technique called the filtered-x LMS method is then developed along with a signal-synthesis method. Both are applied to the problem of active control of acoustic noise. Attention then turns to the problem of identifying nonlinear discrete-time systems using radial basis function or RBF networks. Finally, a GUI module called *g_adapt* is introduced that allows the user to perform system identification without any need for programming. The chapter concludes with an application example, and a summary of adaptive signal processing techniques.

There are a variety of applications of adaptive filters that arise in fields ranging from equalization of telephone channels to geophysical exploration for oil and gas deposits. In this section some general classes of applications of adaptive signal processing are introduced. A brief history of adaptive signal processing and its applications can be found in Haykin (2002).

9.1.1 System Identification

The success enjoyed by engineers in applying analysis and design techniques to practical problems often can be traced to the effective use of mathematical models of physical phenomena. In many instances, a mathematical model can be developed by applying underlying physical principles to each component. However, there are other instances where this bottom-up approach is less effective because the physical system or phenomenon is too complex and is not well understood. In these cases it is often useful to think of the unknown system as a *black box* where measurements can be taken of the input and output, but little is known about the details of what is inside the box (hence the term black). It is assumed that the unknown system can be modeled as a linear discrete-time system. The problem of obtaining a model of the system from input and output measurements is called the *system identification* problem. Adaptive filters are highly effective for performing system identification using the configuration shown in Figure 9.1.

It is standard practice to represent an adaptive filter in a block diagram using a diagonal arrow through the block. The arrow can be thought of as a needle on a dial that is adjusted as the parameters of the adaptive filter are changed. The system identification configuration in Figure 9.1 shows the adaptive filter in parallel with the unknown black box system. Both systems are driven by the same test input $x(k)$. The objective is to adjust the parameters or coefficients of the adaptive filter so that its output mimics the response of the unknown system. Thus the *desired output* $d(k)$ is the output of the unknown system, and the difference between

Black box

System identification

FIGURE 9.1:
System
Identification

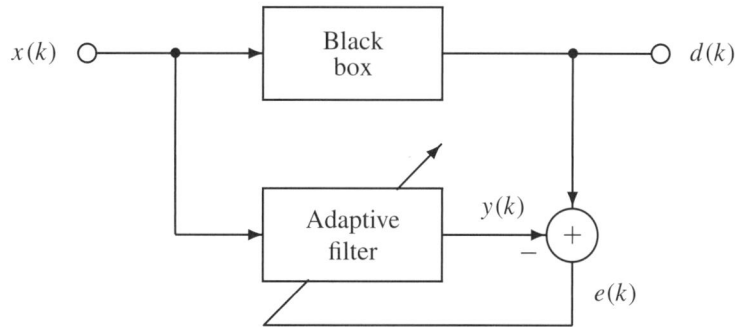

FIGURE 9.1:
System
Identification

Error signal the desired output and the adaptive filter output $y(k)$ is the *error* signal $e(k)$.

$$e(k) \overset{\Delta}{=} d(k) - y(k) \tag{9.1.1}$$

The algorithm for updating the parameters of the adaptive filter uses the error $e(k)$ and the input $x(k)$ to adjust the weights so as to reduce the square of the error. Later the adaptive filter block will be explored in more detail. For now, notice that if the error signal can be made to go to zero, then the adaptive filter output is an exact reproduction of the unknown system output. In this case the adaptive filter becomes an exact model of the unknown black box system. This model can be used in simulation studies, and it also can be used to predict the response of the unknown system to new inputs.

9.1.2 Channel Equalization

Another important class of applications of adaptive filters can be found in the communication industry. Consider the problem of transmitting information over a communication channel. At the receiving end, the signal will be distorted due to the effects of the channel itself. For example, the channel invariably will exhibit some type of frequency response characteristic with some spectral components of the input attenuated more than others. In addition, there will be phase distortion and delay, and the signal may be corrupted with additive noise. To remove, or at least minimize, the detrimental effects of the communication channel, one must pass the received signal through a filter that approximates the inverse of the channel so that the cascade or series connection of the two systems restores the original signal. The technique *Equalization* of inserting an inverse system in series with an original unknown system is called *equalization* because it results in an overall system with a transfer function of one. Equalization, or inverse modeling, can be achieved with an adaptive filter using the configuration shown in Figure 9.2.

Here the black box system, that represents the unknown communication channel, is in series with the adaptive filter. This series combination is in parallel with a delay element corresponding to a delay of M samples. Thus the desired output in this case is simply a delayed version of the transmitted signal.

$$d(k) = x(k - M) \tag{9.1.2}$$

The reason for inserting a delay is that the black box system typically imparts some delay to the signal $x(k)$ as it is processed by the system. Therefore an exact inverse system would have to include a corresponding time advance, something that is not feasible for a causal filter. Furthermore, if the unknown black box system does represent a communication channel, then

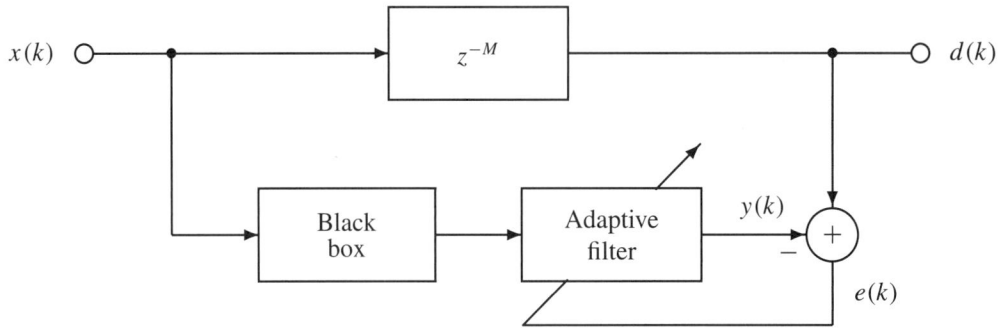

FIGURE 9.2: Channel Equalization

delaying the signal by M samples will not distort the information that arrives at the receiver. Recall that a constant group delay can be achieved by using a linear-phase FIR filter.

9.1.3 Signal Prediction

As an illustration of another class of applications, consider the problem of encoding speech for transmission or storage. The direct technique is to encode the speech samples themselves. An effective alternative is to use the past samples of the speech to predict the values of future samples. Typically, the error in the prediction has a smaller variance that the original speech signal itself. Consequently, the prediction error can be encoded using a smaller number of bits than a direct encoding of the speech itself. In this way an efficient encoding system can be implemented. An adaptive filter can be used to predict future samples of speech, or other signals, by using the configuration shown in Figure 9.3.

In this case the desired output is the input itself. Since the adaptive filter processes a delayed version of the input, the only way the error can be made to go to zero is if the adaptive filter successfully predicts the value of the input M samples into the future. Of course an exact prediction of a completely random input is not possible with a causal system. Typically, the input consists of an underlying deterministic component plus additive noise. In these cases, information from the past samples can be used to minimize the square of the prediction error.

9.1.4 Noise Cancellation

Still another broad class of applications of adaptive filters focuses on the problem of interference or noise cancellation. As an illustration, suppose the driver of a car places a call using a cell phone. The cell phone microphone will pick up both the driver's voice plus ambient road noise

FIGURE 9.3:
Signal
Prediction

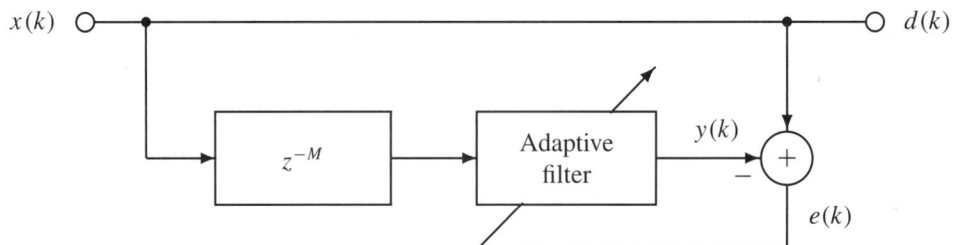

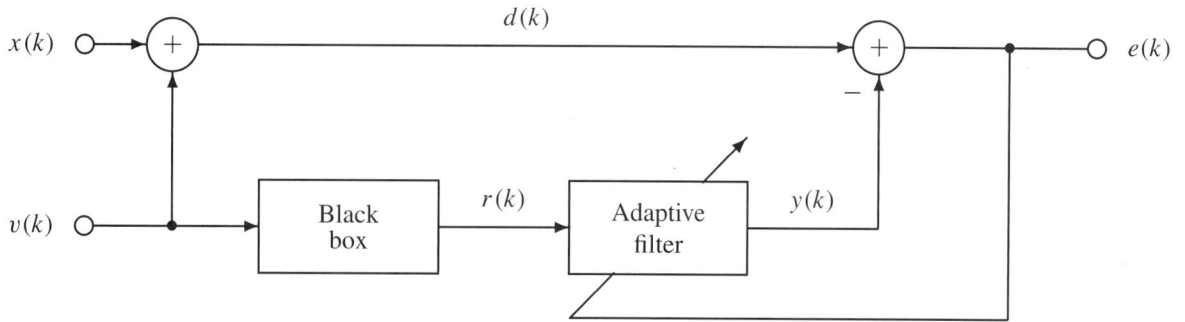

FIGURE 9.4: Noise Cancellation

that varies with the car speed and the driving conditions. To make the speaker's voice more intelligible at the receiving end, a second reference microphone can be placed in the car to measure the ambient road noise. An adaptive filter then can be used to process this reference signal and subtract the result from the signal detected by the primary microphone using the configuration shown in Figure 9.4.

Note that the desired output $d(k) = x(k) + v(k)$ consists of speech plus road noise. The reference signal $r(k)$ is a filtered version of the noise. The presence of an unknown black box system takes into account the fact that the primary microphone and the reference microphone are placed at different locations and therefore the reference signal $r(k)$ is different from, but correlated to, the noise $v(k)$ appearing at primary microphone. The error in this case is

$$e(k) = x(k) + v(k) - y(k) \tag{9.1.3}$$

If the speech $x(k)$ and the additive road noise $v(k)$ are uncorrelated with one another, then the minimum possible value for $e^2(k)$ occurs when $y(k) = v(k)$, which corresponds to the road noise being removed completely from the transmitted speech signal $e(k)$.

● ● ● ● ● ● ● ● ● ● ● ● ● ● ● ● ● ●

9.2 Mean Square Error

9.2.1 Adaptive Transversal Filters

An mth-order adaptive transversal filter is a linear time-varying discrete-time system than can be represented by the following difference equation.

$$y(k) = \sum_{i=0}^{m} w_i(k)x(k-i) \tag{9.2.1}$$

Note that the filter output is a time-varying linear combination of the past inputs. A signal flow graph of an adaptive transversal filter is shown in Figure 9.5 for the case $m = 4$. Given the structure shown in Figure 9.5, a transversal filter is sometimes referred to as a tapped delay line.

A compact input-output formulation of a transversal filter can be obtained by introducing the following pair of $(m + 1) \times 1$ column vectors.

$$u(k) \overset{\Delta}{=} [x(k), x(k-1), \ldots, x(k-m)]^T \tag{9.2.2a}$$

$$w(k) \overset{\Delta}{=} [w_0(k), w_1(k), \ldots, w_m(k)]^T \tag{9.2.2b}$$

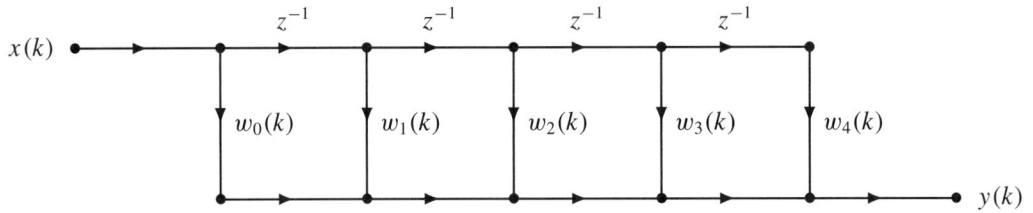

FIGURE 9.5: Signal Flow Graph of an Adaptive Transversal Filter

State, weight vectors

Here $u(k)$ is a vector of past inputs called the *state* vector, and $w(k)$ is the current value of the *weight* vector. If we combine (9.2.1) and (9.2.2), the adaptive filter output can be expressed as a dot product of the two vectors.

$$y(k) = w^T(k)u(k), \quad k \geq 0 \qquad (9.2.3)$$

In view of (9.2.3), an adaptive filter can be thought of has having two inputs, the time-varying weight vector $w(k)$, and the vector of past inputs $u(k)$. The vector $w(k)$ is itself the output of a weight-update algorithm, as shown in Figure 9.6. Recall that $d(k)$ in Figure 9.6 represents the *desired output*, and the difference between $d(k)$ and the filter output $y(k)$ is the *error $e(k)$*. That is,

$$e(k) = d(k) - y(k) \qquad (9.2.4)$$

The details of how the desired output $d(k)$ and the filter input $x(k)$ are generated depend on the type of adaptive filter application. Examples of different classes of applications were presented in Section 9.1.

9.2.2 Cross-correlation Revisited

Typically, the filter input $x(k)$ and the desired output $d(k)$ are modeled as random signals or, more formally, as random processes. This makes $y(k)$ and $e(k)$ random as well. For the purpose of this analysis, let us assume that $x(k)$ and $d(k)$ are *stationary* random signals, meaning that their statistical properties do not change with time. The notion of cross-correlation, first introduced in Section 2.8, can be extended to random signals using the expected value operator.

FIGURE 9.6:
Expanded Adaptive
Filter Block
Showing the
Weight-update
Algorithm

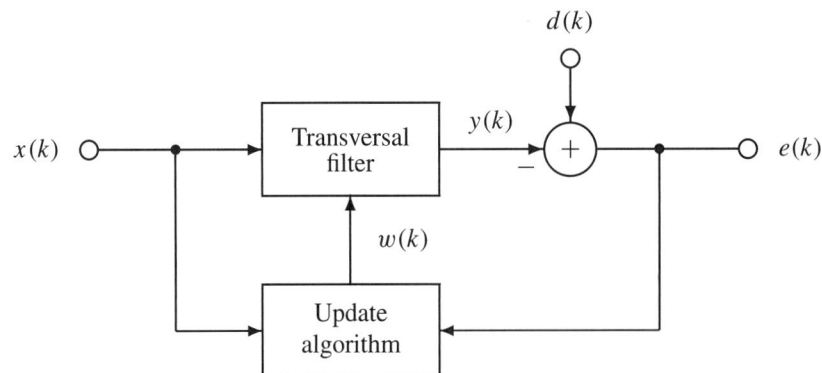

DEFINITION
....
9.1: Random Cross-
correlation

Let $y(k)$ be an L-point random signal and let $x(k)$ be an M-point random signal where $M \leq L$. Then the *cross-correlation* of $y(k)$ with $x(k)$ is denoted as $r_{yx}(i)$ and defined as

$$r_{yx}(i) \triangleq E[y(k)x(k-i)], \quad 0 \leq i < L$$

Recall from (4.7.3) that if a random signal has the property that it is *ergodic*, then the expected value operation can be computed using a time average. Consequently, for a signal $x(k)$, the expected, or mean, value can be approximated as

$$E[x(k)] \approx \frac{1}{N} \sum_{i=0}^{N-1} x(k-i), \quad N \gg 1 \tag{9.2.5}$$

If the signal $x(k)$ is periodic, then the expected value can be determined exactly by computing the average value of over one period.

When the approximation in (9.2.5) is used to evaluate the cross-correlation of two random signals, Definition 9.1 reduces to the deterministic definition of linear cross-correlation introduced previously in Definition 2.5. It is in this sense that Definition 9.1 is a generalization of the notion of cross-correlation to random signals.

There are two properties of cross-correlation that are helpful in the analysis of adaptive *Stationary* filters. Since $x(k)$ is assumed to be *stationary*, the statistical properties of $x(k)$ do not change *signals* with time. Consequently, if $x(k)$ is translated in time, its expected or mean value does not change.

$$E[x(k-i)] = E[x(k)], \quad i \geq 0 \tag{9.2.6}$$

Another fundamental property concerns the expected value of the product of two signals. *Statistically* If two random signals $x(k)$ and $y(k)$ are *statistically independent* of one another, then the *independent signals* expected value of their product is equal to the product of their expected values.

$$E[x(k)y(k)] = E[x(k)]E[y(k)] \tag{9.2.7}$$

Note that if $x(k)$ and $y(k)$ are statistically independent and either $x(k)$ or $y(k)$ has zero mean, then the expected value of their product is zero. Zero-mean signals for which $E[x(k)y(k)] = 0$ *Uncorrelated signals* are said to be *uncorrelated*.

Suppose that $v(k)$ denotes white noise uniformly distributed over $[a, b]$. This particular broadband signal turns out to be an excellent input signal for system identification purposes due to its flat power density spectrum. White noise was examined in detail in Section 4.6. For convenient reference, the following expression denotes the average power of uniform white noise, as developed in Chapter 4 and summarized in Appendix 2.

$$P_v = \frac{b^3 - a^3}{3(b-a)} \tag{9.2.8}$$

9.2.3 Mean Square Error

Mean square error The average of the square of the error of the system in Figure 9.5 is referred to as the *mean square error* of the system. The mean square error $\epsilon(w)$ can be expressed in terms of the expected value operator as follows.

$$\epsilon(w) \triangleq E[e^2(k)] \tag{9.2.9}$$

To see how the mean square error (MSE) is affected by the filter weights, consider the case when the weight vector w is held constant. If we use (9.2.3) and (9.2.4), the square of the error can be expressed as follows.

$$
\begin{aligned}
e^2(k) &= [d(k) - w^T u(k)]^2 \\
&= d^2(k) - 2d(k)w^T u(k) + [w^T u(k)]^2 \\
&= d^2(k) - 2w^T d(k)u(k) + w^T u(k)u^T(k)w
\end{aligned}
\tag{9.2.10}
$$

Since the expected value operation is linear, the expected value of the sum is the sum of the expected values, and scaling a variable simply scales its expected value. Taking the expected value of both sides of (9.2.10) then yields the following expression for the mean square error.

$$
\epsilon(w) = E[d^2(k)] - 2w^T E[d(k)u(k)] + w^T E[u(k)u^T(k)]w
\tag{9.2.11}
$$

To develop a compact formulation of $\epsilon(w)$, consider an $(m+1) \times 1$ column vector p and an $(m+1) \times (m+1)$ matrix R defined as follows.

$$
p \triangleq E[d(k)u(k)]
\tag{9.2.12a}
$$

$$
R \triangleq E[u(k)u^T(k)]
\tag{9.2.12b}
$$

Cross-correlation vector

The vector p is referred to as the *cross-correlation* vector of the desired output $d(k)$ with the vector of past inputs $u(k)$. From (9.2.2a), $p_i = E[d(k)x(k-i)]$. It then follows from Definition 9.1 that

$$
p_i = r_{dx}(i), \quad 0 \le i \le m
\tag{9.2.13}
$$

Auto-correlation matrix

The square matrix R, obtained by taking the expected value of the outer product of $u(k)$ with itself, is referred to as the *auto-correlation* matrix of the past inputs. Again note from (9.2.2a) that

$$
R_{ij} = E[x(k-i)x(k-j)], \quad 0 \le i, j \le m
\tag{9.2.14}
$$

Since $x(k)$ is assumed to be stationary, the signal $x(k-i)x(k-j)$ can be translated in time without changing its expected value. Replacing k with $k+i$ yields $R_{ij} = E[x(k)x(k+i-j)]$. Thus from Definition 9.1

$$
R_{ij} = r_{xx}(j-i), \quad 0 \le i, j \le m
\tag{9.2.15}
$$

The auto-correlation matrix has a number of interesting and useful properties. First, notice from (9.2.14) that since $x(k-i)x(k-j) = x(k-j)x(k-i)$, it follows that R is symmetric. Next, observe from (9.2.15) that $j = i$ yields $R_{ii} = r_{xx}(0)$. But the auto-correlation evaluated at a lag of zero is just the *average power*. That is,

Average power

$$
\begin{aligned}
R_{ii} &= r_{xx}(0) \\
&= E[x^2(k)] \\
&= P_x, \quad 0 \le i \le m
\end{aligned}
\tag{9.2.16}
$$

Consequently, when $x(k)$ is stationary, the diagonal elements of R are all identical and equal to the average power of the input. More generally, it is clear from (9.2.15) that the symmetric auto-correlation matrix R has *diagonal bands*, or stripes of equal elements, above and below

Toeplitz matrix the diagonal, as can be seen in the case $m = 4$ shown in (9.2.17). A matrix with this diagonal striped structure is referred to as a *Toeplitz matrix*.

$$R = \begin{bmatrix} r_{xx}(0) & r_{xx}(1) & r_{xx}(2) & r_{xx}(3) & r_{xx}(4) \\ r_{xx}(1) & r_{xx}(0) & r_{xx}(1) & r_{xx}(2) & r_{xx}(3) \\ r_{xx}(2) & r_{xx}(1) & r_{xx}(0) & r_{xx}(1) & r_{xx}(2) \\ r_{xx}(3) & r_{xx}(2) & r_{xx}(1) & r_{xx}(0) & r_{xx}(1) \\ r_{xx}(4) & r_{xx}(3) & r_{xx}(2) & r_{xx}(1) & r_{xx}(0) \end{bmatrix} \qquad (9.2.17)$$

If we use the definitions of the cross-correlation vector p and the input-correlation matrix R in (9.2.12), the expression for the mean square error performance function in (9.2.11) simplifies to

$$\epsilon(w) = P_d - 2w^T p + w^T R w \qquad (9.2.18)$$

Here $P_d = E[d^2(k)]$ is the average power of the desired output. It is clear from (9.2.18) that the mean square error is a *quadratic* function of the weight vector w. Note that when $m = 1$, the mean square error can be thought of as a surface over the w plane. The objective is to locate the lowest point on this error surface.

To find an optimal value for w, one that minimizes the mean square error, consider the gradient vector $\nabla \epsilon(w)$ of partial derivatives of $\epsilon(w)$ with respect to the elements of w. Taking the derivative of $\epsilon(w)$ in (9.2.18) with respect to w_i and combining the results for $0 \leq i \leq m$, it is possible to show that

$$\nabla \epsilon(w) = 2(Rw - p) \qquad (9.2.19)$$

Consider the case when the input-correlation matrix R is invertible. Setting $\nabla \epsilon(w) = 0$ in (9.2.19) and solving for w, one arrives at the following optimal value for the weight vector.

$$w^* = R^{-1} p \qquad (9.2.20)$$

Wiener solution The optimal weight vector w^* in (9.2.20) is referred to as the *Wiener* solution (Levinson, 1947).

Example 9.1

Optimal Weight Vector

As an illustration of the mean square error and the optimal weight vector, consider the following example adapted from Widrow and Sterns (1985). Suppose that $m = 1$ and the input and desired output are the following periodic functions of period N where $N > 2$.

$$x(k) = 2 \cos\left(\frac{2\pi k}{N}\right)$$

$$d(k) = \sin\left(\frac{2\pi k}{N}\right)$$

Thus the adaptive filter must perform a scaling and a phase shifting operation on the input. First, consider the cross-correlation vector p. Since $d(k)$ and $x(k)$ are periodic, their expected values can be computed by averaging over one period. For convenience, let $\theta = 2\pi / N$.

Using (9.2.13), and the trigonometric identities in Appendix 2, yields

$$
\begin{aligned}
p_i &= E[d(k)x(k-i)] \\
&= E[2\sin(k\theta)\cos\{(k-i)\theta\}] \\
&= 2E[\sin(k\theta)\{\cos(k\theta)\cos(i\theta) + \sin(k\theta)\sin(i\theta)\}] \\
&= 2\cos(i\theta)E[\sin(k\theta)\cos(k\theta)] + 2\sin(i\theta)E[\sin^2(k\theta)] \\
&= \cos(i\theta)E[\sin(2k\theta)] + \sin(i\theta)E[1 - \cos(2k\theta)] \\
&= \sin(i\theta), \quad 0 \le i \le 1
\end{aligned}
$$

Thus the cross-correlation between the desired output and the vector of past inputs is

$$
p = [0,\ \sin(\theta)]^T
$$

Next, consider the auto-correlation matrix of the past inputs. Here

$$
\begin{aligned}
E[x(k)x(k-i)] &= E[4\cos(k\theta)\cos\{(k-i)\theta\}] \\
&= 4E[\cos(k\theta)\{\cos\{k\theta\}\cos(i\theta) + \sin(k\theta)\sin(i\theta)\}] \\
&= 4\cos(i\theta)E[\cos^2(k\theta)] + 4\sin(i\theta)E[\cos(k\theta)\sin(k\theta)] \\
&= 2\cos(i\theta)E[1 + \cos(2k\theta)] + 2\sin(i\theta)E[\sin(2k\theta)] \\
&= 2\cos(i\theta)
\end{aligned}
$$

Thus, from (9.2.17), the auto-correlation matrix is

$$
R = 2\begin{bmatrix} 1 & \cos(\theta) \\ \cos(\theta) & 1 \end{bmatrix}
$$

Since $\theta = 2\pi/N$ and $N > 2$, it is evident that R is symmetric, banded, and nonsingular with

$$
\begin{aligned}
\det(R) &= 4[1 - \cos^2(\theta)] \\
&= 4\sin^2(\theta)
\end{aligned}
$$

From (9.2.20), the optimal value for the weight vector in this case is

$$
\begin{aligned}
w^* &= \begin{bmatrix} 2 & 2\cos(\theta) \\ 2\cos(\theta) & 2 \end{bmatrix}^{-1} \begin{bmatrix} 0 \\ \sin(\theta) \end{bmatrix} \\
&= \frac{1}{4\sin^2(\theta)} \begin{bmatrix} 2 & -2\cos(\theta) \\ -2\cos(\theta) & 2 \end{bmatrix} \begin{bmatrix} 0 \\ \sin(\theta) \end{bmatrix} \\
&= \frac{1}{2\sin^2(\theta)} \begin{bmatrix} -\cos(\theta)\sin(\theta) \\ \sin(\theta) \end{bmatrix} \\
&= .5\begin{bmatrix} -\cot(\theta), & \csc(\theta) \end{bmatrix}^T
\end{aligned}
$$

To make the problem more specific, suppose $N = 4$, in which case $\theta = \pi/2$. A plot of the resulting mean square error surface is shown in Figure 9.7. In this case the optimal weight vector that minimizes the mean square error is

$$
w^* = [0,\ .5]^T
$$

When adaptive filters are used for system identification, the user often has direct control of the input signal $x(k)$. To get reliable results, the spectral content of the input should be sufficiently rich that it excites all of the natural modes of the system being identified. One

FIGURE 9.7: Mean-square-error Surface

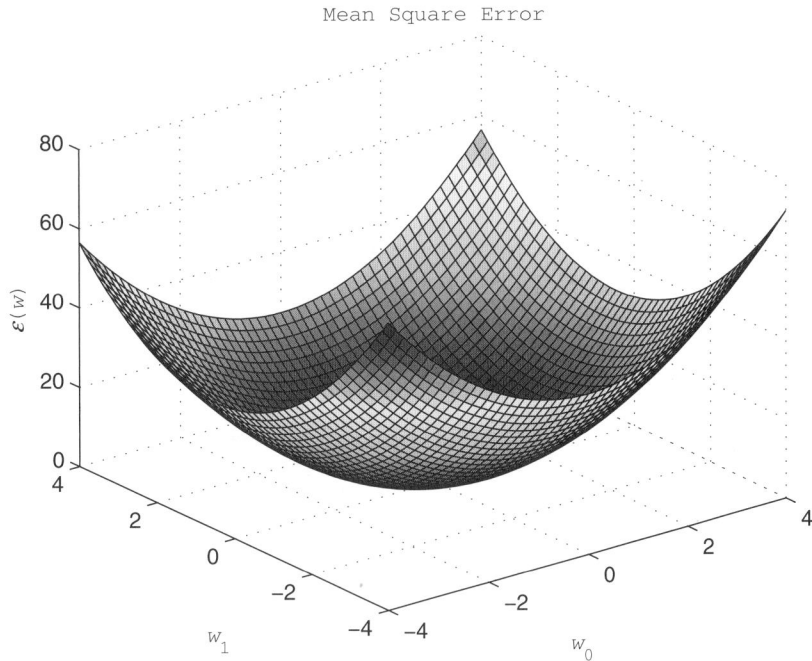

input that is particularly rich in frequency content is a random white noise input, a signal whose power density spectrum is flat.

Example 9.2

White Noise Input

Suppose that $x(k)$ is zero-mean white noise with an average power of

$$P_x = E[x^2(k)]$$

To determine the input-correlation matrix, first note that for white noise the signals $x(k)$ and $x(k - i)$ are statistically independent for $i \neq 0$. Thus the expected value of the product is equal to the product of the expected values. Since $x(k)$ is zero-mean white noise, $E[x(k)] = 0$. Hence for $i \neq 0$

$$E[x(k)x(k - i)] = E[x(k)]E[x(k - i)]$$
$$= 0$$

It then follows from (9.2.16) that for zero-mean white noise with average power P_x, the input-correlation matrix is simply

$$R = P_x I$$

Consequently, zero-mean white noise produces a nonsingular, diagonal input-correlation matrix with the average power of $x(k)$ along the diagonal. It follows from (9.2.20) that the optimal weight in this case is simply

$$w^* = \frac{p}{P_x}$$

●●●●●●●●●●●●● ● ● ● ● ●
9.3 The Least Mean Square (LMS) Method

The optimal weight vector found in Section 9.2 was based on the assumption that the input $x(k)$ and the desired output $d(k)$ are stationary random signals. This assumption is useful for analysis purposes because it allows one to determine, for example, the characteristics of the input that are required to ensure that an optimal weight vector exists and is unique. However, in actual applications of adaptive filters the input and desired output are often not stationary; instead their statistical properties evolve with time. When the input and desired output are not stationary, it is useful to iteratively update an estimate of the optimal weight vector using a numerical search technique.

Suppose that $w(0)$ is an initial guess for the optimal weight vector. For example, in the absence of any specialized knowledge about the application, one might simply take $w(0) = 0$. At subsequent time steps, the new weight is set to the old weight plus a correction term as follows.

$$w(k+1) = w(k) + \Delta w(k), \quad k \geq 0 \tag{9.3.1}$$

A simple way to compute a correction term $\Delta w(k)$ is to use the gradient vector of partial derivatives of the mean square error $\epsilon(w)$ with respect to the elements of w.

$$\nabla \epsilon_i(w) \triangleq \frac{\partial \epsilon(w)}{\partial w_i}, \quad 0 \leq i \leq m \tag{9.3.2}$$

The gradient vector $\nabla \epsilon(w)$ points in the direction of maximum increase of $\epsilon(w)$. For example, when $m = 1$, the mean square error is a surface and $\nabla \epsilon(w)$ is a 2×1 vector that points in the steepest uphill direction, the direction of steepest ascent. Since the objective is to find the minimum point on this surface, consider a step of length $\mu > 0$ in the opposite direction of the gradient. That is, set $\Delta w(k) = -\mu \nabla \epsilon[w(k)]$, in which case the weight-update algorithm becomes

$$w(k+1) = w(k) - \mu \nabla \epsilon[w(k)], \quad k \geq 0 \tag{9.3.3}$$

Steepest descent method   The weight-update formula in (9.3.3) is called the method of *steepest descent*. Note that the *step size* μ must be kept small because as one departs from the point $w(k)$, the direction of steepest descent changes. The main computational difficulty of the steepest-descent method is the need to compute the gradient vector $\nabla \epsilon(w)$ at each discrete time. Since the ith element of the gradient vector represents the slope of $\epsilon(w)$ along the ith dimension, the gradient vector can be approximated numerically using differences. For example, let i^j denote the jth column of the $(m+1) \times (m+1)$ identity matrix $I = [i^1, i^2, \ldots, i^{m+1}]$. If $\delta > 0$ is small, then the jth element of the gradient vector can be approximated with the following forward difference.

$$\nabla \epsilon_j(w) \approx \frac{\epsilon(w + \delta i^{j+1}) - \epsilon(w)}{\delta}, \quad 0 \leq j \leq m \tag{9.3.4}$$

Observe that in the limit as δ approaches zero, the expression in (9.3.4) is the partial derivative of $\epsilon(w)$ with respect to w_j. The approximation of the gradient in (9.3.4) requires $m + 2$ evaluations of the mean square error. Suppose that the mean square error is itself approximated by using a time average of N samples of the square of the error. From (9.2.3), each sample of $e^2(k)$ requires $m + 1$ floating-point multiplications or FLOPs. Thus the total number of FLOPs required to numerically estimate the gradient of the mean square error is

$$r = N(m+1)(m+2) \quad \text{FLOPs} \tag{9.3.5}$$

For a large filter $m \gg 1$, and for an accurate estimate $N \gg 1$. Consequently, implementing the steepest-descent method using a numerical estimate of the gradient vector can be computationally expensive.

There is an alternative approach to estimating the gradient vector that is much more cost effective (Widrow and Sterns, 1985). Suppose that, for the purpose of computing the gradient, the mean square error is approximated using the *instantaneous* value of the square of the error. That is, for the purpose of computing $\nabla \epsilon(w)$, the following approximation is used for the mean square error.

$$\epsilon(w) \approx e^2(k) \tag{9.3.6}$$

This is clearly a rough approximation because it is equivalent to using a single sample to estimate the mean. The approximation in (9.3.6) leads to a substantial simplification in the expression for the gradient. Let $\hat{\nabla} \epsilon(w)$ be the estimate of the gradient using (9.3.6). Then from (9.2.3) and (9.2.4)

$$\hat{\nabla} \epsilon(w) = 2e(k)\frac{\partial e(k)}{\partial w}$$
$$= -2e(k)u(k) \tag{9.3.7}$$

Using this estimate of the gradient, the steepest-descent method in (9.3.3) then reduces to the following simplified weight-update algorithm.

$$w(k+1) = w(k) + 2\mu e(k)u(k), \quad k \geq 0 \tag{9.3.8}$$

LMS Method   The weight-update formula in (9.3.8) is called the *least mean square* or LMS method (Widrow and Hoff, 1960). Note that unlike the traditional steepest-descent method, the LMS method requires only $m + 1$ FLOPs to estimate the gradient at each time step. The LMS is a highly efficient way to update the weight vector. For example, when $N = 10$ and $m = 10$, the LMS method is more than two orders of magnitude faster than the numerical steepest-descent method.

Although the approximation used to estimate the gradient vector in (9.3.6) may appear to be rather crude, experience has shown that the LMS algorithm for updating the weights is quite robust. Indeed, Hassibi et al, (1996) have shown that the LMS algorithm is optimal when a minimax error criterion is used.

The estimate of the gradient in (9.3.7) is itself a random signal. It is instructive to examine the mean or expected value of this random signal. Suppose that the weight w has converged to its steady-state value and is constant. Starting from (9.3.7), and using the definitions of p and R in (9.2.12),

$$E[\hat{\nabla} \epsilon(w)] = -2E[e(k)u(k)]$$
$$= -2E[d(k)u(k) - y(k)u(k)]$$
$$= -2E[d(k)u(k) - u(k)\{u^T(k)w\}]$$
$$= -2\{E[d(k)u(k)] - E[u(k)u^T(k)]w\}$$
$$= 2(Rw - p) \tag{9.3.9}$$

But from (9.2.19), the exact value of the gradient of the mean square error is $\nabla \epsilon(w) = 2(Rw - p)$. Hence

$$E[\hat{\nabla} \epsilon(w)] = \nabla \epsilon(w) \tag{9.3.10}$$

Unbiased estimate   That is, the expected value of the estimate of the gradient of the mean square error is equal to the gradient of the mean square. Therefore, $\hat{\nabla} \epsilon(w)$ is an *unbiased* estimate of $\nabla \epsilon(w)$.

Example 9.3

System Identification

To illustrate the LMS method, consider the system identification problem shown in Figure 9.8. To make the example specific, suppose that the system to be identified has the following transfer function.

$$H(z) = \frac{2 - 3z^{-1} - z^{-2} + 4z^{-4} + 5z^{-5} - 8z^{-6}}{1 - 1.6z^{-1} + 1.75z^{-2} - 1.436z^{-3} + .6814z^{-4} - .1134z^{-5} - .0648z^{-6}}$$

Next, suppose that the input $x(k)$ consists of $N = 1000$ samples of white noise uniformly distributed over $[-1, 1]$. Let the order of the adaptive transversal filter be $m = 50$, and suppose the step size is $\mu = .01$. A plot of the first 500 samples of the square of the error, obtained by running *exam9_3*, is shown in Figure 9.9. It is clear that the square of the error converges close to zero after approximately 400 samples.

One way to assess the effectiveness of the adaptive filter is to compare the magnitude response of the system $H(z)$ with the magnitude response of the adaptive filter $W(z)$, using the final steady-state weight $w(N - 1)$. The two magnitude responses are plotted in Figure 9.10 where it is evident that they are nearly identical. Note that this is true in spite of the fact that $H(z)$ is a an IIR filter with six poles and six zeros, while the steady-state adaptive filter is an

FIGURE 9.8: System Identification

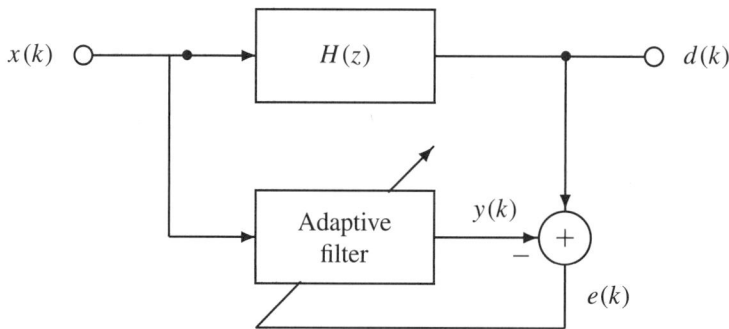

FIGURE 9.9: First 500 Samples of Squared Error During System Identification Using the LMS Method with $m = 50$ and $\mu = .01$

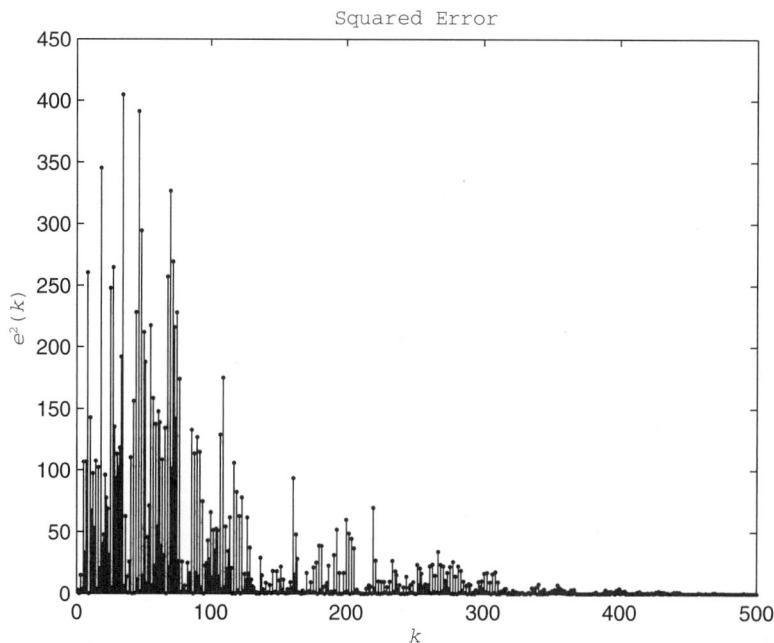

FIGURE 9.10:
Magnitude
Responses of the
Original IIR System
and the Identified
System Using the
LMS Method with
$m = 50$, $\mu = .01$,
and $N = 1000$
Samples

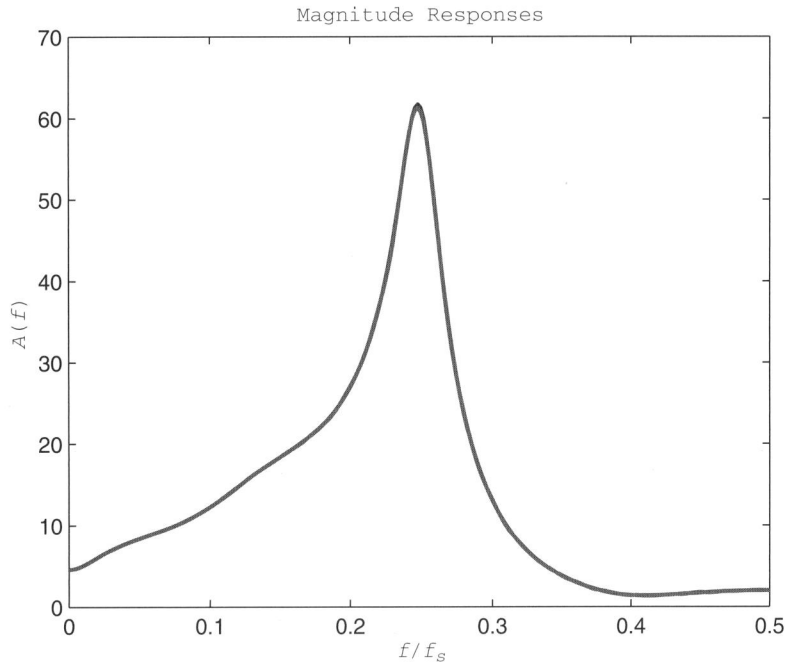

FIR filter of order $m = 50$. By making the order of the adaptive filter sufficiently large, it is apparent that it can model an IIR filter as well. Of course, if the system to be identified is an FIR filter of order p, then an exact fit will be obtained using any adaptive filter of order $m \geq p$, assuming infinite precision arithmetic is used.

FDSP Functions

The FDSP toolbox contains the following function that implements the LMS method.

```
% F_LMS: System identification using least mean square (LMS) method
%
% Usage:
%         [w,e] = f_lms (x,d,m,mu,w)
% Pre:
%      x    = N by 1 vector containing input samples
%      d    = N by 1 vector containing desired output
%             samples
%      m    = order of transversal filter (m >= 0)
%      mu   = step size to use for updating w
%      w    = an optional (m+1) by 1 vector containing
%             the initial values of the weights.
%             Default: w = 0
```

Continued on p. 660

Continued from p. 659

```
% Post:
%        w = (m+1) by 1 weight vector of filter
%            coefficients
%        e = an optional N by 1 vector of errors where
%            e(k) = d(k)-y(k)
% Notes:
%        Typically mu << 1/[(m+1)*P_x] where P_x is the
%        average power of input x.
```

● ● ● ● ● ● ● ● ● ● ● ● ● ● ● ● ●

9.4 Performance Analysis of LMS Method

Although the LMS method is very simple to implement, there remains the question of determining an effective value for the step size μ. The step size should be small enough to guarantee convergence to an acceptable steady-state error, yet large enough to ensure that the convergence is rapid.

9.4.1 Step Size

Recall that the LMS method uses the error $e(k)$ and the vector of past inputs $u(k)$ to update the weight estimate as follows.

$$w(k+1) = w(k) + 2\mu e(k)u(k), \quad k \geq 0 \tag{9.4.1}$$

Since $w(k)$ is a random signal, it is helpful to consider what happens to its mean or expected value as k increases. Taking the expected value of both sides of (9.4.1), and noting that $e(k) = d(k) - u^T(k)w(k)$, this yields

$$
\begin{aligned}
E[w(k+1)] &= E[w(k)] + 2\mu E[e(k)u(k)] \\
&= E[w(k)] + 2\mu E[\{d(k) - u^T(k)w(k)\}u(k)] \\
&= E[w(k)] + 2\mu E[d(k)u(k) - u(k)\{u^T(k)w(k)\}] \\
&= E[w(k)] + 2\mu E[d(k)u(k)] - 2\mu E[u(k)u^T(k)w(k)] \tag{9.4.2}
\end{aligned}
$$

Recall that if two random signals are statistically independent of one another, the expected value of the product is the product of the expected values. For moderate convergence rates, the past inputs $u(k)$ and the weights $w(k)$ can be assumed to be statistically independent. Using the definitions of R and p from (9.2.12), one can rewrite (9.4.2) as

$$
\begin{aligned}
E[w(k+1)] &= E[w(k)] + 2\mu E[d(k)u(k)] - 2\mu E[u(k)u^T(k)]E[w(k)] \\
&= E[w(k)] + 2\mu p - 2\mu R E[w(k)] \\
&= (I - 2\mu R)E[w(k)] + 2\mu p \tag{9.4.3}
\end{aligned}
$$

To further simplify the expression for the expected value of the weight vector, it is helpful to introduce a signal that represents the variation of the weight from its optimal value. The *weight variation* vector is denoted as $\delta w(k)$ and defined as

Weight variation

$$\delta w(k) \overset{\Delta}{=} w(k) - w^* \tag{9.4.4}$$

To reformulate (9.4.3) in terms of the weight variation, one can substitute $w(k) = \delta w(k) + w^*$ and use $E[w^*] = w^*$ and $Rw^* = p$. This yields

$$E[\delta w(k+1)] + w^* = (I - 2\mu R)\{E[\delta w(k)] + w^*\} + 2\mu p$$
$$= (I - 2\mu R)E[\delta w(k)] + (I - 2\mu R)w^* + 2\mu p$$
$$= (I - 2\mu R)E[\delta w(k)] + w^* \qquad (9.4.5)$$

Thus the expected value of the weight variation at iteration $k + 1$ is simply

$$E[\delta w(k+1)] = (I - 2\mu R)E[\delta w(k)] \qquad (9.4.6)$$

The virtue of the formulation in (9.4.6) is that it can be solved directly for $E[\delta w(k)]$ using induction. Note that $E[\delta w(0)] = \delta w(0)$ in which case $E[\delta w(1)] = (I - 2\mu R)\delta w(0)$. More generally,

$$E[\delta w(k)] = (I - 2\mu R)^k \delta w(0), \quad k \geq 0 \qquad (9.4.7)$$

The solution in (9.4.7) can be expressed in terms of the original weight vector $w(k)$ by simply replacing $\delta w(k)$ with $w(k) - w^*$. This yields the following closed-form solution for the expected value of the weight vector at step k.

$$E[w(k)] = w^* + (I - 2\mu R)^k [w(0) - w^*], \quad k \geq 0 \qquad (9.4.8)$$

It is clear from (9.4.8) that $E[w(k)]$ will converge to the optimal weight w^* starting from an arbitrary initial guess if and only if $(I - 2\mu R)^k$ converges to the zero matrix as k approaches infinity. At this point, it is helpful to make use of a result from linear algebra. If A is a square matrix, then

$$A^k \to 0 \quad \text{as} \quad k \to \infty \qquad (9.4.9)$$

if and only if the eigenvalues of A all lie strictly inside the unit circle of the complex plane (Noble, 1969). It can be shown by direct substitution that the ith eigenvalue of $A = I - 2\mu R$ is $r_i = 1 - 2\mu\lambda_i$, where λ_i is the ith eigenvalue of R. Thus $E[w(k)]$ in (9.4.8) converges to w^* as $k \to \infty$ if and only if

$$|1 - 2\mu\lambda_i| < 1 \quad \text{for} \quad 1 \leq i \leq m + 1 \qquad (9.4.10)$$

Since R is symmetric and positive-definite, its eigenvalues λ_i are real and positive. Consequently, (9.4.10) can be rewritten as $-1 < 1 - 2\mu\lambda_i < 1$. Subtracting one from each term, and dividing each term by $-2\lambda_i$ then yields the inequality $1/\lambda_i > \mu > 0$. This must hold for all $m + 1$ eigenvalues of R. Let

$$\lambda_{\max} \overset{\Delta}{=} \max\{\lambda_1, \lambda_2, \dots, \lambda_{m+1}\} \qquad (9.4.11)$$

It then follows that the range of step sizes over which $E[w(k+1)]$ converges, starting from an arbitrary $w(0)$, is $0 < \mu < 1/\lambda_{\max}$.

PROPOSITION

9.1: LMS Convergence

Let $w(0) \in R^{m+1}$ be arbitrary and let $\lambda_{\max}$ be the largest eigenvalue of the auto-correlation matrix R. The LMS method converges in the following statistical sense if and only if $0 < \mu < 1/\lambda_{\max}$.

$$E[w(k)] \to w^* \quad \text{as} \quad k \to \infty$$

Example 9.4

Step Size

As a simple illustration of how to find a range of step sizes for the LMS method, consider an adaptive filter of order $m = 1$. Suppose $N \geq 4$ and the input and desired output are as follows.

$$x(k) = 2\cos\left(\frac{2\pi k}{N}\right)$$

$$d(k) = \sin\left(\frac{2\pi k}{N}\right)$$

For convenience, let $\theta = 2\pi/N$. This two-dimensional adaptive filter was considered previously in Example 9.1 where it was determined that

$$R = 2\begin{bmatrix} 1 & \cos(\theta) \\ \cos(\theta) & 1 \end{bmatrix}$$

Thus the characteristic polynomial of the auto-correlation matrix is

$$
\begin{aligned}
\Delta(\lambda) &= \det\{\lambda I - R\} \\
&= \det\left\{ \begin{bmatrix} \lambda - 2 & -2\cos(\theta) \\ -2\cos(\theta) & \lambda - 2 \end{bmatrix} \right\} \\
&= (\lambda - 2)^2 - 4\cos^2(\theta) \\
&= \lambda^2 - 4\lambda + 4 - 4\cos^2(\theta) \\
&= \lambda^2 - 4\lambda + 4\sin^2(\theta)
\end{aligned}
$$

With the quadratic formula, the eigenvalues of R are

$$
\begin{aligned}
\lambda_{1,2} &= \frac{4 \pm \sqrt{16 - 16\sin^2(\theta)}}{2} \\
&= 2 \pm 2\sqrt{1 - \sin^2(\theta)} \\
&= 2 \pm 2\cos(\theta) \\
&= 2[1 \pm \cos(\theta)]
\end{aligned}
$$

Note that the eigenvalues of R are real and positive. Since $N \geq 4$, it follows that $0 < \theta \leq \pi/2$ and the largest eigenvalue is $\lambda_{\max} = 2[1 + \cos(\theta)]$. Thus from Proposition 9.1 the range of step sizes over which the LMS method converges is Note that the eigenvalues of R are real and positive. Since $N \geq 4, 0 < \theta \leq \pi/2$ and the largest eigenvalue is $\lambda_{\max} = 2[1 + \cos(\theta)]$. Thus from Proposition 9.1 the range of step sizes over which the LMS method converges is

$$0 < \mu < \frac{.5}{1 + \cos(\theta)}$$

A key advantage of the LMS method in (9.4.1) is that it is very simple to implement. For example, there is no need to compute the auto-correlation matrix R or the cross-correlation vector p. Unfortunately, the upper bound on the step size in Proposition 9.1 is not nearly as easy to determine. Not only must the auto-correlation matrix R be constructed, but its eigenvalues must be computed as well. By making use of another result from linear algebra, an alternative step size bound can be developed, one that is more conservative but much easier to compute.

Trace   Recall that the *trace* of a square matrix is just the sum of the diagonal elements. The trace is also equal to the sum of the eigenvalues. That is,

$$\text{trace}(R) = \sum_{i=1}^{m+1} \lambda_i \qquad (9.4.12)$$

Since R is symmetric and positive-definite, $\lambda_i > 0$. It then follows from (9.4.12) that

$$\lambda_{\max} < \text{trace}(R) \qquad (9.4.13)$$

Applying (9.4.13) to Proposition 9.1, a more conservative range for the step size is $0 < \mu < 1/\text{trace}(R)$. This effectively eliminates the need to compute eigenvalues. The requirement to find R itself also can be eliminated by exploiting the special banded structure of R. Recall from (9.2.16) that the diagonal elements of R are all equal to one another with $R_{ii} = P_x$ where $P_x = E[x^2(k)]$ is the average power of the input. Since the trace is just the sum of the $m + 1$ diagonal elements, $\text{trace}(R) = (m + 1)P_x$. This leads to the following smaller, but much simpler, range of values for the step size over which convergence of the LMS method is assured.

LMS step size

$$0 < \mu < \frac{1}{(m+1)P_x} \qquad (9.4.14)$$

Note how the upper bound on μ decreases with the order of the filter and the power of the input signal. In practical applications, it is not uncommon to choose a value for the step size in the range $.01 < (m + 1)P_x\mu < .1$, well below the upper limit (Kuo and Morgan, 1996).

Example 9.5 | **Revised Step Size**

Suppose that the input for an mth-order adaptive filter consists of zero-mean white noise uniformly distributed over an interval $[-c, c]$. From (9.2.8), the average power of $x(k)$ is

$$P_x = \frac{c^2}{3}$$

Applying (9.4.14), yields the following range of step sizes for the LMS method.

$$0 < \mu < \frac{3}{(m+1)c^2}$$

For example, for the system identification application in Example 9.3, the magnitude of the white noise input was $c = 1$, and the filter order was $m = 50$. Thus the range of step sizes for Example 9.3 is

$$0 < \mu < .0588$$

The step size used in Example 9.3, $\mu = .01$, was well within this range.

9.4.2 Convergence Rate

The step size determines not only whether the LMS method converges; it also determines how fast it converges. One way to view convergence is to examine what happens to $w(k)$ in (9.4.8).

Learning curve

An equivalent way to view convergence is to examine the *learning curve* which is a plot of the mean square error as a function of the iteration number. Recall from (9.2.18) that the mean square error can be expressed as follows

$$\epsilon[w(k)] = P_d - 2w^T(k)p + w^T(k)Rw(k) \qquad (9.4.15)$$

To see how fast the mean square error converges, it is helpful to develop an alternative representation for the auto-correlation matrix R. Let λ_i denote the ith eigenvalue of R, and let

Eigenvectors q^i be its associated eigenvector. That is,

$$Rq^i = \lambda_i q^i, \quad 1 \le i \le m+1 \qquad (9.4.16)$$

Suppose that the $m+1$ eigenvectors are arranged as columns of an $(m+1) \times (m+1)$ matrix $Q = [q^1, q^2, \ldots, q^{m+1}]$. Next, let Λ be the diagonal matrix of order $m+1$ with the eigenvalues along the diagonal.

$$\Lambda \triangleq \begin{bmatrix} \lambda_1 & 0 & \cdots & 0 \\ 0 & \lambda_2 & \cdots & 0 \\ \vdots & \vdots & \ddots & \vdots \\ 0 & 0 & \cdots & \lambda_{m+1} \end{bmatrix} \qquad (9.4.17)$$

With Q and Λ, the $m+1$ eigenvector equations in (9.4.16) can be rewritten as a single matrix equation as follows.

$$RQ = \Lambda Q \qquad (9.4.18)$$

Note that the right-hand side of (9.4.18) can be replaced with $Q\Lambda$ because Λ is diagonal. Since the auto-correlation matrix R is symmetric, the eigenvectors of R form a linearly independent set which means the eigenvector matrix Q is invertible. First, commute the matrices on the right-hand side of (9.4.18) and then post-multiply both sides of (9.4.18) by Q^{-1}. This yields the following alternative representation of the auto-correlation matrix in terms of its eigenvectors and eigenvalues.

$$R = Q\Lambda Q^{-1} \qquad (9.4.19)$$

The representation of R in (9.4.19) has a number of useful properties. For example, a direct calculation reveals that $R^2 = Q\Lambda^2 Q^{-1}$. Using induction, it is not difficult to show that

$$R^k = Q\Lambda^k Q^{-1}, \quad k \ge 0 \qquad (9.4.20)$$

This property can be used to evaluate the rate of convergence. Substituting the expression for R from (9.4.19) into (9.4.8) and using (9.4.18) we have

$$\begin{aligned} E[w(k)] &= w^* + (I - 2\mu Q\Lambda Q^{-1})^k [w(0) - w^*] \\ &= w^* + (QIQ^{-1} - 2\mu Q\Lambda Q^{-1})^k [w(0) - w^*] \\ &= w^* + \{Q(I - 2\mu\Lambda)Q^{-1}\}^k [w(0) - w^*] \\ &= w^* + Q(I - 2\mu\Lambda)^k Q^{-1} [w(0) - w^*] \end{aligned} \qquad (9.4.21)$$

The factor $(I - 2\mu\Lambda)^k$ in (9.4.21) consists of the following diagonal matrix where $r_i = 1 - 2\mu\lambda_i$.

$$(I - 2\mu\Lambda)^k = \begin{bmatrix} r_1^k & 0 & \cdots & 0 \\ 0 & r_2^k & \cdots & 0 \\ \vdots & \vdots & \ddots & \vdots \\ 0 & 0 & \cdots & r_{m+1}^k \end{bmatrix} \qquad (9.4.22)$$

Note that this confirms that the LMS method converges if and only if $|1 - 2\mu\lambda_i| < 1$ for $1 \le i \le m + 1$. The speed of convergence is dominated by the factor r_i, whose magnitude is largest because this corresponds to the slowest mode. Let

$$\lambda_{\min} \overset{\Delta}{=} \min\{\lambda_1, \lambda_2, \cdots, \lambda_{m+1}\} \tag{9.4.23}$$

Suppose that the step size is constrained to be at most half of its maximum value, $\mu \le .5/\lambda_{\max}$. Then $0 \le r_i < 1$ and the radius of the slowest or dominant mode is

$$r_{\max} = 1 - 2\mu\lambda_{\min} \tag{9.4.24}$$

The rate of convergence of the LMS method can be characterized by an exponential time constant τ_{mse}. Since the mean square error is a quadratic function of the weights, the mean square error converges at a rate of $r_{\max}^{2k}$. Suppose that the input and desired output are obtained by sampling with an sampling interval of T. Then the exponential rate of convergence is $\exp(-kT/\tau_{\mathrm{mse}})$. Using (9.4.24) results in the following equation for the mean square error time constant.

$$\exp(-kT/\tau_{\mathrm{mse}}) = (1 - 2\mu\lambda_{\min})^{2k} \tag{9.4.25}$$

Taking the log of both sides of (9.4.25) and solving the resulting equation for τ_{mse} yields

$$\tau_{\mathrm{mse}} = \frac{-T}{2\ln(1 - 2\mu\lambda_{\min})} \tag{9.4.26}$$

If μ is sufficiently small, one can use the approximation $\ln(1 + x) \approx x$. This results in the

LMS time constant following simplified approximation for the mean square error *time constant* of the LMS method.

$$\tau_{\mathrm{mse}} \approx \frac{T}{4\mu\lambda_{\min}} \quad \text{sec.} \tag{9.4.27}$$

Note that the time constant can be expressed in units of iterations, rather than seconds, by setting $T = 1$. Furthermore, observe that in order to speed up convergence, one must increase the step size. However, if the step size is made too large, then from Proposition 9.1 the LMS will not converge at all.

Example 9.6 **Time Constant**

Consider the system identification example presented in Example 9.3. There the filter order was $m = 50$, and the input consisted of $N = 1000$ samples of white noise uniformly distributed over $[-1, 1]$. From (9.2.8), the average power of the input is $P_x = 1/3$. Recall that for a zero-mean white noise input, the auto-correlation matrix is very easy to compute. In particular, from Example 9.2,

$$R \approx P_x I$$
$$= \left(\frac{1}{3}\right) I$$

Since R is diagonal, it has a single eigenvalue, $\lambda = 1/3$, repeated $m + 1 = 51$ times. Thus the minimum eigenvalue is

$$\lambda_{\min} = \frac{1}{3}$$

The step size used in Example 9.3 was $\mu = .01$. Applying (9.4.27) results in the following time constant estimate for the system identification example.

$$
\begin{aligned}
\tau_{\text{mse}} &\approx \frac{1}{4\mu\lambda_{\min}} \\
&= \frac{3}{.04} \\
&= 75
\end{aligned}
$$

Here $T = 1$, which yields the time constant in iterations. Since $\exp(-5) = .007$, the mean square error should be reduced to less than one percent of its peak value after five time constants or $M = 375$ iterations. Inspection of the plot of $e^2(k)$ versus k in Figure 9.9 confirms that this is the case, at least approximately. It should be pointed out that the plot of the squared error in Figure 9.9 is a rough approximation to the learning curve. Recall that the learning curve is a plot of the mean square error $\epsilon[w(k)]$, and to obtain a better approximation one would have to perform the system identification many times with different white noise inputs and then average the squares of the errors for each run (See Problem 9.36).

9.4.3 Excess Mean Square Error

The previous analysis of the LMS method suggests that one should choose a step size that is as large as possible, consistent with convergence, in order to reduce the mean square error time constant. As it turns out, there is one more factor called the excess mean square error that mitigates against making the step size too large. Once the excess mean square error is taken into account, one finds that in selecting μ there is a tradeoff between convergence speed and steady-state accuracy.

MSE   Recall that the essential assumption of the LMS method is the approximation of the *mean square error* (MSE) with the squared error for the purpose of estimating the gradient vector. This leads to the gradient approximation, $\hat{\nabla}\epsilon(w) = -2e(k)u(k)$, found in (9.3.7). This estimate differs from the exact value, $\nabla\epsilon(w) = 2(Rw - p)$, in (9.2.19). The error in the estimate of the gradient of the mean square error can be modeled as an additive noise term as follows.

$$\nabla\epsilon[w(k)] = \hat{\nabla}\epsilon[w(k)] + v(k) \tag{9.4.28}$$

Excess MSE   The noise term $v(k)$ causes the steady-state value of the mean square error to be larger than the theoretical minimum value, and the difference is referred to as the *excess mean square error*.

$$\epsilon_{\text{excess}}(k) \overset{\Delta}{=} \epsilon[w(k)] - \epsilon_{\min} \tag{9.4.29}$$

To determine an expression for the minimum mean square error, it is useful to reformulate the mean square error in terms of the weight variation, $\delta w(k) = w(k) - w^*$. Using (9.4.15) and substituting $w(k) = \delta w(k) + w^*$, we have

$$
\begin{aligned}
\epsilon(w) &= P_d - 2p^T(w^* + \delta w) + (w^* + \delta w)^T R(w^* + \delta w) \\
&= P_d - 2p^T w^* - 2p^T\delta w + (w^*)^T Rw^* + (w^*)^T R\delta w + \delta w^T Rw^* + \delta w^T R\delta w \\
&= P_d - 2p^T w^* - 2p^T\delta w + (w^*)^T p + (R^{-1}p)^T R\delta w + \delta w^T p + \delta w^T R\delta w \\
&= P_d - 2p^T w^* - 2p^T\delta w + p^T w^* + p^T(R^{-1})^T R\delta w + p^T\delta w + \delta w^T R\delta w \\
&= P_d - p^T w^* + \delta w^T R\delta w \tag{9.4.30}
\end{aligned}
$$

Here use was made of the following observations: $R^T = R$, the transpose of the inverse is the inverse of the transpose, the transpose of the product is the product of the transposes in reverse order, and the transpose of a scalar is the scalar. From (9.4.30) it is clear that the mean square error or MSE achieves its minimum value at $\delta w = 0$. Using $w^* = R^{-1}p$ in (9.4.30) one arrives at the following expression for the *minimum MSE*.

Minimum MSE

$$\epsilon_{\min} = P_d - p^T R^{-1} p \tag{9.4.31}$$

Combining (9.4.29) through (9.4.31) then results in the following formulation of the excess mean square error in terms of the weight variation.

$$\epsilon_{\text{excess}}(k) = \delta w^T(k) R \delta w(k) \tag{9.4.32}$$

Excess MSE

If we examine the statistical properties of the gradient noise term in (9.4.28), it is possible to develop the following approximation for the *excess mean square error* (Widrow and Sterns, 1985).

$$\epsilon_{\text{excess}}(k) \approx \mu \epsilon_{\min}(m+1) P_x \tag{9.4.33}$$

It is apparent from (9.4.31) that substantial computational effort is required to determine the minimum mean squared error. For this reason, a normalized version of the excess mean square error is often used. The *misadjustment factor* of the LMS method is denoted M_f and defined

Misadjustment factor

$$M_f \overset{\Delta}{=} \frac{\epsilon_{\text{excess}}}{\epsilon_{\min}} \tag{9.4.34}$$

From (9.4.33) the misadjustment factor of the LMS method is simply

$$M_f \approx \mu(m+1) P_x \tag{9.4.35}$$

Note that the normalized excess mean square error increases with the step size, the filter order, and the average power of the input. The dependence on the step size means that in order to reduce M_f, one must decrease μ, but to reduce τ_{mse}, one must increase μ. Thus there is a tradeoff between the convergence speed and the steady-state accuracy of the LMS method.

Example 9.7   Excess Mean Square Error

To illustrate the relationship between excess mean square error and step size, consider an adaptive filter or order $m = 1$ with the following input and desired output.

$$x(k) = 2\cos(.5\pi k) + v(k)$$
$$d(k) = \sin(.5\pi k)$$

Here $v(k)$ represents white noise uniformly distributed over the interval $[-.5, .5]$ that is statistically independent of $2\cos(.5\pi k)$. Thus the average power of the input is

$$
\begin{aligned}
P_x &= E[x^2(k)] \\
&= E[4\cos^2(.5\pi k) + 4v(k)\cos(.5\pi k) + v^2(k)] \\
&= 4E[\cos^2(.5\pi k)] + 4E[v(k)\cos(.5\pi k)] + E[v^2(k)] \\
&= 2E[\cos^2(\pi k) + 1] + 4E[v(k)]E[\cos(.5\pi k)] + P_v \\
&= 2 + P_v
\end{aligned}
$$

FIGURE 9.11: Excess Mean Square Error of an Adaptive Filter of Order $m = 1$ Using Two Step Sizes, (a) $\mu = .1$, (b) $\mu = .01$

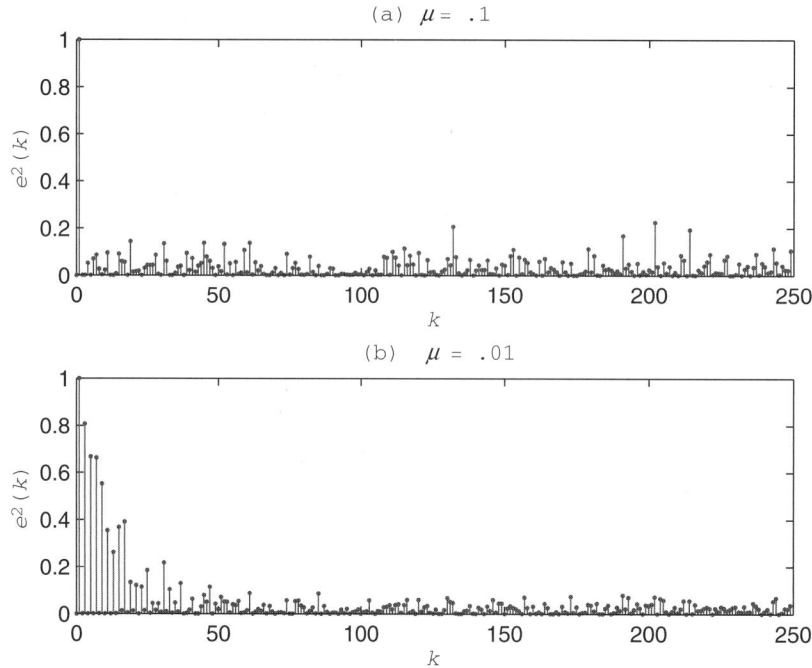

From (9.2.8), the average power of the white noise uniformly distributed over $[-.5, .5]$ is $P_v = (.5)^2/3$. Thus the average power of the input is

$$P_x = \frac{25}{12}$$

If we use (9.4.14) with $m = 1$, the range of step sizes needed for convergence of the LMS method is

$$0 < \mu < .24$$

Next, from (9.4.35), the normalized excess mean square error or misadjustment factor, M_f, is

$$M_f \approx \frac{25\mu}{6}$$

To see the effects of μ, suppose that $w(0) = 0$ and consider two special cases corresponding to $\mu = .1$ and $\mu = .01$. Plots of the squared error can be obtained by running *exam9_7*. Observe from Figure 9.11a that when $\mu = .1$, the squared error converges very rapidly due to the relatively large step size. However, it is apparent that the steady-state error is also relatively large. When $\mu = .01$ in Figure 9.11b, the squared error takes longer to converge, but one is rewarded with a steady-state excess mean square error that is clearly smaller. This illustrates the tradeoff between convergence speed and steady-state accuracy.

The effects of step size, filter order, and input power on the performance characteristics of the LMS method are summarized in Table 9.1.

TABLE 9.1: ▶
Performance
Characteristics of
an Adaptive
Transversal Filter of
Order *m* with Step
Size μ and Input
Power $P_x = E[x^2(k)]$

Property	Value
Convergence range	$0 < \mu < \dfrac{1}{(m+1)P_x}$
Learning curve time constant	$\tau_{\text{mse}} \approx \dfrac{1}{4\mu\lambda_{\min}}$
Misadjustment factor	$M_f \approx \mu(m+1)P_x$

● ● ● ● ● ● ● ● ● ● ● ● ● ● ●●●

9.5 Modified LMS Methods

There are a number of useful modifications that can be made to the LMS method to enhance performance. In this section three variants of the basic LMS method are examined (Kuo and Morgan, 1996).

9.5.1 Normalized LMS Method

Recall from Table 9.1 that the upper bound on μ needed to ensure convergence depends on the filter order m and the input power P_x. A similar observation holds for the step size needed to achieve a given misadjustment factor or excess mean square error. To develop a version of the LMS method that has a step size α that does not depend on the input power or the filter order, the following *normalized LMS* method has been proposed.

$$w(k+1) = w(k) + 2\mu(k)e(k)u(k), \quad k \geq 0 \qquad (9.5.1)$$

Note how the normalized LMS method differs from the basic LMS method in that the step size, $\mu(k)$, is no longer constant. Instead, it varies with time as follows.

$$\mu(k) = \frac{\alpha}{(m+1)\hat{P}_x(k)} \qquad (9.5.2)$$

Here $\hat{P}_x(k)$ is a running estimate of the average power of the input. Observe that if $\hat{P}_x(k)$ is replaced with the exact average power P_x, then from Table 9.1 the range of *constant* step sizes needed to ensure convergence is simply

$$0 < \alpha < 1 \qquad (9.5.3)$$

It is in this sense that the step size has been normalized. The beauty of the normalized approach is that a single value can be used for α, independent of the filter size and the input power.

The simplest way to estimate the average power of the input is to use a rectangular window or running average filter. The following is an Nth-order running-average filter with input $x^2(k)$.

$$\hat{P}_x(k) = \frac{1}{N+1} \sum_{i=0}^{N} x^2(k-i) \qquad (9.5.4)$$

One of the key features the LMS method is its highly efficient implementation. In order to preserve this feature, care must be taken to minimize the number of floating-point operations required at each iteration. The number of multiplications and divisions needed to compute $\hat{P}_x$ in (9.5.4) is $N + 2$. This can be reduced with a recursive formulation of the running-average filter. With the change of variable, $j = i - 1$, (9.5.4) can be rewritten as

$$\hat{P}_x(k) = \frac{1}{N+1} \sum_{j=-1}^{N-1} x^2(k-1-j)$$

$$= \frac{1}{N+1} \left[\sum_{j=0}^{N} x^2(k-1-j) \right] + \frac{x^2(k) - x^2(k-N)}{N+1}$$

$$= \hat{P}_x(k-1) + \frac{x^2(k) - x^2(k-N)}{N+1} \qquad (9.5.5)$$

The recursive formulation in (9.5.5) reduces the number of FLOPs per iteration from $N+2$ to three. Although the implementation in (9.5.5) is faster than the one in (9.5.4), it is not more efficient in terms of memory requirements because $N + 1$ samples of the input still have to be stored. Recall from (9.5.1) that $m + 1$ samples of the input are already being stored in the form of the vector u. If $N = m$, then the following dot product can be used instead to estimate the average power of the input.

$$\hat{P}_x(k) = \frac{u^T(k)u(k)}{m+1} \qquad (9.5.6)$$

This approach has the advantage that no additional values of $x(k)$ need to be stored. Substitution of (9.5.6) into (9.5.2) results in a further simplification because the $m + 1$ factors cancel. This yields the following simplified expression for the time-varying step size (Slock, 1993).

$$\mu(k) = \frac{\alpha}{u^T(k)u(k)} \qquad (9.5.7)$$

There is one additional practical difficulty that can arise when a nonstationary input is used. If the input $x(k)$ is zero for $m + 1$ consecutive samples, then $u(k) = 0$ and the step size in (9.5.7) becomes unbounded. This problem also occurs when the algorithm starts up if $u(0) = 0$. To avoid this numerical difficulty, let δ be a small positive value. Then the step size will never be larger than α/δ if the following modified step size is used for the normalized LMS method.

Normalized LMS method

$$\mu(k) = \frac{\alpha}{\delta + u^T(k)u(k)} \qquad (9.5.8)$$

Example 9.8   **Normalized LMS Method**

To illustrate how the step size changes with the normalized LMS method, suppose the input is the following amplitude-modulated sine wave.

$$x(k) = \cos\left(\frac{\pi k}{100}\right) \sin\left(\frac{\pi k}{5}\right)$$

Next, suppose that the desired output $d(k)$ is produced by applying $x(k)$ to the following second-order IIR resonator filter.

$$H(z) = \frac{1 - z^{-2}}{1 + z^{-1} + .9z^{-2}}$$

FIGURE 9.12:
Normalized LMS
Method with $m = 5$,
$\alpha = .5$, and $\delta = .05$,
(a) Input, (b) Step
Size, (c) Squared
Error

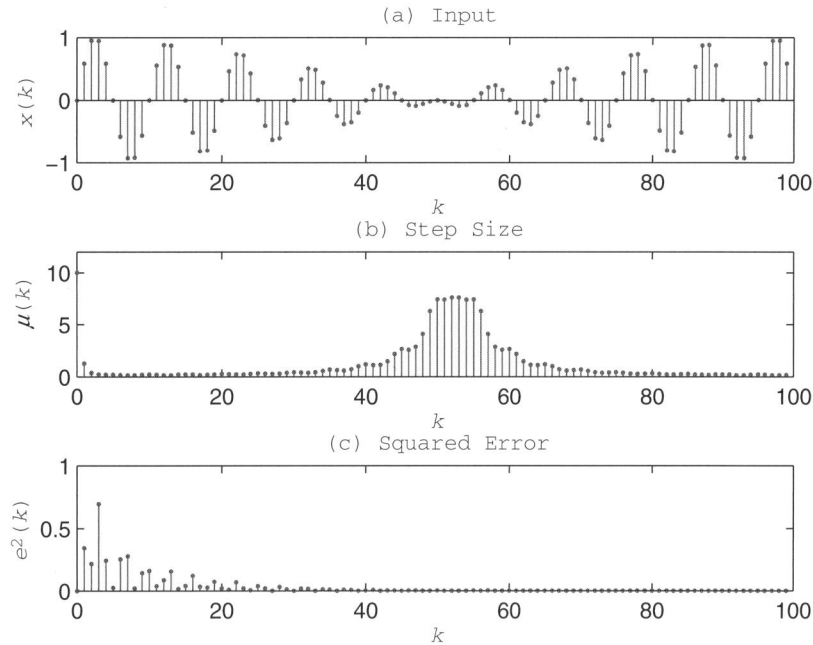

Let the order of the adaptive filter be $m = 5$, and suppose that the normalized step size is $\alpha = .5$. If $\delta = .05$, then the maximum step size is $\alpha/\delta = 10$. Plots of the input $x(k)$, the squared error $e^2(k)$, and the variable step size $\mu(k)$, obtained by running *exam9_8*, are shown in Figure 9.12. Notice that the amplitude-modulated input signal reaches its minimum value at $k = 50$ in Figure 9.12a where the cosine factor is zero. The step size hits its peak value in Figure 9.12b somewhat later due to the delay in the running average estimate of the input power. At $k = 0$ the step size saturates at $\mu(0) = \alpha/\delta$. Observe that in spite of the increase in step size due to a loss of input signal power, the squared error in Figure 9.12c converges and remains small even when the step size increases.

9.5.2 Correlation LMS Method

Recall from Table 9.1 that the learning-curve time constant is inversely proportional to the step size. Consequently, the step size μ should be relatively large to ensure rapid convergence. However, the excess mean square error following convergence is directly proportional to μ which means that μ should be relatively small to improve steady-state accuracy. To avoid this tradeoff, one might use a large step size while convergence is occurring, and then a small step size once convergence has been achieved. The essential task, then, is to detect when convergence has taken place. It is not realistic to use w^* or ϵ_{min} to detect convergence because of the computational burden involved. Instead, a less direct means must be employed. Suppose $w(k)$ has converged to w^*, and consider the product of the error $e(k)$ with the vector of past inputs $u(k)$. If we recall that $u^T(k)w$ is a scalar,

$$e(k)u(k) = [d(k) - y(k)]u(k)$$
$$= [d(k) - u^T(k)w]u(k)$$
$$= d(k)u(k) - u(k)u^T(k)w \qquad (9.5.9)$$

Taking the expected value of both sides of (9.5.9), and evaluating the result at the optimal weight, $w^* = R^{-1}p$, then yields

$$
\begin{aligned}
E[e(k)u(k)] &= E[d(k)u(k)] - E[u(k)u^T(k)]w^* \\
&= p - Rw^* \\
&= 0
\end{aligned}
\tag{9.5.10}
$$

Given the definition of $u(k)$ in (9.2.2), the expected value of $e(k)u(k)$ can be interpreted as a cross-correlation of the error with the input. In particular, using Definition 9.1 and (9.2.2), one can rewrite (9.5.10) as

$$
r_{ex}(i) = 0, \quad 0 \le i \le m
\tag{9.5.11}
$$

From (9.5.11) it is evident that, when the weight vector is optimal, the error is *uncorrelated* with the input. This is an instance of a more general principle that says when an optimal solution is found, the error in the solution is orthogonal to the data on which the solution is based. From the special case $i = 0$ in (9.5.11) one can obtain the following scaler relationship which holds when the LMS method has converged.

$$
E[e(k)x(k)] = 0
\tag{9.5.12}
$$

The basic idea behind the correlation LMS method (Shan and Kailath, 1988) is to choose a step size that is directly proportional to the magnitude of $E[e(k)x(k)]$. This way, the step size will become small when the LMS method has converged, but will be larger during the convergence process. One way to estimate the expected value in (9.5.12) is to use a running-average filter with input $e(k)x(k)$. However, this would mean increased storage requirements because the samples of $e(k)$ are not already stored. A less expensive alternative approach to approximating $E[e(k)x(k)]$ is to use a first-order lowpass IIR filter with the following transfer function.

$$
H(z) = \frac{(1-\beta)z}{z-\beta}
\tag{9.5.13}
$$

The scalar $0 < \beta < 1$ is called a *smoothing* parameter, and typically $\beta \approx 1$. The filter output will be an estimate of $E[e(k)x(k)]$ using an exponentially weighted average. The equivalent width of the exponential window is $N = 1/(1-\beta)$ samples. If $r(k)$ is the filter output, and $e(k)x(k)$ is the filter input, then

$$
r(k+1) = \beta r(k) + (1-\beta)e(k)x(k)
\tag{9.5.14}
$$

Since $r(k)$ becomes small once convergence has taken place, the step size is made proportional to $|r(k)|$ using a proportionality constant or *relative* step size of $\alpha > 0$. This results in the following time-varying step size for the *correlation LMS method*.

Correlation LMS method

$$
\mu(k) = \alpha|r(k)|
\tag{9.5.15}
$$

Sleep mode

The correlation LMS method can be interpreted as having two modes of operation. When convergence has been achieved, the step size becomes small and the algorithm is in the *sleep* mode. Because $\mu(k)$ is small, the excess mean square error is also small. Furthermore, if $y(k)$ contains measurement noise that is uncorrelated with $x(k)$, this noise will not cause an increase in $\mu(k)$. However, if $d(k)$ or $x(k)$ change significantly, this causes $|r(k)|$ to increase and the algorithm then enters the *active* or tracking mode characterized by an increased step size. Once the algorithm converges to the new optimal weight, the step size decreases again and the algorithm reenters the sleep mode.

Example 9.9	**Correlation LMS Method**

To illustrate how the correlation method detects convergence and changes modes of operation, let the input be N samples of white noise uniformly distributed over the interval $[-1, 1]$. Consider the feedback system shown in Figure 9.13. Suppose the open-loop transfer function (when the switch is open) is

$$H_{\text{open}}(z) = \frac{D(z)}{X(z)}$$
$$= G(z)$$
$$= \frac{1.28}{z^2 - .64}$$

The system in Figure 9.13 is a time-varying linear system because the switch in the feedback path starts out closed but opens starting at sample $k = N/2$. This might correspond, for example, to a feedback sensor malfunctioning. When the switch is closed, the Z-transform of the intermediate signal $q(k)$ is

$$Q(z) = X(z) - D(z)$$
$$= X(z) - G(z)Q(z)$$

Solving for $Q(z)$ yields

$$Q(z) = \frac{X(z)}{1 + G(z)}$$

It then follows that, when the switch is closed, the output is

$$D(z) = G(z)Q(z)$$
$$= \frac{G(z)X(z)}{1 + G(z)}$$

FIGURE 9.13: A Time-varying Feedback System Where the Switch Opens at $k = N/2$

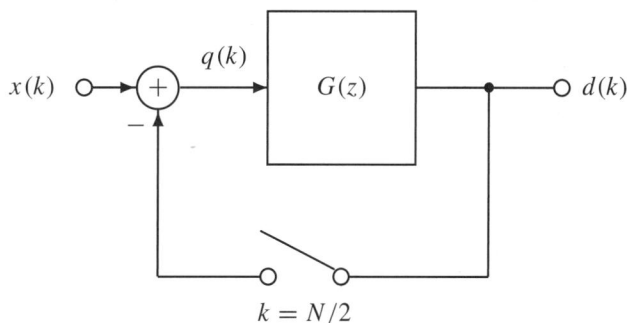

FIGURE 9.14:
Identification of a
Time-varying
Feedback System
Using the
Correlation LMS
Method with
$m = 25$, $\alpha = .5$, and
$\beta = .95$, (a) Squared
Error, (b) Step Size

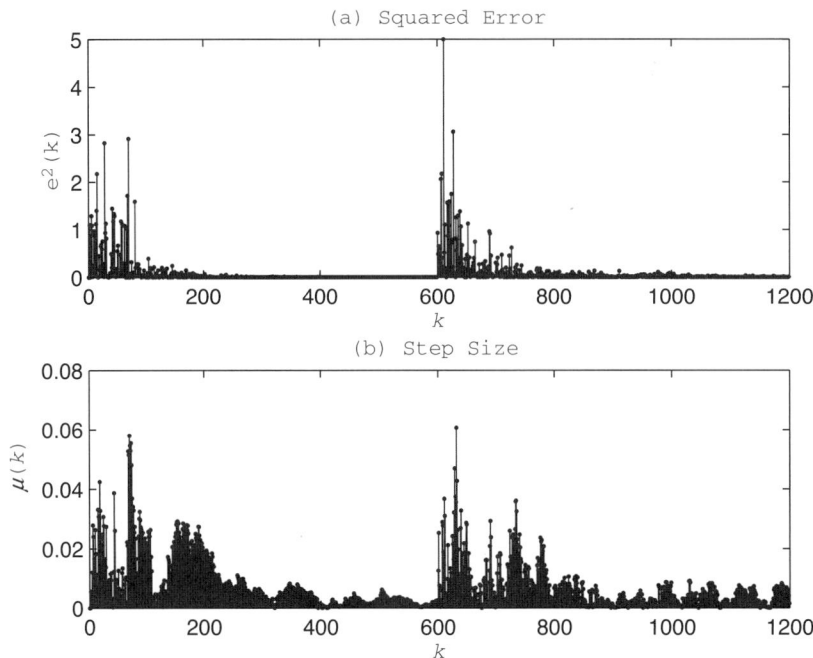

Thus the closed-loop transfer function of the system in Figure 9.13, with the switch closed, is

$$
\begin{aligned}
H_{\text{closed}}(z) &= \frac{D(z)}{X(z)} \\
&= \frac{G(z)}{1 + G(z)} \\
&= \frac{1.28/(z^2 - .64)}{1 + 1.28/(z^2 - .64)} \\
&= \frac{1.28}{z^2 + .64}
\end{aligned}
$$

Consequently, when the switch is closed for samples $0 \le k < N/2$, the system has an imaginary pair of poles at $z = \pm j.8$. At time $k = N/2$, the switch opens and the transfer function reduces to $G(z)$ with real poles $z = \pm.8$. Suppose that this time-varying system is identified with an adaptive filter of order $m = 25$ using the correlation LMS method. Let the relative step size be $\alpha = .5$, and the smoothing factor be $\beta = .95$. A plot of the squared error and the step size, obtained by running *exam9_9*, is shown in Figure 9.14. Observe how the algorithm converges in about 200 samples in Figure 9.14a and enters the sleep mode around 400 samples in Figure 9.14b with a very small step size. At sample $k = 600$, the desired output abruptly changes, and the step size increases, indicating that the active or tracking mode has been entered. The algorithm reconverges around 900 samples and then reenters the sleep mode, this time with a somewhat larger resting step size.

9.5.3 Leaky LMS Method

White noise is a highly effective input for system identification because it has a flat power density spectrum and therefore excites all of the natural modes of the system being identified.

When an input with poor spectral content is used, the auto-correlation matrix R can become singular and the LMS method can diverge with one or more elements of the weight vector growing without bound. An elegant way to guard against this possibility is to introduce a second term in the mean square error objective function. Let $\gamma > 0$, and consider the following *augmented MSE*.

Augmented MSE

$$\epsilon_\gamma[w(k)] \triangleq E[e^2(k)] + \gamma w^T(k)w(k) \tag{9.5.16}$$

Penalty function

The last term in (9.5.16) is called a *penalty function* term because the minimization process tends to penalize any selection of w for which $w^T w$ is large. In this way, the search for a minimum automatically avoids solutions for which $\|w\|$ is large. The parameter $\gamma > 0$ controls how severe the penalty is, and when $\gamma = 0$, the objective function in (9.5.16) reduces to the original mean square error $\epsilon[w(k)]$.

To see what effect the penalty term has on the LMS algorithm, consider the gradient vector of partial derivatives of ϵ_γ with respect to the elements of w. If we use the assumption $E[e^2(k)] \approx e^2(k)$ from (9.3.6) to compute an estimate of the gradient

$$
\begin{aligned}
\hat{\nabla}\epsilon_\gamma(w) &= 2e\frac{\partial e}{\partial w} + 2\gamma w \\
&= -2e\frac{\partial y}{\partial w} + 2\gamma w \\
&= -2eu + 2\gamma w
\end{aligned}
\tag{9.5.17}
$$

Substituting this estimate for the gradient into the steepest-descent method in (9.3.3) then yields the following weight-update formula

$$
\begin{aligned}
w(k+1) &= w(k) - \mu[2\gamma w(k) - 2e(k)u(k)] \\
&= (1 - 2\mu\gamma)w(k) + 2\mu e(k)u(k)
\end{aligned}
\tag{9.5.18}
$$

Leaky LMS method

Finally, define $\nu \triangleq 1 - 2\mu\gamma$. Substituting ν into (9.5.18) then results in the following simplified formulation called the *leaky LMS method*.

$$w(k+1) = \nu w(k) + 2\mu e(k)u(k), \quad i \geq 0 \tag{9.5.19}$$

Leakage factor

The composite parameter ν is called the *leakage* factor. Note that when $\nu = 1$, the leaky LMS method reduces to the basic LMS method. If $u(k) = 0$, then $w(k)$ "leaks" to zero at the rate $w(k) = \nu^k w(0)$. Typically, the leakage factor is in the range $0 < \nu < 1$ with $\nu \approx 1$. It can be shown that including a leakage factor has the same effect as adding low-level white noise to the input (Gitlin et al., 1982). This makes the algorithm more stable for a variety of inputs. However, it also means that there is a corresponding increase in the excess mean square error due to the presence of the penalty term. Bellanger (1987) has shown that the excess mean square error is proportional to $(1 - \nu)^2/\mu^2$, which means that this ratio must be kept small. Since $\nu = 1 - 2\gamma\mu$, this is equivalent to $4\gamma^2 \ll 1$ or

$$\nu = 1 - 2\mu\gamma \tag{9.5.20a}$$

$$\gamma \ll .5 \tag{9.5.20b}$$

It is of interest to note that, because $y(k) = w^T(k)u(k)$, limiting $w^T(k)w(k)$ has the effect of limiting the magnitude of the output. This can be useful in applications such as active noise control where a large $y(k)$ can overdrive a speaker and distort the sound (Elliott et al., 1987).

FIGURE 9.15:
Prediction of the
Input Signal

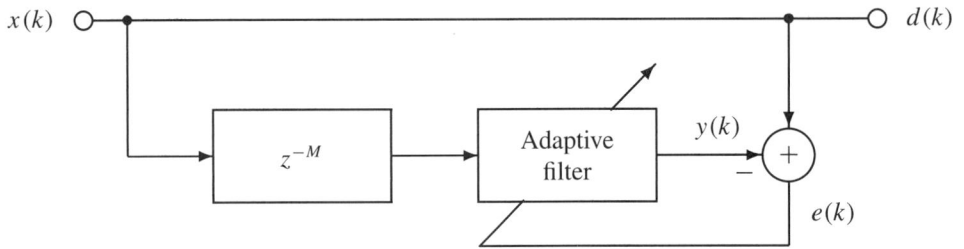

Example 9.10

Leaky LMS Method

To illustrate the use of the leaky LMS method, consider the problem of using an adaptive filter to predict the value the input signal as shown in Figure 9.15. Suppose the input consists of two sinusoids plus noise.

$$x(k) = 2\sin\left(\frac{\pi k}{12}\right) + 3\cos\left(\frac{\pi k}{4}\right) + v(k)$$

Here $v(k)$ is white noise uniformly distributed over the interval $[-.2, .2]$. Next, suppose it is desired to predict the value of the input $M = 10$ samples into the future. Let the order of the adaptive filter be $m = 20$, and the step size be $\mu = .002$. To keep the excess mean square error associated with leakage small, one needs a leakage factor of $v = 1 - 2\mu\gamma$ where $\gamma \ll .5$. Setting $\gamma = .05$ yields $v = .9998$. Running *exam9_10* from *f_dsp* produces the plots shown in Figure 9.16. Observe in Figure 9.16c that after about 40 samples, the algorithm has converged. The filter output $y(k)$ in Figure 9.15b is an effective approximation of the input $x(k)$ in Figure 9.16a, but advanced by $M = 10$ samples.

FIGURE 9.16:
Prediction $M = 10$
Samples Ahead
Using the Leaky
LMS Method with
$m = 20$, $\mu = .002$,
and $v = .9998$, (a)
Input, (b) Output,
(c) Squared Error

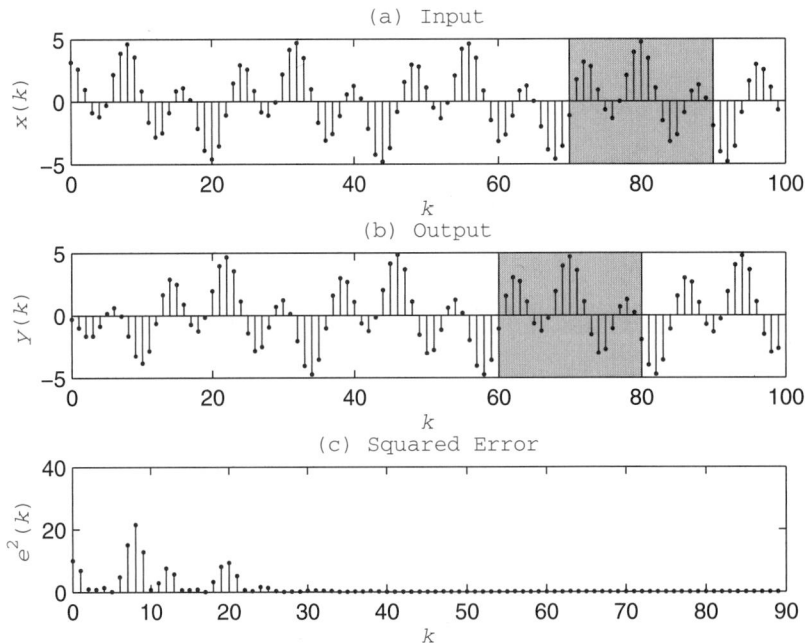

The normalized, correlation, and leaky LMS methods are three popular variants of the basic LMS method. There are other modifications that have been proposed (Kuo and Morgan, 1996). For example, the step size μ can be replaced by a diagonal time-varying step-size matrix $M(k)$ where the ith diagonal element is $\mu_i(k)$. This modification, with each dimension having its own step size, is called the *variable step size* LMS method. Other modifications, intended to increase speed at the expense of accuracy, include the *signed* LMS methods which replace $e(k)$ with sgn$[e(k)]$ or $u(k)$ with sgn$[u(k)]$, where *sgn* is the sign or signum function.

FDSP Functions

The FDSP toolbox contains the following functions which correspond to modified versions of the basic LMS method.

```
% F_LMSNORM: System identification using normalized LMS method
% F_LMSCORR: System identification using correlation LMS method.
% F_LMSLEAK: System identification using leaky LMS method
%
% Usage:
%       [w,e,mu] = f_lmsnorm (x,d,m,alpha,delta,w);
%       [w,e,mu] = f_lmscorr (x,d,m,alpha,beta,w);
%       [w,e]    = f_lmsleak (x,d,m,mu,nu,w);
% Pre:
%       x     = N by 1 vector containing input samples
%       d     = N by 1 vector containing desired output
%               samples
%       m     = order of transversal filter (m >= 0)
%       alpha = normalized step size (0 to 1)
%       delta = an optional positive scalar controlling
%               the maximum step size which is mu  =
%               alpha/delta. Default: alpha/100
%       w     = an optional (m+1) by 1 vector containing
%               the initial values of the weights.
%               Default: w = 0
%       beta  = an scalar containing the smoothing
%               parameter. beta is approximately one
%               with 0 < beta < 1. Default: 1 - 0.5/(m+1)
%       mu    = step size to use for updating w
%       nu    = an optional leakage factor in the range 0 to 1.
%               Pick nu = 1 - 2*mu*gamma where gamma << 0.5.
%               (default: 1 - 0.1*mu).
% Post:
%       w  = (m+1) by 1 weight vector of filter
%            coefficients
%       e  = an optional N by 1 vector of errors where
%            e(k) = d(k)-y(k)
%       mu = an optional N by 1 vector of step sizes
```

Continued on p. 678

Continued from p. 677

```
% Notes:
%          1. When nu = 1, the leaky LMS method reduces to the basic
%             LMS method.
%          2. Typically mu << 1/[(m+1)*P_x] where P_x is the
%             average power of input x.
```

● ● ● ● ● ● ● ● ● ● ● ● ● ● ● ●

9.6   Adaptive FIR Filter Design

9.6.1   Pseudo-filters

Recall from Chapter 6 that most FIR filters are designed to have prescribed magnitude responses and linear-phase responses. Filters with desired magnitude and phase characteristics, plus a group delay, can be designed using the quadrature method in Chapter 6. As an alternative, the LMS method can be used to compute filter coefficients. The basic idea behind this approach is to use a synthetic *pseudo-filter* to generate the desired output as shown in Figure 9.17.

As the name implies, a pseudo-filter is a fictional linear discrete-time system that may or may not have a physical realization. A pseudo-filter is characterized implicitly by a desired relationship between a periodic input and a steady-state output. Let T be the sampling interval, and suppose that the input consists of a sum of N sinusoids as follows.

$$x(k) = \sum_{i=0}^{N-1} C_i \cos(2\pi f_i kT) \qquad (9.6.1)$$

The only constraints on the discrete frequencies $\{f_0, f_1, \cdots, f_{N-1}\}$ are that they be distinct from one another, and that they lie within the Nyquist range $0 \le f_i < f_s/2$. For example, the discrete frequencies are often taken to be uniformly spaced with

$$f_i = \frac{i f_s}{2N}, \quad 0 \le i < N \qquad (9.6.2)$$

Relative weights   The amplitudes or *relative weights*, $C_i > 0$, are selected based on the importance of each frequency in the overall design specification. For example, if $C_k > C_i$ then frequency f_k will

FIGURE 9.17:
Adaptive Filter
Design Using a
Pseudo-Filter

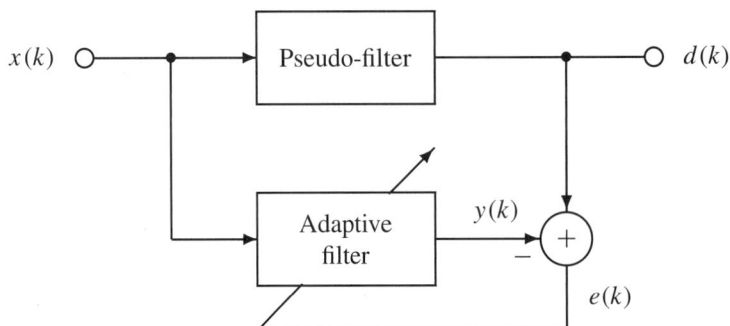

be given more weight than frequency f_i. As a starting point, one can use *uniform* weighting with

$$C_i = 1, \quad 0 \leq i < N \tag{9.6.3}$$

Pseudo-filter output

Once the input is selected, the desired steady-state output is then specified. Since the artificial pseudo-filter is assumed to be linear, the steady-state output will be periodic with the same period as the input.

$$d(k) = \sum_{i=0}^{N-1} A_i C_i \cos(2\pi f_i kT + \phi_i) \tag{9.6.4}$$

Here A_i and ϕ_i denote the desired gain and phase shift, respectively, at frequency f_i. Thus the design specification consists of a set of four $N \times 1$ vectors, $\{f, C, A, \phi\}$. Here f is the frequency vector, C is the relative weight vector, A is the magnitude vector, and ϕ is the phase vector. In this way, N samples of the desired frequency response, both magnitude and phase, can be specified.

It should be emphasized that the implicit relationship between $x(k)$ and $d(k)$ represents a pseudo-filter because the magnitude $A_i = A(f_i)$ and phase $\phi_i = \phi(f_i)$ of a causal discrete-time system are *not* independent of one another. For a causal filter the real and imaginary parts of the frequency response are interdependent, and so are the magnitude and phase (Proakis and Manolakis, 1992). Consequently, one can not independently specify both the magnitude and the phase and expect to obtain an exact fit using a causal linear filter. Instead, an optimal approximation to the pseudo-filter specifications using a causal adaptive transversal filter of order m can be obtained.

When the order of the adaptive filter is relatively small, the $(m + 1) \times 1$ weight vector w can be computed offline by solving $Rw = p$. Given $x(k)$ and $d(k)$, closed form expressions for the input auto-correlation matrix R and the cross-correlation vector p can be obtained (Problems 9.1, 9.16). This direct approach can become computationally expensive (and sensitive to roundoff error) as m becomes large. In these instances it makes more sense to numerically search for an optimal w using the LMS method.

$$w(k + 1) = w(k) + 2\mu e(k)u(k) \tag{9.6.5}$$

Since the objective is to find a fixed FIR filter of order m that best fits the design specifications, the adaptive filter is allowed to run for $M \gg 1$ iterations until the squared error has converged to its steady-state value. The numerator coefficient vector of the fixed FIR filter, $W(z)$, is then set to the final steady-state value of the weights.

$$b = w(M - 1) \tag{9.6.6a}$$

$$W(z) = \sum_{i=0}^{m} b_i z^{-i} \tag{9.6.6b}$$

The frequency response of the FIR filter is then computed and compared with the pseudo-filter specifications. If the fit is acceptable, then the process terminates. Otherwise the order m can be increased or the relative weights C can be changed at those frequencies where the error is largest. The overall design process is summarized in the flowchart in Figure 9.18.

FIGURE 9.18:
Adaptive FIR Filter
Design Process

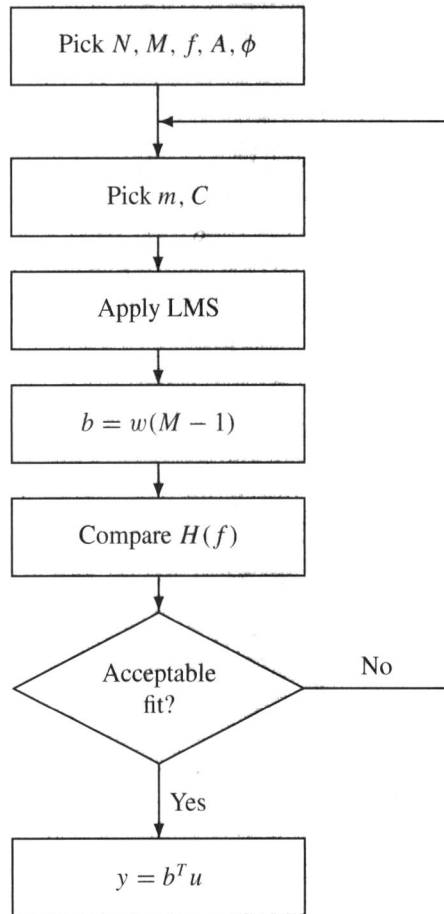

Example 9.11 **Adaptive FIR Filter Design**

To illustrate the use of a pseudo-filter, consider the problem of designing an FIR filter with the following desired magnitude response.

$$A(f) = \begin{cases} 1.5 - \dfrac{6f}{f_s}, & 0 \le f_s < \dfrac{f_s}{6} \\[2mm] .5, & \dfrac{f_s}{6} \le f < \dfrac{f_s}{3} \\[2mm] .5 + .5\sin\left(\dfrac{6\pi f}{f_s}\right), & \dfrac{f_s}{3} \le f < \dfrac{f_s}{2} \end{cases}$$

Suppose that there are $N = 60$ discrete frequencies uniformly distributed as in (9.6.2). Let the relative weights also be uniform as in (9.6.3). Suppose that the order of the adaptive FIR filter is $m = 30$, and the step size is $\mu = .0001$. Let the LMS method run for $M = 2000$ iterations starting from an initial guess of $w(0) = 0$. Two cases are considered. For the first case, the desired phase response of the pseudo-filter is simply

$$\phi(f) = 0$$

A comparison of the desired and actual frequency responses can be obtained by running *exam9_11* with the results shown in Figure 9.19. Notice that the fit is rather poor because of

FIGURE 9.19:
Frequency
Responses of a
Zero-phase
Pseudo-filter and
an FIR Filter of
Order $m = 30$,
(a) Magnitude
Responses,
(b) Phase Responses

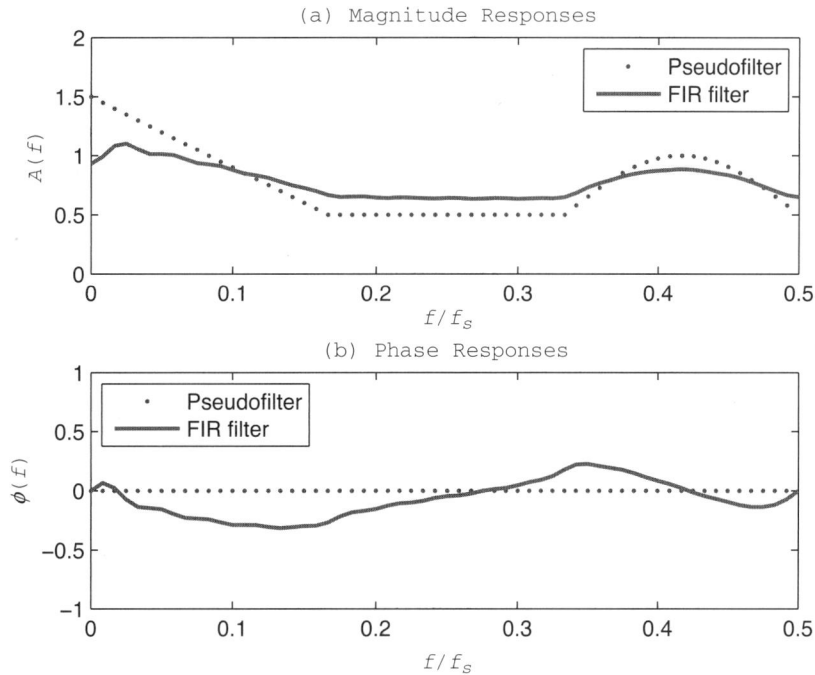

the unrealistic design specification of zero-phase shift. Recall from Section 5.3 that zero-phase filters can be implemented, but only if noncausal filters are employed. The fit can be improved somewhat by increasing m. Alternatively, a linear-phase pseudo-filter with a constant group delay can be specified. If the group delay is set to $\tau = mT/2$, then this corresponds to a phase response of

$$\phi(f) = -m\pi f$$

This combination of magnitude and phase is much easier to synthesize with a causal FIR filter, as can be seen by the results shown in Figure 9.20. Notice that there is a good fit of both the magnitude response in Figure 9.20a and the linear phase response in Figure 9.20b.

9.6.2 Linear-phase Pseudo-filters

The special case of linear-phase filters is important because it corresponds to each spectral component of $x(k)$ being delayed by the same amount as it is processed by the filter. Hence there is no phase distortion of the input, only a delay. Recall from Table 5.1 that an FIR filter of order m with coefficient vector b is a type-1 linear-phase filter if m is even and the filter coefficients satisfy the following even symmetry condition.

$$b_i = b_{m-i}, \quad 0 \le i \le m \tag{9.6.7}$$

A signal flow graph of a transversal filter satisfying this linear-phase symmetry condition is shown in Figure 9.21 for the case $m = 4$. The signal flow graph in Figure 9.21 can be made more efficient by combining branches with identical weights. This results in the equivalent signal flow graph shown in Figure 9.22 which features fewer floating-point multiplications. To

FIGURE 9.20:
Frequency
Responses of a
Linear-phase
Pseudo-filter and
an FIR Filter of
Order $m = 30$,
(a) Magnitude
Responses,
(b) Phase Responses

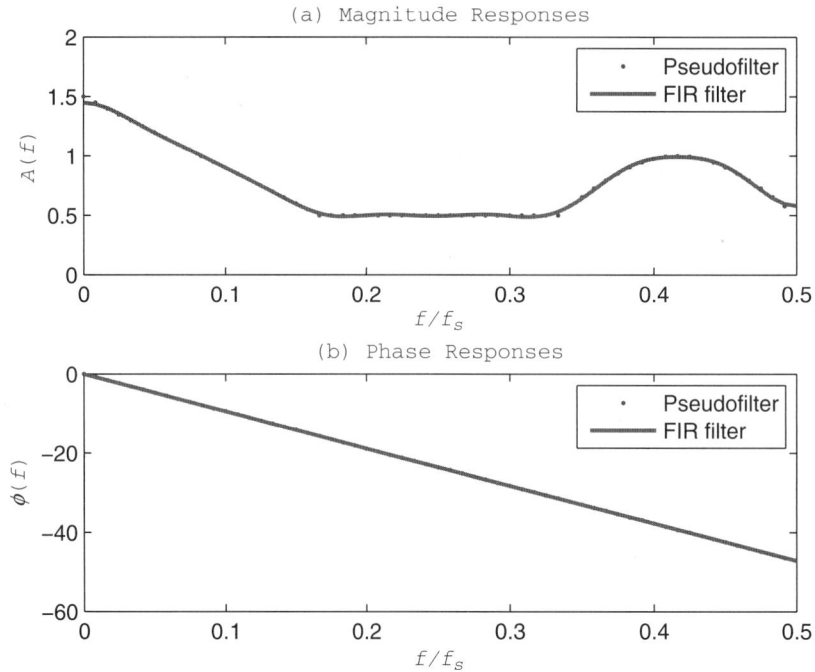

develop a concise formulation of the filter output, consider the following pair of $(m/2+1) \times 1$ vectors.

$$\hat{w}(k) \overset{\Delta}{=} [w_0(k), w_1(k), \cdots, w_{m/2}(k)]^T \tag{9.6.8a}$$

$$\hat{u}(k) \overset{\Delta}{=} [x(k-m/2), x(k-m/2-1) + x(k-m/2+1), \cdots,$$
$$x(k-m) + x(k)]^T \tag{9.6.8b}$$

Here $\hat{w}(k)$ consists of the first $m/2+1$ weights, while $\hat{u}(k)$ is constructed from pairs of the past inputs. If we apply (9.6.8) to Figure 9.22, the linear-phase transversal filter has the following compact representation using the dot product.

$$y(k) = \hat{w}(k)^T \hat{u}(k), \quad k \geq 0 \tag{9.6.9}$$

Note that for large values of m, the linear-phase formulation in (9.6.9) requires approximately half as many floating-point multiplications or FLOPs as the standard representation in (9.2.3). When the LMS method is applied to the linear-phase transversal structure, there is a similar savings in computational effort using

$$\hat{w}(k+1) = \hat{w}(k) + 2\mu e(k)\hat{u}(k), \quad k \geq 0 \tag{9.6.10}$$

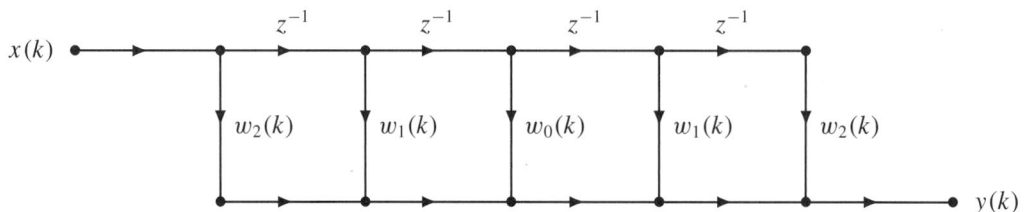

FIGURE 9.21: A Type-1 Linear-phase Transversal Filter of Order $m = 4$

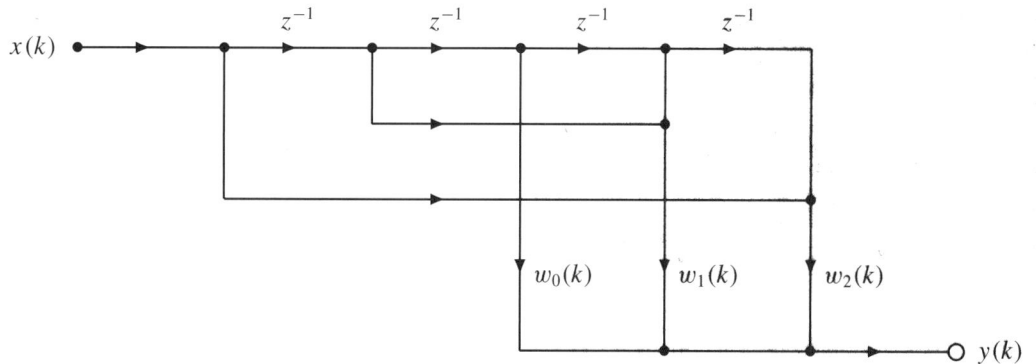

FIGURE 9.22: Equivalent Linear-phase Transversal Filter of Order $m = 4$

Example 9.12 **Adaptive Linear-phase FIR Filter Design**

To illustrate the design of a linear-phase FIR filter, consider a pseudo-filter with the following piecewise-continuous magnitude response specification.

$$A(f) = \begin{cases} \left(\dfrac{6f}{f_s}\right)^2, & 0 \le f_s < \dfrac{f_s}{6} \\ .5, & \dfrac{f_s}{6} \le f < \dfrac{f_s}{3} \\ \left(\dfrac{1 - 2(3f - f_s)}{f_s}\right)^2, & \dfrac{f_s}{3} \le f < \dfrac{f_s}{2} \end{cases}$$

Suppose that there are $N = 90$ discrete frequencies uniformly distributed as in (9.6.2). Let the order of the FIR filter be $m = 45$, and the step size be $\mu = .0001$. Suppose that the LMS method runs for $M = 2000$ iterations starting from an initial guess of $w(0) = 0$. Again two cases are considered. For the first case, the relative weighting of the discrete frequencies is uniform as in (9.6.3). A comparison of magnitude responses, obtained by running *exam9_12*, is shown in Figure 9.23. Observe that the two responses are roughly similar, but that substantial

FIGURE 9.23: Magnitude Responses of a Pseudo-filter and a Linear-phase FIR Filter of Order $m = 45$ Using Uniform Relative Weighting

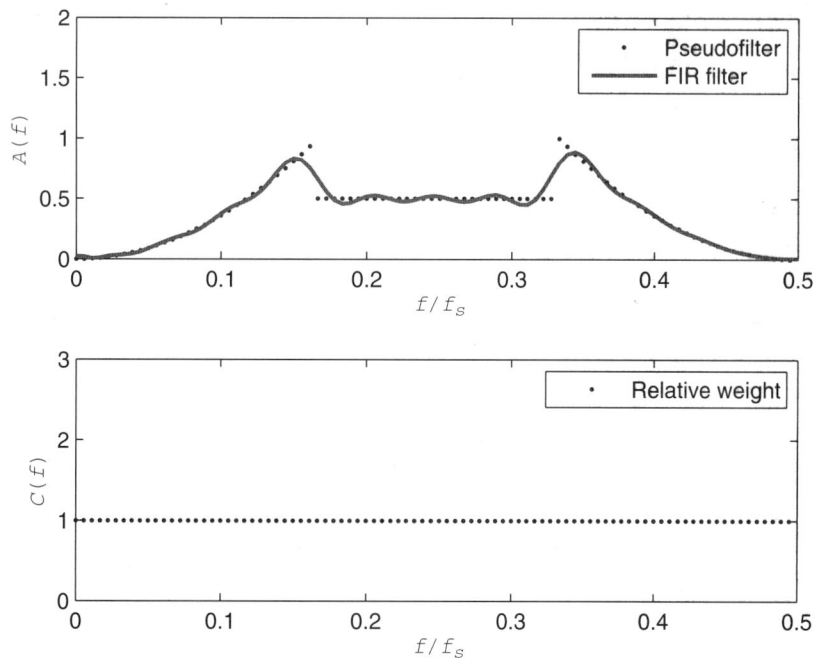

FIGURE 9.24:
Magnitude
Responses of a
Pseudo-filter and a
Linear-phase FIR
Filter of Order
$m = 45$ Using
Nonuniform
Relative Weighting

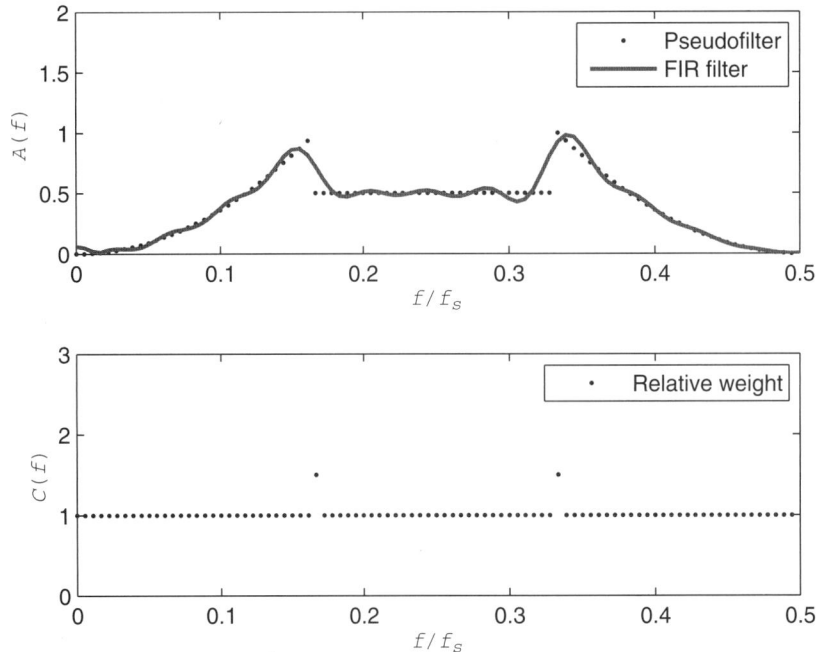

error occurs at multiples of $f_s/6$, where the desired magnitude response in Figure 9.23a has jump discontinuities. The fit in the vicinity of these frequencies can be improved by using a nonuniform relative weighting. Setting $C_{M/6} = C_{M/3} = 1.5$ results in the magnitude response plot shown in Figure 9.24, which is somewhat more accurate near the jump discontinuities.

9.7 The Recursive Least-Squares (RLS) Method

There is a popular alternative to the LMS method called the recursive least-squares or RLS method (Treichler et al., 2001; Haykin, 2002). The RLS method typically converges much faster than the LMS method, but at a cost of more computational effort per iteration.

9.7.1 Performance Criterion

To formulate the RLS method, consider a modification of the least-squares performance criterion introduced in Section 9.2. Suppose the following more general *time-varying* performance criterion is used.

$$\epsilon_k(w) = \sum_{i=1}^{k} \gamma^{k-i} e^2(i) + \delta \gamma^k w^T w, \quad k \geq 1 \tag{9.7.1}$$

Forgetting factor

Regularization parameter

Here the exponential weighting factor, $0 < \gamma \leq 1$, is also called the *forgetting factor* because when $\gamma < 1$, it has the effect of reducing the contributions from errors in the remote past. The second term in (9.7.1) is a regularization term with $\delta > 0$ called the *regularization parameter*. Note that the second term is similar to the penalty function term in the leaky LMS method in that it tends to prevent solutions for which $w^T w$ grows arbitrarily large. Thus it has the effect of making the RLS method more stable. When $\gamma = 1$ (no exponential weighting) and $\delta = 0$ (no regularization), the performance criterion in (9.7.1) is proportional to the mean square error at time k.

To determine a weight vector w that minimizes $\epsilon_k(w)$, substitute $e(i) = d(i) - w^T u(i)$ into (9.7.1) where $d(i)$ is the desired output, and $u(i)$ is the vector of past inputs. This results in the following more detailed expression for the performance index.

$$
\begin{aligned}
\epsilon_k(w) &= \sum_{i=1}^{k} \gamma^{k-i}[d(i) - w^T u(i)]^2 + \delta\gamma^k w^T w \\
&= \sum_{i=1}^{k} \gamma^{k-i}\{d^2(i) - 2d(i)w^T u(i) + [w^T u(i)]^2\} + \delta\gamma^k w^T w \\
&= \sum_{i=1}^{k} \gamma^{k-i}[d^2(i) - 2w^T d(i)u(i)] + \sum_{i=1}^{k} \gamma^{k-i} w^T u(i)[u^T(i)w] + \delta\gamma^k w^T w \\
&= \sum_{i=1}^{k} \gamma^{k-i}[d^2(i) - 2w^T d(i)u(i)] + \\
&\quad w^T \left(\sum_{i=1}^{k} \gamma^{k-i} u(i)u^T(i) + \delta\gamma^k I \right) w
\end{aligned}
\tag{9.7.2}
$$

The expression for $\epsilon_k(w)$ can be made more concise by introducing the following generalized versions of the auto-correlation matrix and cross-correlation vector, respectively, at time k.

$$
R(k) \triangleq \sum_{i=1}^{k} \gamma^{k-i} u(i)u^T(i) + \delta\gamma^k I
\tag{9.7.3a}
$$

$$
p(k) \triangleq \sum_{i=1}^{k} \gamma^{k-i} d(i)u(i)
\tag{9.7.3b}
$$

Substituting (9.7.3) into (9.7.2), the exponentially weighted regularized performance criterion can be expressed as the following quadratic function of the weight vector.

$$
\epsilon_k(w) = \sum_{i=1}^{k} \gamma^{k-i} d^2(i) - 2w^T p(k) + w^T R(k)w
\tag{9.7.4}
$$

Following the same procedure that was used in Section 9.2, the gradient vector of partial derivatives of $\epsilon_k(w)$ with respect to the elements of w can be shown to be $\nabla\epsilon_k(w) = 2[R(k)w - p(k)]$. Setting $\nabla\epsilon_k(w) = 0$ and solving for w, one arrives at the following expression for the optimal weight at time k.

$$
w(k) = R^{-1}(k)p(k)
\tag{9.7.5}
$$

Unlike the LMS method which asymptotically approaches the optimal weight vector using a gradient-based search, the RLS method attempts to find the optimal weight at each iteration.

9.7.2 Recursive Formulation

Although the weight vector in (9.7.5) is optimal in terms of minimizing $\epsilon_k(w)$, it is apparent that the computational effort required to find $w(k)$ is large, and it grows more burdensome as k increases. Fortunately, there is a way to reformulate the required computations to make them more economical. The basic idea is to start with the solution at iteration $k - 1$ and add a correction term to obtain the solution at iteration k. In this way, the required quantities can be computed *recursively*. For example, using (9.7.3a) one can recast the expression for $R(k)$ as follows.

$$R(k) = \gamma \left(\sum_{i=1}^{k} \gamma^{k-i-1} u(i) u^T(i) + \delta \gamma^{k-1} I \right)$$

$$= \gamma \left(\sum_{i=1}^{k-1} \gamma^{k-i-1} u(i) u^T(i) + \delta \gamma^{k-1} I \right) + u(k) u^T(k) \qquad (9.7.6)$$

Observe from (9.7.3a) that the coefficient of γ in (9.7.6) is just $R(k-1)$. Consequently, the exponentially-weighted and regularized auto-correlation matrix can be computed recursively as follows.

$$R(k) = \gamma R(k-1) + u(k) u^T(k), \quad k \geq 1 \qquad (9.7.7)$$

An initial value for $R(k)$ is required to start the recursion process. Using (9.7.3a), and assuming that the input $x(k)$ is causal, one can set $R(0) = \delta I$.

A similar procedure can be used to obtain a recursive formulation for the generalized cross-correlation vector in (9.7.3b) by factoring out a γ and separating the $i = k$ term. This yields the following recursive formulation for $p(k)$ that can be initialized with $p(0) = 0$.

$$p(k) = \gamma p(k-1) + d(k) u(k), \quad k \geq 1 \qquad (9.7.8)$$

Although the recursive computations of $R(k)$ and $p(k)$ greatly simplify the computational effort for these quantities, there remains the problem of inverting $R(k)$ in (9.7.5) to find $w(k)$. It is this step that dominates the computational effort because the number of floating-point operations or FLOPs needed to solve $R(k)w = p(k)$ is proportional to m^3 where m is the order of the transversal filter. As it turns out, it is also possible to compute $R^{-1}(k)$ recursively. To achieve this, one needs to make use of a result from linear algebra. Let A and C be square *Matrix inversion* nonsingular matrices, and let B and D be matrices of appropriate dimensions. Then the *matrix* *lemma* *inversion lemma* can be stated as follows (Woodbury, 1950; Kailath, 1960).

$$(A + BCD)^{-1} = A^{-1} - A^{-1} B(DA^{-1}B + C^{-1})^{-1} DA^{-1} \qquad (9.7.9)$$

This result is precisely what is needed to express $R^{-1}(k)$ in terms of $R^{-1}(k-1)$. Recalling (9.7.7), let $A = \gamma R(k-1)$, $B = u(k)$, $C = 1$, and $D = u^T(k)$. Then applying the matrix inversion lemma in (9.7.9)

$$R^{-1}(k) = \frac{1}{\gamma} \left[R^{-1}(k-1) - \frac{R^{-1}(k-1) u(k) u^T(k) R^{-1}(k-1)}{\gamma + u^T(k) R^{-1}(k-1) u(k)} \right] \qquad (9.7.10)$$

To simplify the final formulation, it is helpful to introduce the following notational quantities.

$$r(k) \triangleq R^{-1}(k-1) u(k) \qquad (9.7.11a)$$
$$c(k) \triangleq \gamma + u^T(k) r(k) \qquad (9.7.11b)$$

From the expression for $R(k)$ in (9.7.3a) it is evident that $R(k)$ is a symmetric matrix. Since the transpose of the inverse equals the inverse of the transpose, this means that $r^T(k) = u^T(k) R^{-1}(k-1)$. If we substitute (9.7.11) into (9.7.10), the inverse of the generalized auto-correlation matrix can be expressed recursively as follows.

$$R^{-1}(k) = \frac{1}{\gamma}\left[R^{-1}(k-1) - \frac{r(k)r^T(k)}{c(k)}\right], \quad k \geq 1 \qquad (9.7.12)$$

The beauty of the formulation in (9.7.12) is that no explicit matrix inversions are required. Instead, the expression for the inverse is updated at each step using dot products and scalar multiplications. To start the process an initial value for the inverse of the auto-correlation matrix is required. Assuming $x(k)$ is causal in (9.7.3a), let

$$R^{-1}(0) = \delta^{-1}I \qquad (9.7.13)$$

Although this initial estimate of the inverse of the auto-correlation matrix is not likely to be accurate (except perhaps for a white noise input), the exponential weighting associated with *Effective window* $\gamma < 1$ tends to minimize the effects of any initial error in the estimate after a sufficient number *length* of iterations. The *effective window length* associated with the exponential weighting is

$$M = \frac{1}{1-\gamma} \qquad (9.7.14)$$

The steps required to compute the optimal weight at each step using the RLS method are summarized in Algorithm 9.1. To emphasize the fact that no inverses are explicitly computed, the notation Q is used for R^{-1}.

ALGORITHM
....

9.1: RLS Method

1. Pick $0 < \gamma \leq 1, \delta > 0, m \geq 0, N \geq 1$.
2. Set $w = 0$, $p = 0$, and $Q = I/\delta$. Here w and p are $(m+1) \times 1$ and Q is $(m+1) \times (m+1)$.
3. For $k = 1$ to N compute
   {

$$u = [x(k), x(k-1), \ldots, x(k-m)]^T$$
$$r = Qu$$
$$c = \gamma + u^T r$$
$$p = \gamma p + d(k)u$$
$$Q = \frac{1}{\gamma}\left[Q - \frac{rr^T}{c}\right]$$
$$w = Qp$$

   }

The RLS method in Algorithm 9.1 typically converges much faster than the LMS method. However, the computational effort per iteration is larger, even with the efficient recursive formulation. For moderate to large values of the transversal filter order m, the computational effort in Algorithm 9.1 is dominated by the computation of r, w, and Q in step 3. The number of FLOPs required to compute r, w, and the symmetric Q is approximately $3(m+1)^2$. Thus the computational effort is proportional to m^2 for large values of m. This makes the RLS method an algorithm of order $O(m^2)$. This is in contrast to the much simpler LMS method that is an algorithm of order $O(m)$ with the computational effort proportional to m. There are faster versions of the RLS method that exploit recursive formulations of $r(k)$ and $w(k)$ (Ljung et al., 1978).

The design parameters associated with the RLS method are the forgetting factor $0 < \gamma \leq 1$, the regularization parameter $\delta > 0$, and the transversal filter order $m \geq 0$. The required filter order depends on the application and is often found empirically. Haykin (2002) has shown that the parameter $\mu = 1 - \gamma$ plays a role similar to the step size in the LMS method. Therefore, γ should be close to unity to keep μ small. If a given effective window length for the exponential weighting is desired, then the forgetting factor can be computed using (9.7.14). The choice for the regularization parameter depends on the signal-to-noise ratio (SNR) of the input $x(k)$. Moustakides (1997) has shown that when the SNR is high (e.g., 30 dB or higher), the RLS method exhibits fast convergence using the following value for the regularization parameter.

$$\delta = P_x \tag{9.7.15}$$

Here $P_x = E[x^2(k)]$ is the average power of $x(k)$ which is assumed to have zero mean, Otherwise the variance of $x(k)$ should be used. As the SNR of $x(k)$ decreases, the value of δ should be increased.

Example 9.13   RLS Method

To compare the performance characteristics of the RLS and LMS methods, again consider the system identification problem posed in Example 9.3. Here the system to be identified was a sixth-order IIR system with the following transfer function.

$$H(z) = \frac{2 - 3z^{-1} - z^{-2} + 4z^{-4} + 5z^{-5} - 8z^{-6}}{1 - 1.6z^{-1} + 1.75z^{-2} - 1.436z^{-3} + .6814z^{-4} - .1134z^{-5} - .0648z^{-6}}$$

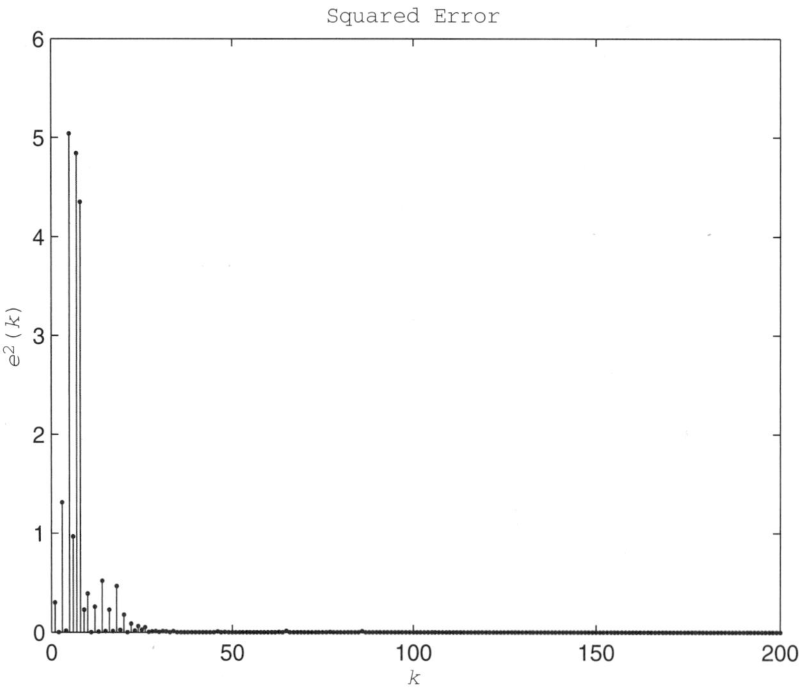

FIGURE 9.25: First 200 Samples of Squared Error During System Identification Using the RLS Method with $m = 50$, $\gamma = .99$, and $\delta = P_x$

FIGURE 9.26:
Magnitude
Responses of the
System, $H(z)$, and
the Identified
Adaptive Model,
$W(z)$, Using the RLS
Method with
$m = 50$, $\gamma = .99$,
and $\delta = P_x$

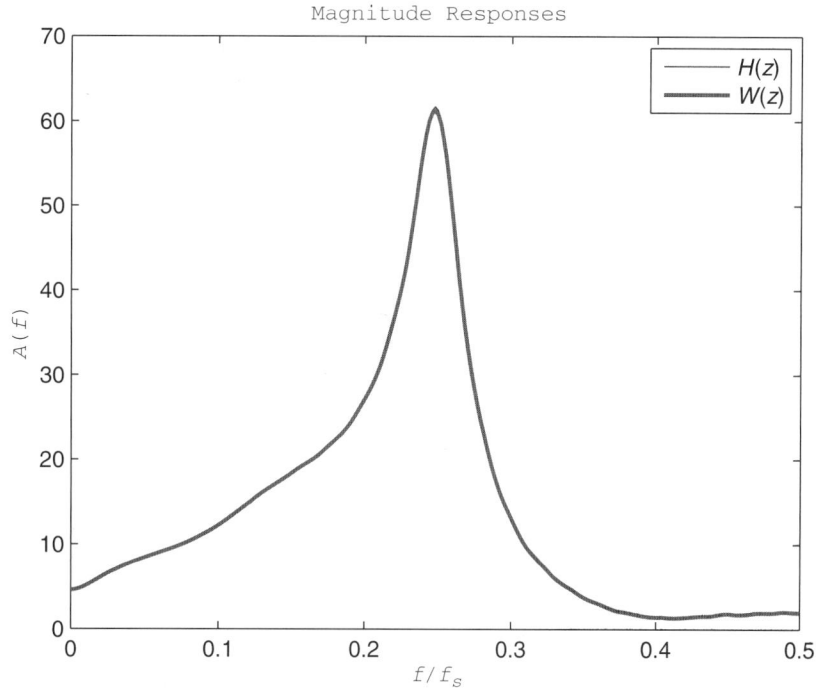

Magnitude Responses

Suppose that the input $x(k)$ consists of $N = 1000$ samples of white noise uniformly distributed over $[-1, 1]$. Let the order of the adaptive transversal filter be $m = 50$, and suppose that a forgetting factor of $\gamma = .98$ is used. The regularization parameter δ is set to the average power of the input as in (9.7.15). A plot of the first 200 samples of the square of the error, obtained by running *exam9_13*, is shown in Figure 9.25. In this case the square of the error converges close to zero after approximately 30 samples. This is in contrast to the LMS method, previously shown in Figure 9.26, which took approximately 400 samples to converge. Thus, when measured in iterations, the RLS method is faster than the LMS method by an order of magnitude. However, it should be kept in mind that the LMS method requires about $m = 50$ FLOPs per iteration, whereas the RLS method requires about $3m^2 = 7500$ FLOPs per iteration. In terms of FLOPs the two methods appear to be roughly equivalent in this case.

Based on (9.7.5), one might expect the RLS method to converge even faster, say, in one iteration. The reason it took approximately 30 iterations to converge was due to the transients associated with the startup of the algorithm. Since $x(k)$ is a causal signal, the vector of past inputs u continues to be populated with zero samples for the first $m = 50$ iterations. The RLS method converges in less than m samples in this instance due to the presence of the forgetting factor $\gamma = .99$, which tends to reduce the influence of samples in the remote past.

An FIR model, $W(z)$, identified with the RLS method is obtained by using the final steady-state estimate for the weight vector, $w(N - 1)$. The magnitude responses of this FIR system and the original system $H(z)$ are plotted in Figure 9.26 where it is evident that they are nearly identical.

FDSP Functions

The FDSP toolbox contains the following function which implements the RLS method in Algorithm 9.1.

```
% F_RLS: System identification using the RLS method
%
% Usage:
%         [w,e] = f_rls (x,d,m,gamma,delta,w)
% Pre:
%         x      = N by 1 vector containing input samples
%         d      = N by 1 vector containing desired output
%                  samples
%         m      = order of transversal filter (m >= 0)
%         gamma  = forgetting factor (0 to 1)
%         delta  = optional regularization parameter
%                  (delta > 0). Default P_x
%         w      = optional initial values of the weights.
%                  Default: w = 0
% Post:
%         w = (m+1) by 1 weight vector of filter
%             coefficients
%         e = an optional N by 1 vector of errors where
%             e(k) = d(k)-y(k)
% Note:
%         As the SNR of x decreases, delta should be
%         increased.
```

• • • • • • • • • • • • • • ••••

9.8 Active Noise Control

One of the emerging application areas of adaptive signal processing is active noise or vibration control. The basic idea behind the active control of acoustic noise is to inject a secondary sound into an environment so as to cancel the primary sound using destructive interference. Application areas include jet engine noise, road noise in automobiles, blower noise in air ducts, transformer noise, and industrial noise from rotating machines (Kuo and Morgan, 1996). Active noise control requires an adaptive filter configuration as shown in Figure 9.27

FIGURE 9.27: Active Control of Acoustic Noise Using an Adaptive Filter

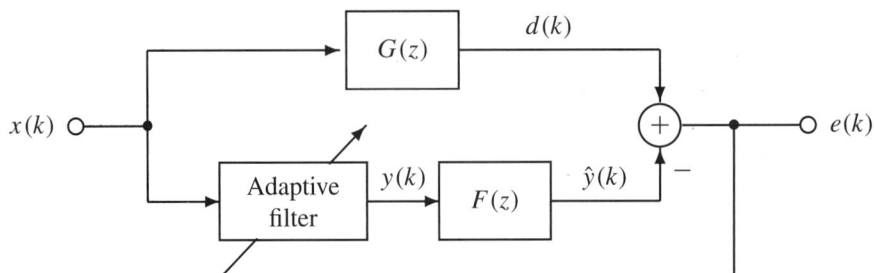

Here the input or *reference signal*, $x(k)$, denotes samples of acoustic noise obtained from a microphone or a non-acoustic sensor. The transfer function $G(z)$ represents the physical characteristics of the air channel over which the noise travels. The primary noise $d(k)$ is combined with secondary noise $y(k)$, produced by the adaptive filter. The secondary noise, also called *anti-noise*, is designed to destructively interfere with the primary noise so as to produce silence at the error microphone $e(k)$. The feature that makes Figure 9.27 different from a standard system identification configuration is the appearance of a secondary path system with transfer function $F(z)$. The secondary system represents the hardware used to produce the secondary sound. It includes such things as a power amplifier, a speaker, the secondary air channel, the error microphone, and a preamp. The system $F(z)$ can be modeled offline using system identification techniques.

9.8.1 The Filtered-*x* LMS Method

The basic LMS method needs to be modified to take into account the presence of the secondary path transfer function $F(z)$ in Figure 9.27. To this end, suppose that the adaptive transversal filter in Figure 9.27 has converged to an FIR system with a constant weight vector w. The transfer function of the resulting FIR filter is then

$$W(z) = \sum_{i=0}^{m} w_i z^{-i} \tag{9.8.1}$$

If we replace the adaptive filter in Figure 9.27 with $W(z)$, the Z-transform of the steady-state error signal is

$$
\begin{aligned}
E(z) &= D(z) - \hat{Y}(z) \\
&= G(z)X(z) - F(z)W(z)X(z) \\
&= [G(z) - F(z)W(z)]X(z)
\end{aligned}
\tag{9.8.2}
$$

For the error to be zero for all inputs $x(k)$, it is required that $G(z) - F(z)W(z) = 0$ or

$$W(z) = F^{-1}(z)G(z) \tag{9.8.3}$$

On the surface, the expression for $W(z)$ in (9.8.3) would appear to provide a simple solution to the problem of finding an optimal active noise control filter. Unfortunately, this solution is not a practical one. Recall that the secondary path transfer function $F(z)$ includes the air channel over which the sound travels from the speaker to the error microphone. The propagation of sound through air introduces a delay that is proportional to the path length. Given the presence of a delay in $F(z)$, the inverse of $F(z)$ must have a corresponding advance, which means that $F^{-1}(z)$ is not causal and therefore not physically realizable in real time.

To obtain a causal approximation to $W(z)$, consider the error signal in the time domain. Recall that multiplication of Z-transforms in the frequency domain corresponds to convolution of the corresponding signals in the time domain. Consequently, from Figure 9.27

$$
\begin{aligned}
e(k) &= d(k) - \hat{y}(k) \\
&= d(k) - f(k) * y(k) \\
&= d(k) - f(k) * [w^T(k)u(k)]
\end{aligned}
\tag{9.8.4}
$$

Here $f(k)$ is the impulse response of the secondary system $F(z)$, and $*$ denotes the linear convolution operation. Recall that $u(k)$ is the $(m+1) \times 1$ vector of past inputs at time k, and $w(k)$ is the $(m+1) \times 1$ weight vector at time k. The mean square error objective function is

$$\epsilon(w) = E[e^2(k)] \tag{9.8.5}$$

For the purpose of estimating the gradient vector of partial derivatives of $\epsilon(w)$ with respect to the elements of w, the LMS approximation, $E[e^2(k)] \approx e^2(k)$, can be used. Combining this with (9.8.4) yields

$$\nabla\epsilon(w) \approx 2e(k)\nabla e(k)$$
$$= -2e(k)[f(k) * u(k)] \tag{9.8.6}$$

To simplify the final result, let $\hat{x}(k)$ denote a filtered version of the input using the filter $F(z)$. That is, $\hat{X}(z) = F(z)X(z)$ or

$$\hat{x}(k) = f(k) * x(k) \tag{9.8.7}$$

Similarly, let $\hat{u}(k)$ denote the $(m+1) \times 1$ vector of filtered past inputs. That is,

$$\hat{u}(k) = [\hat{x}(k), \hat{x}(k-1), \cdots, \hat{x}(k-m)]^T \tag{9.8.8}$$

If we combine (9.8.6) through (9.8.8), it then follows that the gradient of the mean square error can be expressed in terms of the filtered input as

$$\nabla\epsilon(w) \approx -2e(k)\hat{u}(k) \tag{9.8.9}$$

The LMS method uses the steepest descent method as a starting point. If $\mu > 0$ denotes the step length, then the steepest descent method for updating the weights is

$$w(k+1) = w(k) - \mu\nabla\epsilon[w(k)] \tag{9.8.10}$$

FXLMS method   Substituting the approximation for the gradient from (9.8.9) into (9.8.10) results in the following weight update algorithm called the *filtered-x LMS* method or simply the FXLMS method.

$$w(k+1) = w(k) + 2\mu e(k)\hat{u}(k), \quad k \geq 0 \tag{9.8.11}$$

Note that the only difference between the FXLMS method and the LMS method is that the vector of past inputs is first filtered by the secondary-path transfer function, hence the name filtered-x LMS method. In practice, an approximation to the secondary system $\hat{F}(z) \approx F(z)$ is used for the prefiltering of $x(k)$ because an exact model of the secondary path is not available. A block diagram of the FXLMS method is shown in Figure 9.28.

FIGURE 9.28: The Filtered-x LMS Method

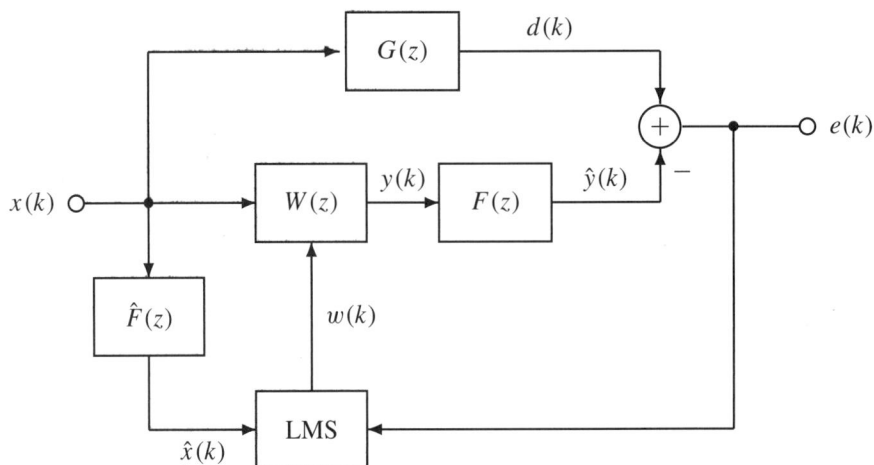

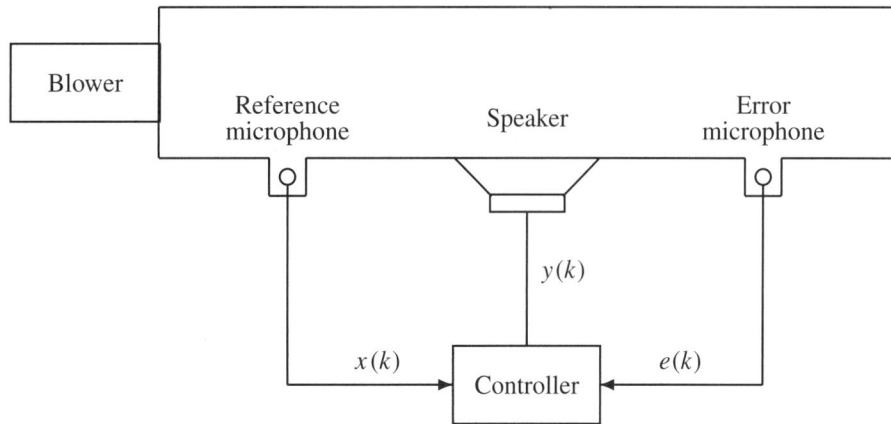

FIGURE 9.29: Active Noise Control in an Air Duct

9.8.2 Secondary Path Identification

The key to implementing the FXLMS method in Figure 9.28 is the identification of a model for the secondary path $F(z)$. To illustrate a secondary path, consider the air duct active noise control system shown in Figure 9.29. The secondary system represents the path traveled by the noise-canceling sound, including the loud speaker and the error microphone. It should be pointed out that there may be some feedback from the speaker to the reference microphone. For the purpose of this analysis, it is assumed that the feedback is negligible. One way to reduce the effects of feedback is to use a directional microphone. Another way to eliminate feedback completely is to use a non-acoustic sensor in place of the reference microphone to produce a signal $x(k)$ that is correlated with the primary noise. Alternatively, the effects of feedback can be taken into account using a feedback neutralization scheme (Warnaka et al., 1984).

The block diagram in Figure 9.29 is a high-level diagram that leaves many of the details implicit. A more detailed representation that shows the measurement scheme used to identify a model for the secondary path is shown in Figure 9.30.

Active control of random broadband noise is a challenging problem. Fortunately, the noise that appears in practice often has a significant narrowband or periodic component. For example, rotating machines generate harmonics whose fundamental frequency varies with the speed of rotation. Similarly, electrical transformers and overhead fluorescent lights emit harmonics with a fundamental frequency of $F_0 = 60$ Hz. Let $f_s = 1/T$ denote the sampling frequency. Then the primary noise can be modeled as follows.

$$x(k) = \sum_{i=0}^{r} c_i \cos(2\pi i F_0 kT) + b_i \sin(2\pi i F_0 kT) + v(k) \quad 0 \le k < N \quad (9.8.12)$$

Here r is the number of harmonics, and F_0 is the frequency of the fundamental harmonic. Since $f_s/2$ is the highest frequency that can be represented without aliasing, the number of harmonics is limited to $r < f_s/(2F_0)$. The broadband term $v(k)$ is additive white noise.

To determine the amount of noise cancellation achieved by active noise control, suppose that the controller is not activated for the first $N/4$ samples. The average power over the first $N/4$ samples of $e(k)$ provides a base line relative to which noise cancellation can be measured.

$$P_u = \frac{4}{N} \sum_{k=0}^{N/4-1} e^2(k) \quad (9.8.13)$$

If the controller is activated at sample $k = N/4$ with the weights updated as in (9.8.11), then the system will undergo a transient segment with the weights converging to their optimal values

FIGURE 9.30:
Identification of
Secondary Path
Model, $\hat{F}(z)$

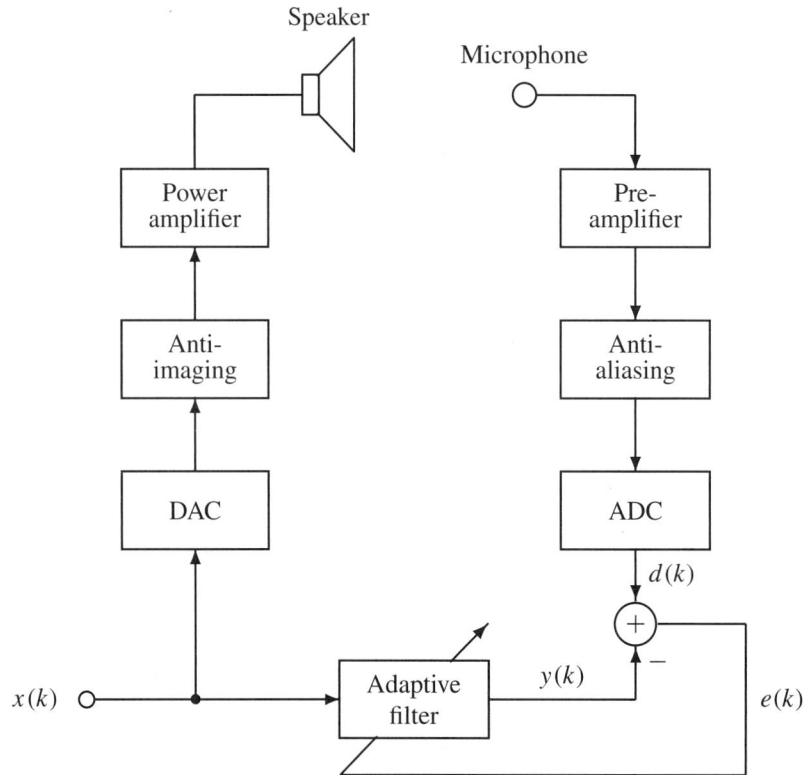

assuming the step size μ is sufficiently small. If the adaptive filter has reached the steady state by sample $3N/4$, then the average power of the error achieved by the controller is

$$P_c = \frac{4}{N} \sum_{k=3N/4}^{N-1} e^2(k) \qquad (9.8.14)$$

Noise reduction   The noise reduction achieved by active noise control can be expressed in decibels as follows.

$$E_{\text{anc}} = 10 \log_{10} \left(\frac{P_c}{P_u} \right) \text{ dB} \qquad (9.8.15)$$

Example 9.14   **FXLMS Method**

To illustrate the use of the FXLMS to achieve active noise control, consider the air duct system shown in Figure 9.29. Suppose that the primary path $G(z)$ and secondary path $F(z)$ are each modeled using FIR filters of order $n = 20$. For the purpose of this simulation, suppose that the coefficients of the filters consist of random numbers uniformly distributed over the interval $[-1, 1]$. Suppose that the noise-corrupted periodic input in (9.8.12) is used, that the sampling frequency is $f_s = 2000$ Hz, and that the fundamental frequency is $F_0 = 100$ Hz. Let the number of harmonics be $r = 5$, and suppose that the additive white noise $v(k)$ is uniformly distributed over $[-.5, .5]$. Finally, suppose that the coefficients for each harmonic are generated randomly in the interval $[-1, 1]$, thereby producing random amplitudes and phases for the r harmonics. The FXLMS method in Figure 9.30 can be applied to this system by running *exam9_14*.

FIGURE 9.31: Active Noise Control Using the FXLMS Method

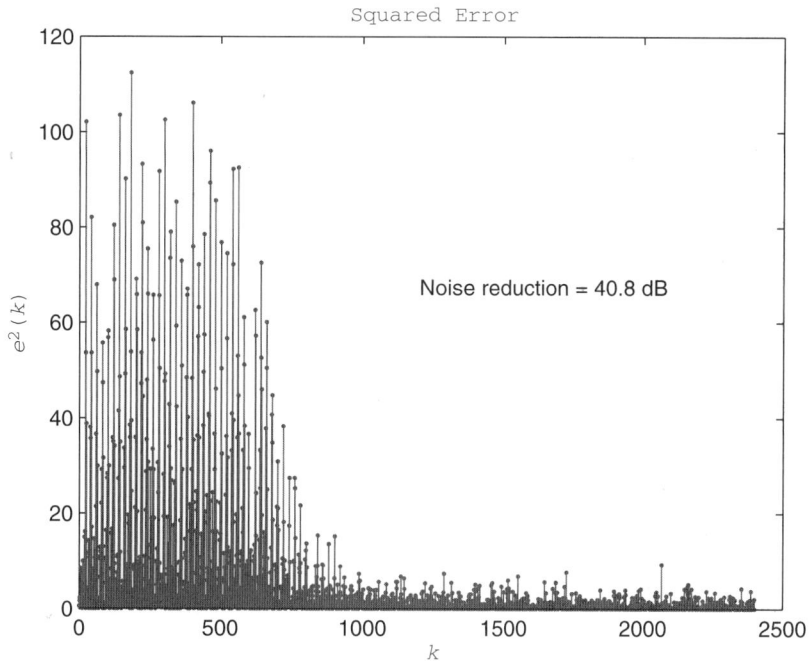

Here $N = 2400$ samples are used with an adaptive filter of order $m = 40$ and a step size of $\mu = .0001$. A plot of the resulting squared error is shown in Figure 9.31. Note that the active noise control is activated at sample $k = 600$. Once the transients have decayed to zero, it is evident that a significant amount of noise reduction is achieved. The noise reduction measured using (9.8.15) is $E_{\text{anc}} = 40.8$ dB.

9.8.3 Signal-synthesis Method

An alternative approach to active noise control that is applicable to narrowband noise is the signal synthesis method. Suppose the input or reference signal is a noise-corrupted periodic signal as in (9.8.12). If the frequency of the fundamental harmonic, F_0, is known or can be measured, then a more direct signal-synthesis approach can be used, as shown in Figure 9.32. For notational convenience, define

$$\theta_0 \triangleq 2\pi F_0 T \tag{9.8.16}$$

FIGURE 9.32: Active Noise Control by the Signal-synthesis Method

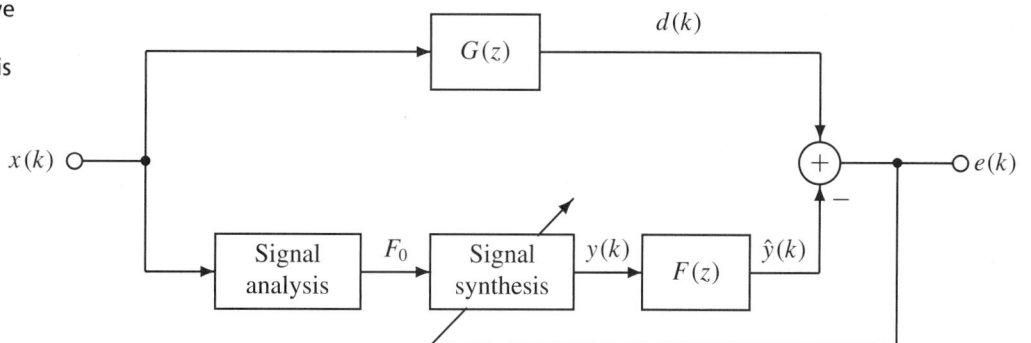

Since the primary and secondary-path models are assumed to be linear, the form of the control signal required to cancel the periodic component of the noise is

$$y(k) = \sum_{i=1}^{r} p_i(k)\cos(ik\theta_0) + q_i(k)\sin(ik\theta_0) \qquad (9.8.17)$$

Next, suppose that the secondary path in Figure 9.28 is modeled with an FIR filter of order m with coefficient vector f. That is,

$$F(z) = \sum_{i=0}^{m} f_i z^{-i} \qquad (9.8.18)$$

To obtain a concise formulation of the secondary noise signal, let $g(k)$ denote the vector of past control signals $y(k)$.

$$g(k) \triangleq [y(k), y(k-1), \cdots, y(k-m)]^T \qquad (9.8.19)$$

It then follows that the secondary noise signal at time k can be expressed with a dot product as

$$\hat{y}(k) = f^T g(k) \qquad (9.8.20)$$

The objective is to choose values for the coefficients $p(k)$ and $q(k)$ in (9.8.17) so as to minimize the mean square error $\epsilon(p, q) = E[e^2(k)]$. As with the LMS method, for the purpose of estimating the gradient, one can approximate the mean square error with $E[e^2(k)] \approx e^2(k)$. From (9.8.17) through (9.8.20), the partial derivative of the mean square error with respect to the ith element of the coefficient vector p is

$$\frac{\partial \epsilon(p, q)}{\partial p_i} \approx 2e(k)\frac{\partial e(k)}{\partial p_i}$$

$$= -2e(k)\frac{\partial f^T g(k)}{\partial p_i}$$

$$= -2e(k)\sum_{j=1}^{m} f_j \frac{\partial y(k-j)}{\partial p_i}$$

$$= -2e(k)\sum_{j=1}^{m} f_j \cos[i(k-j)\theta_0] \qquad (9.8.21)$$

Similarly, the partial derivative of the mean square error with respect to the ith element of the coefficient vector q is

$$\frac{\partial \epsilon(p, q)}{\partial q_i} \approx -2e(k)\sum_{j=1}^{m} f_j \sin[i(k-j)\theta_0] \qquad (9.8.22)$$

To simplify the final result, let $P(k)$ and $Q(k)$ be $r \times 1$ vectors of intermediate variables defined as follows.

$$P_i(k) \triangleq \sum_{j=0}^{m} f_j \cos[i(k-j)\theta_0] \qquad (9.8.23a)$$

$$Q_i(k) \triangleq \sum_{j=0}^{m} f_j \sin[i(k-j)\theta_0] \qquad (9.8.23b)$$

It then follows that the partial derivatives of the mean square error with respect the elements of p and q are

$$\frac{\partial \epsilon(p, q)}{\partial p_i} \approx -2e(k) P_i(k) \tag{9.8.24a}$$

$$\frac{\partial \epsilon(p, q)}{\partial q_i} \approx -2e(k) Q_i(k) \tag{9.8.24b}$$

Signal synthesis method

One can now use the steepest descent method to update the coefficients of the secondary-sound control signal. Let $\mu > 0$ denote the step size. Using (9.8.24) results in the following *signal-synthesis* active noise control method.

$$\begin{bmatrix} p(k+1) \\ q(k+1) \end{bmatrix} = \begin{bmatrix} p(k) \\ q(k) \end{bmatrix} + 2\mu e(k) \begin{bmatrix} P(k) \\ Q(k) \end{bmatrix}, \quad k \geq 0 \tag{9.8.25}$$

One feature that sets the signal-synthesis method apart from the FXLMS method is that it is typically a lower dimensional method. The signal-synthesis method has a weight vector of dimension $n = 2r$, where r is the number of harmonics in the periodic component of the primary noise. Since the maximum number of harmonics that can be accommodated with a sampling frequency of f_s is $r < f_s/(2F_0)$, this means that the maximum dimension of the signal-synthesis method is $n < f_s/F_0$, which is often relatively small.

Although the dimension of the signal synthesis method is small, it is apparent from (9.8.23) that there is considerable computational effort required to compute the intermediate variables $P(k)$ and $Q(k)$. At each time step, a total of $2r(m+1)$ floating-point multiplications or FLOPs are required. Note that the trigonometric function evaluations can be precomputed and stored in a look-up table. Since the $2r(m+1)$ FLOPs must be performed in less than T seconds, this could limit the sampling frequency f_s. Furthermore, for large values of m the computations in (9.8.23) could exhibit significant accumulated roundoff error. Fortunately, these limitations can be minimized by developing a recursive formulation for $P(k)$ and $Q(k)$. Using the cosine of the sum trigonometric identity from Appendix 2

$$P_i(k) = \sum_{j=0}^{m} f_j \cos[i(k-j-1+1)\theta_0]$$

$$= \sum_{j=0}^{m} f_j \cos[i(k-1-j)\theta_0] \cos(i\theta_0) - \sin[i(k-1-j)\theta_0] \sin(i\theta_0)$$

$$= \cos(i\theta_0) P_i(k-1) - \sin(i\theta_0) Q_i(k-1) \tag{9.8.26}$$

Similarly, using the sine of the sum trigonometric identity from Appendix 2, one can show that $Q(k)$ can be expressed recursively as

$$Q_i(k) = \sin(i\theta_0) P_i(k-1) + \cos(i\theta_0) Q_i(k-1) \tag{9.8.27}$$

The recursive update formulas for $P(k)$ and $Q(k)$ can be expressed compactly in vector form as follows.

$$\begin{bmatrix} P_i(k) \\ Q_i(k) \end{bmatrix} = \begin{bmatrix} \cos(i\theta_0) & -\sin(i\theta_0) \\ \sin(i\theta_0) & \cos(i\theta_0) \end{bmatrix} \begin{bmatrix} P_i(k-1) \\ Q_i(k-1) \end{bmatrix}, \quad 1 \leq i \leq r \tag{9.8.28}$$

Note that the number of FLOPs per iteration has been reduced from $2r(m+1)$ to $4r$. Thus the computational effort of the recursive implementation is independent of the size of the filter used to model $F(z)$. To start the recursive computation of $P(k)$ and $Q(k)$, initial values must be used. From (9.8.23), the appropriate starting values are

$$P_i(0) = \sum_{j=0}^{m} f_j \cos(ij\theta_0) \tag{9.8.29a}$$

$$Q_i(0) = -\sum_{j=0}^{m} f_j \sin(ij\theta_0) \tag{9.8.29b}$$

The signal-synthesis method is based on the key assumption that the fundamental frequency of the periodic component of the primary noise is known or can be measured. Examples of applications where F_0 is known include the noise from electrical transformers and from overhead fluorescent lights, where $F_0 = 60$ Hz is a parameter that is regulated carefully by the power company. In those applications where F_0 is not known, it might be measured as shown by the signal analysis block in Figure 9.31. For example, if the primary noise is produced by a rotating machine or motor, then a tachometer might be used to measure F_0. Another approach is to lock onto the periodic component of the noise using a phase-locked loop (Schilling et al., 1998). The virtue of measuring the fundamental frequency is that the signal-synthesis method then can track changes in the period of the periodic component of the primary noise.

Example 9.15

Signal-synthesis Method

To illustrate the use of the signal-synthesis method to achieve active noise control, consider the air duct system shown in Figure 9.29. To facilitate comparison with the FXLMS method, suppose that the input and system parameters are the same as in Example 9.14. The signal-synthesis method in Figure 9.32 can be applied to this system by running *exam9_15*. Again, $N = 2400$ samples are used, but this time the selected step size is $\mu = .001$. A plot of the resulting squared error is shown in Figure 9.33. Note that the active noise control is activated at sample $k = 600$. Once the transients have decayed to zero, it is apparent that a

FIGURE 9.33: Active Noise Control Using the Signal-synthesis Method

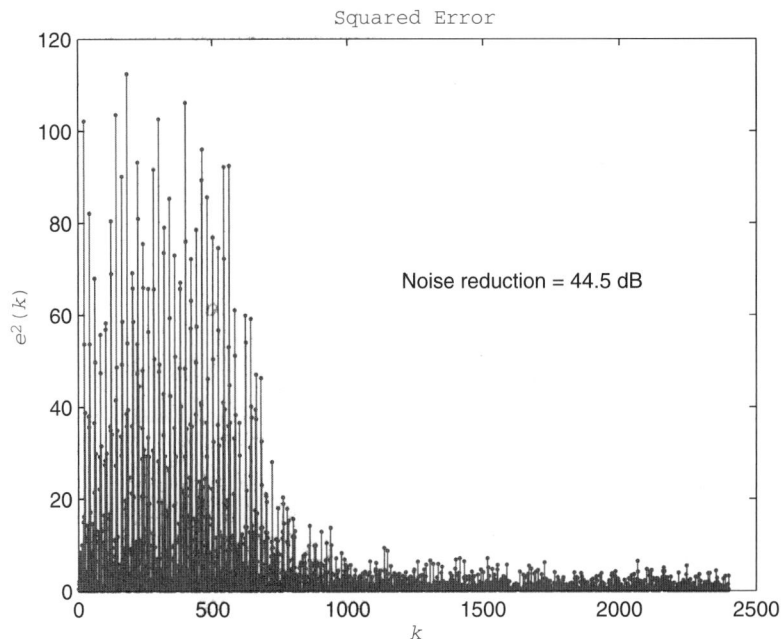

FIGURE 9.34:
Magnitude Spectra
of Error, (a)
Without Active
Noise Control and
(b) with Active
Noise Control Using
the Signal-synthesis
Method

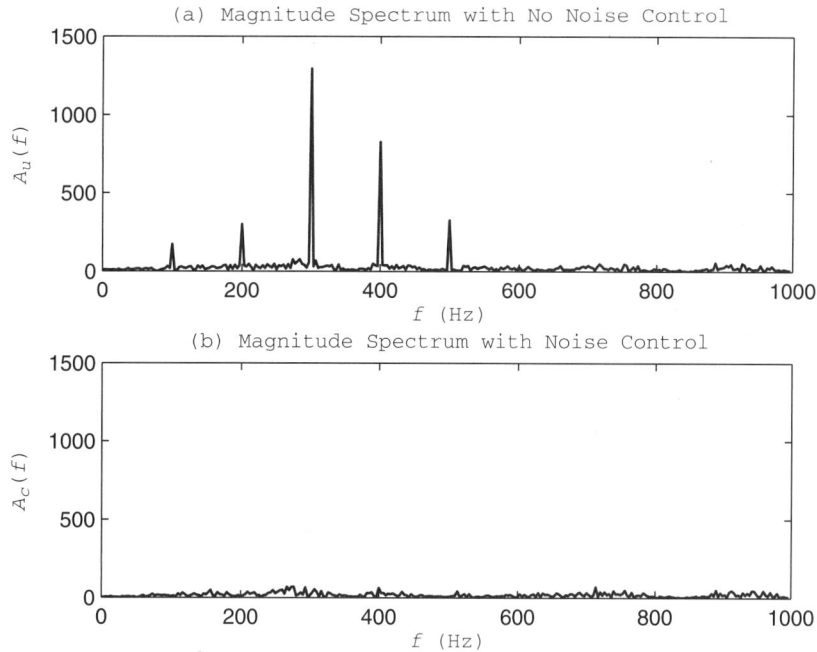

significant amount of noise reduction is achieved. The noise reduction measured using (9.8.15) is $E_{\text{anc}} = 44.5$ dB. In this case the final estimates for the control signal coefficients were as follows.

$$p = [-.0892, -.1159, 1.0692, -.9898, -.1797]^T$$
$$q = [-.1065, .1826, 1.3938, -1.6894, -.3873]^T$$

Another way to look at the performance of an active noise control system is to examine the magnitude spectrum of the error signal with and without the noise control activated. The results are shown in Figure 9.34. The five harmonics are clearly evident in Figure 9.34a, which corresponds to the first $N/4$ samples before the active noise control is activated. The harmonics effectively are eliminated in Figure 9.34b, which corresponds to the last $N/4$ samples after the noise controller has reached a steady state.

FDSP Functions

The FDSP toolbox contains the following functions for implementing active noise control.

```
% F_FXLMS:  Active noise control using the filtered-x LMS method
% F_SIGSYN: Active noise control using the signal synthesis method
%
% Usage:
%        [w,e]   = f_fxlms  (x,g,f,m,mu,w)
%        [p,q,e] = f_sigsyn (x,g,f,f_0,f_s,r,mu)
```

Continued on p. 700

Continued from p. 699

```
% Pre:
%       x  = N by 1 vector containing input samples
%       g  = n by 1 vector containing coefficients of
%            the primary system.  The desired output
%            is D(z) = G(z)X(z)
%       f  = n by 1 vector containing coefficients of
%            the secondary system.
%       m  = order of transversal filter (m >= 0)
%       mu = step size to use for updating w
%       w  = an optional (m+1) by 1 vector containing
%            the initial values of the weights. Default:
%            w = 0
%       f_0 = fundamental frequency of periodic component
%             of the input in Hz
%       f_s = sampling frequency in Hz
%       r  = number of harmonics of x(k) it is desired
%            to cancel (1 to f_s/(2*f_0))
%       mu  = step size to use for updating p and q
% Post:
%       w = (m+1) by 1 weight vector of filter
%           coefficients
%       e = an optional N by 1 vector of errors
%       p = r by 1 vector of cosine coefficients
%       q = r by 1 vector of sine coefficients
% Notes:
%       Typically mu << 1/[(m+1)*P_x'] where P_x' is the
%       average power of filtered input X'(z) = F(z)X(z).
```

9.9 Nonlinear System Identification

9.9.1 Nonlinear Discrete-time Systems

All of the discrete-time systems investigated thus far have been linear systems. One can generalize the notion of a linear discrete-time system to a nonlinear system in the following way.

$$y(k) = f[x(k), \ldots, x(k-m), y(k-1), \ldots, y(k-n)], \quad k \geq 0 \tag{9.9.1}$$

Here f is a real-valued function that is assumed to be continuous, $x(k)$ is the system input at time k, and $y(k)$ is the system output at time k. Thus the present output, $y(k)$, depends on the past inputs and the past outputs in some nonlinear fashion. For convenience, the nonlinear discrete-time system in (9.9.1) will be referred to as the system S_f. A signal flow graph of the system S_f is shown in Figure 9.35 for the case $m = 2$ and $n = 3$.

A more compact formulation of S_f can be obtained by introducing the following generalization of the vector of past inputs called the *state* vector.

State vector

$$u(k) \triangleq [x(k), \ldots, x(k-m), y(k-1), \ldots, y(k-n)]^T \tag{9.9.2}$$

FIGURE 9.35: Signal Flow Graph of Nonlinear Discrete-time System S_f with $m = 2$ and $n = 3$

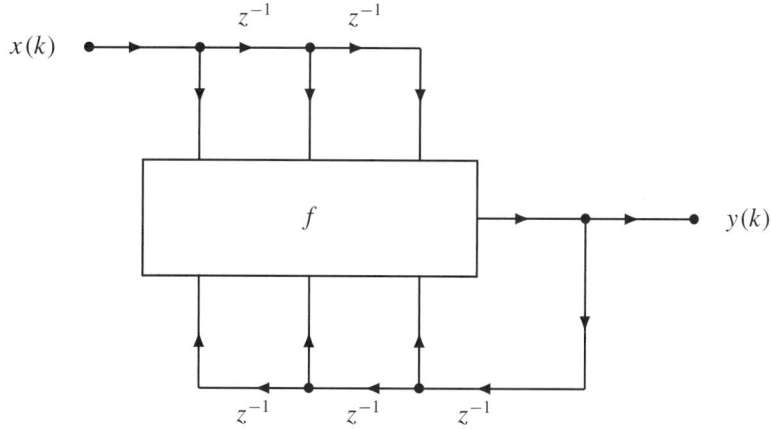

Thus the state vector is a vector containing both the past inputs and the past outputs. The number of elements p in the state vector is called the *dimension* of the system.

Dimension

$$p = m + n + 1 \qquad (9.9.3)$$

Given the $p \times 1$ state vector $u(k)$, the output of the nonlinear system S_f at time k is simply.

$$y(k) = f[u(k)], \quad k \geq 0 \qquad (9.9.4)$$

Suppose the nonlinear system S_f is BIBO stable. Recall from Chapter 2 that a system is BIBO stable if and only if every bounded input $x(k)$ produces a bounded output $y(k)$. Let a be a 2×1 vector, and suppose the input is bounded in the following manner.

$$a_1 \leq x(k) \leq a_2 \qquad (9.9.5)$$

If S_f is BIBO stable, this means there exists an 2×1 vector b such that

$$b_1 \leq y(k) \leq b_2 \qquad (9.9.6)$$

Typically, the vector of input bounds a is selected by the user, and then the vector of output bounds b can be estimated experimentally from measurements. If the input is constrained as in (9.9.5), then the *domain* of the continuous function f is restricted to the following compact subset of R^p.

Domain of S_f

$$U = [a_1, a_2]^{m+1} \times [b_1, b_2]^n \qquad (9.9.7)$$

9.9.2 Grid Points

One approach to approximating the function $f: U \to R$ is to overlay the domain U with a grid of elements, and then develop a *local* representation of f valid over each grid element. To that end, suppose there are $d \geq 2$ grid points equally spaced along each of the p dimensions of U. Then along the ith dimension, the jth *grid value* for $0 \leq j < d$ is

$$U_{ij} = \begin{cases} a_1 + j\Delta x, & 1 \leq i \leq m+1 \\ b_1 + j\Delta y, & m+2 \leq i \leq p \end{cases} \qquad (9.9.8)$$

FIGURE 9.36: Grid Points Over the Domain U with $m = 0$, $n = 1$, and $d = 6$

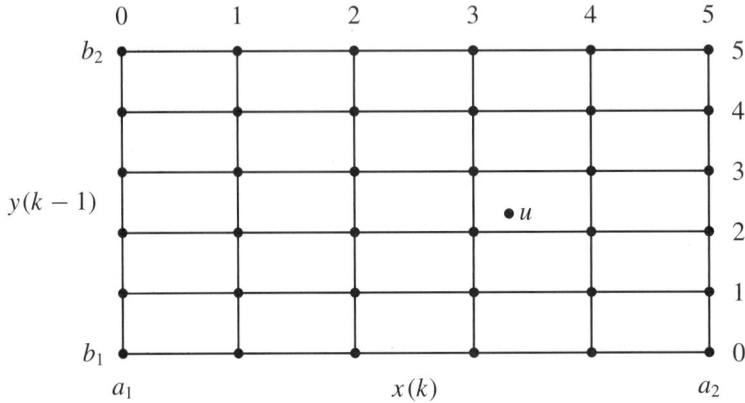

Here Δx and Δy denote the grid point *spacing* along the x and y directions, respectively. That is,

$$\Delta x \triangleq \frac{a_2 - a_1}{d - 1} \tag{9.9.9a}$$

$$\Delta y \triangleq \frac{b_2 - b_1}{d - 1} \tag{9.9.9b}$$

As an illustration, suppose $m = 0$ and $n = 1$, which corresponds to a dimension of $p = 2$. If the number of grid points per dimension is $d = 6$, then the grid consists of 36 grid points, or 25 grid elements, as shown in Figure 9.36. Note that in general the total number of grid points is

$$r = d^p \tag{9.9.10}$$

It is evident from (9.9.10) that as the number of dimensions p and the number of grid points per dimension d grow, the total number of grid points r can be become very large. There are two distinct ways to order this large number of grid points. One is to use a vector subscript q. Suppose q is a $p \times 1$ vector with *integer* elements ranging from 0 to $d - 1$. That is, $0 \le q_i < d$ for $1 \le i \le p$. Here q_i selects the coordinate of the qth grid point along the ith dimension. That is, if $u(q)$ denotes the qth grid point, then using the grid values in (9.9.8)

$$u_i(q) = U_{iq_i}, \quad 1 \le i \le p \tag{9.9.11}$$

Vector subscript Thus q can be thought of as *vector subscript* whose elements select the values of the point $u(q)$ along each of the p dimensions. The virtue of the this vector-subscript approach is that for an arbitrary $u \in U$, it is easy to identify the vertices of the grid element that contains u. For example, the subscript of the *base* vertex of the grid element containing point u is

$$v_i(u) = \begin{cases} \text{floor}\left(\dfrac{u_i - a_1}{\Delta x}\right), & 1 \le i \le m + 1 \\ \text{floor}\left(\dfrac{u_i - b_1}{\Delta y}\right), & m + 2 \le i \le p \end{cases} \tag{9.9.12a}$$

$$v_i(u) = \text{clip}[v_i(u), 0, d - 2] \tag{9.9.12b}$$

Grid element vertices The clipping of the computed subscript to the interval $[0, d-2]$ in (9.9.12b) is included in order to account for the possibility that $u \notin U$. When clipping occurs, the base vertex of the grid element closest to u is obtained. Once the subscript of the base vertex is found, the subscripts of the other vertices are easily determined by adding combinations of 0 and 1 to the elements of $v(u)$. For example, let b^i denote a binary $p \times 1$ vector representing the decimal value i.

Then the subscript of the ith *vertex* of the grid element containing the point u is computed as follows.

$$q^i = v(u) + b^i, \quad 0 \le i < 2^p \tag{9.9.13}$$

The following example illustrates how to find the vertices of the local grid element containing an arbitrary state vector u.

Example 9.16 | **Local Grid Element**

Consider the two-dimensional case $m = 0$ and $n = 1$. Suppose $a = [-10, 10]^T$ and $b = [-5, 5]^T$. Let the number of grid elements per dimension be $d = 6$, as shown in Figure 9.36. In this case the domain of the function f is

$$U = [-10, 10] \times [-5, 5]$$

From (9.9.9), the grid element size is $\Delta x \times \Delta y$ where the grid point spacings are

$$\Delta x = \frac{a_2 - a_1}{d - 1} = 4$$

$$\Delta y = \frac{b_2 - b_1}{d - 1} = 2$$

Suppose $u = [3.2, -.4]^T$ represents an arbitrary point in the domain U. Then from (9.9.12) the subscript of the base, or lower-left, vertex of the grid element containing u is

$$v_1(u) = \text{floor}\left(\frac{u_1 - a_1}{\Delta x}\right) = \text{floor}\left(\frac{13.2}{4}\right) = 3$$

$$v_2(u) = \text{floor}\left(\frac{u_2 - b_1}{\Delta y}\right) = \text{floor}\left(\frac{4.6}{2}\right) = 2$$

There are a total of $2^p = 4$ vertices per grid element. From (9.9.13), the subscripts of the vertices of the grid element containing u are

$$q^0 = v(u) + [0, 0]^T = [3, 2]^T$$
$$q^1 = v(u) + [0, 1]^T = [3, 3]^T$$
$$q^2 = v(u) + [1, 0]^T = [4, 2]^T$$
$$q^3 = v(u) + [1, 1]^T = [4, 3]^T$$

The point u is shown in Figure 9.36. Inspection confirms that $\{q^0, q^1, q^2, q^3\}$ do indeed specify the vertices of the local grid element containing u.

9.9.3 Radial Basis Functions

The overall objective is to model the nonlinear system S_f using input-output measurements by approximating the function f over the domain U. To do this it is helpful to consider a second way to order the r grid points, this time using a scalar. For grid point $u(q)$, the *scalar subscript* i can be computed from the vector subscript q as follows.

Scalar subscript

$$i = q_1 + q_2 d + \cdots + q_p d^{p-1} \tag{9.9.14}$$

As the integer elements of q range from 0 to $d - 1$, the value of i ranges from 0 to $r - 1$ where r is as in (9.9.10). The computation in (9.9.14) can be seen to be a base d-to-decimal

FIGURE 9.37:
Transformations
Between Vector
and Scalar
Subscripts of the
Grid Points

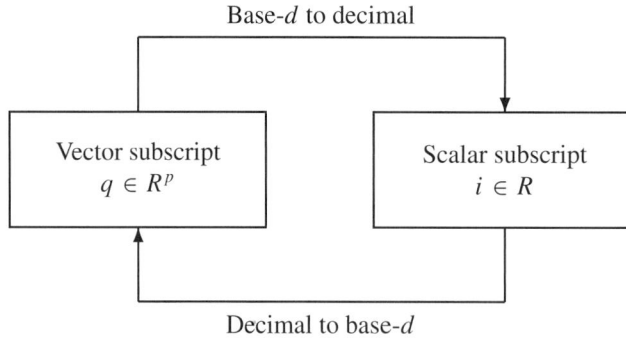

Base-d to decimal

| Vector subscript $q \in R^p$ | | Scalar subscript $i \in R$ |

Decimal to base-d

FIGURE 9.37:
Transformations
Between Vector
and Scalar
Subscripts of the
Grid Points

conversion of q. Consequently, a decimal-to-base d conversion of i can be used to recover q from i. MATLAB functions *base2dec* and *dec2base* are available to perform these conversions. The relationship between the vector subscript q and the scalar subscript i of the grid points is summarized in Figure 9.37.

Using the scalar subscript i, the following one-dimensional ordering of the r grid points is obtained.

$$\Gamma = \{u^0, u^1, \cdots, u^{r-1}\} \tag{9.9.15}$$

Given the one-dimensional ordering of the grid point in (9.9.15), consider the following structure for approximating the function f on the right-hand side of the nonlinear system S_f.

$$f_0(u) = w^T g(u) \tag{9.9.16}$$

Here $w = [w_0, \ldots, w_{r-1}]^T$ is an $r \times 1$ *weight* vector whose elements will be determined from input-output measurements of the system S_f. The function $g : R^p \to R^r$ represents an $r \times 1$ vector of functions called *radial basis* functions with the ith radial basis function centered *RBF*   at grid point u^i. A radial basis function or RBF is a continuous function $g_i : R^p \to R$ such that

$$g_i(u^i) = 1 \tag{9.9.17a}$$
$$|g_i(u)| \to 0 \quad \text{as} \quad \|u - u^i\| \to \infty \tag{9.9.17b}$$

Thus the value of an RBF goes to zero as the radius from the point around which it is centered *Gaussian RBF*   goes to infinity. A popular example of a radial basis function is the *Gaussian RBF*.

$$g_i(u) = \exp\left[\frac{(u - u^i)^T (u - u^i)}{2\sigma^2}\right] \tag{9.9.18}$$

Here the variance σ^2 controls the rate at which the RBF goes to zero as the distance from the center point increases. The Gaussian RBF has a number of useful properties not the least of which is the observation that it has an infinite number of derivatives, all of which are continuous. However, the Gaussian RBF also has some drawbacks. Since $g_i(u) > 0$ for all u, it follows that if σ is too large, there will be many terms in (9.9.16) that contribute to $f_0(k)$. This defeats the local nature of the representation which, ideally, has only a few terms contributing to $f_0(k)$. On the other hand, if σ is too small then the value of $g_i(u)$ will be near zero midway between the grid points which reduces the effectiveness of g_i in approximating f.

FIGURE 9.38:
Comparison of
One-dimensional
Gaussian and
Raised-cosine RBFs

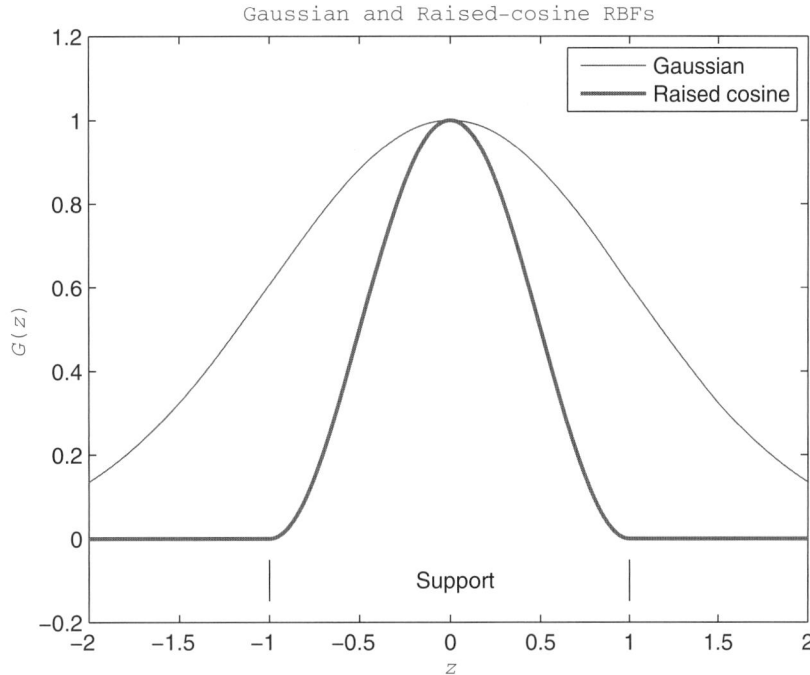
Gaussian and Raised-cosine RBFs

Raised cosine RBF

There are several potential candidates for radial basis functions (Webb and Shannon, 1998). As an alternative to the Gaussian RBF, suppose $p = 1$ and consider the following one-dimensional *raised-cosine RBF* centered about $z = 0$ (Schilling et al., 2001).

$$G(z) = \begin{cases} \dfrac{1 + \cos(\pi z)}{2}, & |z| \le 1 \\ 0, & |z| > 1 \end{cases} \tag{9.9.19}$$

Note that $G(z)$ satisfies the two basic properties in (9.9.17). A plot comparing the one-dimensional Gaussian RBF and the one-dimensional raised-cosine RBF is shown in Figure 9.38 for the case $\sigma = 1$. Observe that the raised-cosine RBF is a continuously differentiable function. Furthermore, the set of z over which the raised-cosine RBF is nonzero is contained in the compact (closed and bounded) set $S = [-1, 1]$. That is, the raised-cosine RBF has *compact support*. This is in contrast to the Gaussian RBF which is not a function of compact support. The compact support property ensures that a representation based on a sum of RBFs will be a *local* representation with only a few of the terms contributing to $f_0(k)$.

Compact support

The raised-cosine RBF can be generalized from one dimension to p dimensions by forming a product of scalar RBFs, one for each dimension. The scale factors Δx and Δy in (9.9.9) can be used to convert the normalized scaler RBF in (9.9.19) into a scalar RBF that takes into account the grid-point spacing. This results in the following p-dimensional raised-cosine RBF centered at u^o. Here the notation $\prod$ denotes the product.

$$g_i(u) = \prod_{j=1}^{m+1} G\left(\frac{u_j - u_j^i}{\Delta x}\right) \prod_{j=m+2}^{p} G\left(\frac{u_j - u_j^i}{\Delta y}\right) \tag{9.9.20}$$

FIGURE 9.39: A
Two-dimensional
Raised-cosine RBF
Centered at $u = 0$

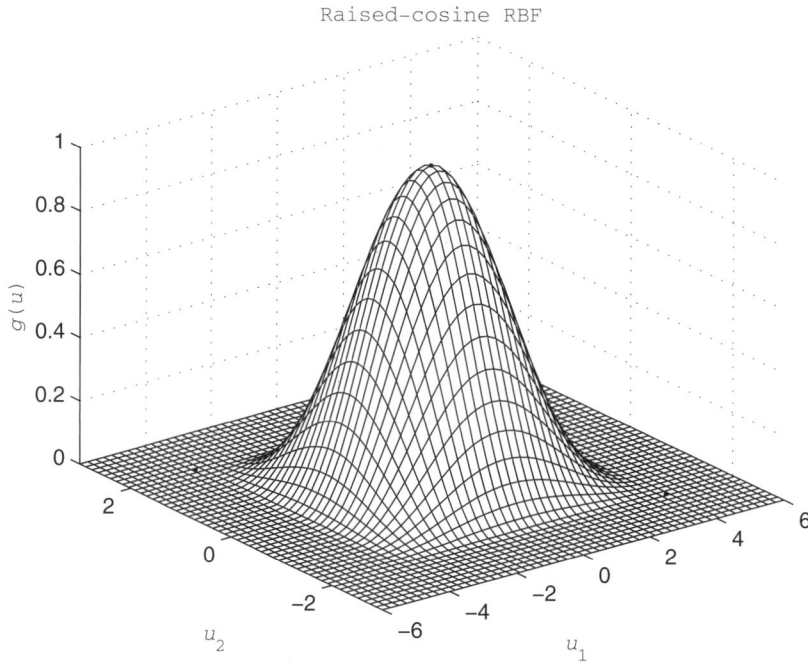

Raised-cosine RBF

Example 9.17

Raised-cosine RBF

Suppose $m = 0$ and $n = 1$. Let the signal bounds be $a = [-4, 4]$ and $b = [-2, 2]$, and suppose the number of grid points per dimension is $d = 3$. In this case the grid-point spacing is $\Delta x = 4$ and $\Delta y = 2$ and the domain of f is

$$U = [-4, 4] \times [-2, 2]$$

There are a total of $r = 9$ grid points with a RBF centered about each one. A plot of the RBF centered about the origin can be obtained by running *exam9_17* with the result shown in Figure 9.39. The continuous differentiability and compact support characteristics of this two-dimensional raised-cosine RBF are evident from inspection.

Properties

Raised-cosine RBFs have a number of interesting and useful properties. To start with, they are continuously differentiable functions of u, which means that the approximation to f in (9.9.16) is also continuously differentiable. The compact support property, evident in Figures 9.38 and 9.39, means that the RBFs do not interact with one another at the grid points. This leads to the following *orthogonality property* on the grid points.

Orthogonal property

$$g_j(u^i) = \begin{cases} 1, & i = j \\ 0, & i \neq j \end{cases} \quad (9.9.21)$$

The orthogonality property suggests an easy way to select the $r \times 1$ vector of weights w. If we apply the orthogonality property to (9.9.16), it follows that $f_0(u^i) = w_i$. Thus the approximation, f_0, to the function f can be made *exact* on the set of r grid points with the following selection of the weights.

$$w_i = f(u^i), \quad 0 \leq i < r \quad (9.9.22)$$

If the number of grid points per dimension d is sufficiently large, then the weights in (9.9.22) will produce an effective model of the nonlinear discrete-time system S_f over the domain U. If the dimension p is too large to permit a relatively large value for d, then the weights in (9.9.22) still constitute an effective initial guess $w(0)$. A procedure for updating the weights to minimize the mean square error starting from this initial guess is discussed in the next subsection.

The orthogonality property is based on a lack of interaction between RBFs at the grid points. Perhaps the most remarkable property of the raised-cosine RBF occurs between the grid points. Suppose the weights in (9.9.16) are all set to unity. It can be shown that the resulting approximation to f has the following property (Schilling et al., 2001).

$$\sum_{i=0}^{r-1} g_i(u) = 1, \quad u \in U \tag{9.9.23}$$

Constant interpolation property

For convenience, (9.9.23) will be referred to as the *constant interpolation* property. It says that if the RBFs are equally weighted, then the surface that they produce when their contributions are combined is perfectly flat over the domain U. This property, which is not shared by Gaussian RBFs, is useful in those instances where the function being approximated is flat over part of its domain. This might occur, for example, if f includes saturation or dead-zone effects.

Example 9.18

Constant Interpolation Property

As an illustration of the constant interpolation property, suppose $m = 0$ and $n = 1$, in which case $p = 2$. Let $a = [-1, 1]$ and $b = [-1, 1]$, and suppose the number of grid points per dimension is $d = 2$. Thus the grid-point spacing is $\Delta x = 2$ and $\Delta y = 2$ and the domain of the function f is

$$U = [-1, 1] \times [-1, 1]$$

In this case there are $r = 4$ grid points located at the corners of the square U. Suppose the weight vector is $w = [1, 1, 1, 1]^T$. A plot of the resulting interpolated surface $f_0(u)$, obtained by running *exam9_18*, is shown in Figure 9.40. It is apparent that on the domain U the interpolated surface is flat.

9.9.4 Adaptive RBF Networks

RBF network

Given the approximation to f in (9.9.16), the output of the raised-cosine *RBF network* at time k can be expressed as follows.

$$y_0(k) = w^T(k)g[u(k)], \quad k \geq 0 \tag{9.9.24}$$

The system in (9.9.24) will be referred to as the RBF network S_0. Recall that as the dimension p and the number of grid points per dimension d grow, the number of terms r in the dot product in (9.9.24) can become very large. Fortunately, for each value of $u(k)$, almost all of these terms are zero. Only the local terms in the neighborhood of $u(k)$ contribute to $y_0(k)$. Because each raised-cosine RBF has compact support and goes to zero at the adjacent grid points, the only terms contributing to $y_0(k)$ are the terms corresponding to the vertices of the grid element containing $u(k)$. Consequently, for each $u(k)$, the number of nonzero terms in $y_0(k)$ is at most

$$M = 2^p \tag{9.9.25}$$

Raised-cosine RBF Network

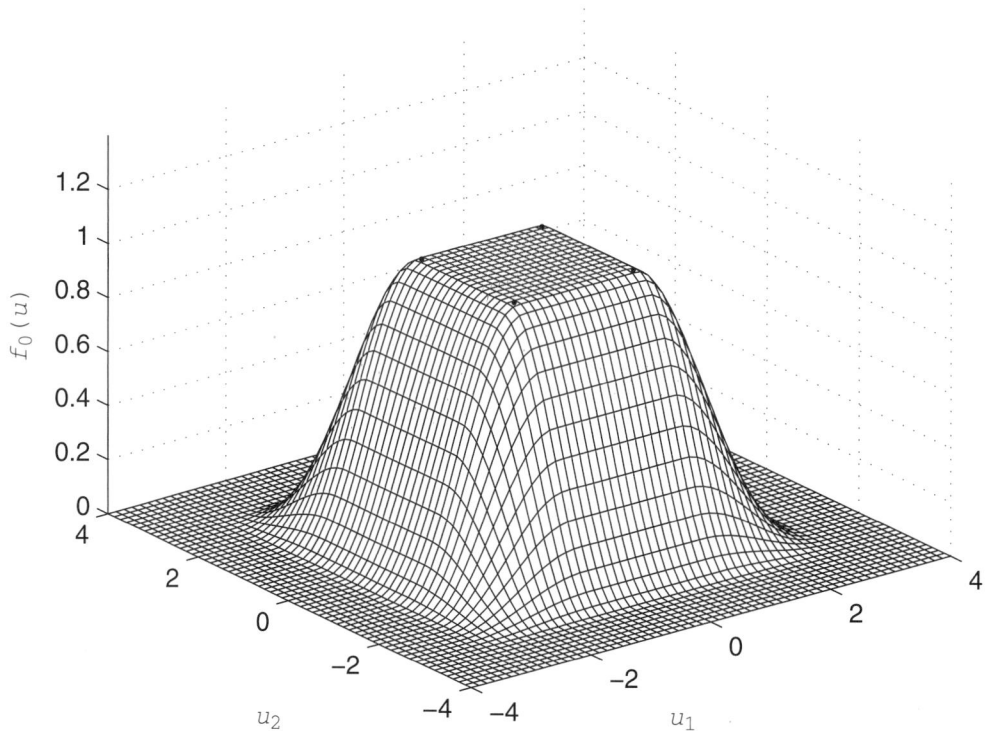

FIGURE 9.40: Constant Interpolation Produced by Four Equally Weighted Raised-cosine Radial Basis Functions.

Recall that the vertices of the local grid element containing u can be found using (9.9.12) and (9.9.13). This is the basis for the following highly efficient algorithm for evaluating the RBF network output.

ALGORITHM

9.2: RBF Network Evaluation

1. Set $y_0 = 0$. Compute $p = m = n + 1$ and $M = 2^p$. Compute Δx and Δy using (9.9.9).
2. Use (9.9.12) and (9.9.13) to compute the vector subscripts of the vertices $\{q^0, \ldots, q^{M-1}\}$ of the grid element containing $u(k)$.
3. For $j = 0$ to $M - 1$ do
   {
   (a) Convert q^j to the scalar subscript i using function *base2dec* or (9.9.14).
   (b) Compute grid point u^i as follows.

   $$u_s^i = \begin{cases} a_1 + q_s^j \Delta x, & 1 \leq s \leq m+1 \\ b_1 + q_s^j \Delta y, & m+2 \leq s \leq p \end{cases}$$

   (c) Using (9.9.20) compute

   $$y_0(k) = y_0(k) + w_i(k)g_i[u(k)]$$

   }

The difference in speed between Algorithm 9.2 and a direct brute-force evaluation of (9.9.24) can be dramatic. For example, suppose $m = 2$, $n = 3$, and $d = 10$. Then $p = 6$ and there are a total of $r = 10^6$ potential terms to evaluate. However, using Algorithm 9.2 requires the evaluation of only $M = 64$ terms. Consequently, apart from the overhead of step 2, Algorithm 9.2 is faster than direct evaluation in this instance by a factor of $r/M = 15625$, or more than four orders of magnitude!

Next, consider the problem of updating the weights of the RBF network so as to achieve optimal performance. The *error* between the system S_f and the RBF network S_0 is

$$e(k) = y(k) - y_0(k) \tag{9.9.26}$$

The objective is to minimize the mean square error $\epsilon(w) = E[e^2(k)]$. For the purpose of estimating the gradient, the fundamental LMS assumption, $E[e^2(k)] \approx e^2(k)$, is used. Then from (9.9.24) the partial derivative of $\epsilon(w)$ with respect to the ith element of w is

$$\begin{aligned}
\frac{\partial \epsilon(w)}{\partial w_i} &\approx 2e(k)\frac{\partial e(k)}{\partial w_i} \\
&= -2e(k)\frac{\partial w^T g[u(k)]}{\partial w_i} \\
&= -2e(k)g_i[u(k)], \quad 0 \le i < r
\end{aligned} \tag{9.9.27}$$

Thus the gradient vector is $\nabla \epsilon(w) = -2e(k)g[u(k)]$. Using this gradient estimate in the steepest-descent method in (9.8.10) then results in the following update formula for the RBF network where $\mu > 0$ is the step size.

$$w(k+1) = w(k) + 2\mu e(k)g[u(k)], \quad k \ge 0 \tag{9.9.28}$$

Observe that the weight-update algorithm for the nonlinear RBF network S_0 is quite simple and is essentially the same as the LMS method for linear systems, but with u replaced by $g(u)$.

Practical Considerations

Before considering an example of an RBF model of a nonlinear discrete-time system, it is useful to briefly consider some practical issues. The first is the question of how to determine an effective set of output bounds b for the nonlinear system S_f. Suppose x is an $P \times 1$ *test* input of white noise uniformly distributed over $[a_1, a_2]$ where $P \gg 1$. Let $y(k)$ be the resulting output of S_f, and let y_m and y_M denote the minimum and maximum of $y(k)$, respectively. Then output bounds can be selected as follows where $\beta \ge 1$ is a *safety factor*.

$$b_1 = \frac{y_m + y_M}{2} - \beta\left(\frac{y_M - y_m}{2}\right) \tag{9.9.29a}$$

$$b_2 = \frac{y_m + y_M}{2} + \beta\left(\frac{y_M - y_m}{2}\right) \tag{9.9.29b}$$

Note that when $\beta = 1$, the bounds in (9.9.29) reduce to $b = [y_m, y_M]$. Typically, $\beta > 1$ is used because this takes into account the fact that $x(k)$ is of finite length. Furthermore, it is prudent to use $\beta > 1$ because $y_0(k)$ of the RBF network may range outside of the interval $[y_m, y_M]$, given that $y_0(k)$ is only an approximation to $y(k)$. Of course, making β too large effectively reduces the precision of the RBF model by making the grid-point spacing Δy large.

It is also important to take some care in selecting the initial condition $u(0)$ when computing the RBF network output. Recall that the RBF network approximation, $f_0(u)$ in (9.9.16) has compact support. One can show that $f_0(u) = 0$ for $u \notin \Omega$ where

$$\Omega = [a_1 - \Delta x, a_2 + \Delta x]^{m+1} \times [b_1 - \Delta y, b_2 + \Delta y]^n \qquad (9.9.30)$$

For example, the support of the RBF network shown in Figure 9.40 is contained in $\Omega = [-3, 3] \times [-3, 3]$. Because of this compact support characteristic, it is important to choose the initial state such that $u(0) \in \Omega$. Otherwise, the RBF network output may start out zero and remain zero for $k \geq 1$.

The final practical issue that needs to be addressed is the question of how to determine whether or not the RBF model represents a good fit to the system S_f. Let $e \in R^N$ and $y \in R^N$ be column vectors containing samples of the error $e(k)$ and system output $y(k)$, respectively. Then the following *normalized mean square* error can be used.

$$E \triangleq \frac{e^T e}{y^T y} \qquad (9.9.31)$$

Note that when the RBF network output is $y_0(k) = 0$, the normalized mean square error reduces to $E = 1$. Thus values of $E \ll 1$ represent a good fit between the system S_f and the RBF network model S_0 with $E = 0$ being a perfect fit.

Example 9.19

Nonlinear System Identification

As practical example of a nonlinear discrete-time system, consider a continuous stirred tank chemical reactor. Suppose $x(k)$ is the reactant concentration at time k, and $y(k)$ is the product concentration at time k. If the input is scaled such that $0 \leq x(k) \leq 1$, then the Van de Vusse reaction can be characterized by the following nonlinear discrete-time system.

$$y(k) = c_1 + c_2 x(k-1) + c_3 y(k-1) + c_4 x^3(k-1) + c_5 y(k-2)x(k-1)x(k-2)$$

The presence of the fourth and fifth terms makes this system nonlinear. If the sampling interval is $T = .04$ hours, then the parameters of the system can be taken to be (Hernandez and Arkun, 1996)

$$c = [.558, .538, .116, -.127, -.034]^T$$

For this system $m = 2$ and $n = 2$. Thus the system dimension is $p = 5$. Since the input is non-negative and has been normalized,

$$a = [0, 1]$$

To determine the vector of output bounds b, consider a test input consisting of $P = 1000$ points and a safety factor of $\beta = 1.5$, as in (9.9.29). This results in $b = [.4310, 1.1930]$ which means the domain of the function f in this case is

$$U = [0, 1]^3 \times [.4310, 1.1930]^2$$

Suppose the initial condition for the RBF network is set to $u(0) = [0, 0, 0, c, c]^T$ where $c = (b_1 + b_2)/2$. Let the number of grid points per dimension be $d = 5$. From (9.9.10) this produces an RBF network with $r = 3125$ terms. However, from (9.9.25) only $M = 32$ of these terms are nonzero for each u. From (9.9.9) the grid-point spacing in the x and y directions is

$$\Delta x = .2500$$
$$\Delta y = .1905$$

9.9 Nonlinear System Identification 711

FIGURE 9.41:
Comparison of
System S_f and RBF
Model with $m = 2$,
$n = 2$, $d = 5$

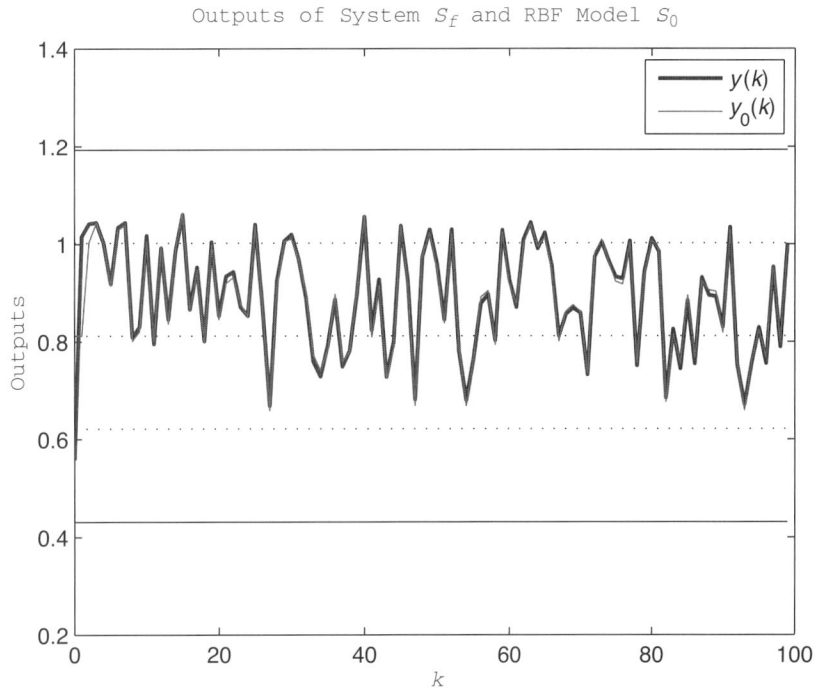

Outputs of System S_f and RBF Model S_0

Suppose the elements of the weight vector are computed using (9.9.22). Finally, the system is tested with $N = 100$ samples of white noise uniformly distributed over $[a_1, a_2]$. A comparison of the two outputs, obtained by running *exam9_19*, is shown in Figure 9.41. Note that the solid horizontal lines indicate the boundaries of the domain U in the y direction, and the dotted lines show where the grid points are located. Careful inspection reveals that there are only minor differences between the two outputs. From (9.9.31), the normalized mean square error in this case was

$$E = .0014$$

First-order RBF Network

The notation $y_0(k)$ used for the output of the RBF network in (9.9.24) is indicative of the fact that the coefficients of the RBF terms are constants, that is zeroth-degree polynomials in the elements of u. In view of the constant interpolation property illustrated in Figure 9.40, it can be shown that $y_0(k)$ will reproduce the system output $y(k)$ exactly, using the minimum of $d = 2$ grid points per dimension, when the nonlinear function $f(u)$ in (9.9.1) is a zeroth-degree polynomial in u. It is also possible to introduce a first-order RBF network that has the following structure.

$$y_1(k) = (Vu + w)^T g(u) \tag{9.9.32}$$

Here the adjustable parameters are a $r \times 1$ weight vector w and the an $r \times p$ *weight matrix* V. This first-order formulation requires $(p + 1)$ times as much storage for the parameters. However, it has the benefit that $y_1(k)$ will reproduce $y(k)$ exactly, using the minimum size network ($d = 2$) when $f(u)$ is a polynomial in u of degree one. Therefore, the first-order RBF network will model linear discrete-time systems exactly using the minimum number of parameters. The update algorithm for the weight vector $w(k)$ and weight matrix $V(k)$ can be found in (Schilling et al, 2001).

Universal approximation property

Suppose the function f in (9.9.1) is not only continuous but also continuously differentiable. It then can be shown (Problem 9.23) that the zeroth-order RBF network converges *uniformly* on the domain U to the system S_f as $d \to \infty$. The same is true of the first-order RBF network in (9.32). This is called the *universal approximation* property.

FDSP Functions

The FDSP toolbox contains the following functions which implement nonlinear system identification using a raised-cosine RBF network.

```
% F_RBFW:  Nonlinear system identification using an RBF network
% F_RBFO:  Compute output of raised-cosine RBF network
% F_STATE: Construct state vector from inputs and outputs
%
% Usage:
%         [w,e] = f_rbfw  (@f,N,a,b,m,n,d,mu,ic,w)
%          y     = f_rbfo  (x,w,a,b,m,n,d)
%          u     = f_state (x,y,k,m,n)
% Pre
%         f  = name of user-supplied function that specifies the
%              right-hand side of the nonlinear discrete-time system.
%
%              y(k) = f(u(k),m,n)
%              u(k) = [x(k),...,x(k-m),y(k-1),...,y(k-n)]'
%
%         N  = number of training samples (N >= 0).
%              If N = 0,the weight returned is the
%              initial weight computed according to
%              input ic.
%         a  = 2 by 1 vector of input bounds
%         b  = 2 by 1 vector of output bounds
%         m  = number of past inputs (m >= 0)
%         n  = number of past outputs (n >= 0)
%         d  = number of grid points per dimension
%         mu = step length for gradient search
%         ic = an initial condition code.  If ic <> 0,
%              compute the initial weights to ensure that
%              the network is exact at the grid points.
%         w  = an optional r by 1 vector containing the
%              initial values of the weights. default:
%                w = 0
%         x  = N by 1 vector of inputs
%         y  = N by 1 output vector
%         k  = current time (1 to N)
```

Continued on p. 713

Continued from p. 712

```
% Post:
%        w = r by 1 weight vector
%        e = an optional N by 1 vector of errors where
%        y = N by 1 vector of outputs.
%            e(k) = y(k)-y_0(k)
%        u = p by 1 state vector at time k. Here
%            u = [x(k),...,x(k-m),y(k-1),...y(k-n)]'
% Notes:
%        1. r = d^p where p = m+n+1
%        2. A good value for the initial w is
%            w(i) = f(u(i)).
```

● ● ● ● ● ● ● ● ● ● ● ● ● ● ● ● ● ●

9.10   GUI Software and Case Study

This section focuses on system identification using adaptive signal processing techniques. A graphical user interface module called *g_adapt* is introduced that allows the user to perform system identification without any need for programming. A case study example is presented and solved using MATLAB.

g_adapt: Perform adaptive signal processing

The FDSP toolbox includes a graphical user interface module called *g_adapt* that allows the user to perform system identification using a variety of adaptive techniques without any need for programming. GUI module *g_adapt* features the display screen with tiled windows shown in Figure 9.42. The upper left-hand *Block diagram* window contains a block diagram of the adaptive system under investigation. It features an mth-order transversal filter characterized by the following time-varying difference equation.

$$y(k) = \sum_{i=0}^{m} w_i(k)x(k-i), \quad 0 \le k < N \tag{9.10.1}$$

The *Parameters* window below the block diagram displays *edit boxes* containing the simulation parameters. The contents of each edit box can be directly modified by the user with the Enter key used to activate the changes. The parameters a and b are the coefficient vectors of the black box system to be identified.

$$\sum_{i=0}^{n} a_i d(k-i) = \sum_{i=0}^{p} b_i x(k-i), \quad 0 \le k < N \tag{9.10.2}$$

It is important to select a such that the resulting black box system is BIBO stable. The parameter fs is the sampling frequency, in Hz, and the parameter $m \ge 0$ is the order of the adaptive filter. The remaining parameters are all real scalars that control the behavior of the adaptive algorithm. Parameter mu is the step length, and it should be chosen to satisfy the following

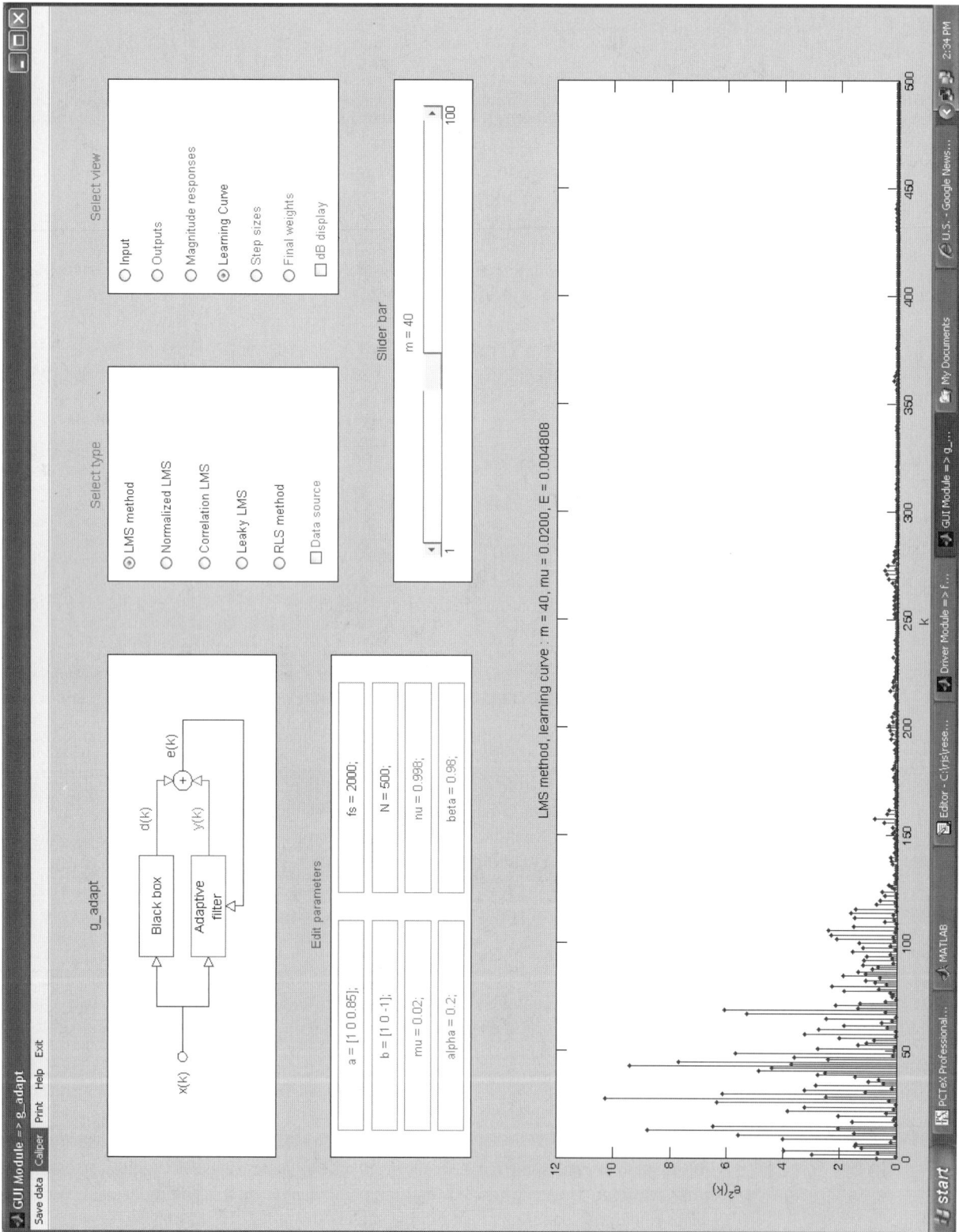

FIGURE 9.42: Display Screen of Chapter GUI Module *g_adapt*

bound where P_x is the average power of the input x.

$$0 < \mu < \frac{1}{(m+1)P_x} \tag{9.10.3}$$

Parameter nu is the leakage factor for the leaky LMS method. Typically, $nu < 1$ with $nu \approx 1$. When $nu = 1$, the leaky LMS method reduces to the LMS method. Parameter $alpha$ is the normalized step size used in the normalized LMS method and the relative step size of the correlation LMS method. The correlation LMS method also uses the smoothing parameter $beta$ where $0 < beta < 1$ with $beta \approx 1$. Inappropriate values for the scalar control parameters can cause some of the methods to diverge.

The *Type* and *View* windows in the upper-right corner of the screen allow the user to select both the type of adaptive algorithm and the viewing mode. The algorithm options include the LMS method, the normalized LMS method, the correlation LMS method, the leaky LMS method, and the RLS method. The *View* options include the input, a comparison of outputs, a comparison of magnitude responses, the learning curve, the step sizes used during learning, and the final weights found. The *Plot* window along the bottom half of the screen shows the selected view.

There are two checkbox controls. The dB checkbox toggles the magnitude response display between linear and logarithmic scales. The Data from file checkbox toggles the source of the input and desired output data. When it is checked, the user is prompted for the name of a user-supplied MAT file that contains the vector of input samples x, the vector of desired output samples d, and the sampling frequency fs. In this way, systems with input-output data generated offline from another source (e.g., from measurements) can be identified. The results of the identification can be saved using the *Save data* menu option at the top of the screen. When the Data from file checkbox is not checked, the input consists of white noise uniformly distributed over $[-1, 1]$, and the desired output is computed using the coefficients a and b as in (9.10.2). The number of samples N is controlled with the horizontal slider bar appearing below the *Type* and *View* windows. This control is active only when the Data from file checkbox is not checked.

Data options 　The *Menu* bar at the top of the screen includes several menu options. The *Caliper* option allows the user to measure any point on the current plot by moving the mouse crosshairs to that point and clicking. The *Save data* option is used to save the current x, y, d, fs, w, a, and b in a user-specified MAT file for future use. Files created in this manner can be later loaded using the Data from file checkbox control. The *Print* option prints the contents of the *Plot* window. Finally, the *Help* option provides the user with some helpful suggestions on how to effectively use module *g_adapt*.

CASE STUDY 9.1 　**Identification of a Chemical Process**

In the chemical process control industry it is common to have dynamic systems with time delays due to transportation lags. This leads to process models that are described by both differential and difference equations. As an illustration, consider the following first-order with dead time system that can be used to model, for example, a heated stirred tank (Bequette, 2003).

$$\frac{dy_a(t)}{dt} + py_a(t) = cx_a(t - \tau) \tag{9.10.4}$$

Here τ is the dead time or delay in the input, and p and c are system parameters. Let T denote the sampling interval. The differential-difference system in (9.10.4) can be converted to a discrete-time system by approximating the derivative using a backwards difference as follows.

$$\frac{dy_a(t)}{dt} \approx \frac{y(k) - y(k-1)}{T} \tag{9.10.5}$$

Here it is understood that $y(k) = y_a(kT)$. The input delay can be modeled exactly in discrete time if some care is taken is choosing the sampling interval. Suppose $T = \tau/M$ for some integer $M \geq 1$. Then $x_a(t-\tau)$ can be replaced by $x(k-M)$, where $x(k) = x_a(kT)$. These substitutions yield the following equivalent discrete-time model.

$$\frac{y(k) - y(k-1)}{T} + py(k) = cx(k-M) \tag{9.10.6}$$

Thus the differential-difference system in (9.10.4) can be approximated by an IIR filter if the sampling interval is chosen to be an integer submultiple of the delay τ. To investigate how well the IIR model can be approximated by an adaptive transversal filter, suppose the delay is $\tau = 5$ sec and the sampling interval is $T = .5$ sec. This yields an input delay of $M = 10$ samples. Let the remaining system parameters be $p = .2$ and $c = 4$. The normalized LMS method can be used to identify this system using *case9_1*.

CASE
STUDY 9.1

```
function case9_1

% Case Study 9.1: Identification of a chemical process

f_header('Case Study 9.1: Identification of a chemical process')
tau = 5
T = 0.5
M = tau/T
p = 0.2;
c = 4;
N = 1000;
rand ('seed',1000)
m = f_prompt ('Enter adaptive filter order',0,80,40);
alpha = f_prompt ('Enter normalized step length',0,1,0.1);

% Compute the coefficients of the IIR model

a = [1+p*T, -1];
b = [zeros(1,M) c*T];

% Compute the input and desired output

x = f_randu (N,1,-1,1);
d = filter (b,a,x);

% Identify a model using normalized LMS method

[w,e] = f_normlms (x,d,m,alpha);

% Learning curve

figure
k = 0 : N-1;
```

Continued on p. 717

Continued from p. 716

```
stem (k,e.^2,'filled','.')
f_labels ('Learning curve','{k}','{e^2(k)}')
box on
f_wait

% Compare responses to new input

P = 200;
x = f_randu (P,1,-1,1);
d = filter (b,a,x);
y = filter (w,1,x);
figure
k = 0 : P-1;
hp = plot (k,d,k,y);
set (hp(1),'LineWidth',1.5)
legend ('{d(k)}','{y(k)}')
e = d - y;
E = sum(e.^2)/sum(d.^2)
caption = sprintf ('Comparison of outputs, {E} = %.4f',E);
f_labels (caption,'{k}','Outputs')
f_wait
```

When *case9_1* is run with an adaptive filter of order $m = 40$ using $N = 1000$ samples and a normalized step size of $\alpha = .1$, it produces the learning curve shown in Figure 9.43. Observe that the normalized LMS method converges in approximately 700 samples. Next the final values for the weights are used and a new white noise test input is generated to compare the responses of the two systems. The results, shown in Figure 9.44, confirm that there is a good fit in this case, with a normalized mean square error of $E = .0029$.

FIGURE 9.43: Learning Curve Using the Normalized LMS Method with $m = 40$ and $\alpha = .1$

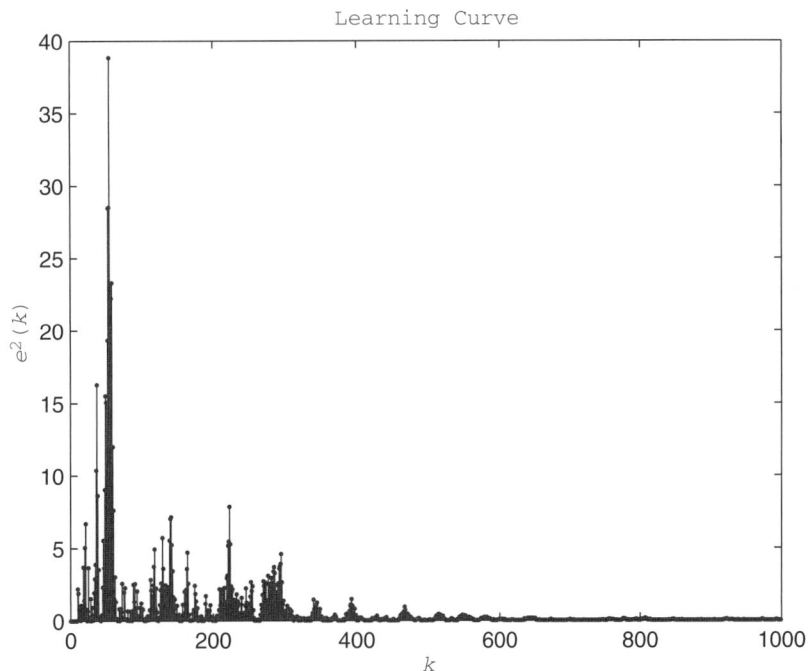

FIGURE 9.44: Comparison of Desired and Actual Outputs Using the Final Values for the Weights

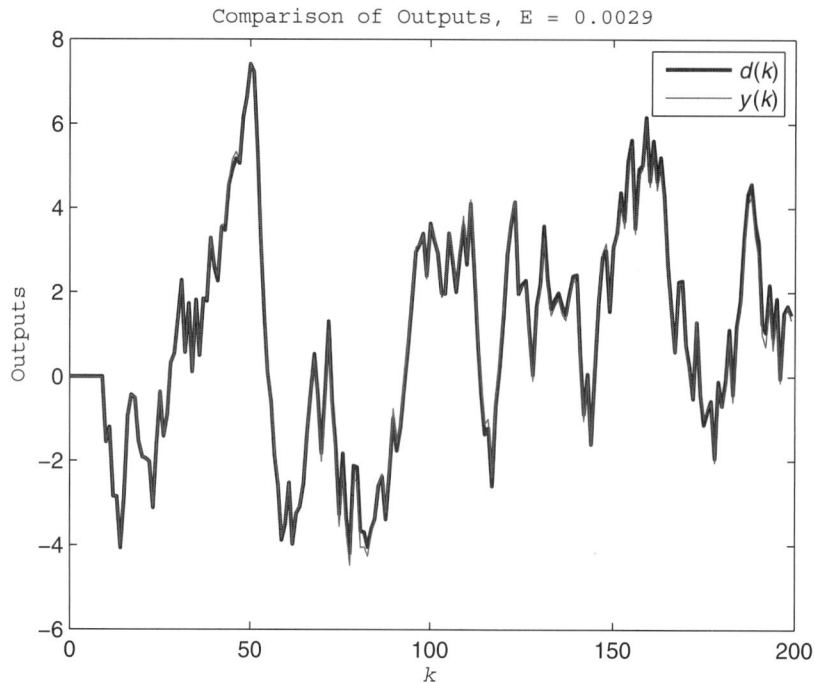

● ● ● ● ● ● ● ● ● ● ● ● ● ● ● ●

9.11 Chapter Summary

Least Mean Square Techniques

This chapter focused on linear and nonlinear adaptive signal processing techniques, and their applications. The adaptive filter structure used was a linear mth-order transversal filter of the following form.

$$y(k) = \sum_{i=0}^{m} w_i(k)x(k-i), \quad k \geq 0 \tag{9.11.1}$$

One important qualitative feature of this structure is that once the weights converge to their steady-state values, the resulting FIR filter is guaranteed to be BIBO stable. Adaptive systems can be configured in a number of ways depending on the application. Examples include system identification, channel equalization, signal prediction, and noise cancellation. Consider the following $(m+1) \times 1$ *state* vector of past inputs.

State vector

$$u(k) = [x(k), x(k-1), \ldots, x(k-m)]^T \tag{9.11.2}$$

A compact expression for the transversal-filter output can be obtained in terms of $u(k)$ using the following dot product formulation.

$$y(k) = w^T(k)u(k), \quad k \geq 0 \tag{9.11.3}$$

Weight vector

The $(m+1) \times 1$ *weight* vector $w(k)$ is adjusted so as to minimize the mean square error $\epsilon(w) = E[e^2(k)]$, where E is the expected value operator. Here the system error is $e(k) = d(k) - y(k)$, where $d(k)$ is the desired output. For example, in the system identification application, $d(k)$ is the output of the system to be identified. When the input $x(k)$ is sufficiently rich in frequency content, the mean square error can be shown to be a positive-definite quadratic function of the weight vector w, which means that a unique optimal weight vector exists. The weights can be adjusted by searching the mean square error function using a steepest-descent search method. For the purpose of computing the gradient of $\epsilon(w)$, one can use the simplifying assumption $E[e^2(k)] \approx e^2(k)$. This leads to the following popular weight-update algorithm called the least mean square or *LMS method*.

LMS method

$$w(k+1) = w(k) + 2\mu e(k)u(k), \quad k \geq 0 \tag{9.11.4}$$

The scalar parameter μ is the *step size*. If $P_x = E[x^2(k)]$ denotes the average power of the input, then the LMS method will *converge* for step lengths in the following range.

Convergence

$$0 < \mu < \frac{1}{(m+1)P_x} \tag{9.11.5}$$

Since the time constant of the LMS method is inversely proportional to the step size, larger step sizes are preferable for faster convergence. However, once convergence has been achieved, excess mean square error is encountered as a result of the approximation used for the gradient of the mean square error. Because excess mean square error is proportional to the step size, there is a tradeoff in choosing μ between convergence speed and steady-state accuracy. In many applications, it is common to choose values for μ that are about an order of magnitude less than the upper bound in (9.11.5).

The basic LMS method in (9.11.4) can be modified in a number of ways to enhance performance. These include the normalized LMS method, the correlation LMS method, and the leaky LMS method. The normalized and correlation methods feature step sizes that vary with time. For example, for the normalized LMS method, the step size is as follows where $0 < \alpha < 1$ is the normalized step size, and $\delta > 0$ is included to ensure that the step size never grows beyond α/δ.

Normalized step size

$$\mu(k) = \frac{\alpha}{\delta + u^T(k)u(k)} \tag{9.11.6}$$

The leaky LMS method tends to make the LMS method more stable by preventing the weights from becoming too large. It is useful for narrowband inputs because it has the effect of adding white noise to the input. However, this tends to increase the excess mean square error once convergence has taken place.

Adaptive Signal Processing Applications

Adaptive signal processing can be used to design FIR filters with prescribed magnitude and phase response characteristics. This is done by applying system identification to a synthetic pseudo-filter. The magnitude and phase responses of a physical system are not completely independent of one another. Consequently, an optimal transversal filter may or may not produce a close fit to the pseudo-filter specifications.

An important alternative to the LMS family of methods for updating the weights is the *recursive least-squares* or RLS method. The RLS method attempts to find the optimal weight at each time step, unlike the LMS method which approaches the optimal weights gradually using a steepest-descent search. As a consequence, the RLS method typically converges much faster than the LMS method, but there is more computational effort per iteration. The computational effort of the RLS method is of order $O(m^2)$ whereas the computational effort for the LMS method is of order $O(m)$ where m is the filter order.

An interesting application area of adaptive signal processing is the active control of acoustic noise. The basic premise is to inject a secondary sound into an environment so as to cancel the

FXLMS method

primary sound using destructive interference. This application requires adaptive techniques because the nature of the unwanted sound and the characteristics of the environment typically are unknown and change with time. The secondary sound must be generated by a speaker, transmitted over an air channel, and detected by a microphone. Consequently, active noise control requires a modification of the LMS method called the filtered-x or FXLMS method.

$$w(k + 1) = w(k) + 2\mu e(k)\hat{u}(k), \quad k \geq 0 \tag{9.11.7}$$

The FXLMS differs from the LMS method in that the state vector of past inputs first must be filtered by a model of the path traveled by the secondary sound. This model typically is obtained offline using standard system identification techniques. Often the primary noise consists of a narrowband periodic component plus white noise. When the fundamental frequency of the periodic component of the noise is known, a more direct signal synthesis approach can be used to cancel the periodic component of the primary noise.

Nonlinear System Identification

Nonlinear system

System identification using adaptive signal processing can be extended from linear systems to *nonlinear systems* of the following form.

$$y(k) = f[u(k)] \tag{9.11.8}$$

Here the state vector includes not only the $(m + 1)$ past inputs but also n past outputs. Thus f is a real-valued continuous nonlinear function of $p = m + n + 1$ variables. If the nonlinear system in (9.11.8) is BIBO stable and the input is bounded, then the domain of the function f is restricted to a closed bounded region $U \subset R^p$. This compact domain can be covered with a grid of r points, and over each grid element a simple local representation of $f(u)$ can be developed. This leads to the following adaptive nonlinear structure called a radial basis function or *RBF network*.

RBF network

$$y_0(k) = w(k)^T g[u(k)] \tag{9.11.9}$$

Here $g : R^r \rightarrow R$ is a $r \times 1$ vector of radial basis functions, one centered about each grid point. If a raised-cosine RBF is used, then the resulting network can be shown to have several useful properties. For example, $g(u)$ is continuously differentiable and if $w_i = f(u^i)$ where u^i is the ith grid point, then the error between the RBF model and the original nonlinear system is zero at the grid points. When the grid is sufficiently fine, this results in a RBF network that is an effective model of the nonlinear system over the domain U.

GUI Module

The FDSP toolbox includes a GUI module called *g_adapt* that allows the user to evaluate and compare several adaptive system identification techniques without any need for programming. The weight-update algorithms featured include the LMS method, the normalized LMS method, the correlation LMS method, the leaky LMS method, and the RLS method. The input and desired output data can be obtained from a user-specified IIR filter, or from a user-supplied MAT file. The latter option allows for identification based on actual physical measurements.

Learning Outcomes

This chapter was designed to provide the student with an opportunity to achieve the learning outcomes summarized in Table 9.2.

TABLE 9.2: ▶ Learning Outcomes for Chapter 9		

Num.	Learning Outcome	Sec.
1	Understand how to use adaptive filters to perform system identification, channel equalization, signal prediction, and noise cancellation.	9.1
2	Know how to compute the mean square error and how to find an optimal weight vector that minimizes the mean square error.	9.2
3	Understand how to implement the least mean square (LMS) method for updating the weight vector.	9.3
4	Know how to find bounds on the step size that ensure steady-state convergence of the LMS method.	9.4
5	Know how to estimate the rate of convergence and the error of the LMS method.	9.4
6	Understand how to modify the basic LMS method to enhance performance using the normalized, correlation, and leaky LMS methods.	9.5
7	Be able to design FIR filters using pseudo-filter input-output specifications.	9.6
8	Know how to apply the recursive least squares (RLS) method.	9.7
9	Know how to apply the filtered-x LMS and signal-synthesis adaptive methods to achieve active control of acoustic noise.	9.8
10	Be able to identify nonlinear discrete-time systems using an radial basis function or RBF network.	9.9
11	Know how to use the GUI module g_adapt to perform system identification	9.10

● ● · ● ● ● ● ● ● ● ● ● ● ● ● ● · ● ·

9.12 Problems

The problems are divided into Analysis and Design problems that can be solved by hand or with a calculator, GUI Simulation problems that are solved using GUI module g_adapt, and MATLAB Computation problems that require a user program.

9.12.1 Analysis and Design

9.1 Consider the following periodic input that is used as part of the input-output specification for a pseudo-filter. Suppose $f_i = i f_s/(2N)$ for $0 \le i < N$. Find the auto-correlation matrix R for this input.

$$x(k) = \sum_{i=0}^{N-1} C_i \cos(2\pi f_i kT)$$

9.2 Find the constant term, $P_d = E[d^2(k)]$, of the mean square error when the desired output is the following signal.

$$d(k) = b + \sin\left(\frac{2\pi k}{N}\right) - \cos\left(\frac{2\pi k}{N}\right)$$

9.3 There is an offline or batch procedure for computing the optimal weight vector called the least-squares method (see Problem 9.35). For large values of m, the least-squares method requires approximately $4(m + 1)^3/3$ FLOPs to find w. How many iterations are required before the computational effort of the LMS method equals or exceeds the computational effort of the least-squares method?

9.4 Suppose the misadjustment factor for the LMS method is $M_f = .4$ when the input is white noise uniformly distributed over $[-2, 2]$.
   (a) Find the average power of the input.
   (b) If the step size is $\mu = .01$, what is the filter order?
   (c) If the filter order is $m = 9$, what is the step size?

9.5 Suppose an input $x(k)$ and a desired output $d(k)$ have the following auto-correlation matrix and cross-correlation vector. Find the optimal weight vector w^*.

$$R = \begin{bmatrix} 5 & 1 \\ 1 & 5 \end{bmatrix}, \quad p = \begin{bmatrix} 3 \\ -2 \end{bmatrix}$$

9.6 Consider a transversal filter of order $m = 1$. Suppose the input and desired output are as follows.

$$x(k) = 2 + \sin(\pi k/2)$$
$$d(k) = 1 - 3\cos(\pi k/2)$$

   (a) Find the cross-correlation vector p.
   (b) Find the input auto-correlation matrix R.
   (c) Find the optimal weight vector w^*.

9.7 Suppose $v(k)$ is white noise uniformly distributed over $[-c, c]$. Consider the following input.

$$x(k) = 2 + \sin(\pi k/2) + v(k)$$

Find the input auto-correlation matrix R. Does your answer reduce to that of Problem 9.6 when $c = 0$?

9.8 Suppose the mean square error is approximated using a running average filter of order $M - 1$ as follows.

$$\epsilon(w) \approx \frac{1}{M} \sum_{i=0}^{M-1} e^2(k - i)$$

   (a) Find an expression for the gradient vector $\nabla \epsilon(w)$ using this approximation for the mean square error.
   (b) Using the steepest-descent method and the results from part (a), find a weight-update formula.
   (c) How many floating-point multiplications (FLOPs) are required per iteration to update the weight vector? You can assume that 2μ is computed ahead of time.
   (d) Verify that when $M = 1$ the weight-update formula reduces to the LMS method.

9.9 Financial considerations dictate that a production system must remain in operation while the system is being identified. During normal operation of the linear system, the input $x(k)$ has relatively poor spectral content.
   (a) Which of the modified LMS methods would appear to be an appropriate choice? Why?
   (b) How might the input be modified slightly to improve identification without significantly affecting the normal operation of the system?

9.10 Consider the normalized LMS method.
   (a) What is the maximum value of the step size?
   (b) Describe an initial condition for the past inputs that will cause the step size to saturate to its maximum value.

9.11 The transversal filter structure used in this chapter is a time-varying FIR filter. One can generalize it by using the following time-varying IIR filter.

$$y(k) = \sum_{i=0}^{m} b_i(k)x(k-i) - \sum_{i=1}^{n} a_i(k)y(k-i)$$

(a) Find suitable definitions for the state vector $u(k)$ and the weight vector $w(k)$ such that the output of the time-varying IIR filter can be expressed as a dot product as in (9.2.3). That is,

$$y(k) = w^T(k)u(k), \quad k \geq 0$$

(b) Suppose the weight vector $w(k)$ converges to a constant. Is the resulting filter guaranteed to be BIBO stable? Why or why not?

9.12 Suppose an input $x(k)$ has the following auto-correlation matrix.

$$R = \begin{bmatrix} 2 & 1 \\ 1 & 2 \end{bmatrix}$$

(a) Using the eigenvalues of R, find a range of step sizes that ensures convergence of the LMS method.
(b) Using the average power of the input, find a more conservative range of step sizes that ensures convergence of the LMS method.
(c) Suppose the step size is one-tenth the maximum in part (b). Find the time constant of the mean square error in units of iterations.
(d) Using the same step size as in part (c), find the misadjustment factor M_f.

9.13 Suppose the LMS learning curve converges to within one percent of its final steady-state value in 200 iterations.
(a) Find the learning-curve time constant τ_{mse} in units of iterations.
(b) If the minimum eigenvalue of R is $\lambda_{min} = .1$, what is the step size?
(c) If the step size is $\mu = .02$, what is the minimum eigenvalue of R?

9.14 Suppose a transversal adaptive filter is of order $m = 2$. Find the input auto-correlation matrix R for the following cases.
(a) The input $x(k)$ consists of white noise uniformly distributed over the interval $[a, b]$.
(b) The input $x(k)$ consists of Gaussian white noise with mean μ_x and variance σ_x^2.

9.15 Suppose the first row of an auto-correlation matrix R is $r = [9, 7, 5, 3, 1]$.
(a) Find R.
(b) What is the average power of the input?
(c) Suppose $x(k)$ is white noise uniformly distributed over the interval $[0, c]$. Find c.

9.16 Consider the following periodic input and desired output that form the input-output specification for a pseudo-filter. Suppose $f_i = if_s/(2N)$ for $0 \leq i < N$. Find the cross-correlation vector p for this input and desired output.

$$x(k) = \sum_{i=0}^{N-1} C_i \cos(2\pi f_i kT)$$

$$d(k) = \sum_{i=0}^{N-1} A_i C_i \cos(2\pi f_i kT + \phi_i)$$

9.17 Consider a raised-cosine RBF network with $m = 0, n = 0, d = 2$, and $a = [0, 1]$. Using the trigonometric identities from Appendix 2, show that the constant interpolation property holds in this case. That is, show that

$$g_0(u) + g_1(u) = 1, \quad a_1 \le u \le a_2$$

9.18 Suppose the nonlinear function in (9.9.4) is $f(u) = h^T u + c$ for some $p \times 1$ vector h and some constant c. Let $d_i = 2$ for $1 \le i \le p$, $w_i = c$ for $0 \le i < r$, and $V_{ij} = h_j$ for $1 \le i \le r$ and $1 \le j \le p$. Show that the first-order RBF network S_1 is exact. That is, show that if $f_1(u) = (Vu + w)^T g(u)$, then

$$f_1(u) = h^T u + c \quad \text{for} \quad u \in U$$

9.19 Suppose the nonlinear function in (9.9.4) is $f(u) = c$ for some constant c. Let $d_i = 2$ for $1 \le i \le p$ and $w_i = c$ for $0 \le i < r$. Show that the zeroth-order RBF network S_0 is exact. That is, show that if $f_0(u) = w^T g(u)$, then

$$f_0(u) = c \quad \text{for} \quad u \in U$$

9.20 Consider the active noise control system shown in Figure 9.45. Suppose the secondary path is modeled as a delay with attenuation. That is, for some delay $\tau > 0$ and some attenuation $0 < \alpha < 1$,

$$\hat{y}(t) = \alpha y(t - \tau)$$

(a) Let the sampling interval be $T = \tau/M$. Find the transfer function, $F(z)$.

(b) Suppose the primary path $G(z)$ is modeled as follows. Find $W(z)$ using (9.8.3).

$$G(z) = \sum_{i=0}^{m} \frac{z^{-i}}{1+i}$$

(c) Is the controller $W(z)$ physically realizable? Why or why not?

FIGURE 9.45: Active Control of Acoustic Noise

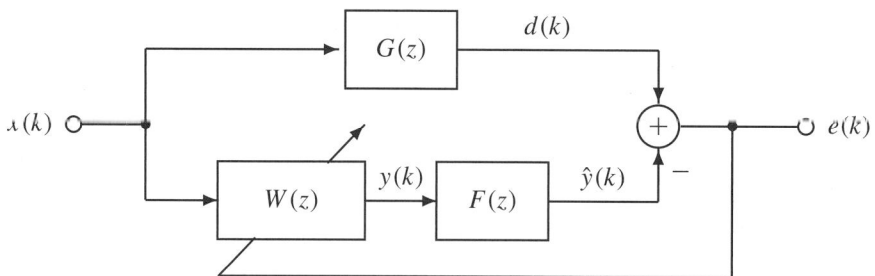

9.21 Consider the following expression for the generalized cross-correlation vector used by the RLS method.

$$p(k) = \sum_{i=1}^{k} \gamma^{k-i} d(i) u(i)$$

Show that $p(k)$ can be expressed recursively in terms of $p(k - 1)$ by deriving the expression for $p(k)$ in (9.7.8).

9.22 Consider the following candidate for a scalar radial basis function.

$$G_i(z) = \begin{cases} \cos^{2i}\left(\dfrac{\pi z}{2}\right), & |z| \leq 1 \\ 0, & |z| > 1 \end{cases}$$

(a) Show that $G_i(z)$ qualifies as an RBF for $i \geq 1$.
(b) Does $G_i(z)$ have compact support?
(c) Show that $G_i(z)$ reduces to the raised-cosine RBF when $i = 1$.

9.23 Suppose the nonlinear function f in (9.9.4) is continuously differentiable. Let $F_0(u) = w^T g(u)$ and consider the following metric for the error between the output of the system S_f and the output of the zeroth-order RBF network S_0.

$$E(d) \stackrel{\Delta}{=} \max_{u \in U}\{|f(u) - f_0(u)|\}$$

Show that the RBF model S_0 converges uniformly to the nonlinear system S_f as d approaches infinity. That is, show that

$$E(d) \to 0 \quad \text{as} \quad d \to \infty$$

9.24 Consider a raised-cosine RBF network with $m = 2$ past inputs and $n = 2$ past outputs. Suppose the range of values for the inputs is $a = [0, 5]$, and the range of values for the outputs is $[-2, 8]$. Let the number of grid points per dimension be $d = 6$.

(a) Find the compact support Ω of the overall network. That is, find the smallest closed, bounded region $\Omega \subset R^p$ such that

$$u \notin \Omega \Rightarrow f_0(u) = 0$$

(b) Show that, in general, $\Omega \to U$ as $d \to \infty$ where $U \in R^p$ is the domain of f.

9.25 Consider the problem of identifying the nonlinear discrete-time system in (9.9.4) using a raised-cosine RBF network. Let the number of past inputs be $m = 1$ and the number of past outputs be $n = 1$. Suppose the range of values for the inputs is $a = [-2, 2]$, and the range of values for the outputs is $b = [-3, 3]$. Let the number of grid points per dimension be $d = 4$.

(a) Find the domain U of the function f.
(b) What is the total number of grid points?
(c) What is the grid point spacing in the x direction and in the y direction?
(d) For each u, what is the maximum number of nonzero terms in the RBF network output?
(e) Consider the following state vector. Find the vector subscripts of the vertices of the grid element containing u.

$$u = [.3, -1.7, 1.1]^T.$$

(f) Find the scalar subscripts of the vertices of the grid element containing the u in part (e).

9.12.2 GUI Simulation

9.26 Using the GUI module g_adapt and the Data source option, load the input and desired output from the MAT-file u_adapt1. Then identify the system that produced these input-output data using the normalized LMS method. Plot the following
(a) The learning curve.
(b) The magnitude responses using the dB scale.
(c) The step sizes. Use the *Caliper* option to mark the largest step size.

9.27 Consider the following FIR black box system. Use the GUI module *g_adapt* to identify this system using the LMS method.

$$H(z) = 1 - 2z^{-1} + 7z^{-2} + 4z^{-4} - 3z^{-5}$$

Save the data in a MAT-file named *prob9_27.mat* and then reload it using the Data source option.
(a) Plot the learning curve when $m = 3$.
(b) Plot the learning curve when $m = 5$.
(c) Plot the learning curve when $m = 7$.
(d) Plot the final weights $m = 7$.

9.28 Use the GUI module *g_adapt* to identify the following black box system using the normalized LMS method.

$$H(z) = \frac{3}{1 - .7z^{-4}}$$

(a) Plot the magnitude responses.
(b) Plot the learning curve.
(c) Plot the step sizes.

9.29 Using the GUI module *g_adapt*, identify the black box system using the LMS method. Set the step size to $\mu = .03$, and then plot the following.
(a) The outputs.
(b) The magnitude responses.
(c) The learning curve.
(d) The final weights.

9.30 Using the GUI module *g_adapt*, identify the black box system using the leaky LMS method. Adjust the number of samples to $N = 500$ and the leakage factor to mu $= .999$. Plot the following.
(a) The outputs.
(b) The magnitude responses.
(c) The learning curve.

9.31 Using the GUI module *g_adapt*, identify the black box system using the following two methods. Plot the learning curve for each case. Observe the scale of the dependent variable.
(a) The LMS method.
(b) The RLS method.

9.32 Using the GUI module *g_adapt*, identify the black box system using the leaky LMS method. Plot the learning curve for the following cases corresponding to different values of the leakage factor.
(a) *nu* = .999.
(b) *nu* = .995.
(c) *nu* = .990.

9.33 Use the GUI module *g_adapt* to identify the following black box system using the correlation LMS method with a filter of order $m = 50$.

$$H(z) = \frac{2}{1 + .8z^{-4}}$$

(a) Plot the magnitude responses.
(b) Plot the learning curve.
(c) Plot the step sizes.

9.12.3 MATLAB Computation

9.34 Consider the problem of designing an adaptive noise-cancellation system as shown in Figure 9.46. Suppose the additive noise $v(k)$ is white noise uniformly distributed over $[-2, 2]$. Let the primary microphone signal be as follows.

$$x(k) = \cos\left(\frac{\pi k}{10}\right) - .5\sin\left(\frac{\pi k}{20}\right) + .25\cos\left(\frac{\pi k}{30}\right)$$

Suppose the path for detecting the noise signal has the following transfer function.

$$H(z) = \frac{.5}{1 + .25z^{-2}}$$

Write a MATLAB program that uses the FDSP toolbox function f_lms to cancel the noise $v(k)$ corrupting the signal $d(k)$. Use an adaptive filter of order $m = 30$, $N = 3000$ samples, and a step size of $\mu = .003$.
(a) Plot the learning curve
(b) Using the final weights, compute $y(k)$ using input $r(k)$. Then plot $x(k)$, $d(k)$, and $e(k)$ for $0 \le k \le N/10$ on the same graph with a legend.

FIGURE 9.46: Noise cancellation

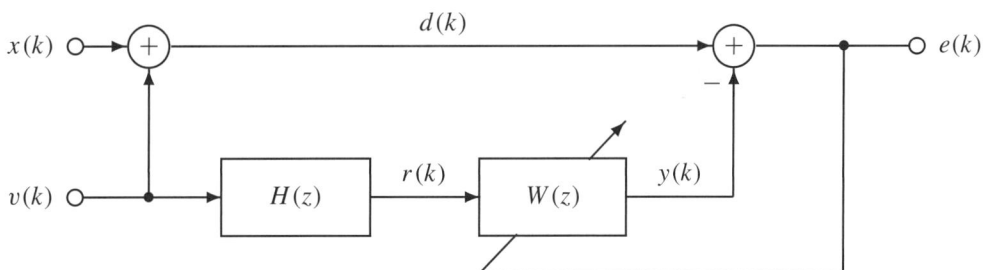

9.35 There is an offline alternative to the LMS method called the *least-squares* method that is available when the entire input signal and desired output signal are available ahead of time. Suppose the weight vector w is constant. Taking the transpose of (9.2.3), and replacing the actual output by the desired output, yields

$$u^T(k)w = d(k), \quad 0 \le k < N$$

Let $d = [d(0), d(1), \ldots, d(N-1)]^T$ and let X be an $N \times (m+1)$ past input matrix whose ith row is $u^T(i)$ for $0 \le i < N$. Then the N equations can be recast as the following vector equation.

$$Xw = d$$

Normal equations When $N > (m+1)$, this constitutes an over-determined linear algebraic system of equations. A weight vector that minimizes the squared error $E = (Xw - d)^T(Xw - d)$ is obtained by premultiplying both sides by X^T. This yields the *normal equations*

$$X^T X w = X^T d$$

The coefficient matrix $X^T X$ is $(m + 1) \times (m + 1)$. If $x(k)$ has adequate spectral content, $X^T X$ will be nonsingular. In this case the optimal weight vector in a least-squares sense can be obtained by premultiplying by the inverse of $X^T X$ which yields

$$w = (X^T X)^{-1} X^T d$$

Write a MATLAB function called f_lsfit that computes the optimal least-squares FIR filter weight vector, $b = w$, by solving the normal equations using the MATLAB left division operator, \. The calling sequence should be as follows.

```
% F_LSFIT: FIR system identification using offline least-squares fit method
%
% Usage:
%          w = f_lsfit (x,d,m)
% Pre:
%          x    = N by 1 vector containing input samples
%          d    = N by 1 vector containing desired output samples
%          m    = order of transversal filter (m < N)
% Post:
%          b = (m+1) by 1 least-squares FIR filter coefficient vector
```

In constructing X, you can assume that $x(k)$ is causal. Test f_lsfit by using $N = 250$ and $m = 30$. Let x be white noise uniformly distributed over $[-1, 1]$, and let d be a filtered version of x using the following IIR filter.

$$H(z) = \frac{1 + z^{-2}}{1 - .1z^{-1} - .72z^{-2}}$$

(a) Use *stem* to plot the least-squares weight vector b.
(b) Compute $y(k)$ using the weight vector b. Then plot $d(k)$ and $y(k)$ for $0 \leq k \leq 50$ on the same graph using a legend.

9.36 Consider the problem of designing an equalizer as shown in Figure 9.47. Suppose the delay is $M = 15$, and $H(z)$ represents a communication channel with the following transfer function.

$$H(z) = \frac{1 + .5z^{-1}}{1 + .4z^{-1} - .32z^{-2}}$$

Write a MATLAB program that uses the FDSP toolbox function f_lms to construct an equalizer of order $m = 30$ for $H(z)$. Suppose $x(k)$ consists of $N = 1000$ samples of white noise uniformly distributed over $[-3, 3]$. Use a step size of $\mu = .002$.

FIGURE 9.47:
Equalization of a
Communication
Channel, $H(z)$

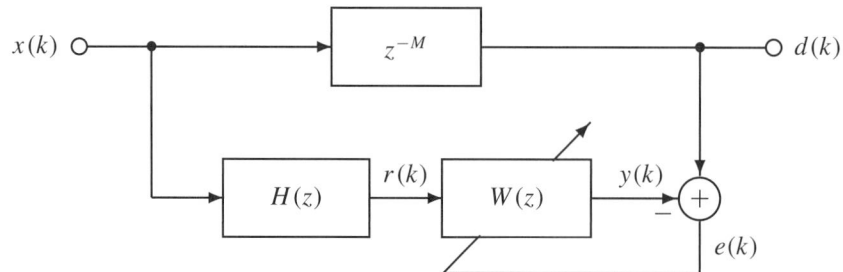

(a) Plot the learning curve.

(b) Using the final weights, compute $y(k)$ using input $r(k)$. Then plot $d(k)$ and $y(k)$ for $0 \leq k \leq N/10$ on the same graph with a legend.

(c) Using the final weights, plot the magnitude responses of $H(z)$, $W(z)$, and $F(z) = H(z)W(z)$ on the same graph using a legend. For the abscissa, use normalized frequency, f/f_s.

9.37 A plot of the squared error is only a rough approximation to the learning curve in the sense that $E[e^2(k)] \approx e^2(k)$. Write a MATLAB program that uses the FDSP toolbox function f_lms to identify the following system. For the input use $N = 500$ samples of white noise uniformly distributed over $[-1, 1]$, and for the filter order use $m = 30$.

$$H(z) = \frac{z}{z^3 + .7z^2 - .8z - .56}$$

(a) Use a step size μ that corresponds to .1 of the upper bound in (9.4.16). Print the step size used.

(b) Compute and print the mean square error time constant in (9.4.29), but in units of iterations.

(c) Construct and plot a learning curve by performing the system identification $M = 50$ times with a different white noise input used each time. Plot the average of the M $e^2(k)$ versus k curves and draw vertical lines at integer multiples of the time constant.

9.38 Consider the following IIR filter.

$$H(z) = \frac{10(z^2 + z + 1)}{z^4 + .2z^2 - .48}$$

Write a MATLAB program that uses the FDSP toolbox function $f\_lmsleak$ to identify a model of order $m = 30$ for this system. Use an input consisting of $N = 120$ samples, a step size of $\mu = .005$, and the following periodic input.

$$x(k) = \cos\left(\frac{\pi k}{5}\right) + \sin\left(\frac{\pi k}{10}\right)$$

(a) Plot the learning curve for $\nu = .99$.

(b) Plot the learning curve for $\nu = .98$.

(c) Plot the learning curve for $\nu = .96$.

(d) Using $\nu = .995$ and the final value for the weights, plot $d(k)$ and $y(k)$ on the same graph with a legend.

9.39 Use the FDSP toolbox to write a MATLAB program that designs an FIR filter to meet the following pseudo-filter design specifications.

$$A(f) = \begin{cases} 2, & 0 \leq f < \dfrac{f_s}{6} \\[2mm] 3, & \dfrac{f_s}{6} \leq f < \dfrac{f_s}{3} \\[2mm] 3 - 24\left(f - \dfrac{f_s}{3}\right), & \dfrac{f_s}{3} \leq f < \dfrac{5f_s}{12} \\[2mm] 1, & \dfrac{5f_s}{12} \leq f \leq \dfrac{f_s}{2} \end{cases}$$

$$\phi(f) = -30\pi f/f_s$$

Suppose there are $N = 80$ discrete frequencies equally spaced over $0 \leq f < f_s/2$, as in (9.6.2). Use f_lms with a step size of $\mu = .0001$ and $M = 2000$ iterations.

(a) Choose an order for the adaptive filter that best fits the phase specification. Print the order m.

(b) Plot the magnitude response of the filter obtained using the final weights. On the same graph, plot the desired magnitude response with isolated plot symbols at each of the N discrete frequencies, and a plot legend.

(c) Plot the phase response of the filter obtained using the final weights. On the same graph, plot the desired phase response with isolated plot symbols at each of the N discrete frequencies, and a plot legend.

9.40 Consider the problem of performing system identification as shown in Figure 9.48. Suppose the system to be identified is the following auto-regressive or all-pole filter.

$$H(z) = \frac{1}{z^4 - .1z^2 - .72}$$

Write a MATLAB program that uses the FDSP toolbox function $f\_lmsnorm$ to identify a model of order $m = 60$ for this system. Use an input consisting of $N = 1200$ samples of white noise uniformly distributed over $[-1, 1]$, a constant step size of $\alpha = .1$, and a maximum step size of $\mu_{max} = 5\alpha$.

(a) Plot the learning curve.

(b) Plot the step sizes.

(c) Plot the magnitude response of $H(z)$ and $W(z)$ on the same graph using a legend where $W(z)$ is the adaptive filter using the final values for the weights.

FIGURE 9.48:
Identification of
Linear Discrete-time
System, $H(z)$

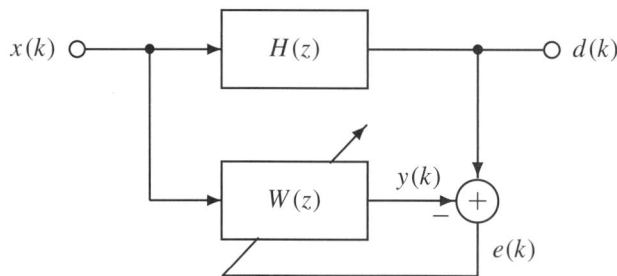

9.41 Consider the problem of performing system identification as shown in Figure 9.48. Suppose the system to be identified is the following IIR filter.

$$H(z) = \frac{z^2}{z^3 + .8z^2 + .25z + .2}$$

Write a MATLAB program that uses the FDSP toolbox function $f\_lmscorr$ to identify a model of order $m = 50$ for this system. Use an input consisting of $N = 2000$ samples of white noise uniformly distributed over $[-1, 1]$, a relative step size of $\alpha = 1$, and the default smoothing parameter β.

(a) Plot the learning curve.

(b) Plot the step sizes.

(c) Plot the magnitude response of $H(z)$ and $W(z)$ on the same graph using a legend where $W(z)$ is the adaptive filter using the final values for the weights.

9.42 Consider the problem of designing a signal predictor as shown in Figure 9.49. Suppose the signal whose value is to be predicted is as follows.

$$x(k) = \sin\left(\frac{\pi k}{5}\right) \cos\left(\frac{\pi k}{10}\right) + v(k), \quad 0 \le k < N$$

Here $N = 200$ and $v(k)$ is white noise uniformly distributed over $[-.05, .05]$. Write a MATLAB program that uses the FDSP toolbox function f_rls to predict the value of this signal $M = 20$ samples into the future. Use a filter of order $m = 40$ and a forgetting factor of $\gamma = .9$.

(a) Plot the learning curve.

(b) Using the final weights, compute the output $y(k)$ corresponding to the input $x(k)$. Then plot $x(k)$ and $y(k)$ on separate graphs above one another using the *subplot* command. Use the *fill* function to shade a section of $x(k)$ of length M starting at $k = 160$. Then shade the corresponding predicted section of $y(k)$ starting at $k = 140$.

FIGURE 9.49: Signal Prediction

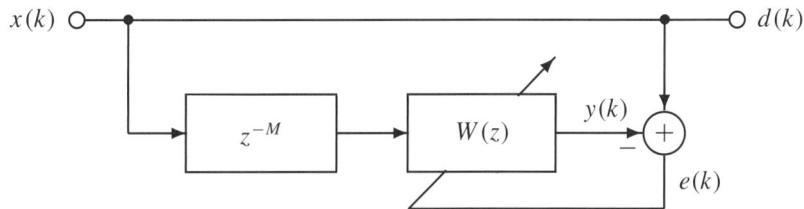

9.43 Consider the active noise control system shown previously in Figure 9.45. Suppose the primary noise $x(k)$ consists of the following noise-corrupted periodic signal.

$$x(k) = 2 \sum_{i=1}^{5} \frac{\sin(2\pi F_0 ikT)}{1 + i} + v(k)$$

Here the fundamental frequency is $F_0 = 100$ Hz and $f_s = 2000$ Hz. The additive noise term $v(k)$ is white noise uniformly distributed over $[-.2, .2]$. Coefficient vectors for FIR models of the secondary path $F(z)$ and the primary path $G(z)$ are contained in MAT file *prob9_44.mat*. The coefficient vectors are f and g. Write a MATLAB program that loads f and g and uses the FDSP toolbox function f_sigsyn to apply active noise control with the signal synthesis method starting at sample $N/4$ where $N = 2000$. Use a step size of $\mu = .04$.

(a) Plot the learning curve. Add a title that displays the amount of noise cancellation in dB using (9.8.15).

(b) Plot the magnitude spectra of the noise without cancellation.

(c) Plot the magnitude spectra of the noise with cancellation.

9.44 Consider the active noise control system shown previously in Figure 9.45. Suppose the primary noise $x(k)$ consists of the following noise-corrupted periodic signal.

$$x(k) = 2 \sum_{i=1}^{5} \frac{\sin(2\pi F_0 ikT)}{1 + i} + v(k)$$

Here the fundamental frequency is $F_0 = 100$ Hz and $f_s = 2000$ Hz. The additive noise term $v(k)$ is white noise uniformly distributed over $[-.2, .2]$. Coefficient vectors for FIR models of the secondary path $F(z)$ and the primary path $G(z)$ are contained in MAT-file *prob9_44.mat*. The coefficient vectors are f and g. Write a MATLAB program that loads f and g and uses the FDSP toolbox function *f_fxlms* to apply active noise control with the filtered-x LMS method starting at sample $N/4$ where $N = 2000$. Use a noise controller of order $m = 30$ and a step size of $\mu = .002$. Plot the learning curve including a title that displays the amount of noise cancellation in dB using (9.8.15).

9.45 Consider the following nonlinear discrete-time system which has $m = 0$ past inputs and $n = 1$ past outputs.

$$y(k) = .8y(k_1) + .3[x(k) - y(k-1)]^3$$

Let the range of inputs be $-1 \le x(k) \le 1$ and the number of grid points per dimension be $d = 8$. Write a MATLAB program that does the following.
(a) Use the FDSP toolbox function *f_state* to compute a set of output bounds b such that $b_1 \le y(k) \le b_2$. Use $P = 1000$ points of white noise uniformly distributed over $[-1, 1]$ for the test input and a safety factor of $\beta = 1.2$ as in (9.9.29). Print a, b, and the total number of grid points r.
(b) Use FDSP toolbox function *f_rbfw* with $N = 0$ and $ic = 1$ to compute a weight vector w that satisfies (9.9.22). Then use *f_rbf0* to compute the output $y_0(k)$ to a white noise input with $M = 100$ points uniformly distributed over $[-1, 1]$. Use *f_state* to compute the nonlinear system response $y(k)$ to the same input. Plot the two outputs on one graph using a legend. Compute the error E using (9.9.31) and add this to the graph title.

9.46 Consider the following nonlinear discrete-time system which has $m = 0$ past inputs and $n = 1$ past outputs.

$$y(k) = .8y(k_1) + .3[x(k) - y(k-1)]^3$$

Suppose the input $x(k)$ consists of $N = 1000$ samples of white noise uniformly distributed over $[-1, 1]$. Let the number of grid points per dimension be $d = 8$. Write a MATLAB program that performs the following tasks.
(a) Use the FDSP toolbox function *f_state* to compute a set of output bounds b such that $b_1 \le y(k) \le b_2$. Use a safety factor of $\beta = 1.2$ as in (9.9.29). Print a, b, Δx, Δy, and the total number of grid points r.
(b) Plot the output $y(k)$ corresponding to the white noise input $x(k)$. Include dashed lines showing the grid values along the y dimension.
(c) Let $f(u)$ denote the right-hand side of the nonlinear difference equation where $u(k) = [x(k), y(k-1)]^T$. Plot the surface $f(u)$ over the domain $[a_1, a_2] \times [b_1, b_2]$.

9.47 Consider the active noise control system shown previously in Figure 9.45. Suppose the secondary path is modeled by the following transfer function which takes into account the delay and attenuation of sound as it travels through air, and the characteristics of the microphones, speaker, amplifiers, and DAC.

$$F(z) = \frac{.2z^{-3}}{1 - 1.4z^{-1} + .48z^{-2}}$$

Suppose the sampling frequency is $f_s = 2000$ Hz. Write a MATLAB program that uses the FDSP toolbox f_lms to identify an FIR model of the secondary path $F(z)$ using an adaptive filter of order $m = 25$. Choose an input and a step size that causes the algorithm to converge.

(a) Plot the learning curve to verify convergence.
(b) Plot the magnitude responses of $F(z)$ and the model $\hat{F}(z)$ on the same graph using a legend.
(c) Plot the phase responses of $F(z)$ and the model $\hat{F}(z)$ on the same graph using a legend.

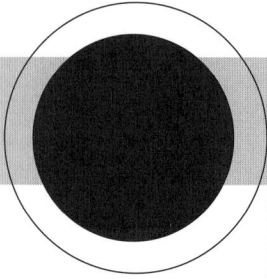

References and Further Reading

1. Ahmed, N., and Natarajan, T., *Discrete-Time Signals and Systems*, Reston: Reston, VA, 1983.
2. Bartlett, M. S., "Smoothing Periodograms from Time Series with Continuous Spectra," *Nature*, Vol. 161, pp. 686–687, May, 1948.
3. Bellanger, M., Bonnerot, G., and Coudreuse, M., "Digital Filtering by Polyphase Network: Application to Sample Rate Alteration and Filter Banks," *IEEE Trans. Acoustics, Speech, and Signal Processing*, Vol. 24, pp. 109–114, 1976.
4. Bellanger, M., *Adaptive Digital Filters and Signal Analysis*, Marcel Dekker: New York, 1987.
5. Bequette, B. W., *Process Control: Modeling, Design, and Simulation*, Prentice Hall: Upper Saddle River, NJ, 2003.
6. Burrrus, C. S., and Guo, H., *Introduction to Wavelets and Wavelet Transforms: A Primer*, Prentice Hall: Upper Saddle River, NJ, 1997.
7. Candy, J. C., and Temes, G. C., *Oversampling Delta-Sigma Data Converters*, IEEE Press: New York, 1992.
8. Chapman, S. J., *MATLAB Programming for Engineers*, Second Edition, Brooks/Cole: Pacific Grove, CA, 2002.
9. Chatfield, C., *The Analysis of Time Series*, Chapman and Hall: London, 1980.
10. Constantinides, A. G., "Spectral Transformations for Digital Filters," *Proc. Inst. Elec. Engr.*, Vol. 117, pp. 1585–1590, 1970.
11. Cook, T. A., *The Curves of Life*, Dover: Mineola, NY, 1979.
12. Cooley, J. W., and Tukey, R. W., "An Algorithm for Machine Computation of Complex Fourier Series," *Mathematics of Computation*, Vol. 19, pp. 297–301, 1965.
13. Crochiere, R. E., and Rabiner, L. R., "Optimum FIR Digital Filter Implementations for Decimation, Interpolation, and Narrow-band Filtering," *IEEE Trans. Acoustics, Speech, and Signal Processing*, Vol. 23, No. 5, pp. 444–456, 1975.
14. Crochiere, R. E., and Rabiner, L. R., "Further Considerations in the Design of Decimators and Interpolators," *IEEE Trans. Acoustics, Speech, and Signal Processing*, Vol. 24, pp. 296–311, 1976.
15. Dorf, R.C., and Svoboda, J.A., *Introduction to Electric Circuits*, Wiley: New York, 2000.
16. Durbin, J., "Efficient Estimation of Parameters in Moving Average Model," *Biometrika*, Vol. 46, pp. 306–316, 1959.
17. Dwight, H. B., *Tables of Integrals and other Mathematical Data*, Fourth Edition, MacMillan: New York, 1961.
18. Elliott, S. J., Stothers, I. M., and Nelson, P. A., "A Multiple Error LMS Algorithm and its Application to the Active Control of Sound and Vibration," *IEEE Trans. Acoustics, Speech and Signal Processing*, Vol. ASSP-35, pp. 1423–1434, 1987.

19. Franklin, G. E., Powell, J. D., and Workman, M. L., *Digital Control of Dynamic Systems*, Second Ed., Addison-Wesley Publishing: Reading, MA, 1990.

20. Gerald, C. F., and Wheatley, P. O., *Applied Numerical Analysis*, Fourth Edition, Addison-Wesley: Reading, MA, 1989.

21. Gitlin, R. D., Meadors, H. C., and Weinstein, S. B., "The Tap-leakage Algorithm: An Algorithm for the Stable Operation of a Digitally Implemented, Fractional Adaptive Spaced Equalizer," *Bell System Tech. J.*, Vol. 61, pp. 1817–1839, 1982.

22. Grover, D., and Deller J. R., *Digital Signal Processing and the Microcontroller*, Prentice Hall: Upper Saddle River, NJ, 1999.

23. Hanselman, D., and Littlefield, B., *Mastering MATLAB 6*, Prentice Hall: Upper Saddle River, NJ, 2001.

24. Hassibi, B. A., Sayed, H., and Kailath, T., "H^∞ Optimality of the LMS Algorithm," *IEEE Trans. on Signal Processing*, Vol. 44, pp. 267–280, 1996.

25. Haykin, S., *Adaptive Filter Theory*, Fourth Edition, Prentice Hall: Upper Saddle River, NJ, 2002.

26. Hernandez, E., and Arkun, Y., "Stability of Nonlinear Polynomial ARMA Models and Their Inverse," *Int. J. Control*, Vol. 63, No. 5, pp. 885–906, 1996.

27. Ifeachor, E. C., and Jervis, B. W., *Digital Signal Processing: A Practical Approach*, Second Edition, Prentice-Hall: Harlow, UK, 2002.

28. Ingle, V. K., and Proakis, J. G., *Digital Signal Processing Using MATLAB*, Brooks/Cole: Pacific Grove, CA, 2000.

29. Jackson, L. B., *Digital Filters and Signal Processing*, Third Edition, Kluwer Academic Publishers: Boston, 1996.

30. Jaffe, D. A., and Smith, J. O., "Extensions of the Karplus-Strong Plucked-string Algorithm," *Computer Music Journal*, Vol. 7, No. 2, pp. 56–69, 1983.

31. Jameco Electronics Catalog, 1355 Shoreway Road, Belmont, CA, 94002–4100, 2004.

32. Jansson, P. A., *Deconvolution*, Academic Press: New York, 1997.

33. Kailath, T., *Estimating Filters for Linear Time-Invariant Channels*, Quarterly Progress Rep., 58, MIT Research Laboratory for Electronics, Cambridge, MA, pp. 185–197, 1960.

34. Kaiser, J. F., "Digital Filters," Chap. 7 of *System Analysis by Digital Computer*, F. F. Kuo and J. F. Kaiser, Eds., Wiley: New York, 1966.

35. Kaiser, J. F., "Nonrecursive Digital Filter Design Using the I_0-sinh Window Function," *Proc. 1974 IEEE Int. Symp. on Circuits and Systems*, San Francisco, CA, pp. 20-23, April, 1974.

36. Kuo, S. M., and Gan, W.-S., *Digital Signal Processing: Architectures, Implementations, and Applications*, Pearson-Prentice Hall: Upper Saddle River, NJ, 2005.

37. Kuo, S. M., and Morgan, D. R., *Active Noise Control Systems: Algorithms and DSP Implementations*, Wiley: New York, 1996.

38. Lam, H. Y.-F., *Analog and Digital Filters*, Prentice-Hall: Englewood Cliffs, NJ, 1979.

39. Levinson, N., "The Wiener RMS Criterion in Filter Design and Prediction," *J. Math. Phys.*, Vol. 25, pp. 261–278, 1947.

40. Ljung, L., Morf, M., and Falconer, D., "Fast Calculation of Gain Matrices for Recursive Estimation Schemes," *Int. J. Control*, Vol. 27, pp 1–19, 1978.

41. Ludeman, L. C., *Fundamentals of Digital Signal Processing*, Harper and Row: New York, 1986.

42. Markel, J. D., and Gray, A. H., Jr., *Linear Prediction of Speech*, Springer-Verlag: New York, 1976.

43. Marwan, N., "Make Install Tool for MATLAB," www.agnld.uni-potsdam.de/ marwan/ 6.download/ whitepaper_makeinstall.html, Potsdam, Germany, 2003.

44. McGillem, C.D., and Cooper, G.R., *Continuous and Discrete Signal and System Analysis*, Holt, Rhinehart and Winston: New York, 1974.

45. Mitra, S. K., *Digital Signal Processing: A Computer-Based Approach*, Second Edition, McGraw-Hill Irwin: Boston, 2001.

46. Moorer, J.A., "About the Reverberation Business," *Computer Music Journal*, Vol. 3, No. 2, pp. 13–28, 1979.

47. Moustakides, G. V., "Study of the Transient Phase of the Forgetting Factor RLS," *IEEE Trans. Signal Processing*, Vol. 45, pp. 2468–2476, 1997.

48. Nilsson, J. W., *Electric Circuits*, Addison-Wesley: Reading, MA, 1982.

49. Noble, B., *Applied Linear Algebra*, Prentice Hall: Englewood Cliffs, NJ, 1969.

50. Oppenheim, A. V., Schafer, R. W., and Buck, J. R., *Discrete-Time Signal Processing*, Prentice Hall: Upper Saddle River, NJ, 1999.

51. Papamichalis, P., *Digital Signal Processing Applications with the TMS320 Family. Theory, Algorithms, and Implementations*, Vol. 3, Texas Instruments: Dallas TX, 1990.

52. Park, S. K., and Miller, K. W., "Random Number Generators: Good Ones are Hard to Find," *Communications of the ACM*, Vol. 31, pp. 1192–1201, 1988.

53. Parks, T. W., and McClellan, J. H., "Chebyshev Approximation for Nonrecursive Digital Filters with Linear Phase," *IEEE Trans. Circuit Theory*, Vol. CT-19, pp. 189–194, Mar., 1972.

54. Parks, T. W., and McClellan, J. H., "A Program for the Design of Linear Phase Finite Impulse Response Filters," *IEEE Trans. Audio Electroacoustics*, Vol. AU-20, No. 3, pp. 195–199, Aug., 1972.

55. Parks, T. W., and Burrus, C. S., *Digital Filter Design*, Wiley: New York, 1987.

56. Porat, B., *A Course in Digital Signal Processing*, Wiley: New York, 1997.

57. Proakis, J. G., and Manolakis, D. G., *Digital Signal Processing: Principles, Algorithms, and Applications*, Second Edition, Macmillan Publishing: New York, 1992.

58. Rabiner, L. R., and Schafer, R. W., *Digital Processing of Speech Signals*, Prentice-Hall: Englewood Cliffs, NJ, 1978.

59. Rabiner, L. R., Gold, B., and McGonegal, C. A., "An Approach to the Approximation Problem for Nonrecursive Digital Filters," *IEEE Trans. Audio and Electroacoustic*, Vol. AU-18, pp. 83–106, June, 1970.

60. Rabiner, L. R., and Crochiere, R. E., "A Novel Implementation for Narrow-band FIR Digital Filters," *IEEE Trans. Acoustics, Speech, and Signal Processing*, Vol. 23, No. 5, pp. 457–464, 1975.

61. Rabiner, L. R., McClellan, J. H., and Parks, T. W., "FIR Digital Filter Design Techniques Using Weighted Chebyshev Approximation," *Proc. IEEE*, Vol. 63, pp. 595–610, 1975.

62. Remez, E. Y., "General Computational Methods of Chebyshev Approximations," *Atomic Energy Translation 4491*, Kiev, USSR, 1957.

63. Roads, C. Pope, S. T., Piccialli, A., and DePolki, G., Editors, *Musical Signal Processing*, Swets & Zeitlinger: Lisse, Netherlands, 1997.

64. Schilling, R. J., Al-Ajlouni, A., Carroll, J. J., and Harris, S. L., "Active Control of Narrow-band Acoustic Noise of Unknown Frequency Using a Phase-locked Loop," *Int. J. Systems Science*, Vol. 29, No. 3, pp. 287–295, 1998.

65. Schilling, R. J., Carroll, J. J., and Al-Ajlouni, A., "Approximation of Nonlinear Systems with Radial Basis Function Neural Networks," *IEEE Trans. Neural Networks*, Vol. 12, No. 1, pp. 1–15, 2001.

66. Schilling, R. J., and Lee, H., *Engineering Analysis: A Vector Space Approach*, Wiley: New York, 1988.

67. Schilling, R. J., and Harris, S. L., *Applied Numerical Methods for Engineers Using MAT-LAB and C*, Brooks-Cole: Pacific Grove, CA, 2000.

68. Slock, T. T. M., "On the Convergence Behavior of the LMS and the Normalized LMS Algorithms," *IEEE Trans. Signal Processing*, Vol. 41, pp. 2811–2825, 1993.

69. Shan, T. J., and Kailath, T., "Adaptive Algorithms with an Automatic Gain Control Feature," *IEEE Trans. Circuits and Systems*, Vol. CAS-35, pp. 122–127, 1988.

70. Shannon, C. E., "Communication in the Process of Noise," *Proc. IRE*, Jan., pp. 10–21, 1949.

71. Steiglitz, K., *A Digital Signal Processing Primer with Applications to Digital Audio and Computer Music*, Addison-Wesley: Menlo Park, CA, 1996.

72. Strum, R. E., and Kirk, D. E., *First Principles of Discrete Systems and Digital Signal Processing*, Addison-Wesley: Reading MA, 1988.

73. Treichler, J. R., Johnson, C. R., Jr., and Larimoore, M. G., *Theory and Design of Adaptive Filters*, Prentice-Hall: Upper Saddle River, NJ, 2001.

74. Tretter, S. A., *Introduction to Discrete-Time Signal Processing*, Wiley: New York, 1976.

75. Warnaka, G. E., Poole, L. A. and Tichy, J., "Active Acoustic Attenuators," U.S. Patent 4,473906, Sept. 25, 1984.

76. Wasserman, P. D., *Neural Computing: Theory and Practice*, Van Nostrand Reinhold: New York, 1989.

77. Webb, A., and Shannon, S., "Shape-adaptive Radial Basis Functions," *IEEE Trans. Neural Networks*, Vol. 9, Nov., 1998.

78. Weiner, N., and Paley, R.E.A.C., *Fourier Transforms in the Complex Domain*, American Mathematical Society: Providence, RI, 1934.

79. Welch, P. D., "The Use of Fast Fourier Transform for the Estimation of Power Spectra: A Method Based on Time Averaging over Short Modified Periodograms," *IEEE Trans. Audio and Electroacoustics*, Vol, AU-15, pp. 70–73, June, 1967.

80. Widrow, B., and Hoff, M. E., Jr., "Adaptive Switching Circuits," *IRE WESCON Conv. Rec., Part 4*, pp. 96–104, 1960.

81. Widrow, B., and Stearns, S. D., *Adaptive Signal Processing*, Prentice Hall: Englewood Cliffs, NJ, 1985.

82. Wilkinson, J. H., *Rounding Error in Algebraic Processes*, Prentice-Hall: Englewood Cliffs, NJ, 1963.

83. Woodbury, M., *Inverting Modified Matrices*, Mem. Rep. 42, Statistical Research Group, Princeton University, Princeton, NJ, 1950.

Appendix 1
Transform Tables

1.1 Fourier Series

Complex Form:

$$x_a(t + T) = x_a(t)$$

$$x_a(t) = \sum_{k=-\infty}^{\infty} c_k \exp\left(\frac{j2\pi kt}{T}\right)$$

$$c_k = \frac{1}{T} \int_T x_a(t) \exp\left(\frac{-j2\pi kt}{T}\right)$$

Trigonometric Form:

$$x_a(t) = \frac{a_0}{2} + \sum_{k=1}^{\infty} a_k \cos\left(\frac{2\pi kt}{T}\right) + b_k \sin\left(\frac{2\pi kt}{T}\right)$$

$$a_k = \frac{2}{T} \int_T x_a(t) \cos\left(\frac{2\pi kt}{T}\right) dt = 2\text{Re}\{c_k\}$$

$$b_k = \frac{2}{T} \int_T x_a(t) \sin\left(\frac{2\pi kt}{T}\right) dt = -2\text{Im}\{c_k\}$$

Cosine Form:

$$x_a(t) = \frac{d_0}{2} + \sum_{k=1}^{\infty} d_k \cos\left(\frac{2\pi kt}{T} + \theta_k\right)$$

$$d_k = \sqrt{a_k^2 + b_k^2} = 2|c_k|$$

$$\theta_k = \tan^{-1}\left(\frac{-b_k}{a_k}\right) = \tan^{-1}\left(\frac{\text{Im}\{c_k\}}{\text{Re}\{c_k\}}\right)$$

Description	$x_a(t)$	Fourier series
Odd square wave	$\operatorname{sgn}\left[\sin\left(\dfrac{2\pi t}{T}\right)\right]$	$\dfrac{4}{\pi}\displaystyle\sum_{k=1}^{\infty}\dfrac{1}{2k-1}\sin\left[\dfrac{2\pi(2k-1)t}{T}\right]$
Even square wave	$\operatorname{sgn}\left[\cos\left(\dfrac{2\pi t}{T}\right)\right]$	$\dfrac{4}{\pi}\displaystyle\sum_{k=1}^{\infty}\dfrac{(-1)^{k-1}}{2k-1}\cos\left[\dfrac{2\pi(2k-1)t}{T}\right]$
Impulse train	$\delta_T(t)$	$\dfrac{1}{T}+\dfrac{2}{T}\displaystyle\sum_{k=1}^{\infty}\cos\left(\dfrac{2\pi kt}{T}\right)$
Even pulse train	$\mu_a\left[\cos\left(\dfrac{2\pi t}{T}\right)-\cos\left(\dfrac{2\pi\tau}{T}\right)\right]$	$\dfrac{2\tau}{T}+\dfrac{4\tau}{T}\displaystyle\sum_{k=1}^{\infty}\operatorname{sinc}\left(\dfrac{2k\tau}{T}\right)\cos\left(\dfrac{2\pi kt}{T}\right)$
Rectified sine wave	$\left\lvert\sin\left(\dfrac{2\pi t}{T}\right)\right\rvert$	$\dfrac{2}{\pi}-\dfrac{4}{\pi}\displaystyle\sum_{k=1}^{\infty}\dfrac{1}{4k^2-1}\cos\left(\dfrac{4\pi kt}{T}\right)$
Sawtooth wave	$\operatorname{mod}(t,T)$	$\dfrac{1}{2}-\dfrac{1}{\pi}\displaystyle\sum_{k=1}^{\infty}\dfrac{1}{k}\sin\left(\dfrac{2\pi kt}{T}\right)$

TABLE A1: ▶ Fourier Series Pairs

1.2 Fourier Transform

● ●

Fourier transform (FT):

$$X_a(f)\triangleq\int_{-\infty}^{\infty}x_a(t)\exp(-j2\pi ft)\,dt,\quad f\in R$$

Inverse Fourier transform (IFT):

$$x_a(t)=\int_{-\infty}^{\infty}X_a(f)\exp(j2\pi ft)\,df,\quad t\in R$$

Entry	$x_a(t)$	$X_a(f)$	Description
1	$\exp(-ct)\mu_a(t)$	$\dfrac{1}{c + j2\pi f}$	Causal exponential
2	$\exp(-c|t|)$	$\dfrac{2c}{c^2 + 4\pi^2 f^2}$	Double exponential
3	$\exp[-(ct)^2]$	$\dfrac{\sqrt{\pi}\,\exp[-(\pi f/c)^2]}{c}$	Gaussian
4	$\exp(j2\pi F_0 t)$	$\delta_a(f - F_0)$	Complex exponential
5	$t\exp(-ct)\mu_a(t)$	$\dfrac{1}{(c + j2\pi f)^2}$	Damped polynomial
6	$\exp(-ct)\cos(2\pi F_0 t)\mu_a(t)$	$\dfrac{c + j2\pi f}{(c + j2\pi f)^2 + (2\pi F_0)^2}$	Damped cosine
7	$\exp(-ct)\sin(2\pi F_0 t)\mu_a(t)$	$\dfrac{2\pi F_0}{(c + j2\pi f)^2 + (2\pi F_0)^2}$	Damped sine
8	$\delta_a(t)$	1	Unit impulse
9	$\mu_a(t)$	$\dfrac{\delta_a(f)}{2} + \dfrac{1}{j2\pi f}$	Unit step
10	1	$\delta_a(f)$	Constant
11	$\mu_a(t + T) - \mu_a(t - T)$	$2\tau\,\mathrm{sinc}(2Tf)$	Pulse
12	$2B\,\mathrm{sinc}(2Bt)$	$\mu_a(f + B) - \mu_a(f - B)$	Sinc function
13	$\mathrm{sgn}(t)$	$\dfrac{1}{j\pi f}$	Signum function
14	$\cos(2\pi F_0 t)$	$\dfrac{\delta_a(f + F_0) + \delta_a(f - F_0)}{2}$	Cosine
15	$\sin(2\pi F_0 t)$	$\dfrac{j[\delta_a(f + F_0) - \delta_a(f - F_0)]}{2}$	Sine

TABLE A3: ▶
Fourier Transform
Properties

Property	$x_a(t)$	$X_a(f)$
Symmetry	Real	$X_a^*(f) = X(-f)$
Even magnitude	Real	$|X_a(-f)| = |X_a(f)|$
Odd phase	Real	$\angle X_a(-f) = -\angle X_a(f)$
Linearity	$ax_1(t) + bx_2(t)$	$aX_1(f) + bX_2(f)$
Time scale	$x_a(at)$	$\dfrac{1}{|a|} X_a\left(\dfrac{f}{a}\right)$
Reflection	$x_a(-t)$	$X_a(-f)$
Duality	$X_a(t)$	$x_a(-f)$
Complex conjugate	$x_a^*(t)$	$X_a^*(-f)$
Time shift	$x_a(t-T)$	$\exp(-j2\pi fT)X_a(f)$
Frequency shift	$\exp(j2\pi F_0 t)x_a(t)$	$X_a(f - F_0)$
Time differentiation	$\dfrac{d^k x_a(t)}{dt^k}$	$(j2\pi f)^k X_a(f)$
Frequency differentiation	$t^k x_a(t)$	$\left(\dfrac{1}{2\pi}\right)^k \dfrac{d^k X_a(f)}{df^k}$
Time convolution	$\displaystyle\int_{-\infty}^{\infty} x_1(\tau)x_2(t-\tau)d\tau$	$X_1(f)X_2(f)$
Frequency convolution	$x_1(t)x_2(t)$	$\displaystyle\int_{-\infty}^{\infty} X_1(\alpha)X_2(f-\alpha)d\alpha$
Cross-correlation	$\displaystyle\int_{-\infty}^{\infty} x_1(\tau)x_2^*(t+\tau)d\tau$	$X_1(f)X_2^*(f)$
Parseval	$\displaystyle\int_{-\infty}^{\infty} x_a(t)y_a^*(t)dt$	$\displaystyle\int_{-\infty}^{\infty} X_a(f)Y_a^*(f)df$
	$\displaystyle\int_{-\infty}^{\infty} |x_a(t)|^2 dt$	$\displaystyle\int_{-\infty}^{\infty} |X_a(f)|^2 df$

1.3 Laplace Transform

Laplace transform (LT):

$$X_a(s) \triangleq \int_0^{\infty} x_a(t)\exp(-st)\,dt, \quad \mathrm{Re}(s) > c$$

Inverse Laplace transform (ILT):

$$x_a(t) = \frac{1}{j2\pi}\int_{c-j\infty}^{c+j\infty} X_a(s)\exp(st)\,ds, \quad t \geq 0$$

TABLE A4: ▶
Laplace Transform Pairs

Entry	$x_a(t)$	$X_a(s)$	Description
1	$\delta_a(t)$	1	Unit impulse
2	$\mu_a(t)$	$\dfrac{1}{s}$	Unit step
3	$t^m \mu_a(t)$	$\dfrac{m!}{s^{m+1}}$	Polynomial
4	$\exp(-ct)\mu_a(t)$	$\dfrac{1}{s+c}$	Exponential
5	$\exp(-ct)t^m \mu_a(t)$	$\dfrac{m!}{(s+c)^{m+1}}$	Damped polynomial
6	$\sin(2\pi F_0 t)\mu_a(t)$	$\dfrac{2\pi F_0}{s^2 + (2\pi F_0)^2}$	Sine
7	$\cos(2\pi F_0 t)\mu_a(t)$	$\dfrac{s}{s^2 + (2\pi F_0)^2}$	Cosine
8	$\exp(-ct)\sin(2\pi F_0 t)\mu_a(t)$	$\dfrac{2\pi F_0}{(s+c)^2 + (2\pi F_0)^2}$	Damped sine
9	$\exp(-ct)\cos(2\pi F_0 t)\mu_a(t)$	$\dfrac{s+c}{(s+c)^2 + (2\pi F_0)^2}$	Damped cosine
10	$t\sin(2\pi F_0 t)\mu_a(t)$	$\dfrac{4\pi F_0}{[s^2 + (2\pi F_0)^2]^2}$	Polynomial sine
11	$t\cos(2\pi F_0 t)\mu_a(t)$	$\dfrac{s^2 - (2\pi F_0)^2}{[s^2 + (2\pi F_0)^2]^2}$	Polynomial cosine

TABLE A5: ▶
Laplace Transform Properties

Property	$x_a(t)$	$X_a(f)$
Linearity	$ax_1(t) + bx_2(t)$	$aX_1(s) + bX_2(s)$
Complex conjugate	$x^*(t)$	$X^*(s^*)$
Time scale	$x_a(at), a > 0$	$\dfrac{1}{a}X_a\left(\dfrac{s}{a}\right)$
Time multiplication	$tx_a(t)$	$-\dfrac{dX_a(s)}{ds}$
Time division	$\dfrac{x_a(t)}{t}$	$\displaystyle\int_s^\infty X_a(\sigma)d\sigma$
Time shift	$x_a(t-T)\mu_a(t-T)$	$\exp(-sT)X_a(s)$
Frequency shift	$\exp(-at)x_a(t)$	$X_a(s+a)$
Derivative	$\dfrac{dx_a(t)}{dt}$	$sX_a(s) - x_a(0^+)$
Integral	$\displaystyle\int_0^t x_a(\tau)d\tau$	$\dfrac{X_a(s)}{s}$
Differentiation	$\dfrac{d^k x_a(t)}{dt^k}$	$s^k X_a(s) - \displaystyle\sum_{i=0}^{k-1} s^{k-i-1}\dfrac{d^i x_a(0^+)}{dt^i}$
Convolution	$\displaystyle\int_0^t x_a(\tau)y_a(t-\tau)d\tau$	$X_a(s)Y_a(s)$
Initial value	$x_a(0^+)$	$\displaystyle\lim_{s\to\infty} sX_a(s)$
Final value	$\displaystyle\lim_{t\to\infty} x_a(t)$	$\displaystyle\lim_{s\to 0} sX_a(s), \ \text{stable}$

1.4 Z-transform

$\bullet \bullet$

Z-transform (ZT):

$$X(z) \triangleq \sum_{k=0}^{\infty} x(k) z^{-k}, \quad r < |z| < R$$

Inverse Z-transform (IZT):

$$x(k) = \frac{1}{j2\pi} \int_C X(z) z^{k-1} dz, \quad |k| = 0, 1, \dots$$

TABLE A6: ▶
Z-transform Pairs

Entry	$x(k)$	$X(z)$	Description
1	$\delta(k)$	1	Unit impulse
2	$\mu(k)$	$\dfrac{z}{z-1}$	Unit step
3	$k\mu(k)$	$\dfrac{z}{(z-1)^2}$	Unit ramp
4	$k^2\mu(k)$	$\dfrac{z(z+1)}{(z-1)^3}$	Unit parabola
5	$a^k\mu(k)$	$\dfrac{z}{z-a}$	Exponential
6	$ka^k\mu(k)$	$\dfrac{az}{(z-a)^2}$	Linear exponential
7	$k^2a^k\mu(k)$	$\dfrac{az(z+a)}{(z-a)^3}$	Quadratic exponential
8	$\sin(bk)\mu(k)$	$\dfrac{z\sin(b)}{z^2 - 2z\cos(b) + 1}$	Sine
9	$\cos(bk)\mu(k)$	$\dfrac{z[z - \cos(b)]}{z^2 - 2z\cos(b) + 1}$	Cosine
10	$a^k\sin(bk)\mu(k)$	$\dfrac{az\sin(b)}{z^2 - 2az\cos(b) + a^2}$	Damped sine
11	$a^k\cos(bk)\mu(k)$	$\dfrac{z[z - a\cos(b)]}{z^2 - 2az\cos(b) + a^2}$	Damped cosine

TABLE A7: ▶
Z-transform
Properties

Property	$x(k)$	$X(z)$
Linearity	$ax(k) + by(k)$	$aX(z) + bY(z)$
Complex conjugate	$x^*(k)$	$X^*(z^*)$
Time reversal	$x(-k)$	$X(1/z)$
Time shift	$x(k-r)$	$z^{-r}X(z)$
Time multiplication	$kx(k)$	$-z\dfrac{dX(z)}{dz}$
Z-scale	$a^kx(k)$	$X(z/a)$
Convolution	$h(k) \star x(k)$	$H(z)X(z)$
Correlation	$r_{yx}(k)$	$\dfrac{Y(z)X(1/z)}{L}$
Initial value	$x(0)$	$\lim\limits_{z\to\infty} X(z)$
Final value	$x(\infty)$	$\lim\limits_{z\to 1}(z-1)X(z)$, stable

1.5 Discrete-time Fourier Transform

● ●

Discrete-time Fourier transform (DTFT):

$$X(f) \triangleq \sum_{k=-\infty}^{\infty} x(k) \exp(-jk2\pi fT), \quad f \in R$$

Inverse discrete-time Fourier transform (IDTFT):

$$x(k) = \frac{1}{f_s} \int_{-f_s/2}^{f_s/2} X(f) \exp(jk2\pi fT)\, df, \quad |k| = 0, 1, 2, \ldots$$

TABLE A8: ▶
Discrete-time
Fourier Transform
Pairs

Entry	$x(k)$	$X(f)$	Description
1	$\delta(k)$	1	Unit impulse
2	$a^k \mu(k), |a| < 1$	$\dfrac{\exp(j2\pi fT)}{\exp(j2\pi fT) - a}$	Exponential
3	$k(a)^k \mu(k), |a| < 1$	$\dfrac{a\exp(j2\pi fT)}{[\exp(j2\pi fT) - a]^2}$	Linear exponential
4	$2F_0 T \operatorname{sinc}(2kF_0 T)$	$\mu(f + F_0) - \mu(f - F_0)$	Sinc function
5	$\mu(k+r) - \mu(k-r-1)$	$\dfrac{\sin[\pi(2r+1)f]}{\sin(\pi f)}$	Pulse function

TABLE A9: ▶
Discrete-time
Fourier Transform
Properties

Property	Time Signal	DTFT
Periodic	General	$X(f + f_s) = X(f)$
Symmetry	Real	$X^*(f) = X(-f)$
Even magnitude	Real	$A_x(-f) = A_x(f)$
Odd phase	Real	$\phi_x(-f) = -\phi_x(f)$
Linearity	$ax(k) + by(k)$	$aX(f) + bY(f)$
Complex conjugate	$x^*(k)$	$X^*(-f)$
Time reversal	$x(-k)$	$X(-f)$
Time shift	$x(k - r)$	$\exp(-j2\pi r fT)X(f)$
Frequency shift	$\exp(jk2\pi F_0 T)x(k)$	$X(f - F_0)$
Multiplication	$x(k)y(k)$	$\int_{-f_s/2}^{f_s/2} X(\alpha)Y(f - \alpha)d\alpha$
Convolution	$h(k) \star x(k)$	$H(f)X(f)$
Correlation	$r_{yx}(k)$	$\dfrac{Y(f)X(-f)}{L}$
Wiener-Khintchine	$r_{xx}(k)$	$\dfrac{S_x(f)}{L}$
Parseval	$\sum_{k=-\infty}^{\infty} x(k)y^*(k)$	$\dfrac{1}{f_s}\int_{-f_s/2}^{f_s/2} X(f)Y^*(f)df$
	$\sum_{k=-\infty}^{\infty} \|x(k)\|^2$	$\dfrac{1}{f_s}\int_{-f_s/2}^{f_s/2} \|X(f)\|^2 df$

1.6 Discrete Fourier Transform (DFT)

Discrete Fourier transform (DFT):

$$X(i) \triangleq \sum_{k=0}^{N-1} x(k)\exp\left(\frac{-jki2\pi}{N}\right), \quad 0 \le i < N$$

Inverse discrete Fourier transform (IDFT):

$$x(k) = \frac{1}{N}\sum_{i=0}^{N-1} X(i)\exp\left(\frac{jki2\pi}{N}\right), \quad 0 \le k < N$$

TABLE A10: ▶
Discrete Fourier
Transform
Properties

Property	Time Signal	DFT	Comments
Periodic	General	$X(i + N) = X(i)$	
Symmetry	Real	$X^*(i) = X(N - i)$	
Even magnitude	Real	$A_x(N/2 + i) = A_x(N/2 - i)$	N even
Odd phase	Real	$\phi_x(N/2 + i) = -\phi_x(N/2 - i)$	N even
Linearity	$ax(k) + by(k)$	$aX(i) + bY(i)$	
Time reversal	$x_p(-k)$	$X^*(i)$	Real x
Circular shift	$x_p(k - r)$	$\exp\left(\dfrac{-j2\pi i r}{N}\right) X(i)$	
Circular convolution	$x(k) \circ y(k)$	$X(i)Y(i)$	
Circular correlation	$c_{yx}(k)$	$\dfrac{Y(i)X^*(i)}{N}$	Real x
Wiener-Khintchine	$c_{xx}(k)$	$S_x(i)$	
Parseval	$\displaystyle\sum_{k=0}^{N-1} x(k)y^*(k)$	$\dfrac{1}{N}\displaystyle\sum_{i=0}^{N-1} X(i)Y^*(k)$	
	$\displaystyle\sum_{k=0}^{N-1} \lvert x(k) \rvert^2$	$\dfrac{1}{N}\displaystyle\sum_{i=0}^{N-1} \lvert X(i) \rvert^2$	

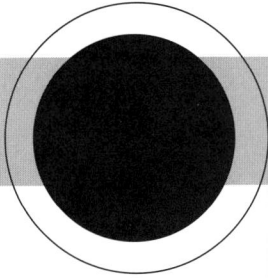

Appendix 2
Mathematical Identities

2.1 Complex Numbers

Rectangular form:

$$j = \sqrt{-1}$$
$$z = x + jy$$
$$z^* = x - jy$$
$$z + z^* = 2\mathrm{Re}(z) = 2x$$
$$z - z^* = j2\mathrm{Im}(z) = j2y$$
$$zz^* = |z|^2 = x^2 + y^2$$

Polar form:

$$z = A\exp(j\phi)$$
$$A = \sqrt{x^2 + y^2}$$
$$\phi = \tan^{-1}\left(\frac{y}{x}\right)$$
$$x = A\cos(\phi)$$
$$y = A\sin(\phi)$$

2.2 Euler's Identity

$$\exp(\pm j\phi) = \cos(\phi) \pm j\sin(\phi)$$
$$\cos(\phi) = \frac{\exp(j\phi) + \exp(-j\phi)}{2}$$
$$\sin(\phi) = \frac{\exp(j\phi) - \exp(-j\phi)}{j2}$$
$$\exp(\pm j\pi/2) = \pm j$$

2.3 Trigonometric Identities

Analysis:

$$\cos^2(a) + \sin^2(a) = 1$$
$$\cos(a \pm b) = \cos(a)\cos(b) \mp \sin(a)\sin(b)$$
$$\sin(a \pm b) = \sin(a)\cos(b) \pm \cos(a)\sin(b)$$
$$\cos(2a) = \cos^2(a) - \sin^2(a)$$
$$\sin(2a) = 2\sin(a)\cos(a)$$

Synthesis:

$$\cos^2(a) = \frac{1 + \cos(2a)}{2}$$
$$\sin^2(a) = \frac{1 - \cos(2a)}{2}$$
$$\cos(a)\cos(b) = \frac{\cos(a+b) + \cos(a-b)}{2}$$
$$\sin(a)\sin(b) = \frac{\cos(a-b) - \cos(a+b)}{2}$$
$$\sin(a)\cos(b) = \frac{\sin(a+b) + \sin(a-b)}{2}$$

2.4 Inequalities

Scalar:

$$|ab| = |a| \cdot |b|$$
$$|a + h| \le |a| + |h|$$

Vector:

$$\|x\|^2 = \sum_{i=1}^{n} x_i^2$$
$$\|A\| = \max_{i=1}^{n}\{|\lambda_i|\}$$
$$\det(\lambda I - A) = \prod_{i=1}^{n}(\lambda - \lambda_i)$$
$$\|x + y\| \le \|x\| + \|y\|$$
$$|x^T y| \le \|x\| \cdot \|y\|$$
$$\|Ax\| \le \|A\| \cdot \|x\|$$

2.5 Uniform White Noise

• •

$$P_v = E[v^2(k)] \approx \frac{1}{N} \sum_{i=0}^{N-1} v^2(k)$$

$$P_v = \frac{b^3 - a^3}{3(b - a)} \quad \text{when} \quad a \le v \le b$$

$$P_v = \frac{c^2}{3} \quad \text{when} \quad -c \le v \le c$$

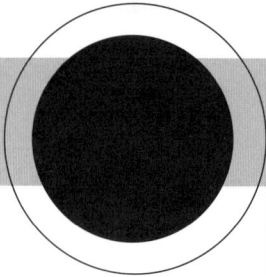

Appendix 3
FDSP Toolbox Functions

The companion Web site of the publisher contains a Fundamentals of Digital Signal Processing (FDSP) toolbox that uses MATLAB to implement the signal processing techniques discussed in the text. This appendix summarizes the contents of the FDSP toolbox. Although MATLAB is platform-independent, the FDSP toolbox was designed primarily for use on a Windows PC with a sound card.

3.1 Installation

The FDSP toolbox was developed to help students solve the GUI Simulation problems and MATLAB Computation problems appearing at the end of each chapter. It also provides the instructor and the student with a convenient way to run all of the computational examples and reproduce all of the MATLAB figures and tables that appear in the text. A novel component of the toolbox is a collection of graphical user interface (GUI) modules that allow the user to interactively explore the signal processing techniques covered in each chapter without any need for programming. The FDSP toolbox is installed using MATLAB itself (Marwan, 2003). For older versions of MATLAB running under Windows Vista, the user may have to right click on the MATLAB icon and select Run as Administrator. Once in MATLAB, issue the following command from within the download folder.

```
>> setup
```

The FDSP folder includes five subfolders. Subfolder *fdsp* contains the FDSP toolbox functions and the chapter GUI modules. The FDSP functions are named using the convention *f_xxx*, and the chapter GUI modules and support functions are named using the convention *g_xxx*. These conventions are adopted in order to ensure compatibility with other MATLAB toolbox boxes, such as the Signal Processing and Filter design toolboxes. The software supplied with this text does not require any optional toolboxes. This was done to keep student expenses to a minimum. Users who do have optional toolboxes can access them without conflict because of the naming conventions. The *examples*, *figures*, and *tables* subfolders contain all of the MATLAB examples, figures, and tables that appear in the text. The *problems* subfolder contains solutions to selected end-of-chapter problems in the form of pdf files. Although most students will download the FDSP toolbox directly from the publisher's companion Web site, it is also possible to download the FDSP toolbox from the following Web site maintained by the authors.

```
www.clarkson.edu/~rschilli/fdsp
```

3.2 Driver Module: *f_dsp*

All of the FDSP software can be conveniently accessed through a driver module called *f_dsp* that is launched by entering the following command from the MATLAB command prompt:

```
>> f_dsp
```

A startup screen for *f_dsp* was shown previously in Figure 1.39 of Chapter 1. Most of options on the menu toolbar at the top of the screen produce submenus of selections. The *GUI Modules* option is used to run the graphical user interface modules. With the *Examples* option, all MATLAB examples appearing in the text can be executed. Similarly, the *Figures* option is used to recreate the MATLAB figures, and the *Tables* option is used to view the tables from the text. The *Help* option provides online help for the GUI modules and the FDSP toolbox functions. The *Web* option connects the user to the companion web site. Using this option, the user can download a zip file that contains the latest version of the FDSP software. The *Exit* option returns control to the MATLAB command window.

3.3 Chapter GUI Modules

When the *GUI Modules* option is selected from the *f_dsp* toolbar, the user is provided with the list of chapter GUI modules summarized in Table A.11.

The chapter GUI modules feature a common user interface that is simple to learn and easy to use. In addition, data can be shared between GUI modules by exporting with the Save option and importing with the User-defined option. Each of the GUI modules is described in detail near the end of the corresponding chapter. The GUI modules are designed to provide the student with a convenient means of interactively exploring the signal processing concepts covered in that chapter without any need for programming. There is also a set of GUI Simulation problems at the end of each chapter that are designed to be solved using the chapter GUI module. Users who are familiar with MATLAB programming can provide optional data files and optional user functions that interact with the GUI modules.

TABLE A11: ▶ Chapter GUI Modules

Module	Description	Chapter
g_sample	Signal sampling	1
g_reconstruct	Signal reconstruction	1
g_systime	Discrete-time systems, time domain	2
g_correlate	Signal correlation and convolution	2
g_sysfreq	Discrete-time systems, frequency domain	3
g_spectra	Signal spectral analysis	4
g_filters	Filter specifications and structures	5
g_fir	FIR filter design	6
g_iir	IIR filter design	7
g_multirate	Multirate signal processing	8
g_adapt	adaptive signal processing	9

3.4 FDSP Toolbox Functions

● ●

The use of the GUI modules is convenient, but it is not as flexible as having users write their own MATLAB programs to perform signal processing tasks. Algorithms developed in the text are implemented as FDSP toolbox functions. These functions fall into two broad categories, main-program support functions and chapter functions. Instructions for usage of any of the FDSP functions and GUI modules can be obtained by using the *helpwin* command with the appropriate argument. Note that the MATLAB *lookfor* command can be used to find a list of the names of functions containing a given key word in the initial comment line.

```
helpwin fdsp            % Help for all FDSP toolbox functions
helpwin f_dsp           % Help for the FDSP driver module
helpwin g_xxx           % Help for GUI module g_xxx
helpwin f_xxx           % Help for FDSP toolbox function f_xxx
```

The *Help* option in the FDSP driver module in Figure 3.2 also provides documentation on all of the chapter GUI modules and the FDSP functions.

The main program support functions consist of general low-level utility functions that are designed to simplify the process of writing MATLAB programs by performing some routine tasks. These functions, in alphabetic order, are summarized in Table A12. Relevant MATLAB functions are also listed.

The second group of toolbox functions includes implementations of algorithms developed in the chapters. Specialized functions are developed in those instances where corresponding MATLAB functions are not available as part of the standard MATLAB interpreter. Summaries of the FDSP functions, organized by chapter, can be found in the following tables. To learn more about the usage of any of these functions simply type *helpwin* followed by the function name.

TABLE A12: ▶
FDSP Main Program
Support Functions

Name	Description
f_caliper	Measure points on plot using mouse crosshairs
f_clip	Clip value to an interval, check calling arguments
f_getsound	Record signal from the PC microphone
f_header	Display headers for an example, figure, or problem
f_labels	Label graphical output
f_prompt	Prompt for a scalar in a specified range
f_randinit	Initialize the random number generator
f_randg	Gaussian random matrix
f_randu	Uniformly distributed random matrix
f_wait	Pause to examine displayed output
soundsc	Play a signal as sound on the PC speakers (MATLAB)

TABLE A13: ▶
Sampling and
Reconstruction,
Chapter 1

Name	Description
f_adc	Perform N-bit analog-to-digital conversion
f_dac	Perform N-bit digital-to-analog conversion
filter	Discrete-time system output (MATLAB)
f_freqs	Frequency response, continuous time
f_quant	Quantization operator

TABLE A14: ▶
Discrete-time
Systems–Time
Domain, Chapter 2

Name	Description
f_blockconv	Fast block cross-convolution
f_conv	Fast convolution
f_corr	Fast cross-correlation
f_corrcoef	Correlation coefficient of two vectors
f_filter0	Filter response with nonzero initial condition
f_impulse	Impulse response

TABLE A15: ▶
Discrete-time
Systems–Frequency
Domain, Chapter 3

Name	Description
f_freqz	Frequency response, discrete time
f_idar	Identify an auto-regressive (AR) filter
f_idarma	Identify an auto-regressive moving-average (ARMA) filter
f_pzplot	Pole-zero plot showing unit circle
f_pzsurf	Surface plot of transfer function magnitude
f_spec	Magnitude, phase, and power density spectra

TABLE A16: ▶
Fourier Transforms
and Signal Spectra,
Chapter 4

Name	Description
fft	Fast Fourier transform (MATLAB)
ifft	Inverse fast Fourier transform (MATLAB)
fftshift	Reorder FFT output (MATLAB)
nextpow2	Next higher power of two (MATLAB)
f_pds	Power density spectrum
f_specgram	Spectrogram
f_window	Data windows

TABLE A17: ▶
Filter Design
Specifications,
Chapter 5

Name	Description
f_chebpoly	Chebyshev polynomials
f_filter1	Filter response using quantized indirect realizations
f_minall	Minimum-phase allpass factorization
f_zerophase	Zero-phase filter

TABLE A18: ▶
FIR Filter Design,
Chapter 6

Name	Description
f_cascade	Find cascade-form realization
f_differentiator	Design FIR differentiator filter
f_firamp	Frequency-selective amplitude response
f_filtcas	Use cascade-form realization
f_filtlat	Use lattice-form realization
f_firideal	Design ideal linear-phase FIR windowed filter
f_lattice	Find lattice-form realization
f_firls	Design linear-phase FIR least-squares filter
f_firquad	Design nonlinear-phase FIR quadrature filter
f_firparks	Design linear-phase FIR equiripple filter
f_firsamp	Design linear-phase FIR frequency-sampled filter
f_firwin	Design general linear-phase FIR windowed filter
f_hilbert	Design FIR Hilbert transformer filter

TABLE A19: ▶
IIR Filter Design,
Chapter 7

Name	Description
f_bilin	Bilinear analog-to-digital filter transformation
f_butters	Design analog Butterworth lowpass filter
f_butterz	Design digital IIR Butterworth filter
f_cheby1s	Design analog Chebyshev-I lowpass filter
f_cheby2s	Design analog Chebyshev-II lowpass filter
f_cheby1z	Design digital IIR Chebyshev-I filter
f_cheby2z	Design digital IIR Chebyshev-II filter
f_elliptics	Design analog elliptic lowpass filter
f_ellipticz	Design digital IIR elliptic filter
f_filtpar	Use parallel-form realization
f_iircomb	Design digital IIR comb filter
f_iirinv	Design digital IIR inverse comb filter
f_iirnotch	Design digital IIR notch filter
f_iirres	Design digital IIR resonator filter
f_low2lows	Lowpass-to-lowpass analog frequency transformation
f_low2highs	Lowpass-to-highpass analog frequency transformation
f_low2bps	Lowpass-to-bandpass analog frequency transformation
f_low2bss	Lowpass-to-bandstop analog frequency transformation
f_orderz	Estimate filter order of classical digital IIR filters
f_parallel	Find parallel-form realization
f_reverb	Compute output of digital IIR reverb filter
f_string	Compute output of digital IIR plucked-string filter

TABLE A20: ▶
Multirate Signal
Processing,
Chapter 8

Name	Description
f_decimate	Integer sampling rate decimator
f_interpol	Integer sampling rate interpolator
f_rateconv	Rational sampling rate converter
f_subsignals	Create examples of subsignals

TABLE A21: ▶
Adaptive Signal
Processing,
Chapter 9

Name	Description
f_base2dec	Convert base d array to a decimal scalar
f_dec2base	Convert decimal scalar to a base d array
f_fxlms	Filtered-x LMS active noise control
f_gridpoint	Find vector subscript of a grid point
f_lms	Least mean square (LMS) method
f_lmscorr	Correlation LMS method
f_lmsleak	Leaky LMS method
f_lmsnorm	Normalized LMS method
f_neighbors	Find scalar subscripts of neighbors of a grid point
f_pll	Estimate frequency using a phase-locked loop (PLL)
f_rbf0	Zeroth-order RBF network evaluation
f_rbf1	First-order RBF network evaluation
f_rbfg	Compute a raised cosine RBF centered at zero
f_rbfv	First-order RBF system identification
f_rbfw	Zeroth-order RBF system identification
f_rls	Recursive least-squares (RLS) method
f_sigsyn	Signal-synthesis active noise control
f_state	Evaluate state of nonlinear discrete-time system

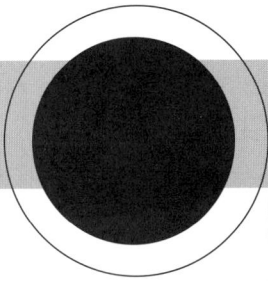

Index

A

Absolutely summable signals, 77
Acoustic (active) noise control,
  690–700, 720
Active noise control, 7–9
Active system, 86
Adaptive filters, 383–386, 394, 645–650,
  678–684
  adaptive signal processing, 645–646,
    649–650
  channel equalization, 647–648
  design specification, 383–386, 394
  error signal, 646–647, 650
  mean square error, 649–650
  noise cancellation, 648–649
  pseudo-filters, 386, 678–684
  signal prediction, 648
  transversal filters, 383–385, 645–646,
    649–650
  FIR filter design, 678–684
Adaptive signal processing, 645–737
  active noise control, 690–700
  adaptive FIR filter design, 678–684,
    719–720
  adaptive transversal filters, 645–650,
    678–684
  black box model for, 646–647
  channel equalization, 647–648
  chemical process identification,
    715–718
  FDSP functions for, 659–660,
    677–678, 690, 699–700, 712–713
  filtered-x LMS (FXLMS) method,
    691–695, 720
  graphical user interface (GUI),
    713–718
  least mean square (LMS) method,
    656–678, 684–695,
    718–719
  mean square error (MSE), 649–655,
    666–669
  noise cancellation, 648–649
  nonlinear systems, 700–713, 720

recursive least mean squares (RLS)
    method, 684–690, 719
  signal prediction, 648
  state vector for, 650, 700–701, 720
  system identification, 646–647,
    700–713, 720
  weight vector for, 650, 720–721
Algorithm order of FFT, 256
Alias-free two-channel QMF bank,
  610–612
Aliasing, 10–11, 23–26, 33–39, 54–59,
  61, 615, 622
  anti-aliasing filters, 33–37,
    54–57, 61
  anti-imaging filter, 37–39
  bandlimited signals for, 24–26
  continuous-time signal sampling,
    23–26, 61
  defined, 10
  error factor, 615, 622
  folding frequency, 26
  formula, 23
  graphical user interface (GUI), 54–59
  oversampling factor, 54, 58
  pixels, 10
  prefilters and postfilters for, 33–39
  sample corruption by, 24–25
  video, 10–11, 57–59
Allpass filters, 362–367, 393
  FDSP functions for, 366–367
  minimum-phase decomposition,
    363–365
  reflective structure, 362
Amplifiers, 6–7, 39–41
  operational (op amp), 39–41
  total harmonic distortion (THD), 6–7
Amplitude modulation, 22
Amplitude response $A_r(f)$, 351–353,
  411–412, 450
Analog filters, *see* Classical analog filters
Analog frequency transformation,
  536–538
Analog signal processing, 4–6, 13–14

digital signal processing (DSP) and,
    4–6
  quantization and, 13–14
Analog-to-digital converters (ADC), 4–5,
    41–45, 612–620, 632
  anti-aliasing filters and, 612–615
  effective precision, 614–615
  FDSP functions for, 45
  flash, 43–45
  multirate signal processing,
    612–620, 632
  oversampling, 612–620, 632
  sigma-delta quantization, 615–620
  signal processing, 4–5, 41–45
  successive-approximation, 41–43
Analysis filter bank, 381–382, 602, 632
Anti-aliasing filters, 33–37, 54–57, 61,
    612–615
  ADC oversampling, 612–615
  Butterworth, 33–37
  classical analog, 37
  cutoff frequency, 33
  first-order, 34–35
  graphical user interface (GUI), 54–57
  multirate signal processing, 612–615
  oversampling, 54, 61, 612–615
  second-order, 35–36
Anti-imaging filters, 37–39, 61, 620–621
  multirate signal processing, 620–621
  oversampling DAC, 620–621
  signal processing, 37–39, 61
Antisound, 8
Aperiodic signals, 75–76
Auto-correlation, 282–290, 652–653
  adaptive signal processing, 652–653
  circular, 282
  mean square error (MSE) and,
    652–653
  noise, periodic signals extracted from,
    286–290
  periodic signal extraction using,
    286–290
  power density spectrum, 284–285

Auto-correlation (*Continued*)
  spectral analysis and, 282–290
  Weiner-Khintchine DFT theorem for, 284–285
  white noise, 282–284
Auto-regressive (AR) model, 183
Auto-regressive moving average (ARMA) model, 183–184
Auto-regressive systems, 149
Average periodogram, 304–308, 311
Average power, 77, 652–653

B

Backward Euler approximation, 367–368, 408
Bandlimited signals, 24–26, 33–39, 60–61
  aliasing and, 24–26
  continuous-time signal sampling, 24–26, 60–61
  defined, 24
Bandpass filters, 421–422, 432–433
  least-squares method for, 432–433
  windowing, 420–421
Bandstop filter design, 479–484
Bandwidth, 375, 420
Bartlett's method, 304–308, 311
Base band, 24
Bessel filters, 351
Bilinear transformations, 529–535, 568–569
  FDSP functions for, 535, 540
  frequency warping, 531–532
  IIR filter design, 529–535, 568–569
  trapezoid integrator, 529–530
Bin frequencies, 241–242, 304
Binary number representation, errors and, 465–466
Bipolar DAC circuits, 39
Black box concept, 198–199, 646–647
Blackman windows, 300–301, 416–417, 419–420
Block diagrams, 94–96
Bounded-input bounded-output (BIBO) systems, 85, 117–119, 130, 185–188
  frequency domain, 185–188
  time domain, 85, 117–119, 130
Bounded signals, 18, 76–77
Butterworth filters, 33–37, 517–522
  cascade connection for, 36
  first-order, 34–35
  frequency transformation, 521–522
  IIR filter design, 517–522
  maximally flat, 519
  normalized, 518–519
  second-order, 35–36

C

Caliper option, 52–53, 123
Cancelled mode, 178–179

Cascade connection, 36
Cascade form, 191, 340–342, 459–461, 547–549, 569
  filter design specifications, 340–342
  FIR filter design, 459–461
  frequency domain stability and, 191
  IIR filter design, 547–549, 569
Cauchy residue theorem, 170–171
Causal exponential, 80, 154–155, 235–237
Causal signals, 15–16, 75, 152–153, 162–163
Causal systems, 83–84
Channel equalization, 647–648
Characteristic polynomial, LTI systems, 87, 130
Chebyshev filters, 338–342, 522–526
  Chebyshev-I, 522–524
  Chebyshev-II, 525–526
  design specifications, 338–342
  equiripple filters, as, 523, 525–526
  IIR filter design, 522–526
  lowpass, 338–342
  ripple factor ε, 522–523, 525
Chebyshev polynomials, 372–373, 434, 522–523
Circular auto-correlation, 282
Circular convolution, 103–107, 131, 252–256
Circular cross-correlation, 114–116, 253
Circular shift property, DFT, 252
Classical analog filters, 37, 517–529, 568
  Butterworth, 517–522
  Chebyshev, 522–526
  elliptic, 526–528
  IIR filter design, 517–529, 568
Clipping, 554
Closed-form expression, inverse Z-transform, 166
Coefficient quantization, 388–392, 470–473, 550–553
  digital filter design specifications, 388–392
  error, 470–473, 550–553
  FIR filter design, 470–473
  graphical user interface (GUI), 388–392
  IIR filter design, 550–553
  pole-zero locations and, 551–552
Colored noise, IIR filters for, 502–504
Comb filters, 180–181, 376, 510–514
  gain factor b_0, 511, 512
  IIR filter design, 510–514
  inverse, 376, 511–514
  notch filter design and, 376–377, 394
  pole-zero placement, 510–514
  transfer functions and, 180–181
Complete response, 92–93
Complex numbers, 474

Complex signal, 371
Computational effort (speed), FFT, 260–262, 265, 271–272
Constant interpolation property, 707
Continuous-time, 3, 11, 16–17, 20–32, 52–54, 60–61
  frequency response, 19, 60
  classification as, 11, 16–17
  FDSP toolbox functions for, 32
  graphical user interface (GUI), 52–54
  impulse response, 20–21
  reconstruction, 26–32
  sampling, 21–26, 52–54, 60–61
  signals, 3, 11, 21–32, 52–54
  system, 16–17
  transfer functions, 29–31
Controller gain, 147
Convergence rate, LMS method, 663–666, 719
Converters, *see* Sampling rate converters
Convolution, 70–71, 100–110, 115–116, 121–123, 130–131, 160–161, 252–256, 263–270
  circular, 103–107, 131, 252–256
  cross-correlation compared to, 71, 115–116
  deconvolution, 108–109, 131
  DFT property of, 252–256
  difference equations for, 70–71, 100–110, 130–131
  discrete-time signals, 70–71, 100–110, 130–131, 160–161
  DSP algorithm use of, 70–71
  fast, 263–266
  fast block, 267–270
  fast Fourier transforms (FFT), 263–270
  FDSP functions for, 107, 270
  GUI modules for, 121–123
  linear, 100–103, 130–131
  MATLAB functions for, 102, 110
  operator, 101–102
  periodic extension for, 103–104
  polynomial arithmetic for, 109–110
  properties of, 102
  spectral analysis and, 252–256, 263–270
  Z transform, 160–161
  zero padding for, 105–107
  zero-state response and, 101–102
Correlation, *see* Auto-correlation; Cross-correlation
Correlation LMS method, 671–674
Cross-correlation, 71, 110–117, 123, 131–132, 161–162, 253, 270–274, 650–652. *See also* Auto–correlation
  adaptive signal processing, 650–652
  circular, 114–116, 253
  convolution compared to, 71, 115–116
  DFT property of, 253

difference equations for, 71, 110–117, 131–132
discrete-time signals, 71, 110–117, 131–132, 161–162
fast, 270–274
fast Fourier transform (FFT), 270–274
FDSP functions for, 117, 274
GUI modules for, 123
lag variable for, 111
linear, 110–114, 116
mean square error (MSE) and, 650–652
normalized, 113–114
signal shape and, 111–113
spectral analysis and, 253, 270–274
symmetry property of, 115
Z transform, 161–162
Cutoff frequency, 33

D

Data windows, 299–301
DC gain, 180–181
Decibel scale (dB), 293–294, 348–349
frequency response, 293–294
frequency-selective filters and, 348–349
logarithmic design specifications, 348–349
zero–padding and, 293–294
Decimation factor, 600–601
Decimation in time, FFT, 256–260
Decimators, 583–584, 587–588, 596–598, 630
integer, 588
multirate filter realization, 596–598
polyphase, 596–598
sampling rate conversion, 583–584, 587–588, 630
Deconvolution, 108–109, 131
Delay block, 94–95
Delay line (τ), 350–351
Delay operator, Z-transform, 158, 169
Delay systems, fractional, 586–587
Design specifications, 337–405
decibel scale (dB), 348–349
digital filters, 337–405
frequency-selective filters, 342–350
linear, 343–348
logarithmic, 348–350
lowpass filters, 338–342
magnitude response $A(f)$, 337–350, 392
passband, 339
phase response $\phi(f)$, 337, 342–343, 350–367, 392–393
realization of filter structures, 339–342
stopband, 339
transition band, 339

Difference equations, 70–74, 86–94, 100–117, 130–132
characteristic polynomial for, 87, 130
complete response, 92–93
convolution of signals using, 70–71, 100–110, 130–131
correlation of signals using, 71, 110–117, 131–132
dimension of the system, 86
discrete-time system analysis, 70–74, 86–94, 100–117, 130–132
DSP applications of, 71–74
FDSP functions for, 93
initial conditions for, 86–87, 130
linear time-invariant (LTI) systems, 86–94
MATLAB functions for, 88, 93
time domain representation by, 70–74
zero-input response, 87–90, 130
zero-state response, 90–94
Differentiators, 408–409, 442–445
Digital-and-aliasing filter, 587–588
Digital filters, 335–580
adaptive, 383–386, 394
allpass, 362–367, 393
design, 335–580
FDSP functions for, 358, 366–367
filter banks, 381–383, 394
finite impulse response (FIR), 353–358, 393, 406–498
frequency response, 337
frequency-selective, 342–350
graphical user interface (GUI), 386–392
infinite impulse response (IIR), 349–350, 499–580
linear-phase, 350–356
lowpass design specification, 338–342
magnitude response $A(f)$, 337–350, 359–361, 392
minimum-phase, 359–362, 366–367, 393
narrowband, 378–381, 394
notch, 374–376, 393–394
passband design specification, 339
phase response $\phi(f)$, 337, 342–343, 350–367, 392–393
quadrature, 367–374, 393
realization structures, 339–342
resonators, 376–378, 393–394
specifications, 337–405
stopband design specification, 339
transition band design specification, 339
zero-phase, 356–358
Digital frequency transformation, 539–540
Digital oscillator, 372–374
Digital signal, 3, 12–13

Digital signal processing (DSP), 3–9, 14–15, 31–32, 37–38, 70–71
active noise control, 7–9
analog signal processing and, 4–6
anti-imaging filters and, 37–38
applications, 3–4
convolution and, 70–71
mathematical model of, 31–32
notch filters, 7
quantization noise and, 14–15
total harmonic distortion (THD), 6–7
zero-order hold, 31–32, 37–38
Digital-to-analog converters (DAC), 5, 37–41, 45, 620–623, 632
anti-imaging filters and, 37–39, 620–621
bipolar circuits, 39
circuits, 39–41
FDSP toolbox functions for, 45
magnitude equalization, 621–622, 632
multirate signal processing, 620–623, 632
operational amplifier (op amp), 39–41
oversamplings, 620–623, 632
passband equalization, 621–623, 632
signal processing, 5, 37–41
unipolar circuits, 39
Dimension, LTI system, 86
Direct current (DC) wall transformer analysis, 230–231
Direct forms, 340, 457–459, 541–544, 569
direct form I, 541
direct form II, 340, 541–542
FIR filter design, 457–459
IIR filter design, 541–544, 569
linear-phase form, 458–459
realization of filter structure, 340, 457–459
tapped delay line, 457
transposed direct form II, 542–544
transposed tapped delay line, 458
Discrete Fourier transform (DFT), 229–230, 241–255, 291–294, 320–321, 745–746
bin frequencies, 241–242
circular convolution of, 252–256
circular correlation of, 253
circular shift property, 252
coefficients, 246
defined, 242
discrete spectrum, 246
FDSP functions for, 247–248
Fourier series and, 230, 245–247
inverse (IDFT), 243, 745
linearity property, 250
matrix formulation, 243–245
orthogonal property, 242
Parseval's identity, 253–255

Discrete Fourier transform (*Continued*)
  periodic property, 248–249
  power density spectrum, 254
  power signals, 245–246
  properties of, 248–255, 745
  signal spectra, 247
  spectral analysis and, 229, 241–255,
    291–294, 320–321
  symmetry property, 249–251
  time reversal property, 251–252
  transform tables, 745–746
  Z-transform and, 229
Discrete (frequency) spectrum, 246
Discrete-time, 3, 11–12, 14–17, 60,
    70–227, 700–701, 720
  adaptive signal processing,
    700–701, 720
  block diagrams for, 94–96
  classification of signals, 11–12, 74–82
  classification of systems, 16–17,
    82–86
  convolution of signals, 70–71,
    100–110, 130–131, 160–161
  correlation of signals, 71, 110–117,
    131–132, 161–162
  difference equations for, 70–74, 86–94,
    100–117, 130–132
  DSP applications of, 71–74, 146–149
  FDSP functions for, 93, 100, 107, 117,
    198, 202–203
  Fibonacci sequence and the golden
    ratio, 210–212
  frequency domain, 145–227
  frequency response, 191–198, 214
  graphical user interface (GUI) for, 71,
    119–129, 132, 203–212, 214
  home mortgage analysis, 71–72,
    123–126
  impulse response, 96–100, 130–131
  MATLAB functions for, 73–74, 88, 93,
    102, 110, 173
  motivation, 70–74, 145–149
  nonlinear systems, 700–701
  poles and zeros, 150, 166–170,
    177–181, 213
  quantization noise and, 14–15
  radar echo detection, 72–73, 127–129
  region of convergence, 150–153, 213
  sample number, 12
  satellite attitude control system,
    146–148, 205–208
  signal flow graphs, 181–184
  signals, 3, 11–12, 14–15, 60, 74–82,
    110–117, 129–132
  speech/vocal tract modeling, 148–149,
    208–210
  stability of systems, 85, 91–92,
    117–119, 146, 184–191, 213–214
  state vector for, 700–701, 720

system identification, 198–203, 214,
    700–701, 720
  systems, 16–17, 82–86, 96–100,
    130–131
  time domain, 70–144
  transfer functions, 174–181, 213
  Z-transform for, 145–146, 149–173,
    213
Discrete-time Fourier transform (DTFT),
    228–229, 233–241, 319–320,
    744–745
  defined, 233–234
  frequency shift property, 237–238
  pairs, 241
  Parseval's identity, 238–239
  periodic property, 234
  properties of, 234–239
  signal spectrum, 234–237
  spectral analysis and, 228–229,
    233–241, 319–320
  symmetry property, 234–235
  time shift property, 237
  transform tables, 744–745
  Wiener-Khintchine theorem, 239
  Z-transform and, 228–229
Discrete wavelet transform (DWT), 302
Discrimination factor, 516
Down-sampling, 587–588, 630

E

Echo, signal transmission, 72–73
Elliptic filters, 526–528
Empty matrix [ ], 49
Energy, discrete-time signals, 77–79. *See
    also* Power
Equalization, 366, 453–456, 621–623,
    632, 647–648
  adaptive signal processing, 647–648
  channel, 647–648
  FIR filter design, 454–456
  inverse systems and, 366
  magnitude, 453, 621–622, 632
  optimal delay, 453–454
  oversampling and, 621–623, 632
  passband, 621–623, 632
  quadrature filter, 453–456
Equiripple filters, 434–442, 485. *See also*
    Chebyshev filters
  FDSP functions for, 442
  minimax error correction, 434–436
  Parks-McClellan algorithm, 436–442
Equivalent convolution, 106–107
Errors, 341–342, 464–476, 486–487,
    550–560, 569, 612–623, 632,
    646–647
  adaptive filter error signal, 646–647
  aliasing error factor, 615, 622
  binary number representation and,
    465–466

clipping, 554
coefficient quantization, 470–473,
    550–553
  FDSP functions for, 559–560
  finite word length effects and,
    341–342, 464–476, 486–487,
    550–560, 569
  IIR filter design, 550–560, 569
  input quantization, 466–469
  limit cycles, 557–559
  linear-phase block, 472–473
  multirate signal processing,
    612–623, 632
  overflow, 474–476, 554–555, 557
  oversampling, 612–623, 632
  precision and, 342, 464–465
  quantization, 341–342, 466–473,
    550–553
  roundoff, 473–474, 553–554
  scaling, 475–476, 554–557
  unit circle zeros, 472
Euler's identity, 89, 747
Excess mean square error and,
    666–669

F

Factored form, transfer functions, 177
Fast Fourier transform (FFT), 229,
    256–274, 321
  algorithm order of, 256
  alternative implementations, 262
  computational effort (speed), 260–262,
    265, 271–272
  decimation in time, 256–260
  fast block convolution, 267–270
  fast convolution, 263–266
  fast correlation, 270–274
  FDSP functions for, 270, 274
  floating-point operations (FLOPs), 256
  MATLAB functions for, 263
  spectral analysis and, 229,
    256–274, 321
  Z-transform, 229
File name conversion, FDSP toolbox, 48
Filter banks, 381–383, 394, 584–586,
    601–612, 631–632
  analysis, 381–382, 602, 634
  filter design specifications,
    381–383, 394
  frequency-division multiplexing,
    383, 601
  multirate signal processing, 584–586,
    601–612, 631–632
  narrowband, 584–586
  Quadrature mirror (QMF),
    607–612, 632
  signal synthesis using, 604–607
  subband processing, 601–607, 632
  synthesis, 383, 602, 632

time-division multiplexing, 383, 607–608
uniform DFT, 603–604
Filtered-*x* LMS (FXLMS) method, 691–695, 720
Filters, 7, 19–21, 33–39, 54–57, 60–61, 335–580. *See also* Digital filters
adaptive, 383–386, 394
allpass, 362–367, 393
anti-aliasing, 33–37, 54–57
anti-imaging, 37–39
Butterworth, 33–37
classical analog, 517–529, 568
cutoff frequency, 33
design specifications, 337–405
digital, design of, 335–580
filter banks, 381–383
FIR design, 406–498
first-order, 34–35
highpass, 389–392
ideal lowpass, 20–21
IIR design, 499–580
lowpass, 20–21, 338–342
narrowband, 378–381
notch, 7, 374–376
parameters for design, 514–517, 568
passband, 339
prototypes, 514–516, 568
quadrature, 367–374
resonators, 376–378
second-order, 35–36
spectrum of signals and, 19, 60
stopband, 339
transition band, 339
Final value theorem, Z-transform, 162–163
Finite impulse response (FIR) systems, 97–98, 130, 187, 353–358, 393, 406–498, 678–684
adaptive filter design, 678–684
bandstop filter design, 479–484
BIBO stability of, 187
cascade-form filters, 459–461
differentiators, 442–445
direct-form filters, 457–459
equiripple filters, 434–442, 485
FDSP functions for, 422–423, 429–430, 433, 442, 447, 456–457
filter design, 353–358, 406–498, 678–684
filter errors, 464–476, 486–487
finite word length effects, 464–476, 486–487
frequency sampling, 423–430, 485
graphical user interface (GUI), 477–484, 487
Hilbert transformers, 445–447
impulse response, 97–98, 130, 412–415

lattice-form filters, 461–463
least-squares method for, 430–433, 485–486
linear-phase, 353–356, 485
MATLAB functions for, 463–464
numerical differentiators, 407–409
pseudo-filters, 678–684
quadrature filters, 442–457, 486
realization structures, 457–464, 486
signal-to-noise ratio, 409–411
symmetry conditions, 353–356, 393
transfer function, 187
windowing, 411–423, 485
zero-phase, 356–358
Finite signals, 74
Finite word length effects, *see* Errors
First-order filters, 34–35
Flash converters, 43–45
Floating-point operations (FLOPs), 256
Folding frequency, 26
Forced mode, 178
Forgetting factor, RLS method, 684
Fourier series, 229–230, 245–247, 738–739
continuous-time signals, 229–230
discrete-time signals, 230
coefficients, 230
discrete Fourier transform (DFT) and, 230, 245–247
transform tables, 738–739
Fourier transforms (FT), 19–20, 28–29, 228–334, 739–741
continuous-time signal analysis and, 19–20, 28–29
discrete (DFT), 229, 241–255, 291–294, 320–321
discrete-time (DTFT), 228–229, 233–241, 319–320
fast (FFT), 229, 256–274, 321
inverse (IFT), 739
pairs, 740
properties, 741
short-term (STFT), 299–300
spectral analysis and, 228–334
transform tables, 739–741
Fractional delay systems, 586–587
Frequency-division multiplexing, 371, 383, 394
Frequency domain, 145–227, 608–609
discrete-time systems in, 145–227
DSP applications of, 146–149
frequency response, 191–198, 214
graphical user interface (GUI) in, 203–212, 214
motivation, 145–149
quadrature mirror filter (QMF) bank, 608–609
rate conversion in, 608–609
region of convergence, 150–153, 213

signal flow graphs for, 181–184
stability of discrete-time systems, 146, 184–191
system identification, 198–203, 214
transfer functions for, 174–181, 213
Z-transform for, 145–146, 149–173, 213
Frequency precision, 295–296
Frequency (spectral) resolution, 296–299, 321
Frequency response, 19, 60, 191–198, 214, 232–233, 291–294, 321, 337
continuous-time systems, 19, 60
decibel scale (dB), 293–294
discrete Fourier transform (DFT) for, 291–293
discrete-time systems, 191–198, 214
FDSP functions for, 198
gain, 19, 194
magnitude response, 19, 60, 193, 214, 337
periodic inputs, 196–197
phase response, 19, 50, 193, 214, 337
phase shift, 19, 194
sinusoidal inputs, 193–195
spectral analysis and, 232–233, 291–294, 321
steady-state response, 193–194
symmetry property, 192
zero padding and, 291–294
Frequency sampling, 423–430, 485
FDSP functions for, 429–430
FIR filter design, 423–430, 485
interpolated response, 424
lowpass filter, 425
transition-band optimization, 425–429
Frequency-selective filters, 342–350, 452–453
decibel scale (dB), 348–349
gain, 343
linear design specifications, 343–348
linear phase response, 343
logarithmic design specifications, 348–350
magnitude response $A(f)$, 342–350
phase response $\phi(f)$, 342–343
phase shift, 343
quadrature filter, 452–453
Frequency shift property, DTFT, 237–238, 603
Frequency transformations, 521–522, 535–540, 568
analog, 536–538
Butterworth filters, 521–522
digital, 539–540
FDSP functions for, 540
IIR filter design, 521–522, 535–540, 568
Frequency warping, 531–532

Full rank, 200
Fundamental frequency (pitch), 148
Fundamentals of Digital Signal
    Processing (FDSP) toolbox, 32,
    45–52, 61–62, 93, 100, 107, 117,
    198, 202–203, 247–248, 270, 274,
    281–282, 294, 303–304, 311, 358,
    366–367, 422–423, 429–430, 433,
    442, 447, 456–457, 514, 528–529,
    535, 540, 549–550, 559–560,
    595–596, 659–660, 677–678,
    690, 699–700, 712–713,
    750–754
  active noise control, 699–700
  adaptive signal processing, 659–660,
    677–678, 690, 699–700, 712–713
  allpass filters, 366–367
  analog-to-digital converters (ADC), 45
  bilinear transformation, 535
  circular convolution, 107
  classical analog filter design, 528–529
  complete responses using, 93
  continuous-time systems, 32
  cross-correlation, 117
  digital filter design, 358, 366–367
  digital-to-analog converters (DAC), 45
  discrete Fourier transform (DFT),
    247–248
  driver module, 46–47, 751
  equiripple filter design, 442
  file name conversion, 48
  finite word length effects (errors),
    559–560
  FIR filter design, 422–423, 429–430,
    433, 442, 447, 456–457
  frequency response, 198
  frequency sampling, 429–430
  frequency transformation, 540
  functions, 46–48, 750–754
  graphical user interface (GUI)
    modules, 49–52, 751
  help, 48–49
  IIR filter design, 514, 528–529, 535,
    540, 549–550, 559–560
  impulse response, 100
  installation, 750
  least-squares method, 433, 659–660,
    677–678
  lookfor command, 48
  minimum-phase filters, 366–367
  multirate signal processing, 595–596
  nonlinear system identification,
    712–713
  pole-zero placement, 514
  power density spectrum
    estimation, 311
  quadrature filter design, 456–457
  radial basis functions (RBF), 712–713
  realization of filter structure, 549–550

  recursive least mean squares (RLS)
    method, 690
  sampling rate converters, 595–596
  spectral analysis, 247–248, 270, 274,
    281–282, 294, 303–304, 311
  spectrograms, 303–304
  system identification, 202–203,
    712–713
  transformation methods, 535, 540
  use of, 46
  white noise, 281–282
  windowing, 422–423
  zero padding, 294
  zero-phase filters, 358

G
Gain, frequency response, 19, 194, 343
Gain factor b_0, 505–506, 508, 511–512
Gaussian radial basis functions (RBF),
    704–705
Gaussian white noise, 278–282
Geometric series, 78–79, 150
Graphical user interface (GUI), 4,
    52–59, 71, 119–129, 132, 203–212,
    214, 311–319, 322, 386–392,
    477–484, 487, 560–567, 570,
    623–630, 632, 713–718, 720
  adaptive signal processing,
    713–718, 720
  anti-aliasing filters, 54–57
  bandstop filter design, 479–484
  chemical process identification,
    715–718
  coefficient quantization, 388–392
  continuous-time signals, 52–59
  convolution, 123
  correlation, 121–123
  digital filter design, 386–392
  discrete-time signals, 312–313
  discrete-time systems, 71, 119–129,
    132, 203–212, 214
  distortion due to chirping, 316–319
  FDSP toolbox modules, 49–52
  Fibonacci sequence and the golden
    ratio, 210–212
  FIR filter design, 477–484, 487
  frequency-domain analysis,
    203–212, 214
  home mortgage analysis, 123–126
  IIR filter design, 560–567, 570
  multirate signal processing,
    623–630, 632
  radar echo detection, 127–129
  reconstruction, 54–55
  reverb filter design, 562–567
  sampling rate converters, 626–630, 632
  sampling, 52–54
  satellite attitude control, 205–208
  signal detection, 314–315

  spectral analysis, 311–319, 322
  speech compression, 208–210
  time-domain analysis, 119–129, 132
  video aliasing, 57–59
Grid points, 701–703

H
Half-band signal, 371
Hamming windows, 300–301, 416–417,
    419–420
Hanning windows, 300–301,
    416–418, 420
Harmonic forcing, 179
Help, FDSP toolbox, 48–49
Highpass filters, 389–392
Hilbert transformer, 369–371, 393,
    445–447

I
Ideal lowpass filter, 20–21,
    240–241
Impulse response, 20–21, 96–100,
    130–131, 165–166, 412–415
  continuous-time systems, 20–21
  discrete-time systems, 96–100,
    130–131
  FDSP functions for, 100
  finite (FIR) systems, 97–98, 130,
    412–415
  infinite (IIR) systems, 97–100, 130
  inverse Z-transform, 165–166
  linear time-invariant (LTI) systems,
    96–100
  sinc function, 21
  truncated, 412–415
  windowing and, 412–415
Indirect forms, *see* Cascade form;
    Parallel form
Inequalities, scalar and vector, 748
Infinite impulse response (IIR) systems,
    97–100, 130, 187, 349–350,
    499–580
  BIBO stability of, 187
  bilinear transformations, 529–535,
    568–569
  Butterworth filters, 517–522
  Chebyshev filters, 522–526
  classical analog filters, 517–529, 568
  colored noise, 502–504
  comb filters, 510–514
  elliptic filters, 526–528
  FDSP functions for, 514, 528–529,
    535, 540, 549–550, 559–560
  filter design, 349–350, 499–580
  filter errors, 550–560, 569–570
  finite word length effects,
    550–560, 569
  frequency transformations, 521–522,
    535–540, 568

graphical user interface (GUI), 560–567, 570
impulse response, 97–100, 130
logarithmic design specifications, 349–350
notch filters, 508–510, 568
parameters for filter design, 514–517, 568
pole-zero placement, 504–514, 551–552, 568
prototype filters, 514–516, 568
realization of filter structures, 541–550, 569
resonators, 504–507, 568
reverb filters, 562–567
tunable plucked-string filter, 500–502
Infinite signals, 74–75
Initial conditions, difference equations, 86–87, 130
Initial value theorem, 162–163, 171
Input-output representations, 184–185
Input polynomial, LTI systems, 90
Input quantization error, 466–469
Integer sampling rate converters, 587–591, 595–596
Interpolated response, 424
Interpolators, 584, 588–591, 598–599, 630
  integer, 584, 588–591
  multirate filter realization, 598–599
  polyphase, 598–599
  sampling rate conversion, 584, 588–591
Intersample delay, 586, 631
Inverse comb filter, 376, 511–514
Inverse discrete Fourier transform (IDFT), 243
Inverse Fourier transform (IFT), 739
Inverse systems, 366
Inverse Z-transform, 146, 164–173, 213, 743
  closed-form expression for, 166
  impulse response method for, 165–166
  MATLAB function for, 173
  noncausal signals and, 164
  partial fraction expansion for, 166–170
  residue method for, 170–173
  synthetic division method for, 164–165
  transform tables, 743

J

Jury test, 188–191

K

Kaiser windows, 420–421

L

Lag variable, 111
Laplace transform, 22–23, 29–31, 741–742

Lattice form, filter realization structure, 461–463
Leakage periodogram, 308–311
Leaky LMS method, 674–676, 719
Least mean square (LMS) method, 385, 430–433, 485–486, 656–678, 684–695, 718–720
  adaptive signal processing, 656–678, 684–695, 718–719
  bandpass filters, 432–433
  convergence rate, 663–666, 719
  correlation, 671–674
  error, 385
  excess mean square error and, 666–669
  FDSP functions for, 659–660, 677–678
  filtered-x (FXLMS) method, 691–695, 720
  FIR systems, 430–433, 485–486
  leaky, 674–676, 719
  misadjustment factor, 667
  modified, 669–678
  normalized, 669–671, 719
  performance analysis of, 660–669
  recursive (RLS) method, 684–690, 719
  steepest-decent method, 656–657
  step size, 660–663, 719
  system identification using, 658–659
Least-squares fit, 199–202
Limit cycles, 557–559
Linear cross-correlation, 110–114, 116
Linear convolution, 100–103, 130–131
Linear design specifications, 343–348
Linear-phase filters, 350–356, 472–473
  amplitude response $A_r(f)$, 351–353
  block, 472–473
  delay line (τ), 350–351
  phase response $\phi(f)$, 350–356
  quantization error, 472–473
  symmetry of, 352–356
Linear-phase form, 458–459
Linear-phase pseudo-filters, 681–684
Linear phase response, 343
Linear systems, 17–18, 82, 86–94, 96–100
  impulse response, 96–100
  difference equations for, 86–94
  time-invariant (LTI), 86–94, 96–100
Linearity property, 157, 250
Logarithmic design specifications, 348–350
lookfor command, FDSP toolbox, 48
Lossless system, 86
Lowpass filters, 20–21, 338–342, 417–419, 425
  cascade form, 340–341
  Chebyshev, 338–342
  design specifications, 338–342
  direct form II, 340
  frequency sampling, 425

ideal, 20–21
  passband, 339
  quantization error, 341–342
  realization structures, 339–342
  stopband, 339
  transition band, 339
  windowed, 417–419

M

Magnitude equalization, 453, 621–622, 632
Magnitude response $A(f)$, 19, 60, 193, 214, 337–350, 359–361, 392, 517, 621–622
  DAC oversampling and, 621–622
  digital filter design, 337–350, 359–361, 392
  frequency response and, 19, 60, 193, 214, 337
  frequency-selective filters, 342–350
  lowpass filters, 338–342
  minimum-phase filters, 359–361
  squared, 517
Magnitude spectrum, 19, 60, 234
MATLAB functions, 73–74, 88, 93, 102, 110, 173, 263, 276, 279, 298–299, 463–464
  deconvolution, 110
  fast Fourier transform, 263
  FIR filter design, 463–464
  frequency (spectral) resolution, 298–299
  inverse Z-transform residue term, 173
  linear convolution, 103
  realization of filter structures, 463–464
  signal creation, 73–74
  spectral analysis, 263, 276, 279, 298–299
  white noise, 73–74, 276, 279
  zero-input response, 88
  zero-state response, 93
Matrix formulation, DFT, 243–245
Maximum-phase filters, 360–362
Mean square error (MSE), 649–655, 666–669. *See also* Least mean square error (LMSE)
  adaptive signal processing, 649–655, 666–669
  adaptive transversal filters, 649–650
  cross-correlation and, 650–652
  excess, 666–669
  optimal weight vector, 653–654
  white noise input, 654–655
Mean square error, 384
Minimax error correction, 434–436
Minimum-phase filters, 359–367, 393
  allpass decomposition, 363–365
  equalization, 366
  FDSP functions for, 366–367

Minimum-phase filters (*Continued*)
  inverse systems, 366
  magnitude response $A(f)$, 359–361
Misadjustment factor, LMS method, 667
Mixed-phase filters, 360–361
Modes, 87, 130, 177–180
  cancelled, 178–179
  forced, 178
  multiple, 178
  natural, 87, 130, 178
  stable, 179–180
  transfer functions, 177–180
  zero-input response, 87, 130
Modulation, 21–23
Moving average (MA) model,
  95–96, 184
Multiple mode, 178
Multirate signal processing, 379–380
  analog-to-digital (ADC) signals,
    612–620, 632
  digital-to-analog (DAC) signals,
    620–623, 632
  FDSP functions for, 595–596
  filter banks, 584–586, 601–612,
    631–632
  fractional delay systems, 586–587
  graphical user interface (GUI),
    623–630, 632
  integer sampling rate converters,
    587–591, 595–596
  narrowband filters, 584–586, 600–601,
    631–632
  oversampling, 612–623, 632
  quadrature mirror filter (QMF) bank,
    607–612
  rational sampling rate converters,
    591–596
  realization of multirate filter structures,
    596–599
  sampling rate converters, 583–584,
    587–596, 626–631
  subband processing, 601–607, 632
Multistage converters, 593–595, 631

N

Narrowband filters, 378–381, 394,
  584–586, 600–601, 631–632
  banks, 584–586
  decimation factor, 600–601
  multirate signal processing, 379–380,
    584–586, 600–601, 631–632
  sampling challenges, 379
  sampling rate converters, 380–381
Natural mode, 87, 130, 178
Noise, 14–15, 274–284, 286–290,
  305–307, 409–411, 500–504,
  562–567, 613, 648–649, 690–700,
  720
  active control, 690–700, 720

adaptive signal processing, 648–649,
  690–700, 720
auto-correlation of, 286–290
cancellation, 648–649
colored, 502–504
FDSP functions for, 699–700
filtered–x LMS (FXLMS) method,
  691–695, 720
FIR filter design, 409–411
IIR filter design, 500–504, 562–567
period estimation, 286–287
periodic signal extraction of, 286–290
quantization, 14–15, 613
reduction, 694
reverb filters, 562–567
secondary path estimation, 693–694
signal estimation, 287–289
signal-synthesis method, 695–699
signal-to-noise ratio, 409–411
spectral analysis of, 274–284,
  286–290, 305–307
tunable plucked–string filter, 500–502
white, 274–284, 305–307, 411,
  502–504
Noncausal filters, 357–358
Noncausal signals, 15, 75, 164
Noncausal systems, 83–84
Nonlinear systems, 17, 82, 700–713, 720
  adaptive signal processing,
    700–713, 720
  discrete-time systems, 700–701
  FDSP functions for, 712–713
  grid points for, 701–703, 720
  identification, 710–713
  radial basis functions (RBF),
    703–713, 720
Normalized cross-correlation, 113–114
Normalized filter, 518–519
Normalized frequency, 348
Normalized LMS method, 669–671, 719
Normalized mean square error, 710
Notch filters, 7, 374–378, 393–394,
  508–510, 568
  bandwidth, 375
  comb filters and, 377, 394
  design of, 374–376, 393–394
  gain factor b_0, 508
  IIR filter design, 508–510, 568
  inverse comb filters, 376
  pole-zero placement, 508–510, 568
  resonators, power-complementary
    relationship of, 376–378, 394
Notch frequency F_0, 508
Numerical differentiators, 407–409

O

Offline processing, 83
Online system identification, 202
Operational amplifier (op amp), 39–41

Operators, 12–13, 101–102
  convolution, 101–102
  quantization, 12–13
Optimal weight vector, 653–654
Orthogonal property, 242, 706–707
Overflow error, 474–476, 554–555, 557
Oversampling, 26–27, 54, 58, 61,
  612–623, 632
  aliasing error factor, 615, 622
  analog-to-digital (ADC) signals,
    612–620, 632
  anti-aliasing filters and, 54, 61,
    612–615
  anti-imaging filters, 620–621
  continuous-time signal reconstruction,
    26–27
  digital-to-analog (DAC) signals,
    620–623, 632
  factor (α), 54, 58
  multirate signal processing, 612–623,
    632
  passband equalization, 621–623, 632
  sigma-delta ADC quantization,
    615–620
  video aliasing and, 58

P

Paley-Wiener theorem, 344
Parallel form, 544–546, 569
Parameters for filter design, 514–517, 568
Parks-McClellan algorithm, 436–441
Parseval's identity, 238–239, 253–255
Partial fraction expansion, inverse
  Z-transform, 166–170
Passband equalization, 621–623, 632
Passband filter specification, 339
Passive system, 86
Period estimation, 286–287
Periodic extension, 78, 103–104
Periodic impulse train, 21–22
Periodic inputs, 196–197, 307–308, 310
  frequency response, 196–197
  power density spectrum, 307–308, 310
Periodic property, 234, 248–249
Periodic signals, 75–76, 286–290
Periodograms, 304–311, 322
  average, 304–308, 311
  leakage, 308–311
  power density spectrum estimation,
    304–311, 322
Persistently exciting inputs, 202
Phase offset, 353
Phase quadrature, 367, 442
Phase response, 19, 60, 193, 214
Phase shift, 19, 194, 343
Phase spectrum, 19, 60, 234
Phonemes, 148
Piecewise-constant approximation, 31
Pitch (fundamental frequency), 148, 501

Pixels, 10
Pole radius, 505
Pole-zero cancellation, 178
Pole-zero placement, 504–514,
    551–552, 568
    comb filters, 510–514
    FDSP functions for, 514
    gain factor b_0, 505–506, 508,
        511–512
    IIR filter design, 504–514, 551–552,
        568
    notch filters, 508–510, 568
    quantization error and, 551–552
    resonators, 504–507, 568
Poles, 150, 166–170, 177–181, 213
    cancelled mode, 178–179
    complex, 169–170
    discrete–time system roots, 150, 213
    factored form of, 177
    inverse Z-transform, 166–170
    multiple, 167–169
    multiple mode, 178
    partial fraction expansion and,
        166–170
    simple, 166–167
    stable mode, 179–180
    transform functions, 177–181
    Z-transform, 150, 213
Polyphase decimator, 596–598
Polyphase decomposition, 597–599
Polyphase interpolator, 598–599
Postfilters, *see* Anti–imaging filters
Power, 77–82, 245–246
    average, 77
    discrete Fourier transform (DFT) and,
        245–246
    discrete-time signals, 77–78, 80–82
    energy and, 77–79
    energy signals, 77
    geometric series, 78–79
    periodic extension for, 78
    spectral analysis and, 245–246
    signals, 78, 80–82, 245–246
Power density spectrum, 247, 254,
    284–285, 304–311, 322
    auto-correlation and, 284–285
    average periodogram, 304–308, 311
    Bartlett's method for, 304–308, 311
    bin frequency, 304
    discrete Fourier transform (DFT) and,
        247, 254
    estimation, 304–311, 322
    FDSP functions for, 311
    leakage periodogram, 308–311
    periodograms, 304–311, 322
    spectral analysis, 247, 254, 284–285,
        304–311, 322
    Welch's method for, 308–311
Power gain, 614

Prefilters, *see* Anti-aliasing filters
Print option, 52–54
Probability density function, 274–275,
    278–279
Prototype filters, 514–516, 568
Pseudo-filters, 386, 678–684
    adaptive filter design, 678–684
    linear-phase, 681–684
Pseudo-inverse, 200, 432

Q
Quadrature filters, 367–374, 393,
    442–457, 486
    amplitude response $A_r(f)$, 450
    backward Euler approximation,
        367–368
    Chebyshev polynomials for,
        372–373
    differentiators, 442–445
    digital oscillator, 372–374
    equalizer filter design, 453–456
    FDSP functions for, 456–457
    FIR design, 442–457, 486
    frequency-selective filter, 452–453
    Hilbert transformer, 369–371, 393,
        445–447
    pair generation, 448–449
    phase quadrature, 442
    residual phase, 450
Quadrature mirror filter (QMF) bank,
    607–612, 632
    alias-free, 610–612
    frequency domain, rate conversion in,
        608–609
    two-channel, 607–612, 632
Quantization, 12–15, 60, 388–392, 613,
    615–620
    coefficients, GUI function for,
        388–392
    expected value (mean), 14
    level, 12, 60
    noise, 14–15, 613
    operator, 12–13
    sigma-delta ADC, 615–620
    signal classification using, 12–15, 60
Quantization error, 341–342, 466–473,
    550–553
    coefficient, 470–473, 550–553
    digital filter design, 341–342
    finite word length effects, 341–342,
        466–473
    FIR filter design, 466–473
    IIR filter design, 550–553
    input, 466–469
    linear-phase blocks and, 472–473
    pole-zero locations and, 551–552
    unit circle zeros and, 472, 552
    white noise modeled as, 466–469
Quantized signal, 12

R
Radial basis functions (RBF),
    703–713, 720
    adaptive networks, 707–709
    constant interpolation property of, 707
    FDSP functions for, 712–713
    first-order network, 711–712
    Gaussian, 704–705
    nonlinear systems, 703–713, 720
    normalized mean square error, 710
    orthogonal property of, 706–707
    raised cosine, 705–706
    safety factor, 709
Raised cosine radial basis functions
    (RBF), 705–706
Rational sampling rate converters,
    591–596, 631
Rayleigh limit, 296–297
Realization of filter structures, 339–342,
    457–464, 486, 541–550, 569,
    596–599
    cascade form, 340–341, 459–461,
        547–549, 569
    direct forms, 340, 457–459,
        541–544, 569
    FDSP functions for, 549–550
    filter design specifications, 339–342
    FIR filter design, 457–464, 486
    IIR filter design, 541–550, 569
    indirect forms, 544–550
    lattice form, 461–463
    linear-phase form, 458–459
    MATLAB functions for, 463–464
    multirate signal processing, 596–599
    parallel form, 544–546, 569
    polyphase decimators, 596–598
    polyphase interpolators, 598–599
    quantization error, 341–342
    tapped delay line, 457
    transposed direct form II, 542–544
    transposed tapped delay line, 458
Real-time signal applications, 5–6
Reconstruction, 26–32, 54–55, 61
    continuous–time signals, 26–32,
        54–55, 61
    formula, 26–29
    Fourier transform for, 28–29
    graphical user interface (GUI), 54–55
    Laplace transform for, 29–31
    oversampling and, 26–27, 61
    transfer function for, 29–30
    zero-order hold, 29–32, 61
Rectangular windows, 300–301, 308,
    416–418, 420
Recursive least mean squares (RLS)
    method, 684–690, 719
    adaptive signal processing,
        684–690, 719
    FDSP functions for, 690

Recursive least mean squares (*Continued*)
  forgetting factor, 684
  performance criterion, 684–685
  recursive formulation for, 685–688
Reflective structure, 362
Region of convergence, 150–153, 213
Relative weights, 678–679
Residual error, 200
Residual phase, 450
Residue, 166
Residue method, inverse Z-transform, 170–173
Resonant frequency F_0, 504–505
Resonators, 376–378, 393–394, 504–507, 568
  filter design, 376–378, 393–394
  gain factor b_0, 505–506
  IIR filter design, 504–507, 568
  notch filters, power-complementary relationship of, 376–378, 394
  pole-zero placement, 504–507, 568
Ripple factor ε, 522–523, 525, 527
Ripple voltage, 231
Roots, 87–90, 150, 166–170, 177–181, 123
  discrete-time systems, 150, 166–170, 177–181, 213
  inverse Z-transform poles, 166–170
  LTI systems, 87–90
  transfer function poles and zeros, 177–181
  Z-transform poles and zeros, 150, 166–170, 213
Rotation matrix, 372
Roundoff error, 473–474, 553–554

S

Safety factor, 709
Sampling, 3, 10–12, 21–27, 52–54, 57–61, 423–430, 485
  aliasing, 23–26, 61
  amplitude modulation, 22
  bandlimited signals, 24–26, 60–61
  continuous-time signals, 21–26, 52–54
  corrupted samples, 24–25
  folding frequency, 26
  frequency f_s, 12, 423–430, 485
  graphical user interface (GUI), 52–54
  imposters, 25
  interval T, 3, 12
  Laplace transform for, 22–23
  modulation, 21–23
  oversampling, 26–27, 54, 58, 61
  periodic impulse train, 21–22
  Shannon theorem, 25, 61
  undersampling, 24
  video aliasing, 10–11, 57–59
Sampling rate converters, 380, 583–584, 587–599, 612–623, 626–632

analog-to-digital (ADC) signals, 612–620, 632
decimators, 583–584, 587–588, 596–598, 630
digital-to-analog (DAC) signals, 620–623, 632
down-sampling, 587–588, 630
FDSP functions for, 595–596
filter design specifications, 380
integer, 587–591, 595–596
interpolators, 584, 588–591, 598–599, 630
multirate signal processing, 583–584, 587–596, 626–631
multistage, 593–595, 631
oversampling, 612–623, 632
rational, 591–596, 631
single-stage, 591–593
up-sampling, 590, 631
Scalar inequalities, 748
Scaling, 475–476, 554–557
Second-order backward differentiator, 408–409
Second-order filters, 35–36
Secondary path estimation, 693–694
Selectivity factor, 516
Shannon sampling theorem, 25, 61
Short-term Fourier transform (STFT), 299–300
Side bands, 24
Sifting property, 16
Signal and system analysis, 1–334
  discrete-time systems, 70–144, 145–227
  Fourier transforms, 228–334
  frequency domain, 145–227
  signal processing, 3–69
  spectral analysis, 228–334
  time domain, 70–144
Signal conditioning circuit, 41
Signal estimation, 287–289
Signal flow graphs, 181–184
Signal prediction, 648
Signal processing, 3–69, 73–82, 110–117, 129–132, 581–737
  active noise control, 7–9
  adaptive, 645–737
  advanced, 581–737
  aliasing, 10–11, 23–26, 33–39, 54–59, 61
  analog, 4–6, 13–14
  analog-to-digital converters (ADC), 4–5, 41–45
  continuous-time, 3, 11, 16–17, 21–32, 52–54, 60–61
  digital (DSP), 3–9, 14
  digital-to-analog converters (DAC), 5, 39–41

discrete–time, 3, 11–12, 16–17, 73–82, 110–117, 129–132
  filters, 7, 19–21, 33–39, 54–57, 60–61
  frequency response, 19, 60
  Fundamentals of Digital Signal Processing (FDSP) toolbox, 32, 45–52, 61–62
  graphical user interface (GUI), 4, 52–59
  impulse response, 20–21
  magnitude spectrum, 19, 60
  MATLAB functions for, 73–74
  motivation, 3–11
  multirate, 583–644
  notch filters, 7
  phase spectrum, 19, 60
  prefilters and postfilters, 33–39
  quantization, 12–15, 60
  reconstruction, 26–32, 54–55, 61
  sampling, 10–12, 21–26, 52–54, 57–61
  signal classification, 11–16, 74–82
  system classification for, 16–21
  total harmonic distortion (THD), 6–7
  transforms, 19–23, 28–31
  video aliasing, 10–11, 57–59
Signal shape, cross-correlation and, 111–113
Signal spectra, 247
Signal synthesis, 604–607, 695–699
Signal-to-noise ratio, 409–411
Sinc function, 21
Single-stage converters, 591–593
Sinusoidal inputs, frequency response, 193–195
Sound, *see* Noise
Spectral analysis, 228–334
  auto-correlation, 282–290
  convolution and, 252–256, 263–270
  correlation and, 253, 270–274
  direct current (DC) wall transformer, 230–231
  discrete Fourier transform (DFT), 229–230, 241–255, 291–294, 320–321
  discrete-time Fourier transform (DTFT), 228–229, 233–241, 319–320
  discrete-time signals, 312–313
  distortion due to chirping, 316–319
  fast convolution, 263–270
  fast correlation, 270–274
  fast Fourier transform (FFT), 229, 256–274, 321
  FDSP functions for, 247–248, 270, 274, 281–282, 294, 303–304, 311
  Fourier series, 229–230, 245–247
  frequency (spectral) resolution, 296–299, 321
  frequency response, 232–233, 291–294, 321

graphical user interface (GUI), 311–319, 322
MATLAB functions for, 263, 276, 279, 298–299
motivation, 228–233
noise, 274–284, 286–290, 306–307
periodic inputs, 307–308, 310
power density spectrum, 247, 254, 284–285, 304–311, 322
signal detection, 314–315
spectrograms, 299–304, 321
white noise, 274–282, 306–307
zero padding and, 291–296, 321
Spectral components, signals, 19–23
Spectral leakage, 300
Spectral (frequency) resolution, 296–299, 321
Spectrograms, 299–304, 321
data windows, 299–301
FDSP functions for, 303–304
spectral analysis using, 299–304, 321
subsignals, 299–300
window functions for, 300–301
Spectrum, defined, 228
Speed (computational effort), FFT, 260–262, 265, 271–272
Square summable signals, 77
Stable mode, 179–180
Stable systems, 18, 85, 91–92, 117–119, 130, 146, 184–191
bounded–input bounded–output (BIBO), 85, 117–119, 130, 185–188
continuous-time systems, 18
discrete-time systems, 85, 117–119, 130, 146, 184–191
frequency domain, 146, 184–191
input-output representations, 184–185
Jury test, 188–191
stability triangle, 190–191
time domain, 85, 117–119, 130
zero-input response and, 91–92
State vector, 384, 650, 700–701, 720
Steepest-decent method, 656–657
Step size, LMS method and, 660–663, 719
Stimulus, 16
Stopband filter specification, 339
Subband processing, 601–607, 632
Subsignals, 299–300, 305
Successive-approximation converters, 41–43
Symmetry property, 115, 192, 234–235, 249–251, 352–356
amplitude response $A_r(f)$, 352–353
cross-correlation, 115
discrete Fourier transform (DFT), 249–251
discrete-time Fourier transform (DTFT), 234–235

even, 352–353
frequency response, 192
linear-phase filters, 352–356
odd, 352–353
reciprocal, 354–355
Synthesis filter bank, 383, 602, 632
Synthetic division method, 164–165
System classification, 16–21, 82–86
continuous-time, 16–21
discrete-time, 82–86
System identification, 198–203, 214, 646–647, 700–713, 720
adaptive signal processing, 646–647, 700–713, 720
black box concept, 198–199, 646–647
discrete-time, 198–203, 214, 700–701
FDSP functions for, 202–203, 712–713
grid points for, 701–703
least-squares fit, 199–202
nonlinear systems, 700–713, 720
persistently exciting inputs, 202
radial basis functions (RBF), 703–713, 720
state vector for, 700–701, 720

T

Tapped delay line, 457
3-dB cutoff frequency, 517–518
Time-division multiplexing, 383, 607–608
Time domain, 70–144
block diagrams for, 94–96
bounded-input bounded–output (BIBO) systems, 85, 117–119, 130
convolution of signals, 70–71, 100–110, 130–131
correlation of signals, 71, 110–117, 131–132
difference equations for, 70–74, 86–94, 100–117, 130–132
discrete-time systems in, 70–144
DSP applications of, 71–74
graphical user interface (GUI) in, 71, 119–121, 132
impulse response, 96–100, 130–131
motivation, 70–74
signal processing in, 74–82, 110–117, 129–132
stability of discrete–time systems, 85, 91–92, 117–119
Time-invariant system, 17, 83
Time-multiplication property, 159–160
Time reversal property, 161, 251–252, 356
Time shift property, 237
Time-varying systems, 17–18, 83
Toeplitz matrix, 653
Total harmonic distortion (THD), 6–7, 231

Transfer functions, 29–31, 174–181, 187–188, 213
BIBO stability and, 187–188
cancelled mode, 178–179
continuous-time signals, 29–31
DC gain, 180–181
discrete-time systems, 174–181, 187–188, 213
factored form of, 177
FIR, 187
frequency-domain representation, 174
multiple mode, 178
poles and zeros, 177–181
stable mode, 179–180
unstable, 187–188
zero-input response, 174
zero-order hold, 29–32
zero-state response, 174, 176–177
Transformation methods, 521–522, 529–540, 568–569
analog frequency, 536–538
bilinear, 529–535, 568–569
digital frequency, 539–540
FDSP functions for, 535, 540
frequency warping, 531–532
frequency, 521–522, 535–540, 568
IIR filter design, 521–522, 529–540, 568–569
trapezoid integrator, 529–530
Transforms, 19–23, 28–31, 145–146, 738–746
continuous-time signals, 19–23, 28–31
reconstruction and, 28–31
sampling, 19–23
Fourier, 19–20, 28–29, 738–741
Laplace, 22–23, 29–31, 741–742
polar form, 19
spectral components determined by, 19–23
tables, 738–746
Z-transform, 145–146, 743
Transition band filter specification, 339
Transition-band optimization, 425–429
Transition bandwidth, 420
Transposed direct form II, 542–544
Transposed tapped delay line, 458
Transversal filters, 383–384, 645. See also Adaptive filters
Trapezoid integrator, 529–530
Trigonometric identities, 748
Tunable plucked-string filter design, 500–502

U

Unbounded signals, 76–77
Uncorrelated signals, 283
Undersampling, 24
Uniform DFT filter bank, 603–604
Uniform weighting, 430

766 Index

Uniform white noise, 274–278, 749
Unipolar DAC circuits, 39
Unit impulse, 15–16, 79, 153
Unit ramp, 160
Unit step, 15, 79–80, 153–154
Unstable systems, 18, 85, 117, 130. *See also* Stable systems
Up-sampling, 590, 631

V

Vector inequalities, 748

W

Weight vector, 383, 650, 720–721
Weighting function, 430
Welch's method, 308–311
White noise, 73–74, 274–284, 305–307, 411, 466–469, 502–504, 654–655, 749
  adaptive signal processing, 654–655
  auto-correlation of, 282–284
  colored noise from, 502–504
  creation of in MATLAB, 73–74
  FDSP functions for, 281–282
  Gaussian, 278–282
  input, 654–655
  MATLAB functions for, 73–74, 276, 279
  mean square error (MSE), 654–655
  power density spectrum of, 305–307
  probability density function, 274–275, 278–279
  quantization error modeled as, 466–469
  spectral analysis of, 274–282, 305–307
  uniform, 274–278, 749
  zero–mean, 411
Wiener-Khintchine theorem, 239, 284–285
Wiener solution, 653
Windowing, 299–301, 411–423, 485
  amplitude response $A_r(f)$, 411–412
  bandpass filter, 421–422
  Blackman windows, 300–301, 416–417, 419–420
  data windows, 299–301

FDSP functions for, 422–423
FIR filter design methods, 411–423, 485
  Hamming windows, 300–301, 416–417, 419–420
  Hanning windows, 300–301, 416–418, 420
  Kaiser windows, 420–421
  lowpass filter, 417–419
  rectangular windows, 300–301, 416–418, 420
  spectral leakage, 300
  spectrograms and, 299–300
  truncated impulse response, 412–415
  window functions, 300–301

Z

z-scale property, 159
Z-transform, 145–146, 149–173, 213, 228–229, 743
  causal signal analysis, 162–163
  convolution of signals using, 160–161
  correlation of signals using, 161–162
  defined, 149
  delay operator, 158, 169
  discrete-time system analysis, 145–146, 149–173, 213
  final value theorem, 162–163
  Fourier transforms and, 228–229
  geometric series for, 150
  initial value theorem, 162–163, 171
  inverse, 146, 164–173, 213, 743
  linearity property, 157
  MATLAB functions for, 173
  operator Z convention, 149–150
  pairs, 149–157
  poles and zeros, 150, 166–170, 213
  properties, 157–163
  region of convergence, 150–153, 213
  signal analysis using, 154–156
  time-multiplication property, 159–160
  time reversal property, 161
  transform tables, 743
  unit impulse, 153
  unit step, 153–154
  z-scale property, 159

Zero-input response, 87–93, 130, 174
  characteristic polynomial for, 87, 130
  complete response using, 92–93
  complex roots, 89–90
  MATLAB functions for, 88
  multiple roots, 88–89
  natural mode, 87, 130
  simple roots, 87–88
  transfer functions, 174
Zero-mean white noise, 411
Zero-order hold, 29–32, 37–38, 61
  anti-imaging filters and, 37–38
  continuous-time signal reconstruction, 29–32, 61
  magnitude of response, 37
  mathematical model of DSP system, 31–32
  transfer functions, 29–31
Zero padding, 105–107, 291–296, 321
  convolution and, 105–107
  decibel scale (dB), 293–294
  discrete Fourier transfer (DFT) for, 219–292
  equivalent convolution by, 106–107
  FDSP functions for, 294
  frequency precision and, 295–296
  frequency response and, 291–294
  spectral analysis and, 291–296, 321
Zero-phase filters, 356–358
Zero-state response, 90–94, 101–102, 174, 176–177
  complete response using, 92–93
  convolution and, 101–102
  MATLAB functions for, 93
  numerical, 93–94
  transfer functions, 174, 176–177
Zeros, 150, 177–181, 213, 354–355, 472. *See also* Poles
  discrete-time system roots, 150, 177–181, 213
  linear-phase filters, 354–355
  quantization error and, 472
  transform functions, 177–181
  unit circle, 472
  Z-transform, 150, 213